D1175654

Introduction to Marine Biogeochemistry

Introduction to Marine Biogeochemistry, Second Edition
by Susan Libes

Introduction to Marine Biogeochemistry

Second Edition

Susan Libes

College of Natural and Applied Sciences
Coastal Carolina University
Conway, South Carolina

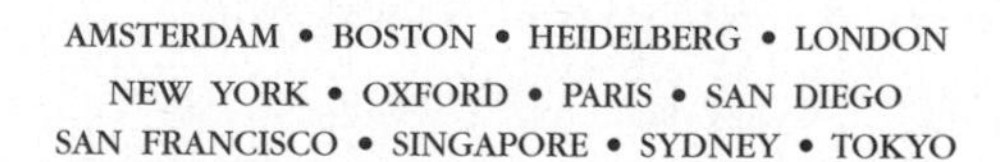
AMSTERDAM • BOSTON • HEIDELBERG • LONDON
NEW YORK • OXFORD • PARIS • SAN DIEGO
SAN FRANCISCO • SINGAPORE • SYDNEY • TOKYO

Academic Press is an imprint of Elsevier

Academic Press is an imprint of Elsevier
30 Corporate Drive, Suite 400, Burlington, MA 01803, USA
525 B Street, Suite 1900, San Diego, California 92101-4495, USA
84 Theobald's Road, London WC1X 8RR, UK

This book is printed on acid-free paper. ♾

Library of Congress Cataloging-in-Publication Data
APPLICATION SUBMITTED

British Library Cataloguing-in-Publication Data
A catalogue record for this book is available from the British Library.

ISBN: 978-0-12-088530-5

For information on all Academic Press publications
visit our Web site at *www.books.elsevier.com*

Printed in the United States of America
09 10 11 12 9 8 7 6 5 4 3 2 1

Contents

Please visit **http://elsevierdirect.com/companions/9780120885305** for supplementary web materials.

Preface

Seawater covers nearly 71% of Earth's surface making the oceans, at least to us humans, the dominant feature of Planet Earth. International interest in utilizing and conserving this vast resource has given rise to undergraduate and graduate degree programs in oceanography. Most rely on curricula built upon a set of core courses that individually provide surveys of marine biology, geology, physics and chemistry. This text is designed for a one-semester survey in marine chemistry at the advanced undergraduate or introductory graduate level. Students are expected to have completed foundational course work in basic chemistry, biology, physics, and calculus along with an introductory course in marine science.

Segregating oceanographic disciplines for educational purposes has grown increasingly confining as research continues to confirm the importance of a multidisciplinary understanding of ocean processes. This has led to an appreciation of the critical role that the oceans play in regulating atmospheric and thereby, terrestrial processes. Of all the subdisciplines of oceanography, marine chemistry relies most heavily on a multidisciplinary approach, now referred to as marine biogeochemistry. This makes formulation of a suitable introductory textbook challenging given the enormous diversity of important topics and complexity of approaches now in use.

To capture the multidisciplinary nature of marine chemistry, this text highlights the ocean's role in the global biogeochemical cycling of elements that are key to regulation of climate and marine biology. The impact of humans on the oceans and climate is given special emphasis, as are some of the practical triumphs of applied marine biogeochemistry, namely petroleum prospecting and the development of marine natural products, such as drugs and food additives. Part I covers the hydrological cycle and basic physico-chemical processes including chemical speciation. Part II provides an introduction to redox chemistry in the context of microbial ecology. Part III considers how marine processes result in the formation of sediments and the role of the sediments in regulating the chemistry of seawater. Part IV constitutes a survey of the field of marine organic chemistry including coverage of the molecular composition, sources and sinks of organic compounds along with a discussion of the elemental cycling of carbon, nitrogen, oxygen, sulphur, and phosphorus. A comprehensive discussion of marine pollution is the subject of Part V. Special features of the text include: (1) a thematic emphasis on the marine cycling of iron to exemplify marine biogeochemical processes that provide feedbacks regulating global climate and that are thereby linked to terrestrial processes, (2) basic details on introductory chemical principles including equilibria, rate laws, redox energetics, and organic chemistry on an as needed basis, (3) advances in paleoceanographic reconstructions of past ocean chemistry, biology, sedimentology and climate, and (4) human impacts on the ocean including climate change and marine pollution.

By focusing on the "hot" areas of research in marine chemistry, I have attempted to communicate the sense of excitement and discovery that is an essential characteristic of this relatively young and growing science. The nature of the "hot" topics has changed a bit since my writing of the first edition. Namely, the singular impacts of marine microbes on elemental cycling have come to be recognized as so powerful as to play a critical role in regulating climate and hence, indirectly, ocean circulation and terrestrial erosion rates. Conversely, physical processes, including oceanic circulation, hydrothermal activity, the production of flood basalts, and mountain building, have come to be recognized as having key impacts on ocean chemistry. These phenomena are now thought to have triggered important evolutionary shifts amongst the marine biota, including the Precambrian explosion of metazoans.

This book also includes a few aspects of marine analytical chemistry to address the operational nature of much of our data collection. This includes the use of remote sensing, such as satellite imagery, and in-situ sensing. Exciting advances in the latter include the use of submerged chemical detectors, such as mass spectrometers, and devices powered by natural redox processes, such as those occurring in marine sediments. Examples are provided of a particularly powerful and widespread approach—the use of naturally occurring stable and radioactive isotopes as tracers of biogeochemical processes. Artificial radionucludes, such as those introduced by bomb testing, have also proven to be excellent tracers of ocean circulation. This text seeks to give the reader enough of a background to pursue a more detailed study of this complicated topic.

Research efforts are now being urgently directed at understanding and mitigating the impacts of humans on the oceans as we have come to appreciate the truly global scope and scale of marine pollution and anthropogenically-driven climate change. The field of marine biogeochemistry is uniquely suited to help in this respect and hence represents a critical frontier of knowledge that can help us sustain ourselves and other life forms on planet Earth.

How to Use this Book

The study of marine chemistry is challenging but highly satisfying as you will use many of the skills and concepts learned in your basic science and math courses. Please be patient - true mastery takes time. Consider this text as a future reference book that you can return to long after graduation.

Useful features in the text include: (1) a glossary that defines technical terms, abbreviations, and acronyms, whose first appearance in the text is shown in *italic font*, (2) appendices containing various constants, equations, and conversion factors, and (3) a thorough index. Because we learn best from doing, this text has a companion website with resources to support a variety of active learning strategies (http://elsevierdirect.com/companions/9780120885305). Online features include: (1) a study guide, (2) homework problems, (3) supplemental content material, (4) a set of lengthy appendices containing geochemical and physical constants and other computational details, (5) color versions of figures as noted in the text's captions, and (6) a list of the full citations for works referenced in the text.

The supplemental content material available at the companion website is briefly described in the text at the relevant locations. For example, information on the evolution of the global carbon cycle over geologic time is briefly presented in Section 25.4 of the text as part of Chapter 25. A lengthier version of Section 25.4 is available online, and is so noted in the text. Text references to figures and tables that are available only online are labeled with a "W". For example, Figure W25.14 is available only in the online supplemental material for Section 25.4.

Acknowledgements

Many others, besides me, spent long hours working on this textbook. My heartfelt thanks go to them and to my students who provided me with the necessary teaching experience as well as to my fellow faculty members who have been enthusiastic supporters and best friends. Coastal Carolina University has provided a stable and supportive environment since my arrival in 1983. Many administrators and staff have been instrumental in providing essential resources, particularly Paul Gayes, Rob Young, Pete and Betsy Barr, and CCU's librarians. My thanks also go to the editors and publisher of the first edition, John Wiley & Sons. The editors of the second edition, Frank Cynar, Philip Bugeau, Laura Kelleher, and Linda Versteeg of Elsevier Press, provided critical support including an unlimited online subscription to Science Direct! The following colleagues served as reviewers: Tom Tisue, Ron Kiene, Erin Burge, Kevin Xu, Brent Lewis, Paul Haberstroh, Margareta Wedborg, and Courtney Burge.

Important motivation for undertaking this second edition came from numerous highly vocal users of the first edition who have variously threatened and cajoled mc sincc 2003. This effort turned out to be far more of an undertaking than anticipated and required drafting various family members into service, namely Lennie, who proofread what she could understand and even stuff she couldn't, Sol, who kept me in functioning computers, Don, who served as my book agent, Prashant, who checked all the math and left me alone for very long periods of time, and last but not least, the three best kitties in the world, Prem, Kali and Moti.

My personal interest in marine biogeochemistry stems from my experiences in the early 1980s as a graduate student in the Massachusetts Institute of Technology/Woods Hole Oceanographic Institution Joint Program in Oceanography and Ocean Engineering where I had first-hand contact with many of the most active researchers in the field. My thanks and admiration goes to them all, as well as to all the researchers and publishers who generously granted permission to use their copyrighted figures and tables herein. It has truly been an honor and a pleasure to summarize and present to the next generation of biogeochemists, the depth and breadth of research now being conducted around the world by an increasingly numerous, diverse, sophisticated and highly dedicated group of marine biogeochemists.

Susan Libes
6/17/08

PART 1

The Physical Chemistry of Seawater

CHAPTER 1

The Crustal-Ocean-Atmosphere Factory

All figures are available on the companion website in color (if applicable).

1.1 INTRODUCTION

The study of marine chemistry encompasses all chemical changes that occur in seawater and the sediments. The ocean is a place where biological, physical, geological, and chemical processes interact, making the study of marine chemistry very interdisciplinary and more appropriately termed *marine biogeochemistry*. Chemical approaches are now commonly used by marine biologists, marine geologists, and physical oceanographers in support of their research efforts. Likewise, oceanographers recognize the interconnectedness of Earth's hydrosphere with its atmosphere and crust, requiring that a true understanding of the ocean include consideration of its interactions with the rest of the planet. Also important are extraterrestrial forces, such as changes in solar energy and meteorites. For these reasons, this textbook covers topics that range far beyond the margins of the seashore and seafloor, as well as the boundaries of a classical study of chemistry.

1.2 WHY THE STUDY OF MARINE BIOGEOCHEMISTRY IS IMPORTANT

Most of the water on Earth's surface is in the ocean; relatively little is present in the atmosphere or on land. Because of its chemical and physical properties, this water has had a great influence on the continuing biogeochemical evolution of our planet. Most notably, water is an excellent solvent. As such, the oceans contain at least a little bit of almost every substance present on this planet. Reaction probability is enhanced if the reactants are in dissolved form as compared with their gaseous or solid phases. Many of the chemical changes that occur in seawater and the sediments are mediated by marine organisms. In some cases, marine organisms have developed unique biosynthetic pathways to help them survive the environmental conditions found only in the oceans. Some of their metabolic products have proven useful to humans as pharmaceuticals, nutraceuticals, food additives, and cosmeceuticals.

Another important characteristic of water is its ability to absorb a great deal of heat without undergoing much of an increase in temperature. This enables the ocean to act as a huge heat absorber, thereby influencing weather and climate.

Thus through many means, water sustains life, both marine and terrestrial. Scientific evidence supports the hypothesis that on Earth, life first evolved in a wet environment, such as an early ocean or submarine hydrothermal system. In turn, biological activity has had important effects on the chemical evolution of the planet. For example, the photosynthetic metabolism of plants is responsible for the relatively high concentration of oxygen gas (O_2) in our present-day atmosphere. Most of this oxygen was originally present as CO_2 emitted onto Earth's surface as part of volcanic gases. Over the millennia, photosynthesizers, such as marine phytoplankton, have converted this CO_2 into O_2 and organic matter (their biomass). Burial of their dead biomass (organic matter) in marine sediments has enabled O_2 to accumulate in the atmosphere. In this way, microscopic organisms have effected a global-scale transformation and transport of chemicals. This in only one example of many in which microscopic organisms serve as global bioengineers.

In studying the ocean, marine biogeochemists focus on exchanges of energy and material between the crust, atmosphere, and ocean. These exchanges exert a central influence on the continuing biogeochemical evolution of Earth. Particular concern is currently focused on the role of the ocean in the uptake and release of greenhouse gases, such as CO_2. As part of the atmosphere, these gases influence solar heat retention and, hence, influence important aspects of climate, such as global temperatures, the hydrological cycle, and weather, including tropical storms. Exchanges of material between the land and sea control the distribution of marine life. For example, transport of nutrients from the nearby continents causes marine organisms to grow in greater abundance in coastal waters than in the open ocean. The exchange rates of many substances have been or are being altered by human activities. Thus, the study of marine chemistry has great practical significance in helping us learn how to use the ocean's vast mineral and biological resources in a sustainable fashion to ensure its health for future generations of humans and other organisms.

1.3 THE CRUSTAL-OCEAN-ATMOSPHERE FACTORY AND GLOBAL BIOGEOCHEMICAL CYCLES

As illustrated in Figure 1.1, the planet can be viewed as a giant chemical factory in which elements are transported from one location to another. Along the way, some undergo chemical transformations. These changes are promoted by the ubiquitous presence of liquid water, which is also the most important transporting agent on Earth's surface. It carries dissolved and particulate chemicals from the land and the inner earth into the ocean via rivers and hydrothermal vents. Chemical changes that occur in the ocean cause most of these materials to eventually become buried as sediments or diffuse across the sea surface to accumulate in the atmosphere. Geological processes uplift marine sediments to locations where terrestrial weathering followed by river transport

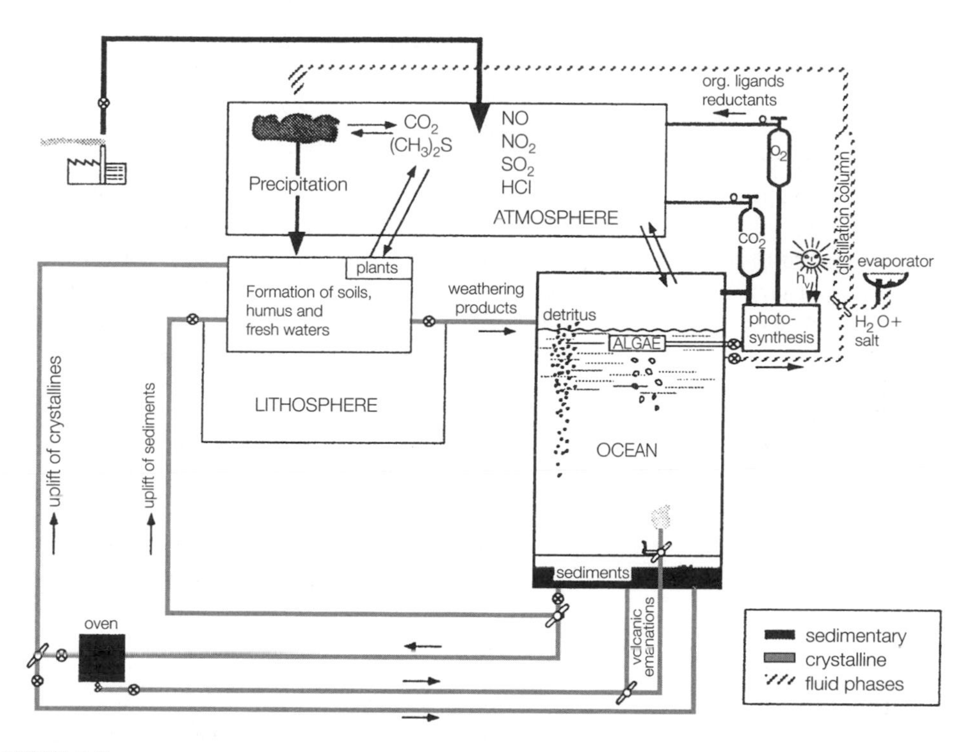

FIGURE 1.1

The crustal-ocean-atmosphere factory. *Source*: Stumm, W. and J. J. Morgan (1996) *Aquatic Chemistry*, 3rd ed. Wiley-Interscience, p. 874.

returns the chemicals to the ocean. The mobility of chemicals within the crustal-ocean-atmosphere factory is strongly affected by partitioning at interfaces. In the ocean, these include the air-sea and sediment-water interfaces, as well as the contact zone between seawater and suspended or sinking particulate matter. Thus, the ocean acts as a giant stirred flow-through reactor in which solutes and solids are added, transformed, and removed.

The representation of the ocean presented in Figure 1.1 is not a complete description of the ocean but serves to illustrate aspects important to the discussion at hand. Scientists refer to these simplified descriptions as a *model*. Models are useful ways of summarizing knowledge and identifying avenues for further study. Those that include mathematical information can be used to make quantitative predictions. The model illustrated in Figure 1.1 is a mechanistic one that emphasizes the flow of materials between various reservoirs. Because most material flows appear to follow closed circuits (if observed for long enough periods of time), the entire loop is referred to as a *biogeochemical cycle*. Such a cycle can be defined for any particular substance, whether it be an element, molecule, or solid. An example of the latter, the rock cycle, is given in Figure 1.2. This type of depiction is called a *box model* because each reservoir, or form that a substance occurs in, is symbolized by a box (e.g., sedimentary rock). The flow

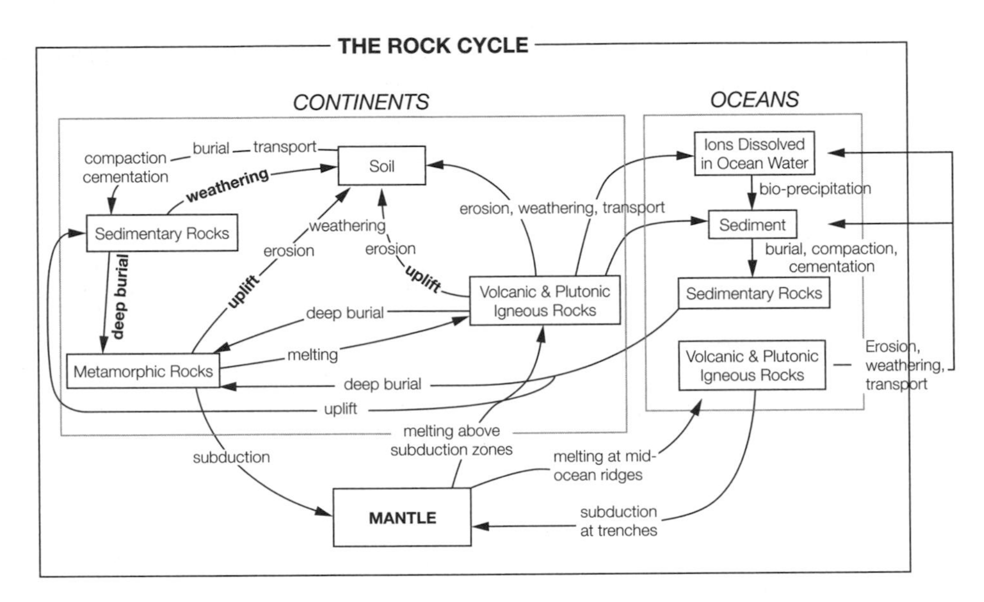

FIGURE 1.2

The global rock cycle. *Source*: After Bice, D. Exploring the Dynamics of Earth Systems: Modeling Earth's Rock Cycle. http://www.carleton.edu/departments/geol/DaveSTELLA/Rock%20Cycle/rock_cycle.htm.

of materials between reservoirs is indicated by arrows that point from the source of a substance to its *sink*. The magnitude of the exchange rates and sizes of the reservoirs are often included in these diagrams. For some substances, such as carbon and nitrogen, humans have significantly altered exchange rates and reservoir sizes. In these cases, the box model approach has proven valuable in assessing current impacts of human activities. These insights are used to predict how the crustal-ocean-atmosphere factory is likely to respond in the future, enabling a cost-benefit analysis of various environmental management strategies.

1.3.1 Steady State, Residence Times, and Turnover Times

If the size of a reservoir remains constant over time, the combined rates of input (J_{in}) to each box must equal the combined rates of output (J_{out}):

$$\frac{dM_C}{dt} = 0 = J_{in} - J_{out} \tag{1.1}$$

where M_C is the amount of material, C, in the reservoir, and J has the units of amount per unit time.[1] This condition is referred to as *steady state*. The average period of time

[1] Typically, M_c has units of mass.

that a specified unit of a substance spends in a particular reservoir is called its *residence time*. This steady-state residence time is given by

$$\text{Residence time} = \frac{\text{Total amount of a substance in a reservoir}}{\text{Total rate of supply to or removal of the substance from the reservoir}} = \frac{M_C}{J_{in} \text{ or } J_{out}} \quad (1.2)$$

In the case of water,

$$\text{Residence time} = \frac{\text{Total volume of water in the ocean}}{\text{Annual volumetric rate of water input}} = \frac{V_{SW}}{Q_{RW}} \quad (1.3)$$

As shown in the next chapter, the average molecule of water spends 3800 years in the ocean before being removed, mostly via the process of evaporation.

The steady-state concentration of a chemical with an oceanic residence time much longer than that of water can be predicted if it is assumed that its removal rate is directly proportional to its abundance in seawater, i.e.,

$$J_{out} = kM = kC_{SSW}V_{SW} \quad (1.4)$$

where k is a removal rate constant and C_{SSW} is the steady-state concentration of C in seawater. Since river input is the major source of most elements to seawater,

$$J_{in} = Q_{RW}C_{RW} \quad (1.5)$$

where C_{RW} is the concentration of C in riverwater. Substituting Eqs. 1.3 and 1.4 into Eq. 1.1 and solving for C_{SSW} yields

$$[C]_{SSW} = \frac{[C]_{RW}}{\text{RT} \times k} \quad (1.6)$$

where RT is the residence time of water in the ocean. Equation 1.5 indicates that the steady-state concentration of a given chemical is dependent on the relative magnitudes of its k and $[C]_{RW}$. Steady-state concentrations can shift given a sustained change in k and/or $[C]_{RW}$. In many biogeochemical cycles, changes in the steady-state concentration are difficult to achieve because natural systems tend to have feedbacks that act to reduce the effects of rate and/or concentration changes and, hence, stabilize biogeochemical cycles against perturbations.

Equation 1.6 is built upon the assumption that each of the removal processes that C undergoes follows first-order behavior. If these are chemical reactions, a first-order rate law can be written for each (individual) process in which

$$\text{Rate of change of } C \text{ due to reaction } i = -\left(\frac{d[C]_{SW}}{dt}\right)_i = k_i [C]_{SW} \quad (1.7)$$

where k_i is the first-order reaction rate constant that has a positive value if C is lost from the ocean through chemical reaction. These rate constants are additive so the k used in Eq. 1.7 can be computed as the sum of the individual reaction rate constants:

$$k = \sum_i^n k_i \quad (1.8)$$

First-order chemical behavior is commonly assumed because reaction rate laws are generally not known. Although this approach is accepted as a reasonable and practical accommodation, marine scientists are careful to acknowledge any computed results as "first approximations" or "back-of-the-envelope estimates."

Equation 1.2 assumes that the concentration of C is constant throughout the ocean, i.e., that the rate of water mixing is much faster than the combined effects of any reaction rates. For chemicals that exhibit this behavior, the ocean can be treated as one well-mixed reservoir. This is generally only true for the six most abundant (major) ions in seawater. For the rest of the chemicals, the open ocean is better modeled as a two-reservoir system (surface and deep water) in which the rate of water exchange between these two boxes is explicitly accounted for.

Another useful measure of biogeochemical processing is the *fractional residence time* or *turnover time* of a material in a reservoir. Computation of this "time" is similar to that of a residence time except that some subset of the input or output processes is substituted into the denominator of Eq. 1.2. The resulting turnover time represents how long it would take for that subset of processes by itself to either supply or remove all of the material from the reservoir. Turnover times can be calculated for reservoirs that are not in steady state. As will be shown in Chapter 21, the residence time can be computed by summing the reciprocals of the turnover times.

Using the rock cycle as an example, we can compute the turnover time of marine sediments with respect to river input of solid particles from: (1) the mass of solids in the marine sediment reservoir (1.0×10^{24} g) and (2) the annual rate of river input of particles (1.4×10^{16} g/y).[2] This yields a turnover time of $(1.0 \times 10^{24}\ \text{g})/(1.4 \times 10^{16}\ \text{g/y}) = 71 \times 10^{6}$ y. On a global basis, riverine input is the major source of solids buried in marine sediments; lesser inputs are contributed by atmospheric fallout, glacial ice debris, hydrothermal processes, and in situ production, primarily by marine plankton. As shown in Figure 1.2, sediments are removed from the ocean by deep burial into the seafloor. The resulting sedimentary rock is either uplifted onto land or subducted into the mantle so the ocean basins never fill up with sediment. As discussed in Chapter 21, if all of the fractional residence times of a substance are known, the sum of their reciprocals provides an estimate of the residence time (Equation 21.17).

1.4 CONSIDERATION OF TIME AND SPACE SCALES

Box models are limited in their ability to show temporal and spatial variability. In the case of the former, rates and reservoir sizes are liable to change over time. For example, plankton distributions tend to fluctuate on a seasonal, and even a daily, basis. Climate change appears to be causing rate and abundance changes over longer time periods, such as decades. This temporal variability is difficult to show in the box model format. One approach is to provide a range of values for the rate or reservoir size. Likewise,

[2] Pre-anthropocene suspended load carried by rivers as estimated by Syvitski, J. P. M, *et al.* (2005). *Science*, 308: 376–380.

spatial variability is also difficult to depict. Reservoirs in box models are assumed to be *homogenous*, i.e., having uniform composition. In reality, most reservoirs have some degree of *heterogeneity* or nonuniformity. For example, surface-water concentrations of nutrients tend to be much lower than deep-water concentrations, and coastal waters tend to have much higher concentrations than open-ocean waters.

One approach to dealing with spatial variability is to partition reservoirs into subreservoirs, such as into surface, deep-water, and coastal-water boxes. Sediments also tend to exhibit great horizontal and vertical variability. For example, most of the solid particles carried by rivers into the ocean are deposited nearshore on the continental margin. In the open ocean, most of the input of particles to the sediments is from atmospheric fallout of dust particles and in situ production of calcareous hard parts by plankton. Thus, calcareous oozes are common on mid-ocean ridges and rare on continental shelves. These examples of temporal and spatial variability highlight the important role of marine organisms in controlling chemical distributions. In turn, their biological activity and spatial distributions are greatly influenced by physical processes such as water movement, gravity, gas diffusion, and heat exchange. In many cases, chemical distributions can be used to trace the pathways and rates of these physical processes.

As illustrated in Figure 1.3, these biogeochemical and physical phenomena occur over a wide range of time and space scales in the crustal-ocean-atmosphere factory. Some are restricted to short time and space scales, whereas others are important only over long time and/or space scales. This requires that oceanographers sample strategically to ensure that their measurements of rates, concentrations, and amounts are truly representative. Because of the complex nature of variability in the marine environment, statistical techniques are now commonly used to design these strategic sampling plans. The goal of these plans is to most effectively target limited resources by adequately covering the temporal and spatial scales over which the processes of interest operate. In some cases, the best approach is to collect large numbers of small samples. In other cases, it is more cost effective to collect very large samples.

Temporal variability in the crustal-ocean factory can disrupt or prevent attainment of steady-state conditions for a given element. Examples of catastrophic events that can perturb global biogeochemical cycles include: (1) meteorite impacts, (2) changes in the rate and pattern of plate tectonic activity, and (3) climate change induced by fluctuations in delivery of solar radiation. Fortunately, many of the biogeochemical cycles seem to have an inherent structure that drives them back toward a steady-state condition. This stabilizing effect is the result of interactions among the transport processes that constitute the biogeochemical cycles. For example, a perturbation that causes an increase in the rate of supply of some element will be countered by an ensuing increase in the rate of its removal. In this way, the steady state is reestablished, although most likely at a new setpoint concentration. This type of interconnected response is termed a *negative feedback loop*.

Unfortunately, some perturbations can induce a *positive feedback* response in which perturbations are amplified. For example, the warming associated with global climate

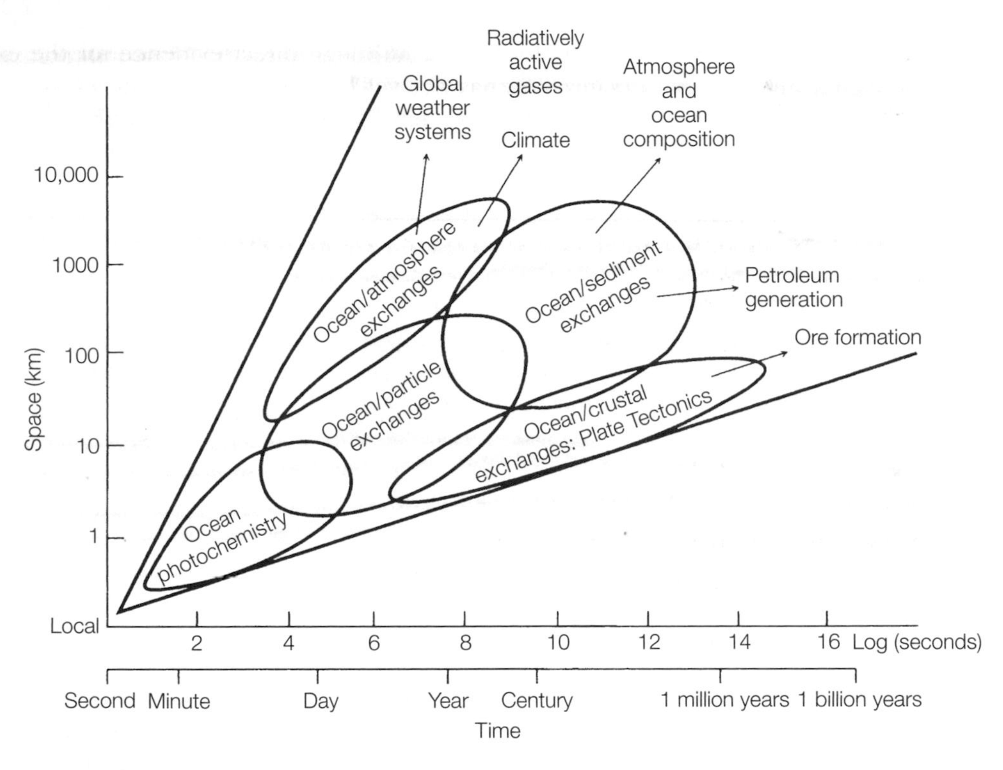

FIGURE 1.3

Time- and space-scales of processes in the crustal-ocean-atmosphere factory.
Source: From Bard, A., *et al*. (1988). *Applied Geochemistry* 3, 5.

change is reducing ice cover, which is in turn reducing Earth's ability to reflect, rather than absorb, incoming solar radiation, thereby enhancing global warming.

Geological evidence documents that Earth has experienced numerous catastrophic changes throughout its history, leading to at least five major extinction events during which a majority of species died off. After each extinction event, a repopulation occurred of life forms that were able to adapt to changed environmental conditions. Over the long term, this has lead to a steady chemical evolution of Earth's surface from a hot, acidic, rocky, airless place to one with a moderate climate, soils, an atmosphere that absorbs UV radiation, and an oxygenated atmosphere and ocean. Much of this evolution is attributable to the effects of marine microbes and algae, some of which have endured for billions of years. Others, such as the diatoms and coccolithophorids, are relative newcomers whose recent evolution has added important stabilizing structure to many of the global biogeochemical cycles. Some scientists consider that these negative feedback loops have conferred upon Earth self-regulatory functions akin to those exhibited by an organism. In this view, called the ***Gaia hypothesis***, Earth's biota and its abiotic environment interact so as to maintain the atmosphere, land, and ocean in a steady state that favors the survival of life. Although direct evidence for the existence of such a high level of organization has not yet been found, a significant body of data support the existence of various negative feedbacks. Some are discussed in this text.

1.5 THE HISTORY OF THE STUDY OF MARINE BIOGEOCHEMISTRY

Marine chemistry became a formal subdiscipline of chemistry in the early 1900s, with the advent of scientists who focused all their research efforts in this field and with the development of doctoral degree programs. Prior to the 1900s, the study of marine chemistry focused on investigations of the composition of the salts in seawater. The first such work was published in 1674 by the English chemist Robert Boyle, the discoverer of Boyle's law, which describes the behavior of ideal gases. Many other notable early chemists chose to focus their efforts on seawater and, in so doing, discovered new elements, established important new principles, and developed new investigative techniques. In 1772, the French chemist Antoine Lavoisier published the first analysis of seawater based on a method of evaporation followed by solvent extraction. Twelve years later, the Swedish chemist Olaf Bergman also published results of the analysis of seawater. To make his measurements, Bergman developed the method of weighing precipitated salts. Through their efforts, the field of analytical chemistry was born. Between 1824 and 1836, the technique of volumetric titrimetry was developed by Joseph Louis Gay-Lussac. Using this method of analysis, Gay-Lussac determined that the salt content of open-ocean seawater is nearly geographically constant. This conclusion was confirmed in 1818 by John Murray and in 1819–1822 by Alexander Marcet, who proposed that seawater contained small quantities of all soluble substances and that the relative abundances of some were constant. This hypothesis is now known as *Marcet's principle*.

The concept of salinity was introduced by Georg Forchhammer in 1865. From extensive analyses of seawater samples, he was able to demonstrate the validity of Marcet's principle for the most abundant of the salt ions: chloride, sodium, calcium, potassium, magnesium, and sulfate. Thus, he recognized that the salinity of seawater could be inferred from the easily measurable chloride concentration or *chlorinity*. The details of this relationship were worked out by Martin Knudsen, Carl Forch, and S. P. L. Sorenson between 1899 and 1902. With the international acceptance of their equation relating salinity to chlorinity ($S‰ = 1.805\ Cl‰ + 0.030$), the standardization necessary for hydrographic research was provided. A slight revision in this equation ($S‰ = 1.80655\ Cl‰$) was made in 1962 by international agreement.

The modern era of oceanography began in 1876 with the Challenger Expedition. This voyage of exploration was the first undertaken for purely scientific reasons. The results from the analysis of 77 seawater samples collected during this cruise were published by William Dittmar in 1884 and supported Marcet's principle. During the remainder of the 19th century, progress was made in the development of analytical methods for the measurement of trace constituents, such as dissolved oxygen (O_2) and nutrients. With this information, attention shifted to the investigation of the chemical controls on marine life.

The study of oceanography grew increasingly more sophisticated during the period from 1925 to 1940, with the initiation of systematic and dynamic surveys. The most famous was performed by the R/V *Meteor*, in which echo sounding was first used to map seafloor topography. Oceanography and the field of marine chemistry entered a new era in the 1940s, primarily as a result of submarine activity during World War II. This was a period of rapid development in technology and instrumentation. Analytical methods were developed for the measurement of trace constituents, such as metals, isotopes and organic matter, with detection levels dropping to subnanomolar levels. The salinity and temperature of seawater became recognized as a powerful tracer of large-scale water movements, including surface and deep ocean currents. Salinity and temperature were also employed to determine the density of seawater for the purposes of correcting sonar and computing geostrophic current velocities.

Modern oceanography is presently characterized by multidisciplinary research projects conducted collaboratively by large groups of scientists often from different research institutions. This approach is necessitated by the complexity of studying marine processes, such as ones that involve global scales, like climate change. This current era was initiated in 1958 with the International Geophysical Year, which was organized by the United Nations' UNESCO General Assembly. For marine chemistry, the first multi-investigator, multi-institution project was the Geochemical Ocean Sections Study (GEOSECS) that ran from 1968 to 1978 during the National Science Foundation's International Decade for Ocean Exploration (IDOE). Its goal was to determine the pathway of deep-ocean circulation using radioisotopes, such as radiocarbon, as tracers of water movement. This work was continued in the Transient Tracers in the Ocean (TTO) program, which ran from 1980 to 1983. Both programs sought to take advantage of the global injection of artificial radionuclides into the ocean from fallout generated during the nuclear weapons testing conducted in the 1950s and 1960s.

The Joint Global Ocean Flux Study,[3] which ran from 1987 to 2003, investigated fluxes of chemicals, primarily carbon and other biogenically controlled elements, to better understand linkages to global climate change. This international program was one of the first core projects of the International Geosphere-Biosphere Programme (IGBP) developed by the Scientific Committee on Oceanic Research (SCOR), a committee of the International Council for Science (ICSU). An important component of JGOFS was the establishment of time-series measurements at two sites, HOTS (Hawaii Ocean Time Series) and BATS (Bermuda Atlantic Time Series Study), to provide interannual and seasonal resolution of biogeochemical variability. Sampling at the BATS site was initiated in 1978 by Dr. Werner Deuser at the Woods Hole Oceanographic Institution as part of the Oceanic Flux Program (OFP) and is the longest time series of its kind; recording

[3] http://www.uib.no/jgofs/

temporal variability in the delivery of sinking biogenic detritus to the seafloor. JGOFS was also notable in its use of remote sensing data collected by satellites.

Data from GEOSECS, TTO, BATS, and HOTS and other major oceanographic research projects, such as the WOCE (World Ocean Circulation Experiment) are available online.[4] The GEOSECS, TTO, and WOCE datasets are part of the *Java Ocean Atlas*, which provides a graphic exploration environment for generating vertical profiles, cross-sections, and property-property plots.[5] Many of the data presented in this text were obtained from this source.

The research ships that supported these major projects were largely managed by the University-National Oceanographic Laboratory System (UNOLS),[6] a consortium of 64 academic institutions established in 1971. UNOLS now coordinates schedules of 28 research vessels ranging in size from 20 to 85 m that are operated by 20 different member organizations, including universities, research institutions, and federal agencies. Ship time is available to all federally funded oceanographers. Deep-sea submersibles and remotely operated vehicles (ROVs) schedules are also coordinated through UNOLS. This technology has played a major role in the study of hydrothermal vents and cold-water seeps. The vents were first discovered in 1977, providing marine chemists with direct observations of large sources and sinks of materials associated with venting along submarine plate boundaries. It also lead to the discovery of a new food web based on chemosynthetic bacteria.

An increasing focus of ongoing work is directed at understanding anthropogenic impacts on the crustal-ocean-atmosphere factory: not just climate change, but also the long-range transport and fate of pollutants. Of particular interest are processes that occur at interfaces, such the fate of river input after it mixes with seawater, the effect of sunlight on the photochemistry of surface water, and the role of organisms in the formation and solubilization of particles. Much of the work involving particles and the fate of pollutants relies on research into very small-scale phenomena, namely the role of phytoplankton and microbes, such as bacteria and viruses, in translocating and transforming chemicals.

1.6 NEW TECHNOLOGIES, NEW APPROACHES

The next step in obtaining a true systems-level understanding of the crustal-ocean-atmosphere factory requires establishment and maintenance of a global-scale, long-term observational program. For marine scientists, this requires switching from short-duration, ship-based expeditions in which discrete samples are collected and brought back to shore for lab-based analysis to one that relies on continuous data collection using

[4] http://whpo.ucsd.edu/index.html
[5] http://odf.ucsd.edu/joa
[6] http://www.unols.org

in situ and remote-sensing technologies. The latter is referred to as *operational oceanography*. In the United States, implementation of these approaches is being directed through the National Aeronautics and Space Administration (NASA), National Oceanic and Atmospheric Administration (NOAA), the Joint Oceanographic Institutions (JOI), and the Consortium for Oceanographic Research and Education (CORE). In 2004, these groups established the Ocean Research Interactive Observatory Networks (ORION) Program to coordinate development and operation of large-scale ocean observatories.[7] Examples include NOAA's seafloor observatories, such as the Aquarius, an underwater laboratory moored at 20 m in the Florida Keys National Marine Sanctuary since 1993, and the Long-term Ecosystem Observatory (LEO-15) established in 1996. LEO-15 is located in 15 m of water over a $30 \times 30\,km^2$ area on the inner continental shelf of New Jersey.[8] In 2001, LEO-15 was expanded into the New Jersey Shelf Observing System (NJSOS), which covers a $300 \times 300\,km^2$ area. More than a dozen different sensors have been deployed at this site, carried by autonomous underwater vehicles (AUVs), ROVs, and human-occupied vehicles (HOVs).

Another example of such a comprehensive approach to ocean monitoring is the High Latitude Time Series Observatory, which is located in the NW Pacific. This observatory was established in 2001 by the Joint North Pacific Research Center to study what appears to be a site of major CO_2 uptake. It is a collaborative effort between the Woods Hole Oceanographic Institution and two Japanese groups, Mutsu Institute for Oceanography and the Japan Marine Science and Technology Center. An innovative technology being used at this site is moored geochemical profilers that shuttle 200 times a year between the mixed layer and deep zone, providing in situ measurements of conductivity, temperature, depth, and 3D current velocity.[9] Also deployed are automated samplers that collect experiments conducted on filtered water, sediment and plankton in automated incubators. In the mixed layer, an optical sensor continuously measures fluorescence, chlorophyll, and particles to depths of 35 m. These measurements are being coordinated with remote sensing obtained from the ADEOS-II, a satellite launched by the National Space Development Agency of Japan (NSDA).

Space-based earth observations began in 1960 with NASA's Television Infrared Observation Satellite (TIROS). In the United States, NOAA and NASA have since developed sensors to measure sea surface temperature, winds, and topography. The first experimental effort to obtain remotely sensed color data was made in 1978 with the launch of the Coastal Zone Color Scanner (CZCS) aboard the Nimbus-7 satellite. The first effort to collect biogeochemical data began in 1997 with NASA's SEAWIFS (Sea-viewing Wide Field of View Sensor) Project, which relies on an ocean color sensor to provide an estimate of phytoplankton production by estimating in vivo fluorescence from chlorophyll. These data were designed to help assess the oceans' role in the global carbon cycle, and were by JGOFS. In 1999, NASA began its Earth Observing System (EOS) program

---

[7] http://www.orionprogram.org.

[8] NURP; http://www.nurp.noaa.gov

[9] http://Jpac.whoi.edu/hilats/strategy/instruments.html

with the launch of the Terra satellite, which contains an upgraded color sensor called the Moderate Resolution Imaging Spectroradiometer (MODIS). MODIS has 36 spectral channels, as compared to SEAWIFS' eight, enabling it to collect information on colored dissolved organic matter (CDOM) and detritus at a resolution of 0.25 to 1 km. A second MODIS satellite, Aqua, was launched in 2002. Near real-time imagery from MODIS sensors is available online.[10] The NOAA Polar Orbiting Environmental Satellites (POES) also carry a multispectral sensor called an Advanced Very High Resolution Radiometer (AVHRR). As shown in Table 1.1, many other countries have launched satellites with ocean color sensors.

Plans have been made to fly a new high-resolution multispectral sensor, Visible/Infrared Imager/Radiometer Suite (VIIRS), aboard the National Polar-Orbiting Operational Environmental Satellite System (NPOESS). The Navy also has plans to send a Coastal Ocean Imaging Spectrometer (COIS) with a resolution of 30 m aboard the Navy Earth Map Observer (NEMO). This sensor is designed to enable detection of oil spills and plankton blooms from spectral signatures. In addition to improving spectral coverage and spatial resolution, future efforts will be directed at increasing temporal resolution. Satellites have been deployed by other countries than the United States. For example, Japan's Advanced Earth Observing Satellite (ADEOS), launched in 2002, carries a Global Imager (GLI) with resolution of 250 m in some of its 36 spectral channels. An international group, the Committee on Earth Observation Satellites (CEOS), was formed in 1984 to coordinate and enhance productivity of space-related earth observation activities. With 100 new satellites expected to be launched over the next decade, this technology can be expected to play an increasingly important role in oceanographic research.

1.7 THE FUTURE OF MARINE BIOGEOCHEMISTRY

Operational oceanography is a first step in the direction of obtaining a systems-level understanding of the crustal-ocean-atmosphere factory. The next step is integrating oceanography with other earth sciences and translating our new understanding into a form that can be used to protect resources and humans. Formal work toward this end began at the First Earth Observation Summit held in July 2003. At its conclusion, thirty countries agreed to support the development of a Global Earth Observation System of Systems (GEOSS). GEOSS currently includes a land-based component, the Global Terrestrial Observing System (GTOS), a satellite-based component (CEOS), and an ocean-based component, the Global Ocean Observing System (GOOS). A systems approach will facilitate integration of data collection with data processing, database management, and data delivery conducted via query-based web pages to provide open access. Forty countries are now participating in GEOSS.

In the United States, GOOS will be implemented through a new Integrated Ocean Observing System (IOOS) run by a new organization, Ocean.US, the National Office

[10] http://rapidfire.sci.gsfc.nasa.gov/

Table 1.1 Summary of Recent and Current Satellite Ocean Color Sensors.

Sensor	Agency	Satellite	Operating Dates	Resolution (m)	Number of Bands	Spectral Coverage (nm)	Ref.
CZCS	NASA (USA)	Nimbus-7 (USA)	24/10/78–22/06/86	825	6	433–12500	a
OCTS	NASDA (Japan)	ADEOS (Japan)	17/08/96–01/07/97	700	12	402–12500	b
POLDER-1	CNES (France)	ADEOS (Japan)	17/08/96–01/07/97	6000	9	443–910	c
MOS	DLR (Germany)	IRS P3 (India)	Launched 21/03/96	500	18	408–1600	d
SeaWiFS	NASA (USA)	OrbView-2 (USA)	Launched 01/08/97	1100	8	402–885	e
OCI	NEC (Japan)	ROCSAT-1 (Taiwan)	Launched Jan 1999	825	6	433–12500	f
OCM	ISRO (India)	IRS-P4 (India)	Launched 26/05/99	350	8	402–885	g
MODIS-Terra	NASA (USA)	Terra (USA)	Launched 18/12/99	1000	36	405–14385	h
OSMI	KARI (Korea)	KOMPSAT (Korea)	Launched 20/12/99	850	6	400–900	i
MERIS	ESA (Europe)	ENVISAT-1 (Europe)	Launched 01/03/02	300/1200	15	412–1050	j
MODIS-Aqua	NASA (USA)	Aqua (EOS-PM1)	Launched 04/05/02	1000	36	405–14385	h
CMODIS	CNSA (China)	Shen Zhou-3 (China)	25/03/02–15/09/02	400	34	403–12500	k
COCTS	CNSA (China)	HaiYang-1 (China)	Launched 15/05/02	1100	10	402–12500	k
CZI	CNSA (China)	HaiYang-1 (China)	15/05/02–01/12/03	250	4	420–890	k
GLI	NASDA (Japan)	ADEOS-II (Japan)	14/12/02–25/10/03	250/1000	36	375–12500	l
POLDER-2	CNES (France)	ADEOS-II (Japan)	14/12/02–25/10/03	6000	9	443–910	m

Data used courtesy of the International Ocean Color Coordinating Group, http://www.ioccg.org/sensorshttp://www.ioccg.org/sensors.
a: http://daac.gsfc.nasa.gov/DATASET_DOCS/czcs_dataset.html. http://daac.gsfc.nasa.gov/DATASET_DOCS/czcs_dataset.html.
b: http://www.eoc.nasda.go.jp/guide/satellite/sendata/octs_e.html. http://www.eoc.nasda.go.jp/guide/satellite/sendata/octs_e.html.
c: http://smsc.cnes.fr/POLDER. http://smsc.cnes.fr/POLDER.
d: http://www.ba.dlr.de/NE-WS/ws5/mos_home.html. http://www.ba.dlr.de/NE-WS/ws5/mos_home.html.
e: http://seawifs.gsfc.nasa.gov. http://seawifs.gsfc.nasa.gov.
f: http://rocsat1.oci.ntou.edu.tw/en/oci/index.htm. http://rocsat1.oci.ntou.edu.tw/en/oci/index.htm.
g: http://www.isro.org/programmes.htm. http://www.isro.org/programmes.htm.
h: http://modis.gsfc.nasa.gov. http://modis.gsfc.nasa.gov.
i: http://kompsat.kari.re.kr/english/index.asp. http://kompsat.kari.re.kr/english/index.asp.
j: http://envisat.esa.int/instruments/meris. http://envisat.esa.int/instruments/meris.
k: http://www.cnsa.gov.cn/main_e.asp. http://www.cnsa.gov.cn/main_e.asp.
l: http://www.eoc.nasda.go.jp/guide/satellite/sendata/gli_e.html. http://www.eoc.nasda.go.jp/guide/satellite/sendata/gli_e.html.
m: http://polder-mission.cnes.fr. http://polder-mission.cnes.fr.
Source: From Pinkerton, M. H. et al. (2005). Remote Sensing of Environment *97, 382–402.*

for Integrated and Sustained Ocean Observations. The goals of IOOS are to benefit humans by (1) improving predictions of climate change and weather and their effects on coastal communities and the nation; (2) improving the safety and efficiency of maritime operations; (3) more effectively mitigating the effects of natural hazards; (4) improving national and homeland security; (5) reducing public health risks; (6) more effectively protecting and restoring healthy coastal ecosystems; and (7) enabling the sustained use of ocean and coastal resources. IOOS is split into an ocean and a coastal component. The coastal component is divided into 10 Regional Coastal Ocean Observing Systems (RCOOS), each run by a Regional Association (RA). The National Federation of Regional Associations (NFRA) is charged with producing an integrated network by coordinating efforts of the RCOOSs. One interesting challenge lies in linking the existing freshwater observational framework, such as gaging stations maintained by the U.S. Geological Survey (USGS), to downstream efforts in estuaries. Several RCOOSs have seafloor observatories, such as LEO-15, which will continue to be coordinated through the ORION. One of the important challenges of these initiatives is in developing strategies for coping with large data streams generated by multiple sensors and instruments, including providing power, high-speed data transmission, and two-way, shore-to-seafloor communications. Another important initiative is the development of telepresence at sea in which an advanced type of videoconferencing is used to transmit video and digital data between ship and shore in near real time using satellite links to the Internet. Through telepresence, the science command center for an expedition can be located on land, thereby reducing costs and scheduling problems amongst the lead scientists.

A notable example of a GOOS program is the broad-scale global array of temperature/salinity profiling floats, known as Argo. Deployments began in 2000, with the final array to be composed of 3000 floats that will generate 100,000 vertical profiles of temperature, salinity and velocity measurements per year at an average 3-degree spacing. Sensor technology is improving rapidly, enabling in situ measurement of other parameters such as dissolved oxygen, nitrate, in vivo chlorophyll fluorescence, CDOM, turbidity, pH, photosynthetically active radiation (PAR), and redox potential (ORP). At present, long-term deployments of these sensors are rare because of biofouling and calibration issues. An interesting short-term deployment technique uses towed undulating ROVs. Depth and GPS sensors provide location information used by an onboard computer to produce horizontal maps, cross-sectional depth profiles, and property-property plots. A new generation of in situ sensors using a wet chemistry approach for measuring nutrients and iron in seawater is now commercially available, but the need for a larger variety of in situ sensors for identifying and quantifying a wide range of gases, organic compounds, and plankton, including microbes, is great. Future approaches will likely seek to create instrument packages that carry sophisticated chemical instrumentation such as high-pressure liquid chromatographs (HPLCs), UV-VIS spectrophotometers, mass and Raman spectrometers, and even DNA analyzers.[11]

---

[11] http://www.whoi.edu/institutes/oli/activities/short_report.pdf

Though great progress has been made in the past four decades, many gaps remain in our understanding of the chemical processes that occur in the sea. There are several reasons for this. First, except for water and the six major ions, all the other substances in seawater are present at very low concentrations. The combination of trying to detect low concentrations in the presence of large amounts of salts makes measurement of the trace constituents in seawater very difficult. To make matters even more complicated, most elements are present in several different forms, or species, in seawater. The speciation of an element determines its reactivity. Thus, the concentration of each species of an element must be known to fully understand the chemical behavior of that element.

Another great challenge in furthering our understanding of the ocean lies in improving our theoretical approach to the ocean. Marine chemists have traditionally resorted to assuming that the chemical reactions of interest attain equilibrium. This greatly simplifies computations, but provides limited insight into the wide variety of biogeochemical processes controlled by marine organisms. Since living organisms are themselves not at equilibrium, neither are the reactions they mediate. Some attempts have been made at kinetic descriptions of marine processes, with most relying on an assumption of first-order rate behavior. Higher-order rate laws are more likely to be the rule and are thought to confer stability on biological systems.

Marine chemistry has traditionally been divided into two fields. One seeks to understand the chemistry of organic substances in the ocean. The other investigates inorganic substances. Because of analytical difficulties, more is known about the latter than the former. Continuing methodological advances are causing this gap to close rapidly. Our growing recognition of the ubiquitous influence of marine organisms has also blurred the distinction between the two fields. This has direct impact on how research is now being conducted to elucidate the controls on ocean fertility, namely assessing the role of trace metals, such as iron, in supporting the growth of phytoplankton. Understanding ocean fertility will help better manage fisheries and cope with pollution problems. Related areas of research include (1) establishing the molecular structure and reactivity of dissolved and particulate organic matter, (2) elucidating the role of marine organisms in packaging materials into solids that settle and become buried in marine sediments, and (3) quantifying material inputs to the ocean from the coastal ocean, atmosphere, and hydrothermal vents. Other efforts are directed at exploring the recovery of mineral resources from the seabed and the discovery of marine natural products. Many of these research areas are characterized by multidisciplinary approaches, making it difficult to separate chemical studies from biological, geological, physical, atmospheric, and even aquatic work. As a result, biogeochemists are being increasingly common and can be found working in laboratories and departments of biology, geology, physics, atmospheric, space, and environmental science!

Biogeochemistry has been particularly useful in efforts to study the ocean's past. This subdiscipline is called *paleoceanography.* Because of the linkages among the crust, ocean, and atmosphere, the field of paleoceanography also provides insight into past climate and terrestrial conditions. Much of the geochemical reconstruction of the ocean's past has relied on compositional analysis of marine microfossils recovered from long sediment cores. These cores are collected by specialized drill ships. The first of these was the *Glomar Challenger*, deployed in 1966 as part of the Deep Sea Drilling Project (DSDP). In 1984, DSDP was transformed into the Ocean Drilling Program (ODP) and acquired a new vessel, the *Resolution*, operated by JOIDES (Joint Oceanographic Institutions for Deep Earth Sampling). In 2003 this program was retooled as the Integrated Ocean Drilling Program (IODP) that now involves 22 countries, including the United States, Japan, and the European Union.[12] IODP has a new drill ship, the Japanese *Chikyu*, and an annual budget of $160 million! IODP's goals include elucidating the history of global climate change and discovering new energy resources and microbes. The ocean covers most of Earth's surface, contains half the planet's biota, and controls our climate. Thus the story of the ocean's past is truly the story of Earth's past. Using information about the causes and behavior of such phenomena as ice ages and plate tectonics, paleoceanographers hope to predict the future of our ocean and planet. This goal is of more than academic interest. Humans have greatly accelerated the transport rates of some materials into the atmosphere and ocean. These changes are so profound that they have arguably launched planet Earth into a new geological epoch, dubbed the Anthropocene.[13] It is critical to our own continued existence on this planet that we predict the effects of our own actions so we can take appropriate actions to protect our home, planet Earth.

---

[12] http://www.iodp-usio.org

[13] Geologists have deemed it necessary to recognize this new epoch because sediments now accumulating on the seafloor are chemically distinct from those whose origins predate human impacts on the crustal-ocean-atmosphere factory.

CHAPTER

The Waters of the Sea 2

All figures are available on the companion website in color (if applicable).

2.1 INTRODUCTION

What is the most abundant substance in the ocean? Water! Not only does water constitute approximately 97 percent of the mass of seawater, but it has some very unusual and important physical characteristics. Because water has a relatively high boiling point, it occurs mostly in the liquid phase. In fact, water is the most common liquid on our planet. Water is essential for life processes largely because of its unique ability to dissolve at least a little bit of virtually every substance. Water is also important because it plays a major role in controlling the distribution of heat on the planet. As water moves through the global hydrological cycle, it transports solutes, gases, and particles, including organisms. In this chapter, the physical and chemical features of water are discussed along with the processes by which this important substance is transported around our globe.

2.2 THE HYDROLOGICAL CYCLE

Among the planets of our solar system, Earth is unique in its great abundance of *free water*.[1] On Earth's surface, most of this water currently resides in the oceans. The origin of this water is still a matter of debate. The favored hypothesis is that most came from the degassing of the planet's interior during the early stages of Earth's formation. Other potential sources that could still be supplying water include (1) radiogenic processes within Earth's mantle followed by volcanic emission and (2) vaporization from small water-rich comets or asteroids as they enter the upper atmosphere. The latter was first observed in the 1980s from satellite imagery. The comets appear to be striking at a rate

[1] Geochemists distinguish free water from bound water based on the degree to which the water is physically or chemically associated with particles, such as mineral or organic surfaces. Examples of binding forces are Van der Waals interactions and hydrogen bonding. They cause bound water to exhibit physical characteristics that are markedly different from those found in free water.

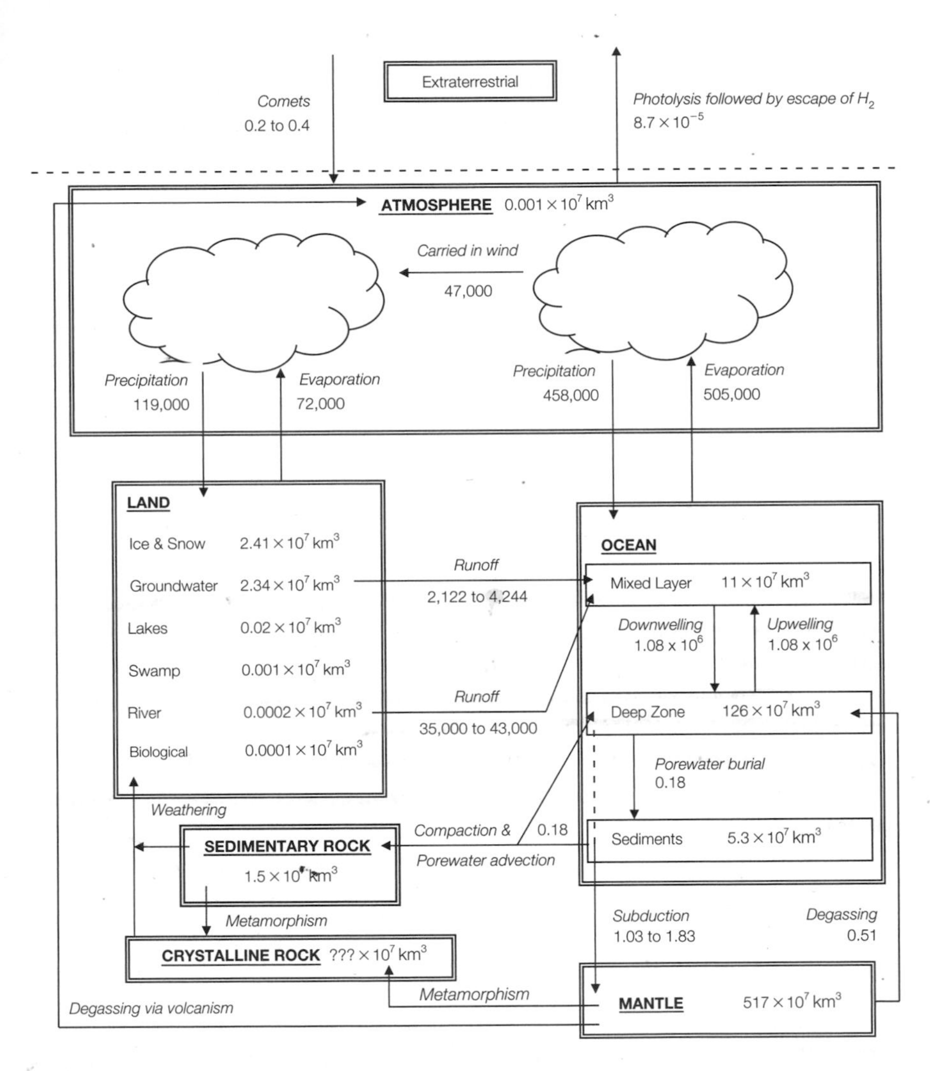

FIGURE 2.1

The global hydrological cycle. Rates are in units of km^3/y and reservoir volumes in km^3. Note that global estimates of rates and reservoirs are still a matter of uncertainty leading to the ranges reported in the figure. *Sources*: (1) Gleick, P. M. (1993). *Water in Crisis*. Oxford University Press, p. 14. (2) Frank, L. A. *Small Comets and Our Origins*. University of Iowa. http://sdrc.lib.uiowa.edu/preslectures/frank99/. (3) Bounama, C., *et al*. (2001). *Hydrology and Earth System Sciences* 5(4), 569–575. (4) Jarrard, R. D. (2003). *Geochemistry, Geophysics Geosystems* 4(5), 15. (5) Burnett, W. C., *et al*. (2003). *Biogeochemistry* 66, 3–33.

of approximately 5 to 30 per minute with each one carrying 20 to 40 tons of water into Earth's upper atmosphere.

As shown in Figure 2.1, the free water on Earth's surface is now transported between the land, atmosphere, ocean, and mantle through a global hydrological cycle. From the perspective of the ocean, water is supplied from direct precipitation, river runoff, groundwater seepage, and mantle degassing. If the volume of the ocean remains constant over time, these inputs must be balanced by an equal amount of output. Water is largely removed from the ocean by evaporation. A small amount is buried as part of the sediments that accumulate on the seafloor. Some of this buried water is subducted into the mantle, where it can be returned to the atmosphere by various geological means, including subaerial volcanism and terrestrial weathering.

Ninety percent of the water that evaporates from the ocean is returned in the form of rainfall. The rest is transported over land, where it is rained out onto the continents. River runoff and groundwater seeps carry this missing 10 percent back to the sea. At the rates that precipitation and runoff (including seeps) deliver water, it would take 3800 y to cycle the amount of water in the ocean through the atmosphere and back into the sea. This is probably a good estimate of the residence time of water in the ocean, assuming the cometary source of water is small and that the ocean's volume is in a steady state. Because rivers and groundwater are the major transport agents of dissolved and solid materials into the ocean, the turnover time of the marine reservoir of water with respect to these processes (30,000 y) is a more geochemically interesting measure than the residence time. In comparison, one mixing cycle of the ocean is approximately 1000 y.

Planet Earth acquired an ocean early in its history, probably by 3.8 billion years before present. Most of the water is thought to have been released during the process of differentiation in which density-driven convection and cooling caused the still-molten planet to separate into layers of decreasing density, i.e., core, mantle, crust, and atmosphere. Once the crust had cooled sufficiently, gaseous water condensed to form a permanent ocean.

Most depictions of the hydrological cycle, such as Figure 2.1, indicate that on time scales experienced by humans, the volume of the ocean remains constant. The most recent significant changes in ocean volume occurred during the Ice Ages of the Pleistocene Epoch. During the last Ice Age, which ended 18,000 y ago, 4.2×10^7 km^3 of seawater was transformed into glacial ice, reducing the ocean's volume by 3% and lowering sea level about 120 m below that of present day.

On longer time scales, continuing mantle and extraterrestrial processes will likely cause shifts in the sizes of the reservoirs. In the case of the former, subduction into the mantle is large enough to be causing a net loss. As the planet continues to cool, this rate will diminish. On the other hand, as the Sun's luminosity increases, the rate of photodissociation of atmospheric water into H_2 and O_2, followed by escape of H_2 to outer space, will increase. In a billion years or so, this process will have stripped all the surface water from Earth. On time scales of greater import to humans is a predicted intensification of the global hydrological cycle associated with global climate change, some of which is natural and some of which is driven by human activities. This is predicted to lead to an increase in the frequency and intensity of floods and droughts that could then alter reservoir sizes in the hydrological cycle, at least regionally. Some evidence already exists for an increase in global runoff rates over the last century.

2.3 WATER: A PHYSICALLY REMARKABLE LIQUID

Water is an unusual liquid. For example, solids tend to be denser than their liquid phase, whereas ice floats in its liquid. Another oceanographically relevant behavior is that water is nearly incompressible. This causes seawater at great depths to have nearly the same density and viscosity as surface seawater of matching temperature and salinity. Water also has a relatively high heat capacity, making it a large heat reservoir that effectively moderates weather and climate. In comparison to other hydrides of the group VI elements, the hydride of oxygen, H_2O, has a relatively high boiling (100°C) and freezing point (0°C). Given the temperature range on Earth's surface, water is commonly found in its liquid form, making it the most common naturally occurring liquid. Another interesting characteristic of liquid water is its high surface tension. This is exploited by aquatic insects to keep them atop water's surface. All of these unusual behaviors of water can be traced back to its tendency to form hydrogen bonds between adjacent water molecules, which is in turn a consequence of water's polar intramolecular bonds.

2.3.1 The Molecular Structure of Water

The molecular structure of water is shown in Figure 2.2. Each atom of hydrogen is covalently bonded to a central oxygen atom, with two electrons shared between the atoms. This sharing is not equal because the eight protons in the nucleus of the oxygen atom exert a stronger force of electrostatic attraction than does the single proton in the hydrogen nucleus. The magnitude of the force of this attraction (F in Newtons [N]) between oppositely charged particles is given by Coulomb's law:

$$F = k\frac{q_1 q_2}{r^2} \tag{2.1}$$

where q_1 is the negative charge on an electron, q_2 is the positive charge on a proton, both 1.602×10^{-12} coulombs (C), r is the distance of separation (in meters) between the charges, and k is a constant (8.99×10^9 N m^{2}C^2).

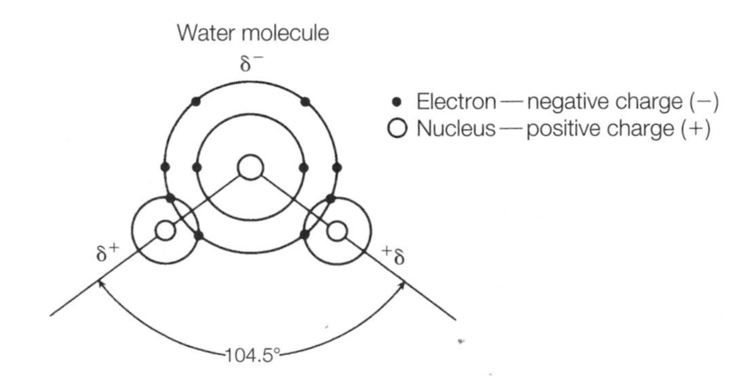

FIGURE 2.2

The Lewis structure and molecular geometry of the water molecule.

Because of the stronger pulling power of the larger atom, the bonding electrons spend more time closer to the oxygen atom. This unequal sharing of electrons is referred to as a *polar covalent bond*. It is characterized by a small net positive charge at the hydrogen end of the molecule and a small net negative charge at the oxygen end. Since these net charges are significantly weaker than those associated with ions and ionic bonds, they are represented by the symbols δ^- and δ^+.

In a water molecule, the oxygen atom shares its valence electrons with two hydrogen atoms. This electron sharing causes the oxygen atom to have four pairs of valence electrons. Two of the pairs form the polar covalent bonds found between oxygen and each hydrogen atom. The other two pairs are nonbonding. If all four pairs were distributed equally through three-dimensional space, the water molecule would exhibit tetrahedral geometry with bonding angles of 109.5°. This is not observed because the two electrons in the nonbonding pairs exert a net repulsive force that reduces the bonding angle between the H–O–H bonds to 104.5°. The two nonbonding pairs also contribute to the small net negative charge that is present on the oxygen end of the water molecule.

2.3.2 The Phases of Water

Water is one of the few substances on the planet that naturally occurs in three phases. Gaseous water is usually referred to as *steam* or water vapor. This phase is characterized by a relatively random arrangement of molecules. Like any gas, a quantity of steam has no definite shape or size. For example, one can put some gas in a balloon and then change the size and shape of the gas just by manipulating the size and shape of the balloon. Some gases, such as steam and oxygen (O_2), are composed of molecules, while others, such as the noble gases, are composed of separate atoms. In the gas phase, these particles of matter are less tightly packed together than in either the liquid or solid phases. The relative compactness of the phases of matter is shown in Figure 2.3.

The degree of compactness can be expressed as the density of a substance, which is defined as

$$\text{Density} = \frac{\text{Mass}}{\text{Volume}} \tag{2.2}$$

The SI unit of density is kg/m^3. Oceanographers more commonly use units of g/cm^3 and kg/L. The density of pure liquid water at 4°C is exactly $1\,g/cm^3$. Thus, a cube of liquid water, measuring exactly 1 cm on all sides, has a mass of exactly 1 g. This is how the unit of a gram was originally defined. Density is an intrinsic property of matter. It remains constant regardless of the amount of substance being measured. For example, at 4°C both 1000 kg and 10 mg of pure water have a density of exactly $1\,g/cm^3$. The density of a substance gives important information on its behavior. For example, oil floats on liquid water because oil has a lower density than water. A rock will sink in liquid water because the rock has the higher density.

The liquid phase of matter is denser than the gaseous phase and has a more orderly arrangement of particles. A liquid has a definite volume, but no definite shape. So a cup of liquid water can take on the shape of its container, whether it be a cylinder or a box. Water in the solid phase is referred to as *ice*. Solids possess the most orderly arrangement

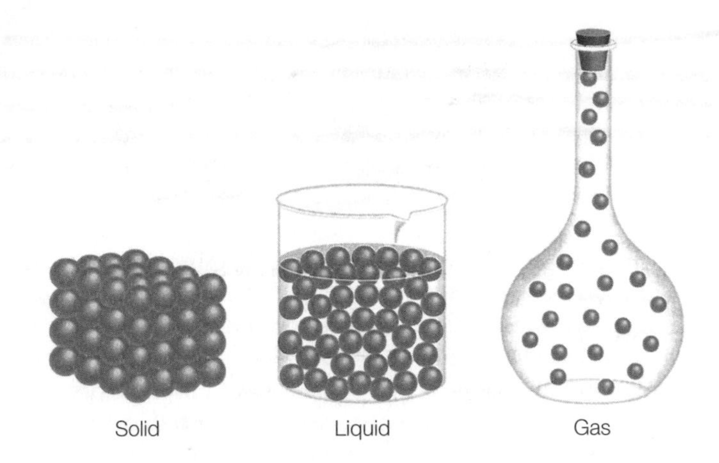

FIGURE 2.3

Particle distributions in the solid, liquid, and gaseous phases of matter. *Source*: From Chang, R. (1994). *Chemistry*, 5th ed. McGraw-Hill, Inc, 994 pp.

of particles. As shown in Figure 2.4, which uses a grain of sodium chloride salt as an example, crystalline solids possess such an orderly arrangement that the positions of the particles can be predicted over long distances. Because of this long-range order, solids are mechanically rigid and thus have a size and shape that is independent of any container.

The dimensions of any given chunk of crystalline solid are determined by the environmental conditions under which it solidifies or is mechanically fractured. In the case of table salt, an average grain ($0.1\ mm^3$) contains about 10^{13} atoms of Na and Cl. Thus, it is not possible to write one molecular formula that describes all grains of crystalline sodium chloride. Instead, chemists use an empirical formula that indicates the combining ratios of the atoms. For crystalline sodium chloride, this empirical formula is NaCl(s).

If the pressure on a substance is kept constant, its phase can be changed simply by adding or removing heat. For water, specific names are given to each phase change. The transition from solid to liquid state is termed *melting* and its reverse is *freezing*. If the water temperature is held at 0°C in a closed container held at 1 atm pressure,[2] the two phases will coexist and interconvert as represented by the following equation:

$$H_2O(s) \rightleftharpoons H_2O(l) \quad (2.3)$$

The two phases are said to be in equilibrium when the rate at which water molecules entering the solid state is exactly matched by rate entering the liquid state. The temperature at which this occurs is called the *melting point*, or *freezing point*, of water. Note that true phase changes are not considered to be chemical reactions as no intramolecular bonds are broken or formed.

[2] SI units for atmospheres are pascals where 1 atm = 101325 Pa = 1.01325 bar. Most oceanographers use bars when referring to pressures at depth below the sea surface.

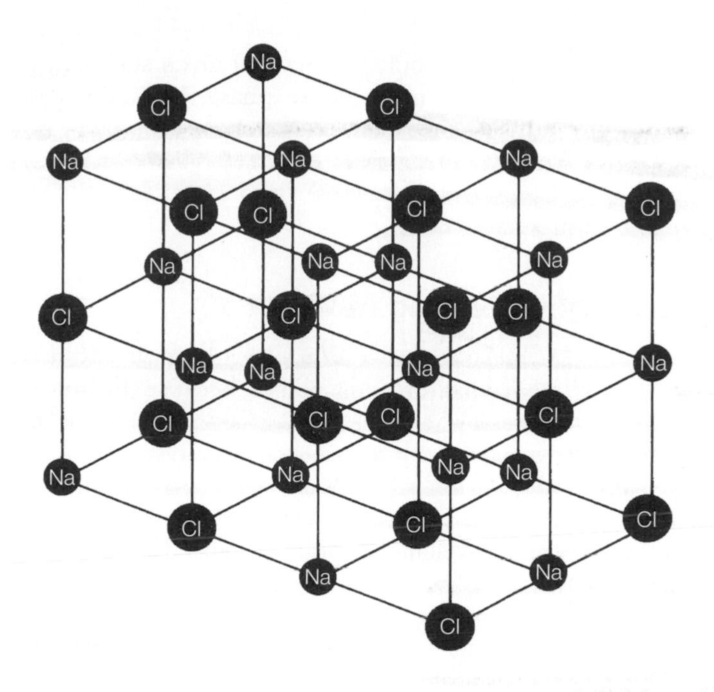

FIGURE 2.4

Crystal lattice of NaCl.

The transition from the liquid to the gaseous state is called *evaporation* or *vaporization*. The reverse is referred to as *condensation* or, in terms of rainfall, *precipitation*. If heated to 100°C in a closed container at 1 atm pressure, the two phases of water will coexist in the equilibrium given in Eq. 2.4.

$$H_2O(l) \rightleftharpoons H_2O(g) \tag{2.4}$$

This temperature is called the *normal boiling point* of water. If the container were to be opened, some of the gas molecules would escape. To replace the missing water, the phase change represented by Eq. 2.4 would be driven toward the products until all of the liquid water evaporated. The direct transition from the solid to the gaseous phase is termed *sublimation*. Ice will sublime under arid conditions, especially in polar climates.

Heat transfer causes phase transitions by changing the average *kinetic energy* of the particles.[3] When heated, particles move faster and, if unconfined, farther apart. In so doing, thermal energy (heat) is transformed into kinetic energy. By driving the particles apart, the density of a substance is lowered. When heat is removed from a substance, the particles slow down. In this lower energy state, they come closer together, causing an increase in density.

[3] The generic term *particle* is used to refer to either atoms or molecules.

From this discussion, we would predict that given sufficient cooling, a liquid should be transformed into a solid, more dense phase. Why, then, does ice float in liquid water? Some force must keep the water molecules far enough apart in ice so as to cause its density to be lower than that of liquid water. It is somewhat ironic that the most abundant and important of liquids on our planet is the only one to exhibit this anomalous density behavior.

2.3.3 Hydrogen Bonding in Water

The force that influences the orientation of water molecules in ice is called *hydrogen bonding*. This is somewhat of a misnomer because hydrogen bonding is an *intermolecular force* rather than a true chemical bond, which is an intramolecular force. Hydrogen bonding is caused by the electrostatic attraction of the negatively charged end of a water molecule for the positively charged end of a neighboring molecule. As shown in Figure 2.5, this attraction causes the unshared electron pairs on the oxygen end of each water molecule to orient themselves toward the hydrogen atoms of neighboring water molecules. The strength of a hydrogen "bond" is on the order of 5 kcal/mol. In comparison, the energy of a typical single covalent bond ranges from 50 to 110 kcal/mol (depending on the molecular setting). So hydrogen bonds are weaker than true intramolecular bonds.

In ice, all of the water molecules have formed the maximum number of hydrogen bonds, which is four per molecule. This creates the hexagonal pattern illustrated in Figure 2.6.

As shown in Figure 2.7, liquid water also contains some degree of hydrogen bonding. Although the details of the structure of liquid water are not well understood, it is thought to be composed of transitory clusters of four to five molecules held together by multiple hydrogen bonds. Since the molecules have a high kinetic energy in the

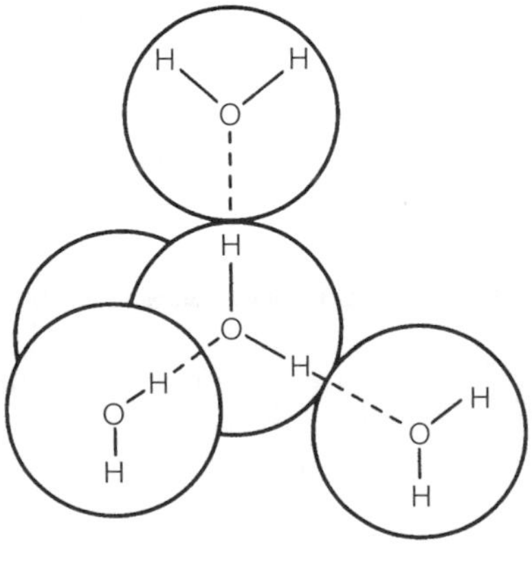

FIGURE 2.5

Hydrogen bonding between water molecules. Hydrogen bonds are represented by dashed lines.

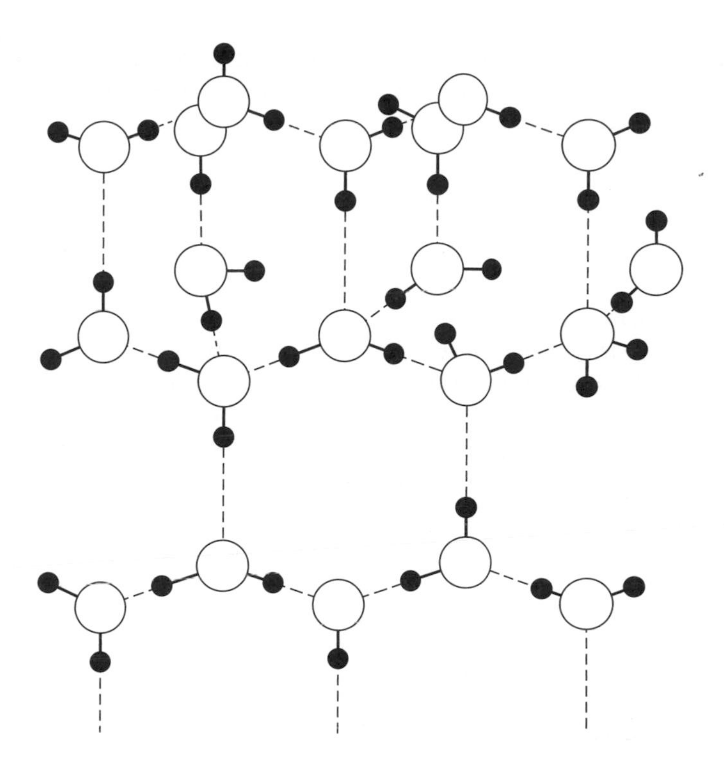

FIGURE 2.6

The crystalline structure of fully hydrogen-bonded water in ice. Hydrogen bonds are represented by dashed lines.

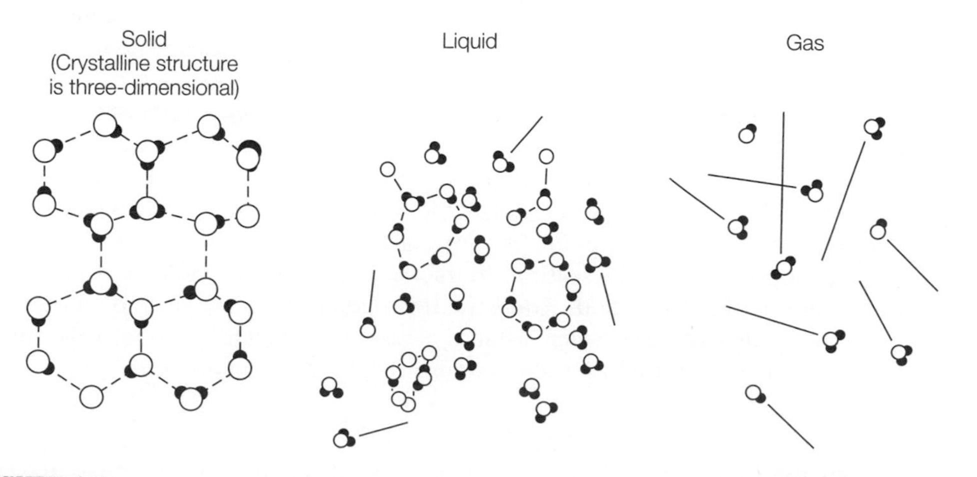

FIGURE 2.7

A comparison of hydrogen bonding in the solid, liquid, and gaseous phases of water. *Source*: From Thurman, H. V., (1988). *Introductory Oceanography*, 5th ed., Merrill Publishing Company, p. 150.

liquid state, these intermolecular "bonds," and, hence, clusters, are rapidly broken and reformed.[4] This results in regions of varying density in liquid water, with some greater than that found in ice.

2.3.4 The Effect of Hydrogen Bonding on the Physical Behavior of Water

Hydrogen bonding is not restricted to water. A few other hydrides, such as NH_3 and HF, have polar covalent bonds with large enough charge differences to support hydrogen bonding. But these substances are gases at the temperatures and pressures usually encountered on this planet. Therefore hydrogen bonding is of little importance to their environmental chemistry, except when they are dissolved in water. Hydrogen bonding also occurs between biochemicals, such as proteins and DNA, and helps define their three-dimensional molecular structure, which in turn affects their chemical stability and reactivity. In water, hydrogen bonding plays a large role in determining a variety of unusual physical and chemical properties as summarized in Table 2.1 and discussed next.

First, water has a relatively high boiling and freezing point. As illustrated in Figure 2.8, extrapolation of the molecular weight trends established by the Group VIA hydrides suggests that the boiling and freezing points of H_2O should be −68°C and −90°C, respectively. Instead, water has a boiling point of 100°C. A higher temperature is needed to give water enough kinetic energy to overcome the hydrogen bonds and thus enable the water molecules to separate and escape into the gas phase. The anomalously high freezing point (0°C) is caused by the formation of hydrogen bonds as water cools. This extra force helps organize the molecules into the long-range order necessary to produce a solid. Thus, less heat removal is required to freeze water.

Second, water has a relatively high *heat capacity*, which is a measure of how much heat can be absorbed per unit of temperature increase. As shown in Figure 2.9, the temperature of 1 g of liquid water is increased by 1°C for every calorie of heat energy added. In other words, the heat capacity of liquid water is $1\,cal\,°C^{-1}\,g^{-1}$. The heat capacities of ice and steam are 0.51 and $0.48\,cal\,°C^{-1}\,g^{-1}$, respectively.[5] The cause of the relatively high heat capacity of liquid water is similar to that which produces the anomalously high boiling point. Because of the presence of hydrogen bonds, heat that would otherwise go to increasing the motion of the water molecules instead goes into breaking the hydrogen bonds. Once the hydrogen bonds have been disrupted, the added heat energy is expressed solely as an increase in molecular motion. It is this increased motion that is measured as a temperature rise by a thermometer.

---

[4] Experimental evidence and theoretical modeling suggest that these clusters involve several to several hundred water molecules.

[5] These heat capacities vary slightly with temperature and pressure. For example, the heat capacity of liquid water increases from $1.000\,cal\,°C^{-1}\,g^{-1}$ at 14°C to $1.007\,cal\,°C^{-1}\,g^{-1}$ at 100°C under 1 atm pressure.

Table 2.1 Notable Physical Properties of Liquid Water.

Property	Comparison with Other Substances	Importance in Physical-Biological Environment
Heat capacity	Highest of all solids and liquids except liquid NH_3	Prevents extreme ranges in temperature Heat transfer by water movements is very large Tends to maintain uniform body temperatures
Latent heat of fusion	Highest except NH_3	Thermostatic effect at freezing point owing to absorption or release of latent heat
Latent heat of evaporation	Highest of all substances	Large latent heat of evaporation extremely important for heat and water transfer in atmosphere
Thermal expansion	Temperature of maximum density decreases with increasing salinity; for pure water it is at 4°C	Fresh water and dilute seawater have their maximum density at temperatures above the freezing point; this property plays an important part in controlling temperature distribution and vertical circulation in lakes
Surface tension	Highest of all liquids	Important in physiology of the cell Controls certain surface phenomena and drop formation and behavior
Dissolving power	In general dissolves more substances and in greater quantities than any other liquid	Obvious implications in both physical and biological phenomena
Dielectric constant	Pure water has the highest of all liquids	Of utmost importance in behavior of inorganic dissolved substances because of resulting high dissociation
Electrolytic dissociation	Very small	A neutral substance, yet contains both H^+ and OH^- ions
Transparency	Relatively great	Absorption of radiant energy is large in infrared and ultraviolet; in visible portion of energy spectrum there is relatively little selective absorption, hence is "colorless"; characteristic absorption important in physical and biological phenomena
Conduction of heat	Highest of all liquids	Although important on small scale, as in living cells, the molecular processes are far outweighed by eddy diffusion

Source: From Sverdrup, H. U., et al. (1941). The Ocean. *Prentice Hall, p. 48.*

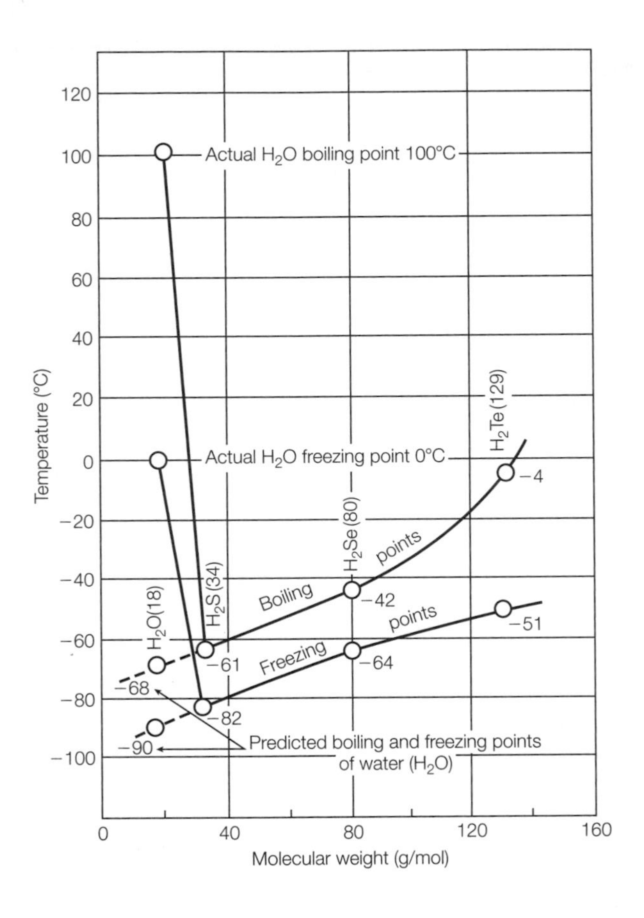

FIGURE 2.8

Molecular weight versus freezing- and boiling-point temperatures of the group VIA hydrides. *Source*: From Thurman, H. V., (1988). *Introductory Oceanography*, 5th ed., Merrill Publishing Company, p. 148.

You have probably experienced the high heat capacity of water for yourself during your last trip to the beach. Standing on the water's edge on a hot, sunny day, you can have one foot in the pleasantly cool water of the ocean and the other foot, just a few inches away, in the painfully hot sand. How can this be? Both the sand and the ocean have received the same amount of solar energy. The explanation, of course, is that the water has absorbed the solar energy without experiencing as large a rise in temperature as the sand. The heat capacity of some common materials is given in Table 2.2.[6]

[6] Because of its high heat capacity, ammonia is used as the working fluid in Ocean Thermal Energy Conversion (OTEC) units. See http://www.nrel.gov/otec/for more information.

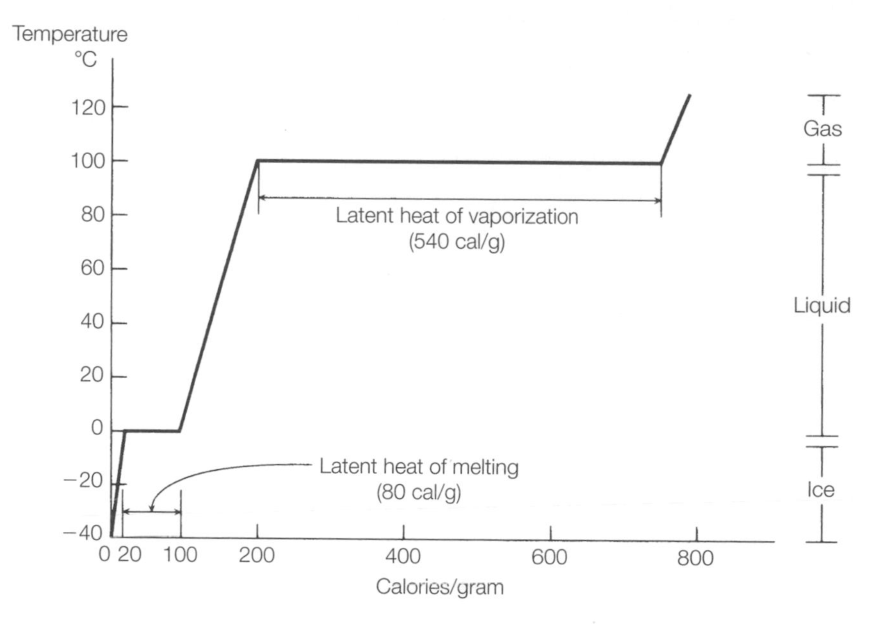

FIGURE 2.9

The phase transitions of water as caused by changing heat content. Slopes of the lines indicate heat capacity. Note that the latent heats are slightly temperature dependent, i.e., the latent heat of vaporization is 540 cal/g at 100°C and 533 cal/g at 110°C.

The relatively high heat capacity of water has important consequences for climate and life on Earth. For example, seasonal changes in atmospheric temperatures are moderated at mid-latitudes via adsorption of heat by the ocean's surface waters during the summer and release of this heat during the winter. Thus, mid-latitude coastal zones experience much smaller seasonal atmospheric temperature fluctuations than occur inland.

Third, water has a relatively large *latent heat of fusion* and *latent heat of vaporization*. The former is the amount of heat required to transform 1 g of ice into liquid water or the amount of heat that must be removed to transform 1 g of liquid water into ice. The latent heat of evaporation is analogous to the latent heat of fusion, but refers to the liquid-gas phase transition. These relatively high latent heats are another consequence of hydrogen bonding. Before water can undergo these phase transitions, heat is needed to disrupt the hydrogen bonds. More heat is required for the liquid-to-gas transition than for the solid-to-liquid transition because almost all the hydrogen bonds must be broken to reach the gaseous state. As shown in Figure 2.9, water does not experience a temperature change during these phase transitions.

Fourth, hydrogen bonding causes the density of ice to be lower than that of liquid water. The process by which water freezes is illustrated in Figure 2.10. Panels above the graph depict the arrangement of water molecules during various stages of cooling. Decreasing the temperature of water by removing heat slows the water molecules, bringing them closer together. In this way, the density of water increases

Table 2.2 Heat Capacity of Common Materials.[a]

Material	Heat Capacity (cal $°C^{-1}$ g^{-1})
Air	0.23
Aluminum	0.22
Ammonia	1.13
Copper	0.09
Grain alcohol	0.23
Lead	0.03
Iron	0.20
Mercury	0.03
Quartz sand	0.22 to 0.25
Plastic	0.57
Human tissue	0.85
Steam	0.48
Ice	0.51
Liquid water	1.00

[a] *Heat capacity defined on a per unit mass basis is also called "specific heat".*

from 0.9982 g/cm^3 at 20°C to 0.9991 g/cm^3 at 15°C. This trend continues to 4°C. At this temperature, pure water reaches its maximum density, exactly 1 g/cm^3. Further cooling causes the density to decrease. At these low temperatures, molecular motion has been slowed such that hydrogen bonds form between enough molecules to create some hexagonal clusters. At 0°C, the water molecules are completely hydrogen bonded, forming the hexagonal crystal lattice that is ice. Because this arrangement is retained at temperatures below the freezing point, it must be more energetically favored than any denser arrangement of water molecules. In summary, ice floats in its liquid because the average distance between water molecules in the crystal lattice is greater than found in liquid water at temperatures greater than 4°C.

Because of the increasing number of hydrogen bonds, water expands as it freezes. You've probably experienced this expansion. Think what happens when you put an unopened can of soda in the freezer. As the water freezes, its volume increases, making the can bulge.

The anomalous density behavior of water has important consequences for the survival of aquatic organisms at mid-latitudes. As winter approaches, the surface waters of ponds and lakes cool. The ensuing increase in density causes this water to sink to the bottom of the water body. This process continues until water temperatures drop

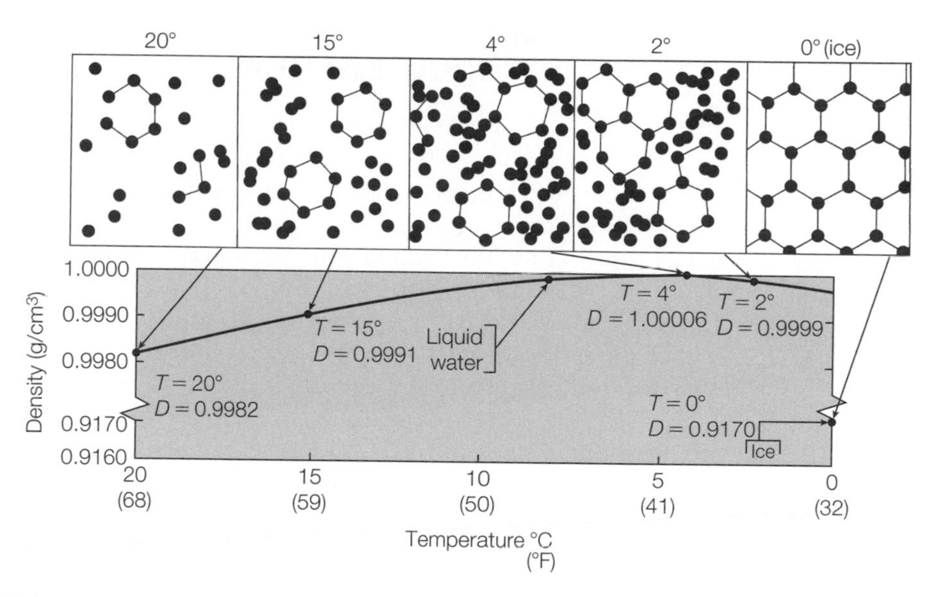

FIGURE 2.10

The influence of temperature on hydrogen bonding and water density. *Source*: From Thurman, H. V. (1988). *Introductory Oceanography*, 5th ed., Merrill Publishing Company, p. 158.

below 4°C. At lower temperatures, further cooling produces water of a lower density, so the sinking stops. If the atmospheric temperature reaches 0°C, ice will form at the surface and act as an insulating layer that isolates the underlying water from further atmospheric cooling. Thus, ponds and lakes freeze from the top down. Since very long and cold winters are required to complete this process, frogs, fish, and other creatures that hibernate in the mud are protected from freezing. In the ocean, cooling of surface seawater at polar latitudes also causes an increase in density that leads to sinking of these waters. This process is one of the forces driving deep-water circulation.

Three other physical characteristics of water are affected by the presence of hydrogen bonds. These are: (1) an unusually high *surface tension*, (2) high *viscosity*, and (3) relatively low *compressibility*. The surface tension effect results from interconnections between water molecules causing the surface of the liquid to behave as if covered by a thin skin. Because of this surface tension, a carefully filled cup of water forms a dome of the liquid above its rim. Viscosity is a measure of how much a fluid will resist changing its form when a force is exerted on it. Because hydrogen bonds act to orient neighboring water molecules, they can not flow past each other as easily as molecules in fluids without hydrogen bonds leading to a relatively high viscosity. Finally, hydrogen bonds prevent water molecules from being pushed too far together, even at high pressures. Pressure increases by approximately 1 atm (1.01 bar) for every 10 m increase in water depth. Thus, at the average depth of the ocean, approximately 4000 m, the pressure is 400 times greater than at the sea surface. Due to the low compressibility of water, this great increase in pressure causes only a slight increase in density.

2.4 WATER AS THE UNIVERSAL SOLVENT

Water is called the *universal solvent* because of its ability to dissolve at least a little of virtually every substance. Water is a particularly good solvent for substances held together by polar or ionic bonds. Indeed, the most abundant substance dissolved in seawater is an ionic solid, sodium chloride. In comparison, only small amounts of nonpolar substances, such as hydrocarbon oils, will dissolve in water.

Ionic solids are also called *salts*. Salts are composed of atoms held together by ionic bonds. These bonds are the result of electrostatic attractions between positively charged ions (*cations*) and negatively charged ions (*anions*). The force of electrostatic attraction is inversely related to the square of the distance of separation of the ions (Eq. 2.1).

When placed in water, salts dissolve because the cations and anions are electrostatically attracted to the water molecules. The cations attract the oxygen ends of the water molecules, and the anions attract the hydrogen ends. When surrounded by water molecules, the ions are too far apart to exert a significant force of attraction on each other. Thus, the ionic bond is broken and the ions are said to be dissolved or *hydrated*. Once the surface ions have become hydrated, the underlying salt ions are exposed to water and eventually become hydrated as well. This process is illustrated in Figure 2.11.

The chemical equation that describes the dissolution of NaCl is given in Eq. 2.5, where n and m equal the number of waters of hydration in direct contact with each ion.

$$NaCl(s) + (n + m)H_2O(l) \longrightarrow Na(H_2O)_n^+ + Cl(H_2O)_m^- \qquad (2.5)$$

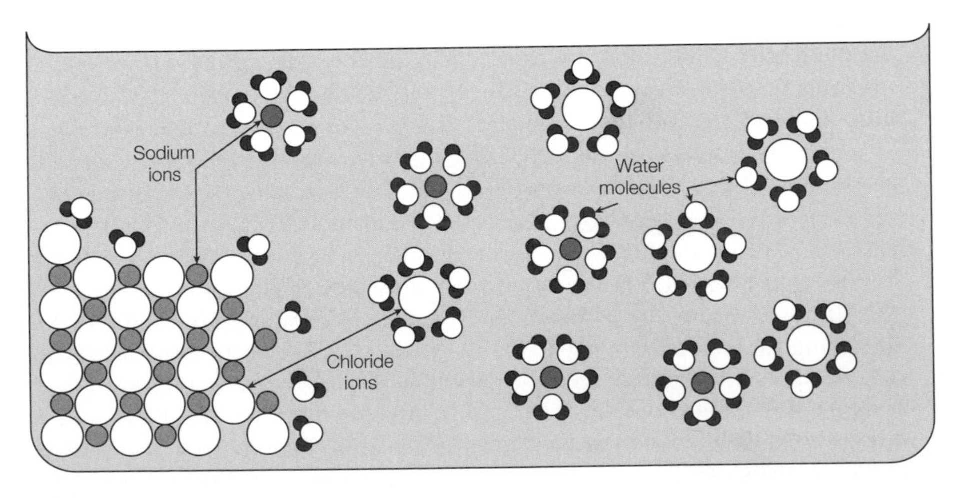

FIGURE 2.11

Dissolution of sodium chloride in water. *Source*: From Kotz, J. C., and K. F. Purcell. (1987). *Chemistry and Chemical Reactivity*, Saunders College Publishing Company, p. 85.

The number of waters of hydration is determined by an ion's charge and radius, as well as the presence of other solutes. In the case of Na^+ and Cl^-, this number ranges from four to six.

Since the only chemical bonds broken during the dissolution of salt are the ionic bonds, water is not truly a reactant in Eq. 2.5. Thus, Eq. 2.5 is more commonly written as

$$NaCl(s) \rightarrow Na^+(aq) + Cl^-(aq) \tag{2.6}$$

where the term *(aq)* indicates that the ions are hydrated, i.e., in aqueous form.

The ability of an ionic solid to dissolve depends on its lattice energy, as well as the degree to which its ions can become hydrated. The lattice energy of an ionic crystal is a measure of the strength of its three-dimensional network of bonds. If these interactions are weaker than the solute-solvent attractions, the ionic bonds will be easily disrupted by water molecules.

In the case of NaCl(s), the dissolution products are strongly favored at equilibrium. In other words, NaCl(s) is very soluble in water, dissociating into $Na^+(aq)$ and $Cl^-(aq)$. Such salts are termed strong *electrolytes*. If an electromagnetic field is applied to a solution of strong elcctrolytes, the ions will migrate, producing an electric current. The ability of a solution to conduct an electrical current increases with increasing electrolyte concentration. As described in the next chapter, this characteristic has been used to develop a very precise method for measuring the saltiness of seawater.

2.5 THE EFFECT OF SALT ON THE PHYSICAL PROPERTIES OF WATER

Adding salt to water increases the density of a solution. This effect is discussed in greater detail in the next chapter. As shown in Table 2.3, the presence of salt alters other physical properties.

For example, adding salt to water lowers the freezing point of the resulting solution. Thus, the freezing point of seawater is a function of its salt concentration, or *salinity*, as illustrated in Figure 2.12c. (A rigorous definition of salinity is given in Chapter 3.) At low enough temperatures, the water in the seawater will freeze, forming sea ice. For average seawater (salinity of 35‰), the freezing point is 1.9°C. Seawater is said to have a sliding freezing point because the formation of sea ice causes the salt content of the remaining seawater to increase and, hence, its freezing point to decline.

The temperature at which seawater reaches its maximum density also decreases with increasing salinity. Most seawater in the ocean has a salinity between 33% and 37%. At salinities greater than 26%, the freezing point of seawater is higher than the temperature at which it reaches its maximum density. Thus, seawater never undergoes the anomalous density behavior of pure water. Instead, sea ice floats because it is mostly pure water (some pockets of brine are often occluded into the crystal structure).

As shown in Figure 2.12b, the vapor pressure of seawater decreases with increasing salinity. Since more heat is required to raise the vapor pressure to atmospheric pressure, the normal boiling point of seawater increases with increasing salinity. There are few

Table 2.3 Comparison of Pure Water and Seawater Properties.

Property	Seawater, 35‰ S	Pure Water
Density (g/cm^3) at 25°C	1.02412	1.0029
Specific conductivity (ohm^{-1} cm^{-1}) at 25°C	0.0532	—
Viscosity (millipoise) at 25°C	9.02	8.90
Vapor pressure (mm Hg) at 20°C	17.4	17.34
Isothermal compressibility (vol/atm) at 0°C	46.4×10^{-6}	50.3×10^{-6}
Temperature of maximum density (°C)	−3.52	+3.98
Freezing point (°C)	−1.91	0.00
Surface tension (dyne/cm) at 25°C	72.74	71.97
Velocity of sound (m/s) at 0°C	1450	1407
Specific heat (jg^{-1} $°C^{-1}$) 17.5°C	3.898	4.182

Source: After Horne, R. A. (1969). Marine Chemistry, *John Wiley & Sons, Inc., p. 57.*

practical applications of this, because seawater is rarely close to its boiling point. The hottest seawater is found in hydrothermal systems as a result of close contact with magma. This water reaches temperatures in excess of 400°C, but the high hydrostatic pressure lowers its vapor pressure and prevents boiling. This can lead to the attainment of a *supercritical* condition in which water no longer behaves strictly as a liquid or a gas. Instead it takes on some gas-like characteristics, such as low viscosity, and some liquid-like characteristics, such as high density. The contact of supercritical seawater with hot basalt is thought to enhance the solubility of certain elements, causing elevated concentrations in the *hydrothermal fluids*. Lab simulations have also demonstrated the abiotic synthesis of amino acids in supercritical seawater from small inorganic precursor molecules, lending support to the hypothesis that life may have evolved in hydrothermal systems. At high temperatures and pressures, hydrothermal fluids and seawater can also undergo complicated phase separations that generate a vapor and brine.

The high concentrations of ions in seawater cause it to have a greater *osmotic pressure* than pure water. Because of this osmotic pressure, water molecules will spontaneously diffuse across semipermeable membranes from regions of low salt concentration to regions of higher concentration. Net diffusion ceases when the water transfer has equalized the salt concentrations across the membrane. The most common and important examples of natural semipermeable membranes are cell membranes. The salt content in the intracellular fluid of most lower order marine organisms is quite close to that of seawater. This minimizes the amount of energy needed to maintain an internal salt content different from ambient seawater. In higher order organisms, such as fish, turtles, and seabirds, excess salt is excreted through specialized organs.

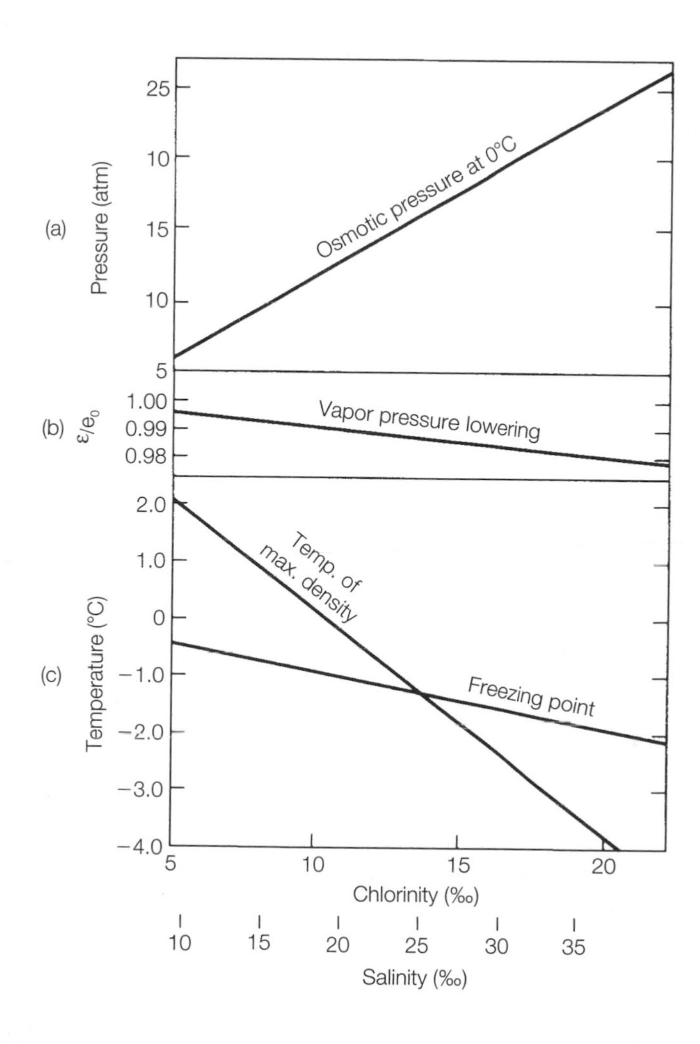

FIGURE 2.12

(a) Osmotic pressure, (b) vapor pressure relative to that of pure water, (c) freezing point and temperature of maximum density as a function of salinity. *Source*: From Sverdrup, H. U., *et al.* (1941). *The Oceans*, Prentice Hall, Inc., p. 66.

The addition of salt to water increases the viscosity of water. This is caused by the electrostatic attraction between the solutes and water. Because of slight spatial differences in salinity, the viscosity, and, hence, the speed of sound in seawater is also geographically variable. This is of practical consequence because the operation of SONAR (Sound Navigation Ranging) depends on a precise knowledge of the speed of sound in seawater. As described in Chapter 1, World War II made the need for accurate SONAR essential. This demand motivated the first detailed studies of the distribution of salinity and temperature in the ocean, which marked the beginning of the modern age of oceanography.

CHAPTER 3
Seasalt Is More Than NaCl

All figures are available on the companion website in color (if applicable).

3.1 INTRODUCTION

Water moving through the hydrological cycle interacts with rocks, soil, sediment, organisms, and the atmosphere. Along the way, water can acquire and lose solutes, gases, and particles. This makes the chemical composition of most natural waters highly variable across space and time. A notable exception to this is seawater, which has a nearly constant composition of the salt ions Cl^-, Na^+, SO_4^{2-}, Mg^{2+}, Ca^{2+}, and K^+. These ions are present in relatively high concentrations compared to the natural waters found on land in rivers, lakes, and ponds. They constitute the bulk of solutes in seawater and, hence, are referred to as *major ions*. In this chapter, some of the physical and chemical aspects of the major ions are discussed.

3.2 CLASSIFICATIONS OF SUBSTANCES IN SEAWATER

Because water is a universal solvent, at least some of virtually every element is present as a solute in seawater. As shown in Table 3.1, the most abundant substances in seawater are the *major ions* (Cl^-, Na^+, SO_4^{2-}, Mg^{2+}, Ca^{2+}, and K^+). They are present in nearly constant proportions in the open ocean because their concentrations are largely controlled by physical processes associated with water movement, such as transport by currents, mixing via turbulence, evaporation, and rainfall. These solutes are also referred to as *conservative ions*. Most of the rest of the solutes in seawater are not present in constant proportions because their concentrations are altered by chemical reactions that occur faster than the physical processes responsible for water movement. These chemicals are said to be *nonconservative*. Though most substances in seawater are nonconservative, they collectively comprise only a small fraction of the total mass of solutes and solids in the ocean.

The concentration ranges for specific elements are provided in Figure 3.1, ranked from the highest (Cl at 0.6 mol/L to Au and Bi at 10^{-13} mol/L).

The last two entries in Table 3.1 are not solutes. From a theoretical perspective, true solutes have diameters less than 0.001 μm (10 Å).True solids have a particle

Table 3.1 Classifications of Materials Present in Seawater and Approximate Concentration Ranges.

Phase	Category	Concentration Range	Examples
Solutes			
	Major Elements/Ions	>50 mM	Cl^-, Na^+
		10 to 50 mM	Mg^{2+}, SO_4^{2-}, Ca^{2+}, K^+
	Minor Elements	0.1 to 10 mM	C*, Br^-, N_2(g), B**, O_2(g)
		0.1 to 100 μM	Si, Sr^{2+}, F^-, NO_3^-, Li^+, Ar(g), PO_4^{3-}, Rb^+, I^-, Ba^{2+}
	Trace Elements	1 to 100 nM	V, As, U, H^+, Ne(g), Ni, Kr(g), Zn, Cr, Cu, He(g), Se, Sb, Al
		<1 nM	Ti, Be^{2+}, Mn, Fe, Co, Ga, Ge, Y, Zr, Cd, Xe(g)
	Dissolved Organic Matter	ng/L to mg/L	Amino acids, lipids, humic materials, organometallic compounds
	Molecular Gases	$<$mM	CH_4, CO_2, CO, N_2O, Freon
Solids			
	Organic Particulate Matter	ng/L to mg/L	Plankton biomass, fecal pellets, molts, feeding nets
	Inorganic Particulate Matter	ng/L to mg/L	Atmospheric dust, riverborne clay minerals, iron oxyhydroxides, micrometeorites
Colloids			
	Organic Colloids		
	Inorganic Colloids		iron oxyhydroxides

C = inorganic carbon including HCO_3^- and CO_3^{2-}*
*B** = inorganic boron including $B(OH)_3^0$ and $B(OH)_4^-$*
Metals and metalloids listed as elements are present in several forms in seawater. See Chapters 5 and 11 for further detail.

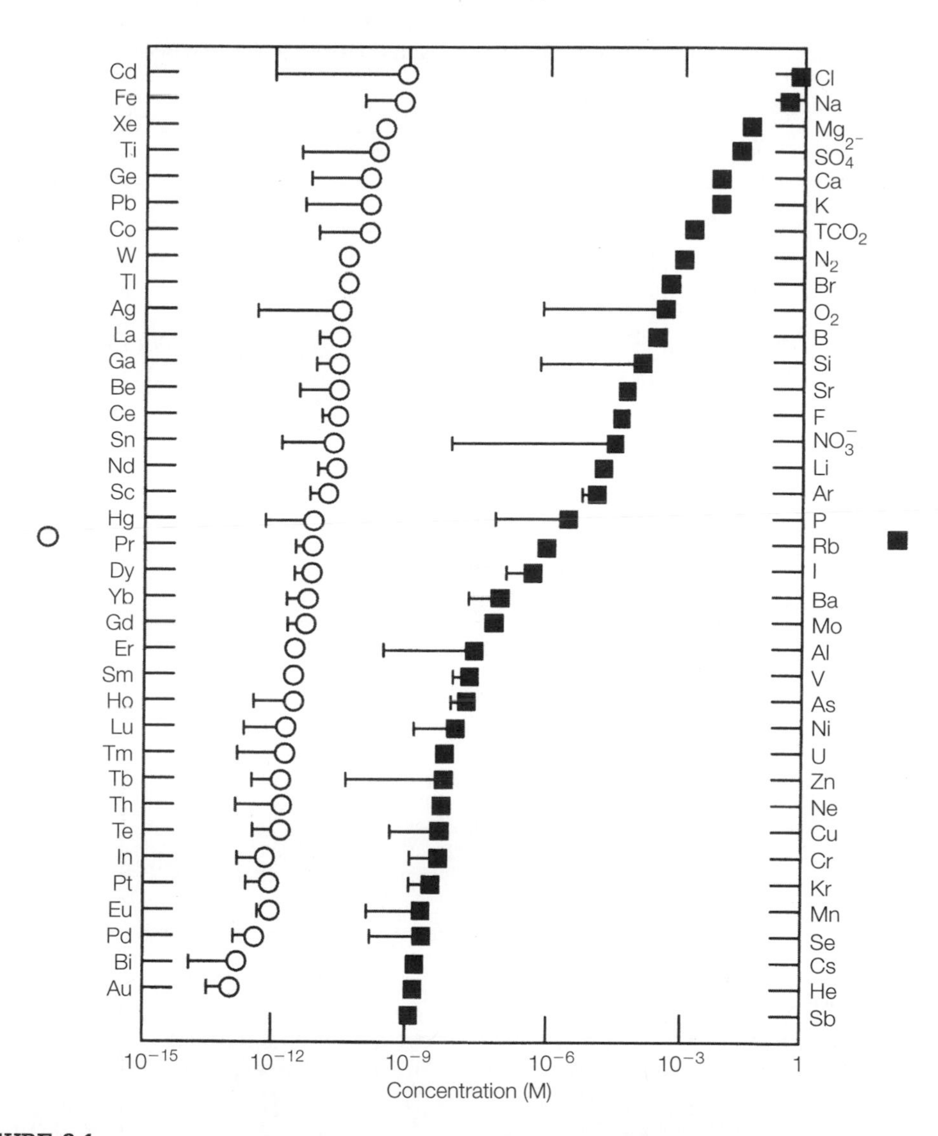

FIGURE 3.1

Concentrations of solutes in seawater. TCO_2 is the sum of all inorganic carbon compounds (CO_2, H_2CO_3, HCO_3^-, and CO_3^{2-}). *Source*: From Johnson, K. S., *et al.* (1992). *Analytical Chemistry* 64, 1065A–1075A.

diameter greater than 0.1 μm (1000 Å). An intermediate class of particles ranging from 0.001 to 10 μm (10 to 100,000 Å) are termed *colloids*.[1] In practice, solutes, colloids,

[1] Some scientists recognize the upper size limit for colloids as 0.1 μm (1000 Å) as shown in Figure 5.3 or 0.5 μm (5000 Å) as shown in Figure 22.1.

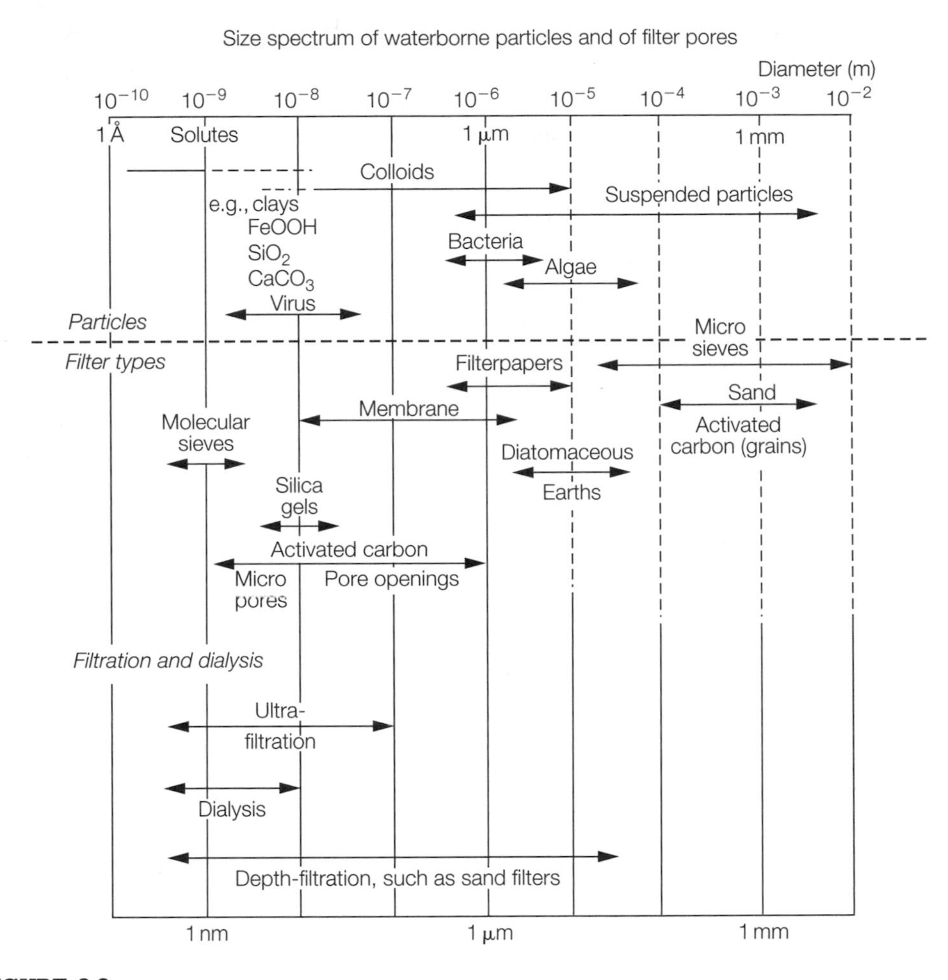

FIGURE 3.2

Size spectrum of particles and cutoffs for various separation techniques. *Source*: After Stumm, W., and J. J. Morgan (1996). *Aquatic Chemistry*, 3rd ed. Wiley Interscience, p. 821.

and solids are defined by the technique used to isolate them for measurement. Some of the techniques used involve filtration, dialysis, centrifugation, and adsorption. As shown in Figure 3.2, these techniques do not have size cutoffs that follow the theoretical definitions. In the case of filters, a variety of nominal pore widths are possible, as filters are made out of many materials, including polycarbonate, glass fibers, and silver. In practice, the substances retained by the filter are referred to as solids or *particulate matter*. The substances that pass through the filter are referred to as solutes or *dissolved matter*. These are examples of *operational definitions*, as the separation into size categories is dependent upon the conditions under which the filtration is performed. For example, during the filtration process, solids clog the filter's pores, decreasing the effective pore size and thereby causing the retention of increasingly smaller particles including colloids. Size separation of particles is also complicated by the ability of the filter and retained solids to adsorb solutes and colloids during the filtration process.

Marine colloids can be thought of as gels or sols. Their composition is highly variable. Some are inorganic substances, such as amorphous silica and polymetallic oxyhydroxides, whereas others are composed of organic compounds. Many are composites of organic and inorganic materials. Some organic colloids form via the aggregation of biopolymers exuded or excreted by marine organisms. The most abundant organic colloids are aggregations of high-molecular-weight molecules called humic substances. These are discussed in Chapter 23. Colloids have negatively charged surfaces and readily adsorb dissolved trace metals. While some colloids remain suspended in seawater, others continue to grow in size by clumping together until they become dense enough to sink. Such particles continue to adsorb metals during their transit through the water column. This process represents a major sink for dissolved trace metals. Colloids are separated from seawater via tangential flow ultrafiltration, a technique that enables large volumes of water to be filtered through membrane filters with a 0.1 μm particle size cutoff.

Fully hydrated gas molecules are solutes. Some gases, such as the noble gases, are conservative, whereas others, such as CO_2, O_2, and CH_4, are nonconservative. The chemistry of CO_2 is further complicated by its spontaneous reaction with water. In some cases, gases in seawater and the sediments can collect into bubbles. In this form, the gases are not truly dissolved.

3.3 MEASURING SALT CONTENT

The major ions constitute about 99.8% of the mass of solutes dissolved in seawater. Sodium and chloride alone account for 86%. Thus, seawater is a very salty solution. Early oceanographers invented the term *salinity* to refer to the mass of dissolved salts in a given mass of seawater. The mathematical form of this theoretical definition is:

$$S(‰) = \frac{\text{g of inorganic dissolved ions}}{\text{1 kg seawater}} \times 1000 \tag{3.1}$$

Since the mass ratio is multiplied by 1000, the units of salinity are parts per thousand, symbolized by ‰. The average salinity of seawater is 35‰, which is equivalent to a 3.5% salt solution. As shown in Figure 3.3, 99% of the seawater in the ocean has a salinity between 33‰ and 37‰. Note that the temperature of seawater is far more variable, ranging from −2° to 30°C with an average of 3.5°C. Fifty percent of the water has a salinity between 34.6‰ and 34.8‰ and a temperature between 1.3°C and 3.8°C.

As with solids, colloids, and solutes, marine scientists have developed practical definitions of salinity that are based on measurement technique. Those in current use are listed in Table 3.2 with their typical measurement precision. Each method has its advantages and disadvantages. For example, refractometers are small, lightweight, and low cost, but have low precision, so they are used for field work in which only a rough

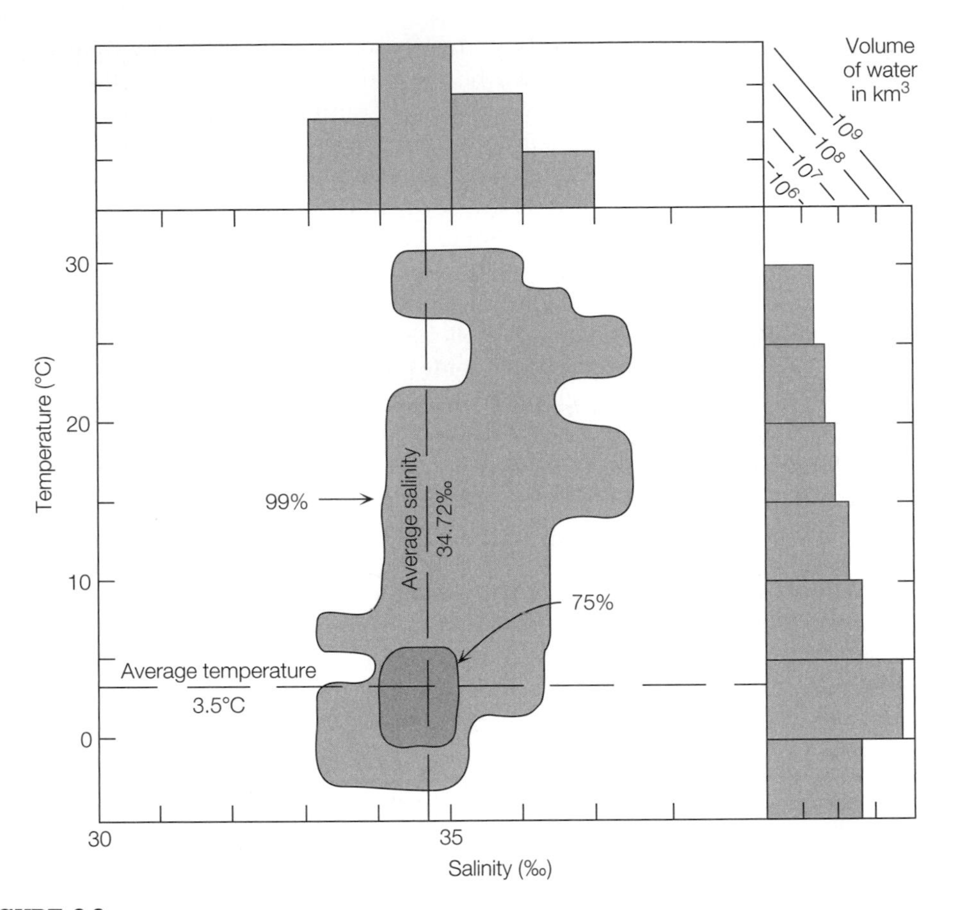

FIGURE 3.3

The temperature and salinity of 99% of ocean water are represented by points with the stippled area enclosed by the 99% contour. The shaded area represents the range of temperature and salinity of 75% of the water in the world ocean. *Source*: From Gross, M. G. (1987). *Oceanography: A View of the Earth*, 4th ed. Prentice Hall, Inc., p. 164.

Table 3.2 Precision of Methods Used to Measure Salinity.

Method	**Precision (‰)**
Field refractometer	1
Knudsen titration	0.02
Inductive salinometer	0.001

Precision = ± 1 *SD.*

estimate of salinity is required. Because of their high precision, inductive salinometers are the technique of choice when portability and cost are not an issue. By international agreement, salinity is now defined in terms of *conductivity* as measured by an inductive salinometer. As described later, this definition has its historical roots in the first techniques used to quantify the salt content of seawater, i.e., gravimetry and the titration of halides with a silver nitrate solution (Knudsen titration).

3.3.1 Gravimetric Salinity

Early oceanographers first attempted to measure the salt content of seawater by evaporating the water and weighing the residue. This turned out to be a most difficult endeavor because the required heating (480°C) causes bicarbonate (HCO_3^-) and carbonate (CO_3^{2-}) to decompose into oxides. The halides, bromide, chloride, and iodide, are partially vaporized and the organic matter is converted to $CO_2(g)$. Recognizing these limitations, an operational definition was developed that considered "salinity" to be "the total amount of solid material in grams contained in one kilogram of seawater when all carbonate has been converted to oxide, all the bromine and iodine replaced by chlorine, and all the organic material oxidized." To use this definition, the amount of halides lost was determined by comparing the results of silver nitrate titrations performed before and after heating. By agreement, the halide loss was compensated for by adding to the gravimetric mass of salts a weight of chloride equivalent to the loss of chloride, bromide, and iodide. This, along with a few other issues, makes the operational definition a few hundred milligrams less than the theoretical salinity. Despite these accommodations, the gravimetric technique is still a tedious and, hence, impractical approach for quantifying salinity.

3.3.2 Chlorinity

Because the major ions are present in nearly constant proportions, the salinity of seawater can be inferred from any of their individual concentrations. The easiest concentration to measure is that of the chloride ion, which is also the most abundant. In practice, this concentration is determined by titrating a sample of seawater with a standardized solution of silver nitrate. The reactions that take place are:

$$\begin{aligned} Ag^+(aq) + Cl^-(aq) &\rightleftharpoons AgCl(s) \\ Ag^+(aq) + Br^-(aq) &\rightleftharpoons AgBr(s) \\ Ag^+(aq) + I^-(aq) &\rightleftharpoons AgI(s) \end{aligned} \tag{3.2}$$

and the specific standard technique is referred to as a *Knudsen titration*. The silver ion reacts with all the halides, so bromine and iodine are titrated along with the chloride. Thus, the total amount of silver ion consumed can only be used to estimate the total molar concentration of halides. Recognizing this limitation, oceanographers defined an operational definition of the *chlorinity* of seawater as "the weight in grams (in vacuo) of the chlorides contained in one gram of seawater (in vacuo) when all the bromides and iodides have been replaced by chlorides." This is expressed mathematically as:

$$\text{Chlorinity} = \left[\frac{\begin{matrix}\text{Atomic weight} \\ \text{of } Cl^-\end{matrix} \times \begin{matrix}\text{moles of } Ag^+ \text{ required to} \\ \text{precipitate } Cl^-, \text{ Br, and } I^-\end{matrix}}{1000\,\text{g seawater}} \right] \times 1000 \qquad (3.3)$$

Chlorinity was first defined in 1902 and, hence, was affected by subsequent refinements in measurement of the atomic weight of chlorine. To make chlorinity independent of any such future changes, a new definition of chlorinity was adopted in 1937, e.g., "the mass of silver required to precipitate completely the halogens in 0.3285234 kg of sample seawater."

By careful comparison of the chlorinity of seawater to its gravimetric salinity, the following empirical relationship was developed and adopted internationally in 1962:

$$\text{Salinity} = 1.80655 \times \text{Chlorinity} \qquad (3.4)$$

Minor deviations from this relationship can occur due to naturally occurring variations in the ion ratios as discussed later in this chapter.

3.3.3 Conductivity

Since the invention of highly precise inductive salinometers in the 1960s, salinity has come to be defined in terms of the conductivity of seawater. The ability of ions in seawater to conduct electrical charge is directly proportional to their concentrations. The constant of proportionality for each ion is dependent on that ion's charge and ionic radius, i.e., its charge density. Because of their high concentrations, the major ions are responsible for most of the conductivity of seawater. Thus, the higher their concentrations, the greater the conductivity of the seawater sample. Other electrolytes, such as the minor ions, trace metals, nutrients, and dissolved organic matter, are present at such low concentrations that they usually make an insignificant contribution to the total conductivity of seawater.

The inductive salinometer has become the measurement technique of choice because of its ease of use, speed, and precision. The reason for this high precision lies in the inductive salinometer's ability to detect very small differences in the ratio of the conductivity of a seawater sample as compared to that of a "standard." By international agreement, this standard is prepared by diluting filtered Atlantic ocean water to produce seawater of chlorinity 19.374‰. The International Association for Physical Sciences of the Ocean (IAPSO) has designated Ocean Scientific International Ltd. (OSIL) in Wormley, UK, as the sole preparer and distributor of this standard, which is called "IAPSO Standard Seawater."

This standard seawater has since proven problematic because it is based on real seawater, whose conductivity is influenced by concentration variations in the nonconservative ions and subtle fluctuations in the ratios of the major ions. To eliminate these issues, a practical salinity scale (PSS-78) was adopted by international agreement in 1978. As a

Table 3.3 Equation Relating the Conductivity Ratio Measured by an Inductive Salinity to Practical Salinity, PSS 1978.

$$S = a_0 + a_1 R_t^{\frac{1}{2}} + a_2 R_t + a_3 R_t^{\frac{3}{2}} + a_4 R_t^2 + a_5 R_t^{\frac{5}{2}}$$

$$+ \frac{t - 15}{1 + k(t - 15)} \left(b_0 + b_1 R_t^{\frac{1}{2}} + b_2 R_t + b_3 R_t^{\frac{3}{2}} + b_4 R_t^2 + b_5 R_t^{\frac{5}{2}} \right)$$

where

S = practical salinity, PSS 1978. Conventional units for S include psu and ‰.

t = temperature in °C

R_t = the ratio of conductivities $= \frac{C(S,t,0)}{C(35,t,0)}$

C = conductivity at salinity S, temperature t, and 0 bar applied pressure

$k = +0.0162$

$a_0 = +0.0080$	$b_0 = +0.0005$
$a_1 = -0.1692$	$b_1 = -0.0056$
$a_2 = +25.3851$	$b_2 = -0.0066$
$a_3 = +14.0941$	$b_3 = -0.0375$
$a_4 = -7.0261$	$b_4 = +0.0636$
$a_5 = +2.7081$	$b_5 = -0.0144$

result, salinity is now defined as the ratio of the electrical conductivity of a seawater sample to that of a potassium chloride solution at both 15°C and 1 atm. The mass fraction of KCl in the potassium chloride solution is 32.4356×10^{-3}, which is equivalent to a chlorinity of 19.374‰. This solution has been assigned a practical salinity value of 35.000.

Recognizing that IAPSO Standard Seawater was used to calibrate inductive salinometers prior to 1978, oceanographers agreed to continue its use by backcalibrating to the primary KCl standard upon which PSS-78 is based. As a result, the conductivity measurements of seawater are now traceable to a primary standard that is well defined and, thus, can be reproducibly created. Since IAPSO water is prepared by diluting North Atlantic seawater to a chlorinity of 19.374‰, this solution also serves as a suitable standard for measuring chlorinity via the Knudsen titration. The equation used to relate the conductivity ratio measurements generated by an inductive salinometer to a PSS-78 value is given in Table 3.3. This equation is valid for salinities of 2 to 42 and temperatures of −2 to 35°C. At sea, salinity is now commonly measured in situ with CTD's (conductivity, temperature, and depth) sensors. These can be deployed at great depths in the ocean. In this case, additional factors are included to the conversion equation to address the effects of pressure on conductivity.

The PSS-78 is based on the measurement of a conductivity ratio and, hence, is technically unitless. Nevertheless, some oceanographers use a "psu" designation to represent a *practical salinity unit* and others report salinity in units of parts per thousand (‰). The latter convention has been adopted in this text. In any event, it is important to appreciate that the practical salinity is no longer directly traceable to the theoretical definition given in Eq. 3.1.

3.3.4 Density

The in situ density of seawater is a measure of the total mass of seawater in a given unit volume. In situ density increases with decreasing temperature and increasing salinity and pressure. Density is difficult to measure, so it is usually computed from salinity, temperature, and pressure (depth) data using an empirical equation, called the *equation of state of seawater*. This equation was developed by careful, direct observations based on experiments performed in the 1970s. The internationally accepted version is shown in Table 3.4. The observed effects of temperature, salinity, and pressure on density are slightly nonlinear. This behavior is mathematically described by the polynomial expansions in The Equation of State of Seawater. In practice, oceanographers do not hand calculate density, they use computer programs. Calculators of seawater density are now available online[2] and provide a quick means for obtaining density from temperature, salinity, and pressure (depth) data. Another approach is to use tabulated values (see online appendix on companion website).

The in situ density of seawater ranges from 1.02 to 1.07 g/cm^3. In other words, the in situ density of seawater exhibits a narrow range of values under the salinity, temperature, and pressures found in the ocean. Nevertheless, these small differences in density are significant drivers of physical oceanic processes, such as water transport in currents. Thus, oceanographers must focus their attention on these small density differences. To do this (and to save space and time), density is usually reported in a shorthand form called *sigma* (σ), which is defined as

$$\sigma = (\rho - 1) \times 1000 \tag{3.5}$$

where ρ is density in units of g/cm^3 (or kg/m^3). Values of $\sigma_{\text{in situ}}$ for seawater range from 20 to 70. Sigma is technically unitless but often reported in terms of the SI unit of ρ, i.e., kg/m^3. The accuracy of the observational techniques used to measure salinity, temperature, and pressure enable σ to be computed to four significant figures, i.e., $\sigma_{\text{in situ}}$ for seawater of 35‰, 0°C at 0 bar (sea level) is 28.13. Variations in the last two significant digits are physically meaningful and used to characterize physical processes, such as water transport.

Unfortunately, seawater is slightly compressible, so in situ density increases with increasing pressure. The rate at which pressure increases with increasing depth below the sea surface is nearly equal to 1 dbar per m. At 45° latitude, the actual rate is 1.01 dbar/m from 0 to 2500 m. Below 2500 m, it increases to 1.02 dbar/m. This rate increase is due to an increasing resistance to further compression. Because Earth is not a perfect sphere, the rate at which pressure increases with depth also varies with latitude. Thus, the depth at which 4500 dbar of pressure is attained is 4428 m at the equator and 23 m shallower at the poles (4405 m). Tabulated values of pressure as a

[2] Three examples are: (1) http://www.es.flinders.edu.au/~mattom/Utilities/density.html, (2) http://www.phys.ocean.dal.ca/~kelley/seawater/WaterProperties.html, and (3) http://ioc.unesco.org/Oceanteacher/oceanteacher2/01_GlobOcToday/02_CollDta/02_OcDtaFunda/02_OcMeasUnits/SWequationofstatecalculator.htm.

Table 3.4 The International Equation of State of Seawater, 1980, Definition.

I. The One Atmosphere International Equation of State of Seawater formulated to compute $\rho(S, t, 0)$. This can be converted into σ_t using Eq. 3.5

The density $(\rho, \mathrm{kgm}^{-3})$ of seawater at one standard atmosphere ($p = 0$) is to be computed from the practical salinity (S) and the temperature (T, °C) with the following equation:

$$\rho(S,t,0) = \rho_\omega + \left(8.24493 \times 10^{-1} - 4.0899 \times 10^{-3}t + 7.6438 \times 10^{-5}t^2 - 8.2468 \times 10^{-7}t^3 + 5.3875 \times 10^{-9}t^4\right)S + \left(-5.72466 \times 10^{-3} + 1.0227 \times 10^{-4}t - 1.6546 \times 10^{-6}t^2\right)S^{\frac{3}{2}} + 4.8314 \times 10^{-4}S^2$$

where ρ_ω. the pure water term, is given by:

$$\rho_\omega = 999.842594 + 6.793952 \times 10^{-2}t - 9.095290 \times 10^{-3}t^2 + 1.001685 \times 10^{-4}t^3 - 1.120083 \times 10^{-6}t^4 + 6.536332 \times 10^{-9}t^5$$

This equation is valid for practical salinity from 0 to 42 and temperatures from −2 to 40°C.

II. The International Equation of State of Seawater formulated to compute $\rho(S, t, p)$. This can be converted into σ_θ using Eq. 3.5.

The in situ density $(\rho, \mathrm{kg\,m}^{-3})$ of seawater at pressure (p, bar) is to be computed from the practical salinity (S) and the temperature (T, °C) with the following equation:

$$\rho(S,t,p) = \frac{\rho(S,t,0)}{1 - p/K(S,t,p)}$$

where $K(S,t,p)$ is the secant bulk modus given by:

$$K(S,t,p) = K(S,t,0) + Ap + Bp^2$$

where

$$K(S,t,0) = K_\omega + \left(54.6746 - 0.603159t + 1.09987 \times 10^{-2t^2} - 6.1670 \times 10^{-5}t^3\right)S + \left(7.944 \times 10^{-2} + 1.6483 \times 10^{-2}t + 5.3009 \times 10^{-4}t^2\right)^{\frac{3}{2}}$$

$$A = A_\omega + (2.2838 \times 10^{-3} - 1.0981 \times 10^{-5}t - 1.6078 \times 10^{-6}t^2)S1.91075 \times 10^{-4}pS^{\frac{3}{2}}$$

$$B = B_\omega + (-9.9348 \times 10^{-7} + 2.0816 \times 10^{-8}t + 9.1697 \times 10^{-10}t^2)S$$

and the pure water terms are given by:

$$K_\omega = 19.65221 + 148.4203t - 2.327105t^2 + 1.360477 \times 10^{-2}t^3 - 5.155288 \times 10^{-5}t^4$$

$$A_\omega = 3.239 + 1.43716 \times 10^{-3}1.16092 \times 10^{-4}t^2 - 5.77908 \times 10^{-7}t^3$$

$$B_\omega = 8.50935 \times 10^{-3} - 6.12293 \times 10^{-6}t + 5.2787 \times 10^{-8}t^2$$

Source: *From Fofonoff, N. (1985).* Journal of Geophysical Research *90 (C2), 3332–3342.*

Table 3.5 Effect of Pressure on the In Situ Density of Seawater Having a Salinity of 35‰ and Temperature of 0°C.

Depth (m)	Pressure (bar)	Sigma	% Increase in Density Relative to 0 m
0	0	28.11	
100	10	28.60	0.05%
200	21	29.08	0.10%
500	51	30.54	0.24%
1000	103	32.95	0.47%
2000	206	37.68	0.93%
3000	308	42.32	1.38%
4000	411	46.85	1.82%
5000	514	51.28	2.25%
6000	617	55.61	2.68%
7000	720	59.85	3.09%
8000	822	64.00	3.49%
9000	925	68.06	3.89%
10,000	1028	72.04	4.27%

function of depth and latitude are provided in the online appendix on the companion website. Online calculators for converting pressure to depth are now available.[3] These get heavy use because pressure sensors are now routinely employed to establish sampling locations beneath the sea surface. Alternatively, sample locations are reported in dbar beneath the sea surface. Since pressure is reported relative to the sea surface, it is assigned a value of 0 dbar at 0 m water depth. This convention is used in the equation of state of seawater (Table 3.4) where p represents the applied pressure in excess of 1 atm. Thus at the sea surface, $p = 0$ bar.

Because seawater is slightly compressible, the in-situ density of a parcel of seawater will increase as it is lowered into the deep ocean. As shown in Table 3.5, this effect is small, causing only a 4% increase if the seawater parcel is lowered from 0 to 4000 m (in the absence of any exchange of heat or salt with adjacent parcels). The hydrostatic

[3] An example can be found at http://ioc.unesco.org/Oceanteacher/oceanteacher2/01_GlobOc Today/02_CollDta/02_OcDtaFunda/02_OcMeasUnits/SWequationofstatecalculator.htm.

pressure exerted by the overlying water column is responsible for this compression. Some of the energy imparted to the water parcel by the increased hydrostatic pressure raises the kinetic energy of the water molecules. This leads to a slight increase in temperature, which in very deep waters can be as much as 0.45°C.

The force of gravity acts to spontaneously relocate surface water parcels to subsurface depths of matching in situ density. For example, cooling and evaporation at the sea surface can increase the density of a seawater parcel, causing it to spontaneously sink. As the parcel sinks, its density increases slightly because of the increase in pressure. The water stops sinking when its in situ density matches that of the surrounding water. Thus, the denser the surface water, the deeper it will dive. In water columns that are not subject to strong turbulent mixing, this spontaneous density sorting leads to the maintenance of a vertical gradient in which density increases with increasing depth as illustrated in Figure 3.4.

Water columns with large density gradients are said to exhibit ***density stratification***. When the gradient is such that density increases with increasing depth, this is termed a ***stable density stratification***. The larger the gradient, the greater the stability. Water columns with large (steep) density gradients are said to be highly stratified. When a water column exhibits a stable density gradient, its stratification can only be disturbed if another force, such as winds, tides, or geothermal heating, is imposed.

Since the degree to which seawater is compressed is largely determined by in situ pressure, the reason why some surface seawater parcels sink deeper than others is wholly due to their temperature and salinity. Oceanographers are most interested in

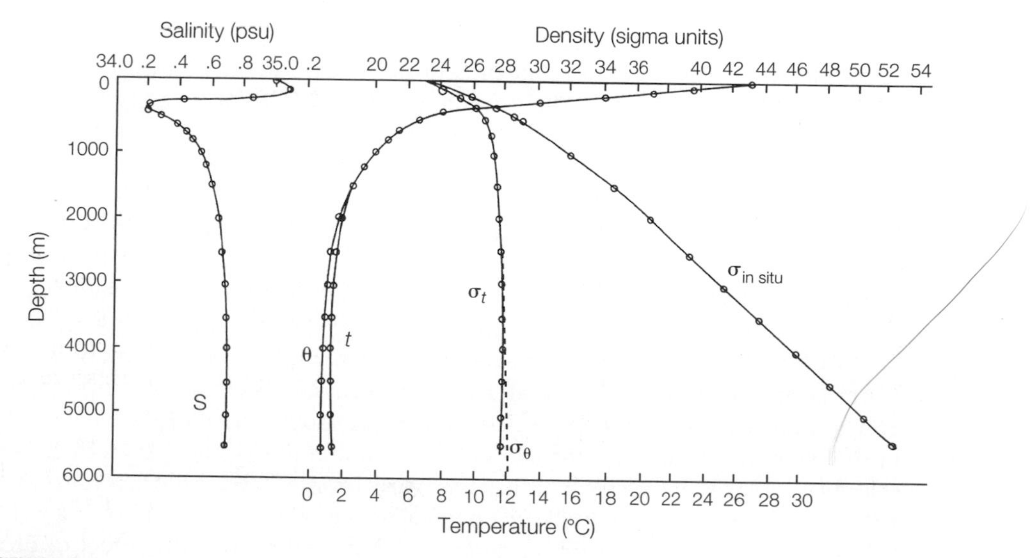

FIGURE 3.4

Vertical profiles of salinity, potential temperature (θ), in situ temperature (t), σ_θ, σ_t, and $\sigma_{\text{in situ}}$ for a station in the North Pacific at 17°N and 162°W. At 6000 m, $\sigma_\theta - \sigma_t \approx 0.05$. *Source*: After Knauss, J. A. (1997). *Introduction to Physical Oceanography*, 2nd ed. Prentice Hall, p. 31.

tracking water parcel movements and, hence, have defined a property that reflects just the temperature and salinity effects on density. This property is called *potential density* (σ_θ). It is the density that a water parcel would have if it were raised to the sea surface without undergoing any change in heat, water or salt content. Because of the effect of increasing hydrostatic pressure on temperature, an adiabatic relocation of a deep-water parcel to the sea surface should cause its temperature to decline. (An adiabatic relocation is one in which no heat is exchanged.) This operationally defined lower temperature is called *potential temperature* (θ).

Potential density is computed from measurements of salinity (S), in situ temperature (t), and pressure (p) using the equations provided in Table 3.4 to generate a value of $\rho(S, t, p)$, which is then transformed into σ format using Eq. 3.5. As shown in Figure 3.4, σ_θ exhibits a much smaller range in values (20 to 30) than $\sigma_{\text{in situ}}$. Although the vertical profile of σ_θ is stably stratified, the gradient is much smaller than that exhibited by $\sigma_{\text{in situ}}$. This illustrates that most of the vertical increase in in-situ density is caused by increased pressure.

Unfortunately, the compressibility of seawater is not just a function of pressure. It is also dependent (slightly and nonlinearly) on temperature and salinity. In very deep waters, this can cause σ_θ to exhibit a very slight decrease with increasing depth although $\sigma_{\text{in situ}}$ increases with depth. Although this effect is minor, it can impair the use of σ_θ in tracking physical processes within the deep ocean, such as the pathway of horizontal deep-water currents. To reduce this problem, oceanographers will define density with respect to a particular depth (pressure) close to that of their observations. For example, σ_4 is the density a water parcel would have if it were relocated to 4000 m (about 4000 dbar) without exchanging salinity or heat.

In shallow water depths, this pressure effect is too minimal to have a significant impact on density. In this setting, oceanographers use sigma-t (σ_t), which is the density a water parcel would have if it were brought to the sea surface while retaining its in situ temperature. At the sea surface, $\sigma_{\text{in situ}} = \sigma_t = \sigma_\theta$. σ_t is computed from measurements of salinity (S) and in situ temperature (t) using the first equation in Table 3.4 to generate a value of $\rho(S, t, 0)$, which is then transformed into σ format using Eq. 3.5. As illustrated in Figure 3.4, the vertical ranges exhibited by σ_t and σ_θ in a highly stratified stable water column are so large that any depth-specific differences are insignificant. Thus, in practice, both σ_t and σ_θ are used for quantifying density in surface waters.

As noted earlier, the effect of salinity and temperature on the compressibility of seawater is slightly nonlinear. Even at a constant pressure, salinity and temperature interact in a nonlinear way to influence density. This is shown in Figure 3.5 for σ_t. The curves in the diagram are lines of constant σ_t. As temperatures decline, the effect of increasing salinity on density increases. This is particularly pronounced at the low temperatures characteristic of the deep sea and surface polar waters. For seawater at 0°C, a rise in salinity from 35 to 36‰ increases the σ_t density 15 times more than the effect of dropping the temperature by 1°C.

As demonstrated by the polynomials in the equation of state of seawater, density is not linearly related to temperature or salinity and does not exhibit conservative behavior. One of the interesting consequences of this nonconservative behavior is that an

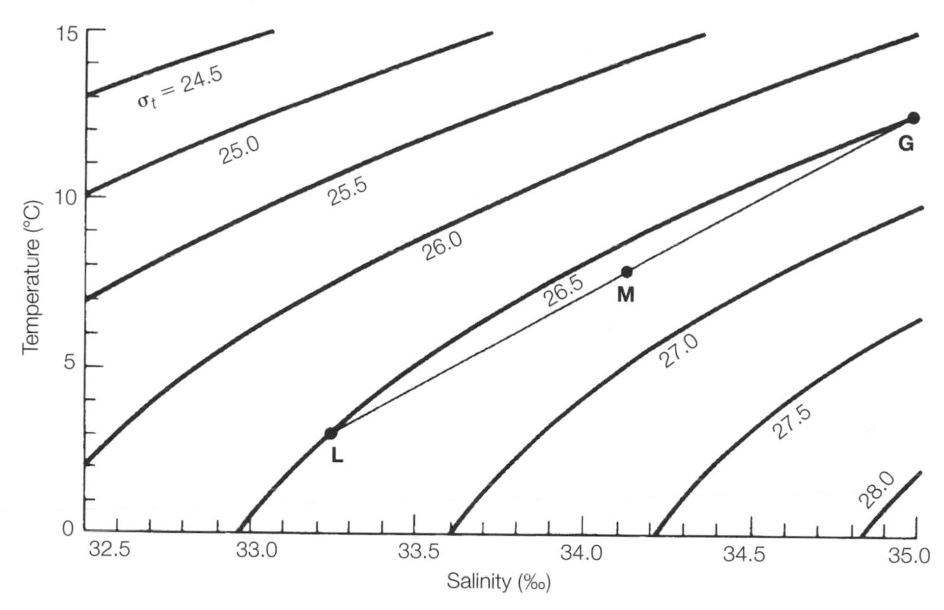

FIGURE 3.5

Sigma-t of seawater at 0 dbar pressure as a function of temperature and salinity. The straight line drawn between points L and G represents the mixing of two waters with the same σ_t but different combinations of temperature and salinity. Because of the nonlinear behavior of the equation of state of seawater, the mixture (M) has a higher density than the contributing waters and should spontaneously sink. This phenomenon is called caballing or densification. In this example, surface water carried southward by the Labrador Current (L) is mixing with the northern most edge of the Gulf Stream (G). This leads to the formation of subtropical mode water in the northwestern Atlantic Ocean. (See Figure 4.16.) *Source*: From Tolmazin, D. (1985). *Elements of Dynamic Oceanography*. Allen & Unwin, p. 13.

admixture of two masses of seawater of equal density, but different temperature and salinity, can generate a new water mass of higher density. This is referred to as ***caballing*** or ***densification***. The effect of mixing two water masses on the density of an admixture is illustrated by the straight line in Figure 3.5, which represents the mixing pathway of water parcels L and G. The admixture (M) is denser and should spontaneously sink.

The nonconservative behavior of seawater density and compressibility is caused partly by H bonding and partly by the electrostatic attractions exerted by the salt ions on their neighboring water molecules. The effect of these attractions can be estimated by trying to compute the density of seawater as a simple sum of the volumes of water and salt present in 1 kg of seawater (S = 35‰ and $t = 4^{\circ}$C). As shown in Table 3.6, the actual density, as tabulated in the online appendix on the companion website ($\sigma_t = 27.81$, so $\rho = 1.02781\ \text{g/cm}^3$), is about 1% higher than that predicted from summing the volumes of salt and water (1.0192 g/cm^3).

Table 3.6 Theoretical Calculation of the Density of 1000 g of Seawater with a Salinity 35‰ at a Temperature of 4°C.

Substance	Mass (g)	Density (g/cm³)	Computed Volume (mass/density, cm³)
Water	965.00	1.0000	965.00
Salt	35.00	2.165[a]	16.17
Seawater	1000.00	1.0192[b]	981.17[c]

[a] *For NaCl from 4 to 25°C*

[b] $\frac{\text{Mass}}{\text{Computed volume}} = \left(\frac{1000\,g}{981.17\,cm^3}\right)$

[c] $\text{Volume}_{water} + \text{Volume}_{salt} = \text{Volume}_{seawater}$

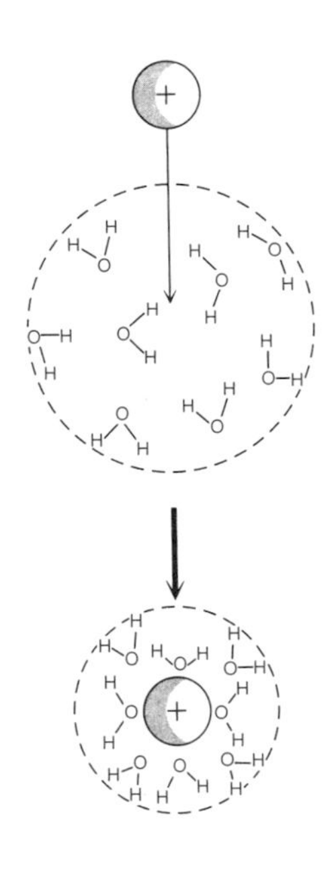

FIGURE 3.6

Electrostriction around a cation. *Source*: From Horne, R. A. (1968). *Survey of Progress in Chemistry*, vol. 4. Academic Press, p. 15.

The increased density is caused by the clustering of the polar water molecules around the salt ions as illustrated in Figure 3.6. This process is called *electrostriction*. It is enhanced at lower temperatures, increasing the nonlinear behavior of density as a function of temperature and salinity as illustrated in Figure 3.4. As we will see in Chapter 6, this is one example of several nonideal thermodynamic behaviors that seawater exhibits as a consequence of its high concentration of dissolved salts.

3.4 THE CONSERVATIVE NATURE OF THE MAJOR IONS

The ions that compose the bulk of the dissolved salts are present in constant proportions to each other and to the total salt content of seawater. Their abundances relative to salinity and chloride are given in Table 3.7 in a variety of commonly used concentration units. Also shown is the concentration of charge supplied by each ion demonstrating that seawater is electrically neutral. Note that for seawater of salinity 35.000 by the PSS-78, the mass of salt is a few hundred milligrams above 35.000 g. Likewise, the total grams of ions per gram of Cl is a bit higher than the constant of proportionality in Eq. 3.4. Both reflect the historical connection that was established between the original gravimetric definition of salinity and the PSS-78.

This constancy in relative ion concentration was first postulated by Alexander Marcet in 1819 and, hence, is known as *Marcet's Principle* or the *Rule of Constant Proportions*. Formally stated, it says that "regardless of how the salinity may vary from place to place, the ratios between the amounts of the major ions in the waters of the open ocean are nearly constant."

These proportions are constant because the rate at which water is moved through and within the ocean is much faster than any of the chemical processes that act to remove or supply the major ions. Since adding or removing water does not change the total amount of salt in the ocean, only the concentrations of the ions, and, hence, salinity, are altered. This is termed *conservative behavior*. The relatively slow biogeochemical processes that determine the ion proportions and the total amount of salt in the ocean are the subject of Part III and summarized in Chapter 21. For example, the residence time of water in the ocean is about 2700 y, the mixing time is 1000 y, and the residence time of the major ions is on the order of 10 to 100 million years. Thus, the average atom of sodium circulates around the ocean many times before being removed by various chemical means.

The types of water transport that can alter the major ion concentrations without significantly affecting their relative abundances are listed in Table 3.8. Note that these are all physical phenomena. In other words, the spatial gradients in concentrations of conservative substances are largely controlled by physical, rather than chemical, processes.

Water is added to the ocean through rainfall (precipitation) and the melting of ice. The addition of water to seawater lowers its salinity. Since the concentrations of all the solutes are decreased by the same amount, the ion ratios remain constant. Likewise, the removal of water from the ocean, through the processes of evaporation and freezing, increases the salinity of seawater without altering the ion ratios. This simple description is complicated by the presence of small amounts of salts in precipitation and ice melt, which are not necessarily present in the same ratios as seawater. Much larger amounts of salts are carried into the ocean as part of river runoff, groundwater seeps, and hydrothermal venting. These waters generally have ion ratios significantly different from those in seawater and, hence, as discussed later, cause deviations from the Rule of Constant Proportions in coastal waters and near hydrothermal vents.

Table 3.7 Composition of Seawater of Salinity = 35.000 psu and Chlorinity = 19.374‰ at 25°C and 1 atm.

	Ion	Atomic Weight (amu)	g ion per kg Seawater	g ion per Cl (‰)	Mole Ion per kg Seawater	Mole Ion per kg Water	Mole Ion per L Seawater	Equivalents of Charge per Ion	Equivalents of Charge per kg Seawater
Cations	Na^{+}	22.9898	10.7838	0.556614	0.46907	0.48617	0.48002	1	0.46907
	Mg^{2+}	24.3051	1.2837	0.066260	0.05282	0.05474	0.05405	2	0.10563
	Ca^{2+}	40.0784	0.4121	0.021270	0.01028	0.01066	0.01052	2	0.02056
	K^{+}	39.0983	0.3991	0.020600	0.01021	0.01058	0.01045	1	0.01021
	Sr^{2+}	87.6210	0.0079	0.000410	0.00009	0.00009	0.00009	2	0.00018
Anions	Cl^{-}	35.4532	19.3529	0.998910	0.54587	0.56577	0.55861	−1	−0.54587
	SO_4^{2-}	96.0632	2.7124	0.140000	0.02824	0.02926	0.02889	−2	−0.05647
	HCO_3^{-}	61.0170	0.1070	0.005524	0.00175	0.00182	0.00179	−1	−0.00175
	Br^{-}	79.9041	0.0672	0.003470	0.00084	0.00087	0.00086	−1	−0.00084
	CO_3^{2-}	60.0091	0.0161	0.000830	0.00027	0.00028	0.00027	−2	−0.00054
	$B(OH)_4^{-}$	78.8412	0.0080	0.000415	0.00010	0.00011	0.00010	−1	−0.00010
	F^{-}	18.9984	0.0013	0.000067	0.00007	0.00007	0.00007	−1	−0.00007
	OH^{-}	17.0074	0.0001	0.000007	0.00001	0.00001	0.00001	−1	−0.00001
	$B(OH)_3$	61.8338	0.0194	0.001002	0.00031	0.00033	0.00032	0	0.00000
Total			35.171	1.81538	1.1199	1.1608	1.1461		0.0000

To convert mol ion per kg seawater to units of molarity (mol ion per L seawater), multiply by the density of seawater (kg seawater per L seawater). At 25°C, 1 atm, and S = 35, density = 1.023343 g/cm³.
To convert mol ion per kg seawater to units of molality (mol ion per kg water), multiply by kg water per kg seawater. For S = 35, kg water per kg seawater = 0.96483.
Data from Millero, F. T. (1996) Chemical Oceanography. *CRC Press.*

Table 3.8 Physical Processes That Can Alter the Salinity of Seawater without Significantly Affecting Ion Ratios.

Evaporation
Precipitation
Freezing
Thawing
Molecular diffusion of ions between water masses of different salinity
Turbulent mixing between water masses of different salinity
Water-mass advection

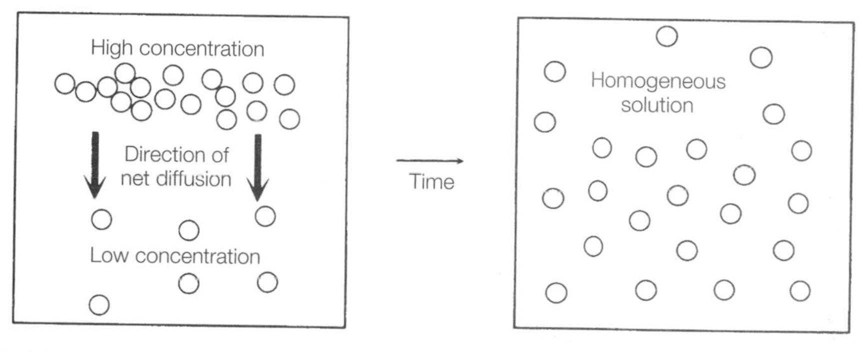

FIGURE 3.7
The process of molecular diffusion.

Table 3.8 contains one physical process that involves the movement of ions rather than water (i.e., *molecular diffusion*). As mentioned in Chapter 2, all particles experience some degree of motion. In the absence of any other influences, the random motion of solutes will eventually lead them to become evenly dispersed throughout a solution. As shown in Figure 3.7, molecular diffusion leads to the destruction of concentration gradients. This requires a net transport of solutes from regions of higher concentration to regions of lower concentration. Once the solution concentration is homogenous, the random motion of the solute particles continues, but results in no further net molecular diffusion and, hence, no further concentration change over time.

The net transfer of solutes along a concentration gradient via molecular diffusion is an example of a mass flux. As illustrated in Figure 3.8, a mass flux of solutes can be thought of as the amount of particles, expresses in mass units or moles, that move through a unit area (1 m^2) in a unit time (1 s) giving SI units of kg m^{-2} s^{-1} or mol m^{-2} s^{-1}.

The flux caused by molecular diffusion follows Fick's first law, which states that the diffusive flux (F_c^{diff}) is directly proportional to the concentration gradient of the chemical undergoing net diffusive transport ($d[C]_z/dz$). Thus, the larger the concentration difference, or gradient, the larger the flux. For a diffusive flux occurring in the vertical

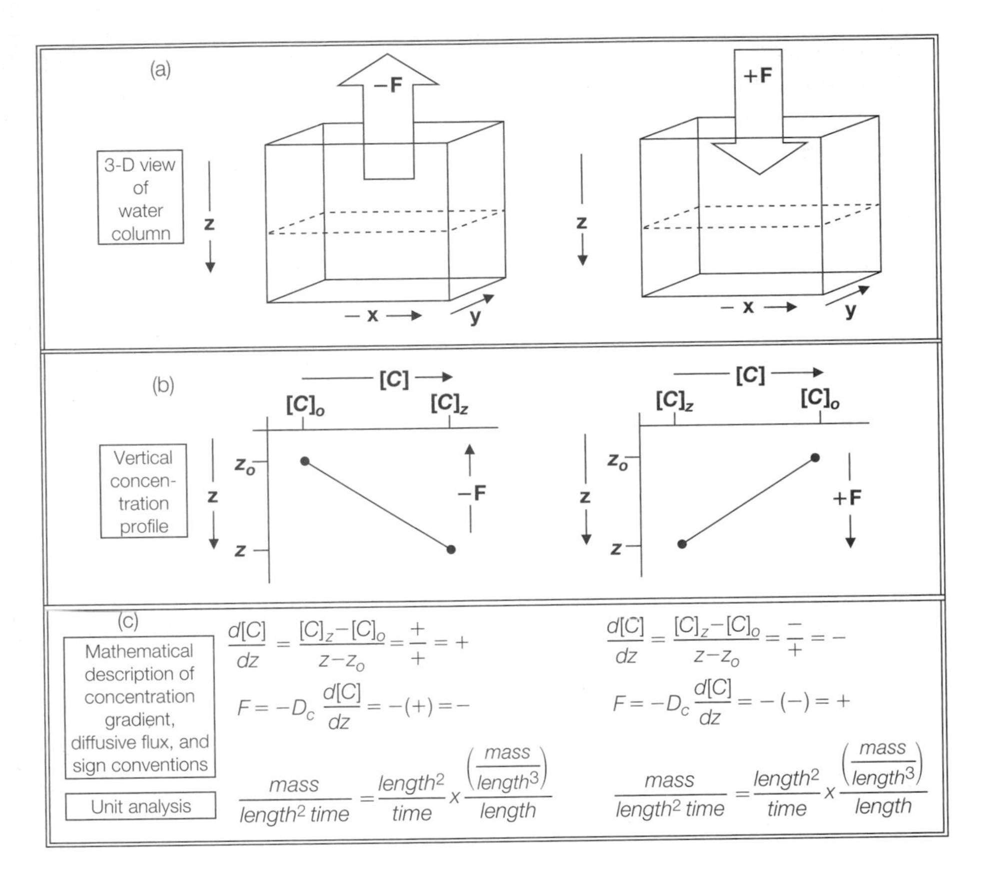

FIGURE 3.8

Diffusive flux: (a) Three-dimensional view of a vertical diffusive flux showing the oceanographic sign convention, (b) vertical concentration profile of the diffusing solute (*C*), and (c) mathematical description of the concentration gradient of *C* and its diffusive flux.

dimension, Fick's first law takes the following form:

$$F_c^{\text{diff}} = -D_c \frac{d[C]_z}{dz} \tag{3.6}$$

where $[C]_z$ is the concentration of a solute C at depth z, z is depth beneath the sea surface, and D_c, the proportionality constant, is called the molecular diffusivity coefficient. The SI units for D_c are m^2/s. If C has units of kg/m^3, then the units of its diffusive flux (F_c^{diff}) are $kg\,m^{-2}\,s^{-1}$. If the concentration is expressed as mol/m^3, the units of F_c^{diff} are $mol/(m^2\ s)$. In water, D_c ranges from 10^{-5} to $10^{-6}\ cm^2/s$. In other words, D_c is not a constant. Its value depends on various solution characteristics, such as temperature, and various solute characteristics, such as an ion's charge and radius.

Because of the negative sign in Eq. 3.6, a positive concentration gradient supports a negative flux as illustrated in Figure 3.8. Positive gradients arise if concentrations increase with increasing distance along the length scale, whether it be the x, y, or z dimensions. By oceanographic convention, z (water depth) increases with increasing distance below the sea surface. Thus, for vertical diffusive fluxes, a positive F_c^{diff} results from concentrations that decrease with increasing depth supporting a net downward transport. A negative diffusive flux arises when concentrations increase with increasing depth, leading to a net upward transport.

In practice, the concentration gradient is estimated from observational data by computing the concentration difference ($C_{z_z} - C_{z_o}$) between two samples collected from two different depths ($z_z - z_o$). Real concentration gradients are not usually linear because of the effects of other processes that concurrently affect solute distributions, such as advection and biological activity. Thus, estimating fluxes from the concentration differences of discrete samples is best done over small depth intervals.

$$F_c^{diff} = -D_c \frac{dC}{dz} \approx -D_c \frac{(C_{z_z} - C_{z_o})}{z_z - z_o} \qquad (3.7)$$

In seawater, physical processes that transport water can also cause mass fluxes and, hence, are another means by which the salinity of seawater can be conservatively altered. The physical processes responsible for water movement within the ocean are turbulent mixing and water-mass advection. Turbulent mixing has been observed to follow Fick's first law and, hence, is also known as ***eddy diffusion***. The rate at which solutes are transported by turbulent mixing and advection is usually much faster than that of molecular diffusion. Exceptions to this occur in locations where water motion is relatively slow, such as the pore waters of marine sediments. The effects of advection and turbulent mixing on the transport of chemicals are discussed further in Chapter 4.

3.5 EXCEPTIONS TO THE RULE OF CONSTANT PROPORTIONS

As noted earlier and shown in Table 3.9, small deviations in the proportions of the major ions have been observed particularly in coastal and pore waters. Reasons for these exceptions to the rule of constant proportions are described briefly next and addressed at length in later chapters.

3.5.1 Marginal Seas and Estuaries

The ion proportions in most river water is significantly different from that in seawater. As a result, river runoff can have a local impact on the ion ratios of coastal waters. This effect is most pronounced in marginal seas and estuaries where mixing with the open ocean is restricted and river input is relatively large. The variable composition of river water and its impact on the chemical composition of seawater are discussed further in Chapter 21.

Table 3.9 Major Constituent-to-Chlorinity Ratios for Various Oceans and Seas.

Ocean or Sea	Na^+ / ‰ Cl	Mg^{2+} / ‰ Cl	K^+ / ‰ Cl	Ca^{2+} / ‰ Cl	Sr^{2+} / ‰ Cl	SO_4^{2-} / ‰ Cl	Br^- / ‰ Cl
N. Atlantic	—	—	0.02026	—	—	—	0.00337–0.00341
Atlantic	0.5544–0.5567	0.0667	0.01953–0.0263	0.02122–0.02126	0.000420	0.1393	0.00325–0.0038
N. Pacific	0.5553	0.06632–0.06695	0.02096	0.02154	—	0.1396–0.1397	0.00348
W. Pacific	0.5497–0.5561	0.06627–0.0676	0.02125	0.025058–0.02128	0.000413–0.000420	0.1399	0.0033
Indian	—	—	—	0.02099	0.000445	0.1399	0.0038
Mediterranean	0.5310–0.5528	0.06785	0.02008	—	—	0.1396	0.0034–0.0038
Baltic	0.5536	0.06693	—	0.02156	—	0.1414	0.00316–0.00344
Black	0.55184	—	0.0210	—	—	—	—
Irish	0.5573	—	—	—	—	0.1397	0.0033
Puget Sound	0.5495–0.5562	—	0.0191	—	—	—	—
Siberian	0.5484	—	0.0211	—	—	—	—
Antarctic	—	—	—	0.02120	0.000467	—	0.00347
Tokyo Bay	—	0.0676	—	0.02130	—	0.1394	—
Barents	—	0.06742	—	0.02085	—	—	—
Arctic	—	—	—	—	0.000424	—	—
Red	—	—	—	—	—	0.1395	0.0043
Japan	—	—	—	—	—	—	0.00327–0.00347
Bering	—	—	—	—	—	—	0.00341
Adriatic	—	—	—	—	—	—	0.00341

Source: From Culkin, F. and R. A. Cox (1966). Deep-Sea Research, *13, 789–804.*

The impact of river input on the major ion ratios in seawater was not well understood at the time that the first operational definitions of salinity were developed. Of the nine seawater samples used to initially define the relationship between gravimetric salinity and chlorinity in 1902, only two were from the open ocean (Atlantic); the others were from marginal seas including the Baltic, North, and Red Seas. These samples generated the salinity to chlorinity relationship presented in Chapter 1 ($S‰ = 1.805\ Cl‰ + 0.030$). As shown in Table 3.9, seawater from the Baltic and Red Seas deviates significantly from the average values given in Table 3.7. By the time this was recognized, 60 years of salinity data had been computed from chlorinity measurements. In 1962, an international agreement was reached to replace this flawed relationship with Eq. 3.4, which was formulated using a more diverse set of samples from the open ocean.

3.5.2 Anoxic Basins

Deviations in the SO_4^{2-} ion ratios have also been observed in coastal areas, particularly in the sediments. This effect is due to bacterial reduction of sulfate to sulfide, which occurs in waters devoid of dissolved oxygen. Environmental conditions that contribute to the depletion of dissolved oxygen include restricted water circulation and high rates of organic matter supply. This subject is discussed in Chapters 8 and 12.

3.5.3 Sea Ice

Small amounts of salt are commonly occluded in sea ice. Not all of the ions are incorporated to the same degree. This alters the ion ratios in the remaining brine, leading to deviations from the Rule of Constant Proportions during freezing. Likewise, meltwater from sea ice can contain ions in ratios that deviate from average open ocean water.

3.5.4 Calcareous Shells and Coral Skeletons

Small deviations in the Ca^{2+} ion ratios in seawater are caused by the formation and dissolution of calcareous ($CaCO_3$) hard parts deposited by marine organisms. Despite the great abundance of these organisms, their activities produce a maximum variation of only 1% in Ca^{2+} ion ratios. The marine chemistry of carbonates is discussed in Chapter 15.

3.5.5 Hydrothermal Vents

Hydrothermal vents are another source of water entering the ocean. These vents are submarine hot-water geysers that are part of seafloor spreading centers. The hydrothermal fluids contain some major ions, such as magnesium and sulfate, in significantly different ratios than found in seawater. The importance of hydrothermal venting in determining the chemical composition of seawater is described in Chapters 19 and 21.

3.5.6 Evaporites

Salt is removed from the sea when evaporation raises the major ion concentrations to levels that exceed the solubility of minerals such as $CaCO_3$ (limestone), $CaMg(CO_3)_2$ (dolomite), gypsum ($CaSO_4 \cdot 2H_2O$) and $CaSO_4$ (anhydrite), and NaCl (halite). These precipitates are deposited sequentially, so the ions are removed at different times and rates. This causes the ion ratios in the remaining brine to deviate from those of average seawater. The formation of evaporites requires climatological conditions supporting high evaporation rates and restricted mixing with the open ocean. Evaporites currently form only in shallow-water environments. Past formation rates were much larger, causing evaporites to constitute a very large sedimentary reservoir of the major ions. Thus, over geologic time scales, evaporite formation and dissolution play a major role in the global cycling of the major ions. This subject is discussed in Chapters 17 and 21.

3.5.7 Bursting Bubbles

The major ions are transported across the air-sea interface by the ejection of water droplets from the sea surface. These droplets result from water turbulence at the sea surface that causes microscopic bubbling. Some of these bubbles burst, ejecting seawater into the atmosphere. Since not all of the salt ions are ejected to the same degree, bursting bubbles can alter the ion ratios in the remaining water.

Once in the atmosphere, the water evaporates and some of the sea salt falls back to the sea surface. The rest is transported considerable distances by winds until it is washed out of the atmosphere by rainfall. The salts that are transported back to the continents by this process are termed *cyclic salts*. After having been rained out onto the continents, the salts are carried back into the ocean by river runoff. On short time scales, the global cycling of chlorine and sodium are dominated by this process. The cyclic salts are discussed further in Chapter 21.

3.5.8 Interstitial Waters

A significant amount of seawater is trapped in the open spaces that exist between the particles in marine sediments. This fluid is termed *pore water* or *interstitial water*. Marine sediments are the site of many chemical reactions, such as sulfate reduction, as well as mineral precipitation and dissolution. These sedimentary reactions can alter the major ion ratios. As a result, the chemical composition of pore water is usually quite different from that of seawater. The chemistry of marine sediments is the subject of Part III.

CHAPTER 4
Salinity as a Conservative Tracer

All figures are available on the companion website in color (if applicable).

4.1 INTRODUCTION

In Chapter 3, the conservative nature of salinity was discussed. Because of its conservative behavior, salinity makes an excellent tracer of water movement in the ocean. Hence, it is routinely measured, along with another conservative tracer, temperature. These two conservative tracers have been used to ascertain the patterns of surface and deepwater circulation. They also determine the degree of density stratification in a water column, which controls the degree to which water can undergo turbulent mixing. Thus, knowledge of the processes that control the heat and salt content of seawater are key to understanding much of the physical dynamics of seawater.

In this chapter, we discuss the factors that control the global heat and hydrological cycles. This information is used to explain the spatial and temporal variations in temperature, salinity, and density of the modern-day ocean. Oceanographers use the conservative properties of seawater to provide insight into the physical processes in operation. These physical processes also act on the nonconservative substances in seawater. The type and intensity of the physical effects can be assessed from concurrent measurement of the conservative tracers. This enables identification of the biogeochemical processes acting on the nonconservative substances. Mathematical models are used to infer the rates of some biogeochemical processes by effectively subtracting off the effects of the physical processes as estimated from the conservative tracers.

4.2 GLOBAL HEAT AND WATER BALANCE

The global heat cycle drives the hydrological cycle, which in turn controls the salinity of seawater. The most important contributor of heat to the crustal-ocean-factory is solar radiation. The flux of solar radiation that reaches Earth is termed *insolation*. Only a fraction of the incoming solar radiation reaches Earth's surface, because a large portion is either reflected or absorbed by the atmosphere. That which reaches Earth's surface is also either reflected or absorbed. In the end, about half of the incoming radiation is absorbed by the rocks and water on Earth's surface. (A detailed heat budget is provided

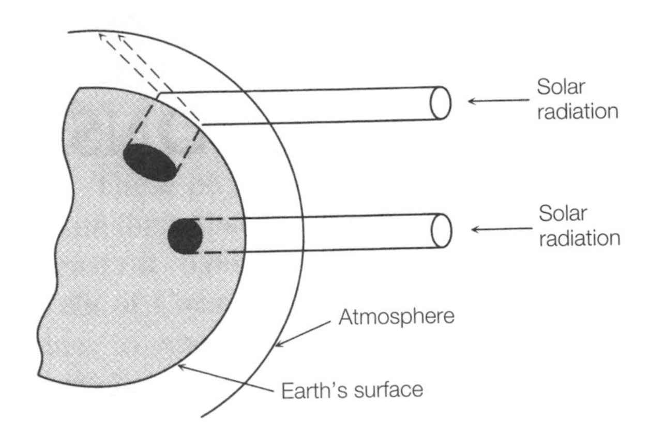

FIGURE 4.1

Latitudinal variations in insolation caused by differences in the angle at which the sun's radiation strikes Earth's surface. *Source*: From Thurman, H. V. (1988). *Introductory Oceanography*, 5th ed. Merrill Publishing Company, p. 170.

in Figure 25.6.) Water has a higher heat capacity than crustal rock, so it is able to absorb more solar radiation without undergoing a temperature increase (Table 2.2).

Insolation decreases with increasing latitude due to the decreasing angle of the sun's rays (Figure 4.1). Since the sun is at its highest angle (90°) over the equator, this location receives the most insolation. At Earth's surface, the effect of the sun's angle on insolation is threefold. This effect is illustrated at low latitudes where: (1) the higher incident angle causes a ray of solar radiation to be spread over a smaller surface area; (2) the ray passes through a smaller thickness of atmosphere; and (3) the higher incident angle causes less insolation to be reflected.

Insolation absorbed by the ocean influences water temperatures and rates of evaporation. Because insolation varies temporally and spatially, so does the temperature of surface seawater. As shown in Figure 4.2, low latitudes are characterized by warmer surface water, with sea surface temperatures generally declining with increasing latitude. Seasonal shifts in sea surface temperature are caused by changes in insolation that are a consequence of Earth's revolution around the sun and the tilt of its orbital axis (Figure 4.3). Shorter time scale changes include: (1) diurnal changes in insolation caused by Earth's rotation, (2) changes in cloud cover, and (3) changes in sea state. In the case of the last, the choppier the water's surface, the greater the surface area for reflection and, hence, the larger the percentage of insolation reflected.

The latitudinal heat gradient in the atmosphere and ocean remains relatively constant over time despite the short-term and spatial variations in insolation. This steady state is maintained by the net transport of heat from low to high latitudes where it is radiated back into space. Atmospheric currents (winds) are responsible for about half of this meridional net transport of heat.[1] The rest is accomplished by water movement in the

[1] Meridional flows cross latitudes and, hence, follow meridians of longitude. In contrast, zonal flows cross longitude and, hence, follow parallels of latitude.

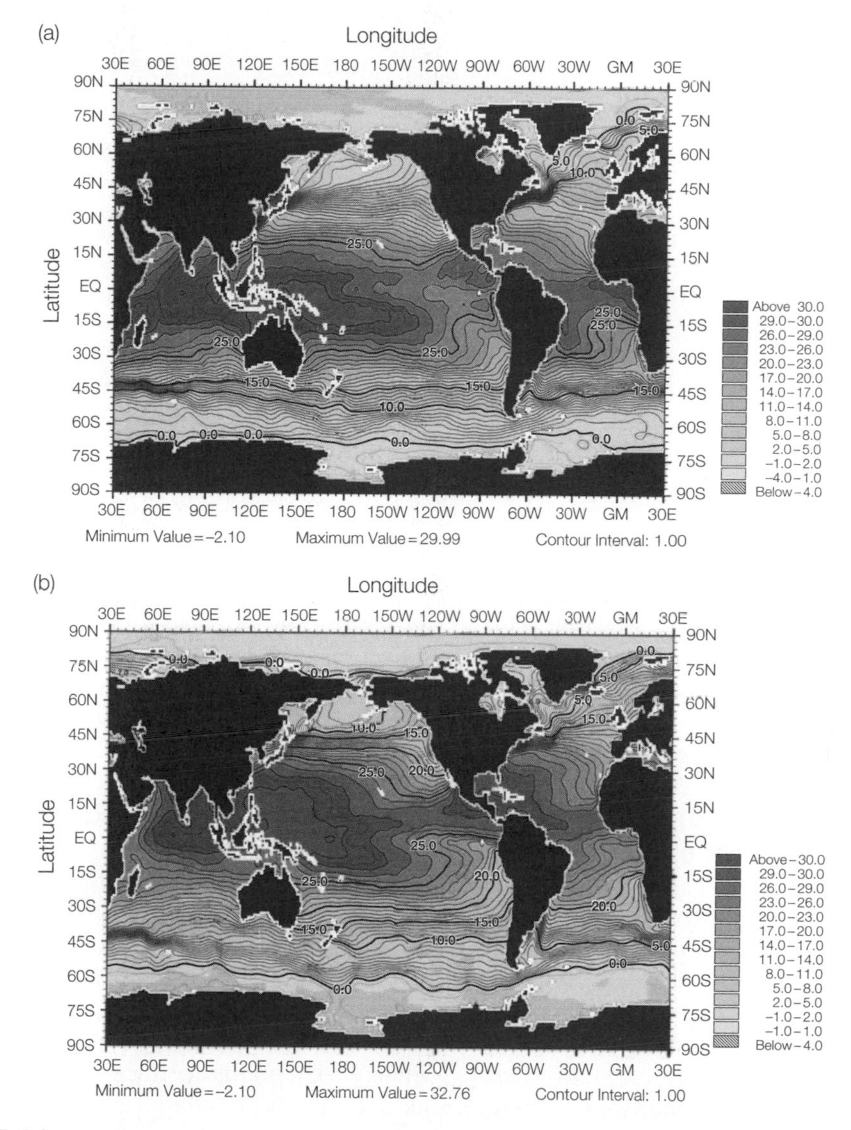

FIGURE 4.2

Average sea-surface temperature (°C) (a) in winter (Jan–Mar) and (b) in summer (Jul–Sep). *Source*: After Reynolds, R. W., and T. M. Smith (1995). *Journal of Climate* 8, 1571–1583. (See companion website for color version.)

surface and deep-water currents. In all but the Southern Ocean, oceanic circulation causes a net transport poleward.

Winds play a central role in inducing ocean currents and turbulent mixing. Winds themselves are a type of convection current generated by spatial and temporal variations in atmospheric heating. In the atmosphere, these air currents form the major wind

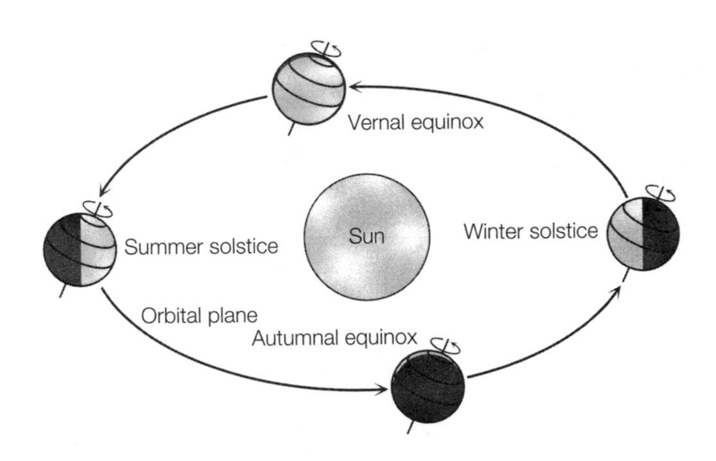

FIGURE 4.3

Earth's annual revolution around the sun.

bands, the Trade Winds, Westerlies, and Polar Easterlies (Figure 4.4a). As winds blown across the sea surface they transfer their kinetic energy to the ocean. Some of the energy ends up as surface waves. The rest generates advection movements collectively called *Ekman transport*. The direction of transport is heavily influenced by the Coriolis effect. As shown in Figures 4.5a and b, the Trade Winds cause upwelling due to divergence of surface flows. A similar divergence caused by the Westerlies also generates upwelling. The upwelling waters heading equatorward from the Westerlies meet the upwelling waters heading poleward from the Trades. The water converges to form a small hill centered at 30°. Gravity acts to pull water down the hill. The water flows down the hill in a circular pattern due to the Coriolis effect. This water flow is called a *geostrophic* current.

These currents cause surface waters to flow in approximately cyclonic gyres in the northern hemisphere and anticyclonic gyres in the southern hemisphere as shown in Figure 4.4b.[2] The geostrophic currents that move water poleward are called Western Boundary Currents. They are notable for their fast velocities—up to 1 m/s. In contrast, the currents that move water equatorward, the Eastern Boundary Currents, are much slower—on the order of 10 cm/s. Notable exceptions to the gyre pattern of flow are the zonal Equatorial Countercurrents and the West Wind Drift. Although the winds are the driving force behind the geostrophic currents, their circulation patterns are also influenced by the Coriolis effect and the physical barriers presented by the land masses.

Ekman transport can also cause coastal upwelling and downwelling as shown in Figures 4.5c and d for the southern hemisphere. The patterns would be reversed for the northern hemisphere; southerly winds induce downwelling and northerly winds lead to upwelling.

[2] Currents in cyclonic gyres move water in a clockwise direction and counterclockwise in anticyclonic gyres.

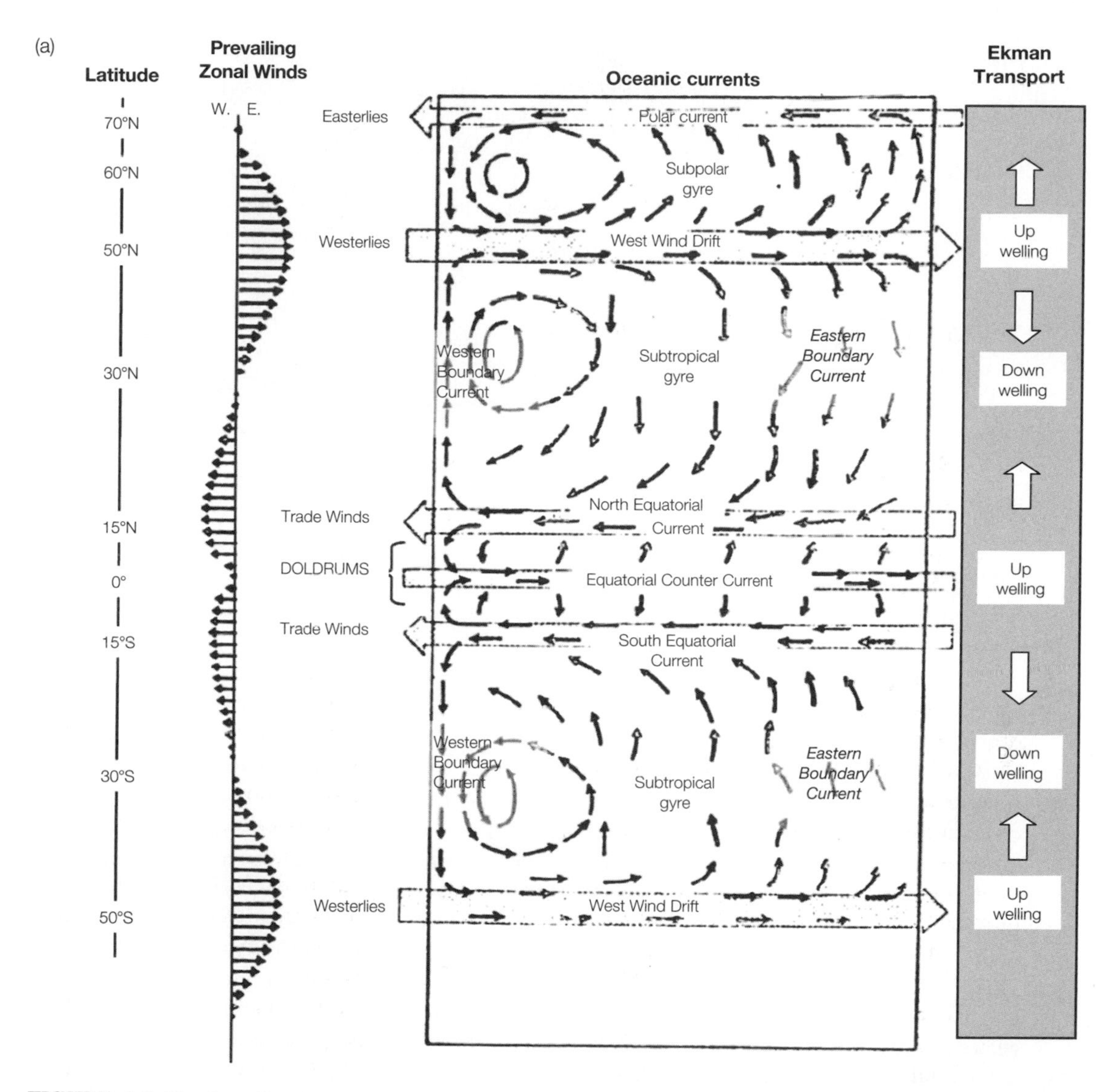

FIGURE 4.4 *(Continued)*

The winds also play an important role in driving deep-water currents. The pathway of this deep circulation is shown in Figure 4.6. It is often described as resembling a conveyor belt that transfers surface water to the deep ocean via downwelling at selected sites, causes it to travel through the ocean basins, and then returns it back to the sea surface via upwelling. Because this deep-water circulation tends to transport water across latitudes, it is termed *meridional overturning circulation* (MOC). As described later in this chapter, thermohaline convection currents play an important role in determining

(b)

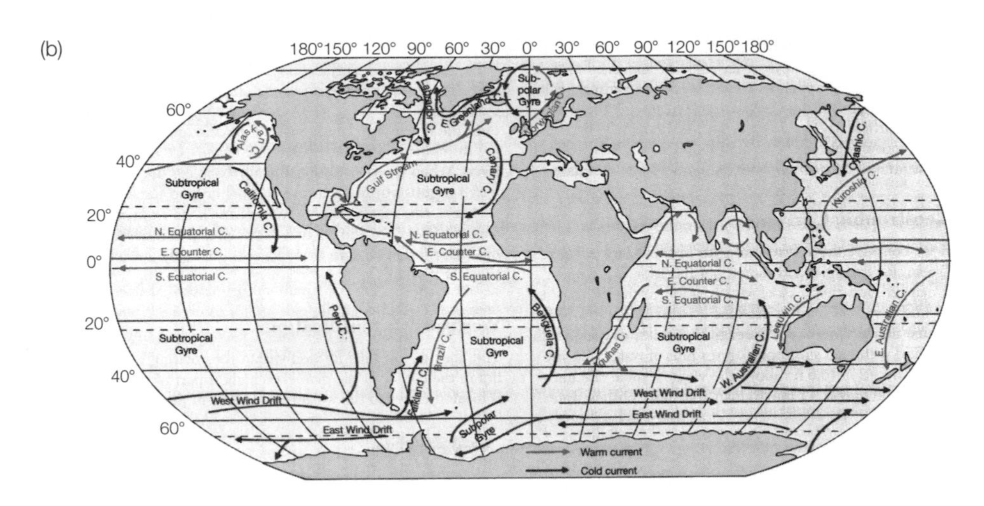

FIGURE 4.4

(a) Generalized schematic of the major wind bands and the underlying geostrophic gyres. (b) Global map of geostrophic ocean currents. *Source*: After (a) Apel, J. (1987) *Principles of Ocean Physics*. Academic Press (Figure 6.36 on p. 310), and (b) Mellor, G. L. (1996). *Introduction to Physical Oceanography*. Springer, AIP Press (Figure 1–3, p. 4). (See companion website for color version.)

where and how surface water is downwelled, but winds and tides are thought to be the primary drivers of MOC.

Since the oceans can absorb and release heat, the circulation of the ocean is intimately tied to that of the atmosphere. Changes in climate can induce changes in oceanic circulation and vice versa. Most of these feedbacks are complicated and nonlinear. For example, a change in the hydrological cycle can alter the salinity and, hence, the heat capacity of seawater. Thus, mathematical models of climate require the coupling of global ocean and atmospheric circulation models. Considerable attention is now directed at refining these models in an effort to improve predictions of global climate change. Recent research suggests that even changes in the abundance of phytoplankton can significantly alter the ability of surface seawater to absorb solar radiation. This is related to the ability of chlorophyll (a photosynthetic pigment) to absorb a significant amount of the incoming shortwave solar radiation.

4.2.1 Temperature Distributions

In Figure 4.2, we saw that the temperature of surface seawater is temporally and spatially variable, exhibiting approximately meridional gradients. As a rule, the temperature of seawater located at depths greater than 1000 m is uniformly low, with values less than 5°C. This causes the vertical profiles of temperature to exhibit predictable variations with latitude as shown in Figure 4.7. At high latitudes, the deep waters are overlain by surface waters of low temperature, making this water column nearly isothermal. In the mid- and low latitudes, where insolation is higher, depth profiles in the open ocean

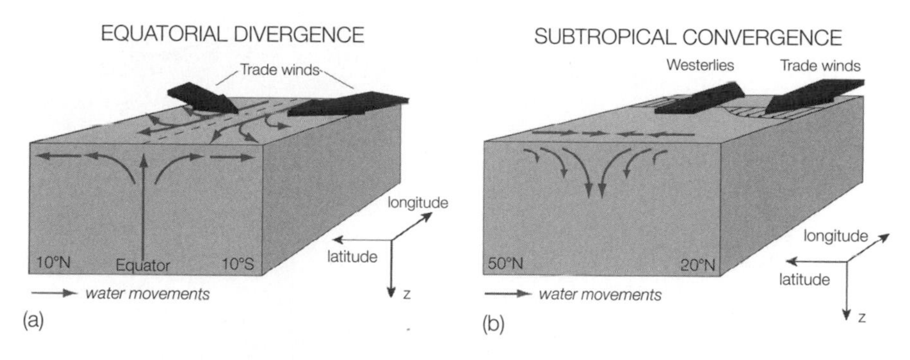

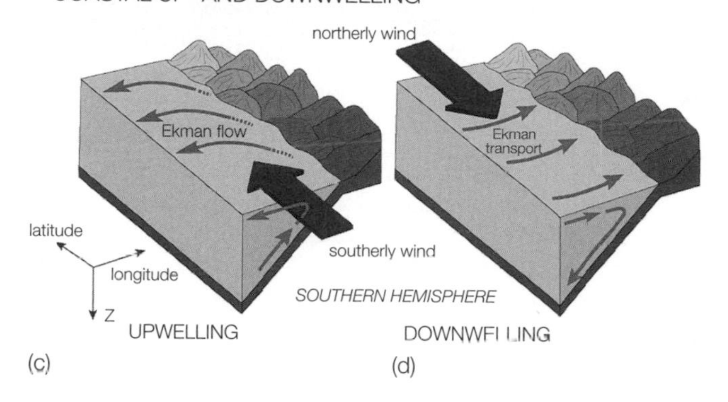

FIGURE 4.5

Examples of Ekman transport resulting from wind-driven currents (a) upwelling from divergence at equatorial latitudes, (b) downwelling from convergence in subtropical latitudes of the northern hemisphere, (c) coastal upwelling in southern hemisphere, and (d) coastal downwelling in the southern hemisphere. Wind directions are shown by large arrows.

are characterized by three temperature regimes. At the sea surface lies a warm homogeneous pool of water called the *mixed layer*. The temperature and depth of this zone are controlled by local insolation and wind mixing. Mixed layer depths (MLDs) range from 20 to 500 m.[3] Large seasonal shifts occur at mid and subpolar latitudes as winter cooling causes the MLD to deepen and summer heating to rise (Figure 4.8). Below the mixed layer, temperature decreases with increasing depth. This zone of temperature decline is called the *thermocline*. Below approximately 1000 m, water temperatures are nearly constant with increasing depth. This thermally homogeneous region is called the *deep zone*.

Technically the thermocline is the depth zone over which the vertical temperature gradient reaches maximal values. This depth zone varies with season, latitude and longitude, and local environmental conditions. The top of the thermocline is defined by

[3] A common definition used to establish the bottom of the mixed layer is the depth at which temperature has decreased more than 0.2°C from that in the overlying water located at 10 m.

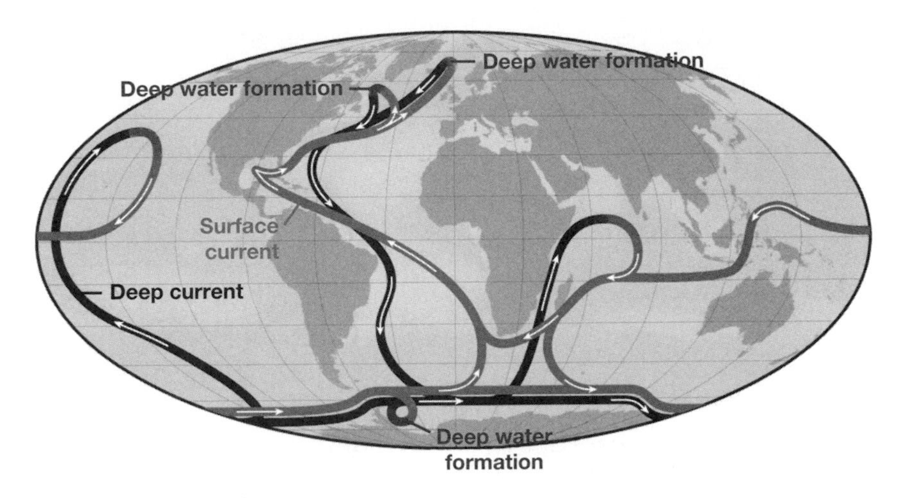

FIGURE 4.6

Schematic map of the ocean's meridional overturning circulation. Deep and bottom water currents are shown in blue. These currents are fueled by the cooling of warm salty water in the North Atlantic and Wedell Sea in Antarctica. These waters flow into all of the oceans and slowly ascend throughout them, becoming warmer and less salty through admixture with intermediate and surface waters. The surface water currents are shown in red. They represent a return flow that resupplies water to the sites of deep and bottom water formation. Also shown is the distribution of annually averaged surface salinity. See Figures W10.1 and W10.2 for more detailed views. *Source*: After R. Simmon, Explaining Rapid Climate Change: Tales from the Ice, Earth Observatory Features, NASA, http://earthobservatory.nasa.gov/Study/Paleoclimatology_Evidence/paleoclimatology evidence 2.html, accessed May 2008. (See companion website for color version.)

the bottom of the mixed layer. At mid-latitudes, a permanent thermocline is present and typically extends down to depths of 1000 to 2000 m. During the summer, increased insolation elevates the surface-water temperatures, producing a seasonal thermocline as the MLD rises. Winter storms destroy the seasonal thermocline by cooling the surface water and lowering the MLD via wind mixing. These seasonal shifts are very important in controlling the transport of nonconservative chemicals, such as nutrients, between the mixed layer and upper thermocline. At low latitudes, the temperature gradient between the mixed layer and deep zone is very large because of permanently high surface-water temperatures. In this region, the permanent thermocline is very steep. At high latitudes, cold surface water prevent the maintenance of a permanent thermocline although a seasonal thermocline can occur. At mid-latitudes, seasonal variations in insolation cause seasonal variations in the depth and steepness of the upper thermocline.

4.2.2 Salinity Distributions

In the open ocean, surface water salinities exhibit meridional gradients largely controlled by the local balance between water loss through evaporation and water gain through

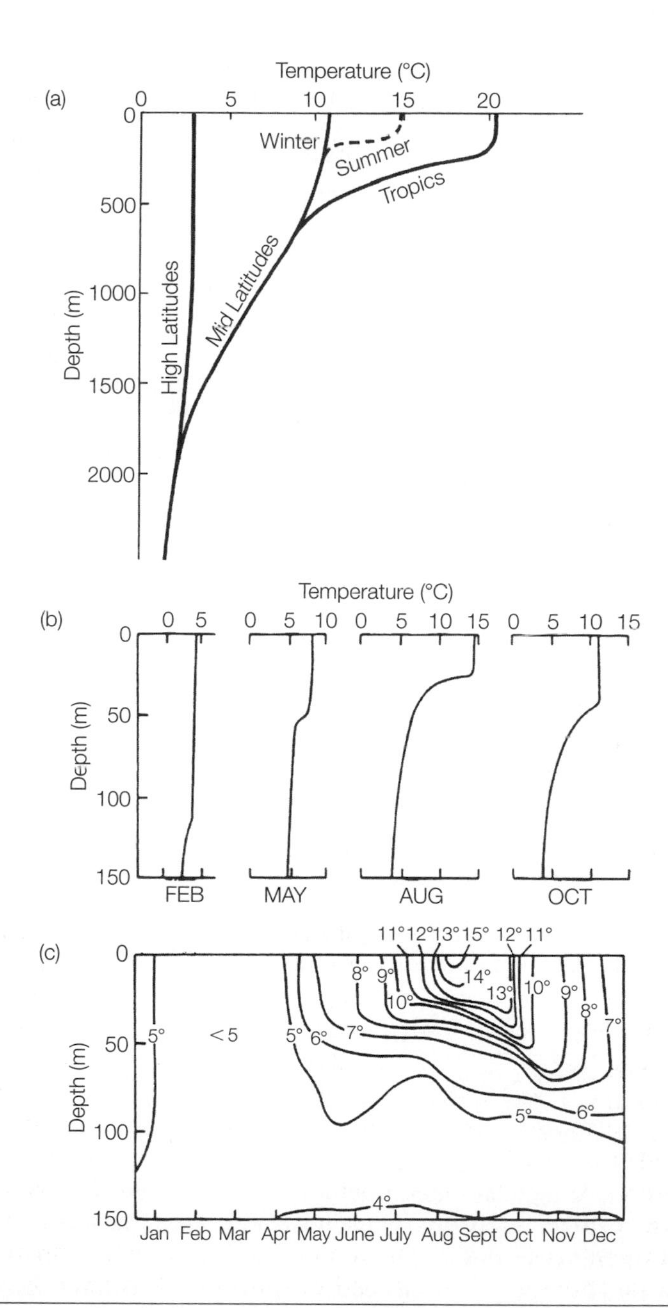

FIGURE 4.7

(a) Typical temperature profiles in the open ocean at low, mid, and high latitudes. A summer thermocline is found at mid-latitudes. (b) Typical seasonal changes in temperature structure in the top 150 m at a northern hemisphere mid-latitude site (50°N, 145°W in the Eastern North Pacific). (c) Data from (b) plotted versus time. Note the summer buildup and winter destruction of the thermocline. *Source*: From Knauss, J. A. (1997). *Introduction to Physical Oceanography*, 2nd ed. Prentice Hall, p. 2 and p. 7.

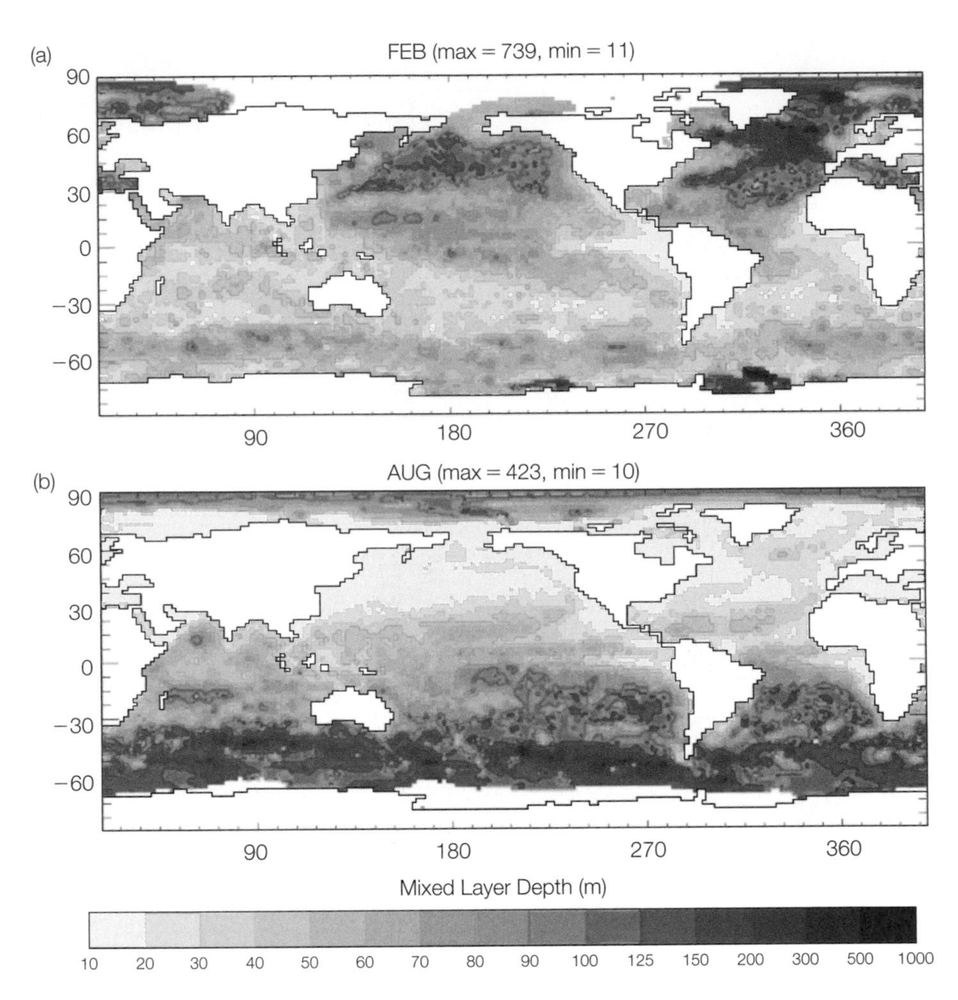

FIGURE 4.8

Global map of average mixed layer depths in (a) February and (b) August based on hydrographic data collected from 1941 through 2002. *Source*: After de Boyer Montégut, C., *et al*. (2004). *Journal of Geophysical Research* 109, C12003. (See companion website for color version.)

precipitation. At high latitudes, the formation of sea ice involves brine exclusion leading to local elevations in surface salinity. Generally, salinities are lower in coastal regions than the open ocean due to closer proximity to freshwater sources such as rivers and groundwater. Exceptions to this occur in some semienclosed marginal seas, such as the Mediterranean Sea, and small lagoons trapped behind barrier islands where evaporation rates are high and freshwater inputs are absent. Other marginal seas, such as the Red Sea, have high salinities due to admixture with hydrothermal brines.

The meridional variations in evaporation and precipitation rates that control surface salinity in the open ocean are shown in Figure 4.9. These variations are caused

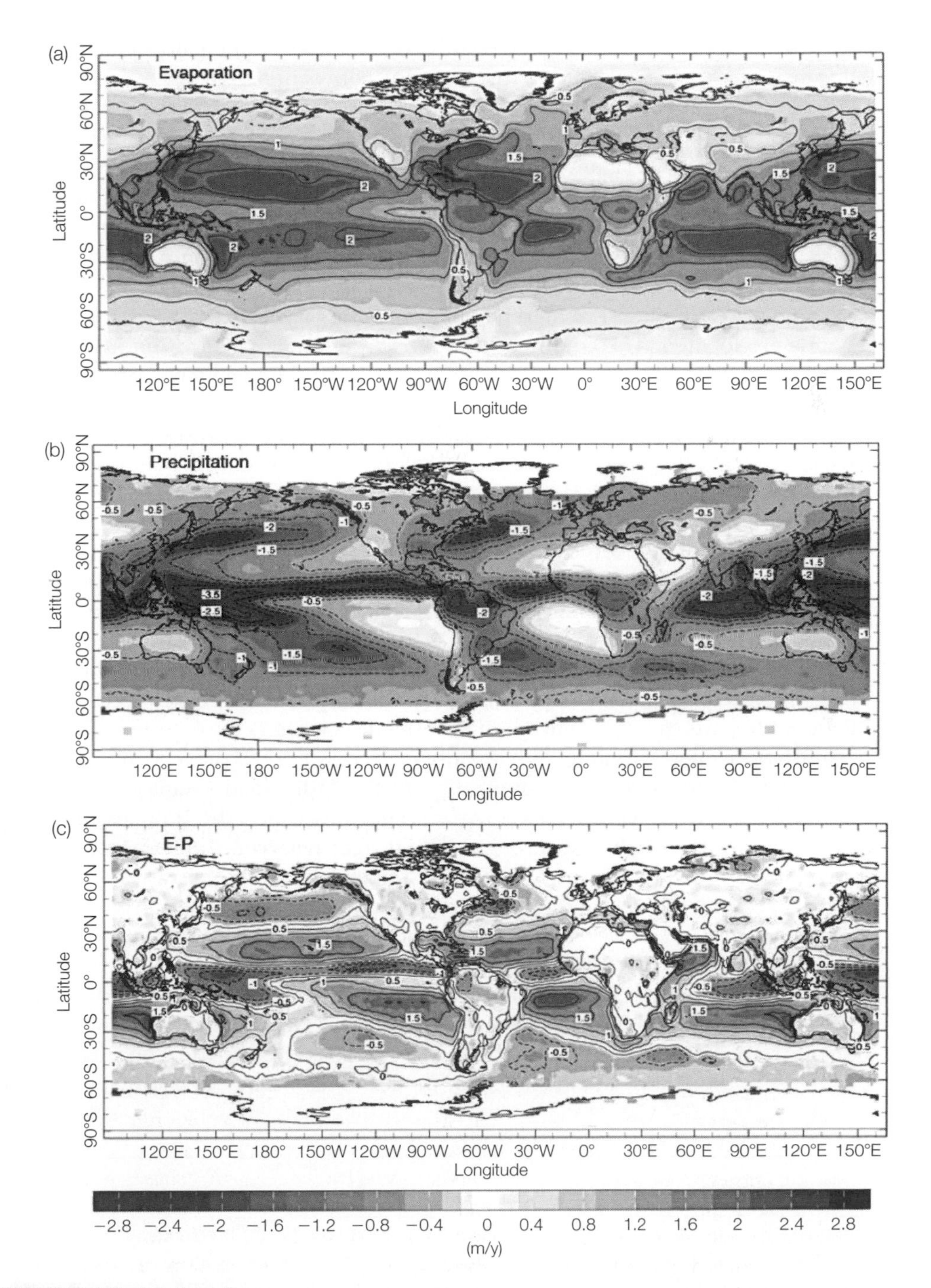

FIGURE 4.9 *(Continued)*

by atmospheric convection currents driven by the latitudinal gradient in insolation. Because of the Coriolis effect, three sets of convection cells are present in the northern and southern hemispheres. The bottom legs of the convection cells are the winds shown in Figure 4.4a. Between these wind bands lie regions of rising air (~0° and ~60°) and sinking air (~30° and ~90°). The regions of rising air are characterized by high precipitation rates due to condensation of water vapor as the rising air moves into the upper atmosphere and cools. The dried air returns as a downward, or sinking, flow at ~30 and ~90°, causing these regions to have low precipitation rates. Evaporation rates are highest in regions with strong winds, dry air, and high surface water temperatures. This causes evaporation rates to be highest at mid-latitudes (20 to 25°) where the Trade Winds are located. The combined effect of the meridional variation in evaporation and precipitation results in net evaporation at mid-latitudes and net precipitation at low and high latitudes. Regions characterized by net evaporation have higher salinities than those experiencing net precipitation. The variations in surface salinity of the subpolar waters reflects the degree of sea ice formation and, hence, brine exclusion. The relative lack of sea ice in the North Pacific leads to the relatively low salinity of its subpolar surface water.

A map of surface seawater salinity is presented in Figure 4.10, illustrating that mid-latitude regions of high salinity are centered over the subtropical geostrophic gyres with net evaporation rates reaching 150 cm/y. This figure also shows that the surface salinity is higher in the Atlantic than the Pacific despite its numerous larger rivers and smaller surface area. This is apparently due to a persistent net export, via the Trade Winds, of evaporated water from the Atlantic to the Pacific and Indian Oceans. As shown in Figure 4.9c, a greater percentage of the Atlantic Ocean is covered by regions where net evaporation rates are high. This export of freshwater is balanced by a return flow in the form of surface seawater origination in the Indian and Pacific Oceans. This return flow leaves the Pacific Ocean via surface currents, called the Indonesian Throughflow, that move water southwest through the passages between the Indonesian Islands. This flow feeds the Agulhas Current, which transports surface seawater from the Indian Ocean around the southern tip of Africa and into the Atlantic (Figure 4.4b). In the South Atlantic, the water is then entrained in the Benguela Current, which is an Eastern Boundary Current. This current brings the return flow into the equatorial current systems where it is ultimately transferred to the Gulf Stream. This surface water transport from the Pacific to the Indian to the Atlantic Ocean constitutes an important part of the return pathway of the meridional overturning circulation because, as shown in Figure 4.6, the North Atlantic is one of the few sites where seawater sinks into the deep zone.

FIGURE 4.9

Annual mean rates in the surface ocean of (a) evaporation (E), (b) precipitation (P), and (c) evaporation minus precipitation (E – P). In the bottom map, areas where E > P are colored green and areas where P > E are colored brown. The contour interval is 0.5 m/y. (See companion website for color version.) Data after Kalnay, E., *et al.* (1996): *Bullet. Amer. Meteorol. Soc.*, 77, 437–471.

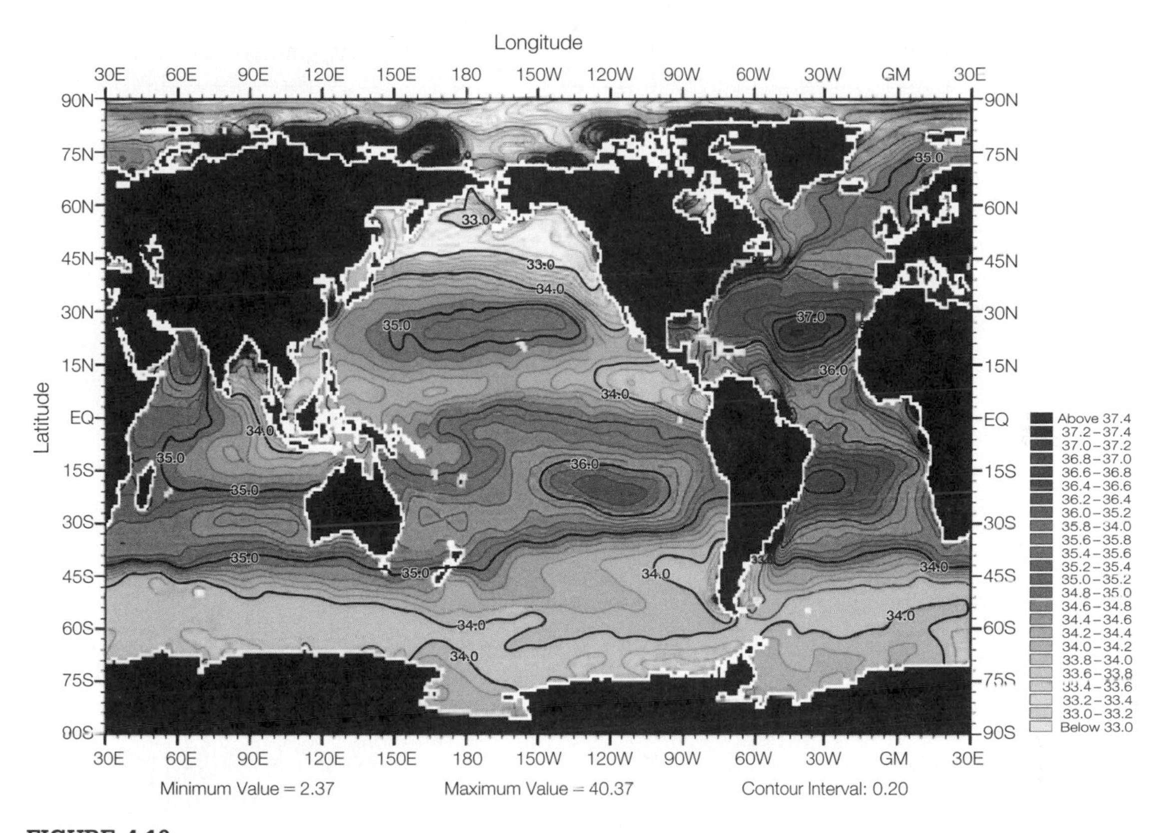

FIGURE 4.10

Annually averaged salinity of surface seawater in the world's oceans. *Source*: After Levitus, S. (1982). Climatological atlas of the world ocean. NOAA Professional Paper 13, U.S. Government Printing Office, Washington, DC. (See companion website for color version.)

As with temperature, the vertical profiles of salinity in the open ocean vary with latitude. Figure 4.11 compares vertical profiles from high, mid-, and low latitudes in the Atlantic and Pacific Oceans. Although the Atlantic Ocean surface salinities are higher than those of the Pacific, the vertical salinity gradients are largest at low and mid-latitudes because of net evaporation. This salinity gradient is called the ***halocline***. At very low latitudes in the tropics, excess precipitation dilutes the surface salinity, generating a pool of low-density (low-salinity and high-temperature) water that floats on the sea surface. At high latitudes, excess precipitation also lowers surface salinities, but wind mixing penetrates as deep as 1000 m, making the water column nearly isohaline. (The low salinity water at mid-depths in the mid-latitude Atlantic Ocean profile reflects the presence of a low-salinity water mass, Antarctic Intermediate Water.)

Global climate change is having an impact on sea surface temperatures and salinities. Recent research has identified a systematic decline in mixed layer salinities at high latitudes (−0.03 to −0.2‰) and an increase at low latitudes (+0.1 to +0.4‰) between the

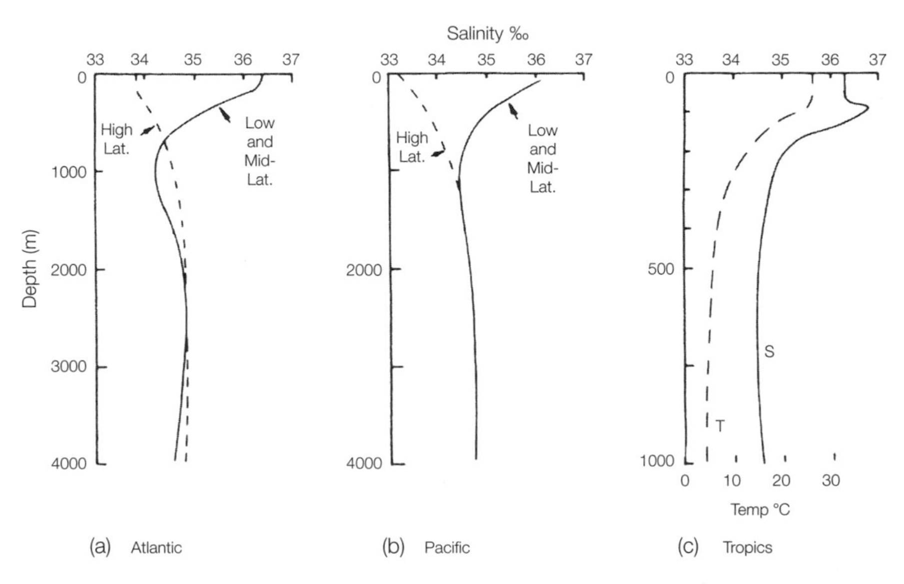

FIGURE 4.11

Latitudinal variations in depth profiles of salinity in the (a) Atlantic, (b) Pacific, and (c) tropical oceans: High-latitude salinities are given by the dashed lines. *Source*: After Pickard, G. L., and W. J. Emery (1999). *Descriptive Physical Oceanography: An Introduction*, 5th ed. Butterworth-Heinemann, p. 52.

1950s and the 1990s in the Atlantic Ocean. This suggests that the meridional patterns of net evaporation and precipitation are shifting. Increases in sea surface temperature have also been observed. These changes alter the density of seawater and, hence, have the potential to affect oceanic circulation patterns.

4.2.3 Density Distributions

The in situ density of seawater is determined by its temperature, salinity and depth (overlying pressure). In Figure 3.4, we saw that the vertical profiles of in situ density are largely determined by pressure. In water columns that are stably stratified, density increases with increasing depth. This stable state is also referred to as a type of equilibrium, since any relocation of a water parcel will result in a spontaneous movement of that parcel back to its energetically neutral location. As noted in Chapter 3, the depth of this energetically neutral location is determined by the water parcel's temperature and salinity. This is why oceanographers use potential density (σ_θ) or σ_t for identifying and tracking water parcels. As shown in Figure 4.12, vertical profiles of potential density exhibit meridional variations mostly caused by differences in sea-surface temperatures. Surface densities are lowest at low latitudes due to high surface temperatures. As with temperature and salinity, vertical gradients of density are nearly absent at high latitudes due to the great depth of mixed layer. At mid- and low latitudes, a density gradient lies between the mixed layer and the deep zone. This gradient is called the *pycnocline* and is largely the result of the thermocline.

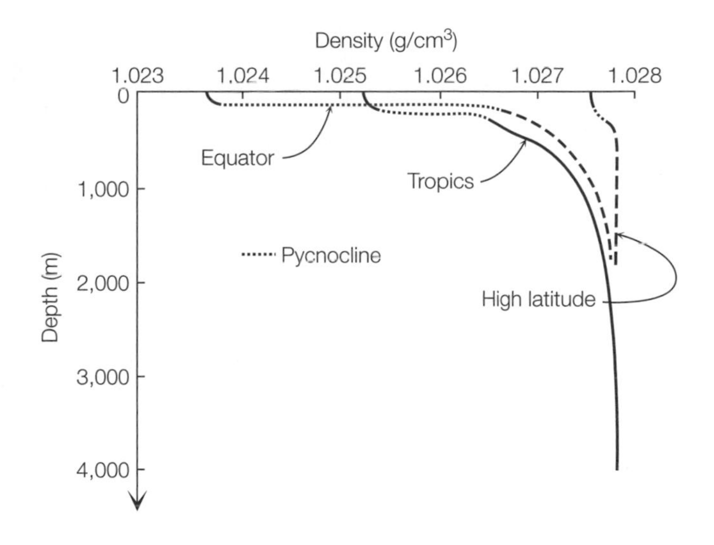

FIGURE 4.12

Typical potential density profiles of ocean water. *Source*: From Chester, R. (2003). *Marine Geochemistry*, 2nd ed. Blackwell Publishing, p. 143.

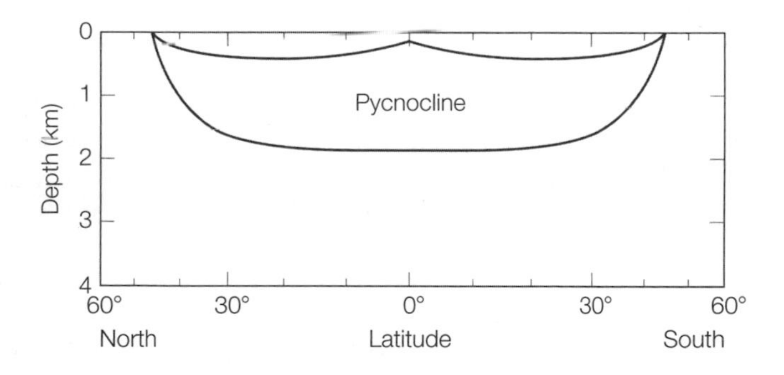

FIGURE 4.13

Density layering through a longitudinal cross section of an idealized ocean basin. *Source*: From Chester, R. (2003). *Marine Geochemistry*, 2nd ed. Blackwell Publishing, p. 143.

These latitudinal variations in potential density are depicted along a longitudinal cross section of an idealized ocean basin in Figure 4.13. At high latitudes, winter mixing essentially merges the mixed layer with the deep zone by preventing the formation of a vertical density gradient. In these regions, the cold surface waters are in direct contact with the deep zone. At latitudes lower than 45°, the presence of a pycnocline separates the mixed layer from the deep zone. The mixed layer represents only 20% of the ocean. Most of the ocean's water lies in the deep zone out of direct contact with humans.

An example of a longitudinal profile of potential density is presented in Figures 4.14c and d for the Atlantic Ocean. Note that the water column at any latitude exhibits stable

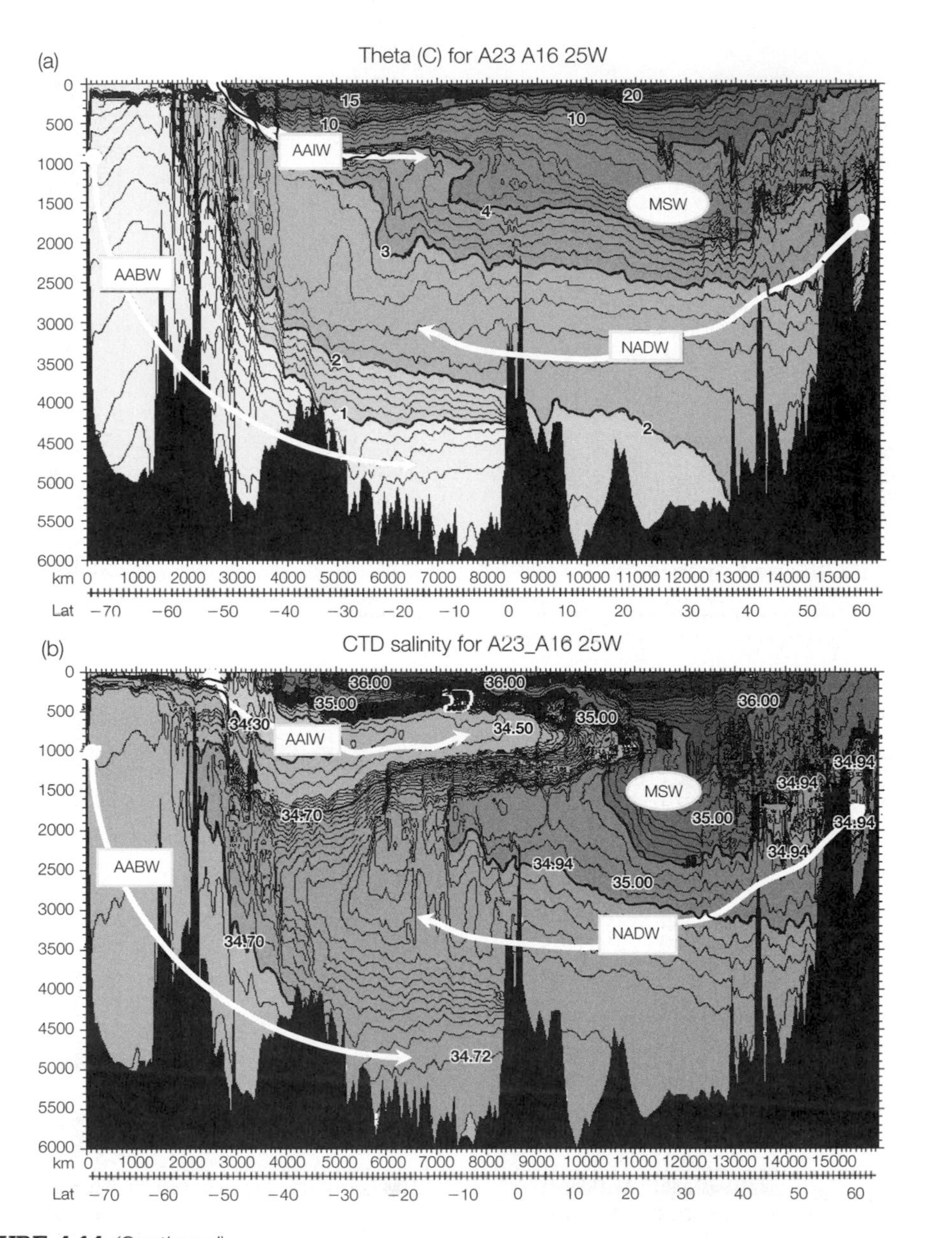

FIGURE 4.14 *(Continued)*

density stratification. In these diagrams, the lines of constant density, called isopycnals, lie parallel to each other. To maintain stable stratification, water movement occurs along isopycnals. The tight clustering of isopycnals in mid-depths marks the pycnocline. In this region, the gravitational stability imparted by strong density stratification inhibits vertical

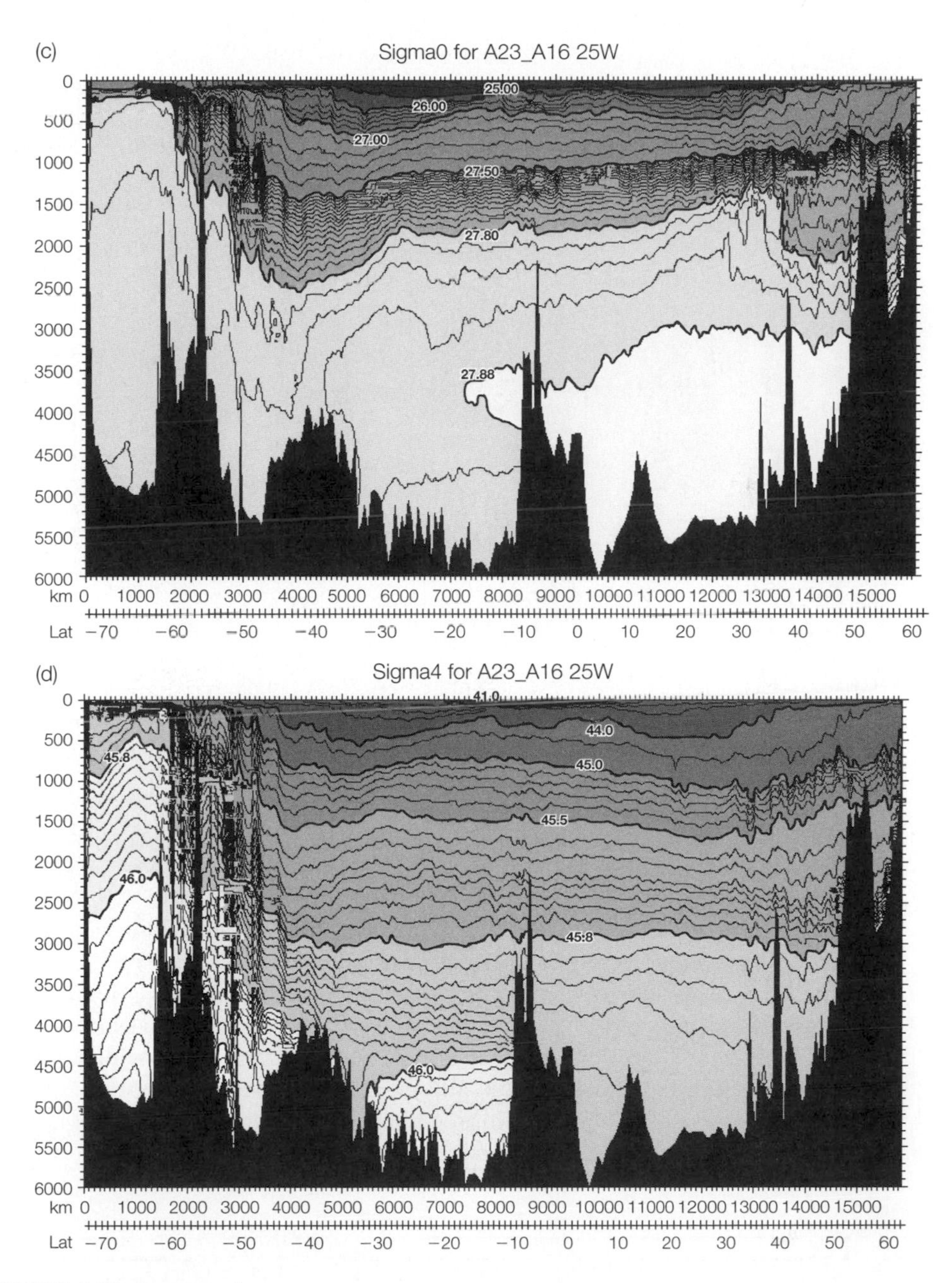

FIGURE 4.14 *(Continued)*

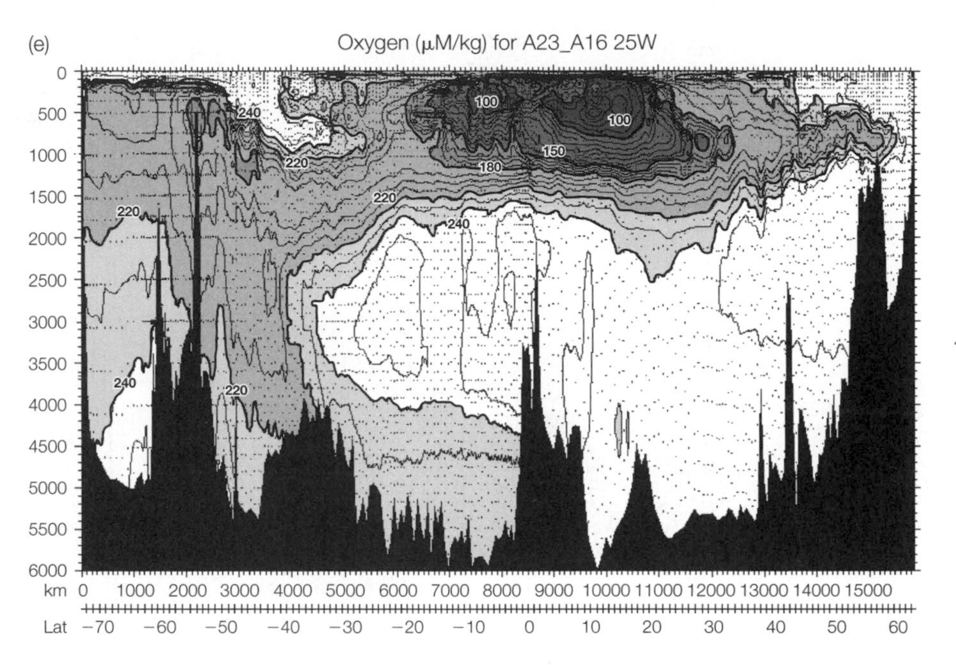

FIGURE 4.14

Longitudinal profiles in the Atlantic Ocean at about 25°W. (a) Potential temperature (°C), (b) salinity, (c) potential density (0 dbar), (d) potential density (4000 dbar), and (e) dissolved oxygen (μmol/kg). *Source*: After Talley, L. (1996). Atlantic Ocean: Vertical Sections and datasets for selected lines. http:/sam.ucsd.edu/vertical_sections/Atlantic.html. Scripps Institute of Oceanography, University of California – San Diego. Data are from WOCE hydrographic program. (See companion website for color version.)

mixing. In comparison, little energy is required to induce vertical mixing where density stratification is absent or minimal, such as at high latitudes, where the isopycnals are nearly vertical. This illustrates that the isopycnals are not always oriented horizontally. The degree to which the isopycnals are inclined reflects the local balance in physical forces acting on seawater. The most important are gravity, the winds, and the tides. Note the difference in profiles of σ_t and σ_4. As discussed in Chapter 3, the effects of temperature and salinity on the compressibility of seawater at great pressures make the use of σ_4 more appropriate for characterizing potential densities in the deep sea.

4.3 TRANSPORT OF HEAT AND SALT VIA WATER MOVEMENT

Water motion in the ocean is the result of two general phenomena, advection and turbulence. *Advection* causes water to experience large-scale net displacement (directed transport), whereas *turbulent mixing* involves the random motion of water molecules (random transport). Both types of water motion are important in transporting chemicals and marine organisms, namely the plankton. Most advective transport in the ocean is the result of currents caused by winds, tides, and thermohaline gradients. The geostrophic currents move surface water at velocities ranging from a few centimeters to a few meters per second. (The velocity of the Gulf Stream is 2 m/s.) In comparison, the currents that transport deep water as part of MOC have horizontal velocities on the order of a few centimeters per second or less. The upwelling of these waters to the sea surface is much slower, with velocities of a few meters per year.

Thermohaline currents are driven by convection caused by a loss of vertical density stability, specifically one in which the buoyancy of a water parcel is lost. This can result from evaporative cooling, brine exclusion, caballing, and double diffusion. The first two are restricted to the sea surface. Double diffusion is caused by a hundred-fold faster molecular diffusion of heat (conduction) as compared to the molecular diffuse of salt ions. This enables heat transfer to occur between adjacent water masses before any significant transport of salinity can take place via molecular diffusion. Thus, a warm salty water can become cooler and denser by losing heat to an underlying cooler and fresher water mass. The salty water then sinks. This process leads to the formation of microscale vertical layers called salt fingers. Because of the small spatial scales involved, this water movement is considered to be a type of turbulent mixing process. It is important in the Arctic where cold freshwater can overlie warm salt and in western tropical oceans where warm salt can overlie cold freshwater. Since it requires substantial salt and temperature gradients between water masses, the largest effects of double diffusion are seen in the thermocline. In the most extreme cases, the vertical temperature and salinity profiles take on a staircase appearance (Figure 4.15).

In the surface waters, evaporation, cooling, and brine exclusion can increase surface-water densities such that a water parcel will spontaneously sink. This commonly occurs at high latitudes or in restricted basins, such as the Mediterranean Sea. The depth to which a water parcel sinks is defined by its potential density, which is in turn determined by its temperature and salinity. Therefore, water parcels that share a common mode of formation generally exhibit a narrow range of temperature and salinity. This characteristic range of temperature and salinity has been used to classify water parcels. For example, Antarctic Intermediate Water (AAIW), which is formed by sinking at 50 to 60°S, has a temperature of 3 to 7°C and a salinity of 34.2 to 34.44‰. Although its salinity is rather low, its low temperature produces a σ_t (26.82 to 27.43) high enough to cause it to sink to depths of 500 to 1000 m. As shown in Figure 4.13, it is clearly recognizable as a tongue of low-salinity water.

Antarctic Bottomwater, which is formed by sinking in the Weddell Sea, is defined by a single salinity and temperature (−0.4°C, 34.66‰). This is referred to as a *water type*, whereas water parcels of common original that exhibit a range of temperature and salinity values are said to constitute a *water mass*. The variability in temperature and salinity within a particular water mass is due to spatial and temperature variations in the processes responsible for their formation, i.e., cooling, evaporation, sea ice formation, etc. In general, the deeper water masses exhibit less variability than the shallower water masses. The most common are listed in Table 4.1. The deepest and, hence, the densest

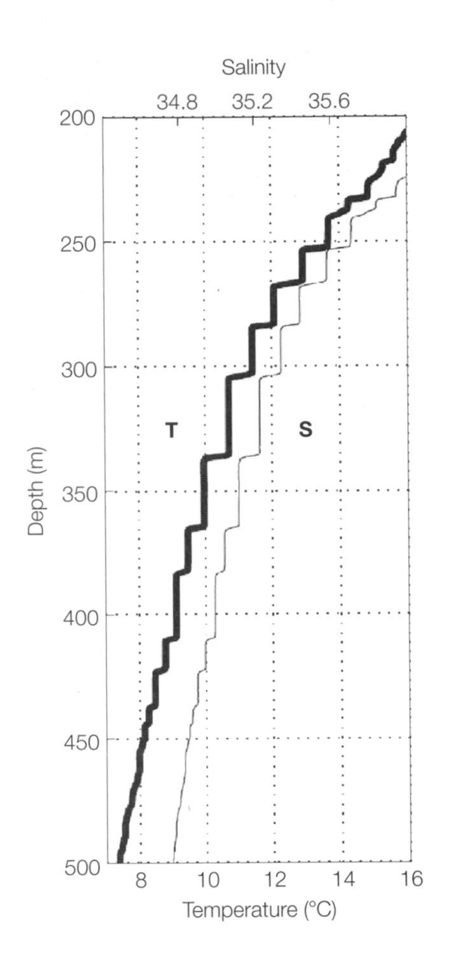

FIGURE 4.15

Profiles of potential temperature and salinity in the thermocline of the western tropical north Atlantic Ocean. (12° 24.7′N, 53° 40.0′W on 7 November 2001.) *Source*: From Schmitt, R. W., *et al.* (2005). *Science* 308, 685–688.

are the bottom waters. Of decreasing depth and density are the deep water, intermediate waters, and central waters.

4.3.1 Advection

In the open ocean, the major advective water motion is associated with the surface-water geostrophic currents and meridional overturning circulation. These flow paths are shown in Figures 4.4b and 4.6. Advection is much faster than molecular diffusion and turbulence. This enables water masses to retain their original temperatures and salinities as they are advected away from their sites of formation. Slow turbulent mixing with adjacent water masses eventually alters this temperature and salinity signal beyond

Table 4.1 Major Water Masses of the World Ocean.

Water Mass	Salinity (‰)	Temperature (°C)	Potential Density (g/cm^3)	Depth Range (m)
Antarctic Bottom (AABW)	34.66	–0.4	1.02786	4000–bottom
Antarctic Circumpolar	34.7	0–2	1.02775–1.02789	100–4000
Antarctic Intermediate (AAIW)	34.2–34.4	3–7	1.02682–1.02743	500–1000
Arctic Intermediate	34.8–34.9	3–4	1.02768–1.02783	200–1000
Mediterranean (MIW)	36.5	8–17	1.02592–1.02690	1400–1600
North Atlantic Central	35.1–36.7	8–19	1.02630–1.02737	100–500
North Atlantic Intermediate	34.73	4–8	1.02716–1.02765	300–1000
North Atlantic Deep and Bottom (NADW)	34.9	2.5–3.1	1.02781–1.02788	1300–bottom
South Atlantic Central	34.5–36.0	6–18	1.02606–1.02719	100–300
North Pacific Central	34.2–34.9	10–18	1.02521–1.02634	100–800
North Pacific Intermediate (NPIW)	34.0–34.5	4–10	1.02619–1.02741	300–800
Red Sea	40.0–41.00	18	1.02746–1.02790	2900–3100

recognition. Because of the sharp distinctions in temperature and salinity among the water masses, the pathway of meridional overturning circulation can be traced for long distances. This is called the *core technique of water mass tracing*. The application of this technique to Atlantic Ocean is shown in Figure 4.14 where the pathway of North Atlantic Deep Water (NADW), Antarctic Bottom Water (AABW), and AAIW are identified by their signature temperatures and salinities.

Although meridional overturning circulation moves water far more slowly than the geostrophic currents, it is important because it is responsible for water movement in the deep zone. As noted earlier, its return flow drives a net transport of surface water from the Pacific to the Indian to the Atlantic Ocean. The major features of MOC are a seasonal sinking of water masses in very limited geographic areas to form deep, bottom, and intermediate waters followed by their lateral transport throughout the ocean basins. These waters are returned to the sea surface by vertical advection and turbulent mixing. The upward advective flow is very slow but seems to be enhanced in some areas, such as the equatorial waters and the Southern Ocean. This advection appears to be wind-driven, drawing the deep, bottom, and intermediate waters toward the upwelling sites. Thus, the currents in the deep zone and the permanent thermocline are thought to be driven ultimately by the winds.

A key process in MOC is the formation and sinking of the bottom, deep, and intermediate water masses. This is thought to involve thermohaline currents induced by seasonal increases in surface water density. In the Greenland and Labrador Seas, NADW is formed by evaporative cooling and brine rejection associated with sea ice formation. The water sinks in narrow (1-km) cells called chimneys. In the Mediterranean Sea, evaporation alone is responsible for elevating density sufficiently to create Mediterranean Intermediate Water. Because of its high temperature, this water is not as dense as NADW. The densest water, AABW, forms in the Weddell Sea as a result of evaporative cooling. Another important water mass is AAIW, whose formation was described in the previous section. While intermediate waters are formed in the Pacific and Indian oceans, the only ocean where deep and bottom waters form is the Atlantic (with a small amount of AABW forming in the Pacific Ocean). Some water mass formation is also caused by caballing (Figure 3.5). Seasonal thermohaline convection is also responsible for the formation of shallow water masses, called mode waters. These form at latitudes of 40° and define the top of the permanent thermocline. The rest of the thermocline is defined by the various intermediate waters. The locations of formation of the bottom, deep, intermediate, and mode waters are shown in Figure 4.16. Note that the depiction of MOC in Figure 4.6 includes only the sites of formation of deep and bottom waters. A more detailed depiction of MOC showing the flow paths of the intermediate and mode waters is presented in Figure W10.1.

NADW flows south from its site of formation until it reaches the Southern Ocean where it joins up with AABW. The water masses then flow eastward under the influence of the Westerlies. A branch heads off into the Indian Ocean and the rest enters the South Pacific. All along these flow paths, upward advection and turbulent mixing slowly return the water to the surface where the geostrophic currents eventually carry it back to the Atlantic Ocean. Because a major feature of the flow paths is transport across latitudes,

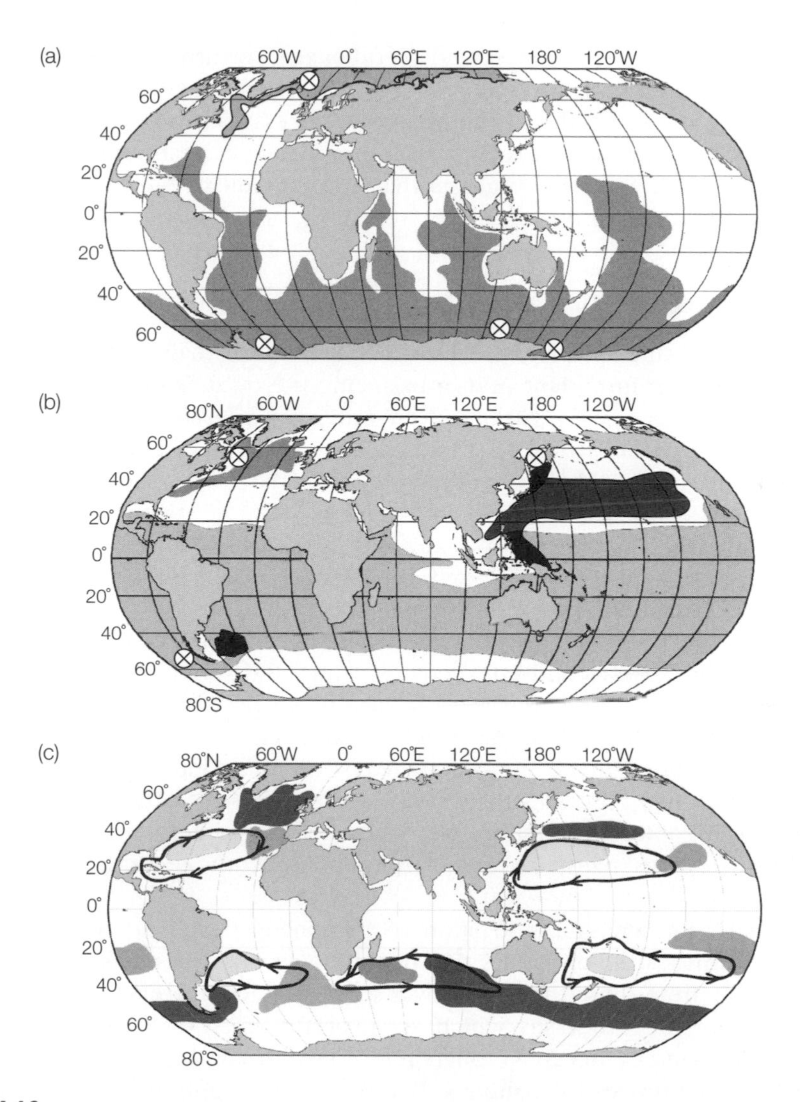

FIGURE 4.16

Sites of water mass formation. Shading shows the lateral extent of each water as far as it can be followed by the core technique of water mass tracing: (a) Bottom waters with X marking the spots of NADW (medium gray) and AABW (dark gray) formation. Shading shows the location of $\sigma_4 = 45.92\,\mathrm{kg/m^3}$. (b) Intermediate water with X marking the spots of AAIW (light gray), LSW (medium gray), and NPIW (dark gray) formation. Black areas are regions where mixing is essential for setting the temperature and salinity properties of the water masses. (c) Subtropical mode waters. Dark shading is subpolar mode waters. Medium and light shading is subtropical mode waters. The subtropical geostrophic gyre paths are superimposed as black arrows. *Source*: After Talley, L. D. (1999). *Mechanisms of Global Climate Change at Millennial Time Scales*, Geophys. Mono. Ser. 112. American Geophysical Union, pp. 1–22. (See companion website for color version.)

the entire loop has been termed Meridional Overturning Circulation. The formation of NADW plays an important role in global climate as it drives a net heat transport northward at all latitudes in the Atlantic Ocean. This heat is carried by the water flows that feed the formation of NADW. In contrast, the net heat transport in the other oceans is always poleward. This suggests that a disruption in the formation of NADW could affect global climate.

4.3.2 Turbulent Mixing

Turbulent mixing is typically 10^3 to 10^5 times slower than advective transport. Nonetheless, it plays an important role in balancing the ocean's energy budget, influencing its density structure, and returning deep waters to the sea surface. The major features of the ocean's density structure, a homogenous deep zone separated by a shallow thermocline, is the result of how the ocean dissipates the energy input its receives from winds, tides, and other sources. [The other sources are relatively negligible. They include: (1) heating and cooling by the atmosphere (thermohaline convection), (2) exchange of freshwater with the atmosphere (evaporation/precipitation), (3) geothermal heating, and (4) atmospheric pressure loading.]

Some of these energy inputs to the ocean fuel advective flows, such as the geostrophic currents and convective sinking, and some generate surface wind waves. All of the energy input is eventually dissipated by molecular diffusion. This proceeds by way of an energy cascade in which large-scale phenomena, such as the advective geostrophic flows that occur on the gyre scale (thousands of kilometers) transfer energy to nonadvective processes that occur on the mesoscale. The latter then transfer energy down to the microscale (submeter) where molecular diffusion can act to dissipate heat. How this energy transfer occurs is not well understood and hence is an active area of research.

What is known is that the nature of the nonadvective fluid motion is scale dependent, with the mesoscale processes that operate over hundreds of kilometers behaving quite differently from those that occur over scales of a few meters to tens of kilometers. Both differ from the processes which act over the microscale (submeter). To distinguish these phenomena, oceanographers refer to the larger scale processes as stirring and the smaller scale processes as mixing. Because oceanographers are limited in their ability to distinguish stirring from mixing, the term *turbulent mixing* is commonly used to refer collectively to all of the nonadvective processes.

Of particular interest are the mesoscale processes generated by internal waves. The two that appear to be most important are Rossby (planetary) waves and baroclinic mesoscale eddies. Rossby waves have wavelengths that range from a few hundred to several thousand kilometers. They move westward with internal heights of 10 to 100 m. They can cause substantial vertical water displacements, which is referred to as *eddy pumping*. For waves passing through the top of the thermocline, this displacement can bring nutrient-rich subsurface water into the euphotic zone, leading to the fertilization of phytoplankton. Rossby waves can travel across an entire ocean basin over a period of months to years. Their basin-scale movements are well correlated with phytoplankton pigments (chlorophyll) distributions, suggesting that they play a major role in controlling surface water productivity. Global warming is expected to cause an increase in their speed.

Most of the energy transmitted into the ocean appears to be dissipated through the formation of baroclinic mesoscale eddies (also known as meddies). They are closed vortices with length scales of a few hundred kilometers. They can extend several hundred meters and, hence, reach into the permanent thermocline. The most well known are the Gulf Stream Rings. In the stratified mid-latitude ocean, cyclonic mesoscale eddies play an important role in driving an eddy-pumped flux of nutrients. They tend to form at oceanic fronts, such as between the Gulf Stream and the North Atlantic subtropical gyre.

At the horizontal scale of a few kilometers, vertical shear is an important agent in generating turbulent mixing.[4] Shear occurs at the boundary between isopycnals, especially between two adjacent advective flows, such as the deepwater currents associated with MOC. Shear is also generated by interactions between internal waves of different frequencies and by water movement against the seafloor, especially in regions with rugged, steeply sloping topography, such as fracture zones, mid-ocean ridges, seamounts, and any structures that serve as sills separating basins. In the case of the last, turbulence is also produced when deep-water flows are forced to flow through narrow restrictions such as sills and channels. This is due to the increased shear generated by higher current velocities.

Tides appear to be the most important source of energy fueling turbulent mixing because of the limited depth penetration of the wind effects. The tides transmit energy into the deep sea in the form of internal waves with lengths on the order of a few hundred kilometers. An important source of tidally generated turbulence is drag along the seafloor. This process seems to be very important in the Southern Ocean because of its rugged topography. The Southern Ocean is also the place where the Westerlies have an unrestricted fetch, leading to very strong upwelling. The wind forcing induces such strong upwelling that the isopycnals located at 1300 m water depth immediately north of this divergence zone rise and outcrop at the sea surface between 50 and 60°S (Figure 4.14c). For these two reasons, the Southern Ocean is thought to play an essential role in determining the pattern and speed of MOC. In later chapters, we will see how this control on MOC causes the Southern Ocean to play a pivotal role in determining global nutrient distributions and the air-sea exchange of CO_2. These in turn influence biological productivity and potentially, climate.

Because of the stabilizing effect of density stratification, most turbulent mixing occurs laterally along isopycnals. Outside of the polar and subpolar zones, isopycnals tend to follow horizontal pathways (Figures 4.14c and d). Thus, lateral mixing does not involve movement across isopycnals, whereas vertical mixing does. Although suppressed by density stratification and, hence, very slow, diapycnal mixing is important for

[4] Shear stresses develop in fluids when adjacent particles have different velocities. This causes the fluid to deform and undergo turbulent mixing.

returning deep water to the surface ocean. Evidence for this is seen in Figures 4.14c and d, which illustrates that most of the bottom-water isopycnals (namely those of AABW) do not rise and outcrop at the sea surface at any location other than the sites of their formation. This requires that the mechanism by which the bottom waters are returned to the sea surface maintain the horizontal isopycnals. This mechanism is diapycnal mixing. Over long vertical space scales, a component of this mixing leads to a net very slow upward advection that balances out the downwelling of water into the deep zone. (In the thermocline, wind forcing in the Southern Ocean contributes to the upward return flow.)

Diapycnal mixing is also required to maintain the ocean's energy balance, enabling the dissipation of energy input to the deep sea. Because of the dampening effect of density stratification, diapycnal mixing occurs over relatively short space scales via turbulent eddies on the 1 to 100 cm scale and, if the heat and salt gradients permit, by double diffusion (Figure 4.15). In the deep sea, the turbulence that fuels diapycnal mixing is generated by (1) shear between adjacent advection flows, (2) tidal drag, (3) interactions between internal waves of different frequencies, and (4) the molecular diffusion that leads to double diffusion. At the sea surface, additional mechanisms involve small-scale density instabilities associated with water mass sinking, including loss of buoyancy from evaporative cooling and caballing. Although slow, the rates of vertical upwelling due to diapycnal mixing and slow upward advection are geographically variable. The slowest rates are found in mid-latitude thermoclines due to the stabilizing effect of strong density stratification. Along the equator, rates are enhanced due to horizontal shear generated by the strong wind-driven equatorial currents and countercurrent. Diapycnal mixing is responsible for the concave upward curvature of the isopycnals in the upper thermocline at the equator (Figures 4.14c and d). Similar curvatures are also seen in the temperature and salinity cross sections (Figures 4.14a and b) and in Figure 4.13.

The effect of turbulent mixing has been shown to follow the same behavior as molecular diffusion as previously shown in Eq. 3.6 (Fick's first law), where the diffusive flux, $F_{\text{diff}}(\text{mol m}^{-2}\,\text{s}^{-1})$, of a solute, C (mol/m^3), is given by:

$$F_c^{\text{diff}} = -D_z \frac{d[C]_z}{dz} = -D_z \frac{([C]_{z_z} - [C]_{z_o})}{z_z - z_o} \tag{4.1}$$

In this case, D_z is the *turbulent mixing coefficient*, which has units of m^2/s. Because of this mathematical similarity to molecular diffusion, turbulent mixing is also called *eddy diffusion*. Equation 4.1 has been written for turbulent flow occurring across depth (z). Similar equations can be written for turbulent mixing in the horizontal dimensions of x or y. As shown in Table 4.2, turbulent mixing coefficients are much larger than those of molecular diffusion (D_c). This causes turbulence to have a much larger effect on the distribution of a solute in solution than molecular diffusion. The most important exception to this are locations where water flow is restricted, such as in the pore waters of marine sediments, and on the short space scales over which double diffusion occur. Note that the $D_{x,y}$ for lateral mixing (10^2 to 10^8) is much larger than that for vertical mixing (10^0 to 10^1). This means that lateral mixing is faster than vertical mixing and

Table 4.2 Range of Molecular Diffusivity and Turbulent Mixing Coefficients in Natural Environments.

System	Diffusivity Coefficient (cm^2/s)[a]
Molecular	
In water	10^{-6}–10^{-5}
In air	10^{-1}
Turbulent in ocean	
Vertical, mixed layer[b]	0.1–10^4
Vertical, deep sea	0.1–10
Horizontal[c]	10^2–10^8
Turbulent in lakes	
Vertical, mixed layer[b]	0.1–10^4
Vertical, deep water	10^{-3}–10^{-1}
Horizontal[c]	10^1–10^7
Turbulent in atmosphere	
Vertical	10^4–10^{-5}
Note: Horizontal transport mainly by advection (wind)	
Mixing in rivers	
Turbulent vertical	1–10
Turbulent lateral	10–10^3
Longitudinal by dispersion	10^{-5}–10^6

[a] *1 cm^2/s = 8.64 m^2/d.*
[b] *Maximum numbers for storm conditions.*
[c] *Horizontal diffusivity is scale dependent.*
Source: From Lerman, A. (1979). Geochemical Processes: Water and Sediment Environments. *John Wiley & Sons, Inc., p. 57.*

exhibits a large degree of scale dependence. In other words, lateral mixing does not follow Fick's first law very well. The reason for this is that horizontal mixing occurs over very large spatial scales and, hence, involves stirring as well as mixing, whereas the spatial scale of vertical mixing is too small, due to suppression by density stratification, to support the stirring processes.

4.3.3 Conservative Mixing

Turbulence and advection can lead to the mixing of adjacent water masses (or types). These water motions create horizontal and vertical gradients in temperature and salinity. As illustrated in Figures 4.17a and 4.17b, vertical mixing at the boundary between two water types produces waters of intermediate temperature and salinity. Since mixing does not alter the ratios of the conservative ions, the water in the mixing zone acquires a salinity intermediate between that of the two water types. The salinity of the admixture is a direct function of the proportions of water mixing. This behavior is termed *conservative mixing*. Temperature also exhibits conservative mixing behavior if mixing is fast enough to swamp the effects of conduction.

Conservative mixing produces nonlinear vertical gradients in temperature and salinity as illustrated by the curved depth profiles shown in Figures 4.17a and 4.17b. This curvature is a consequence of the combined effects of turbulent mixing and advection as described in Chapter 4.3.4. In contrast, an *xy* plot of temperature versus salinity generates a straight line (Figures 4.17c, d, e). This type of plot is called a *T-S diagram*, and the linear relationship is called a *conservative mixing line*. The ends of this line are defined by the temperature and salinity of the original water masses (or types). The latter are also referred to as *mixing end members*.

A water parcel sampled from anywhere in the mixing zone should have a *T-S* signature that plots on the conservative mixing line established by the mixing end members. Thus, a 50-50 mixture of two adjacent water masses (or types) generates a temperature and salinity signature that is midway between that of the end members. Because of this linear relationship, the relative proportions of these end members in any admixture can be calculated from a system of two simultaneous equations:

$$x_a + y_b = 1 \tag{4.2}$$

$$x_a S_a + y_b S_b = S_{\text{mix}} \text{ or}$$

$$x_a T_a + y_b T_b = T_{\text{mix}}$$

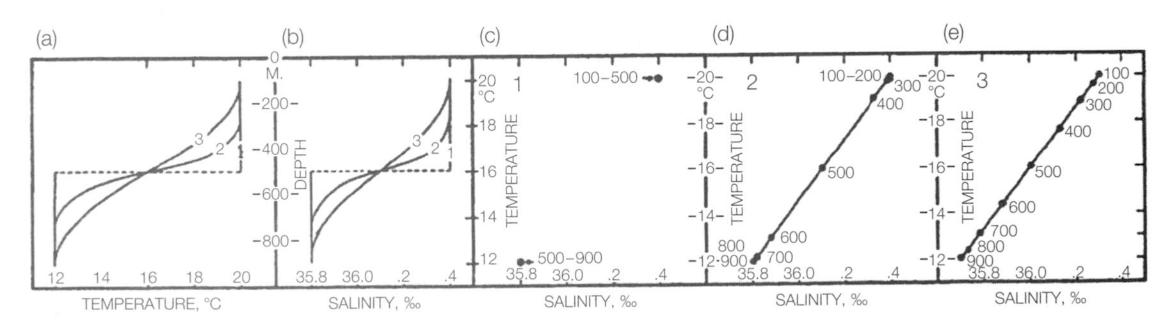

FIGURE 4.17

Effects of the progressive conservative mixing of two water types on the vertical profiles of (a) temperature, (b) salinity, and (c, d, e) *T–S* diagrams. Progressive mixing is represented by stages 1, 2, and 3 where stage 1 is the condition prior to initiation of mixing in which two water types are present. During Stage 2, mixing begins to blend the temperature and salinity characteristics of the two water types, leading to development of curvature in the depth profiles. In Stage 3, mixing has progressed sufficiently to induce curvature throughout the entire depth ranges of both water types. These are idealized curves; the exact forms found in the ocean will depend on the relative strengths and duration of the mixing processes. As shown in the progression of *T–S* diagrams (c, d, e), the resultant admixture generates a *T–S* signature that always plots on the conservative mixing line. *Source*: From Sverdrup, H. U., *et al.* (1942) *The Oceans*: *Their Physics*, *Chemistry*, and *General Biology*, Prentice-Hall, p. 144.

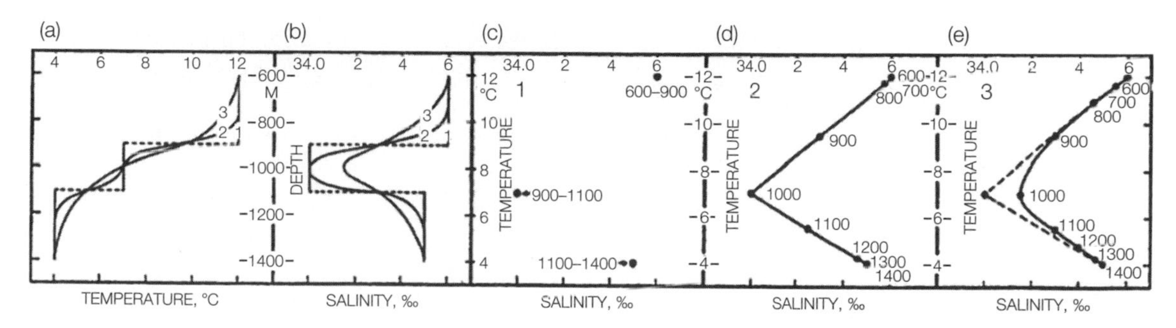

FIGURE 4.18

Effects of progressive conservative mixing of three water types. This diagram is analogous to Figure 4.17 except that three water types are mixing. Note that in mixing stage 3, the *T–S* diagram exhibits nonlinear behavior between 900 and 1100 m. This reflects the mixing of three end members over this depth range. *Source*: From Sverdrup, H. U., *et al.* (1942) *The Oceans: Their Physics, Chemistry, and General Biology*, Prentice-Hall, p. 144.

in which x_a = fraction of water mass a and y_b = fraction of water mass b contributing to the mixture. The second equation is based on the salinity (or temperature) of water mass a (S_a or T_a), water mass b (S_b or T_b), and the mixture (S_{mix} or T_{mix}).

Figure 4.18 illustrates the effects of mixing when three water masses (or types) are present. In this case, mixing occurs at the upper boundary between the surface and intermediate depth layers and at the lower boundary between the intermediate and bottom depth layers. Mixing at the upper boundary establishes a conservative mixing line that is distinct from the line established by mixing at the lower boundary. Given sufficient mixing (labeled stage 3 in Figure 4.18e), some degree of admixture of all three waters can occur. The resultant family of T-S points will lie within a triangular region defined by the T-S signatures of the three end members.

T–S diagrams are commonly used to identify water masses. An example is presented in Figure 4.19 for the South Atlantic, showing the presence of four water masses: central water (< 800 m), AAIW (800 to 1000 m), NADW (1500 to 3000 m), and AABW (> 4000 m).[5] Because *T–S* diagrams are constructed from observational data, the mixing end members are not necessarily included. To establish the identities of these end members, the *T–S* values of the water masses and types given in Table 4.1 can be plotted onto the *T–S* diagrams. By doing this, each of the inflections in the *T–S* diagram can be attributed to the contribution of a unique end member.

T–S diagrams based on ocean data typically exhibit curvature. Causes for this nonlinearity include: (1) mixing of more than two end members (Figure 4.18e), (2) temporal variations in environmental conditions during formation of the end members, and (3)

[5] These data are from latitudes that are too far south to be affected by admixture with Mediterranean seawater.

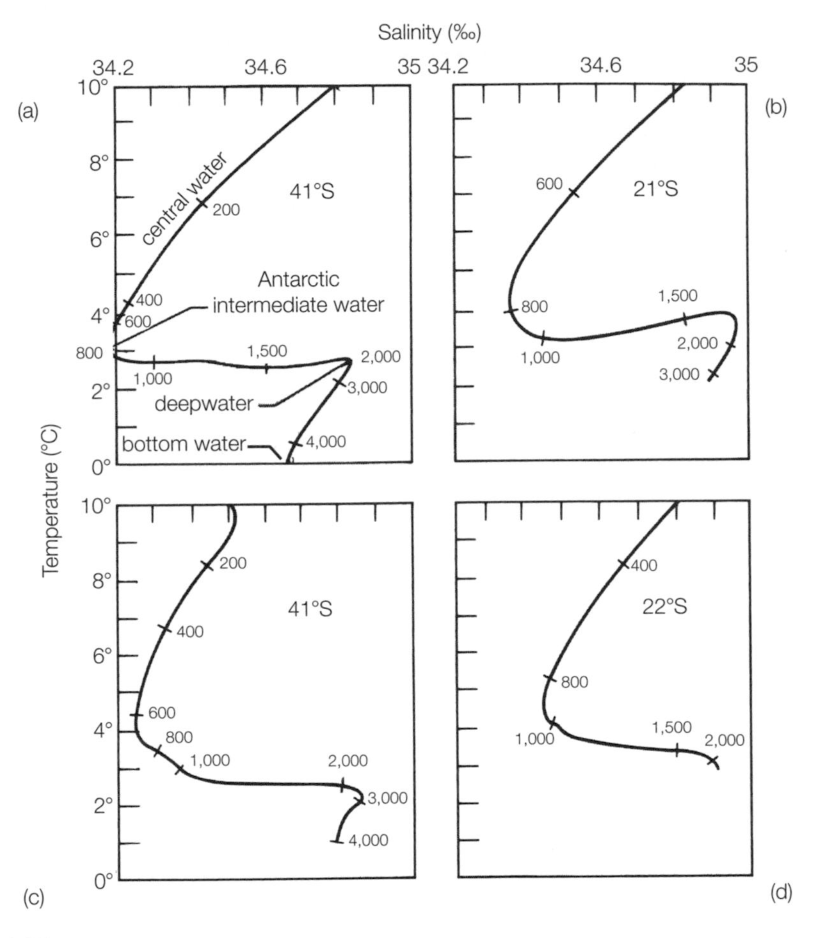

FIGURE 4.19

T–S diagrams for the western and eastern basins of the South Atlantic. (a) and (b) represent two locales in the western basin of the South Atlantic on the same line of longitude but different by 20° of latitude. (c) and (d) are in the eastern basin of the South Atlantic, again about the same 20° of latitude apart from one another but on the same longitude. The small numbers on each graph line represent depths in meters. The horizontal portion of each diagram indicates mixing.

nonconservative behavior caused by double diffusion (Figure 4.15) and geothermal heating. The latter is associated with production of oceanic crust at spreading centers and mantle hot spots.

4.3.4 One-Dimensional Advection-Diffusion Model

Oceanographers follow three basic approaches to collecting data for the study of biogeochemical processes: (1) they measure concentrations over time and space, (2) they make direct measurements of chemical fluxes, and (3) they perform laboratory experiments that simulate in situ processes. In situ measurement of fluxes and laboratory simulations are difficult and expensive, so most of what we know about biogeochemical processes is based on inferences from vertical and horizontal concentration profiles. In the preceding section, we saw how mixing alters vertical temperature and salinity profiles and how to use *T–S* diagrams to identify water masses. By employing mathematical models, these same data can be used to infer rates of water motion. This is commonly done because direct measurements of water movement are also technically difficult and expensive. Other conservative tracers used to estimate rates of water movement include ^3He and freons.

Whereas the concentration profiles of the conservative elements are shaped primarily by physical processes, those of the nonconservative ones are shaped by both physical and biogeochemical processes. The effects of the latter can be isolated from the former using site-specific salinity and temperature information to identify water masses and rates of movement. This is why marine chemists rely on temperature and salinity profiles to aid in their interpretation of the concentration distributions of the nonconservative solutes. Once the effects of the physical processes are recognized, the effects of the biogeochemical processes can be identified and investigated. The comparative inspection of concentration profiles to identify physical and biogeochemical processes is an essential skill for a marine scientist. Therefore many examples are provided in this text. As with the conservative properties, mathematical models can be used to infer rates of nonconservative behavior, which are either in situ production or consumption. An example of a mathematical approach to computing biogeochemical processing rates is provided later but first requires obtaining estimates of the rates of water movement using the temperature and salinity data.

In-depth profiles, the combined effects of advection and turbulent mixing can cause the concentrations of conservative solutes to exhibit curvature. The degree of this curvature reflects the relative rates of vertical advection and turbulent mixing and tends to be largest in the pycnocline. Downward vertical advection is limited to the few sites where water masses form and subsequently sink. The return flow occurs through upward vertical advection and is thought to occur over most of the ocean outside of the sites of water mass formation. Velocities are thought to be slow, ranging from 4 to 10 m/y depending on location, with the exception of the Southern Ocean where the Westerlies greatly enhance Ekman transport, supporting more rapid rates of upward vertical advection.

The mathematical models used to infer rates of water motion from the conservative properties and biogeochemical rates from nonconservative ones were first developed in the 1960s. Although they require acceptance of several assumptions, these models represent an elegant approach to obtaining rate information from easily measured constituents in seawater, such as salinity and the concentrations of the nonconservative chemical of interest. These models use an Eulerian approach. That is, they look at how a conservative property, such as the concentration of a conservative solute C, varies over time in an infinitesimally small volume of the ocean. Since C is conservative, its concentrations can only be altered by water transport, either via advection and/or turbulent mixing. Both processes can move water through any or all of the three dimensions

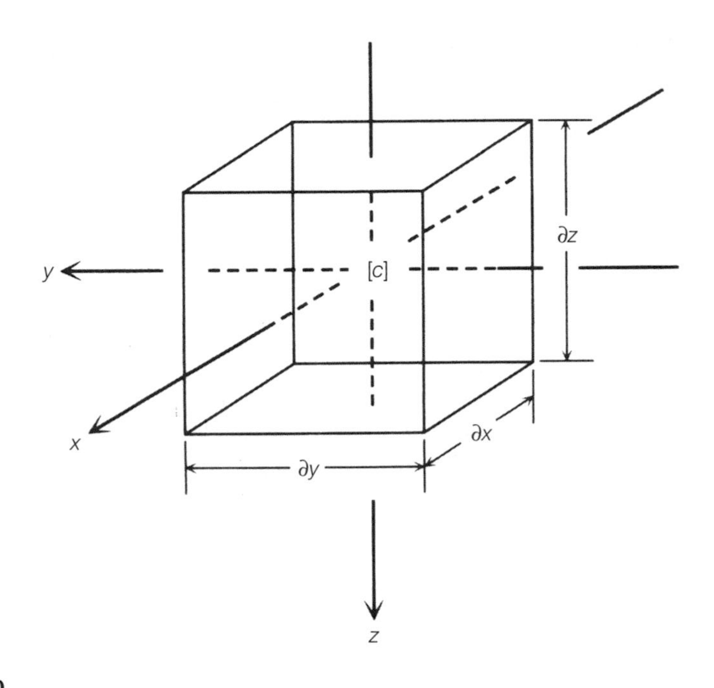

FIGURE 4.20

Directions of seawater transport through an infinitesimally small volume of the ocean of a conservative solute, C. The spatial frame of reference is fixed in place, while advection and turbulence move seawater through the box.

(x, y, and z) as shown in Figure 4.20. This can lead to concentration changes over time and/or space within the box in any or all of the three dimensions.

The total flux of solute C is the sum of the diffusive and advective fluxes

$$F_C^{\text{tot}} = F_C^{\text{diff}} + F_C^{\text{adv}} \tag{4.3}$$

F_C^{diff} can be computed from Fick's first law (Eq. 4.1) and has units of Mass/(Length^2 Time). The advective flux of C is caused by water transport, so F_C^{adv} is given by

$$F_C^{\text{adv}} = v_z\,[C] \tag{4.4}$$

where v_z is the vertical advective water velocity and has units of Length/Time. [Recall that the units of concentration for $[C]$ are Mass/(Vol seawater) = Mass/Length^3. If concentration is given as (Mass solute)/(Mass seawater), the density of the seawater can be used to convert to Mass/(Vol seawater)]. Note that F_C^{adv} also has units of mass/(length^2 time). By oceanographic convention, v_z is assigned a negative value for upward transport and positive for downward transport.

To construct a model that can describe changes in $[C]$ over time and space, we must have a way of expressing the mass fluxes as functions of time. This is done with

Gauss's theorem, which states that the rate of change in $[C]$ with respect to time (t) and at some depth (z) is equal to the negative spatial gradient of the mass flux (F),

$$\left.\frac{\partial[C]}{\partial t}\right|_{z=\text{constant}} = -\left.\frac{\partial F}{\partial z}\right|_{t=\text{constant}} \tag{4.5}$$

By rearranging, we see that Gauss's theorem obeys the law of conservation of mass

$$\left.\frac{\partial[C]}{\partial t}\right|_{z=\text{constant}} + \left.\frac{\partial F}{\partial z}\right|_{t=\text{constant}} = 0 \tag{4.6}$$

Equations 4.5 and 4.6 are examples of partial differential equations because they contain partial derivatives, i.e., $\frac{\partial[C]}{\partial t}$ and $\frac{\partial F}{\partial z}$. The ∂ symbol indicates that $[C]$ and F are functions of several variables. In this case, the variables are time (t) and depth (z), respectively. To evaluate a partial derivative, all but one of the variables must be held constant; in the case of $\frac{\partial[C]}{\partial t}$, depth ($z$) is held constant and only time (t) is considered a variable. In the case of $\frac{\partial F}{\partial z}$, time ($t$) is held constant and only depth (z) is considered a variable.

We can rewrite Gauss's theorem (Eq. 4.5) in terms of F_C^{diff} and F_C^{adv},

$$\left.\frac{\partial[C]}{\partial t}\right|_{z=\text{constant}} = -\left.\frac{\partial F_C^{\text{diff}}}{\partial z}\right|_{t=\text{constant}} + -\left.\frac{\partial F_C^{\text{adv}}}{\partial z}\right|_{t=\text{constant}} \tag{4.7}$$

Substituting in the expressions for F_C^{adv} and F_C^{diff} from Eqs. 4.1 and 4.4 yields:

$$\left.\frac{\partial[C]}{\partial t}\right|_{z=\text{constant}} = \left(-\frac{\partial\left(-D_z\frac{\partial[C]_z}{\partial z}\right)}{\partial z}\right) + \left(-\frac{\partial\left(v_z[C]_z\right)}{\partial z}\right) = D_z\frac{\partial^2[C]_z}{\partial z^2} - v_z\frac{\partial[C]_z}{\partial z} \tag{4.8}$$

in which D_z and v_z are assumed to be constant and the units of $\frac{\partial[C]}{\partial t}$ are mass/(length^3 time). By analogy, similar terms can be included to cover the other two dimensions. But since horizontal gradients tend to be much smaller than the vertical gradients, they make a negligible contribution to the concentration change in the small box (Figure 4.20).[6] Therefore, we do not include them, and, hence, Eq. 4.8 is referred to as the *one-dimensional advection-diffusion model.*

In practice, C is assumed to be present in steady state $\left(\frac{\partial[C]_z}{\partial t} = 0\right)$, so

$$0 = D_z\frac{\partial^2[C]_z}{\partial z^2} - v_z\frac{\partial[C]_z}{\partial z} \tag{4.9}$$

Since C is only a function of z, Eq. 4.9 can now take the form of an ordinary differential equation, i.e.,

$$0 = D_z\left(\frac{d^2[C]_z}{dz^2}\right) - v_z\left(\frac{d[C]_z}{dz}\right) \tag{4.10}$$

[6] Important exceptions to this are the strong horizontal gradients in chemical composition found in hydrothermal plumes (Chapter 19.4).

In the case where no turbulent mixing is occurring and only vertical advection is transporting water and C, then $[C]$ does not vary with depth. The depth profile will have the shape of a vertical straight line. In other words, vertical advection alone cannot explain the presence of a vertical salinity gradient. Turbulent mixing must be acting for a gradient to form and be maintained.

In the case where no advection is occurring and only turbulent mixing is acting on C, Eq. 4.8 is reduced to:

$$\left.\frac{\partial C}{\partial t}\right|_{z=\text{constant}} = -\left.\frac{\partial F_C^{\text{diff}}}{\partial z}\right|_{t=\text{constant}} = \left(-\frac{\partial\left(-D_z\frac{\partial[C]_z}{\partial z}\right)}{\partial z}\right) = D_z\left(\frac{\partial^2[C]_z}{\partial z^2}\right) \tag{4.11}$$

which is called *Fick's second law*. If a steady state exists and D_z is constant over time and depth, the solution to Eq. 4.11 is

$$[C]_Z = [C]_o + ([C]_L - [C]_o)\frac{Z}{L} \tag{4.12}$$

where L is the thickness of the mixing zone, Z is the depth below the top of the mixing zone, and C_L and C_o are the salinity at the bottom and top of the mixing zone, respectively. This relationship gives rise to linear depth profiles. Since this is a diffusive flux, the two possible profile shapes are the same as was illustrated in Figure 3.8 for molecular diffusion.

If both advection and turbulent mixing are acting on C, the solution to the one-dimensional advection-diffusion model at steady state (Eq. 4.10) is

$$[C]_Z = [C]_o + ([C]_L - [C]_o)\left(\frac{1 - e^{\frac{v_z}{D_z}Z}}{1 - e^{\frac{v_z}{D_z}L}}\right) \tag{4.13}$$

where L is the thickness of the mixing zone, Z is the depth below the top of the mixing zone, and $[C]_L$ and $[C]_o$ are the salinity at the bottom and top of the mixing zone, respectively. This equation gives rise to nonlinear distributions of $[C]_Z$. In other words, the interaction of turbulent mixing with advection can cause curvature in the vertical and horizontal profiles of solute concentrations and temperature. (In the case of the nonconservative chemicals, biogeochemical processes can also contribute to the observed curvature.)

If neither D_z nor v_z is known, depth profiles of salinity are used to estimate the value of the ratio, v_z/D_z. This is done using curve-fitting computer programs that find the best match between the observed salinity profile and one generated using physically reasonable values for v_z and D_z. Specifically, v_z is thought to be on the order of a few meters per year (4 to 10 m/y) and D_z to have values ranging from 0.1 to 10 cm^2/s. Note that this model can only be applied over depth intervals where a salinity gradient exists, i.e., an interval in which water masses are mixing. Thus $[C]_L$ and $[C]_o$ represent the cores of the water masses that are mixing over the depth interval. Since these water masses are usually far from their sites of formation, their T–S signatures have been somewhat modified by turbulent mixing. Therefore the top and bottom of the mixing zone is usually identified from T–S diagrams. This then determines which samples to use for $[C]_L$ and $[C]_o$.

An example of how well the one-dimensional advection-diffusion model fits real data is given in Figure 4.21a using depth profiles of salinity from the southeastern Atlantic Ocean. Two vertical mixing zones were modeled in this profile. The upper one covers the admixture of South Atlantic Central Water with underlying AAIW. The lower one covers the admixture of AAIW with underlying NADW. In the upper zone $[C]_o > [C]_L$, and in the lower zone, $[C]_o < [C]_L$, which serves to illustrate the effect of the relative sizes of $[C]_L$ and $[C]_o$ on the shape of the profile. In both zones, $|D_z/v_z| >> 1$ and salinity profile exhibits curvature. Larger values of $|D_z/v_z|$ intensify this curvature and lower ratios diminish it. A reversal in the direction of vertical advection to downward transport would alter the curvature direction, e.g., from concave upward (∪) to concave downward (∩).

The ratio v_z/D_z can then be used to calculate a chemical reaction rate for a nonconservative solute, S. To do this, the one-dimensional advection-diffusion model is modified to include a chemical reaction term, J. This new equation is called the one-dimensional advection-diffusion-reaction model and has the following form:

$$0 = D_z\left(\frac{d^2[S]_z}{dz^2}\right) - v_z\left(\frac{d[S]_z}{dz}\right) + J \qquad (4.14)$$

This model was first applied to dissolved oxygen gas (O_2) profiles to estimate the rate of aerobic respiration. This biological process is responsible for the presence of a pronounced mid-depth O_2 concentration minimum in the mid- and low latitudes throughout all the ocean basins. The concentration minimum in the Atlantic can be seen in Figure 4.14e. The solution to Eq. 4.14, in the presence of an upward vertical advection, is

$$[S]_Z = [S]_o + ([S]_L - [S]_o)\left(\frac{1 - e^{\frac{v_z}{D_z}Z}}{1 - e^{\frac{v_z}{D_z}L}}\right) + \frac{J}{v_z}Z - \frac{J}{v_z}L\left(\frac{1 - e^{\frac{v_z}{D_z}Z}}{1 - e^{\frac{v_z}{D_z}L}}\right) \qquad (4.15)$$

An estimate of J is obtained via curve fitting using an approach analogous to the one described earlier for the salinity profiles where D_z/v_z is obtained from the salinity profile curve fitting.

An example of the use of the one-dimensional advection-diffusion-reaction model is given in Figure 4.21b. The estimated O_2 consumption rate in the upper mixing zone is -0.023 $mLO_2L^{-1}\,y^{-1}$.[7] The computed O_2 consumption rate in the lower mixing zone is -0.0026 $mLO_2\,L^{-1}\,y^{-1}$. A plausible explanation for the decrease in oxygen consumption rate with depth is that the concentration and lability of the sinking detritus, which fuels aerobic respiration, decreases with depth. Note that the J term in the model represents a net reaction rate to which several processes are likely contributing, including

[7] This unusual concentration unit will be discussed in Chapter 6.

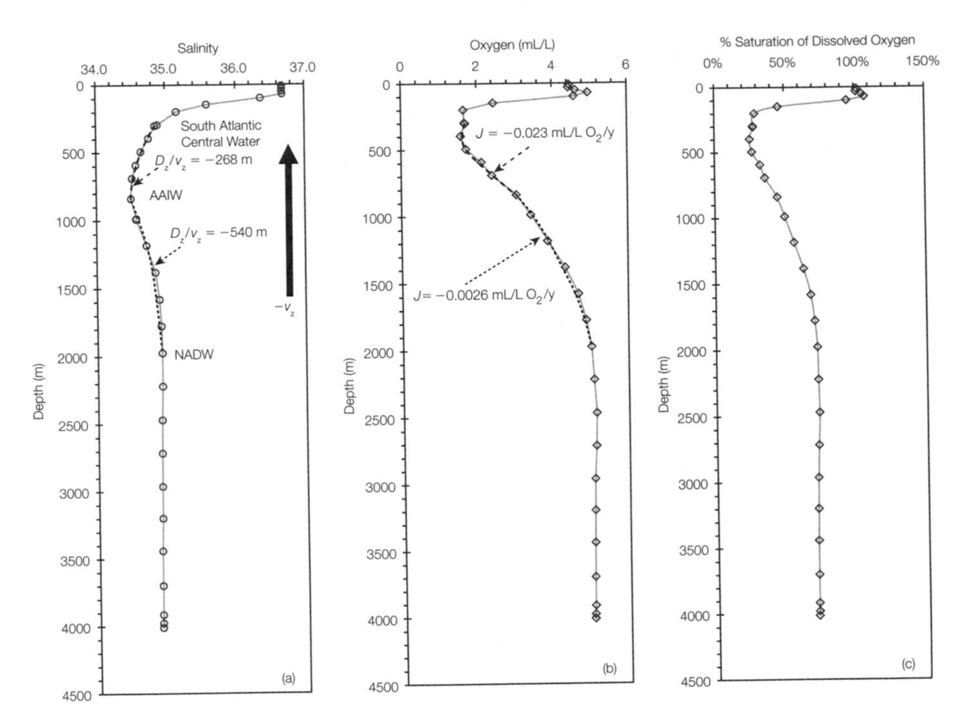

FIGURE 4.21

Depth profiles of (a) salinity (‰), (b) dissolved oxygen (mL/L), and (c) percent saturation of dissolved oxygen in the Southeastern Atlantic Ocean (9°30'W 11°20'S). Samples were collected in March 1994. Dotted lines represent the curves generated by the one-dimensional advection-diffusion model (see text for details). The values of D_z, v_z, and J are the ones that best fit the data. Data are from Java Ocean Atlas (http:/odf.ucsd.edu/joa). Values of percent saturation of oxygen less than 100 reflect the effects of aerobic respiration. Values greater than 100 indicate a net input, such as from photosynthesis. (See companion website for color version.)

aerobic respiration of bacteria, zooplankton, fish, and even mammals! The high O_2 concentrations in the surface water are due to in situ production from photosynthesis and physical exchange with the atmosphere across the air-sea interface.

For some depth profiles, the dissolved oxygen data are best fit with a J term that is first order, i.e., $J = -k[O_2]_z$, or exponential, i.e., $J = [O_2]_{z_o} e^{-kz}$, where k is the reaction rate constant. No mechanistic understanding of the chemical reaction is implied by the mathematical form of J. It is no more than a function that provides the best empirical fit to the data. Despite the assumptions and limitations of the one-dimensional advection-diffusion model, it has provided important information on the rates of physical and chemical processes in the ocean and pore waters of marine sediments. An example of the latter is presented in Chapter 12.

CHAPTER 5

The Nature of Chemical Transformations in the Ocean

All figures are available on the companion website in color (if applicable).

5.1 INTRODUCTION

In Chapter 4, we saw how conservative chemicals are used to trace the pathway and rates of water motion in the ocean. True conservative behavior is exhibited by a relatively small number of chemicals, such as the major ions and, hence, salinity. In contrast, most of the minor and trace elements display nonconservative behavior because they readily undergo chemical reactions under the environmental conditions found in seawater. The rates of these reactions are enhanced by the involvement of marine organisms, particularly microorganisms, as their enzymes serve as catalysts. Rates are also enhanced at particle interfaces for several reasons. First, microbes tend to have higher growth rates on particle surfaces. Second, the solution in direct contact with the particles tends to be highly enriched in reactants, thereby increasing reaction probabilities. Third, adsorption of solutes onto particle surfaces can create favorable spatial orientations between reactants that also increases reaction probabilities.

Some chemical reactions can cause elements to switch phases. For example, photosynthesis by marine phytoplankton transforms the oxygen in liquid water ($H_2O(l)$) and carbon dioxide ($CO_2(g)$) into algal biomass (organic matter abbreviated as $CH_2O(s)$ and oxygen gas ($O_2(g)$). This example illustrates that an element can be present in many chemical forms, which are called *species*. In this case, oxygen is present in four species, i.e., H_2O, CO_2, $C(H_2O)$, and O_2. Because of their unique molecular forms, each of these species can undergo further chemical reactions. For example, CO_2 can react with H_2O to produce H_2CO_3 (carbonic acid). Organic matter can be consumed by heterotrophs, such as zooplankton. In this chapter, we will see that O_2 can react with dissolved iron (Fe^{2+}) to create several solid phases, namely the minerals Fe_2O_3 (hematite), Fe_3O_4 (magnetite), $FeO(OH)$ (goethite), and $Fe(OH)_3$ (ferrihydrite).

Chemical transformations play a controlling role in determining the growth and distribution of marine organisms. Some marine organisms are able to control the speciation of chemicals in their surrounding seawater, which enables them to enhance food availability, reduce the effects of toxins, fend off predators, and improve their reproductive success rates. For example, some phytoplankton have been shown to detoxify seawater by releasing dissolved organic molecules that bind with, and thereby inactivate, harmful chemicals such as copper. Other organic exudates stimulate plankton growth by forming dissolved organometallic complexes with sparingly soluble micronutrients, such as iron. The formation of these complexes increases the solubility of the micronutrients and, hence, their bioavailability to the plankton. These examples demonstrate the important roles of chemical transformations and speciation in marine biological and geological phenomena. They also highlight the critical role of chemical transformations in determining transports within the crustal-ocean-atmosphere factory. In this chapter, the ways and means by which marine scientists study chemical transformations are described using the story of one element, iron, as a case study.

5.2 TRACKING NONCONSERVATIVE BEHAVIOR

The easiest technique for establishing the nonconservative behavior of an element, or one of its chemical species, is to compare its concentration to that of a conservative tracer. Salinity is typically used because it is a standard measurement performed on most seawater samples.

Figure 5.1 illustrates how the total dissolved iron concentration in an estuary changes as a function of salinity. If conservative mixing was the sole process controlling the iron concentration, then all of the data would fall on a straight line connecting the pure freshwater end member (S = 0‰) and the pure marine end member (S = coastal ocean value, i.e., approximately 35‰). This straight line is referred to as the *conservative mixing line*. Iron is plotted on the y-axis because its concentration is considered to be "dependent" on salinity. Assuming that this estuarine profile remains constant over time, the iron concentrations are less than what would be predicted from the conservative mixing line. This can be attributed to the effect of some kind of net chemical removal.

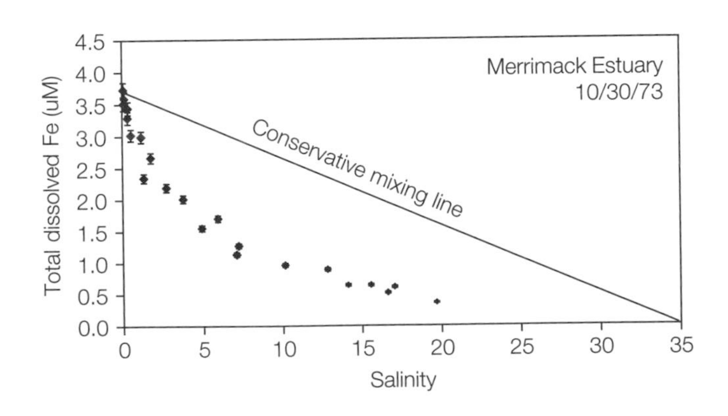

FIGURE 5.1

Total dissolved iron versus salinity in the Merrimack Estuary (Massachusetts). Data from Table 2 in Boyle, E., *et al*. (1974). *Geochimica et Cosmochimica Acta* 38, 1719–1728.

Additional information is required to determine what is causing this net removal. In the case of iron, research has demonstrated that its solubility decreases with increasing salinity leading to the formation of two types of solids: (1) iron oxide minerals, and (2) organic flocs. Some iron is also removed by uptake as a micronutrient by plankton. The flocs form from the co-precipitation of iron with the high-molecular-weight dissolved organic compounds naturally present in river water.

Note that plots, such as Figure 5.1, provide information only on the net outcome of chemical reactions. In the case of iron, a small addition does take place in estuaries as a result of desorption of Fe^{2+} from the surfaces of riverine particles. As these solids move through the estuarine salinity gradient, the major cation concentrations increase and effectively displace the iron ions from the particle surfaces. Since this release of iron is much smaller than the removal processes, the net effect is a chemical removal of iron. Sedimentation of these iron-enriched particles serves to trap within estuaries most of the riverine transport of reactive iron, thereby preventing its entry into the oceans.

In the case of iron, the river water concentration is greater than that in seawater. This relationship is generally seen for solutes with terrestrial sources, such as those released into river water as a result of chemical weathering and pollution. Some of these solutes exhibit a net addition in estuaries, so their concentrations lie above the conservative mixing line (Figure 5.2b). Although the major ions have terrestrial inputs from chemical weathering, their slow chemical removal times in the ocean cause their seawater concentrations to be greater than their riverine concentrations. Thus, their conservative mixing plots take the general form shown in Figure 5.2a. The

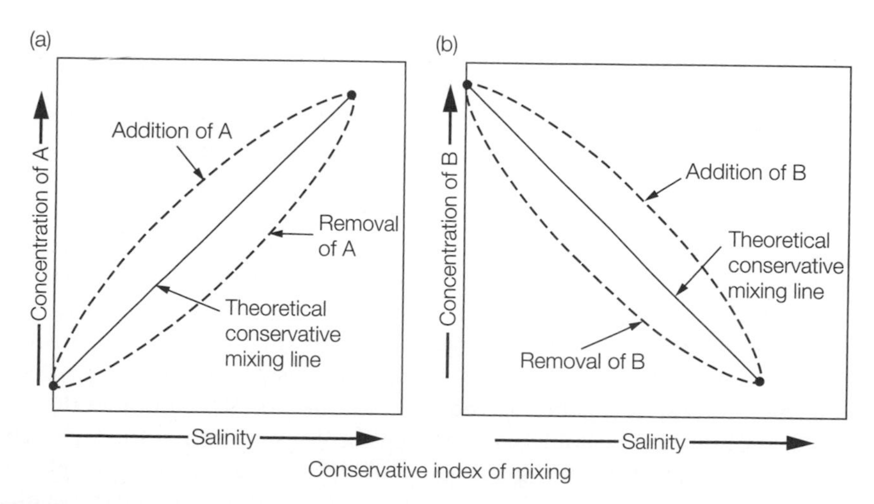

FIGURE 5.2

Possible salinity relationships resulting from conservative and nonconservative behavior when (a) the concentration of a chemical A in the lower salinity end member is lower than that of the higher salinity end member and (b) the concentration of a chemical B in the lower salinity end member is greater than that of the higher salinity end member.

approach illustrated in Figure 5.2 is also used to evaluate the degree of nonconservative behavior exhibited by chemicals during the vertical and lateral mixing of open ocean water masses.

5.3 DEFINING CHEMICAL SPECIES

In Figure 5.1, the fraction of iron whose concentration is being reported is identified as the "total dissolved iron" concentration. In practice, this fraction is operationally defined by the analytical method used in its measurement. For the data in Figure 5.1, the total dissolved iron concentration was determined by filtration to remove the solid iron, followed by *colorimetric analysis* to quantify the solutes. Another analytical technique, such as filtration followed by atomic absorption spectrophotometry, might yield a different total dissolved concentration, so it is important to be aware of the analytical methods used. To address this issue, marine chemists engage in intercalibration experiments to assess differences in results from various analytical methods.

Why do different measurement methods sometimes yield quite different results? To answer this question, we must recognize that a dissolved chemical can be present in several species. In the case of iron, the major species are the aquo ions, Fe^{2+}, Fe^{3+}, $Fe(OH)^{2+}$, $Fe(OH)_2^+$, $Fe(OH)_2^{4+}$, $Fe(OH)_3^0$, $Fe(OH)_4^-$, and dissolved organo-iron complexes. Each analytical method detects these species to varying degrees. The detectability of each species depends on how reactive a given species is to the physicochemical processes inherent in the measurement technique. The analytical methods that generate the least ambiguous results are the ones that come closest to detecting all of the species that constitute the specified fraction, such as total dissolved iron. Difficulties also arise during the separation of fractions into the dissolved, colloidal, and particulate phases. As mentioned in Chapter 3, filtration is commonly used with the particle size cutoffs for each fraction determined by the pore sizes of the filtration medium. For example, total organic carbon (TOC) can be fractionated into dissolved organic carbon (DOC), particulate organic carbon (POC), and colloidal organic carbon pools. DOC contains a multitude of molecules ranging from small compounds, like glucose, to very large polymers, such as alginic acid. Conversely, DOC is one component of dissolved organic matter (DOM).

As shown in Figure 5.3, chemicals, such as iron, can be present in a variety of species and phases that span a large size spectrum. The dissolved fraction can include inorganic complexes, organometallic molecules, and the uncomplexed ions. In the case of iron, two oxidation states are possible, so the free ion can be in the form of Fe^{2+}(aq) or Fe^{3+}(aq). In the colloidal and particulate phases, iron can be present as part of a mineral (inorganic) or an organic molecule. Within the particulate phase, a distinction is often made between the fraction that is adsorbed, usually electrostatically as an ion, onto the surface and the fraction that is covalently bound into the crystal lattice.

The relative abundance of these species is commonly presented as a pie chart or as a phase-style diagram. In the case of the latter, the concentrations of each species is plotted as a function of a master variable such as pH or salinity. Examples are shown in Figure 5.4.

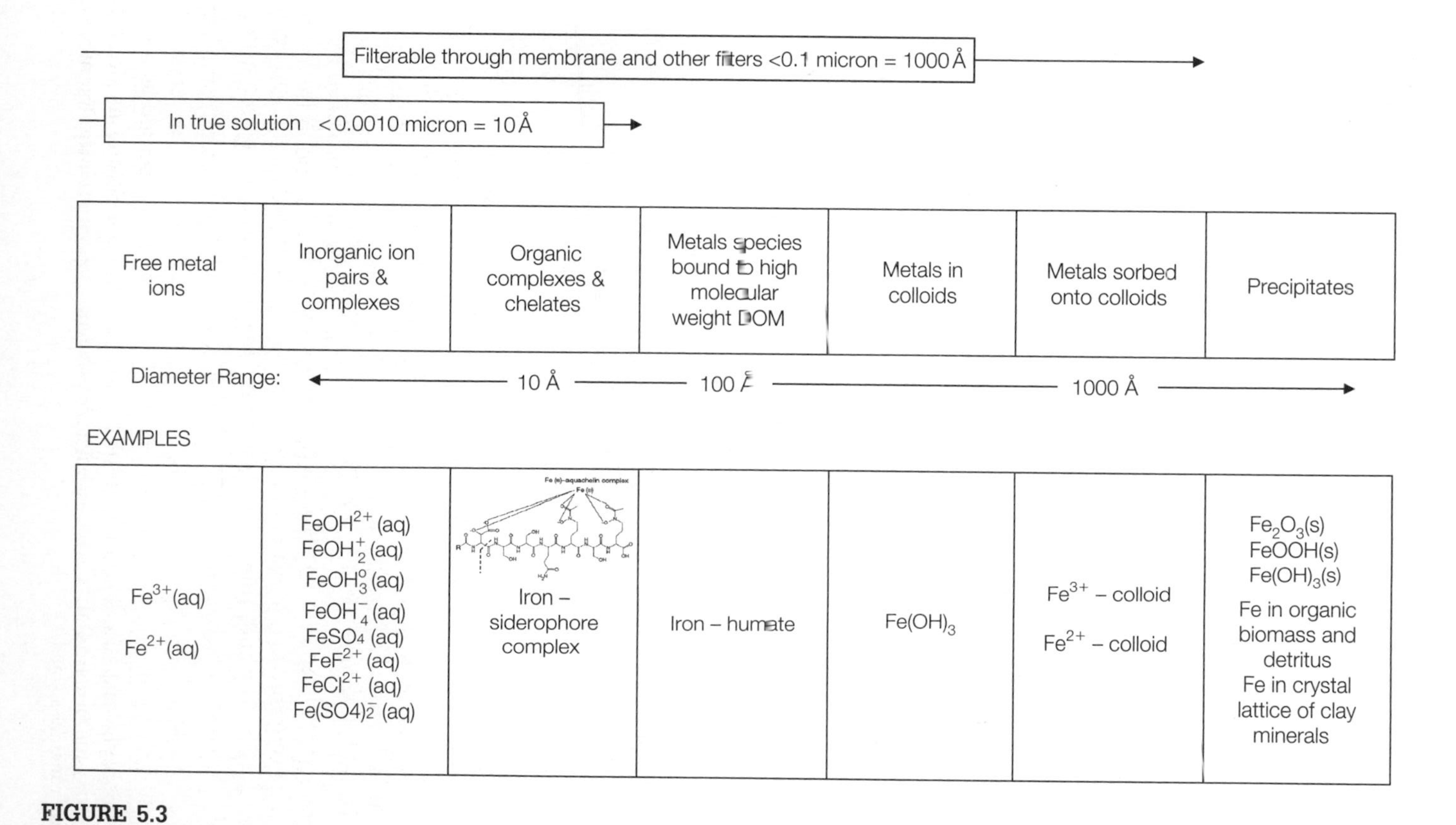

FIGURE 5.3

Chemical species in seawater.

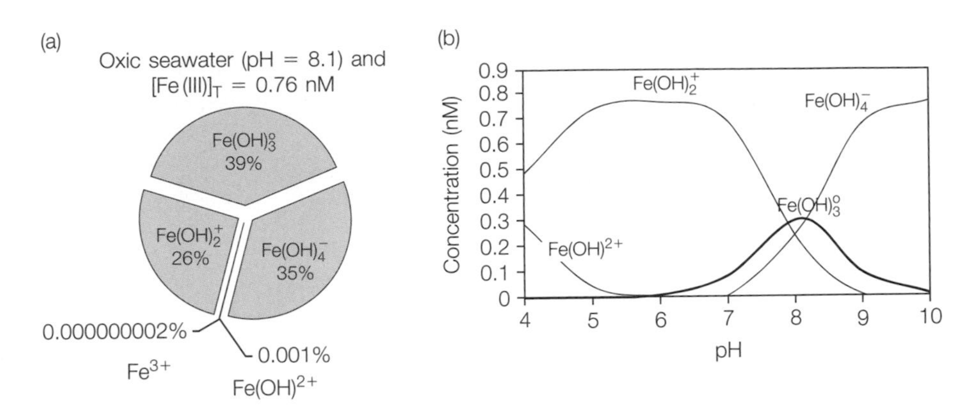

FIGURE 5.4

Speciation of Fe(III) in oxic seawater in the absence of organic matter. (a) Pie chart and (b) concentration as a function of pH.

Information on speciation is also commonly shown on x–y plots that display the results of an experiment. An example is given in Figure 5.5 from growth bioassays performed on marine phytoplankton that are coastal red tide species. The growth of these toxic algae is hypothesized to be stimulated by exposure to elevated iron concentrations. Like most organisms, marine algae have metabolic iron requirements associated with biochemical processes including electron transfer, nitrogen assimilation, and the synthesis of DNA, RNA, and chlorophyll. In some locations, iron availability is more important than the macronutrients, nitrogen, phosphorus, and silicon in limiting algal growth. As the bioassay results show, algal growth in laboratory cultures was stimulated by exposure to organically bound iron (Fe–EDTA)[1] and highly soluble $FeCl_3$. A lesser degree of enhancement was generated by exposure to two particulate forms of iron ($FePO_4$ and FeS). Growth was not enhanced by the other two particulate forms, Fe_2O_3 and FeO(OH).

For iron to stimulate algal growth, it must first cross the cell membrane so it can enter the cell matrix. The species that can cross the cell membrane are thought to be Fe^{2+}(aq) or Fe^{3+}(aq). As algal withdraw these species from seawater, their continued growth will depend on a resupply of Fe^{2+}(aq) or Fe^{3+}(aq). This can be achieved through the breakdown of other species, such as the dissociation of Fe–EDTA and the dissolution of highly soluble $FeCl_3$. The stimulatory effects of $FePO_4$ and FeS on algal growth suggests that these two solids undergo enough dissolution to resupply the Fe^{2+}(aq) or Fe^{3+}(aq) required to stimulate algal growth. The solubility of Fe_2O_3 and FeO(OH) in seawater is very low and, thus, not likely to be a good source of soluble iron as confirmed by the bioassay results. Other recent research has shown that marine algal are able to

[1] EDTA is a low-molecular-weight organic molecule, ethylenediaminetetraacetate, that is commonly used as a proxy for marine DOM. It is acts as buffer, maintaining constant pH and/or metal ion concentrations in experimental solutions.

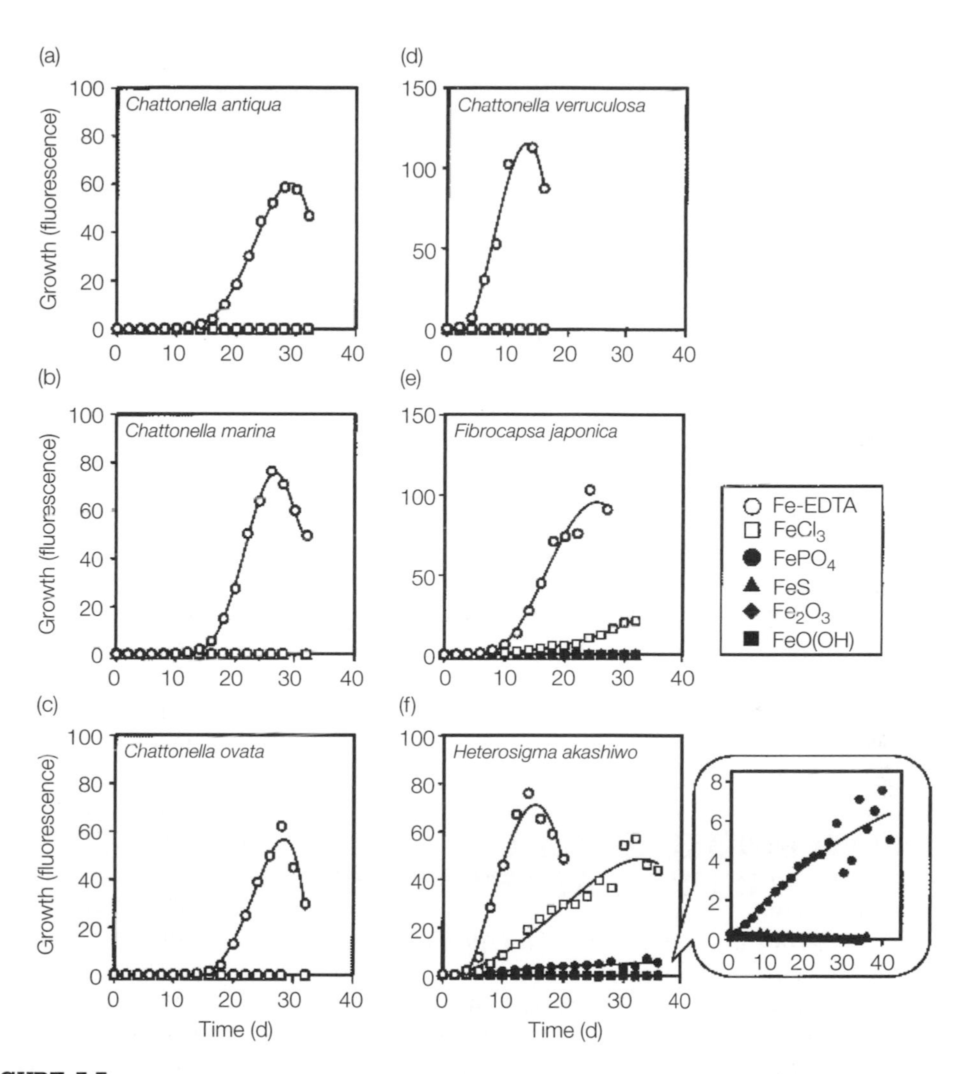

FIGURE 5.5

Growth of *Raphidophyceae* (red tide microalgae) as measured by change in fluorescence. These bioassays were conducted under conditions of iron limitation and emendation with four insoluble iron species [$FeO(OH)$, Fe_2O_3, FeS, and $FePO_4$ $4H_2O$], soluble inorganic iron ($FeCl_3$ $6H_2O$), and an artificial organic ion species (Fe-EDTA). Growth curves are for (A) *Chattonella antiqua*, (B) *Chattonella marina*, (C) *Chattonella ovata*, (D) *Chattonella verruculosa*, (E) *Fibrocapsa japonica*, and (F) *Heterosigma akashiwo*. *Source*: From Naito, K., *et al.* (2005). *Harmful Algae* 4, 1021–1032.

enhance the bioavailability of iron in seawater by exuding organic molecules, called siderophores, that bind iron into soluble compounds. The organically bound iron, like Fe–EDTA, can undergo dissociation to provide $Fe^{2+}(aq)$ or $Fe^{3+}(aq)$ and, hence, serves as a reservoir of iron that is readily accessible to the algae. In the following sections, we discuss how the concentrations of each of these iron species can be predicted from the thermodynamic principles governing chemical equilibrium.

The practical consequence of the bioassay results presented in Figure 5.5 is that the growth of coastal red tide plankton are potentially controlled through contact with sedimentary $FeCl_3$ and $FePO_4$. The red tide plankton engage in diurnal migrations deep enough (8 m) to bring them into contact with the seafloor in shallow waters where $FePO_4$ and FeS are common components of the sediments. Thus, diurnal migration could help these phytoplankton meet their iron needs by enabling uptake from marine sediments. It also suggests that storm-driven mixing could lead to the formation of red tides by enhancing contact of the plankton with sedimentary iron.

5.4 CONCENTRATION UNITS

To perform speciation calculations, chemical distributions must be in the form of molar $\left(\frac{\text{mol solute}}{\text{solution volume}}\right)$ or molal $\left(\frac{\text{mol solute}}{\text{solvent mass}}\right)$ concentrations. Like most scientists, oceanographers use a variety of units to express chemical concentrations. The aqueous concentration units in common use, along with their base SI units, which are rarely used, are shown in Table 5.1. Unit conversions are provided in Appendix 3. In practice, seawater samples are usually measured out by their volumes and density is used to convert from liters of seawater to kilograms of seawater. Note that $1\,g/cm^3 = 1\,kg/L$. Concentration units for chemicals in the solid phase are presented in Chapter 12.

5.5 MODELING CHEMICAL REACTIONS: EQUILIBRIUM VERSUS KINETIC APPROACHES

For a chemical, C, whose concentration is in steady state, the rate of supply equals the rate of its loss. This can be represented by the simple box model shown in Figure 5.6 for a chemical, C, which is present in three reservoirs, and by the following chemical equation:

$$C_1 \xrightarrow{k_i} C_2 \xrightarrow{k_o} C_3 \qquad (5.1)$$

where k_i is the rate constant for the input process and k_o is the rate constant for the output. If the rates of supply and removal to reservoir 2 are both first-order processes with respect to C, the rate law for the input is $\frac{d[C]_2}{dt} = k_i\,[C]_1$ and $-\frac{d[C]_2}{dt} = k_o\,[C]_2$. At a steady state, $\frac{d[C]_2}{dt} = 0$ and $k_i\,[C]_1 = k_o\,[C]_2$.

In the special case, where the process supplying the chemical, C, is the mechanistic reverse of the process removing it, *chemical equilibrium* is said to exist. This condition is illustrated in Figure 5.7 for two reservoirs, C_1 and C_2, and by the following chemical equation:

$$C_1 \underset{k_b}{\overset{k_f}{\rightleftharpoons}} C_2 \qquad (5.2)$$

Table 5.1 Common Aqueous Concentration Units (where SW = seawater).

Base SI Unit	Non-SI Units in Common Use	Common Name and Abbreviation
kg/kg	g solute per kg SW	parts per thousand (‰)
	mg solute per kg SW	parts per million (ppm)
	μg solute per kg SW	parts per billion (ppb)
	ng solute per kg SW	parts per trillion (ppt)
	pg solute per kg SW	parts per quadrillion (ppq)
kg/m^3	g solute per L SW	parts per thousand (‰)[a]
	mg solute per L SW	parts per million (ppm)[a]
	μg solute per L SW	parts per billion (ppb)[a]
	ng solute per L SW	parts per trillion (ppt)[a]
	pg solute per liter SW	parts per quadrillion (ppq)[a]
mol/kg	mol solute per kg water	Molality (m)
	mmol solute per kg of water	Millimolal (mm)
	mol solute per kg SW	Molinity
	mmol solute per kg SW	Molinity
	μmol solute per kg SW	Molinity
	nmol solute per kg SW	Molinity
	pmol solute per kg SW	Molinity
	equivalents (eq) of charge per kg SW[b]	Molinity
mol/m^3	mmol solute per m^3 SW	
	μmol solute per m^3 SW	
	nmol solute per m^3 SW	
	pmol solute per m^3 SW	
	mol solute per L SW	Molarity (M)
	mmol solute per L SW	Millimolar (mM)
	μmol solute per L SW	Micromolar (μM)
	nmol solute per L SW	Nanomolar (nM)
	pmol solute per L SW	Picomolar (pM)
	g-atom of an element per L SW[c]	
	mg-atom of an element per L SW[c]	
	μg-atom of an element per L SW	
	equivalents (eq) of charge per L SW[b]	
	milliequivalents (meq) of charge per L SW[b]	
m^3 per m^3	ml gas per L of SW	
	L gas per L atmosphere	

[a] *Although technically not correct, wt/vol concentrations are commonly denoted with the wt/wt abbreviations. This introduces only a minor error as the density of seawater is approximately 1.02 to 1.03 g/cm³.*

[b] *An equivalent of charge = 1 mol of charge.*

[c] *For example, seawater with a nitrate concentration of 1 mM NO_3^- has a nitrogen concentration expressed as 1 mg-atom N/L seawater because each mol of NO_3^- contains 1 mol of N.*

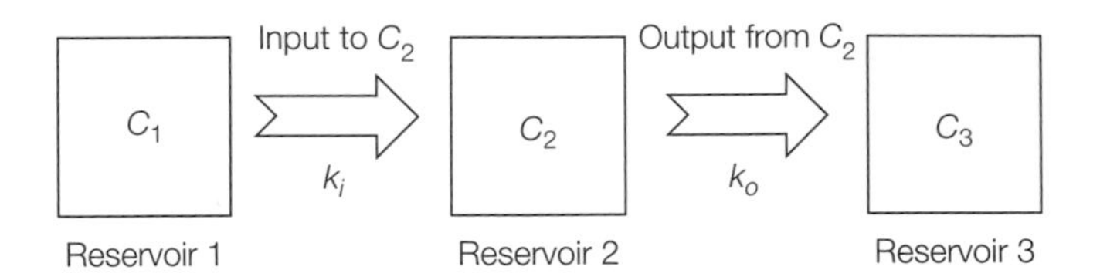

FIGURE 5.6

Condition of steady state.

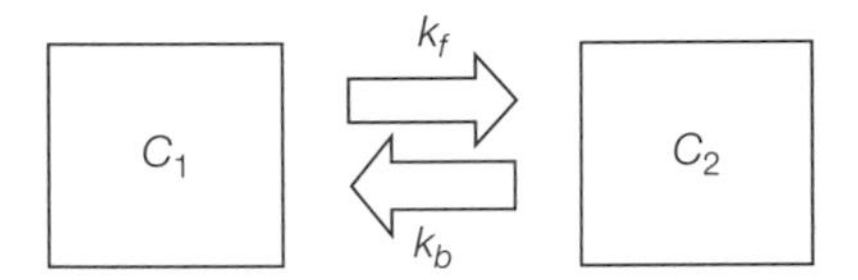

FIGURE 5.7

Condition of chemical equilibrium.

where k_f and k_b are the rate constants for the forward and back reactions. If the rate laws for both reactions are first order with respect to C, $\frac{d[C]_2}{dt} = k_f [C]_1$ and $-\frac{d[C]_2}{dt} = k_b [C]_2$. At equilibrium, $\frac{d[C]_1}{dt} = \frac{d[C]_2}{dt} = 0$, and $k_f [C]_1 = k_b [C]_2$. Thus, $\frac{[C]_2}{[C]_1}$ is a constant whose value is determined by the relative magnitudes of the rate constants. This is also true for higher order rate laws as the existence of equilibrium requires that the forward reaction be the mechanistic reverse of the back reaction.

At equilibrium, the reactant concentrations and products can be used to define a mass ratio called an *equilibrium constant* (K). This constant can then be used to predict the equilibrium concentrations of the reactants and products from the total amount of C or from either the equilibrium concentration of the products or the reactants. Although K is referred to as an equilibrium constant, it is a function of salinity, temperature, and pressure. With the appropriate value of K, calculations can be made to predict the equilibrium speciation of elements in seawater. The procedure for doing this is provided in the next section along with an expansion of K to multicomponent chemical systems.

Many chemical reactions in seawater do not achieve equilibrium. The most notable are ones that involve marine organisms. Since organisms require energy, they cannot survive if their constituent biochemicals are at equilibrium. Equilibrium is also not likely to be achieved if some other process is adding or removing a chemical faster than equilibrium can be reattained. For example, calcium carbonate shells should spontaneously dissolve in deep ocean water, but some sink so fast that they can reach the sediments where they eventually become buried and, hence, preserved. In other words, the equilibrium approach is most applicable to reactions that attain equilibrium faster than any other competing processes acting on the chemical of interest.

Even when equilibrium is not attained, knowledge of theoretical equilibrium concentrations provides information on the direction and driving force (or intensity) of the most likely spontaneous reactions. The other approach to predicting chemical behavior is through the use of reaction rate laws. Development of these kinetic models is difficult because we do not know the rate laws of most reactions in seawater. This is why first-order behavior is often assumed. Furthermore, the mathematics of dealing with higher order rate laws and competing reactions is complicated. To simplify these calculations, equilibrium speciation calculations can first be performed to identify reactions that are unlikely to occur and, hence, need not be included in kinetic models. Commercially available computer programs, such as STELLA,[2] are now commonly used to develop and implement kinetic models.

5.6 USING EQUILIBRIUM MODELS IN SEAWATER

Chemical changes proceed toward a condition called chemical equilibrium, which represents the lowest energy state for a reactant/product system. In seawater, an important class of reactions that achieve equilibrium are those that can be represented by Eq. 5.3,

$$aM^+ + bL^- \rightleftharpoons M_aL_b \tag{5.3}$$

which describes the formation of an ion pair, M_aL_b, from the combination of a metal cation, M^+, with an electron donor (ligand), L^-. The molar reaction stoichiometry is given by the coefficients, a and b.

Formation reactions can also involve ligands that are not anions. Water is technically a ligand and quite effective as it is able to share its nonbonded electron pair with the cation. If the interaction between the metal and the ligand has significant covalent character, the resulting molecule is termed an *ion complex* or *coordination complex*. Some examples are shown in Figure 5.8 including their waters of hydration. The number of ligands or waters of hydration that a metal ion can accommodate depends on how many coordination sites it possesses. This is determined by its electronic configuration.

At equilibrium, the rate of the forward reaction (formation) is equal to that of the reverse (dissociation), causing the concentrations of all the chemical species to remain constant over time. The following *equilibrium constant expression* can be written for this reaction:

$$K^o_{eq} = \frac{[M_aL_b]}{[M^+]^a\,[L^-]^b} \tag{5.4}$$

in which K^o_{eq} is the equilibrium constant observed when solute concentrations are low. Solutions are said to exhibit "ideal behavior" if their equilibrium constant is accurately described by the simple form given in Eq. 5.4. In ideal solutions, the solvent has a negligible effect on the reaction of interest and the only interactions exhibited by the solutes are those of the reaction of interest, i.e., Equation 5.3. Since solvents exhibit

[2] http://www.iseesystems.com/softwares/Education/StellaSoftware.aspx

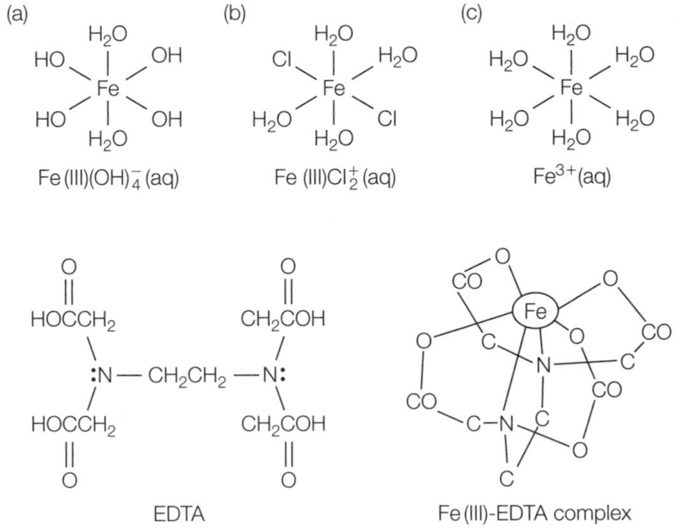

FIGURE 5.8

Examples of coordination complexes for Fe(III) with coordination number = 6: (a) $Fe(OH)_4^-(aq)$, (b) $FeCl_2^+(aq)$ and (c) $Fe^{3+}(aq)$.

negligible effects on reaction chemistry, their concentrations are not included in equilibrium constant expressions for ideal solutions. Nor are those of any other pure liquids or solids. By convention, gas concentrations are included as partial pressures in units of atmospheres.

Seawater has high concentrations of solutes and, hence, does not exhibit ideal solution behavior. Most of this nonideal behavior is a consequence of the major and minor ions in seawater exerting forces on each other, on water, and on the reactants and products in the chemical reaction of interest. Since most of the *nonideal* behavior is caused by electrostatic interactions, it is largely a function of the total charge concentration, or ionic strength of the solution. Thus, the effect of nonideal behavior can be accounted for in the equilibrium model by adding terms that reflect the ionic strength of seawater as described later.

Equilibrium constants are also dependent on temperature and pressure. The temperature functionality can be predicted from a reaction's enthalpy and entropy changes. The effect of pressure can be significant when comparing speciation at the sea surface to that in the deep sea. Empirical equations are used to adapt equilibrium constants measured at 1 atm for high-pressure conditions. Equilibrium constants can be formulated from solute concentrations in units of molarity, molality, or even moles per kilogram of seawater.

Equilibrium speciation calculations have been performed for all of the varied types of chemical transformations that occur in seawater as listed in Table 5.2. Note that some involve only a phase change and, thus, are not chemical reactions. Equilibrium

Table 5.2 Types of Equilibrium Processes.

Types of Chemical Process	Description
Hydration	Incorporation of water molecule(s) into a molecule or crystal lattice of a mineral without hydrolysis
Hydrolysis	Reaction with water molecule, which acts as an acid or a base
Acid/base	Reaction with a proton donor, such as H^+, or a proton acceptor, such as OH^-
Precipitation/dissolution	Formation or dissolution of a solid; this usually involves hydration
Adsorption/desorption	Reactions that involve solute becoming chemically bonded to the surface of a solid. The reverse process releases solutes from the surface of a solid
Complexation	Reactions involving a cation and an anion that produces a ion complex
Redox	Reactions that involve electron transfers between reactants causing a change in the oxidation state
Photochemistry	Reactions that require sunlight as an energy source
Gas exchange	Dissolution of gases into seawater or degassing from seawater to the atmosphere or into occluded gas bubbles

speciation results for each of the processes listed in Table 5.2 are presented in context throughout this text, reflecting the importance of this approach in understanding marine biogeochemical processes.

5.6.1 Nonspecific versus Specific Effects

Special mathematical techniques have been developed to adapt the ideal equilibrium model for use in seawater. They rely on partitioning solution behavior into two categories: specific and nonspecific effects. *Specific effects* are ones that can be described as the result of a chemical reaction for which an equation similar to Eq. 5.3 can be written and a limited type of equilibrium constant expression, such as Eq. 5.4, can be defined. *Nonspecific effects* are longer range interactions that do not result in formation of a chemical bond. In seawater, most nonspecific effects are electrostatic in nature. For example, oppositely charged ions will attract one another and ions of similar charge will repel each other, even when separated by many layers of water molecules. The intensity of these interactions is dependent on the distance of separation of the ions and on their charge densities. The latter are usually estimated by the ratio, z/r^2, where z is an ion's net charge and r is its ionic radius. Thus, nonspecific effects vary from ion to ion in a predictable way.

Nonspecific interactions also include interactions between ions and their solvent, especially one that it is as polar as water. As already illustrated in Figure 3.6, high concentrations of ions can impose some degree of order in the arrangement of nearby water molecules leading to electrostriction. This affects solution density and, hence, solute concentrations. The impact of this nonspecific effect is greatest for water molecules in direct contact with the ion. As shown in Figure 5.9, this region is termed the primary solvation shell. Water molecules in the secondary solvation shell are still close enough to experience some electrostatic attraction, but the degree of ordering is much less than in the primary shell. At greater distances, the water molecules are too far from the ion to be significantly influenced by electrostatic attraction. This region is termed the bulk solution.

Nonspecific effects can also limit the availability of solutes to undergo chemical reactions. The usual case is that the long-range electrostatic interactions between ions leads to a reduction in their availability to form ionic or covalent bonds. As with solvent-ion

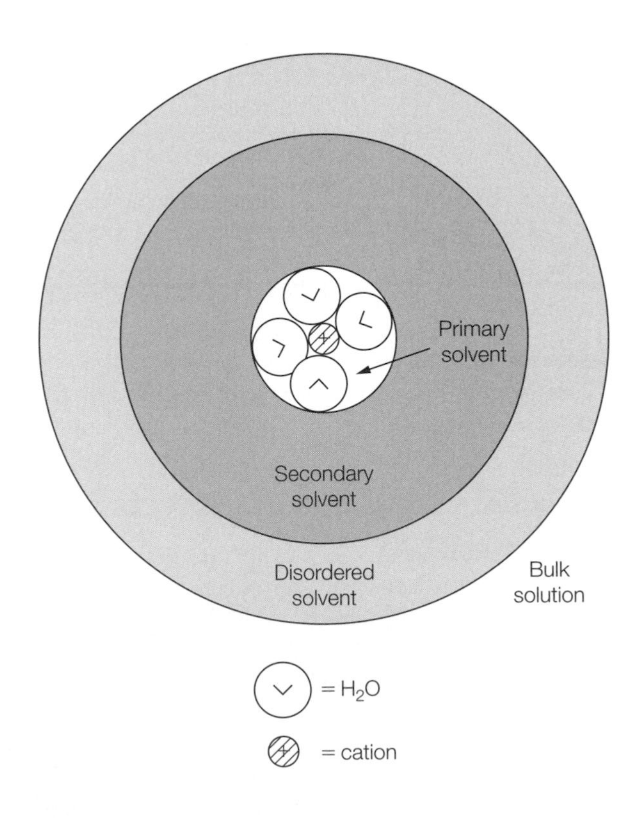

FIGURE 5.9

The various solvent regions around a metal ion. *Source*: From Burgess, J., (1978). *Metal Ions in Solution*, John Wiley & Sons, Inc., p. 20.

interactions, a continuum of solute-ion-to-solute-ion interactions exists whose intensity depends on the separation distance of the ions. At one extreme, i.e., very short separation distances, ionic and covalent bonds forms. At the other, i.e., very long separation distances, no effects are exerted. Thus, the nonspecific effects occur over a middle spatial range across which no theoretical threshold exists to differentiate a very weak specific effect from a very strong nonspecific one. Rather, marine chemists differentiate the two operationally. The specific effects are considered to be those chemical behaviors that can be adequately described by equilibrium constants. Nonspecific effects are parameterized in the equilibrium model through the use of "effective concentrations" in place of the total solute concentrations. The effect concentrations are also called *activities*.

An example of this continuum of interactions is given by the dissolution of NaCl(*s*). In the process illustrated in Eqs. 2.5, 2.6, and Figure 2.11, full dissolution is achieved when the ionic bonds between Na and Cl are broken and the ions are completely surrounded by water molecules. Most of these ions are so well hydrated that they experience only nonspecific interactions and, hence, are termed *free ions*. They are symbolically represented as $Na^+(aq)$ and $Cl^-(aq)$. Ions that are not as well hydrated can come into closer contact with each other and experience specific interactions. In cases where the ions are close enough to share some of their primary solvation shells, the resulting strong electrostatic attractions are referred to as *ion pairing*. Various degrees of ion pairing are possible as shown in Figure 5.10.

One possibility is the formation of an ion pair in which one sodium ion is in direct contact with one chloride ion. Its formation is represented by the following reaction,

$$Na^+(aq) + Cl^-(aq) \rightleftharpoons NaCl^0(aq) \tag{5.5}$$

where the relevant equilibrium expression is:

$$K_f = \frac{[NaCl^0(aq)]}{[Na^+(aq)]\,[Cl^-(aq)]} \tag{5.6}$$

K_f is called a *formation or stability constant*. Note that the formula for the ion pair, $NaCl^0(aq)$, symbolizes the interaction of 1 atom of Na with 1 atom of Cl, whereas the representation of crystalline halite, NaCl(s), is an empirical formula in which an unspecified number of Na and Cl atoms are present in a 1:1 stoichiometric ratio.

Ion pairing can occur among groups of ions to create ternary ($n = 3$), quaternary ($n = 4$), and other higher order associations. While ion pairing is an example of a specific interaction, it also represents a continuum of interactions that range in strength from relatively weak associations, as in solvent-separated ion pairs, to strong ones, as in ion or coordination complexes.

5.6.2 **Handling Nonideal Solution Behavior**

Marine chemists have developed two approaches to handling *nonspecific* effects. The easiest one, which we will adopt in this book, is to use equilibrium constants appropriate for the temperature, pressure, and salinity of seawater. (Since most of the ionic

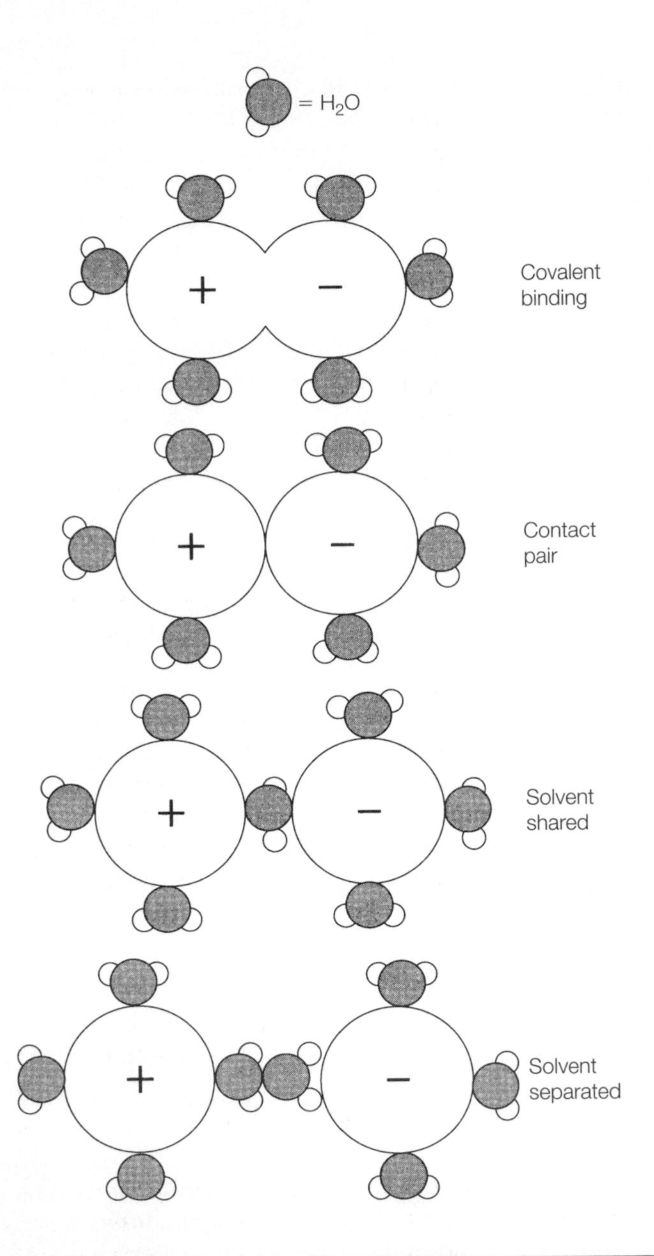

FIGURE 5.10

Types of ion pairs. *Source*: From Millero, F., and D. Pierrot (2002). *Chemistry of Marine Water and Sediment*, Springer-Verlag, pp. 193–220.

strength of seawater is contributed by the major ions, salinity makes a good proxy for ionic strength.) In the second approach, equilibrium expressions are redefined in terms of a type of solute concentration that obeys ideal solution behavior. This special solute concentration is referred to as an *activity* (a) or *effective concentration*. It yields an equilibrium expression that has a constant value regardless of equilibrium solute concentrations or ionic strength.

The activity of an ion, *i*, is defined as the product of its total, or stoichiometric concentration, and an "activity coefficient" (γ_i):

$$a_i = \{i\} = \gamma_i m_i \tag{5.7}$$

where { } is used to denote activity and m_i is the total *molality* of the solute. Because γ_i is dimensionless, a_i has the units of molality. In most cases, the activity coefficient of ionic solutes is less than 1, making the activity of an ion less than its stoichiometric concentration. In other words, the solute is less available to participate in chemical reactions due to nonspecific interactions exerted by the solution. You have probably already worked with activities as pH is defined as the $-\log\{H^+\}$. In other words, pH is an operational definition; it is the activity of H^+ sensed by a pH electrode.

For the reaction in Eq. 5.3, the equilibrium constant expression (Eq. 5.4) can be rewritten in terms of activities as

$$K_f = \frac{a_{M_aL_b}}{a^a_{M^+}a^b_{L^-}} = \frac{\{M_aL_b\}}{\{M^+\}^a\{L^-\}^b} = \frac{\gamma_{ML}\,[M_aL_b]}{\gamma^a_{M^+}\,[M^+]^a\,\gamma^b_{L^-}\,[L^-]^b} \tag{5.8}$$

where K_f is called the formation or stability constant (as in Eq. 5.6). At very low ionic strengths and solute concentrations, the values of the activity coefficients are typically close to 1 and the equilibrium constant obeys ideal behavior as described by Eq. 5.4.

Activity coefficients can be determined by experimental observations. Since they are functions of ionic strength, temperature and pressure, marine scientists typically estimate values at the environmental conditions of interest from semi-empirical equations. In dilute solutions, the activity coefficient of a monoatomic ion can be calculated from the Debye-Hückel equation:

$$\log \gamma_i = -Az_i^2\sqrt{I} \tag{5.9}$$

where A is a temperature dependent term, z_i the charge of the ion, and I is the *ionic strength* of the solution. The ionic strength is computed as:

$$I = \frac{1}{2}\sum_i m_i z_i^2 \tag{5.10}$$

where m_i is the molality of each ion and z_i is the ion's charge. Equation 5.9 is valid only for $I < 0.01\,m$. Many natural waters, such as seawater ($0.7\,m$), have much higher ionic strengths and require the use of more complex equations, such as those listed in Table 5.3, to estimate activity coefficients.

Despite the additional complexity, all the equations in Table 5.3 are functionally equivalent. That is, the activity coefficients approach a value of 1 as the ionic strength of the solution is decreased to $0\,m$. Thus, in dilute solutions, $a_i \approx m_i$. In other words, the effective concentration of an ion decreases with increasing ionic strength. In contrast, the activity coefficients of uncharged solutes can be greater than 1 in solutions of high ionic strength, such as seawater.

Table 5.3 Various Expressions for the Calculation of Activity Coefficients for Monoatomic Ions.

Approximation	Equation	Approximate Applicability [ionic strength (M)]
Debye–Hückel	$\log \gamma = -Az^2\sqrt{I}$	$<10^{-2.3}$
Extended Debye–Hückel	$= -Az^2 \dfrac{\sqrt{I}}{1 + Ba\sqrt{I}}$	$<10^{-1}$
Güntelberg	$= -Az^2 \dfrac{\sqrt{I}}{1 + \sqrt{I}}$	$<10^{-1}$ useful in solutions of several electrolytes
Davies	$= -Az^2 \left(\dfrac{\sqrt{I}}{1 + \sqrt{I}} - 0.2I\right)$	<0.5
Brönsted–Guggenheim	$\ln \gamma_S = \ln \gamma_{DHS} + \sum_j As_j(C_j) + \sum_j \sum_k B_{s_{jk}}(C_j)(C_k) + \cdots$	≤ 4

Where $A = 1.82 \times 10^6(\epsilon T)^{-3/2}$, $B = 50.3(\epsilon T)^{-1/2}$ *and a is an ion size parameter which ranges in value from 0.10 to 0.99 depending on ion size and I. The dielectric constant (ε) of water is a function of temperature and pressure. At 25°C and 1 atm,* $\epsilon_{H_2O} = 78.54$.
Source: After Stumm, W. W., and J. J. Morgan (1981). Aquatic Chemistry. *John Wiley and Sons, Inc., p. 135.*

The activity coefficients can be used to compute a stoichiometric equilibrium constant, K_c, from K^o_{eq} as follows:

$$K_c = \frac{[M_aL_b]}{[M^+]^a\,[L^-]^b} = \frac{(\gamma_{M^+})^a(\gamma_{L^-})^b}{\gamma_{M_aL_b}} K^o_{eq} \tag{5.11}$$

K_c is sometimes referred to as a conditional formation or stability constant reflecting its validity only at a specified set of environmental conditions. With values of K_c, equilibrium speciation calculations can be performed using stoichiometric concentrations. This is the approach used throughout this text. Since K^o_{eq} is defined under standard conditions of 25°C and 1 atm, additional corrections are needed to adapt K_c to other temperatures and pressures.

For some reactions, K_c has been determined by direct measurement over a broad range of temperature, pressure, and salinities. Enough data exist to formulate empirical equations that enable extrapolation to the exact temperature, salinity, and pressure of interest. This has been done for the chemical reactions in the carbonate system, for the dissociation of water (K_w), and for the dissolution of gases. These equations have been used to formulate look-up tables, such as those presented in the online appendix on the companion website.

More commonly, K_c's are computed on the fly by computer programs as part of an equilibrium speciation calculation. These programs employ various numerical approaches for estimating K_c, such as the one shown in Eq. 5.11. In this case, the programs rely on large internal databases of K^o_{eq}'s and various numerical assumptions to estimate γ_i. Marine scientists have not reached agreement on acceptable values for many of the K^o_{eq}'s or how to extrapolate them to the environmental conditions present in the ocean. Thus, the computer programs are inherently limited by the quality of their databases and their computational strategies.

5.7 COMPETITIVE COMPLEXING: CALCULATING SPECIATION IN A MULTI-ION SOLUTION

Most chemicals in seawater have the potential to concurrently engage in multiple reactions whose products are a variety of ion pairs involving the chemical of interest. These reactions can be viewed as a competition among the ligands for one or more of the coordination sites on the metal ions. The winner ligands are the ones with the highest affinity for the metal (high K_c) and the highest concentration. The ion pairs that are most favored at equilibrium (high K_c) are the ones with the most covalent character, i.e., the ion or coordination complexes. Some ligands are particularly good at binding metals because they possess multiple sites at which electrons can be donated to a metal ion (Figure 5.11). This type of bonding is referred to as *chelation* and is common in organic ligands making them very effective at binding metal ions. Some examples are provided in Figure 5.11 for a class of naturally occurring ligands, called *siderophores*, that have very high K_c's for $Fe^{3+}(aq)$ (10^{10} to 10^{13}). These ligands are produced by microorganisms, such as marine phytoplankton and bacteria. Following exudation into seawater, the siderophores form soluble organo-iron complexes that enhance the bioavailability of the metal. Most of the binding affinity is contributed by the multiple oxygen atoms that serve as electron donors.

The chemical speciation of iron also includes many inorganic ion complexes. Iron is also present in seawater in the solid, colloidal, and dissolved phases. Its chemistry is further complicated by oxidation-reduction reactions in which iron cycles between two oxidation states, Fe(II) and Fe(III). (The redox chemistry of seawater is the subject of Chapter 7.) The consequences of this complicated speciation behavior appear to have global significance as iron is a micronutrient that controls phytoplankton growth in large areas of the oceans. The production of algal biomass is a significant sink for CO_2 and, hence, provides a feedback link to global climate. For this reason, a great deal of attention has been focused on understanding the biogeochemical cycling of iron in the ocean. Thus, iron makes an excellent case study for illustrating how chemical speciation aids in the investigation of chemical transport pathways in the crustal-ocean-atmosphere factory. Our current understanding of the crustal-ocean-atmosphere cycling of iron is shown in Figure 5.12.

5.7.1 The "Iron Hypothesis"

In the late 1980s, improvements in the measurement methods for trace metals in seawater lead to the observation that dissolved iron concentrations in the open ocean are

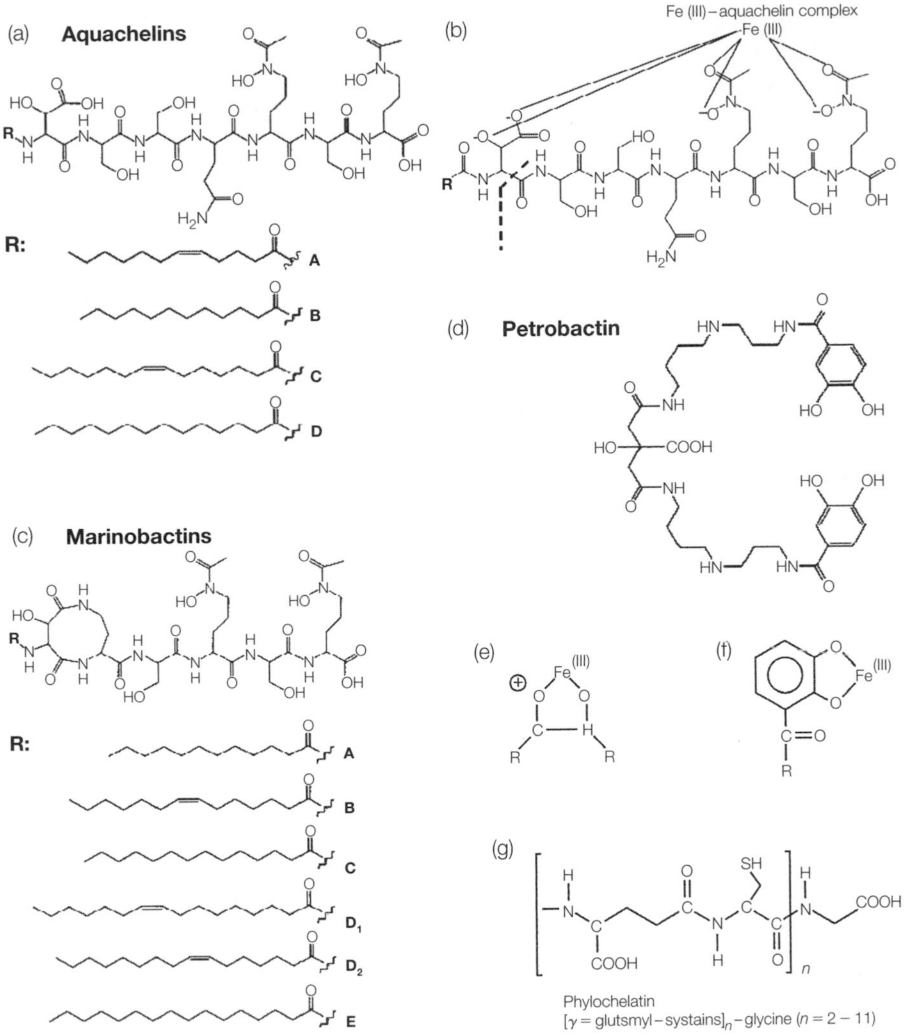

FIGURE 5.11

Structures of siderophores synthesized by marine heterotrophic bacteria: (a) aquachelins from *Halomonas aquamarina*, (b) iron binding to multiple oxygen atoms in an aquachelin, (c) marinobactins from a *Marinobacter* species, and (d) petrobactin from *Marinobacter hydrocarbonoclasticus*. Partial structures of siderophores synthesized by marine phytoplankton bacteria: (a) hydroxamate group, (b) catechol group, and (c) phytochelatines. *Sources*: From (b) Barbeau, K., *et al.* (2001). *Nature* 413, 409–413. (a,c,d) Barbeau, K., *et al.* (2003). *Limnology and Oceanography* 48, 1069–1078. (e,f,g) Stumm, W., and J. J. Morgan (1996). *Aquatic Chemistry*, 3rd ed. Wiley-Interscience, p. 306.

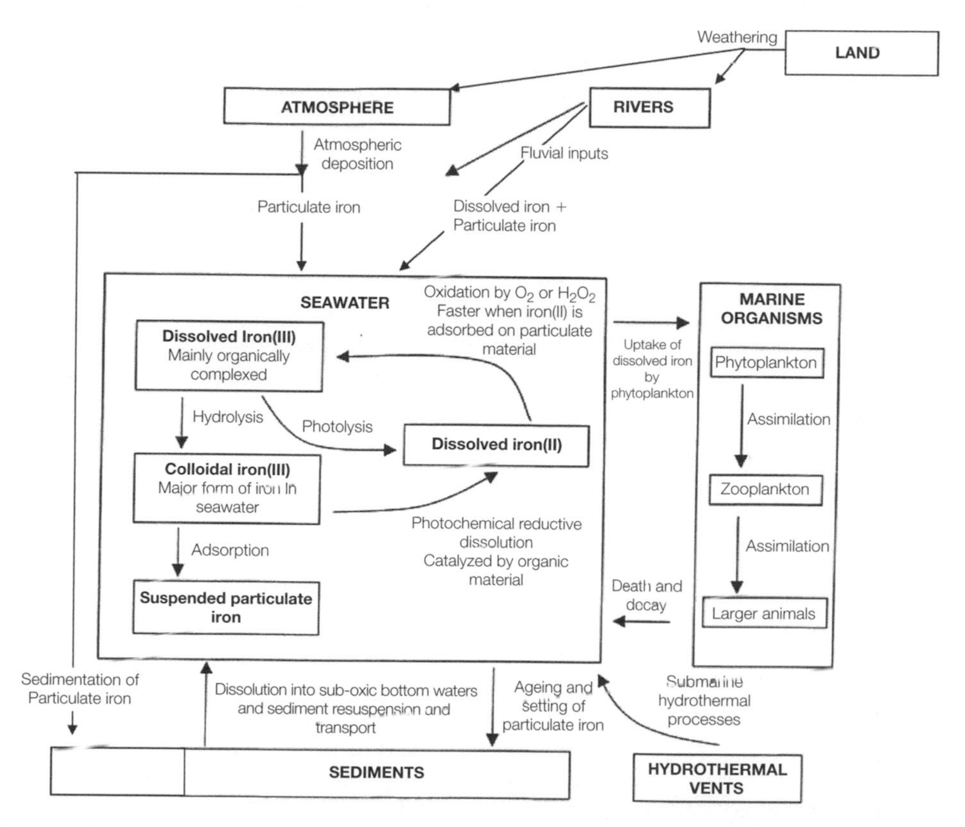

FIGURE 5.12

Biogeochemical cycle of iron in the crustal-ocean-atmosphere factory. *Source*: After Achterberg, E. P., *et al.* (2001). *Analytica Chemica Acta* 442, 1–14.

very low, ranging from 0.1 to 1 (nmol Fe)/(kg SW) at the surface, Below the mixed layer, concentrations are generally higher, but still low ranging from 0.3 to 1.3 (nmol Fe)/(kg SW) [average of 0.76 ± 0.25 (nmol Fe)/(kg SW)]. As we shall see in Chapter 9, this type of concentration behavior is characteristic of the nutrients nitrate, phosphate, and silicon in which low surface water concentrations are caused by nearly complete removal by phytoplankton followed by release at depth from decomposition of sinking dead biomass. Subsequent work has demonstrated that iron is likely limiting phytoplankton growth rates and biomass production in a significant fraction (40%) of the open ocean, particularly the equatorial Pacific, subpolar regions, and some of the coastal ocean. Some of the evidence for this comes from in situ fertilization experiments in which the deliberate addition of dissolved iron has produced algal blooms, some large enough to be detected from space using the color sensors on satellites.

Given this state of iron limitation, inputs of iron to the open ocean could result in enough new primary production to lower atmospheric carbon dioxide levels. Because CO_2 is a greenhouse gas, cooling of the global atmosphere could result. This potential

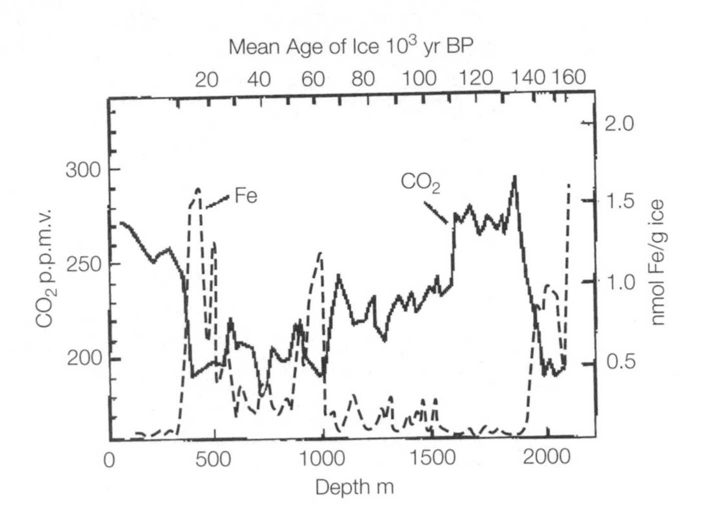

FIGURE 5.13

Iron concentrations as a function of depth in the Antarctic Vostok ice core, together with mean CO_2 concentrations in air trapped in ice, versus mean age of air. *Source*: From Martin, J. (1990). *Paleoceanography* 5, 1–13.

linkage between iron supply and climate led to formulation of the "Iron Hypothesis" by John Martin in 1986. He proposed that changes in the supply of iron to the ocean over geologic time have led to fluctuations in atmospheric CO_2 levels and, hence, to global climate change. As shown in the beginning of this chapter, most of the iron delivered by rivers is trapped in estuarine sediments. Thus, the major source of "new" iron to the open surface ocean is via airborne dust. Although wind-driven upwelling is another potential source of iron to the surface waters, this represents a recycled input supplied by release from sinking dead biomass. Carbon is concurrently recycled back into CO_2, so the uptake of upwelled iron does not result in a net removal of CO_2. To reduce atmospheric CO_2 levels, the new biomass carbon must be retained by the ocean. This occurs mostly through burial in the sediments.

Evidence in support of the Iron Hypothesis has been obtained from ice cores that contain records of past dust deposition, atmospheric CO_2 levels and global temperatures. As shown in Figure 5.13, during the past 160,000 years, periods during which dustborne iron levels have been high coincide with lower atmospheric CO_2 levels and global temperatures, i.e., the most recent Ice Ages.

A tremendous amount of research has since been conducted to investigate various components of the Iron Hypothesis because of the potential linkage to climate control. This connection has even led to the proposal that we "iron" our way out of our current greenhouse gas problem![3] Particular attention has been paid to determining

[3] See Chapter 25.6.

whether the dust flux of iron is significant enough to affect marine production and if so, what controls this flux. Ironically, we may have inadvertently reduced some of the dust flux through successful soil conservation efforts. On the other hand, climate change is expanding the deserts in the Sahel region of Africa, providing new dust sources. Another episodic input of iron-rich dust is probably associated with subaerial volcanism. Submarine volcanism introduces iron-rich hydrothermal fluids into the deep sea. The importance of this source of "new" iron into the deep sea to marine production is not known, but if significant, it could result in episodic impacts on phytoplankton production, because hydrothermal venting tends to be highly variable over time and space and in fluid composition. Although we are far from knowing all the details, scientists agree that the complex and interconnected processes in iron's global biogeochemical cycle provide significant linkages between crustal weathering, marine productivity, and climate. These linkages are likely to involve positive and negative feedbacks.

One of the major unknowns in this story is whether the natural inputs of iron to the surface ocean are large enough to have a significant impact on biological production. This requires understanding how plankton meet their iron needs, including what species the iron must be present in and how the formation of these chemical species is regulated. Our ability to directly measure individual chemical species is limited, so their concentrations are usually predicted from equilibrium models developed for a given set of environmental conditions, i.e., temperature, pressure, salinity, and ligand concentrations. Based on these predictions and a limited number of direct measurements, marine scientists have come to appreciate that iron is rapidly cycled in seawater through a large number of chemical species, including dissolved, colloidal, and solid phases. As shown in Figure 5.14, some of the chemical transformations are mediated by sunlight. Phytoplankton and bacteria appear to engage in a variety of strategies for converting particulate iron into a form that can be assimilated into their cells. One biological approach involves the excretion of siderophores that selectively bind with iron and thereby increase its solubility in seawater. If the particulate iron is not solubilized, it will eventually sink below the euphotic zone and deposit on the seafloor. Interestingly, dust from polluted regions appears to have enhanced solubility, probably due to elevated levels of acid contributed by SO_2, a fossil fuel combustion by-product.

5.7.2 The Chemical Speciation of Iron: Ion Pairing

The iron story is just one example of the critical importance of speciation in determining biogeochemical pathways through the crustal-ocean-atmosphere factory. As a result, many computer software programs are now available for predicting speciation. See Table 5.4 for some examples; some are commercial products and others were developed with government funding and are freeware. As noted earlier, these programs are limited by the thermodynamic databases from which they compute environmentally appropriate equilibrium constants. Especially problematic is the lack of information on organic ligands. And of course, all these programs assume that equilibrium has been achieved. In practice, marine scientists do not attempt to include all the chemical species in

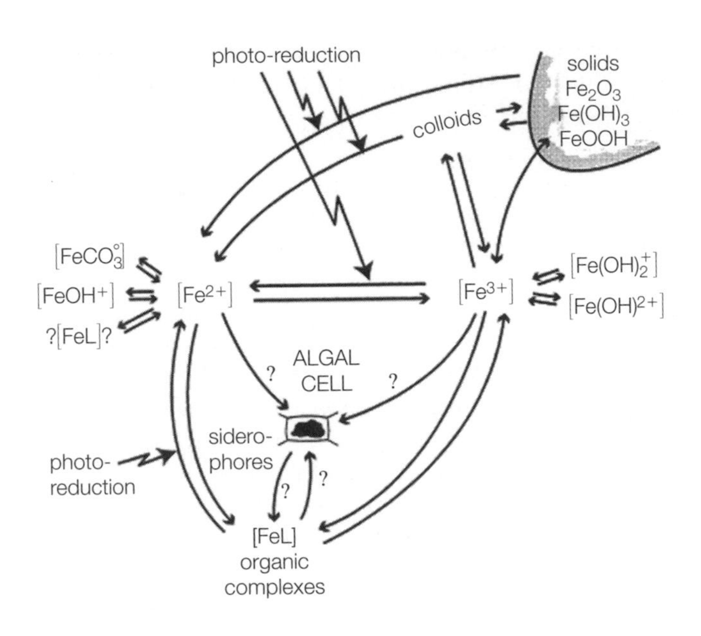

FIGURE 5.14

Iron assimilation by phytoplankton. *Source*: From Gerringa, L. J. A., *et al*. (2000). *Marine Chemistry* 68, 235–346.

seawater when they develop speciation models; instead they focus computations on the chemicals and reactions of interest.

An example of the results of such a calculation is presented in Figure 5.15 for two scenarios: (1) iron undergoes hydrolysis in a 0.7 molal NaCl solution that has the ionic strength of seawater and (2) iron undergoes hydrolysis in organic-free seawater. These calculations have been performed with the assumptions that the seawater is oxic, and that no dissolved organic matter is present. In oxic seawater, most of the dissolved iron in seawater is in its higher oxidation state (Fe(III)) rather than as Fe(II) due to the ready oxidation of Fe(II) to Fe(III) by $O_2(g)$. The speciation calculation results for these two scenarios predict that at pH 8, which is typical for seawater, four species of inorganic dissolved iron should be present, $Fe(OH)^{2+}(aq)$, $Fe(OH)_2^+(aq)$, $Fe(OH)_3^o(aq)$, and $Fe(OH)_4^-(aq)$. Using the percentages presented in Figure 5.15, the concentration of each species can be calculated from any possible total dissolved iron concentration. The difference in results between the two scenarios is due to the more complicated ion composition of seawater as compared to the NaCl solution. In seawater, a greater diversity of ion pairing and nonspecific interactions are possible. The latter affects the values of the K_c's.

Although computer programs are now used to perform speciation calculations, examining how these calculations are performed provides important insights into the limitations of the model predictions. Thus, we will step through a small part of the calculation used to generate the results presented in Figure 5.4, which represents the iron

Table 5.4 Examples of Commercially Available and Free Software for Computing Aqueous Chemical Speciation in Natural Waters.

MINTEQA2 *http://www.epa.gov/ceampubl/mmedia/minteq/index.htm*
MINTEQA2 is an equilibrium speciation model that can be used to calculate the equilibrium composition of dilute aqueous solutions in the laboratory or in natural aqueous systems. The model is useful for calculating the equilibrium mass distribution among dissolved species, adsorbed species, and multiple solid phases under a variety of conditions including a gas phase with constant partial pressures. A comprehensive database is included that is adequate for solving a broad range of problems without need for additional user-supplied equilibrium constants.

MINEQL+ *http://www.mineql.com/index.html*
A chemical equilibrium modeling system that can be used to perform calculations on low temperature (0–50°C), low to moderate ionic strength (<0.5 M) aqueous systems. MINEQL+ uses a thermodynamic database that contains the entire USEPA MINTEQA2 database plus data for chemical components that the EPA did not include, so all calculations will produce results compatible with EPA specifications.

The Geochemist's Workbench *http://hercules.geology.uiuc.edu/~bethke/hydro_gwb.htm*
A set of software tools for manipulating chemical reactions, calculating stability diagrams and the equilibrium states of natural waters, tracing reaction processes, modeling reactive transport, and plotting the results of these calculations. These include: (1) RXN balances chemical reactions and calculates equilibrium constants, temperatures, and equations. (2) ACT2 generates stability diagrams on activity, E_h, $p\varepsilon$, pH, and fugacity axes. (3) TACT generates temperature-activity and temperature-fugacity diagrams. (4) SPECE8 computes the distribution of species, sorption onto surfaces, mineral saturation, and gas fugacity in aqueous solutions. (5) Aqplot projects SPECE8 calculations in a variety of ways, including ternary, Piper, Stiff, Durov, Schoeller, and ion balance diagrams. (6) REACT traces reaction paths involving fluids, minerals, and gases. REACT can account for kinetic rate laws, isotope fractionation, microbial metabolism and growth, and much more. (7) X1t and X2t models reactive transport in groundwater flows in one and two dimensions, including the effects of injecting and producing wells. (8) XTPLOT renders the results of X1t and X2t simulations as x-y graphs, color maps, and contour plots.

WATEQ4F *http://water.usgs.gov/software/wateq4f.html*
A USGS model for computing the major and trace element speciation and mineral saturation for natural waters.

PHREEQE *http://water.usgs.gov/software/phreeqe.html*
A USGS model for computing the chemical evolution of the aqueous phase as the result of specified geochemical reactions. Based on an ion-pairing aqueous model, PHREEQE can calculate pH, redox potential, concentration of elements, molalities and activities of aqueous species, and mineral or gas mass transfer as a function of reaction progress. The program is capable of simulating reactions due to mixing, titrating, net irreversible reaction, temperature changes, and mineral- or gas- phase equilibration.

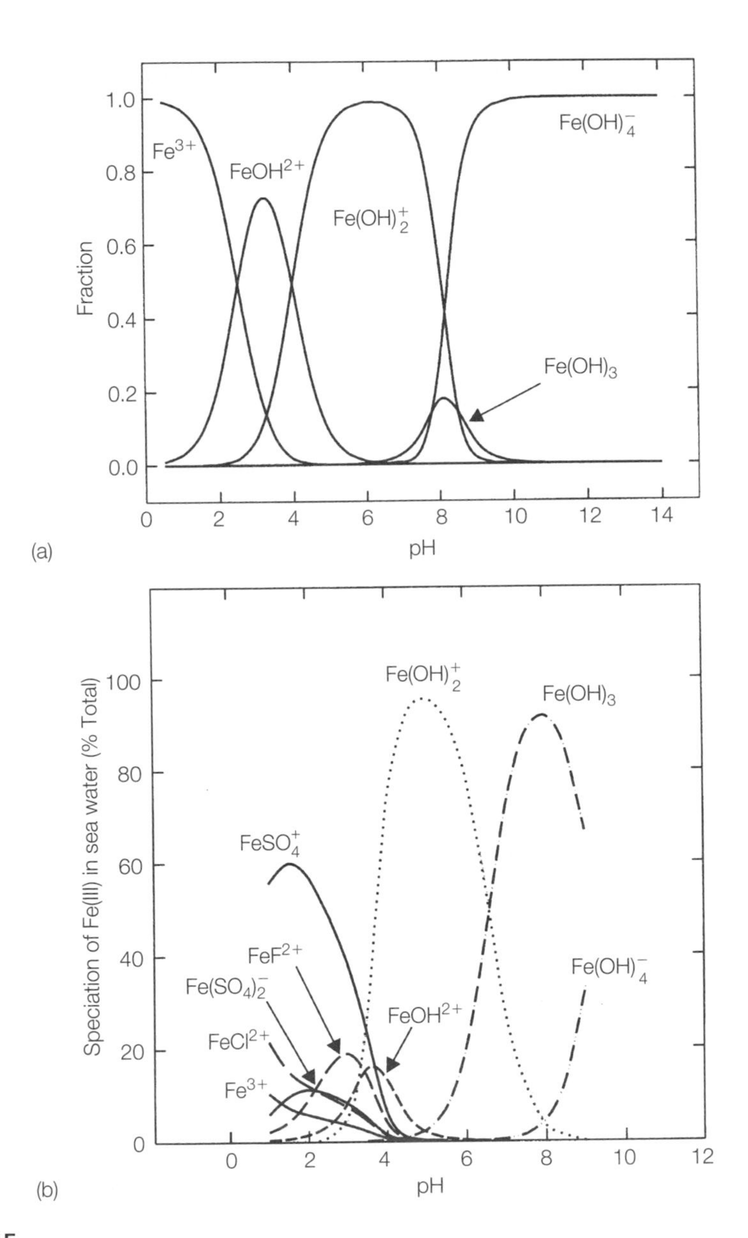

FIGURE 5.15

Iron speciation shown as % total Fe present in a particular species as a function of pH. (a) In 0.7 *m* NaCl and (b) for organic-free seawater $S = 35$, both at 25°C. *Source*: From Millero, F., and D. Pierrot (2002). *Chemistry of Marine Water and Sediment*, Springer-Verlag, pp. 193–220.

speciation in a 0.7 *m* NaCl solution, whose total dissolved iron concentration is equal to that of average deepwater, 0.76 (nmol Fe)/(kg SW).

The first step is to identify the chemical reactions of interest, in this case, the hydrolysis reactions for Fe(III):

$$Fe^{3+}(aq) + H_2O(l) \rightleftharpoons Fe(OH)^{2+}(aq) + H^+(aq) \quad (5.12)$$

$$Fe^{3+}(aq) + 2H_2O(l) \rightleftharpoons Fe(OH)_2^+(aq) + 2H^+(aq) \quad (5.13)$$

$$Fe^{3+}(aq) + 3H_2O(l) \rightleftharpoons Fe(OH)_3^o(aq) + 3H^+(aq) \quad (5.14)$$

$$Fe^{3+}(aq) + 4H_2O(l) \rightleftharpoons Fe(OH)_4^-(aq) + 4H^+(aq) \quad (5.15)$$

The following stoichiometric equilibrium expressions can be written for each of these species:

$$K_{c_{Fe(OH)^{2+}}} = \frac{[Fe(OH)^{2+}][H^+]}{[Fe^{3+}]} \quad (5.16)$$

$$K_{c_{Fe(OH)_2^+}} = \frac{[Fe(OH)_2^+][H^+]^2}{[Fe^{3+}]} \quad (5.17)$$

$$K_{c_{Fe(OH)_3^o}} = \frac{[Fe(OH)_3^o][H^+]^3}{[Fe^{3+}]} \quad (5.18)$$

$$K_{c_{Fe(OH)_4^-}} = \frac{[Fe(OH)_4^-][H^+]^4}{[Fe^{3+}]} \quad (5.19)$$

A mass balance equation can be written to show that Fe(III) is present in five dissolved species,

$$\begin{aligned}[Fe(III)]_{Total} &= [Fe(III)]_{Fe^{3+}} + [Fe(III)]_{Fe(OH)^{2+}} \\ &+ [Fe(III)]_{Fe(OH)_2^+} + [Fe(III)]_{Fe(OH)_3^o} + [Fe(III)]_{Fe(OH)_4^-}\end{aligned} \quad (5.20)$$

Since each species has one iron atom per molecule, Eq. 5.20 can be rewritten in terms of the species concentrations as:

$$[Fe(III)]_{Total} = [Fe^{3+}] + [Fe(OH)^{2+}] + [Fe(OH)_2^+] + [Fe(OH)_3^o] + [Fe(OH)_4^-] \quad (5.21)$$

The fraction of Fe(III) present as free iron, i.e., $[Fe(III)]_{Fe^{3+}}/[Fe(III)]_{Total}$ can computed by substituting in definitions for the concentrations of the molecular species, $Fe(OH)^{2+}$, $Fe(OH)_2^+$, $Fe(OH)_3^0$, $Fe(OH)_4^-$, based on their stoichiometric formation constants (K_c) and $[Fe^{3+}]$ as follows:

$$[Fe(OH)^{2+}] = \frac{K_{c_{Fe(OH)^{2+}}}[Fe^{3+}]}{[H^+]} \quad (5.22)$$

$$[Fe(OH)_2^+] = \frac{K_{c_{Fe(OH)_2^+}}[Fe^{3+}]}{[H^+]^2} \quad (5.23)$$

$$[Fe(OH)_3^o] = \frac{K_{c_{Fe(OH)_3^o}}[Fe^{3+}]}{[H^+]^3} \tag{5.24}$$

$$[Fe(OH)_4^-] = \frac{K_{c_{Fe(OH)_4^-}}[Fe^{3+}]}{[H^+]^4} \tag{5.25}$$

where the value of K_c is appropriate for the salinity, temperature, and pressure of interest.

Substituting these expressions into the mass balance equation yields

$$[Fe(III)]_{Total} = [Fe^{3+}] + \frac{K_{c_{Fe(OH)^{2+}}}[Fe^{3+}]}{[H^+]} + \frac{K_{c_{Fe(OH)_2^+}}[Fe^{3+}]}{[H^+]^2} + \frac{K_{c_{Fe(OH)_3^o}}[Fe^{3+}]}{[H^+]^3} + \frac{K_{c_{Fe(OH)_4^-}}[Fe^{3+}]}{[H^+]^4} \tag{5.26}$$

This equation can be simplified by rearranging to isolate $[Fe^{3+}]$,

$$[Fe(III)]_{Total} = [Fe^{3+}]\left(1 + \frac{K_{c_{Fe(OH)^{2+}}}}{[H^+]} + \frac{K_{c_{Fe(OH)_2^+}}}{[H^+]^2} + \frac{K_{c_{Fe(OH)_3^o}}}{[H^+]^3} + \frac{K_{c_{Fe(OH)_4^-}}}{[H^+]^4}\right) \tag{5.27}$$

and solving for the fraction of $[Fe(III)]_{Total}$ present as $[Fe^{3+}]$;

$$\frac{1}{\left(1 + \frac{K_{c_{Fe(OH)^{2+}}}}{[H^+]} + \frac{K_{c_{Fe(OH)_2^+}}}{[H^+]^2} + \frac{K_{c_{Fe(OH)_3^o}}}{[H^+]^3} + \frac{K_{c_{Fe(OH)_4^-}}}{[H^+]^4}\right)} = \frac{[Fe^{3+}]}{[Fe(III)]_{Total}} \tag{5.28}$$

$[H^+]$ can be computed from the pH and an appropriate activity coefficient. Values for this activity coefficient and appropriate K_c's are provided in Table 5.5. Thus, for a given pH, a value for $[Fe^{3+}]/[Fe(III)]_{Total}$ can be obtained. At a pH of 8.1, $[Fe^{3+}]/[Fe(III)]_{Total} = 1.87 \times 10^{-11}$. This is shown in Figure 5.4a in terms of a percentage, i.e., only 0.000000002% of the Fe(III) is present as the free ion, $Fe^{3+}(aq)$.

The fraction of the free Fe(III) obtained from Eq. 5.28 can then be used to compute the fractions of the other iron species as per the following example:

$$\left(\frac{[Fe^{3+}]}{[Fe(III)]_{Total}}\right) \times \frac{K_{c_{Fe(OH)^{2+}}}}{[H^+]} = \frac{[Fe(OH)^{2+}]}{[Fe(III)]_{Total}} \tag{5.29}$$

The results of this multicomponent equilibrium calculation showing the speciation of Fe(III) as a function of pH are presented in Figure 5.4b:

Table 5.5 Equilibrium Constants for Fe(III) and γ_{H^+} at 25°C and 0 bar Determined in 0.7 m NaCl.

Ion	log K_c	γ
$Fe(OH)^{2+}(aq)$	−2.52	NA
$Fe(OH)_2^+(aq)$	−6.52	NA
$Fe(OH)_3^0(aq)$	−14.67	NA
$Fe(OH)_4^-(aq)$	−23.04	NA
$H^+(aq)$	NA	0.59

These K_c's are defined in terms of molarity. Also provided is γ_{H^+} for these environmental conditions. Data sources: Liu, X. and F. J. Millero (1999). Geochimica et Cosmochimica Acta *63, 3487–3497 (Fig. 3) and Millero, F., and D. Pierrot (2002).* Chemistry of Marine Water and Sediment, *Springer-Verlag, pp. 193–220.*

If the concentration of $[Fe(III)]_{Total}$ is known, Eq. 5.27 can be rearranged to solve for $[Fe^{3+}]$:

$$\frac{[Fe(III)]_{Total}}{\left(1+\frac{K_{c_{Fe(OH)^{2+}}}}{[H^+]}+\frac{K_{c_{Fe(OH)_2^+}}}{[H^+]^2}+\frac{K_{c_{Fe(OH)_3^0}}}{[H^+]^3}+\frac{K_{c_{Fe(OH)_4^-}}}{[H^+]^4}\right)}=[Fe^{3+}] \quad (5.30)$$

Note that since the K_c's in Table 5.5 are defined in molarity units, the $[Fe(III)]_{Total}$ should be in similar units. (The density of seawater is used to convert kg SW to L SW.) The resulting $[Fe^{3+}]$ can then be substituted into Eqs. 5.22 through 5.25 to compute the concentrations of each of the molecular species. These results are shown in Figure 5.4b. They are somewhat different than those in Figure 5.15a because of slight differences in the values of K_c used. Because the solubility of iron is a matter of great interest, marine scientists are continuing to improve their measurements of the relevant K_c s and, thus, no single set of values has obtained universal acceptance.

A full understanding of the speciation of dissolved iron requires consideration of ligands other than water and hydroxide. The most important ones are listed in Table 5.6 along with their concentration ranges in seawater and freshwater. For Fe(III) in seawater at pH > 4, the formation of complexes with hydroxide is most important, but at pH < 4, sulfate, chloride, and fluoride pairing predominates (Figure 5.15b). To predict the equilibrium speciation at low pH, these anions need to be added to the mass balance equation for Fe(III) (Eq. 5.20). Seawater with low pH tends to have low O_2 concentrations. Under these conditions, most of the dissolved iron is present as Fe(II), which undergoes complexation with sulfide and carbonate.

Including multiple anions into speciation calculations greatly increases their complexity because the anions also undergo multiple ion pairing reactions. The most important involve the formation of ion pairs with the major ions, Na^+, Ca^{2+}, Mg^{2+},

Table 5.6 Concentrations of Ligands in Seawater and River Water. (log Conc.(M)).

	Freshwater	Seawater
HCO_3^-	−4 to −2.3	−2.6
OH^-	−4 to −8	−5.5 to −6.5
CO_3^{2-}	−6 to −4	−4.5
Cl^-	−5 to −3	−0.26
SO_4^{2-}	−5 to −3	−1.55
F^-	−6 to −4	−4.2
HS^-/S^{2-} (anoxic conditions)	−6 to −3	—
Amino acids	−7 to −5	−7 to −6
Organic acids	−6 to −4	−6 to −5
Particle surface groups	−8 to −4	−9 to −6

Source: After Stumm, W., and J. J. Morgan (1996). Aquatic Chemistry, *3rd ed. Wiley-Interscience.*

and K^+. This ion pairing significantly lowers the free ion concentrations of the anions. To compute the degree of anion binding, a calculation similar to that outlined earlier for Fe(III) must be performed for each of the anions. This requires development of a mass balance equation, analogous to Eq. 5.20, for each anion. To obtain the free anion concentrations from these mass balance equations requires that each of the free cation concentrations be known. To obtain this information requires generation of a set of mass balance equations for the cations. The cation mass balance equations cannot be solved without knowledge of the free anion concentrations. In other words, the resulting system of equations for the anions and cations has too many unknowns to be explicitly solved.

Various numerical techniques are used to indirectly obtain solutions to large systems of equations with too many unknowns to solve explicitly. One approach is to solve the equations iteratively. This is done by first assuming that all of the anions are unbound and, hence, their free ion concentrations are equal to their total (stoichiometric) concentrations. By substituting these assumed anion concentrations into the cation mass balance equations, an initial estimate is obtained for the free cation concentrations. These cation concentrations are substituted into the anion mass balance equations to obtain a first estimate of the free anion concentrations. These free anion concentrations are then used to recompute the free cation concentrations. The recalculations are continued until the resulting free ion concentrations exhibit little change with further iterations. The computer programs used to perform speciation calculations perform these iterations in a matter of seconds.

5.7.3 The Chemical Speciation of Iron: Mineral Precipitation and Dissolution

Some solutes in seawater participate in equilibria that involve minerals. These reactions can be represented by

$$AB(s) \rightleftharpoons A^+(aq) + B^-(aq) \tag{5.31}$$

where the thermodynamic equilibrium constant is

$$K_{eq} = \frac{\{A^+(aq)\}\{B^-(aq)\}}{\{AB(s)\}} \tag{5.32}$$

For pure mineral phases in dilute solutions,[4] $\{AB(s)\} = 1$ and Eq. 5.32 becomes

$$K_{sp} = \{A^+(aq)\}\{B^-(aq)\} \tag{5.33}$$

where K_{sp} is called the ***solubility product***. (Technically K_{sp}'s are defined only for sparingly soluble salts.) As with any equilibrium, mineral solubility is a function of temperature, pressure, and the ionic strength of the solution.

To avoid the necessity of using activity coefficients, mineral solubility can be defined in terms of stoichiometric concentrations where

$$K^*_{sp} = [A^+(aq)]\,[B^-(aq)] \tag{5.34}$$

For the remainder of this text, we will use the stoichiometric K^*_{sp} defined for the temperature, pressure, and salinity of interest rather than the thermodynamic K_{sp}.

Solubility products can be used to predict the stability of a mineral by comparing the observed ***ion product***, $[A^+(aq)][B^-(aq)]$, to the mineral's K^*_{sp}. If the ion product is greater than K^*_{sp}, the solution is supersaturated with respect to that mineral. In this case, precipitation should proceed spontaneously until the ion concentrations are decreased to levels that lower the ion product to the value dictated by the K^*_{sp}. Conversely, if the ion product is less than the K^*_{sp}, the solution is undersaturated with respect to that mineral. Dissolution should proceed spontaneously until the ion concentrations are increased to levels that raise the ion product to a value equal to the K^*_{sp}. At equilibrium, where the ion product has a value equal to that of the K^*_{sp}, the rate of mineral dissolution is equal to the rate of precipitation, so the ion concentrations remain constant over time.

To return to our case study of iron, the equilibrium concentration of Fe(III) is ultimately controlled by its mineral solubility. Since atmospheric dust is a major source of "new" iron to the ocean, its solubility is a matter of hot debate. If the solubility is low, the particulate iron is likely to settle out of the euphotic zone before it can be assimilated by plankton. Iron is one of the most abundant elements in Earth's crust, so it is not surprising that concentrations in dust are high, ranging from 3 to 5% dry weight.

---

[4] This is true only in dilute solutions, because at high ionic strengths, the adsorption of ions alters the mineralogy of the solid's surface, thereby affecting its activity.

The three most abundant minerals forms are Fe_2O_3 (hematite), $Fe(OH)_3$ (hydrous ferric oxide or ferrihydrite), and FeO(OH) (goethite). The chemical reactions describing their dissolution and K^*_{sp} in slightly acidic water at 25°C, 1 atm, and the ionic strength of seawater are:

$$Fe_2O_3(s) + 6H^+(aq) \rightleftharpoons 2Fe^{3+}(aq) + 3H_2O(l) \quad \log K^*_{sp} = -1.76 \tag{5.35}$$

$$Fe(OH)_3(s) + 3H^+(aq) \rightleftharpoons Fe^{3+}(aq) + 3H_2O(l) \quad \log K^*_{sp} = 4.2 \tag{5.36}$$

$$FeOOH(s) + 3H^+(aq) \rightleftharpoons Fe^{3+}(aq) + 2H_2O(l) \quad \log K^*_{sp} = -0.648 \tag{5.37}$$

Since the hydrous ferric oxide has the largest K^*_{sp}, it should be the most soluble. The equilibrium concentration of Fe^{3+} can be predicted for this mineral at a given pH, as follows:

$$K^*_{sp} = \frac{[Fe^{3+}]}{[H^+]^3} = \frac{[Fe^{3+}]}{\left[\frac{10^{-pH}}{\gamma_{H^+}}\right]^3} \tag{5.38}$$

$$[Fe^{3+}] = K^*_{sp}\left[\frac{10^{-pH}}{\gamma_{H^+}}\right]^3 \tag{5.39}$$

For both organic-free seawater and 0.7 m NaCl, $K^*_{sp} \approx 4.2$ at pH 8.1 and 25°C, 1 atm. Using a value of $\gamma_{H^+} = 0.59$,

$$[Fe^{3+}] = 10^{4.2}\left[\frac{10^{-8}}{0.59}\right]^3 = 3.91 \times 10^{-20}\,M \tag{5.40}$$

As shown in the preceding section, very little of the Fe(III) dissolved in seawater is predicted to be in the form of Fe^{3+} at equilibrium. Using the data in Table 5.5 for 0.7 m NaCl, we computed that $[Fe(III)]_{Fe^{3+}}/[Fe(III)]_{Total} = 1.87 \times 10^{-11}$ at pH = 8.1. This can be used to compute the equilibrium concentration of $Fe(III)_{Total}$:

$$[Fe(III)]_{Total} = \frac{[Fe^{3+}]}{\frac{[Fe(III)]_{Fe^{3+}}}{[Fe(III)]_{Total}}} = \frac{3.91 \times 10^{-20}}{1.87 \times 10^{-11}} = 2.09 \times 10^{-9} = 2.09\,nM \tag{5.41}$$

Since $[Fe(III)]_{Total} >> [Fe^{3+}]$, the formation of ion pairs and complexes is greatly enhancing the equilibrium solubility of ferrihydrite. This is called the salting-in effect and illustrates why mineral solubility calculations in seawater must take ion speciation into consideration.

This simplified calculation is used to illustrate basic computational techniques. It assumes that all of the $Fe(OH)^0_3(aq)$ is a true solute. The quality of this assumption is a matter of debate as at pH 8, $Fe(OH)^0_3(aq)$, tends to form colloids. Thus, laboratory measurements of ferrihydrite solubility yield results highly dependent on the method by which $[Fe(III)]_{Total}$ is isolated. Ultrafiltration techniques that exclude colloids from the $[Fe(III)]_{Total}$ pool produce very low equilibrium solubility concentrations, on the order of 0.01 nM. This is an important issue because a significant fraction of the iron in seawater is likely colloidal, some of which is inorganic and some organic. In oxic seawater, inorganic colloidal iron is thought to be composed of ferrihydrite produced by the rapid oxidation of $Fe^{2+}(aq)$. As shown in Figure 5.14, photochemical reduction reactions appear to play an important role in the production of $Fe^{2+}(aq)$ and, hence, that of inorganic colloidal iron.

How do these equilibrium speciation predictions compare to observed iron concentrations in seawater (0.1 to 1.3 (nmol Fe)/(kg SW))? The computational results are similar, suggesting the importance of hydrolysis and solubility equilibria in controlling seawater's $[Fe(III)]_{Total}$. The general increase in $[Fe(III)]_{Total}$ with depth is also consistent with equilibrium control as the solubility of Fe(III) increases with decreasing temperature. On the other hand, the observed seawater concentrations are far higher than those obtained from the laboratory measurements of ferrihydrite solubility (0.01 nM). These equilibrium solubility measurements were made in organic-free solutions suggesting that seawater contains organic ligands that greatly enhance iron solubility.

5.7.4 The Chemical Speciation of Iron: Cation Exchange

The effect of solids on ion speciation is not limited to precipitation/dissolution reactions. Most solid surfaces in seawater possess a net negative charge that enables them to electrostatically attract cations (M^+). This attraction can be represented as:

$$M_1^+(aq) + \{M_2^+ - R(s)\} \rightleftharpoons M_2^+(aq) + \{M_1^+ - R(s)\} \tag{5.42}$$

where R represents a mineral with a negatively charged surface and M_1 and M_2 are metals. Particulate organic matter and colloids tend to have negatively charged surfaces and, thus, also adsorb cations. These adsorbed ions are not part of the crystal lattice or molecular structure of the minerals. Thus, their adsorption and desorption reactions are readily reversible. An equilibrium constant can be defined for this process. Its magnitude reflects the relative affinity of a cation for adsorption onto the minerals surface. This is largely a result of an ion's charge density (z/r^2).

The equilibrium speciation of a metal ion influenced by cation exchange is dependent on the relative concentrations of the cations competing for the negatively charged sites on the particle's surface and their relative affinities for adsorption. Since one cation displaces another from the negatively charged sites, this process is termed *cation exchange*.

Cation exchange is particularly important in estuaries because particles transported in river flows experience increasingly higher major cation concentrations as the freshwater mixes with seawater. This process is represented by

$$M_1^+(aq) + \underbrace{\{Fe^{3+} - R(s)\}}_{\text{riverine particle}} \rightleftharpoons Fe^{3+}(aq) + \{M_1^+ - R(s)\} \tag{5.43}$$

where $M_1^+ = Na^+(aq)$, $Ca^{2+}(aq)$, $K^+(aq)$, or $Mg^{2+}(aq)$.

This leads to desorption of the minor and trace metals because the major cation ion concentrations are so high as to swamp any differences in relative adsorption affinity. As shown in Figure 5.1, desorption must play a minor role in the estuarine chemistry of iron because the net effect is a chemical removal. In the open ocean, adsorption of dissolved metals onto sinking particles can represent a significant sink as these particles eventually settle onto the seafloor and become buried in the sediments.

5.7.5 The Chemical Speciation of Iron: Dissolved Organic Matter as a Ligand

The solubility concentration of $Fe(III)_{Total}$ predicted earlier is based on equilibrium constants measured in organic-free seawater and NaCl solutions ($I = 0.7$ m). $Fe(OH)_3(s)$ has been observed to be far more soluble in seawater than in the organic-free solutions. In some cases, the solubility in seawater can be decreased by exposure to UV radiation. These observations suggest that the enhanced solubility of Fe(III) is due to the formation of soluble organometallic complexes, which are subject to degradation by UV radiation. Direct measurements using voltammetry provide estimates of the concentrations of these organic ligands and their formation constants. As shown in the examples provided in Table 5.7, the concentrations of these ligands tend to be greater than the concentration of $Fe(III)_{Total}$ and their formation constants are quite high.

These high concentrations and binding strengths support the hypothesis that a significant fraction of the Fe(III) is present as an organometallic complex. We can estimate this fraction by adding the following reaction:

$$Fe^{3+}(aq) + L^{-}(aq) \rightleftharpoons FeL^{2+}(aq) \quad (5.44)$$

to our $Fe(III)_{Total}$ mass balance equation:

$$[Fe(III)]_{Total} = [Fe^{3+}] + [Fe(OH)^{2+}] + [Fe(OH)_2^{+}] + [Fe(OH)_3^{o}] + [Fe(OH)_4^{-}] + [FeL^{2+}] \quad (5.45)$$

A test calculation, using the data from the Atlantic Ocean where $[L]_T = 4$ nM and $\log K_{FeL} = 19.3$ and the equilibrium constants from Table 5.5, yields a solubility $[Fe(III)]_{Total} = 5.2$ nM, close to a threefold increase! This estimate is substantially higher

Table 5.7 Concentrations and Formation Constants for Fe(III) Organic Complexes in SeaWater.

Region	$[L]_T$(nM)	$[Fe]_T$(nM)	$\log K_{FeL}$
Menai Straits	4–10	3.5–8.5	21.3
Atlantic	3–5	0.8–1.8	19.3
Atlantic	0.4–0.6	0.17–0.27	23.2
Pacific	0.4	0.2–0.8	23.1
Mediterranean	4–13	3.1	21.8

Data sources provided in Millero, F., and D. Pierrot (2002). Chemistry of Marine Water and Sediment, *Springer-Verlag, pp. 193–220.*

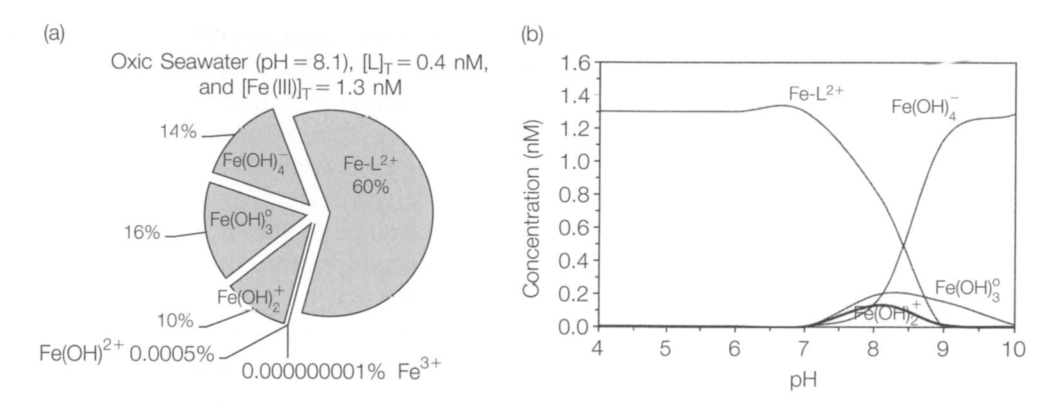

FIGURE 5.16

Iron speciation in seawater with ligands using data from Atlantic Ocean as per Table 5.7, where $[L]_T$ = 4 nM, $[Fe(III)]_T$ = 1.3 nM, and log $K_{FeL^{2+}}$ = 19.3 (a) Pie chart and (b) Concentration as a function of pH.

than the observed concentrations (0.8 to 1.8 nM) suggesting that $[Fe(III)]_{Total}$ in seawater is not wholly controlled by equilibrium processes. Or it could simply reflect uncertainties in the equilibrium model, such as those associated with values for the K_c's and colloid formation.

Although the details of the equilibrium model are still uncertain, the general trends are likely reliable. As shown in Figure 5.16, most of the Fe(III) in seawater is predicted to be in the form of the FeL complex. The equilibrium model also predicts that this degree of complexation should enhance iron solubility such that 10 to 50% of the iron delivered to the ocean as dust will eventually become dissolved if equilibrium is attained. If this model is a reasonable representation for iron speciation in seawater, uptake of $[Fe(III)]_{Total}$ by phytoplankton should induce a spontaneous dissolution of additional particulate iron so as to drive the dissolved iron concentrations back toward their equilibrium values.

Why Are Organic Ligands Such Good Binding Agents?

Dissolved organic matter contains negatively charged functional groups that can form coordination complexes with dissolved metals. The degree to which an organic compound complexes with a metal depends on its concentration and affinity for that metal. For example, in estuarine waters where DOM concentrations can exceed 10 mg/L, a significant amount of some trace metals are present as organic complexes. A similar effect is not seen on the major cations because their seawater concentrations are much higher than the organic ligand concentrations.

Though a great variety of dissolved organic compounds are present in seawater, the functional groups responsible for metal complexation are primarily the anions $R{-}O^-$, $R{-}COO^-$, $R{-}S^-$, and $R_2{-}\ddot{N}H$, $R{-}\ddot{N}H_2$, where R represents any organic structure. Nonpolar organic matter is not expected to function well as a ligand. As illustrated in

Figure 5.11, many organic compounds contain more than one of these functional groups and, thus, are able to form multiple bonds or associations with metals.

The organic compounds most effective at binding with trace metals are the ones with very high and specific affinities for these ions. The most extreme examples are the biomolecules present in marine organisms. These compounds are so effective that they enable organisms to selectively concentrate certain trace metals in their tissues up to a millionfold over the dissolved metal concentrations in seawater. This great specificity is probably related to the three-dimensional geometry of the biomolecules that directs the binding in a manner analogous that which enables enzymes to bind only certain substrates. As discussed later, organisms seem to be responsible for the production of the organic ligands with high binding affinities.

Where Do Organic Ligands Come From?

The sources and sinks of specific dissolved organic compounds are the subject of Chapter 23. Thus, we will here just briefly summarize the major sources: (1) in situ production by marine organisms, (2) input via river runoff of terrestrial organic matter, and (3) decomposition of sinking or sedimentary particulate organic matter. Dissolution of organic gases from the atmosphere and hydrothermal systems is of lesser importance. Most of the DOM supplied by rivers is operationally defined as humic substances. These are high-molecular-weight compounds that are variable in composition and relatively resistant to degradation. They are likely formed by abiogenic reactions among biomolecules and fragments of biomolecules, such as the nutritionally poor compounds characteristic of terrestrial plants, i.e., cellulose and lignin. Some humic substances are similarly created by abiogenic reactions between organic compounds released in seawater during decomposition processes. As shown in Figure 5.17, the affinity of metals for humic substances is highest for copper and mercury.

These affinities are generally less than observed for DOM excreted directly into seawater by marine organisms. These biomolecules are relatively low in molecular weight. The siderophores are a well-studied example of such ligands. As shown in Figure 5.11, they contain atoms of nitrogen, oxygen, and sulfur atoms that serve as effective chelating agents. These biomolecules are ubiquitous in marine bacteria, blue-green algae, and fungi. Within the cells, the siderophores have extraordinarily high affinities for Fe^{3+} ($K \approx 10^{20}$ to 10^{60}). These high affinities demonstrate the importance of iron to the metabolic processes of this diverse group of organisms. These high affinities also reflect the prevalence of iron limitation, making biosynthesis of siderophores an evolutionary advantage in meeting nutritional iron needs from oxic seawater.

How Can Organic Complexation Affect Metal Bioavailability?

Complexation by organic matter causes a wide variety of effects on the bioavailability of metals. As noted earlier, the toxicity of dissolved copper can be reduced by complexation with organic matter. In this case, complexed copper is not biologically available to phytoplankton. The other extreme is exhibited by iron, in which the metal's bioavailability is enhanced by complexation with organic matter. There are two likely

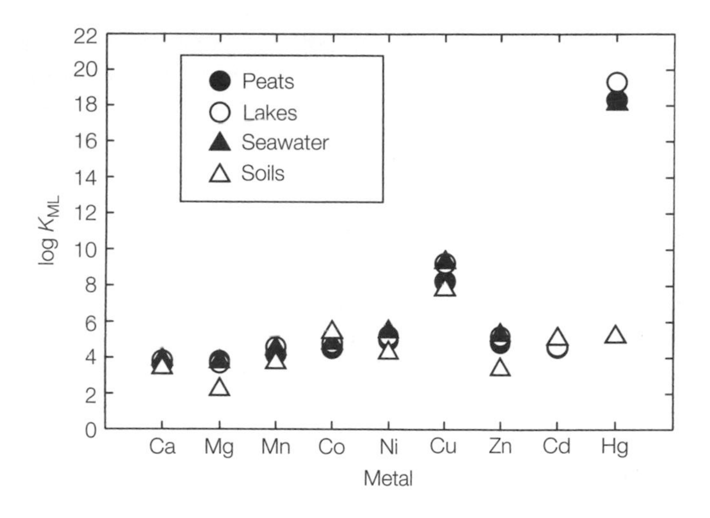

FIGURE 5.17

Formation constants for metals complexing with humic materials for marine and terrestrial sources. *Source*: From Millero, F., and D. Pierrot (2002). *Chemistry of Marine Water and Sediment,* Springer-Verlag, pp. 193–220.

mechanisms by which this occurs. The first is that cells have a metabolic pathway for transporting the complex across their cell membranes. In the case of siderophores, this is likely because these organic compounds are synthesized and excreted by the marine microbes that presumably are assimilating the organo-iron complexes from seawater. Photochemical reactions appear to play an important role in converting iron-siderophore complexes into a more biologically available form. Second, a ready supply of $Fe^{3+}(aq)$ for uptake across the cell membrane can be supported by the breakdown of the organometallic complexes. For decomplexation to be a significant supply of biologically active iron, this reaction must occur at a fairly rapid rate in near proximity to the cells.

Not all phytoplankton species seem to be similarly equipped to excrete siderophores. As illustrated in Figure 5.5, several species that contribute to red tides are so well equipped that their growth can be enhanced by exposure to particulate iron. This may be achieved by adhesion of the particulate iron to the cell membrane where solubilization occurs. Open-ocean fertilization experiments have also demonstrated that diatoms are more likely to experience growth stimulation following iron additions. This suggests that iron availability controls not just plankton growth, but also the biological species composition.

Complexation of metals with organic compounds can also increase the toxicity of metals. This is the case with mercury, in which the organo-Hg species, methyl- and dimethylmercury, are far more toxic than elemental or ionic mercury ($Hg^{2+}(aq)$). The enhanced toxicity is caused by the increased tendency of the organo species to be retained, and therefore concentrated within, organisms. As discussed in Chapter 28.6.8, mercury is naturally biomethylated by bacteria under conditions of low pH and low

O_2 concentrations, such as found in marine wetlands. High biomethylation rates have also been observed in coastal sediments. Because methylmercury is transferred up the food chain, the marine fish that occupy high trophic levels have very high mercury concentrations. In some cases, such as for tuna and swordfish, concentrations are high enough to pose human health risks.

5.8 ION SPECIATION IN SEAWATER

5.8.1 Major and Minor Ion Speciation in Seawater

The speciation of the major and minor ions in seawater for 25°C, 1 atm, and 35‰ is shown in Figure 5.18. These pie charts were constructed from multi-component competitive complexing calculations using appropriate equilibrium constants. The major cations are largely present as their free ions, with sulfate being the most important ligand. In comparison, only about one third of the sulfate is present as the free ion. In these calculations, chloride is assumed to be wholly unpaired. This is supported by direct observations. Significant amounts of the minor anions (bicarbonate, carbonate, hydroxide, borate, and fluoride) are also paired, predominantly with sodium and magnesium. For the major ions, the cation concentrations are much larger than the anion concentrations (with the exception of Cl^-). This causes ion pairing to influence the speciation of the anions much more than that of the cations.

The results of these calculations are dependent on salinity, temperature, and pressure. As noted earlier, semiempirical formulas have been developed to extrapolate equilibrium constants to the temperature, salinity, and pressure of interest. Using sulfate as an example (Table 5.8), we see that these effects can be significant, demonstrating the need to use the appropriate equilibrium constants.

5.8.2 Trace Metal Speciation in Seawater

As illustrated by iron, the equilibrium speciation of trace metals is characterized by ion pairing. The major inorganic species for each metal are listed in Table 5.9 in order of relative abundance. The free ions are relatively important for Mn, Co, Ni, Ag, and Cd. In oxic water, the most important inorganic ion pairs are those with hydroxide, chloride, sulfate, carbonate, and dissolved organic matter. In anoxic waters, sulfide complexes are important. The unique speciation of each metal is due to its electronic characters, particularly its ionic charge density. This subject is discussed further in Chapter 11. The metals that exhibit the highest degree of organic complexing are Cu, Zn, Fe, Pb, and Hg. This complexing is highly variable due to large vertical and horizontal gradients in DOM concentrations and structural composition that leads to differences in binding affinities.

In most cases, the ligands are present at far greater concentrations than the trace metals, (i.e., $L^-(aq) \gg M^+(aq)$). Thus, ion associations between the trace metals and ligands influence the speciation of the former to a far greater extent than that of the latter.

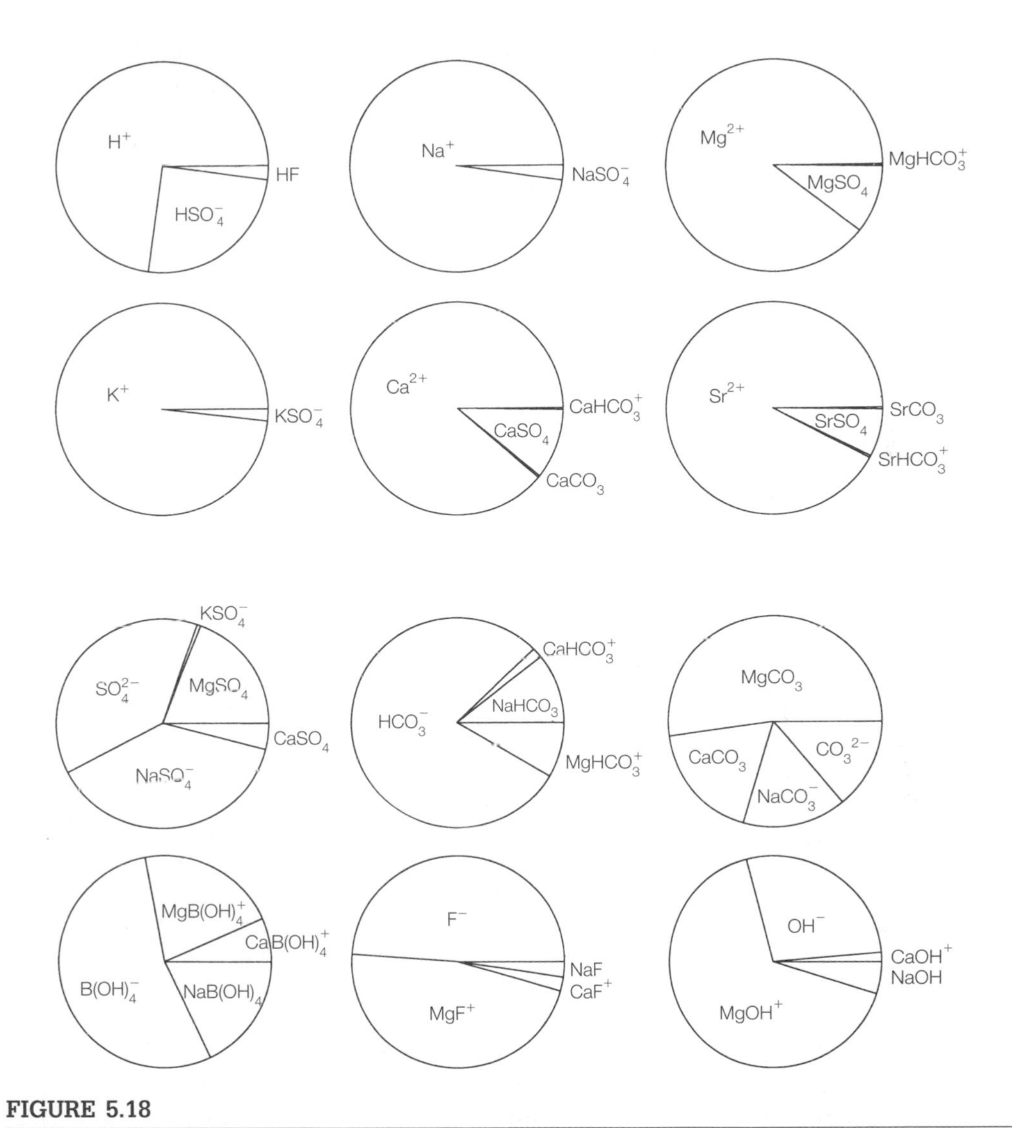

FIGURE 5.18

Major Ion Speciation. *Source*: From Millero, F. (1996). *Chemical Oceanography*, 2nd ed. CRC Press, p. 157.

Since the metals can be considered as the limiting reactants, the effects of metal-metal competition for the ligands can be neglected. Thus, speciation calculations for each trace metal can be performed independently.

As with all speciation calculations, the results for the trace metals depend on the values used for the thermodynamic equilibrium constants and activity coefficients. Determination of these constants is very difficult because direct detection of trace metal species is complicated by their low concentrations and analytical interferences

Table 5.8 Effect of Temperature and Pressure on Sulfate Species Speciation in Seawater.

T (°C)	*P* (atm)	Free SO_4^{2-} (%)	$NaSO_4^-$ (%)	$MgSO_4^0$ (%)	$CaSO_4^0$ (%)
25 (1)	1	39.0	38.0	19.0	4.0
2 (1)	1	28.0	47.0	21.0	4.0
2 (1)	1000	39.0	32.0	24.0	5.0
(2)	1000	42.0	35.0	19.0	4.0

Data from: (1) Millero, F. J. (1971). Geochimica et Cosmochimica Acta, 35, 1089–1098; and (2) Kester, D. R. and R. M. Pytkowicz (1970). Geochimica et Cosmochimica Acta, 34, 1039–1051.

Table 5.9 Major Inorganic Species in Natural Waters.

Condition	Element	Major Species	Freshwater $[M^{n+}]/M_T$	Seawater $[M^{n+}]/M_T$
Hydrolyzed, anionic	B(III)	H_3BO_3, $B(OH)_4^-$		
	V(V)	HVO_4^{2-}, $H_2\ VO_4^-$		
	Cr(VI)	CrO_4^{2-}		
	As(V)	$HASO_4^{2-}$		
	Se(VI)	SeO_4^{2-}		
	Mo(VI)	MoO_4^{2-}		
Hydrolyzed	Si(IV)	$Si(OH)_4$		
Predominantly free aquo ions	Li	Li^+	1.00	1.0
	Na	Na^+	1.00	1.0
	Mg	Mg^{2+} (Mg^{2+}, $MgCO_3$)	0.94	0.84
	K	K^+	1.00	0.98
	Ca	Ca^{2+} (Ca^{2+}, $CaSO_4$)	0.94	0.83
	Sr	Sr^{2+}	0.94	0.71
	Cs	Cs^+	1.00	0.93
	Ba	Ba^{2+}	0.95	0.86
Complexation with OH^-, CO_3^{2-}, HCO_3^-, Cl^-	Be(II)	$BeOH^+$, $Be(OH)_2^0$	1.5×10^{-3}	1.8×10^{-3}
	Al(III)	$Al(OH)_3(s)$, $Al(OH)_2^+$, $Al(OH)_4^-$	1×10^{-9}	6×10^{-10}
	Ti(IV)	$TiO_2(s)$, $Ti(OH)_4^0$		
	Mn(IV)	$MnO_2(s)$		
	Fe(III)	$Fe(OH)_3(s)$, $Fe(OH)_2^+$, $Fe(OH)_4^-$	2×10^{-11}	1×10^{-12}
	Co(II)	Co^{2+}, $CoCO_3^0$	0.5	0.58
	Ni(II)	Ni^{2+}, $NiCO_3^0$ (Ni^{2+}, NiCl)	0.4	0.47
	Cu(II)	$CuCO_3^0$, $Cu(OH)_2^0$	0.01	9.3×10^{-2}

(Continued)

Table 5.9 *(Continued)*

Condition	Element	Major Species	Freshwater $[M^{n+}]/M_T$	Seawater $[M^{n+}]/M_T$
	Zn(II)	Zn^{2+}, $ZnCO_3^0$ (Zn^{2+}, ZnCl)	0.4	0.45
	Ag(I)	Ag^+, $AgCl^0$ ($AgCl_2^-$, AgCl)	0.6	5.5×10^{-6}
	Cd(II)	Cd^{2+}, $CdCO_3^0$ ($CdCl_2$)	0.5	2.7×10^{-2}
	La(III)[a]	$LaCO_3^+$, $La(CO_3)_2^-$	8×10^{-3}	0.38
	Hg(II)	$Hg(OH)_2^0$, ($HgCl_4^{2-}$)	1×10^{-10}	6×10^{-15}
	Tl(I), (III)	Tl^+, $Tl(OH)_3$, $Tl(OH)_4^-$	2×10^{-21b}	3×10^{-21b}
	Pb(II)	$PbCO_3^0$ ($PbCl^+$, $PbCO_3$)	5×10^{-2}	3×10^{-2}
	Bi(III)	$Bi(OH)_3^0$	7×10^{-16}	1.6×10^{-15}
	Th(IV)	$Th(OH)_4^0$		2×10^{-16}
	U(VI)	$UO_2(CO_3)_2^{2-}$, $UO_2(CO_3)_3^{4-}$	1×10^{-7c}	1×10^{-7c}

[a]La(III) *is representative of the lanthanides.*
[b]*Redox state of* Tl(I) *under natural conditions is uncertain; ratio is for* Tl(III).
[c]*As* UO_2^{2+}.
Source: From Stumm, W., and J. J. Morgan (1996). Aquatic Chemistry, 3rd ed. Wiley-Interscience, p. 293.

caused by the high concentrations of the major ions. For many species, either no thermodynamic data exist or a wide range in values has been reported by various authors. Thus, the calculated speciation of the trace metals is much more tentative than for the major ions.

Trace metal speciation is a subject of great interest due to its impact on marine organisms. Some of the metal species appear to be micronutrients, potentially as important as those of iron. Most of these metals, which include Ni, Co, Zn, and Mn, have been demonstrated to play key roles in enzyme systems. Others, like Cu^{2+}, exert toxic effects. Marine microbes have been observed to alter the speciation of a variety of trace metals in seawater by excreting organic ligands. In the case of Cu^{2+}, this binding appears to make the metal biologically unavailable and, hence, lowers the toxicity of seawater. In an effort to predict what life forms may be present on other planets, biologists are now exploring extreme environments on Earth to obtain a better appreciation for the full range of microbial metabolic strategies. They are discovering organisms able to metabolically benefit from metals, such as arsenic and cadmium, formerly thought to be toxic in all forms and concentrations. This ability is likely due to biochemical pathways that involve complexation reactions that serve to reduce the toxicity of these metals.

5.8.3 Acid and Base Speciation in Seawater

Some equilibria in seawater involve weak acids (proton donors) and bases (proton acceptors) that represent potential pH buffers. The most abundant ones in seawater are listed in Table 5.10.

Table 5.10 Weak Acids and Bases in Natural Waters.

		Seawater		
	Freshwater Mean	**Warm Surface**	**Deep Atlantic**	**Deep Pacific**
Carbonate	0.97 mM	2.1 mM	2.3 mM	2.5mM
Silicate	220 μM	<3 μM	30 μM	150 μM
Ammonia	0–10 μM	<0.5 μM	<0.5 μM	<0.5 μM
Phosphate	0.7 μM	<0.2 μM	1.7 μM	2.5 μM
Borate	1 μM	0.4 mM	0.4 mM	0.4 mM
	Typical Anoxic Hypolimnion	**Black Sea (Deep Water)**	**Cariaco Trench**	
Sulfide	50–150 μM	330 μM	20 μM	
Ammonia	10–40 μM	53 μM	10 μM	

Source: *From Morel, F. M. M., and J. G. Hering (1993).* Principles and Applications of Aquatic Chemistry. *Wiley Interscience, p. 159.*

Their speciation is dominated by hydrolysis reactions. For acids, this is typically represented as a dissociation reaction,

$$HA(aq) \rightleftharpoons H^+(aq) + A^-(aq) \tag{5.46}$$

where A^- is the conjugate base of the acid HA. For bases, the hydrolysis reaction is given by

$$B(aq) + H_2O(1) \rightleftharpoons BH^+(aq) + OH^-(aq) \tag{5.47}$$

where B is a base. The thermodynamic equilibrium constant. for the acid hydrolysis is given by

$$K_a = \frac{\{H^+(aq)\}\{A^-(aq)\}}{\{HA(aq)\}} \tag{5.48}$$

and for bases by

$$K_b = \frac{\{BH^+(aq)\}\{OH^-(aq)\}}{\{B(aq)\}} \tag{5.49}$$

This assumes $\{H_2O\} = 1$, which is nearly true, even in seawater. For example, $\{H_2O\} = 0.98$ at 35‰, 25°C, and 1 atm. As with other equilibrium expressions, K_a and K_b can be rewritten as stoichiometric constants that are specific for a particular temperature, pressure, and ionic strength.

The larger the value of K_a and K_b, the greater the strength of the acid or base. Very strong acids, such as HCl and HNO_3, can be considered as completely dissociated in aqueous solutions. Thus, the dissolution of 1 mol of HCl in 1 L of water will produce a solution that is 1 M in H^+. Dissolution of 1 mol of a weak acid in 1 L of water will produce a solution that has a $H^+ < 1$ M because a significant amount of HA will be present at equilibrium. In the case of the di- and triprotic strong acids, such as H_2SO_4 and H_3PO_4, dissolution generates H^+ and a mixture of the conjugate bases, (i.e., HSO_4^-, and SO_4^{2-}, or $H_2PO_4^-$, HPO_4^{2-}, and PO_4^{3-}). At equilibrium, the relative abundances of the conjugate bases decrease with increasing ionic charge.

An increase in $[H^+]$ produced by the dissociation of an acid causes the following reaction to shift in favor of the reactants:

$$H_2O \rightleftharpoons H^+(aq) + OH^-(aq) \tag{5.50}$$

thereby causing a decrease in $[OH^-]$, since

$$K_w = \{H^+(aq)\}\{OH^-(aq)\} \tag{5.51}$$

is a constant. As with any equilibrium constant, K_w is a function of temperature and pressure. The effect of the former can be significant, as shown in Table 5.11. Thus, the degree to which the chemical reaction in Eq. 5.50 is driven toward the products depends on ambient temperature and pressures. The effect of pressure also becomes significant at values greater than 1000 dbar, which is equivalent to water depths exceeding 1000 m.

As noted earlier, H^+ activity is usually reported as *pH*, which is defined as

$$pH = -\log\{H^+\} \tag{5.52}$$

Thus, at 25°C and 1 atm, a solution with a pH which is less than 7, has $\{H^+(aq)\} > \{OH^-(aq)\}$ and is said to be *acidic*. For pH greater than 7, $\{H^+(aq)\} < \{OH^-(aq)\}$, and the solution is said to be *basic*, or *alkaline*.

The speciation of polyprotic acids is complicated by the occurrence of multiple hydrolyses. For example, the dissolution of carbonic acid involves the following equilibria:

$$CO_2(aq) + H_2O(l) \rightleftharpoons H_2CO_3(aq) \tag{5.53}$$

$$H_2CO_3(aq) \rightleftharpoons H^+(aq) + HCO_3^-(aq) \tag{5.54}$$

$$HCO_3^-(aq) \rightleftharpoons H^+(aq) + CO_3^{2-}(aq) \tag{5.55}$$

As a result of these reactions, the carbonate system can buffer against changes in pH caused by addition of acid via two reactions:

$$H^+(aq) + HCO_3^-(aq) \rightleftharpoons H_2CO_3(aq) \tag{5.56}$$

$$H^+(aq) + CO_3^{2-}(aq) \rightleftharpoons HCO_3^-(aq) \tag{5.57}$$

Table 5.11 Equilibrium Constants for the Dissociation of Water, 0 to 60°C, and 1 atm.

T(°C)	$K_W \times 10^{-14}$
0	0.1139
5	0.1846
10	0.2920
15	0.4505
20	0.6809
25	1.008
30	1.469
35	2.089
40	2.919
45	4.018
50	5.474
55	7.297
60	9.614

Source: *From Harned, H. S., and B. B. Owen (1958).* The Physical Chemistry of Electrolytic Solutions, *Reinhold, p. 638.*

Buffering against pH changes caused by the addition of a base occurs via

$$OH^-(aq) + HCO_3^-(aq) \rightleftharpoons H_2O(l) + CO_3^{2-}(aq) \quad (5.58)$$

$$OH^-(aq) + H_2CO_3(aq) \rightleftharpoons H_2O(l) + HCO_3^-(aq) \quad (5.59)$$

On time scales of less than a few million years, these buffering reactions keep the pH of seawater within a narrow range of 7.9 to 8.4. The average pH of seawater is generally assumed to be 8.1, but rising atmospheric CO_2 concentrations are causing a significant decline. (This is discussed in detail in Chapter 15.) The other weak acids, except for borate, have a negligible impact on the pH and buffering ability of seawater due to their relatively low concentrations.

Speciation calculations can be performed for the weak acids and bases in a fashion similar to that presented earlier for Fe(III). The results of these calculations as a function of pH are shown in Figure 5.19. At the pH of seawater, the dominant species are carbonate, bicarbonate, ammonium, hydrogen phosphate, dihydrogen phosphate, and boric and silicic acid. In waters with low O_2 concentrations, significant concentrations of HS^- can be present.

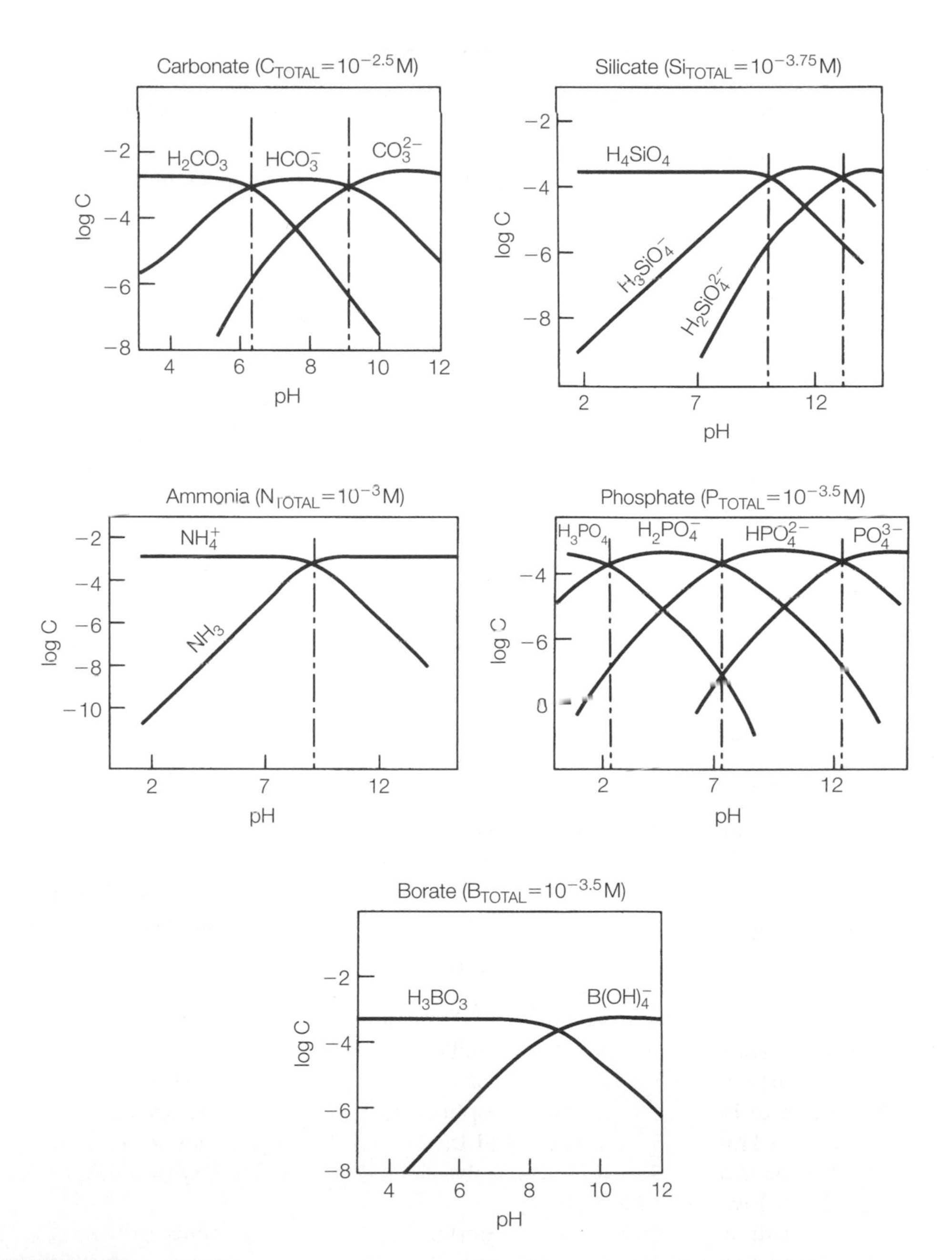

FIGURE 5.19

Concentration versus pH diagrams for weak acids and bases that are most common in natural waters. Note that the concentration axis is in log form. *Source*: From Morel, F. M. M., and J. G. Hering (1993). *Principles and Applications of Aquatic Chemistry*. Wiley Interscience, p. 160.

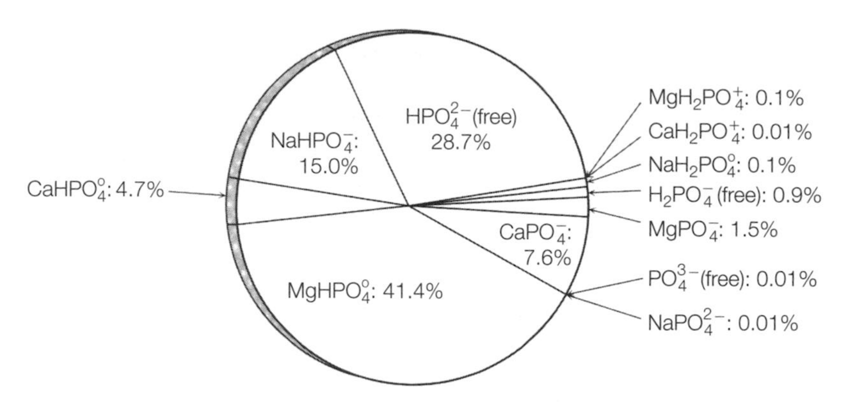

FIGURE 5.20

Ion pairing of phosphate species. *Source*: From Atlas, E., *et al.* (1976). *Marine Chemistry*, 4, 243–254.

As noted already for carbonate and sulfate, some conjugate bases are subject to a significant amount of ion pairing. This is also seen in phosphate, whose speciation is strongly influenced by ion pairing and acid-base reactions (Figure 5.20).

5.8.4 Other Speciation Reactions in Seawater

Some of the chemical reactions that control speciation involve gases and electron transfers. The latter are referred to as redox reactions. Most of these are controlled by the biochemical activities of living marine organisms, so they rarely attain equilibrium. In these cases, theoretical equilibrium calculations are used to predict the spontaneous direction of reaction and their relative rates. Gases are the subject of Chapter 6 and redox reactions are covered in Chapter 7. Another category of speciation reactions involves changes in the isotopic composition of ions and molecules. For example, radioactive decay transforms $^{14}CO_2$ into $^{12}CO_2$. Even the stable isotopes of elements can engage in speciation transformations. For example, plankton selectively assimilate $^{12}CO_2$ over $^{13}CO_2$ causing the relative abundance of ^{12}C in algal biomass to be higher than that of ^{13}C.

CHAPTER 6

Gas Solubility and Exchange across the Air-Sea Interface

All figures are available on the companion website in color (if applicable).

6.1 INTRODUCTION

Gases are involved in many biogeochemical cycles. For example, the marine biogeochemical cycle of organic matter is largely controlled by the competing processes of photosynthesis and respiration. During photosynthesis, plants consume carbon dioxide gas (CO_2) and produce oxygen gas (O_2). Animals, bacteria, and other heterotrophic microorganisms use O_2 to respire organic matter. Without O_2, fish and benthic animals die.

Some parts of the marine environment, such as salt marshes and the mid-waters of the thermocline, are naturally subject to low dissolved oxygen concentrations. Human activities are leading to an overall decline in dissolved oxygen concentrations. In estuaries and nearshore waters, the decline is related to an overfertilization of native algae by nutrient pollution. On a global scale, climate change also appears to be lowering oxygen concentrations as a result of higher water temperatures, which reduce solubility and enhance respiration rates.

O_2 is supplied to the surface waters of the ocean through two processes: photosynthesis and the dissolution of atmospheric O_2 across the air-sea interface. Because both processes are restricted to the surface waters, the only source of O_2 to the deep sea is through the sinking of surface water masses. If the rate of deepwater formation was to slow or stop, so would the transport of O_2 to the deep sea, with potentially fatal consequences for deep-dwelling aerobic organisms.

Many of the atmospheric greenhouse gases have oceanic sources and/or sinks and, thus, are involved in global climate regulation. These include CO_2, N_2O, and CH_4, whose marine cycling is regulated by plankton, including algae, bacteria, and archaea. Since water vapor is a greenhouse gas, a net transfer of water from the ocean to the atmosphere, will also have climate consequences. Biogeochemical cycling in the ocean also controls the atmospheric sources and sinks of other gases that play a role in climate regulation, such as dimethylsulfide.

Other gases with significant anthropogenic sources contribute to acid rain (NO_x and SO_x), the reduction of the ozone layer (chlorofluorocarbons), nutrient transport (NH_3

and NO_x), and the input of toxic materials. The last include some volatile heavy metals, such as Hg, and volatile organic compounds, such as PCBs. Thus, our understanding of gas behavior in the crustal-ocean-atmosphere factory is essential to assessing and controlling pollution.

In this chapter, the fundamentals of aqueous gas chemistry are described, such as the factors that determine gas solubility and the rate of gas exchange across the air-sea interface. This background information is important to understanding the biogeochemical processes discussed through the rest of the text, because many involve gaseous species.

6.2 DALTON'S LAW OF PARTIAL PRESSURES

The most abundant gases present in the ocean and atmosphere are listed in Table 6.1. In the atmosphere, each gas exerts a pressure, called a *partial pressure*. Because of their relatively low concentrations, most atmospheric gases exhibit near ideal behavior such that their partial pressures are independent of each other. In this case, the total atmospheric pressure (P_T) is the sum of the partial pressures of the component gases. This is termed *Dalton's Law of Partial Pressures*. For air:

$$P_T = P_{N_2} + P_{O_2} + P_{Ar} + P_{H_2O} + P_{CO_2} + \cdots. \tag{6.1}$$

Since each gas exerts its own partial pressure, each gas also follows the *ideal gas law,* which is the equation of state for the gas phase,

$$P_A V_A = n_A RT \tag{6.2}$$

where P_A is the pressure of a gas A, V_A is its volume (L), n_A is the number of moles of the gas, T is its temperature in kelvins (K = 273.15 + °C) and R is the ideal gas constant (0.082053 L atm K^{-1} mol^{-1}). At conditions of standard temperature and atmospheric pressure (STP = 0°C and 1 atm), 1 mol of an ideal gas occupies 22.414 L. Deviations from ideal behavior are observed at high gas pressures and with molecules that have high molecular weights or complex structures. To accommodate these behaviors an Equation of State for non-ideal gases has been developed, but is not necessary for most oceanographic applications.

The atmospheric composition information in Table 6.1 is presented as the volume mixing ratio for each gas (V_A/V_{AIR}). At a constant temperature and pressure, the ideal gas law can be used to show that

$$\frac{PV_A}{PV_{AIR}} = \frac{n_A RT}{n_{AIR} RT} \tag{6.3}$$

$$\left[\frac{V_A}{V_{AIR}}\right]_{T,P} = \left[\frac{n_A}{n_{AIR}}\right]_{T,P}$$

demonstrating that the volume mixing ratio is equal to the mole fraction (χ_A). The ideal gas law also states that at constant temperature and volume, pressure is directly

Table 6.1 Composition of the Atmosphere Given as the Molar or Volume Mixing Ratio in Dry Air for $P_T = 1$ atm.

	Constituent	Chemical Symbol	Concentration: Part per hundred (%) or Part per million (ppm)	Concentration: Atmospheres
Gases of Homogenous Atmospheric Composition	Nitrogen	N_2	78.084%	0.78084
	Oxygen	O_2	20.947%	0.20947
	Argon	Ar	0.934%	0.00937
	Neon	Ne	18.18 ppm	0.00001818
	Helium	He	5.24 ppm	0.00000524
	Krypton	Kr	1.14 ppm	0.00000114
	Xenon	Xe	0.087 ppm	0.000000087
Gases of Variable Atmospheric Composition	Carbon dioxide	CO_2	385 ppm[a]	0.000385
	Methane	CH_4	~1.745 ppm	~0.000001745
	Water[b]	H_2O	0.1 ppm to 4%	0.0000001 to 0.04
	Hydrogen	H_2	~0.55 ppm	~0.00000055
	Nitrous Oxide	N_2O	~0.314 ppm	~0.000000314
	Ozone	O_3	0.01 to 0.5 ppm	0.00000001 to 0.0000005
	Carbon Monoxide	CO	~0.05 to 0.200 ppm	~0.00000005 to 0.0000002
	Sulfur Dioxide	SO_2	~0.100 ppm	~0.0000001
	Iodine	I_2	~0.010 ppm	~0.00000001
	Nitrogen Dioxide	NO_2	~0.002 ppm	~0.000000002
	Ammonia	NH_3	~0.003 ppm	~0.000000003
	CFC-12	CCl_2F_2	~0.0005 ppm	~0.0000000005
	CFC-11	CCl_3F	~0.0003 ppm	~0.0000000003

[a] *Circa 2008 Annual Mean at Mauna Loa (19°32'N. 155°34'W. 3397 m above MSL).*
[b] *Range of values exhbited for very dry to very moist air.*

proportional to the amount of gas ($P = kn$). This leads to the relationship

$$P_A = \chi_A P_T \quad (6.4)$$

that can be used to determine P_A for any P_T. In the case where $P_T = 1$ atm, the mole fractions presented in Table 6.1 are numerically equivalent to the partial pressures of the gases (when pressures are in units of atmospheres).

Because water vapor makes up a variable and not insignificant portion of air (as much as 4%), the gas composition data in Table 6.1 are given for dry air. P_A^{dry} can be used to obtain P_A^{moist} from the following:

$$P_{A_{moist}} = P_{A_{dry}} x \left(1 - \frac{P_{H_2O}}{P_T}\right) \quad (6.5)$$

where P_{H_2O} is the vapor pressure of seawater. This relationship assumes that the air is at saturation with respect to water vapor. At $P_T = 1$ atm, P_{H_2O}/P_T ranges from an insignificant 0.5% at 0°C to a significant 4% at 30°C. An empirical relationship relating the saturation P_{H_2O} to temperature at $P_T = 1$ is provided in the online appendix on the companion website.

At sea level, P_T is approximately 1 atm, but exhibits some temporal and spatial variability. For example, the annual mean pressure in the northern hemisphere is 0.969 atm and in the southern hemisphere is 0.974 atm, with monthly averages varying by as much as 0.0001 atm, i.e., about 1 mbar (1 atm = 1013.25 mbar). These fluctuations are caused by spatial and temporal variations in atmospheric temperature and water vapor content associated with weather, and seasonal and longer-term climate shifts. P_T is also affected by diurnal atmospheric tides, and it decreases with increasing altitude above sea level. Some gases, such CO_2 and O_2, exhibit seasonal variability that is caused in part by seasonal variability in plant and animal activity (see Figures 25.4 and 6.7).

6.3 GAS SOLUBILITY

Gases can be present in the ocean in three phases: (1) in the gas phase as part of a bubble in which gas molecules are surrounded by other gas molecules, (2) as a true solute in which the gas molecule is completely surrounded by water molecules, and (3) as a solid in which the gas molecule is trapped within water ice or clathrate hydrates. The most important phase transformation involves the continual passage of gas molecules across the air-sea interface. The *dissolution* of atmospheric gases into seawater is termed *ingassing*. The reverse process in which gases are released from the ocean to the atmosphere is termed *degassing*. When the rates of ingassing and degassing are equal, the gas is said to be at equilibrium. At gaseous equilibrium, the aqueous concentration is constant over time. This equilibrium concentration is a function of the water's temperature and salinity as well as the atmospheric partial pressure of the gas.

A net transfer of gas across the air-sea interface will spontaneously take place if the gas is not at equilibrium. In the case where the aqueous gas concentration exceeds the equilibrium value, a spontaneous net transport from the ocean to the atmosphere will occur. The transport will continue until the gas concentration has been lowered to the equilibrium level. Likewise, a gas will undergo net transport into the ocean if its aqueous concentration is less than the equilibrium level.

Gaseous equilibrium can be represented by the following expression:

$$A(g) \rightleftharpoons A(aq) \tag{6.6}$$

where A(g) and A(aq) represent gas A in the gaseous state and aqueous solution, respectively. For a dilute solution in which concentrations are approximately equal to activities, the thermodynamic equilibrium constant for the equilibrium gas exchange represented by Eq. 6.7 is given by

$$K_{eq} = \frac{[A(aq)]}{[A(g)]} \tag{6.7}$$

A(g) is usually expressed as a partial pressure, P_A. Since $P_A V_A = n_A RT$, P_A is related to A(g) as follows:

$$[A(g)] = \frac{n_A}{V_A} = \frac{P_A}{RT} \tag{6.8}$$

Substituting P_A/RT into Eq. 6.7 and solving for A(aq) yields

$$[A(aq)] = \frac{K_{eq}}{RT} P_A \tag{6.9}$$

This is known as *Henry's Law*, and is usually written as

$$[A(aq)] = K_H \times P_A \tag{6.10}$$

where Henry's law constant $K_H = K_{eq}/RT$.

In 2004, the partial pressure of CO_2 in our atmosphere was 377 ppm or 377 μatm. The K_H for CO_2 at 0°C and 35‰ is 0.06582 (mol CO_2)/(kg SW atm). Thus, the concentration of CO_2 that would be achieved if seawater reached gaseous equilibrium with the atmosphere (dry air) would be 24.8 (μmol CO_2)/(kg SW). This concentration unit can be converted to (μmol CO_2)/(L SW) using the density of seawater (at 0°C, 35‰ and 1 atm, $\rho = 1.028\,g/cm^3$, so $[CO_2(aq)] = 25.5$ (μmol CO_2)/(L SW) = 25.5 (mmol CO_2)/(m^3 SW)). Once CO_2(aq) dissolves, it will spontaneously react with H_2O to form the inorganic carbon species: H_2CO_3(aq), HCO_3^-(aq) and CO_3^{2-}(aq). Thus, 24.8 (μmol CO_2)/(kg SW) is functionally the concentration of the total dissolved inorganic carbon, i.e., $TDIC = [CO_2(aq)] + [H_2CO_3(aq)] + [HCO_3^-(aq)] + [CO_3^{2-}(aq)]$.

Oceanographers use a specialized form of Henry's law that has the following form:

$$[A(aq)] = \beta_A \times P_A^{moist} \tag{6.11}$$

β_A is called the *Bunsen solubility coefficient*. It is the volume of gas (in mL) that will dissolve in 1 L of seawater when equilibrium has been reached at STP where

$P_A = 1$ atm. The units of β_A are (mL gas)/(L SW atm), giving $[A(aq)]$ units of (mL gas)/(L SW). In this concentration unit, the gas abundance is expressed as the volume it would occupy if extracted from the seawater and subjected to STP. Under these conditions, the gas's molar volume is 22,414 mL (assuming ideal gas behavior). Thus, gas concentrations expressed in units of $\frac{\text{mL gas}}{\text{L SW}}$ can be converted to molarity and molinity (mol/kg) using the following:

$$\frac{\text{mol gas}}{\text{L SW}} = \frac{\text{mL gas}}{\text{L SW}} \times \left(\frac{1 \text{ mol gas}}{22{,}414 \text{ mL gas}}\right)_{\text{STP}} \tag{6.12}$$

and

$$\frac{\text{mol gas}}{\text{kg SW}} = \frac{\text{mL gas}}{\text{L SW}} \times \left(\frac{1 \text{ mol gas}}{22{,}414 \text{ mL gas}}\right)_{\text{STP}} \times \frac{\text{L SW}}{\text{kg SW}} \tag{6.13}$$

Similarly, K_H and β_A are related by

$$\frac{\text{mol gas}}{\text{L SW} \cdot \text{atm}} \times \left(\frac{22{,}414 \text{ mL gas}}{1 \text{ mol gas}}\right)_{\text{STP}} = K_H \times 22{,}414 = \beta_A \tag{6.14}$$

K_H and β_A are functions of seawater's temperature and salinity and the gas's partial pressure and molecular composition. As shown in Figure 6.1, solubility increases with decreasing temperature.

Figure 6.1 also indicates that for the monoatomic gases, solubility increases with increasing molecular weight. In the case of the more complex gases, such as CO_2 and

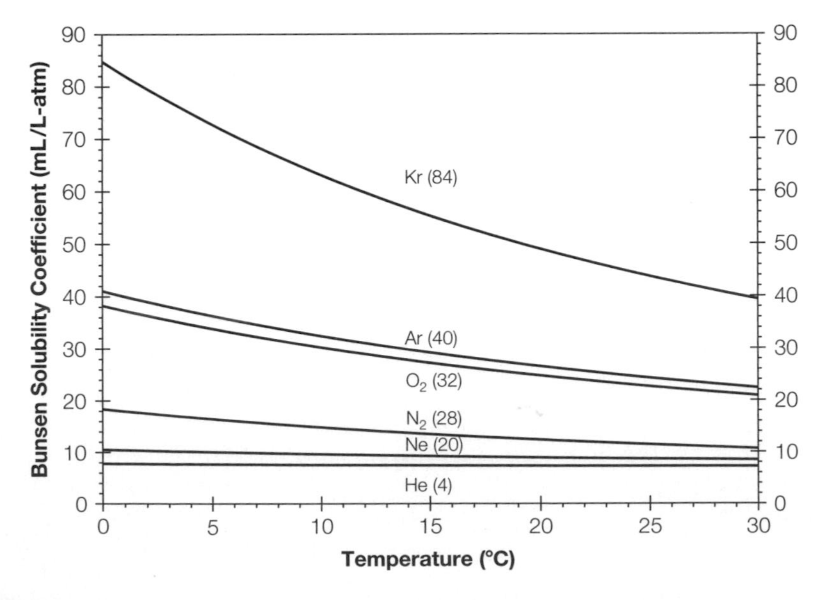

FIGURE 6.1

The effect of temperature on the Bunsen solubility coefficients of the monoatomic atmospheric gases for seawater of 35‰. Molecular weights are shown in parentheses.

CH_4, solubility is generally much greater than predicted by a simple interpolation from the monoatomic gas's molecular weight trends.

Gas solubility decreases with increasing salinity. This phenomenon is referred to as *salting out*. It is caused by the electrostatic forces exerted by the salt ions. These forces have to be overcome to create spaces between water molecules to accommodate a gas atom or molecule. So higher salinities lead to less favorable energetics for gas dissolution. The high salt content of seawater also leads to nonspecific interactions that cause gases to have activity coefficients on the order of 1.1 to 1.2 at a salinity of 35‰ and temperature of 25°C.

The solubility of some gases, such as CO_2, is complicated by their chemical reaction with water. In the case of CO_2, these reactions tend to reach equilibrium rapidly, making the equilibrium concentration of CO_2 a function of temperature, salinity, the atmospheric partial pressure of CO_2, and pH. The last determines the degree to which CO_2 is converted into the other inorganic species, H_2CO_3(aq), HCO_3^-(aq), and CO_3^{2-}(aq). Other gases that react with water include H_2S and NH_3. Note that these hydrolysis and acid dissociation reactions act to enhance the solubility of the parent gas.

Empirical equations have been formulated to enable calculation of the Bunsen solubility coefficient for any given temperature and salinity at P_T = 1 atm. These empirical equations are presented in the online appendix on the companion website for the most common gases found in seawater but being empirical, they are still subject to refinement. The equilibrium gas concentrations computed from the Bunsen solubility coefficient should be thought of as the gas concentration that a water mass would attain if it were allowed to equilibrate with the atmosphere at its in situ salinity and potential temperature.

Because P_T = 1 atm, these concentrations are called *normal atmospheric equilibrium concentrations*, or *NAEC*. NAECs for the most common gases in seawater (35‰) over the range of temperatures encountered in the surface ocean are shown in Table 6.2.

6.3.1 Deviations from NAEC

If the in situ gas concentration is less than its NAEC, the seawater is said to be *undersaturated* with respect to that gas. If the observed in situ gas concentration exceeds the NAEC, the seawater is said to be *supersaturated*. If the in situ concentration is equal to the NAEC, the seawater is said to be *saturated*. The degree of gas saturation is usually expressed as

$$\% \text{ saturation} = \frac{[\text{A(aq)}]_{\text{in situ}}}{\text{NAEC for A}} \times 100 \qquad (6.15)$$

At equilibrium, seawater is 100% saturated. If the percentage of saturation is greater than 100, the seawater is *supersaturated*. If the percentage of saturation is less than 100, the seawater is undersaturated.

Most deviations from NAECs are the result of nonconservative behavior involving chemical processes that remove or supply the gas faster than the water mass can reequilibrate with the atmosphere. As shown in Figure 6.2, large areas of the surface ocean

Table 6.2 NAECs of Gases in Seawater at a Salinity of 35.0.

T(°C)	0	5	10	15	20	25	30
	Saturation Water Vapor Pressure $\left(\frac{P_{H_2O}}{P_T} \times 100\right)$						
	0.6%	0.8%	1.2%	1.6%	2.3%	3.1%	4.1%
	Concentrations in mmol/m^3						
N_2	635.7	565.6	508.3	460.7	420.4	385.7	355.1
O_2	355.6	313.2	278.7	250.0	225.9	205.1	186.9
CO_2	23.37	19.26	16.09	13.6	11.61	10.00	8.66
Ar	17.01	14.98	13.33	11.96	10.81	9.81	8.93
	Concentrations in μmol/m^3						
N_2O	14.84	12.16	10.09	8.46	7.16	6.10	5.23
Ne	8.45	8.03	7.66	7.33	7.04	6.79	6.56
Kr	4.31	3.68	3.18	2.78	2.44	2.16	1.93
CH_4	3.44	3.00	2.64	2.35	2.12	1.92	1.76
He	1.81	1.76	1.73	1.68	1.67	1.62	1.62

N_2, O_2, Ar, Ne, Kr, and He concentrations computed from equations in the online appendix on the companion website.
CO_2 from Weiss, R. F. (1974). Marine Chemistry *2, 203–215.*
CH_4 from Wisenburg, D. A., and N. L. Guinasso (1979). J. Chem. Engr. Data *24, 356–360.*
H_2O and N_2O from Weiss, R. F. and B. A. Price (1980). Marine Chemistry *8, 347–359.*

are supersaturated with respect to O_2 whereas some, such as the Southern Ocean, are undersaturated. In the case of O_2, these deviations from saturation are ultimately due to the relative rates of aerobic respiration and photosynthesis as compared to those of gas exchange across the air-sea interface and water movement via turbulent mixing and advection.

The balance between relative rates of aerobic respiration and water movement were considered in Section 4.3.4. We saw that a subsurface concentration minimum, the oxygen minimum zone (OMZ), is a common characteristic of vertical profiles of dissolved oxygen and is produced by in situ respiration. Waters with O_2 concentrations less than 2.0 ppm are termed *hypoxic*[1]. The term *anoxic* is applied to conditions when O_2 is absent. (Some oceanographers use the term *suboxic* to refer to conditions where O_2 concentrations fall below 0.2 ppm but are still detectable.) As illustrated by Figure 4.21b, this water column is hypoxic in the OMZ. The dissolved oxygen concentrations are presented as % saturations in Figure 4.21c. With the exception of the mixed layer, the water column is undersaturated with respect to dissolved oxygen with the most intense undersaturations present in mid-depths. Surface supersaturations are the result of O_2 input from photosynthesis and bubble injection.

[1] Some scientists define hypoxic waters as having O_2 concentrations less than 2 mg/L.

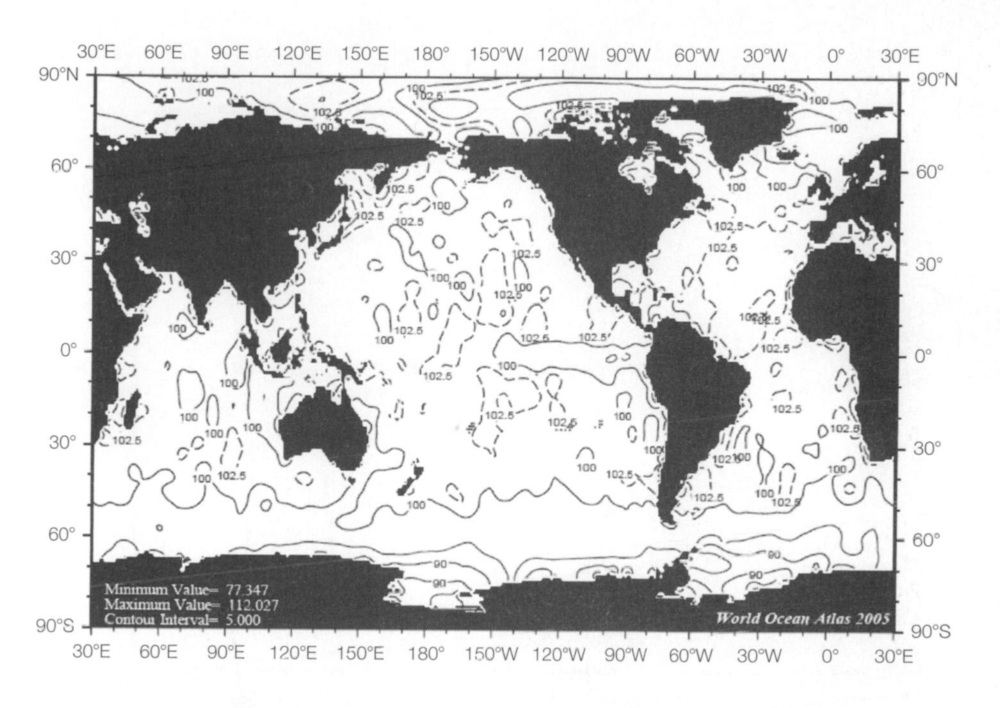

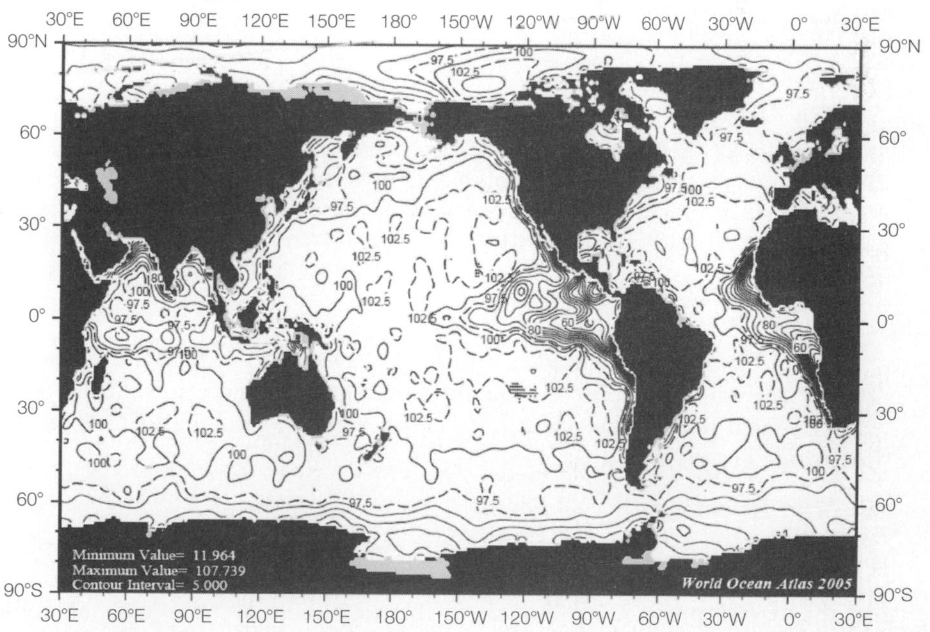

FIGURE 6.2

Annual percent saturation of dissolved oxygen in (a) surface waters (0 m) and (b) 50 m water depth. *Source*: From Garcia, H. E., *et al.* (2006). *World Ocean Atlas 2005, Volume 3: Dissolved Oxygen, Apparent Oxygen Utilization, and Oxygen Saturation*. NOAA Atlas NESDIS 63, U.S. Government Printing Office, p. 342.

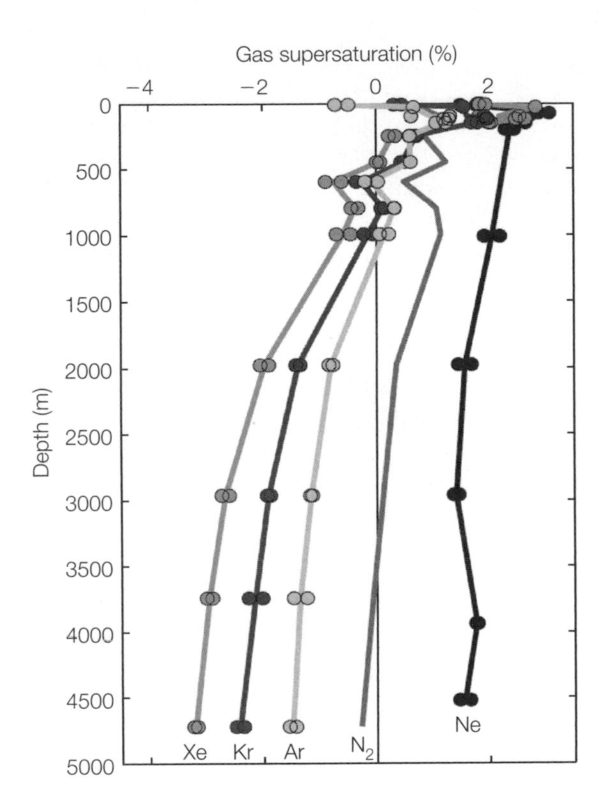

FIGURE 6.3

Depth profiles of Ne, N_2, Ar, Kr, and Xe supersaturations (in %) measured at station ALOHA (22°45'N 158°W) in August 2004. Station ALOHA is located in the central region of the North Pacific Subtropical Gyre, about 100 km north of Oahu. Equilibrium with the atmosphere is indicated by the vertical line at 0%. Points indicate individual samples while the lines are the average of duplicates. *Source*: After Hamme, R. C., and J. P. Severinghaus (2007). Deep Sea Research Part I: Oceanographic Research Papers 54, 939–950.

A variety of physical effects can lead seawater to exhibit deviations from NAEC. This is illustrated in Figure 6.3 for the noble gases, which presumably do not undergo any chemical reactions in seawater. The physical phenomena that can give rise to such deviations are discussed later.

Bubble Injection

At high enough wind speeds, bubbles of air are injected into the sea surface. The gases in these bubbles dissolve under some pressure. The extent of dissolution is dependent on the partial pressure of the gas in the air bubble and the gas's solubility. This can create supersaturated conditions to a degree that varies from gas to gas.

Kinetic Considerations

The establishment of equilibrium takes time. If environmental conditions change faster than equilibrium can be attained, deviations from NAECs result. For example, water masses that form at the sea surface and sink have not necessarily attained their saturation concentration. This will occur when atmospheric temperature and pressure are changing faster than gaseous equilibrium can be reestablished. The attainment of equilibrium can also be prevented if barriers to gas exchange are present, such as sea-surface slicks. For example, slicks left by oil spills can reduce the rate of influx of O_2 from the atmosphere such that fish and other animals will eventually suffocate.

Even if equilibrium had been achieved before the water mass sank, the observed water mass concentrations could exhibit apparent deviations from equilibrium, if the atmospheric pressures have changed or if the water mass' temperature has been altered. Thus, deepwater that was last at the sea surface hundreds of years ago could potentially have equilibrated with the atmosphere at much different partial pressures of gases such as methane and carbon dioxide whose atmospheric levels have risen over time. For example, the partial pressure of CO_2 in dry air has risen from about 280 ppm in the 1860s to 385 ppm in 2008.

Postequilibration Temperature Changes

If the temperature of a water mass is altered after having been isolated from the sea surface, deviations from saturation can result. Such alterations are a consequence of heat transport via either conduction (molecular diffusion of heat) or turbulent mixing of adjacent water masses. In the case of the former, hydrothermal systems are important subsurface heat sources. If two water masses of different temperature undergo turbulent mixing, the temperature of the admixture will differ from that of the source waters.

The effect of this mixing on gas saturation is shown in Figure 6.4. Using the simple situation where equal volumes of two water masses undergo complete mixing, a new water mass is formed that has a temperature intermediate between the two source waters. If the original water masses contained the NAEC of a conservative gas, such as N_2, the gas concentration in the admixture will be halfway between the NAEC's of the source waters. In this example, mixing produces a water mass with a gas concentration of 11.10 mL N_2/L and temperature of 15°C. But the online appendix on the companion website lists the NAEC for N_2 at 15°C and 35‰ as 10.33 mL N_2/L. Therefore, the mixing has caused an apparent supersaturation of 108%.

How can this be? No additional gas was added to the water. The answer lies in the nonlinear temperature effect on the Bunsen solubility coefficient (Figure 6.1). Because of the concave nature of the curves relating the Bunsen solubility coefficient to temperature, the result of this type of postequilibration temperature change is always supersaturation.

Subsurface Sources

Some gases have subsurface sources that are related to physical phenomena, such as inputs from the introduction of hydrothermal fluids in bottom waters or release from warming sediments. The latter is a source of methane, which can occur in sediments in a solid phase called a clathrate hydrate. Biogeochemical reactions in sediments can also produce gases that diffuse from the pore waters into the deep sea.

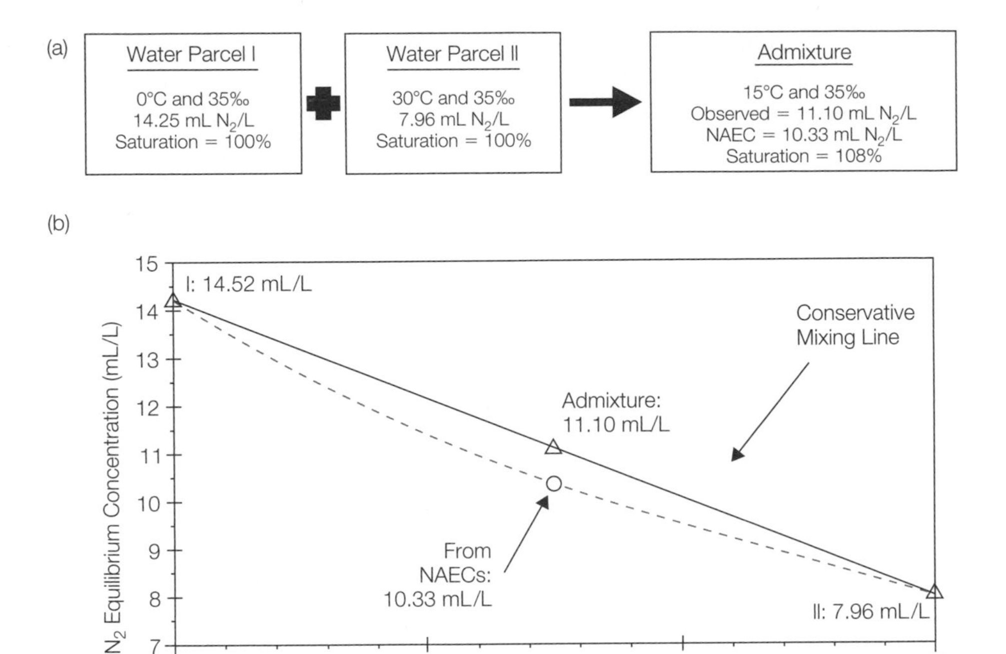

FIGURE 6.4

Changes in percent saturation caused by the mixing of two water masses. (a) Mixing scenario and (b) nonlinear temperature effect.

6.4 RATES OF GAS EXCHANGE

The exchange of gases across the air-sea interface is critical to the function of the crustal-ocean-atmosphere factory particularly in mediating the ocean's role in climate change. For example, the rate of CO_2 uptake and release from the sea surface and the rate of its transfer to deep waters determines the degree to which the ocean functions as a sink for this greenhouse gas. The ocean is a net sink for other greenhouse gases such as ozone and methane. The ocean is a net source of organosulfur gases, such as dimethylsulfide, that could play an important role in influencing climate by acting as cloud condensation nuclei. The ocean is also a sink for anthropogenic gases that contribute to acid rain and for those that damage the ozone layer.

The first part of this chapter used the thermodynamic concept of equilibrium to predict the direction of gas exchange, either into or out of the ocean. This does not provide information on the rates of gas exchange. Despite the importance of gas exchange rates, few direct measurements have been made because of the technical difficulties associated with working at the sea surface. Instead, rates are generally inferred from mathematical models. The most commonly used ones are described below. These models focus on the physical phenomenon of gas exchange across the air-sea interface. Approaches for estimating rates of gas exchange across the sediment-water interface are described in Chapter 12.

6.4.1 Thin-Film Model

The thin-film model is the simplest and, therefore, most commonly used approach to estimate air-sea fluxes of gases. In this model, molecular diffusion is assumed to present a barrier to gas exchange in each of two layers. As illustrated in Figure 6.5, one layer is composed of a shallow region of the atmosphere that lies in direct contact with the sea surface. The second is a shallow layer of seawater that lies at the sea surface. These layers have depths less than 100 μm and, hence, are referred to as ***thin films***.

Above the atmospheric film lies the bulk atmosphere, which is well mixed by turbulence and advection and, hence, is homogeneous in gas composition. Below the sea surface film lies the bulk seawater, which is also well mixed by turbulence and advection and is consequently homogeneous in gas composition. The thin films are regions in which turbulence and advection play minor roles, such that molecular diffusion controls the movement of gases. Because of the limited degree of air and water motion in

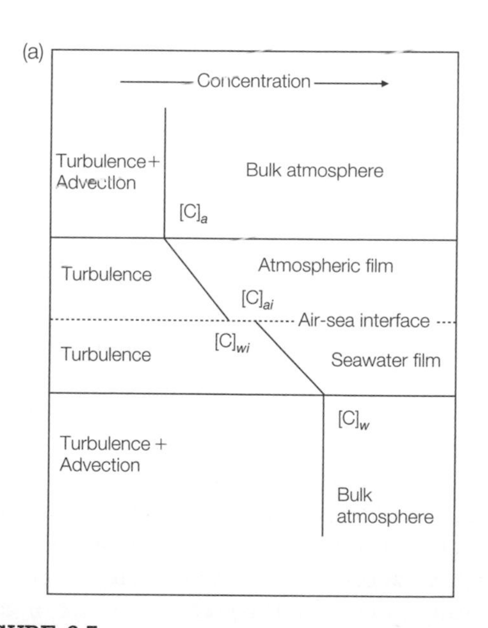

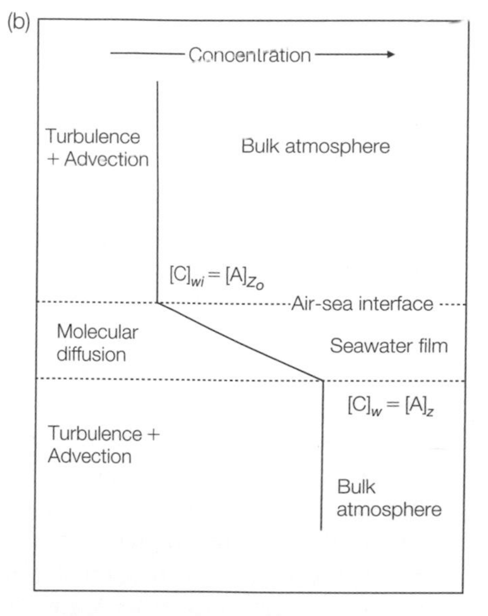

FIGURE 6.5

(a) Two-film gas model. C_a is the concentration of the gas in the bulk atmosphere, C_{ai} is the concentration of the gas at the bottom of the atmospheric thin film, C_w is the bulk concentration of the gas in seawater, and C_{wi} is the concentration of the gas at the top of the sea surface. (b) Single-film gas model. $C_w = [A(aq)]_z$ is the bulk concentration of the gas in seawater and $C_{wi} = [A(aq)]_{z_o}$ is the concentration of the gas at the top of the sea surface.

these films, they can be thought of as regions of relatively stagnant air and water. In the case of nonconservative gases, the rates of nonconservative processes within the thin film are assumed to be relatively slow as compared to molecular diffusion.

If a gas in the bulk atmosphere is not at equilibrium with the gas in the bulk seawater, a concentration gradient exists as shown in Figure 6.5. Molecular diffusion of the gas through the thin films will spontaneously occur in an effort to destroy this gradient. The movement of gas causes a net flux through the air-sea interface whose magnitude is directly related to the size of the concentration gradient. For gases that are sparingly soluble, diffusion through the seawater film is rate limiting. In this case, the model can be simplified to a single film in which the net gas flux is given by Fick's first law:

$$F_A = -D_A \frac{d\,[A]}{dz} = -D_A \frac{[A(aq)]_z - [A(aq)]_{z_o}}{z} \tag{6.16}$$

where D_A is the molecular diffusivity coefficient for the gas. These coefficients are functions of molecular weight, temperature, and salinity. Because the seawater film is very thin, special sampling devices are required to measure $[A(aq)]_z$ because even during calm seas, surface grab samples will typically contain water from the top meter of the ocean.

The concentration at the top of the film is approximated from the atmospheric concentration of the gas because the single film model assumes that the atmosphere is at equilibrium with the top of the film. Thus, Eq. 6.16 can be rewritten as

$$F_A = D_A \frac{[A(aq)]_{z_o} - [A(aq)]_z}{z} = \frac{D_A}{z}\left[(K_H P_A) - [A(aq)]_z\right] \tag{6.17}$$

where D_A/z has units of cm/s and is referred to as the *transfer velocity* or *piston velocity*. F_A is typically reported in units of mol/(m^2 y). Positive fluxes arise when $[A(aq)]_{z_o} > [A(aq)]_z$ and represent ingassing or a net gain of gas by the ocean. Negative fluxes arise when $[A(aq)]_{z_o} < [A(aq)]_z$ and represent degassing or a net loss of gas from the ocean. Gas fluxes tend to be small; for example, the net average helium flux out of the sea is 1.8×10^{-6} mol/(m^2 y) at 0°C. In the absence of a concentration gradient, gas exchange can occur, but the rate of gas leaving the sea surface must equal the rate entering the sea surface so no net flux of gas results.

As per Eq. 6.16, the net diffusive flux is directly proportional to the magnitude of the concentration gradient. The larger the difference in concentration between the top and bottom of the stagnant film, the greater the flux of gas across the air-sea interface. Thus, the larger the degree of supersaturation, the greater the flux of gas out of the ocean, and the larger the degree of undersaturation, the greater the flux of gas into the ocean. These rates can also be affected by reactions that enhance the solubility of a gas such as the reaction of CO_2 with water.

The air-sea fluxes are inversely proportional to the thickness of the thin film. From the measurement of gas fluxes and concentration gradients, oceanographers have inferred that the stagnant-film thickness ranges from 10 to 60 μm. As mentioned earlier, direct flux measurements are difficult to make at sea, especially in the presence of waves. Indirect methods of estimating fluxes include the use of natural tracers, such as radiocarbon and radon, and artificial tracers intentionally injected into the sea surface, such as SF_6. Some experimental work has been done in laboratory tanks.

Since the conceptual development of the thin-film model, marine scientists have observed that a true microlayer does exist at the sea surface. These microlayers contain high concentrations of DOM and are enriched in metals probably as a result of complexation with the organic matter. The films range in thickness from 50 to 100 μm. Bacteria and plankton are also present at elevated concentrations and appear to be responsible for a significant fraction of the DOM in the microlayer. The microlayers inhibit the effects of winds and waves on water motion and, hence, act to increase the depth of the thin film, thereby decreasing gas fluxes. When turbulence is high, such as in the presence of breaking waves, the microlayers are dispersed.

The diffusive flux of gas across the air-sea interface is also directly proportional to the molecular diffusivity coefficient. Molecular diffusivity coefficients range in magnitude from 1×10^{-5} to 4×10^{-5} cm^2/s. As shown in Table 6.3, D_A's increase with increasing temperature and decreasing molecular weight. This causes gas fluxes to increase with increasing temperature and decreasing molecular weight.

The combined effects of D_A and z on the net diffusive gas flux is given by the piston velocity (D_A/z). This can be thought of as the rate at which a column of gas is pushed through the sea surface. For D_A's of approximately 2×10^{-5} cm^2/s and z's that average 40 μm, piston velocities are approximately 18 cm/h or 1600 m/y. This is

Table 6.3 Molecular Diffusivity Coefficients of Various Gases in Seawater.

Gas	Molecular Weight, (g/mol)	Diffusion Coefficient ($\times 10^{-5} cm^2/s$)	
		0°C	24°C
H_2	2	2.0	4.9
He	4	3.0	5.8
Ne	20	1.4	2.8
N_2	28	1.1	2.1
O_2	32	1.2	2.3
Ar	40	0.8	1.5
CO_2	44	1.0	1.9
Rn	222	0.7	1.4

Source: *From Broecher, W. S. (1974).* Chemical Oceanography, *Harcourt, Brace, and Jovanovich Publishers, p. 127.*

equivalent to pushing a 4-m high column of gas through the sea surface every day. Piston velocities can be used to compute the time required for gaseous equilibration with the atmosphere. If the mixed layer ranges in depth from 20 to 100 m, the length of time required to push all of the gas through this zone is given by

$$20 \text{ to } 100\,\text{m} \times \frac{\text{d}}{4\,\text{m}} = 5 \text{ to } 25\,\text{d}$$

As discussed in Chapter 4, Fick's first law also describes solute fluxes driven by turbulent mixing (aka eddy diffusion). The eddy diffusivity coefficient (D_z) for vertical mixing in the ocean ranges in magnitude from 0.1 to 10 cm^2/s (Table 4.2). Thus, when vertical turbulence is present, as in the water below the thin film, transport due to mixing overwhelms that from molecular diffusion. In other words, molecular diffusion is a significant transport mechanism only under stagnant conditions. When vertical turbulence is strong, such as under windy conditions, the thin film is greatly reduced in thickness and the gas flux is enhanced (Figure 6.6).

At high enough wind speeds, gas exchange is enhanced by bubble injection, in which atmospheric gases are entrained into the sea surface by breaking waves. High sea states also provide greater surface area for gas exchange, providing another mechanism for enhancing fluxes, which is undoubtedly part of the cause for the relationship shown in Figure 6.6. Various types of heat and water transfer across the air-sea interface also influence gas exchange rates including rainfall and evaporation. The latter involves heat conduction that causes the sea surface to have a "cool skin" 0.1 to 0.3°C lower than the bulk seawater of the mixed layer. This lower temperature enhances gas solubility and, hence, affects concentration gradients at the air-sea interface.

6.4.2 Surface Renewal and Boundary Layer Models

Clearly, the real ocean does not have a stagnant film at the sea surface. A more realistic approach views the sea surface as a region occupied by slabs or cells of moving water. These slabs are transported from the mixed layer to the sea surface as a result of turbulent mixing. While at the sea surface, the top of the slab reaches gaseous equilibrium with the atmosphere, while its bottom retains the chemical composition of the underlying mixed layer. Thus, a gas concentration gradient is set up across the slab. Turbulent mixing eventually returns the slab to the mixed layer. Its site at the sea surface is immediately occupied by another slab. If the rate of slab exchange is fast, the mixed layer will attain gaseous equilibrium with the atmosphere.

In this surface-renewal model of gas exchange, the gas flux across the air-sea interface is determined by the frequency at which the slab is replaced or "renewed." Various parameterizations have been developed for this model. One example relates the net diffusive flux (F_A) to the frequency of slab renewal (θ) as follows

$$F_A = \sqrt{\frac{D_A}{\theta}}\left([\text{A(aq)}]_{z_o} - [\text{A(aq)}]_z\right) \tag{6.18}$$

where $[\text{A(aq)}]_{z_o}$ is the gas concentration at the top of the slab when it is at the sea surface and $[\text{A(aq)}]_z$ is the concentration at the bottom of the slab. If a concentration

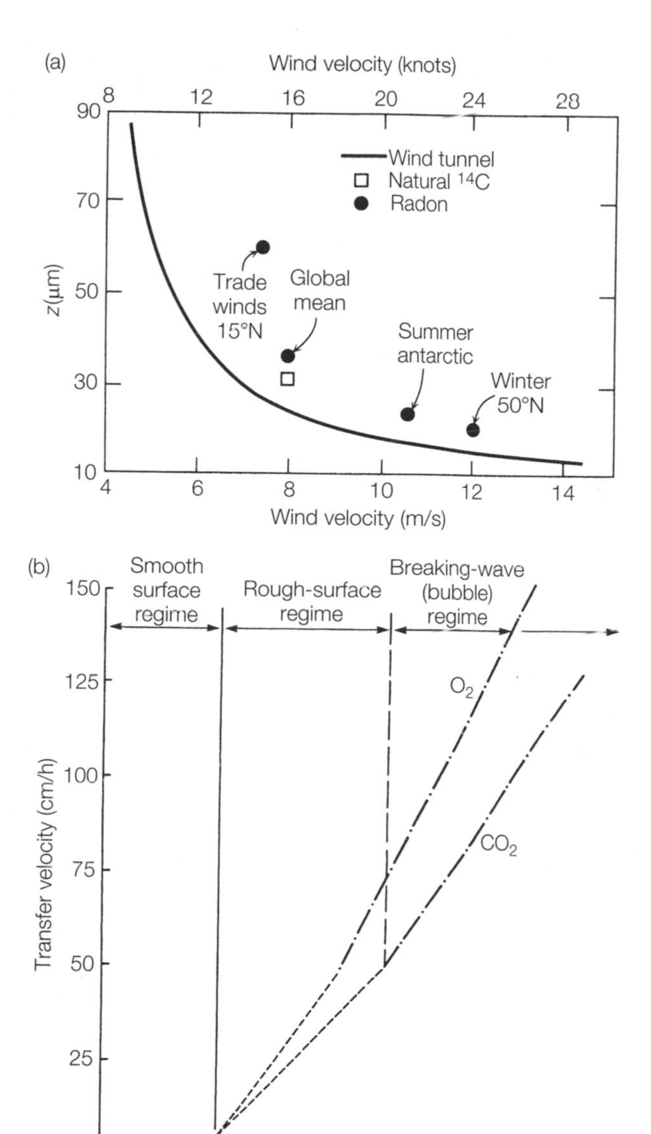

FIGURE 6.6

The effect of wind velocity on (a) thin-film thickness and (b) piston velocity. The solid line represents results obtained from measurements made in wind tunnels. In situ measurements were made from distributions of the naturally occurring radioisotopes of carbon and radon. *Source*: From (a) Broecker, W. S., and T.-H. Peng (1982). *Tracers in the Sea*. Lamont–Doherty Geological Observatory, p. 128, and (b) Bigg, G. R. (1996). *The Oceans and Climate*. Cambridge University Press, p. 85.

gradient exists, the less time the slab spends at the sea surface, the greater the net flux. This replacement time is functionally equivalent to the film thickness in the thin-film model. Though the gas flux in the surface-renewal model is proportional to the square root of D_A, the results from the two models are similar because D_A is small. The frequency of slab renewal is a function of turbulence and is typically parameterized as a function of wind speed.

Boundary layer models take a similar approach but attempt to extend the parameterization of gas exchange to individual micrometeorological processes including transfer of heat (solar radiation effects including the cool skin), momentum (friction, waves, bubble injection, current shear), and other effects such as rainfall and chemical enhancements arising from reaction with water.

6.5 NONCONSERVATIVE GASES

Gases can exhibit nonconservative behavior if their rates of production and/or consumption are faster than water transport. Nonconservative gases typically exhibit percent saturations that deviate from 100%. The extent to which a gas is nonconservative is often reflected in its degree of deviation from saturation. For example, O_2 supersaturations as high as 120% have been observed in the euphotic zone. Such high deviations are possible because the rate of O_2 supply from photosynthesis is faster than the rate at which exchange across the air-sea interface can reestablish gaseous equilibrium. O_2 undersaturations as low as 0% have been observed in subsurface waters of some coastal upwelling areas. In these regions, the rate of O_2 uptake via aerobic respiration exceeds the rate at which water motion can resupply the gas.

Not all chemical reactions proceed at rates fast enough to create significant deviations from NAECs. For example, although N_2 gas is consumed by nitrogen-fixing plankton and produced by denitrifying bacteria, these processes are too slow to affect the relatively high seawater concentrations established by equilibration with the atmosphere.

The greater the deviation from saturation, the greater the flux of the gas across the air-sea interface. As described later, the ocean is a net source of some gases to the atmosphere and a net sink of others. The direction of gas exchange can vary spatially and temporally. For example, gases controlled by phytoplankton production can exhibit diurnal variations as a function of light/dark cycles. Their fluxes are also likely to vary seasonally. This is seen in the behavior of O_2 as illustrated in Figure 6.7. These fluxes are driven by the disequilibria illustrated in Figure 6.2. Oxygen gas fluxes also exhibit variability over longer time scales as a result of changes in biological productivity and solubility caused by shifts in nutrient supply and climate.

The ocean plays an important role in determining atmospheric composition because it is net source of some gases and a net sink of others. As described later, most of the gases that undergo net degassing are produced by phytoplankton, bacteria, or the photochemically mediated oxidation of DOM. Many of the gases for which the ocean

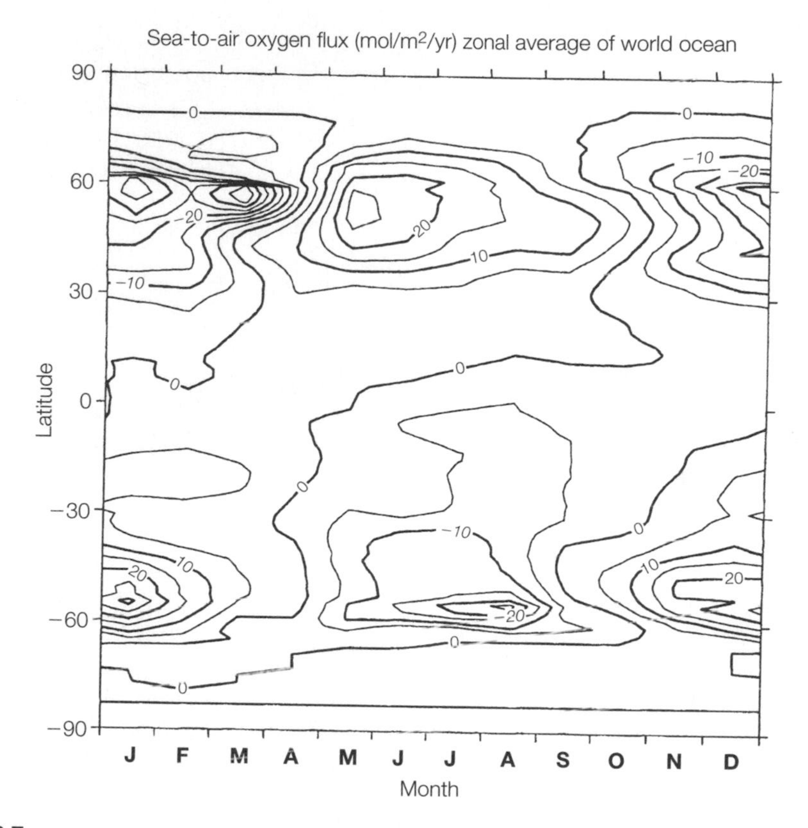

FIGURE 6.7

Zonal average of the monthly sea-to-air O_2 oxygen flux for the world ocean (mol m^{-2} y^{-1}). Outgassing (negative) and ingassing (positive) fluxes are concentrated at mid- and high latitudes. Outgassing prevails when photosynthetic rates are highest (hemispheric summer) and ingassing when net respiration is at a maximum (hemispheric winter). These patterns oscillate seasonally between the hemispheres. *Source*: Najjar, R. G. and R. F. Keeling (2000). *Global Biogeochemical Cycles* 14, 573–584.

serves as a net sink are anthropogenic in origin, so human activity has had a significant impact on their rates of ingassing and seawater concentrations. A summary of the gases in seawater and their roles in determining atmospheric composition is provided in Table 6.4.

The ocean is a source to the atmospheric of the following: (1) sulfur gases, such as dimethylsulfide, methyl mercaptan, carbonyl sulfide (COS), and carbon disulfide (CS_2); (2) halogen gases, such as methyl chloride, chloroform ($CHCl_3$), methyl iodide (CH_3I), ethyl iodide, propyl iodide, bromoiodomethane, chloroiodomethane, and diiodomethane; (3) nitrogen gases, such as nitrous oxide (N_2O), and the alkyl (methyl, ethyl, and propyl) nitrates; (4) carbon gases, such as methane (CH_4) and carbon monoxide (CO); (5) volatile metals, such as mercury; and (6) hydrogen (H_2).

Table 6.4 Summary of Gases in Seawater and the Importance of Their Global Atmosphere/Ocean Flux.

Gas	Atmospheric Role	Main Production Mechanism	Net Annual Flux to the Atmosphere (+) or to the Ocean (−)	% of Atmospheric Source (+) or Sink (−)
DMS (dimethyl sulfide)	Cloud formation and acidity	Phytoplankton	15 to 33 Tg S	80
CH_3SH	Cloud formation and acidity	Phytoplankton	15 to 33 Tg S	?
COS	Cloud formation and acidity	Photochemistry	−0.1 to 0.3 Tg	40
CS_2	Cloud formation and acidity	Photochemistry	0.13 to 0.24 Tg	20 to 35
H_2S	?	Hydrolysis	1.8 Tg	25
CH_3I	Oxidation capacity	Phytoplankton? Photochemistry?	0.13 to 0.36 Tg	?
CH_3Cl	Ozone depletion	Phytoplankton? Chemical?	0.2 to 0.4 Tg	7 to 14
$CHCl_3$	?	?	0.1 to 0.35 Tg	25 to 70
N_2O	Greenhouse gas, ozone depletion	Microbial (de)nitrification	11 to 17 Tg N	60 to 90
CH_4	Greenhouse gas, oxidation capacity	Bacteria	15 to 24 Tg	3 to 5
CO	Oxidation capacity	Photochemistry	10 to 650 Tg C	3 to 20
NMHC (nonmethane hydrocarbons)	Oxidation capacity	Photochemistry?	2.1 Tg	0.2

Table 6.4 (*Continued*)

Gas	Atmospheric Role	Main Production Mechanism	Net Annual Flux to the Atmosphere (+) or to the Ocean (−)	% of Atmospheric Source (+) or Sink (−)
Alkyl nitrates	Oxidation capacity	Photochemistry?	−30 Gg	Significant
Oxygenated organics	Oxidation capacity	Photochemistry?	?	?
H_2	Oxidation capacity	Biological photochemistry	1 Tg	5
Mercury	Pollution	Biological	1 Gg[a]	20[a]
Se	Geochemical cycling	Biological	5 to 8 Gg	45 to 77
Po	?	Biological?	?	?
CO_2	Greenhouse gas	Respiration	-1.7 ± 0.5 Pg C[b]	−30[b]
CH_3Br	Ozone depletion	Phytoplankton?	−11 to −20 Gg	−9 to −17%
$CHBr_3$	Oxidation capacity	Macroalgae	0.22 Tg	70
NH_4	Aerosol formation	Biological	20 Tg	Significant
O_3	Oxidation capacity	NA	−500 Tg	?
SO_2	Aerosol formation	NA	?	?
HCN + CH_3CN	Oxidation capacity	NA	−1.3 Tg N	−80
CFCs (chlorofluorocarbons)	Ozone depletion	NA	?	?

[a]*See Figure 28.27 for another estimate.*
[b]*See Table 25.5 for another estimate.*
Source: After Nightingale, P. D., and P. S. Liss (2003) Treatise in Geochemistry, pp. 49–81. NA, not applicable.

Other gases have temporally and geographically variable fluxes. These include carbon dioxide (CO_2), hydrogen sulfide (H_2S), gaseous Hg, the bromine gases (methyl bromide (CH_3Br) and bromoform ($CHBr_3$)), and the nitrogen gases (ammonia (NH_3) and methylamines). The organobromine gases are of interest because of their role in destroying stratospheric ozone. As a result, anthropogenic production is currently limited by international agreement. Thus it is of considerable interest that the ocean appears to be a significant source of these gases, which are thought to be of biogenic origin. Likewise, anthropogenic sources of volatile nitrogen gases that dissolve into coastal surface waters are large enough to cause algal blooms. Like DMS, ammonia gas plays an important role in the formation of cloud condensation nuclei and is likely to have some influence on climate.

The ocean is a significant sink for atmospheric ozone (O_3), which is likely consumed via oxidation reactions involving DOM. Other anthropogenic gases for which the ocean is a net sink include sulfur dioxide (SO_2), hydrogen cyanide (HCN), acetonitrile (CH_3CN), chlorofluorocarbons (freons and CCl_4), and synthetic organic compounds, such as the polychlorinated biphenyls (PCBs) and chlorinated pesticides (such as DDT, chlordane, and dieldrin). These organochlorine compounds are of environmental concern because they are known carcinogens and endocrine disruptors. They have no known natural sources. Because they are lipophilic (highly soluble in lipids) and relatively inert, they are readily moved through food webs. Their volatility has also enabled them to become widely dispersed over the world's oceans.

PART 2

The Redox Chemistry of Seawater

CHAPTER 7

The Importance of Oxygen

All figures are available on the companion website in color (if applicable).

7.1 INTRODUCTION

An important class of chemical reactions, called *redox* reactions, involves the transfer of electrons between reactants. Redox reactions play a major role in the biogeochemical cycles of the minor and trace elements. Most of these reactions are mediated by marine organisms because their enzymes serve as potent catalysts. For marine organisms, redox reactions are a means of obtaining energy and essential elements. The ultimate source of the energy fueling most redox reactions is derived from solar radiation harvested by photosynthetic organisms. These organisms transform the solar energy into a chemical form, the organic compounds that constitute their biomass, and generate O_2 as a by-product. The chemical energy stored in the organic compounds is then transferred through the marine food web as phytoplankton, the primary producers, are either eaten by primary consumers or have their dead tissues decomposed by microbes. The consumers and decomposers use planktonic organic matter to fuel their metabolic processes through a series of redox reactions that are collectively termed *respiration*. The energy yields of the various types of respiration reactions and the energy requirements of the primary producers can be predicted from thermodynamic principles.

Thermodynamic calculations indicate that the redox chemistry of aerobic seawater should be controlled by the spontaneous reduction of O_2 and oxidation of organic matter, because these two chemicals are the most abundant of the strong oxidizing and reducing agents. Where O_2 concentrations are low, microbes engage in metabolic strategies that involve reduction of oxidized solutes, such as nitrate, Mn(IV), Fe(III), sulfate, and even organic matter. And in environments where organic matter is scarce, microbes can obtain chemical energy from oxidizing inorganic compounds, such as CH_4, sulfide, Fe(II), Mn(II), and H_2.

In this chapter, a thermodynamic approach is presented to enable prediction of energy yields and redox speciation. Reactions whose energy yields are favorable would be expected to proceed spontaneously. Field work has demonstrated that for every energetically favorable redox reaction, there exists some marine organism, or community of organisms, capable of exploiting that energy resource. This has been made possible through the evolutionary development of enzymes that serve as catalysts, which enhance reaction rates. Most of the metabolic diversity is found in microbes, such that any environment in which oxidants and reductants are present will have a characteristic microbial community adapted to exploiting this chemical resource. Examples are provided in this chapter and even include microbes capable of using redox reactants that are in particulate form!

Because all metabolism relies on redox reactions, biological productivity tends to be highest in regions where redox species are furthest from chemical equilibrium, such as the euphotic zone, around hydrothermal vents, and at the sediment-water interface. These biologically mediated redox reactions act to control the marine chemistry of many minor and trace elements. This influence arises from the important roles of these elements in intracellular processes, such as energy transfer and storage, and in enzyme structure and function. Some of these elements are so crucial to biological processes that they are strongly concentrated in living tissues, shell, and bone. All of these biotic materials are eventually converted to nonliving or detrital forms that collectively represent a significant elemental reservoir in the crustal-ocean-atmosphere factory, particularly in the sediments. Some of the chemical waste products of redox reactions are gases, such as O_2, CO_2, CH_4, and N_2O, which are readily transported across the air-sea interface.

Over geologic time, the cumulative effects of all this biological redox has caused a chemical evolution of Earth's crust, ocean, and atmosphere. Future changes are likely in store as global climate change and other anthropogenic impacts are altering microbial food webs. Hence, marine chemists and biologists are currently working hard to elucidate the roles of microbes and their redox reactions in the crustal-ocean-atmosphere factory. These subjects are considered in Part II of this text, starting first with the basic electrochemical concepts necessary for understanding redox reactions.

7.2 BASIC CONCEPTS IN ELECTROCHEMISTRY

7.2.1 Half-Cell Reactions

Redox reactions occur when electrons are transferred between atoms or molecules. Most first-year chemistry students have performed the redox reaction that occurs spontaneously when metallic zinc is placed in a beaker containing an aqueous solution of copper sulfate. A vigorous exothermic reaction ensues and at its conclusion, the zinc has dissolved, the solution has lost its blue tint, and an orange solid has formed. The reaction that occurs is the following:

$$Zn(s) + Cu^{2+}(aq) \rightleftharpoons Zn^{2+}(aq) + Cu(s) \tag{7.1}$$

during which electrons have been transferred from the zinc to the copper. This electron flow can be monitored by conducting the reaction in an electrochemical cell outfitted with a voltmeter, as illustrated in Figure 7.1, in which a bar of solid zinc is connected

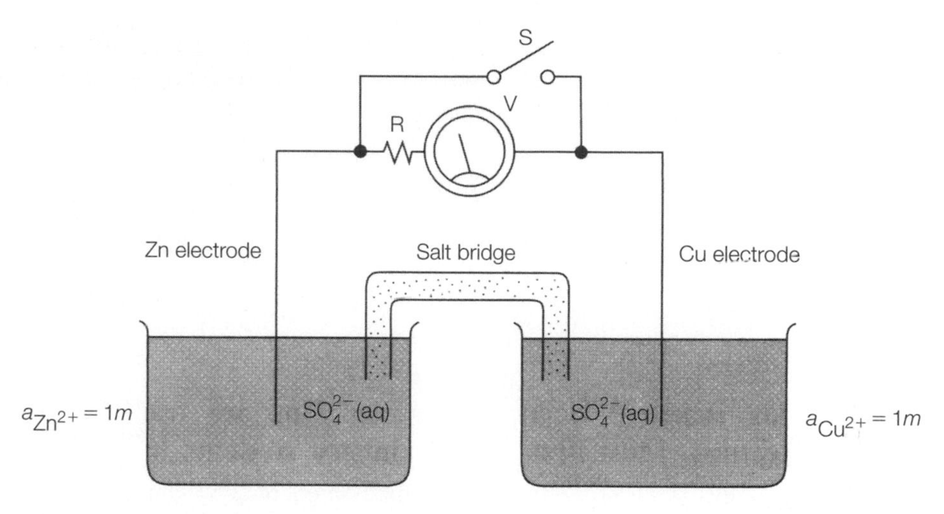

FIGURE 7.1

A galvanic cell. S = switch, R = resistor, V = voltage.

via a wire to a bar of solid copper. Because the electron flow from the zinc electrode to the copper electrode is spontaneous, this is termed a *galvanic* cell.

At the start of the reaction, the reactants are present at nonequilibrium concentrations, so the following reaction occurs spontaneously at the zinc electrode:

$$Zn(s) \rightarrow Zn^{2+}(aq) + 2e^- \quad (7.2)$$

and at the copper electrode:

$$Cu^{2+}(aq) + 2e^- \rightarrow Cu(s) \quad (7.3)$$

As the zinc dissolves, more positive charge [i.e., Zn^{2+}(aq)] is introduced into this side of the cell. As the copper precipitates, the concentration of positive charge declines on its side of the cell. To maintain electroneutrality throughout the cell, sulfate ions diffuse across a salt bridge from the copper to the zinc side of the cell.

Electrons spontaneously flow from the zinc to the copper electrode because copper has a greater affinity for electrons than does zinc. We will discuss a procedure for quantifying these affinities in a few pages. At this point, we recognize that each reactant has a characteristic affinity and that this difference in affinity creates an electron pressure differential across the wire. This differential is detected as a voltage.

Equations 7.2 and 7.3 are examples of electrochemical *half-cell reactions*. Since free electrons are not found in nature, half-cell reactions always occur in pairs such that the electrons generated by one are consumed by the other. The half-cell reaction that releases electrons is referred to as an *oxidation reaction*. The half-cell reaction that consumes electrons is referred to as a *reduction reaction*. For the redox reaction shown in Eq. 7.1, the oxidation and reduction half-cell reactions are given by Eqs. 7.2 and 7.3, respectively. The reactant that gets reduced is called an *oxidizing agent*; in our example, this is Cu. The reactant that gets oxidized is called a *reducing agent*; in our example this is Zn.

Please visit **http://elsevierdirect.com/companions/9780120885305** for brief tutorials on assignment of oxidation numbers and methods for balancing redox reactions. Both skills are required for mastery of the remainder of the materials presented in this chapter.

7.2.2 Energetics of Redox Reactions

Reactions proceed spontaneously in the direction that minimizes the energy of the reaction system (reactants and products). When the system reaches a minimal energy level, no driving force remains to cause further changes in the concentrations of the chemicals. The resulting state is referred to as *chemical equilibrium*. The further the energy of a reaction system is from that minimal level, the stronger is the chemical drive to undergo the reactions required to reach equilibrium.

Gibbs Free Energy

Because energy can be transformed or transferred in several forms during a chemical reaction, the total energy of a chemical system must be considered when predicting the direction and extent of spontaneous reactions. This total energy is referred to as the *Gibbs free energy*, G. Chemical systems that have large free energies will spontaneously react until the equilibrium state is achieved. In doing so, energy is often transferred to the surroundings and can be used to do work. The resulting *free energy change*, ΔG, is a measure of how far the original reaction mixture was from equilibrium.

By convention, ΔG is negative for a spontaneous reaction. The SI units are joule per mole (J/mol). Other common units are calories per mole (cal/mol, where $1\,\text{cal} = 4184\,\text{J}$) and kilocalories per mole (kcal/mol), where $1000\,\text{cal} = 1\,\text{kcal}$. As shown next, ΔG of a reaction can be calculated from various types of thermodynamic data. The more negative the ΔG, the more favored the reaction and the greater the amount of energy that can be released to the surroundings.

Standard Cell Potentials

The galvanic cell pictured in Figure 7.1 is not at equilibrium. If switch S is closed, electrons will spontaneously flow from the zinc (anode) to the copper (cathode) electrode. This flow will continue until the reactants and products attain their equilibrium concentrations. If switch S is opened before the cell reaches equilibrium, the electron flow will be interrupted. The voltmeter would register a positive voltage, which is a measure of the degree to which the redox reaction drives electrons from the anode to the cathode. Since this voltage is a type of energy that has the potential to do work, it is referred to as a redox potential or *cell potential*, denoted as E_{cell}.

Like ΔG, the magnitude of E_{cell} is a measure of how far a reaction mixture is from equilibrium. It is related to ΔG by

$$\Delta G = -nF(E_{cell}) \tag{7.4}$$

where F, Faraday's constant = 23.062 kcal/(V · mol electrons transferred) and n = mol electrons transferred. For the Cu/Zn cell illustrated in Figure 7.1, E_{cell} = +1.10 V; thus

$$\Delta G = -(2\,\text{mol electrons transferred})\left(23.062\,\frac{\text{kcal}}{\text{V mol electrons transferred}}\right)(+1.10\,\text{V})$$
$$= -50.7\,\text{kcal} \tag{7.5}$$

Since ΔG is negative, this reaction is spontaneous.

For redox reactions conducted at standard conditions,[1] such as those shown in Figure 7.1, Eq. 7.4 can be written as

$$\Delta G^\circ = -nF(E^\circ_{cell}) \tag{7.6}$$

where ΔG° is the standard free energy change and E°_{cell} is the standard redox potential.

The degree to which electrons are driven from the anode to the cathode can also be thought of as an *electron activity*, $\{e^-\}$. This activity is analogous to an electron pressure, rather than to a concentration, and is usually expressed as $p\varepsilon$ where

$$p\varepsilon = -\log\{e^-\} \tag{7.7}$$

The $p\varepsilon$ in a galvanic cell is related to the redox potential by

$$p\varepsilon_{cell} = \frac{F}{2.303\,RT}\,E_{cell} \tag{7.8}$$

where $R = 1.987 \times 10^{-3}$ kcal/(K mol) and T is temperature in kelvins.

E_{cell} is the sum of the energies, or half-cell potentials, contributed by the half-cell reactions. Thus, E_{cell} for any redox reaction can be calculated by summing its half-cell potentials. These half-cell potentials cannot be measured directly because oxidations and reductions always occur in pairs. Instead, half-cell potentials are determined indirectly by measuring the redox potential of a galvanic cell in which one half-cell reaction is provided by a *standard hydrogen electrode (SHE)*. Such a cell is shown in Figure 7.2 for the Zn/Zn^{2+} couple.

In this cell, the following reaction occurs spontaneously at the anode:

$$Zn(s) \rightarrow Zn^{2+}(aq) + 2e^- \tag{7.9}$$

and at the cathode:

$$2H^+(aq) + 2e^- \rightarrow H_2(g) \tag{7.10}$$

Equation 7.10 refers to the standard hydrogen electrode in which a piece of platinum as a reaction catalyst.

---

[1] Standard electrochemical conditions are (1) temperature = 25°C; (2) atmospheric pressure = 1 atm; and (3) all solutes are maintained at unit activity (i.e., $a = 1\,m$).

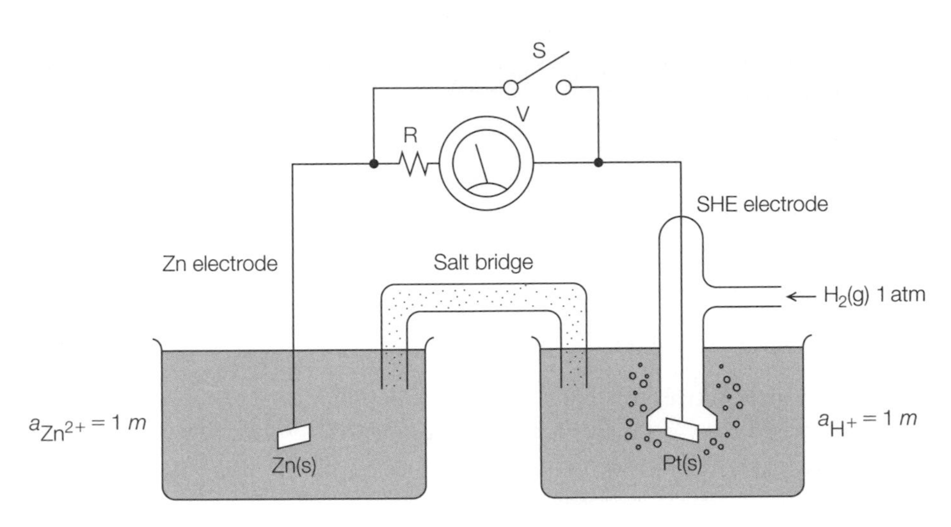

FIGURE 7.2

Galvanic cell used for measuring E_h°. S = switch, R = resistor, V = voltage.

Summing Eqs. 7.9 and 7.10 yields the redox reaction

$$Zn(s) + 2H^+(aq) \rightarrow Zn^{2+}(aq) + H_2(g) \quad (7.11)$$

The E_{cell} of this standard cell is +0.76 V. By international convention, the half-cell potential of the hydrogen reduction is assigned a value of exactly 0 V. Thus, the half-cell potential of the zinc oxidation is equal to E°_{cell} (i.e., +0.76 V). This voltage is called the standard half-cell potential and is represented by the symbol E_h°, to indicate that it was determined against a standard hydrogen electrode.

As shown in Table 7.1, E_h° values are usually listed for half-cell reactions written as reductions. Thus, the standard half-cell reduction potential for the Zn^{2+}/Zn couple is −0.76 V. Half-cell reductions that are "strong" enough to spontaneously oxidize $H_2(g)$ have a positive E_h°. Conversely, half-cell reductions that have a negative E_h° are not "strong" enough to oxidize $H_2(g)$. Instead, they spontaneously proceed as oxidations, causing the reduction of $H^+(aq)$.

Tabulated E_h° values can be used to calculate the E°_{cell} for any reaction, as illustrated in Table 7.2 for the Zn/Cu galvanic cell. The redox reaction is spontaneous when the half-reaction (Cu^{2+}/Cu) with the larger reduction E_h° (+0.34 V) acts as the oxidizing agent. In this case, the other half-reaction (Zn^{2+}/Zn) proceeds as an oxidation. The half-cell potential for this reduction is +0.76 V as it represents the reverse of the half-cell reduction potential as listed in Table 7.2. The sum of the oxidation and reduction half reactions is +0.34 V + 0.76 V = +1.10 V. Thus E°_{cell} for the galvanic Zn/Cu cell is +1.10 V.

The most energetic galvanic cell (highest E°_{cell}) is created by pairing the half-cell reaction, which has the largest reduction E_h°, with the one that has the smallest. Using the entries in Table 7.1, this would involve $Co^{3+}(aq)$ as the oxidant and Na(s) as the

Table 7.1 Standard Electrode Potentials and Equilibrium Constants for Some Reduction Half-Reactions.

Reaction	Log K at 25°C	E_h°(v)	pε°
$Co^{3+} + e^- = Co^{2+}$	+31	+1.82	−46
$Cl_2(g) + 2e^- = 2C1^-$	+46	+1.36	−39.7
$MnO_2(s) + 4H^+ + 2e^- = Mn^{2+} + 2H_2O$	+43.6	+1.29	−13
$IO_3^- + 6H^+ + 5e^- = \frac{1}{2}I_2(s) + 3H_2O$	+104	+1.23	−2.45
$Fe(OH)_3(s) + 3H^+ + e^- = Fe^{2+} + 3H_2O$	+17.1	+1.01	−4.75
$Ag^+ + e^- = Ag(s)$	+13.5	+0.80	−4.30
$Fe^{3+} + e^- = Fe^{2+}$	+13.0	+0.77	0
$Cu^+ + e^- = Cu(s)$	+8.8	+0.52	2.4
$Cu^{2+} + 2e^- = Cu(s)$	+11.4	+0.34	2.7
$AgCl(s) + e^- = Ag(s) + Cl^-$	+3.7	+0.22	3.7
$Cu^{2+} + e^- = Cu^+$	+2.7	+0.16	5.7
$S(s) + 2H^+ + 2e^- = H_2S$	+4.8	+0.14	8.8
$2H^+ + 2e^- = H_2(g)$	0.0	0.00	13.0
$V^{3+} + e^- = V^{2+}$	−4.3	−0.26	13.5
$Co^{2+} + 2e^- = Co(s)$	−9.5	−0.28	17.1
$Fe^{2+} + 2e^- = Fe(s)$	−14.9	−0.44	20.8
$Zn^{2+} + 2e^- = Zn(s)$	−26	−0.76	21.8
$Mg^{2+} + 2e^- = Mg(s)$	−79.7	−2.35	23
$Na^+ + e^- = Na(s)$	−46	−2.71	31

Source: After Stumm, W., and J. J. Morgan (1996). Aquatic Chemistry, *3rd ed. Wiley-Interscience, p. 445.*

Table 7.2 Computing E_{cell}° from Half-Cell Potentials: The Zn/Cu Cell.

Location	Reaction	E_h°(v)
Cathode	$Cu^{2+}(aq) + 2e^- \rightarrow Cu(s)$	+0.34
Anode	$Zn(s) \rightarrow Zn^{2+}(aq) + 2e^-$	+0.76
Redox reaction	$Zn(s) + Cu^{2+}(aq) \rightarrow Cu(s) + Zn^{2+}(aq)$	+1.10

reductant. Conversely, the closer in value the reduction $E_h^{\circ'}$s are, the smaller the resulting E_{cell}°. Using the entries in Table 7.1, this would involve Cu^{2+}(aq) as the oxidant and H_2S as the reductant.

Nonstandard Cell Potentials: The Nernst Equation

The results obtained under standard conditions can be used to predict thermodynamic behavior at other concentrations and temperatures. To derive the necessary equations, consider the general redox reaction:

$$a\mathrm{OX}_1 + b\mathrm{RED}_2 \rightleftharpoons c\mathrm{RED}_1 + d\mathrm{OX}_2 \tag{7.12}$$

which can be viewed as the sum of the reduction half-reaction:

$$a\mathrm{OX}_1 + ne^- \rightleftharpoons c\mathrm{RED}_1 \qquad E_{h_1}^{\circ} \tag{7.13}$$

and oxidation half-reaction:

$$d\mathrm{OX}_2 + ne^- \rightleftharpoons b\mathrm{RED}_2 \qquad E_{h_2}^{\circ} \tag{7.14}$$

The oxidation half-reaction (Eq. 7.14) has been written as a reduction, such as it would appear in a standard table of half-cell reductions (E_h°). Thus $E_{cell}^{\circ} = E_{h_1}^{\circ} - E_{h_2}^{\circ}$. If the redox reaction (Eq. 7.12) proceeds spontaneously to the right, the half-cell reduction potential, $E_{h_1}^{\circ}$, is greater than the half-cell reduction potential, $E_{h_2}^{\circ}$. In other words, species 1 is a stronger oxidizing agent than species 2.

From thermodynamic principles, chemists have demonstrated that the free energy change at nonstandard conditions, ΔG, is related to the free energy change under standard conditions, ΔG°, by

$$\Delta G = \Delta G^{\circ} + RT \ln\ Q \tag{7.15}$$

where Q is the reactant quotient and is similar in form to the equilibrium constant, K, that is,

$$Q = \frac{\{\mathrm{RED}_1\}^c \{\mathrm{OX}_2\}^d}{\{\mathrm{OX}_1\}^a \{\mathrm{RED}_2\}^b} \tag{7.16}$$

Substituting Eqs. 7.4 and 7.6 into Eq. 7.15 yields the *Nernst equation*:

$$E_{cell} = E_{cell}^{\circ} - \frac{RT}{n\mathrm{F}} \ln Q \tag{7.17}$$

Since $2.303 \log x = \ln x$, the Nernst equation at 25°C becomes

$$E_{cell} = E_{cell}^{\circ} - \frac{0.0592}{n} \log Q \tag{7.18}$$

At equilibrium, $E_{cell} = 0$ V. Substituting this condition into Eq. 7.18 yields

$$E_{cell}^{\circ} = \frac{0.0592}{n} \log Q = \frac{0.0592}{n} \log K_{cell} \tag{7.19}$$

where $Q = K_{cell}$ and K_{cell} is the thermodynamic equilibrium constant for the redox reaction. Note that E°_{cell} is measured under standard conditions in which all solutes are present at concentrations of 1 molal. Since this represents a nonequilibrium state, $E^{\circ}_{cell} \neq 0$ V.

Electrochemical Expressions in Terms of pε

By substituting in terms of $p\varepsilon_{cell}$ (as per the definition given in Eq. 7.8), the Nernst equation becomes

$$p\varepsilon_{cell} = p\varepsilon^{\circ}_{cell} - \frac{\log Q}{n} \tag{7.20}$$

and at equilibrium

$$p\varepsilon^{\circ}_{cell} = \frac{\log K_{cell}}{n} \tag{7.21}$$

Thus, for a one-electron transfer, $p\varepsilon^{\circ}_{cell} = \log K_{cell}$.

The pε and E_h of a half-cell reaction can similarly be related to the ΔG and K of that half-reaction as follows. For the half-reaction given in Eq. 7.13,

$$K_1 = \frac{\{RED_1\}^c}{\{OX_1\}^a\{e^-\}^n} \tag{7.22}$$

Rearranging,

$$\frac{1}{\{e^-\}^n} = K_1 \frac{\{OX_1\}^a}{\{RED_1\}^c} \tag{7.23}$$

and then taking the log of all the terms,

$$p\varepsilon_1 = \frac{1}{n}\left(\log K_1 + \log \frac{\{OX_1\}^a}{\{RED_1\}^c}\right) \tag{7.24}$$

and substituting $p\varepsilon^{\circ}_1 = \left(\frac{1}{n}\right)\log K_1$ (obtained by analogy from Eq. 7.21) yields

$$p\varepsilon_1 = p\varepsilon^{\circ}_1 + \frac{1}{n}\left(\log \frac{\{OX_1\}^a}{\{RED_1\}^c}\right) \tag{7.25}$$

At equilibrium $\{e^-\} = 1$, making $p\varepsilon_1 = 0$. For the reaction in Eq. 7.13, $K_1 = \{RED_1\}^c/\{OX_1\}^a$. Substituting these expressions into Eq. 7.25 yields

$$p\varepsilon^{\circ}_1 = \frac{1}{n}\log K_1 \tag{7.26}$$

A similar treatment of the half-cell reaction given in Eq. 7.14 yields

$$p\varepsilon^{\circ}_2 = \frac{1}{n}\log K_2 \tag{7.27}$$

Since $K_{cell} = K_1/K_2$,

$$p\varepsilon^{\circ}_1 - p\varepsilon^{\circ}_2 = \frac{1}{n}\log K_{cell} \tag{7.28}$$

Values for K_1 are provided in Table 7.1 for biogeochemically significant half reactions along with their $E^{\circ}_{h_1}$, $p\varepsilon^{\circ}_1$.

The reaction with the greatest tendency to proceed spontaneously will be the one with the largest equilibrium constant. This is achieved by pairing the oxidizing agent (OX_1) with the largest pε to the reducing agent (RED_2) whose half-cell reduction has the most negative pε. In seawater, these chemicals are O_2 and organic matter, respectively.

Finally, since at equilibrium, $\Delta G^\circ = -RT \ln K$,

$$\Delta G^\circ = 2.303\ nRT(p\varepsilon_2^\circ - p\varepsilon_1^\circ) \tag{7.29}$$

It can also be shown that

$$\Delta G = 2.303\ nRT(p\varepsilon_2 - p\varepsilon_1) \tag{7.30}$$

A summary of the electrochemical formulae developed above is provided in Table 7.3. ΔG, pε, E_h, and K contain virtually the same thermodynamic information. While E_h is the quantity that is analytically measured, pε is preferred by marine chemists as it is temperature independent and numerically easier to work with.[2] ΔG is often used to compare the relative stability of species because it provides a measure of energy yields in units of calories or joules. A comparison of the three electrochemical scales at 25°C is given in Figure 7.3. The merits of each thermodynamic parameter will become evident in the next section of the chapter where the energetics of some marine redox processes are considered.

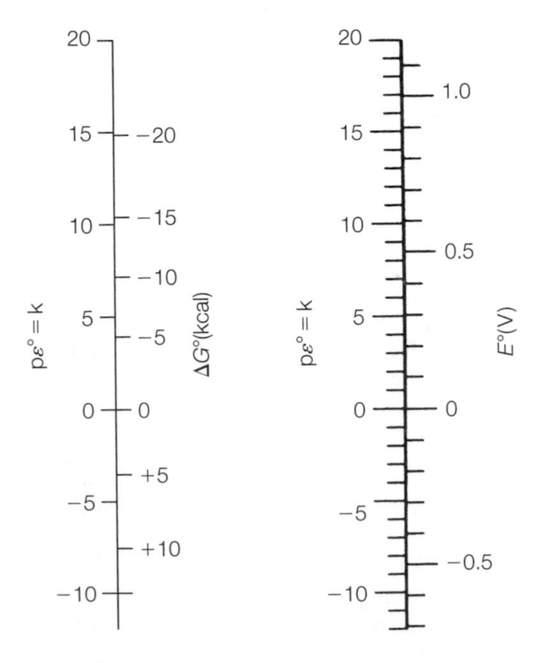

FIGURE 7.3

A comparison of the pε°, ΔG°, and E° scales using a 1-electron transfer at 25°C.

[2] pε is temperature independent in the sense that a condition of 25°C is assumed.

Table 7.3 Electrochemistry Relationships.

I. Types of Energy (for interconverting)

Constants:

F, Faraday's constant $= 23.062 \frac{\text{kcal}}{\text{V mol electrons transferred}}$

$R = 1.987 \times 10^{-3} \frac{\text{kcal}}{\text{K mol}}$

T is temperature in K.

	Standard State	Nonstandard State
A.	$\Delta G^{\circ}_{cell} = -nFE^{\circ}_{cell}$	$\Delta G_{cell} = -nFE_{cell}$
B.	$E^{\circ}_{cell} = \frac{\Delta G^{\circ}_{cell}}{-nF} = p\varepsilon^{\circ}_{cell}\frac{2.303RT}{F}$	$E_{cell} = \frac{\Delta G_{cell}}{-nF} = p\varepsilon_{cell}\frac{2.303RT}{F}$
C.	$p\varepsilon^{\circ}_{cell} = \frac{F}{2.303RT}E^{\circ}_{cell}$	$p\varepsilon^{\circ}_{cell} = \frac{F}{2.303RT}E_{cell}$

where $p\varepsilon = -\log\{e^-\}$

II. Nonstandard States (for computing concentrations)

Cell reaction:	$a\text{OX}_1 + b\text{RED}_2 \rightleftharpoons c\text{RED}_1 + d\text{OX}_2$	E°_{cell}
Half reaction 1:	$a\text{OX}_1 + ne^- \rightleftharpoons c\text{RED}_1$	$E^{\circ}_{h_1}$
Half reaction 2:	$d\text{OX}_2 + ne^- \rightleftharpoons b\text{RED}_2$	$E^{\circ}_{h_2}$

For $E^{\circ}_{h_1} > E^{\circ}_{h_2}$, $Q = \frac{\{\text{RED}_1\}^c\{\text{OX}_2\}^d}{\{\text{OX}_1\}^a\{\text{RED}_2\}^b}$

A. $\Delta G_{cell} = \Delta G^{\circ}_{cell} + RT\ \ln Q$

B. $E_{cell} = E^{\circ}_{cell} - \frac{RT}{nF}\ln Q$ (Nernst Equation)

C. $p\varepsilon_{cell} = p\varepsilon^{\circ}_{cell} - \frac{\log Q}{n}$

III. At Equilibrium (for computing K and concentrations)

$Q = K$

A.	$\Delta G = 0$	$\Delta G^{\circ} = -RT\ \ln K$	
B.	$E_{cell} = 0$	$E^{\circ}_{cell} = \frac{RT}{nF}\ln K = \frac{0.0592}{n}\log K$	at 25°C
C.	$p\varepsilon_{cell} = 0$	$p\varepsilon^{\circ}_{cell} = \frac{\log K}{n}$	at 25°C

IV. Half Reactions

A.	$E^{\circ}_{cell} = E^{\circ}_{h_1} - E^{\circ}_{h_2} = \frac{RT}{nF}\ln K = \frac{0.0592}{n}\log K$	at 25°C
B.	$p\varepsilon^{\circ}_1 - p\varepsilon^{\circ}_2 = \frac{\log K}{n}$	at 25°C

(Continued)

Table 7.3 *(Continued)*

V. Types of Standard States		
A.	Classical :	T = 25°C partial pressure of each gas = 1 atm activity of each solute = 1 molal Standard Hydrogen Electrode (SHE)
B.	Aquatic:	same as classical with additional constraint of pH = 7. Sometimes solute concentrations are specified.

7.3 THE REDOX CHEMISTRY OF SEAWATER

Aquatic chemists have defined their own electrochemical standard state to facilitate calculation of redox speciation in aqueous solutions. In this standard state, all reactions are conducted at pH 7.0, 25°C, and 1 atm. The concentrations of all other solutes are 1 molal (unless otherwise specifically noted). Values so obtained are designated with the subscript "w." The $p\varepsilon^\circ_w$'s for the most important redox couples in seawater are given in Table 7.4.

These values can be used to predict redox speciation in seawater as illustrated by the following example. Consider the half-reactions

$$\frac{1}{4}O_2(g) + H^+ + e^- \rightleftharpoons \frac{1}{2}H_2O \qquad p\varepsilon^\circ_w = +13.75 \tag{7.31}$$

$$\frac{1}{8}SO_4^{2-} + \frac{5}{4}H^+ + e^- \rightleftharpoons \frac{1}{8}H_2S(g) + \frac{5}{4}H_2O \quad p\varepsilon^\circ_w = -3.50 \tag{7.32}$$

Since the $p\varepsilon^\circ_w$ for Eq. 7.31 is larger (+13.75) than that of Eq. 7.32 (−3.50), the former proceeds as the reduction and the latter as the oxidation. This yields the following redox reaction in which H_2S is spontaneously oxidized by O_2 to SO_4^{2-} via the transfer of one electron:

$$\frac{1}{8}H_2S(g) + \frac{1}{4}O_2(g) \rightleftharpoons \frac{1}{8}SO_4^{2-} + \frac{1}{4}H^+ \tag{7.33}$$

This reaction is mediated by bacteria called sulfide oxidizers.

How favored is this reaction? How much H_2S should be present at equilibrium? These are important questions because H_2S appears to be a major source of chemical energy supporting the hydrothermal marine food web.

The relative abundances of SO_4^{2-} and H_2S at equilibrium can be calculated from Eq. 7.28 using the stoichiometry given in Eq. 7.33:

$$13.75 - (-3.50) = \frac{1}{1}\log\frac{\{SO_4^{2-}\}^{1/8}\{H^+\}^{1/4}}{P_{O_2}^{1/4}P_{H_2S}^{1/8}} \tag{7.34}$$

Table 7.4 Electrochemical Energies of Aquatic Redox Half Reactions.[a]

Reaction			$p\varepsilon^\circ$ (= log K)	$p\varepsilon^\circ_w$	$E^\circ_h(w)$
(1)	$\frac{1}{2}H_2O_2 + H^+ + e^-$	$= H_2O$	+30.0	+23.0	+1.36
(2)	$\frac{1}{4}O_2(g) + H^+ + e^-$	$= \frac{1}{2}H_2O$	+20.75	+13.75	+0.81
(3)	$\frac{1}{5}NO_3^- + \frac{6}{5}H^+ + e^-$	$= \frac{1}{10}N_2(g) + \frac{3}{5}H_2O$	+21.05	+12.65	+0.75
(4)	$\frac{1}{2}MnO_2(s) + 2H^+ + e^-$	$= \frac{1}{2}Mn^{2+} + H_2O$	+20.8	+9.8[b]	+0.58
(5)	$\frac{1}{2}NO_3^- + H^+ + e^-$	$= \frac{1}{2}NO_2^- + \frac{1}{2}H_2O$	+14.15	+7.15	+0.42
(6)	$\frac{1}{8}NO_3^- + \frac{5}{4}H^+ + e^-$	$= \frac{1}{8}NH_4^+ + \frac{3}{8}H_2O$	+14.9	+6.15	+0.36
(7)	$\frac{1}{6}NO_2^- + \frac{4}{3}H^+ + e^-$	$= \frac{1}{6}NH_4^+ + \frac{1}{3}H_2O$	+15.2	+5.82	+0.34
(8)	$\frac{1}{2}O_2(g) + H^+ + e^-$	$= \frac{1}{2}H_2O_2$	+11.5	+4.5	+0.27
(9)	$Fe(OH)_3(am) + 3H^+ + e^-$	$= Fe^{2+} + 3H_2O$	+16.0	+1.0[b]	+0.06
(10)	$\frac{1}{2}CH_3OH + H^+ + e^-$	$= \frac{1}{2}CH_4(g) + \frac{1}{2}H_2O$	+9.88	+2.88	+0.17
(11)	$\frac{1}{4}$"CH_2O"$+ H^+ + e^-$	$= \frac{1}{4}CH_4(g) + \frac{1}{4}H_2O$	+6.94	−0.06	+0.00
(12)	$\frac{1}{6}SO_4^{2-} + \frac{4}{3}H^+ + e^-$	$= \frac{1}{48}S_8(col) + \frac{2}{3}H_2O$	+5.9	−3.4	−0.20
(13)	$\frac{1}{8}SO_4^{2-} + \frac{5}{4}H^+ + e^-$	$= \frac{1}{8}H_2S(g) + \frac{1}{2}H_2O$	+5.25	−3.50	−0.21
(14)	$\frac{1}{8}SO_4^{2-} + \frac{9}{8}H^+ + e^-$	$= \frac{1}{8}HS^- + \frac{1}{2}H_2O$	+4.25	−3.75	−0.22
(15)	$\frac{1}{2}$"CH_2O"$+ H^+ + e^-$	$= \frac{1}{2}CH_3OH$	+3.99	−3.01	−0.18
(16)	$\frac{1}{8}HCO_3^- + \frac{9}{8}H^+ + e^-$	$= \frac{1}{8}CH_4(g) + \frac{3}{8}H_2O$	+3.8	−4.0	−0.24
(17)	$\frac{1}{8}CO_2 + H^+ + e^-$	$= \frac{1}{8}CH_4(g) + \frac{1}{4}H_2O$	+2.9	−4.13	−0.24
(18)	$\frac{1}{16}S_8(col) + H^+ + e^-$	$= \frac{1}{2}H_2S(g)$	+3.2	−3.8[c]	−0.22
(19)	$\frac{1}{16}S_8(col) + \frac{1}{2}H^+ + e^-$	$= \frac{1}{2}HS^-$	−0.8	−4.3	−0.25
(20)	$\frac{1}{6}N_2 + \frac{4}{3}H^+ + e^-$	$= \frac{1}{3}NH_4^+$	+4.65	−4.7	−0.28
(21)	$H^+ + e^-$	$= \frac{1}{2}H_2(g)$	+0.00	−7.0	−0.41
(22)	$\frac{1}{4}HCO_3^- + \frac{5}{4}H^+ + e^-$	$= \frac{1}{4}$"CH_2O"$+ \frac{1}{2}H_2O$	+1.8	−7.0	−0.41
(23)	$\frac{1}{2}HCOO^- + \frac{3}{2}H^+ + e^-$	$= \frac{1}{2}$"CH_2O"$+ \frac{1}{2}H_2O$	+2.82	−7.68	−0.45
(24)	$\frac{1}{4}CO_2(g) + H^+ + e^-$	$= \frac{1}{4}$"CH_2O"$+ \frac{1}{4}H_2O$	−1.2	−8.20	−0.49
(25)	$\frac{1}{2}HCO_3^- + \frac{3}{2}H^+ + e^-$	$= \frac{1}{2}CO(g) + H_2O$	+2.2	−8.3	−0.49
(26)	$\frac{1}{2}CO_2(g) + \frac{1}{2}H^+ + e^-$	$= \frac{1}{2}HCOO^-$	−4.83	−8.33	−0.49
(27)	$\frac{1}{2}CO_2(g) + H^+ + e^-$	$= \frac{1}{2}CO(g) + \frac{1}{2}H_2O$	−1.7	−8.7	−.051

am = amorphous; col = colloidal.

[a] *Values for $p\varepsilon^\circ(w)$ and $E^\circ_h(w)$ reflect pH 7.0 and temperature 25°C.*

[b] *In the reaction of reductive dissolution of metal oxides, the concentrations of the dissolved metals (Mn^{2+} and Fe^{2+}) have been fixed at 1 μM to more accurately reflect their relative redox properties.*

[c] *This reaction is listed out of order so as not to separate it from the reaction of formation of the bisulfide ion, HS^-.*

Sources: After Stumm, W., and J. J. Morgan, (1996). Aquatic Chemistry, *3rd edition, Wiley-Interscience, p. 465. Morel, F. M. M., and J. G. Herring, (1993).* Principles and Applications of Aquatic Chemistry, *Wiley-Interscience, p. 439.*

The superscripts in the equilibrium constant become coefficients since $\log x^a = a \log x$. The log $\{H^+\}$ term can be replaced by −pH to yield

$$17.25 = \frac{1}{8}\log\frac{\{SO_4^{2-}\}}{P_{H_2S}} - \frac{1}{4}\mathrm{pH} - \frac{1}{4}\log P_{O_2} \tag{7.35}$$

Assuming the solution has a pH = 8 and is in gaseous equilibrium with the atmosphere (i.e., $P_{O_2} = 0.21$ atm),

$$\frac{\{SO_4^{2-}\}}{P_{H_2S}} = 10^{155.4} = 2 \times 10^{155} \tag{7.36}$$

In other words, virtually no H_2S should be present at equilibrium.

The equilibrium constant, which can be calculated by substituting into Eq. 7.28, is

$$13.75 - (-3.50) = \frac{1}{1}\log K = 17.25 \tag{7.37}$$

or $K = 10^{17.50}$. ΔG_w° can be calculated from Eq. 7.29 as

$$\begin{aligned}\Delta G_w^\circ &= (2.303)(1\ \text{mol})(1.987 \times 10^{-3}\ \text{kcal K}^{-1}\ \text{mol}^{-1})(298.15\ \text{K}) \\ &\quad (-3.50 - 13.75) \\ &= -23.5\ \text{kcal}\end{aligned} \tag{7.38}$$

In this calculation, −23.5 kcal of energy is produced per $\frac{1}{8}$ mol of H_2S and $\frac{1}{4}$ mol O_2 consumed while $\frac{1}{8}$ mol SO_4^{2-} and $\frac{1}{4}$ mol H^+ are produced. These calculations predict that sulfide-oxidizing bacteria should obtain large amounts of energy from oxidizing hydrogen sulfide gas. Or to state this a different way, since K is very large, the thermodynamic driving force is large.

Biologically mediated redox reactions tend to occur as a series of sequential subreactions, each of which is catalyzed by a specific enzyme and is potentially reversible. But despite favorable thermodynamics, kinetic constraints can slow down or prevent attainment of equilibrium. Since the subreactions generally proceed at unequal rates, the net effect is to make the overall redox reaction function as a unidirectional process that does not reach equilibrium. Since no net energy is produced under conditions of equilibrium, organisms at equilibrium are by definition dead. Thus, redox disequilibrium is an opportunity to obtain energy as a reaction proceeds toward, but ideally for the sake of the organism does not reach, equilibrium.

7.3.1 **Relative Redox Intensity**

The half-cell reduction reaction with the highest pε will force all other half-cell reactions to proceed as oxidations. Although seawater contains stronger oxidizing agents than O_2, such as H_2O_2, these others do not exert a controlling influence on the redox chemistry of the ocean. This is due to their relatively low concentrations and/or slow reaction rates. In comparison, nearly all reactions that involve O_2 proceed relatively rapidly because they are mediated by enzymes produced by a wide variety of marine organisms. Because of O_2's great oxidizing power, equilibrium thermodynamics predicts that all biochemically active elements should exist primarily in their highest oxidation states in oxic seawater.

The greater the difference in pε between the oxidizing and reducing agents, the greater the free energy yield of the resulting redox reaction. Since organisms depend on this energy to fuel their metabolic processes, the redox reaction that produces the most energy is of greatest biological benefit. In seawater, the redox reaction that yields the most energy is the aerobic oxidation of organic matter. (As noted earlier, seawater contains trace amounts of oxidants stronger than O_2, but these are relatively unimportant due to their limited abundance and reactivity.)

The chemical equation that describes the aerobic oxidation of organic matter is:

$$\frac{1}{4}CH_2O + \frac{1}{4}O_2 \rightleftharpoons \frac{1}{4}CO_2 + \frac{1}{4}H_2O \quad (7.39)$$

where organic matter is represented generically by the empirical formula "CH_2O" in which the oxidation number of C is 0. (In Chapter 8, we will consider the other elements in organic matter, such as nitrogen, phosphorus, sulfur, and trace metals.) Note that the oxidation number of carbon in CO_2 is IV so in this reaction, carbon is oxidized and oxygen is reduced. (See the online appendix on the companion website for rules used in assigning oxidation numbers to carbon.)

The free energy yield (ΔG°_w) per mole "CH_2O" oxidized can be computed for this reaction by substituting into Eq. 7.29. as follows:

$$\begin{aligned}\Delta G^\circ_w &= (2.303)(1\ \text{mol})(1.987 \times 10^{-3}\ \text{kcal K}^{-1}\ \text{mol}^{-1})(298.15\ \text{K})\\ &\quad(-8.20 - 13.75)\\ &= -29.95\ \text{kcal per mole "CH}_2\text{O" oxidized}\end{aligned} \quad (7.40)$$

where the values of $p\varepsilon_1^0$ and $p\varepsilon_2^0$ are for the half-cell reactions 2 and 24, as listed in Table 7.4, respectively. The aerobic respiration of organic matter is the most important metabolic reaction by which animals obtain energy from their food.

Reducing Agents in Seawater

If one of these reactants is depleted, the next most energetic redox reaction will occur in its stead. In the absence of organic matter, the next most energetic electron donor is hydrogen gas (H_2). This reaction produces only slightly less energy (−28.4 kcal per mole "CH_2O" oxidized) than does the aerobic oxidation of organic matter. The other half-cell oxidations that proceed in the presence of O_2 are listed in order of their relative redox intensity at the bottom of Figure 7.4. This diagram is designed to illustrate that the greatest energy is provided by matching half reactions that have the greatest different in pε's (represented diagrammatically as the greatest distance between the heads of the two arrows). Thus, the oxidation half-cell reactions (L through T) occur most energetically when paired with half-reaction A (O_2 reduction). In order of decreasing energy yield, they include the aerobic oxidation of CH_2O, H_2, CH_4, H_2S, Fe(II), ammonium (NH_4^+), and Mn(II). The stoichiometry and energy yields of some of these reactions

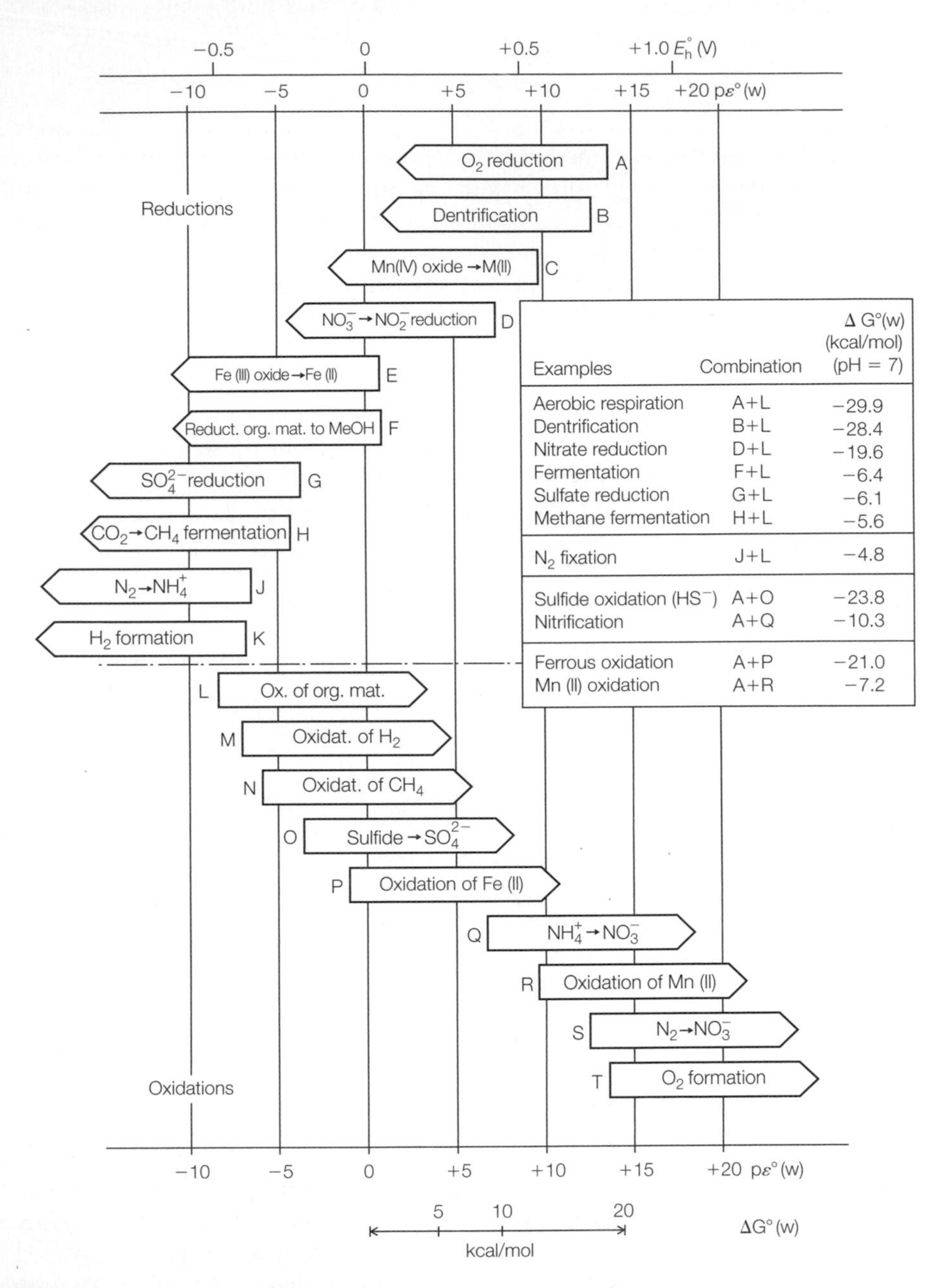

Examples	Combination	Δ G°(w) (kcal/mol) (pH = 7)
Aerobic respiration	A+L	−29.9
Dentrification	B+L	−28.4
Nitrate reduction	D+L	−19.6
Fermentation	F+L	−6.4
Sulfate reduction	G+L	−6.1
Methane fermentation	H+L	−5.6
N_2 fixation	J+L	−4.8
Sulfide oxidation (HS^-)	A+O	−23.8
Nitrification	A+Q	−10.3
Ferrous oxidation	A+P	−21.0
Mn (II) oxidation	A+R	−7.2

FIGURE 7.4

Relative energetics of microbially mediated redox reactions. Inset box lists redox reactions in terms of decreasing ΔG_w^o yields. Relative energetic of the component half reactions are shown by arrows that point in the direction of the spontaneous redox reaction. The originating point of the arrows indicates the redox energy associated with each half-cell reaction. The distance between the originating points of two half reactions provides an estimate of ΔG_w^o for the redox reaction as per scale on the bottom of the figure. No information is contained in the relative length of the arrows. MeOH = methanol. *Source*: After Stumm, W. S., and J. J. Morgan (1981). *Aquatic Chemistry*, John Wiley & Sons, Inc., p. 460.

Table 7.5 Oxidation of Reduced Inorganic Compounds.[a]

Hydrogen oxidation
$\frac{1}{2}H_2(g) + \frac{1}{4}O_2(g) = \frac{1}{2}H_2O$; $\Delta G_w^0 = -28.4$ kcal
Methane oxidation
$\frac{1}{8}CH_4 + \frac{1}{4}O_2 = \frac{1}{8}CO_2 + \frac{1}{4}H_2O$; $\Delta G_{(w)}^0 = -24.4$ kcal
Sulfide oxidation
$\frac{1}{8}H_2S(g) + \frac{1}{4}O_2(g) = \frac{1}{8}SO_4^{2-} + \frac{1}{4}H^+$; $\Delta G_w^0 = -23.5$ kcal
Iron oxidation
$Fe^{2+} + \frac{1}{4}O_2(g) + \frac{5}{2}H_2O = Fe(OH)_3(s) + 2H^+$; $\Delta G_w^0 = -21.0$ kcal
Nitrification
$\frac{1}{6}NH_4^+ + \frac{1}{4}O_2(g) = \frac{1}{6}NO_2^- + \frac{1}{3}H^+ + \frac{1}{6}H_2O$; $\Delta G_w^0 = -10.8$ kcal
$\frac{1}{2}NO_2^- + \frac{1}{4}O_2(g) = \frac{1}{2}NO_3^-$; $\Delta G_w^0 = -9.0$ kcal
Manganese oxidation
$\frac{1}{2}Mn^{2+} + \frac{1}{4}O_2(g) + \frac{1}{2}H_2O = \frac{1}{2}MnO_2(s) + H^+$; $\Delta G_w^0 = -5.4$ kcal

[a] *These are accomplished by chemolithotrophs, mostly autotrophs. Source: After Morel, F. M. M. (1983)* Principles of Aquatic Chemistry. *John Wiley & Sons, p. 330.*

are presented in Table 7.5. The oxidation of ammonium is called nitrification. Only prokaryotes (bacteria and archaea)[3] are capable of exploiting the energy of the reactions listed in Table 7.5. In most cases, a given species is capable of using only one kind of electron donor, so these metabolisms tend to be species specific.

Oxidizing Agents in Seawater

If, during the oxidation of organic matter, O_2 is depleted first, the next-strongest oxidizing agent, nitrate, will take its place (half-reaction B paired with half-reaction L in Figure 7.4). Nitrate is reduced in a stepwise fashion to $N_2(g)$. This process is termed denitrification and is also mediated by certain types of marine microbes.[4] As shown in Table 7.6, the energy yield is only slightly less than that of aerobic respiration. If a large amount of organic matter is present, the nitrate will be depleted and the next reducing agent of choice is Mn(IV) (half reaction C paired with half reaction L). The relative

[3] Archaea are microbes that include many species capable of survival in extreme environments.

[4] Recent research suggests that denitrification may involve the reduction of nitrate to nitrite followed by the reaction of nitrite with ammonium in which the nitrogen in nitrite oxidizes the nitrogen in ammonium thereby generating a molecule of $N_2(g)$. This is termed the anammox reaction.

Table 7.6 Oxidation of Organic Compounds Represented Generically as "CH_2O".[a]

Aerobic respiration
$\frac{1}{4}$"CH_2O" + $\frac{1}{4}O_2(g) = \frac{1}{4}CO_2(g) + \frac{1}{4}H_2O$; $\Delta G^0_w = -29.9$ kcal
Denitrification
$\frac{1}{4}$"CH_2O" + $\frac{1}{5}NO_3^- + \frac{1}{5}H^+ = \frac{1}{4}CO_2(g) + \frac{1}{10}N_2(g) + \frac{7}{20}H_2O$; $\Delta G^0_w = -28.4$ kcal
Manganese respiration
$\frac{1}{4}$"CH_2O" + $\frac{1}{2}MnO_2(s) + H^+ = \frac{1}{4}CO_2(g) + \frac{1}{2}Mn^{2+} + \frac{3}{4}H_2O$; $\Delta G^0_w = -24.6$ kcal
Iron respiration
$\frac{1}{4}$"CH_2O" + $Fe(OH)_3(s) + 2H^+ = \frac{1}{4}CO_2(g) + Fe^{2+} + \frac{11}{4}H_2O$; $\Delta G^0_w = -12.6$ kcal
Sulfate reduction
$\frac{1}{4}$"CH_2O" + $\frac{1}{8}SO_4^{2-} + \frac{1}{8}H^+ = \frac{1}{4}CO_2(g) + \frac{1}{8}HS^- + \frac{1}{4}H_2O$; $\Delta G^0_w = -6.1$ kcal
Methane fermentation
$\frac{1}{4}$"CH_2O" = $\frac{1}{8}CO_2(g) + \frac{1}{8}CH_4(g)$; $\Delta G^0_w = -5.6$ kcal
Hydrogen fermentation
$\frac{1}{4}$"CH_2O" + $\frac{1}{4}H_2O = \frac{1}{4}CO_2(g) + \frac{1}{2}H_2(g)$; $\Delta G^0_w = -1.6$ kcal

[a]*These are accomplished by chemoorganotrophs, all heterotrophs.*
Source: After Morel, F. M. M. (1983). Principles of Aquatic Chemistry. *John Wiley and Sons, p. 330.*

strength of the other reducing agents in seawater is shown in the top half of Figure 7.4 and their reaction stoichiometry is given in Table 7.6. In order of decreasing energy yield, they include the reduction of O_2, NO_3^- to N_2, Mn(IV), NO_3^- to NO_2^-, Fe(III), "CH_2O" to methanol, and SO_4^{2-} and "CH_2O" to CH_4. Reduction of "CH_2O" (fermentation) by organic matter produces a variety of products, including CH_4, H_2, and some low-molecular-weight organic compounds such as methanol.

Biological Nitrogen and Carbon Fixation

Because of its relative abundance and low $p\varepsilon^\circ_w$, organic matter is the most important reducing agent in the ocean. Other reductants paired with reduction reactions in the top half of Figure 7.4 produce less energy than organic matter. All of this organic matter is ultimately derived from biological "fixation" reactions in which oxidized inorganic carbon is converted to a reduced organic form. The most important carbon fixation reactions are listed in Table 7.7.

These reactions require energy to proceed. Sunlight is the energy source for some, but not all, of these carbon fixation pathways. The sunlight-fueled fixation of carbon, i.e., photosynthesis, is responsible for most of the organic matter production on Earth.

Table 7.7 Carbon and Nitrogen Fixation Reactions.

Reaction
Carbon fixation[a]
$\frac{1}{4}CO_2(g) + \frac{1}{4}H_2O = \frac{1}{4}\text{"}CH_2O\text{"} + \frac{1}{4}O_2(g);\ \Delta G^0_w = +29.9\text{ kcal}$
$\frac{1}{4}CO_2(g) + \frac{1}{2}H_2S(g) = \frac{1}{4}\text{"}CH_2O\text{"} + \frac{1}{16}S_8(col)^b + \frac{1}{4}H_2O;\ \Delta G^0_w = +4.7\text{ kcal}$
$\frac{1}{4}CO_2(g) + \frac{1}{6}NH_4^+ + \frac{1}{12}H_2O = \frac{1}{4}\text{"}CH_2O\text{"} + \frac{1}{6}NO_2^- + \frac{1}{3}H^+;\ \Delta G^0_w = +17.8\text{ kcal}$
Nitrogen fixation[c]
$\frac{1}{6}N_2(g) + \frac{1}{3}H^+ + \frac{1}{4}\text{"}CH_2O\text{"} + \frac{1}{4}H_2O = \frac{1}{3}NH_4^+ + \frac{1}{4}CO_2(g);\ \Delta G^0_w = -4.8\text{ kcal}$

[a] *These are accomplished by autotrophs.*
[b] *col = colloidal.*
[c] *This is accomplished by nitrogen (N_2) fixers.*
Source: After Morel, F. M. M. (1983). Principles of Aquatic Chemistry. *John Wiley and Sons, p. 330.*

In this redox reaction, plants use solar energy to transform inorganic carbon into organic carbon, $\left(\text{i.e., } \frac{1}{4}CO_2 + \frac{1}{4}H_2O \rightarrow \frac{1}{4}\text{"}CH_2O\text{"} + \frac{1}{4}O_2\right)$. The energy required to fix 0.25 mol of inorganic carbon into organic form is given by

$$\begin{aligned}\Delta G^\circ_w &= (2.303)\,(1\text{ mol})\,(1.987 \times 10^{-3}\text{ kcal K}^{-1}\text{ mol}^{-1})\\ &\quad (298.15\text{ K})\,(13.75 - (-8.20))\\ &= +29.95\text{ kcal}\end{aligned} \tag{7.41}$$

This reaction is not spontaneous. To form 0.25 mol of organic matter requires the input of solar energy in an amount equal to the energy obtained from the aerobic oxidation of 0.25 mol of organic matter. Photosynthesis is listed in Figure 7.4 as "O_2 formation." The O_2 produced by photosynthesis is available to engage in the oxidation reactions listed in the bottom half of Figure 7.4. Carbon fixation can also be driven by energy obtained from the oxidation of reduced inorganic compounds, such as sulfide oxidation.

Also included in Table 7.7 are the nitrogen fixation reactions. These are similar to the carbon fixation reactions in that they involve the conversion of an oxidized inorganic species (N_2) to a reduced form, such as ammonium. The fixed forms of nitrogen can be taken up by plants. As with carbon fixation, this process requires an energy source in order to proceed. Some N_2 fixers are photosynthetic and others use energy obtained from the oxidation of reduced inorganic compounds.

7.3.2 Metabolic Classifications of Organisms

Biologists use a taxonomic classification system for organisms that recognizes three domains: the archaea, bacteria, and eukarya. Eukaryotes include all multicelled and some single-celled life forms. Marine biologists use the term *protist* to refer to eukaryotic microorganisms that are neither animal, fungi, plant, or archaean. Unicellular forms include the protozoans and algae. Some algae are either multicellular or colonial. The protists are characterized by relatively large cells that have flexible walls and a nucleus.

In contrast, the archaea and bacteria, which are *prokaryotes*, are all unicellular and have rigid cell walls. They do not possess a nucleus and are small (generally 1 to 2 μm in size). In this text, the term *microbe* is used to collectively refer to all of the single-celled organisms. In marine systems, microbes exhibit three lifestyles: (1) floating and drifting in seawater, (2) attached to surfaces, or (3) living within a host. The free-living microbes are considered to be members of the plankton, i.e., weak swimmers. They are typically classified into size categories as shown in Table 7.8. The bacteria and archaea are also referred to as bacterioplankton, the photosynthetic microbes as phytoplankton, and the animals as zooplankton.

The biological classification schemes for bacteria and archaea are still being developed because of the rapid pace of new discoveries in genomics. The two most important phyla of marine bacteria are the cyanobacteria, which are photosynthetic, and the proteobacteria. The latter include some photosynthetic species, such as the purple photosynthetic bacteria and N_2 fixers. Other members of this diverse phylum are the methanotrophs, nitrifiers, hydrogen, sulfur and iron oxidizers, sulfate and sulfur reducers, and various bioluminescent species.

Table 7.8 Plankton Size Classifications.

Size Class	Diameter	Examples	Examples
Femtoplankton	<0.2 μm	Viroplankton	Viruses
Picoplankton	0.2 to 2.0 μm	Proteobacteria Cyanobacteria	*Pelagibacter*, *Roseobacter* *Prochlorococcus*, *Synechococcus*, *Trichodesmium*
Nanoplankton	2.0 to 20 μm	Protists Mycoplankton Phytoplankton	Amoebae, flagellates, Euglenozoa, dinoflagellates[a] Yeasts and fungi Coccolithophorids, *Phaeocystis*, dinoflagellates[a]
Microplankton	20 to 200 μm	Protists Phytoplankton	Radiolarians, foraminiferans Diatoms, raphidophytes, *Phaeocystis*[b]
Mesoplankton	0.2 to 20 mm	Zooplankton	Copepods, amphipods, appendicularians
Macroplankton	2 to 20 cm	Zooplankton	Chaetognaths
Megaplankton	20 to 200 cm	Zooplankton	Larvaceans, euphausiids, salps, pteropods, jellyfish

[a] *Some species of dinoflagellates are not photosynthetic. Some species are mixotrophic.*
[b] *Phaeocystis blooms form large colonies.*

The flexible cell wall of the protists enables them to obtain fuel via phagocytosis in which a food particle is engulfed by the cell membrane and then enzymatically decomposed. Thus, metabolically they are capable of consuming organic matter (heterotrophy), although not all do, such as the photosynthetic eukaryotes. Because of their rigid cell wall, the prokaryotes cannot engage in phagocytosis and, thus, must obtain their fuel and nutrients via transport across their cell membranes. The prokaryotes exhibit a great diversity of metabolic strategies that take advantage of a wide variety of energy sources, electron donors, and carbon sources. Most of the prokaryotes engage in only one type of metabolism (obligate) although some can alternate depending on environmental conditions (facultative). Surprisingly, microbial species that seem to be very similar genetically can exhibit diverse metabolic strategies. Another important characteristic of the microbes is that some metabolic strategies are functional only in a symbiotic setting, either intracellular or extracellular. In the case of the latter, the symbiosis involves mixed species assemblages, or consortia, of microbes that grow best as a group rather than by themselves. This phenomenon is called *syntrophy*.

Microbes are of central importance in the ocean. Most of the marine biomass is microbial, with the majority contributed by photosynthesizers. Because of their abundance, high growth rates, and metabolic diversity, marine microbes play a key role in the transformation and transport of chemicals in the crustal-ocean-atmosphere factory. To aid in discussions of their metabolic strategies, which are all based on redox reactions, microorganisms are classified in terms of their energy, carbon, and electron sources as shown in Table 7.9. The major metabolic strategies are listed in Table 7.10 along with examples of organisms that use these pathways.

Phototrophy

Phytoplankton are *photoautolithotrophs* as their energy source is sunlight (photo), their carbon source is inorganic carbon (auto), and their electron donor is inorganic (litho). Representative members of the phytoplankton include the diatoms, some dinoflagellates, the coccolithophorids, phytoflagellates, and photosynthetic bacteria. The most well known of the phytoplankton are the oxygenic forms that use H_2O as their electron donor. Less well known are anoxygenic phototrophs, which, depending on their species, can use H_2, sulfur compounds (H_2S and S°), Fe(II), and organic compounds (sugars, amino acids, and organic acids) as their electron donor. The anoxygenic phototrophs are bacteria. The first species discovered were the purple and green sulfur

Table 7.9 Trophic Prefix Naming Scheme for Specifying Metabolic Pathways.

Energy Source	Carbon Source	Electron Source
Light → "Photo-"	CO_2 → "Auto-"	Inorganic → "Litho-"
Chemical → "Chemo-"	Organic C → "Hetero-"	Organic → "Organo-"

Examples: Photoautolithotrophy, chemoautolithotrophy, chemoheteroorganotrophy.

Table 7.10 Main Types of Energy Metabolisms with Examples of Representative Organisms.

Metabolism		Energy Source	Carbon Source	Electron Donor	Organisms
Photoautotroph		Light	CO_2	H_2O	Green plants, algae, cyanobacteria
				H_2S, S^0, Fe^{2+}	Purple and green sulfur bact. (*Chromatium*, *Chlorobium*), Cyanobacteria
Photoheterotroph			Org. C $\pm$ CO_2		Purple and green nonsulfur bact. (*Rhodospirillum*, *Chloroflexus*)
Chemolithotroph	Chemolithoautotroph	Oxidation of inorganic compounds	CO_2		*Aerobic*:
				H_2S, S^0, $S_2O_3^{2-}$, FeS_2	Colorless sulfur bact. (*Thiobacillus*, *Beggiatoa*)
				NH_4^+, NO_2^-	Nitrifying bact. (*Thiobacillus*, *Nitro- bacter*)
				H_2	Hydrogen bact. (*Hydrogenomonas*)
				Fe^{2+}, Mn^{2+}	Iron bact. (*Ferrobacillus*, *Shewanella*)
					Anaerobic:
				$H_2 + SO_4^{2-}$	Some sulfate reducing bact. (*Desulfovibrio spp.*)
				H_2 $S/S^0/S_2O_3^{2-} + NO_3^-$	Denitrifying sulfur bact. (*Thiobac. denitrificans*)
				$H_2+CO_2 \rightarrow CH_4$	Methanogenic bact.
				$H_2+CO_2 \rightarrow$ acetate	Acetogenic bact.
	Mixotroph		CO_2 or Org. C	H_2S, S°, $S_2O_3^{2-}$	Colorless sulfur bact. (some Thiobacillus)
	Chemolithoheterotroph		Org. C	H_2S, S°, $S_2O_3^{2-}$	Colorless sulfur bact. (some Thiobacillus, Baggiatoa)
				H_2	Some sulfate reducing bact.
Heterotroph (=chemoorganotroph)		Oxidation of organic compounds	Org. C (max. 30% CO_2)	Org. C	Aerobic: Animals, fungi, many bacteria
					Anaerobic: Denitrifying bacteria
					Mn- or Fe-reducing bacteria
					Sulfate reducing bacteria
					Sulfate reducing bacteria
					Fermenting bacteria
			Org. C (30–90% CO_2)	CH_4	Methane oxidizing bact.

Source: From Jorgensen, B. B. (2000). Marine Geochemistry. *Springer, p. 188.*

bacteria that inhabit anoxic environments, such as salt marshes and deep waters in marginal seas, e.g., the Red, Black, and Mediterranean Seas.

More recently, marine biologists have discovered that aerobic anoxygenic bacteria, such as *Roseobacter*, are a widespread and abundant component of the picoplankton. These organisms use a protein called rhodopsin for light harvesting and do not appear to be strict lithotrophs as they use organic matter as an energy source. This phenomenon, where an organism uses both energy sources listed in Table 7.9, is termed mixotrophy. Rhodopsins have also been found in other bacteria and archaea, suggesting that chemoautotrophy augmented by some degree of phototrophy might be a widespread strategy. These light-harvesting pigments are highly efficient, enabling the anoxygenic bacteria to live as far as 200 m below the surface. At even greater depths, a bacterium has recently been isolated from hydrothermal vent fluids that requires light, sulfur, and CO_2 to grow. Hydrothermal vents emit a very dim light due to sonoluminesence. It is thought that this newly discovered bacterium is using the vents' luminescence as an energy source!

Oxygenic photosynthetic bacteria include the nitrogen fixer *Trichodesmium* and the cyanobacteria *Synechococcus* and *Prochlorococcus*. The latter represent a major fraction of the microbial biomass in the ocean and are probably the most abundant primary producers in the ocean. They were not discovered until the late 1980s, probably because of their small size as they are members of the picoplankton. These cyanobacteria are photosynthetic. They use bacteriochlorophyll *a* and plastocyanin, a blue Cu-based pigment, for light harvesting.

Heterotrophy

Animals (multicellular eukaryotes) are chemoorganoheterotrophs, since they use organic matter as a source of carbon, energy, and electrons. Most fungi, bacteria, and some protists do the same. This metabolism is usually referred to simply as *heterotrophy*. The most abundant of the heterotrophs is the proteobacterium *Pelagibacter ubique*, which was discovered in the early 1990s. This organism is very small ($<1\ \mu m$), having about half the volume of a typical marine bacterium. Nonetheless, it is so abundant and widespread that *Pelagibacter* appears to constitute about 25% of the microbial biomass in the ocean. It may be the most numerous bacterium on the planet. Like *Roseobacter*, *Pelagibacter ubique* seems to engage in photoheterotrophy using a type of rhodopsin. As noted earlier, photoassisted heterotrophy appears to be a common metabolic strategy, especially for very small microbes living in nutrient-poor (*oligotrophic*) waters.

Some prokaryotes are anaerobic heterotrophs. These include the denitrifiers, sulfate reducers, and fermenters, as well as the bacteria capable of reducing metals, such as Fe(III) to Fe(II) and Mn(IV) to Mn(II). Because the oxidized metals are present as solids, e.g., FeOOH(s), Fe_2O_3(s), and MnO_2(s), these bacteria must be in direct contact with the mineral surface and have a mechanism for transferring electrons across their cell membranes. One bacterium that appears to have such a mechanism is the facultative anaerobe *Shewanella oneidensis*, which produces a specific protein on its outer membrane only under anaerobic conditions when it is in direct contact with a suitable mineral surface. This protein is thought to effect the transfer of electrons from the mineral across the bacterium's cell wall. In other words, *Shewanella* is able to respire particulate metals!

Some heterotrophic eukaryotes are anaerobes. These include some fungi and protists. The latter were formerly called amoeboid and ciliated protozoans. They have a special organelle, called a hydrogenosome, instead of a mitochondrion. The hydrogenosome plays the role of a mitochondrion in transferring the chemical energy of organic matter into the ATP-driven electron transport chain. But in this case, the products are H_2, CO_2, and acetate, rather than H_2O and CO_2. These anaerobic protozoans have an archaean endosymbiont that uses the hydrogenosome's metabolic waste products (H_2 and CO_2) to generate methane. Microbes that generate methane are called *methanogens*.

Chemoautolithotrophy

The methanogens that use H_2 as a source of energy and electrons and CO_2 as their carbon source are chemoautolithotrophs. All methanogens are archaea and strict anaerobes, but not all methanogens are chemoautolithotrophs. Some use methanol as their energy source and CO_2 as their carbon source, making them chemoorganoautotrophs. Others disproportionate acetate into methane and CO_2, making them chemoorganoheterotrophs.

Methanogens appear to be generating methane at great depth beneath the crust (from 300 to 500 m) using H_2 of radiogenic origin. Supersaturations of methane in surface waters suggest that methanogenesis is also occurring in the anaerobic interiors of the detrital POM (particulate organic matter), which is most abundant in the euphotic zone. High methane concentrations have also been detected in the mid-depth suboxic zone characteristic of upwelling areas. All this methane is of great concern, because it is a potent greenhouse gas. Thus, the degree to which the ocean is a source of atmospheric methane is critical to climate control in the crustal-ocean-atmosphere factory. Fortunately, a variety of microbes, called *methanotrophs*, oxidize methane. These include aerobic proteobacteria and anaerobic archaeans. Anaerobic methane oxidation seems to require that the methanotrophs grow syntropically with sulfate-reducing bacteria. Some methane oxidizers, called *methylotrophs*, also utilize other single-carbon compounds, such as methanol and formate.

Other *chemoautolithotrophic* metabolisms involve the oxidation of (1) sulfur, (2) reduced nitrogen (ammonium, nitrite, and N_2O), (3) H_2, and (4) reduced metals, such as Fe(II) and Mn(II). Chemoautolithotrophy appears to be a varied and important biological adaptation exhibited by many types of archaeans and bacteria. The sulfide oxidizers, which were first discovered in the late 1970s living on H_2S emanating from hydrothermal vents, have since been observed to inhabit less exotic settings, such as salt marshes. Recent research suggests that some sulfide oxidizers also reduce Fe(III) to Fe(II). Marine nitrifiers are also widely distributed and important in transforming ammonium, which is toxic to eukaryotes at high concentrations, into nitrate.

Mixotrophy

As noted earlier, some marine protists are strictly phototrophic and some are strictly phagotrophic. Others engage in both strategies, making them mixotrophic. Examples include some phytoflagellates that are photosynthetic and ingest particulate prey, usually bacteria. Others include the dinoflagellates and ciliates. Some ciliate species are considered to be members of the microzooplankton because they consume small phytoplankton, copepod eggs, bacteria, and smaller protists. Even the true protistan phototrophs have some metabolic wrinkles to them. For example, some species of diatoms harbor endosymbiotic N_2-fixing bacteria.

Marine biogeochemists have increasingly come to realize that a continuum of metabolic strategies exists ranging from pure photoautolithotrophy to pure chemoorganoheterotrophy with varying degrees of mixotrophy between. This is illustrated in Figure 7.5. The most important consequence of this new understanding is that considerably more primary production could be occurring than previously estimated,

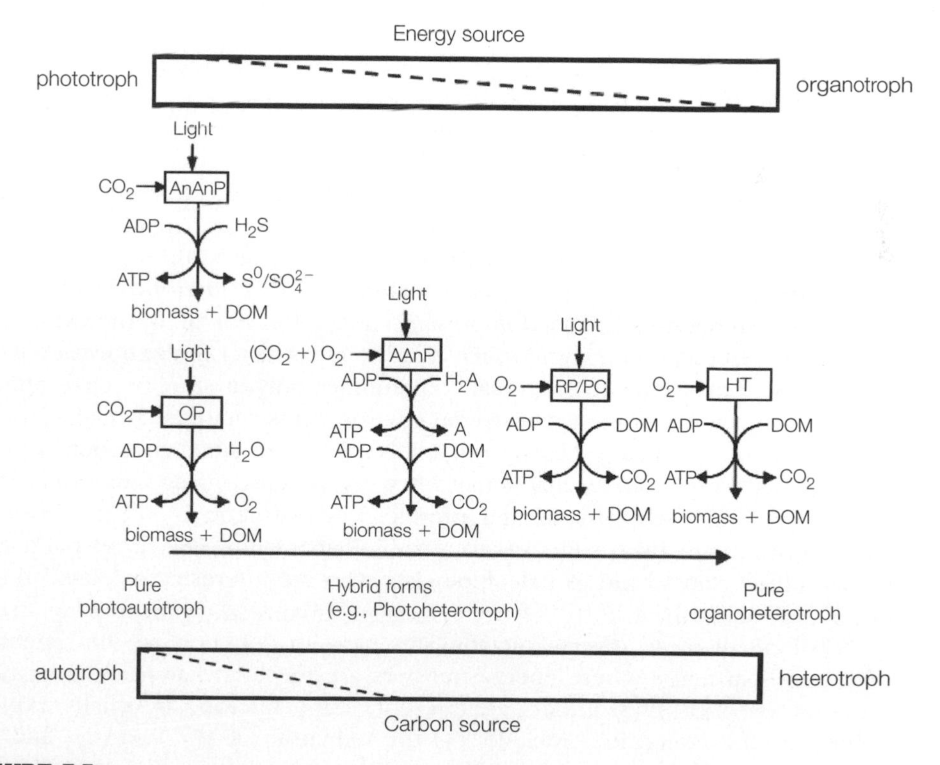

FIGURE 7.5

Continuum of metabolic strategies. AnAnP, anaerobic anoxygenic photosynthesis; OP, oxygenic photosynthesis; AAnP, aerobic anoxygenic photosynthesis; RP/PC, rhodopsin and other pigments; HT, heterotroph; DOM, dissolved organic matter. *Source*: From Eiler, A. (2006). *Applied Environmental Microbiology*, 42(12), 7431–7437.

especially if aerobic anoxygenic photosynthesis is as important as the widespread abundance of *Roseobacter* and *Pelagibacter* suggests.

Other Life Forms: Viruses, Fungi, and Yeasts

Fungi and yeasts are also members of the marine heterotrophic eukaryotes. They are generally found living on or within tissues of other organisms or on detrital POM. Fungi are important primarily in coastal water where they serve as decomposers of terrestrial vascular plant detritus. Yeasts occur as parasites of copepods.

Viruses are also present in the sea and are so abundant that they are probably the major life form in the ocean. They infect all kinds of microbes and are responsible for a significant amount (10 to 40%) of bacterial mortality primarily through cell lysis. This process has two important effects: (1) it releases the bacteria's cytoplasmic DOM (dissolved organic matter) into seawater, where it is consumed by other bacteria, thereby boosting microbial productivity, and (2) viruses acquire genetic material from their host and transmit it into the next host cell that they infect. This leads to a transmission of genetic material whose consequence is as yet unknown.

Extremophiles

Some marine microbes live under very adverse environmental conditions, such as very high or very low temperatures, high pressures, high salt concentration, very low and very high pH, and high radiation. These microbes are termed *extremophiles* and are classified into the following categories: thermophiles (temperature $>40^\circ C$), acidophiles ($pH<2$), alkalophiles ($pH>11$), halophiles (salt $>20\%$ w/v), barophiles (pressure >100 atm), and psycrophiles (temperature $<20^\circ C$). Microbes that are adapted to moderate conditions are termed *mesophiles*. Most, but not all, of the extremophiles are archaeans. Although the extremophiles tend to grow faster under normal environmental conditions, they dominate microbial communities only in extreme environments.

The survival of life in such extreme environments requires special adaptations that are largely biochemically based and involve novel organic compounds. Since these compounds are physiologically active, they are potential drug candidates and, hence, are undergoing testing by natural product chemists and biotechnology companies. Extremophiles are also of interest to astrobiologists who seek terrestrial analogues for life on other planets and to paleobiologists who are interested in elucidating how life originated on Earth.

What all of these diverse metabolisms have in common are the immutable laws of thermodynamics. Where energy transfers are thermodynamically favorable and the required chemical constituents are present, some microbe is usually exploiting this niche. Notable exceptions include: (1) the oxidation of Mn^{2+} and N_2 and (2) anoxygenic photosynthesis using NH_3, PH_3 or CH_4 as the electron donor. Also notable is the fact that no microbes seem to have developed a mechanism for harvesting energy from ocean currents or heat. Thus, all microbial energy traces back to either reduced inorganic chemicals or sunlight. Another interesting characteristic of the prokaryotes is that most have metabolisms that are highly specialized. For example, denitrifiers cannot reduce sulfate or oxidize methane. The reason for this is thought to reflect the biochemical demands of dealing with toxic waste products and intermediates. Syntrophic relationships help with the waste removal but require proximity to the microbes in the consortia. Likewise, detoxifying toxic intermediates requires deployment of specialized enzymes to speed degradation.

More novel metabolic pathways are likely to be discovered. It has been estimated that biologists have characterized only 1 to 5% of the bacteria on the planet. In 1987, the number of known bacterial phyla was 12. This had increased to 37 by 2000. The archaea were not even recognized as a separate domain until the 1970s. The relatively new field of genomics is likely to increase the pace of discovery of new microbes. Thus, we can expect to see significant changes in our understanding of metabolic strategies and community energetics.

7.3.3 Electron Transfer on the Intracellular Level

Within an organism, the energy provided by spontaneous redox reactions is used to fuel nonspontaneous metabolic processes to support growth, reproduction, and motion. Energy is transported around the cell for these various uses via redox reactions involving the half reactions listed in Table 7.11. Some are electron transport processes and others rely on transfer of phosphate groups. All of these reactions involve complex organic compounds. Virtually all organisms uses the adenosine diphosphate—adenosine triphosphate couple in which phosphate is transferred between the oxidized (ADP) and reduced (ATP) forms by a process termed phosphorylation. Other redox couples serve as electron carriers. They include nicotinamide adenine dinucleotide phosphate (NADPH/NADP), nicotinamide adenine dinucleotide (NADH/NAD), flavin adenine dinucleotide (FADH/FAD), the ferredoxins, and the cytochromes.

The roles of these electron transfer processes in photosynthesis and aerobic respiration are summarized in Figure 7.6. Photosynthetic eukaryotes have chloroplasts in which the light reactions and Calvin cycle take place. During the light reactions, solar energy, in the form of photons, is absorbed by a pigment molecule, such as chlorophyll, resulting in the production of a high energy (excited state) electron. This electron is sequentially transferred, via redox reactions, among a variety of molecules, including ferredoxin, ubiquinone, and the cytochromes, with the end result being the reduction of $NADP^+$ to NADPH and the generation of H^+. The oxidation of NADPH back to $NADP^+$ provides the energy to convert ADP to ATP.

Photosynthetic prokaryotes do not have chloroplasts. Their photosynthetic pigments are embedded in their cell walls. Some use bacteriochlorophyll for light harvesting. In the proteobacteria and archaea, light harvesting is accomplished by the protein rhodopsin, which acts as a photo-driven proton pump that fuels phosphorylation of ADP.

The metabolic machinery responsible for the heterotrophic respiration reactions is contained in specialized organelles called mitochondria. These reactions occur in three stages: (1) glycolysis, (2) the Krebs or tricarboxylic acid cycle, and (3) the process of oxidative phosphorylation also known as the electron transport chain. As illustrated in

Table 7.11 Some Cellular Energy Transfer Reactions.

Redox Half-Reactions (Reduction)	**$p\varepsilon_W^\circ$**
$NAD^+ + 2H^+ + 2e^- = NADH^+ + H^+$	−5.4
$NADP^+ + 2H^+ + 2e^- = NADPH + H^+$	−5.5
2 Ferredoxin(Fe^{3+}) + $2e^-$ = 2 ferredoxin(Fe^{2+})	−7.1
Ubiquinone + $2H^+$ + $2e^-$ = ubiquinol	+1.7
2 Cytochrome C(Fe^{3+}) + $2e^-$ = 2 cytochrome C(Fe^{2+})	+4.3
Phosphate Exchange Half-Reactions (Hydrolysis)	**ΔG_W° (kcal/mol)**
Phosphoenol pyruvate = pyruvate + P_i	−14.8
Phosphocreatinine = creatinine + P_i	−10.3
Acetylphosphate = acetate + P_i	−10.1
Adenosine triphosphate (ATP) = ADP + P_i	
37°C, pH = 7.0, excess Mg^{2+}	−7.3
25°C, pH = 7.4, 10^{-3} M Mg^{2+}	−8.8
25°C, pH = 7.4, no Mg^{2+}	−9.6
Adenosine diphosphate (ADP) = AMP + P_i	−7.3
Glucose 1-phosphate = glucose + P_i	−5.0
Glucose 6-phosphate = glucose + P_i	−3.3
Glycerol 1-phosphate = glycerol + P_i	−2.2

P_i = inorganic phosphate
Source: After Morel, F. M. M. (1983). Principles of Aquatic Chemistry. *John Wiley & Sons, p. 338.*

Figure 7.6b, the transformation of organic compounds, such as sugars, into electron energy involves the half-cell reactions shown in Table 7.11. Biologists hypothesize that both the mitochondria and chloroplasts originated as endosymbiotic prokaryotes, which through time evolved into organelles.

7.3.4 Redox Speciation

In the search for energetically favorable reactions, marine scientists often predict redox speciation over a range of environmental conditions using the thermodynamic relationships and E° data presented earlier in this chapter. The results are usually depicted as *Pourbaix diagrams*, which summarize the relative abundance of the various redox species of an element over a range of E_{cell} (or $p\varepsilon_{cell}$) and pH conditions assuming that equilibrium has been achieved. While measurement of pH is relatively straightforward, measurement of E_{cell} in natural water is complicated by fouling of electrodes,

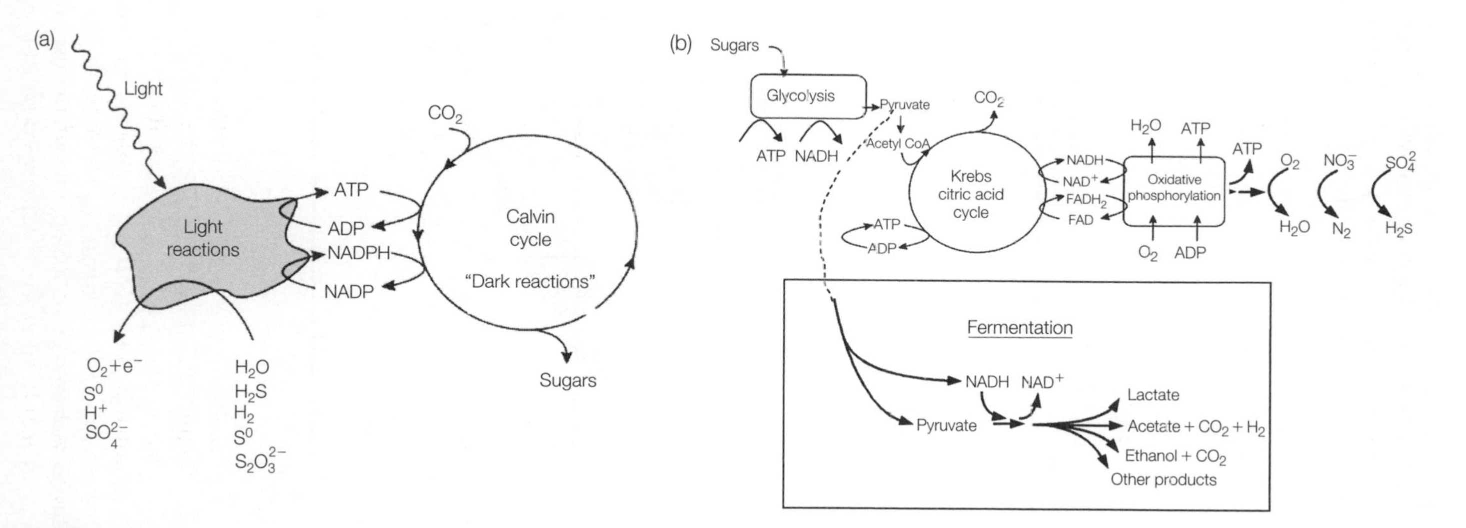

FIGURE 7.6

Intracellular redox processes. (a) The two stages of photosynthesis: light reactions and the Calvin cycle. During oxygenic photosynthesis, H_2O is used as an electron source. Organisms capable of anoxygenic photosynthesis can use a variety of other electron sources (H_2S, H_2, S°, $S_2O_3^{2-}$) during the light reactions and do not liberate free O_2. Energy in the form of ATP and reducing power in the form of NADPH are produced by the light reactions and subsequently used in the Calvin cycle to deliver electrons to CO_2 to fuel the production of sugars. (b) The three components of aerobic respiration: glycolysis, the Krebs cycle, and oxidative phosphorylation. Sugars are used to generate energy in the form of ATP during glycolysis. The product of glycolysis, pyruvate, is converted to acetyl–CoA, which enters the Krebs cycle. The Krebs cycle produces CO_2, stores energy as ATP, and stores reducing power as NADH and $FADH_2$ (the reduced form of flavin adenine dinucleotide [FAD]). O_2 is only directly consumed during oxidative phosphorylation to generate ATP as the final product of aerobic respiration. Dashed lines show the fermentative pathway. *Source*: After Petsch, S. T. (2003) *Treatise on Geochemistry*, Elsevier, pp. 515–555.

sluggish reaction kinetics, and the presence of numerous redox couples, which are poorly coupled with one another. Thus, the measured E_{cell} is a function of environmental and operational conditions, such as how long the electrode was permitted to equilibrate in the seawater.

The Pourbaix diagram for the O_2-H_2O couple is presented in Figure 7.7, along with the E_{cell}-pH conditions characteristic of various natural and polluted waters. The equations for the boundary lines are calculated as follows. The redox half reaction that defines the upper boundary is given by Eq. 7.31. Its equilibrium constant is

$$K = \frac{1}{P_{O_2}^{1/4}\{H^+\}\{e^-\}} \tag{7.42}$$

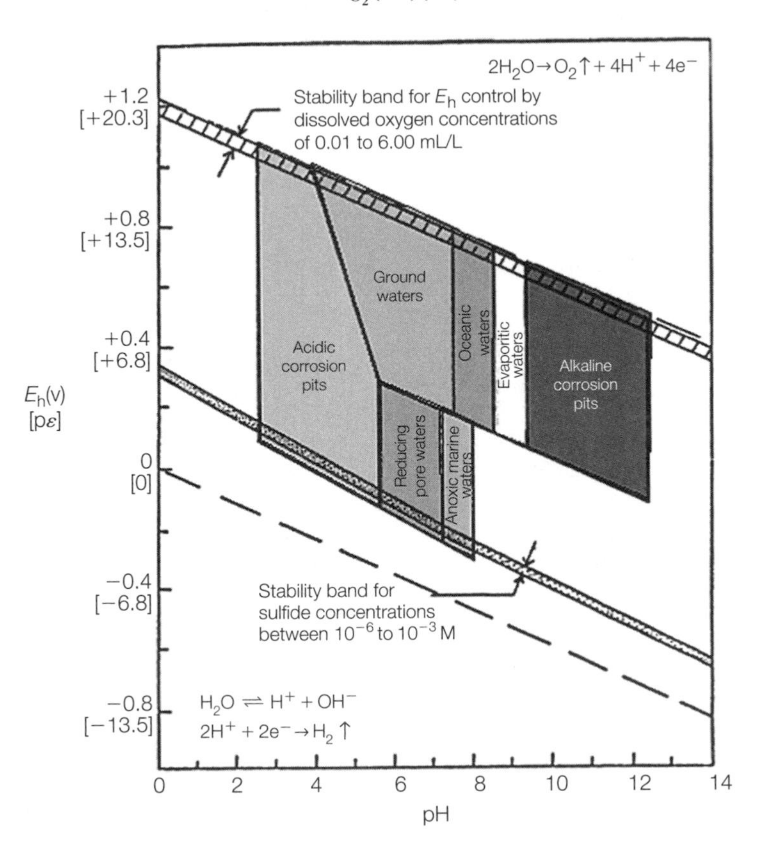

FIGURE 7.7

Pourbaix diagram for O_2-H_2O couple showing conditions characteristic of various natural and polluted waters. The upper hatched zone represents the stability band for the O_2-H_2O couple. The dashed line represents the stability boundary for the H^+/H_2 couple. The line defined by the sulfate-sulfide couple is included as description of the lowest E_{cell}-pH characteristics exhibited by natural waters. *Source*: After Kester, D. Retrieved July 2005 from http://www.gso.uri.edu/~dkester/eh/ehph2r.htm.

Inverting and then taking the log of both sides yields

$$-\log K = \log P_{O_2}^{1/4} + \log\{H^+\} + \log\{e^-\} \tag{7.43}$$

or,

$$\log K = -\frac{1}{4}\log P_{O_2} + pH + p\varepsilon \tag{7.44}$$

At equilibrium, $\log K = p\varepsilon^\circ$ for this one-electron transfer. Since $p\varepsilon^\circ$ for this one-electron reaction is +20.75, Eq. 7.44 becomes

$$20.75 + \frac{1}{4}\log P_{O_2} = pH + p\varepsilon \tag{7.45}$$

For $P_{O_2} = 1$ atm,

$$p\varepsilon = -pH + 20.75 \tag{7.46}$$

At pH 7, by definition $p\varepsilon = p\varepsilon_w^\circ$, which for this half-reaction is +13.75 or +0.81 V. The choice of $P_{O_2} = 1$ atm is arbitrary; it represents the extreme case where O_2 comprises the bulk of the atmosphere. The boundary line equation can be redefined for other partial pressures of O_2, but this causes only minor changes. For example, the lowest partial pressure at which seawater is considered to be oxic is $P_{O_2} = 0.001$ atm. At this P_{O_2}, Eq. 7.45 becomes

$$p\varepsilon = -pH + 20.75 \tag{7.47}$$

In Figure 7.7, the upper bound of the hatched region depicting the equilibrium O_2-H_2O couple reflects $P_{O_2} = 1$ atm. The lower bound reflects $P_{O_2} = 0.001$ atm. An environment in which the E_{cell} and pH lie above this hatched region is a condition under which H_2O should spontaneously decompose into O_2 and H^+.

Since water dissociates into H^+ and OH^-, the redox half reaction that defines the lower set of boundary lines is

$$H^+ + e^- \rightleftharpoons \frac{1}{2}H_2 \tag{7.48}$$

Its equilibrium constant is

$$K = \frac{P_{H_2}^{1/2}}{\{H^+\}\{e^-\}} \tag{7.49}$$

Taking the log of both sides and rearranging:

$$\log K = -\frac{1}{2}\log P_{H_2} - \log\{H^+\} - \log\{e^-\} \tag{7.50}$$

or,

$$\log K = \frac{1}{2}\log P_{H_2} + pH + p\varepsilon \tag{7.51}$$

By convention, log $K = p\varepsilon^\circ = 0$ for the H^+/H_2 couple. Substituting this value into Eq. 7.51 yields

$$-\frac{1}{2}\log P_{H_2} = pH + p\varepsilon \quad (7.52)$$

For the extreme case where $P_{H_2} = 1$ atm,

$$p\varepsilon = -pH \quad (7.53)$$

This line defines the lowest boundary above which water is stable. An environment in which the E_{cell} and pH plot below this line is a condition under which H^+ should spontaneously become reduced to H_2. The line defined by the sulfate-sulfide couple is included in Figure 7.7, because it serves as a more practical description of the lowest E_{cell}-pH characteristics exhibited by natural waters.

As shown in Figure 7.7, seawater is characterized by E_{cell} and pH values that lie within the stability field for water. Oxic seawater has a relatively high pH and E_{cell} as compared to anoxic marine water. Conditions of low pH and low E_{cell} are caused by the aerobic oxidation of large amounts of organic matter. The removal of O_2 lowers the E_{cell} of the system, whereas the concurrent production of CO_2 lowers the pH. The observed E_{cell} of oxic seawater ranges from +0.1 to +0.8 V with an average of +0.4 V. Much of this range falls below the hatched zone of the upper boundary line reflecting the operational limitations inherent in the measurement of seawater's E_{cell}.

In a similar fashion, E_{cell}-pH diagrams can be constructed for other redox half reactions. Some examples are given in Figure 7.8. These diagrams suggest that in oxic seawater (E_{cell} = +0.4 V and pH = 8), the stable form of iron is $Fe(OH)_3^0$, nitrogen is stable as nitrate, sulfur as sulfate, and carbon as bicarbonate, if each of these species reaches redox equilibrium.

In reality, the E_{cell} of seawater, like its pH, is the result of many competitive and interactive reactions, not all of which attain equilibrium. Nevertheless, the use of the equilibrium approach is not totally unreasonable as the redox reactions proceed somewhat independently. Furthermore, most achieve a type of steady state that approximates equilibrium. Redox reactions tend to proceed independently for two reasons. First, reactions tend to occur in a sequence dictated by their relative energy yields. Thus, a closed container of seawater will first undergo aerobic redox reactions, then anaerobic ones, if sufficient organic matter is present to deplete the O_2. Similarly, sediments tend to be aerobic at the surface and anaerobic at depth. This separation is also enhanced by the poisoning effect some redox chemicals have on the enzymes of competing processes. Without the catalytic effect of enzymes, many redox reactions proceed at very slow rates. In syntrophic microbial assemblages, one group consumes the waste products of the other, thereby overcoming this poisoning effect and permitting several redox processes to co-occur.

Redox processes also tend to be separated in time and space due to the relative sluggishness of solute transport. For example, molecular diffusion is the major mechanism by which solutes can be transported through the pore waters of sediments. In many cases this process is slower than the chemical reaction rates and, thus, prevents

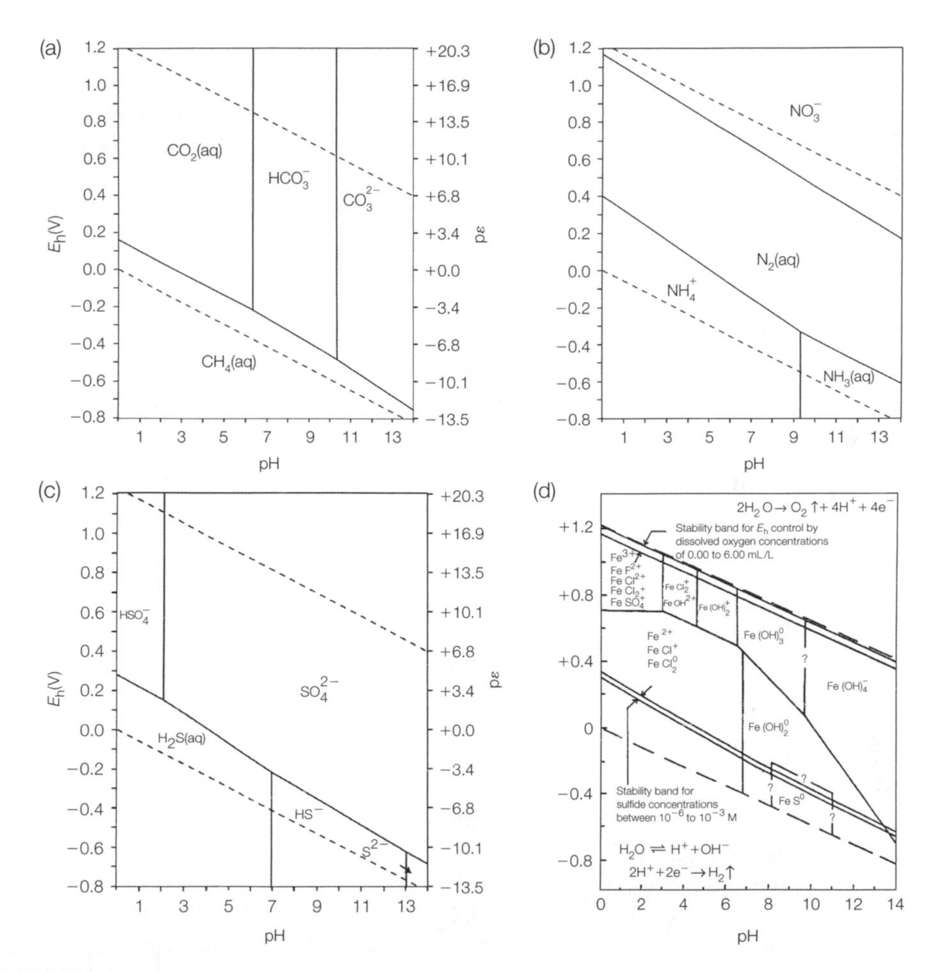

FIGURE 7.8

Pourbaix diagrams for (a) carbon, (b) nitrogen, (c) sulfur, and (d) iron. In (a)–(c), total concentration of each elements is 10^{-10} mol/kg. Such a low concentration makes the activity coefficient = 1 and impedes precipitation of the solid phases that hide the area of dominant aqueous species in the E_{cell}-pH diagrams. The upper and lower stability boundaries for water are shown as dashed lines. Computed for 25°C and 1 atm. *Source*: After Takeno, N. (2005). Atlas of E_{cell}-pH diagrams: Intercomparison of thermodynamic databases. Geological Survey of Japan Open File Report No. 419. (d) Redox speciation of iron in seawater. *Source*: After Kester, D. Retrieved July 2005 from http://www.gso.uri.edu/~ dkester/eh/ehph2r.htm.

the attainment of redox equilibrium. Likewise, gas exchange across the air-sea interface can be inhibited by the presence of dense algal mats. In this case, the underlying water tends to go anaerobic. As shown in Figure 7.8b, nitrate should be the dominant species at the $p\varepsilon_{cell}$ and pH of seawater. The large amounts of N_2 that are actually present in seawater suggest that redox equilibrium is not attained. This is likely due to the kinetic problems associated with breaking the triple bond in N_2 during nitrogen fixation. In comparison, sulfate dominates at the E_{cell} and pH of aerobic seawater, as predicted in Figure 7.8c, whereas sulfide is favored under the reducing conditions commonly caused by the decomposition of large amounts of organic matter. As shown in Figure 7.8a, all forms of dissolved organic matter are thermodynamically unstable in aerobic seawater. The presence of substantial amounts of DOM demonstrates that the speciation of carbon is not regulated by thermodynamic equilibrium. This is due in part of the low reactivity of the high molecular weight compounds in DOM.

7.3.5 The Global Biogeochemical Redox Cycle

The transfer of redox energy can be represented as a global biogeochemical cycle as shown in Figure 7.9. This cycle is largely driven by solar energy. Through the process of photosynthesis, solar energy is converted into a variety of thermodynamically unstable chemical species. This is initiated by the transfer of electrons from water to carbon dioxide, creating electron-rich organic matter and electron-poor O_2. This organic matter is thermodynamically unstable in the presence of O_2 thereby serving as a reductant for heterotrophs. (A small amount of inorganic carbon is fixed into organic form by chemoautolithotrophs whose energy is derived by reductants emitted from Earth's interior, e.g., H_2, H_2S, S°, and CH_4. Some of these reductants are primordial and some are the product of ongoing nuclear reactions. Collectively, chemoautolithotrophy contributes less than 1% of the total annual primary production in the sea and, thus, has not been

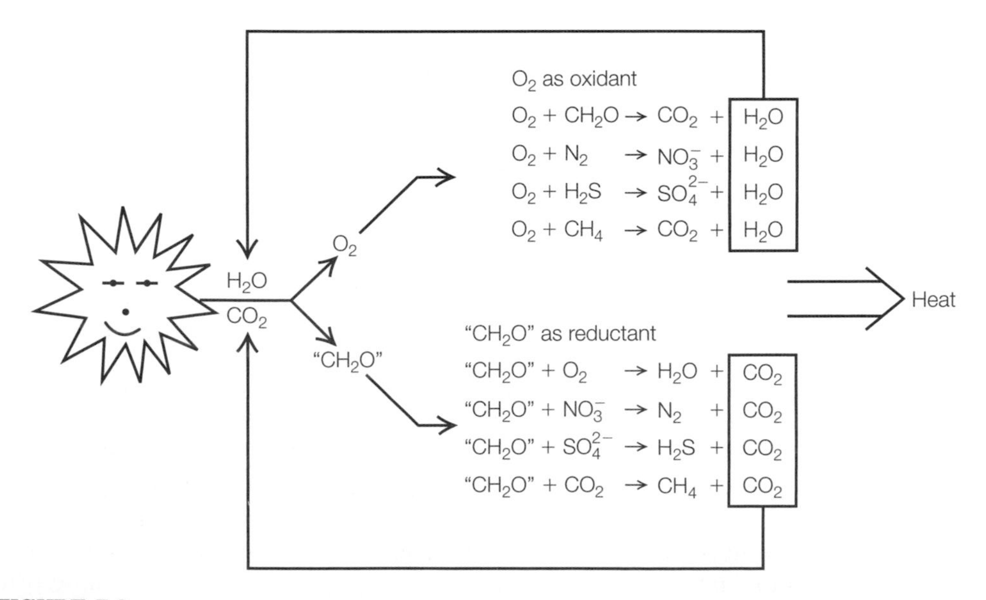

FIGURE 7.9

Global geochemical cycle of solar redox energy. Note that no reaction stoichiometry is shown in this schematic depiction.

included in Figure 7.9. At present, the eruption of new mantle material is not thought to represent a significant electron sink. This topic is considered further in the next chapter as part of a description of the global biogeochemical cycle of O_2.)

By catalyzing the oxidation of organic matter, the redox reactions conducted by the heterotrophs serve to restore the reduced atoms in the organic compounds back to their oxidized forms. During these reactions, electrons supplied by the organic matter cause O_2, NO_3^-, SO_4^{2-}, and CO_2 to be reduced to H_2O, N_2, H_2S, and CH_4, respectively. The oxidation of organic matter regenerates the CO_2 required for photosynthesis.

The reduced compounds produced from the oxidation of organic matter are thermodynamically unstable in the presence of O_2. Their ensuing oxidation by O_2 regenerates the oxidized species (NO_3^-, SO_4^{2-}, and CO_2) needed to reinitiate the cycle. Water is also regenerated and, hence, is made available to participate in photosynthesis.

The solar energy that provided the original source of energy fueling this redox cascade of electrons has two possible fates: (1) it becomes buried as organic matter in the sediments or soil, or (2) it is dissipated as heat. The latter is one of the products of metabolism. If insolation and the concentrations of the redox species remain constant over time, this heat would have to be radiated from the planet to maintain a steady-state energy balance with respect to incoming solar radiation. The biological processes that participate in the global cycle of electrons also affect Earth's heat budget by serving as sources and sinks of important greenhouse gases, such as CO_2, CH_4, and N_2O.

Figure 7.9 shows only the major features of the global biogeochemical electron cycle. For example, it does not show anoxygenic photosynthesis, the role of Fe and Mn, or the chemoautolithotrophy that is occurring independent of solar energy, such as the production of methane from CO_2 and H_2 by the archaean methanogens living deep within the crust on H_2 generated during the crystallization and cooling of magma. (An example of such a reaction is given in Eq. 19.5).

7.4 PHOTOCHEMICAL REDOX REACTIONS

Photochemical reactions are ones in which photons are absorbed by an atom or molecule called a *chromophore*. The chromophore transforms the solar energy into a high energy (excited state) electron capable of conducting chemical work. Photosynthesis is an example of a photochemical reaction in which the chromophore is usually a pigment molecule. Photochemical reactions also occur directly in seawater with the most important chromophores being high-molecular-weight DOM collectively referred to as *humic materials*. Solar energy is also probably absorbed by metal oxide minerals, such as Fe_2O_3. Most of the photochemical reactions in seawater are probably powered by ultraviolet radiation, because its wavelengths ($\lambda < 450$ nm) are more energetic than those of visible light.

The high energy electrons are emitted by the chromophore and then are free to undergo reactions with molecules such as $^{\bullet}O_2$, thereby generating highly reactive and unstable species, such as the free radicals O_2^- (superoxide), $^{\bullet}HO_2$ (hydroperoxyl), $^{\bullet}OH$

(hydroxyl), and $^1O_2$ (singlet oxygen). These free radicals are stronger oxidizing agents than O_2, although some can also act as reductants as noted in the next paragraph. All are highly reactive and hence, occur at very low concentrations. Other reactive photochemical species include H_2O_2 (hydrogen peroxide), $^{\bullet}Br$ and $^{\bullet}Br_2^-$ (bromine and dibromide radicals), and excited states of DOM. Because of their high reactivity, the influence of these chemicals is restricted to the surface waters (<100 m) where their production rates are highest.

Probably the most important photochemical reactions in seawater are the ones that involve humic materials. The humic substances carried by rivers into the ocean seem especially photochemically labile. These compounds are colored, usually brown or yellow, and their photochemical reactions cause them to become oxidized or "bleached." The oxidation reactions degrade the molecular structure and in so doing create smaller molecules such as CO and CO_2 and lower weight molecular DOM. In some situations, inorganic nitrogen (ammonium) and phosphate appear to be released. Photochemical reactions also seem to enhance the bioavailability of the remaining DOM and associated trace metals. As noted in Chapter 5, the uptake rate of iron by marine plankton appears to be increased by the action of sunlight on Fe(III) complexes, partly because some of the iron is reduced and partly because of changes in the structure of the organic-complexing compound. Humic materials also seem to promote the photoreduction of iron, copper, and manganese. O_2^- ,$^{\bullet}OH$, and H_2O_2 can also variously cause the oxidation or reduction of these metals, e.g.,

$$Fe^{3+} + O_2^- \longrightarrow Fe^{2+} + O_2 \quad (7.54)$$

$$Fe^{2+} + O_2^- \longrightarrow Fe^{3+} + 2H^+ + H_2O_2 \quad (7.55)$$ [5]

$$Fe^{2+} + H_2O_2 \longrightarrow Fe^{3+} + {}^{\bullet}OH + OH^- \quad (7.56)$$

$$Fe^{2+} + {}^{\bullet}OH \longrightarrow Fe^{3+} + OH^- \quad (7.57)$$

Other photochemically sensitive elements include chromium, mercury, and the halides: bromine, chlorine, and iodine. Photocatalytic halogenation of organic compounds appears to naturally produce organohalogens that were thought (until recently) to have only a human origin. Photochemical reactions also appear to be important in regulating the concentrations of volatile sulfur compounds, such as dimethyl sulfide, carbonyl sulfide, and carbon disulfide. These gases play an important role in regulating climate because they act as cloud condensation nuclei once they diffuse across the air-sea interface to become part of the atmosphere. Because photochemical reactions absorb solar radiation, they have a role in controlling the depth of the euphotic zone.

[5] In this reaction, Fe^{2+} is present as an organometallic complex that donates the hydrogen found in the products.

CHAPTER

Organic Matter: Production and Destruction

8

All figures are available on the companion website in color (if applicable).

8.1 INTRODUCTION

In the previous chapter, organic matter was identified as the most important electron donor in the marine environment. As such, it provides the energy needed to drive most of the biologically mediated redox reactions. Marine chemists use the term *organic matter* to refer collectively to any and all organic compounds. The compounds that constitute organic matter are diverse in structure, molecular weight, and elemental composition. Some organic compounds, called *biomolecules*, are created directly by organisms. Others, such as humic materials, are thought to be amalgamations of fragments of biomolecules produced abiotically, such as through photochemical reactions. The most abundant elements in these organic compounds are carbon, hydrogen, nitrogen, oxygen, phosphorus, and sulfur. Small amounts of trace metals are also present.

Compared to carbon, hydrogen, oxygen, and sulfur, the oceanic reservoirs of biologically available nitrogen and phosphorus are small. The relative scarcity of these nutrients causes them to limit biological production. For this reason, nitrogen and phosphorus are classified as *biolimiting elements*. An important consequence of this nutrient limitation is that biological processes govern the seawater concentrations of these nutrients. Most of these processes involve redox reactions associated with the biogeochemical cycling of organic matter. Although this cycling involves the other elements present in high abundance in organic matter, namely carbon, hydrogen, oxygen and sulfur, only in the case of carbon are seawater concentrations influenced. In the case of hydrogen and oxygen, the major elemental reservoir is water and whose size is not affected by chemical processes. Most of the sulfur in seawater is present as sulfate, which is one of the six major ions. Given the large size of this reservoir, biological impacts to the sulfate concentration of seawater are limited to regions in which sulfate reduction occurs. This requires anoxic conditions, which are rarely found in the water column of the present-day ocean.

In this chapter, the biogeochemical cycling of organic matter is discussed from the perspective of its carbon, hydrogen, nitrogen, oxygen, phosphorus, and sulfur content.

To do this, a model compound is defined to represent an average molecule of organic matter. This compound is then used in chemical equations to illustrate how metabolic processes act on organic matter and its constituent elements.

8.2 THE PRODUCTION OF ORGANIC MATTER

Most of the organic matter in the ocean is produced in situ by phytoplankton. Other sources are relatively minor. They include inputs from rivers, the atmosphere, hydrothermal emissions, and extraterrestrial sources, such as meteorite impacts. Chemoautolithotrophic production is also thought to be small compared to that of phototrophy. Photosynthetic organisms require visible light. Solar radiation is the source of this light. It does not penetrate far beneath the sea surface because of absorption and reflection by water molecules, dissolved organic matter, and particulate matter. Thus, the photosynthesizers are limited to a relatively shallow surface layer called the *euphotic zone*. The bottom of the euphotic zone is defined as the depth at which 1% of the insolation remains unabsorbed. The greater the insolation and the lower the *turbidity* of the water, the deeper the *euphotic* zone.[1] The deepest euphotic zone is found in the open ocean at low latitudes, but even here, very little light penetrates deeper than 200 m below the sea surface.

Although phototrophs are restricted to the surface ocean, photosynthesis is the ultimate source of almost all the organic matter that supports heterotrophic activity in the sea. As shown in Figure 8.1, this organic matter is transferred through the marine food web by the feeding activities of heterotrophs including bacteria, protozoans, and animals.

Some of this organic matter is transferred up the food chain in the form of POM when phytoplankton are consumed by herbivorous zooplankton, which are in turn eaten by higher-order consumers. When these organisms die, their tissues degrade, first through lysis of cell membranes, causing DOM to be released into seawater. DOM is also supplied to seawater from the breakdown of other forms of detrital POM such as: (1) fecal pellets; (2) mucous feeding nets; (3) structural materials, such as chitin that makes the exoskeletons of crustacea; (4) dead tissues; and (5) amorphous fragments termed *marine snow*. Some of the DOM is consumed by bacteria and other protists. These organisms are consumed by other members of the plankton, such as bacterivorous protists, which are in turn eaten by zooplankton. As shown in Figure 8.1, not all of the organic matter is retained in euphotic zone. Some of the POM sinks out of the euphotic zone and some of the DOM is transported into the deep zone by physical water motion. Sinking POM is the main source of food to the microbial heterotrophs of the deep sea and sediments. Only around hydrothermal vents, cold seeps, and deeply

[1] High concentrations of particulate matter cause water to appear cloudy due to the absorption and scattering of light. The cloudiness of water is operationally measured as the amount of light transmitted through water. This quantity is termed *turbidity* and is reported in nephelometric turbidity units (NTU).

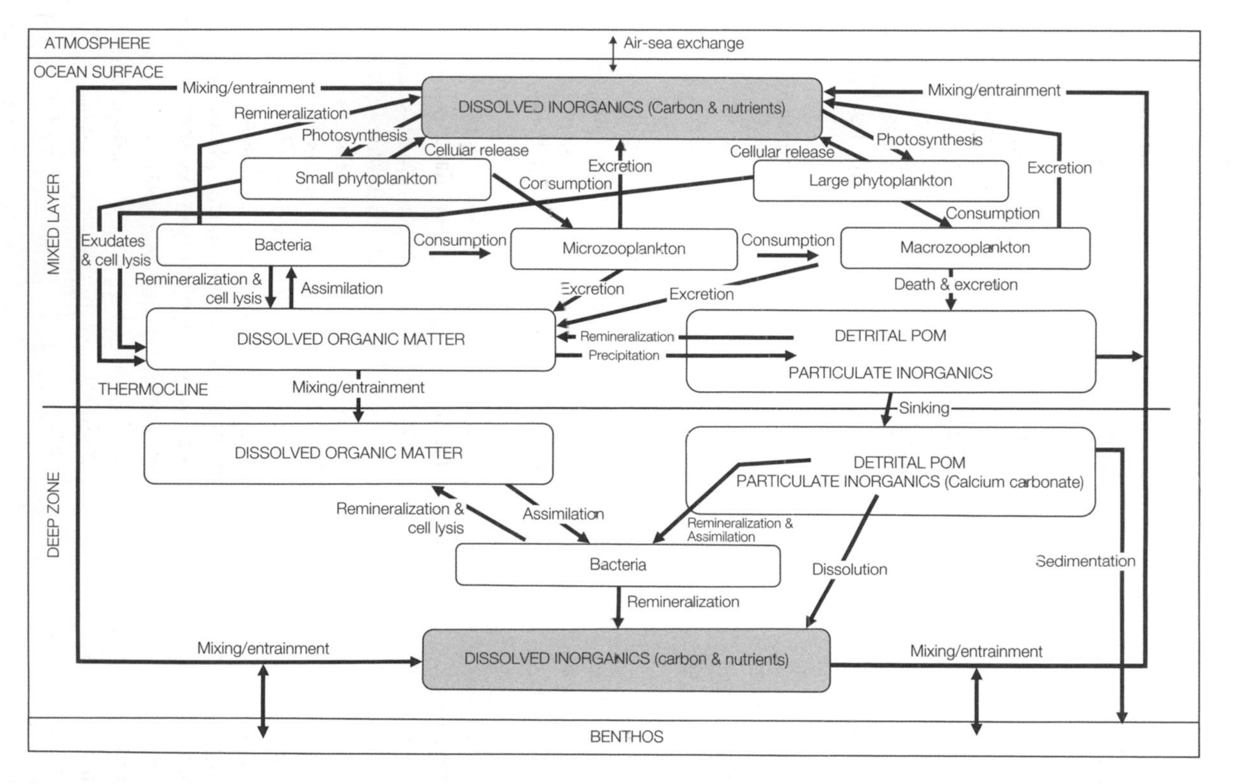

FIGURE 8.1

The biogeochemical cycling of POM in the oceans. All of the trophic levels contribute to the POM pool. For the sake of clarity, only the macrozooplankton source is connected to this pool via an arrow. Also shown are the pools of: (1) dissolved organic matter (DOM), (2) dissolved inorganics (carbon and nutrients), and (3) particulate inorganic matter (calcium carbonate).

buried sediments is chemoautotrophy an important source of POM. Therefore the chemical composition of POM sinking out of the euphotic zone is of critical importance as it determines the nutritive quality of the primary food source to the deep-sea biota.

From a geochemical perspective, sinking POM is an important mechanism by which carbon and other elements are transferred from the sea surface into the deep sea and onto the sediments. This transport is termed the *biological pump* and includes the sinking of inorganic particles that are of biogenic origin, namely calcium carbonate and silicate shells.

The elemental composition of marine plankton, as collected by net tows, is remarkably constant, with an average atomic ratio of C to N to P of 106 to 16 to 1 through all the world's oceans. This is called the *Redfield-Richards ratio.* It is named after the scientists who formulated this relationship in the 1960s from analysis of plankton collected in net tows. Using this elemental ratio, Redfield and Richards developed a model empirical formula ($C_{106}(H_2O)_{106}(NH_3)_{16}PO_4$) to represent average planktonic organic matter. The oxygen and hydrogen composition is based on an assumption that the organic matter is composed of a hypothetical biomolecule, a "phosphorylated amino-carbohydrate," formed as a consequence of photosynthesis via the following stoichiometry:

$$106\ CO_2 + 122\ H_2O + 16\ HNO_3 + H_3PO_4 \rightarrow (CH_2O)_{106}(NH_3)_{16}H_3PO_4 + 138\ O_2 \qquad (8.1)$$

This reaction illustrates that, in addition to carbon dioxide and water, phytoplankton also require dissolved inorganic nitrogen and phosphorus. These elements are usually present in relatively low concentrations in seawater and, thus, are often the limiting reactants in photosynthesis. (We'll consider the biolimiting effects of trace elements in Chapter 11.) Any chemicals that when added to seawater stimulate plant growth are termed "biolimiting" elements or *nutrients*. Nitrogen and phosphorus are also called *macronutrients* because they are needed in much higher amounts than *micronutrients*. Examples of the latter include iron, zinc, molybdenum, manganese, and magnesium.

Equation 8.1 illustrates only the stoichiometry of the process of photosynthesis. The actual chemical species that are taken up, or *assimilated*, by the phytoplankton are somewhat variable, depending the biological species, their physiological state, and environmental conditions. The process of assimilation occurs via passage of ions through the cell membrane. Inorganic carbon is taken up as either CO_2 or bicarbonate (HCO_3^-). Some phytoplankton secrete extracellular enzymes that convert bicarbonate to $CO_2(aq)$, which is then transported across the cell membrane. Nitrogen can be assimilated as nitrate (NO_3^-), nitrite (NO_2^-), ammonium (NH_4^+), and as an organic species, urea ($(NH_2)_2CO$). Phosphorus is probably assimilated as orthophosphate (PO_4^{3-}) and sulfur as sulfate (SO_4^{2-}). Sulfur is not included in the Redfield-Richards plankton model compound because the production and consumption of organic matter has little impact on sulfate ion concentrations.

The oxygen yield of Eq. 8.1, 138 mol O_2, was developed from a two-step model in which 106 mol O_2 are considered to be generated from the production of 1 mol of glucose (the base unit of the model carbohydrate) via

$$106\ CO_2 + 106\ H_2O \rightarrow (CH_2O)_{106} + 106\ O_2 \quad (8.2)$$

In the second step, 32 mol O_2 are generated from the assimilatory reduction of nitrate, in the form of HNO_3, to amine nitrogen, represented in the organic molecule as $(NH_3)_{16}$,

$$16\ HNO_3 + 16\ H_2O \rightarrow (NH_3)_{16} + 32\ O_2 \quad (8.3)$$

The biomolecules that constitute planktonic organic matter also include protein, lipids, DNA, and RNA. These contain much less oxygen and hydrogen than carbohydrates. Thus, the oxygen and hydrogen composition of the model Redfield-Richards organic matter is somewhat higher than observed in real plankton. Various alternative elemental ratios have been proposed to more accurately reflect the average composition of marine plankton. In this text, we use the Redfield-Richards ratio, as the proposed alternatives are still under debate and are not significantly different from C:N:P::106:16:1.

8.3 THE AEROBIC DESTRUCTION OF ORGANIC MATTER

Heterotrophic microbes consume organic matter to fuel respiration metabolisms that provide energy. Aerobic respiration of the Redfield-Richards planktonic organic matter can be represented stoichiometrically as

$$(CH_2O)_{106}(NH_3)_{16}H_3PO_4 + 138\ O_2 \rightarrow 106\ CO_2 + 122\ H_2O + 16\ HNO_3 + H_3PO_4 \quad (8.4)$$

during which nitrogen, phosphorus, and carbon are converted into inorganic species. The products of this reaction can be assimilated by phytoplankton. Anaerobic respiration metabolisms also transform nutrients back into inorganic form. These transformations are collectively termed *nutrient regeneration* or *remineralization*.

Microbes can only consume DOM. Most of this DOM comes from planktonic exudations or from the degradation of detrital POM. Since the organic compounds in detrital POM are out of redox equilibrium in oxic seawater, they are very reactive and quickly undergo degradation reactions that reduce them in size. Bacteria are known to excrete enzymes, such as proteases, which facilitate the degradation of organic matter. Degradation of detrital POM is also facilitated by viruses, which lyse cell membranes. Once the particles are small enough, these organic compounds become part of the DOM pool and can be assimilated by microbes, including bacteria, archaea, and some protistan heterotrophs. Because this organic matter is the sole source of chemical energy for heterotrophs, it is rapidly consumed. This happens throughout the entire water column, so little of the sinking detrital POM survives to settle onto the seafloor. What survives represents less than 1% of the primary production.

Even as part of the sediments, POM is subject to respiration. The effects of the benthos are greatest at the sediment-water interface, but remineralization due to respiration is still a significant sink of POM through the top 10 cm of sediment. Thus, the organic carbon content of marine sediments is usually less than 1% by mass, except in coastal areas, where values as high as 10% can occur. In Chapter 23, we'll take a closer look at the fate of DOM. As with POM, a small fraction of the DOM escapes degradation, probably because it is of relatively poor nutritional quality, so it remains dissolved in seawater as a rather inert pool of organic matter.

Equation 8.4 illustrates only the stoichiometry of aerobic respiration. The actual metabolic processes usually occur as a series of coupled reactions. For example, during the initial stages of remineralization, nitrogen is released from organic matter in the form of ammonium. If O_2 is present, the ammonium is rapidly oxidized by certain species of bacteria to nitrite, and by others to nitrate. This is a chemoautolithotrophic process called *nitrification*. While it is important in the nitrogen cycle, the amount of carbon fixed is less than 1% of that fixed by phytoplankton. In the euphotic zone, phytoplankton are such effective competitors for ammonium that a significant fraction of primary production is supported by this regenerated nitrogen. Below the euphotic zone, most of the ammonium is oxidized to nitrate by nitrifying bacteria. Thus, nitrogen carried back to the surface waters via vertical eddy diffusion and the return flow of meridional overturning circulation is primarily in the form of nitrate.

8.4 IMPACTS ON O_2 AND NUTRIENT CONCENTRATIONS

Since respiration consumes O_2, subsurface water masses at mid- and low latitudes are generally undersaturated with respect to this gas. As illustrated in Figure 4.21, the largest undersaturations occur within the thermocline. This oxygen minimum zone (OMZ) is a consequence of several physical and biogeochemical phenomena. First, the strong density stratification in the thermocline inhibits vertical mixing. This isolates the thermocline from the mixed layer and, hence, the atmospheric and photosynthetic pools of O_2. Molecular diffusion is too slow to transport a significant amount of O_2 into the thermocline or deep zone. Second, the flux of sinking POM declines with depth because particles are continuously decomposing as they settle. Thus, the supply of oxygen-demanding substances decreases with depth. Third, water circulation is less effective at transporting O_2 into the intermediate waters as compared to the deep waters. Recall from Chapter 4 that water masses form via sinking from the sea surface. The intermediate waters are warmer than the deep waters and, hence, have a lower NAEC for O_2.

8.4.1 Apparent Oxygen Utilization

Since detrital POM is continuously settling out of the surface waters of all the world's ocean, water masses moving laterally through the ocean basins are continuously receiving a rain of detrital POM. Aerobic respiration of this detrital POM causes the O_2 concentration in a water mass to decrease as it travels through the deep sea. The amount of O_2 consumed since a water mass was last at the sea surface can be calculated if its original gas concentration is known. Assuming that the water mass was in gaseous equilibrium with the atmosphere at the time it sank from the sea surface, the original concentration is equal to the NAEC.[2] The difference between the in situ O_2 concentration and the NAEC is called the *AOU*, or *apparent oxygen utilization*, of the water mass. This is defined mathematically as

$$AOU = NAEC - [O_2]_{in\ situ} \tag{8.5}$$

AOU is usually expressed in units of μmol O_2/kg SW. This called an "apparent" utilization because it is a measure of the net difference between the AOU and NAEC. The latter is inferred from the in situ temperature and salinity. Thus, the AOU calculation assumes that (1) the water mass had attained gaseous equilibrium with the atmosphere, (2) the partial pressure of O_2 in the atmosphere has not changed since the water mass was last at the sea surface, and (3) neither have the in situ temperature and salinity. As described in Chapter 6, this last assumption can be violated by post-equilibration heating, such as via conduction from a hydrothermal system. This leads to an underestimate of the true NAEC, and, hence, an underestimate of the AOU.

The higher the AOU, the greater the amount of O_2 removed since the water mass was last at the sea surface. Thus, AOU increases with increasing distance from the site at which the subsurface water mass was formed. Since the AOU increases with the age of the water mass, the pathway of deep-water circulation can be traced from the distribution of AOU in the deep zone. As shown in Figure 8.2, the AOU at 4000 m is lowest in polar regions, indicating these areas are the sites of deep-water formation.

An estimate of the amount of organic matter respired since a water mass was last at the sea surface can be inferred from its AOU and the stoichiometry given in Eq. 8.4. The respiration of 1 mol of POM comprised of average detrital plankton biomass requires the oxidation of 106 mol organic carbon. As per Eq. 8.6, this requires 106 mol O_2:

$$106\ CH_2O + 106\ O_2 \rightarrow 106\ CO_2 + 106\ H_2O \tag{8.6}$$

Similarly, 16 mol of organic nitrogen must be oxidized to nitrate. As shown in Eq. 8.7, this requires 32 mol O_2:

$$16\ NH_3 + 32\ O_2 \rightarrow 16\ HNO_3 + 16\ H_2O \tag{8.7}$$

Since phosphorus is not oxidized during the respiration of organic matter, it does not contribute to the O_2 uptake. Thus, 138 mol O_2 is consumed during the respiration of 1 mol of average plankton detritus, making the molar ratio of organic carbon respired to O_2 consumed equal to 106:138.

The concentration of POM remineralized since a deep-water mass was last at the sea surface can be calculated using this ratio. For example, the highest AOU shown in Figure 8.2 occurs in the North Pacific. The amount of organic carbon that must be

[2] In reality, the O_2 saturation level of surface seawater is geographically and temporally variable, ranging from 60 to 110% with a mean value of 103% (Figure 6.2).

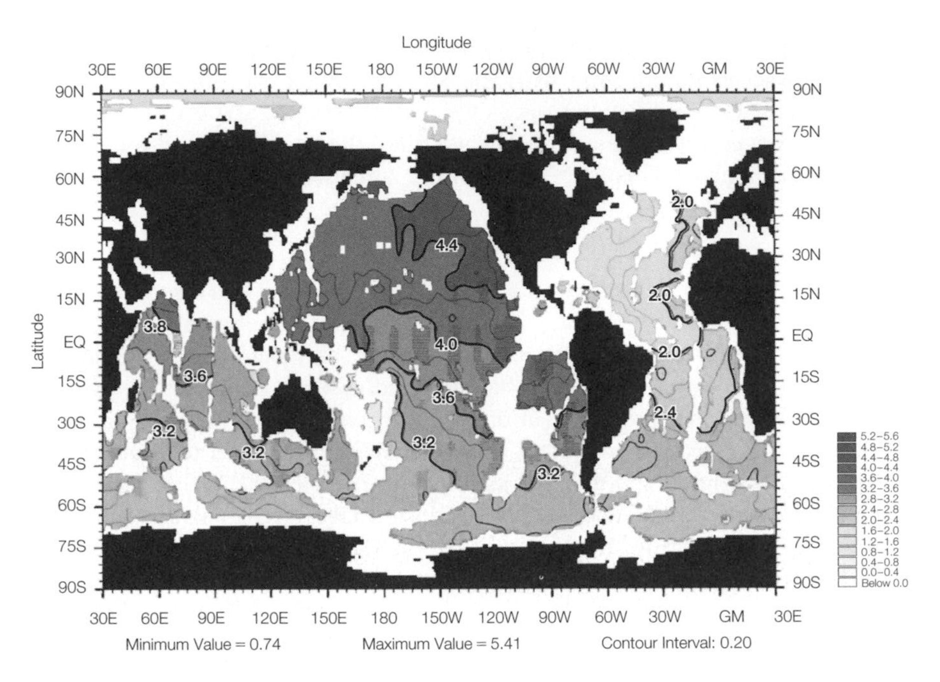

FIGURE 8.2

Distribution of AOU at a depth of 4000 m. Units are mL O_2/L seawater. To convert to μmol O_2/L seawater, multiply by 44.6. *Source*: After Conkright, M. E., *et al.* (2002). *World Ocean Atlas 2001, Volume 4: Nutrients*. NOAA Atlas NESDIS 52, U.S. Government Printing Office. (See companion website for color version.)

oxidized to cause this AOU is

$$190\ \frac{\mu\text{mol O}_2}{\text{kg SW}} \times \frac{106\,\text{mol C}}{138\,\text{mol O}_2} = 146\ \frac{\mu\text{mol C}}{\text{kg SW}} \tag{8.8}$$

In other words, this process causes the inorganic carbon content of the seawater to increase by 146 (μmol C)/(kg SW). Thus, 146 (μmol C)/(kg SW) of organic carbon has been remineralized. Using the Redfield-Richards stoichiometry, the nitrate concentration is predicted to rise by

$$190\ \frac{\mu\text{mol O}_2}{\text{kg SW}} \times \frac{16\,\text{mol N}}{138\,\text{mol O}_2} = 22\ \frac{\mu\text{mol N}}{\text{kg SW}} \tag{8.9}$$

and the phosphate concentration by

$$190\ \frac{\mu\text{mol O}_2}{\text{kg SW}} \times \frac{1\ \text{mol P}}{138\ \text{mol O}_2} = 1.4\ \frac{\mu\text{mol P}}{\text{kg SW}} \tag{8.10}$$

Deviations from the stoichiometry given in Eq. 8.4 have been observed. They are the result of several phenomena, such as variability in the elemental composition of plankton. Furthermore, not all of the POM that is respired is plankton. Other types of

POM, such as fecal pellets, have higher C-to-N ratios than plankton. Equation 8.4 also assumes that all of the organic carbon and nitrogen are completely oxidized to their highest oxidation states, which is not always the case. Finally, detrital POM tends to undergo a sequential degradation in which nitrogen is remineralized faster than carbon, so the C-to-N ratio of organic matter increases as it decomposes. Nitrogen can also be lost from the water column by denitrification.

8.4.2 Why Is the N-to-P Ratio Equal to 16?

Equation 8.4 predicts that aerobic respiration should release dissolved inorganic nitrogen and phosphorus into seawater in the same ratio that is present in plankton, i.e., 16:1. As shown in Figure 8.3, a plot of nitrate versus phosphate for seawater taken from all depths through all the ocean basins has a slope close to 16:1. Why do both plankton and seawater have an N-to-P ratio of 16:1? Does the ratio in seawater determine the ratio in the plankton or vice versa? Current thinking is that the N-to-P ratio of seawater reflects a quasi steady state that has been established and stabilized by the collective impacts of several biological processes controlled by marine plankton.

One important determinant in setting the Redfield-Richards ratio is the species composition of the plankton. For example, green algae on average tend to have much higher N-to-P ratios (27) than the red algae (10). Even within a species, these ratios are somewhat variable as they depend on nutrient availability and physiological

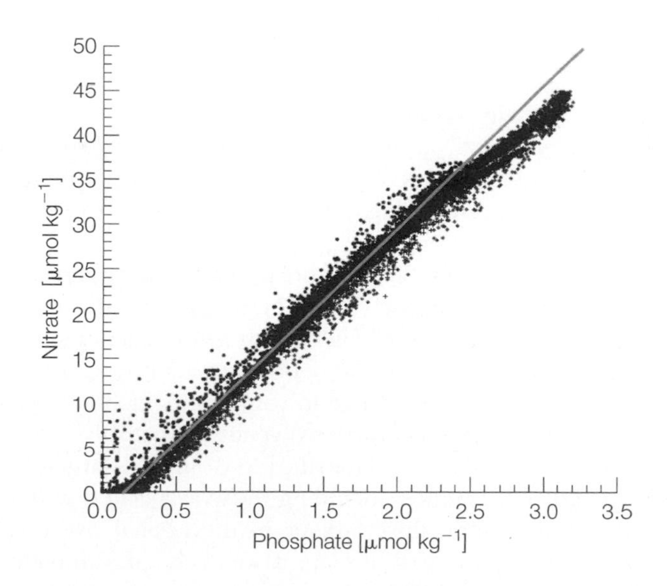

FIGURE 8.3

Plot of nitrate versus phosphate from all depths and from selected WOCE cruises in all ocean basins. Straight line represents the mean oceanic trend in the data, which has a slope of about 16:1. *Source*: From Gruber, N. (2004). *Carbon-Climate Interactions*, NATO ASI Series, p. 102.

state. For phosphorus, nutrient availability is controlled by river inputs, whereas the ocean acquires nitrogen from atmospheric deposition and via the microbially mediated processes of N_2 fixation. Most marine nitrogen fixers are phototrophs.

Nitrogen also differs from phosphorus in that it has an additional oceanic loss mechanism, which is conversion to $N_2(g)$ via denitrification. Water-column denitrification occurs only under conditions of low O_2, such as in the OMZ. It causes the average N-to-P ratio of deep water to be somewhat lower than 16:1, e.g., 14.7:1. (This deviation from the Redfield-Richards ratio is seen in the high concentration end of the curve in Figure 8.3.) When the return flow of meridional overturning circulation transports this water back to the euphotic zone, the missing nitrogen can be resupplied by phototrophic nitrogen fixation. Although denitrification and N_2 fixation occur at different times and places in the ocean, the two processes seem to be coupled over time scales that enable them to stabilize the N-to-P ratio of seawater.

Since nitrogen fixation can supply new assimilable nitrogen to the ocean, phosphorus appears to be the ultimate element limiting plankton growth. On the other hand, N_2 fixation has a large iron requirement because the necessary enzyme, nitrogenase and its cofactors, requires seven atoms of Fe per enzyme-cofactor unit. This suggests that the atmospheric dust transport of iron (as described in Chapter 5.7.1) could be an equally important control on plankton growth and, hence, the Redfield-Richards ratio. (This hypothesis is complicated by the tendency of phosphate to adsorb onto the surfaces of particulate iron oxides and oxyhydroxides. So colimitation by nitrogen, phosphorus, and iron is more likely.)

As illustrated in Figure 8.3, considerable scatter exists in the N-to-P ratio of nitrate and phosphate. From the preceding discussion, this scatter is understood to reflect differences in plankton species composition and physiological state, both of which are temporally and spatially variable. Thus, the convergence of the N-to-P ratio at a value of 16:1 reflects an overall balance in the world ocean between plankton production of POM, denitrification, and N_2 fixation. Some oceanographers view the nearly zero x and y intercepts in Figure 8.3 as further evidence for the tight coupling of N and P biogeochemical cycling.

The residence times of nitrogen and phosphorus in the ocean are 1500 to 5000 y and 10,000 to 20,000 y, respectively. The mixing of the ocean is then fast enough (1000 y) to homogenize any local and short-term excursions from the Redfield-Richards ratio. Can the Redfield-Richards ratio be permanently changed? Yes, but only if the essential biological processes change in some fundamental way. One such example is a large-scale change in plankton species diversity, especially if this involves the microbes responsible for N_2 fixation or denitrification. Such a large-scale change could result from environmental alterations, including those associated with climate change, such as a shift in the atmospheric dust flux or in meridional overturning circulation. These kinds of environmental alterations are also likely to stimulate evolutionary shifts in plankton composition. Such shifts have the potential to act as feedbacks in the crustal-ocean-atmosphere factory. For example, changes in plankton production could affect the ocean's role in the global biogeochemical cycle of carbon, leading to changes in atmospheric CO_2 levels and, hence, global climate.

8.5 THE ANAEROBIC DESTRUCTION OF ORGANIC MATTER

If the rate of O_2 removal at a particular location exceeds its rate of supply, O_2 concentrations will be undersaturated. The degree of undersaturation can be extreme, resulting in suboxic and even anoxic conditions. Such oxygen-deficient zones occur in areas where the overlying primary productivity is high and the rate of deepwater motion is relatively slow. The former ensures a large flux of organic matter and the latter a slow supply of O_2. As shown in Figure 8.4, these conditions are found in coastal upwelling areas, some marginal seas, and in the eastern tropical Pacific Ocean. About 2% of the continental margin seafloor lies below suboxic or anoxic waters.

In most of these regions, oxygen deficiency has been caused by high primary production that supplies a large rain of POM into the subsurface waters. Aerobic respiration has depleted the O_2 concentrations to suboxic levels. Under these conditions, anaerobic microorganisms can metabolize any remaining organic matter. As described in Chapter 7, manganese(IV), nitrate, Fe(III), sulfate, and organic matter (via fermentation reactions) can function as electron acceptors in the absence of O_2, thus enabling the oxidation of organic matter to continue. The concentrations of iron and manganese are generally too low for them to play a significant role as an oxidizing agent in seawater, but they are important in marine sediments and, hence, are considered further in Chapter 12.

The most common anaerobic heterotrophic metabolisms found in the water column are denitrification and sulfate reduction. Their stoichiometries are given in Eqs. 8.11

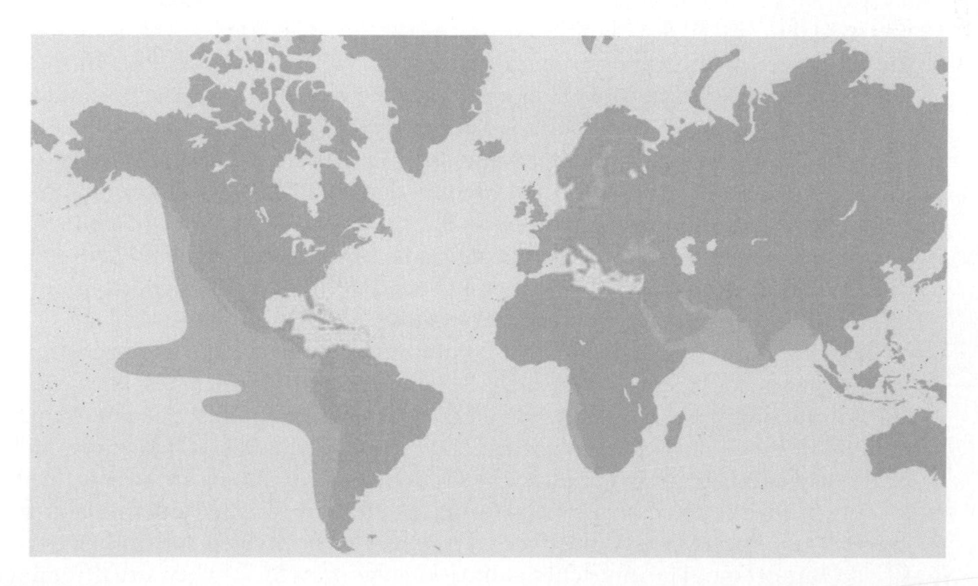

FIGURE 8.4

Oxygen minimum zones (gray) where intermediate and deep water have concentrations <1 mL O_2/L. *Source*: From Levin, L. A. (2002). *American Scientist* 90, 436–444.

and 8.12.

$$(CH_2O)_{106}(NH_3)_{16}H_3PO_4 + 84.8\ HNO_3 \rightarrow 106\ CO_2 + 148.8\ H_2O + 42.4\ N_2 + 16\ NH_3 + H_3PO_4 \quad (8.11)$$

$$(CH_2O)_{106}(NH_3)_{16}H_3PO_4 + 53\ SO_4^{2-} \rightarrow 106\ CO_2 + 106\ H_2O + 16\ NH_3 + 53\ S^{2-} + H_3PO_4 \quad (8.12)$$

In the absence of O_2, the remineralization of nitrogen and sulfur generates NH_3 and S^{2-}. These weak bases react with the acid generated by the hydrolysis of CO_2 ($H_2O + CO_2 \rightarrow H_2CO_3 \rightarrow H^+ + HCO_3^-$). This titration transforms the remineralized nitrogen and sulfur into NH_4^+(aq), HS^-(aq) and H_2S (g). In both remineralization reactions (Eqs. 8.11 and 8.12), nitrogen and phosphorus are released from organic matter in the Redfield-Richards ratio. Since much of the nitrogen remineralized during denitrification is converted to N_2, the remaining fixed inorganic nitrogen-to-phosphorus ratio is less than 16:1.

Because of its higher free energy yield, denitrification is thermodynamically favored over sulfate reduction and will proceed until the nitrate is depleted. If organic matter is still present, sulfate reduction will then occur. These conditions are not found in open ocean waters, but have been observed within the water columns of coastal upwelling zones and in the underlying sediments. Periodic eruptions of H_2S(g) out of the sediments have been observed via SeaWIFs satellite imagery in the upwelling zone off Walvis Bay along the west coast of Africa as milky white clouds of particulate elemental sulfur that float to the sea surface. (The elemental sulfur is produced during the oxidation of H_2S(g) which occurs as the gas rises out of the sediments and into the oxic part of the water column.) Elevated CH_4(g) concentrations in the subsurface waters of some coastal upwelling areas provide evidence for ongoing methanogenesis. As noted in Chapter 7.3.2, surface waters tend to be supersaturated with respect to CH_4(g). Likely sources of this methane are: (1) methanogenesis that takes place within the anoxic microenvironments harbored by detrital POM and (2) from microbial decomposition of certain types of biomolecules. Evidence of the latter has been observed in incubation experiments where large amounts of methane are generated during the aerobic bacterial degradation of methylphosphonate, a naturally occurring component of the dissolved organic phosphorus pool (Table 22.8).

Low O_2 concentrations are also common to shallow-water environments with high organic matter concentrations, such as estuaries and salt marshes. Hypoxic conditions are not uncommon and seem to be increasing in geographic scope and frequency, most notably in the Gulf of Mexico around the mouth of the Mississippi River, and in Chesapeake Bay and Long Island Sound. The O_2-deficient conditions are largely due to nutrient loading from the excessive use of fertilizers that are washed off the land by rain and carried into the rivers. Atmospheric transport of volatilized nitrogen (NH_3 and NOx) is also important. The nutrients stimulate algal growth. The eventual remineralization of the detrital algal remains via microbial aerobic respiration consumes O_2, leading to hypoxic conditions. This process is termed *eutrophication* and is discussed further in Chapter 28.6.2 and 28.6.3. Other marginal seas with very restricted water circulation, such as the Black Sea, are naturally O_2 deficient. These deficiencies have been increased as a result of human impacts.

8.6 THE INFLUENCE OF ORGANIC MATTER ON THE GLOBAL BIOGEOCHEMICAL OXYGEN CYCLE

The marine cycle of organic matter has played an important role in the development and stabilization of atmospheric O_2 levels and, hence, seawater concentrations. Climate change has the potential to overwhelm these stabilizing loops. Evidence already supports a decline oxygen concentrations arising from warming of the oceans. This is likely to have profound consequences on fish distributions. Paleoceanographic records indicate that the global ocean has become anoxic for extended periods. These ocean anoxic events coincide with mass extinction events and periods of rapid climate change. Can we expect this to happen in the future?

A review of this subject is provided in the supplemental information for Chapter 8.6 that is available online at **http://elsevierdirect.com/companions/9780120885305**. It includes a discussion of how Earth's atmosphere became oxic and how marine plankton and the cycling of organic matter have contributed to setting and stabilizing the O_2 concentrations in seawater and the atmosphere. This subject is of great importance as the development of an oxic atmosphere has had a profound effect in directing the evolutionary trajectories of marine and terrestrial microbes and animals.

CHAPTER 9

Vertical Segregation of the Biolimiting Elements

All figures are available on the companion website in color (if applicable).

9.1 INTRODUCTION

Marine organisms have had a large impact on the chemical evolution of the planet. Burial of their biomass in the sediments has, over the millennia, helped make Earth's atmosphere O_2-rich. Although a slow process, the deposition and burial of this organic matter has sequestered a large amount of carbon in the ocean's sediments. In this chapter, we consider the effects of marine biomass formation and burial on the other major constituents of organic matter, namely nitrogen, phosphorus, and silicon. Nitrogen and phosphorus are required for formation of tissues (soft parts). Silicon is an example of an element that is required for the formation of plankton hard parts. For example, diatoms require silicon to form their shells, called *frustules* or *tests*.

Plankton require so much nitrogen, phosphorus, and silicon to meet their biological needs that surface-water concentrations are usually low enough to limit growth. Hence, these elements are considered to be "biolimiting." Bacterial remineralization of detrital plankton biomass and other organic remains releases the biolimiting elements back into seawater. This release occurs throughout the water column. Thus, delivery of the biolimiting elements to surface waters is dependent on physical processes, including advection, turbulence, river runoff, and atmospheric deposition. The interaction of these physical and biological processes causes the biolimiting elements to have much lower concentrations in the surface waters as compared to the deep zone. The reasons for this are discussed herein using a two-box model for the world's ocean. This quantitative approach enables estimation of the oceanic residence times and recycling efficiencies of these elements.

9.2 SURFACE-WATER DEPLETIONS, BOTTOM-WATER ENRICHMENTS

The biolimiting elements (also called nutrients or micronutrients) tend to be present at low, and often undetectable, levels in the surface waters at mid- and low latitudes. Examples of nutrient depth profiles are provided in Figure 9.1 for a mid-latitude site in

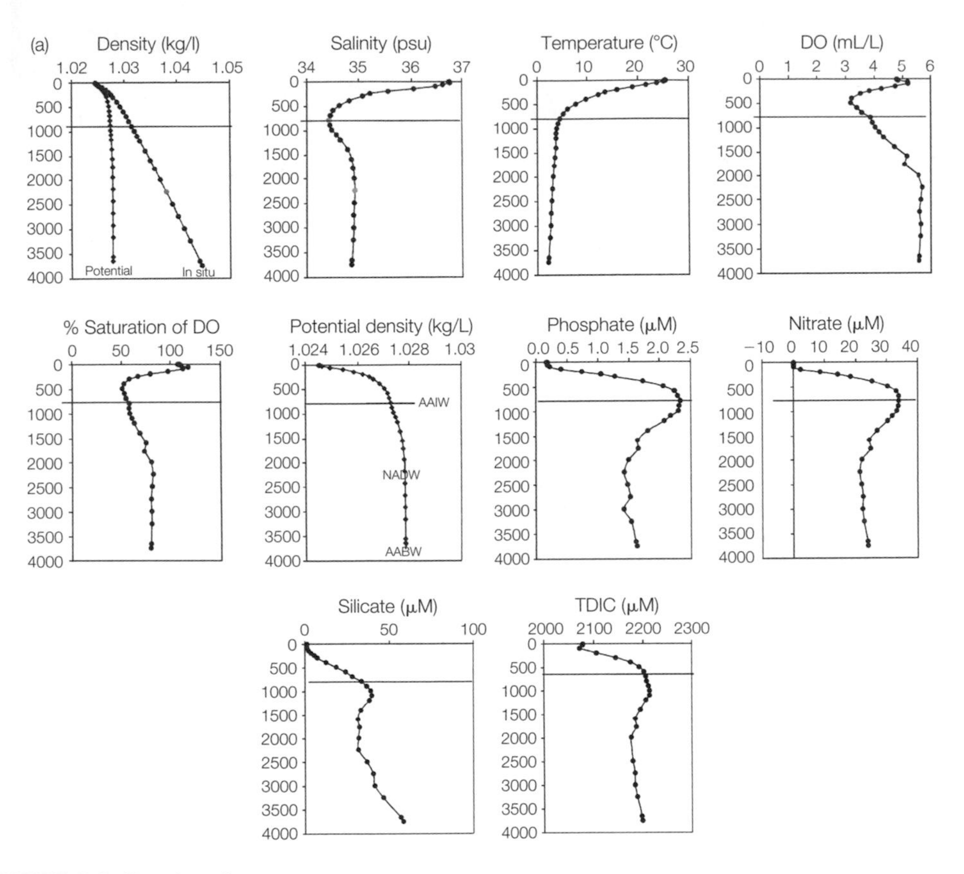

FIGURE 9.1 *(Continued)*

the southern Atlantic Ocean. This figure also includes depth profiles of temperature, salinity, density, dissolved oxygen (O_2), and inorganic carbon (TDIC).[1] The shapes of these profiles can be used to infer the biogeochemical processes that are influencing the depth distributions of the biolimiting elements as well as those of O_2 and TDIC. Note that the nitrogen and phosphorus species plotted are nitrate and phosphate. These are the biologically usable species present at highest concentration in oxic seawater.

The nutrient profiles are characterized by much higher concentrations in the deep-waters than in the surface. In some locations, such as shown in Figure 9.1, mid-water concentration maxima are present. The depth region over which concentrations exhibit the largest vertical gradients is usually defined by the thermocline. All biolimiting elements have similar depth profiles, having surface-water depletions and deep-water enrichments.

[1] In Chapter 5, we noted that CO_2 readily dissolves in seawater to form several inorganic carbon species. TDIC is defined as the sum of the concentration of those species, i.e., the sum of the carbonate, bicarbonate, carbonic acid, and carbon dioxide concentrations.

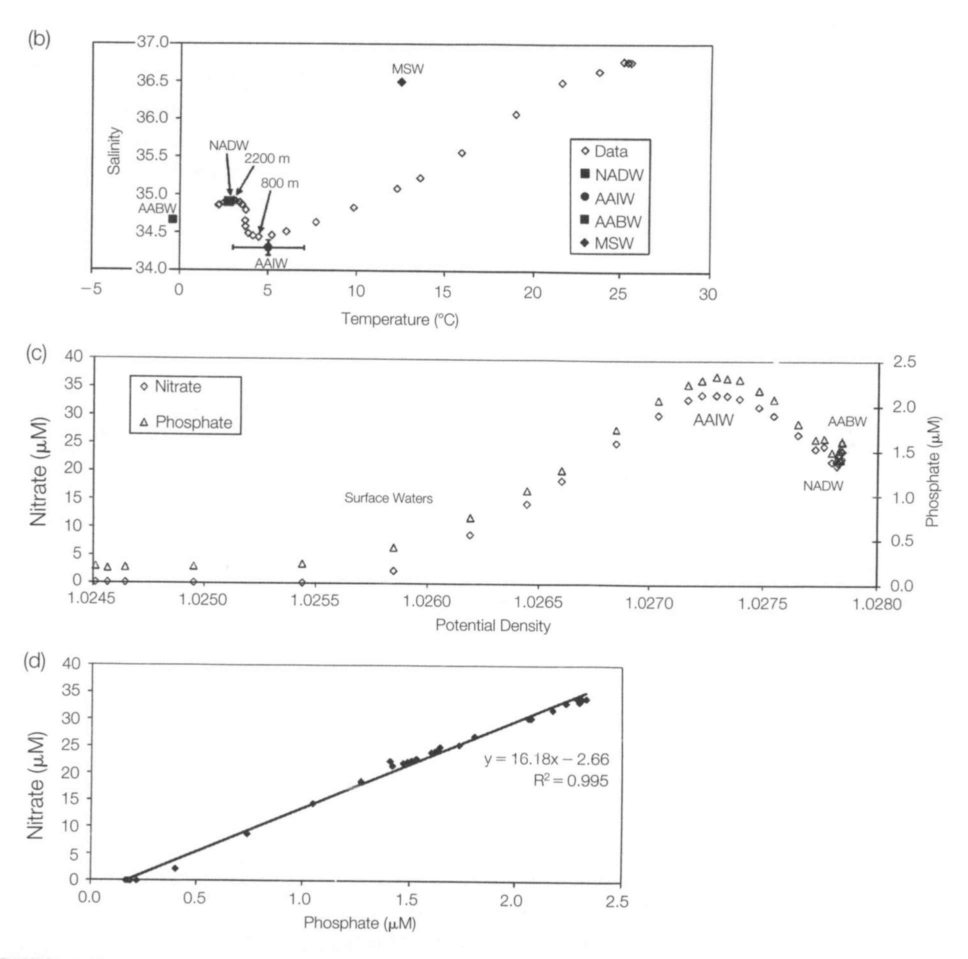

FIGURE 9.1

(a) Depth profiles of density, salinity, temperature, dissolved oxygen, percent saturation of dissolved oxygen, phosphate, nitrate, silicate, and total dissolved inorganic carbon in the southern Atlantic (19.0°S and 16.08°W sampled 2/24/91 at 1454 hours as part of the Transient Tracers in the Ocean program). This hydrographic station is located in the middle of the Atlantic Ocean. The horizontal line at 800 m marks the location of the core of AAIW. Also provided are *xy* plots of (b) salinity versus temperature for the data in Figure 9.1a and the core water mass T-S values from Table 4.1. (c) nitrate versus phosphate concentrations, and (d) nitrate and phosphate concentrations versus potential density. Data from: JAVA Ocean Atlas Section A09.

Elements that are not biolimiting have quite different vertical concentration profiles. Thus, the shapes of vertical concentration profiles can be used to infer the most important biogeochemical processes acting on the chemical of interest. In this chapter and the next, we will explore several sets of vertical profiles for nitrogen, phosphorus, and silicon, obtained from different parts of the world's ocean. In Chapter 11, we will investigate the vertical profiles of the micronutrients, such as iron and zinc.

The deep-water enrichments and surface-water depletions seen in the vertical concentration profiles of the biolimiting elements are a consequence of how biogenic particles are produced and destroyed in the ocean. The availability of these elements in surface seawater limits the growth of phytoplankton such that these microbes extract nearly all of the dissolved inorganic species out of the waters of the euphotic zone. (The depth of the euphotic zone is roughly coincident with the depth of the mixed layer, around 100 to 200 m.) Some of these elements, such as nitrogen and phosphorus, are incorporated into the tissues, or soft parts, of the phytoplankton. Others, such as silicon, are deposited into hard parts that constitute shells or skeletal material, such as the siliceous frustules created by diatoms. (The other type of common planktonic test is composed of calcite [$CaCO_3$]. Carbon and calcium are present in such high concentrations in seawater that they do not limit plankton growth.)

In Chapter 8, we saw how planktonic organic matter can be passed up through the food chain giving rise to detrital soft parts, such as dead tissues, fecal pellets, molts, and feeding nets. These organic materials are eventually remineralized by bacteria. This process returns the nitrogen and phosphorus to dissolved inorganic form. In oxic seawater, these forms are mostly nitrate and phosphate. The hard parts also undergo remineralization. Once an organism dies or molts, its hard parts are subject to dissolution. For example, the dissolution of the siliceous frustules returns dissolved silicon to seawater in the form of silicate.

When remineralization occurs in the surface waters, the regenerated nutrients are rapidly reassimilated by plankton and returned to particulate form. Because some of the detrital hard and soft parts sink out of the surface waters, remineralization also occurs in the thermocline and deep zone. Sinking particles can include living organisms, such as bacteria attached to decomposing particles and phytoplankton incorporated in fecal pellets. Because photosynthesis cannot occur below the euphotic zone, the regenerated nutrients remain in dissolved form unless consumed by chemoautolithotrophs. (As discussed in Chapter 7, some bacteria and archaea are chemoautolithotrophs, but their contribution to global marine primary production is minor.)

Most of the regenerated nutrients remain trapped below the mixed layer because strong density stratification inhibits vertical mixing. The ocean has not become a biological desert, so some of these nutrients must eventually be returned to the surface ocean. The return flow of nutrient-enriched subsurface waters is associated with Ekman-driven upwelling (Figure 4.5a) and meridional overturning circulation (Figure 4.6). The relative rates of biological uptake, POM sinking, and water mass circulation are presently such that the biolimiting elements are concentrated in the deepwaters as compared to the surface. This effect is commonly referred to as the *biological pump*. The biological pump also serves to transport carbon into the deep sea.

9.3 INTERPRETING DEPTH PROFILES

Vertical concentration and temperature profiles are commonly used to assess the nature and relative rates of biogeochemical processes. An example was provided in Chapter 4 using the one-dimensional advection-diffusion model to obtain in situ O_2 uptake rates. In Chapter 5, we saw how salinity can be used to assess the degree of nonconservative behavior in any element or species of interest. In this chapter, we introduce the essential skill of vertical profile interpretation, which is used to identify specific biogeochemical processes responsible for the profile curve shapes. To do this, oceanographers routinely measure a fixed set of parameters at a sampling site.

Sampling sites are also referred to as station locations. For water column work, depth profiles are constructed from seawater samples collected at representative depths. Temperature and salinity are measured in situ with sensors. Remote-closing sampling bottles deployed from a hydrowire are used to collect water for later chemical analysis, either on the ship or in a land-based laboratory. The standard chemical measurements made on the water samples include nutrients (nitrate, phosphate, and silicate), dissolved O_2, and total dissolved inorganic carbon (TDIC) concentrations.

Usually, several sampling bottles are deployed at one time so that multiple depths can be sampled during deployment of the hydrowire. A single deployment is called a hydrocast. It typically takes several hydrocasts to obtain enough samples to construct a detailed depth profile, especially in the deepwater columns of the open ocean. An example of a standard set of hydrographic profiles is shown in Figure 9.1.

When oceanographers start reviewing the standard hydrographic profiles for any given station, they usually first focus on the temperature, salinity, and density data. This helps determine the density stratification of the water column and what water masses are present. The latter provide clues as to the water masses' origin and relative age (time elapsed since they were last at the sea surface). For the sampling site profiles in Figure 9.1, the mixed layer appears to be quite shallow, being less than 50 m in depth, and the thermocline zone extends from the bottom of the mixed layer to about 1000 m. *T-S* plots, such as shown in Figure 9.1b, are used to identify water masses. At this sampling location, three subsurface water masses are present, namely AAIW, NADW, and AABW. The core of the AAIW appears to be located at 800 m and the core of NADW at about 2000 m. The bottom of the thermocline is formed by the mixing of AAIW with NADW. The top of the thermocline is formed by the mixing of AAIW with the overlying surface water masses, namely South Atlantic Central Water (Table 4.1).

The dissolved oxygen data follow depth trends that are nearly a mirror image of the nutrients. The OMZ lies at depths slightly above the core of the AAIW. Why is the OMZ located at these depths? To answer this question, oceanographers use the vertical concentration profiles of O_2, nutrients, and TDIC to assess the relative rates of aerobic respiration and photosynthesis as a function of depth. (The TDIC concentration is used as a measure of how much CO_2 has been taken up from or released into the water.)

The effects of photosynthesis are clearly seen in the low TDIC and nutrient concentrations of the surface water. The O_2 concentrations are high because of contact with the sea surface and production by phytoplankton. The temperature and O_2 concentration data have been used to compute the percent saturation with respect to O_2. The high degree of supersaturation in the surface water suggests that the rate of O_2 supply via photosynthesis is exceeding its removal via the dual processes of aerobic respiration and degassing across the air-sea interface.

In Figure 9.1, the seawater is undersaturated with respect to O_2 at depths greater than 100 m because of increasing distance from the air-sea interface and increasing light limitation of the phytoplankton. With such a shallow mixed layer (< 50 m), the physical resupply of O_2 from turbulence cannot penetrate far. Nor is molecular diffusion across the air-sea interface a significant resupply mechanism, because it is too slow to provide much transport. Since light levels are too low to support photosynthesis, there is no in situ O_2 production. So at this site, waters depths greater than 100 m are undersaturated with respect to O_2 because its consumption via aerobic respiration is faster than its resupply.

The low nutrient and inorganic carbon concentrations in the surface waters reflect efficient removal by the plankton.[2] Any nutrients regenerated by aerobic respiration in the euphotic zone appear to be readily assimilated, keeping surface water concentrations low. Since photosynthesis does not occur below the euphotic zone and aerobic respiration does, the products of POM remineralization accumulate in the waters lying below 100 m. This causes vertical profiles to exhibit trends of increasing nutrient and TDIC concentration with depth. The concomitant decline in O_2 concentration supports the hypothesis that nutrient remineralization occurs via the process of aerobic respiration. As shown in Figure 9.1d, the nutrients are released in the Redfield–Richards ratio. The rate of aerobic respiration beneath the mixed layer can be estimated from the vertical oxygen concentrations profiles using the one-dimensional advection-diffusion model described in Chapter 4 (Figure 4.21).

Below the top of the thermocline, O_2 concentrations continuing declining with increasing depth until a minimum value is reached at 500 m. Nutrient and TDIC concentrations do the reverse; they increase with increasing depth, reaching a maximum concentration at slightly deeper depths, i.e., 800 m for nitrate and phosphate and 1100 m for silicate and TDIC. Below these depths, all the concentration trends reverse; O_2 concentrations increase and nutrient and TDIC decrease until 2000 m is reached, where the core of the NADW is located.

These concentration trend reversals produce an O_2 minimum and nutrient maximum at mid-depths that nearly coincide in location with the core of the AAIW. The locations of these mid-depth concentration maxima and minima are defined by the supply of biogenic hard and soft parts sinking out of the mixed layer, which are then remineralized at mid-depths via aerobic respiration and in so doing consume O_2. The bottom limbs of the mid-depth nutrient maxima and oxygen minimum are defined by two processes: (1) the supply of sinking particles declines with depth and (2) the underlying water mass, NADW, has a relatively higher O_2 and lower nutrient and TDIC concentrations. The latter reflects the effect of meridional overturning circulation on water mass history. Once a water mass sinks below the sea surface, it begins accumulating remineralization products, such as nitrate, phosphate, and TDIC. The greater the time since the water mass

[2] The surface waters in Figure 9.1 appear to have higher phosphate than nitrate concentrations, suggesting the plankton are nitrogen limited. This may not be the case as the measurement detection limit for nitrogen is usually an order of magnitude higher than that for phosphorus.

was last at the sea surface, the greater the accumulation of remineralization products. Thus, the profiles in Figure 9.1 are from a station location where the NADW is still "young" enough to have a lower nutrient (and higher O_2) content than the overlying AAIW. The slight increase in nutrient and TDIC concentrations below 3000 m marks the start of the mixing zone between NADW and the underlying AABW. This is particularly pronounced in the silicate profile and is partly due to diffusion of this remineralized nutrient out of the underlying sediments.

As exemplified by the silicate profile, all biolimiting elements do not behave identically. In the case of dissolved silicon and TDIC, their concentration maxima lie below the nitrate and phosphate maxima. This reflects the different mechanisms by which the elements are resolubilized. Nitrate and phosphate are regenerated from soft parts. This process seems to occur more readily than the dissolution of hard parts, which releases silicate causing the nitrate and phosphate concentration maxima to lie at shallower depths. Since TDIC is released in nearly equal amounts from soft parts as CO_2 and the dissolution of calcareous hard parts as CO_3^{2-}, the resulting concentration maximum lies below that of nitrate and phosphate.

9.4 BROECKER BOX MODEL

The vertical distribution of biolimiting elements is characterized by deep-water enrichments and surface-water depletions. As described above, this *vertical segregation* is caused by the remineralization of biogenic particles in the deep sea. Not all particulate matter that sinks into the deep zone is remineralized. Some survives to become buried in the sediments. How much of the biogenic particle flux escapes from surface waters? How much of this particle flux is remineralized in the deep zone? How much is lost from the ocean by burial in the sediments? What effect does this have on the concentrations of the biolimiting elements?

Answers to these questions are important as they provide insight on the availability of the elements that ultimately limit global marine productivity. The burial of carbon in the sediments is also important in regulation of global climate by sequestering greenhouse gases (CO_2 and CH_4) and enabling the buildup of atmospheric O_2. This same burial of carbon has, over the millennia, contributed to the formation of petroleum reserves. Thus, considerable attention and interest has been focused on finding out what controls marine POM production, destruction, and its preservation through burial in the sediments.

The degree to which a biolimiting element is remineralized can be thought of as its *recycling efficiency*. Using a simple mathematical model devised by Dr. W. S. Broecker, a rough estimate of the recycling efficiencies of the biolimiting elements can be obtained from their average surface-, deep-, and river-water concentrations. In this model, the waters of the ocean are divided into two reservoirs, a surface and a deep layer. The former represents the warm waters of the mixed layer and the top half of the thermocline; the latter is composed of the relatively cold waters of the bottom half of the thermocline and the deep zone.

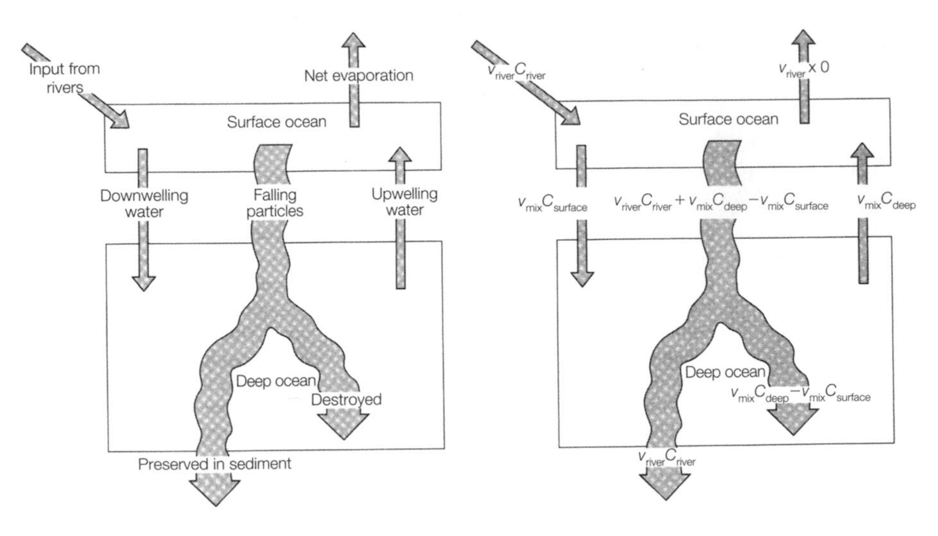

FIGURE 9.2

Box model for the biolimiting elements. *Source:* From Broecker, W. S. (1974). *Chemical Oceanography*, Harcourt, Brace and Jovanovich Publishers, pp. 14–15.

As shown in Figure 9.2, the only communication between the two reservoirs is assumed to occur through the upwelling and downwelling associated with meridional overturning circulation. The thermocline, which is present at mid- and low latitudes, inhibits vertical circulation. Thus, downwelling, achieved by the sinking of deep-water masses, occurs primarily in polar regions where the thermocline is absent. As described in Chapter 4, the deepwaters are returned to the sea surface rather slowly by vertical advection and turbulent mixing. This return flow is not well understood; some upward transport seems to occur along the entire pathway of meridional overturning circulation, with some areas, such as the Southern Ocean, having locally higher rates.

In the Broecker Box model, the total amount of water in the ocean is assumed to remain constant over time. In other words, the evaporation rate and burial of water in the sediments is equal to the rate of water input from river runoff and precipitation. The sizes of the surface- and deep-water reservoirs are also assumed to remain constant over time. This requires the global rate of upwelling to equal the global rate of downwelling.

Since the continental margins are relatively small compared to the total surface area of the ocean, the surface areas of the two reservoirs are also considered to be equal in size. From this perspective, the global rate of water transport between the boxes can be expressed as the annual exchange of a water layer (v_{mix}) that is 300 cm thick. This value of v_{mix} (300 cm/y) is obtained by guesstimating that it takes the ocean 1000 y to mix the ocean through an average depth of 3000 m.

The Broecker Box model also assumes that materials enter the ocean only through river and groundwater runoff. This is a realistic assumption as all the other sources, such as atmospheric fallout and hydrothermal emissions, supply negligible amounts of the biolimiting elements from a global perspective. The model also assumes that the only pathway by which the biolimiting elements are removed from the ocean is via burial in the sediments as unremineralized biogenic particles. These particles are considered to be composed of both hard and soft parts created by marine organisms.

From the perspective of the surface box, the biolimiting elements are supplied via river runoff and from upwelling. The elements are removed via the sinking of biogenic particles and downwelling. Since this model considers only the transport of materials into and out of the ocean and between the two reservoirs, details as to what happens to the elements while they reside in the boxes are not needed other than that they are present in a steady state.[3] In such a case, the input rate of a biolimiting element will equal its output rate. For the surface-water reservoir, the mass balance that describes this steady state is given by

$$\text{Upwelling flux} + \text{River runoff flux} = \text{Downwelling flux} + \text{Particle flux} \tag{9.1}$$

The first three fluxes can be calculated by multiplying the biolimiting element concentration (C) by the annual water transport (v); that is,

$$\text{Flux} = C \times v = \frac{\text{mol}}{\text{L}} \times \frac{\text{cm}}{\text{y}} \times \frac{\text{L}}{1000\ \text{cm}^3} = \frac{\text{mol}}{\text{cm}^2\ \text{y}} \tag{9.2}$$

Therefore, the upwelling flux is computed as $v_{mix} \times C_{deep}$, the river runoff flux as $v_{river} \times C_{river}$, and the downwelling flux as $v_{mix} \times C_{surface}$.

In this model, the rate of river runoff (v_{river}) is expressed as the depth of a layer of water produced by spreading the annual river-water input across the entire surface area of the ocean. The annual amount of river water entering the ocean is 47,000 km^3/y (Figure 2.1). Assuming that the average area of the ocean is equal to that at the sea surface ($3.6 \times 10^{18}\ cm^2$),[4] the river input represents the annual addition of a layer of water approximately 10 cm deep, making $v_{river} = 10$ cm/y.

The surface-ocean mass balance equation (Eq. 9.1) can be rewritten as

$$v_{river}\ C_{river} + v_{mix}\ C_{deep} = v_{mix}\ C_{surface} + P \tag{9.3}$$

where P is the biogenic particle flux. Solving for P yields

$$P = v_{mix}\ C_{deep} + v_{river}\ C_{river} - v_{mix}\ C_{surface} \tag{9.4}$$

The surface-ocean recycling efficiency is calculated as the fraction of the element in the surface box that is removed in particulate form. This fraction, g, is given by

$$g = \frac{\text{Amount of particles leaving the surface ocean}}{\text{Total input to the surface ocean}} \tag{9.5}$$

$$= \frac{v_{mix}\ C_{deep} + v_{river} C_{river} - v_{mix} C_{surface}}{v_{mix} C_{deep} + v_{river} C_{river}} \tag{9.6}$$

[3] The rates of input of nitrogen and phosphorus into the ocean have likely increased as a result of human activities. This is discussed further in Sections 24.9 and 28.6.2.

[4] From the online appendix on the companion website.

This equation can be converted algebraically by dividing each of the products on the right-hand side by the term, $v_{mix}C_{river}$. Cancelling common terms and regrouping yields

$$g = 1 - \left(\frac{\frac{C_{surface}}{C_{river}}}{\frac{C_{deep}}{C_{river}} + \frac{v_{river}}{v_{mix}}} \right) \tag{9.7}$$

One final transformation is generated by multiplying each of the ratios in Eq. 9.7 by v_{mix}/v_{river}. This produces a mathematical definition of g that is solely based on the values of three ratios, v_{mix}/v_{river}, $C_{surface}/C_{river}$, and C_{deep}/C_{river}:

$$g = 1 - \left(\frac{\frac{v_{mix}}{v_{river}} \frac{C_{surface}}{C_{river}}}{1 + \frac{v_{mix}}{\text{river}} \frac{C_{deep}}{C_{river}}} \right) \tag{9.8}$$

Since v_{mix} is 30 cm/y and v_{river} is 10 cm/y, $v_{mix}/v_{river} = 30$,

$$g = 1 - \left(\frac{30 \frac{C_{surface}}{C_{river}}}{\left(30 \frac{C_{deep}}{C_{river}} \right) + 1} \right) \tag{9.9}$$

Thus, to compute the surface-water recycling efficiency of an element, all that must be known are its average surface-, deep-, and river-water concentrations.

For phosphorus, $C_{deep}/C_{river} = 3.0$ and $C_{surface}/C_{river} = 0.15$, so $g = 0.95$. This means that if enough time has elapsed for the complete exchange of water between the two reservoirs, then 95% of the phosphorus that enters the surface box is removed in particulate form. Detailed studies of nutrient dynamics in the mixed layer indicate that the average atom is recycled 10 times before escaping as a sinking particle into the deep sea.

If the ocean is in steady state with respect to a biolimiting element, then its flux into the ocean must equal its flux out; that is,

$$v_{river}C_{river} = \text{Particle flux to the sediments} \tag{9.10}$$

The particle flux to the sediments can be calculated by first defining f as

$$f = \frac{\text{Particle flux to the sediments}}{\text{Total particle flux}} \tag{9.11}$$

This is the fraction of the total particle flux, P, that survives the descent through the deep-water box to become buried in the sediments.

The particle flux to the sediments can be calculated by multiplying f and P, since

$$\frac{\text{Particle flux to the sediments}}{\text{Total particle flux}} \times P = \text{Particle flux to the sediments} \tag{9.12}$$

The particle flux to the sediments represents the sole route by which the element is lost from the ocean, so

$$\frac{\text{Particle flux to the sediments}}{\text{Total particle flux}} \times P = v_{\text{river}} C_{\text{river}} \tag{9.13}$$

or

$$f \times P = v_{\text{river}} C_{\text{river}} \tag{9.14}$$

Substituting the definition of P given in Eq. 9.4 and solving for f yields

$$f = \frac{v_{\text{river}} C_{\text{river}}}{v_{\text{river}} C_{\text{river}} + v_{\text{mix}} C_{\text{deep}} - v_{\text{mix}} C_{\text{surface}}} \tag{9.15}$$

This is simplified algebraically to

$$f = \frac{1}{1 + \frac{v_{\text{mix}}}{v_{\text{river}}} \left(\frac{C_{\text{deep}}}{C_{\text{river}}} - \frac{C_{\text{surface}}}{C_{\text{river}}} \right)} \tag{9.16}$$

Substituting $v_{\text{mix}}/v_{\text{river}} = 30$ gives

$$f = \frac{1}{1 + 30 \left(\frac{C_{\text{deep}}}{C_{\text{river}}} - \frac{C_{\text{surface}}}{C_{\text{river}}} \right)} \tag{9.17}$$

For phosphorus, $f \approx 0.01$. This means that only 1% of the particle flux that enters the deep-water box during any given mixing cycle survives to become buried in the sediments. Ninety-nine percent is remineralized in the deepwater.

The overall oceanic recycling efficiency of a biolimiting element is given by the fraction of the river input that is buried in the sediments during one complete mixing cycle. This is calculated as

$$f \times g = \frac{\text{Particle flux to the sediments}}{\text{Total particle flux}} \times \frac{\text{Total particle flux}}{\text{Total input to surface seawater}} \tag{9.18}$$

$$= \frac{\text{Particle flux to the sediments}}{\text{Total input to surface seawater}} \tag{9.19}$$

For phosphorus, $f \times g = 0.01 \times 0.95 \approx 0.01$. This means that only 1 percent of the phosphorus introduced into the ocean by river runoff is removed to the sediments during each mixing cycle.

The residence time of a biolimiting element can be calculated from $f \times g$ and the average mixing time of the ocean (1000 y) as follows:

$$\tau = \frac{\text{total amount of an element in the ocean}}{\text{removal rate}} = \frac{1}{\left(\frac{f \times g}{1000\ y}\right)} \tag{9.20}$$

where the total amount of an element in the ocean is represented as unity since $f \times g$ is the fraction of element removed. Since $f \times g$, or 1% of the phosphorus, is removed from the ocean in 1000 y, removal of 100% requires 100,000 y. This means that on average, each atom spends 100 mixing cycles in the ocean, getting recycled in the surface waters and exchanged between the surface and deep-water reservoirs, before becoming buried in the sediments. Recent research suggests that phosphate is rapidly cycled across the sediment-water interface of the continental shelves, causing the actual residence time of phosphorus to be somewhat shorter (10,000 to 17,000 y) than predicted by the Broecker Box model.

Regardless of the degree of its accuracy, the Broecker Box model can be used to obtain further insights into the relative importance of the biological pump. For example, we note that the surface-water volume is 10 times smaller than the deepwater. Since the rates of downwelling and upwelling are equal, seawater must spend one-tenth of its time in the surface-water box. The residence time of surface water is then 1/10 × 1000, or 100 y. Since the volume of surface water is small, as is $C_{surface}$ for the biolimiting elements, downwelling cannot be a significant removal mechanism. Thus, the particle flux, P, must be responsible for most of the removal of the biolimiting elements from the surface box. To maintain a 20-fold deepwater enrichment for phosphorus (and neglecting any sediment inputs), the particle flux must operate 20 times faster than the downwelling flux. This also requires the residence time of phosphorus in the surface-water box to be 20 times smaller than that in the deep water.

Thus, during one mixing cycle of the ocean, the simple Broecker Box model predicts that an atom of phosphorus spends, on average, only 5 y (1/20 of 100 y) in the surface box. Direct measurements suggest this time may be as short as 1 to 20 d. The average atom spends the rest of the mixing cycle in the deep-water. After somewhere between 10^4 to 10^5 y, the average atom escapes to the sediments. In comparison to the ocean, transport rates in the rock cycle are slow. The average atom of phosphorus spends 2×10^8 or so years buried in the sediments before geologic processes cause it to be uplifted, eroded, and carried back into the ocean.

9.5 BIOLIMITING VERSUS BIOINTERMEDIATE VERSUS BIOUNLIMITED

Table 9.1 contains a summary of recycling efficiencies for a variety of representative elements. The results fall into three groups, which have been termed *biolimiting, biointermediate*, and *biounlimited*. The biolimiting elements have $g \approx 1$ and, thus, are almost

Table 9.1 Broecker Box Model Ratios and Recycling Efficiencies for Representative Biolimiting, Biointermediate, and Biounlimited Elements.

Category	Element	$C_{deep}/C_{surface}$	C_{deep}/C_{river}	C_{deep}/C_{river}	g	f	$f \times g$	τ(y)
Biolimiting	P	20	0.15	3.0	0.95	0.01	0.01	1×10^5
	Si	35	0.05	0.7	0.97	0.05	0.05	2×10^4
	N*	14	0.056	0.78	0.93	0.04	0.04	2×10^4
Biointermediate	Ba	3	0.1	0.3	0.70	0.14	0.10	1×10^4
	Ca	1.01	24.8	25	0.01	0.14	0.0013	8×10^5
Biounlimited	S	1	5000	5000			0.0001	2×10^7
	Na	1	50,000	50,000			0.00001	2×10^8

*Assumes N_2 fixation is negligible. If N_2 fixation is supplying nitrogen to the ocean at a rate equivalent to river input, then $g = 0.93$, $f = 0.08$, $f \times g = 0.08$, and $\tau = 1 \times 10^4$ y.

Source: *After Broecker, W. S. (1974)* Chemical Oceanography. *Harcourt, Brace, and Jovanovich Publishers, p. 21.*

completely removed from the surface seawater in particulate form. They also have an $f \ll 1$, so the particles that settle out of the surface are almost completely remineralized in the deep-water. Thus, a large deepwater enrichment is established without much loss of the element from the ocean during a mixing cycle.

In comparison to phosphorus, silicon exhibits a much greater deepwater enrichment ($C_{deep}/C_{surface} = 35$) reflecting a smaller recycling efficiency in the surface ocean. Thus, a greater fraction of the silicon is transported to the deep sea. The fraction leaving the ocean in particulate form ($f = 0.05$) is also larger. So siliceous hard parts experience less remineralization as they sink through the water column as compared to the elements in soft parts. This causes a greater fraction of the silicon to become buried in the sediments. This difference in behavior among the biolimiting elements is due to the different mechanisms by which phosphorus and silicon are remineralized. Phosphorus is resolubilized during the aerobic respiration of POM, whereas silicon is released by the dissolution of siliceous frustules. This dissolution proceeds at a slower rate than the remineralization of POM, so a greater fraction of the siliceous hard parts survives the trip to the seafloor.

For elements whose concentrations do not vary with depth, $g \ll 1$ and f is approximately equal 1. These chemicals generally are present at high concentrations in seawater, so they are not growth limiting. An insignificant fraction of these chemicals is incorporated into biogenic particles and so they are termed biounlimited elements. Sodium and sulfur are examples of biounlimited elements. Most of their removal from the ocean is accomplished by other means. In the case of sodium, this is through the deposition of evaporites and for sulfur, by incorporation into evaporite, pyrite, and hydrothermal minerals. The biounlimited elements are characterized by relatively long residence times. Though biounlimited elements can be a significant component of biogenic particles, such as is the case with sulfur, either the particle flux is too small to have much of an impact on their seawater concentrations, or most of the sinking particles are buried in the sediment and, hence, cannot contribute to a deep-water enrichment.

Elements such as barium and calcium have intermediate values of f and g. These are classified as biointermediate elements. In the case of barium, the fraction removed as particles from the surface water is similar to that of a biolimiting element. But a much greater proportion of the particulate barium escapes from the ocean, causing its $C_{deep}/C_{surface}$ to be relatively small ($\sim$3). This modest degree of vertical segregation suggests some degree of cycling in biogenic particles. For barium, this cycling appears to involve abiotic formation of a particulate phase comprised of the mineral barite ($BaSO_4$). Precipitation of this mineral phase is thought to occur in the surface ocean from barium remineralized from POM and hard parts. Most of the remineralized barium comes from the dissolution of siliceous and celestite ($SrSO_4$) tests deposited by radiolarians, a very abundant type of amoeboid protist. Some bacteria also generate intracellular barite precipitates. (In the sediments, barite is precipitated by benthic foraminiferans.) Because a large fraction of the barite sinks out of the ocean to become buried in the sediments, the residence time of barium is shorter than that of the biolimiting elements.

Some biointermediate elements, such as calcium, have residence times longer than the biolimiting elements. In the case of calcium, this element is a major component of

Table 9.2 The Role of Elements in Life Processes.

General Function	Elements	Chemical Form	Examples of Specific Functions
Major structural components of soft parts	C, H, N, O, P, S	Organic compounds	Components of biomolecules that comprise tissues, membranes, organelles, organic exoskeletons
Major structural components of hard parts	Ca, C, Si, O, P, F, Sr, S	Calcite, aragonite, opaline silica, celestite, apatite, fluoroapatite	Components of frustules and tests, bone, teeth
No known role but present in organisms	Li, Ba, Rb, Cs, Pt, Al, Ti, Sn, W, Ge, B, Br		NA
Electrochemical functions	H, Na, K, Cl, Mg, Ca, HPO_4^{2-}	Free ions	Osmotic regulation; transmission of chemical signals; energy production including light harvesting
Redox catalysis	Zn, Fe, Cu, Mn, Mo, Co, V, Se, Cd, Ni	Enzymes (see Table 11.4 for more information)	Reactions with oxygen (Fe, Cu) Oxygen evolution (Mn) Nitrogen fixation (Fe, Mo) Inhibition of lipid peroxidation (Se) Carbonic anhydrase (Cd) Reduction of nucleotides (Co) Reactions with H_2 (Ni) Bromoperoxidase activity (V)
Oxygen transport	Fe, Cu	Proteins	Hemoglobin
Electron donors or acceptors used by extremophiles	As, Hg, Pb, U, Cr		
Thyroid function in animals	I	Thyroxine	

calcareous ($CaCO_3$) hard parts deposited by phytoplankton, such as coccolithophorids, and protists, such as foraminiferans. Because the seawater concentration of calcium is so high (it is a major ion), only a small fraction ($g = 0.01$) is removed from the surface water in particulate form. Calcareous hard parts are remineralized by dissolution. A significant fraction of the particulate flux of detrital calcite sinks fast enough to become buried in the sediments ($f = 0.14$). Because g is small, $C_{deep}/C_{surface}$ is nearly equal to 1 for calcium and its residence time is intermediate between the biolimiting and biointermediate elements.

In Chapter 11, we will discuss a fourth category of elements, one with vertical profiles nearly opposite to the biolimiting elements. These elements have surface-water enrichments and bottom-water depletions. Most are trace metals that adsorb onto sinking particles enabling their transport to the sediments. These elements tend to have shorter residence times than the biolimiting elements because they lack the remineralization step. Still other elements have a foot in both camps; in some locations, they exhibit biolimiting behavior and have profiles with surface-water depletion and bottom-water enrichments, and in other locations, the profiles appear to be controlled by particle adsorption. Iron is an example of such an element.

Table 9.1 provided only two examples each of elements exhibiting biolimiting and biointermediate behavior. Other elements that fall into these classification categories have the following characteristics in common. First, their seawater concentrations are in the trace to minor range. Second, they are involved in essential metabolic processes or are a major constituent of a structural component, such as a cell membrane, exoskeleton, bone, or shell. As illustrated in Table 9.2, many of these biologically essential elements, are trace metals. Some, such as Cu, Zn, Cd, and Co, have known metabolic roles in enzymes but are generally toxic at high concentrations. Others are used only by certain extremophiles as electron donors or acceptors, such as As, Hg, Pb, and U.

CHAPTER 10
Horizontal Segregation of the Biolimiting Elements

All figures are available on the companion website in color (if applicable).

10.1 INTRODUCTION

As discussed in the previous chapter, the biological pump transfers nutrients and micronutrients to the deep sea, leaving surface-water concentrations of the biolimiting elements uniformly low. The degree of this vertical segregation is geographically variable. The most important trend is an increase in deep-water concentrations along the pathway of deep-water circulation. This phenomenon is termed *horizontal segregation*.

The patterns and causes of horizontal segregation are the subject of this chapter. This matter is of considerable interest because it plays an important role in determining the locations and rates at which nutrients are returned to the sea surface. This, in turn, controls primary production. Because phytoplankton fix CO_2 into organic matter, changes in nutrient supply to the sea surface have the potential to affect P_{CO_2} and, hence, global climate. Changes in global climate have the potential to alter the rates and pathways of deep-water circulation. Since this water motion influences horizontal segregation, oceanographers hypothesize that a negative feedback loop exists in which nutrient cycling and deep-water circulation interact so as to stabilize marine productivity and possibly global climate.

10.2 GEOGRAPHIC DIFFERENCES IN DEPTH PROFILES

The water-column profiles in Figure 10.1 illustrate that the degree of vertical segregation for nitrate, phosphate, and silicate varies geographically. Since the surface-water concentrations of the biolimiting elements are uniformly low, most of the variability in vertical segregation reflects geographic differences in deep-water concentrations. In general, the deep-water nutrient concentrations are highest in the Pacific Ocean and lowest in the Atlantic. As shown in Figure 10.2, a similar pattern is seen in the TDIC concentrations. This pattern is reversed for the O_2 profiles, with highest deepwater

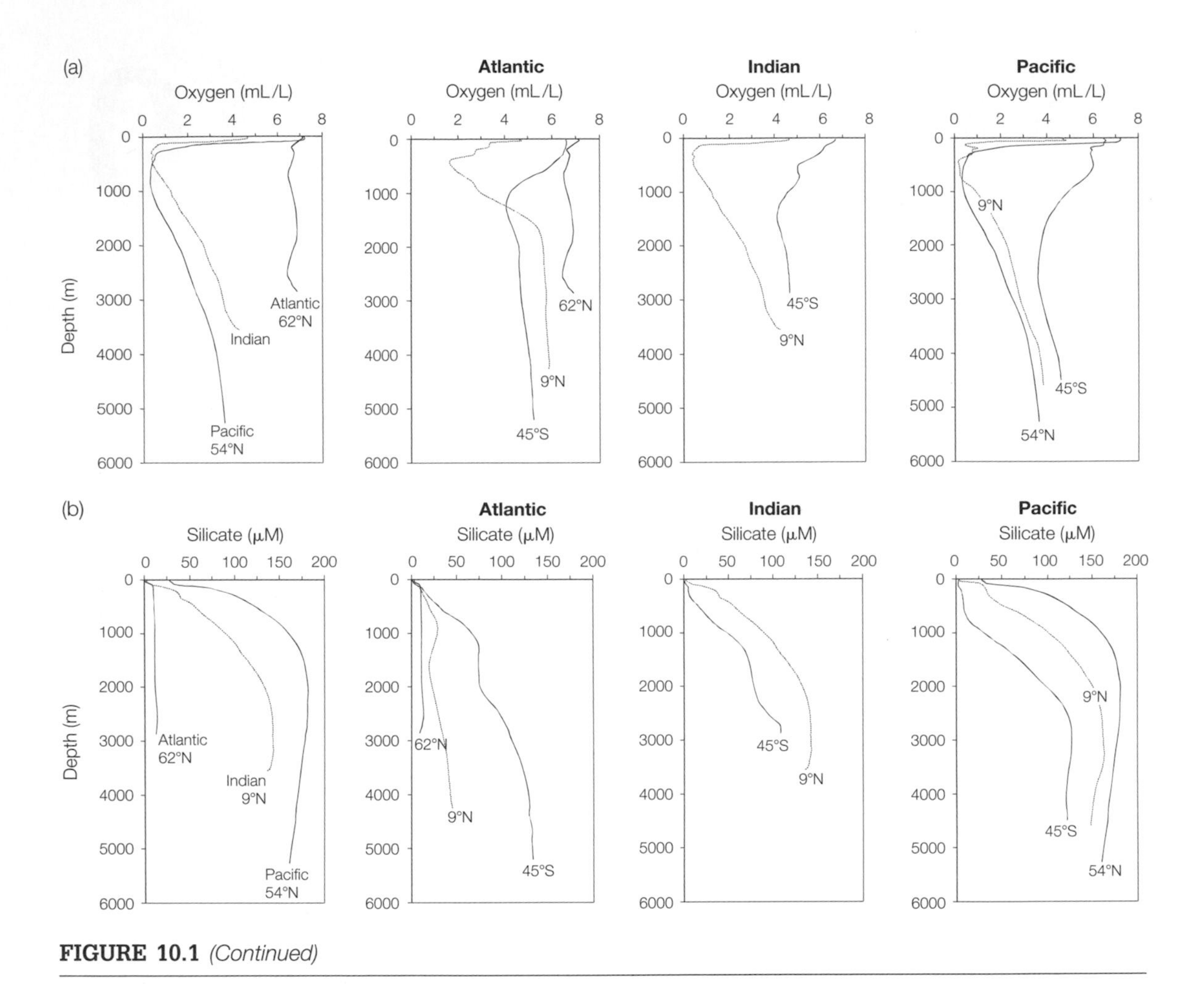

FIGURE 10.1 *(Continued)*

concentrations in the Atlantic and the lowest in the Pacific. Deepwater circulation is a major factor determining these concentration trends.

As noted in Chapter 4, the term *meridional overturning circulation*, abbreviated MOC, is used to collectively describe all the water movements associated with the creation, lateral flows, and upwelling of the subsurface water masses.

> To understand segregation, you will need a detailed understanding of these water currents. This information is provided on the companion website **(http://elsevierdirect.com/companions/9780120885305)**.

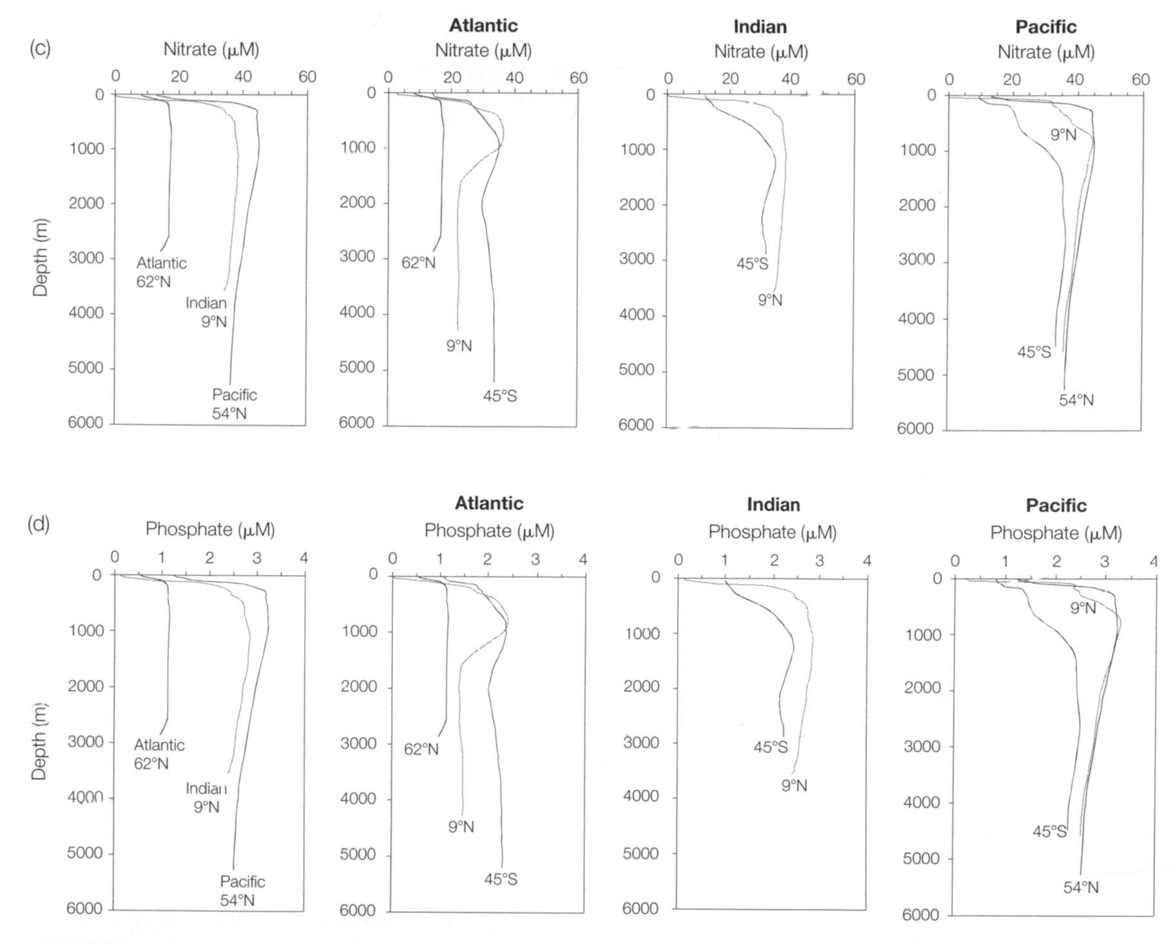

FIGURE 10.1

Vertical distributions of (a) oxygen, (b) dissolved silicon (silicate is the primary chemical species), (c) nitrate, and (d) phosphate at selected latitudes within the Atlantic (20 to 30°W), Indian (91 to 95°W), and Pacific (135 to 158°W) oceans. These are WOCE data obtained from Java Ocean Atlas.

10.3 HORIZONTAL SEGREGATION OF NUTRIENTS IN THE DEEP ZONE

Plankton produce biogenic particles in the surface waters of all the ocean basins. Most of these particles sink into the deep sea and are then remineralized. The rain of biogenic particles causes the nutrient concentration of the deep-water masses to increase as they move through the ocean basins for two reasons. First, the further a deep-water mass has traveled from its site of formation, the greater the amount of particles it will

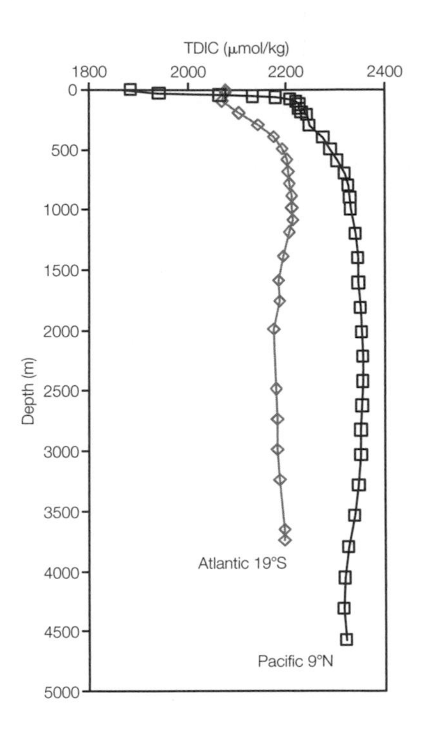

FIGURE 10.2

Vertical distributions of total dissolved inorganic carbon (TDIC) in the subtropical South Atlantic (19°S 16°W) and equatorial North Pacific (9°N 135°W) oceans. These are WOCE data obtained from Java Ocean Atlas.

have accumulated. And second, the more time a biogenic particle has spent in the deepwater, the greater the likelihood that it has been remineralized. Thus, the nutrient concentrations of deep-water masses tend to increase as they "age." Since the oldest deep waters are located in the North Pacific, the overall impact has been to push the nutrients into the deep zone and toward the Pacific Ocean. The latter effect is termed *horizontal segregation*.

Nutrients are carried back to the sea surface by the return flow of deep-water circulation. The degree of horizontal segregation exhibited by a biolimiting element is thus determined by the rates of water motion to and from the deep sea, the flux of biogenic particles, and the element's recycling efficiency (f and g from the Broecker Box model). If a steady state exists, the deep-water concentration gradient must be the result of a balance between the rates of nutrient supply and removal via the physical return of water to the sea surface.

The present-day deep-water concentration gradients of O_2,phosphate, nitrate, and dissolved silicon are illustrated in Figure 10.3 as a global map at 4000 m and in Figure 10.4 as a longitudinal profile running down the middle of the Atlantic Ocean.

(a)

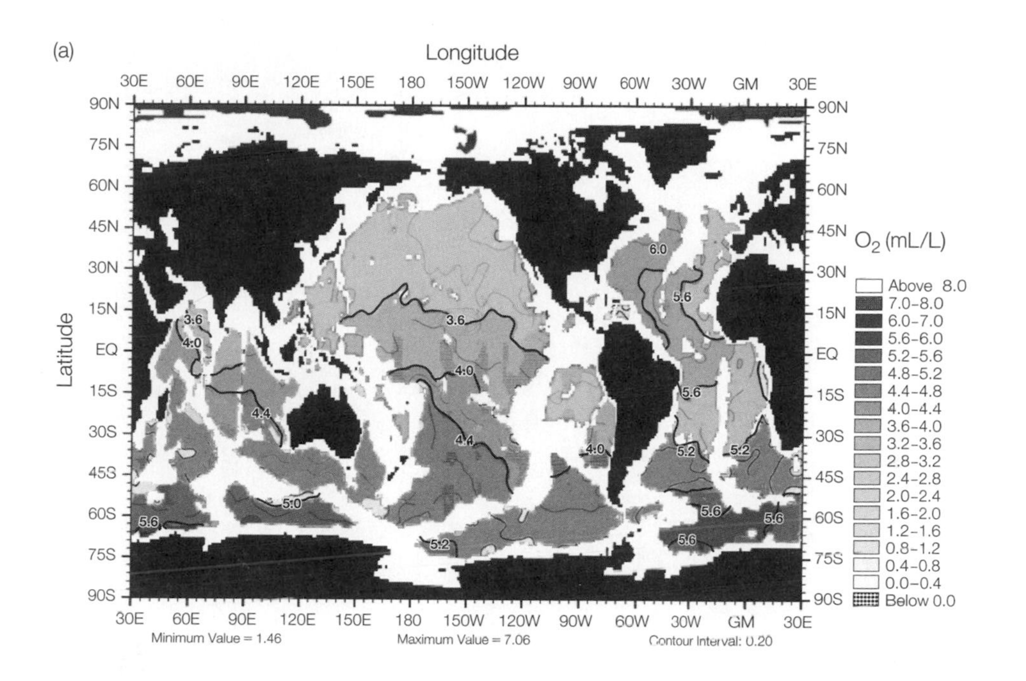

(b)

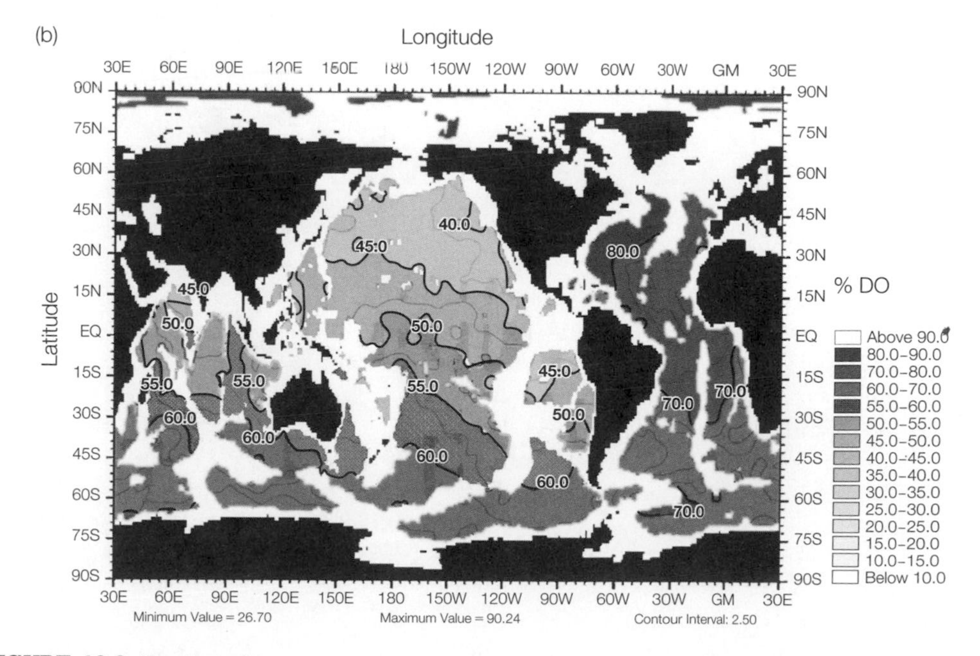

FIGURE 10.3 *(Continued)*

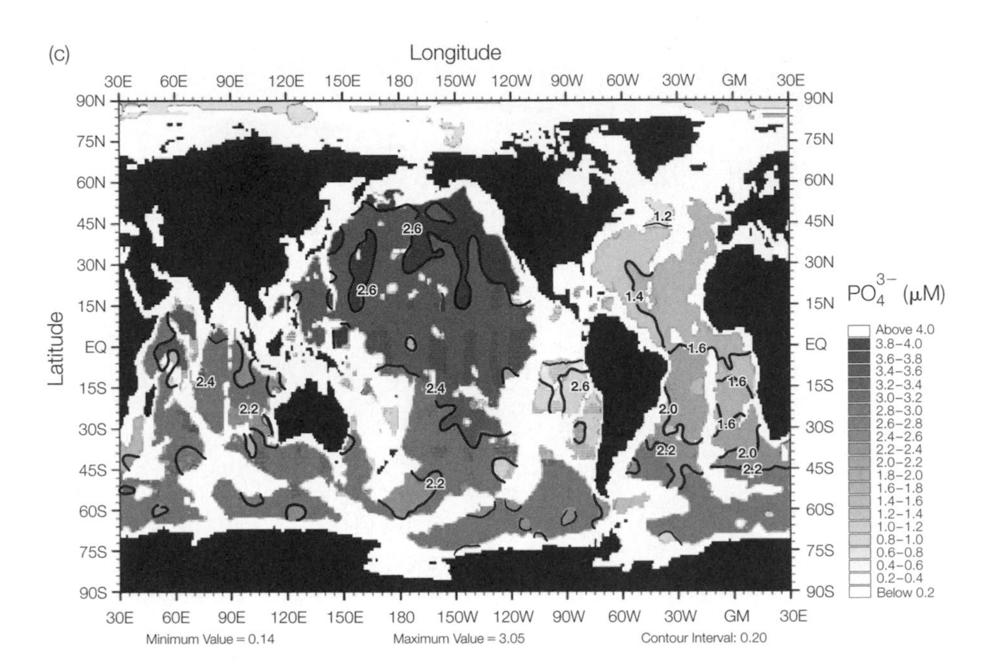

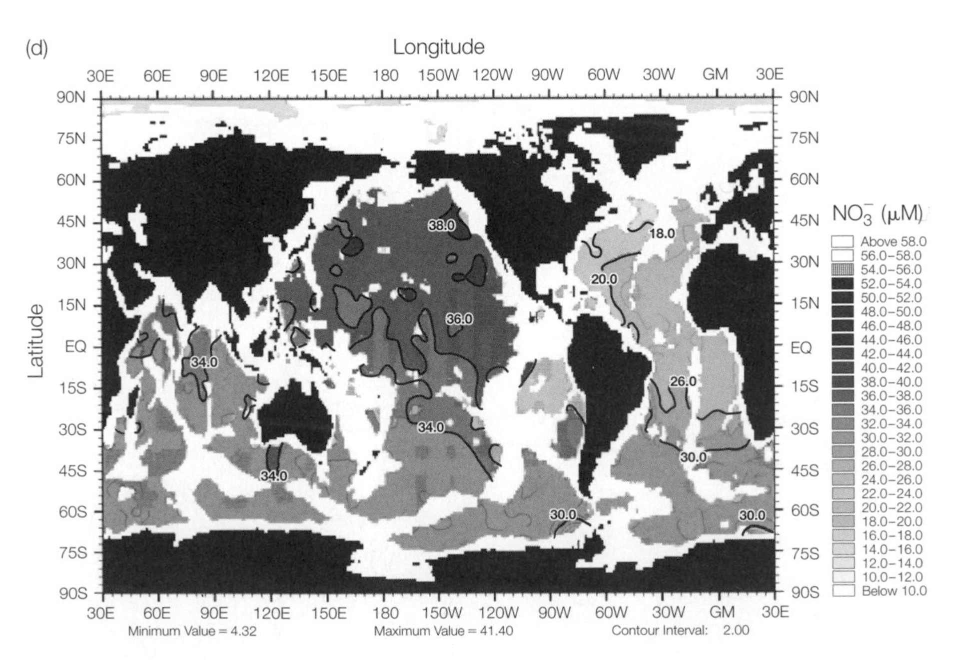

FIGURE 10.3 *(Continued)*

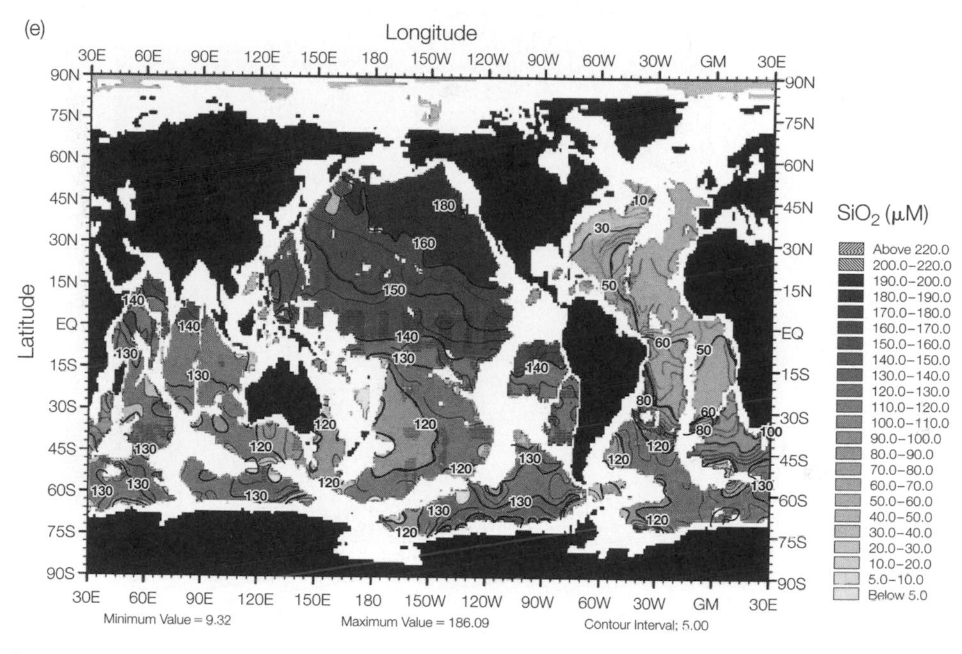

FIGURE 10.3

Horizontal distributions of (a) O_2 (mL/L), (b) percent saturation of O_2 (%), (c) phosphate (μmol/L), (d) nitrate (μmol/L), and (e) silicate (μmol/L) at 4000 m depth in the world's major ocean basins. The horizontal distribution of AOU was presented in Figure 8.2. *Source*: After Conkright, M. E., *et al.* (2002). *World Ocean Atlas 2001, Volume 4: Nutrients*, NOAA Atlas NESDIS 52, U.S. Government Printing Office (See companion website for color version.)

The horizontal gradients are the cause of the geographic variations depicted in Figure 10.1. Note that in Figures 10.1a and 10.3a, the O_2 depth profiles also reflect the influence of water temperature on gas solubility and density stratification. The uniformly high O_2 concentrations observed at high latitudes are due to intense cooling, which increases O_2 solubility and eliminates density stratification. At these locations, sinking and vertical mixing occur throughout the water column, thereby transporting O_2 to the deep sea. The effect of temperature is eliminated by viewing oxygen as its percent saturation (Figure 10.3b) or AOU (Figure 8.2). Subsurface concentrations decrease with increasing age of the water mass because O_2 is removed during the remineralization of POM.

The O_2 content of the surface waters is lower at mid-latitudes because of higher temperatures, which lead to lower gas solubility. As shown in Figure 10.1a, the thermocline is characterized by a concentration minimum that increases in intensity from the Atlantic to the North Pacific. Note that the O_2 minimum is less pronounced in the vertical profile from 45°S as compared to 9°N in the Atlantic Ocean because of close proximity to the site of AABW formation. Mid-water phosphate and nitrate maxima

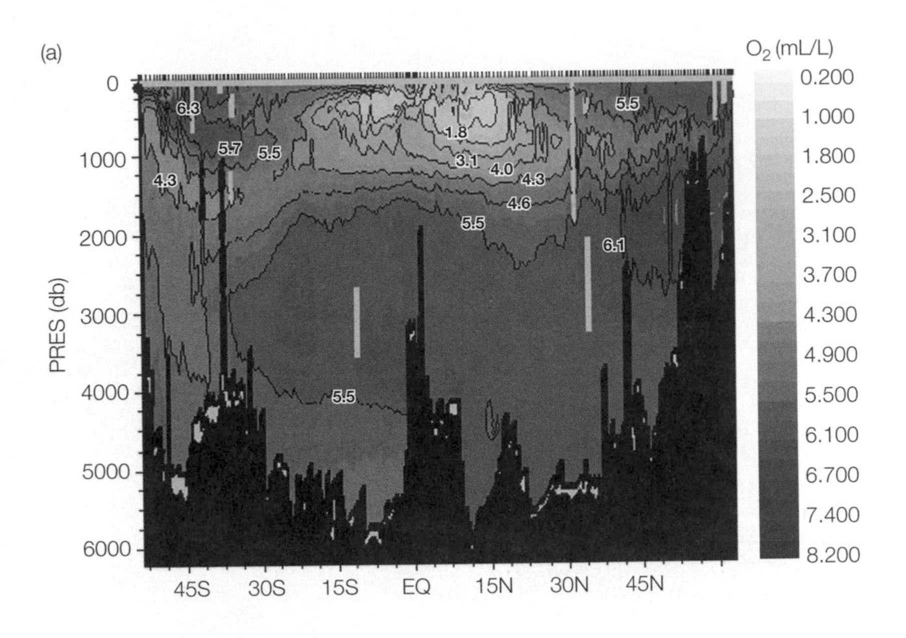

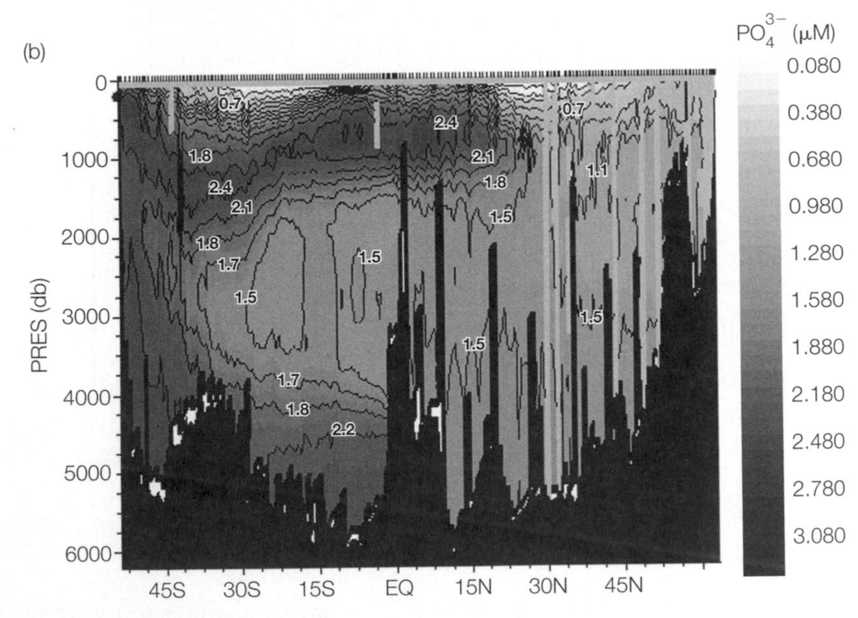

FIGURE 10.4 *(Continued)*

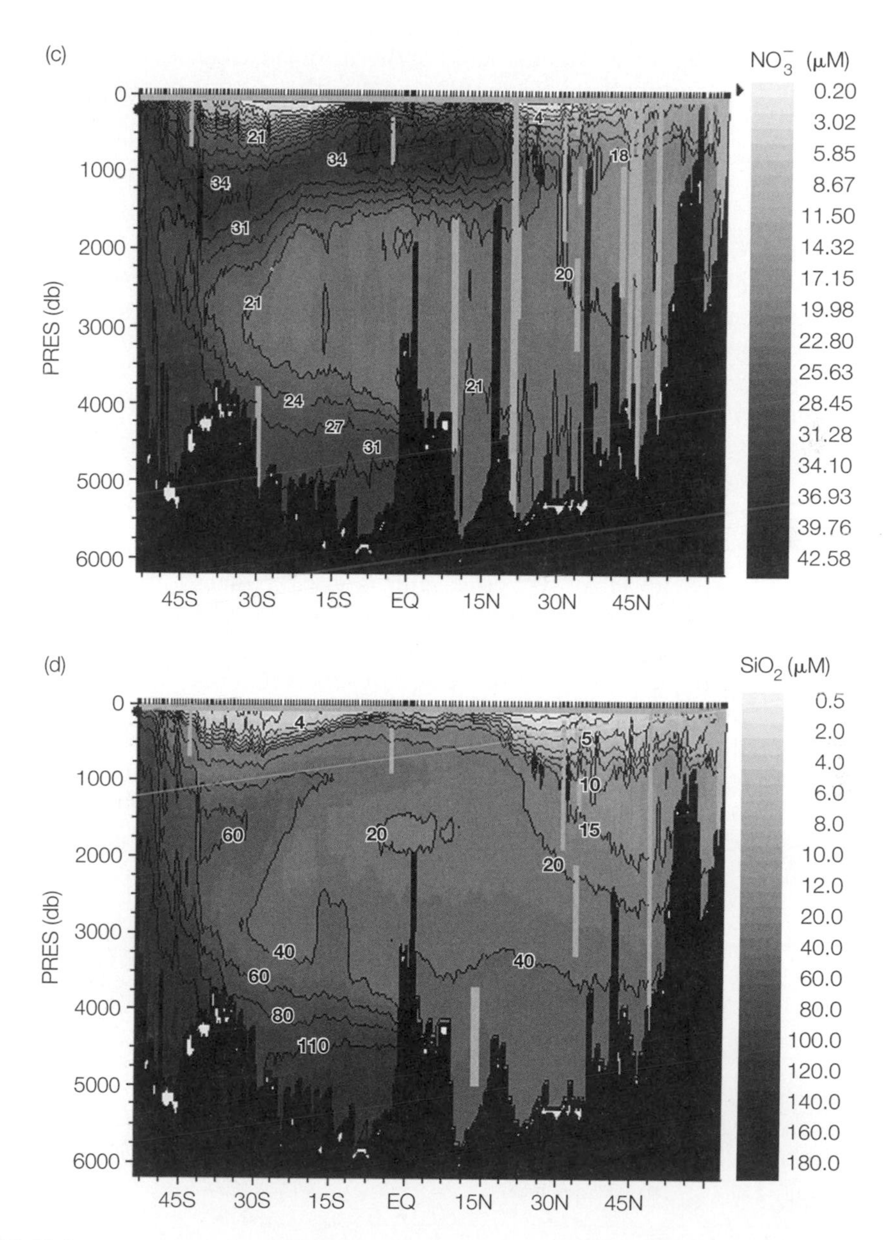

FIGURE 10.4

Longitudinal cross sections of through the Atlantic Ocean at approximately 25°W of (a) O_2 (mL/L), (b) phosphate (μmol/L), (c) nitrate (μmol/L), and (d) silicate concentrations (μmol/L). These are WOCE data (Cruise Track AO16) from the Java Ocean Atlas. Salinity and temperature cross sections are presented in Figure 4.14. *Source*: After Conkright, M. E., *et al*. (2002). *World Ocean Atlas 2001, Volume 4: Nutrients*, NOAA Atlas NESDIS 52, U.S. Government Printing Office. (See companion website for color version.)

are present only in the Atlantic and Southern Indian Oceans. This maximum coincides with the OMZ. By the time deep-water reaches the Northern Indian and Pacific Oceans, its nutrient concentrations have increased to a level that matches the bottom of the thermocline, thus subsuming the mid-water maximum.

The silicate depth profiles are notably different from those of nitrate and phosphate. Because this element is remineralized by the relatively slow dissolution of hard parts, it has a low surface-water recycling efficiency. This leads to a higher degree of both vertical and horizontal segregation. In Chapter 9, $C_{deep}/C_{surface}$ was used as a measure of the intensity of vertical segregation. The concentration difference, $C_{deep} - C_{surface}$, can also be used as a measure of vertical segregation. This concentration difference is greater in the Pacific than in the Atlantic because of horizontal segregation and can be quantified as the ratio:

$$\frac{(C_{deep} - C_{surface})_{Pacific}}{(C_{deep} - C_{surface})_{Atlantic}}.$$

As indicated in Table 10.1, this ratio is greater than 1 for all the biolimiting elements, demonstrating that their deepwater concentrations are greater in the Pacific than in the Atlantic Ocean since at most locations, $C_{surface}$ is much less than C_{deep}. The higher the ratio, the greater the degree of horizontal segregation. The relatively slow remineralization of silicate is responsible for its greater enrichment in the deep waters of the Pacific. The high degree of horizontal segregation exhibited by barium suggests that this element is also transported to the deep sea as a component of a hard part. As noted in Chapter 9, some plankton appear to precipitate barite ($BaSO_4$) and some barite seems to be formed abiotically in the water column from precipitation of remineralized barium.

Table 10.1 Horizontal Segregation of Biologically Utilized Elements between the Deep Atlantic and the Deep Pacific.

Element	$\frac{(C_{deep} - C_{surface})_{Pacific}}{(C_{deep} - C_{surface})_{Atlantic}}$
Nitrogen (as NO_3^-)	2
Phosphorus	2
Carbon	3
Silicon	5
Barium	4

Source: From Broecker, W. S. (1974). Chemical Oceanography, *Harcourt, Brace and Jovanovich Publishers, p. 23.*

10.4 HORIZONTAL SEGREGATION AND THE REDFIELD–RICHARDS RATIO

As a deepwater mass ages, it accumulates the products of biogenic particle remineralization. For nitrogen and phosphorus, the resulting changes in water chemistry can be predicted by assuming that (1) the degrading POM has the Redfield–Richards ratio of C to N to P and (2) the POM is completely remineralized by the process of aerobic respiration as presented in Eq. 8.4. If so, the regeneration of 1 mol of phosphate would require the removal of 138 mol of O_2 while producing 16 mol of nitrate.[1] Thus, deepwater nitrogen concentrations should increase 16 times faster than those of phosphorus. Likewise, the O_2 concentrations should decline 138 times faster than the phosphorus increases. To determine whether in situ nutrient regeneration conforms to this ideal behavior, the deep-water concentrations shown in Figure 10.1 (from 2500 m, the core of NADW) are replotted in Figure 10.5. If the nutrients have been regenerated by the ideal behavior just described, the deep-water nitrate and phosphate concentrations should generate a best-fit line with a slope equal to 16. The slope of the best-fit regression line of O_2 versus phosphate should be −138.

Figure 10.5 demonstrates that considerable deviations from the ideal behavior exist. Some of the reasons for this were described in Chapter 9. For example, the average oxygen content of plankton is likely much lower than the Redfield-Richards ratio, requiring that more O_2 be consumed per mole of P remineralized such that the $\Delta O_2/\Delta PO_4^{3-}$ is more likely to be closer to 150 than 138. Some of the deviations reflect decomposition of POM that is not plankton biomass, such as feces, molts, and animal tissues, and, hence, does not necessarily have the Redfield-Richards elemental ratios. Incomplete oxidation of plankton detritus can also lead to deviations if the organic compounds that are preferentially degraded do not have the Redfield-Richards ratio. For example, the preferential decomposition of nitrogen-rich compounds, such as proteins, regenerates more nitrate than would be predicted from the accompanying changes in O_2 or phosphate. Mixing with adjacent water masses can also alter nutrient and O_2 ratios. For example, AABW has a much higher phosphate concentration than NADW relative to their O_2 concentrations. As discussed later, this is a consequence of how the water masses formed.

Deviations from the Redfield-Richards ratio can also be caused by remineralization that proceeds by processes other than aerobic respiration. As shown in Figure 10.6, water samples obtained from depths where denitrification has occurred have lower dissolved inorganic nitrogen concentrations ($[NO_3^-] + [NO_2^-]$) than would be predicted from their AOU. During denitrification, nitrate is first reduced to nitrite and then to N_2,

[1] In oxic seawater, ammonium released during the remineralization of organic matter is readily oxidized to nitrite and then to nitrate by the actions of nitrifying bacteria. Thus, ammonium and nitrite concentrations in oxic seawater tend to be much lower than nitrate, so tracking the nitrate concentration is an effective measure of how much organic remineralization has occurred.

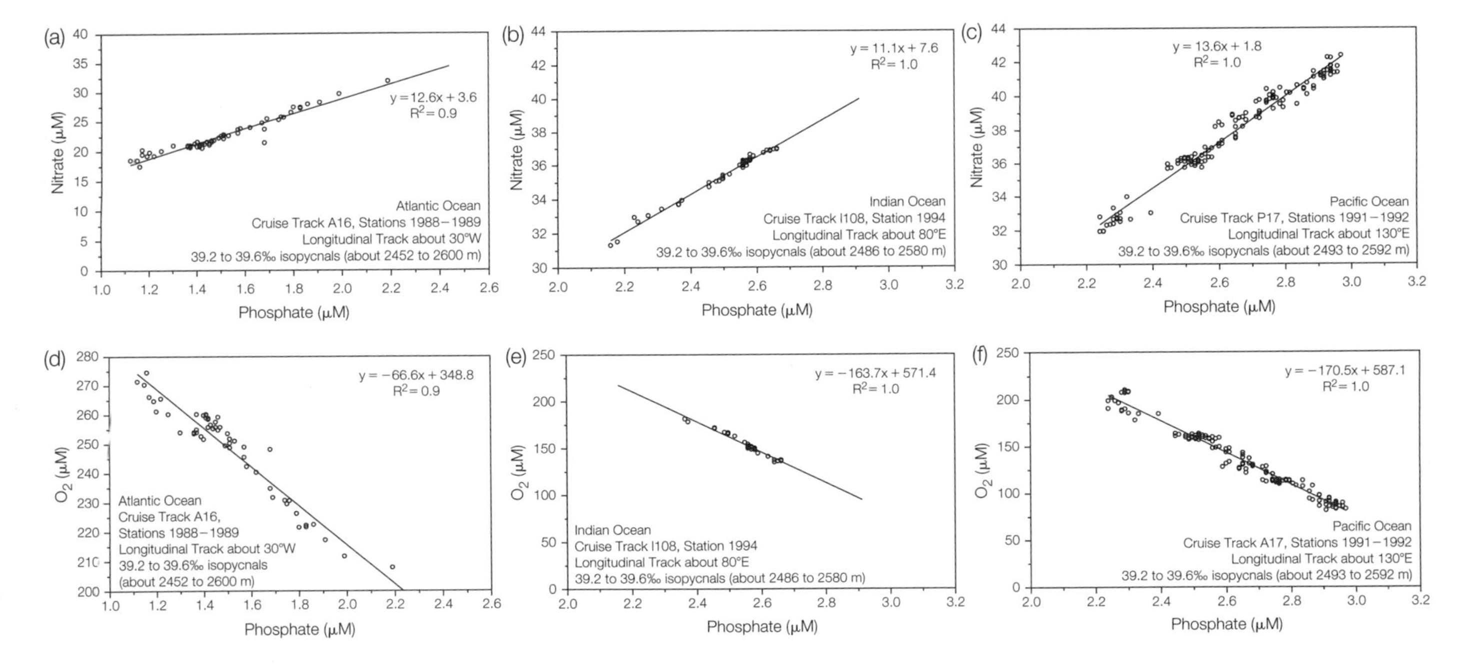

FIGURE 10.5

Nitrate versus phosphate concentrations at 2500 m in the (a) Atlantic, (b) Indian, and (c) Pacific Oceans. Dissolved oxygen versus phosphate concentrations at 2500 m in the (d) Atlantic, (e) Indian, and (f) Pacific Oceans. The slopes of these lines represent the proportions by which these constituent concentrations are altered by the remineralization of POM in the deep sea. These data are replotted from Figure 10.1. *Source*: From Conkright, M. E., *et al.* (2002). *World Ocean Atlas 2001, Volume 4: Nutrients*, NOAA Atlas NESDIS 52, U.S. Government Printing Office.

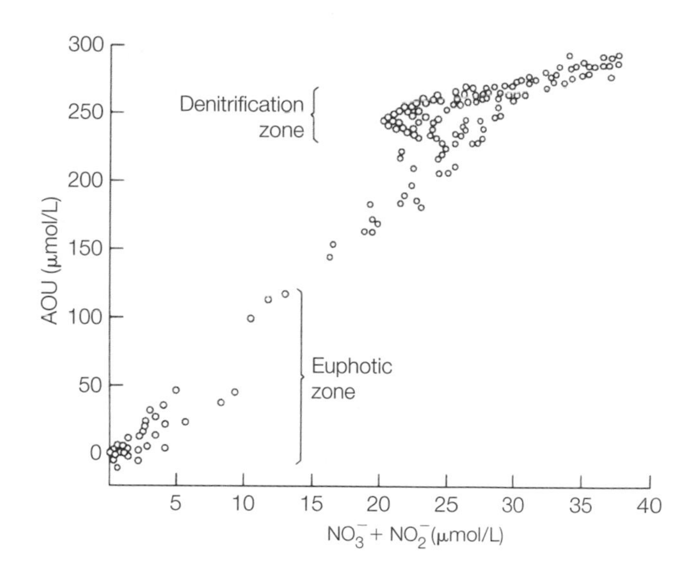

FIGURE 10.6

AOU versus NO_3^- + NO_2^- in the waters of the Arabian Sea. *Source*: From Deuser, W. G., *et al.* (1978). *Deep Sea Research* 25, 431–445.

so the sum of the concentrations of both species, $[NO_3^-] + [NO_2^-]$, must be considered when assessing the regeneration of dissolved inorganic nitrogen in O_2-deficient waters.[2] In such waters, dissolved inorganic fixed nitrogen appears to be "missing," because denitrification converts it to unfixed nitrogen, namely N_2.

Because phosphate is released during remineralization with no decrease in O_2, the $\Delta O_2/\Delta PO_4^{3-}$ produced via denitrification should be lower than that predicted by the aerobic respiration of Redfield-Richards planktonic detritus. To reach the suboxic conditions required for denitrification requires the aerobic respiration of a considerable amount of POM and, hence, release of phosphate. Thus, $\Delta O_2/\Delta PO_4^{3-}$ ratios less than 138 are most likely to be found in waters with high phosphate concentrations. The prevalence of denitrification in deep waters is suggested by their low (14.7) average N-to-P ratio (Figure 8.3). Areas where the OMZ are pronounced, such as coastal upwelling areas, have particularly low N-to-P ratios as shown in Figure 10.7.

10.5 PREFORMED NUTRIENTS

If the plankton assimilation of nutrients was complete, a newly formed deepwater mass should have no nutrients. Such a water mass should also have an O_2 concentration

[2] Some of the missing nitrogen is also likely due to the anammox reaction, in which ammonia is oxidized to N_2 by anaerobic microbes.

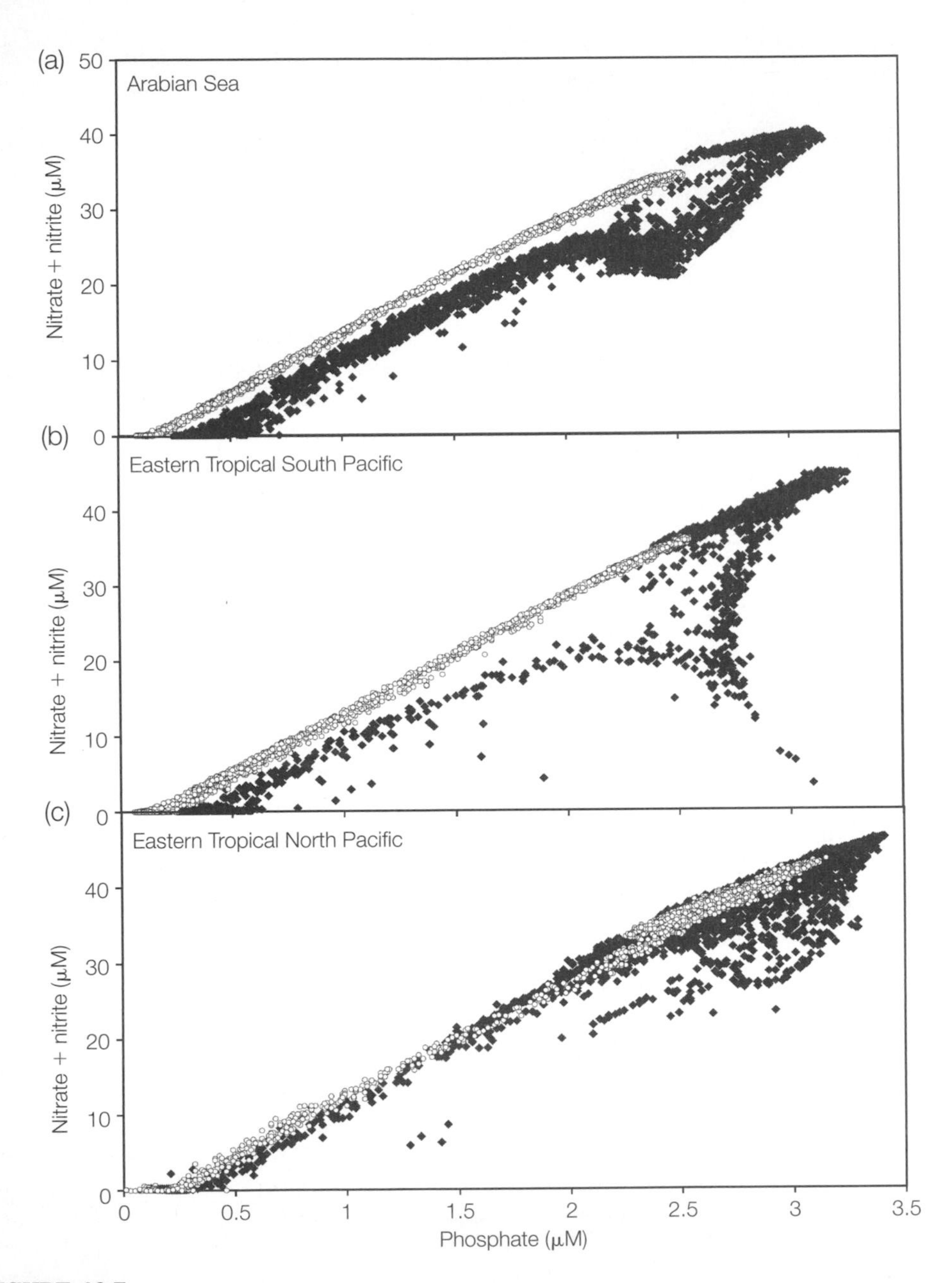

FIGURE 10.7

Nutrient chemistry of coastal upwelling regions supporting anoxic water columns and influenced by N_2 production (black diamonds), compared to the nutrient chemistry of adjacent water bodies unaffected by water-column anoxic conditions (open circles). *Source*: From Canfield, D. E. (2006). *Geochimica et Cosmochimica Acta* 70, 5753–5765.

equal to its NAEC. Extrapolating the best-fit lines in Figures 10.5d through 10.7f back to an O_2 concentration equivalent to the deep waters' NAEC suggests that a nonzero level of phosphate was present in these deep waters at the time of their formation. The same can be demonstrated for nitrate. Thus, phytoplankton growth in the source surface waters was likely limited by some other factor, such as temperature, silica, or the availability of essential trace metals, like iron. When a nutrient is not limiting, it is not completely extracted from the surface waters by the plankton and, hence, sinks as a deep-water mass is formed. This background concentration is termed the *preformed* nutrient concentration.

As illustrated in Figure 10.8, surface-water phosphate and nitrate concentrations are uniformly low at all latitudes below 45°. In contrast, high-latitude surface waters are characterized by much higher concentrations. This causes the phosphate concentration in new NADW to range from 0.7 to 0.9 μM and 1.8 to 2.0 μM in AABW. As the water masses sink and travel through the ocean basins, their phosphate concentrations increase as sinking POM is remineralized. Thus, the phosphate concentration at any location can be thought of as the sum of the preformed, phosphate $\left(\left[PO_4^{3-}\right]^\circ\right)$, and the phosphate added via remineralization.

Assuming Redfield-Richards behavior, i.e., the aerobic respiration of planktonic detritus, the concentration of remineralized phosphate can be estimated from AOU/138. Thus, the preformed phosphate concentration $\left(\left[PO_4^{3-}\right]^\circ\right)$ is given by the following mass balance equation:

$$\left[PO_4^{3-}\right]_{observed} = \left[PO_4^{3-}\right]^\circ + \frac{AOU}{138} \qquad (10.1)$$

(This assumes that the newly formed deep waters were at 100% saturation with respect to O_2 when they were last at the sea surface. As noted in Chapter 6, the percent saturation of surface water is geographically and temporally variable with an average value of 103%.)[3]

Longitudinal cross sections of preformed phosphate concentrations in the Atlantic and Pacific Oceans are presented in Figure 10.9. The presence of NADW moving through the Atlantic centered at about 2500 m is clearly present as a tongue of lower preformed phosphate concentration relative to that of the overlying AAIW and underlying AABW. The higher preformed phosphate concentrations found in AAIW and AABW are thought to be caused by relatively rapid sinking of newly upwelled CDW, giving plankton insufficient time to extract nutrients. In contrast, the waters that feed the formation of NADW have spent much longer periods at the sea surface, enabling plankton to reduce nutrient concentrations to somewhat lower levels than in the AABW. Another possibility is that plankton at the sites of AABW formation (> 70°S) are far more iron limited so that although time, nitrogen, and phosphorus have been sufficient, growth cannot proceed due to lack of iron.

[3] A similar procedure has been used for nitrogen and silica to isolate preformed concentrations from the remineralized concentrations. These quantities are called N* and Si*.

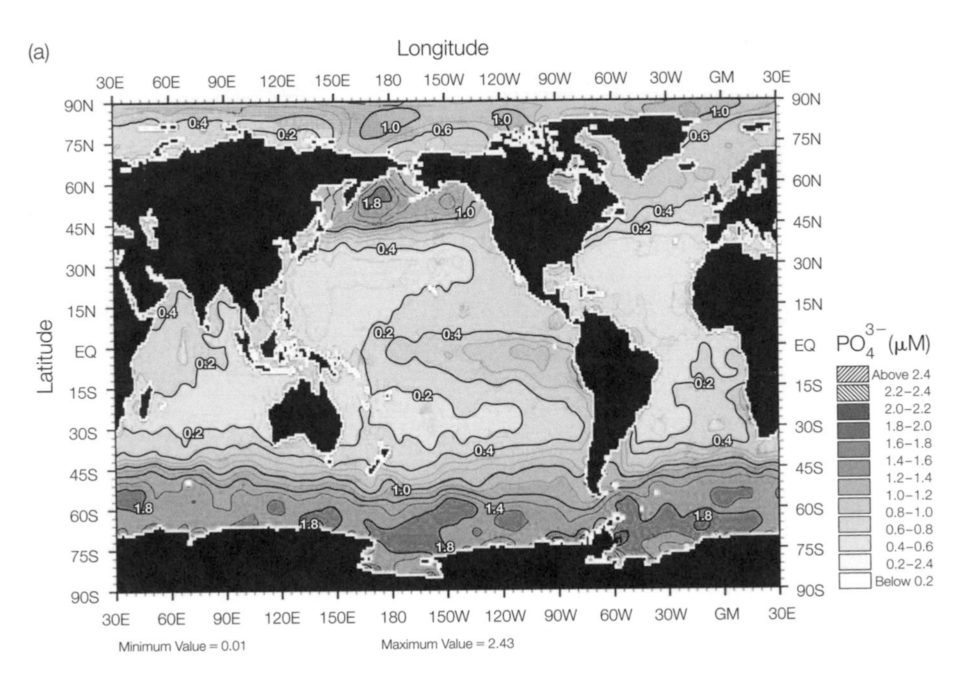

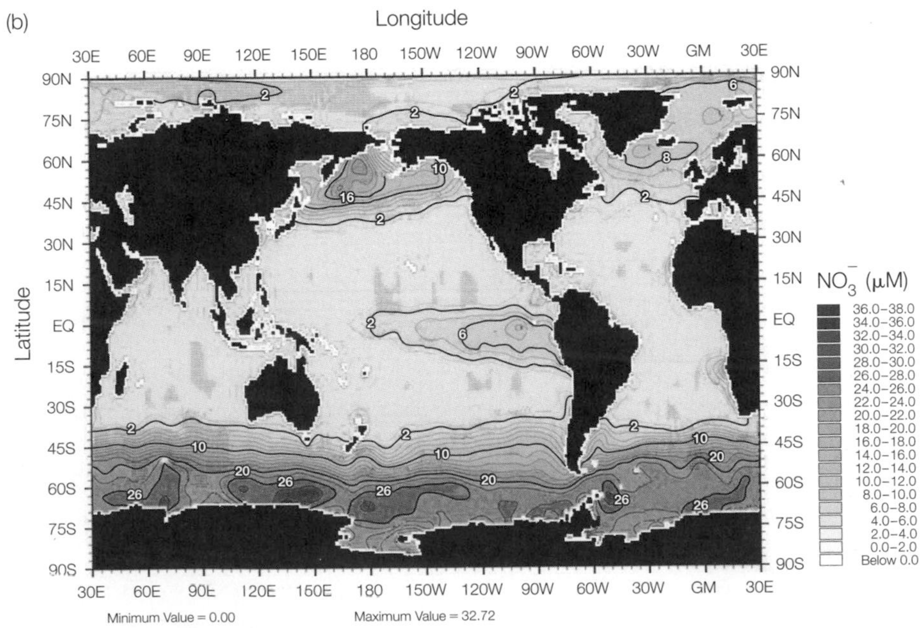

FIGURE 10.8 *(Continued)*

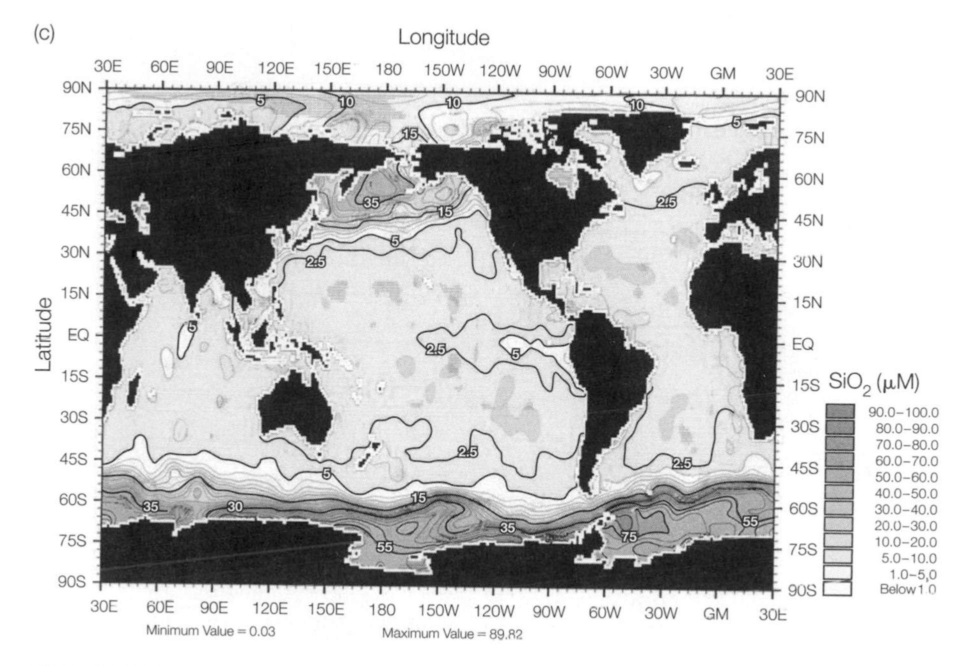

FIGURE 10.8

Annual mean surface water nutrient concentrations (μM) of (a) phosphate, (b) nitrate, and (c) silicate. *Source*: After Conkright, M. E., *et al.* (2002). *World Ocean Atlas 2001, Volume 4: Nutrients*, NOAA Atlas NESDIS 52, U.S. Government Printing Office (See companion website for color version.)

In the Pacific Ocean, most of the waters at 2500 m have a preformed phosphate concentration intermediate between NADW and AABW. Because preformed phosphate is a conservative tracer, it can be used to estimate the proportions of NADW and AABW present in the deep zones of the ocean basins. The average deep-water preformed phosphate concentration is 1.4 μM. This concentration would result from an equal-volume admixture of NADW and AABW. This conservative mixing estimate is based on the assumption that the preformed phosphate concentrations of the end-member water masses have remained constant over time scales at least as long as the mixing time of the ocean.

Preformed phosphate represents an important component of the dissolved phosphate reservoir. This is seen in the relatively low ratio of remineralized to preformed phosphate, which ranges from 0.36 to 0.70. Based on the Broecker Box model presented in Chapter 9, only about 1% of the phosphate escapes from the ocean on any given mixing cycle to become buried in the sediments. So to a first approximation, only 36 to 70% of the phosphate in a deepwater mass originates from remineralization. (The wide range in percentage reflects geographic variability related to the age of the water

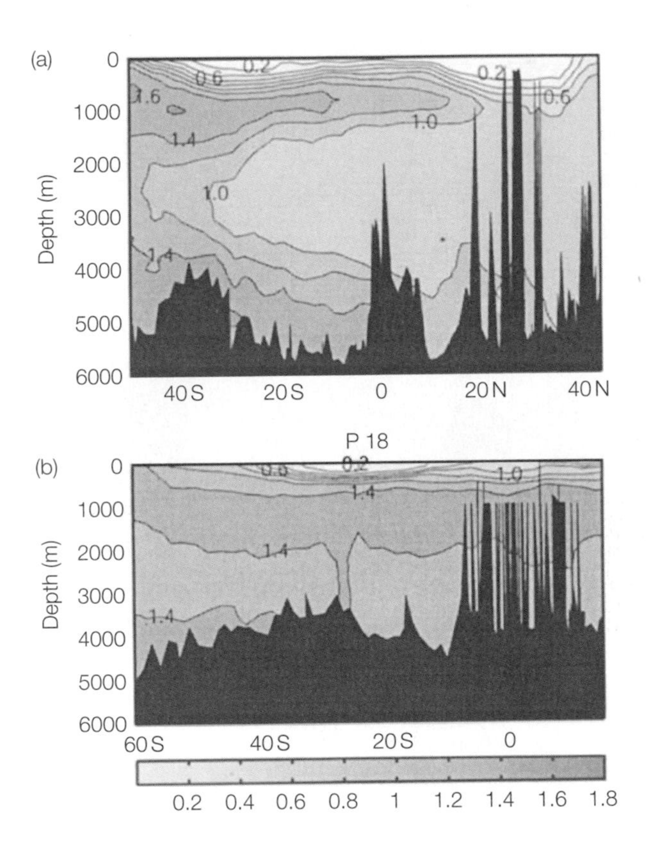

FIGURE 10.9

Longitudinal cross section of preformed phosphate concentrations (μM) in the (a) Atlantic and (b) Pacific Oceans. Constructed from WOCE-JGOFS data, cruises A20, A17, and P18.
Source: From Ito, T. and M. J. Follows (2005). *Journal of Marine Research* 63, 813–839.

mass.) Some oceanographers have interpreted this as evidence that nitrogen, rather than phosphorus, is the nutrient limiting plankton production.

Some of the preformed nutrients carried by the intermediate waters are transported into the thermocline by turbulent mixing and vertical advection. The latter is most intense in the equatorial divergence zones where the Trade Winds induce upwelling. This flow path is shown in Figure W10.2 illustrating how SAMW supplies nutrients to the equatorial thermocline. In the North Pacific, nutrients also appear to be resupplied to the thermocline by enhanced turbulence and vertical advection, possibly related to the effect of the tides as NPIW moves over the rugged topography around the Kurile Islands.

The preceding discussion suggests that the Southern Ocean plays a key role in regulating nutrient availability throughout the world ocean. The mechanisms by which nutrient concentrations are set in AAIW, SAMW, and AABW are complex. A schematic model is shown in Figure 10.10. Upwelling of CDW brings nutrient-rich water to the

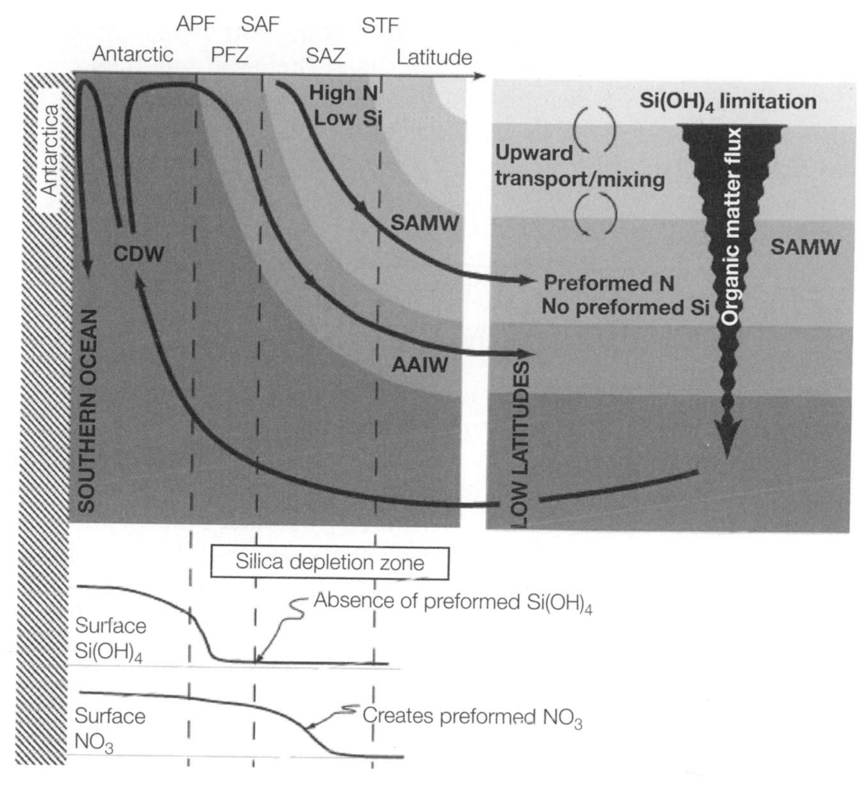

FIGURE 10.10

Southern Ocean control on thermocline nutrient concentrations. APF = Antarctic polar front, PFZ = polar frontal zone, SAF = subantarctic front, SAZ = subantarctic zone, STF = subtropical front. *Source*: From Sarmiento, J. L., *et al.* (2003). *Nature* 427, 56–60.

surface in the Southern Ocean. These waters are relatively rich in phosphate and silica, but depleted in iron. Under silica-rich conditions, diatoms dominant. At very high latitudes, where AABW forms, dust deposition is low, so plankton growth is iron limited during polar summers. Under iron-limited conditions, diatoms are less efficient with their silica and, hence, their silica requirements increase. The net result is that silica is stripped out the surface waters and returned to the deep sea in the form of sinking hard parts, leaving nitrate and phosphate behind. Eventually the hard parts dissolve supplying remineralized silica to CDW. Thus, the Southern Ocean acts as a silica trap.

The southern flank of the upwelling CDW sinks around the continent of Antarctica to become AABW. The northern flank of upwelling CDW is transported by surface currents, first into the polar frontal zone (PFZ) where AAIW forms, and then into the subantarctic zone (SAZ) where SAMW forms. At these latitudes, dust deposition is high enough to reduce iron limitation of the diatoms. As the surface waters move northward, they become increasingly depleted in silica. Thus, AAIW and SAMW both contain preformed phosphate and nitrate but little silica. The most important consequence of this silica trapping is to largely restrict diatom growth to the Southern Ocean. The only other regions where diatoms are the dominant plankton are the subpolar North Pacific and North Atlantic, where surface silica concentrations are high.

10.6 FEEDBACK RELATIONS IN THE MARINE BIOGEOCHEMICAL NUTRIENT CYCLE

Oceanographers think that the nutrients are present in steady-state concentrations—at least over time scales equal to their oceanic residence times. These concentrations are thought to hold within a narrow range by a set of negative feedbacks that provide a "dynamic homeostasis." This is fortunate because the biogeochemical cycles of the nutrients have the potential to be perturbed by several physical and biological phenomena, including changes in climate, plate tectonics, and plankton species composition, not to mention changes in extraterrestrial processes such as insolation and delivery of extraterrestrial materials. These perturbations affect the global nutrient cycles because they alter ocean circulation, sea level, erosion rates, and plankton species composition.

Fortunately, the ocean appears to have a set of negative feedback loops that minimize the impacts of such perturbations. A simple example is the hypothesis that increases in the rates of river input of nutrients will be balanced by an increased production of plankton biomass. This should increase the flux of sinking biogenic particles and, hence, their burial rates in marine sediments. Some scientists liken the negative feedbacks to the self-regulating mechanisms that stabilize the biochemistry of individual organisms. The current biogeochemical state of the planet is considered to be the result of, as well as stabilized by, these negative feedbacks. This concept is termed the *Gaia hypothesis*. Much oceanographic research is directed at understanding how these feedbacks operate so that the impact of human activities can be predicted.

The importance of the biogeochemical cycle of organic matter in regulating atmospheric O_2 levels was described in Chapter 8. Organic matter production and destruction, and, hence, nutrient cycling, are also important in stabilizing atmospheric CO_2 levels and, hence, play a role in determining global climate. In other words, the biogeochemical cycles of O_2, carbon, nitrogen, phosphorus, silica, and, to some degree, iron are interlinked. Some of these interlinks function as negative feedbacks and others as positive feedbacks. The nature of these feedbacks will be discussed in Chapter 24 from the perspective of shifts in nutrient limitation (N versus P versus Fe). The consequences of this nutrient cycling for carbon storage and, hence, atmospheric CO_2 levels are covered in Chapter 25.

Considerable evidence exists that human activities have already perturbed parts of the interlinked global biogeochemical cycles of the nutrients, micronutrients, carbon, and O_2. Some of these perturbations are the consequence of climate change and others are associated with changes in the rates of input of nutrients and iron to the sea. As discussed in Chapter 8, a slight decrease in P_{O_2} has been attributed to the burning of fossil fuels. Slight decreases in the O_2 concentrations in the ocean over the past several decades have been attributed to global warming. Some oceanographers report that nutrient concentrations in parts of the ocean are changing and some claim to have detected changes in the deep-water nutrient N-to-P ratios. Marine biologists have observed shifts in plankton species. The degree to which these changes are "natural" versus directly related to human activities is presently unknown. Thus, it is critical that we obtain an understanding of what the "natural" scales of variability are in the elemental biogeochemical cycles, climate, and plankton community. Given the growing scale of human perturbations, we are not likely to have the opportunity to easily define natural variability. In other words, human-driven perturbations have engaged the crustal-ocean-atmosphere factory in a global experiment that will test the degree to which negative feedbacks can stabilize the biogeochemistry of the oceans and, by connection, global climate and, hence, terrestrial life.

CHAPTER 11
Trace Elements in Seawater

All figures are available on the companion website in color (if applicable).

11.1 INTRODUCTION

Virtually every naturally occurring element has been detected in seawater. Those that are present at concentrations less than 100 μmol/kg are termed *trace elements*. As shown in Figure 11.1, the trace elements are mostly metals and metalloids. (Oceanographers tend to use the terms *trace metals* and *trace elements* interchangeably.)

In contrast to their rather low dissolved concentrations in seawater, some of the trace metals, e.g., iron and aluminum, along with oxygen and silicon, comprise the bulk of Earth's crust. Some trace elements are micronutrients and, hence, have the potential to control plankton species composition and productivity. This provides a connection in the crustal-ocean-atmosphere factory to the carbon cycle and global climate.

Most trace elements are removed from the ocean by incorporation into settling particles that are subsequently buried in the sediments. The depositional histories of some of these trace elements represent paleoceanographic records that are used to deduce past changes in circulation and biological productivity. Many of the trace metals undergo biogeochemical cycling in seawater and the sediments that is largely controlled by the biological processes that mediate redox reactions and the formation and aggregation of particulate matter.

Most of our understanding of the marine chemistry of trace metals rests on research done since 1970. Prior to this, the accuracy of concentration measurements was limited by lack of instrumental sensitivity and contamination problems. The latter is a consequence of the ubiquitous presence of metal in the hulls of research vessels, paint, hydrowires, sampling bottles, and laboratories. To surmount these problems, ultra-clean sampling and analysis techniques have been developed. New methods such as anodic stripping voltammetry are providing a means by which concentration measurements can be made directly in seawater and pore waters. Most other methods require the laborious isolation of the trace metals from the sample prior to analysis to eliminate interferences caused by the highly concentrated major ions.

The role of the ocean in the global biogeochemical cycling of trace elements is the subject of this chapter. The processes that introduce and remove these elements from the ocean are summarized in Figure 11.2 and discussed in detail next.

Major Elements	M	> 50 mM
	m	10 to 50 mM
Minor Elements	m	0.1 to 10 mM
	mm	0.1 to 100μM
Trace Elements	t	1 to 100 nM
	tt	<1 nM

Dominant Oxidation number

Alkali Metals	Alkaline Earth Metals			Transition Metals									Metalloids	Nonmetals			Halogens	Noble Gases
1 H I **t**																		2 He 0 **t**
3 Li I **mm**	4 Be II **tt**												5 B III **m**	6 C IV **m**	7 N V **m**	8 O 0 **m**	9 F -I **m**	10 Ne 0 **t**
11 Na I **M**	12 Mg II **m**												13 Al III **mm**	14 Si IV **mm**	15 P V **mm**	16 S VI **m**	17 Cl -I **M**	18 Ar 0 **mm**
19 K I **m**	20 Ca II **m**	21 Sc III **tt**		22 Ti IV ? **tt**	23 V V **t**	24 Cr III **t**	25 Mn II **tt**	26 Fe III **tt**	27 Co II **tt**	28 Ni II **t**	29 Cu II **t**	30 Zn II **t**	31 Ga III **tt**	32 Ge IV **tt**	33 As III **t**	34 Se IV,VI **t**	35 Br -I **m**	36 Kr 0 **t**
37 Rb I **mm**	38 Sr II **mm**	39 Y III **tt**		40 Zr IV **tt**	41 Nb V **tt**	42 Mo VI **tt**	43 (Tc) VII **tt**	44 Ru IV **tt**	45 Rh III **tt**	46 Pd II **tt**	47 Ag I **tt**	48 Cd II **tt**	49 In III **tt**	50 Sn IV **tt**	51 Sb V **mm**	52 Te VI **tt**	53 I V **mm**	54 Xe 0 **tt**
55 Cs I **tt**	56 Ba II **mm**	57 La III **tt**	1	72 H IV **tt**	73 Ta V **tt**	74 W VI ? **tt**	75 Re II **tt**	76 Os (III, IV) **tt**	77 Ir III **t**	78 Pt II **tt**	79 Au III **tt**	80 Hg II **tt**	81 Tl I **tt**	82 Pb II **tt**	83 Bi III **tt**	84 Po VI **tt**	85 At **tt**	86 Rn 0 **tt**
87 Fr I **tt**	88 Ra II **tt**	89 Ac III **tt**	2	104 (Rf) ? ?	105 (Db) ? ?	106 (Dg) ? ?	107 (Bh) ? ?	108 (Hs) ? ?	109 (Mt) ? ?	110 (Uun) ? ?	111 (Uuu) ? ?	112 (Uub) ? ?		114 (Uuq) ? ?		116 (Uuh) ? ?		118 (Uuo) 0 ?

Rare Earth Elements															
Lanthanides	1	58 Ce III **tt**	59 Pr III **tt**	60 Nd III **tt**	61 (Pm) III **tt**	62 Sm III **tt**	63 Eu III **tt**	64 Gd III **tt**	65 Tb III **tt**	66 Dy III **tt**	67 Ho III **tt**	68 Er III **tt**	69 Tm III **tt**	70 Yb III **tt**	71 Lu III **tt**
Actinides	2	90 Th IV,VI **tt**	91 Pa V **tt**	92 U VI **t**	93 (Np) ? ?	94 (Pu) **tt**	95 (Am) ? ?	96 (Cm) ? ?	97 (Bk) ? ?	98 (Cf) ? ?	99 (Es) ? ?	100 (Fm) ? ?	101 (Md) ? ?	102 (No) ? ?	103 (Lr) ? ?

FIGURE 11.1

Periodic Table of Elements showing dominant oxidation states of each element and concentration classifications. Oxidation Numbers from Whitfield, M. (2001). *Advances in Marine Biology* 41, 3–128; and Bruland, K. W. (1983). *Chemical Oceangraphy*, Academic Press. pp. 157–220. Concentrations from MBARI Chemical Sensor Program, Monterey Bay Aquarium Research Institute, Periodic Table of Elements in the Ocean, (http://www.mbari.org/chemsensor/summary.html) accessed July 2008.

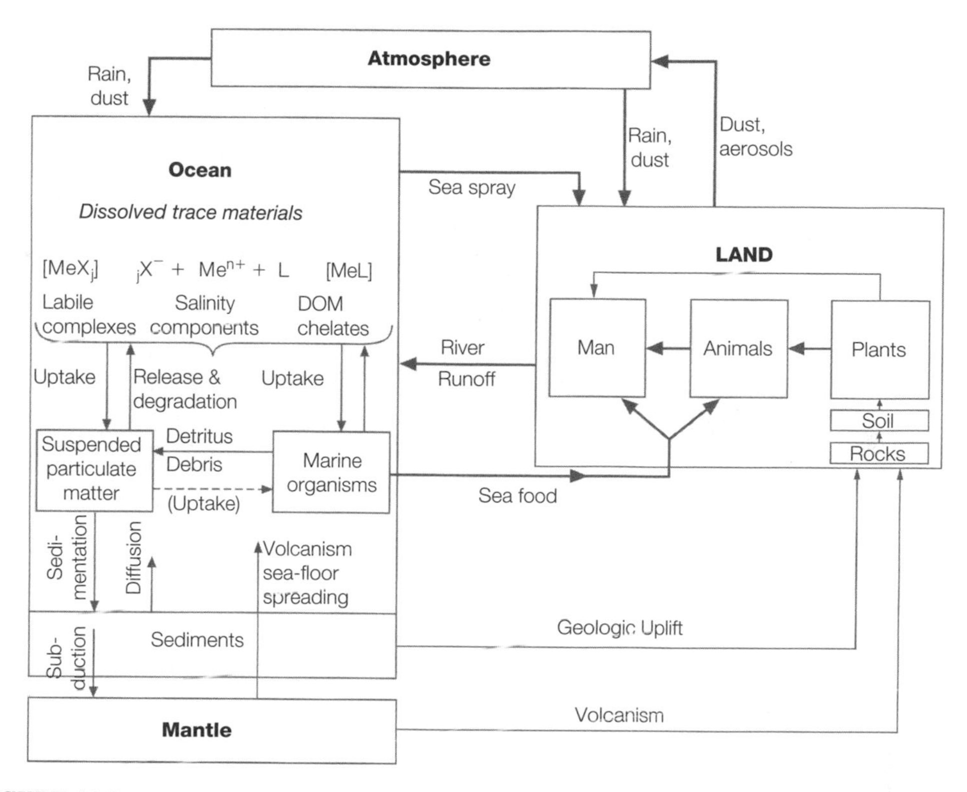

FIGURE 11.2

General box model summarizing the transport pathways and reservoirs for trace elements in the crustal-ocean-atmosphere factory.

11.2 SOURCES OF TRACE ELEMENTS TO THE OCEAN

The trace elements are introduced into seawater by river runoff, atmospheric transport, hydrothermal venting, groundwater seeps, diffusion from the sediments, and transport from outer space, usually as micrometeorites. The magnitudes of the first three of these fluxes, which are considered to be the major ones, are given in Table 11.1. Anthropogenic activities have significantly increased some of these fluxes, as discussed later.

11.2.1 River Runoff

Trace elements are discharged into the ocean in particulate and dissolved form as a component of river runoff and groundwater seeps. They are introduced into these waters during the chemical and mechanical weathering of crustal rocks. Thus, the chemical composition of river water is dependent on the composition of the rocks in the

Table 11.1 Elemental Fluxes to the World Ocean from Major Sources Excluding Modern Anthropogenic Inputs.

Element	Fluvial gross flux: total (particulate + dissolved)	Atmospheric flux: total (particulate + soluble)	Fluvial net dissolved flux	Hydrothermal axial dissolved flux	Atmospheric soluble flux	Fluvial gross particulate flux	Fluvial net particulate flux	Atmospheric particulate flux
Al	54×10^{12}	0.25×10^{12}	$3.5–6.0 \times 10^{10}$	$1.2–6.0 \times 10^{8}$	1.2×10^{10}	54×10^{12}	5.4×10^{12}	0.24×10^{12}
Fe	13.3×10^{12}	0.065×10^{12}	$0.54–2.3 \times 10^{10}$	$2.3–19 \times 10^{10}$	0.49×10^{10}	13×10^{12}	1.3×10^{12}	0.06×10^{12}
Mn	30×10^{10}	0.14×10^{10}	$0.49–0.55 \times 10^{10}$	$1.1–3.4 \times 10^{10}$	0.05×10^{10}	0.29×10^{12}	0.029×10^{12}	0.001×10^{12}
Ni	2.4×10^{10}	$<0.12 \times 10^{10}$	0.05×10^{10}	—	$<0.05 \times 10^{10}$	2×10^{10}	0.23×10^{10}	$<0.075 \times 10^{10}$
Co	0.56×10^{10}	0.004×10^{10}	$0.011–0.013 \times 10^{10}$	$6.6–68 \times 10^{5}$	1×10^{7}	0.52×10^{10}	0.05×10^{10}	0.003×10^{10}
Cr	3.1×10^{10}	$<0.10 \times 10^{10}$	0.036×10^{10}	—	$<0.01 \times 10^{10}$	3×10^{10}	0.3×10^{10}	$<0.09 \times 10^{10}$
V	5.2×10^{10}	$<0.07 \times 10^{10}$	0.07×10^{10}	—	$<0.02 \times 10^{10}$	5×10^{10}	0.5×10^{10}	$<0.05 \times 10^{10}$
Cu	2.5×10^{10}	0.06×10^{10}	$0.1–0.5 \times 10^{10}$	$3–13 \times 10^{8}$	2×10^{8}	2.4×10^{10}	0.24×10^{10}	0.04×10^{10}
Pb	0.75×10^{10}	0.13×10^{10}	$0.2–1.5 \times 10^{8}$	$2.7–110 \times 10^{5}$	4×10^{8}	0.75×10^{10}	0.075×10^{10}	0.09×10^{10}
Zn	6.0×10^{10}	0.31×10^{10}	$0.04–1.4 \times 10^{10}$	$0.12–0.32 \times 10^{10}$	0.14×10^{10}	6×10^{10}	0.6×10^{10}	0.17×10^{10}
Cd	0.15×10^{9}	$<0.03 \times 10^{9}$	3×10^{7}	—	$<2 \times 10^{7}$	0.14×10^{9}	0.014×10^{9}	$<0.7 \times 10^{9}$

Units are in mol/y. *Source: From Chester, R. (2003).* Marine Geochemistry, *2nd edition, Blackwell Publishing, p. 127.*

river drainage basin and local environmental conditions, such as rainfall and temperature, both of which influence the rate of chemical weathering. All of these factors are temporally and spatially variable, making it very difficult to estimate global river fluxes. This is unfortunate given the important role that river transport plays in the crustal-ocean-atmosphere factory. This subject is considered further in Chapter 21.

The data presented in Table 11.1 indicate that the fluvial gross river flux is the major source of trace metals to the oceans and that most of this flux is in particulate form (fluvial gross particulate flux). But the majority of this particulate flux is trapped within estuaries, primarily via settling, and, hence, is not released into the open ocean. As a result, the fluvial net particulate flux is only about 10% of the fluvial gross particulate flux. In seawater, most of this particulate metal remains in solid form due to low solubilities. The particulate metals eventually settle to the seafloor and are subsequently buried in the sediments. In the case of iron, a small fraction of the particulate pool does dissolve. In the surface waters, solubilization of particulate iron can provide a significant amount of this micronutrient to the phytoplankton.

As river water moves downstream into estuaries, mixing with seawater can promote processes that alter the dissolved metal concentrations. An example of the nonconservative behavior of iron in estuaries was given in Figure 5.1. Some of these processes are abiotic in nature, including cation exchange and precipitation, and are promoted by exposure to increasing pH, salinity, and ionic strength as river water mixes with ocean water. Some of these reactions cause the removal of dissolved metals from seawater and others result in their solubilization. Biogeochemical processes in estuaries alter dissolved metal concentrations as a result of uptake into hard and soft parts. These processes are particularly important in estuaries because of locally high biological productivity. In summary, extensive estuarine processing of the riverine transport of particulate and dissolved trace metals makes it essential that a "net" fluvial dissolved flux to the ocean be identified. Since the estuarine processes are temporally and spatially variable, estimating the global "net" fluvial dissolved flux is very difficult. Based on the estimates provided in Table 11.1, the net fluvial dissolved flux appears to be similar to that of the atmospheric and hydrothermal fluxes.

11.2.2 Groundwater Seeps

The contribution of chemicals to the ocean via groundwater seeps has recently been recognized as a potentially significant transport process that, in some cases, rivals river input. For example, nutrient inputs via groundwater can be large enough in some locations to significantly enhance local marine primary production. The groundwater transport of trace metals is likely to be complicated by changes in speciation that occur when the seeping fluids first comes into contact with oxic estuarine or coastal ocean waters.

As illustrated in Figure 11.3, when hydraulic conditions are favorable, freshwaters flow below the land surface toward the ocean. Subsurface flows are only possible if the water has voids to move through, such as cracks in crustal rocks or the spaces (pores) between sediment particles. Subsurface deposits with a high degree of pore spaces are

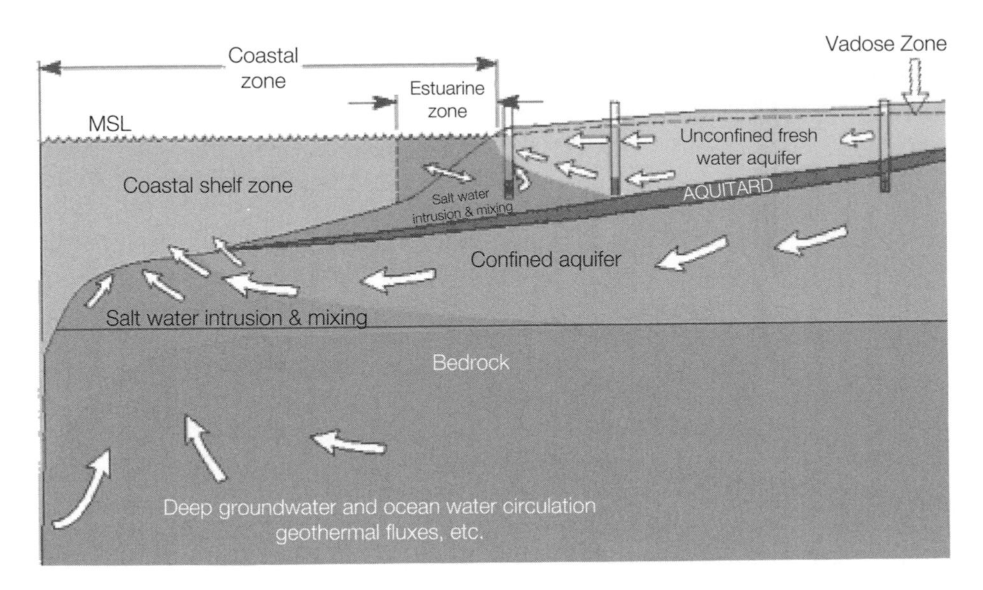

FIGURE 11.3

Movement of groundwater into the ocean illustrating that mixing with seawater can occur in marine sediments and soils located landward of the coastline. MSL = mean sea level.

very permeable to groundwater flow. These are referred to as *aquifers*. Most aquifers are ancient sediments that were deposited as a uniform layer that geological processes generally deform over time. Layers that are not very permeable are termed *aquitards*. Thus, groundwater flows tend to be confined to the aquifer layers.

Freshwaters enter aquifers via the percolation of surface waters, such as rain water, through soils and downward until the groundwater reaches an outcropping of the aquifer. This process is called aquifer recharge. If the hydrostatic pressures are strong, the freshwater flows through the aquifer and discharges into the subsurface waters of the coastal ocean or estuary. In some cases, hydrostatic pressures exerted by the ocean or estuary exceed those of the freshwater. These high pressures cause salty water to infiltrate into the inland aquifers. In this case, the mixing zone between the salty water and the groundwater lies in the porewaters some distance inland. These mixing zones are referred to as *subterranean estuaries*.

The degree of mixing within these subterranean estuaries varies over time in response to storm surges, rainfall, and the tides. In coastal areas with significant tidal ranges, tidal pumping can occur. In these locations, groundwater discharge into the ocean can be salty. In some locations, removal of groundwater for human use has lowered the freshwater hydrostatic head such that saltwater intrusions are now permanent features. In these cases, the aquifer can no longer provide usable groundwater.

Groundwater discharges into the coastal ocean are found emanating from sediments that lie on the continental shelves. The flows are very geographically variable as they depend on the geologic bedforms present within the sediments. For example, groundwater flows are enhanced by the presence of paleochannels. These are former riverbeds, now submerged, in which aquifers outcrop because riverwater had cut channels into the bedforms. In the outer part of the continental shelf, groundwater circulation can also be driven by convection currents that occur in deeply buried sediments as a result of geothermal gradients.

Groundwaters tend to be anoxic and, hence, contain trace metals in reduced form. For example, most groundwaters have high ferrous iron concentrations ($Fe^{2+}(aq)$). When the reduced metals come in contact with oxic seawater, they are likely to undergo oxidation. In the case of $Fe^{2+}(aq)$, reaction with O_2 should lead to the formation of solid phase iron oxyhydroxides. Many other trace metals and phosphate tend to become incorporated into these oxyhydroxides through co-precipitation and electrostatic absorption. Thus, the degree to which metals and phosphate are transported via groundwater into the ocean is likely dependent on their solubility in the presence of O_2 and iron oxyhydroxides. Iron-rich precipitates have been observed in subsurface coastal soils and nearshore sediments, suggesting that these oxidation reactions are likely rapid and effective in reducing the groundwater discharge of metals and phosphate into the ocean.

11.2.3 Atmospheric Input

Trace elements are delivered to the ocean by atmospheric, or *aeolian*, processes in both particulate and soluble forms. Most of the aeolian particles entering the ocean are less than 10 μm in size and are referred to as *aerosols*. Aeolian transport of particles occurs when winds, such as the Trades, pick up small particles from the land's surface and carry them over the ocean. Some trace elements, such as mercury, have a high enough vapor pressure that they are present as atmospheric gases. Still others are ejected during volcanic eruptions in either particulate or gaseous form.

The atmospheric load is delivered to the sea surface via two processes: dry and wet deposition. Dry deposition is mostly gravitational; when winds weaken, solid particles fall to the sea surface. Wet deposition involves incorporation into a raindrop that falls to the sea surface. Particulate trace elements can become entrained in raindrops and thereby carried to the sea surface. Along the way, some of the trace elements can dissolve. Gaseous trace elements, such as mercury, can also be dissolved directly into raindrops. Thus, rainwater can contain dissolved trace metals.

The composition of the aeolian particles is temporally and spatially variable. These particles are typically fragments of weathered rocks, soil, or biogenic detritus, such as terrestrial plant fragments. Other biogenic particles include bacteria, phytoplankton, mold, fungal spores, seeds, and even insects.

Most of the aeolian input of trace metals is in particulate form as weathered rocks and soil. The deposition patterns on the sea surface tend to follow those of the major winds bands, with deserts being the primary source regions (Figure 11.4a). As shown in Figure 11.4b, this results in high deposition to the surface waters immediately downwind of the arid regions of Africa, the Arabian Peninsula, Central Asia, Australia, and

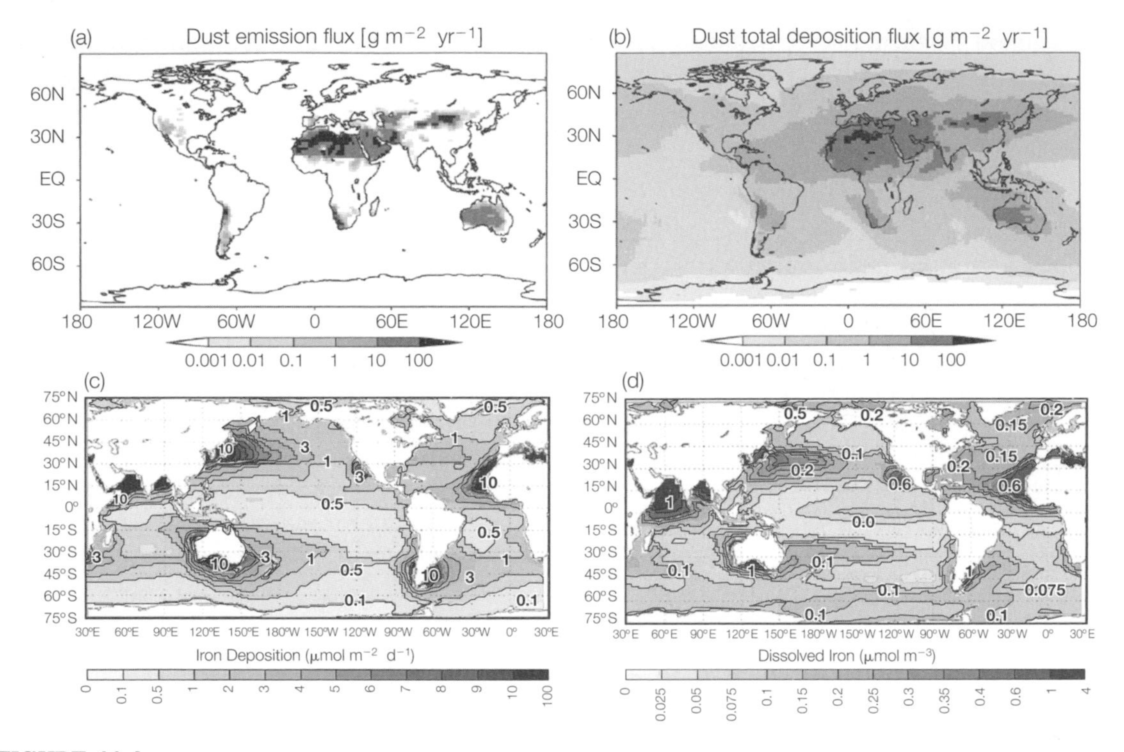

FIGURE 11.4

Simulated annually averaged (a) dust emission flux ($g\,m^{-2}\,y^{-1}$) and (b) total deposition flux ($g\,m^{-2}\,y^{-1}$). Annual mean (c) iron dust deposition rate ($\mu mol\,m^{-2}\,d^{-1}$) and (d) surface dissolved Fe concentration ($\mu mol\,m^{-3}$). *Sources*: After (a–b) Tanaka, T. Y., and M. Chiba (2006). *Global and Planetary Change*, 52, 88–104. (c–d) From Tegen, I., and I. Fung (1994). *Journal of Geophysical Research*, [Atmosphere] 99, 22897–22914, and (c) and (d) Vichi, M., *et al.* (2007). *Journal of Marine Systems*, 64, 110–134. (See companion website for color version.)

South America. The influence of aeolian dust transport on the surface ocean iron deposition rates and concentrations is shown in Figures 11.4c and 11.4d.

As with riverine fluxes, global atmospheric fluxes are difficult to estimate because of spatial and temporal variability. Some of the variability occurs on short time scales as a result of large dust storms and volcanic eruptions. The summary data in Table 11.1 suggests that atmospheric transport is as important as the river and hydrothermal fluxes for the soluble metals. Although the aeolian particulate flux is lower than the fluvial net particulate flux, this belies its relative importance to the open ocean for two reasons. First, the aeolian input is introduced across a much larger surface area than is the river discharge. Second, the solubility of aeolian particulate metals is considered to be higher than that of the fluvial particulate metals. Thus, from the perspective of a phytoplankton cell living in an oceanic gyre, wind transport is far more likely to be a source of iron than any riverine input. This is especially true in the Pacific Ocean, which has fewer large rivers and the greatest surface-area-to-coastline ratio.

11.2.4 Diagenetic Remobilization from Nearshore Sediments

Sinking particles transport trace elements to the sediments. Once in the sediments, chemical reactions can resolubilize a significant fraction of the particulate metals. This process is termed *diagenetic remobilization* and is the subject of the next chapter. The resolubilized elements can diffuse across the sediment-water interface into the deep zone.

Organic-rich sediments, such as those in the coastal zone, are particularly good sources of remobilized trace metals for two reasons. First the organic matter helps create a reducing environment, which leads to redox reactions that solubilize metals. Second, the organic matter is largely derived from the remains of plankton that tend to be enriched in trace metals. Since most of the metals released from the sediments were introduced into the ocean via other processes, such as river and aeolian input, diagenetic remobilization is considered to be a recycling process rather than a "new" source of metals to the ocean.

11.2.5 Hydrothermal Activity

Some trace metals are transported into the ocean as a component of hydrothermal fluids. This process is discussed further in Chapter 19. To briefly summarize, hydrothermal fluids are produced when seawater penetrates into cracks in the crust near tectonic spreading centers. The seawater is heated as it comes into contact with magma. The hot seawater leaches a number of trace metals from the magma. The resulting hydrothermal fluids are acidic and do not contain O_2, so most of the metals are present in reduced form. Because of their high temperatures, the hydrothermal fluids have a lower density than cold seawater. Their increased buoyancy causes them to rise until they are emitted into the deep sea. Admixture with cold, oxic, alkaline seawater causes the hydrothermal metals to undergo various redox and precipitation reactions.

As shown in Table 11.1, hydrothermal emissions are a major source of soluble iron, manganese, and zinc and a minor source of aluminum, cobalt, copper, and lead. Other elements with significant hydrothermal inputs include lithium, rubidium, cesium, and potassium. Considerable uncertainty also surrounds these flux estimates because they are the result of extrapolations from measurements made at a small number of hydrothermal systems at single points in time. These fluxes appear to vary significantly over short time scales as tectonic activity abruptly opens and closes cracks in the oceanic crust.

In the case of iron and manganese, most of these metals are removed from the hydrothermal fluids and converted to particulate form close to their point of entry. Some of these removals are in the form of sulfides, which form as the fluids emerge into the deep sea. The rest occurs as the fluids mix with cold, oxic, alkaline seawater, which promotes the oxidation of reduced metals. Thus, $Fe^{2+}(aq)$ and $Mn^{2+}(aq)$ are transformed into insoluble iron and manganese oxides, forming colloids and particles, the latter of which eventually settle onto the sediments. As described in the next chapter, at least some of these oxidation reactions are biologically mediated. Some of the iron-manganese precipitates form nodules and pavements whose formation is further discussed in Chapter 18. As a result of these precipitations, most of the hydrothermal iron and manganese is removed from seawater before it can be transported into surface waters via ocean currents and mixing.

11.2.6 Anthropogenic Input

Trace elements are also introduced into the ocean as a result of human activities. As shown in Table 11.2, the major pathway by which anthropogenic metals enter the ocean is via aeolian input. Most of the anthropogenic emissions of trace metals to the atmosphere is from fossil-fuel burning by coal-fired power plants and automobiles. Fossil fuels contain trace metals because they form from organic matter whose ultimate source is biogenic soft parts. As will be discussed in Chapter 26, these biogenic soft parts are transformed into fossil fuels by chemical reactions that occur after the soft parts are buried beneath thick piles of sediment. Some of these deposits are marine and some terrestrial in origin. When fossil fuels are burned, the trace metals are converted into aerosols or gaseous form depending on the element. Some of the metals, such as lead, have been intentionally added to refined fossil fuels to enhance their performance as fuel.

Humans also mobilize metals from the earth's crust by mining, which exposes fresh rock surfaces to the atmosphere and to rain water. Some metals are mobilized by leaching into rain water that eventually flows into the ocean. Others, such as mercury, have high vapor pressure and, hence, volatilize. Other anthropogenic transport pathways include ocean dumping of wastes, nuclear bomb explosions, and leaching from metallic structures such as oil-drilling platforms and marine paints used to prevent biofouling.

The metals shown in Table 11.2 are ones whose anthropogenic atmospheric emissions are similar to or greater than their natural atmospheric input. They tend to be toxic to organisms at relatively low concentrations and are usually referred to as *heavy metals*. Not included in Table 11.2 is an especially problematic metal, tin. Tin-based marine paints have been used since the 1950s to prevent biofouling of ship hulls. Leaching of the tin into seawater creates toxic conditions in seawater and sediments. As a result, many countries have banned or strictly limited the use of these paints. Note that some of the metals listed in Table 11.2, such as iron and zinc, are micronutrients for plants and animals. But they are required only in small quantities; high concentrations can have negative and even toxic effects.

11.3 OCEANIC SINKS OF TRACE ELEMENTS

Most of the trace elements are present in concentrations well below those that should result from the maintenance of simple equilibria with mineral phases. Clearly some other processes are effectively removing these elements from seawater. As it turns out, all of these processes involve incorporation of the trace metals into sinking particles.

Table 11.2 Comparison of Natural and Anthropogenic Fluxes of Heavy Metals Circa 1983 and 1995.

Trace Metal Sources	Fluxes (10^3 tonnes per year)							
	Cd	Hg	Pb	Cu	Zn	As	Cr	Ni
Natural emission to atmosphere	1.4	2.2	12	28	45	12	43	29
Anthropogenic emissions to atmosphere	7.6	3.6	332	35	132	19	31	52
Anthropogenic emissions to atmosphere circa 1995	3.0	2.2	119	26	57	5.0	15	95
Total atmospheric emissions	9.0	5.8	344	63	177	31	74	81
% anthropogenic	84%	62%	97%	56%	75%	61%	42%	64%
Atmospheric inputs to ocean[a]	3.2	1.7	88.0	34.0	136.0	5.8	NA	25.0
Riverine inputs to ocean[a]	0.07	0.03	0.3	4.0	5.6	1.6	5.0	4.6
Atmospheric + riverine inputs to ocean[a]	3.27	1.73	88.3	38.0	141.6	7.4	NA	29.6
% atmospheric	98%	98%	100%	89%	96%	78%	NA	84%

[a] *Including natural and anthropogenic inputs.*

NA = data not available.

Because of uncertainties in global estimates, these values differ somewhat from the estimates in Table 1.1. Data from 1983 are from Nriagu, J. O. (1990). Environment *32, 8–11 and 28–33; and Nriagu, J. O. (1992). Proc. Symp. on the Deposition and Fate of Trace Metals in Our Environment, pp. 9–21. U.S. Dept. Agriculture, Forest Service. Gen. Tech. Rep. NC-150. Data from mid-1990s are from Pacyna, J. M., and E. G. Pacyna (2001).* Environmental Reviews *9, 269–298.*

This has been whimsically described as "the Great Particle Conspiracy." An important set of these processes involves the incorporation of trace metals into biogenic hard and soft parts. Another involves the adsorption of metals onto the surfaces of particles. Although some metals can theoretically be removed by abiotic precipitation into mineral phases, continuing research confirms that most of the trace metal removal is biologically mediated.

These removal processes are relatively rapid, causing most of the trace elements to have residence times less than 50,000 y. As shown in Figure 11.5, their residences time follow periodic trends that reflect the unique chemical reactivity of each element as determined by its electronic configuration. In Chapter 5, the chemical reactions that occur in aqueous solutions, such as seawater, were described as encompassing a range of interactions. These include the relatively weak electrostatic attractions involved in ion pairs and hydrogen bonds and the relatively strong ones found in ionic and covalent bonds. The likelihood for an element to engage in a particular interaction is largely determined by its electronic configuration as the latter determines an element's fundamental atomic characteristics. Since these fundamental characteristics, such as the ratio of an element's ionic charge to its ionic diameter (z/r) were used to develop the structure (families and groups) of the periodic table, the speciation of trace metals in seawater tends to follow periodic trends.

11.3.1 Scavenging

Most particulate matter in the ocean, such as clay minerals, metal oxyhydroxides, and POM, possess a small net negative charge at the pH of seawater. Hence, metal cations are electrostatically attracted to the particle surfaces. As shown in Figure 11.6, these

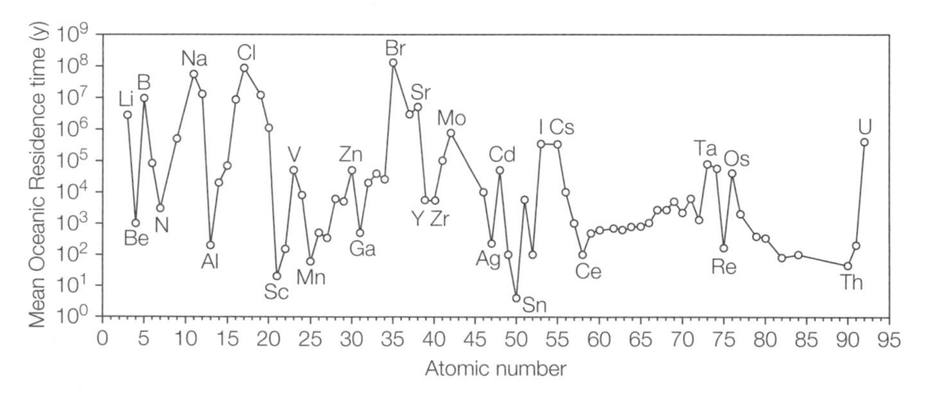

FIGURE 11.5

Relationship of mean oceanic residence time to atomic number. Residence times from MBARI PTEO http://www.mbari.org/chemsensor/summary.html. Residence times do not include seasalt cycling.

attractions range in strength from weak outer-sphere attractions, to ion pairing, up to ones in which covalent bonds are formed.

Some of these particles eventually sink to the seafloor, thus removing metals from the ocean. This process of surface adsorption followed by settling is referred to as *particle scavenging*. The rate and degree to which a dissolved metal is scavenged from the ocean depends on (1) its elemental nature, (2) the abundance of particulate matter, (3) the concentrations of other solutes that can compete for adsorption sites, and (4) the depth of the water column. Metal scavenging rates have been inferred from the concentrations of naturally occurring radionuclides, such as ^{234}Th, ^{230}Th, and ^{228}Th.

In the deep sea, the sole process responsible for the removal of metals from the ocean is via scavenging. Thus, a residence time for trace metals in the deep sea can be computed from their scavenging rates and concentrations if a steady-state condition is assumed. As indicated in Table 11.3, the deepwater residence times of many trace metals is close to or less than the mixing time of the ocean (1000 y). Thus, areas with high rates of scavenging can have local dissolved metal concentrations considerably below that of average seawater. This can occur in regions with large particle fluxes. Some of these sinking particles are created in situ through the aggregation of smaller particles.

For some metals, scavenging is reversible, making the trip down to the sediments a stepwise process in which metals are desorbed for short periods before hitching another

Table 11.3 Deepwater Scavenging Residence Times of Some Trace Elements in the Ocean.

Element	Scavenging Residence Time (years)	Element	Scavenging Residence Time (years)
Sn	10	Mn	51–65
Th	22–33	Al	50–150
Fe	40–77	Sc	230
Co	40	Cu	385–650
Po	27–40	Be	3700
Ce	50	Ni	15,850
Pa	31–67	Cd	177,800
Pb	47–54	Particles	0.365
Fe	270 ± 140*		

Data after Balistrieri, L., et al. (1981). Deep Sea Research, *28, 101–121; Orians, K. J., and K. W. Bruland (1986).* Earth and Planetary Science Letters, *78, 397–410; and Whitfield, M. and D. R. Turner (1987).* Aquatic Surface Chemistry: Chemical Processes at the Particle-Water Interface, *John Wiley & Sons.*

**After B. Berquist and E. A. Poyle (2006).* Earth and Planetary Science Letters, *248, 54–68.*

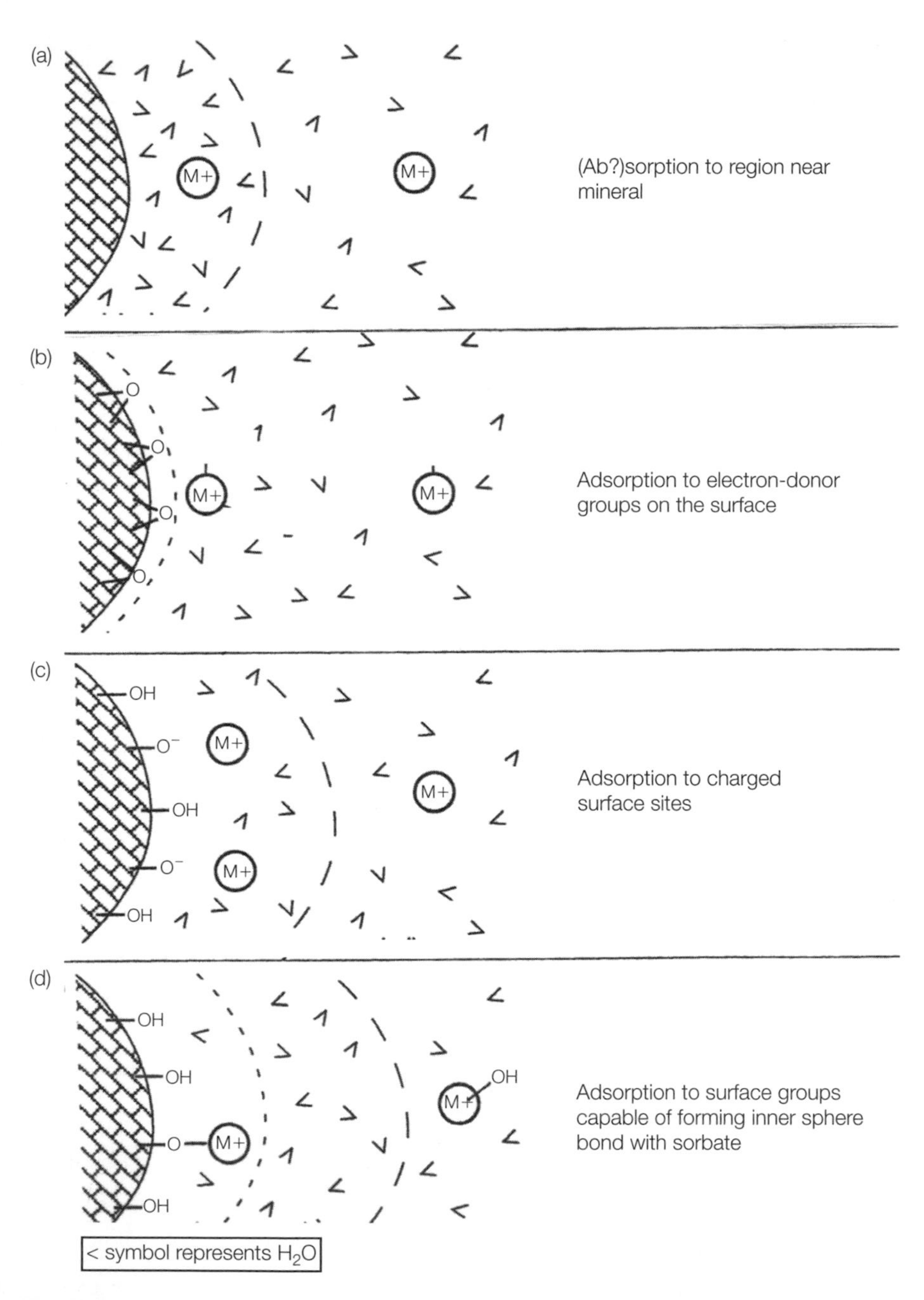

FIGURE 11.6

Types of metal adsorption to mineral and organic surfaces: (a) outer sphere interactions, (b) adsorption by electron-donor groups, (c) adsorption by negatively charged surface sites, and (d) adsorption to surface groups capable of forming metal-covalent bonds. The only parts of the molecular structure of the particle shown are the atoms engaged in metal adsorption.

ride downward by adsorbing onto another sinking particle until the seafloor is reached. Metals that participate in these reversible adsorption reactions have longer deepwater scavenging times.

Some metals are irreversibly adsorbed, probably via incorporation into the mineral phases, such as amorphous iron oxyhydroxides, as shown in Figure 11.6d. Some of these amorphous phases form by direct precipitation from seawater. As noted earlier, hydrothermal fluids are an important source of iron and manganese, both of which subsequently precipitate from seawater to form colloidal and particulate oxyhydroxides. Other metals tend to coprecipitate with the iron and manganese, creating a polymetallic oxyhydroxide. It is not clear the degree to which biological processes mediate the formation of such precipitates. Since the metals are incorporated into a mineral phase, this type of scavenging is better referred to as an absorption process.

Minerals formed by abiotic precipitation from seawater are commonly called *hydrogenous precipitates*. These are discussed further in Chapter 18. They play a role in the removal of some trace elements. For example, trace metals are removed by precipitation into sulfide minerals under anoxic conditions commonly found in nearshore sediments. In the absence of oxygen, sulfate reduction generates sulfide. Anoxic conditions also promote diagenetic remobilization leading to elevated concentrations of dissolved metals. The resulting levels of sulfide and dissolved metals are often high enough to exceed the solubility product of some *metal sulfides*. As a result, metal sulfides, such as pyrite (FeS_2), are commonly found in organic-rich anoxic sediments.

Other resolubilized trace metals precipitate as replacement ions in existing solids such as fecal pellets and bone. Examples of these fossilized materials include *barite*, *phosphorite*, and *glauconite*. These precipitates contain small amounts of a variety of trace metals as well as other elements. As a result, their chemical composition is variable and their structure is usually amorphous, making it difficult to assign them an empirical formula.

11.3.2 Incorporation into Biogenic Materials

Marine organisms concentrate metals in their tissues and skeletal materials. Many of these trace metals are classified as micronutrients because they are required, albeit in small amounts, for essential metabolic functions. Some are listed in Table 11.4, illustrating the role of metals in the enzyme systems involved in glycolysis, the tricarboxylic acid cycle, the electron-transport chain, photosynthesis, and protein metabolism. These micronutrients are also referred to as *essential metals* and, as discussed later, have the potential to be biolimiting.

Trace Metal Enrichment Factors

The degree to which a metal is concentrated in a marine organism is given by an enrichment factor (EF), which is defined as

$$EF = \frac{\text{Metal concentration in biogenic material}}{\text{Metal concentration in seawater}} \tag{11.1}$$

Table 11.4 Role of Trace Metals in Enzyme Systems.

Metal Ion	Examples
Co	Vitamin B_{12}: rearrangements, reduction, and C and H transfer reactions with glycols and ribose
Cu	Laccase, oxidases Plastocyanin: photosynthetic electron transport Cytochrome *c* oxidase: mitochondrial electron transport
Fe	Cytochrome oxidase: reduction of oxygen to water Cytochrome P-450: O-insertion from O_2, and detoxification Cytochromes *b* and *c*: electron transport in respiration and photosynthesis Cytochrome *f* : photosynthetic electron transport Ferredoxin: electron transport in photosynthesis and nitrogen fixation Iron-sulfur proteins: electron transport in respiration and photosynthesis Nitrate and nitrite reductases: reduction to ammonium
Fe (Se) (V) (Mn)	Catalase Peroxidases in H_2O_2 breakdown and in reactions involving halogens
Fe, Mn, Cu	Superoxide dismutases: disproportionation of O_2^- radicals to O_2 and H_2O_2
Fe, Mn	Acid phosphatase
Mn	Oxygen-generating system of photosynthesis
Mo	Nitrate and nitrite reductases: reduction to ammonium Sulfate reductase
Mo (Fe) (V)	Nitrogenase: nitrogen fixation
Ni	Urease: hydrolysis of urea
Ni (Fe)	Methanogenesis
Zn (Cd)	Carbonic anhydrase: hydration and dehydration of CO_2
Zn	Alkaline phosphatase: hydrolysis of phosphate esters Peptidases: hormone control DNA and RNA polymerases: nucleic acid replication and polymerization

Source: *From Whitfield, M. (2001).* Advances in Marine Biology, *41, 3–128.*

Table 11.5 Enrichment Factors for Metals in the Tissues of Phytoplankton and Brown Algae.

Element	Plankton	Brown Algae
Al	25,000	1550
Cd	910	890
Co	4600	650
Cr	17,000	6500
Cu	17,000	920
Fe	87,000	17,000
I	1200	6200
Mg	0.59	0.96
Mn	9400	6500
Mo	25	11
N	19,000	7500
Na	0.14	0.78
Ni	1700	140
P	15,000	10,000
Pb	41,000	70,000
S	1.7	3.4
Si	17,000	120
Sn	2900	92
V	620	250
Zn	65,000	3400

Source: After Bowen, H. J. M. (1966). Trace Elements in Biochemistry, *Academic Press, pp. 87–88.*

EFs as high as 10^5 have been observed. The major cations have the lowest values. As shown in Table 11.5, iron has the highest EF. This is likely due to its binding with biomolecules, such as ferredoxin, which have extraordinarily high affinities for iron. Tables 11.5 and 11.6 also demonstrate that organisms differ in their degree of trace-metal enrichment. In general, organisms lower on the evolutionary ladder tend to have higher EFs. Soft parts also tend to have different degrees of enrichments as compared to hard parts. As shown in Table 11.6, the enrichments vary among the different mineral

Table 11.6 Concentration of Metals in Calcareous Shells of Formanifera and Siliceous Shells of Radiolaria in ppm of Ash.

Element	Foraminifera	Radiolaria
Ba	700	5400
Sr	400	Major
Cu	25	750
Ag	3	0
Zn	0	600
Pb	10	105
Ti	15	30
V	trace	0
Cr	8	90
Mn	300	0
Fe	9	300
Ni	0	57
B	0	0

Source: From Pytkowicz, R. M. (1983). Equilibria, Nonequilibria and Natural Waters. *John Wiley & Sons, p. 300.*

phases of hard parts, e.g., calcite versus silica. Biogenic calcite tends to have significant concentrations of magnesium, strontium, and barium. Species differences are also important. For example, the spicules of the siliceous protozoan *Acantharia* are particularly notable for their large strontium enrichments. Some of these trace metal impurities have been used as paleoceanographic records of changes in alkalinity (Ba in benthic forams), pH (B in corals),[1] and seawater phosphate concentration (Cd in forams).

Trace metals are resolubilized from the biogenic hard and soft parts in much the same way as the macronutrients. Thus, the depth profiles of the trace metals with high EFs tend to be similar in shape to those of the nutrients. Efforts have been made to develop a Redfield-Richards type ratio for the trace metals in marine plankton. Surprisingly, field and lab work[2] suggests that a relatively constant composition can be defined for whole

[1] B is a metalloid element.
[2] As proposed by Ho, T-Y., *et al.* (2003). *Journal of Phycology* 39, 1145–1159.

cells (including hard parts) as

$$(C_{147}N_{16}P_1S_{1.3}K_{1.7}Mg_{0.56}Ca_{23})_{1000}Sr_{54}Fe_{7.5}Zn_{0.80}Cu_{0.38}Co_{0.19}Cd_{0.21}Mo_{0.03} \quad (11.2)$$

and for organic biomass as

$$(C_{124}N_{16}P_1S_{1.3}K_{1.7}Mg_{0.56}Ca_{0.5})_{1000}Sr_{5.0}Fe_{7.5}Zn_{0.80}Cu_{0.38}Co_{0.19}Cd_{0.21}Mo_{0.03} \quad (11.3)$$

(These formulations also include an update to the original Redfield-Richards ratio in the carbon content. Mn is omitted due to its variable stoichiometry.)

Unlike the case of nitrogen and phosphorus, the trace metal stoichiometry of plankton does not resemble that of seawater. On the other hand, as shown in Figure 11.7, the relative abundances of the trace metals in plankton are well correlated with the relative abundances in Earth's crust. A similar relationship has been observed with higher-order terrestrial plants suggesting that some fundamental metal requirement for photosynthesis has been retained over the millennia. For example, the high iron and manganese requirements of plants may reflect the origins of photosynthesis as an anoxygenic process in which Fe and Mn, rather than H_2O, were used as an electron donor.

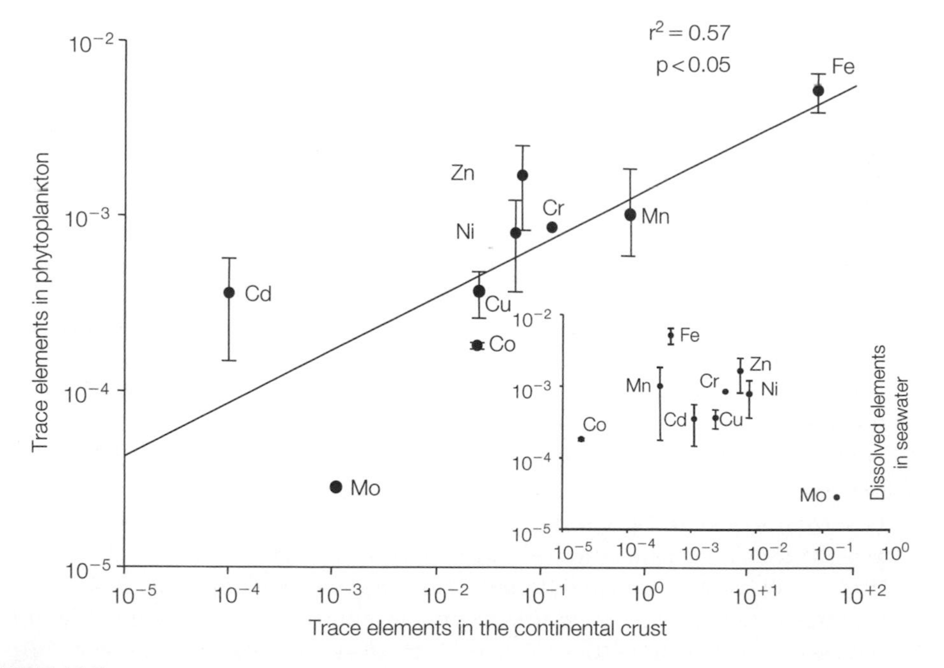

FIGURE 11.7

Relationship between the trace elemental composition of phytoplankton, continental crust and seawater. Phytoplankton and crustal abundances are normalized to phosphorus (ppm trace metal: ppm P). Seawater trace elements abundances are normalized to phosphate (ppb trace metal: ppb P as phosphate). *Source*: From Quigg, A., *et al.* (2003). *Nature* 425, 291–294.

Role of Trace Metals in the Evolution of Oxygenic Photosynthesis

As discussed in Section 8.6.1, the initiation of photosynthesis as an anoxygenic process is thought to be a consequence of the absence of O_2 and, hence, the presence of reducing conditions on the early Earth. Such conditions would have supported high concentrations of dissolved iron and manganese, as $Fe^{2+}(aq)$ and $Mn^{2+}(aq)$. Depletion of these reduced materials is thought to have stimulated evolution of oxygenic photosynthesis in which the highly abundant H_2O is used as the electron donor. Oxygenic photosynthesis appears to use some of the same enzymatic pathways as the anoxygenic modes in which trace metals, such as iron and manganese, play central roles. This suggests that the oxygenic photosynthesizers are descendants of the first anoxygenic forms.

As oxygenic photosynthesis lead to higher atmospheric O_2 levels, the ensuing increase in oxidizing conditions further limited the availability of dissolved iron and manganese. This necessitated development of complicated strategies to ensure that plankton were able to meet their biochemical needs for the essential trace metals. Shifting trace metal availability also appears to have spurred a shift in ecological dominance of the phytoplankton from the green superfamily, which gave rise to the dinoflagellates, to those from the red superfamily. The latter have given rise to the coccolithophorids and diatoms. The red algae have lower Fe and Mn and higher Cd, Co, and Mo requirements than the green algae.

Nutrient Colimitation

Plankton appear to have evolved several biogeochemical strategies for managing their micronutrient limitations. As noted in Chapter 5, the bioavailability of iron appears to be enhanced by complexation with ligands composed of dissolved organic compounds released into seawater by some species of plankton. Similar phenomena are likely to be occurring for the other essential trace elements. For this process to be successful, in terms of ensuring bioavailability, complexation in seawater must be followed by uptake across the cell membrane. This is not a trivial process given the relatively large sizes of the organometallic compounds.

Indirect evidence suggests that plankton growth is controlled by, and is controlling, the dissolved concentrations of the essential trace metals in the euphotic zone. As shown in Figure 11.8, a negative correlation exists between the free ion concentrations of these metals in surface seawater and the rate at which they bind with ligands. This correlation is considered circumstantial evidence that the removal of these metals, and, hence, their surface water concentrations, are controlled by rapid metal-ligand complexation reactions that are likely part of the metabolic machinery responsible for metal transport across the cell membranes of plankton. If this is the case, plankton seem to be meeting their micronutrient needs by rapidly binding metals that are present at very low (subnanomolar) concentrations into an organometallic complex small enough to pass across their cell membranes.

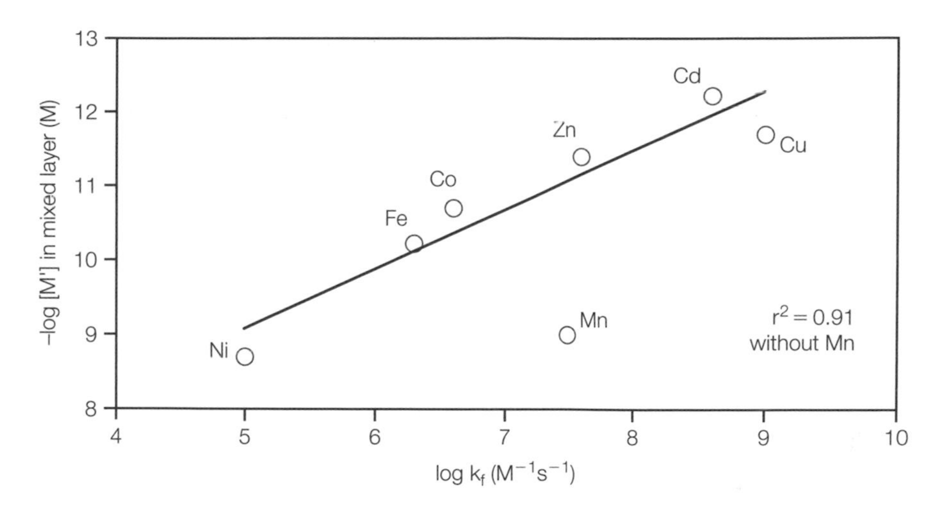

FIGURE 11.8

The relationship between the rate constant for ligand-metal binding (k_f) and the concentration of free metal ion [M^{n+}] in oceanic surface waters for the biologically essential trace metals. Data plotted from Table 1 in Hudson, J. M., and F. M. M. Morel (1993). *Deep Sea Research*, 40, 129–150.

Many laboratory and field studies have shown that micronutrients can limit plankton growth, but in complicated ways that generally involve colimitation by more than one essential metal and/or macronutrient. This is not surprising given the variety of trace metals required by the enzymes involved in nitrogen and phosphorus assimilation and metabolism. It also appears that the ligands responsible for metal transport across the cell membranes of plankton can be metal-specific, providing a mechanism for finer-scale management of micronutrient colimitation. As noted earlier, some micronutrient management seems to be occurring extracellularly via excretion of complexing ligands into seawater. The ultimate limit to all these strategies is thought to rest on the physical processes that control the rates at which micronutrients are brought sufficiently close to the outer surface of the plankton cells. These physical processes are turbulent mixing and molecular diffusion. A further level of complexity to the planktonic management of micronutrient limitation is provided by the extracellular excretion of ligands that act to decrease the bioavailability of some metals, such as Cu^{2+}(aq), which are toxic at high concentrations.

Role of Micronutrients in Global Climate Control

Much of the trace metal demand of plankton in the open ocean may be supplied by the aeolian transport of dust. If trace metals limit plankton growth, the factors that control the transport of dust will also ultimately control the ocean's biological pump and, hence, atmospheric CO_2 levels. This suggests the existence of powerful feedbacks among the processes that control global climate. For example, shifts in climate that lead to regional changes in desertification and wind patterns will affect dust fluxes and thereby alter the flux of trace metals to the ocean. Changes in trace metal inputs have the potential to affect plankton productivity and, hence, the activity of the biological pump. The burial rate of biogenic materials in marine sediments is also influenced by deepwater circulation rates because these affect the oxygenation of the deep sea and, hence, the rates of aerobic respiration of sinking POM. Still other feedbacks exist that involve the production of biogenic greenhouse gases by marine plankton. Thus, the biogeochemical cycling of trace elements through the crustal-ocean-atmosphere factory looks to be intimately linked with the cycling of carbon and global climate. These feedbacks are schematically illustrated in Figure 11.9. A particular concern is that anthropogenically driven changes in land use can lead to changes in dust fluxes and, hence, have the potential to impact these interlinked processes in ways that we, as yet, are not able to predict.

11.3.3 Hydrothermal Activity

Hydrothermal activity is a major net sink for magnesium, but is of unknown importance to the trace metals. Low-temperature weathering of cooled basalt is thought to be a significant sink for some trace metals. As described in Chapter 19, this process is not well understood.

11.4 TYPES OF TRACE ELEMENT DISTRIBUTIONS

The horizontal and vertical distributions of trace metals in seawater are determined by their relative rates of supply and removal. In this section, we inspect a variety of concentration profiles and assess the processes responsible for determining their shapes. In the case of the vertical profiles, the trace metals can be classified into one of the following types: (1) nutrient, (2) conservative, and (3) scavenged, with some elements exhibiting a mixture of these types.

The vertical dissolved concentration profiles characteristic of each element are given in Figure 11.10 in the format of the periodic table to reinforce the connection between electronic configuration and chemical reactivity. As noted in Section 5.6.1, these electronic configurations, as reflected in an element's electronegativity and its z/r, can be used to predict speciation, including the tendency to be present as a hydrated (free), hydrolyzed, or complex ion. Some of these species are more likely than others to become incorporated in sinking particles, both biogenic and abiogenic, and, hence, removed from the ocean. For example, trace metals present as free cations ($M^{n+}(aq)$) are more likely to become electrostatically absorbed onto particles. Redox speciation is also important in determining the concentration distributions of the trace metals as some of these elements are concurrently present in multiple oxidation states. All of these speciation-driven chemical processes influence the shape of an element's vertical concentration profile and its residence time. This leads to the periodic trends in residence times seen in Figure 11.5 and the trends in profile shape seen in Figure 11.10.

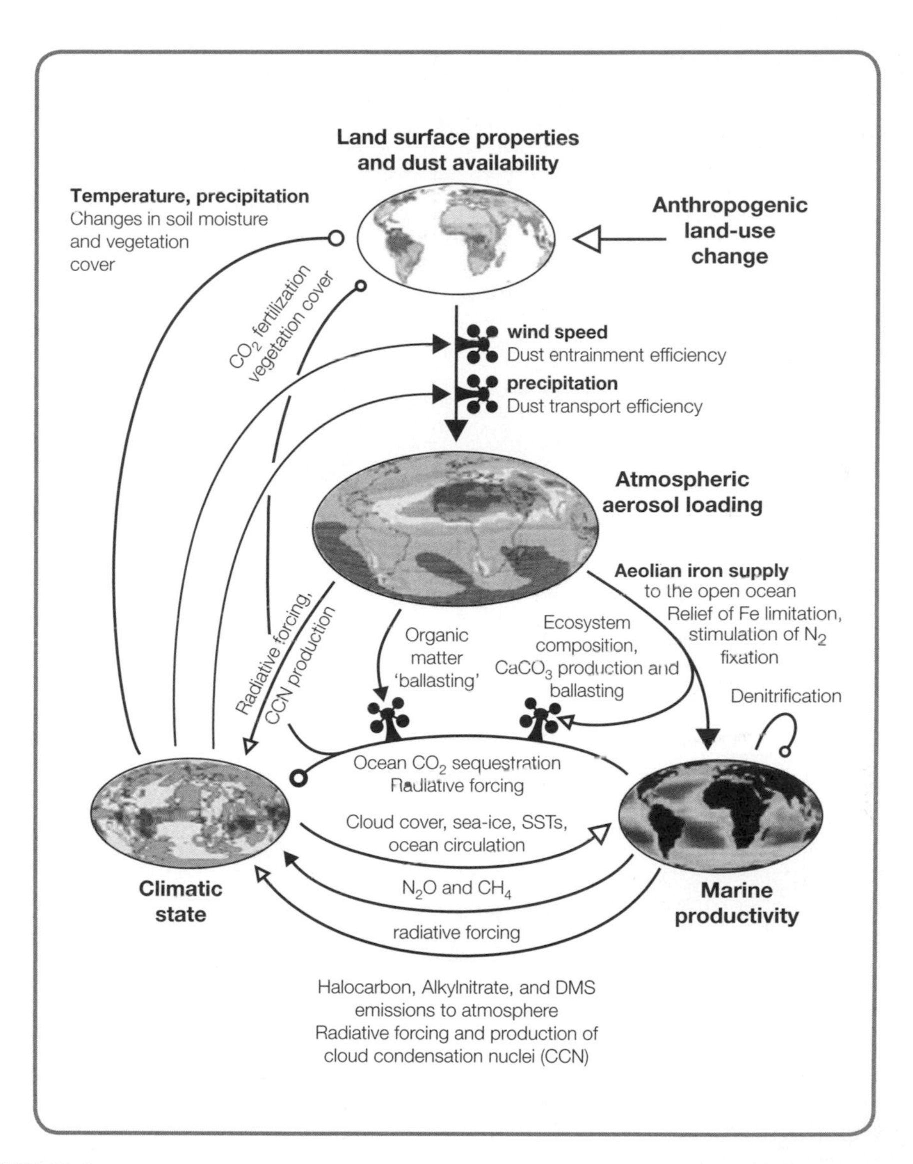

FIGURE 11.9

Schematic representation of global iron connections linked through the state of the land surface, aerosol loading, marine productivity, and climate. Positive feedbacks are represented by solid arrowheads and negative feedbacks by open circles. Feedbacks of uncertain sign are represented by open arrowheads. Linking mechanisms are in italics. Taps represent mechanisms that modulate primary processes. *Source*: From Jickells, T. D., *et al.* (2005). *Science*, 308, 67–71. (See companion website for color version.)

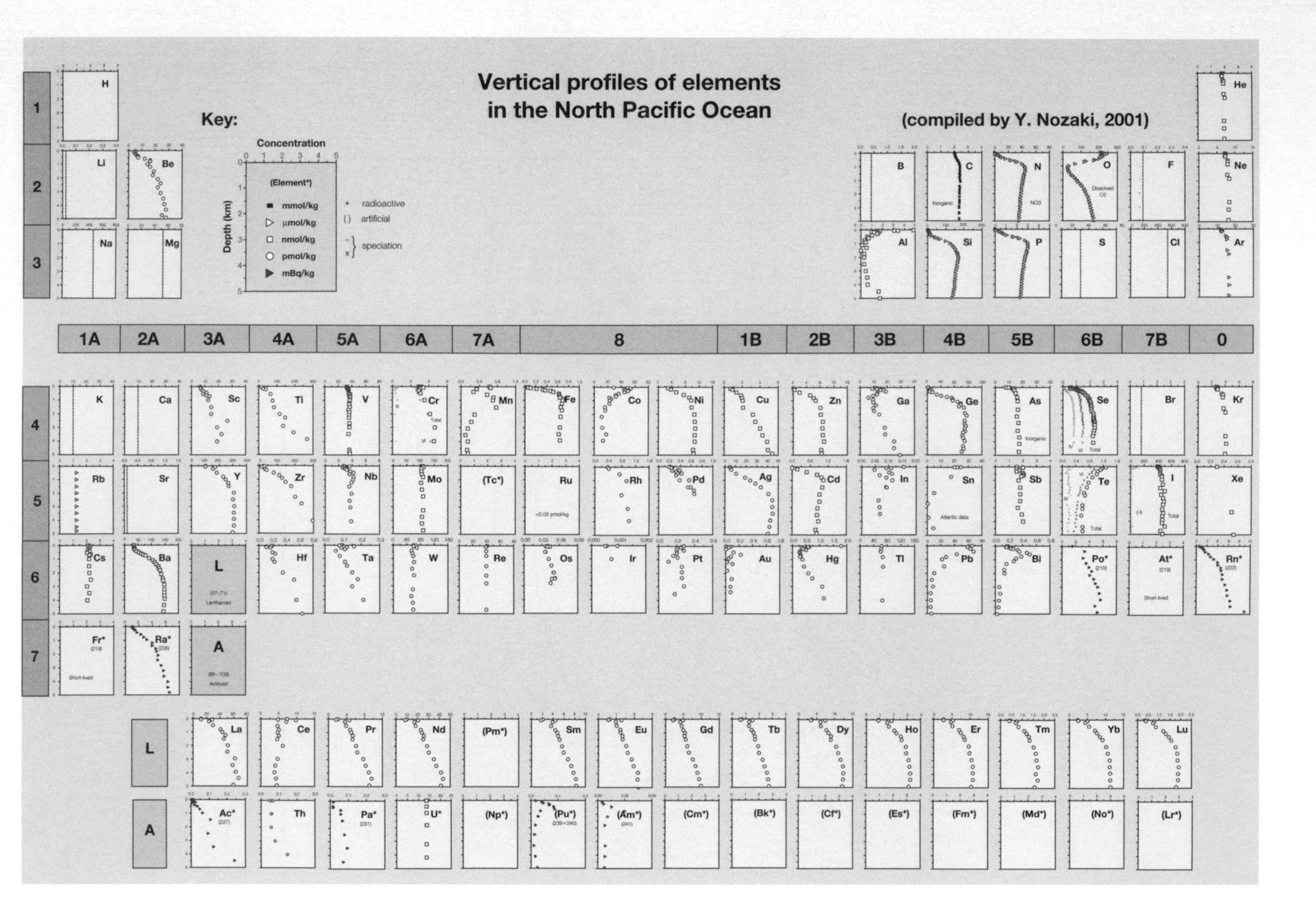

FIGURE 11.10

Vertical profiles of elements in the North Pacific Ocean. These profiles are somewhat subject to geographic variability, particularly proximity to land and upwelling areas. The North Pacific Ocean was selected as a representative region to minimize these effects. Note that concentration units are defined by symbol type: filled square (mmol/kg), open triangle (μmol/kg), open square (nmol/kg), open circle (pmol/kg), and, for the radioisotopes, closed triangle (mBq/kg). The sources of the artificial radioisotopes are largely nuclear bomb testing and discharge from nuclear fuel reprocessing plants. The redox speciation is shown for three elements (Cr, Se, and Te). Concentration information is not yet available for some elements, such as Ru. *Source*: From Nozaki, Y. (1997). *Eos*, May 27, 1997, p. 221.

11.4.1 **Importance of Speciation**

Because speciation determines the reactivity of an element, insights into the biogeochemistry of an element are best obtained by measuring concentrations of the major species, rather than total elemental concentrations. This approach is illustrated for the element selenium, which exists in several species, some dissolved and some particulate, some inorganic and some organic. The two inorganic dissolved species are selenite $Se(IV)O_3^{2-}(aq)$ and selenate $Se(VI)O_4^{2-}(aq)$. As shown in Figure 11.11, these two species are present in similar concentrations although the reduced form, Se(IV), is thermodynamically unstable in oxic seawater. The relatively high concentration of Se(IV) can be interpreted as the result of a kinetic barrier to its oxidation to Se(VI). Redox disequilibria have also been observed for other trace elements such as Cr(III)/Cr(II), Mn(IV)/Mn(II), Cu(II)/Cu(I), I(V)/I(−I), Sb(V)/Sb(III), and As(V)/As(III). Another feature common to the trace metals is the importance of complexation with dissolved organic compounds. In many cases, the resulting organometallic complexes are present in far higher concentrations than either the inorganic complexes or the free ion.

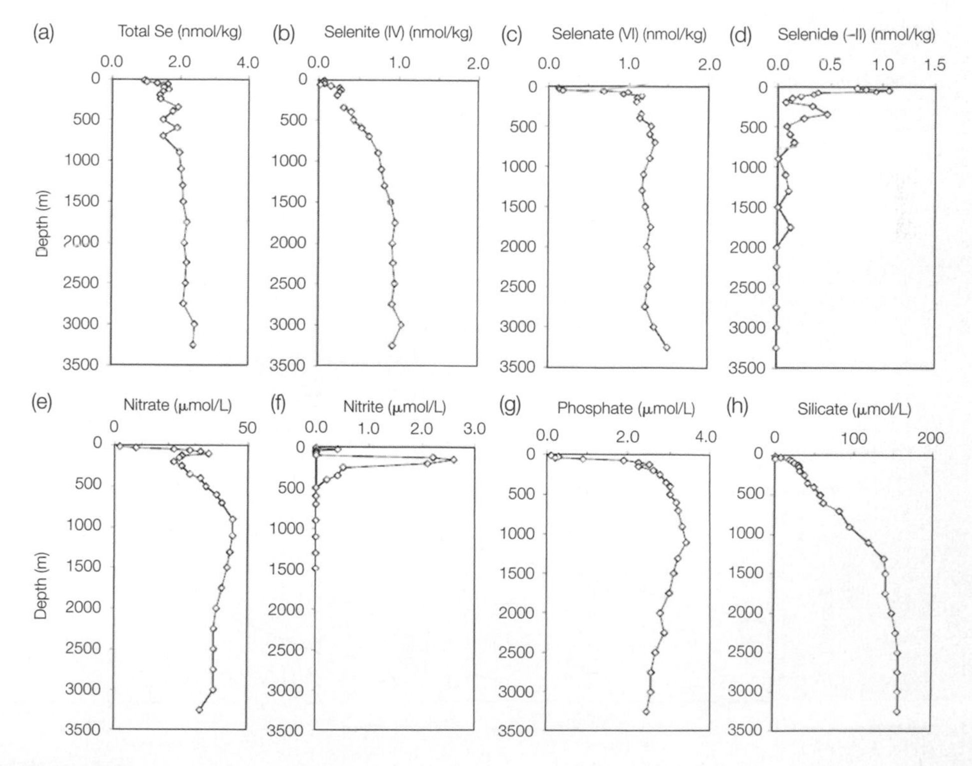

FIGURE 11.11

Vertical concentration profiles of (a) total dissolved Se, (b) selenite $Se(IV)O_3^{2-}$, (c) selenate $Se(VI)O_4^{2-}$, (d) selenide (−II), (e) nitrate, (f) nitrite, (g) phosphate, and (h) silicate in the North Pacific. *Source*: From Cutter, G. A., and K. W. Bruland (1984). *Limnology and Oceanography*, 29 1179–1192.

11.4.2 Nutrient-Type Distributions

The inorganic selenium profiles in Figure 11.11 look much that those of a biolimiting element, having low surface-water and high deep-water concentrations. This biolimiting, or nutrient-type, behavior is exhibited by many trace elements, including most of the micronutrients, suggesting that their seawater concentrations are controlled by the same mechanisms acting on nitrogen, phosphorus, and silica, i.e., the biological pump. Further circumstantial evidence for the importance of incorporation and release of trace elements into and from biogenic particles can be obtained by examining correlations with the nutrients. As shown in Figure 11.12, the total selenium concentration is highly correlated with silica and to a lesser extent with phosphorus. This suggests that some of the selenium is incorporated into hard parts and some into the soft parts of plankton. As the detrital biogenic particles sink, they are remineralized, causing selenium to be released into seawater in dissolved form. As with the nutrients, this causes the total dissolved selenium concentration to increase with depth.

During the remineralization process, selenium is released from the soft parts as a dissolved organometallic compound in the reduced (−II) oxidation state (Figure 11.13). When this organoselenide (−II) comes into contact with the O_2 in seawater, it is oxidized to selenite (IV) and then to selenate (VI). From Figure 11.11, we see that the organoselenide concentration maximum is located in the surface waters where plankton production and remineralization rates are highest. That organoselenide is the dominant

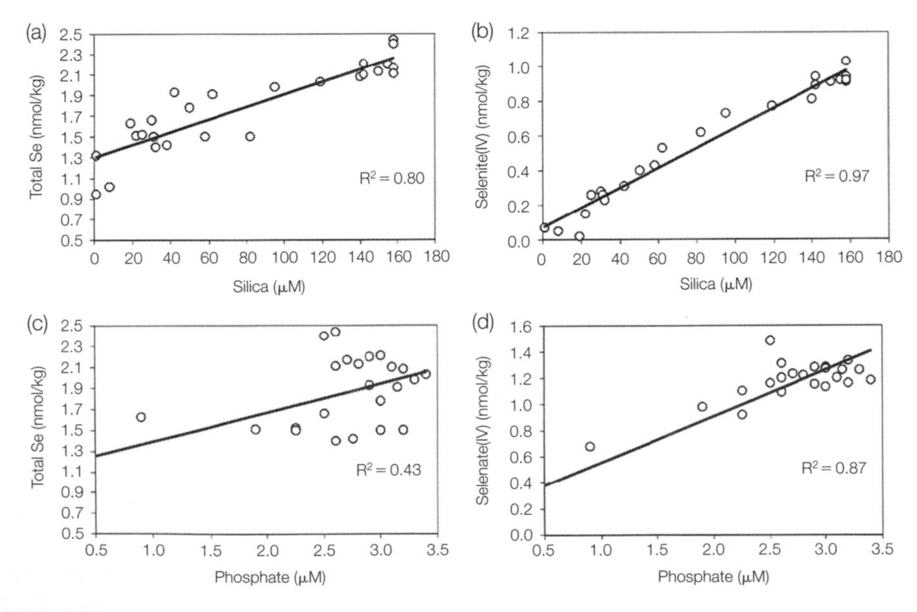

FIGURE 11.12

Correlation plots for concentration data from Figure 11.11. (a) Total Se versus silica, (b) total Se versus phosphate, (c) selenite versus silica, and (d) selenate versus phosphate.

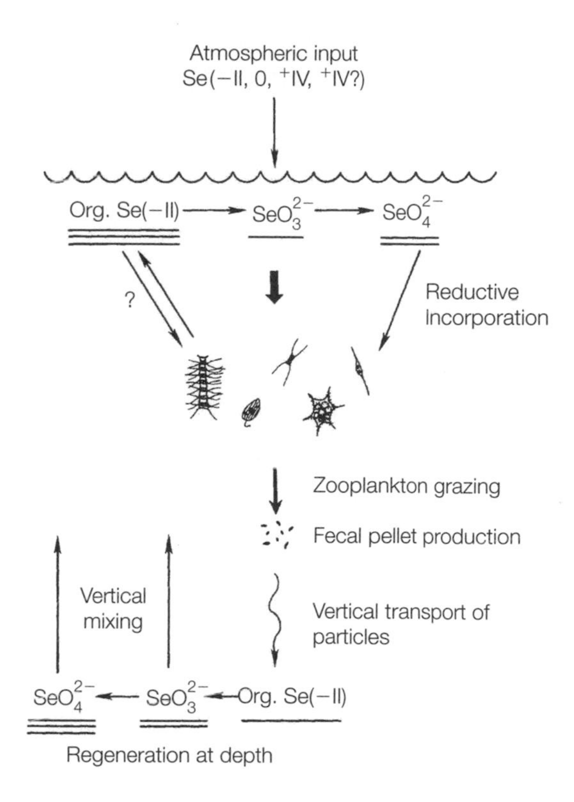

FIGURE 11.13

The redox cycling of selenium. Underlining reflects the relative concentrations of selenium species in surface and deep waters. The preferential uptake of selenite in surface waters is indicated by a larger dissolved-to-particulate arrow. *Source*: After Cutter, G. A., and K. W. Bruland (1984). *Limnology and Oceanography*, 29, 1179–1192.

species in the surface water attests to the importance of remineralization in supplying the surface waters with dissolved Se. The rate of oxidation of selenide (−II) to selenite (IV) is evidently much faster than the rate of oxidation of selenite (IV) to selenate (VI), giving rise to the redox disequilibrium mentioned earlier.

Sediment trap data have demonstrated that particulate selenide (−II) is delivered to the deep water where it undergoes remineralization and stepwise oxidation to selenate (VI). This stepwise oxidation appears to be very efficient, causing the remineralized selenate (VI) to be highly correlated with phosphate (Figure 11.12d). When O_2 levels are low, oxidation to selenate is not as efficient. Evidence for this is seen in the secondary selenide maximum present in the subsurface OMZ located between 350 and 400 m. (Evidence for an OMZ in this depth range is provided by the concurrent presence of a nitrate minimum and nitrite maximum, both of which are diagnostic for suboxic respiration via denitrification.) The correlation shown in Figure 11.12b between selenite (IV) and silica suggests that remineralization from the siliceous hard parts seems to solubilize selenium in a form that is less readily oxidized.

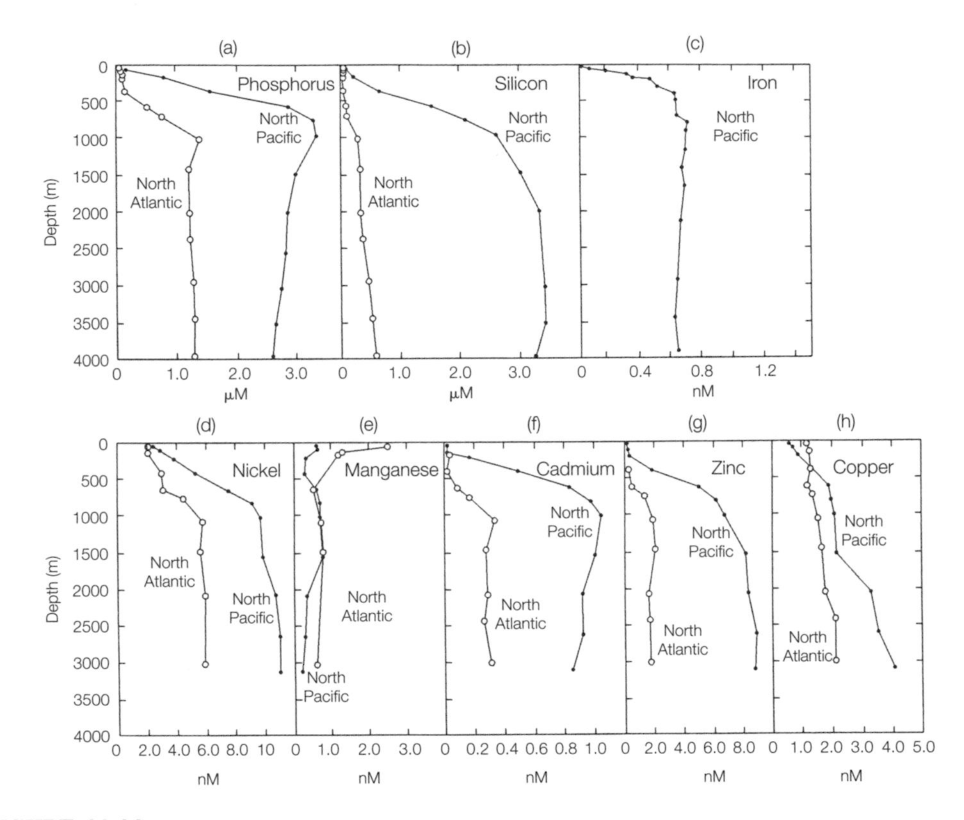

FIGURE 11.14

Concentration profiles from the North Atlantic and North Pacific: (a) phosphorus, (b) silicon, (c) iron, (d) nickel, (e) manganese, (f) cadmium, (g) zinc, and (h) copper. *Source*: From Morel, F. M. M., and J. G. Hering (1993) *Principles and Applications of Aquatic Chemistry*. John Wiley & Sons, p. 406. Data sources: Bruland, K. W., and R. P. Franks (1983). *Trace Metals in Seawater* pp. 395–414, C. S., Wong, *et al.* Plenum Press and Bruland, K. W. (1980). *Earth and Planetary Sciences Letters*, 47, 176–198.

Other examples of nutrient-type profiles exhibited by the essential trace metals are given in Figure 11.14. Some of these profiles, such as Cd and Fe, mimic the shapes of the nitrate and phosphate profiles. These profiles are characterized by concentrations that increase rapidly with increasing depth below the euphotic zone, reaching subsurface maxima near the top of the thermocline. Others, such as Zn, Ni, and Cu, more closely resemble the shape of the silica profile with concentrations increasing over a deeper portion of the water column. As illustrated earlier with selenium, these similarities can be quantified by plotting the trace element concentrations as a function of the nutrient concentrations. High correlations provide circumstantial evidence for incorporation of the trace element in the biogenic part containing the nutrient. These plots are presented in Figure 11.15 and demonstrate that Zn and Ni appear to be associated with siliceous

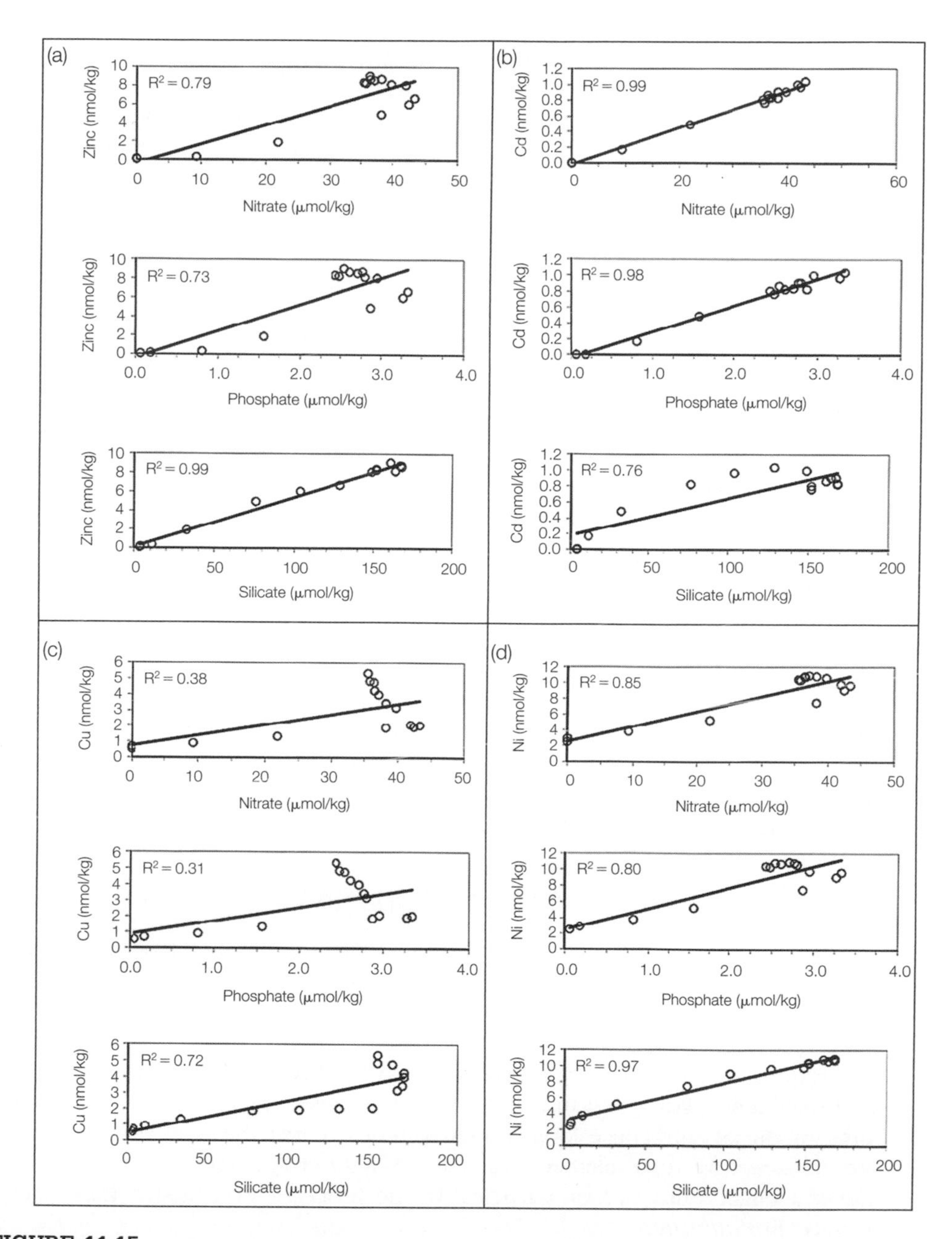

FIGURE 11.15

Correlation plots for the North Pacific Ocean concentration data from Figure 11.14: (a) zinc versus nitrate, phosphate, and silica, (b) cadmium versus nitrate, phosphate, and silica, (c) copper versus nitrate, phosphate, and silica, and (d) nickel versus nitrate, phosphate, and silica.

hard parts, whereas Cd is associated with the soft parts. Copper does not exhibit a strong correlation with any of the nutrients. This suggests that particle scavenging is more likely to be controlling the deepwater concentrations of copper.

The degree of vertical segregation exhibited by a trace element reflects the extent to which it is biolimiting. This biolimitation is a consequence of the nutritional requirement for the trace element as exhibited by the resident plankton community and the bioavailability of this element in the surface seawater. The trace elements with the highest nutrient requirements are the ones with the highest EFs. Trace elements with modest EFs and relatively high seawater concentrations tend to have a weaker degree of vertical segregation, such as is characteristic of a biointermediate element. Indeed, the Broecker Box model can be used to estimate the fraction of a micronutrient element removed from the surface water by sinking biogenic particles (g) and the fraction removed by burial in the sediments (f). As with the nutrients, values of g for the micronutrients range from 0.70 to 0.99, demonstrating efficient removal from the surface waters as sinking biogenic particles.

The micronutrients are also horizontally segregated as shown in Figure 11.14, with the highest concentrations being present in the deep waters of the North Pacific Ocean. Thus, the biological pump has had a marked impact on these elements, such that vertical segregation has reduced their surface water concentrations to low levels and horizontal segregation has established a geographic gradient of deep-water concentrations. Because of remineralization in the deep sea, the micronutrient trace elements have residence times similar to that of the nutrients, ranging from 10,000 to 100,000 y. Important exceptions are iron and manganese, whose residence times are much shorter, being on the order of a few hundred years. In addition to their biogeochemical cycle, these elements are subject to precipitation as oxyhydroxides and particle scavenging. These two process contribute to their relatively rapid removal from the ocean.

11.4.3 Conservative-Type Distributions

A few minor and trace elements, such as Rb^+, Cs^+, and MoO_4^-, have depth distributions that are linearly related to salinity and, hence, are said to exhibit conservative behavior. As illustrated in Figure 11.10, these profiles are characterized by little or no gradients such that concentrations are nearly homogeneous with depth. As with salinity, the concentrations of these elements are controlled by physical processes, such as advection and turbulent mixing, because their residence times are much longer than the mixing time of the ocean. These long residence times reflect the relatively slow reactivity of the conservative trace elements as they are neither concentrated in biogenic particles nor scavenged onto sinking particles. In the Broecker Box model, these elements are termed *biounlimited*.

11.4.4 Scavenged-Type Distributions

Trace elements that undergo scavenging are characterized by vertical profiles in which dissolved concentrations decrease with increasing depth. These include Al, Mn, Pb, Ce,

Th, Co, and, in some locations, Fe. Surface-water enrichments are usually caused by rapid rates of supply to the mixed layer via atmospheric deposition or river runoff. Removal usually occurs through relatively rapid precipitation into or adsorption onto sinking particles. Trace elements controlled by scavenging tend to have short (100 to 1000 y) residence times. Since these residence times are less than the mixing time of the ocean, significant geographic gradients are common.

Lead is an example of a trace element whose surface-water concentrations are elevated as the result of a large atmospheric flux. As shown in Table 11.2, most of this flux is anthropogenic in origin. It is associated with the use of lead-based gasoline additives, such as tetraethyl lead. The Atlantic Ocean has higher concentrations than the Pacific Ocean because it has a smaller surface area and more highly concentrated industrial activity around its coastline. The decreasing use of leaded gasoline in North America since the 1970s has led to a significant and continuing decline in surface-water concentrations (Figure 11.16b).

The influence of river water inputs on trace metal distributions is illustrated in Figure 11.17c, which shows that the surface-water concentration of dissolved Mn in the Pacific Ocean decreases with increasing distance from the California coast. The vertical profile measured in the coastal zone (Figure 11.17b) exhibits a strong surface enrichment characteristic of scavenged trace elements. A similar vertical gradient is seen in the

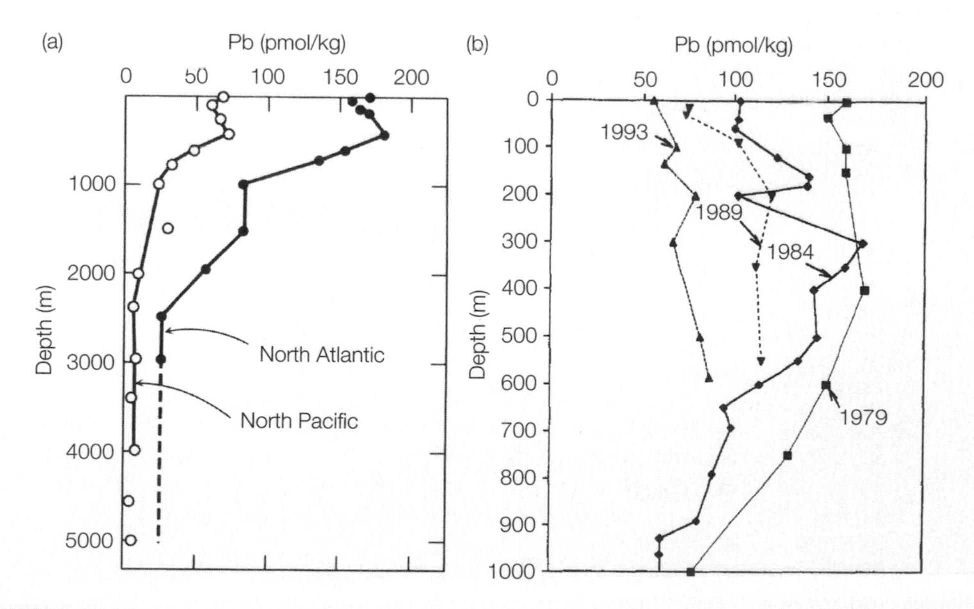

FIGURE 11.16

(a) Vertical profiles of dissolved lead concentrations in the central North Pacific and North Atlantic. *Source*: From Schaule, B. K., and C. C. Patterson (1983). *Trace Metals in Seawater*, Plenum Press, pp. 487–503. (b) Lead profiles near Bermuda. *Source*: From Wu, J., and E. A. Boyle (1997). *Geochimica et Cosmochimica Acta*, 61, 3279–3283.

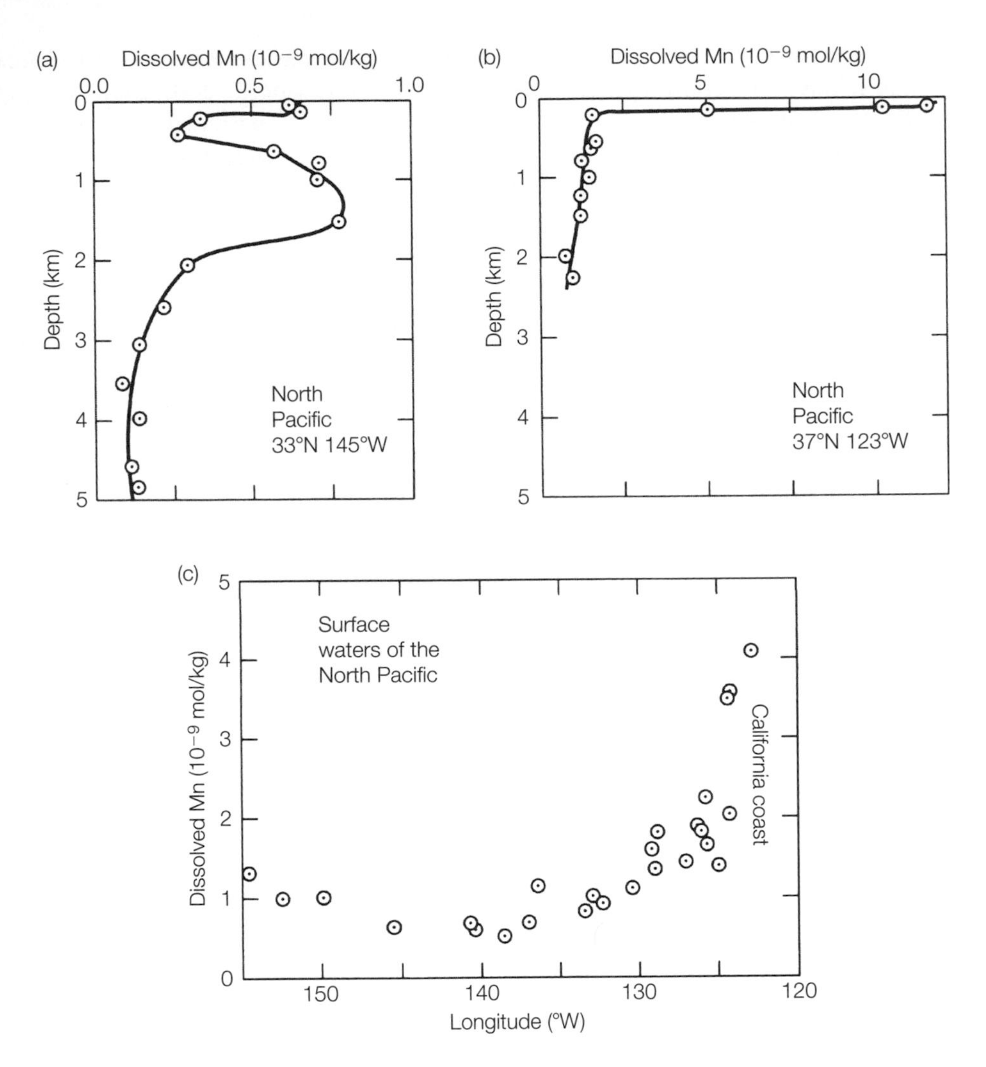

FIGURE 11.17

Upper panels show vertical profiles of manganese in the North Pacific Ocean at (a) an open-ocean station, (b) a coastal station, and (c) the Mn content of surface water with increasing distance from the California coast. Note the tenfold scale difference in concentration between these diagrams. *Source*: From Landing, W. M., and K. W. Bruland (1980). *Earth and Planetary Sciences Letters*, 49, 45–56.

open ocean vertical profile (Figure 11.17a), except that the surface concentrations are lower due to the increased distance from land and a mid-water concentration maximum is present. The source of the mid-water maximum is thought to be lateral transport of dissolved Mn that diffuses out of continental shelf sediments following diagenetic remobilization under suboxic conditions. The lateral transport is associated with geostrophic currents that move the coastal waters offshore, thereby transporting dissolved Mn to the open ocean. As these currents move across the continental shelf, they can cause resuspension of sediments. These sediments are also transported laterally into the open ocean. Subsequent remineralization of Mn from these resuspended particles is another likely contributor to the mid-water concentration maximum. A similar lateral transport of iron has been observed and is thought to represent a significant input of this biolimiting element to the oceans, particularly in regions where the dust flux is low.

11.4.5 Other Vertical Profile Shapes

Some vertical profiles do not fit into the nutrient, conservative, or scavenged categories. Examples and causes for the most common of these are given next.

Mid-Depth Minima

The aluminum profile shown in Figure 11.18 is an example of a mid-depth minimum created by two inputs, one at the surface and one from the sediments. The high surface-water concentrations are due to a large atmospheric dust flux in the North Atlantic. As

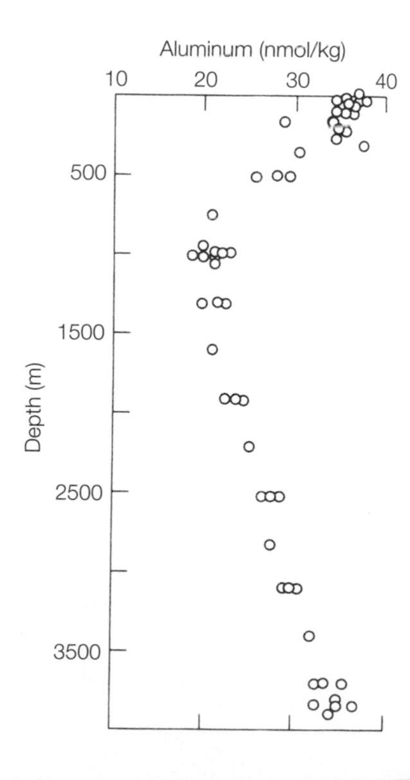

FIGURE 11.18

Vertical profile of dissolved aluminum in the North Atlantic (40°51′N, 64°10′W). *Source*: From Hydes, D. J. (1979). *Science*, 205, 1260–1262.

with Mn, the surface concentrations are geographically variable and generally reflect proximity to land as this determines the magnitude of the aeolian and riverine fluxes. Most of the atmospheric and riverine inputs of Al are rapidly removed by scavenging, causing Al to have a short residence time (~200 y). The high dissolved concentrations in the deep waters near the seafloor are due to an upward flux supported by diffusion of resolubilized Al from the sediments. The low concentrations at mid-depth reflect both the greater distance from the surface and sedimentary sources of aluminum, as well as mid-water removal via particle scavenging. For Al, particle scavenging seems to involve adsorption onto sinking detrital siliceous shells.

Mid-Depth Maxima

Mid-depth maxima are produced by mid-depth sources of metals. Some of these maxima are created by remineralization of detrital biogenic particles, such as seen in Figure 11.4f for cadmium. Others are caused by lateral transport of metals mobilized from coastal sediments as illustrated in Figure 11.17(a) for manganese. Mid-depth maxima can also result from hydrothermal emissions as shown in Figure 11.19 for $Mn^{2+}(aq)$ and $^3He(g)$ at a site in the Eastern North Pacific Ocean. Hydrothermal fluids are emitted into the ocean from chimneys located atop the East Pacific Rise at water depths of about 2500 m. After entering the ocean, the Mn and ^3He are entrained in subsurface currents and

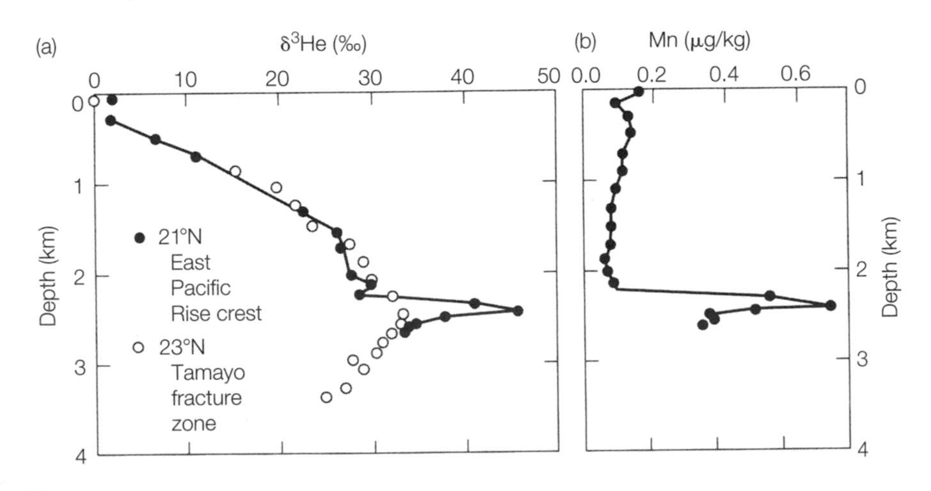

FIGURE 11.19

Vertical profiles of (a) excess ^3He and (b) dissolved manganese at two sites in the North Pacific Ocean. The data represented by the solid circles were obtained from water located directly over the crest of the East Pacific Rise at 21°N. *Source*: From Broecker, W. S., and T.-H. Peng (1982). *Tracers in the Sea*, Lamont-Doherty Geological Observatory, p. 229. See Broecker and Peng (1982) for data sources.

advected horizontally through the Pacific Ocean. Since the concentrations of $Mn^{2+}(aq)$ and $^3He(g)$ are ordinarily very low, this input can be used as a tracer of the pathway of water transport in the deep sea. As the $Mn^{2+}(aq)$ is transported through the deep sea, it is slowly oxidized to particulate $Mn(IV)O_2$, which then settles to the sediments. This input exceeds that of river runoff, making hydrothermal emissions the single largest source of Mn to the oceans.

Mid-Depth Maxima or Minima within a Suboxic or Anoxic Zone

Suboxic and even anoxic conditions are found in oceanic waters that have been subject to a large flux of sinking detrital POM. Suboxic conditions are common in the thermocline at mid- and low latitudes and are recognized as a geographically widespread oxygen minimum zone (OMZ). Horizontal and vertical segregation causes O_2 to be undersaturated in these waters and the oldest deep waters, e.g., those in the North Pacific Ocean. Extreme O_2 depletions also occur in the subsurface waters of coastal upwelling areas, at the bottom of deep-sea trenches, and in some marginal seas and inlets. In these locations, redox cycling of trace elements is observed, particularly near the boundaries that mark the transition from oxic to suboxic or anoxic conditions. This was previously illustrated for selenium in Figure 11.11.

More dramatic examples are found in marginal seas, such as the Black Sea, where permanent anoxic zones are present. This produces very complex redox cycling as illustrated in Figure 11.20. The major source of metals to the waters of the Black Sea are through atmospheric and riverine input. In the case of iron and manganese, their inputs are predominantly as the oxidized species, $Mn(IV)O_2$ and $Fe(III)OOH$, both of which have a low solubility in oxic alkaline seawater. When these particles settle into the subsurface anoxic waters, anaerobic bacteria use them to engage in iron and manganese respiration via the reactions presented in Table 7.6. In so doing, the particulate manganese and iron are reduced, to $Mn^{2+}(aq)$ and $Fe^{2+}(aq)$, respectively. Some of the dissolved metals diffuse upward until they come back into contact with O_2. Their ensuing reoxidation causes them to be converted back into particulate form. These particles eventually settle into the anoxic waters, where the metals are once again resolubilized. This cycling effectively traps the metals at the top of the anoxic zone, creating a particulate metal concentration maximum just above the redox boundary and a dissolved metal concentration maximum just below it. Co exhibits a similar redox cycling to Fe and Mn.

The other metals exhibit different vertical profiles. The dissolved concentrations of Pb and Cu do not exhibit subsurface concentration maxima and, hence, do not appear to undergo any redox reactions. Their dissolved concentrations decline with increasing depth and are likely controlled by precipitation into sulfide minerals as the particulate concentrations increase rapidly with depth in the anoxic zone. In the anoxic waters, sulfide is supplied by in situ sulfate reduction.

In the case of Zn and Cd, a subsurface dissolved concentration maximum in the oxic zone suggests supply via remineralization of sinking detrital biogenic particles. As with Pb and Cu, the dissolved concentrations of Zn and Cd decline as the anoxic zone is

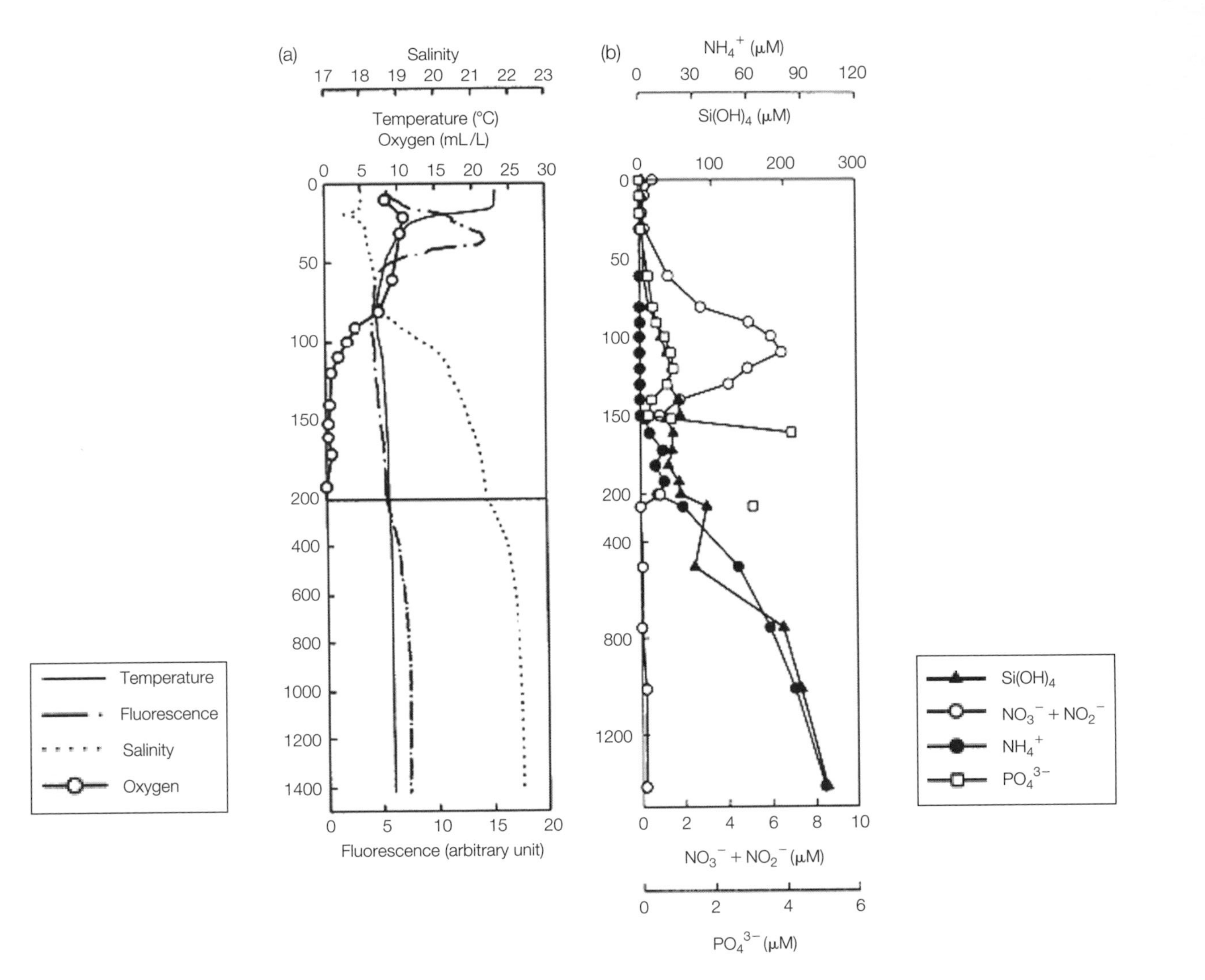

FIGURE 11.20
(Continued)

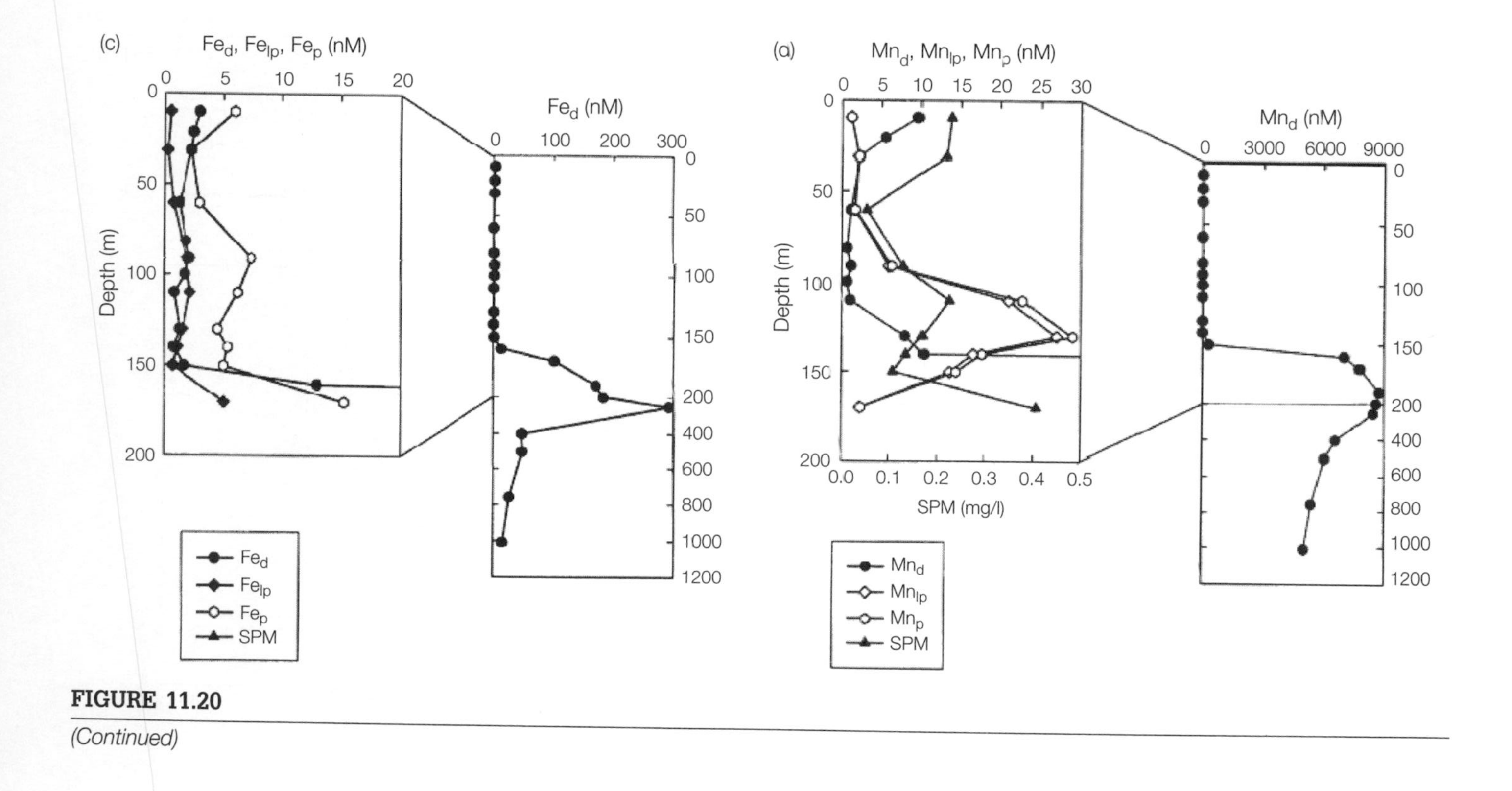

FIGURE 11.20
(Continued)

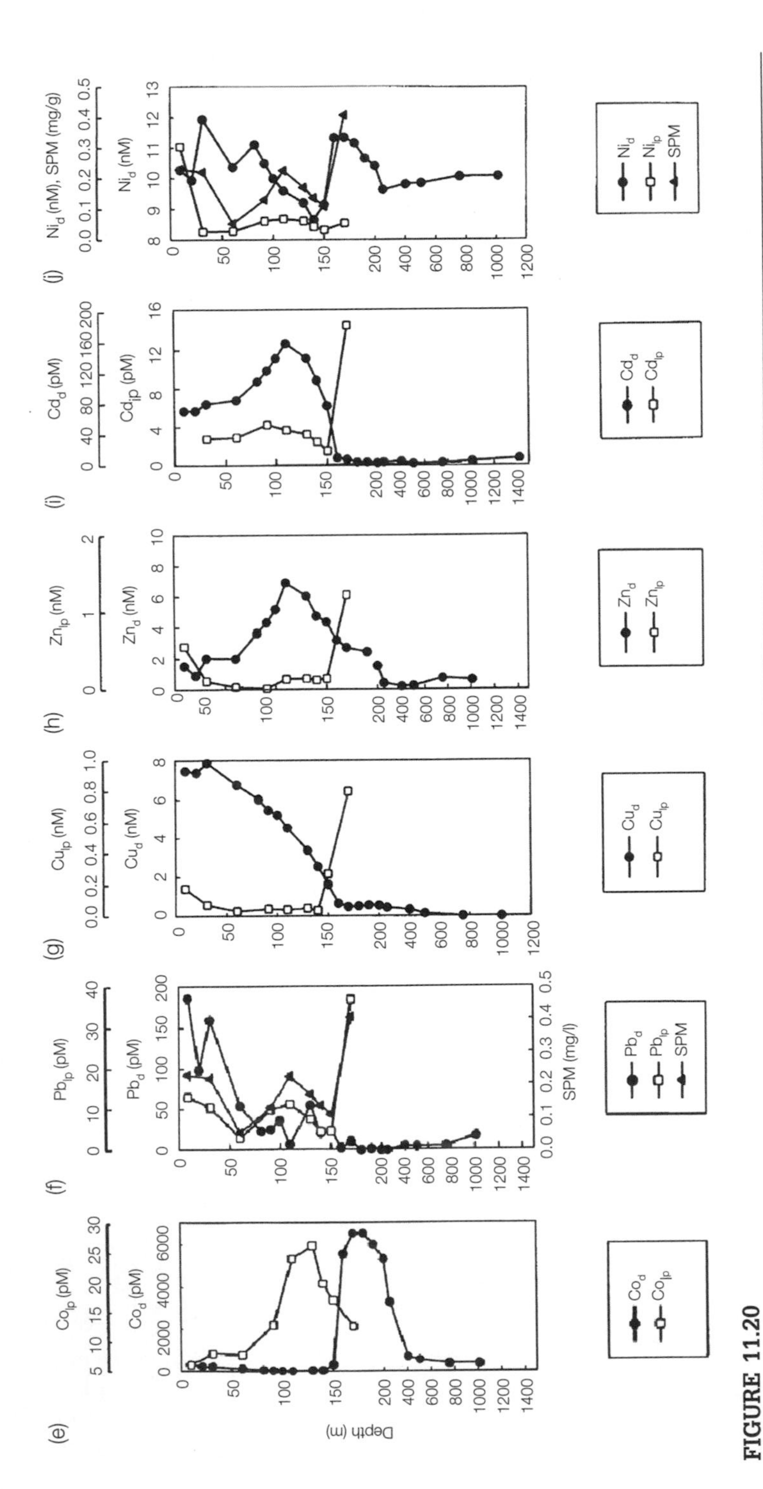

FIGURE 11.20

Vertical profiles of O_2 and particulate and dissolved trace metal concentrations at 32.5°E and 44.5°N in the Black Sea. (a) Temperature, salinity, fluorescence, and O_2; (b) ammonium, silica, nitrate+nitrite, and phosphate; (c) Fe; (d) Mn; (e) Co; (f) Pb; (g) Cu; (h) Zn; (i) Cd; and (j) Ni. In the trace metal profiles, the dissolved concentrations are represented as solid circles, the total particulate concentrations by open circles, the acid-leachable particulate concentrations by open squares, and the suspended particulate matter concentrations by the solid triangles. *Source:* After Tankéré, S. P. C., *et al.* (2001). *Continental Shelf Research*, 21, 1501–1532.

approached, probably as a result of precipitation into sulfide minerals. This is further supported by the sharp increase in particulate Zn and Cd concentrations at the top of the anoxic zone.

A third type of profile is exhibited by Ni in which a dissolved concentration maximum is positioned at the top of the anoxic layer with no overlying particulate concentration maximum. It is not clear what process(es) are supporting this.

CHAPTER 12
Diagenesis

All figures are available on the companion website in color (if applicable).

12.1 INTRODUCTION

As particulate matter settles through the water column, it is subject to physical and biogeochemical processes that can alter its size and chemical composition. These processes continue as the particles accumulate on the seafloor and eventually become buried. Most of the changes that particulate matter undergoes once it reaches the seafloor occur within the first thousand years and are collectively referred to as *diagenesis*. Most of the diagenetic changes take place while the particulate matter is still lying on the surface of the seafloor just prior to burial.

As a layer of sediment becomes buried, its temperature and overlying pressure increase. At burial depths of a few hundred meters, temperatures exceed 50°C. The physical and chemical changes that occur under these conditions are referred to as *catagenesis*. Below a few thousand meters, temperatures exceed 150°C and sediments undergo metamorphosis. The chemical changes that occur during this stage of sediment burial are termed *metagenesis*. As we shall see in Chapter 27, petroleum is the result of the cumulative actions of diagenesis, catagenesis, and metagenesis on particulate organic matter created millions of years ago by marine and aquatic organisms.

Diagenetic changes occur during the earliest stages of burial in the topmost layer of the sediments under environmental conditions close to those present at the time of sedimentation. Some of these changes lead to remobilization of trace elements, nutrients, and carbon. Thus, settling onto the seafloor in particulate form is no guarantee of burial and, hence, removal from the ocean! This makes diagenesis an important process in the crustal-ocean-atmosphere factory as it determines the degree to which the sediments act as a sink for any particular element.

Many of the chemical reactions that occur in sediments during diagenesis are mediated by marine organisms or are a consequence of biotic activities. Most are energy-yielding redox reactions driven by the oxidation of organic matter and, hence, represent a critical metabolic resource to benthic organisms.

Postdepositional changes can obscure the sedimentary paleoceanographic record. Fortunately, these alterations follow predictable geochemical principles, making it possible to glean information from ancient sediments on past changes in the rates and pathways of processes in the crustal-ocean-atmosphere factory. In this chapter, we consider the basic principles of diagenesis as they affect the organic matter, nutrient, and trace metal content of the sediments. These principles are used in Part III of this text to help explain the patterns of sediment accumulation on the ocean floor.

12.2 PHYSICAL PROCESSES: PARTICLE AND WATER TRANSPORT

12.2.1 Sedimentation

Most of the solid matter found in the sediments of the open ocean was transported to the seafloor via the slow sinking of small particles through the water column. This process is termed *pelagic sedimentation*. Other types of sedimentation are discussed in the next chapter and include turbidity flows, hydrothermal deposits, and deposition of large animal carcasses, e.g. whales, squid, and fish.

Pelagic sedimentation brings both inorganic and organic particles to the seafloor. The inorganics include clay minerals, hydrogenous precipitates, and biogenic hard parts, such as calcite and opaline silica. The organic particles are mostly detrital biogenic soft parts, such as dead tissues, fecal pellets, mucous feeding nets, and molts. Living organisms, such as bacteria and phytoplankton, can also be present as a component of sinking particles, particularly fecal pellets. Since biogenic particles contain carbon, nitrogen, phosphorus, and many trace metals, pelagic sedimentation is an important transport mechanism for these elements. The adsorption of dissolved organic matter and trace elements onto settling particles provides another mechanism for transporting chemicals to the seafloor.

Particulate matter that reaches the seafloor becomes part of the blanket of sediments that lie atop the crust. If bottom currents are strong, some of these particles can become resuspended and transported laterally until the currents weaken and the particles settle back out onto the seafloor. The sedimentary blanket ranges in thickness from 500 m at the foot of the continental rise to 0 m at the top of the mid-ocean ridges and rises. Marine scientists refer to this blanket as the *sedimentary column*. Like the water column, the sediments contain vertical gradients in their physical and chemical characteristics. Similar to the vertical profile convention used in the water column, depth in the sediments is expressed as an increasing distance beneath the seafloor.

When particles first become incorporated into the sediments, quite a bit of seawater is usually present between adjacent grains. This is termed *pore water* or *interstitial water*. In some cases, it is difficult to define exactly where the bottom of the ocean stops and the seafloor begins, especially if bottom currents are resuspending a lot of particles. As pelagic sedimentation adds particles to the sediment, layers deposited at an earlier time are eventually buried. This produces distinct horizontal layers if the types of particles collecting on the seafloor vary over time.

12.2.2 **Pore-Water Advection and Diffusion**

As continuing sedimentation increases the depth of a sedimentary layer relative to the seafloor, the overlying pressure increases because of the increased weight of the additional particles. The increased pressure leads to particle compaction if the pore waters can escape upward. Under these conditions, sedimentation generates an upward advective flow of pore water. This flow has the potential to transport solutes.

Compaction reduces the distance between the adjacent particles and thereby reduces sediment porosity. The latter is defined as the ratio of the volume of pore space to the total volume (wet+dry) of the sediment. As shown in Figure 12.1, a reduction in porosity creates a very tortuous path for the water to move through, with some paths being more available than others.

Compaction causes the density of sediments to increase with increasing depth through the top meter or so as water is squeezed out, leaving the denser solid particles behind. Grain size plays an important role in determining the degree to which a sediment can become compacted. Through the top meter or so, wet densities range from 1.3 to 2.1 g/cm^3 and porosities from 50 to 90%. As illustrated by a typical sediment core (Figure 12.2), most of the effects of compaction on density and porosity occur in the top meter. Variations below this depth can be attributed to changes in the relative sedimentation rates of particles with differing mineralogies and grain sizes. The dry density of marine sediments ranges from 2.6 to 3.0 g/cm^3 depending on the mineral content.

At greater depths, under which catagenesis and metagenesis occur, pressures are high enough to cause dewatering, mechanical deformation, and recrystallization of minerals. As shown in Figure 12.3, this reduces porosity to values under 10%, thereby converting clay mineral sediments to sandstones and shales.

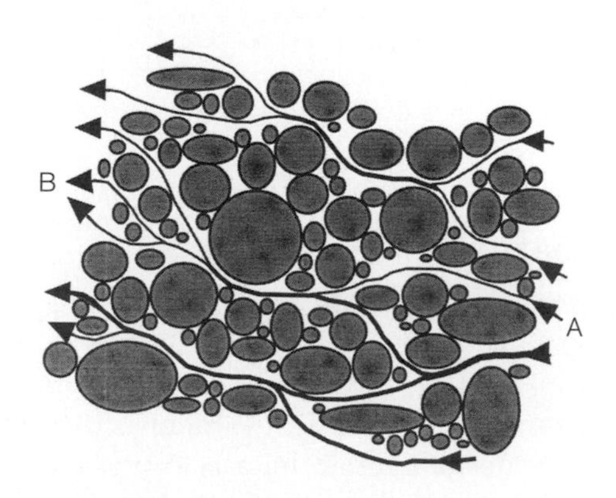

FIGURE 12.1

Movement of pore water in permeable sediments. A solute trying to move from point A to point B cannot pass through the solid sediment particles and, hence, must travel around them following a tortuous path.

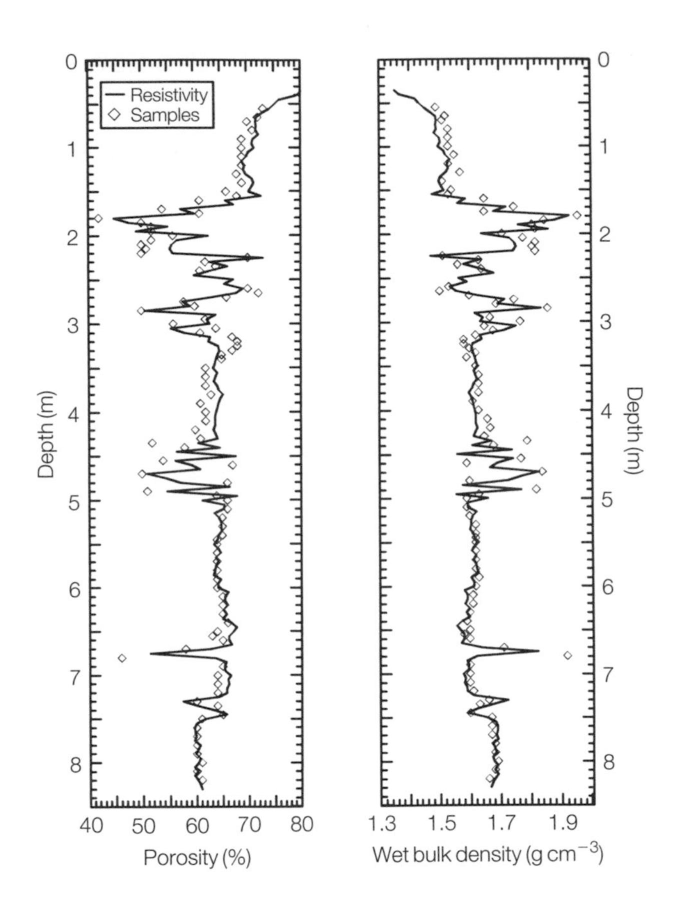

FIGURE 12.2

Downcore variations in porosity (%) and wet bulk densities. Open symbols represent results of direct measurements and lines represent estimates based on resistivity measurements. *Source*: From Breitzke, M. (2000). *Marine Geochemistry*, Springer, p. 38.

12.2.3 **Bioturbation**

Other advective flows occur in marine sediments in addition to those driven by compaction. These include flows caused by terrestrial groundwater movements (Figure 11.3), hydrothermal circulation (described in Chapter 19), waves that reach the seafloor, and *bioirrigation*. The latter is a downward advective flow into the sediments caused by the pumping activities of benthic animals, especially ones that excavate burrows. All of these advective flows have the potential to transport solutes. Solutes are also moved through pore waters by molecular diffusion, transport in rising gas bubbles (*ebullition*), and by the physical mixing of sediments by benthic animals (*bioturbation*).

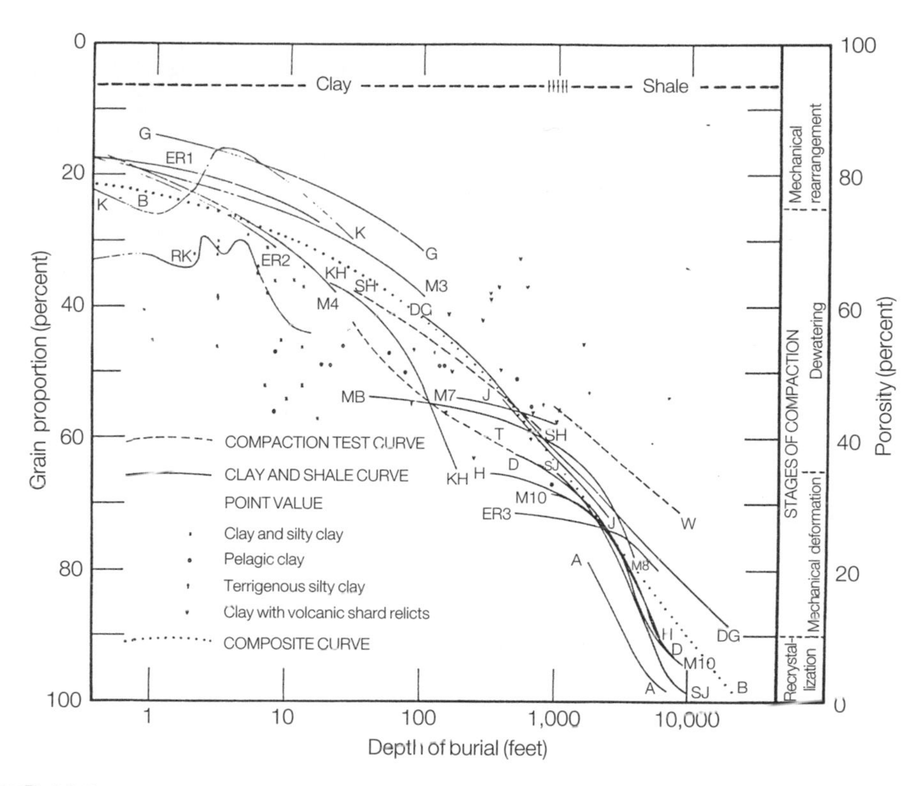

FIGURE 12.3

Porosity versus depth. Grain proportion is the complement of porosity, i.e., the ratio of volume of solid grains to bulk volume data. *Source*: From Baldwin, B. (1971). *Journal of Sedimentary Petrology*, 41, 293–301. See Baldwin (1971) for core sample site information (K, B, G, ER1, RK, etc.).

The feeding and burrowing activities of benthic epifauna and infauna cause bioirrigation and bioturbation as illustrated in Figure 12.4.

A particularly important consequence of bioirrigation and bioturbation is the introduction of relatively O_2-rich bottom water into the sediments. This enhancement in O_2 supply is analogous to the aeration of soil by earthworms. Bioturbation can occur as deeply as 1 m below the sediment surface, but is most intense in the top 10 cm. The depth of O_2 penetration is also strongly influenced by the flux of sedimenting POM. High accumulation rates of organic-rich particles can fuel bacterially mediated aerobic respiration supporting rates of O_2 removal that exceed the benthic animals' abilities to reaerate the sediments. In this case, anoxic conditions result. Since animals require O_2, bioturbation does not occur in anoxic sediments. Thus, the effects of bioturbation are limited to the oxic portion of the sediments.

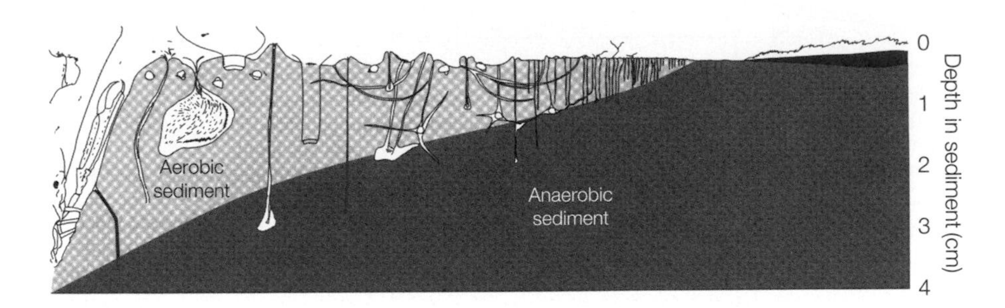

FIGURE 12.4

Bioturbating activities of some benthic marine organisms. *Source*: From Pearson, T. H., and R. Rosenberg (1976). *Ambio*, 5, 79.

Bioturbation can mix out, or homogenize, chemical gradients that would otherwise have been generated by ongoing chemical reactions or changes in sedimenting particle composition. Once the sediments are buried below the bioturbation zone, existing and developing chemical gradients are largely left intact because the remaining physical processes, namely pore-water advection and molecular diffusion, are slow compared to sedimentation and reaction rates.

The marine benthos has other important impacts on marine sediments. For example, deposit feeders ingest the sediment whole and then expel in their fecal pellets any undigested particles. This changes the size and chemical composition of the sediment grains. Organisms that construct burrows alter the particle size and mechanical strength of the sediments by reinforcing the walls of their burrows with mucus-like exudates that "glue" the sedimentary grains together. This increases the mechanical strength of the sediments and lowers their tendency to become resuspended by bottom currents. Burrowing organisms are also responsible for bioirrigation.

12.3 INTERPRETING DEPTH PROFILES IN MARINE SEDIMENTS

Marine scientists obtain information about diagenesis via three strategies: (1) measuring chemical constituent concentrations within the sediment and its pore waters, (2) measuring the flux of chemical constituents across the sediment-water interface, and (3) computing reaction rates from mathematical models. The most common approach for obtaining sediments is via collection of vertical cores. Sediment cores are sectioned into layers whose solids and pore water are chemically analyzed. Alternatively, pore water can be obtained in situ by probes that are inserted into the sediments. Pore water is collected via suction using a syringe or via diffusion across semipermeable membranes. Direct measurements of fluxes across the sediment-water interface are made with chambers installed over a small area of the seafloor from which water is withdrawn periodically for chemical analysis. Changes in concentration over time are used to compute fluxes. Schematics of some benthic sampling equipment are provided in Figure 12.5.

The results of concentration measurements are presented as vertical profiles similar to those for the water column, with the vertical axis representing increasing depth below the sediment-water interface. Depth profiles of concentrations can be used to illustrate downcore variations in the chemical composition of pore waters or in the solid particles. Dissolved concentrations are typically reported in units of moles of solute per liter of pore water. Solid concentrations are reported in mass/mass units, such as grams of carbon per 100 grams of dry sediment (%C) or mg of manganese per kg of dry sediment (ppm Mn).

Some chemicals behave conservatively in the sediments, undergoing no chemical reactions after burial. In this case, downcore concentration variations are a record of past changes in the concentration of that constituent in the sedimenting particles. This is illustrated in Figure 12.6a, which depicts the burial of a layer of sediment over time. For nonreactive chemicals delivered to the sediments as a component of solid particles,

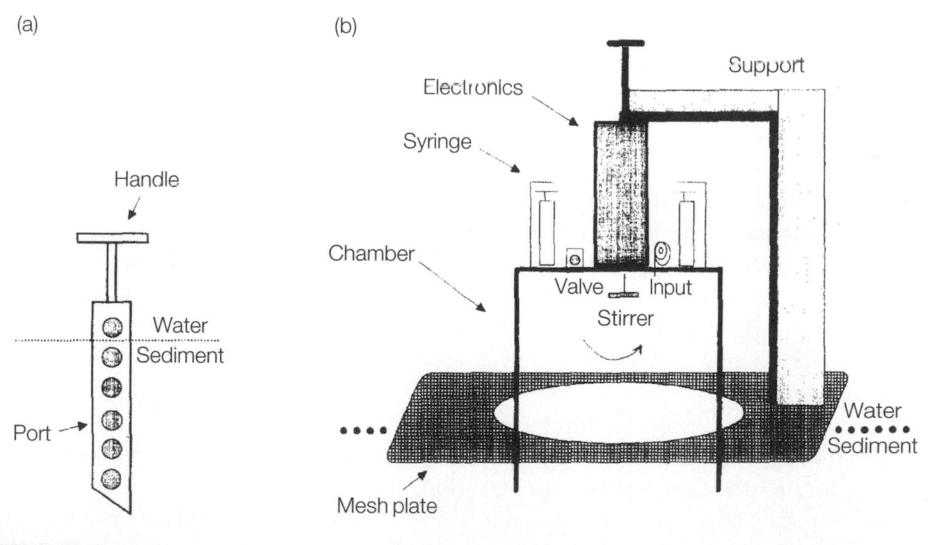

FIGURE 12.5

Diagrams of benthic sampling equipment. (a) Pore water peepers used for collecting pore water in situ via molecular diffusion across a 0.4-μm polycarbonate membrane. These probes were 36 inches long. Each of the six ports on the peeper held 6 mL of equilibration fluid. The probes were deployed for 1 to 2.5 y to ensure equilibrium had been reached with the pore waters. (b) Benthic flux chamber. This chamber covered a bottom area of 368 cm^2. *Source*: From Aller, R. C., *et al*. (1998). *Deep-Sea Research,* 45, 133–165.

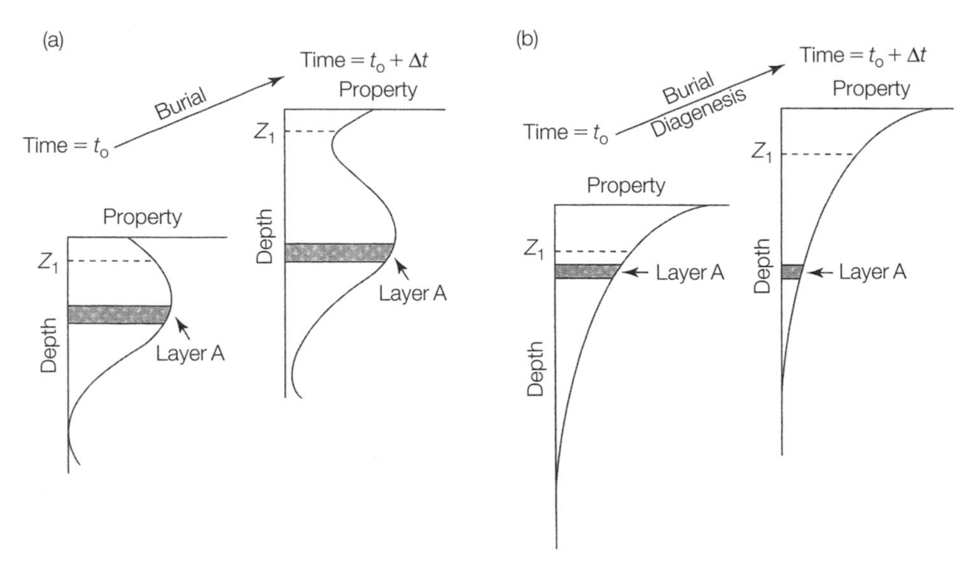

FIGURE 12.6

Burial of a sediment layer as recorded by a chemical constituent present in sedimenting particles. Z_1 is a reference depth located at a fixed distance below the sediment-water interface. (a) Nonreactive case where delivery of the chemical constituent to the surface sediment changes over time, and (b) reactive case for a first-order loss rate where delivery of the chemical constituent to the surface sediment is constant over time. *Source*: From Berner, R. A. (1972). *Journal of Geoscience Education*, 20, 267–272.

downcore concentration minima and maxima can be interpreted as a record of changes in its relative depositional flux, or accumulation rate.[1]

In the surface sediments, the physical processes of compaction, bioirrigation, and bioturbation also influence downcore concentration profiles. The concentration minima and maxima tend to be broadened by the action of benthic animals, whereas compaction has a sharpening effect on the depth profiles of the solids.

In the case of reactive chemicals, downcore variations in concentrations also reflect reaction rates. The resulting depth profile for a chemical that is delivered to the sediments at a constant rate and subsequently undergoes a reaction leading to an exponential decline in its concentration over time is illustrated in Figure 12.6b. If the chemical continues to be delivered to the sediments at a constant rate (as a component of sedimenting particles), its concentration at a given depth below the sediment surface does not change over time. In other words, this profile represents a steady-state

[1] We use the term *relative* here to emphasize that changes in sediment concentrations of nonreactive chemicals can be caused by changes in the rates of accumulation of particles that lack the chemical of interest. For example, a large increase in the flux of such a particle, without an equal increase in the flux of particles containing the chemical of interest, would generate a lower concentration of the chemical in the overall collection of sedimenting particles.

condition. This type of concentration profile is exhibited by organic matter in the solid particles and by dissolved oxygen in the pore waters.

12.4 THE ONE-DIMENSIONAL ADVECTION-DIFFUSION MODEL

Reaction rates of nonconservative chemicals in marine sediments can be estimated from porewater concentration profiles using a mathematical model similar to the one-dimensional advection-diffusion model for the water column presented in Section 4.3.4. As with the water column, horizontal concentration gradients are assumed to be negligible as compared to the vertical gradients. In contrast to the water column, solute transport in the pore waters is controlled by molecular diffusion and advection, with the effects of turbulent mixing being negligible.

The one-dimensional advection-diffusion model has proven useful for studying biogeochemical processes in marine sediments for several reasons. First, the vertical concentration profile of a given chemical constituent is often the result of several physical and chemical processes. Modeling can help identify which processes are most important. Second, some of these processes occur over long time scales and, hence, cannot be directly observed during sampling efforts. Third, most sediments are located at water depths that make direct observations difficult, so sampling is usually restricted to collection of sediment cores. And last, modeling enables marine scientists to predict impacts of future possible changes in fluxes of sedimenting particles. The latter is of particular interest for the elements involved in the global climate change story, i.e., carbon, nitrogen, phosphorus, and iron.

Molecular diffusion follows Fick's second law (Eq. 4.11),

$$\frac{\partial [C]}{\partial t} = D_z \frac{\partial^2 [C]}{\partial z^2} \tag{12.1}$$

where $[C]$ is the solute concentration, t is time, z is depth beneath the sediment-water interface, and D_z is the vertical diffusivity coefficient for solute C. As shown in Figure 12.1, pore water follows a tortuous path through the sediments. The molecular diffusion of solutes through pore water is similarly restricted such that over short space scales, solutes released into the pore waters tend to disperse in a cloudlike fashion away from their entry point. These interactions cause the value of D_z for molecular diffusion in the sediments (0.3×10^{-5} to 1×10^{-5} cm^2/s) to be somewhat less than that for molecular diffusion in free solution (1×10^{-5} to 10×10^{-5} cm^2/s). Also note that these molecular diffusivity coefficients are temperature dependent.

The effect of pore-water advection on solute concentrations is given by

$$\frac{\partial [C]}{\partial t} = -u_z \frac{\partial [C]}{\partial z} \tag{12.2}$$

where u is the rate of vertical pore-water advection. This is analogous to the water velocity term in Eq. 4.8. If the sedimentation rate remains constant over time and the vertical porosity gradient is in steady state, the rate of water advection (upward) is equal to rate of particle deposition (downward). The latter is given by the sedimentation rate (s), so Eq. 12.2 can be approximated by

$$\frac{\partial [C]}{\partial t} = -s\frac{\partial [C]}{\partial z} \tag{12.3}$$

Sedimentation rates range from centimeters per year in coastal sediments down to millimeters per thousand years on the deep-sea floor. Thus, the effects of molecular diffusion are generally greater than that of advection in shaping pore-water concentration profiles.

Combining the effects of molecular diffusion and pore-water advection yields a one-dimensional advection-diffusion equation for a conservative solute:

$$\frac{\partial [C]}{\partial t} = D_z\frac{\partial^2 [C]}{\partial z^2} - s\frac{\partial [C]}{\partial z} \tag{12.4}$$

If the solute undergoes any chemical changes, a reaction term must be added to Eq. 12.4. In the absence of specific rate law information, diagenetic reactions are generally assumed to be first-order with respect to the solute concentration. Thus, the one-dimensional advection-diffusion equation for a nonconservative solute is given by

$$\frac{\partial [S]}{\partial t} = D_z\frac{\partial^2 [S]}{\partial z^2} - s\frac{\partial [S]}{\partial z} - k\,[S] \tag{12.5}$$

where k is the first-order rate constant for the nonconservative solute, S. Note that k represents the net effect of all chemical changes. The negative sign on the reaction term makes k a positive number for reactions that cause $[S]$ to decrease with increasing depth, i.e., diagenesis is resulting in the removal of this solute.

If the pore water concentration profile of the solute does not change over time, $\partial [S]/\partial t = 0$ (which implies that k, D_z, and s are constants), Eq. 12.5 becomes an ordinary differential equation. For a nonconservative solute whose concentration decreases (exponentially) with depth such that $[S] \rightarrow 0$ as $z \rightarrow \infty$, the solution to this equation is given by

$$[S] = [S_o]\, e^{\frac{\left(s-\left[s^2+4kD_z\right]^{1/2}\right)}{2D_z}z} \tag{12.6}$$

where $[S_o]$ is the solute concentration at the sediment surface ($z = 0$). Similar solutions can be developed for solutes whose concentrations increase with increasing depth.

These solutions to the one-dimensional advection-diffusion model can be used to estimate reaction rate constants (k) from the pore-water concentrations of S, if D_z and s are known. More sophisticated approaches have been used to define the reaction rate term as the sum of multiple removals and additions whose functionalities are not necessarily first-order. Information on the reaction kinetics is empirically obtained by determining which algorithmic representation of the rate law best fits the vertical depth concentration data. The best-fit rate law can then be used to provide some insight into potential reaction mechanisms. Models have also been developed to include the effects of bioturbation, bioirrigation, and varying sedimentation rates. In practice, sedimentation rates are usually determined from concentration profiles of naturally occurring radioisotopes that are present in the solid phase, such as ^{210}Pb.

As we saw with the steady-state water-column application of the one-dimensional advection-diffusion-reaction equation (Eq. 4.14), the basic shapes of the vertical concentration profiles can be predicted from the relative rates of the chemical and physical processes. Figure 4.21 provided examples of profiles that exhibit curvatures whose shapes reflected differences in the direction and relative rates of these processes. Some generalized scenarios for sedimentary pore water profiles are presented in Figure 12.7 for the most commonly observed shapes.

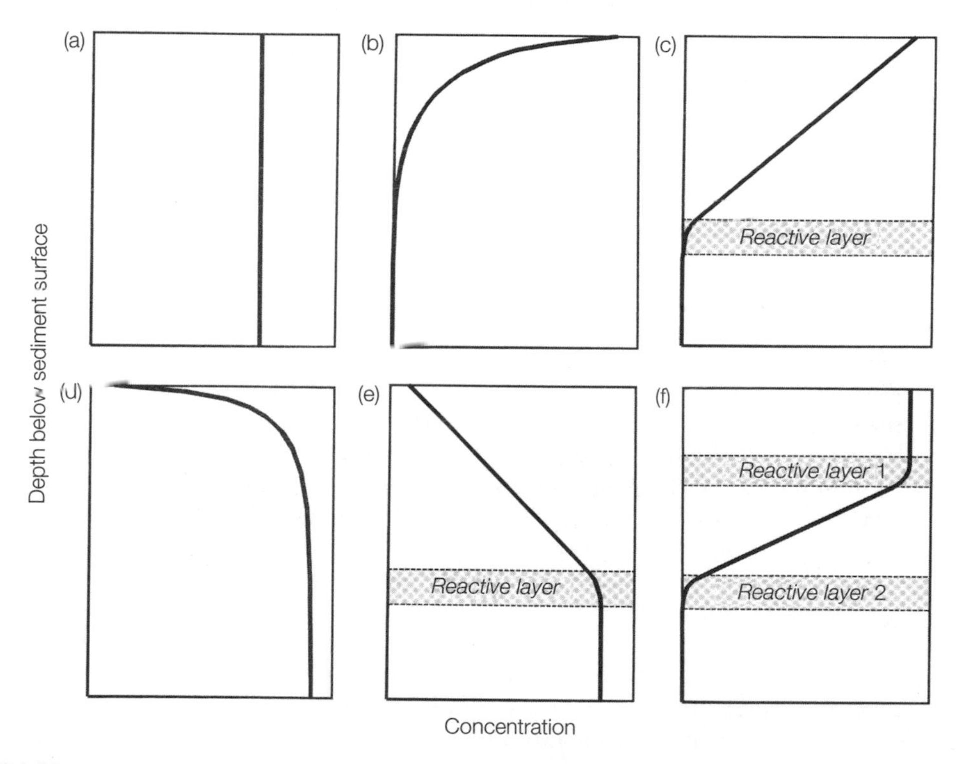

FIGURE 12.7

Most commonly observed pore-water concentration profiles. (a) A nonreactive substance, such as chloride; (b) a chemical, such as O_2, which undergoes removal in the surface sediment as a result of aerobic respiration; (c) a chemical that is consumed by a reaction that occurs in a subsurface layer, such as $Fe^{2+}(aq)$ precipitating with $S^{2-}(aq)$ to form $FeS_2(s)$; (d) a chemical released in surface sediments, such as silica via dissolution of siliceous hard parts; (e) a chemical released into pore waters from a subsurface layer, such as $Mn^{2+}(aq)$ by the reduction of $MnO_2(s)$; and (f) a chemical released at one depth (reactive layer 1), such as $Fe^{2+}(aq)$ by reduction of FeOOH(s), and removal at another depth (reactive layer 2), such as $Fe^{2+}(aq)$ precipitating as $FeS_2(s)$. *Source*: From Schulz, H. D. (2000). *Marine Geochemistry*, Springer Verlag, pp. 87–128.

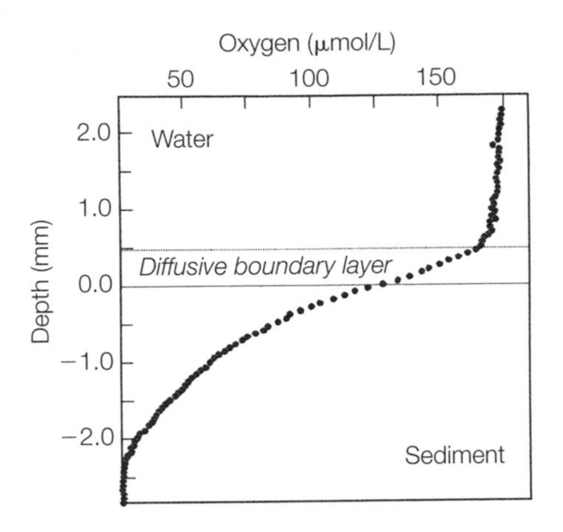

FIGURE 12.8

Diffusive boundary layer at the sediment-water interface measured with oxygen microelectrodes in Danish coastal sediments in a total water depth of 15 m. *Source*: After Gundersen, J. K., and B. B. Jorgensen (1990). *Nature* 345, 604–607.

If the pore-water concentrations are significantly different from the overlying bottom-water concentrations, Fick's first law (Eq. 3.6) can be used to compute the diffusive flux of the solute across the sediment-water interface. Two scenarios are possible: (1) the pore-water concentration exceeds the bottom-water concentration, supporting a net diffusive flux of the solute into the bottom ocean and (2) the pore-water concentration is less than the bottom-water concentration, supporting a net diffusive flux of the solute into the surface sediments. The latter case is illustrated in Figure 12.8 for dissolved oxygen, illustrating the existence of a diffusive boundary layer in the bottom waters whose thickness is highly variable due to disruption by turbulence and advection arising from storm waves, tidal pumping, groundwater discharge and bioirrigation.

In some cases, diagenesis produces enough solutes to thermodynamically favor the precipitation of mineral phases. Sand grains act as crystallization nuclei, so mineral precipitates tend to fill the sedimentary pores. This cementation process is mostly due to the deposition of calcium carbonate and quartz but can also include iron, phosphate, and manganese minerals such as glauconite, fluoroapatites, carbonates, and sulfides. Deep burial causes the dissolution of some of these precipitates because mineral solubility increases with increasing pressure. This effect is called *pressure solution*. At depths greater than several thousand meters below the sediment-water interface, pressures are high enough to crush and recrystallize these particles, thus producing sedimentary rock (Figure 12.3).

12.5 REDOX CONDITIONS IN MARINE SEDIMENTS

The chemical reactions that occur in the sediments fall into three categories: (1) adsorption onto or desorption from sedimentary particles, (2) dissolution and precipitation of solids, and (3) redox reactions, with some involving phase changes. Acid-base changes are a common feature of these reactions. As in the water column, redox reactions in the sediment are mediated by microbes in search of energy. In Chapter 7, we saw that energy can be obtained through heterotrophic metabolisms based on oxidation of organic matter. Other metabolic strategies involve transforming solar energy into chemical energy (photosynthesis) and chemoautolithotrophy during which reduced inorganic compounds are oxidized. Each of these strategies is exploited by various benthic marine microbes. For example, benthic photoautotrophs live in coastal environments where the water is shallow enough for light to reach the sediments. Benthic chemoautolithotrophy is practiced by microbes living in sulfide-rich environments, such as salt marshes and hydrothermal vents where O_2 is used as the oxidant. Another interesting example of benthic chemoautolithotrophy involves sulfate-reducing bacteria that use methane as an electron donor. The methane is a product of methanogenesis conducted by archaea. Marine scientists speculate that this anaerobic oxidation of methane plays a significant role in controlling global climate because it greatly reduces the supply of methane, a more potent greenhouse gas than either CO_2 or H_2O, to the atmosphere.

12.5.1 Oxic Zone Diagenesis

The most common metabolic strategy exhibited by marine benthic microbes is heterotrophy driven by organic matter supplied by pelagic sedimentation of POM generated by plankton in the euphotic zone. The critical role of the biological pump in supporting the benthos is illustrated in Figure 12.9. This pump is not very efficient; most of the vertical flux of POM generated in the euphotic zone never reaches the sediment due to remineralization en route by microbes, primarily via aerobic respiration. Only 10% of the flux of POM from the euphotic zone survives to settle onto the sediment surface. Approximately 90% of this POM is remineralized by benthic microbes, so only 1% of the euphotic zone flux survives to become buried in the sediments. Even after burial, organic matter continues to be consumed via microbial metabolisms at rates that depend on the availability of electron acceptors and the lability of the remaining organic molecules. As discussed in Chapter 7, heterotrophic microbes preferentially utilize electron acceptors so as to maximize free energy yield.

Because of its high free energy yield, the aerobic respiration of sedimentary organic matter is the dominant metabolism when O_2 is present. Thus, the rate of O_2 supply to the sediments is critical to determining the rate at which sedimentary organic carbon is remineralized. Remineralization via anaerobic metabolisms has a lesser impact in reducing the organic carbon content of the sediments. As a result, most of the organic carbon remineralization takes place in the surface sediments. Once the sediments are buried below a few meters, the remaining organic carbon undergoes little or no further reaction and, hence, is "preserved."

Since oceanic sediments are isolated from the atmosphere, O_2 is supplied only through contact with the bottom waters via downward diffusion or through the effects of bioturbation. The depth of O_2 penetration into the sediments reflects the relative rates of O_2 supply from the bottom water and O_2 removal via aerobic respiration. The

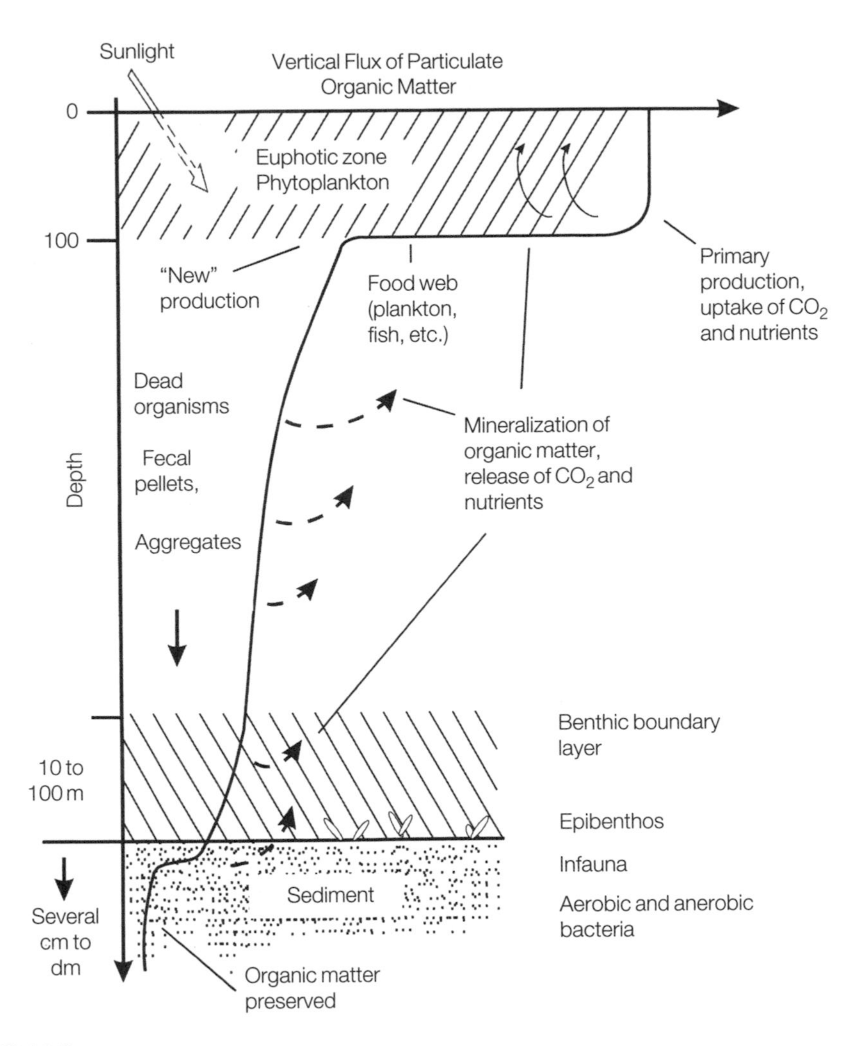

FIGURE 12.9

Importance of the biological pump in supplying the benthos with POM. *Source*: From Rullkotter, J. (2000). *Marine Geochemistry*, Springer Verlag, pp. 125–126.

rate of O_2 supply to the sediments is dependent on the concentration of O_2 in the bottom water and the rates of bioirrigation and bioturbation. Since POM fuels aerobic respiration, organic carbon remineralization rates are directly correlated with the POM flux reaching the seafloor (Figure 12.10). Thus, sediments receiving high POM fluxes will have high rates of aerobic respiration and, hence, rapid O_2 uptake rates. Under these conditions, O_2 is the limiting reactant, making the depth to which O_2 can penetrate into the sediments dependent on the flux of POM to the seafloor.

Areas with high POM fluxes tend to be located in coastal waters due to high production in the overlying euphotic zone and a short water column. The latter ensures

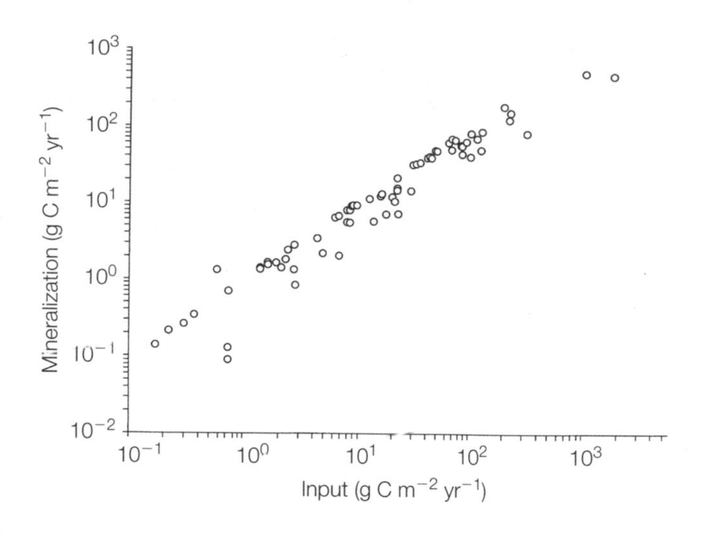

FIGURE 12.10

Relationship between flux of organic carbon to the sea floor and organic carbon remineralization rate. Data are a compilation of measurements made globally. Redrawn by Wakeham, S. (2002). *Chemistry of Marine Water and Sediments*, A. Gianguzza, E. Pelizzetti, and S. Sammartano, eds. Springer, pp. 147–164. From Heinrichs, S. (1993). *Organic Geochemistry—Principles and applications*, Plenum Press, pp. 101–115.

a quick trip to the seafloor, minimizing the loss of sinking detrital POM to aerobic respiration. In the open ocean, the combined effects of lower euphotic zone productivity and the longer water column result in a much smaller flux of detrital POM to the sediments. Thus, the depth of O_2 penetration into the sediments tends to increase with increasing water depth as shown in Figure 12.11. For sediments in the open ocean, the oxic zone generally extends 8 to 15 cm below the seafloor. In coastal sediments, the oxic zone is typically less than 10 cm deep. Beneath coastal upwelling areas, productivity can be so high that the supply of POM is large enough to cause the oxic-anoxic boundary to lie above the sediment surface. This is also seen in estuarine sediments. The position of the redox boundary can shift over time in response to changes in POM flux, temperature, and water circulation rates. In some estuarine sediments, the depth of the redox boundary has been observed to vary seasonally and diurnally.

The aerobic respiration of organic matter of planktonic origin resolubilizes nitrogen (as nitrate), phosphorus (as phosphate), carbon (as CO_2), and sulfur (as sulfate) in ratios that follow the Redfield-Richards stoichiometry (Eq. 8.4). Thus, aerobic respiration elevates the pore-water concentrations of nitrate, phosphate, CO_2, and sulfate. The complete oxidation of organic matter generates CO_2, which reacts with water to generate carbonic acid, thereby enhancing the dissolution of calcareous shells and increasing the pore-water concentrations of carbonate and bicarbonate. Siliceous shells also tend to dissolve, producing pore waters enriched in dissolved silica.

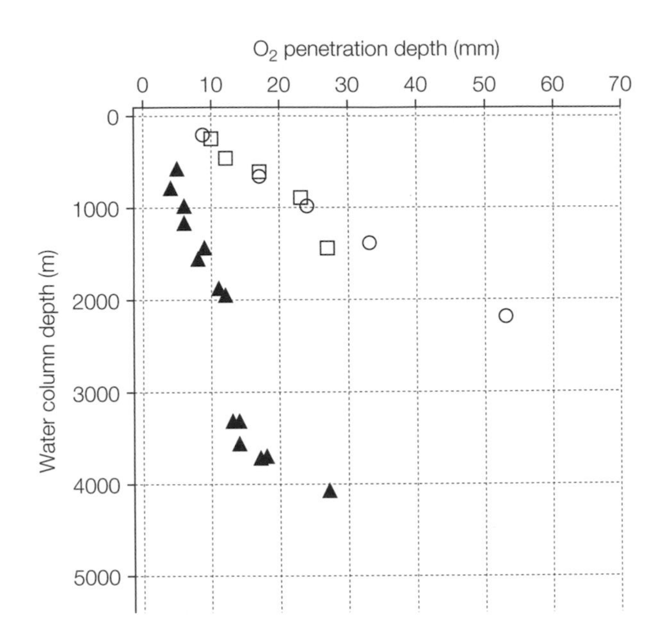

FIGURE 12.11

The sediment oxygen penetration depth as a function of water column depth in the Atlantic Ocean (open symbols) and Pacific Ocean (closed symbols). *Source*: From Martin, W. R., and F. L. Sayles (2003). *Treatise on Geochemistry*, Elsevier.

12.5.2 Redox Zonation

In sediments where O_2 concentrations are low (suboxic) or zero (anoxic), bacteria oxidize organic matter using other electron acceptors including nitrate, manganese, iron, sulfate, or carbon dioxide. The stoichiometries of the most well known of these anaerobic metabolisms are given in Table 12.1 along with an estimate of their free energy yields per mole of organic matter. As noted in the footnote, the number of kilojoules produced by the oxidation of 1 mol of planktonic organic matter of Redfield-Richards composition can be obtained by multiplying the tabulated values by 17.67. These reactions are similar to the ones presented in Table 7.6.

Aerobic respiration has the highest free energy yield, so it is favored over all other organic matter oxidations in the presence of O_2. As a result, this reaction dominates the oxic zone. At some depth below the sediment-water interface, the rate at which O_2 can be resupplied through diffusion, bioirrigation, and bioturbation is less than the rate at which it is consumed by aerobic respiration. This causes O_2 concentrations to decrease with increasing depth. Below this oxic zone, bacteria engage in anaerobic metabolisms that occur at depths dictated by their relative energy yields. This produces a vertical zonation of metabolisms as illustrated in Figure 12.12.

This vertical zonation is a consequence of microbes acting on sedimentary organic matter as it accumulates on the seafloor and eventually becomes buried. In the case

Table 12.1 Oxidation Reactions of Sedimentary Organic Matter.

1. By organic matter (aerobic respiration)

$$(CH_2O)_{106}(NH_3)_{16}(H_3PO_4) + 138O_2 \rightarrow 106CO_2 + 16HNO_3 + H_3PO_4 + 122H_2O$$

$\Delta G^0_w = -3190$ kJ/mol of glucose

2. By MnO_2

$$(CH_2O)_{106}(NH_3)_{16}(H_3PO_4) + 236MnO_2 + 472H^+ \rightarrow 236Mn^{2+} + 106CO_2 + 8N_2 + H_3PO_4 + 366H_2O$$

$\Delta G^0_w = -3090$ kJ/mol (Birnessite)

-3050 kJ/mol (Nsutite)

-2920 kJ/mol (Pyrolusite)

3. By nitrate (denitrification)

$$(CH_2O)_{106}(NH_3)_{16}(H_3PO_4) + 94.4HNO_3 \rightarrow 106CO_2 + 55.2N_2 + H_3PO_4 + 177.2H_2O$$

$\Delta G^0_w = -3030$ kJ/mol

$$(CH_2O)_{106}(NH_3)_{16}(H_3PO_4) + 84.8HNO_3 \rightarrow 106CO_2 + 42.4N_2 + 16NH_3 + H_3PO_4 + 148.4H_2O$$

$\Delta G^0_w = -2750$ kJ/mol

4. By Fe_2O_3

$$(CH_2O)_{106}(NH_3)_{16}(H_3PO_4) + 212Fe_2O_3 \text{ (or } 424FeOOH) + 848H^+ \rightarrow$$

$$424Fe^{2+} + 106CO_2 + 16NH_3 + H_3PO_4 + 530H_2O \text{ (or } 742H_2O)$$

$\Delta G^0_w = -1410$ kJ/mol (Hematite, Fe_2O_3)

-1330 kJ/mol (Limonitic goethite, FeOOH)

5. By sulfate (sulfate reduction)

$$(CH_2O)_{106}(NH_3)_{16}(H_3PO_4) + 53SO_4^{2-} \rightarrow 106CO_2 + 16NH_3 + 53S^{2-} + H_3PO_4 + 106H_2O$$

$\Delta G^0_w = -380$ kJ/mol

6. By fermentation

$$(CH_2O)_{106}(NH_3)_{16}(H_3PO_4) \rightarrow 53CO_2 + 53CH_4 + 16NH_3 + H_3PO_4$$

$\Delta G^0_w = -350$ kJ/mol

The free energy yield of these reactions has been estimated by assuming the organic matter is structurally equivalent to glucose with respect to the carbon, a primary amine with respect to the nitrogen, and glucose-1-phosphate with respect to phosphorus. Multiplication by 17.67 will convert these values to kilojoule per Redfield-molecule mole.
Source: From Froelich, P. N., et al. (1979). Geochimica et Cosmochimica Acta, *43, 1076.*

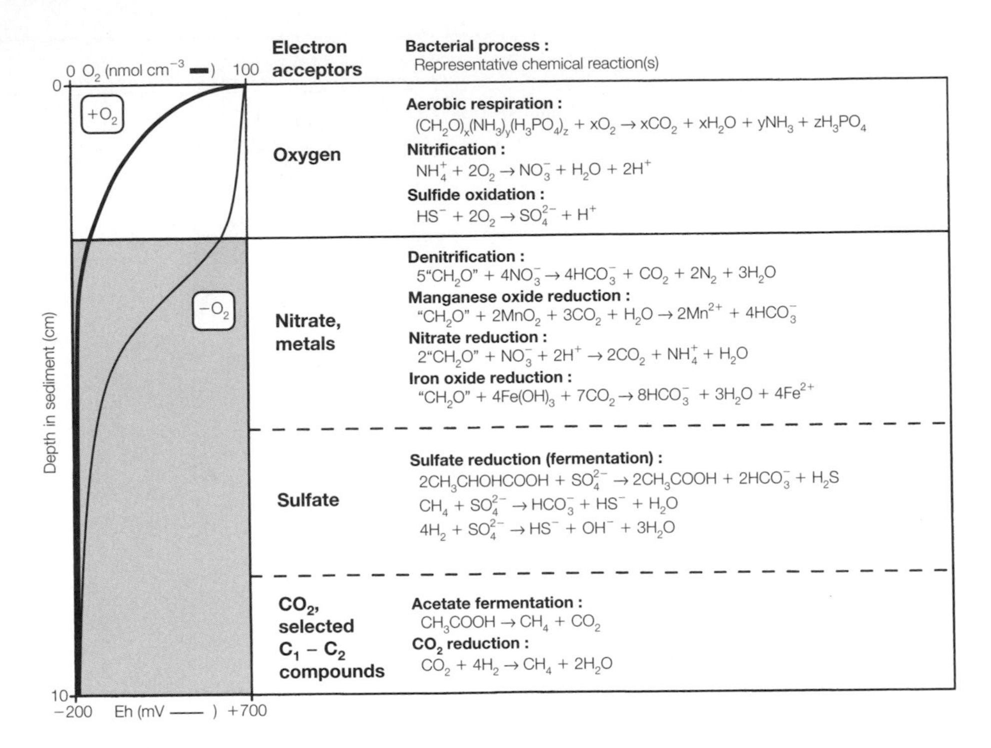

FIGURE 12.12

A schematic of zones of organic matter degradation and respiratory functions of bacteria that occur during diagenesis of organic matter in sediments. From Deming, J. W., and J. A. Baross (1993). *Organic Geochemistry—Principles and Applications*, Plenum Press, pp. 101–115.

where the rate of organic matter deposition and deepwater oxygen concentrations are constant over time, the depth of the oxic zone remains at fixed relative to the sediment-water interface. As a sedimentary layer becomes buried, it is eventually transported below the oxic zone into the suboxic and then anoxic zones, where it is subject to anaerobic metabolisms. Thus, as the depth of burial increases, so has the length of time over which diagenesis has occurred. This means that the deeper a layer lies below the sediment surface, the greater the degree of diagenetic change it has experienced. A layer that has passed through the oxic zone has had its organic content reduced and its pore-water concentrations of nitrate, phosphate, and inorganic carbon increased. As noted earlier, some of this carbon is supplied by the dissolution of calcareous hard parts.

Following aerobic respiration, manganese oxidation is the second most energy-yielding reaction (Table 12.1). An idealized set of vertical concentration profiles are presented in Figure 12.13a depicting typical chemical trends within the oxic and suboxic zones, such as would be observed in an open-ocean setting. The depth of the bioturbated sediments is marked by the surface zone where O_2 concentrations are high

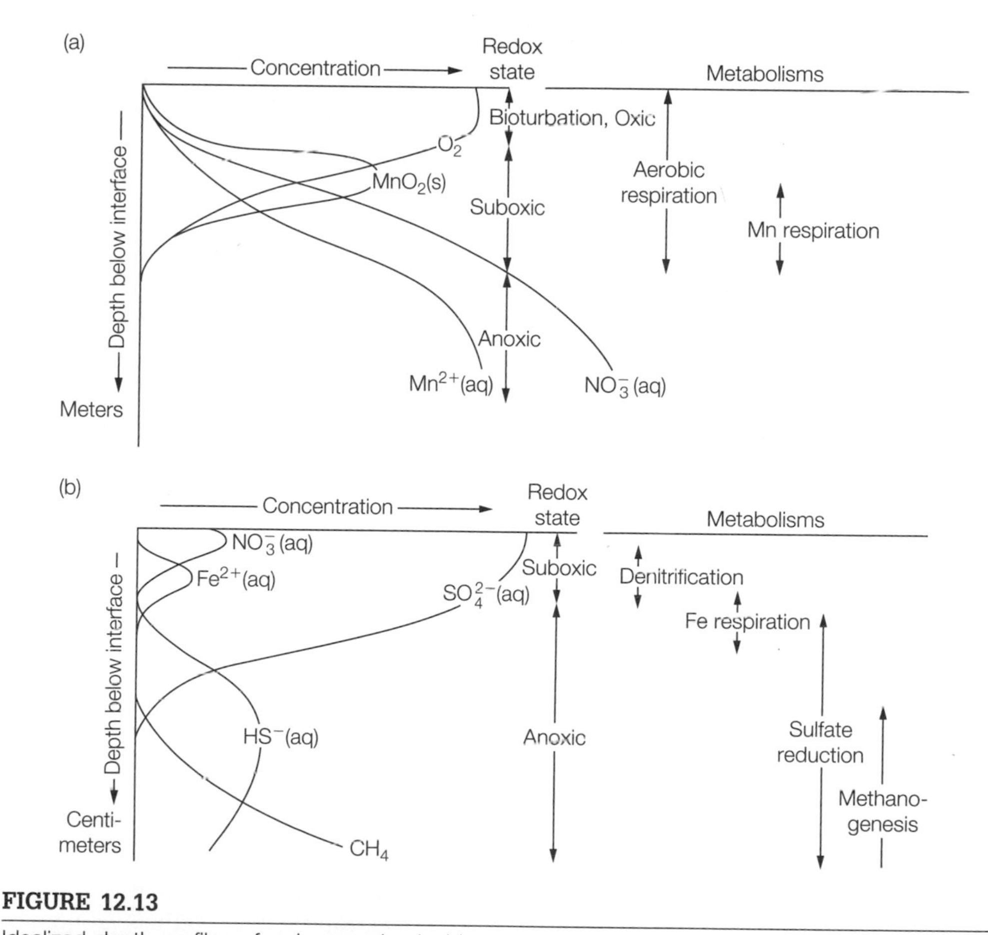

FIGURE 12.13

Idealized depth profiles of redox species in (a) open-ocean sediments (water depth >1000 m) and (b) coastal sediments (water depth <1000 m).

and no gradient is present. Evidence for manganese respiration is seen in the decreasing content of the reactant, Mn(IV)O_2, and increasing pore-water concentration of the product, Mn^{2+}, with increasing depth. As with aerobic respiration, nitrogen, phosphorus, and carbon are remineralized. Once O_2 has been depleted, remineralized nitrogen cannot be oxidized to nitrate and, thus, remains in reduced form as ammonium. As shown in Table 12.1, this ammonia is oxidized to N_2 during manganese respiration. The organic content of open-ocean sediments is generally insufficient to support the other less energy-yielding anaerobic metabolisms.

In sediments that lie in coastal waters, organic carbon levels are high enough to support denitrification, iron respiration, sulfate reduction and methanogenesis. As shown in the idealized profile presented in Figure 12.3b, the depth of O_2 penetration in organic-rich sediments is typically so shallow as to make the zones of aerobic respiration, manganese respiration, and denitrification very thin and appear to overlap. In these sediments, Mn respiration has proceeded such that all of the MnO_2 in the suboxic zone has been converted to Mn^{2+}. This leaves denitrification as the next most energy-yielding heterotrophic metabolism. Denitrification consumes nitrate (NO_3^-) and produces N_2 gas with nitrite (NO_2^-) being generated as an intermediate.

Recent research has identified some other microbial routes for denitrification that are not heterotrophic. One, called the anammox reaction, involves the oxidation of ammonium to N_2 using either nitrite or nitrate as the electron donor. The second has bacteria using Mn^{2+} to reduce nitrate to N_2. As noted earlier, N_2 is generated by the oxidation of ammonium using MnO_2 as the electron acceptor. [Denitrification may also be supported by Fe^{2+}(aq) oxidation.] These reactions are summarized in Table 12.2. The overall consequence of these reactions is that ammonium does not accumulate in the pore waters where Mn respiration and denitrification are occurring.

At depths below which nitrate has been depleted, iron respiration becomes energetically favorable. In this metabolism Fe(III), in the form of particulate oxyhydroxides, is reduced to Fe^{2+}, leading to an increase in pore-water iron concentrations. At depths below which the reactive particulate iron has been depleted, sulfate reduction is thermodynamically favored. In this metabolism sulfate is converted to sulfide. The latter can be present as the free sulfide ion; as hydrogen sulfide gas, which gives organic-rich sediments their rotten-egg smell; or as bound sulfide. Metal sulfides have very low solubilities, so the precipitation of these minerals [FeS (pyrite), FeS_2 (marcasite) and MnS (alabanite)] provides an upper limit to pore-water concentrations of iron and manganese. Other minerals produced as a consequence of anaerobic diagenesis of trace metals include the iron and manganese carbonates (siderite and rhodochrosite), iron phosphate (vivianite), and glauconite, an iron-rich silicate.

At depths below which sulfate is absent, the least energetic of the heterotrophic metabolisms is now the only choice, namely methanogenesis, in which organic matter undergoes a disproportionation reaction causing some of the carbon to be oxidized to CO_2 and some to be reduced to CH_4. This provides an upper limit to the pore-water concentration of dissolved inorganic carbon along with the precipitation of carbonate minerals. As noted earlier, the archaea appear to be responsible for methanogenesis with a significant portion of the resulting methane that diffuses upward being oxidized via nonheterotrophic sulfate reduction (anaerobic oxidation of methane).

Table 12.2 Anaerobic Chemoautolithotrophic Metabolisms That Prevent the Accumulation of Ammonium in Pore Waters.

Annamox	$5\,NH_4^+(aq) + 3\,NO_3^-(aq) \rightarrow 4\,N_2(g) + 9\,H_2O(l) + 2\,H^+(aq)$
Denitrification	$10\,Fe^{2+}(aq) + 2\,NO_3^-(aq) + 24\,H_2O(l) \rightarrow N_2(g) + 10\,Fe(OH)_3(s) + 18\,H^+(aq)$ $5\,Mn^{2+}(aq) + 2\,NO_3^-(aq) + 4\,H_2O(l) \rightarrow N_2(g) + 5\,MnO_2(s) + 8\,H^+(aq)$
NH_3 respiration	$2\,NH_4^+(aq) + 3\,MnO_2(s) + 4\,H^+(aq) \rightarrow N_2(g) + 3\,Mn^{2+}(aq) + 6\,H_2O(l)$

Phosphate is remineralized during the oxidation of organic matter and dissolution of hard parts, such as bones and teeth, that are composed of the minerals hydroxyapatite and fluoroapatite. Unlike the other products of remineralization, pore-water phosphate concentrations are regulated only by mineral solubility, such as through vivianite (iron phosphate) and francolite (carbonate fluoroapatite). Redox reactions are not significant because phosphorus exists nearly entirely in the +5 oxidation state.

In this section, idealized profiles have been used to illustrate redox zonation such as would be predicted to occur if thermodynamic constraints controlled the vertical concentration profiles. Development of this degree of zonation is rarely observed for a variety of reasons. For example, bioturbation and bioirrigation can cause localized and temporary redox disequilibria. This is commonly observed in worm burrows, where deep oxic zones can be established as a result of relatively high rate of O_2 supply from irrigation. The delivery of O_2 promotes the oxidation of Fe^{2+} and Mn^{2+}, which precipitate as oxides, leaving dark-colored mineral deposits marking the location of old burrows. Anoxic microzones can also exist within and around organic-rich particles such as fecal pellets and decaying biomass. For these reasons, surface sediments tend to have considerable horizontal and vertical redox inhomogeneity. Similarly, large spatial gradients in pH have also been observed in close proximity to burrows and detrital organic particles.

Departures from idealized redox zonation can also result from temporal shifts in sedimentation rates and in the depth of the oxic zone. Some examples are provided in the next section using iron and manganese as case studies. These elements are particularly useful as records of past changes in sedimentation rates because they respond to local redox conditions by undergoing postdepositional migration.

12.5.3 Postdepositional Migration of Metals

The redox chemistry of iron and manganese in marine sediments is accompanied by phase changes, which cause particulate Fe and Mn to become concentrated at the redox boundary that marks the transition from oxic to suboxic and anoxic conditions. This trapping process has been likened to a type of zone refining of these metals. It is very similar to the cycling of iron and manganese that occurs in the water column of the Black Sea (Figure 11.20). An example of iron and manganese trapping in surficial sediments is presented in Figure 12.14.

Iron and manganese are initially supplied to the sediments as a component of the sinking flux of POM and particulate oxyhydroxides. Remineralization of the POM releases iron and manganese to the pore waters. In the presence of O_2, the solubilized metals are oxidized and precipitate as oxyhydroxides, thereby increasing the inorganic particulate phase in the oxic layer. Continuing sedimentation eventually carries this particulate Mn and Fe below the oxic zone.

In the anoxic zone, heterotrophic respiration of particulate MnO_2 and Fe_2O_3 or FeOOH causes manganese and iron to be reduced to $Mn^{2+}(aq)$ and $Fe^{2+}(aq)$. As dissolved ions, these trace metals diffuse through the pore waters. The ions that diffuse upwards will reenter the oxic zone, where they react with O_2 to reform the oxyhydroxides. This produces a metal-enriched layer that lies just above the redox

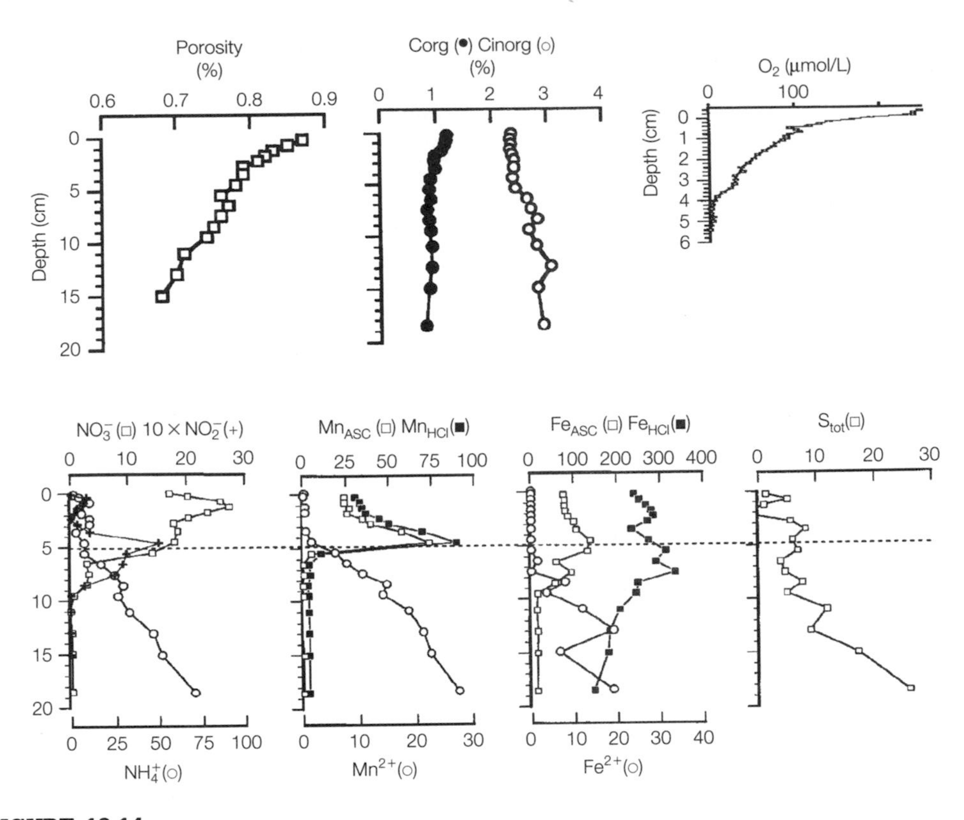

FIGURE 12.14

Vertical profiles of redox sensitive species in modern sediments of the Bay of Biscay in the Northeastern Atlantic Ocean in 2800 m water depth off the coast of France. Carbon content is reported in dry weight percentages (% C_{org}). Note depth scale in O_2 profile. The horizontal bar in the nitrogen, manganese, iron, and sulfur profiles represents the oxygen penetration depth. Pore-water concentrations of the solutes are reported in units of µmol/L and the reactive particulate phases are in units of µmol/g dry sediments. For the solid phases, two fractions were isolated, a reducible fraction (ASC, ascorbate extractable) that reflects the metal oxide and sulfide content and an acid-soluble fraction (HCl, hydrochloric acid extractable fraction) that reflects the metal sulfide and carbonate content. S_{tot} = total particulate sulfur. Sulfate concentrations were constant with depth and no free sulfide was detected. *Source*: From Hyacinthe, C., (2001). *Marine Geology* 177, 111–128.

boundary. Continuing sedimentation acts to drive this layer back down below the oxic zone, thereby repeating the cycle of metal reduction, solubilization, diffusion, and reprecipitation. If the sedimentation rate of particles and metal remains constant over time, iron and manganese will become trapped at the redox boundary, producing the concentration profiles shown in Figures 12.13 and 12.14.

The particulate iron(III) maximum is deeper than that of the particulate Mn maximum. This reflects the lower energy yield of iron respiration and its faster rate of

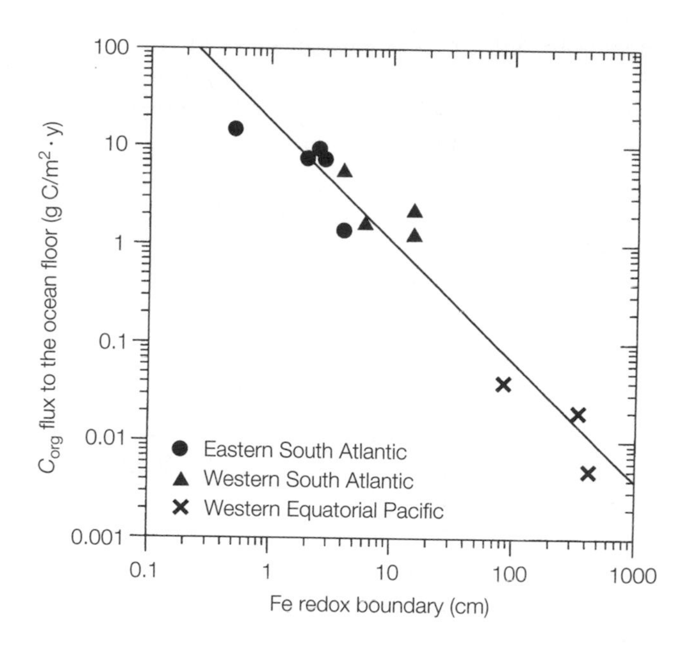

FIGURE 12.15

Subsurface depth of the iron redox boundary versus organic carbon flux to the ocean floor. Source: From Bleil, U. (2000). *Marine Geochemistry*, Springer Verlag, p. 80. See Bleil (2000) for data sources.

oxidation by O_2. Other metals, such as cobalt, copper, and nickel, tend to co-precipitate in the iron and manganese oxide phases and, hence, also exhibit some trapping at the redox boundary.

The importance of sedimenting POM in determining the depth of the redox boundary and, hence, the depth at which iron and manganese are trapped is illustrated in Figure 12.15. This graph shows that the depth of the redox boundary increases with decreasing supply of POM to the sediments. As discussed in the next section, postdepositional shifts in the depth of the redox boundary can alter the diagenetic features of the sedimentary iron and manganese profiles.

The Fe^{2+} and Mn^{2+} that diffuse downward are subject to precipitation as carbonate and sulfide minerals in which the metals are present in reduced form. These minerals do not undergo any further chemical changes unless tectonic processes (uplift) cause them to come into contact with O_2. As with the oxide phase, other metals tend to coprecipitate into the sulfide minerals, such as cadmium, silver, molybdenum, zinc, vanadium, copper, nickel, and uranium.

12.5.4 Diagenetic Overprinting of Fossil Oxidation Fronts

For chemicals that do not undergo diagenetic reactions, concentration profiles in the sediment can be used as records of changes in accumulation rate (Figure 12.6a). In some cases, similar information can be gleaned from the concentration profiles of reactive chemicals despite the altering effects of diagenesis. Since the sediments represent our primary source of information regarding past oceanic conditions, considerable effort has been focused on understanding how diagenesis can alter the sedimentary record.

An example is provided online at **http://elsevierdirect.com/companions/9780120885305**. This material focuses on the redox dynamics of manganese in the sediments of the Mediterranean Sea.

12.6 THE RELATIVE IMPORTANCE OF AEROBIC VERSUS ANAEROBIC DIAGENESIS

The degree to which organic matter can be buried in the sediment has important consequences for the crustal-ocean-atmosphere factory in terms of carbon storage. As noted earlier, only about 10% of the organic matter that makes it to the seafloor survives microbial degradation to become buried in the sediments. For water depths greater than 1000 m, 95% of the sedimentary organic matter oxidation is a consequence of aerobic respiration (Table 12.3a). But 80 to 90% of the organic matter reaching the sediments is deposited on the continental margins in water depths less than 1000 m. In these settings, anoxic respiration is responsible for approximately one third of the sedimentary organic matter oxidation (Table 12.3b).

In Figure 12.10, we saw that sediments in coastal zones have fast remineralization rates. Nevertheless, these sediments tend to have very high organic carbon contents because sedimentation rates are also fast. Rapid burial promotes preservation of organic matter by quickly moving it down the sedimentary column below the redox boundary. The rate of loss of organic matter is much slower under anaerobic conditions. Thus, the organic content of the sediment reflects the time required for burial below the oxic zone.

Two reasons have been proposed to explain why degradation under anaerobic conditions seems to be less effective than under oxic conditions in remineralization of organic matter. First, POM is thought to degrade in a stepwise fashion, producing organic molecules that are increasingly less reactive. This is attributed to a preferential use by microbes of the most reactive and energy-rich compounds first, such as proteins and carbohydrates, leaving behind the less reactive and less nutrient-rich compounds, such as lipids and lignin. Second, microbial biomass is not grazed under anaerobic conditions because the consumers, protozoans and meiofauna, require O_2. Under oxic conditions, these grazers consume the microbial biomass. Because organic matter is lost during the trophic transfer, this feeding activity enhances remineralization rates. This enhancement does not take place in anoxic sediments, enabling the preservation of particulate organic matter.

Table 12.3 Electron Acceptors in (a) Pelagic Sediments and (b) Continental Margin Sediments.

(a) Electron Acceptors in Pelagic Sediments

Site	Region	C_{org} Oxidation Rate (μmol cm^{-2} yr^{-1})	% of Organic *C* Oxidation by Different Electron Acceptors				
			O_2	NO_3^-	Mn(IV)	Fe(III)	SO_4^{2-}
MANOP H	E. eq. Pacific	12.0	99.2	0.8	0.4		
MANOP C	Central eq. Pacific	20.4	98.1	1.6	0.4		
E. eq. Atlantic	0–3°N, 6–16°W	12.4	93.8	4.4	0.1		1.8

(b) Electron Acceptors in Continental Margin Sediments

Location	Water Depths	Total C_{org} Oxidation (μmol cm^{-2} yr^{-1})	% of Organic *C* Oxidation by Different Electron Acceptors				
			O_2	NO_3^-	Mn(IV)	Fe(III)	SO_4^{2-}
NE Atlantic	208–4500	36–158	67–97	1–8.5	0–2.1	0–1.7	1–20
NW Atlantic	260–2510	36–52	74–90	1.8–6.0		8–20	
NE Pac.: O_2 < 50 μM	780–1440	66–75	5.0–46	41–69	0.1	0.7–1.3	5.7–25
NE Pac.: O_2 = 73–145	1900–1070	36–74	69–75	11–18	0.1–6.9	0.3–0.7	5.6–18

Source: From Martin, W. R., and F. L. Sayles (2003). Treatise on Geochemistry, *Elsevier, pp. 37–65.*

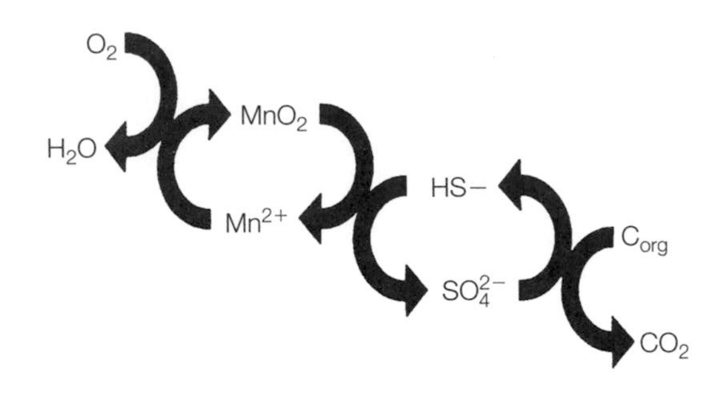

FIGURE 12.16

A proposed electron shuttle in which the oxidative power of O_2 can be conveyed below the oxic zone in marine sediments.

12.7 CHEMOAUTOLITHOTROPHY

A few examples of chemoautolithotrophic processes have been mentioned in this chapter, namely anaerobic methane oxidation coupled to sulfate reduction and the ones listed in Table 12.2 involving manganese, iron, and nitrogen. Another example are the microbial metabolisms that rely on sulfide oxidation. Since sulfide oxidation is a source of electrons, it is a likely source of energy that could be driving denitrification, and manganese and iron reduction where organic matter is scarce.

Less is known about these chemoautolithotrophic processes in comparison to the heterotrophic ones. They are interesting because their overall effect is to convey part of the oxidizing power of O_2 below the oxic zone. This is achieved by a kind of electron "shuttle" as illustrated in Figure 12.16. If this transfer of oxidizing energy is common in marine sediments, then the redox cycling of carbon, O_2, nitrogen, manganese, iron, and sulfur is likely more interconnected and, hence, complex than presented in Figure 12.12.

12.8 CLAY MINERALS AND DIAGENESIS

Clay minerals are the most abundant type of particles in typical marine sediments. They are present primarily as small grains and, thus, represent a very large surface area across which chemical reactions can occur. Clay minerals are discussed in detail in Chapter 14 but we will note here that these particles play three very important roles in diagenetic reactions. First, their surfaces act as cation exchange sites and, thus, are responsible for the uptake and release of cations in pore waters. Second, clay minerals act to enhance the preservation of sedimentary organic matter. This process is not well understood but appears to involve a physical association that protects the organic matter from microbial attack. Third, over time scales of thousands of years, the cations in pore water can react with the clay minerals and thereby alter their mineral composition. These are called reverse weathering reactions and appear to be significant sinks for some ions, such as K^+ and H^+. This process is discussed at length in Chapter 21.4.6.

PART 3

The Chemistry of Marine Sediments

CHAPTER 13
Classification of Sediments

All figures are available on the companion website in color (if applicable).

13.1 INTRODUCTION

The chemical composition of seawater is largely regulated by biogeochemical processes that cause dissolved materials to be converted into solid forms. These solids are then deposited on the seafloor, making the sediments a very important reservoir in the crustal-ocean-atmosphere factory. Marine sediments are also important because they contain our only record of past conditions in the ocean.

The composition of marine sediments is heterogeneous, comprising many particle types whose origins vary from biogenic to cosmogenic. The particle composition of the sedimentary column varies horizontally and vertically. Vertical variations reflect changes in particle supply over time and the effects of diagenesis. Horizontal variations reflect the combined effects of phenomena that control the production of particles, their transport to the seafloor, and preservation in the sediments. The complex interplay between these phenomena are explored for each of the major sediment types within Part III of this text. This provides an understanding of the global distribution of sediment types on the seafloor and how to relate their variations with depth to past changes in the crustal-ocean-atmosphere factory. At the end of Part III, several elemental mass-balance models are presented to quantify the role of marine sediments in controlling the chemical composition of seawater. These models also illustrate the important role of the sediments in feedback mechanisms that regulate climate and marine biological productivity.

In this first chapter of Part III, we discuss the schemes most commonly used to classify marine sediments.

13.2 SEDIMENTS: CLASSIFICATION SCHEMES

Marine sediments are composed of *unconsolidated* particles that blanket the bedrock of the seafloor. They vary greatly in chemical composition, mineralogy, particle size, origin, sedimentation rate, and geographic distribution. These characteristics are commonly used to classify marine sediments and are briefly discussed next. Other classification schemes include color, texture, and degree of bioturbation.

13.2.1 Geographic Distribution

Several naming schemes are currently in use to describe the location of a marine sediment. For example, deposits on the continental margin are termed *neritic* and those that overlie oceanic crust, such as on the abyssal plains and mid-ocean ridges, are referred to as *oceanic*. Sediments have also been classified by water depth: (1) *pelagic* sediments lie beneath a water column greater than 3000 m; (2) *hemipelagic* sediments lie in water depths of 200 to 3000 m (roughly encompassing the continental slope and upper part of the rise); and (3) the continental shelf and coastal sediments lie in water depths less than 200 m.

13.2.2 Grain Size

Sediments can also be categorized on the basis of grain diameter. Since grains can be irregularly shaped, the longest diameter is generally used to classify the particle as either a clay, silt, sand, granule, pebble, cobble, or boulder using the criteria given in Table 13.1. The types of particles that fall into each of these size classes are shown in Figure 13.1. Most sedimentary particles are *sand-*, *silt-*, or *clay-sized*. The term *mud* is used to refer to a mixture of silts and clays. Note that this size classification scheme conveys no information regarding the mineral composition of the particles. So a sand-sized grain can include a typical beach sand as well as the small shells, or tests, of radiolarians and foraminiferans.

The most common grain size in pelagic sediments are the clays and silts. The smaller grains are derived primarily from aeolian dust. The larger grains are detrital fragments of the hard parts deposited by calcareous and siliceous plankton. In contrast, hemipelagic sediments are characterized by larger particles, being composed mostly of silts, sands, and even gravels.

13.2.3 Origin, Chemical Composition, and Mineralogy

The particles that comprise marine sediments have two basic origins. As shown in Figure 13.2, either they are created in situ from the precipitation of dissolved chemicals (*authigenic* or *autochthonous*) or they are carried to the ocean in solid form (*allochthonous*). The major sources of allocthonous particles are Earth's crust (*lithogenous* or *clastic*) and extraterrestrial materials (*cosmogenous*). Some oceanographers also recognize humans as a source of particles (*anthropogenic*). The authigenic sediments are subdivided into two categories: those precipitated by abiogenic means (*hydrogenous*) and those formed as a consequence of biological activity (*biogenous*).

Table 13.1 Udden-Wentworth Grain-Size Scale.

Particle Diameter				Grade	Class	Fraction	
Millimeters (mm)		Micrometers (μm)	Phi(Φ)[a]			Unlithified	Lithified
	4096		−12				
					Boulder	Gravel	Conglomerate
	256		−8				
					Cobble		
	64		−6				
					Pebble		
	4.00		−2				
					Granule		
	2.00		−1				
				Very coarse	**Sand**	Sand	Sandstone
	1.00		0				
				Coarse			
1/2	0.50	500.0	1				
				Medium			
1/4	0.25	250.0	2				
				Fine			
1/8	0.125	125.0	3				
				Very fine			
1/16	0.0625	62.5	4				
				Coarse	Silt	Mud	Mudstone or Shale
1/32	0.0313	31.3	5				
				Medium			
1/64	0.0156	15.6	6				
				Fine			
1/128	0.0078	7.8	7				
				Very fine			
1/256	0.0039	3.9	8				
				Coarse	Clay		
1/512	0.0020	2.0	9				
				Medium			
1/1024	0.0010	1.0	10				
				Fine			
1/2048	0.00049	0.49	11				
				Very fine			
1/4096	0.00024	0.24	12				
1/8192	0.00012	0.12	13				
				Colloid			
1/16,384	0.00006	00.6	14				

[a] *Phi(Φ) = $-\log_2 D/D_0$, where D = particle diameter in millimeters and D_0 is the diameter of a reference particle, chosen to be1 mm. The use of a ratio makes Phi dimensionless.*

Source: After Wentworth, C.K. (1922). Journal of Geology, 30, 377–392.

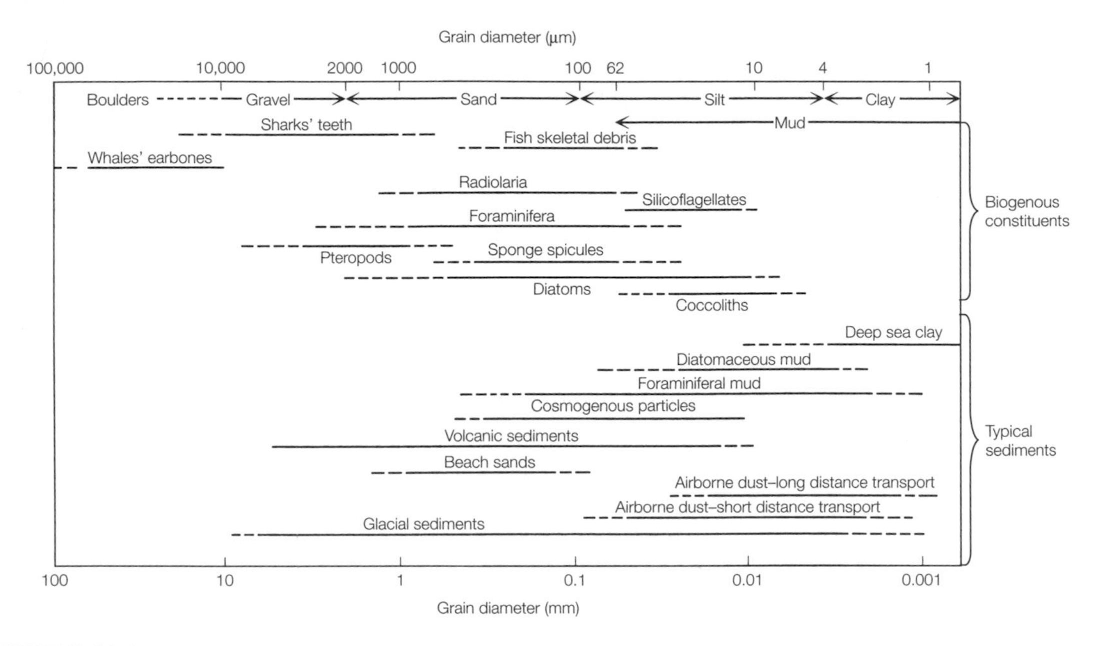

FIGURE 13.1

Grain-size distribution in common marine sediments and in various sediment sources. *Source*: From Gross, M. G. (1987). *Oceanography: A View of the Earth*, 4th ed. Prentice Hall, Inc., p. 81.

Each of these solid phases can be described in terms of their mineralogy. This classification scheme is based on crystal structure and chemical composition. The most common minerals found in marine sediments are listed in Table 13.2. Most are silicates in which Si and O form a repeating tetrahedral base unit. Other minerals common to marine sediments are carbonates, sulfates, and oxyhydroxides. Less common are the hydrogenous minerals as they form only in restricted settings. These include the evaporite minerals (halides, borates, and sulfates), hydrothermal minerals (sulfides, oxides, and native elements, such as gold), and phosphorites.

13.2.4 Sedimentation Rate and Thickness

Marine sediments range in thickness from 0 km (on top of volcanically active mid-ocean ridges and rises) to 10 km (on the continental slopes and rises). The average thickness of the sediments lying on the abyssal plains is 0.5 km. The thickness of a deposit depends on (1) the local rate of particle supply to the seafloor, (2) the degree to which the particles are preserved following sedimentation, and (3) the age of the underlying crust. The latter determines the length of the time over which sedimentation has taken place. Thus, the thinnest layers overlie the youngest oceanic crust and, hence, are located at the mid-ocean ridges and rises. As shown in Figure 13.3, the thickest layers have accumulated at the foot of the continental shelves that lie on passive continental margins.

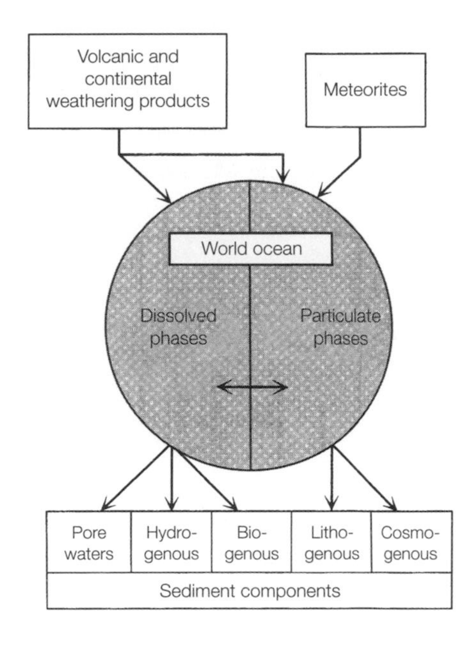

FIGURE 13.2

The classification of the components of marine sediments. *Source*: From Goldberg, E. D. (1964). *Transactions of the New York Academy of Sciences*, 27, 7–19.

This reflects the combined effects of high sedimentation rates and the advanced age of the underlying crust.

Sediments that form at rates exceeding 1 cm/1000 y are termed *nonpelagic*. Those that are deposited at rates slower than 1 cm/1000 y (on the order of millimeters per 1000 y) are termed *pelagic* sediments. As shown in Figure 13.4 and Table 13.3, most oceanic sediments fall into this category. These slow rates reflect the great vertical and horizontal distance of pelagic sediments from any particle source. Sedimentary deposits with sedimentation rates of 0 or less are termed *relict*. Changes in sea level can cause a deposit to be cut off from its particle source and, thus, cease to accumulate. Relict sediments are subject to erosion (negative sedimentation rates) if bottom or contour currents are strong enough to resuspend particles.

13.3 PARTICLE SINKING RATES AND PELAGIC SEDIMENTATION

Many particles reach the seafloor by sinking through the water column. This particle-by-particle accumulation is termed *pelagic sedimentation*. Sinking rates depend on particle size, shape, and density. The sinking rate of a particle in a fluid experiencing laminar flow can be estimated from Ostwald's modification of Stokes' law,

Table 13.2 Mineral Constituents of Marine Sediments Classified by Origin.

Cosmogenous		**Biogenous**	
Spherules		Calcite	$CaCO_3$; $(Ca_{1-x}Mg_x)CO_3$
Iron	FeNi	Aragonite	$CaCO_3$
Olivine	$(Mg,Fe)_2SiO_4$	Opal	$SiO_2\ nH_2O$
Pyroxene	$(Mg,Fe)_2Si_2O_6$	Francolite	$Ca_{10-x-y}Na_xMg_y(PO_4)_{6-z}(Co_3)_zF_{0,4z}F_2$
		Barite	$BaSO_4$
		Celestite	$SrSO_4$
Lithogenous		**Hydrogenous**	
Quartz	SiO_2	FeMn oxides/ oxyhydroxides	Nodules and crusts
Feldspars			
Plagioclase	$(Na,Ca)[Al(Si,Al)Si_2O_8]$		
Orthoclase	$K[AlSi_3O_8]$		
Clay minerals		Francolite	$Ca_{10-x-y}Na_xMg_y(PO_4)_{6-z}(CO_3)_zF_{0,4z}F_2$
Illite	$K_xAl_2(Si_{4-x}Al_x)O_{10}(OH)_2$	Barite	$BaSO_4$
Chlorite	$(Mg,Fe)_5(Al,Fe)_2Si_3O_{10}(OH)_8$	Celestite	$SrSO_4$
Kaolinite	$Al_2Si_2O_5(OH)_4$	Montmorillonite	$(Na,K)_x(Al_{2-x}R_x)Si_4O_{10}(OH)_2$
Montmorillonite	$(M)_{x-y}(R^{3+}_{2-y}R^{2+}_y)(Si_{4-x}Al_x)$ $O_{10}(OH)_2$	Nontronite	$(Na,K)_xFe_2(Al_xSi_{4-x})O_{10}(OH)_2$
		Glauconite	$K_{0.85}(Fe,Al)_{1.34}(Mg,Fe)_{0.66}$ $(Si_{3.76}Al_{0.24})O_{10}(OH)_2$
Volcanic glass			
Amphiboles	$Ca_2(Mg,Fe)_5Si_8O_{22}(OH)_2$	Zeolites	
Pyroxene	$(Mg,Fe)_2Si_2O_6$	Phillipsite	$K_{2.8}Na_{1.6}Al_{4.4}Si_{11.6}O_{32}\ 10H_2O$
Olivine	$(Mg,Fe)_2SiO_4$	Clinoptilolite	$K_{2.3}Na_{0.8}Al_{3.1}Si_{14.9}O_{36}\ 12H_2O$
		Geothite	FeOOH
		Palygorskite	$(OH_2)_4Mg_5Si_8O_{20}(OH)_2\ 4H_2O$
		Sepiolite	$(OH_2)_4Mg_8Si_{12}O_{30}(OH)_2\ 4H_2O$

x,y,z denotes variable stoichiometry in montmorillonite, R^{3+} = Fe,Cr and R^{2+} = Mg, Zn, Li. M = exchangeable ions (See Figure 14.4).
Source: After Li, Y.-H. and J.E. Schoonmaker (2005). Treatise on Geochemistry, Elsevier, Ltd., p. 2.

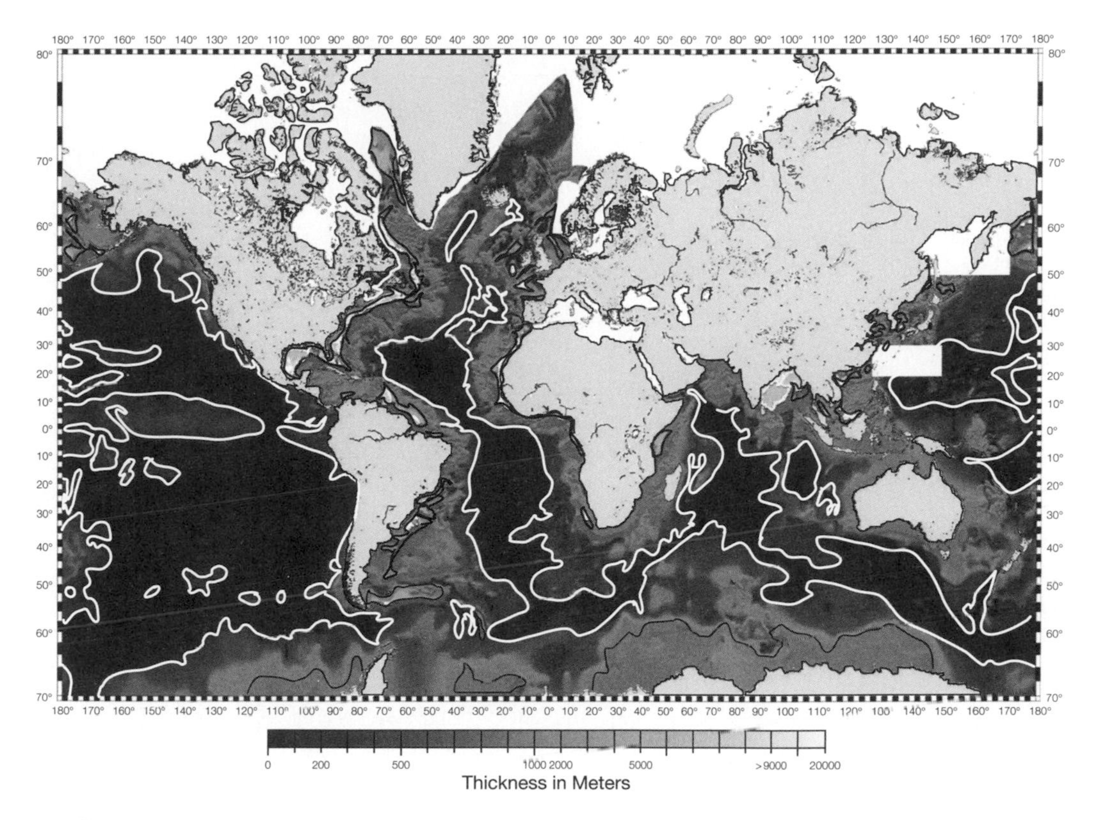

FIGURE 13.3

Sediment thickness in meters to acoustic basement in the world ocean. *Source*: After Divins, D. L. Sediment thickness of the world's oceans & marginal seas. Retrieved September 15, 2007, from: http://www.ngdc.noaa.gov/mgg/sedthick/sedthick.html. (See companion website for color version.)

$$V_s = 2gr^2(\rho' - \rho)/9\eta\Phi_r \tag{13.1}$$

where

V_s = settling velocity (m/s)
g = gravitational acceleration = 9.8 m/s^2
r = radius of settling sphere (m)
ρ' = density of the particle (kg/m^3). Typical densities are shown in Table 13.4
ρ = density of seawater at a given temperature and salinity (kg/m^3)
η = dynamic viscosity of seawater at a given temperature and salinity = 0.9×10^{-3} kg m^{-1} s^{-1} at 30°C and 1.4×10^{-3} kg m^{-1} s^{-1} at 10°C
Φ_r = form resistance = 1 (by definition for a sphere and usually $\geqslant 1$ for a nonspherical particle)

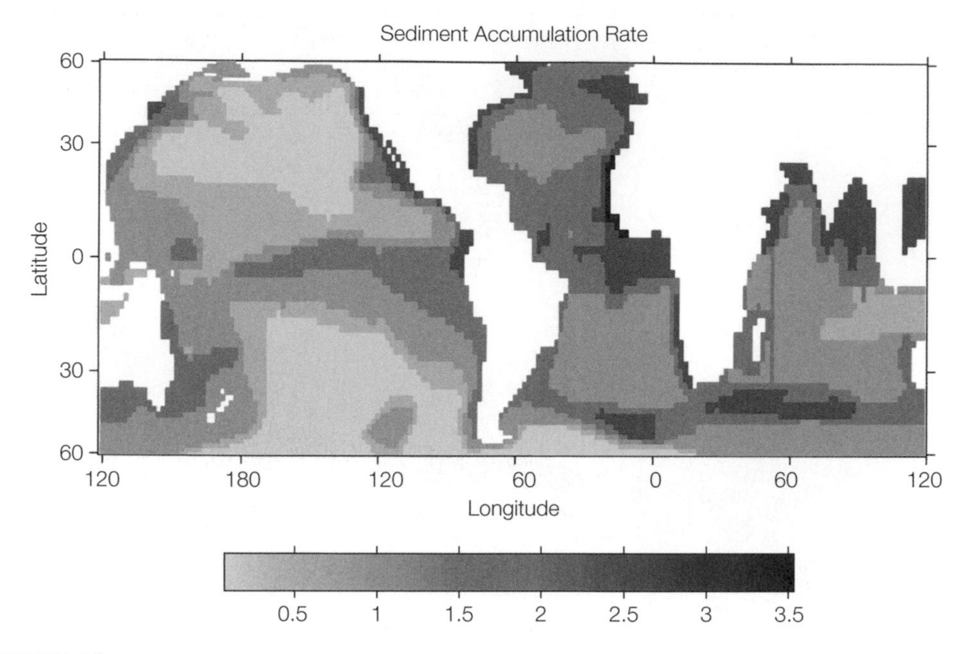

FIGURE 13.4

Global ocean map of accumulation rates ($g/cm^3/10^3$ y) for pelagic sediments. These values can be converted into sedimentation rates ($cm/10^3$ y) by dividing by the bulk sediment density. The latter can be obtained by correcting the dry particle density (Table 13.4) for the sediment's water content. This correction is performed by multiplying the dry particle density by $(1 - \Phi)$ where Φ is the sediment porosity (Figure 12.3). For a dry particle density of 2.5 g/cm^3 and sediment porosity of 0.8, the mapped accumulation rates are, thus, multiplied by 2 to convert to units of $cm/10^3$ y. *Source*: Redrawn from Jahnke, R. A. (1996). *Global Biogeochemical Cycles* 10, 71–88.

Thus, for particles with a similar density and shape, sinking rates increase with increasing diameter because of the decrease in surface area to volume ratio. As shown in Table 13.5, sand-sized grains reach the sediments in a few days, whereas clays can take centuries. If entrained in a current, sinking particles can also experience significant horizontal transport.

Particle sinking rates are of considerable interest because the faster a particle can make the trip to the seafloor, the shorter the time it is subject to decomposition or dissolution and, hence, the greater its chances for burial in the sediments. The length of the trip is dictated by the depth to the seafloor, the horizontal current velocity, and the particle sinking rates. As shown in Figure 13.5, sedimentation rates decrease with increasing water depth. This relationship reflects the preservation issue and the fact that coastal waters tend to have larger sources of particles to the surface zone.

Most of the mass transport to the seafloor is in the form of large and, hence, rapidly sinking particles. Most of these are biogenic in origin, such as fecal pellets, cast-off

Table 13.3 Typical Sedimentation Rates.

Area	Average Sedimentation Rate (cm/1000 y)
Continental margin	
Continental shelf	30 (15–40)[a]
Continental slope	20
Fjord (Saanich Inlet, British Columbia)	400
Fraser River delta (British Columbia)	700,000
Upper Gulf of Thailand	400–1100
Marginal ocean basins	
Black Sea	30
Gulf of California	100
Gulf of Mexico	10
Clyde Sea	500
Deep-ocean sediments	
Coccolith muds	1 (0.2–3)[a]
Deep-sea muds	0.1 (0.03–0.8)[a]

[a]*Observed range is given in parentheses.*
Source: From Gross, M. G. (1987). Oceanography: A View of the Earth, *4th ed. Prentice Hall, p. 89.*

larvacean houses, and clumps of diatoms. Although the flux decreases exponentially with depth, particle sinking rates appear to increase. This trend suggests that particle size increases with increasing depth. Some likely mechanisms for this increase include: (1) zooplankton feeding during which small plankton are consumed and larger fecal pellets are produced; (2) aggregation and flocculation of small particles into larger ones, which appears to be facilitated by microbial production of sticky exopolymers that coat surfaces; and (3) use by some zooplankton, such as larvaceans, of mucous feeding nets, which are discarded once they become clogged with particles. The discarded nets collapse because water is no longer being pumped through them and, hence, sink rapidly. The cast-off nets of giant larvaceans studied in the coastal waters of California generate a POC flux to the sediments about equal to that of particles sampled via sediment traps and, thus, are thought to account for most of the flux to the sediments.

Because the density of POM is close to that of seawater, particle repackaging acts to increase sinking rates by increasing particle density and radius. POM fluxes in the deep sea are also positively correlated with those of inorganic particles, namely biogenic calcite, silica, and clay minerals. Microscopic inspection of the sinking particles has shown that the POM is physically associated with the inorganic particles. The positive effect of this association on POM sinking rates is called *particle ballasting*. It is thought to be a consequence of: (1) increased particle density because the inorganics are denser than the organics and (2) some kind of protection afforded to the POM against decomposition. The mechanism of this protection is unknown.

Table 13.4 Grain Densities of Common Sedimentary Minerals and Bulk Grain Densities of Marine Sediments.

	Density (g/cm^3)	% Sand	% Silt	% Clay
Minerals				
Quartz	2.65–2.66			
Calcite	2.71			
Aragonite	2.95			
Opal	1.90–2.30			
Orthoclase & Plagioclase feldspars	2.55–2.76			
Illite	2.60–2.90			
Montmorillonite	2.00–2.70			
Kaolinite	2.63			
Chlorite	2.60–3.30			
Goethite	4.18			
Hematite	5.26			
Magnetite	5.18			
Pyrite	4.80–5.00			
Deep-Sea Sediments				
Calcareous Ooze	2.61–2.68	3–27	40–76	8–56
Diatomaceour Ooze	2.45–2.47	3–8	37–76	17–60
Red Clay	2.68–2.78	0.1–3.9	19–59	37–81
Abyssal Plains	2.61–2.66	0–35	22–78	19–78
Continental Shelf and Slope Sediments				
Sandy	2.69–2.71	64–100	0–23	0–13
Slity	2.65–2.67	6–26	61–81	13
Clayey	2.66–2.71	5–32	41–60	27–54

Mineral data from Mindat.org (www.mindat.org/index.php). Sediments data from Hamilton, E. L. (1976). Journal of Sedimentary Petrology, 46, 2, 280–300.

As noted in Chapter 12, the rapid burial of particles helps ensure their preservation following deposition on the seafloor. High sedimentation rates are required to achieve rapid burial. If the particle being buried is compositionally different from the bulk of the sedimenting particles, high sedimentation rates can preserve the particle while lowering its mass concentration in the sediment. Thus, the formation of sediments enriched in a given particle type is dependent on burial by sedimentation of like particles. For example, calcareous sediments are rarely found in coastal sediments, even though fluxes of biogenic calcite are high, due to the diluting effect of a large supply of lithogenous particles supported by the close proximity to land.

Table 13.5 Sinking Rates of Marine Particles.

Particle Type	Particle Diameter (micron) or Source	Sinking Rate		Time to Settle through 4 km Water Depth		Horizontal Distance (km) Traveled in Sinking 1000 m through a Current of 1 cm/s	Reference
		m/d	km/yr	y	d		
Sand	100	353	129	0.031	11	2.4	1
Silt	50	7.4	2.7	1.5	538	116	1
Clay	1	0.035	0.013	311	113,379	24,490	1
Particles photographed at 100 to 500 m	500 to >5000	16 to 25	5.8 to 9.1	0.44 to 0.68	160 to 250	35 to 54	2
Particles collected from 1000 to 3000 m	sediment trap	100 to 325	37 to 119	0.034 to 0.11	12 to 40	2.7 to 8.6	3
Particles collected from 2800 and 4000 m	sediment trap	200	73	0.014	5.0	4.3	4
Giant larvacean houses	60,000	800	292	0.0034	1.3	1.1	5

(Continued)

Table 13.5 *(Continued)*

Particle Type	Particle Diameter (micron) or Source	Sinking Rate		Time to Settle through 4 km Water Depth		Horizontal Distance (km) Traveled in Sinking 1000 m through a current of 1cm/s	Reference
		m/d	km/yr	y	d		
Marine snow	500	10 to 100	3.7 to 37	0.11 to 1.1	40 to 400	8.6 to 86	6
Marine snow		16 to 368	5.8 to 134	0.030 to 0.68	11 to 250	2.3 to 54	7
Phytodetritus		100 to 150	37 to 55	0.073 to 0.11	27 to 40	5.8 to 8.6	7
Fecal Pellet	Copepods	5.0 to 22	1.8 to 8	0.498 to 2.2	182 to 800	39 to 173	7
Fecal Pellet	Euphausiids	16 to 862	5.8 to 315	0.013 to 0.68	4.6 to 250	1.0 to 54	7
Fecal Pellet	Doliolids	41 to 504	15 to 184	0.022 to 0.27	7.9 to 98	1.7 to 21	7
Fecal Pellet	Appendicularians	25 to 166	9.1 to 61	0.066 to 0.44	24 to 160	5.2 to 35	7
Fecal Pellet	Chaetognaths	27 to 1313	10 to 479	0.008 to 0.41	3.0 to 148	0.7 to 32	7
Fecal Pellet	Pteropods	120 to 1800	44 to 657	0.006 to 0.09	2.2 to 33	0.5 to 7.2	7
Fecal Pellet	Heteropods	120 to 646	44 to 236	0.017 to 0.09	6.2 to 33	1.3 to 7.2	7
Fecal Pellet	Salps	43 to 2700	16 to 986	0.004 to 0.25	1.5 to 93	0.3 to 20	7

References:

1, *Anikouchine, W. A., and Sternberg, R. A. (1981).* The World Ocean: An Introduction to Oceanography, *Prentice-Hall, Inc., p. 513.*
2, *Pilskaln, C. H., et al. (1998).* Deep-Sea Research, *45, 1803–1837;*
3, *Berelson, William M. (2002).* Deep-Sea Research II, *49, 237–251;*
4, *HOTS at Station ALOHA (http://hahana.soest.hawaii.edu/hot/methods/traps.html);*
5, *Robison, B. H., et al. (2005).* Science, *308, 1609–1611;*
6, *Shanks, A. L. (2002).* Continental Shelf Research, *22, 2045–2064;*
7, *References in Turner, J. T. (2002).* Aquatic Microbial Ecology, *27, 57–102.*

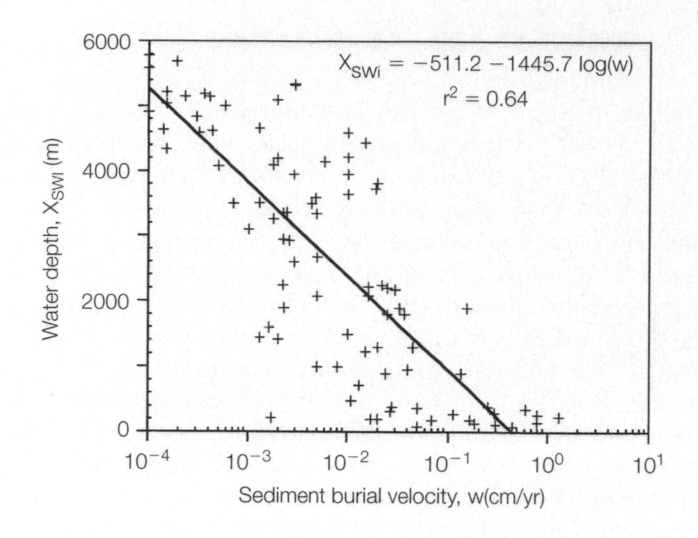

FIGURE 13.5

Plot of oceanic water depth versus sediment burial velocity (burial velocity is equal to sedimentation rate when the latter is constant over time and diagenetic loss is minimal). *Source*: From Boudreau, B. P. (1997). *Diagenetic Models and Their Implementation*. Springer-Verlag, p. 164.

13.4 PELAGIC SEDIMENTS

Sediments are classified as true pelagic deposits if (1) the median grain size is less than 5 μm (excepting authigenic or pelagic biogenous particles) and (2) less than 25% of the particles larger than 5 μm is of terrigenous, volcanogenic, and/or neritic origin. The most abundant particle type in pelagic deposits is clay-sized and lithogenous in origin. These deposits are termed *pelagic clays* if the lithogenous clay content constituents more than 70% of the dry sediment mass. Pelagic clays are the dominant sediment type in the North Pacific ocean. These sediments tend to be red-brown in color due to the presence of oxidized iron. If the biogenous component exceeds 30% by mass, the sediment is termed a *biogenous ooze*. Hydrogenous particles can be locally important, especially near mid-ocean ridges and rises. A brief discussion of the formation and mineralogy of the particle types common to pelagic sediments is provided next.

13.4.1 Lithogenous Sediments

Earth's crust is a source of particles produced as a consequence of weathering and volcanic activity. Weathering of continental rocks generates *terrigenous* particles that are carried into ocean via rivers, glaciers, and winds. As shown in Table 13.2, the most abundant mineral types are quartz, plagioclase, and clay minerals. The most abundant particle size in deep-sea sediments are the small (<3 μm) clays whose major source is aeolian (wind) transport.

The global patterns of mineral dust transport to the world ocean are illustrated in Figure 11.4. These patterns follows the major wind bands (Figure 4.4a) as the mineral dust is lifted directly from soils by the winds. Other windborne particles (collectively termed *aerosols*) include sea salt, soot (black carbon), terrestrial plant fragments, and biota such as insects and bacteria. The soot is a product of fossil fuel burning and forest fires. Aerosols are returned to Earth's surface via gravitational settling and incorporation in rainfall (washout). Most of the windborne particles that fall onto the sea surface are less than a few microns in diameter as the larger size fractions fall back to Earth's surface close to their sites of origin. Aeolian fluxes to the sea surface exhibit large seasonal and interannual variations due to the effect of weather, climate, and volcanism on mineral dust production. Anthropogenic sources of soot now exceed natural sources. The resulting brown and black clouds are visible from satellite imagery and are a recognized agent of climate change as they absorb and scatter insolation.

Some of the clays that enter the ocean are transported by river input, but the vast majority of the riverine particles are too large to travel far and, hence, settle to the seafloor close to their point of entry on the continental margins. The most abundant clay minerals are illite, kaolinite, montmorillonite, and chlorite. Their formation, geographic source distribution and fate in the oceans is the subject of Chapter 14. In general, these minerals tend to undergo little alteration until they are deeply buried in the sediments and subject to metagenesis.

Lithogenous particles created by volcanic activity tend to be larger than clays. They are produced during volcanic eruptions as ejected fragments of volcanic glass. Their mineralogy is primarily that of amphiboles, pyroxene and olivine. Eruptions can occur above and below the sea surface. Once in the ocean, these particles undergo weathering as a result of chemical attack by seawater. The reaction products are listed in Table 13.2 in the hydrogenous particle category, with the most abundant being the zeolite phillipsite and the clay mineral montmorillonite.

As shown in Table 13.6, the sedimentation rate of the clay mineral fraction of pelagic deposits is very slow, being on the order of millimeters per thousand years. Spatial variations reflect geographic variability in the supply rate of these lithogenous particles. This is generally related to distance from land, causing the South Pacific to have sedimentation rates less than 1 mm per thousand years.

13.4.2 **Biogenous Sediments**

Biogenous sediments are composed of detrital (nonliving) hard and soft parts formed by marine organisms. The hard parts are better preserved, so most sedimentary biogenic particles are fragments of structural components, including shell, endoskeletons, and exoskeletons. Soft parts composed of organic matter are present to a minor degree and include tissues, organic exoskeletons, and excretions, such as fecal pellets. The biogenic minerals deposited by plankton as part of their hard parts are listed in Table 13.2 and include calcite, aragonite, opal, barite ($BaSO_4$), and celestite

Table 13.6 Sedimentation Rates of the Noncarbonate Fraction of Pelagic Sediments.

Location	Rate ($mm/10^3$ y)	
	Mean	Range
South Pacific	0.45	0.3–0.6
North Pacific	1.5	0.4–6.0
South Atlantic	1.9	0.2–7.5
North Atlantic	1.8	0.5–6.2

Source: *From Riley, J. P., and R. Chester (1971).* Introduction to Marine Chemistry. *Academic Press, p. 289.*

($SrSO_4$). Carbonate-rich fluoroapatite minerals, such as francolite, are the major component of bones. As noted in Chapter 11, hard and soft parts can contain significant amounts of trace metals.

Most of the calcite ($CaCO_3$) hard parts are deposited by coccolithophorids (<60 μm) and foraminiferans (<2 mm) in which magnesium can be present in significant amounts. Most of the opaline (SiO_2) hard parts are deposited by diatoms (mostly <50 μm) and radiolarians (20–300 μm). If these planktonic hard parts comprise more than 30% by mass of the sediments, the pelagic deposit is termed a biogenous *ooze*. If one type or species of marine organism, or mineral, has contributed most of the biogenous particles, the deposit is named accordingly (e.g., a radiolarian, *globigerina*, or *siliceous ooze*, respectively).

Oceanic sediments contain less than 1% organic matter by mass. On the continental margins, contributions range from 1 to 10%. Higher concentrations can be found in sediments in estuaries and coastal upwelling areas. Although organic detritus accounts for a relatively insignificant fraction of the sediment mass, it is an important reservoir of carbon, nitrogen, phosphorus, and trace metals.

13.4.3 Hydrogenous Sediments

Sediments formed by the abiogenic precipitation of solutes from seawater are termed *hydrogenous*. Unequivocal examples of hydrogenous sediments are ones formed from the evaporation of seawater. The minerals deposited are collectively called *evaporites* and are the subject of Chapter 17.[1] Others form with the assistance, to varying degrees, of marine microbes. For example, bacteria seem to play a role in the formation of Fe–Mn nodules and crusts. Some hydrogenous minerals, such as barite, celestite, glauconite, and francolite, are produced from the precipitation of elements

[1] As described in Chapter 17.4, evaporites can contain biogenic detritus produced by the benthos.

remineralized from hard and soft parts. As noted earlier, some hydrogenous particles, such as montmorillonite and phillipsite, are secondary minerals produced by chemical weathering of volcanic rock. During this process, cations from seawater replace some of those originally present in the volcanic rock.

13.4.4 Cosmogenous Sediments

Extraterrestrial material is constantly entering Earth's atmosphere. The fraction of this flux that is less than 1 mm in diameter is termed *cosmic dust* or *micrometeroids*. It is composed primarily of fragments of comets and asteroids in the 5 to 50 μm size range. Because of their small size, some of these particles are able to pass through Earth's atmosphere and reach the sea surface without melting. These are termed *micrometeorites* and are of considerable interest because their chemical composition provides information on extraterrestrial geochemistry. Current rates of deposition are on the scale of 1 micrometeorite per square meter per day, supporting an annual flux of 40,000 to 60,000 tons.

The micrometeorites that melt during passage through Earth's atmosphere tend to solidify as spheres. These are termed *cosmic spherules*. The mineralogy of these spherules is given in Table 13.2. Their high iron and nickel content make them much denser (3 to 6 g/cm^3) than continental rock (~2.7 g/cm^3). Like aeolian particles, cosmic dust deposited on the sea surface eventually settles to the seafloor via pelagic sedimentation.

Cosmogenous particles are also formed indirectly as a result of the impact energy associated with large meteorites. When a large meteorite enters the atmosphere and strikes Earth, its kinetic energy is dissipated as heat and as pressure waves. If the meteorite strikes terrestrial rock, it can induce melting and/or shock-related changes in mineralogy. These particles are termed *tektites*, if centimeter sized, and *microtektites*, if millimeter sized. Their unusual shapes (Figure 13.6) suggest they were formed by the ejection of terrestrial rock into space following meteorite impact. As the ejecta flew upward, they underwent melting due to frictional heating. They then cooled into the teardrop and button-like shapes shown in Figure 13.6 during their descent back to Earth. Tektites and microtektites are generally concentrated in areas called strew fields.

The unique chemical composition of cosmogenous debris has provided some insight into why approximately 70% of the species of organisms on Earth were driven extinct over a relatively short time interval approximately 66 million years ago. Evidence for this mass extinction has been observed in marine sediments throughout all the ocean basins. In a contemporaneous layer deposited at the end of the Cretaceous period, the hard parts of many species of marine plankton abruptly vanished from the sedimentary record. This sedimentary layer is also characterized by a large enrichment in the rare element iridium.

Since meteorites are known to have high levels of iridium, some scientists have suggested that the sediment enrichment was produced by the impact of a very large comet, which either disintegrated in Earth's atmosphere or exploded upon impact. This hypothesis assumes that a huge amount of airborne debris was produced and distributed

FIGURE 13.6

Tektites. Courtesy of V. E. Barnes, University of Texas at Austin, Austin, TX.

globally by the winds. If so, some would eventually have fallen onto the sea surface and settled onto the seafloor. Such a large quantity of atmospheric fallout would also have caused a temporary decrease in insolation, altering the climate and causing a decline in plant growth. Since plants form the base of the food web, all organisms would have been negatively affected.

Evidence also exists for a terrestrial source of the iridium enrichment as volcanic ejecta is enriched in this rare element. Thus, the enriched sediment layer could also have been caused by an abrupt and large increase in volcanic activity. Evidence for this is suggested by high levels of volcanic ash, soot, and shocked minerals in the iridium-enriched layer. Other geochemical characteristics of this sediment layer appear to have been caused by acid rain and tsunamis, both of which are by-products of volcanic activity.

A coincident reversal in magnetic field orientations at the time of deposition of the iridium-enriched layer suggests that this sediment was deposited during a period of intense geological activity. This activity could have been triggered by a global-scale catastrophic event, such as the impact of a huge comet. In 1991, geologists identified a 180-km-wide impact crater on the northern part of the Yucatan peninsula in Mexico that they think is responsible for the end-Cretaceous extinction event. To create a crater of this size requires the impact of a meteorite with a diameter on the order of 10 to 15 km.

If Earth's crust was penetrated by impact of this meteorite, great quantities of magma would have been released rapidly and explosively, producing massive flood basalts. The geologic record indicates extensive flood basalts were formed at the end of the Cretaceous period. Flood basalts have been deposited during other periods of Earth's history and coincide with other mass extinction events. Some evidence exists for a periodicity on the order of 30 million years. A likely forcing function are disturbances of the Oort Cloud, a group of comets orbiting our solar system, that arise from dynamical interactions with the rest of the Milky Way galaxy.

Mass extinction events accelerate evolution by opening up biological niches on a global scale. Thus, such catastrophes may act as a driving force for biological evolution that periodically supersedes the slow pace of Darwinian-style selection of the fittest. If so, the biogeochemical development of Earth must be greatly influenced by astrophysical processes that operate on a galactic scale.

This bit of paleoceanographic detective work is of more than academic interest. Scientists have used it as an analogy for what might occur globally if a number of nuclear bombs were detonated. Since the explosion of such bombs could ignite fires, the ensuing combustion might produce enough airborne particles to cause a global change in insolation and climate. The resulting "nuclear winter" could have more far-reaching effects than blast damage or radioactive fallout.

13.4.5 Anthropogenic Sediments

Humans have greatly accelerated the rate of river transport of sediment. Most of this material is soil that is being eroded as a result of deforestation and agriculture. At the same time, damming of rivers has halted the input of lithogenous particles to the coastal ocean, thereby creating relict sediments and erosional shorelines. On a global basis, the net effect has been to reduce the riverine sediment load by 10%, which will have important consequences for survival of coastal communities as sea level is rising.

The ocean is receiving anthropogenic particles via piped discharges and stormwater and groundwater flows and via barged garbage dumping. The organic component of these particles, along with runoff of solubilized nutrients, fuels microbial aerobic respiration, contributing to periodic episodes of hypoxia and anoxia in some coastal waters.

13.5 NONPELAGIC SEDIMENTS

Nonpelagic sediments accumulate at rates exceeding 1 cm/1000 y, so they are formed by processes that move a lot of particles quickly. These include *turbidity currents, contour currents, volcanic eruptions, ice rafting*, river discharge, and locally high rates of plankton production. The last can be enhanced by aggregation or clumping of plankton following blooms. Clumping increases the sinking rate of the biogenic particles and, hence, the likelihood of accumulation on the seafloor. Because nonpelagic sediments are produced by relatively rapid and energetic transport processes, they tend to contain a wide range of particle sizes and, hence, are classified as *poorly sorted* or *unsorted* deposits. In comparison, pelagic sedimentation produces well-sorted deposits because they contain only a few size classes of particles, mostly clays and some sands.

13.5.1 Gravity-Driven Sediment Transport

Most nonpelagic sediments lie on the continental margins. They are categorized into hemipelagic, shelf, and coastal deposits. The hemipelagic sediments found on the slope and rise collect at rates of 0.5 to 500 cm/1000 y with most in the range of 10 to

30 cm/1000 y. Hemipelagic deposits are technically defined as having: (1) a median grain size greater than 5 μm (excepting authigenic and pelagic biogenous particles) with (2) more than 25% of its fraction larger than 5 μm being of terrigenous, volcanogenic, and/or neritic origin. Hemipelagic sediments are often referred to as muds because they contain a high percentage of silts and clays. Most have a dark color due to the large terrigenous contribution.

On the continental shelf and closer inshore, such as in estuaries and deltaic environments, sedimentation rates typically range from 10 to 5000 m/1000 y. High accumulation rates are found in areas where river input supports large fluxes of terrigenous particles. Deltaic deposits are common in the mouths of rivers with large particle fluxes, such as the Ganges, Amazon, Yangtze, and Mississippi rivers. Riverine sediments that deposit in the nearshore and shelf can be transported offshore by gravity-driven mechanisms. These include underwater slides, slumps, and gravity-driven sediment flows. This type of sediment transport is generally initiated by a pulse of energy, such as from an earthquake, a hurricane, or an internal wave.

Gravity-driven sediment flows are a major contributor to the transport of terrigenous sediments onto the abyssal plains where they form deposits at the foot of the continental slope called *deep sea fans* (Figure 13.7). These flows are characterized by sediment motion that entrains seawater, thereby forming a moving mixture of particles and water. Gravity-driven sediment flows are classified into four categories, i.e., grain flows, fluidized flows, debris flows, and turbidity currents. They can be erosive enough to carve out submarine canyons as they flow across the continental margins.

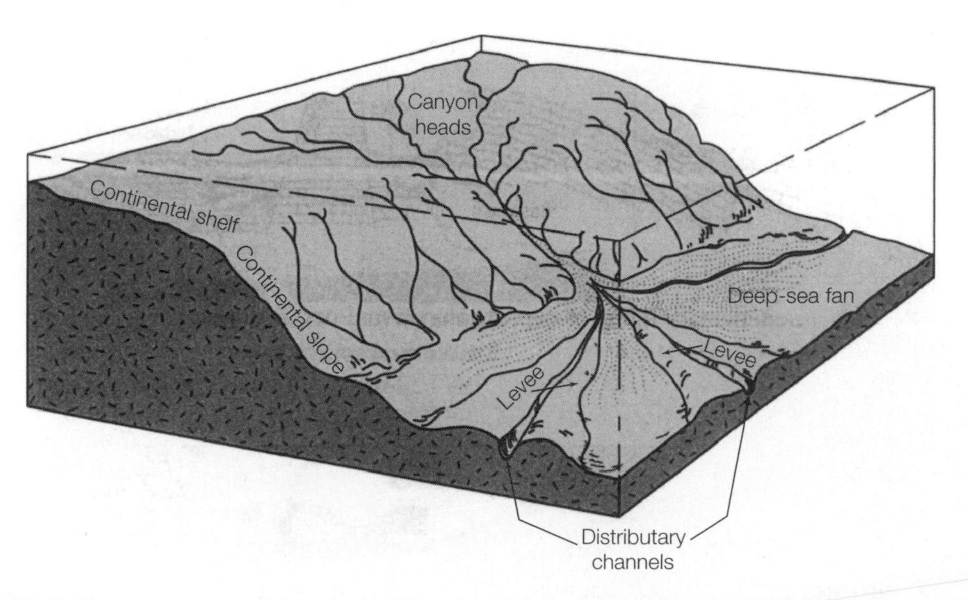

FIGURE 13.7

A submarine canyon and deep-sea fan.

The deposits formed after the sediments settle back onto the seafloor exhibit varying degrees of sorting depending on the nature of the flow. The most common are the grain and fluidized flows, which produce well-sorted deposits of fine and medium grain size. Less frequent are the debris flows and turbidity currents. These are higher in energy and shorter-lived. The debris flows entrain very large particles and produce very poorly sorted deposits. They commonly contain terrestrial debris, such as wood and leaves. Turbidity currents are highly dilute sediment suspensions. They produce sediment deposits called *turbidites*. The particle size sorting in a turbidite can range from poor to moderate depending on how far the turbidity current has traveled and the particle size ranges in the sediment source.

Many turbidites found on the continental margins are characterized by *graded bedding* (Figure 13.8). This type of particle sorting is produced by the sequential deposition of larger grains followed by increasingly finer grain sizes as the smaller particles remain suspended longest. Turbidites can be as much as 1 km thick. At some locations, several sequences of graded bedding have accumulated. These are produced by repeated cycles of deposition driven by turbidity currents that periodically flow over the site. Turbidites are such a common component of the continental rise sediments that they are responsible for most of the mass transport of terrigenous materials from the continental shelf to the abyssal plain.

Sediments also slump, slide, and flow down the slopes of mid-ocean ridges and rises. This phenomenon is called *ponding* and is responsible for smoothing out the jagged relief of the mid-ocean ridges and rises by filling in crevices and abysses (Figure 13.9).

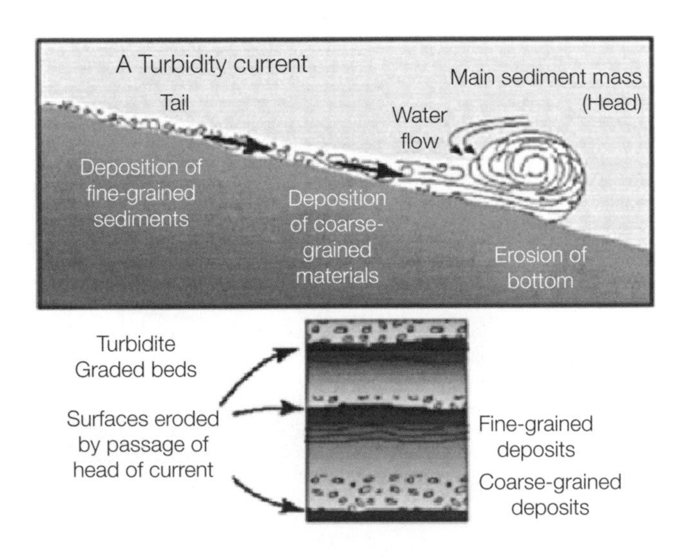

FIGURE 13.8

Graded bedding in turbidites. *Source*: From Morelock, J. and Wilson R. (2005). Marine geology: Terrigenous sediments. http://geology.uprm.edu/Morelock/GEOLOCN/dpseaterrig.htm.

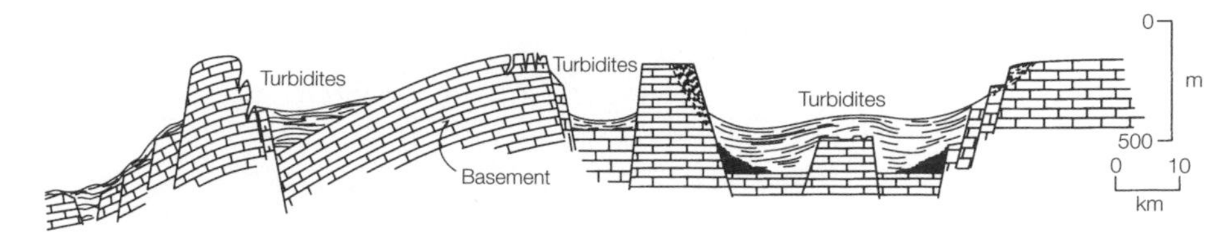

FIGURE 13.9

Sedimentary filling of depressions on mid-ocean ridges.

Since these sediments contain only clay- and silt-sized particles, the graded bedding in these deposits is not as pronounced as in the nearshore turbidites.

13.5.2 Sediment Transport Driven by Bottom-Water Currents

Particles are resuspended from the sediments by the actions of surface gravity waves, such as tides and internal waves, and by bottom currents that flow along the bathymetric contours of many ocean basins. As illustrated in Figure 13.10, a strong contour current flows along the western boundary of the Atlantic Ocean. This current maintains a cloud of resuspended particles, called a *bottom nepheloid layer*, that extends 600 m above the seafloor. Although the cloud fluctuates in particle density and location, it appears to be a permanent feature in the deep zone of the North Atlantic, causing particle concentrations to decrease with increasing vertical distance above the seafloor. Tidal currents are particularly effective in resuspending sediments from continental slopes, thereby creating a mid-depth nepheloid layer as the "dirty" water is advected horizontal offshore.

Where the waves and currents weaken, resuspended sediment settles back down to the seafloor. Given the small particle sizes of the suspended material (mostly 3 to 10 μm), redeposition can take many years. The resulting redistribution of sediments creates patches of clay, mud, and exposed rock on the continental margins. In other words, resuspension from waves and currents can cause some sediments to become relict deposits. Hard bottoms can serve as good habitats for some members of the benthos as they promote the formation of coral reefs. For paleoceanographers, relict deposits are problematic because they represent gaps, or unconformities, in the sedimentary record.

13.5.3 Volcanism

Volcanic eruptions on land spew ash and glassy sand-sized fragments into the atmosphere. Ash can be transported long distances before settling onto the sea surface, but the sand-sized fragments are too heavy to be carried far. If the eruption occurs close to the coast, black sand beaches are often produced. These beaches are extremely hot, as black-colored objects absorb all wavelengths of sunlight. Because of the generally abrasive nature of sandy deposits and their high temperatures, black sand beaches are one of the most inhospitable habitats on the planet. Volcanic eruptions that occur beneath

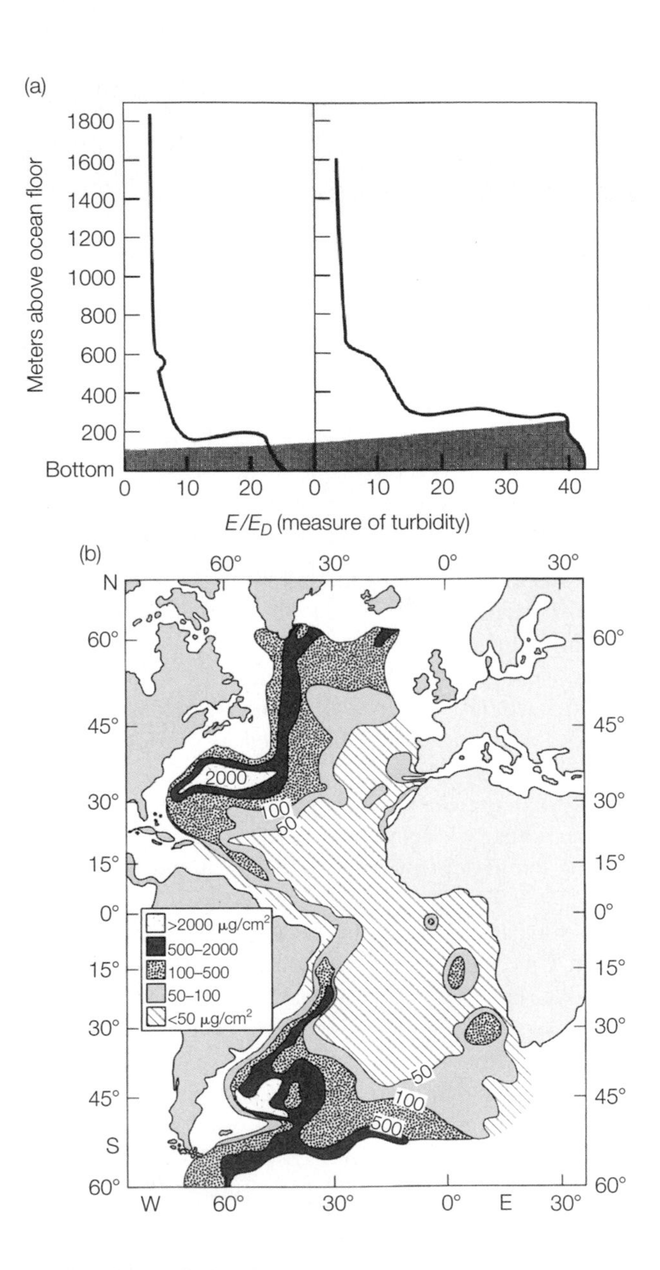

FIGURE 13.10

(a) Vertical distribution of particle concentrations at a site in the North Atlantic as measured by water turbidity. (b) Horizontal distribution of particle concentrations in the deep-water nepheloid layer of the Atlantic ocean. *Source*: From Thurman, H. V. (1988) *Introductory Oceanography*, Merrill Publishing Company, p. 60. Data from Biscaye, P. E. and S. L. Eittreim, (1977) *Marine Geology*, 23, 155–172.

the sea surface also eject particles into the ocean, although most is extruded as lava flows called pillow basalts.

13.5.4 Ice-Driven Transport

Glaciers are very erosive, so as ice moves downslope to the coastline, it cuts a V-shaped channel in the crustal rock. The eroded material, which includes all particle sizes from clays to boulders, is incorporated into the glacier. When the glaciers flow into the ocean, chunks break off, forming icebergs that are then pushed by the winds and currents into the open ocean. The icebergs carry particles with them that sink to the seafloor once the ice melts. This process is called *ice rafting*. The deposits thus formed are called glacial marine sediments and are generally poorly sorted with particle sizes ranging from 180 μm to 3 mm in diameter.

Some glaciers melt before reaching the coastline, leaving a pile of rubble called a *terminal moraine*. If the sea level rises, these deposits often become islands or peninsulas. Cape Cod, Massachusetts, and Long Island, New York, are examples of terminal moraines. Icebergs are presently restricted to latitudes greater than 40°N and 50°S, so actively forming glacial marine deposits are limited to subpolar and cooler temperate regions. Approximately 20% of modern-day sediments are glacial marine deposits.

Sediments deposited in the northeastern Atlantic Ocean during the end of the last ice age (10,000 to 70,000 ybp) contain eight distinct layers in which relatively unweathered coarse lithic particles are the dominant particle type. The lithic particles include *dropstones* that range in size from small pebbles to boulders. The coarse-grained layers alternate with fine-grained sediments dominated by foraminiferan shells. The episodes of increased lithic particle sedimentation are termed *Heinrich Events*. They coincide with rapid oscillations in global climate known as *Dansgaard-Oeschger* events whose occurrence has been documented from oxygen isotope analysis of Greenland ice cores (Chapter 25.4.4). The increase in relative abundance of lithic particles is attributed to increased transport via ice rafting during short-lived (500 ± 200 y) surges in iceberg and sea ice production that occurred during the cold spells along with a decrease in local plankton production. Other likely associated transport mechanisms of the lithic particles include turbidity and debris flows. The reasons for the periodic surges in iceberg production are not known because their timing does not coincide with fluctuations in ice sheet volume, climate, or meridional overturning circulation. Sediments deposited concurrently in other ocean basins contain biogeochemical evidence of global-scale impacts of the Heinrich events. The short-lived nature of these events provides evidence for the relative instability of global climate over short time scales—at least during the end of the last ice age—and provides a cautionary warning regarding the potential for future climate instability.

CHAPTER

Clay Minerals and Other Detrital Silicates 14

All figures are available on the companion website in color (if applicable).

14.1 INTRODUCTION

Weathered fragments of continental crust comprise the bulk of marine sediments. These particles are primarily detrital silicates, with clay minerals being the most abundant mineral type. Clay minerals are transported into the ocean by river runoff, winds, and ice rafting. Some are authigenic, being produced on and in the seafloor as a consequence of volcanic activity, diagenesis and metagenesis.

Clay minerals are important to the crustal-ocean-atmosphere factory, not just for their abundance, but because they participate in several biogeochemical processes. For example, the chemical weathering reactions responsible for their formation are accompanied by the uptake and release of cations and, thus, have a large impact on the chemical composition of river and seawater. This includes acid/base buffering reactions, making clay minerals responsible for the long-term control of the pH of seawater and, hence, of importance in regulating atmospheric CO_2 levels.

Because of their net negative surface charge, clay minerals can adsorb cations and dissolved organic matter. Being particulate matter, the clay minerals eventually settle onto the seafloor, carrying with them a significant amount of adsorbed materials. Clay minerals adsorb organic matter and in so doing appear to afford protection against degradation. This effect enhances the flux of organic matter to the sediments. The ability of clay minerals to adsorb organic matter has led to the hypothesis that they might have played a role in the genesis of life on Earth by catalyzing the formation of biomolecules.

Weathering and authigenic reactions produce a variety of clay minerals. The most abundant marine forms are *illite*, *kaolinite*, *montmorillonite*, and *chlorite*. Their distributions in marine sediments are spatially variable, reflecting regional patterns in the sites of their production and in their transport mechanisms. Downcore changes in their relative abundances provide pale oceanographers with a record of temporal changes in clay mineral production and transport. These fluctuations are largely driven by changes in climate and topography. Thus, clay minerals represent a central player in the crustal-ocean-atmosphere factory and a valuable tracer of past conditions. In this chapter, we discuss the crystal structures, production, transport, and distributions of the most abundant clay minerals and other detrital silicates found in marine sediments.

14.2 THE STRUCTURE OF CRYSTALLINE SILICATE MINERALS

From the perspective of the global rock cycle (Figure 1.2), volcanic activity is the ultimate source of minerals comprising the crust. The crust is 27.7% by mass silicon and 46.6% oxygen, so it is not surprising that silicates are the dominant mineral type. Weathering of these minerals generates *siliclastic particles*. These are also referred to as *detrital silicates*.

Silicon undergoes internal cycling within the crustal-ocean-atmosphere factory during which mineral silicates are produced via precipitation from seawater. These include biotic and abiotic precipitates. As compared to the crystalline silicates that ultimately originate from the igneous rocks, the precipitates are amorphous in structure. To distinguish them from the crystalline silicates, the amorphous forms are collectively referred to as *opaline silica*.

The fundamental structural unit of all crystalline silicates is a tetrahedron that has an O^{2-} anion at each of its four corners and a Si^{4+} cation in the center. The silicate tetrahedra can be arranged in a variety of ways giving rise to a great variety of crystalline silicate minerals. Most of these arrangements involve some degree of sharing of the corner O^{2-} anions between adjacent silicate tetrahedra.

Geologists have classified the crystalline silicate minerals into six types based on the degree of sharing between adjacent tetrahedra. As shown in Figure 14.1, the degree of sharing ranges from none (the neosilicates, such as olivine) to chains (the inosilicates, such as augite and hornblende), to sheets (the phyllosilicates, such as the clay minerals) to framework silicates (the tectosilicates, such as quartz and feldspar). In framework silicates, every corner O^{2-} anion is shared by an adjacent silicate tetrahedron.

Following silicon and oxygen, the next most abundant elements in the crust are Al (8.1%), Fe (5.0%), Ca (3.6%), Na (2.8%), K (2.6%), and Mg (2.1%). These elements are commonly found as substituents for Si^{4+} in silicate tetrahedra. The chemical composition and crystal structure of a given igneous mineral is determined by its source material (magma) and the temperature at which it crystallized (Figure 14.2). Note that the resulting mineralogy also determines an igneous rock's relative stability to weathering. Of the major igneous minerals shown in Figure 14.2, feldspar and quartz comprise 95% of the continental crust. Both are also common components of marine sediments having been transformed into small particles via mechanical weathering. As discussed below, quartz is also a product of intense chemical weathering in which all of the substituent cations are removed from the parent mineral.

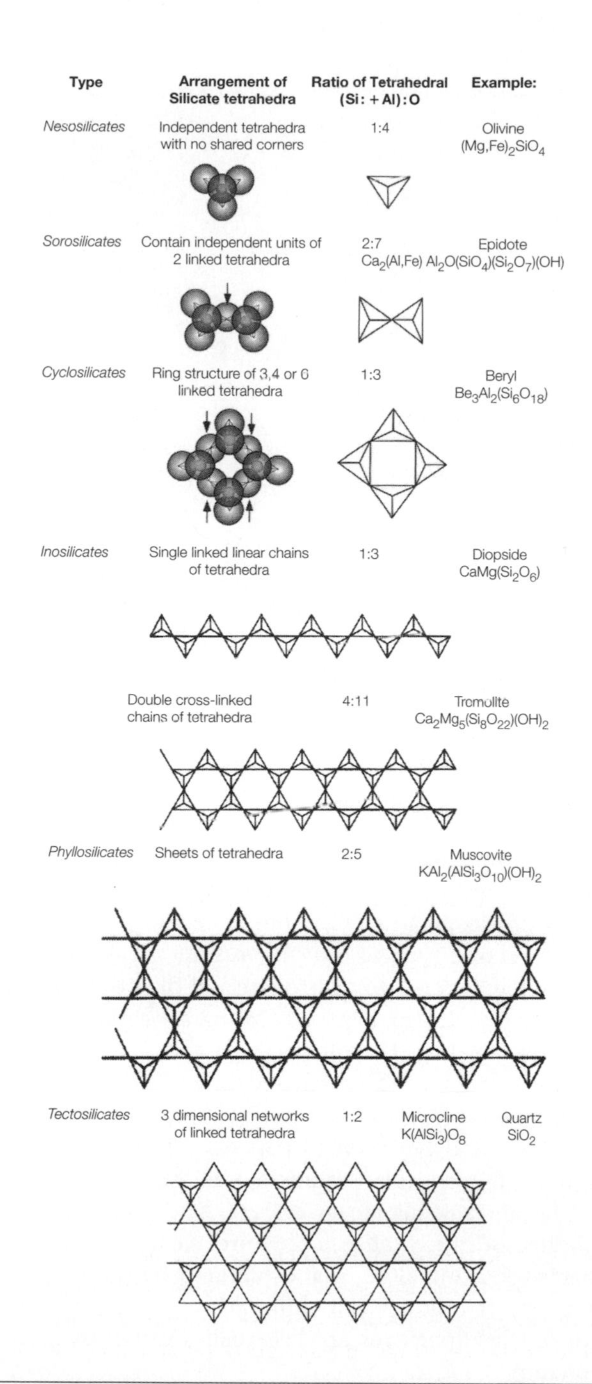

Type
Arrangement of Silicate tetrahedra
Ratio of Tetrahedral (Si: + Al):O
Example:
Nesosilicates
Independent tetrahedra with no shared corners
1:4
Olivine
$(Mg,Fe)_2SiO_4$
Sorosilicates
Contain independent units of 2 linked tetrahedra
2:7
Epidote
$Ca_2(Al,Fe)\ Al_2O(SiO_4)(Si_2O_7)(OH)$
Cyclosilicates
Ring structure of 3,4 or 6 linked tetrahedra
1:3
Beryl
$Be_3Al_2(Si_6O_{18})$
Inosilicates
Single linked linear chains of tetrahedra
1:3
Diopside
$CaMg(Si_2O_6)$
Double cross-linked chains of tetrahedra
4:11
Tremolite
$Ca_2Mg_5(Si_8O_{22})(OH)_2$
Phyllosilicates
Sheets of tetrahedra
2:5
Muscovite
$KAl_2(AlSi_3O_{10})(OH)_2$
Tectosilicates
3 dimensional networks of linked tetrahedra
1:2
Microcline
$K(AlSi_3)O_8$
Quartz
SiO_2

FIGURE 14.1

Arrangement of silica tetrahedra in various classes of silicate minerals. These are top-down views looking at the apex of the tetrahedra.

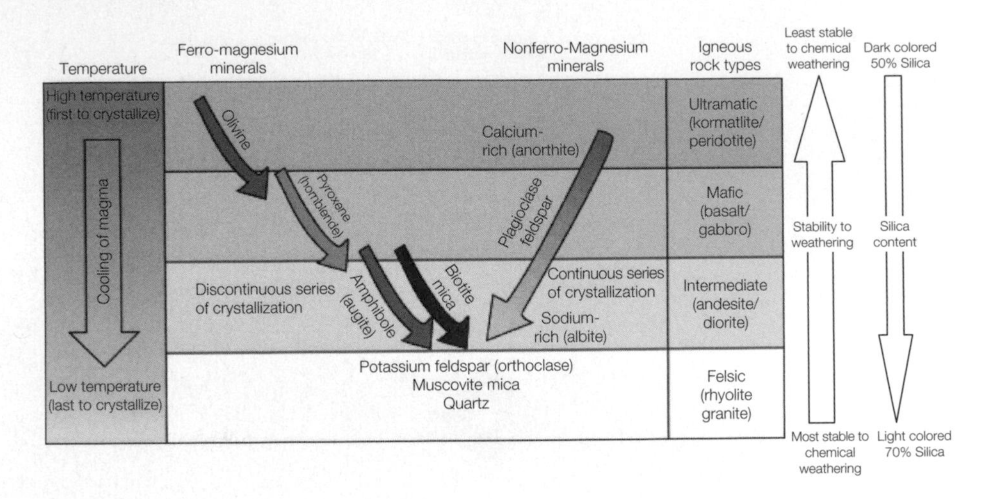

FIGURE 14.2

Bowen's reaction series representing the relationship of igneous silicate mineral composition to crystallization temperature. The plagioclase feldspars represent a continuous series grading from the calcium-rich to sodium-rich to potassium-rich forms. The rest of the minerals constitute a discontinuous series in which distinct crystal structures are characteristic of a particular temperature range under which the magma solidified.

14.2.1 Crystalline Structure of Clay Minerals

When the SiO_2 tetrahedra are arranged so that only the O^{2-} in the bases are shared, a thin layer results (Figure 14.3c). Silicate minerals constituted of these layers are termed sheet silicates (phyllosilicates). Clay minerals and mica are examples of sheet silicates. In both cases, a second type of layer is present whose fundamental structural unit is an octahedron in which each corner is occupied by a OH^- and the centers by either Al or Mg (Figure 14.3b). Linking of these octahedra via sharing of hydroxide groups creates a continuous layer (Figure 14.3d).

Because clay minerals are composed of the two types of sheets, they are termed *layered aluminosilicates*. Different arrangements of these sheets and degrees of cation substitution within and between the sheets give rise to a variety of clay minerals. As listed in Figure 14.4, the clay minerals can be classified into two-, three-, and four-layer clays. The aluminosilicate layers are held together by hydrogen bonds and van der Waals forces. Because these forces are weak, the layered silicates tend to be soft and deformable, cleaving along planes created by the sheets. Mechanical and chemical weathering generally reduces clay minerals to small fragments such that in pelagic sediments, most of these particles are less than 4 μm in diameter and, hence, fall into the clay-sized fraction.

In terms of composition, the simplest of the marine clay minerals is kaolinite in which tetrahedral and octahedral layers alternate (Figure 14.5a) creating a two-layer repeating unit. In three-layer clays, the repeating unit is composed of an octahedral

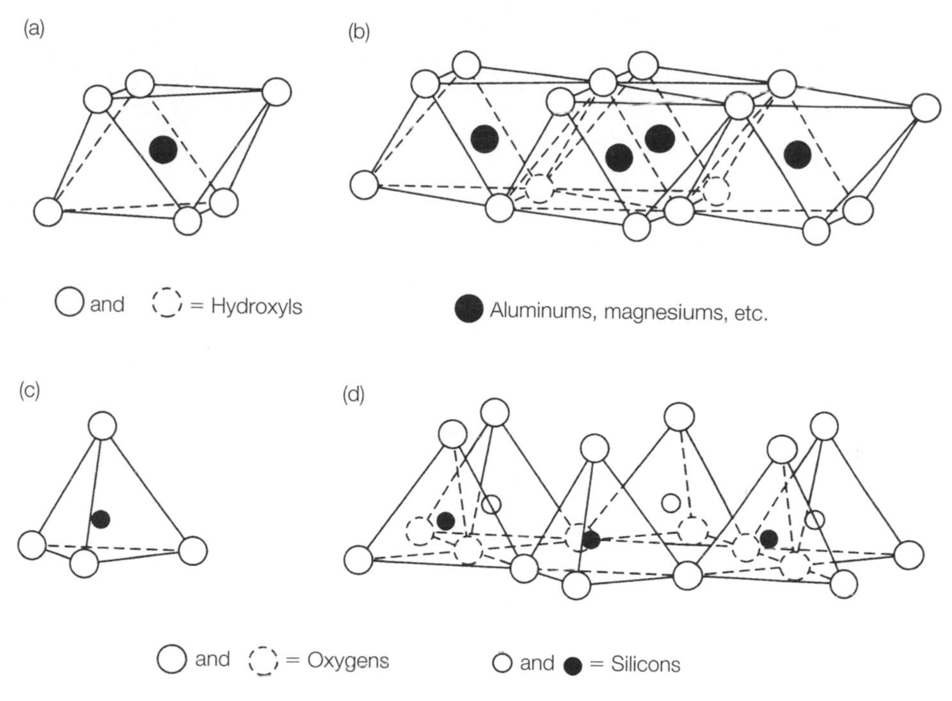

FIGURE 14.3

Fundamental structural units of detrital silicates. (a) octahedron, (b) octahedral layer found in sheet silicates, (c) tetrahedron, and (d) tetrahedral layer found in sheet silicates. *Source*: From Grim, R. E. (1968). *Clay Mineralogy*, 2nd ed., McGraw-Hill Publishing Company, p. 52.

layer sandwiched between two tetrahedral layers. The most common three-layer clays found in marine sediments are montmorillonite, a member of the smectite group, and illite. Illite is similar to mica and, hence, is considered to be part of a clay-mica group (Figure 14.5b). Chlorite is an example of a four-layer clay in which the three-layer sandwich alternates with a fourth layer, called brucite. The brucite layer is composed of octahedra that have Mg^{2+} in their centers and OH^- in their corners (Figure 14.5c).

14.2.2 Cation Exchange Capacity

The substitution of cations into and between the tetrahedral and octahedral layers within clay minerals gives rise to charge imbalances throughout the crystal structure. This causes the exterior surfaces to have a net negative charge as well as the corners and edges. Because of these negative charges, minerals electrostatically attract and adsorb cationic solutes onto their surfaces. If the surface site is already occupied, *cation exchange* can occur as follows:

$$A^+(aq) + B\text{–clay mineral} \rightleftharpoons A\text{–clay mineral} + B^+(aq) \quad (14.1)$$

Type of Clay	Group	Mineral	Structure	Composition (idealized)	Cation exchange capacity (meq/100g)	Remarks	Exchangeable Interlayer cations (Ex)
Two layer	Kaolin	Kaolinite		$Al_2Si_2O_5(OH)_4$	3–15	Little isomorphous substitution small cation exchange capacities Nonexpanding	-none-
Three layer	Smectite	Expanding Montmorillonites	H_2O Ex H_2O	$Ex_x[Al_{2-x}Mg_x]<Si_4>O_{10}(OH)_2$	80–150	Substitution of a small amount of Al for Si in Td-sheet and of Mg, Fe, Cr, Zn, Li for Al or Mg in Oh-sheet Large CEC	Ca^{2+}, Na^+, K^+, Li^+
	Vermiculites	Vermiculites	H_2O Ex H_2O	$Ex_x[Mg_3]<Al_xSi_{4-x}>O_{10}(OH)_2$	100–150	Swell in water or polar organic compounds	Mg^{2+}
	Mica	Nonexpanding (illites)	K^+	$K_{1-x}[Al_2]<Al_{1-x}Si_{3+x}>O_{10}(OH)_2$	10–40	About $\frac{1}{4}$ of Si in Td-sheet replaced by Al, similar Oh-sheet substitutions Small CEC	K^+
Four layer	Chlorites	Chlorite Brucite		$[Mg, Al]_3(OH)_6[Mg, Al]_3<Si, Al>_4O_{10}(OH)_2$	5–30	Three-layer alternating with brucite Brucite layer positively charged [some Al(III) replacing M(II)], partially balances negative charge on Td-Oh-Td (mica) layer Low CEC, nonswelling	-none-

Tetrahedral layer (Td) < >
Octahedral layer (Oh) [ ]
Ex = exchangeable interlayer cations

FIGURE 14.4

The principal species of clay minerals. *Source*: After Stumm, W. S. and J. J. Morgan (1981). *Aquatic Chemistry*, John Wiley & Sons, Inc., p. 442; and Grim, R. E. (1968). *Clay Mineralogy*, 2nd ed., McGraw-Hill Publishing Company, p. 159.

in which cations A and B compete for the negatively charged surface sites on the clay mineral. (See Eq. 5.43 for a more general form of this equilibrium process.) The speciation at equilibrium is determined by: (1) the relative ion concentrations, (2) the ability of a particular cation to compete for an adsorption site, and (3) the *cation exchange capacity (CEC)* of the clay mineral. The latter is a measure of the amount of cations adsorbed at equilibrium. The CEC is commonly reported as the milliequivalents of positive charge that are adsorbed by 100 g of clay mineral.

The CEC of a clay mineral is directly related to its crystalline structure because this determines the density of negative charge on its surfaces. The crystalline structure is in turn a consequence of the types and degree of cation substitution. Since cations similar in ionic radius and charge are most likely to replace the aluminum and silicon within the crystal lattice, this phenomenon is termed *isomorphic substitution*. Some isomorphic substitution occurs during crystallization of the igneous silicates and some occurs during chemical weathering. The larger the degree of isomorphic substitution, the larger the charge imbalances and the larger the CECs.

As shown in Figure 14.4, each clay mineral exhibits a large range in the type and degree of isomorphic substitution. The central silicon atom in the tetrahedral layers can be replaced by aluminum, alkali, alkaline earth, and trace metal atoms. In the octahedral layers, the central Al and Mg atoms can be similarly replaced. The large range in composition within each mineral type reflects variability in the environmental conditions under which crystallization and chemical weathering occur. Thus, the

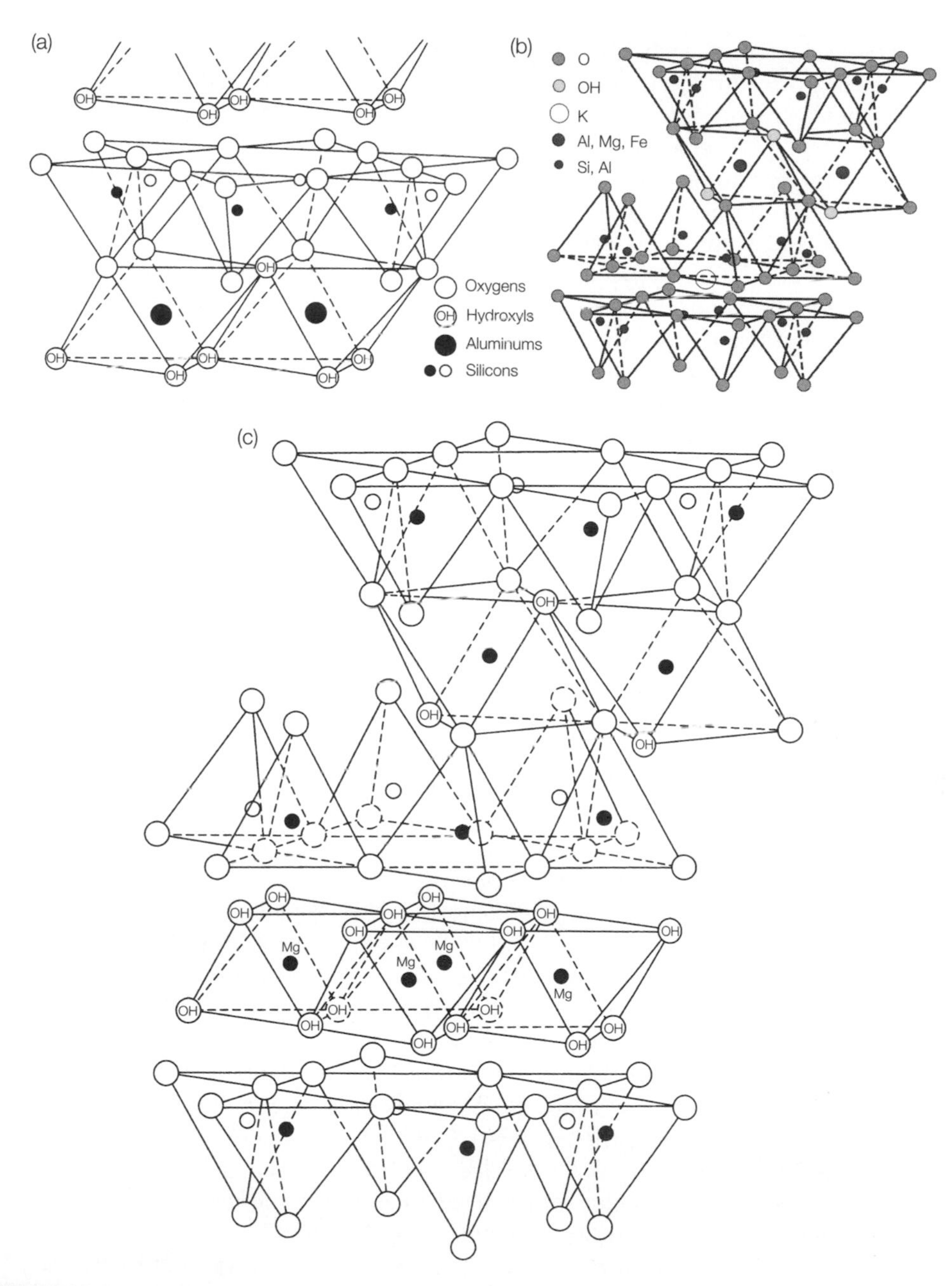

FIGURE 14.5

Crystal structure of (a) kaolinite, (b) illite, and (c) chlorite. *Source*: From Grim, R. E. (1968). *Clay Mineralogy*, 2nd ed., McGraw-Hill Publishing Company, p. 58. Structure of illite is modified from Grim (1962) by Poppe, L. J., *et al.* (2001). A Laboratory Manual for X-Ray Powder Diffraction. U. S. Geological Survey Open-File Report 01–041, http://pubs.usgs.gov/of/2001/of01-041/htmldocs/images/illstruc.jpg.

stoichiometry of each clay mineral type must be represented by ranges in elemental composition. In Figure 14.4, these ranges in isomorphic substitution are given by the elements in parentheses (<Tetrahedral> and [Octahedral]).

The CEC of clay minerals is partly the result of adsorption in the interlayer space between repeating layer units. This effect is greatest in the three-layer clays. In the case of montmorillonite, the interlayer space can expand to accommodate a variety of cations and water. This causes montmorillonite to have a very high CEC and to swell when wetted. This process is reversible; the removal of the water molecules causes these clays to contract. In illite, some exchangeable potassium is present in the interlayer space. Because the interlayer potassium ions are rather tightly held, the CEC of this illite is similar to that of kaolinite, which has no interlayer space. Chlorite's CEC is similar to that of kaolinite and illite because the brucite layer restricts adsorption between the three-layer sandwiches.

14.3 THE PRODUCTION OF CLAY MINERALS FROM TERRESTRIAL WEATHERING

The chemical and physical processes that occur on land convert igneous silicates into a wide variety of clay minerals, most of which are present as sand, silt, and clay-sized particles. These processes are collectively referred to as *terrestrial weathering*. As shown in Figure 14.6, weathering is initiated by physical processes that break the parent rock into small fragments. This increases the surface area, thereby facilitating further chemical weathering. Water and organisms cause much of the physical weathering of terrestrial rocks. For example, the expansion of water when it freezes in cracks and crevices fractures rock. Plants also cause rock fragmentation when their roots grow down and expand in cracks and crevices. This is an example of *biological weathering*.

The newly exposed surfaces undergo chemical weathering, during which bonds are broken and new ones formed. Weathering reactions can yield soluble products (*congruent dissolution reactions*) or a mixture of solid and soluble products (*incongruent dissolution reactions*). Both cations and anions are produced during chemical weathering. In the case of the igneous rocks, the solid products include clay minerals, quartz, and gibbsite. The last is composed of sheets of octahedra that have Al atoms at the center and OH^- groups at the corners (Figure 14.3d). Chemical weathering is initiated by hydration of the parent rock's surface, as illustrated in Figure 14.7. The extent to which a rock is hydrated depends on its surface-area-to-volume ratio, as well as the interatomic spacing and charge distribution in the crystal lattice.

The second stage of chemical weathering involves a type of hydrolysis. During this reaction, hydrogen and hydroxide ions derived from water bond with the crystal lattice. As shown in Figure 14.7b, this causes the breakage of some of the cation-oxygen bonds. The susceptibility of a rock to hydrolysis is determined by the degree to which it can become hydrated and the relative strength of the cation-oxygen bond. The latter is a function of the relative charge density (z/r) of the cation.

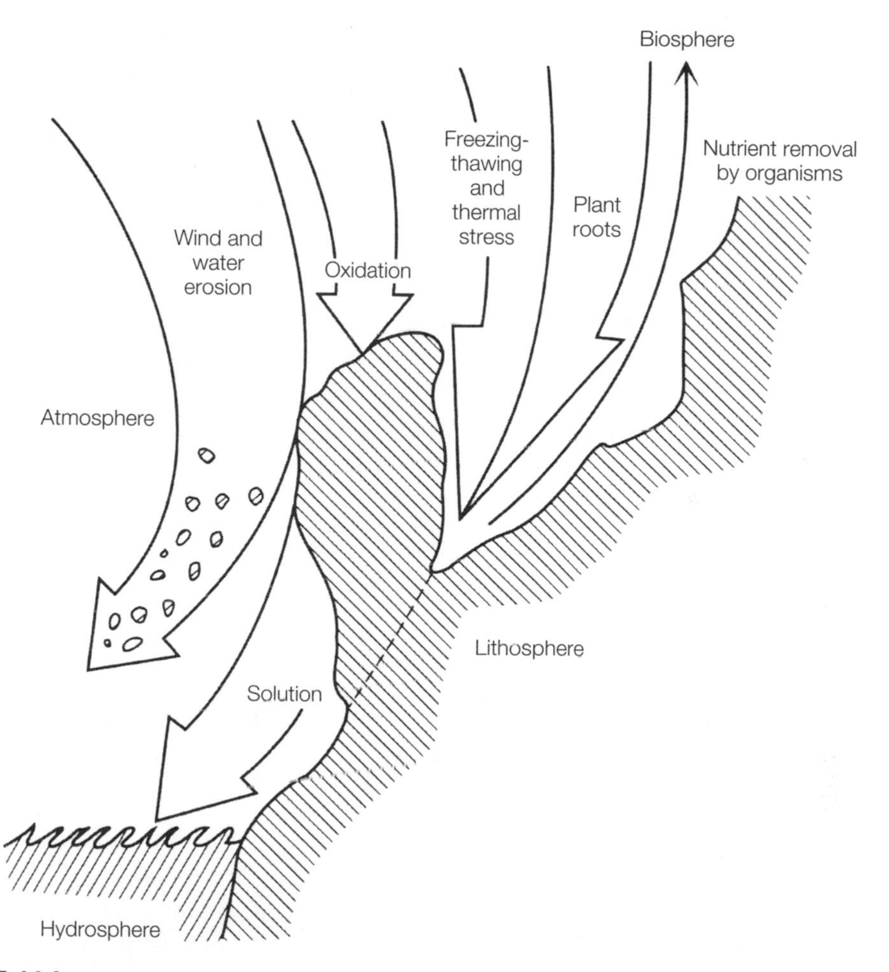

FIGURE 14.6

Weathering attack. *Source*: From Horne, R. A. (1978). *The Chemistry of Our Environment*, John Wiley & Sons, Inc., p. 534.

The breakage of these bonds proceeds more rapidly and to a greater extent in the presence of a stronger acid, such as carbonic acid. As shown in Eq. 5.53, H_2CO_3 is a product of the hydrolysis of $CO_2(g)$, so this acid is present in waters that are in gaseous equilibrium with the atmosphere. Pure water in equilibrium with atmospheric levels of $CO_2(g)$ has a pH of 5.6. Thus, rain water is naturally acidic, although its exact pH is determined by the dissolution of variable amounts of other acids and bases, such as H_2SO_4, HNO_3, and NH_3 and temperature.

Carbonic acid reacts with the parent or primary rock to solubilize cations as follows:

$$\text{Igneous rock(s)} + H_2CO_3(aq) + H_2O(l) \rightarrow HCO_3^-(aq) + H_4SiO_4(aq) + \text{Cations(aq)} + \text{Clay minerals(s)} \quad (14.2)$$

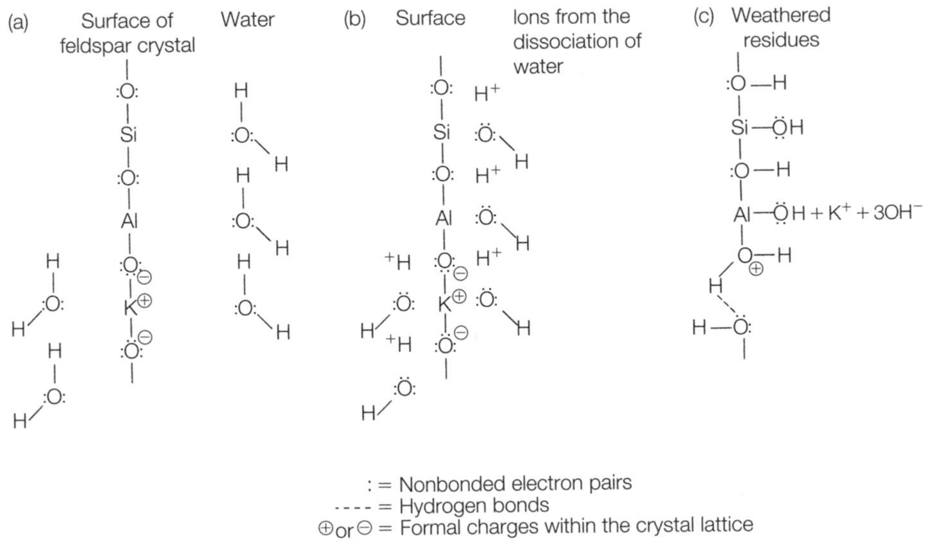

FIGURE 14.7

Decomposition of feldspar lattice by water: (a) hydration, (b) hydrolysis, and (c) weathered residues. Note that the Al–O and Al–Si bonds are mixtures of pure covalent and ionic forms. The bond between K and O is also a hybrid but is much weaker than the other types of O bonds. *Source*: After Friedman, G. M. and J. E. Sanders (1978). *Principles of Sedimentology*, John Wiley & Sons, Inc., 792 p.

The cations become a component of river water and are eventually transported to the sea. About 45% of the dissolved solids entering the ocean are derived from the weathering of detrital silicates. Feldspars are the most important source rock for terrigenous clays as illustrated by the following reaction

$$\underset{\text{(plagioclase feldspar)}}{4Na_{0.5}Ca_{0.5}Al_{1.5}Si_{2.5}O_8(s)} + 6CO_2(g) + 17H_2O(l) \rightarrow 2Na^+(aq) + 2Ca^{2+}(aq) + 4H_4SiO_4(aq) + 6HCO_3^-(aq) + \underset{\text{(kaolinite)}}{3Al_2Si_2O_5(OH)_4(s)} \quad (14.3)$$

Other examples of weathering reactions involving igneous silicates are provided in Table 14.1. In some cases, chemical weathering proceeds in a stepwise fashion in which one clay mineral can be transformed into another given favorable environmental conditions.

The amount and type of chemical weathering that occurs depends on the chemical composition of the parent rock and the weathering solution, as well as the environmental conditions. As shown in Figure 14.2, the high-temperature igneous silicates are more prone to chemical weathering than the low-temperature minerals. Climate is important because it determines temperature and water availability, both of which control

Table 14.1 Examples of Detrital Silicate Weathering Reactions.

Reaction
$3KAlSi_3O_8(s) + 2CO_2(g) + 14H_2O(l) \rightarrow KAl_3Si_3O_{10}(OH)_2(s) + 2K^+(aq)$ (orthoclase feldspar) (illite) $6H_4SiO_4(aq) + 2HCO_3^-(aq)$
$CaAl_2Si_2O_8(s) + 2CO_2(g) + 3H_2O(l) \rightarrow Al_2Si_2O_5(OH)_4(s) + Ca^{2+}(aq) + 2HCO_3^-(aq)$ (anorthite) (kaolinite)
$2NaAlSi_3O_8(s) + 2CO_2(g) + 6H_2O(l) \rightarrow$ (albite) $Al_2Si_4O_{10}(OH)_2(s) + 2Na^+(aq) + 2HCO_3^-(aq) + 2H_4SiO_4(aq)$ (montmorillonite)
$NaAlSi_3O_8(s) + CO_2(g) + \frac{11}{2}H_2O(l) \rightarrow$ (albite) $\frac{1}{2}Al_2Si_2O_5(OH)_4(s) + Na^+(aq) + HCO_3^-(aq) + 2H_4SiO_4(aq)$ (kaolinite)
$KMgFe_2AlSi_3O_{10}(OH)_2(s) + \frac{1}{2}O_2(g) + 3CO_2(g) + 11H_2O(l) \rightarrow$ (biotite) $Al_2O_3 \cdot H_2O(s) + K^+(aq) + Mg^{2+}(aq) + 3HCO_3^-(aq) + 3H_4SiO_4(aq) + 2Fe(OH)_3(s)$ (gibbsite)
$Al_2Si_2O_5(OH)_4(s) + 5H_2O(l) \rightarrow Al_2O_3 \cdot 3H_2O(s) + 2HCO_3^-(aq) + 2H_4SiO_4(aq)$ (kaolinite) (gibbsite)

the amount of vegetation. Plant growth increases the rate and extent of chemical and mechanical weathering because the decomposition of organic detritus produces CO_2, which enhances the dissolution and fragmentation of rocks. Topography is another important environmental control, as it determines water drainage (i.e., the length of time that water is in contact with the rock). This also affects any weathering reactions that involve redox chemistry, because the relative stagnancy of the water influences its O_2 content. Redox reactions are important for detrital rocks with high iron contents.

The weathering of igneous silicates under temperate conditions yields a suite of clay minerals that include, in order of decreasing abundance: illite, vermiculite, kaolinite, smectite, and chlorite. Vermiculite is an expanding three-layer clay (Figure 14.4). Under tropical conditions, weathering is more intense, leading to the removal of more cations, with kaolinite being the most abundant clay mineral product. If the kaolinite remains on land, continued weathering will remove all of its cations and separate the solid residues into silica (quartz) and aluminum (gibbsite) minerals. The iron in igneous silicates is usually oxidized converted into solid oxides, such as goethite (FeOOH) and hematite (Fe_2O_3), which are recognizable by their red-orange coloration. At high latitudes, chemical weathering is minimal due to low temperatures. The dominant clay mineral product is generally chlorite. In contrast to illite, kaolinite, and chlorite, most of the montmorillonite found in pelagic marine sediments is the result of authigenic production.

Because chemical weathering consumes CO_2 and supplies HCO_3^- to river water, the formation of clay minerals plays a role in regulating the CO_2 content of the atmosphere and stabilizing the pH of the oceans. This subject is considered further in Chapter 21. Chemical weathering also solubilizes silicon as silicic acid. The riverine input of this dissolved silicon into the oceans plays a significant role in meeting the nutritional needs of organisms that deposit siliceous tests, namely diatoms and radiolarians.

14.4 THE PRODUCTION OF CLAY MINERALS FROM AUTHIGENIC PROCESSES

After delivery to the ocean, clay minerals react with seawater. The processes that alter the chemical composition of the terrigenous clay minerals during the first few months of exposure are termed *halmyrolysis*. These include: (1) cation exchange, (2) fixation of ions into inaccessible sites, and (3) some isomorphic substitutions. Another important transformation is flocculation of very small (colloidal-size) clay particles into larger ones.

Most cation exchange occurs in estuaries and the coastal ocean due to the large difference in cation concentrations between river and seawater. As riverborne clay minerals enter seawater, exchangeable potassium and calcium are displaced by sodium and magnesium because the Na^+/K^+ and Mg^{2+}/Ca^{2+} ratios are higher in seawater than in river water. Trace metals are similarly displaced.

Over longer time scales, clay minerals can undergo more extensive reactions. For example, fossilization of fecal pellets in contact with a mixture of clay minerals and iron oxides produces an iron- and potassium-rich, mixed-layer clay called *glauconite*. This mineral is a common component of continental shelf sediments. Another example of an authigenic reaction is called *reverse weathering*. In this process, clay minerals react with seawater or porewater via the following general scheme:

$$\begin{aligned} &H_4SiO_4(aq) + \text{degraded clay from land} + \text{ions} \\ &\quad (\text{primarily } Mg^{2+}(aq), K^+(aq), Na^+(aq) + HCO_3^-(aq)) \rightarrow \\ &\quad\quad \text{ion-rich authigenic clay} + H_2O(l) + CO_2(g) \end{aligned} \tag{14.4}$$

leading to the formation of a secondary clay mineral. During this process, major ions are removed along with bicarbonate and dissolved silicon. Reverse weathering releases acid, which then titrates bicarbonate. This generates CO_2, which is eventually released to the atmosphere via air-sea exchange. Thus, the balance between weathering and reverse weathering could play a critical role in controlling atmospheric CO_2 levels.

Direct evidence supporting the occurrence of reverse weathering has proven difficult to obtain for two reasons. First, the same kinds of clay minerals produced by this process are also transported to the ocean as part of the suspended load in river runoff. Second, the rate of reverse weathering is so slow that laboratory studies of this process are difficult to conduct.

In some restricted locations, such as deltaic sediments, reverse weathering reactions appear to be occurring at relatively fast rates. For example, in sediments of the Amazon river delta, secondary clays enriched in Fe, K, and Mg have been observed to form over time scales on the order of months to years. Because the Amazon River is such a large source of particulate matter to the oceans, the removal rate of K^+ via reverse weathering in its deltaic sediments is estimated to be on the order of 10% of the total oceanic sink for this element. These secondary clays exhibit some similarity to glauconite, having chemical enrichments in iron and potassium and an origin in diagenetic reactions that occur in organic-rich sediments.

The authigenic clays that are thought to be formed via the slower type of reverse weathering are likely to be products of late-stage diagenesis and metagenesis. In this case, the sediments have been buried to depths greater than 1.5 km and, hence, have been exposed to high pressures and temperatures ranging from 50 to 300°C. During these reverse weathering reactions, which occur over time scales of millions of years, the sediments are in the process of being transformed into rocks, primarily sandstones and shales. Evidence for chemical transformations are seen in downcore decreases in the relative abundances of smectites and kaolinite, while illite and chlorite increase. Another frequently observed trend is a decrease in the relative abundance of primary igneous silicates, such as plagioclase feldspar, with increasing depth, such that these minerals are rarely found in shales. The following reaction has been proposed to explain this commonly observed downcore trend:

$$\underset{\text{(K-feldspar)}}{KAlSi_3O_8(s)} + \underset{\text{(montmorillonite)}}{Al_2Si_4O_{10}(OH)_2 \cdot nH_2O(s)} \rightarrow \underset{\text{(illite)}}{2KAl_3Si_3O_{10}(OH)_2(s)} + nH_2O(l) \quad (14.5)$$

In addition to reverse weathering and dissolution of feldspars, authigenic clays are thought to form within deeply buried sediments via precipitation of solutes present in pore fluids. Precipitation of calcium carbonate and silicate also occur, generating authigenic calcite and quartz. Because of the large volumes of sediment and rock involved, these diagenetic and metagenetic reactions likely play a significant role in the cycling of elements through the crustal-ocean-atmosphere factory. If some of these reactions are of the reverse-weathering type, they could also have a large impact on maintaining the acid-base balance of the oceans and atmospheric CO_2 levels.

Finally, some authigenic clay minerals are produced by the reaction of seawater with fresh volcanic glass. This commonly occurs near mid-ocean ridges and rises or where lava from coastal volcanoes flows into the sea. Clay minerals produced by this process are primarily smectites, such as montmorillonite, and a type of framework silicate called *zeolites* of which phillipsite and clinoptilite are the most common marine examples. Zeolites are characterized by three-dimensional frameworks with large cavities that can accommodate exchangeable cations and water. The CEC of zeolites is very high (clinoptilite = 220 meq/100 g and phillipsite = 450 meq/100 g).

14.5 TRANSPORT PATHWAYS

14.5.1 Rivers and Oceanic Currents

Rivers transport clay minerals primarily as part of their suspended load (silts and clays). The silt-size fraction is composed of quartz, feldspars, carbonates, and polycrystalline rocks. The clay-sized fraction is dominated by the clay minerals illite, kaolinite, chlorite, and montmorillonite. In addition to suspended particles, rivers carry as a bed load larger size fractions. The bed load constitutes only 10% of the total river load of particles and is predominantly quartz and feldspar sands.

River transport is responsible for 84% of the terrestrial input of clay minerals to the ocean but most is trapped in the coastal zone, accumulating either in river deltas or on the continental shelf and slope. Much of this deposition occurs in marginal seas. Sedimentation of the fine-grain fraction is enhanced by various processes, including biological uptake and colloidal flocculation, both of which increase sinking rates by incorporating the clays into larger particles. Thus, only a small fraction of the riverine sediment input is carried by currents into the open ocean. The small fraction that is delivered to the open ocean eventually undergoes pelagic sedimentation.

River transport of clay minerals into the ocean is spatially and temporally variable. The global annual suspended load of river sediment into coastal waters currently averages 12.6×10^9 ton.[1] This flux is approximately 10% less than was delivered before humans began damming rivers. (One notable exception is the Mississippi River, whose sediment load has increased due to very high rates of soil erosion. The riverine sediments deposited in the mouth of the Mississippi River form one of the world's largest deltas.)

At present count, more than 45,000 dams greater than 15-m high are now collectively holding back about 15% of the total annual river runoff.[2] The largest of the rivers are listed in Table 14.2 along with their modern-day and prehuman annual sediment loading rates. These 25 rivers collectively provide 40% of the annual sediment load to the ocean. They are concentrated in Southeast Asia, the Pacific Ocean Islands, and South America, where the river drainage basins lie in warm climate zones and have headwaters in mountainous regions. Each of the oceans receives about the same load: 3.4×10^9 ton/y in the Atlantic, 3.3×10^9 ton/y in the Indian, and 4.9×10^9 ton/y in the Pacific Ocean.

Turbidity and contour currents resuspend and transport the sediments that lie on the continental margin. These sediments are redeposited when the currents weaken.

---

[1] Suspended load carried by rivers during the Anthropocene as estimated by Syvitski, J. P. M., *et al.* (2005) *Science*, 308, 376–380.

[2] Nilsson, C., *et al.* (2005). *Science*, 308, 405–408.

Table 14.2 Discharge and Sediment Loads into the Ocean from the World's Twenty-Five Largest Rivers.

Rank (Prehuman Sediment Load)	River Name	Continent	Receiving Ocean	Average annual discharge (km^3/yr)	Prehuman Sediment Load (MT/y)	Modern Sediment Load (MT/y)	Sediment Load Difference (Modern-Pre-human)
1	Ganges	Asia	Indian Ocean	40,026	1278	1278	0
2	Amazon	South America	Atlantic Ocean	207,682	1155	1199	44
3	Irrawaddy	Asia	Indian Ocean	20,502	515	260	−255
4	Chang Jiang (Yangtze)	Asia	Pacific Ocean	29,583	345	156	−189
5	Indus	Asia	Indian Ocean	3,333	303	116	−186
6	Mississippi	North America	Atlantic Ocean	19,396	178	288	111
7	Shatt el Arab	Asia	Indian Ocean	2,418	161	25	−136
8	Salween	Asia	Indian Ocean	3,153	150	124	−26
9	Yukon	North America	Pacific Ocean	6,426	147	60	−87
10	Rio Grande (U.S.)	North America	Atlantic Ocean	120	105	0	−105
11	Mackenzie	North America	Arctic Ocean	9,192	102	107	5
12	Huang He (Yellow)	Asia	Pacific Ocean	1,438	102	29	−73
13	Yenisei	Asia	Arctic Ocean	19,655	94	4	−91
14	Irharhar	Africa	Mediterranean + Black Sea	26	92	4	−89

15	Parana	South America	Atlantic Ocean	16,937	91	26	−65
16	Nile	Africa	Mediterranean + Black Sea	1,251	89	12	−77
17	Colorado (Arizona)	North America	Pacific Ocean	21	85	58	−27
18	Mekong	Asia	Pacific Ocean	15,029	80	80	−1
19	Zaire	Africa	Atlantic Ocean	40,546	73	23	−50
20	Orinoco	South America	Atlantic Ocean	35,402	60	30	−30
21	Amur	Asia	Pacific Ocean	10,839	60	22	−37
22	Orange	Africa	Atlantic Ocean	146	59	2	−57
23	Danube	Europe	Mediterranean + Black Sea	6,488	56	23	−32
24	Grande deSantiago	North America	Pacific Ocean	343	53	1	−52
25	Colorado (Argentina)	South America	Atlantic Ocean	356	51	6	−44

MT = $10^{12}g$
Source: After Syvitski, J. P. M, et al. (2005). Science, *308, 376–380. Data downloaded from: INSTAAR, Univ. of Colorado (2007). http://instaar.colorado.edu/deltaforce/papers/global_sediment_flux.html*

In the case of the turbidity currents, this redistribution usually occurs along the foot of the continental slope and is largely responsible for the accumulation of sediments in the continental rise. The resuspension of particles by contour currents can also maintain permanent nepheloid layers as shown in Figure 13.10.

14.5.2 **Winds**

Winds are responsible for the transport of 7% of the terrigenous sediment input to the oceans. Most of the marine aerosols have diameters less than 10 μm, of which a significant percentage is mineral dust. The constituent minerals are similar to those in river runoff, including quartz, feldspar, carbonates, and the clay minerals, plus sulfates (gypsum). The aeolian load of clay minerals is derived from arid and semiarid regions whose soils are made airborne by the major wind bands (Figure 4.4a). This produces the dust belt illustrated in Figure 11.4. Grain size decreases with increasing distance from the coast such that pelagic sediments lying downwind of these source regions receive aeolian dust mostly in the clay-size fraction. For example, the trade winds that blow across the Sahara Desert are responsible for the high concentrations of kaolinite in the sediments of the subtropical South Atlantic.

The clay minerals of aeolian origin comprise 25 to 75% of the mass of pelagic sediments. The large range in composition reflects the latitudinal nature of the dust belt as well as dilution by other locally important particle types such as clay minerals of volcanogenic origin and biogenic hard parts (calcite and opaline silica).

Global estimates of annual dust deposition rates are on the order of 0.4×10^9 ton/y with the highest regional rates in the North Atlantic and Indian Oceans (0.2 and 0.1×10^9 ton/y, respectively). Large dust storms and volcanic activity can cause short-term spikes in the rates of dust supply. Thus, annual emission rates are highly variable. They also appear to be particularly susceptible to alteration by human activities due to the combined effects of changes in land use and climate. Anthropogenic impacts on local and global dust supply are likely to have significant impacts on the crustal-ocean-atmosphere factory. These effects include: (1) the degree to which insolation is reflected or absorbed by atmospheric particles, (2) how heterogeneous reactions proceed in the atmosphere in which mineral dust serves as a catalytic surface, (3) the rate at which iron is delivered to the surface waters of the open ocean, (4) the rate at which particle scavenging removes trace metals from seawater, and (5) the degree to which sinking and sedimentary POM is protected from degradation.

14.5.3 **Ice**

Ice rafting is responsible for 7% of the terrigenous input of siliclastic particles to the ocean. When the ice melts, the particles settle to the seafloor to form glacial marine deposits. These are currently forming at latitudes greater than 40°N and 50°S. Most of the glacial marine sediments are poorly sorted deposits composed of relatively unweathered materials with chlorite being the dominant clay mineral. In the North Atlantic, layers enriched in coarse-grained terrigenous minerals are evidence of mass ice-rafting events[3] that occurred during the Ice Ages of the Pleistocene. Paleoceanographers hypothesize that swarms of icebergs flowed off land during these very cold periods and, hence, greatly increased the flux of glacial marine particles to the sediments.

14.5.4 Organisms

Filter-feeding marine organisms increase the sinking rates of clay minerals by incorporating them into fecal pellets. Clay minerals, in turn, afford protection against degradation to closely associated organic matter during the trip to the seafloor and after deposition onto the sediments. Due to its net negative surface charge, clay minerals are effective scavenging agents that act to transport adsorbed materials, including trace metals and organic matter, to the seafloor.

14.6 GLOBAL PATTERNS OF QUARTZ AND CLAY MINERAL DISTRIBUTIONS

The global distribution patterns of kaolinite, chlorite, montmorillonite, and illite in pelagic sediments are listed in Table 14.3 and illustrated in Figures 14.8 through 14.11.

Table 14.3 Average % Composition of <2 μm Carbonate-Free Size Fraction in Sediments of the Abyssal Plains.

Oceanic Area	Chlorite	Montmorillonite	Kaolinite	Illite
North Atlantic	10	16	20	56
Gulf of Mexico	18	45	12	25
Caribbean Sea	11	27	24	36
South Atlantic	11	26	17	47
North Pacific	18	35	8	40
South Pacific	13	53	8	26
Indian	10	47	16	30
Bay of Bengal	14	45	12	29
Arabian Sea	18	28	9	46

Source: From Griffin, J. J., et al. (1968). Deep-Sea Research, *15, 433–459.*

[3] These are called Heinrich events after the marine scientist who discovered them in 1988 (see Chapter 13.5.4).

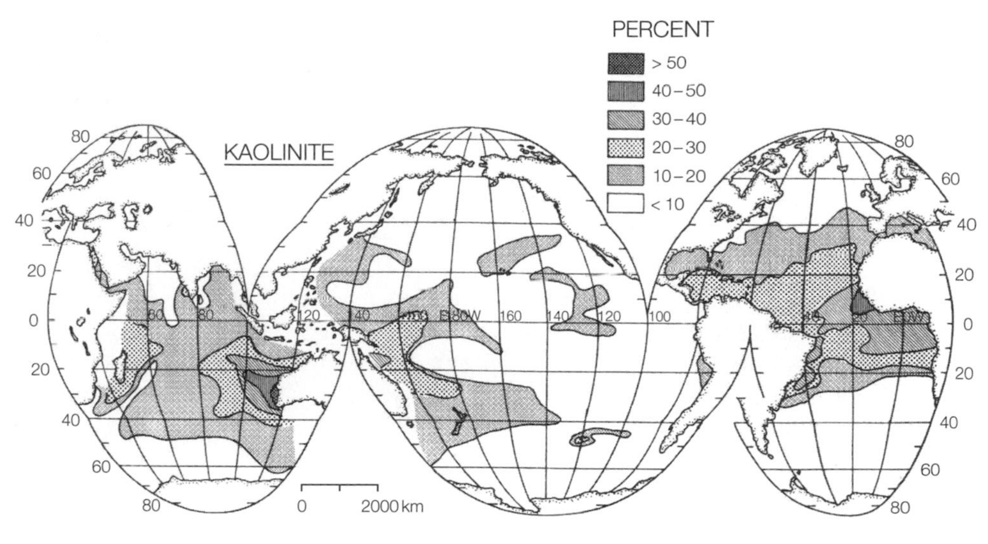

FIGURE 14.8

Distribution of kaolinite (as a percentage of the four major clay minerals) in the < 2 μm noncarbonate fraction of deep-sea sediments. *Source*: From Griffin, J. J., *et al.* (1968). *Deep-Sea Research*, 15, 433–459.

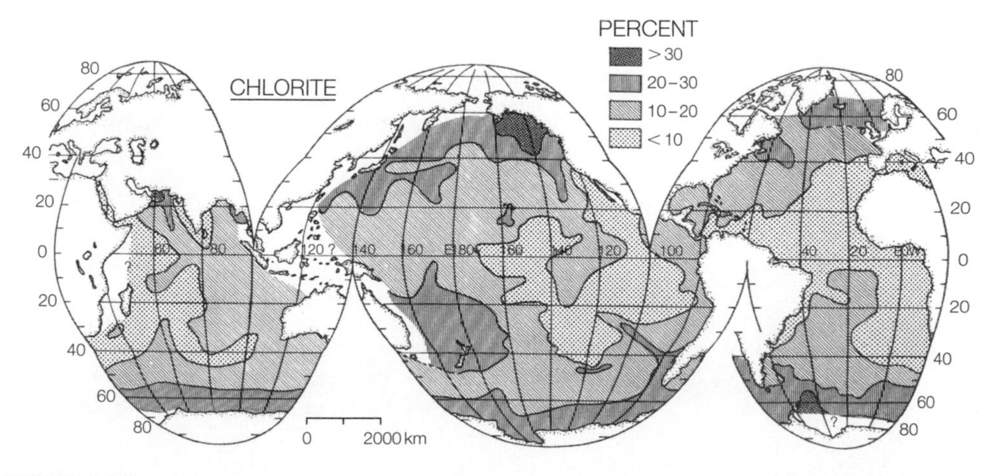

FIGURE 14.9

Distribution of chlorite (as a percentage of the four major clay minerals) in the < 2 μm noncarbonate fraction of deep-sea sediments. *Source*: From Griffin, J. J., *et al.* (1968). *Deep-Sea Research*, 15, 433–459.

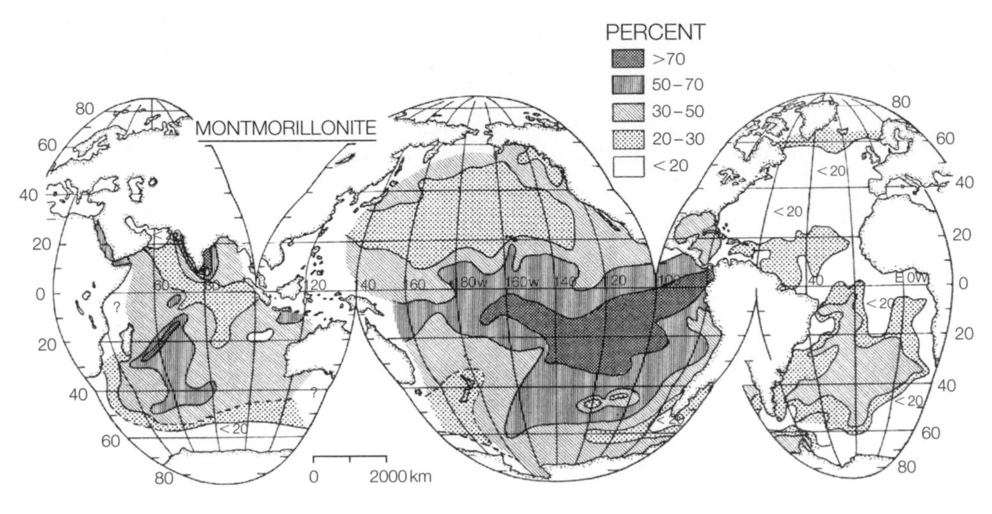

FIGURE 14.10

Distribution of montmorillonite (as a percentage of the four major clay minerals) in the < 2 μm noncarbonate fraction of deep-sea sediments. *Source*: From Griffin, J. J., *et al.* (1968). *Deep-Sea Research*, 15, 433–459.

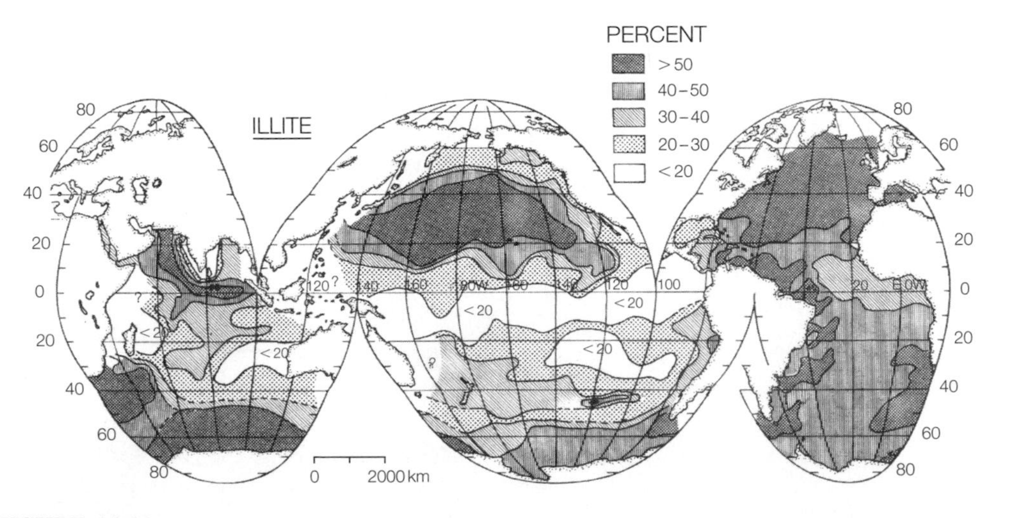

FIGURE 14.11

Distribution of illite (as a percentage of the four major clay minerals) in the < 2 μm noncarbonate fraction of deep-sea sediments. *Source*: From Griffin, J. J., *et al.* (1968). *Deep-Sea Research*, 15, 433–459.

This information is reported as the percentage that each of the clay mineral type contributes to total identifiable clay mineral content of the noncarbonate clay-sized fraction of the surface sediments. These percentages were determined by x-ray diffraction, which is unable to identify noncrystalline solids. Using this technique, clay minerals were found to comprise about 60% of the mass of carbonate-free fine-grained fraction. Most of the noncrystalline solids are probably mixed-layer clay minerals. Carbonate was removed to facilitate the x-ray diffraction characterization of the clay minerals. In some cases, round off errors cause the sum of the percentages of kaolinite, illite, montmorillonite, and chlorite to deviate slightly from 100%.

The global distribution patterns shown in these maps reflects the locations where the clay minerals were formed and the pathways by which they were transported to the open ocean. In contrast to the other particle types found in pelagic sediments, namely biogenous calcite and silica, the clay minerals undergo little chemical alteration during their trip to the sediments. Thus, the sediments with the highest clay mineral contents are located in the North Pacific Ocean, not because the input rate is unusually high to this region, but because the sedimentation rate of other particles is very low.

Table 14.3 indicates that illite is the dominant clay mineral type in all of the ocean basins with the exception of the South Pacific and Indian Oceans, where montmorillonite concentrations are highest. In these locations, montmorillonite is being supplied by volcanism associated with seafloor spreading at the mid-ocean ridges. This explains the very high montmorillonite concentrations in sediments located at the East Pacific Rise (>70%) and the 90° East Ridge (>50%) (Figure 14.10). Since the Atlantic Ocean has a smaller surface area, the contribution of terrigenous clay minerals is very high and dilutes the contribution of montmorillonite produced at the Mid-Atlantic Ridge.

The largest concentrations of chlorite are located at high latitudes (>20%), reflecting the region of its production, as well as the limit of its transport by icebergs (Figure 14.9). Chlorite is also carried to the coastal sediments via river runoff. The region of chlorite-enriched sediment immediately off the coast of East Asia suggests transport by the Yangtze and Yellow Rivers. Some chlorite appears to be of hydrothermal origin, leading to the area of high concentration around the Hawaiian islands.

As shown in Figure 14.8, kaolinite concentrations are highest in tropical and equatorial latitudes, particularly off the western coasts of North Africa and Australia (>40%) and the northeastern coasts of Australia and South America (30%). The first two are the result of aeolian transport by the Trade Winds from the Saharan and Australian deserts, respectively. The other two are the result of river input from the eastern Australian continent and the Amazon River.

The production of illite from chemical weathering occurs at all latitudes. It dominates the clay mineral assemblage in the North Atlantic and North Pacific Ocean, particularly at 40° reflecting aeolian transport by the westerlies (Figure 14.11). In the southern hemisphere, the input of illite by the westerlies is diluted by a large input of authigenic montmorillonite in the South Pacific and Indian Oceans and in the South Atlantic by a large input of kaolinite.

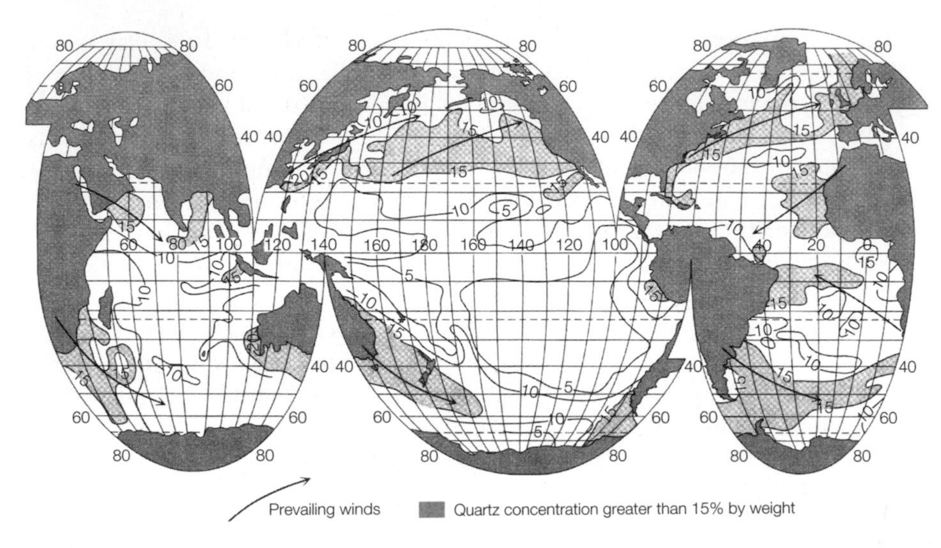

FIGURE 14.12

Carbonate and opal-free weight percent of quartz in marine sediments. *Source*: From Leinen, M., D. Cwienk, G. R. Heath, P. E. Biscaye, V. Kolla, J. Thiede, and J. P. Dauphin (1986). Distribution of biogenic silica and quartz in recent deep-sea sediments, *Geology*, 14, 199–203.

A global map of quartz abundance is given in Figure 14.12. In this case, the contribution of quartz is presented as the contribution to the bulk sediment from which biogenic carbonate and silica have been removed. This map is very similar to the global distribution of dust presented in Figure 11.4, reflecting the importance of aeolian transport for this detrital silicate.

CHAPTER

Calcite, Alkalinity, and the pH of Seawater 15

All figures are available on the companion website in color (if applicable).

15.1 INTRODUCTION

Marine organisms have had a large impact on the distribution of carbon in the crustal-ocean-atmosphere factory. This has been achieved through the burial, after death, of their organic soft parts and inorganic hard parts in marine sediments. As described in Chapter 8, burial of POC created by plankton has lead to the development and maintenance of an oxic atmosphere on Earth. In this chapter, we consider the fate of the particulate inorganic carbon (PIC). Many marine organisms create shells and skeletons out of calcium carbonate, making this mineral the primary biogenic hard part in the modern-day ocean. In comparison to detrital POC, a much larger percentage of the detrital particulate inorganic carbon (PIC) survives the trip to the seafloor to become buried in the sediments. As a result, calcium carbonate comprises the largest crustal reservoir of carbon, with more than half (55%) of the oceanic and continental slope deposits classified as calcareous oozes.

The global distribution of calcareous oozes is a consequence of the relative rates of production and preservation of biogenic calcium carbonate. Another important factor is the relative accumulation rate of other particle types because they cumulatively dilute, on a mass basis, the sedimenting calcium carbonate. This effect is seen on the continental margins where the accumulation rate of inorganic carbon is equal to that of the pelagic ocean, but calcareous oozes do not form because of dilution by large quantities of lithogenous materials.

Buried calcium carbonate is eventually transformed into sedimentary rocks, primarily limestone and dolomite. In the sediments and sedimentary rocks, downcore variations in carbonate mineral abundance have been interpreted as records of change in biogenic calcite production and burial over time. Paleoceanographers also use the naturally occurring stable isotopes of C and O in sedimentary calcite to reconstruct paleotemperatures, glacial ice volumes, and the sizes of the carbon reservoirs through geologic time. The trace metal content of detrital biogenic calcite also contains important information on past changes in the chemical composition of seawater with cadmium, barium, and boron serving as proxies for seawater phosphate, alkalinity, and pH.

Production and burial of biogenic calcite are affected by several important forces in the crustal-ocean-atmosphere factory, including climate, oceanic circulation, and atmospheric CO_2 levels. Because sedimentary biogenic calcite is such a large carbon reservoir, fluctuations in its production and burial are thought to exert important stabilizing feedbacks in the crustal-ocean-atmosphere factory. The first biomineralizing marine organisms evolved early in the Phanerozoic eon, but calcifiers did not become abundant until about 150 to 200 mybp at the end of the Permian period. This means that the stabilizing feedbacks in the modern-day crustal-ocean-atmosphere factory are a relatively recent phenomenon.

In this chapter, we consider the biogeochemical controls on the production of biogenic calcite and its burial in the sediments. This knowledge is critical to understanding how the global carbon cycle affects and is affected by global climate. Anthropogenic injection of CO_2 into Earth's atmosphere has altered some features of the oceanic carbonate system, such as seawater pH. Marine chemists and biologists are now watching closely to see how the the crustal-ocean-atmosphere factory will respond. Particular concern exists regarding the fate of surface-dwelling marine calcifiers, as observed declines in surface-water pH suggest that many species, including tropical coral reef communities, will be unable to survive a continued acidification of seawater.

15.2 THE BIOGENIC INORGANIC CARBON PUMP

Most of the calcite buried in marine sediments is of biogenic origin, having been deposited as a shell or skeletal material by organisms living in seawater (pelagic) or on the seafloor (benthic). In the modern-day ocean, about one fourth of the biogenic carbonate production occurs in the neritic environment where the most important calcifiers are benthic, e.g., coralline algae, corals, foraminiferans, mollusks, and bryozoans. The majority of the biogenic carbonate production takes place in the open ocean. The major calcifiers are planktonic protozoans (foraminferans), phytoplankton (coccolithophorids), and mollusks (pteropods).

This collective production is about 4 times larger than the river input of dissolved calcium and bicarbonate. The ocean is not running out of calcium and bicarbonate because most of the biogenic calcium carbonate dissolves following death of the calcifying organism that created it. A small fraction (25%) escapes dissolution through burial in the sediments with approximately equal amounts accumulating in the neritic and oceanic environments. The sinking of calcareous hard parts into the deep sea and sediments is such an important part of the biological pump, it has been termed the *inorganic carbon* or *carbonate pump*. The component of the biological pump caused by the sinking of particulate organic carbon is called the *soft tissue pump*.

The sedimentary sink for inorganic carbon is about equal to the river input of calcium and bicarbonate, suggesting the cycling of inorganic carbon in the modern-day ocean is in steady state. This statement will likely need to be modified at some point, because human impacts appear to be increasing the dissolution rate of biogenic carbonate. The ocean is thought to have a feedback to compensate against such perturbations. To understand how this feedback works, we must examine how calcification and dissolution alter the inorganic carbon chemistry of seawater and conversely, how the inorganic carbon chemistry of seawater controls biogenic carbonate formation and dissolution.

15.2.1 Planktonic Calcifiers

In the present-day ocean, about half of the PIC exported to oceanic sediments consists of the remains of *foraminiferans*. These microorganisms are protozoans. They are widespread in the marine environment with some species having a pelagic lifestyle and others benthic. As shown in Figure 15.1a, their calcareous structures have the appearance of a chambered snail shell and are composed of the mineral *calcite*. Since this hard part is covered by tissue, it is technically a type of skeleton. These detrital remains are referred to as *tests* or *forams*. Among the present-day and extinct species of foraminiferans, considerable variation exists in the size, shape, and density of their tests.

Second in importance to the sedimentary PIC flux are the detrital remains of *coccolithophorids*, a genus of phytoplankton. As shown in Figure 15.1b, these plants deposit calcium carbonate in plates (about 50 per cell) that overlap to create an external shell. An individual coccolithophorid will create and shed these plates on a continual basis at rate of about 1 per hour. The plates also separate from each other after death of the plant, especially if the detrital remains fall into waters that promote dissolution. These plates are referred to as *coccoliths* and have the crystalline structure of the mineral *calcite*.

Coccolithophorids are geographically widespread. They form monospecific blooms, primarily in subpolar waters, that can extend over thousands of square kilometers (Figure 15.2a). These plankton blooms are readily observable because the coccoliths reflect light and cause the surface waters to have a milky turquoise color (Figure 15.2b). Satellite imagery is now being used to detect coccolithophorid blooms and thereby estimate global production from these phytoplankton. Coccolithophorids are of particular interest as they are a significant source of gaseous DMS, which plays an important role in global climate control (Figure W23.1).

Other pelagic calcifiers include the *pteropods*, which are gastropods whose hard part is a pen-shaped internal structure made of the mineral *aragonite* (Figure 15.1c). Due to the difference in crystalline structure (Figure 15.3), aragonite is more soluble than calcite, making it a minor component of sedimentary carbonates. Some dinoflagellates deposit calcareous cysts during dormant life stages (Figure 15.1d). In terms of size, the pteropods are largest, averaging about 10 mm in length and 2 mm in width. Forams are sand-sized, with diameters on the order of 25 to 100 μm. Coccoliths and dinoflagellate cysts are part of the silt-size class with diameters of about 10 to 20 μm.

Planktonic calcifiers were not always a major component of the marine plankton. They first rose to prominence about 250 mybp at the beginning of the Mesozoic era along with the diatoms and dinoflagellates. Thus, our modern-day assemblage of marine plankton species is a relatively "new" phenomenon. The evolution and succession of these groups is thought to have been a consequence of the end-Permian mass extinction events that occurred just prior to the beginning of the Mesozoic era. As described in

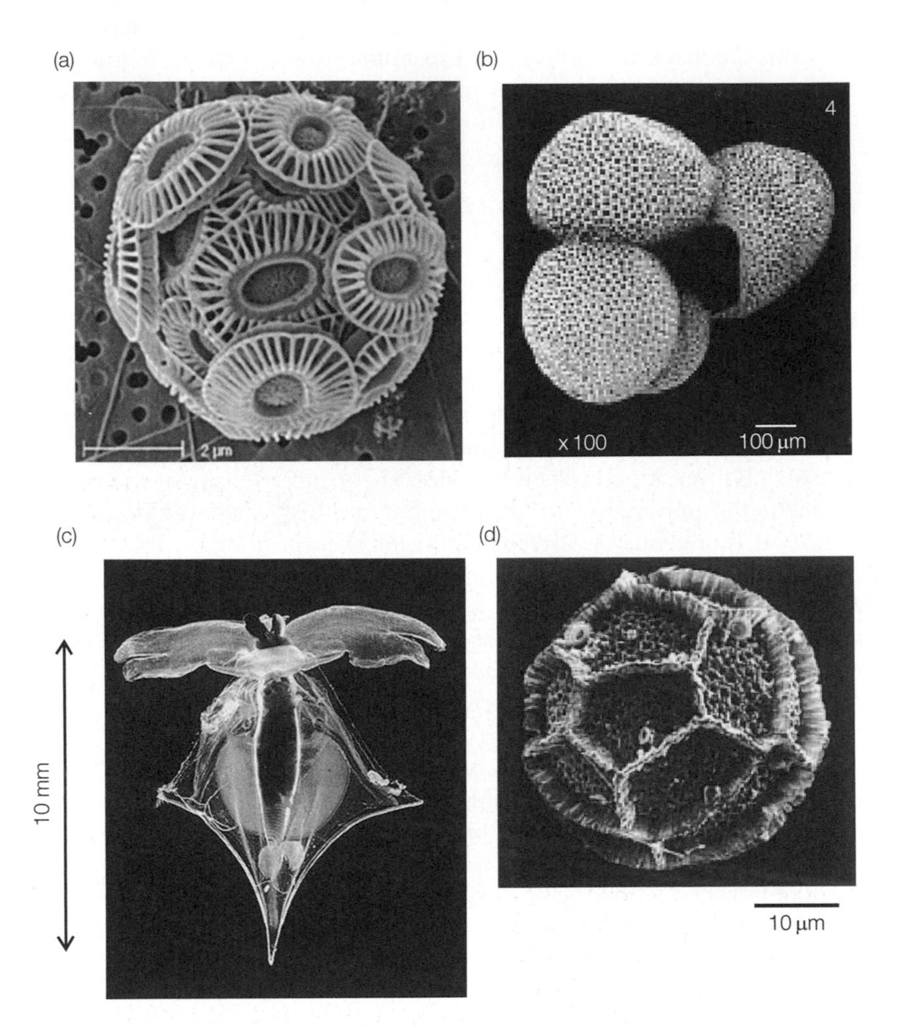

FIGURE 15.1

Planktonic calcifiers: (a) Photograph of coccolithophorid *Emiliania huxleyi*, http://earthguide.ucsd.edu/earthguide/imagelibrary/emilianiahuxleyi.html. (b) Photograph of foram of *Globigerinoides sacculifer*, © 1999–2002 by the Regents of the University of California. http://www.soton.ac.uk/~bam2/col-index/fossi-lindex/Forams/Eelco/Mediterranean/pages/sacculifer2.htm. (c) Pteropod, *Clio pyramidata*, R. W. Gilmer and G. R. Harbison, *Science* (16 July 2004) 305 (5682), cover (http://www.ipsl.jussieu.fr/~jomce/acidification/Clio_pyramidata_images.html). (d) Dinoflagellate cyst, *Calciodinellum albatrosianum*. *Source*: From: Vink, A. (2004). *Marine Micropaleontology* 50, 43–88. (See companion website for color version.)

Section 8.6.2, the Permian period ended with the largest mass extinction event that has yet occurred on planet Earth. As the ocean began a sustained recovery at the beginning of the Mesozoic era, opportunities likely abounded for the survivors to take over empty ecological niches through evolutionary adaptation. Prior to the advent of planktonic

(a)

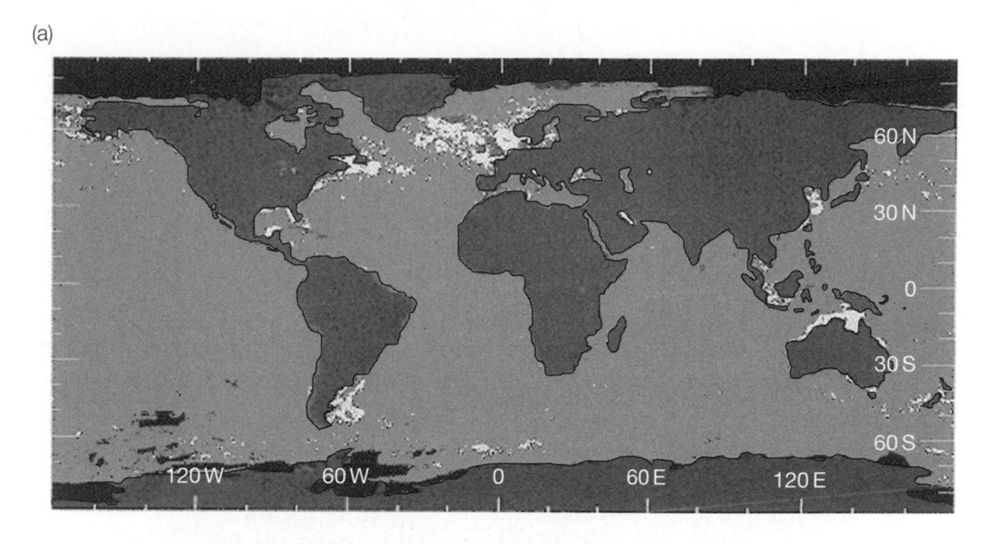

(b)

FIGURE 15.2

(a) Mission climatology of classified coccolithophorid blooms in the world's oceans. The maximum spatial extent of blooms detected during this period are displayed. Coccolithophorid bloom class = white, nonbloom class = blue, land = green, black = lack of data. *Source*: From Brown, C. W., and J. A. Yoder (1994). *Journal of Geophysical Research*, 99(C), 7467–7482. http://www.noc.soton.ac.uk/soes/staff/tt/eh/pics/cbrown.gif (b) SeaWIFS satellite image of a bloom off Newfoundland in the western Atlantic on 21 July 1999. *Source*: Provided by the SeaWIFS Project, NASA/Goddard Space Flight Center and ORBIMAGE. http://www.noc.soton.ac.uk/soes/staff/tt/eh/watl.html. (See companion website for color version.)

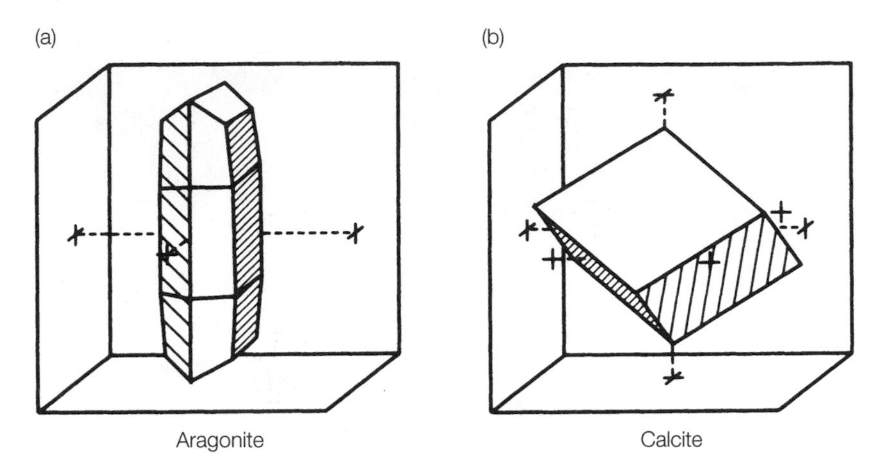

FIGURE 15.3

Crystalline forms of (a) aragonite and (b) calcite. *Source*: From Horne, R. A. (1969). *Marine Chemistry*, John Wiley & Sons, Inc., p. 214. After Berry, L. G., *et al*. (1983). *Mineralogy*, 2nd ed., W. H. Freeman and Co., pp. 330, 340.

calcifiers, calcium carbonate deposition was restricted to the neritic zone, where benthic calcifiers have been present since the beginning of the Phanerozoic eon.

15.2.2 Benthic Calcifiers

In the neritic zone, benthic deposition of biogenic calcium carbonate has led to the formation of massive calcareous deposits called *platform carbonates*. In shallow tropical waters (<50 m), hermatypic corals and green algae are responsible for the formation of reefs and platforms that collectively represent about half of the sedimentary carbonates currently forming on the continental shelves. In cold waters, massive banks of biogenous carbonate are formed by communities of mollusks, foraminiferans, echinoderms, byrozoans, barnacles, ostracods, sponges, worms, ahermatypic corals, and coralline algae. The areal extent of these deposits is shown in Figure 15.4.

About 25% of the carbonates deposited in shallow water are eventually eroded and carried downslope by bottom and turbidity currents to become part of the shelf and pelagic sediments. Shallow-water carbonates are also notable for their mineral composition. In addition to calcite and aragonite, some shallow-water calcifiers deposit hard parts containing high percentages of magnesium. These are referred to as magnesium-rich calcites.

These shallow-water deposits were the sole source of biogenic carbonate until the evolution and proliferation of planktonic calcifiers, namely the coccolithophorids and the foraminiferans, around 250 mybp. This enabled a shift in the site of sedimentary carbonate accumulation from the shallow waters to the deep sea. At present, about half of the sedimentary carbonates are being buried on the shelves and the other half in the deep sea and slopes.

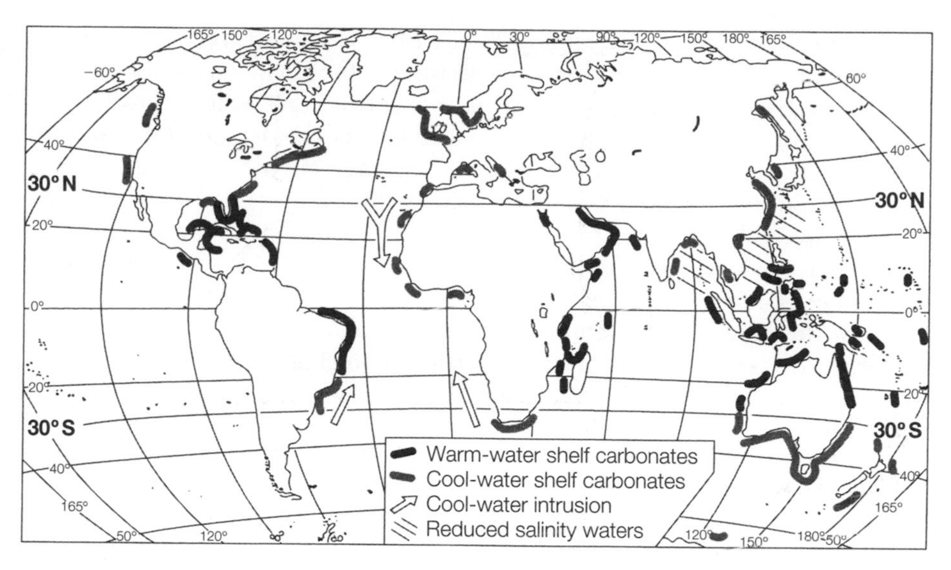

FIGURE 15.4

Distribution of warm- and cool-water shelf carbonate in the modern-day ocean. *Source*: From Mackenzie, F. T., and A. Lerman (2006). *Carbon in the Geobiosphere: Earth's Outer Shell.* Springer-Verlag, p. 283.

15.2.3 Biomineralization of Calcium Carbonate

In terms of organic carbon generation, the coccolithophorids are a minor player, representing only 6 to 8% of global marine primary production. But their detrital remains contribute disproportionately to the burial of carbon in marine sediments. This is due to near complete loss of POC via remineralization as the detrital hard and soft parts settle to the seafloor. As estimated from Broecker's Box model in Chapter 9, only about 1% of the POM that sinks out of the surface water is buried in marine sediments. In comparison, about 20% of the biogenic PIC survives to become buried in the sediments.

Coccolithophorids generate a lot of calcite because they create POC and PIC in a 1-to-1 ratio. The overall stoichiometry for these concurrent processes is shown next, illustrating that calcite deposition generates CO_2, which is then used to create organic matter.[1]

$$\text{Biogenic Calcite Deposition: } Ca^{2+}(aq) + 2HCO_3^-(aq) \rightleftharpoons CO_2(aq) + CaCO_3(s) + H_2O(l) \quad (15.1)$$

$$\text{Organic Matter Formation: } CO_2(aq) + H_2O(l) \rightleftharpoons \text{“}CH_2O\text{”}(s) + O_2(g) \quad (15.2)$$

$$\text{Overall: } Ca^{2+}(aq) + 2HCO_3^-(aq) \rightleftharpoons \text{“}CH_2O\text{”}(s) + CaCO_3(s) + O_2(g) \quad (15.3)$$

[1] Ambient concentrations of CO_2 are very low and usually biolimiting. Hence phytoplankton generally rely on bicarbonate as their carbon source. Phytoplankton must convert this bicarbonate to CO_2 to enable production of organic matter. This conversion is facilitated by the Zn-containing enzyme, carbonic anhydrase (Table 11.4). Some phytoplankton release carbonic anhydrase into seawater with the resulting CO_2 then transported across their cell membrane.

In the surface waters the ratio of POC to PIC is 12 to 1, reflecting the overall production ratio in marine plankton. This illustrates the relatively minor contribution to the POC by coccolithophorids, which generate POC to PIC in a 1-to-1 ratio. By 1000 m, the sinking detritus has a ratio very close to that of coccolithophorids. This is also the same ratio observed in the very surface sediments. This leveling effect, in which the POC is burned off until a constant POC-to-PIC ratio is reached, is thought to reflect a protective role provided by the biogenic carbonate. The mode of protection appears twofold. First, organic residues attached or adsorbed to the sinking detrital tests, skeletons, or shells get the benefit of a fast trip to the seafloor. The faster the particle sinks, the greater the likelihood that the organic matter is not remineralized prior to burial. The second seems to be mechanical shielding against microbial attack. This is a two-way street, as adsorbed materials can also protect the sinking carbonates from dissolution.

Given this relationship between POC and PIC, any change in their relative abundance would be expected to affect their export efficiency, i.e., the percentage of the production flux that reaches the seafloor. For example, an increase in the coccolithophorid population relative to other phytoplankton would be expected to lead to an increased export efficiency of POC and PIC. Such changes have the potential to affect the CO_2 content of the ocean and hence the atmosphere and thereby alter climate.

The degree to which detrital biogenic carbonate is preserved during its trip to the seafloor is therefore a particularly important matter. The factors that control this are complicated and powerful as reflected in the global distribution of calcareous oozes. As shown in Figure 15.5, calcite-rich sediments are not found beneath surface waters where overlying production rates are highest. Rather, the distribution of calcareous oozes bears a greater relationship to the topography of the seafloor with the highest calcite concentrations occurring in the sediments that lie on topographic highs, such as the mid-ocean ridges and rises. To explain this, we must consider all the factors that control the preservation of calcium carbonate during its trip to the seafloor and subsequent burial in the sediments.

15.3 CALCIUM CARBONATE DISSOLUTION

The PIC sinking out of the surface waters fall into three categories: (1) hard parts shed by marine organisms, (2) hard parts left after death, and (3) living organisms. In the case of the latter, living plankton have been observed in sinking fecal pellets and marine snow. All of this PIC, as well as the PIC lying on top of and buried within the sediments, is subject to dissolution if the ambient seawater is undersaturated with respect to whatever mineral phase of the calcium carbonate is present. In this section, the thermodynamic and kinetic controls on PIC dissolution (and formation) are discussed. We start with thermodynamic considerations because they determine the overall tendency of PIC toward dissolution or precipitation. This is followed by a discussion of kinetic constraints that, in some cases, prevent attainment of equilibrium.

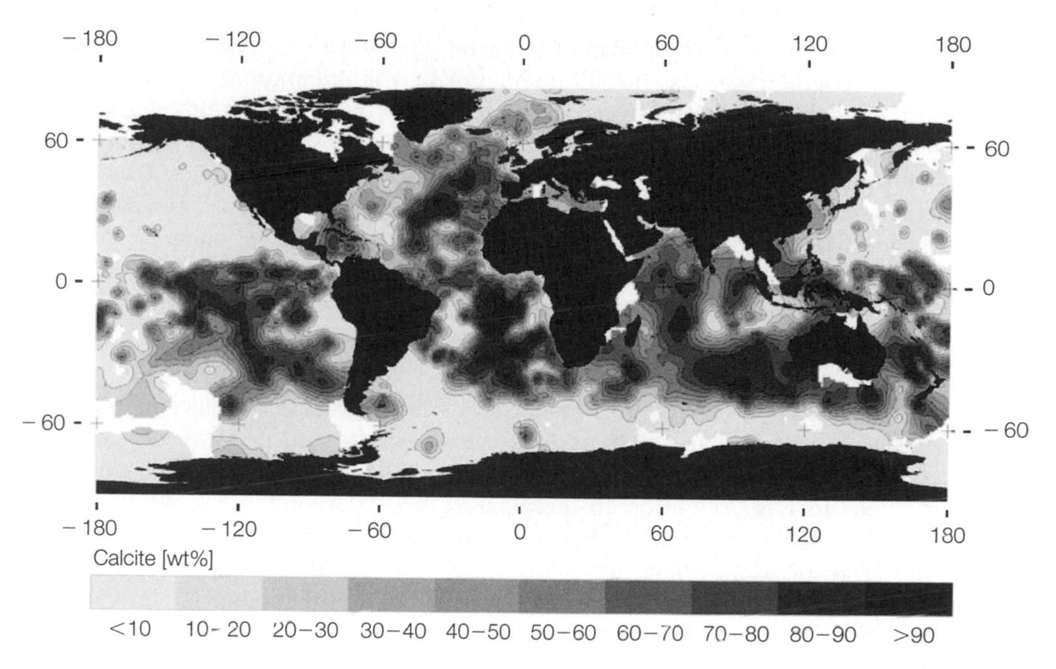

FIGURE 15.5

Calcite content of surface sediments in wt%. *Source*: From Seiter, K., *et al.* (2004). *Deep-Sea Research*, 52, 2001–2026. (See companion website for color version.)

15.3.1 Thermodynamic Controls on Solubility

The amount of calcium carbonate that will spontaneously dissolve in water if thermodynamic equilibrium is attained is governed by the following reaction:

$$CaCO_3(s) \rightleftharpoons Ca^{2+}(aq) + CO_3^{2-}(aq) \tag{15.4}$$

At equilibrium, the rate of calcium carbonate dissolution is equal to the rate of its precipitation. The concentrations of the reactants and products remain constant over time, so no further net dissolution occurs. Since the solution can dissolve no more calcium carbonate, it is said to be saturated. The K_{sp} for this reaction is given by

$$K_{sp} = \{Ca^{2+}\}_{saturated} \times \{CO_3^{2-}\}_{saturated} \tag{15.5}$$

where $\{Ca^{2+}\}_{saturated}$ and $\{CO_3^{2-}\}_{saturated}$ are the activities of Ca^{2+} and CO_3^{2-} at equilibrium.

Marine chemists measure ion concentrations rather than activities, so the solubility product for calcium carbonate is usually defined as

$$K_{sp}^{*} = [Ca^{2+}]_{saturated} \times [CO_3^{2-}]_{saturated} \tag{15.6}$$

This stoichiometric solubility product, or K_{sp}^{*}, incorporates specific and nonspecific effects resulting from solute-solute and solute-solvent interactions. As noted in

Chapter 5, nonspecific interactions can be separately accounted for through the use of activity coefficients and specific interactions via multicomponent competitive complexation calculations. Both effects are important to the solubility of calcite and aragonite, acting to greatly enhance the solubility in seawater as compared to pure water. In the case of the specific effects, this enhancement arises from the significant degree of ion pairing exhibited by carbonate and to a lesser extent by calcium. At a salinity of 35, 86% of the carbonate is ion paired, mostly with magnesium and secondarily with calcium and sodium. For calcium, 11% is ion paired, primarily with sulfate. The combined result of the specific and nonspecific effects causes calcite to be 100 times more soluble in seawater than in freshwater. (The K^*_{sp} of calcite in seawater of 35‰ at 25°C and 1 atm is $10^{-6.4}$, whereas the solubility product in pure water at the same temperature and pressure is $10^{-8.4}$.) For every 1 unit (ppt) increase in salinity, calcite solubility increases 3%.

Calcium carbonate solubility is also temperature and pressure dependent. Pressure is a far more important factor than temperature in influencing solubility. As illustrated in Table 15.1, a 20°C drop in temperature boosts solubility 4%, whereas the pressure increase associated with a 4-km increase in water depth increases solubility 200-fold. The large pressure effect arises from the susceptibility of the fully hydrated divalent Ca^{2+} and CO_3^{2-} ions to electrostriction. Calcite and aragonite are examples of minerals whose solubility increases with decreasing temperature. This unusual behavior is referred to as *retrograde solubility*. Because of the pressure and temperature effects, calcium carbonate is far more soluble in the deep sea than in the surface waters (See the online appendix on the companion website).

Table 15.1 Stoichiometric Equilibrium Constants ($-\log K^*_{sp}$) for Calcite and Aragonite Solubility as a Function of Temperature and Pressure Where K^*_{sp} Has Units of $(mol/kg)^2$ (S = 35)

Calcite	Temperature (°C)				
	24°C	**10°C**	**4°C**	**2°C**	**0°C**
0 m	6.369	6.365	6.366	6.367	6.368
1000 m	6.300	6.288	6.285	6.283	6.282
6000 m	5.983	5.937	5.912	5.903	5.734
Aragonite	**Temperature (°C)**				
	24°C	**10°C**	**4°C**	**2°C**	**0°C**
0 m	6.187	6.169	6.166	6.166	6.166
1000 m	6.123	6.097	6.090	6.088	6.086
6000 m	5.830	5.771	5.743	5.734	5.724

Source: *After Sarmiento, J. L., and N. Gruber (2006).* Ocean Biogeochemical Dynamics. *Princeton University Press.*

Most water masses are not at equilibrium with respect to either calcite or aragonite. The degree to which a water mass deviates from equilibrium for a particular mineral type can be expressed as its degree of saturation (Ω), which is defined as:

$$\Omega = \frac{[Ca^{2+}]_{observed} \times [CO_3^{2-}]_{observed}}{[Ca^{2+}]_{saturation} \times [CO_3^{2-}]_{saturation}} = \frac{\text{Ion product}}{K_{sp}^*} \quad (15.7)$$

where $[Ca^{2+}]_{observed}$ and $[CO_3^{2-}]_{observed}$ are the in situ concentrations in the water mass of interest. Since aragonite is more soluble than calcite, $\Omega_{calcite}$ is always greater than $\Omega_{aragonite}$ for a given water mass.

If Ω is greater than 1, the water mass is supersaturated and calcium carbonate will spontaneously precipitate until the ion concentrations decrease to saturation levels. When Ω is less than 1, the water mass is undersaturated. If calcium carbonate is present, it will spontaneously dissolve until the ion product rises to the appropriate saturation value. Although calcium is a biointermediate element, it is present at such high concentrations that PIC formation and dissolution causes its concentration to vary by less than 1%. Thus, $[Ca^{2+}]_{observed} \approx [Ca^{2+}]_{saturation}$ and Eq. 15.7 can be simplified to

$$\Omega = \frac{[CO_3^{2-}]_{observed}}{[CO_3^{2-}]_{saturation}} \quad (15.8)$$

The degree of saturation is also commonly expressed as a saturation index $(SI) = \log \Omega$ and as excess carbonate ion concentration, $\Delta CO_3^{2-} = [CO_3^{2-}]_{observed} - [CO_3^{2-}]_{saturation}$. Supersaturated solutions have an SI and ΔCO_3^{2-} greater than 0. Tabulated values of the carbonate saturation concentrations are provided in Table 15.2. Because of the physicochemical effects of temperature and pressure, the carbonate saturation concentrations increase with increasing depth. Because aragonite is more soluble, its saturation $[CO_3^{2-}]$ is always higher than that of calcite. These data are plotted in Figure 15.6, which clearly illustrates the greater importance of pressure over temperature in determining calcium carbonate solubility for both calcite and aragonite.

All of these measures of saturation are very useful in predicting the geographic distribution of sedimentary carbonates. For example, sediments lying below waters that are undersaturated with respect to calcite should be devoid of calcareous oozes. Since direct measurement of $[CO_3^{2-}]_{observed}$ is difficult, its concentration is usually computed from two more easily measured parameters, the carbonate alkalinity and pH of a seawater sample.

Calculation of Carbonate Ion Concentrations from Measurements of pH and Alkalinity

Dissolved inorganic carbon (DIC) is present in seawater as the following species: carbon dioxide (CO_2), carbonic acid (H_2CO_3), carbonate (CO_3^{2-}), and bicarbonate (HCO_3^-). Since each of these molecules contains one atom of carbon, the *total DIC concentration*

Table 15.2 Saturation Concentrations of Carbonate Ion in Seawater (μmol/kg) as a Function of Temperature and Pressure. (S = 35)

Calcite	**Temperature (°C)**				
	24°C	**10°C**	**4°C**	**2°C**	**0°C**
0 m	41.5	41.9	41.8	41.7	41.6
1000 m	48.7	50.0	50.5	50.6	50.7
6000 m	101.2	112.4	119.0	121.5	124.2
Aragonite	**Temperature (°C)**				
	24°C	**10°C**	**4°C**	**2°C**	**0°C**
0 m	63.2	65.9	66.3	66.3	66.3
1000 m	73.3	77.8	79.1	79.4	79.8
6000 m	143.9	164.6	175.5	179.5	183.7

Source: After Sarmiento, J. L., and N. Gruber (2006). Ocean Biogeochemical Dynamics. *Princeton University Press.*

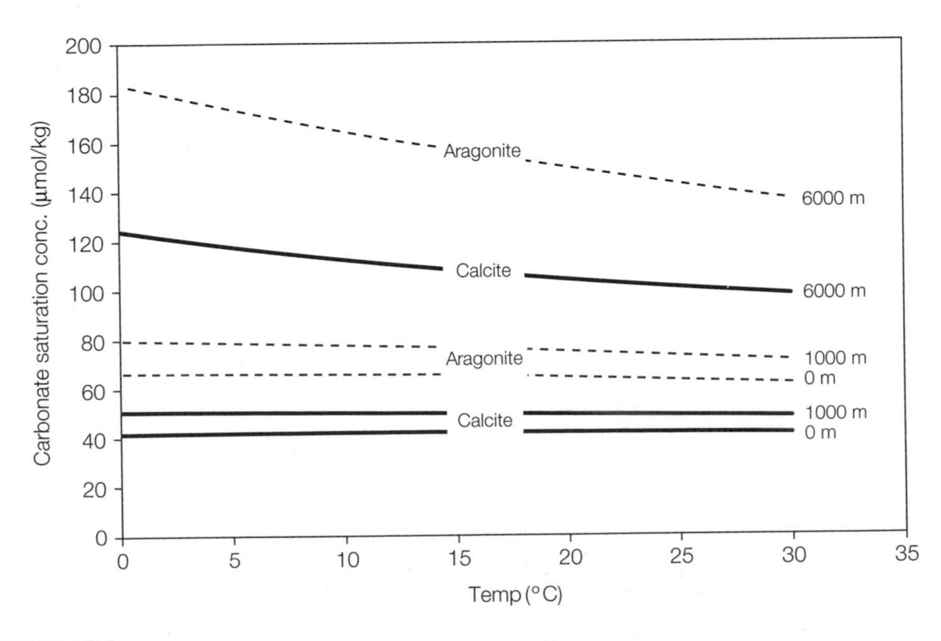

FIGURE 15.6

Saturation concentrations of carbonate ion in seawater as a function of temperature and pressure.

Table 15.3 Oceanic Carbon Inventory.

Species	Pg C	% Inventory	References
Bicarbonate	33,803	87.1%	1
Carbonate	4,056	10.5%	1
Carbonic Acid + Carbon Dioxide	193	0.5%	1
DOC	725	1.9%	2
Detrital POC	30	0.1%	3
Biota POC	3	0.01%	2
PIC	0.2	0.001%	4

1 Pg = 10^{15} g.
[1]*Computed from mean global TDIC concentration and ion speciation ratios. Source: After Sarmiento, J.L. and Gruber, N. (2006) Ocean Biogeochemical Dynamics. Princeton University Press.*
[2]*Source: After Pacific Marine Environmental Laboratory, PMEL Carbon Group, Global Carbon Cycle, http://www.pmel.noaa.gov/co2/gcc.html, accessed July 2008.*
[3]*Source: After Mackenzie, F.T. and A. Lerman (2006)* Topics in Geobiology, *Volume 25, Springer.*
[4]*Computed from biota POC assuming that 6 to 8% of plankton are calcifiers and produce PIC in a 1:1 ratio with POC.*

(symbolized variously as TDIC, ΣCO_2, or C_T) is given by the following mass balance equation:

$$\sum CO_2 = [CO_2] + [H_2CO_3] + [HCO_3^-] + [CO_3^{2-}] \tag{15.9}$$

As shown in Table 15.3, most of the inorganic carbon in seawater is in the forms of bicarbonate and carbonate, so Eq. 15.9 can be simplified to

$$\sum CO_2 = [HCO_3^-] + [CO_3^{2-}] \tag{15.10}$$

ΣCO_2 is measured by acidifying a seawater sample. This converts all the bicarbonate and carbonate ion into CO_2. The CO_2 gas is stripped out of the water sample by bubbling with an inert gas and then swept through either a coulometer or infrared analyzer for quantification. Alternatively, ΣCO_2 can be calculated from two easily measurable concentrations, pH and alkalinity. (See the online appendix on the companion website for the equations.)

Alkalinity is measured by titrating a seawater sample with a monoprotic strong acid, such as HCl. The amount of acid (H^+) required to consume all the titratable charge is operationally defined as a sample's *total alkalinity (T.A.)*. This is reported as a concentration in units of moles of charge per kg (or L) water. A mol of charge is also referred to as an *equivalent of charge* and thus alkalinity is often expressed in units of milliequivalents (meq) per kg (or L) seawater. The total alkalinity of seawater ranges from 2.3 to 2.6 mmol charge per kg (or L) seawater and can be determined to a precision of 0.1%.

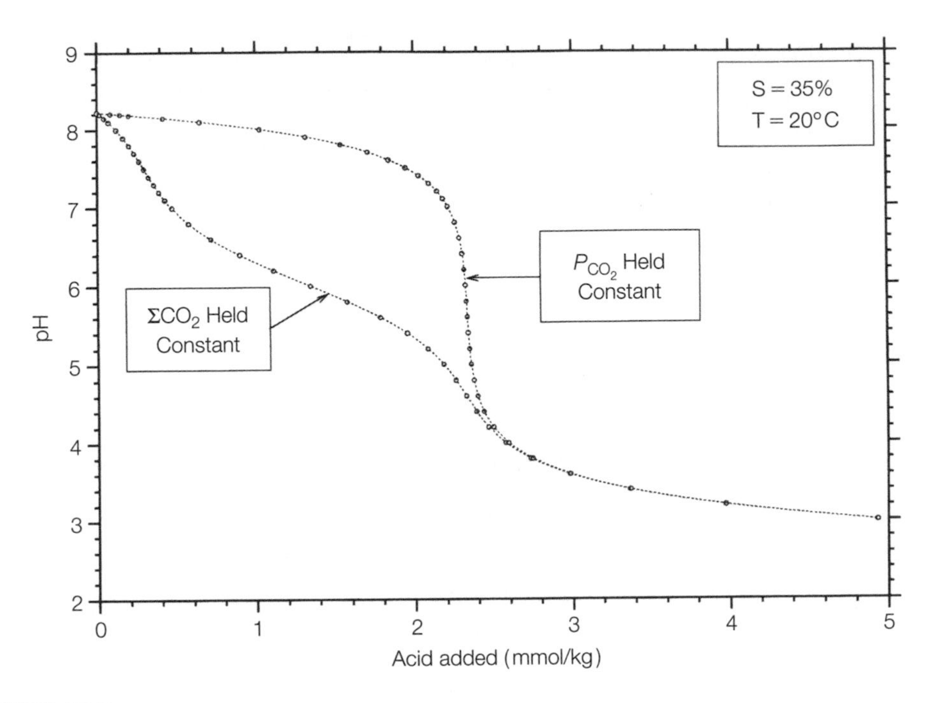

FIGURE 15.7

Titration curve for seawater. The shape of the curve is dependent upon experimental conditions. The top curve is produced when seawater is titrated in an open container so that CO_2 generated after incremental acid addition can escape into the atmosphere. The bottom curve is generated when seawater is titrated in a closed container. In this case, the pH drops faster during the initial part of the titration because of the build-up of CO_2 as acid is added. Once the carbonate/carbonic acid equivalence point is reached, both curves converge upon the same pH for the same volume of acid added, but extensive laboratory work has demonstrated that better accuracy is achieved with the closed container method. *Source*: From Pilson, M. E. Q. (1998). *An Introduction to the Chemistry of the Sea*. Prentice-Hall, p. 119.

Most of the titratable charge in seawater is supplied by bicarbonate because its concentration is much greater than that of carbonate or any of the other weak bases in seawater, such as $B(OH)_4^-$. A typical acid titration curve for a seawater sample is shown in Figure 15.7. If the titration is performed in an open container, initial addition of acid does not cause much of a drop in pH. During this phase of the titration, H^+ is readily consumed, first by carbonate (Eq. 5.57) and then by bicarbonate (Eq. 5.56). Most of the buffering is provided by bicarbonate because of its high concentration. Once most of the bicarbonate has been consumed, further addition of acid causes a rapid decline in pH.

In practice, the alkalinity of seawater is determined by titrating a sample until the pH drops to 3, well below the bicarbonate equivalence point (pH ~4.5). Although most of the titratable negative charge is contributed by bicarbonate and carbonate, the other weak bases present in seawater do consume some acid above and below the bicarbonate equivalence point. The portion of the total alkalinity that is attributable to carbonate and bicarbonate is termed the *carbonate alkalinity*. It is defined as

$$\text{Carbonate alkalinity} = \text{C.A.} = 2\left[CO_3^{2-}\right] + \left[HCO_3^-\right] \quad (15.11)$$

where each mole of carbonate ion contributes two units of titratable charge.

The carbonate ion concentration can be estimated from the following expression

$$\left[CO_3^{2-}\right] = \text{C.A.} - \sum CO_2 \quad (15.12)$$

which is derived by combining Eqs. 15.10 and 15.11. This expression is based on the assumption that $\left[CO_2\right] + \left[H_2CO_3\right] \approx 0$. Because a small amount of CO_2 and H_2CO_3 are present in seawater, this assumption introduces a small (<10%) error in the computed CO_3^{2-}. Thus, Eq. 15.12 is used to provide a quick estimate of $\left[CO_3^{2-}\right]$ from C.A. and ΣCO_2 data.

A more accurate computation of $\left[CO_3^{2-}\right]$ can be obtained using the competitive equilibrium complexing approach described in Chapter 5.7. The details of this calculation for the carbonate system are provided in the online appendix on the companion website including semiempirical equations for estimating equilibrium constants at the temperature, salinity, and pressure of interest. As noted earlier, these speciation calculations also enable computation of ΣCO_2 from a sample's pH and C.A. The effects of temperature, salinity, and pressure on the equilibrium speciation of carbonate, bicarbonate, and CO_2 are illustrated in Figure 15.8. Note the decline in carbonate ion concentration with decreasing temperature. This is due to the effect of temperature on the equilibrium constants for Eqs. 5.56 and 5.57 as given by Eqs. A14.21 and A14.22. This temperature effect is partly responsible for the retrograde solubility behavior of calcite and aragonite. In practice, computer programs, such as MINEQL, are used to compute the inorganic carbon speciation and mineral saturation indices of a seawater sample from its pH and total alkalinity.

More about Alkalinity

As mentioned above, several other weak bases can consume acid during the alkalinity titration of seawater. In order of decreasing ability to react with H^+, they include: OH^-, PO_4^{3-}, $H_3SiO_4^-$, $B(OH)_4^-$, NH_3, $H_2SiO_4^{2-}$, and HPO_4^{2-}. Organic acids can also contribute alkalinity, but are usually present in very low concentrations. By including the contributions of all these species, the total alkalinity of oxic seawater is defined as:

$$\begin{aligned}\textit{Total alkalinity} = \text{T.A.} = {} & \left[HCO_3^-\right] + 2\left[CO_3^{2-}\right] + \left[B(OH)_4^-\right] + \left[HPO_4^{2-}\right] + 2\left[PO_4^{3-}\right] + \\ & \left[H_3SiO_4^-\right] + 2\left[H_2SiO_4^{2-}\right] + \left[NH_3\right] - \left[H^+\right] + \left[OH^-\right] + \\ & \left[\text{other conjugate bases of weak acids}\right] \end{aligned} \quad (15.13)$$

The contribution of a species to T.A. reflects both its concentration in seawater and its relative reactivity towards H^+. The contribution of the conjugate bases that have two moles of titratable negative charge, such as PO_4^{3-} and $H_2SiO_4^{2-}$, is twice their ion

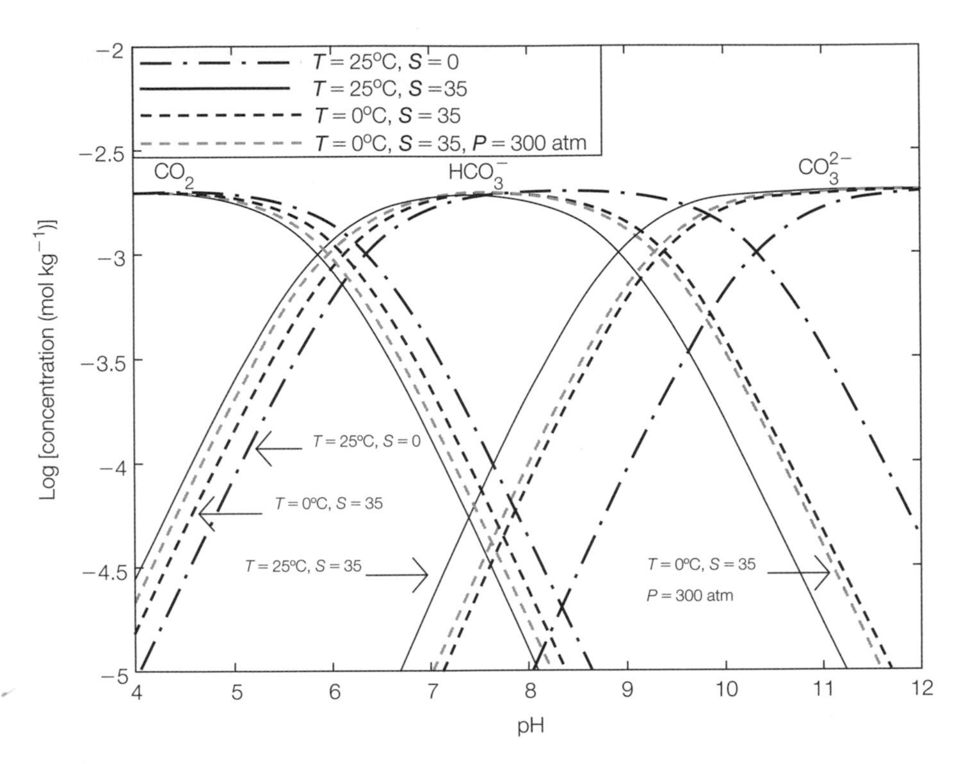

FIGURE 15.8

Effect of temperature, pressure, and salinity on speciation of the dissolved inorganic carbon for ΣCO_2 = 2 mmol/kg. *Source*: After Zeebe, R.E. and D. Wolf-Gladrow (2001) Elsevier Oceanography Series, 65, Elsevier, p. 10.

concentration.[2] In Eq. 15.13, the ambient H^+ concentration is subtracted as its positive charge masks a portion of the sample's titration alkalinity.

Over 96% of the T.A. is contributed by the C.A. Most of the difference between T.A. and C.A. is caused by $B(OH)_4^-$, which contributes 3% of the T.A. In practice, T.A. is converted to C.A. by simply subtracting off $[B(OH)_4^-]$, which, being a nearly conservative ion, is estimated from the sample's salinity. (See the online appendix on the companion website for details.)

Changes in phosphate, nitrate, ammonia, and silicate concentrations associated with the biogenic production and destruction of POM can alter seawater alkalinities. These effects are usually so small in scale that they can be ignored. Since the largest biotic impact on alkalinity in oxic seawater is exerted by the formation and dissolution of

---

[2] During the standard technique used to measure T.A., the addition of acid is halted at a pH that is too high to drive the titration of $H_2PO_4^-$ to H_3PO_4. Therefore, during an alkalinity titration, any HPO_4^{2-} present in the sample will consume only 1 mol of acid whereas PO_4^{3-} will consume 2 mol of acid.

calcium carbonate, we will assume that only these processes are responsible for the geographic and temporal variations in alkalinity.

More detailed information on alkalinity is provided in the supplemental content for Chapter 15 at **http:/elsevierdirect.com/companions/9780120885305**.

15.3.2 Kinetic Controls on Solubility

Observations from deep-water sediment traps have demonstrated that PIC is present in waters that are undersaturated with respect to this mineral. Thus, thermodynamic considerations are not a perfect predictor of the presence of PIC. In other words, some PIC is present out of equilibrium with the seawater it is in. This is largely a result of kinetics in which dissolution is slow enough to enable PIC to persist for some time. Much effort has been applied to determining the factors that control the rate of PIC dissolution. Marine scientists have reached agreement that the rate law for $CaCO_3$ dissolution in undersaturated waters ($\Omega < 1$) can be represented as:

$$\frac{d[CaCO_3]}{dt} = -[CaCO_3]\,k_{CaCO_3}(1-\Omega)^n \qquad (15.14)$$

where $[CaCO_3]$ is the concentration of $CaCO_3$ in units of mmol/m^3, k_{CaCO_3} is the dissolution rate constant in units of d^{-1}, and n is an apparent reaction order. Current thinking is that $n = 1$, making the dissolution rate directly proportional to the degree of undersaturation. Values for k_{CaCO_3} as observed in laboratory experiments are on the order of $0.38\,d^{-1}$ for PIC in seawater and $0.0001\,d^{-1}$ for sedimentary calcium carbonate. In situ values are likely to depend on temperature, particle diameter (wall thickness), mineral composition, and coatings.

This rate law suggests that dissolution should be relatively slow compared to the time that detrital PIC takes to settle to the seafloor. But recent observations indicate that a significant amount of dissolution occurs high in the water column, even in saturated waters. Dissolution under saturated conditions is thought to be a consequence of PIC exposure to metabolic CO_2 in acidic microenvironments such as found in zooplankton guts and feces and within aggregations of marine snow.

15.4 THE EFFECT OF POC AND PIC FORMATION AND DEGRADATION ON THE pH, CARBONATE ALKALINITY, AND $\sum CO_2$ OF SEAWATER

Temporal and geographic variations in C.A. and ΣCO_2 are largely controlled by the formation and degradation of planktonic POC and PIC. ΣCO_2 is supplied to seawater during the remineralization of planktonic POC and dissolution of PIC. Formation of biomass and calcareous hard parts requires the removal of ΣCO_2. Thus, vertical profiles of ΣCO_2 exhibit vertical segregation characteristic of biolimiting elements with low concentrations in the surface waters and high concentrations at depth (Figure 10.2).

The formation and degradation of planktonic POC and PIC influence pH and C.A. as follows. The remineralization of POC produces CO_2, which is rapidly hydrolyzed to carbonic acid, bicarbonate, and carbonate via the reactions given in Eqs. 5.53 through 5.57. Carbonic acid and bicarbonate are both weak acids, so their dissociation generates H^+. This acid enhances the dissolution of PIC through the following reaction:

$$CO_2(aq) + CaCO_3(s) + H_2O(1) \rightleftharpoons Ca^{2+}(aq) + 2HCO_3^-(aq) \qquad (15.15)$$

in which carbonate alkalinity is produced. As illustrated in Eq. 15.3, the formation of planktonic PIC reduces the alkalinity and ΣCO_2 of seawater.

In the case where POC is remineralized in the absence of PIC, the C.A. does not change because the hydrolysis of CO_2 produces one unit of positive charge (H^+) for each unit of negative charge (HCO_3^-). Likewise the formation of planktonic POC reduces the ΣCO_2 of seawater but has no effect on C.A.

Over time scales of thousands of years, the carbonate-bicarbonate equilibria (Eqs. 5.53 through 5.59) maintain the pH of oxic seawater within a narrow range, i.e., 7.4 to 8.2. Geographic and temporal variations in pH are largely a consequence of CO_2 removal by photosynthesis and CO_2 supply by POC remineralization. For example, the pH of seawater in the euphotic zone undergoes a diurnal cycle in which CO_2 uptake by phytoplankton causes an increase in pH during the day. At night, photosynthesis stops, while dark respiration continues, leading to release of CO_2 and an ensuing decline in pH. At depths immediately below the euphotic zone, remineralization of sinking POC releases acid in quantities that exceed the buffering capacity of the water, leading to lowered pHs.

15.4.1 Vertical Segregation of ΣCO_2 and Alkalinity

ΣCO_2 and alkalinity exhibit vertical segregation similar to that of nitrate and phosphate. An example from the North Pacific is provided in Figure 15.9. Concentrations of ΣCO_2 and alkalinity are low in surface waters because of uptake of DIC by phytoplankton, some of which is converted to POC and some to PIC. The biogenic detritus that sinks out of the surface waters is remineralized in the deep zone, causing ΣCO_2 and alkalinity to increase with depth. Because of density stratification, the resolubilized DIC and alkalinity are trapped below the thermocline.

The increase in ΣCO_2 within the thermocline is accompanied by a decrease in pH caused by the hydrolysis of CO_2 produced by the remineralization of POC. This supply is so large that significant amounts of CO_2 and H_2CO_3 are present at these depths. The $[CO_3^{2-}]$ reaches minimum values in the depths over which POC remineralization is most

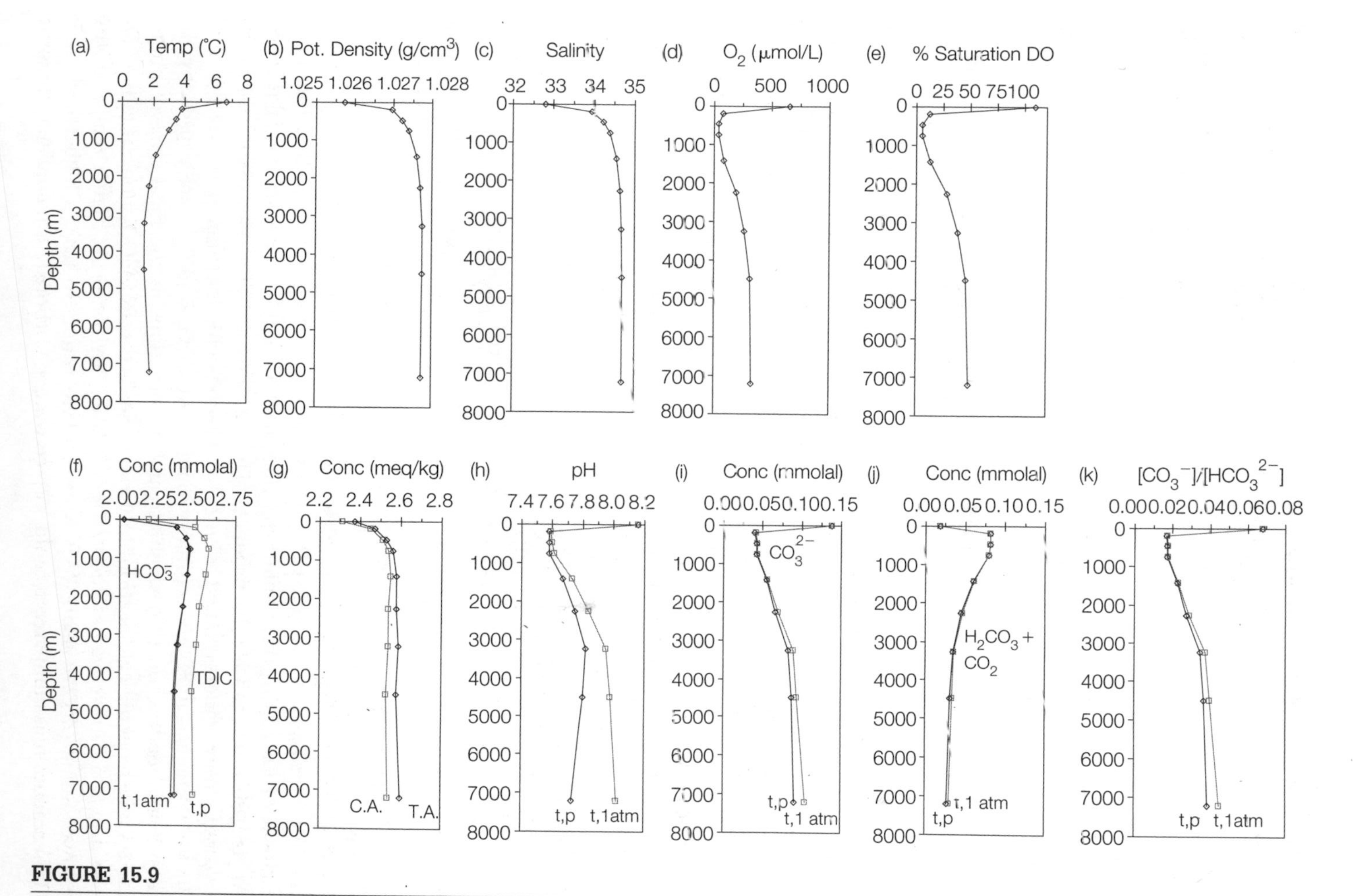

FIGURE 15.9

Vertical concentration profiles of (a) temperature, (b) potential density, (c) salinity, (d) O_2, (e) % saturation of O_2, (f) bicarbonate and TDIC, (g) carbonate alkalinity and total alkalinity, (h) pH, (i) carbonate, (j) carbon dioxide and carbonic acid concentrations, and (k) carbonate-to-bicarbonate ion concentration ratio. Curves labeled *t,p* have been corrected for the effects of in-situ temperature and pressure on equilibrium speciation. Curves labeled t, 1 atm have been corrected for the in-situ temperature effect, but not for that caused by pressure. Data from 50°27.5'N, 176°13.8'W in the North Pacific Ocean on June 1966. *Source*: From Culberson, C., and R. M. Pytkowicz (1968). *Limnology and Oceanography*, 13, 403–417.

intense because the organic carbon is transformed into CO_2 that reacts with H_2O. Acid is generated via:

$$CO_2(aq) + H_2O(l) \rightleftharpoons H^+(aq) + HCO_3^-(aq) \tag{15.16}$$

which then reacts with CO_3^{2-}:

$$H^+(aq) + CO_3^{2-}(aq) \rightleftharpoons HCO_3^-(aq) \tag{15.17}$$

The net combined stoichiometry of these two reactions is:

$$CO_2(aq) + CO_3^{2-}(aq) + H_2O(1) \rightleftharpoons HCO_3^-(aq) \tag{15.18}$$

Thus, $[HCO_3^-]$, rather than $[CO_3^{2-}]$, increases with increasing depth. The concomitant increase in alkalinity with depth indicates that some PIC has also dissolved via the chemical reaction depicted in Eq. 15.18.

In Figure 15.9f, the ΣCO_2 concentration in the deep zone is seen to be less than that in the thermocline. This suggests that the waters in the deep zone have accumulated a smaller amount of the products of biogenic particle decomposition than have the waters in the overlying thermocline. The cause for this is twofold: (1) a smaller flux of biogenic particles reaches the deeper depths as most is remineralized in the overlying waters and (2) the water in the deep zone is "younger" than the water in the thermocline, having more recently been at the sea surface as a consequence of meridional overturning circulation. Since less remineralized CO_2 has been added to the waters of the deep zone, the ratio of $[CO_3^{2-}]$ to $[HCO_3^-]$ is higher here than in the thermocline. Eq.15.18 is shifted in favor of the reactants. Nevertheless, seawater becomes increasingly more undersaturated with respect to biogenic calcium carbonate with increasing depth due to the very large effect of pressure on the K_{sp} of calcite and aragonite.

15.4.2 Horizontal Segregation of ΣCO_2 and Alkalinity

As illustrated in Figure 15.10, both ΣCO_2 and T.A. exhibit horizontal segregation similar to that of nitrate and phosphate. As a water mass moves along the pathway of meridional overturning circulation, it accumulates biogenic particles. As the particles remineralize, ΣCO_2 and T.A is released into the water mass. Thus, the deep waters with the highest T.A. and ΣCO_2 are the oldest, i.e., those in the North Pacific ocean.

As a water mass ages, the ratio of $[CO_3^{2-}]$ to $[HCO_3^-]$ declines because the continuing generation of CO_2 from the remineralization of POC pushes the equilibrium reaction in Eq. 15.18 further toward the products. Thus, as a deep water mass ages, it becomes increasingly more undersaturated with respect to biogenic calcium carbonate.

In the surface waters, geographic variability in ΣCO_2 and T.A. are caused by the effects of temperature on CO_2 solubility and by variations in the local rates of photosynthesis and biogenic calcification. In general, surface water ΣCO_2 concentrations are lowest in warm surface waters due to the low solubility of CO_2 at higher temperatures. The lower influx of CO_2 also causes warm surface waters to have a higher carbonate ion concentration as compared to cold surface waters. Carbonate ion concentrations are

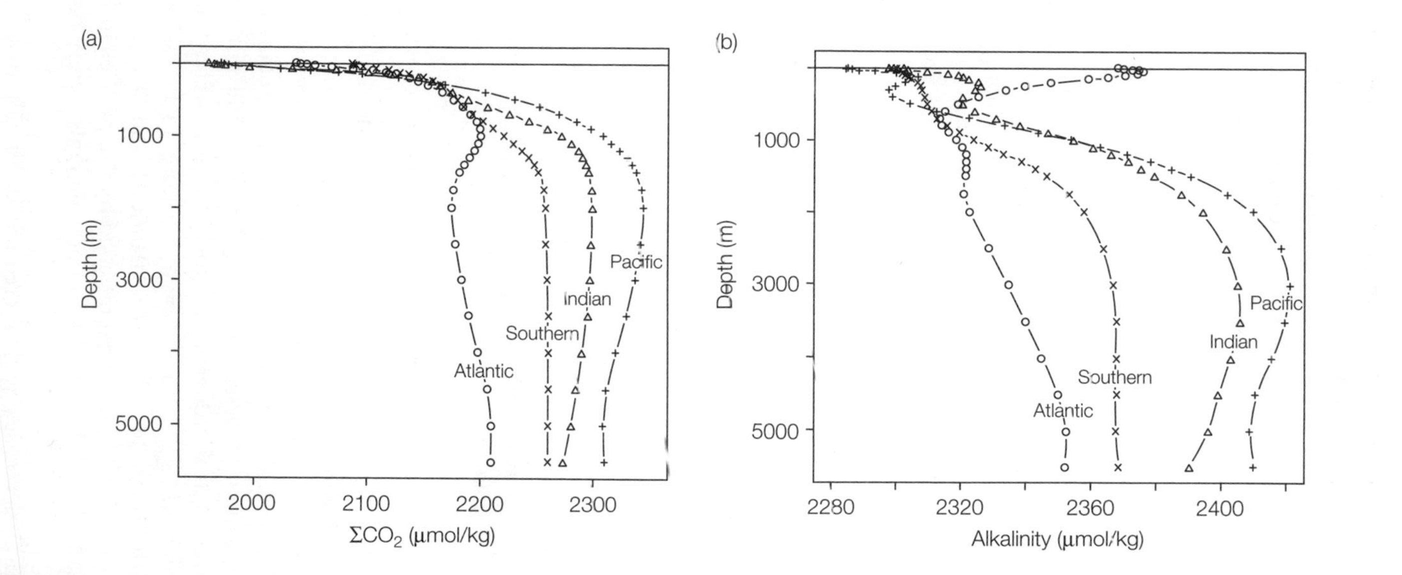

FIGURE 15.10

Geographic variations in (a) $\sum CO_2$ and (b) total alkalinity as reflected by average vertical profiles for each ocean. The averages were computed from data collected during GEOSECS, TTO, SAVE, and INDIGO expeditions. The global dataset includes results from 12,011 hydrographic stations sampled during 116 cruises that were conducted between 1972 and 1999. *Source*: After Key, R. M., *et al.* (2004) *Global Biogeochemical Cycles* 18, GB4031, p. 23.

also higher in warm waters because the K_a for bicarbonate increases with increasing temperature. As a result of these combined effects, warm surface waters have a greater degree of supersaturation with respect to biogenic calcium carbonate than do cold surface waters. Under these conditions, abiogenic calcium carbonate precipitation occurs. This leads to a phenomenon known as whitings that is described in Chapter 18.3.3.

15.5 THE PRESERVATION OF CALCIUM CARBONATE IN MARINE SEDIMENTS

In the preceding sections, we have discussed the marine processes that control calcium carbonate's formation, dissolution, and delivery to the seafloor. Their combined effects determine the geographic distribution of calcium carbonate in marine sediments seen in Figure 15.5. As noted earlier, the global distribution of calcareous sediments does not seem to follow that of plankton production. This points to the overriding importance of the processes that control the dissolution and sedimentation of calcium carbonate.

An equally important consideration in the formation of calcareous oozes is the rate of accumulation of other particles because sedimentary calcium carbonate content is expressed as $\%CaCO_3$ (w/w). Thus, geographic variability in the burial rate of noncalcareous particles can lead to geographic variability in $\%CaCO_3$. For example, continental margin sediments tend to have low $\%CaCO_3$ although biogenic calcite production is high in the overlying waters. This is a consequence of dilution by clay minerals whose source is the nearby land.

Away from the influence of the continents, the $\%CaCO_3$ in the sediments is largely controlled by the processes that determine whether sinking detrital calcium carbonate survives the trip to the seafloor. Some of these processes are related to the thermodynamic controls on calcium carbonate solubility and others are a consequence of the relative rates of particle sinking and dissolution.

15.5.1 Thermodynamic Considerations

The saturation state of seawater can be used to predict whether detrital calcite and aragonite are thermodynamically favored to survive the trip to the seafloor and accumulate in surface sediments. Any PIC or sedimentary calcium carbonate exposed to undersaturated waters should spontaneously dissolve. Conversely, PIC and sedimentary calcium carbonate in contact with saturated or supersaturated waters will not spontaneously dissolve. Typical vertical trends in the degree of saturation of seawater with respect to calcite ($\Omega_{calcite}$) and aragonite ($\Omega_{aragonite}$) are shown in Figure 15.11 for two sites, one in the North Pacific and the other in the North Atlantic Ocean. The depth at which $\Omega = 1$ is termed the *saturation horizon*. As exemplified by the profiles in Figure 15.11, surface waters are supersaturated with respect to both minerals, while bottom waters are undersaturated.

The degree of saturation decreases with increasing depth for two reasons. First, the solubility of biogenic calcite and aragonite increases with depth due to increasing

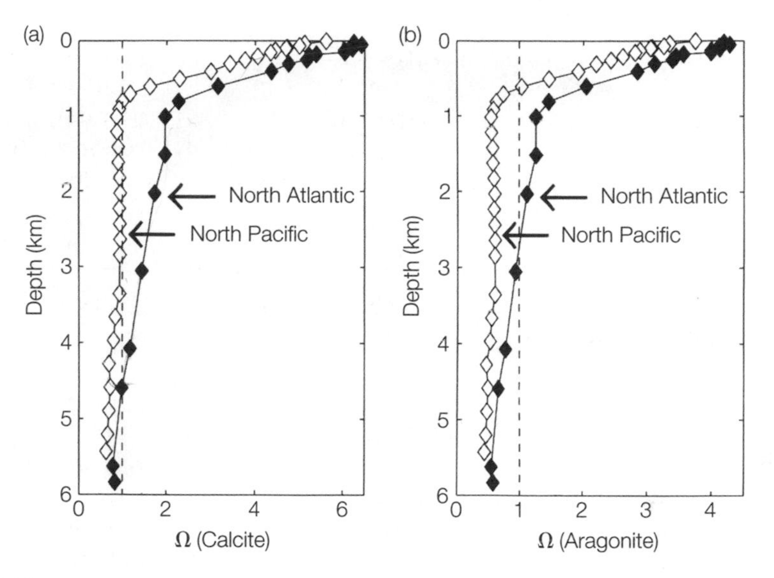

FIGURE 15.11

Saturation state of seawater, Ω, with respect to (a) calcite and (b) aragonite as a function of depth. The dashed vertical line marks the saturation horizon. North Pacific profile is from 27.5°N 179.0°E (July 1993) and North Atlantic profile is from 24.5°N 66.0°W (August 1982) from CDIAC/WOCE database (*http://cdiac.esd.ornl.gov/oceans/CDIACmap.html*) Section P14N, Stn 70 and Section A05, Stn 84. *Source*: From Zeebe, R.E. and D. Wolf-Gladrow (2001) Elsevier Oceanography Series, 65, Elsevier, p. 26.

pressure and decreasing temperature. This causes $[CO_3^{2-}]_{saturation}$ to increase with depth. Second, CO_3^{2-} is removed from seawater as a result of reaction with biogenic CO_2 via Eq. 15.18. This biogenic CO_2 is released as sinking POM undergoes microbially mediated remineralization.

All surface seawater is presently supersaturated with respect to biogenic calcite and aragonite with Ω ranging from 2.5 at high latitudes and 6.0 at low latitudes. The elevated supersaturations at low latitude reflect higher $[CO_3^{2-}]$ due to: (1) the effect of temperature on CO_2 solubility and the K_a for HCO_3^-, and (2) density stratification. At low latitudes, enhanced stratification prevents the upwelling of CO_2-rich deep waters.

Although surface waters are supersaturated with respect to calcium carbonate, abiogenic precipitation is uncommon, probably because of unfavorable kinetics. (The relatively rare formation of abiogenic calcite is discussed further in Chapter 18.) Marine organisms are able to overcome this kinetic barrier because they have enzymes that catalyze the precipitation reaction. Because Ω declines with depth, organisms that deposit calcareous shells in deep waters, such as benthic foraminiferans, must expend more energy to create their hard parts as compared to surface dwellers.

More detail on the horizontal variations in Ω is provided in Figure 15.12, which illustrates that the depth of the saturation horizon for calcite rises from 4500 m in the

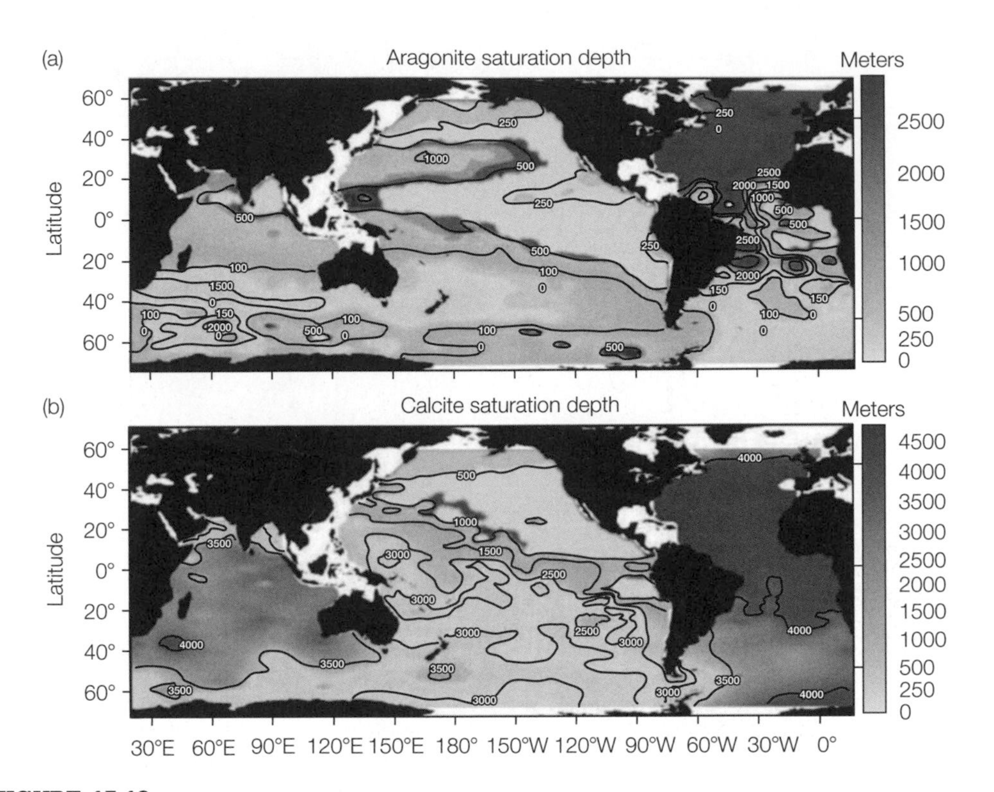

FIGURE 15.12

Depth in meters of the (a) aragonite and (b) calcite saturation horizons ($\Omega = 1$) in the global oceans. *Source*: After Feeley, R. A., *et al.* (2004). *Science* 305(5682), 362–366. (See companion website for color version.)

North Atlantic to 500 m in the North Pacific. This reflects an increasing addition of CO_2 to deep waters as meridional overturning circulation moves them from the Atlantic to the Indian and then to the Pacific Ocean. Thus, as a water mass ages, it becomes more corrosive to calcium carbonate. Since aragonite is more soluble than calcite, its saturation horizon lies at shallower depths, rising from 3000 m in the North Atlantic to 200 m in the North Pacific.

Based on thermodynamic considerations, sediments that lie at depths below the saturation horizon should have 0% $CaCO_3$. This then explains why calcareous oozes are restricted to sediments lying on top of the mid-ocean ridges and rises and why the sediments of the North Pacific are nearly devoid of calcite and aragonite. (The low %$CaCO_3$ in the sediments of the continental margin is a result of dilution by terrestrial clay minerals.)

A significant fraction of the CO_2 injected into the atmosphere as a result of fossil fuel burning is now dissolving into the surface ocean. This is causing a decline in seawater pH and Ω. A recent modeling effort, shown in Figure 15.13, predicts a precipitous rise in the aragonite saturation horizon by the year 2100, with surface waters in

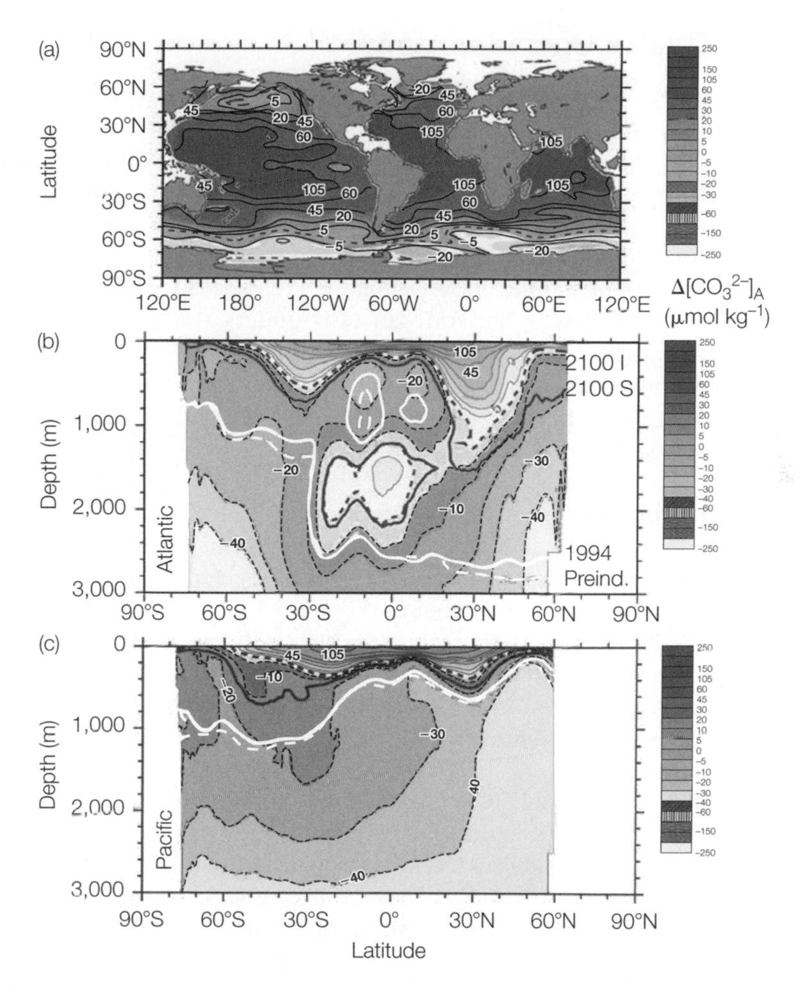

FIGURE 15.13

Aragonite saturation horizon predictions for the year 2100. Values mapped are $\Delta[CO_3^{2-}]_A = [CO_3^{2-}]_{in\ situ} - [CO_3^{2-}]_{aragonite\ saturation}$, where positive $\Delta[CO_3^{2-}]_A$ indicates supersaturation and negative $[\Delta CO_3^{2-}]_A$ indicates undersaturation. The saturation horizon is $\Delta[CO_3^{2-}]_A = 0$. (a) Surface map with saturation horizon marked by black dashed line. Longitudinal cross sections showing zonal averages for the (b) Atlantic Ocean and (c) Pacific Ocean. White dashed line marks the position of the aragonite saturation horizon in 1765 (preindustrial); the white solid line, the position in 1994. The black solid and dashed lines are predictions for 2100. *Source*: After Orr, J. C., *et al.* (2005). *Nature*, 437, 681–686. (See companion website for color version.)

the Southern Ocean becoming undersaturated. This suggests that calcifying organisms will be increasingly limited in their ability to deposit hard parts. This represents a serious perturbation to the biological carbon pump and, hence, the degree to which the sediments can act as a carbon sink in the crustal-ocean-atmosphere factory.

15.5.2 **Kinetic Considerations**

Thermodynamics provides insight into where sedimentary calcium carbonate should and should not be present. This approach is based on the assumption that the mineral phase reaches equilibrium prior to burial. In some cases, thermodynamic predictions appear to be violated. For example, a significant amount of sinking PIC appears to dissolve in waters that lie above the saturation horizon. As noted earlier, this is attributed to dissolution within acidic microenvironments present in zooplankton guts and feces and in aggregates of marine snow. Similarly, PIC deposited in sediments whose surface lies above the saturation horizon can also undergo a significant amount of dissolution. This is attributed to reaction with biogenic CO_2 produced postburial by the remineralization of sedimentary POC. In these two examples, the thermodynamic predictions were faulty because they were based on bulk water column chemistry and ignored the existence of undersaturated conditions in microenvironments and pore waters.

On the opposite end of the spectrum, thermodynamics cannot explain why some PIC can sink through undersaturated waters without dissolving to accumulate on the seafloor. This is a widespread phenomenon as evidenced by the spatial %$CaCO_3$ gradients seen in the surface sediments (Figure 15.5). If the saturation horizon dictated the survival of sinking and accumulating PIC, a sharp depth cutoff should exist below which calcium carbonate is absent from the surface sediment. The importance of this kinetic barrier to dissolution is also seen in the relatively high fraction of surface-water PIC (20 to 25%) that accumulates in the sediments as compared to the low fraction of surface-water POC (1%).

The ability of sinking PIC to survive its trip to the seafloor is dependent on the particle dissolution rates and travel time. The latter is a function of sinking rate and depth to the seafloor. As discussed in Section 15.3.2, PIC dissolution rates are dependent on Ω, water temperature, and various shell characteristics including thickness, shape, mineralogy, and organic coatings. Many of these shell characteristics also determine sinking rate. Since shell characteristics are species specific, so are dissolution rates, leading to selective preservation in the sediments.

Once the PIC reaches the seafloor, it is still subject to dissolution if the bottom waters are undersaturated. Protection against postdepositional dissolution is provided by rapid burial. Calcareous oozes are created when burial is achieved by sedimenting particles that are predominantly PIC.

At some depth below the saturation horizon, the rate of delivery of sinking PIC is equal to its dissolution rate. This is called the *calcium carbonate compensation depth* (CCD). Below the CCD, surface sediments have a negligible (<10%) calcium carbonate content. When considered three-dimensionally, the CCD can be thought of as the bottom edge of a giant bathtub ring of calcium carbonate that encircles the ocean basins. As illustrated in Figure 15.14 for the Equatorial Pacific, the CCD can lie at depths considerably greater than the saturation horizon.

The calcium carbonate content of surface sediments is not a very sensitive indicator of the degree of PIC dissolution. Thus, a significant percentage of the sinking PIC must dissolve before an appreciable impact is seen in the %$CaCO_3$ of the surface

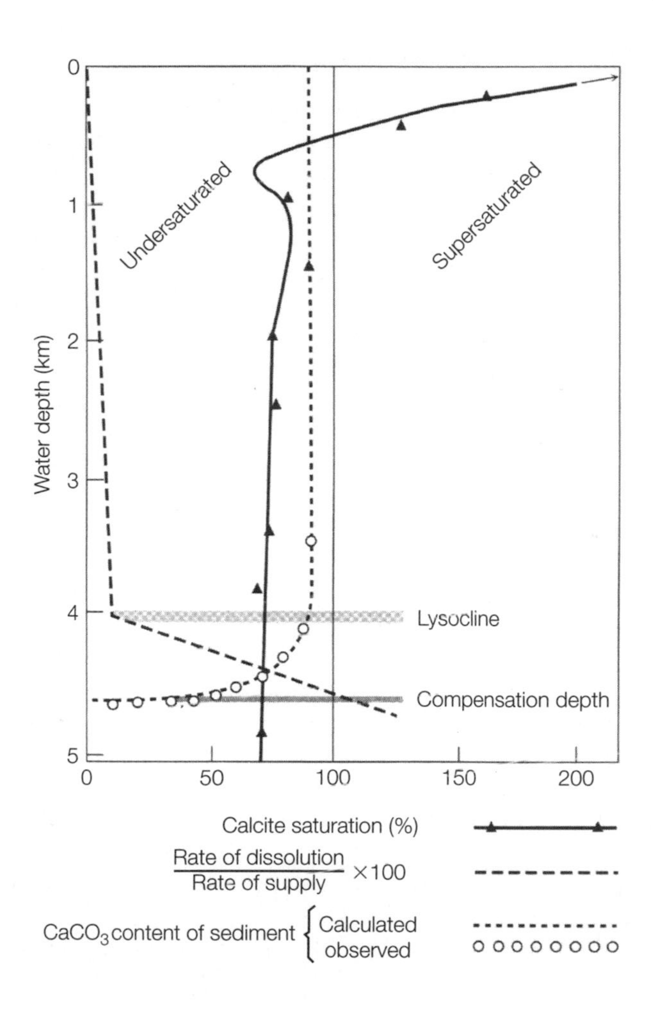

FIGURE 15.14

Parameters influencing the distribution of calcium carbonate with increasing water depth in equatorial Pacific sediment. Note that Ω is reported as a percentage (%). *Source*: From van Andel, Tj. H., *et al.* (1975). *Cenozoic History and Paleoceanography of the Central Equatorial Pacific Ocean*, Geological Society of America, Boulder, CO, p. 40.

sediments. This is illustrated in Figure 15.15, which shows that about half of the sinking PIC flux must dissolve before a significant decline in $\%CaCO_3$ occurs. The depth at which dissolution starts to have a significant impact on the sedimentary $\%CaCO_3$ is called the *lysocline*. As illustrated in Figure 15.14, because half of the sinking PIC flux must first dissolve, the lysocline can also lie quite far below the saturation horizon. This also applies to diagenetic dissolution; a large percentage of the accumulated PIC must dissolve to significantly lower the $\%CaCO_3$ of the sediments post burial.

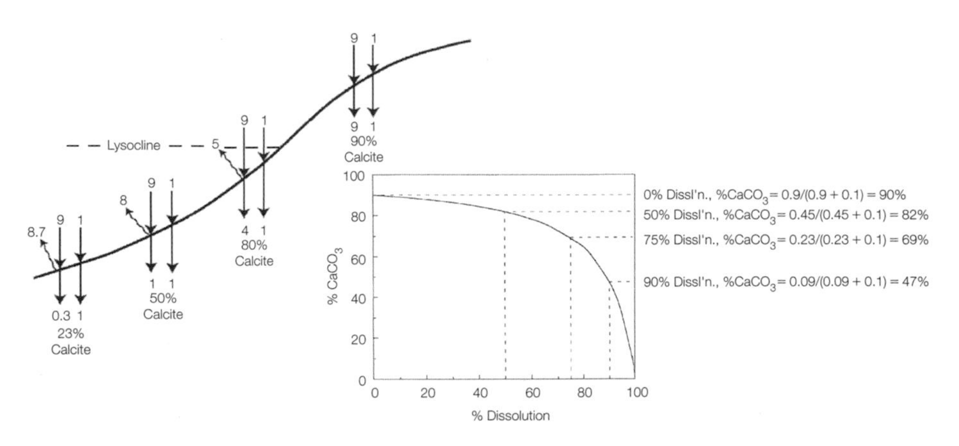

FIGURE 15.15

A diagrammatic view of how the extent of dissolution affects the percent calcite in the sediment. In each example, the right-hand vertical arrows give the sedimenting flux and accumulation rate of noncalcareous particles and the left-hand vertical arrows the sedimenting flux and accumulation rate of calcite. In all cases, the sedimenting flux is 90% $CaCO_3$ and 10% noncarbonate. The wavy arrows represent the dissolution rates of calcite. The inset graph shows that large amounts of dissolution are required before the %$CaCO_3$ (as calcite) of the sediment drops significantly. *Source*: After Broecker, W. S. (2005). *Treatise on Geochemistry*, Elsevier, p. 530.

15.6 THE ROLE OF MARINE CARBONATES IN THE CRUSTAL-OCEAN-ATMOSPHERE FACTORY'S GEOCHEMICAL EVOLUTION, STABILIZATION, AND FUTURE

The marine carbonate system appears to play a critical role in the crustal-ocean-atmosphere factory by participating in two feedback loops in which biogenic calcite burial serves as the primary oceanic sink for carbon. Both loops influence atmospheric CO_2 levels and seawater chemistry, including pH and alkalinity levels, but operate on much different time scales. One loop, called the *deep carbon cycle*, operates over timescales involving eons (million of years). In this loop, marine carbonates participate in the global rock cycle by undergoing metamorphosis into limestone, chalk, and marble, followed by either subduction or uplift. Both subduction and uplift return the carbon back to the part of the crustal-ocean-atmosphere factory where chemical weathering remobilizes the carbon. The deep carbon cycle also involves the weathering of silicate rocks as illustrated in Eq. 14.2. The deep carbon cycle is considered further in Chapter 21.

The marine carbonate system is thought to participate in another set of interlinked processes that acts to regulate atmospheric CO_2 levels over time scales of 5 to 10 ky. This shorter cycle is commonly referred to as *carbonate compensation* and appears to explain a significant component of the swings in P_{CO_2} associated with transitions between glacial and interglacial conditions. In this context, *compensation* refers to the ocean's response to P_{CO_2} perturbations through shifts in its carbonate chemistry. These shifts require changes in the carbonate ion concentration that change the depth of the CCD and, hence, lead to changes in the burial rate of carbon as biogenic calcium carbonate.

If the carbonate compensation model is truly at work, the depth of the calcite and aragonite CCDs must have changed over time. These changes should be recorded as downcore variations in the calcium carbonate content of marine sediments. Marine scientists are presently hard at work trying to parse out this part of the sedimentary record. Their task is complicated by the confounding effects of changes in the rain rate of other particle types and changes in the depth of the sediment surface over time. An example of the latter is illustrated in the supplemental material for Chapter 15 available on the companion website, which shows that seafloor spreading eventually carries sediments into deeper waters. This is caused by the subsidence of oceanic crust that occurs as old basalt is pushed away from active spreading centers. In this way, the sediment surface is eventually transported below the CCD, thereby halting any further accumulation of calcium carbonate. In this case, the downcore depth at which calcium carbonate disappears from the sediments can be used to reconstruct the configuration of the ancient ocean basins if the position of the CCD (relative to the sea surface) has remained constant over time and the sedimentation rate is known.

A description of the carbonate compensation feedback system and its hypothesized role in regulating atmospheric CO_2 levels is provided in the supplemental information for Chapter 15.6 that is available at **http://elsevierdirect.com/companions/9780120885305**.

CHAPTER 16
Biogenic Silica

All figures are available on the companion website in color (if applicable).

16.1 INTRODUCTION

Silicon is the second most abundant element in earth's crust, present primarily as mineral silicates. In Chapter 14, we saw that chemical weathering of terrestrial rocks solubilizes silicon, which is then carried by river runoff into the ocean. This is the major source of dissolved silicon to the ocean. Marine plankton that form siliceous hard parts are extraordinarily effective at removing dissolved silicon from seawater. Like all biolimited elements, dissolved silicon concentrations exhibit classical patterns of vertical and horizontal segregation. In contrast to the other biolimiting elements, nitrogen and phosphorus, a significant fraction of the siliceous hard parts survive the trip to the seafloor to become buried in the sediments. Siliceous oozes, in which *biogenic silica* constitutes more than one third of the sediment mass, are common to high-latitude and equatorial regions, making this the second most important type of biogenic marine sediment. The geographic distribution of siliceous oozes reflects the biogeochemical processes that control the production of biogenic silica and the degree to which these particles are preserved prior to burial in the sediments. Unlike calcium carbonate, all of seawater is undersaturated with respect to biogenic silica, making it unlikely that equilibrium-driven processes control the formation of siliceous oozes.

The availability of dissolved silicon controls the growth of a major class of phytoplankton, diatoms, which are presently responsible for about half of the global ocean's net primary production. The relative dominance of diatoms in the phytoplankton community has shifted over time. This has had consequences in the crustal-ocean-atmosphere factory. For example, diatoms compete with coccolithophorids for nutrients, so their dominance leads to a reduced production of calcareous hard parts and, hence, burial rate of inorganic carbon in marine sediments. Impacts to the burial rate of organic carbon are likely, because the fraction of diatom biomass exported to the sediments as detrital POC is not likely to be the same as for coccolithophorids. Diatoms have evolved a set of adaptations that give them a competitive edge over other phytoplankton when silicon is present. For example, some species have endosymbiotic nitrogen-fixing bacteria. Thus, silicon availability plays an important role in controlling the biogeochemical links in the crustal-ocean-atmosphere factory that regulate primary production, the CO_2 content of the atmosphere, and, hence, climate and ocean circulation. Research in this area is currently focused on investigating: (1) the role that iron plays in limiting diatom growth, (2) what controls the dissolution rates of sedimentary biogenic silica, and (3) how to use downcore variations in the biogenic silica content of marine sediments as a paleoceanographic record.

Like the coccolithophorids, diatoms are a relative newcomer to the oceans, having risen to prominence over the past 100 million years, as part of the evolutionary recovery of the marine biota from the mass extinctions of the Permian. Prior to the advent of the diatoms, seawater dissolved-silicon concentrations were likely much higher and probably regulated by abiotic precipitation. In the present-day ocean, diatom uptake keeps the dissolved silicon concentrations far below mineral saturation concentrations. Given the current important role of diatoms in the plankton community and, hence, the marine carbon cycle, it is of note that human activities are likely reducing the terrestrial fluxes of dissolved silicon to the ocean. This is a consequence of the excessive use of fertilizers that ultimately end up in lakes and streams. The mobilized nitrogen and phosphorus enhance freshwater diatom growth, thereby reducing dissolved silicon concentrations in the river waters that flow into the ocean.

16.2 THE PRODUCTION OF BIOGENIC SILICA

At the pH and ionic strength of seawater, the dominant dissolved species of silicon is *orthosilicic acid* [$H_4SiO_4(aq)$ or $Si(OH)_4(aq)$]. The speciation of silicic acid is shown in Figure 5.19. At the pH of seawater, a minor amount of dissociation occurs, such that about 5% of the dissolved silicon is in the form of $H_3SiO_4^-(aq)$. Dissolved organic complexes of silicon do not occur naturally.

Biogeochemists use the terms *dissolved silica* (DSi) or *dissolved silicate* to collectively refer to all of the dissolved silicon. Silicic acid exhibits tetrahedral geometry with the silicon atom at the center and a hydroxyl group occupying each of the four corners. This structure is similar that of the mineral silicate tetrahedra (Figure 14.3c). Chemical weathering of the silicate minerals is the major source of DSi to the ocean, giving rise to the term *dissolved silicate*, which is usually abbreviated to just "silicate."

The silicate mineral that is composed solely of silicon and oxygen is quartz. As noted in Chapter 14, quartz is a framework silicate whose empirical formula is SiO_2 (*silica*). Silica also occurs in hydrated form ($SiO_2 \cdot nH_2O$), which unlike quartz is not crystalline and, hence, is commonly termed *amorphous silica*. The siliceous hard parts deposited by marine plankton are a type of amorphous silica, variously referred to as *biogenic silica* (BSi), *opal*, *biogenic opal*, and *opaline silica*. Remineralization of BSi resolubilizes silicon; hence DSi is often referred to as "dissolved silica" or simply "silica." In the sediments, some of the remineralized DSi is reprecipitated by benthic bacteria. Opal can also form abiogenically, via precipitation reactions that occur at high temperatures in hydrothermal systems and in the pore waters of sediments.

The burial of BSi is the most important mechanism by which dissolved silicon is removed from the ocean. Most of the BSi is deposited by surface-dwelling plankton whose actions collectively keep surface-water DSi concentrations very low (<5 μM). Remineralization of sinking detrital BSi leads to vertical segregation of DSi as illustrated in Figure 10.1, with average deepwater concentrations around 100 μM. In the present-day ocean, abiogenic precipitation is important only in locations, such as pore waters and estuaries, where DSi concentrations are very high.

BSi is produced primarily by planktonic organisms, namely diatoms and *silicoflagellates*, which are phytoplankton, and *radiolaria*, which are protozoans. Most of radiolarians are surface dwellers, living in the top 100 m of the water column. Some sponges form siliceous spires or spicules, but these constitute a small fraction of the BSi. The modern-day planktonic silicifiers exhibit a high degree of species diversity. In the case of the diatoms, 1800 marine species now exist with many more having gone extinct. Examples of the siliceous skeletal structures of the modern-day diatoms and radiolarians are illustrated in Figures 16.1 and 16.2, respectively. Although the BSi deposited by each species has its own characteristic shape, these hard parts are all generally less than 100 μm in length or diameter.

Since all seawater is undersaturated with respect to BSi, plankton that deposit siliceous hard parts must fight unfavorable thermodynamics. In the case of the diatoms, the phytoplankton have specialized proteins that help concentrate the DSi intracellularly

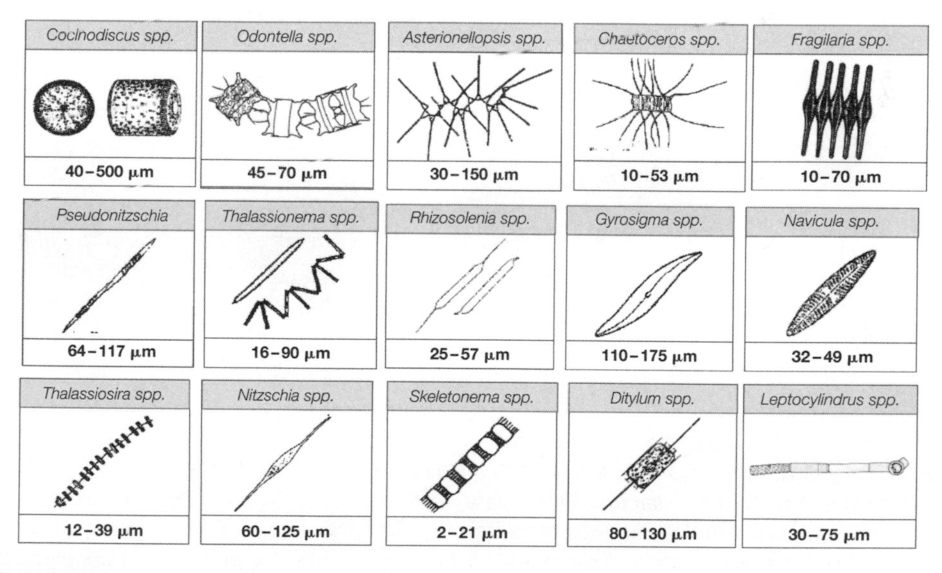

FIGURE 16.1

Common species of modern marine diatoms and their approximate dimensions. *Source*: From The National Ocean Service/National/Center for Coastal Ocean Science Program (2008) Phytoplankton Monitoring Network. Common Phytoplankton Key, Identification Sheet. http://www.chbr.noaa.gov/PMN/downloads/IDsheet_CommonPhytoKey.pdf. Accessed June 2008.

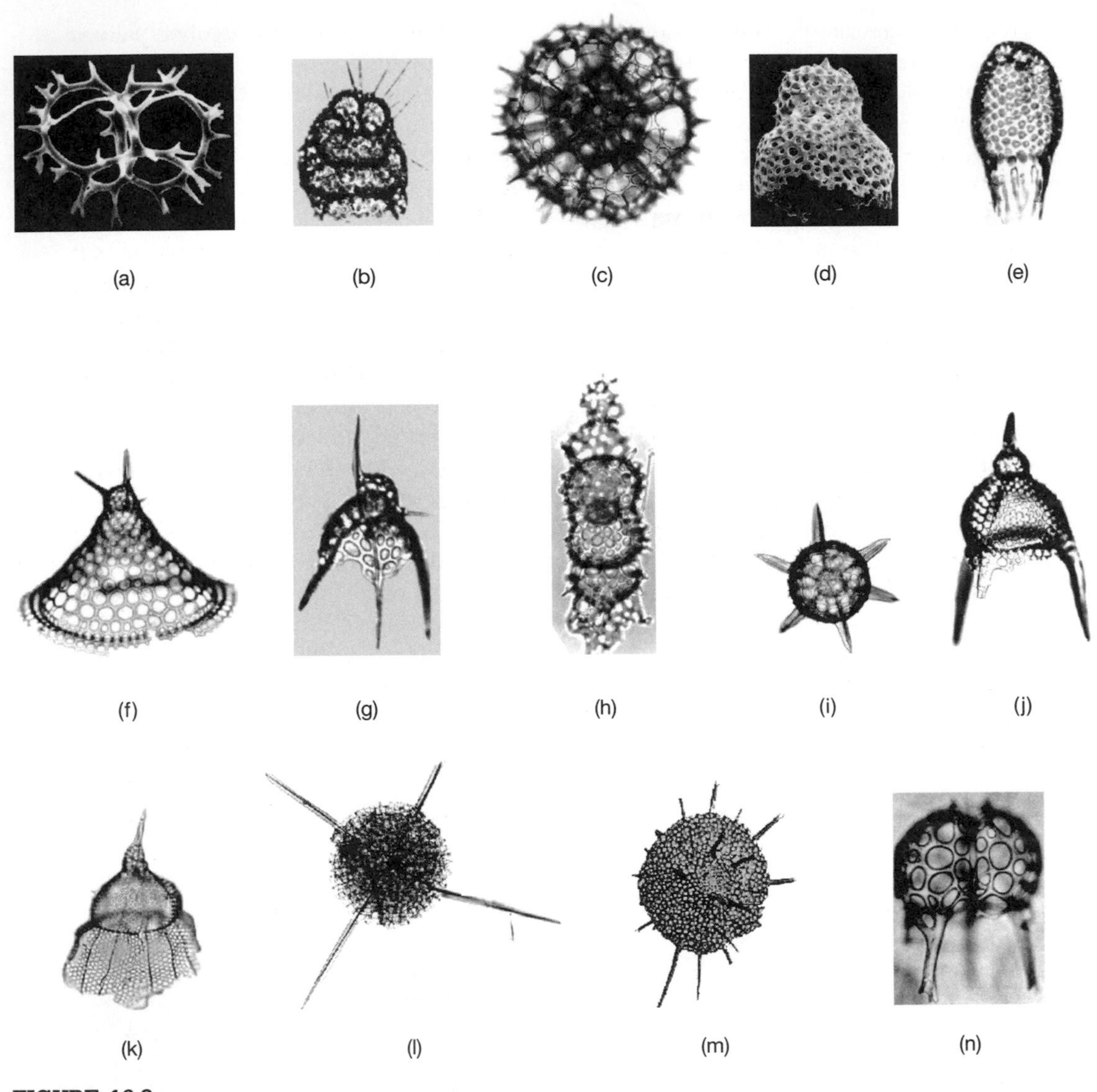

FIGURE 16.2

Common species of modern marine radiolaria: (a) *Acanthodesmia viniculata*, (b) *Botryocyrtis scutum*, (c) *Actinomma antarcticum*, (d) *Antarctissa denticulata*, (e) *Carpocanium* spp. (f) *Corocalyptra cervus*, (g) *Dictyophimus gracilipes*, (h) *Didymocyrtis tetrathalamus*, (i) *Hexacontium armatum/hostile* group, (j) *Pterocanium trilobum*, (k) *Pterocorys herwigii*, (l) *Spongosphaera streptacantha*, (m) *Spongotrochus glacialis*, and (n) *Triceraspyris antarctica*. *Source*: Boltovskoy, D. (1998). *Palaeo-Electronica*, 1, http://palaeo-electronica.org/1998_2/boltovskoy/issue2.htm.

where precipitation occurs. The hard parts deposited by diatoms are called *frustules*. They are thought to serve several functions: (1) protection against predators; (2) pH buffering that shifts the inorganic carbon species from bicarbonate to CO_2, the chemical form that can then be incorporated into organic matter; (3) enhanced diffusion of nutrients enabling rapid uptake into the cell; and (4) buoyancy control.

The ecological advantages conferred by their frustules are part of the reason why diatoms dominate the modern-day phytoplankton. Other reasons include their ability to store nutrients in vacuoles, which enables them to hoard large amounts of nitrate and phosphate and, hence, survive periods of low nutrient availability. This ability also reduces growth of other phytoplankton by depriving them of biolimiting nutrients. Thus, diatoms tend to be the superior competitors in regions where nutrients (DSi, nitrate, and phosphate) are supplied in pulses, such as in coastal upwelling zones and in highly turbulent areas, like high-latitude seas. Some diatom species also have endosymbiotic cyanobacteria that fix N_2 (diazotrophs) and provide some of this fixed nitrogen to their host. These diatoms tend to be iron as well as silica limited due to the trace metal requirement of the diazotrophs. Iron limitation also seems to lead diatoms to deposit thicker-walled frustules, such that a much larger ration of DSi relative to nitrate is required for growth. Thus, diatoms growing under conditions of iron limitation tend to deplete seawater of DSi before exhausting the nitrate.

DSi concentrations are highest in regions subject to wind-driven upwelling because at these locations, nutrient-rich deep water is transported to the sea surface. As illustrated in Figures 4.4a and 4.5, upwelling occurs in the open ocean and in coastal waters as a result of wind-driven Ekman transport of water. Areas of persistent and intense upwelling are located at equatorial and subpolar latitudes as a result of divergence caused by the Trades Wind and Westerlies (Figure 16.3). Coastal upwelling is relatively episodic due to seasonal and longer-term fluctuations in local patterns of wind and surface circulation.

Diatoms are the dominant siliceous plankton at high latitudes, with radiolarians forming a significant component of the BSi production in the equatorial zones of divergence. The planktonic silicifiers are so effective in removing DSi that surface concentrations are virtually zero ($<5\ \mu M$) except in upwelling regions (Figure 10.8c). After death, the detrital remains of the siliceous organisms sink toward the seafloor. As noted in Chapters 9 and 10, a greater fraction of the detrital BSi escapes remineralization in the surface waters as compared with nitrogen and phosphorus. This is a consequence of detrital BSi being remineralized via dissolution, which takes place more slowly, and, hence, at greater depths, than microbial decomposition of POM. Thus, the biological pump tends to create and maintain conditions of DSi limitation in the surface waters. Because a significant fraction of the sinking BSi does dissolve, DSi exhibits vertical and horizontal segregation as illustrated in Figures 10.1b and 10.5e with bottom water concentrations increasing from the Atlantic to the Indian to the Pacific Ocean.

In upwelling areas, the resupply of DSi via Ekman transport leads to a silica trapping effect as illustrated in Figure 16.4. Under these conditions, growth rates in the surface waters are controlled by the supply rate of DSi via upwelling. An example of nutrient

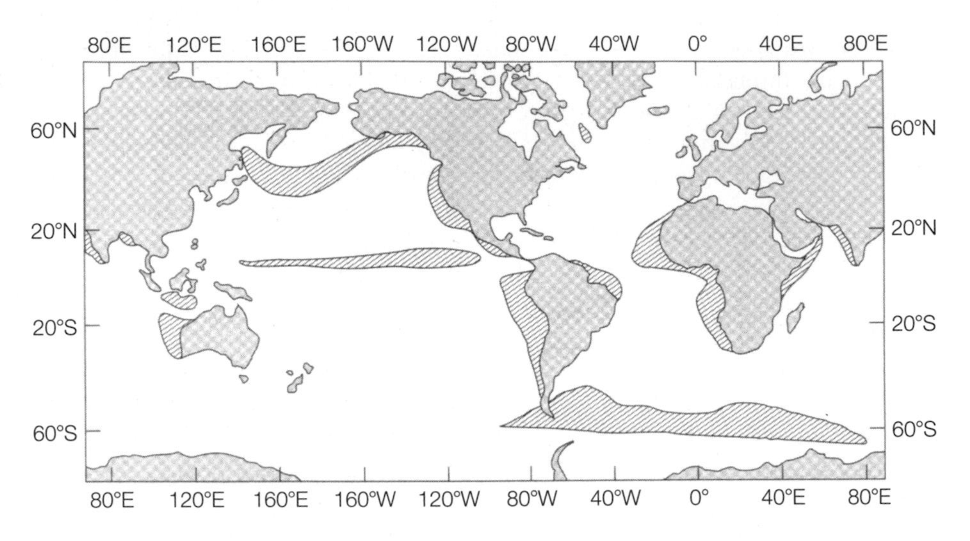

FIGURE 16.3

Regions of offshore upwelling driven by Ekman transport.

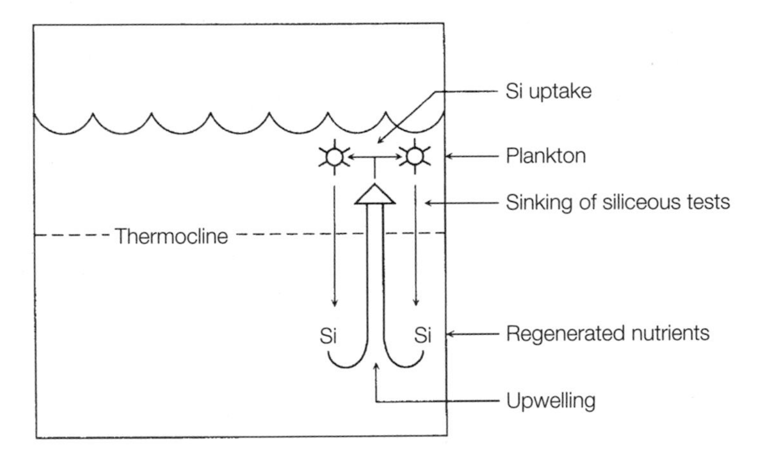

FIGURE 16.4

The silica trap in upwelling areas.

concentrations under these conditions is shown in Figure 16.5a, which demonstrates that DSi rather than nitrate is the limiting nutrient.

The DSi trapping effect is particularly strong in the Southern Ocean because of the upwelling associated with meridional overturning circulation. As discussed in Chapter 10.6, the upwelling of Circumpolar Deep Water (CDW) brings nitrogen and DSi to the sea surface. As Ekman transport drives this water northward, diatoms strip out the nutrients. Because of iron limitation at these latitudes, the DSi uptake is high relative to

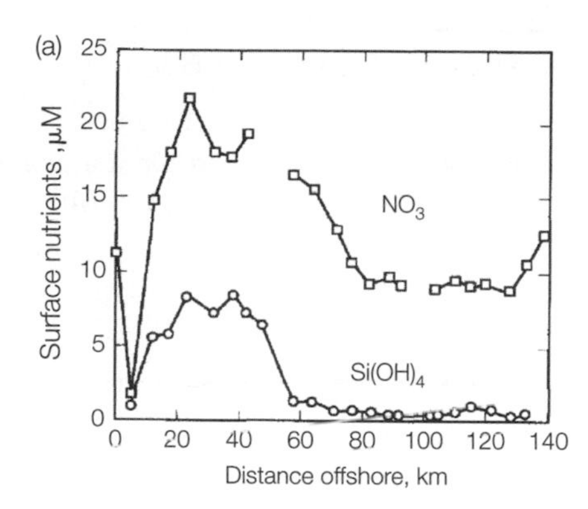

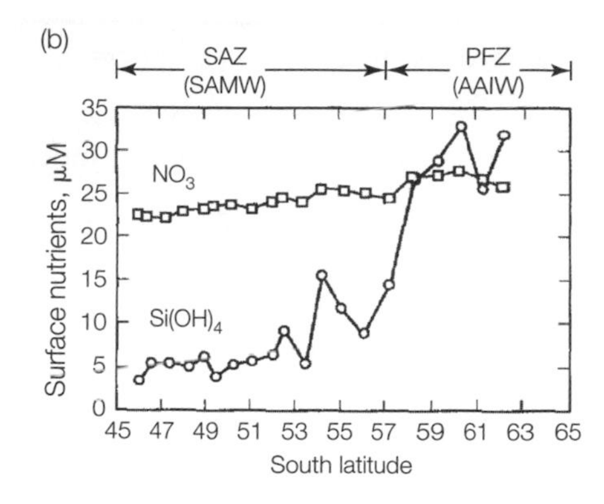

FIGURE 16.5

Surface concentrations of nitrate and DSi (μM) (a) in the upwelling area located offshore of West Africa (18°N 16°W) during March–April 1971 and (b) in the Southern Ocean (45°S to 65°S, along 65°W) during March 1977. *Source*: After Dugdale, R. C., *et al.* (1995). *Deep-Sea Research* I 42, 697–719.

nitrate, so the diatoms deplete the surface waters of DSi, leaving a considerable amount of nitrate unused. As shown in Figures 10.4d and 10.10, this happens around 55°S in the regions where Subantarctic Mode Water (SAMW) is created. When this water mass forms by sinking during the southern polar winter, it carries a [DSi] of about 0 μM and $[NO_3^-]$ of 10 to 15 μM (Figure 16.5b). SAMW is an important water mass as it flows northward, thereby feeding the thermocline. As a result, a significant fraction of the nitrate (and by analogy, phosphate; see Figure 10.9) in the thermocline is preformed, rather than supplied by remineralization of sinking POM.

The DSi stripped out of the Southern Ocean's surface waters at the site of SAMW formation is converted to BSi. This BSi eventually sinks into the deep waters (CDW), where it is remineralized to DSi and driven back south to be upwelled again into the surface waters. This trapping effect is a large part of why half of the global marine sedimentary sink of BSi is located in the Southern Ocean. South of 55°S, iron limitation is so severe, as compared to the rate of upwelling supply of DSi, that the diatoms are not able to reduce silicic acid concentrations to zero.

16.3 PRESERVATION VERSUS DISSOLUTION OF SINKING DETRITAL BIOGENIC SILICA

In contrast to calcium carbonate, all seawater is undersaturated with respect to BSi. As shown in Table 16.1, the undersaturation is very large and increases with depth because the solubility of BSi increases with pressure. Thus, all siliceous hard parts are subject to dissolution. Nevertheless, about 25% of the BSi created in the surface waters survives the trip to the seafloor via pelagic sedimentation. Direct observations of this transport

Table 16.1 Saturation Concentrations of DSi with Respect to BSi and Observed DSi Concentrations.

Depth (m)	BSi Saturation Concentration (μM)		Observed Concentration Range (μM)
	3°C	23°C	
0	800 to 1000	1500 to 1700	0 to 20 (80 to 100 in Antarctica)
10,000	1000 to 1200	1700 to 1900	10 to 180

have been obtained through the use of sediment traps that intercept sinking particles as they settle through the water column.

As with the calcareous tests, BSi dissolution rates depend on (1) the susceptibility of a particular shell type to dissolution and (2) the degree to which a water mass is undersaturated with respect to opaline silica. Susceptibility to dissolution is related to chemical and physical factors. For example, various trace metals lower the solubility of BSi. (See Table 11.6 for the trace metal composition of siliceous shells.) From the physical perspective, denser shells sink faster. They also tend to have thicker walls and lower surface-area-to-volume ratios, all of which contribute to slower dissolution rates. As with calcium carbonate, the degree of saturation of seawater with respect to BSi decreases with depth. The greater the thermodynamic driving force for dissolution, the faster the dissolution rate. As shown in Table 16.1, vertical and horizontal segregation of DSi does not significantly counter the effect of pressure in increasing the saturation concentration DSi. Thus, unlike calcite, there is no deep water that is more thermodynamically favorable for BSi preservation; they are all corrosive to BSi.

The fate of BSi as it sinks out of the surface waters has been studied with sediment traps (Figure 16.6). The traps are deployed at specific depths for controlled lengths of time. The model illustrated in Figure 16.6 is equipped with collection cups programmed to open and close sequentially so that temporal changes in the flux of sinking particles can be studied. The traps are usually attached to a mooring array that is recovered by activating an acoustic release.

Observations of BSi fluxes obtained from sediment traps moored in a range of settings are presented Figure 16.7a in order of high to low BSi productivity. The sites with the highest BSi production are from the Southern Ocean (POOZ, SACC, APFP, NACC, APFA), the next highest are from the North Pacific (OSP) and the Equatorial Pacific. The lowest are from the Atlantic Ocean (BATS and PAP). At each of these locations, sediment-trap samples were collected at three depths: (1) the bottom of the mixed layer (~100 to 200 m), (2) the base of the thermocline (~1000 m), and (3) in the deep zone near the sediment surface. The last is a measure of how much BSi is "raining" onto the sediment surface. Also provided is the rate at which BSi is accumulating in the sediments. Because of postdepositional dissolution, only about 20% of the BSi that "rains" onto the sediments is buried.

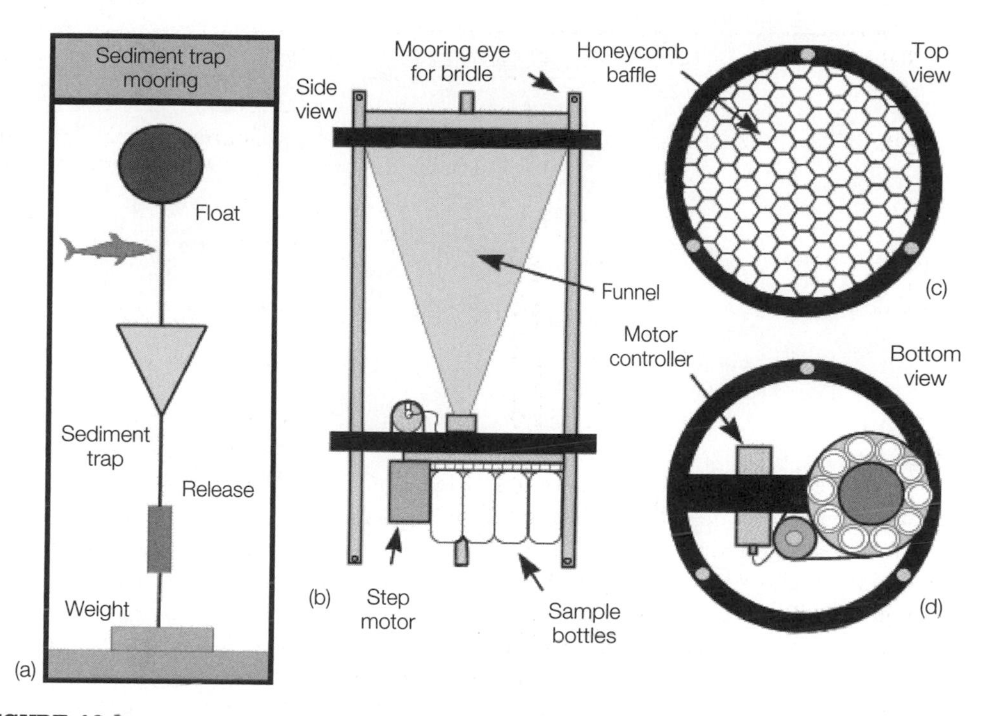

FIGURE 16.6

An example of a time-series sediment trap. This one is commercially available from McClane Research Labs, Inc. (*http://www.mclanelabs.com/*). (a) Trap shown deployed on a moored hydrowire. (b) Schematic of the trap which has a height of 164 cm and a diameter at the top of the funnel of 91 cm providing an interception surface area of 0.66 m^2. (c) A top view of the baffles (cell diameter is 2.5 cm) at the top of the funnel, and (d) a top view of the motor assembly at the bottom of the funnel which rotates the sample bottles. It can hold 21 sample bottles with volumes of 250 or 500 mL and be deployed a maximum of 18 month. *Source*: From Poppe, L. J., *et al.* (2000). East-Coast Sediment Analysis: Procedures, Database, and Georeferenced Displays, U.S. Geological Survey Open-file Report 00-358.

The rain rate of BSi is dependent on (1) the rate of its production by marine organisms, (2) shell dissolution rates, and (3) the time required for a shell to reach the seafloor. High rates of production by siliceous plankton ensure a large supply of opal to the water column. The fraction reaching the seafloor is largest when transit times are shortest. Thus, shells that sink fastest will be preferentially preserved and a greater fraction of the particulate silica flux reaches sediments that lie in shallow waters.

The average percentage of BSi production that makes it out of the surface waters is about 50%. Half of this flux is remineralized in the deep zone by 1000 m, leaving 25% of the original BSi production. Little further dissolution takes place prior to the BSi reaching the sediment, i.e., the percentage of the original BSi production raining onto the sediment is about 25%. Thus, most of the dissolution take place in the upper part of

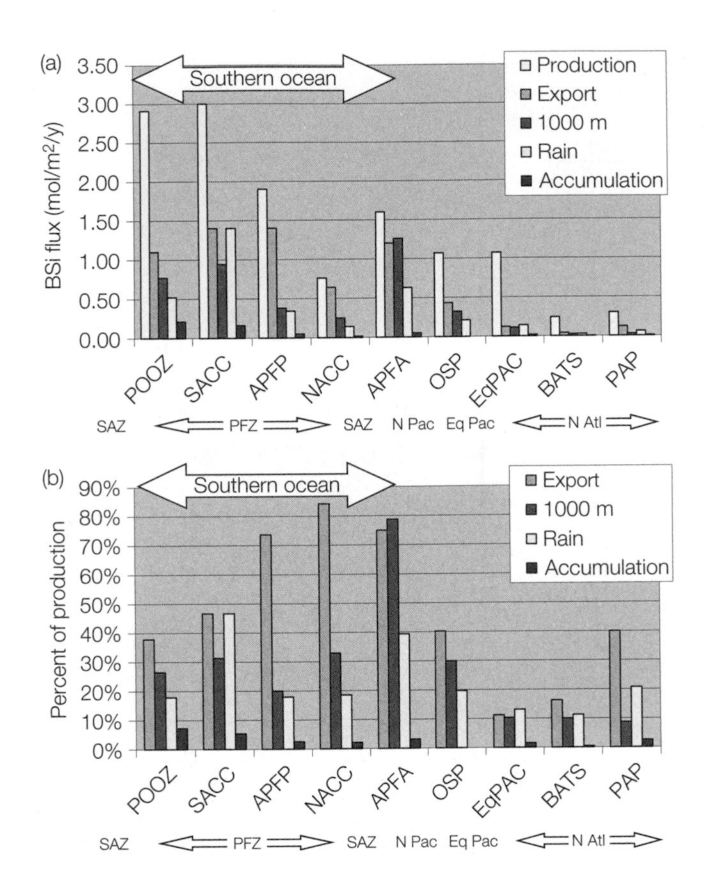

FIGURE 16.7

(a) BSi fluxes in mol m^{-2} y^{-1} and (b) BSi as a percent of BSi production. Sampling locations are as follows: (1) POOZ: 52°S 61°E (SAZ), (2) SACC, APFP, NACC: 55 to 65.5°S 170°W (PFZ), (3) APFA: 50 to 52°S 6 to 11°E (SAZ), (4) OSP: 50°N 145°W (N Pac.), (5) EqPAC: 0°N 140°W (Eq Pac), (6) BATS: 31.8°N 61.4°W (N Atl), and (7) PAP: 48°N 16 to 20°W (N Atl). Regional locations of the sites are shown as SAZ, sub-Antarctic zone; PFZ, polar front zone; N.P., North Pacific; Eq Pac, Equatorial Pacific; N Atl, North Atlantic. Data from: Ragueneau, O., *et al.* (2002). *Deep-Sea Research* II 49, 3127–3154.

the water column. This lack of dissolution in the deep waters is particularly impressive given the increase in undersaturation of BSi with increasing depth. Also, recall that only 1% of detrital sinking POM survives to reach the sediments.

Why does such a high percentage of sinking detrital BSi survive dissolution? The answer is that several factors enable rapid sinking and slow dissolution rates. Rapid sinking of diatoms appears to be promoted by two processes: (1) incorporation into fecal pellets that are relatively large and, hence, sink fast and (2) the formation of dense aggregates of cells following blooms. The most dramatic example of the rapid delivery of BSi to the sediments has been observed to occur when large diatoms that live at the base of the seasonal thermocline clump together to form dense mats. These large species are adapted to low-light conditions and, hence, benefit from the nutrients that are regenerated at these depths. When density stratification is broken down toward the end of the summer by early winter mixing, these diatoms aggregate into dense mats that sink rapidly. This "fall dump" is thought to export more biogenic production to the sediments than plankton blooms that occur in the spring or from upwelling. Because of the rapid delivery of these large diatom frustules, the "fall dump" is recorded in the sediments as a siliceous-rich layer (Figure 16.8).

Until recently, biogeochemists thought that BSi dissolution rates were controlled purely by physical and chemical considerations, such as the degree of undersaturation of the water with respect to BSi, or the surface-to-volume ratio of the siliceous hard part. Biological activity is now understood to play an important role. For example, grazers and bacteria can enhance the dissolution of BSi by removing organic coatings from BSi surfaces or mechanically breaking the hard parts, thereby enhancing the amount of "clean" surface area in contact with the surrounding undersaturated waters. Since most

FIGURE 16.8

Evidence of the "fall dump" of giant diatoms. Sediment layers from (a) the Gulf of California dominated by cell walls of *Stephanopyxis palmeriana* (diameter <70 μm); (b) the Southern Ocean, showing a tangled mass of the rod-shaped species *Thalassiothrix antarctica*, which grows up to 4 mm in length. *Source*: From Kemp, A. E. S., *et al.* (2000). *Deep-Sea Research* II, 47, 2129–2154.

of these biological effects occur in the surface waters, the surfaces of BSi do not undergo much further cleaning with depth. This explains why most of the BSi dissolution occurs above 1000 m. Solubility is also decreased by uptake of dissolved metals, such as aluminum. The lower solubility of BSi in coastal waters as compared to the open ocean is thought to be related to the higher Al concentration in the coastal waters, which leads to the deposition of an Al-rich layer on the BSi. Similar changes in the mineralogy of BSi occur after burial in the sediments and are discussed further in the next section.

The BSi flux is sometimes observed to increase with increasing depth in the water column (Figure 16.7a, SACC and APFA). Several processes are likely explanations: (1) lateral advection of water masses with high BSi, (2) sinking of deep-dwelling radiolarians, (3) resuspension events that affect sediment traps deployed within the benthic nepheloid layer, (4) problems with the trapping efficiency of traps located in shallow water depths, and (5) grazing of diatoms by swimming zooplankton. Temporal changes in BSi production in the surface layer can also theoretically give cause to apparently high BSi at depth if the particles collected at that depth were generated in the overlying waters during a time of unusually high BSi production.

16.4 ACCUMULATION AND PRESERVATION IN THE SEDIMENTS

On average about 20% of the BSi arriving at the sediment-water interface dissolves prior to burial. Thus, only 3 to 5% of the surface production survives to become buried in the sediments. At some locations, where the BSi rain rate is high, a larger percentage (around 30%) is preserved. In comparison, 90% of the POC that rains onto the sediments is remineralized by the benthos, leaving sediment concentrations on the order of 1% w/w organic carbon.

A global map of the BSi content in surface sediments is shown in Figure 16.9 and regional averages in Table 16.2. Even in the regions with highest concentrations, opal comprises on average only 10 to 15% w/w of marine sediments, whereas calcite averages 30 to 50% w/w. Siliceous oozes are restricted to narrow zonal bands in the equatorial Pacific and Indian Ocean, the Southern Ocean, the Bering Sea, and in the immediate vicinity of coastal upwelling areas (Peru and West Africa). Relative high average concentrations are also observed around the continental margin of India and the Middle East.

The geographic distribution of opal in the surface sediments is controlled by (1) the local rain rate of biogenic silica, (2) the degree of its preservation in the sediments, and (3) the relative rate of accumulation of other types of particles. Preservation is promoted by rapid burial as this isolates BSi from seawater. But if the BSi is buried by other particle types, the relative contribution of BSi to the sediment is diluted. This dilution effect causes the BSi content of most continental margin sediments to be low despite high rain rates. Preservation efficiency is also dependent on: (1) the intensity of bioturbation and suspension feeding and (2) the various factors that control

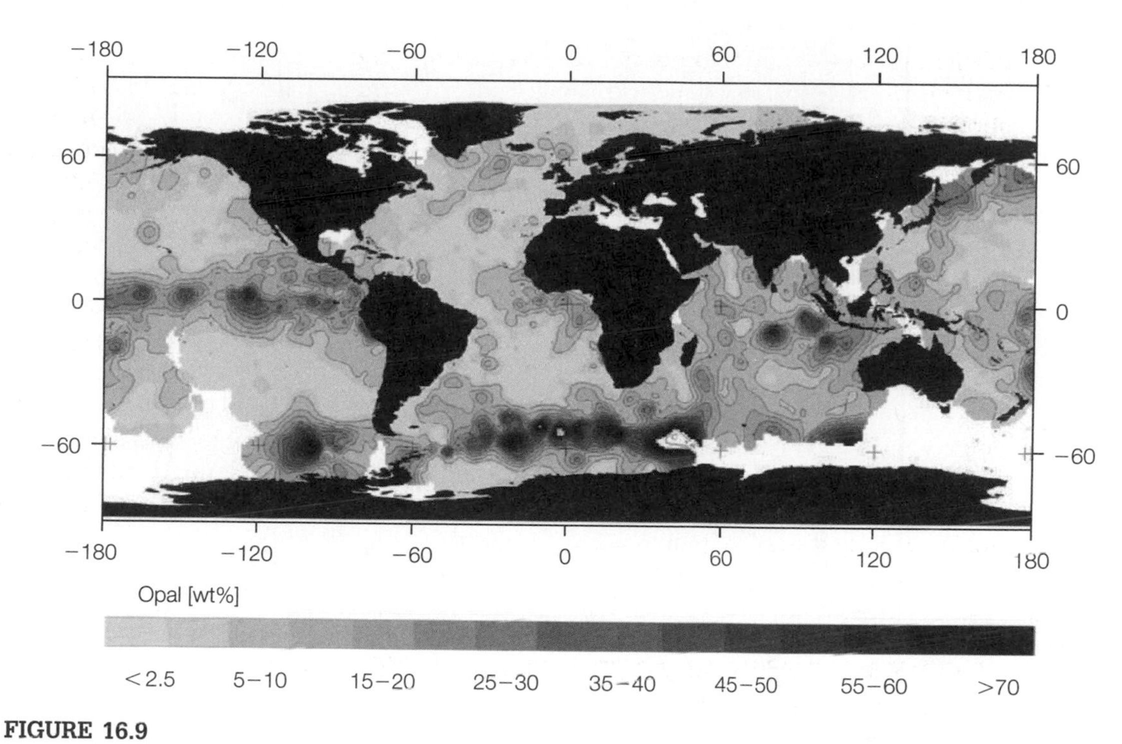

FIGURE 16.9

Weight percent of opal in surface marine sediments (generally 0 to 5 cm). White areas indicate no data. *Source*: From Seiter, K., *et al*. (2004). *Deep-Sea Research* I 51, 2001–2026.

post-burial dissolution, such as the degree of saturation of the pore waters with respect to BSi.

Evidence for the dissolution of BSi following burial in the sediments is seen in pore water DSi depth profiles that typically exhibit downcore increases in concentration. As shown in Figure 16.10, the DSi concentration gradient is exponential with maximal values being attained within 5 to 10 cm beneath the sediment-water interface. Below these depths, DSi concentrations exhibit no further vertical gradient, suggesting thermodynamic control via precipitation into some mineral phase. Interestingly, the maximal values observed in these depth profiles (50 to almost 900 μM) are well below saturation values with respect to BSi, at least as determined from laboratory equilibration measurements. This suggests that the precipitation of amorphous silica is not controlling the pore-water DSi concentrations. Alternatively, other mineral phases could be responsible for setting the upper limit on pore-water DSi concentrations. As noted in Chapter 14.4, evidence has been observed in deltaic sediments for a relatively rapid reverse weathering of clay minerals. As per Eq. 14.4, this reaction consumes DSi.

As in the water column, the question arises as to why any BSi is preserved after its burial in the sediments. Evidently some kinetic inhibition exists. Various hypotheses for the cause of this inhibition include: (1) the buildup of organic or inorganic coatings on

Table 16.2 Biogenic Hard Part Composition of Surface Sediments by Region.

Region	Location	Division	Opal (wt%)		Calcite (wt%)		Total Organic Carbon (wt%)	
			Mean	Max	Mean	Max	Mean	Max
Southern hemisphere	Pacific Ocean	East	7	67	38	93	0.5	3.2
		West	7	39	51	92	0.8	1.3
	Atlantic Ocean		**11**	87	43	100	0.4	1.7
	Indian Ocean		**14**	85	53	97	0.4	2
Northern hemisphere	Pacific Ocean	West	6	33	11	80	0.6	1.9
		East	3	18	3	82	0.4	3.8
	Atlantic Ocean		4	28	44	93	0.4	7.1
Continental shelves[a]	Antarctica	Circumpolar	**10**	49	6	63	0.3	1.4
	India	East	**15**	25	46	90	1.0	1.5
		West	**14**	24	32	57	1.2	3.1
	Middle East		7	25	44	89	1.5	5.5
Coastal upwelling	West Africa	Namibia	6	41	21	85	2.7	9
	West Africa	Southwest	7	40	34	86	1.5	9
	Peru		**24**	44	2	8	4.8	16.2
Equatorial upwelling	Atlantic	East	**13**	27	45	95	0.7	3.4
	Pacific	East	**17**	72	44	96	0.8	1.7
		West	8	34	42	87	1.2	4.9

[a]*Only opal-rich regions are shown. Regions with elevated mean opal contents are in boldface.*
Source: From Seiter, K., et al. (2004). Deep-Sea Research *I 51, 2001–2026.*

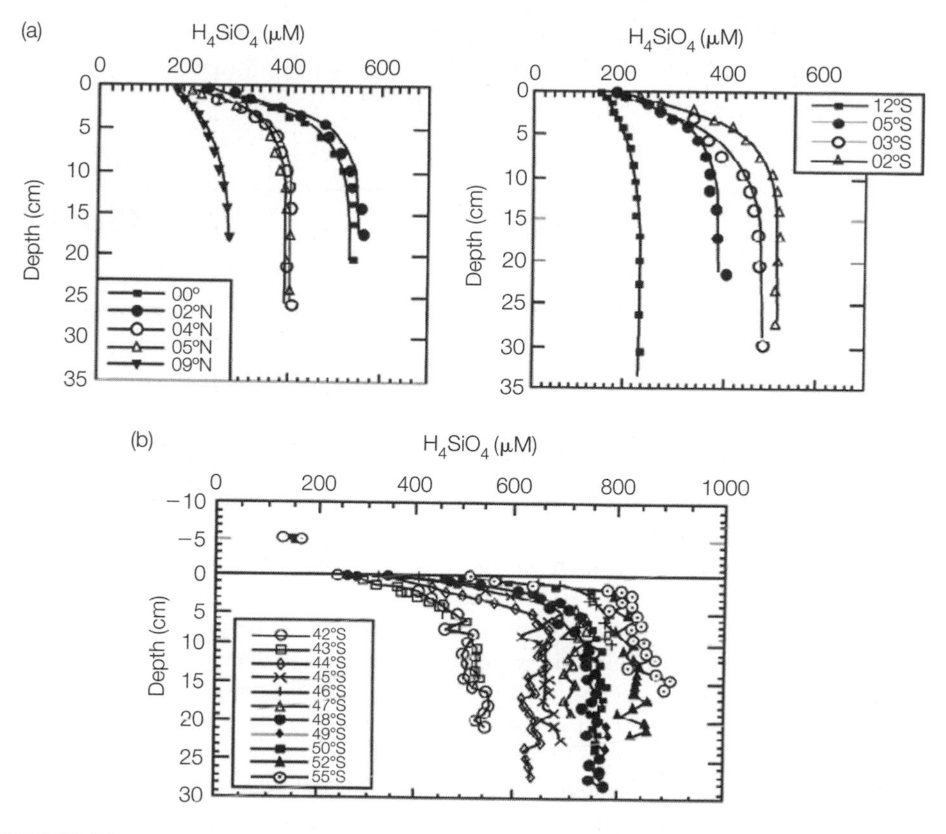

FIGURE 16.10

Pore-water DSi concentrations as a function of depth in sediment on (a) a North-South transect along 140°W in the equatorial Pacific and (b) a North-South transect through the Indian sector of the Polar Front. *Source*: From Emerson, S., and J. Hedges (2003). *Treatise on Geochemistry*, Elsevier, pp. 293–319.

BSi, (2) blocking of reactive surface sites on BSi through adsorption of inhibitors, such as trace metals like Al, (3) loss of reactive sites on BSi as a result of partial dissolution and reprecipitation, and (4) differential dissolution rates across a given siliceous particle due to shell geometry.

16.5 THE FORMATION OF CHERT

The pressure and temperature associated with long-term burial in the sediments eventually converts BSi into chert and quartz, both of which are crystalline. This conversion process involves the partial dissolution of BSi followed by its reprecipitation. Reprecipitation occurs within and upon BSi and probably on clay minerals.

At the higher temperatures and pressures of catagenesis, fossilized BSi is eventually metamorphosed into chert and quartz. Quartz is also produced at high temperatures by direct precipitation from hydrothermal fluids.

16.6 THE CRUSTAL-OCEAN-ATMOSPHERE PERSPECTIVE

Because diatoms play such a large role in the biogeochemistry of silicon, changes in their productivity are thought to have the potential to affect other aspects of the crustal-ocean-atmosphere factory. To consider this future, we first look at the marine silica budget.

16.6.1 Global Marine Silica Budget

A box model for the marine silica cycle is presented in Figure 6.11 with respect to the processes that control DSi and BSi. An oceanic budget is provided in Table 16.3 in which site-specific contributions to oceanic outputs are given. This table illustrates that considerable uncertainty still exists in estimating the burial rate of BSi. Regardless, burial of BSi is responsible for most of the removal of the oceanic inputs of DSi, with the latter being predominantly delivered via river runoff. This demonstrates the importance of the biological silica pump in the crustal-ocean-atmosphere factory.

Assuming that the marine silica cycle is in a steady state, the oceanic residence time of DSi can be estimated from its average seawater concentration (70.6 μM). Using the volume of seawater provided in Figure 2.1, the total amount of DSi in the ocean is 0.97×10^5 Tmol Si. Using the range of output rates provided in Figure 16.11 and Table 16.3, the residence time of silica is estimated to be on the order of 15,000 y. The turnover time relative to biological uptake from the surface waters is about 400 y. We can also estimate that the DSi delivered to the ocean must pass through the internal cycle of biological uptake and dissolution somewhere between 30 to 40 times ($\frac{\text{production}}{\text{accumulation}} = \frac{240\ \text{Tmol Si/y}}{5.5\ \text{to}\ 7.5\ \text{Tmol Si/y}}$) before being removed to the sediments.

Table 16.3 indicates that most of the burial of BSi takes place in the sediments of the Southern Ocean. The importance of burial in estuarine and continental margin sediments is still a matter of debate as shown by the difference in estimates provided in Table 16.3. These regions are likely to be considerable sinks for silica due to rapid sedimentation rates, shallow water depths, and high rates of biological productivity in the overlying waters. Diatom biological productivity is enhanced in coastal waters due to proximity to river runoff and turbulent conditions. Preservation of BSi in the sediments is enhanced by the relatively high aluminum content in coastal diatom frustules. Finally, the role of coastal sediments in removing DSi has likely been underestimated as it is based on measurements of BSi burial and does not include uptake into clay minerals via reverse weathering. The latter has been observed to be occurring during early diagenesis in the deltaic sediments of the Amazon River and is likely to be of importance in other coastal regions.

Table 16.3 Geochemical Balance of Dissolved Silicon in the Modern Ocean.

	DeMaster (1981); Treguer *et al.* (1995)[a]	**Revised Budget**[a]
Sources of silicate		
Rivers	5.6	5.6
Hydrothermal emanations + weathering of marine basalt	0.6	0.6
Eolian	0.5	0.5
Total supply rate	**6.7**	**6.7**
Sedimentary repositories for biogenic silica		
Deep sea total	5.1–6.0	4.1–4.3
Antarctic	4.1–4.8	3.1
Polar Front	2.7–3.4	0.3
Nonpolar Front	1.4	2.8
Bering Sea	0.5	0.5
North Pacific	0.3	0.3
Sea of Okhotsk	0.2	0.2
Low Si Sediments	<0.2	<0.2
Eq. Pacific	0.02	0.02
Continental margins total	0.4–1.5	2.4–3.1
Estuaries	0.2–0.6	<0.6
Gulf of California	0.2	0.2
Walvis Bay	<0.2	0.2
Peru/Chile	<0.1	<0.1
Antarctic margin	0.2	0.2
Other margins	<0.2	1.8
Total removal rates	**5.5–7.5**	**6.5–7.4**

[a] *All fluxes are expressed in units of* $\times 10^{12}$ mol y^{-1}r.
Two sets of estimates for the marine sink are provided. The first set are from: (1) Tréguer, P., et al. (1995). Science *268, 375–379; and (2) DeMaster, D. J. (1981).* Geochimica et Cosmochimica Acta *45, 1715–1732. The second set are from DeMaster, D. J. (2002).* Deep-Sea Research *II 49, 3155–3167.*

16.6.2 Feedbacks in the Crustal-Ocean-Atmosphere Factory for Silica?

Given the large role of BSi production in the oceanic silica cycle, it is not surprising that burial of siliceous hard parts in the sediments is the major regulator of DSi concentrations. An interesting question is whether a negative feedback exists to stabilize the DSi concentration at its current level. If such a feedback exists, it would probably function to decrease the burial rate of BSi if DSi gets too low or conversely, increase the burial rate if DSi gets too high. For example, a decline in DSi can eventually lead to Si limitation in diatoms, thereby lowering the production of BSi and, hence, its burial rate.

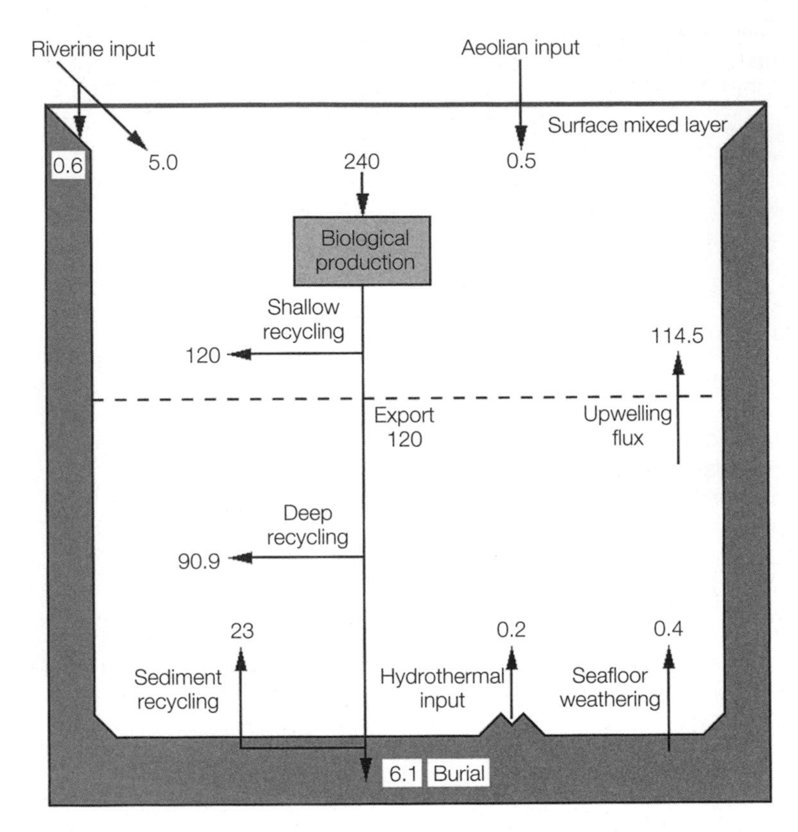

FIGURE 16.11

Biogeochemical cycle of Si in the world ocean at steady state. Fluxes are in Tmol (10^{12}) Si/y. *Source*: After Tréguer, P., *et al*. (1995). *Science* 268, 375–379.

Conversely, perturbations in the burial rate of BSi have the potential to alter the marine silica cycle. For example, changes in sea level affect the expanse of continental shelf. Since BSi burial is more efficient in shelf sediments (because of better preservation), a topographic change that alters the spatial extent of this depositional environment has the potential to alter the size of the DSi reservoir.

Although diatoms appear to have evolved sometime during the Triassic or Jurassic, they did not begin to increase in abundance and diversity until the Cenozoic (65 mybp). Some of this growth spurt is attributed to an increase in turbulent conditions in the ocean associated with greater wind mixing and more intense meridional overturning circulation. This decrease in hydrographic stability is thought to reflect changes in climate caused by the onset of the polar glacial-interglacial cycles and the large meteorite impact that occurred 65 mybp at the end of the Cretaceous period. A very sharp increase in diatom species diversity, and presumably abundance, occurred around 34 mybp at the

Eocene/Oligocene boundary. This shift is thought to reflect an increase in DSi delivery to the ocean associated with the evolution of grasses and grazing ungulates.

Like all plants, the root structure of the grasses acts to mobilize silica from the soils. Grasses are unusual in that they deposit a significant amount of silica in their cell walls as opal phytoliths. This biomineral is highly soluble, at least as compared to the crustal aluminosilicate minerals. As grasses spread in response to a favorable shift in climate at the Eocene/Oligocene boundary, animals evolved to exploit this expanding food resource. These grass grazers served to further mobilize crustal silica by their feeding activities, which moved the phytoliths back into the soils where rain and groundwater could leach and transport DSi into the rivers. If the rise of the diatoms over the Cenozoic era lead to an increased sinking flux of POC to the sediments, the effect on P_{CO_2} and, hence, climate would have been such as to ensure continued turbulent conditions, further favoring the success of the diatoms.

As noted earlier, biogenic hard parts play an important role in determining the degree to which detrital sinking POC survives the trip to the seafloor. Since the hard parts are denser than the POC, aggregates of hard parts and soft parts sink faster than just POC alone. The hard parts also appear to confer a protection against microbial degradation. This suggests that the POC in sinking particles is embedded in the mineral crevices and/or inner chambers of shells, tests, and frustules. Although it is not clear which type of hard part is most effective in improving the export efficiency of POC to the sediments, shifts in plankton community composition are likely to cause changes in the export efficiency of POC by altering the relative magnitudes of the sinking fluxes of detrital BSi and calcite.

One hypothesis for how such shifts are thought to occur is through a "silicate switch" in which increased DSi availability enables diatoms to "outcompete" other phytoplankton species (such as the coccolithophorids). The "switch" could be triggered by an increase in DSi availability caused by an increase in aeolian dust fluxes or continental rock weathering, or by decreased rates of Si removal as a consequence of changes in sea level. If the aeolian dust flux enhances delivery of iron, the efficiency of Si utilization by diatoms increases. Diatoms growing where iron is abundant leave DSi in seawater, having first exhausted the nitrate. If this enhanced iron supply is provided to the Southern Ocean, the DSi left in the surface waters is then available to sink as SAMW forms (Figure 10.10). Meridional overturning circulation carries this water mass to higher latitudes, where it feeds the thermocline leading to the transport of DSi into the overlying mixed layer. This lateral transport of DSi would support an expansion in the growth range of the diatoms, furthering the dominance of the silicifiers over the calcifiers. If such a "silicate" switch does exist, diatom growth could be a important control on the burial rate of POC in the sediments and, hence, atmospheric P_{CO_2} and global climate.

CHAPTER 17 Evaporites

All figures are available on the companion website in color (if applicable).

17.1 INTRODUCTION

On the early Earth, ions were mobilized from volcanic rocks by chemical weathering. Rivers and hydrothermal emissions transported these chemicals into the ocean, making seawater salty. These salts are now recycled within the crustal-ocean-atmosphere factory via incorporation into sediments followed by deep burial, metamorphosis into sedimentary rock, uplift, and weathering. The last process remobilizes the salts, enabling their redelivery to the ocean via river runoff and aeolian transport. In the case of sodium and chlorine, evaporites are the single most important sedimentary sink. This sedimentary rock is also a significant sink for magnesium, sulfate, potassium, and calcium.

Because of their role as an elemental sink, the formation and weathering of evaporites has the potential to affect the salinity of seawater. This can in turn alter climate, because the heat capacity of seawater is a function of its salt content. Changes in the salt content of seawater also have the potential to affect survival of marine biota, particularly the calcifiers.

Although evaporite deposition is currently not very common, evaporite rocks are widespread geographically and temporally. For example, evaporite rocks underlie about 35 to 40% of the surface area of the United States, being distributed through 32 of the 48 contiguous states. Some of these deposits are enormous, containing several million cubic kilometers of salt. Obviously, large fluctuations in evaporite deposition rates have taken place over time. These fluctuations are thought to have resulted from changes in climate, sea level, and plate tectonics. Paleoceanographers use the mineralogy of evaporites to help reconstruct these changes. Evaporite mineralogy also provides insight into past changes in seawater composition.

Evaporites are important sources of economic minerals that have been exploited for at least the past 6000 y. For example, the evaporite mineral trona ($NaHCO_3 \cdot Na_2CO_3 \cdot 2H_2O$) was used by the ancient Egyptians to preserve mummies. Evaporite salts continue to be used for food preservation, construction, road deicing, and in industrial processes. The marine evaporites of Saskatchewan (Canada) are the world's largest source of potash (KCl), which is used as an agricultural fertilizer. In the United States,

15 to 20 million tons of rock salt is mined annually. Petroleum geochemists use sedimentary evaporites as indicators of the likely presence of recoverable petroleum because similar depositional environments are required for their formation. These salt deposits also act to concentrate oil and gas, as evaporite minerals present an impermeable barrier behind which migrating petroleum tends to pool.

The mineralogy, distribution, and formation of evaporites are the subjects of this chapter. The role of evaporite formation and dissolution in determining the salinity of seawater is discussed in Chapter 21.

17.2 FORMATION OF EVAPORITES BY EVAPORATION OF A FIXED VOLUME OF SEAWATER

Evaporites are formed by the precipitation of salts from seawater as water evaporates. For a fixed volume of seawater, the progressive evaporation of water results in a sequence of mineral precipitations determined by the relative solubility of the salts. The sequence illustrated in Figure 17.1 assumes that equilibrium is achieved between the mineral and the brine solution from which the minerals are precipitating. The first precipitates are the carbonates, namely calcite and aragonite. Recall that surface seawater is generally saturated with respect to these minerals, but abiogenic precipitation is rare because of kinetic barriers. The special settings where abiogenic precipitation occurs are ones where large supersaturations present.

As shown in Figure 17.1, about half the water must first be removed before the carbonate minerals begin to precipitate. Precipitation of these minerals removes all the TDIC because Ca^{2+} is present in great excess. Continued evaporation leads to formation of *gypsum* ($CaSO_4 \cdot 2H_2O$), followed by *anhydrite* ($CaSO_4$). Once 90‰ of the water has been removed, *halite* (NaCl) precipitates, along with some magnesium salts [$MgSO_4 \cdot nH_2O$ ($n = 1$, 6, or 7) and $MgCl_2 \cdot 6H_2O$]. The potassium salts (*potash*) deposit last, with the most abundant being *sylvite* or *sylvinite* (KCl). Small amounts of mixed minerals, such as langbeinite ($K_2Mg_2(SO_4)_3$), carnallite ($KMgCl_3 \cdot 6H_2O$), polyhalite ($K_2MgCa_2(SO_4)_4 \cdot 2H_2O$), and kainite ($KCl \cdot MgSO_4 \cdot 3H_2O$), are also formed. As indicated in Figure 17.1, most of the mass of the evaporite minerals is contributed primarily by halite and secondarily by gypsum and anhydrite.

The precipitation sequence presented in Figure 17.1 is a function of temperature and the extent to which equilibrium is maintained with the increasingly saltier brine. Each of these precipitations alters the ion ratios in the remaining seawater. Since the rule of constant proportions is violated, density, rather than salinity, is used to monitor the increasing saltiness of the brine.

The depositional sequences observed in evaporites do not generally follow the predictions obtained from a fixed-volume evaporation. This is largely due to natural processes that act to replenish the evaporated water with fresh seawater. The geohydrological mechanisms by which this occurs are discussed in the next sections of this chapter.

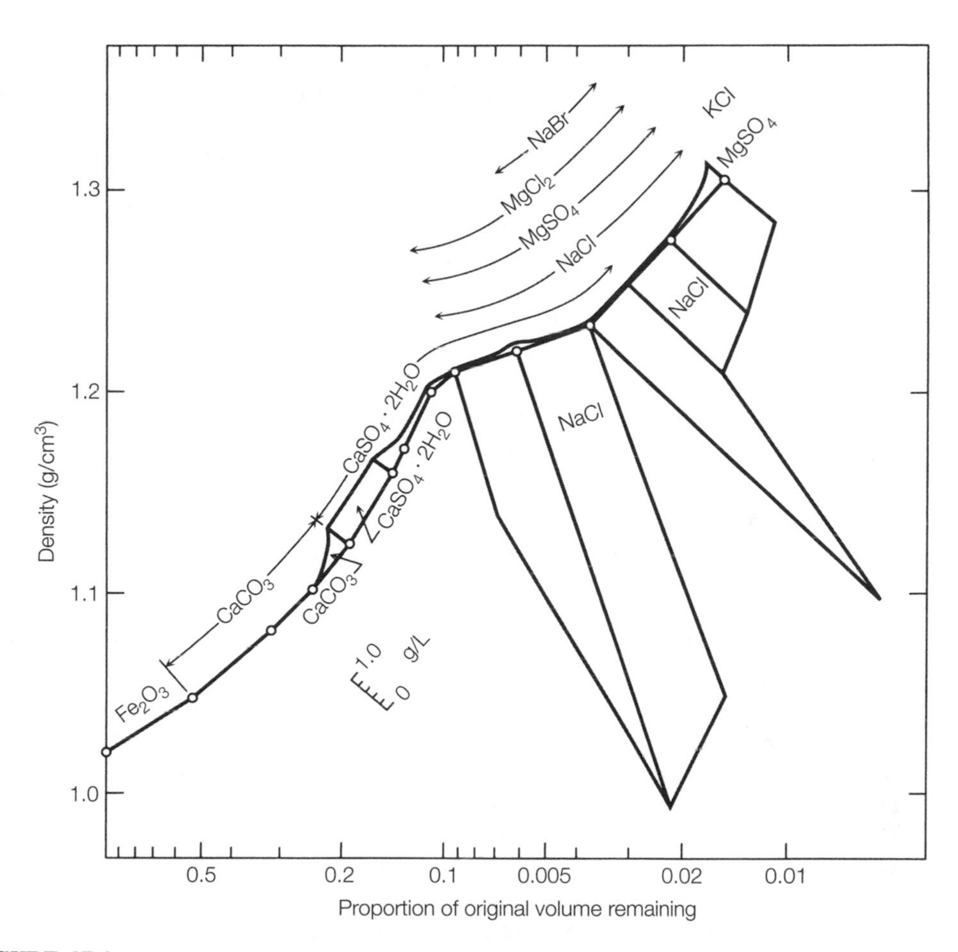

FIGURE 17.1

Brine density versus proportion of original volume remaining in the closed-system evaporation of Mediterranean seawater (initial salinity = 35‰). The concentration of minerals produced (in grams per liter of water) is indicated by the lengths of the lines drawn perpendicular to the curve. *Source*: From Friedman, G. M., and J. E. Sanders (1978). *Principles of Sedimentology*, John Wiley & Sons, Inc., p. 527. See Friedman and Sanders (1978) for data sources.

Diagenesis and catagenesis can alter the evaporite minerals after burial. For example, high temperatures, pressures, and pore-water salinities characteristic of deep burial lead to the conversion of gypsum into anhydrite. Thus, evaporite mineralogy reflects not only the environmental conditions under which the evaporite was formed, but also those under which diagenesis and catagenesis occurred.

Some evaporites form from the evaporation of terrestrial waters. The ions present in these waters are supplied by the chemical weathering of continental rock. Terrestrial waters are more variable in composition than seawater, causing greater variability in the mineralogy of the nonmarine evaporites. Because of relatively high carbonate concentrations, nonmarine evaporites are composed primarily of carbonate minerals. In contrast, the marine forms are mostly chloride and sulfate salts.

Evaporites are very soluble and, hence, are readily weathered. This leads to their poor preservation in the sedimentary rock record, with only a few deposits being greater than 800 million years old. This record has been used to reconstruct the geologic history of seawater over the past 600 million years by assuming that the elemental composition of an evaporite is determined by the chemical composition of its parent seawater. The observation that ancient evaporites vary substantially in their elemental composition suggests that significant shifts in the major ion chemistry of seawater have occurred. As discussed in Chapter 21, the major causes of these shifts are related to changes in the rates of seafloor spreading and associated hydrothermal activity.

17.3 METEOROLOGICAL AND GEOLOGICAL SETTINGS

The formation of substantial evaporite deposits requires two conditions: (1) some mechanism by which salt ion concentrations are kept at supersaturated levels, and (2) a steady resupply of salt ions. The hydrogeologic setting that is most likely to meet these two criteria are shallow-water embayments located in arid climates where sea level is relatively stable and terrestrial runoff is very low or absent.

Shallow-water embayments provide a mechanism to isolate seawater so that evaporation can raise salt ion concentrations. Arid climates are required to ensure that the rate of water loss from evaporation exceeds the rate of water supply by rainfall, groundwater seeps, or river runoff. Seawater can be resupplied continuously via a type of antiestuarine circulation as illustrated in Figure 17.2 or episodically as a result of sea level change, plate tectonics, or very high tides and storm surges.

Because climate, circulation, sea level, and basin morphology are not constant over time, evaporite formation tends to be episodic. This leads to the formation of layers of evaporite minerals interbedded between other sediment types, such as biogenous oozes and lithogenous clays. In some cases, conditions remained stable long enough for thick layers of evaporite minerals to deposit in sequences reflecting their relative solubility. An example is shown in Figure 17.2 illustrating the sequential deposition of evaporite layers in a shallow-water embayment. Formation of such layering requires seawater to have been resupplied at a rate that kept the parent brine supersaturated with respect to the carbonate and sulfate minerals, but not so concentrated as to precipitate halite until near the end of the sequence's deposition. This resupply is thought to be provided by antiestuarine circulation in which evaporation raises the density of the brine, causing it to sink to the bottom and thereby pull fresh seawater from the adjacent ocean into the embayment. As the surface seawater is drawn toward the head of the embayment, evaporation causes its salinity to progressively increase. Since carbonate minerals have the lowest solubility, their saturation concentrations are exceeded first over time and space, followed by the sulfate minerals. A continuing supply of seawater

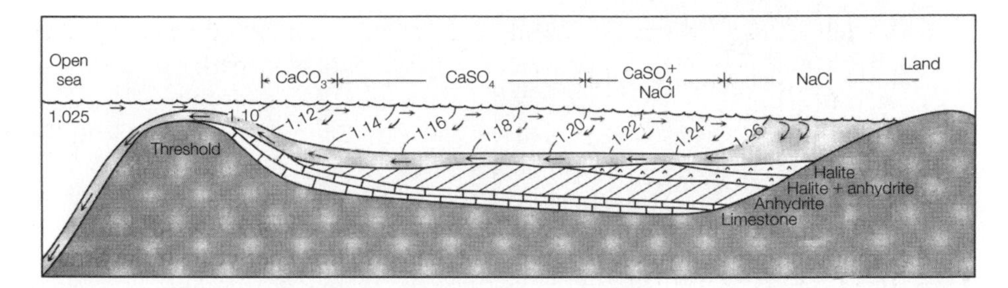

FIGURE 17.2

Schematic longitudinal profile through a semi-isolated basin located in a hot, arid climate and separated from the open sea by a narrow portal. The sill depth, although shallow, is still great enough to permit some two-way flow of surface water. The lines show inferred seawater density (g/cm^3) and the arrows show current directions. The pattern of evaporite deposition is based on the relationships between brine density and precipitate composition as shown in Figure 17.1, assuming that salt particles accumulate on the seafloor through the process of pelagic sedimentation. *Source*: From Scruton, P. C. (1953). *American Association of Petroleum Geologists Bulletin*, 37, 2498–2512.

into the embayment enables formation of thick layers of limestone and anhydrite. The most soluble minerals (NaCl and bittern salts) deposit only toward the head of the embayment where the most highly concentrated brine is present.

The deposition of thick evaporite deposits takes long periods of time. Thus, extended periods of stability in sea level are necessary. This has not been the case over the past 1 million years because of frequent glacial-interglacial cycles. As a result, evaporite deposition is currently restricted to tidal mud flats (*sabkhat*) and coastal salt lakes (*salinas*). Ancient evaporites are comparative giants, with thicknesses of several kilometers and surface areas of hundreds of square kilometers. The mechanisms by which these ancient salt giants formed are not well understood because no modern-day analogue exists. Figure 17.2 illustrates one proposed mechanism. Some others are presented later, following a description of modern evaporites.

17.4 MODERN EVAPORITES

As shown in Figure 17.3, modern evaporites are currently forming in two latitudinal zones, between 15° and 40°N and between 0° and 35°S. These are the latitudinal zones where net evaporation in the open ocean is occurring (Figure 4.9) and, hence, the saltiest surface seawater is found (Figure 4.10). Coastal upwelling also acts to produce arid conditions along coastlines. Cold water rising to the sea surface cools the overlying air mass, causing any water vapor to condense. The resulting air mass is too dry to supply rainwater to the adjacent coastal land areas. As a result, many coastal areas located near upwelling zones, such as those in Peru, Chile, northwestern Africa, and the Baja Peninsula, have large deserts and evaporitic sediments.

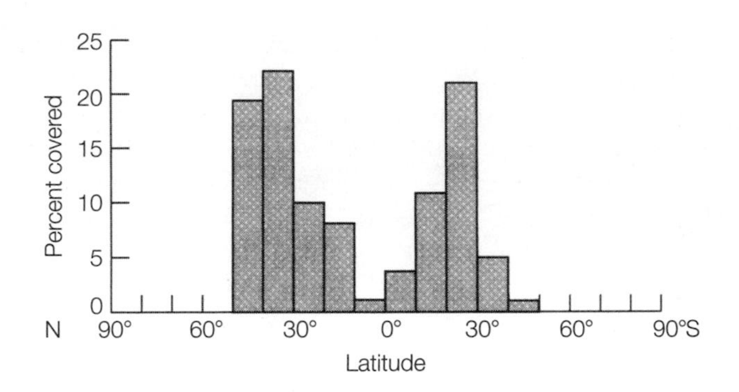

FIGURE 17.3

Latitudinal distribution of the world's modern evaporitic sediments as a percentage of total sediment. *Source*: From Warren, J. K. (1989). *Evaporite Sedimentology*, Prentice Hall, Inc., p. 15.

Table 17.1 Documented Modern Sea-Margin Evaporites.

Sabkha
Abu Dhabi, Arabian Gulf
Sabkha Matti, Arabian Gulf
Bardawil Lagoon, North Sinai coast
Western Nile Delta
Gulf of Suez coast
Tunisian coast
Northwest Australian coast
Spencer Gulf, Australia
Baja California, Mexico
Laguna Madre, Texas
Ojo de Liebre, Baja
Bocana de Virrila, Peru
Salina
South Coast, Australia
Hutt and Leeman Lagoons, Western Australia
Lake MacLeod, Western Australia
Solar Lake, Gulf of Elat
Ras Muhammad, South Sinai coast
Tunisian coast
Pleistocene coast, Sicily
Pekelmeer, Netherland Antilles

Source: After Warren, J. K. (1989). Evaporite Sedimentology. Prentice-Hall, p. 41. See Warren (1989) for data sources.

As noted earlier, modern evaporites are currently forming in only two depositional environments, sabkhat and salinas. These settings are illustrated schematically in Figure 17.4. The locations of specific examples are given in Table 17.1.

A marine sabkha is an intertidal mud flat common in highly arid regions such as the Arabian Gulf and the Baja Peninsula. Sabkhat are commonly located on broad, flat coastal plains where the subtidal zone is populated by coral reefs. This barrier creates shallow-water lagoons that restrict water flow in the intertidal zone. Evaporation causes the salinity of this trapped water to rise. In the Arabian Sea, which is the world's warmest (20° to 34°C), the salinity of the lagoon seawater ranges from 54‰ to 67‰. Because of high biological activity, the sediments are dominated by algal mats, fecal pellets, shells, and other biogenic materials.

Evaporites form throughout the supratidal zone. The lower supratidal receives seawater only during spring high tides and is characterized by gypsum deposits. In some parts of the Arabian Gulf, this region extends 2.5 km inland. The middle supratidal, which is flooded by seawater less than once a month, is 1.5 km wide. Here, gypsum and aragonite are deposited and diagenetically altered to anhydrite and dolomite, respectively. This is one of the few locations where active dolomite formation has been observed.

Periodic drying causes the surface of the tidal mudflats to break into leathery polygonal chips that are tinged with whitish halite crusts. The upper supratidal receives

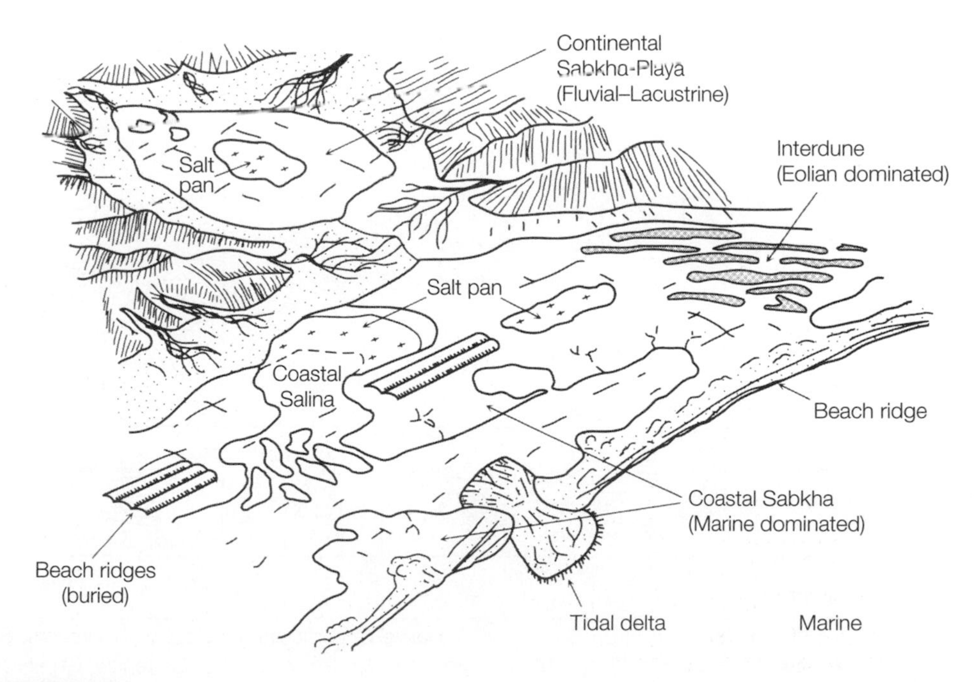

FIGURE 17.4

Modern evaporite settings. *Source*: From Kendall, A. C. (1984), Facies Models, 2nd ed., Geoscience Canada Reprint Series, pp. 259–296.

seawater only from aeolian transport and rare catastrophic floods. This zone is about 4.8 km wide and is covered mostly by anhydrite and halite. Drying is so intense that the winds remove the upper surface. Aeolian transport of this material can produce offshore deposits of evaporites if the relocated materials are rapidly buried.

The deposits in each zone vary in texture, as well as mineralogy, as shown in Figure 17.5. Some of these unique characteristics are preserved after burial and can

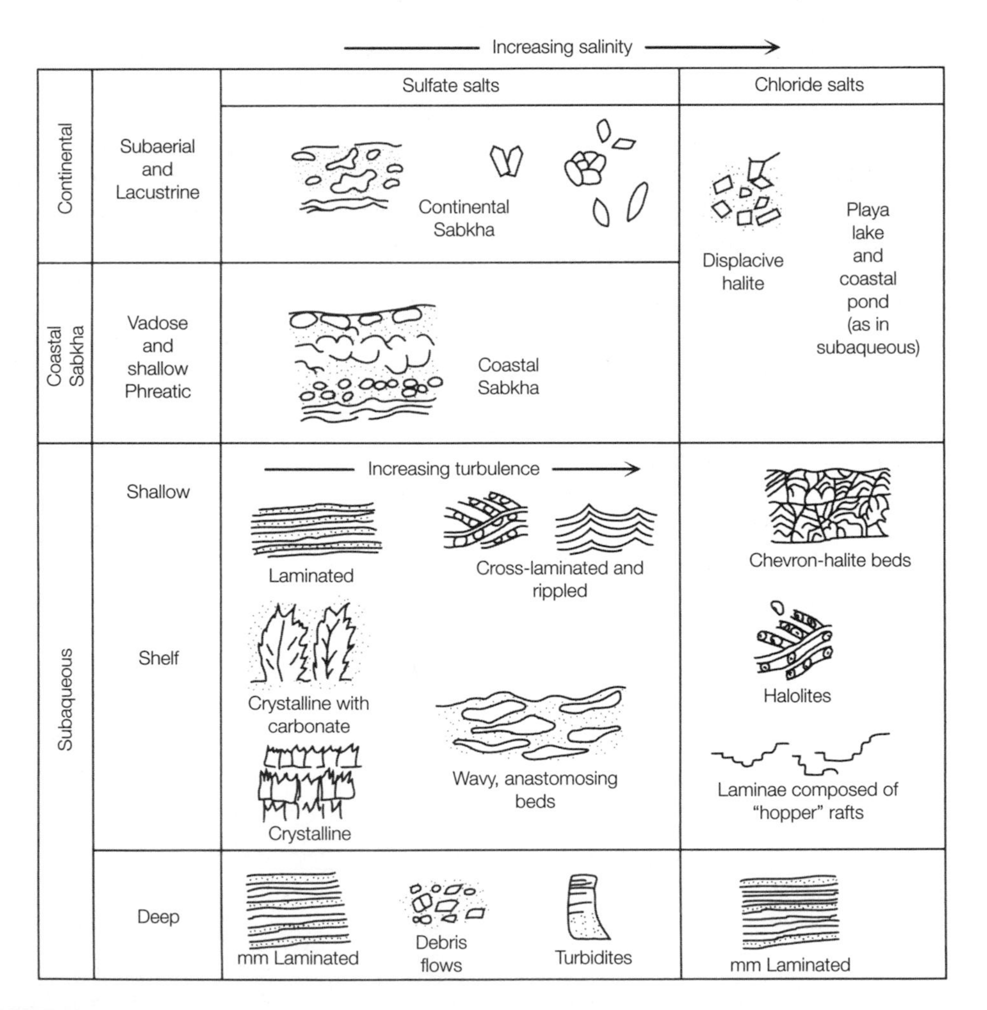

FIGURE 17.5

Summary of the various evaporitic textures indicative of particular physical environments. *Source*: From Kendall, A. C. (1984). Facies Models, 2nd ed., Geoscience Canada Reprint Series, Walter, R. G., ed. pp. 259–296.

be used to determine the conditions under which an evaporite was deposited. Usually diagenesis and relocation by winds, waves, and turbidity currents make such interpretations difficult.

In the event of a decline in sea level, the supratidal zone will move seaward, causing evaporites to deposit on top of old lagoonal sediments. When sea level rises back, it "drowns" these evaporites. These low-amplitude fluctuations in sea level build up laminated sediments in which layers of biogenic oozes and organic-rich muds alternate with evaporites.

Sabkhat also form inland, where river input and saline groundwater seeps contribute salt and water, forming an evaporitic pan. As illustrated in Figure 17.6, these continental sabkhat are far more isolated from the ocean than a marine sabkha. They also contain far less biogenic detritus.

Saline lakes, or salinas, are another type of continental evaporite. Most are located in depressions behind calcareous dunes along the margins of the Mediterranean Sea, the Arabian Gulf, and the western coast of Australia. Seeps of saline groundwater supply salt to these brine pools. In the salinas of southwestern Australia, the evaporites deposit in a bull's-eye pattern. As illustrated in Figure 17.6, stromatolites and carbonates form along the rim with gypsum depositing in the lake's center. These are the only locations where gypsum is presently depositing subaqueously. In some of these salinas, as much as 10 m of gypsum has been deposited over the past 6000 y.

Stromatolites are domal, pillar-like structures composed of fossilized layers of algal mats. Although the organic detritus has been largely replaced by calcium carbonate, enough of the original texture remains to be identifiable as a biogenic deposit. The oldest known fossils are stromatolites, which were deposited approximately 3.45 bybp in coastal marine settings, probably by blue-green algae or cyanobacteria.

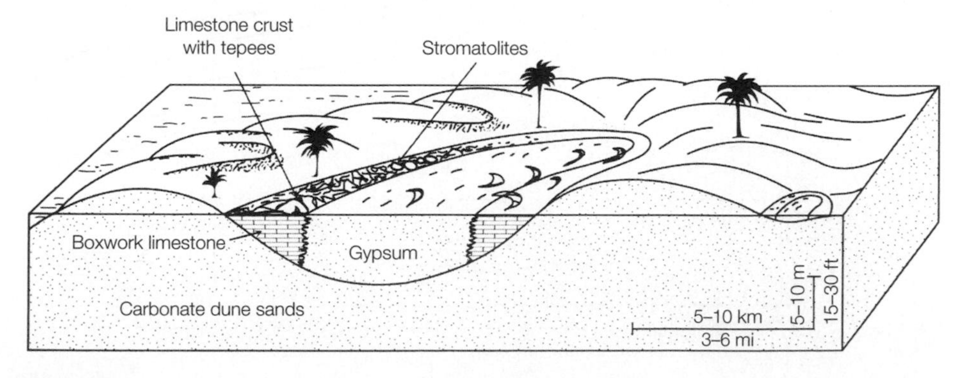

FIGURE 17.6

Depositional facies in the salinas of Southwestern Australia. *Source*: From Warren, J. K., and G. C. St. C. Kendall (1985). *American Association of Petroleum Geologists Bulletin*, 69, 1013–1023.

17.5 ANCIENT EVAPORITES

17.5.1 Geologic Variations in the Rate of Evaporite Deposition

Only a few evaporites have been found that are more than 800 million years old, indicating that most of the salt formed prior to this period has been recycled via uplift and weathering. No evaporites of Archean age have as yet been discovered. The oldest known chemical sediments were deposited 3.45 bybp in what is now western Australia. They appear to have precipitated as shallow-water carbonates. This suggests that sulfate concentrations during the Archean were much lower than present day, probably because of limited oxygenation of the atmosphere and ocean.

As shown in Figure 17.7, the volume of evaporites in the rock record increases from the Proterozoic (2.6 to 0.6 bybp) into the Phanerozoic (0.6 bybp to present), although

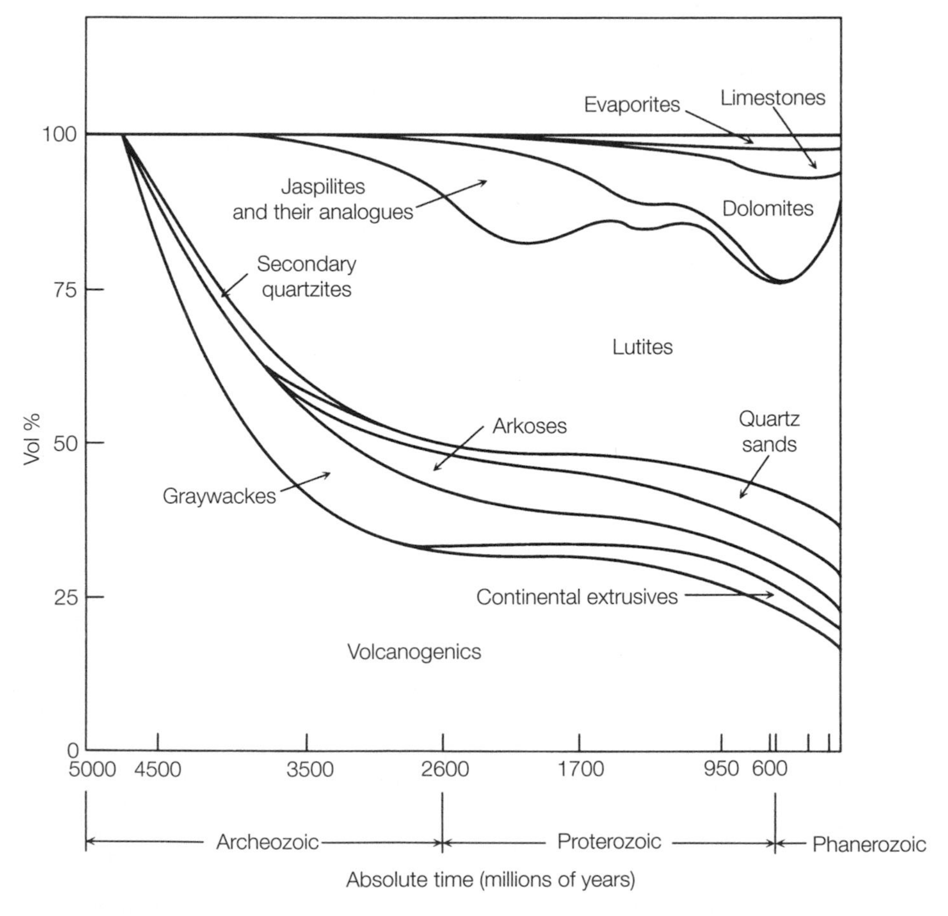

FIGURE 17.7

Volume percent of sedimentary rock as a function of age. *Source*: From Ronov, A. B. (1964). *Geochemistry International* 1, 713–737. Note: Eon timescales have since been redefined.

deposition rates are currently abnormally low. This general rise from the Archean to the Proterozoic to the Phanerozoic could be the result of better preservation, but most geochemists think a change in hydrogeochemical conditions occurred that increased the rate of evaporite formation. This conclusion is supported by the evaporite deposition record during the Phanerozoic. As shown in Figure 17.8, evaporite deposition rates have fluctuated wildly over the past 600 million years, arguing against any wholesale leveling effect from poor preservation.

Calcium sulfate minerals begin appearing in the evaporite record after 1600 to 1700 mybp at which time the oceans become oxic (Figure W8.1). The first episode of massive evaporite formation occurred toward the end of the Proterozoic, about 600 mybp. This seems to be related to the dispersal of a supercontinent, Rodinia, which formed over the late Proterozoic, 1200 to 750 mybp.

During the Phanerozoic, a period of massive evaporite formation occurred around 250 mybp during the Permian and appears to be associated with the dispersal of the Pangean supercontinent. Supercontinents form from the collision of continental plates on time scales of 300- to 500-million-year cycles. Dispersal is achieved by seafloor spreading during which new oceans are formed. Thus, periods of dispersal are characterized by rapid rates of seafloor spreading and enhanced hydrothermal activity. As discussed in Chapter 19, hydrothermal activity is a sink for Mg^{2+} and SO_4^{2-} and a source of Ca^{2+}. Evaporite mineralogy and brine occlusions confirm that Mg^{2+}, SO_4^{2-}, and Ca^{2+} concentrations have changed over the Phanerozoic in patterns that reflect the supercontinent building and dispersal cycles. In the case of the other major ions, modern seawater has a somewhat higher Na^+ and lower Cl^- concentration than the rest of the Phanerozoic, whereas K^+ seems to have remained constant.

Rapid spreading gives rise to new oceans that tend to be relatively shallow, creating excellent conditions for evaporite formation. For example, the dispersal of Pangea during the Permian created a shallow inland sea that extended from present-day West Texas into northwestern Kansas. This sea enabled the formation of a massive evaporite that is 500 to 1500 m thick and contains $10 \times 10^6\ km^3$ of salt.

Other periods during the Phanerozoic in which large volumes of evaporites were deposited include the Devonian, Carboniferous, Triassic, Jurassic, and Cretaceous. The expanse of these deposits in the United States is shown in Figure 17.9. Evaporite rocks underlie about 35 to 40% of the surface area and are found in 32 of the 48 contiguous states. During a short interval in the Messinian age of the Neogene period, massive evaporites were deposited in the Mediterranean Sea. Deposition of these large volumes of salts likely altered the salinity of seawater. In contrast, very little evaporite deposition has occurred during the Quaternary period. At present, evaporites, account for less than 0.1% of the actively forming continental shelf sediments. This scarcity is due to the rapid, high-amplitude changes in sea level caused by frequent glacial-interglacial cycles, particularly during the Pleistocene epoch.

17.5.2 **Modes of Formation**

The ancient massive salt deposits were laid down under geological and climatological settings that do not presently exist. Climate was warmer, so arid zones were more

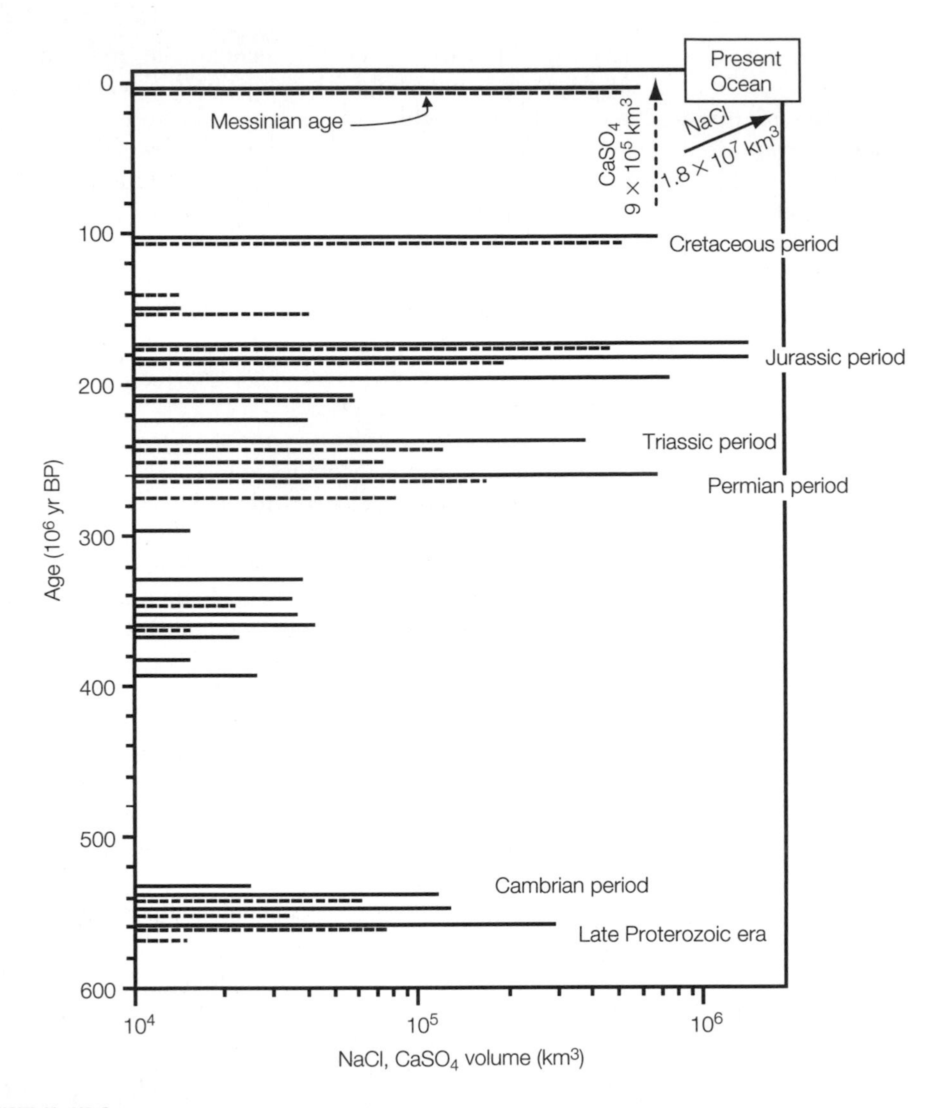

FIGURE 17.8

Timeline of marine evaporite deposition during the Phanerozoic. Shown are the volumes of NaCl (halite, dark line) and $CaSO_4$ (gypsum and anhydrite, dashed line) deposited over time in km^3. The arrows mark the current volumes of NaCl and $CaSO_4$ contained in modern ocean water. These are approximately 1.8×10^7 and 9×10^5 km^3, respectively. *Source*: After Holser, W. T. (1984). *Patterns of Change in Earth Evolution*, Springer, pp. 123–143.

intense and widespread. Sea level was relatively stable. The continental margins were broader and shallower, forming large shallow seas with restricted circulation. Tectonic collisions between lithospheric plates had created several large and shallow marine basins with restricted access to the open ocean. Carbonate reefs were common to both the shelves and inland seas, further restricting seawater circulation.

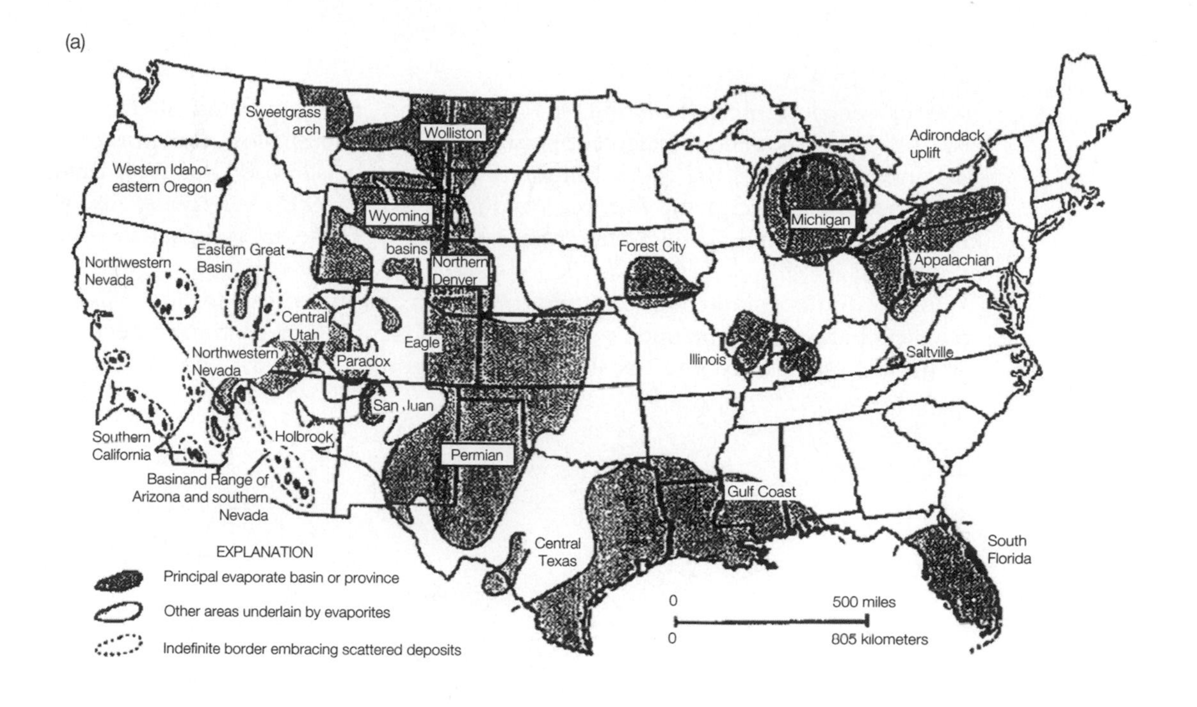

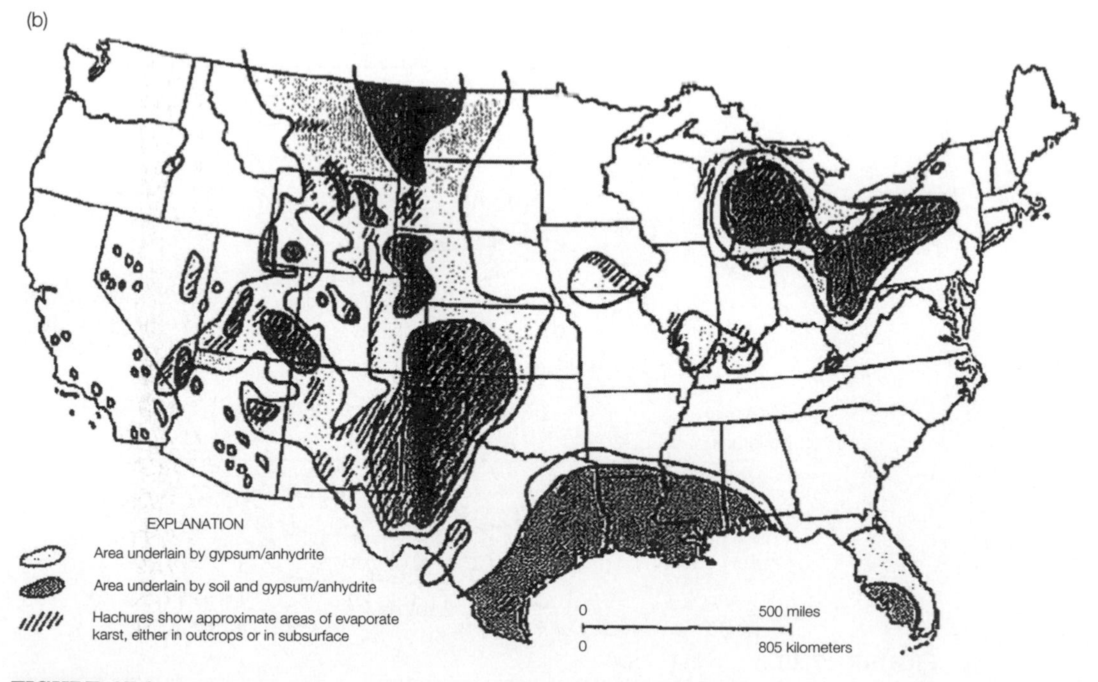

FIGURE 17.9

Maps of the conterminous United States: (a) evaporite basins and districts and (b) distribution of gypsum/anhydrite, salt, and evaporite karst. Karst is limestone or dolomite that has been eroded by dissolution. *Source*: From Johnson, K. S. (1997). *Carbonates and Evaporites* 12, 2–13.

Over very long periods during the Paleozoic era, the continental shelves were covered by only a few centimeters of water. This was too shallow for currents or tidal exchange to circulate the water. Evaporation increased salinities enough to cause halite and gypsum to precipitate, forming evaporites that extend over thousands of square kilometers and range in thickness from 5 to 10 km. These are referred to as *platform evaporites*.

Evaporite deposition also occurred in "barred" basins whose shallow sills restricted water exchange with the open ocean. Several modes of formation have been proposed for these basinal evaporites as illustrated in Figure 17.10. First, if sea level dropped

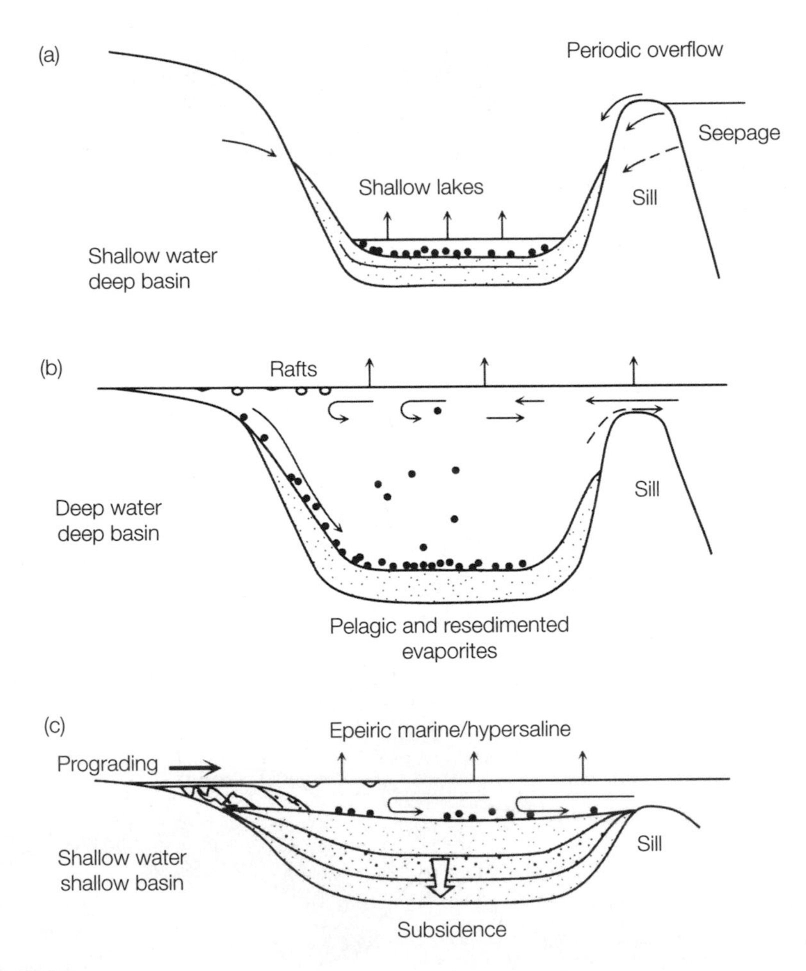

FIGURE 17.10

Basinal settings: (a) shallow-water, deep basin; (b) deep-water, deep basin; (c) shallow-water, shallow basin (epicontinental or epeiric sea). *Source*: From Kendall, A. C. (1984). Facies Models: 2nd ed., Geoscience Canada Reprint Series, Walker, R. G., ed. pp. 259–296.

below the depth of the sill, the basin was completely isolated from the ocean. As the trapped seawater evaporated, basin-wide evaporites should have deposited. Most of the Messinian evaporites in the Mediterranean Sea are thought to have formed in such a shallow-water, deep-basin setting.

Basinal evaporites should also form when the sill is below sea level, as illustrated in Figure 17.10b. In this setting, antiestuarine circulation similar to that illustrated in Figure 17.2 would be present. Salts that precipitated in the surface waters were carried to the basin floor by pelagic sedimentation. The stagnancy of the bottom waters prevented flushing of the particles, thereby ensuring their sedimentation. This type of setting would produce a sequential mineral assemblage. Since the salt content of the basin water should increase with increasing distance from the sill, deposits grade diagonally, as well as vertically, with the most soluble mineral (halite) precipitating last and furthest landward.

Some basinal evaporites appear to have been deposited by turbidity currents as shown in Figure 17.10c. In this mode of formation, salts originally deposited in shallow waters are transported into the basin by turbidity currents. These salts fill in the basin. The decline in water depth then promotes in situ precipitation of evaporites. Subsidence would have kept the basin floor below the sill.

17.5.3 Diapirs

Halite buried beneath heavy loads of sediment will eventually undergo a type of fluidized "creep." Because halite is incompressible, it retains a lower density than the surrounding sediments as the latter become compressed after deep burial. This differential response to increasing pressure causes the sediments to sink through the salt, displacing the halite upward in tall pillars, upwards of 10 km, called *diapirs*.

Diapirs rise episodically and slowly at rates of 0.1 to 1 mm/y for periods of several million years. This is roughly 100 times slower than the lateral speed of the earth's crustal plates. Some diapirs are eventually squeezed out onto the surface of the crust where they form salt islands. Some are protruding through the sides of mountains and are now flowing down slope as a type of salt glacier, called a *namakier*, that moves on the order of a few meters per year.

Contact with rain water causes the exposed evaporites to dissolve, mobilizing salt ions for their return to the ocean. Diapirs are of considerable interest to petroleum geologists because about half of the recoverable petroleum deposits are associated with this salt deposit.

17.5.4 The "Dolomite Problem"

Dolomites compose a significant portion of ancient evaporites. This mineral is essentially a Mg-rich calcite, having variable Mg content. Its empirical formula is commonly given as $MgCa(CO_3)_2$. Having no modern examples of dolomite precipitation beyond those of a few salinas and sabkhat, geochemists have been unable to determine conclusively how the ancient deposits must have formed.

Though the dolomites of Australian salinas are currently being produced by direct precipitation, this mechanism is not thought to have played a major role in formation of the ancient massive dolomite deposits. The current consensus is that the ancient dolomites were produced by some diagenetic process that involved the circulation of hypersaline brines through evaporites prior to their deep burial. The ultimate source of this dolomitizing solution was probably seawater that had been concentrated into a dense, Mg-rich solution. Because of its density, this brine should have sunk through the sediments, thereby displacing the pore waters and seeping through the carbonate deposits.

To convert calcium carbonate to dolomite, some of the calcium must have been replaced by magnesium, requiring the partial dissolution of the carbonate. This process is promoted by contact with acidic pore water, such as occurs in organic-rich sediments because remineralization produces carbon dioxide. This is probably why dolomites are presently forming in detrital algal mats buried beneath sabkhat. The restricted extent of these modern dolomites reflects a kinetic hindrance to precipitation. Apparently dolomite precipitation in this setting is too slow to form substantial deposits when sea level is rapidly fluctuating.

17.6 THE GREAT SALINITY CRISIS OF THE MEDITERRANEAN SEA

The ancient evaporites of the Phanerozoic eon were deposited at rates as fast as 100 m per 1000 y. These rapid rates are thought to have been caused by a lowering of sea level associated with tectonic activity and glaciation. Some of the largest of the salt giants are the Messinian evaporites that formed in the Mediterranean Sea during the late Miocene epoch, 5.5 to 6.5 mybp.

During this time, global cooling led to an intense ice age that caused sea level to drop below the level of the Gibraltar Sill, isolating the Mediterranean Sea from the Atlantic Ocean. Arid conditions promoted evaporation, transforming the Mediterranean into a series of large inland lakes. As these lakes dried up, an enormous volume ($1 \times 10^6\ km^3$) of evaporites was deposited.

The modern Mediterranean Sea has a water volume of $3.7 \times 10^6\ km^3$. Net evaporation presently occurs at a rate of $3.3 \times 10^3\ km^3/y$. If the Straits of Gibraltar were exposed today, the present-day Mediterranean Sea would dry up in about 1000 y. Assuming an average salinity of 37‰ and a seasalt density of $2.165\ g/cm^3$, the Mediterranean would have had to been refilled and evaporated at least 15 times to account for the volume of salt in the Messinian evaporites ($1 \times 10^6\ km^3$). The evaporites on the margins of the Mediterranean Sea tend to be thinner than the deposits in the middle and deepest parts of the basin, which are up to 1600 m in thickness. In some limited areas on the eastern side of the basin, thicknesses up to 3500 m have been observed!

The presence of halite in the Messinian evaporites suggests that Mediterranean seawater must have been very salty during the late Miocene. The absence of fossils indicates that marine organisms could not survive under these conditions. The underlying cause for this "salinity crisis" appears to have been the compounded effects of tectonic activity and a glacioeustatic change in sea level. The presence of erosional surfaces in deposits now deeply buried in the center of the Mediterranean basin suggests that evaporite deposition occurred under shallow-water conditions. The erosional surfaces are probably ancient river valleys cut during a time of dramatically lower sea level. Plate tectonics is thought to have played an important role in creating these shallow-water conditions in two ways: (1) the collision of Africa with Europe closed all connections to the ocean except for shallow sills and (2) uplift kept the basin shallow.

Thus, the Mediterranean Sea must have been refilled in an episodic fashion such that conditions favoring shallow-water evaporite deposition were rapidly reattained. Some geologists have proposed that this was achieved via periodic inflows of seawater from the Atlantic Ocean over the exposed Gibraltar Sill into a nearly dry Mediterranean Sea basin. This must have taken the form of a waterfall hundreds of meters in height! The episodic nature of this process is reflected in the repeating evaporite sequences found throughout the Messinian deposits.

The amount of salt sequestered in the Messinian evaporites represents about 6% of all the salt presently dissolved in the ocean. Its removal must have caused an average reduction of 2‰ in the salinity of the remaining ocean water. This decrease could have had a significant effect on ancient climate as the colligative properties of seawater would have been altered. For example, a decreased salinity would have increased the freezing point of seawater, leading to an increased amount of ice formation, thereby intensifying global cooling because ice reflects insolation. In fact, some scientists hypothesize that the formation of evaporites, as a result of the initial isolation of the Mediterranean Sea, caused the global cooling that occurred during the late Miocene.

CHAPTER

Iron-Manganese Nodules and Other Hydrogenous Minerals 18

All figures are available on the companion website in color (if applicable).

18.1 INTRODUCTION

Hydrogenous minerals are classified into two categories: primary or secondary precipitates. The former are the product of solutes that have precipitated out of seawater, such as the evaporite minerals. The secondary precipitates are the result of reactions between solutes and a preexisting solid. Examples of solids that give rise to the secondary precipitates include clay minerals and biogenic hard and soft parts. In most cases, precipitation is promoted by the presence of solutes that are supersaturated with respect to the hydrogenous mineral. Supersaturations are usually the consequence of some remineralization processes or of redox reactions. In the case of the latter, these redox reactions generally involve a change in oxidation state that converts an element into a species of lower solubility. Most hydrogenous minerals form slowly, at rates less than 1 mm per 1000 years. This restricts their formation to areas on the seafloor where sedimentation rates are nearly zero. Such conditions are common in areas where bottom currents prevent sediment accumulation, such as the tops of seamounts.

Many of the hydrogenous minerals are of commercial interest. As noted in Chapter 17, evaporite minerals have been mined for thousands of years. Other hydrogenous minerals contain precious metals. Indeed, many of the largest metal ore deposits were created in marine settings. Uplift and weathering have brought some of these deposits within easy reach of humans, enabling mining. Another important mineral resource obtained from hydrogenous precipitates is phosphorus. As humans have depleted easily accessible terrestrial deposits, consideration is now being given to mining hydrogenous minerals at their sites of formation on the seafloor.

Two types of metal-rich hydrogenous deposits are formed on the seafloor: *iron-manganese oxides* and *polymetallic sulfides*. The iron-manganese oxides have been deposited as nodules, sediments, and crusts. They are enriched in various trace elements, such as manganese, iron, copper, cobalt, nickel, and zinc, making them a significant repository for some of these metals. Most of the metals in the polymetallic sulfides are of hydrothermal origin. These sulfides have been deposited as *metalliferous sediments* around hydrothermal systems and as rocks that infill cracks within former spreading centers. Other hydrogenous minerals discussed in this chapter are *phosphorite*, *oolitic calcite*, *aragonitic needles*, *barite*, and various clay minerals including *glauconite*, *phillipsite*, *clinoptilite*, and *montmorillonite*.

The formation of hydrogenous minerals requires specific environmental conditions, such as redox level. Thus, their presence in the sedimentary record provides information on past oceanographic conditions. For example, oxic conditions are required for the formation of iron-manganese oxides, so these minerals are not likely to be found in sediments deposited during OAEs. Some hydrogenous minerals participate in the feedback loops that constitute the crustal-ocean-atmosphere factory. For example, phosphorite deposition is thought to play a role in regulating primary production and, hence, climate and atmospheric O_2 levels.

In this chapter, we consider the mineral composition of the hydrogenous minerals and how they form. The evaporite minerals have already been covered in Chapter 17. The hydrothermal minerals (polymetallic sulfides) are discussed further in Chapter 19.

18.2 IRON-MANGANESE OXIDES

Trace metals are introduced to the ocean by atmospheric fallout, river runoff, and hydrothermal activity. The latter two are sources of soluble metals, which are primarily reduced species. Upon introduction into seawater, these metals react with O_2 and are converted to insoluble oxides. Some of these precipitates settle to the seafloor to become part of the sediments; others adsorb onto surfaces of sinking and sedimentary particles to form crusts, nodules, and thin coatings. Since reaction rates are slow, the metals can be transported considerable distances before becoming part of the sediments. In the case of the metals carried into the ocean by river runoff, a significant fraction is deposited on the outer continental shelf and slope. Hydrothermal emissions constitute most of the source of the metals in the hydrogenous precipitates that form in the open ocean.

Not surprisingly, iron and manganese are the most abundant metals in the iron-manganese oxides. Although iron is far more abundant than manganese in Earth's crust (4.17% versus 0.077%), iron and manganese are equally abundant in the oxides. This fractionation of iron and manganese is a result of the more rapid oxidation of iron, causing it to precipitate closer to its point of entry into seawater than does manganese. In contrast, Mn^{2+} is carried long distances through the ocean via advective transport and through pore waters via molecular diffusion. The rate of Mn^{2+} oxidation to Mn^{4+} is enhanced by bacteria, such as *Bacillus* and *Leptothrix*, and by surface catalysis. In particular, iron-manganese oxides are very effective at catalyzing the oxidation of Mn^{2+}.

Reducing conditions also promote the segregation of Fe from Mn as FeS is very insoluble and MnS is soluble. As a result, Mn^{2+} is able to migrate through both oxic and anoxic pore waters, with its aqueous concentration ultimately being controlled by precipitation of $MnCO_3$ (rhodochrosite). Recall from Chapter 7.3.2 that under reducing conditions, bacteria, such as *Shewanella*, solubilize Mn^{2+} and Fe^{2+} using manganese and iron oxides as electron acceptors. These differences in the biogeochemical behavior of iron and manganese collectively cause the deepwater [Fe]/[Mn] to be much lower than is found in crustal rocks, i.e., $([Fe]/[Mn])_{deep\ water} \approx 3.5$ versus $([Fe]/[Mn])_{crustal\ rocks} \approx 54$.

Trace metals, such as copper, nickel, cobalt, zinc, and various rare earth elements, tend to coprecipitate with or adsorb onto Fe-Mn oxides. As shown in Table 18.1, this causes these elements to be highly enriched in the hydrogenous deposits as compared to their concentrations in seawater. The degree of enrichment is dependent on various environmental factors, such as the redox history of the underlying sediments and hydrothermal activity. This makes the composition of the oxides geographically variable.

Fe-Mn oxides are difficult to define stoichiometrically because the process of coprecipitation produces solids of poor crystallinity. The manganese is present in silicate-like octahedral sheets $[(Mn(IV)O_6)^{8-}]$ between which interlayer cations, such as K^+, Ca^{2+}, and Ba^{2+}, can be trapped. Isomorphic substitution of cations into the octahedral sheets is also common. Iron can be present as a substituent or as a distinct mineral layer alternating with the manganese layers. In the nodules, the most commonly observed crystalline structures for manganese are vernadite (δMnO_2), birnessite, and todorokite. The empirical formulae for these minerals are given in Table 18.2, illustrating that they differ in degree of oxidation and metal enrichment. This is thought to reflect differing conditions of formation.

Iron-manganese oxides that form close to the source of hydrothermal emissions, i.e., around mid-ocean ridges and rises, accrete at relatively fast rates on the order of thousands of millimeters per million years. The vent fluids are the primary source of metals to these deposits. The hydrothermal iron-manganese oxides tend to form crusts on boulders that can be dense enough to appear as a pavement. These deposits constitute the outermost ring of a halo of hydrothermal precipitates that lie around active spreading centers. The inner ring is generally covered by polymetallic sulfides (also called metalliferous sediments). These deposits are discussed in Chapter 19.

The Fe-Mn oxides that deposit far from the sources of hydrothermal emissions grow at much slower rates on the order of 1 to a few hundred millimeters per million years. Their formation occurs via two processes: (1) precipitation of trace metals directly from seawater and (2) precipitation of trace metals supplied by postdepositional mobilization from the underlying sediments. These two processes proceed at different rates and produce different patterns of metal enrichment. Generally both processes contribute to the formation of any given iron-manganese concretion, although their relative contributions vary spatially and temporally. This leads to regional variations in the sizes, shapes, and trace metal composition of the iron-manganese oxides. Concretions take the form of crusts, slabs, and nodules. Gradients in chemical composition are commonly found within individual concretions, reflecting changes in growth rate and environmental conditions over time.

The slowest growth rates are found in the Fe-Mn oxides that have formed predominantly by precipitation of solutes from seawater, being on the order of 1 to a few millimeters per million years. Because of slow formation rates, these hydrogenous precipitates tend to form only in areas where sedimentation rates are slow, such as the abyssal plains of the mid-Pacific Ocean, or where bottom currents are strong enough to prevent sediment accumulation, such as on submarine seamounts and plateaus.

Table 18.1 Average Compositions of the Earth's Upper Continental Crust, Shale, Iron-Manganese Oxides, Phosphorite, and Various Types of Marine Sediments (All in Units of ppm, Unless Noted otherwise), along with Seawater and a Hydrothermal Vent Solution from the East Pacific Rise (both in Units of 10^{-9} g L^{-1}).

		Particulate Concentrations in ppm (wt/wt) Unless Otherwise Noted								Dissolved Concentrations (ng/L)	
Element	**Z (atomic #)**	**Upper Crust**	**Shale**	**Pelagic Clay**	**Phosphorite**	**Fe-Mn Nodule**	**Fe-Mn Crust**	**MOR Basal Sediment**	**MOR Ridge Sediment**	**Seawater**	**Hydro-thermal Fluid**
Li	3	23	66	57	5	80		125		1.8×10^5	94×10^5
B	5	12	100	230	16	300		123	500	4.5×10^6	6.0×10^6
F	9	700	740	1,300	31,000	200		466		1.3×10^6	0.14×10^6
Mg (%)	12	1.64	1.5	2.1	0.18	1.6	0.88	2.08		1.3×10^9	0
Al (%)	13	7.83	8.8	8.4	0.91	2.7	0.41	2.73	0.5	300	120,000
Si (%)	14	30	27.5	25	5.6	7.7	2.2	10.8	6.1	2.5×10^6	4.5×10^8
P	15	860	700	1,500	138,000	2,500	3,900		9,000	65,000	18,000
S	16	530	2400	2000	7200	4700				8.98×10^8	1.3×10^7
K (%)	19	2.56	2.66	2.5	0.42	0.7	0.38	1.15		3.9×10^8	9.5×10^8
Ti	22	3,300	4,600	4,600	640	6,700	7,700		240	10	
V	23	140	130	120	100	500	500		450	2,150	
Cr	24	69	90	90	125	35	9.1	15	55	252.6	

Mn (%)	25	0.077	0.085	0.67	0.12	18.6	20.4	6.1	6	72	4.9×10^7
Fe (%)	26	4.17	4.72	6.5	0.77	12.5	12.3	20	18	250	1.39×10^8
Co	27	17	19	74	7	2,700	8,400	82	105	1.2	13,000
Ni	28	55	50	230	53	6,600	3,900	460	430	530	
Cu	29	39	45	250	75	4,500	380	790	730	210	2.8×10^6
Zn	30	67	95	170	200	1,200	540	470	380	320	6.9×10^6
As	33	1.6	13	20	23	140	230		145	1,700	35,000
Se	34	0.14	0.6	0.2	4.6	0.6		2.6		145	4,800
Sr	38	350	170	180	750	830	1,200	351		7.8×10^6	5.8×10^6
Mo	42	1.6	2.6	27	9	400	370		30	10,300	
Ag	47	0.06	0.07	0.11	2		0.09	0.18	6.2	2.5	4,000
Cd	48	0.1	0.3	0.42	18	10	3	0.4 4	76	20,000	
Sn	50	3.3	3	4	3	2		0.6		0.6	
Ba	56	570	580	2,300	350	2,300	1,000	6,230	6,000	1.5×10^4	1.5×10^6
Au (ppb)	79	2.3	2.5	2	1.4	2	250		16	0.03	
Pb	82	17	20	80	50	900	1,400	100	152	2.7	75,000

MOR = mid-ocean ridge.
Source: After Li, K. Y.-H., and J. E. Schoonmaker (2003). Treatise on Geochemistry. Elsevier Ltd.

Table 18.2 Manganese Minerals in Marine Mn Nodules and Crusts.

Name	Other Names	Empirical Formula	Notes
Tetravalent oxides			
Pyrolusite		β-MnO_2	Single chains of $(MnO_6)^{-8}$ octahedra
Ramsdellite		MnO_2	Double chains of $(MnO_6)^{-8}$ octahedra
Nsutite	Vernadite	δ-MnO_2	Intergrowths of single and double chains of $(MnO_6)^{-8}$ octahedra
		$MnO_2\,nH_2O$	
		$MnO_2\,nH_2O{\cdot}m(R_2O, RO, R_2O_3)$ where R = Na, Ca, Co, Fe, Mn	
Hollandite		$(Ba,K)_{1-2}Mn_8O_{16} \cdot xH_2O$	Open tunnels permit large cation incorporation
Cryptomelane		$K_{1\text{-}2}Mn_8O_{16} \cdot xH_2O$	
Psilomelane	Romanechite	$(Ba,K,Mn,Co)_3(O,OH)_6Mn_8O_{16}$	
Todorokite	10 Å Manganite	$(Na,Ca,K,Ba,Mn,)Mn_3O_7 \cdot xH_2O$	
Birnessite	7 Å Manganite	$(Ca,Na)(Mn^{2+}Mn^{4+})_7O_{14} \cdot xH_2O$	
Lithiophorite		$(Al,Li)(OH)_2MnO_2$	Common in the supergene zone
Trivalent oxides			
Bixbyite		α-$(Mn,Fe)_2O_3$	0.3% Fe_2O_3
Spinel structures			
Hausmannite		Mn_3O_4	Up to 7 mol% Fe_3O_4
Jacobsite		Mn_3O_4	0.45 mol% Fe_3O_4
Hydroxides			
Manganite		γ-$MnOOH$	
Pyrochroite		$Mn(OH)_2$	Brucite structure
Silicates			
Braunite		Mn_7SiO_{12}	
Carbonates			
Kutnahorite		$CaMn(CO_3)_2$	
Rhodochrosite		$MnCO_3$	

x = Variable moles of water.
Source: From Maynard, J. B. (1983). Geochemistry of Sedimentary Ore Deposits, *Springer.*

The Fe-Mn oxides that form from the diagenetic remobilization of sedimentary metals accrete at faster rates, on the order of hundreds of millimeters per million years.

18.2.1 **Iron-Manganese Nodules**

The most well studied of the Fe-Mn oxides are the nodules. These deposits are found on the seafloor throughout the world's ocean. Nodule densities are highest in regions of low sedimentation rate, such as beneath the Southern Ocean and the mid-ocean gyres, and where bottom currents prevent sediment from accumulating, such as the tops of seamounts and mid-ocean ridges and rises. As shown in Figure 18.1, the largest expanses of high seafloor with high nodule densities are found in the open ocean of the Pacific. In some regions, nodule densities are so high that pavements have formed, such as illustrated in Figure 18.2.

Fe-Mn nodules represent a very large metal ore resource because they contain significant amounts of precious metals, having as much as 1% w/w Ni, Cu and Co. For example, within the Cook Islands' Exclusive Economic Zone (EEZ), which covers 2 million km^2, nodule densities are on the order of 60 kg/m^2, translating into a total mineable weight of 7.5 million metric tons. The nodules from this region have the highest Co content yet found, collectively representing a potential ore deposit of 32 million

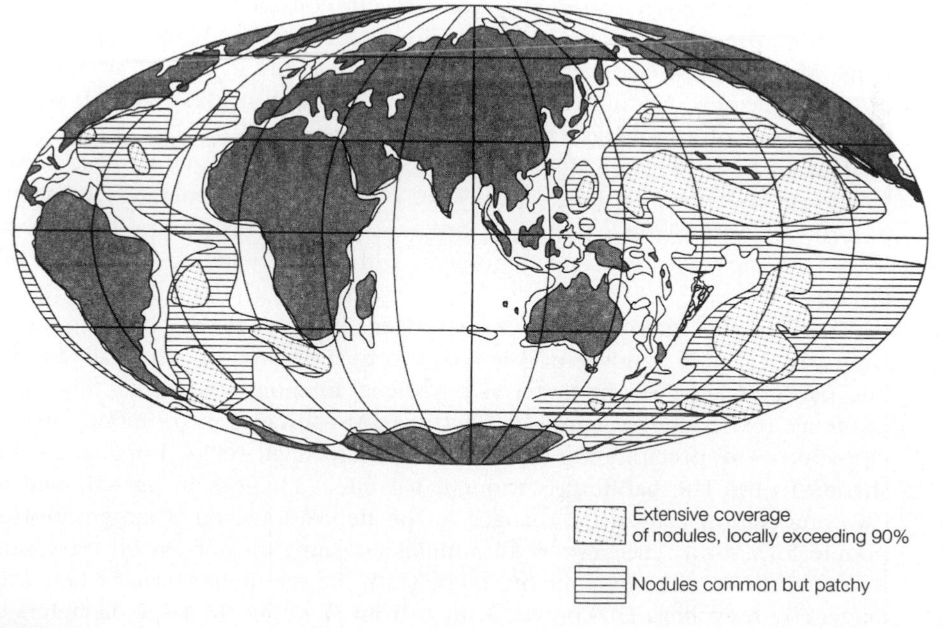

FIGURE 18.1

General distribution of Fe-Mn nodules in the Pacific and Atlantic oceans. *Source*: From Kennett, J. (1982). *Marine Geology*, Prentice Hall, Inc., p. 500. After Cronan, D. S. (1977). *Marine Manganese Deposits*, Elsevier Science Publishers, pp. 11–44.

FIGURE 18.2

Photograph showing a field of unusually closely spaced iron-manganese nodules on the seafloor. Courtesy of the Woods Hole Oceanographic Institution, Woods Hole, MA.

tons. Unfortunately, they are located in water depths of 5000 m, making mining technically challenging and expensive. Nodule pavements are also found in shallower depths (around 1000 m), such as on the Blake Plateau in the western North Atlantic and the Agulhas Plateau, south of South Africa. The first trial of a prototype nodule-mining system took place in 1970 on the Blake Plateau, off Florida.

As shown in Table 18.3, most of the precious metals found in Fe-Mn nodules are currently used for steel production. As a result, considerable effort has been expended in locating the densest concentrations of nodules with the highest precious metals contents. Most of this scientific information was collected in the 1970s and 1980s as part of the Manganese Nodule Project (MANOP). The goal of this research was to facilitate development of a profitable mining industry.

The nodules range in diameter from 20 μm to 15 cm, with most between 1 and 10 cm. As shown in Table 18.4, they are composed predominantly of manganese and iron oxides (about 20% each). They are generally dark brown and have the appearance of dirty, lumpy potatoes or hamburgers, with some being spherical and others more discoidal.

The nodule interiors are characterized by concentric banding, as shown in Figure 18.3. These bands generally conform to the shape of a centrally located particle. The particles are thought to act as a nucleus, initiating and promoting the deposition of metals reflecting the catalytic effect of Fe-Mn surfaces in promoting manganese precipitation. The precipitation nuclei are typically small rocks, bone fragments, or even shark's teeth. The banding is thought to reflect changes in growth rate and chemical composition caused by changes in the depositional environment and/or mode of nodule formation. The very small nodules (<1 mm) do not usually have nuclei. These micronodules tend to become buried beneath the sediment surface where they undergo diagenetic recycling. This prevents them from attaining the large diameters seen in the nodules that have precipitation nuclei.

From a mineralogical perspective, Fe-Mn nodules are composed of layers of manganese oxides (vernadite, todorokite, and birnessite), hydrous Fe oxide $FeOOH \cdot H_2O$), aluminosilicates, quartz, and feldspar. Like the rest of the Fe-Mn oxides, the nodules

Table 18.3 Mineral Consumption, Resources, and Value of Iron, Manganese, Nickel, Copper, Cobalt, Phosphorus and Barium.

Element	2005			Reserves,[a] million metric tonnes	Years of Reserves at Current Global Production Rate	$ (U.S.) per metric tonne	Total Value (Million $ US)		Primary Uses
	U.S. consumption, million metric tonnes	U.S. production, million metric tonnes	Global Production, million metric tonnes				U.S. consumption	Global Production	
Mn	0.76	0	9.7	430	44	$ 471	$ 358	$ 4,569	Steel production, directly in pig iron manufacture and indirectly through upgrading ore to ferroalloys, dry cell batteries, plant fertilizers and animal feed, and brick colorant.
Fe	57	0.055	1520	160,000	105	$ 44	$ 2,508	$ 66,880	Steel production, directly in pig iron manufacture.
Ni	0.221	0	1.5	62	41	$ 14,538	$ 3,213	$ 21,807	45% nonferrous alloy and superalloy production, 36% stainless and alloy steels, 14% electroplating.
Cu	2.29	1.15	14.9	470	32	$ 3,638	$ 8,330	$ 54,200	Building construction, 49%; electric and electronic products, 21%; transportation equipment, 11%; industrial machinery and equipment, 9%; and consumer and general products, 10%.
Co	0.011	0	0.0524	7	134	$ 34,833	$ 383	$ 1,825	43% superalloys, which are used mainly in aircraft gas turbine engines; 9% was in cemented carbides for cutting and wear-resistant applications; 22% was in various other metallic uses.
P	40.5	38.3	148	18,000	122	$ 28	$ 1,130	$ 4,128	45% fertilizer, rest is used for food-additive and industrial applications.

(Continued)

Table 18.3 *(Continued)*

Element	2005						Total Value (Million $ US)		Primary Uses
	U.S. consumption, million metric tonnes	U.S. production, million metric tonnes	Global Production, million metric tonnes	Reserves,[a] million metric tonnes	Years of Reserves at Current Global Production Rate	$ (U.S.) per metric tonne	U.S. consumption	Global Production	
Ba	2.8	0.5	7.6	200	26	$ 36	$ 100	$ 271	Filler, extender, or weighting agent in paints, plastics, and rubber. Others: aggregate in high-density concrete used for radiation shielding, contrast medium in medical X-rays, and ingredient in faceplate glass of cathode ray tubes used in televisions and computer monitors.

[a] *That part of the reserve base that could be economically extracted or produced at the time of determination. The term* reserves *does not signify that extraction facilities are in place and operative. Reserves include only recoverable materials.*

Source: From U.S. Geological Survey, Mineral Commodity Summaries, January 2006, http://minerals.er.usgs.gov/minerals/pubs/mcs/2006/mcs2006.pdf.

Table 18.4 Average Percent Composition of Iron-Manganese Nodules from the Different Oceans.

Minerals	South Pacific[a]	North Pacific[a]	West Indian[a]	Atlantic[b]	Favorable North Pacific Area[c]	
					Red Clay	Siliceous Oozes
Manganese	16.61	12.29	13.56	16.1	17.43	22.36
Iron	13.92	12.00	15.75	21.82	11.45	8.15
Nickel	0.433	0.422	0.322	0.297	0.76	1.16
Copper	0.185	0.294	0.102	0.109	0.50	1.02
Cobalt	0.595	0.144	0.358	0.309	0.28	0.25

Data from: [a]*Cronan, D. S. (1967).* The geochemistry of some manganese nodules and associated pelagic deposits, *Doctoral dissertation, Imperial College, University of London; Cronan, D. S., and J. S. Tooms (1969).* Deep-Sea Research, *16, 335–359.*
[b]*Cronan, D. S. (1972).* Nature Physical Science, *235, 171–172.*
[c]*Horn, D. R., et al. (1972). Ferromanganese Deposits of the North Pacific, National Science Foundation, pp. 40–54.*
Source: From Ross, D. A. (1982). Introduction to Oceanography, *Prentice Hall, p. 411*

exhibit regional variations in trace metal composition and gradients within individual nodules. These differences are a consequence of temporal and spatial differences in their mode of formation. For example, the tops of the nodules form from precipitation of dissolved metals from seawater. The bottoms tend to grow from precipitation of metals remobilized by diagenesis in the underlying sediments. As shown in Figure 18.4, the hydrogenous mode produces more spheroidal nodules and the diagenetic mode, more discoidal nodules. Both of these formation modes are discussed further in the next section.

One of the most curious aspects of the nodules is their very slow growth rates. Radiochemical analyses indicate that the nodules grow, or accrete, at rates ranging from 1 to a few hundred millimeters per million years. Thus, a nodule that is 1 cm in diameter is 10,000 to 1 million years old! The average rate of sedimentation on the abyssal plains, the location of most of the larger nodules, is approximately 1 mm per thousand years. Thus, the sediments are accumulating much faster than the nodules are growing, suggesting that the nodules should become buried. Instead, 75 % of the nodules lie on the sediment surface. The rest are located within the top 4 m of the sedimentary column. Hypotheses to resolve this paradox are discussed later.

Deep-Sea Nodule Formation Processes

Several geochemical processes are responsible for the accretion of metals onto Fe-Mn nodules. In the case of hydrogenous precipitation, the metal ions are supplied by

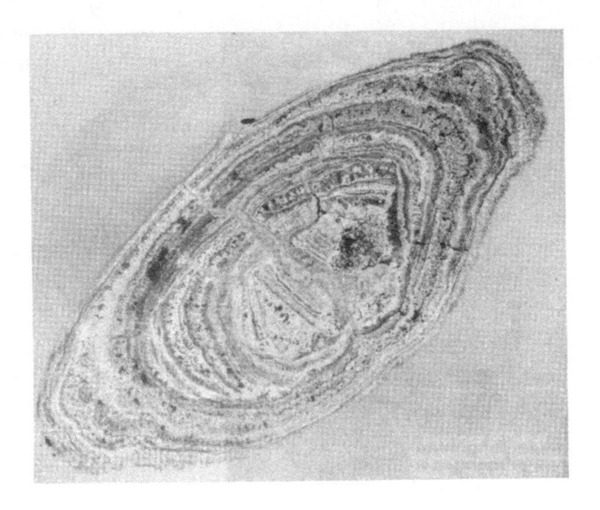

FIGURE 18.3

A polished cross-section (5.3 cm) of a manganese nodule photographed in vertical illumination. The nodule is from the North Pacific and illustrates concentric banding as well as the presence of a distinct central nucleus. Bright areas are opaque oxides; dark areas are mostly clay-sized silicate particles. The core, which nucleated the nodule's growth, consists of broken older nodule fragments. The thin smooth white layer along the left surface marks the top of the nodule as it rested on the seabed. It is enriched in Fe and depleted in Mn, Ni, and Cu compared to the bottom of the nodule. Other layers of different reflectivity, grading from bright to dark, indicate that the nature and rate of mineral deposition varied over time. As nodules increase in size, transverse as well as concentric fractures develop and enlarge, so that the nodules eventually fall apart. *Source*: From Sorem, R. K., and R. H. Fewkes (1979). *Manganese Nodules: Research Data and Methods of Investigation*, Plenum Press, p. 521.

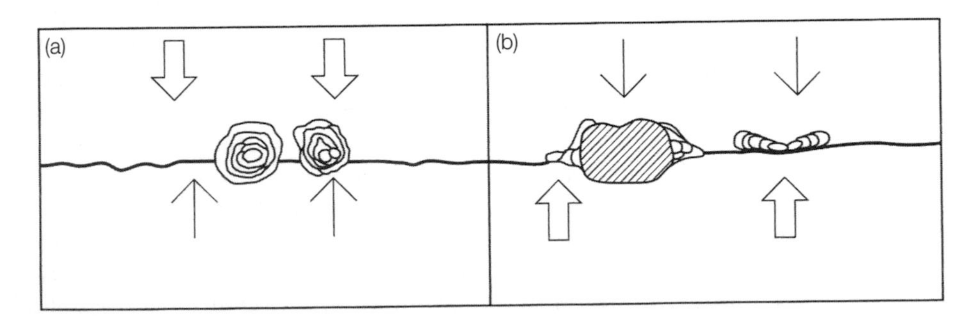

FIGURE 18.4

Schematic representation of manganese nodule end-member morphologies. The size of the arrows indicates the proportion and direction of metal supply. (a) Typical situation in the open ocean with the nodules lying on an oxidized sediment substrate; dominant mode of formation is hydrogenous. (b) Typical situation in nearshore and freshwater environments with nodules lying on a sediment substrate that is partly reducing in character. Dominant supply of metals is via interstitial waters from below the substrate surface. *Source*: From Chester, R. (2003). *Marine Geochemistry*, 2nd ed. Blackwell, p. 425.

seawater. In contrast, the diagenetic processes involve mobilization of metals from the surrounding sediments. As summarized in Table 18.5, these processes occur at different rates and lead to the deposition of Fe-Mn oxides that differ in mineralogy and metal content. Some nodules have been formed by the concurrent operation of more than one of these processes. Many also contain evidence of episodic changes in their modes of deposition.

Hydrogenous Precipitation

The hydrogenous mode of nodule formation involves the direct precipitation of solutes from seawater via the following reactions:

$$4Fe^{2+}(aq) + O_2(aq) + 8OH^-(aq) \rightarrow 2Fe_2O_3(s) + 4H_2O(l) \quad (18.1)$$

$$6Mn^{2+}(aq) + O_2(aq) + 12OH^-(aq) \rightarrow 2Mn_3O_4(s) + 4H_2O(l) \quad (18.2)$$

$$3Mn_3O_4(s) + 12H^+(aq) \rightarrow 3\ \delta\text{-}MnO_2(s) + 6Mn^{2+}(aq) + 4H_2O(l) \quad (18.3)$$

The products are thermodynamically favored under the oxic alkaline conditions that are characteristic of most of the ocean. Reaction rates are slow, so metal oxides tend to precipitate onto detritus or preexisting nodules because of the catalytic effect of the surfaces. The Fe and Mn are supplied by both river and hydrothermal sources. For Mn, these two sources are about equal.

This mode of formation dominates in red clay sediments and produces rather smooth, spherical nodules with lower Mn/Fe ratios than those produced via the diagenetic mode. These nodules tend to be enriched in Fe, Co, and Pb. Their accretion rates are slow—on the order of millimeters per million years and even less in the southwest Pacific.

Formation by Diagenesis

The diagenetic mode of nodule growth dominates in sediments lying beneath waters of moderate productivity, such as the Equatorial Pacific. As discussed in Chapter 12, Fe and Mn are remineralized from suboxic sediments as a result of heterotrophic microbial metabolic activity. The remineralized metals diffuse upward through the pore water until they encounter a nodule in the oxic zone and precipitate onto its surface. The resulting nodules are pebbly and discoidal with growth rates on the order of hundreds of millimeters per million years. As noted in Chapter 12, postdepositional diagenetic remobilization drives Fe deeper into the sediments than Mn, so the Mn/Fe ratio tends to be higher in these nodules than the hydrogenous ones. The source of metals to sediments to fuel diagenetic nodule growth includes biogenic hard and soft parts, with Fe delivered as part of calcareous remains and Ni, Cu, and Zn as part of BSi. Thus, the diagenetic mode of growth enriches nodules in Mn, Ni, Cu, and Zn. (Little of the soft parts survives the trip to the deep sea and, thus, is a smaller source of metals to the nodules than the hard parts.) Some of the diagenetic remobilization appears to occur under oxic as well as suboxic conditions as described next.

Table 18.5 Deep-Sea Iron-Manganese Nodule Modes of Formation.

Mode of Accretion	Source of Metals	Nodule Growth Rate (mm/10^6 y)	Dominant Mn Oxide Mineralogy	Mn Content	Mn/Fe	Mn/Ni
Hydrogenous precipitation	Direct precipitation or accumulation from seawater. Nodule surfaces undergoing accretion are not in direct contact with sediment.	1–2	δ-MnO_2	22%	~1	30–50
Oxic Diagenesis	Metals remobilized from sediments lying in the oxic zone. Remobilization likely occurs in anoxic microzones adjacent to nodules. Bioturbation is an important metal transport agent. Some nodules now found in oxic sediments were likely formed during times when the redox boundary was closer to the seafloor.	10–50	Todorokite (high Cu and Ni content)	32%	5–10	15–20
Suboxic Diagenesis	Metals remobilized from reducing sediments. Upward diffusive transport through pore waters supplies metals to nodule bottoms. Accretion is episodic, occurring only when the depth of the redox boundary rises close to the sediment-water interface.	100–200	Todorokite/Birnessite (low Cu and Ni content)	48%	20–70	60–200

Source: After Table 11 in Dymond, J., et al. (1984). Geochimica et Cosmochimica Acta *48, 931–949.*

Suboxic Diagenesis

Nodule growth under suboxic conditions dominates in sediments underlying areas of high biological productivity where the sedimentation rate of POM is large enough to bring the reducing oxic-anoxic boundary close to the sediment surface. As illustrated in Chapter 12.5.3, the continuing accumulation of sediment eventually transports the former surface layer through the redox boundary, causing it to enter the reducing zone. If sufficient organic matter is present, any Fe-Mn oxides in the layer will remineralized via heterotrophic microbial activity as per the equations in Table 12.1. Some of iron and manganese resolubilized in the sediments comes from the remineralization of biogenic hard and soft parts as plankton naturally concentrate both metals in their tissues and shells (Tables 11.5 and 11.6).

The resolubilized metals that diffuse upward precipitate at the redox boundary as insoluble oxides. If the redox boundary lies close to the sediment surface, precipitation can occur onto the surfaces of nodules lying at or near the sediment-water interface. Nodules formed under these conditions are notable for their high manganese contents ([Mn]/[Fe] > 20) and rapid accretion rates (100 to 200 mm/10^6 y). Accretion is thought to occur episodically rather than at a uniform pace. Nodules exhibiting the highest degree of formation via suboxic diagenesis are found in marginal and hemipelagic sediments because high sedimentation rates of POM are required. Very high sedimentation rates bury the nodules and prevent them from growing. Thus, nodules are rarely found in estuaries and upwelling areas.

Nodules formed via suboxic diagenesis tend to have a low iron content because most is precipitated as FeS below the redox boundary. This is true for other trace metals, such as Cu and Ni, which form highly insoluble sulfides. Because biogenic grains tend to be incorporated in these nodules, all metal concentrations tend to be low, making these nodules economically unattractive.

Oxic Diagenesis

Most Fe-Mn nodules are located in areas where the POM flux is not high enough to keep the redox boundary close to the sediment surface. The oxic conditions in these sediments inhibit postdepositional remobilization. Nevertheless, evidence exists that nodules can grow under oxic conditions. This is seen in the nodules collected from oxic siliceous oozes in the tropical North Pacific. The nodule bottoms are relatively enriched in Cu, Ni, and Zn. These metals are thought to be derived from BSi, suggesting they are somehow being mobilized under oxic conditions. Nodule growth via oxic diagenesis occurs at rates of 10 to 50 mm per million years, i.e., somewhat more slowly than growth via suboxic diagenesis.

Three mechanisms have been proposed to explain how particulate metals could be transported within such sediments so as to support the growth of Fe-Mn nodules: (1) anoxic microzones, (2) bioturbation, and (3) shifts in the depth of the redox boundary over time. Anoxic microzones are present within fecal pellets and the interiors of radiolarian shells where detrital POM is still present. Metals mobilized within these microzones should be able to diffuse through the sediments for substantial distances before becoming oxidized. Because of the catalytic effect of surfaces, once these metals encounter a nodule they should readily precipitate. In the equatorial Pacific, where the sediments are radiolarian oozes, such metal mobilization and precipitation appears to occur in two stages. First remineralized Mn^{2+} precipitates as MnO_2 onto the surface of the radiolarian shells, while the iron precipitates as FeOOH within the shells' interiors. As the BSi dissolves, the iron oxides react with the remineralized Si and volcanic ash, which is common to these sediments, to form a clay mineral called nontronite. As the BSi dissolves, the manganese is remineralized and, hence, available for precipitation onto nodules. This results in a relatively Mn-rich nodule.

Bioturbation is another likely agent facilitating metal transport in oxic sediments by ensuring that iron- and manganese-rich particles are brought into close contact with the nodules. Once this is achieved, only a short-term mobilization is required to transfer the metals to the nodules. As illustrated in Figure 18.5, the source of the metals involved in this transfer are thought to be ions loosely adsorbed to the sediments and in equilibrium with their pore water reservoir. Removal of the ions from pore water via precipitation onto the nodule surface drives an equal amount of desorption from the sediments in an effort to reachieve equilibrium pore water concentrations. Eventually bioturbation is required to move a fresh supply of sorbed material into the reaction zone around the nodule. This mechanism requires that only 1% of the Mn, Ni, and Co present in the sediments located within a 5-cm radius of the nodule be mobilized and precipitated to account for the metal content found in the typical "oxic" nodule.

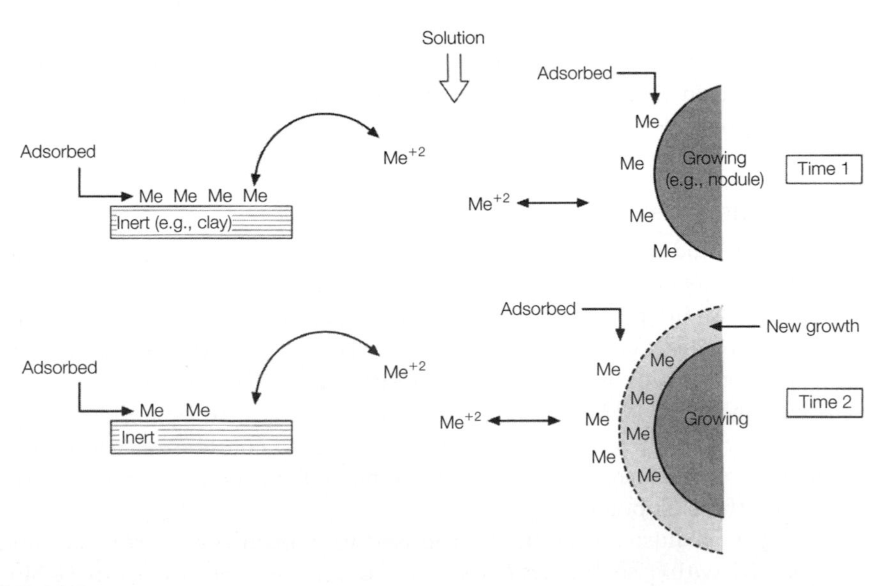

FIGURE 18.5

Model for the formation of manganese nodules under oxic conditions by means of adsorption onto sediment grains Me = metal atom. *Source*: From Lyle, M., *et al*. (1984). *Geochimica et Cosmochimica Acta*, 48, 1705–1715.

Finally, nodules may not be growing under oxic conditions. The nodules found in oxic sediments may have formed at some earlier time when the redox boundary was closer to the sediment-water interface. Changes in the position of the redox boundary are a consequence of changes in the flux of POM and bottom-water O_2 concentrations.

Shallow-Water Diagenesis

Mn nodules and crusts also form in shallow waters within marginal seas, such as the Baltic and Black Seas. In some locations, such as the Gulf of Bothnia, these Fe-Mn concretions are very abundant, reaching surface densities of 15 to 40 kg/m^2. These concretions are similar to ones formed in the open ocean, having concentric layers deposited around a nucleus, but growth rates are three to four orders of magnitude higher. Some elements, such as Zn, that have been mobilized by anthropogenic activity are enriched in the outer layers of these nodules. Thus, the metal content of these nodules makes them useful for monitoring metal pollution.

Formation of the shallow-water concretions is associated with anoxic conditions that range in duration from seasonal to nearly continuous. For example, in the Baltic Sea, nodules and crusts are mostly found around the margins of the deep anoxic basins. They form from Mn and Fe that accumulates from the reduction of Mn and Fe oxides in the anoxic deep waters. When the basin is periodically flushed with oxic water from the North Sea, about once a decade, the concretions undergo growth as the fresh supply of metals is oxidized.

During the time when the basin is anoxic, $MnCO_3$ precipitates at depths where the oxycline intercepts the seafloor, producing a kind of bathtub-like ring as shown in Figure 18.6. It forms as a consequence of the MnO_2 trapping that occurs at the oxic-anoxic interface as illustrated in Figure 11.20 for the Black Sea. Any MnO_2 that sinks below the interface is converted to Mn^{2+}, so sediments that lie below the redox

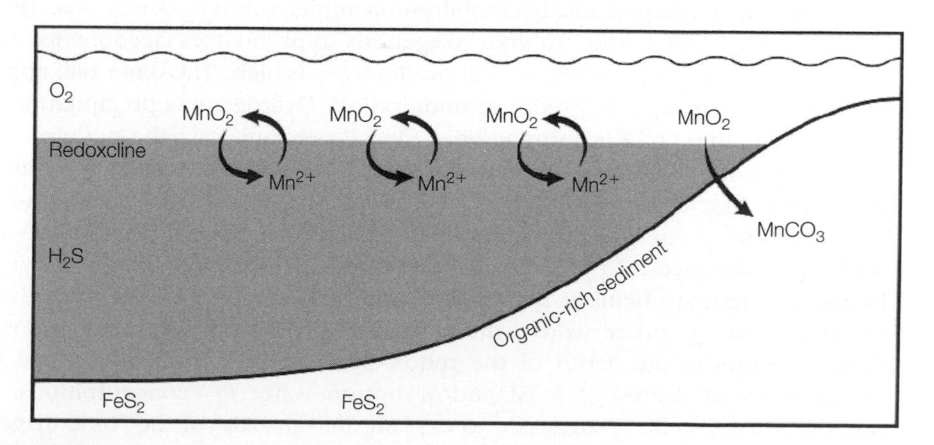

FIGURE 18.6

General model for Mn mineralization in euxinic (sulfate-reducing) basins. *Source*: Maynard, J. B. (2003). *Treatise on Geochemistry*, Elsevier Ltd., pp. 289–308.

interface have a low manganese content. The sediments that lie above the oxycline are also low in manganese because the oxygenated surface waters contain relatively little dissolved manganese—at least as compared to the highly enriched waters at the redox boundary. At some critical depth, which is located near where the oxic-anoxic interface intercepts the seafloor, sinking MnO_2 can reach the sediments before it dissolves. MnO_2 is converted to $MnCO_3$ via the following reaction, which is fueled by the copious flux of POM to the sediments:

$$2MnO_2 + CH_2O + HCO_3^- \longrightarrow 2MnCO_3 + H_2O + OH^- \quad (18.4)$$

If pyrite (FeS) is present, the MnO_2 will also react as follows:

$$FeS + 4.5MnO_2 + 4H_2O \longrightarrow FeOOH + 4.5Mn^{2+} + SO_4^{2-} + 7OH^- \quad (18.5)$$

supplying more Mn^{2+} for the formation of $MnCO_3$ and removing sulfur from the deposit.

Accretion Rates

The observation that nodules grow at widely varying rates provides further support for the existence of multiple formation mechanisms. The nodules that accrete most slowly (1 mm per million years) appear to have formed primarily by the process of hydrogenous precipitation. This accretion rate is equivalent to the annual deposition of a layer that is only one atom deep. These slow rates cause a significant amount of metal-rich seawater to become occluded between the Fe-Mn oxide layers.

The nodules that form at rates on the order of tens of millimeters per million years appear to have been produced primarily by postdepositional remobilization under oxic conditions. These nodules have relatively high copper and nickel contents. The nodules that accrete at the fastest rates (200 mm per million years) appear to have formed primarily via postdepositional remobilization under suboxic conditions. Despite these rapid accretion rates, the "suboxic diagenesis"-type nodules account for only half of those found in areas where biological productivity is high. The other half appear to have been formed primarily by "oxic" remobilization. Hydrogenous precipitation appears to play the dominant role in forming only a small percentage of the nodules.

The "suboxic diagenesis"-type nodules have been found in surface sediments where the redox barrier is located several centimeters below the sediment-water interface. Obviously, the suboxic mechanism for nodule growth cannot operate across such large distances. To explain the presence of these nodules far above the redox boundary, marine chemists hypothesize that the position of the boundary has varied over time. In other words, these nodules grow only when the redox barrier is shallow. Shifts in the depth of the redox boundary are probably related to changes in the rates of supply of POM and/or bottom-water O_2 concentrations. (This phenomenon has also been suggested to explain the "growth" of the "oxic diagenesis"-type nodules.)

If temporal variations in the depth of the redox boundary are important, climatological events, such as ice ages and even seasons, should affect nodule growth rates.

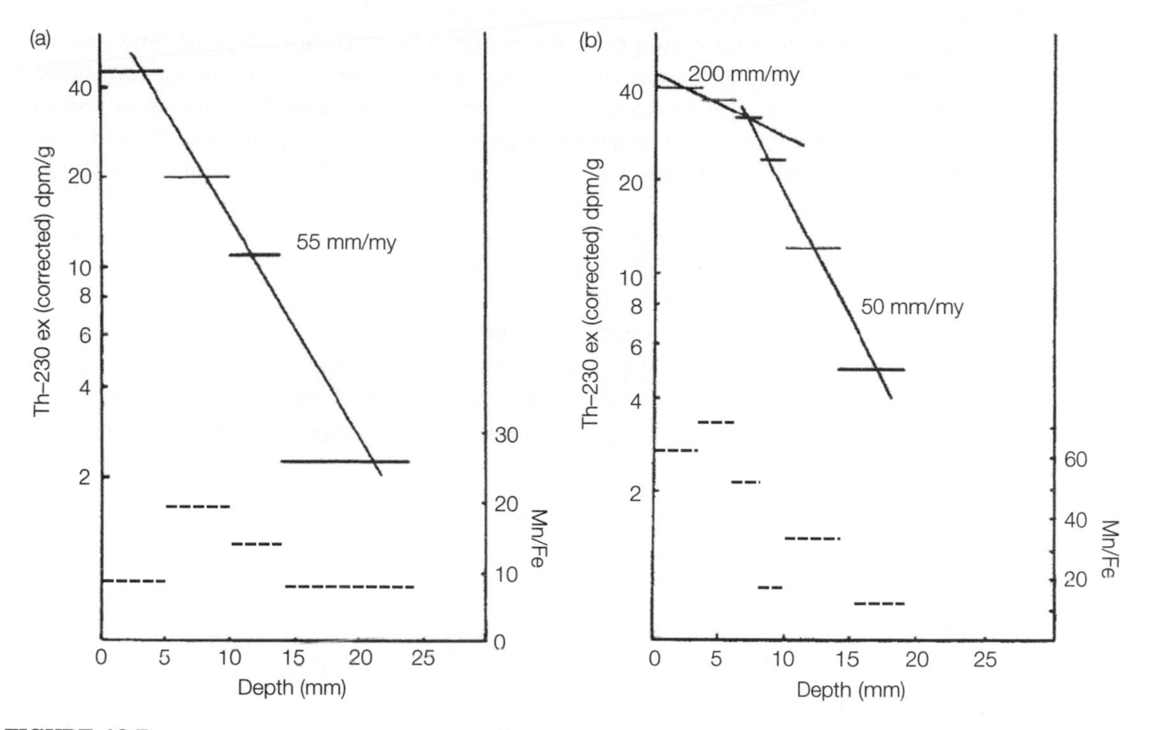

FIGURE 18.7

Corrected excess ^{230}Th concentration (solid lines) and Mn/Fe ratios (dashed lines) versus depth in an Fe-Mn nodule collected from MANOP Site H (6°33'N, 92°49'W in the Guatemala Basin in a water depth of 3600 m). Nodule diameters average about 4 cm at this site, but range up to 20 cm. Nodule density averages 300 nodules/m^2, which is equivalent to 30 kg/m^2. (a) Nodule top and (b) nodule bottom. *Source*: From Finney, B., *et al.* (1984). *Geochimica et Cosmochimica Acta*, 48, 911–919.

Support for this comes from radiochemical dating of "suboxic diagenesis"-type nodules using the naturally occurring isotopes of the ^{238}U decay series. As shown in Figure 18.7, accretion rates on the bottom of the nodule increased from 50 to 200 mm per million years at a depth in the nodule corresponding to 40,000 ybp. This rate shift coincides with a sharp increase in the organic matter content of the sediments. This suggests that the increase in accretion rate was supported by an increased supply of trace metals derived from a higher POM flux to the sediments and more reducing conditions. These nodules have concentric bands and color variations that are also likely records of changes in paleoceanographic conditions.

Even the nodules growing at the fastest rates ought to be buried but are not. Several theories have been advanced to explain how the nodules maintain their position on the sediment surface. For example, earthquakes could rock the nodules frequently enough to keep their tops clear of sediment. Bottom currents could be strong enough to sweep the nodules free of sediment. This explanation is supported by the presence of Fe-Mn nodule pavements in areas subject to fast bottom currents.

The effects of wind winnowing by bottom currents could also explain why 25% of the nodules are buried in the sediments. Variations in local rates of sediment deposition and erosion could cause the periodic burial and exposure of nodules. Growth would be expected only during periods of exposure. This would also explain why buried nodules appear unaffected by diagenesis (i.e., they are not buried long enough to undergo extensive alteration).

Another likely explanation for the concentration of nodules at the sediment surface is linked to bioturbation. If benthic animals are active enough, their crawling and burrowing could jostle the nodules, knocking off enough sediment to prevent burial. Or some combination of animals wedging between a nodule and backfilling cumulatively acts to shuffle sediment beneath the nodule. Bottom photography of the feeding tracks left by benthic worms has given rise to the analogy that the surface sediments are much like a "well-plowed Iowa cornfield," giving support to the hypothesis that bioturbation keeps the large nodules "afloat."

Geomicrobiology of Iron-Manganese Nodules

Bacteria and protozoans are thought to play an important role in the formation and diagenetic recycling of Fe-Mn nodules. The presence of bacteria and protozoans on nodules was first reported in the 1870s as part of the published records of the Challenger Expedition. These observations are not particularly surprising given that, upon submergence in seawater, all surfaces are rapidly covered with a coating of adsorbed organic matter. This organic matter represents a concentrated food source to microbes. Hence, surfaces are rapidly colonized by bacteria, followed by protozoa that feed on the bacteria.

The nodule bacteria synthesize negatively charged extracellular polymers that act as ion exchangers, adsorbing cations such as Mn^{2+}. These polymers are exuded as microfibrils, which significantly extend the surface area available for metal adsorption. The adsorbed ions are eventually converted to MnO_2 as a result of extended contact with oxic seawater. These bacteria are consumed by protozoans, including foraminiferans that tend to grow in clusters called *agglutinations*. As protozoa feed on the bacteria, they "browse" across the nodules. This activity could help keep the nodules free of sediment. The manganese sequestered by the bacteria is passed through the digestive tract of the foraminifera and deposited in their fecal material. This manganese-rich detritus is called a *stercome*. The stercomata become part of the nodule, as do the calcareous tests of the foraminiferans after their death. Eventually, these tests dissolve, leaving Fe-Mn oxides as a type of fossilized remain.

Rhizopod protozoans have also been observed living on nodules. These organisms incorporate manganese into their shells and protoplasm. Their stercomata are also enriched in manganese. Though deposition of manganese oxides by all of these fauna is a possible mechanism by which nodules could form under oxic conditions, no direct observations of this have been reported. In fact, microbes are just as likely to be removing as depositing manganese because their fecal deposits are rich in organic matter.

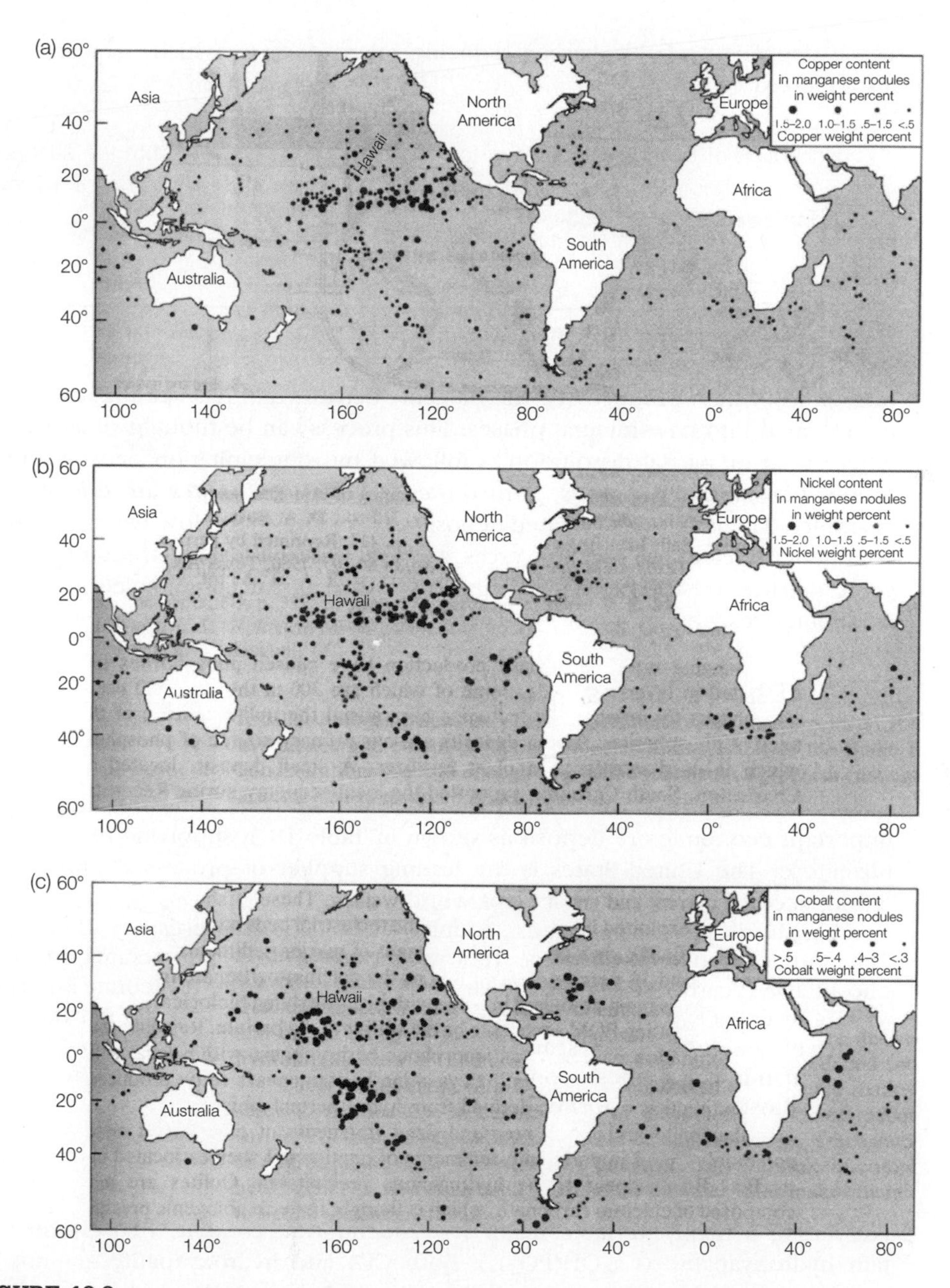

FIGURE 18.8

Concentration of (a) copper, (b) nickel, and (c) cobalt in iron-manganese nodules found on the seafloor. *Source*: From Ross, D. A. (1982). *Introduction to Oceanography*, 3rd ed., Prentice Hall, pp. 411–412. Data from: D. R. Horn, *et al.* (1972). *Ferromanganese Deposits of the North Pacific*, National Science Foundation, pp. 40–54.

These deposits probably harbor reducing microzones in which manganese could be resolubilized.

A description of the status of efforts to mine manganese nodules from the seabed is provided in supplemental material for Chapter 18 available at **http://elsevierdirect.com/companions/97801230885305**.

18.3 OTHER HYDROGENOUS MINERALS

In regions of high productivity, the biogenic detritus that reaches the sediment is often transformed into new mineral phases. This process can be thought of as a fossilization, where total or partial dissolution is followed by reprecipitation or isomorphic substitution. *Phosphorites*, *barites*, *glauconites*, and *calcitic oolites* are the most common examples. The most concentrated deposits are formed in shallow waters, particularly on continental shelves in upwelling areas, as this is where the rates of sedimentation of biogenic detritus are highest. These minerals usually occur commingled as noncrystalline conglomerates.

18.3.1 Phosphorites

Phosphorites are hydrogenous precipitates with phosphorus concentrations greater than 5% w/w P_2O_5. Concentrations as high as 40% have been observed. In comparison the phosphorus content of most sediments is around 0.3%. Phosphorites represent an important economic ore deposit as shown in Table 18.3, supplying phosphorus for fertilizer use. The United States is the leading supplier of processed phosphates in the world, accounting for about 45% of world trade.

Phosphorites occur as sands, nodules, pebbles, slabs, crusts, and conglomerates. The nodules and crusts average 18% w/w P_2O_5. The conglomerates are mixtures of phosphatized carbonate grains and large fossils in a matrix of glauconite and can have up to 15% w/w P_2O_5. The nodules grow slowly, at rates of 1 to 10 mm per thousand years. They have been found on seamounts, plateaus, ridges, banks, atolls, and shelves. The dominant mineral in these concretions is francolite, also called carbonate fluorapatite (CFA), whose empirical formula is

$$(Ca_{10-a-b}Na_aMg_b(PO_4)_{6-x}(CO_3)_{x-y-z}(CO_{3z} \cdot F)_y(SO_4)_zF_2$$

illustrating a highly variable major ion and fluoride content. Phosphorites also contain hydroxyapatite ($Ca_5OH(PO_4)_3$). Both CFA and hydroxyapatite are not restricted to phosphorite deposits and are widely distributed in marine sediments, albeit at low concentrations. Some hydroxyapatite is of direct biogenic origin, because it is the primary mineral in fish bones, scales, and teeth. A significant fraction (22 to 27%) of the phosphorus in marine sediments is sorbed to iron-rich oxides. Particulate organic phosphorus constitutes only 25 to 30% of the sedimentary phosphorus reservoir.

Phosphorite Formation Process

CFA forms as a secondary precipitate as shown in Figure 18.9. In oxic sediments, phosphate remineralized from POM is scavenged from pore waters by adsorption and co-precipitation into Fe(III)OOH. Some fluoride is similarly scavenged. Burial of this precipitate carries it into the reducing zone of the sediments. Under these redox conditions, Fe(III) is reduced and resolubilized to Fe^{2+}(aq), thereby remineralizing phosphate and fluoride. The concurrent dissolution of $CaCO_3$ supplies Ca^{2+} and CO_3^{2-}. Once the concentrations of phosphate, fluoride, calcium, and carbonate exceed the solubility of CFA, francolite precipitates. Microbes may be involved in this precipitation process. Other diagenetic minerals that are thought to form under reducing conditions, include vivianite ($Fe_3(PO_4)_2 \cdot 8H_2O$) and various aluminophosphates.

Not all of the remobilized phosphate is reprecipitated below the redox boundary. Some escapes by diffusing upward through the pore waters. Once this phosphate enters the oxic zone, it is readsorbed by Fe(III)OOH along with any Fe^{2+} that has similarly diffused upward. The Fe^{2+} that diffuses downward into the sulfate-reducing zone precipitates sulfide to form pyrite (FeS).

Phosphate removal by sorption on Fe(III)OOH is particularly important in the vicinity of mid-ocean ridges and rises because hydrothermal activity supports a large flux of

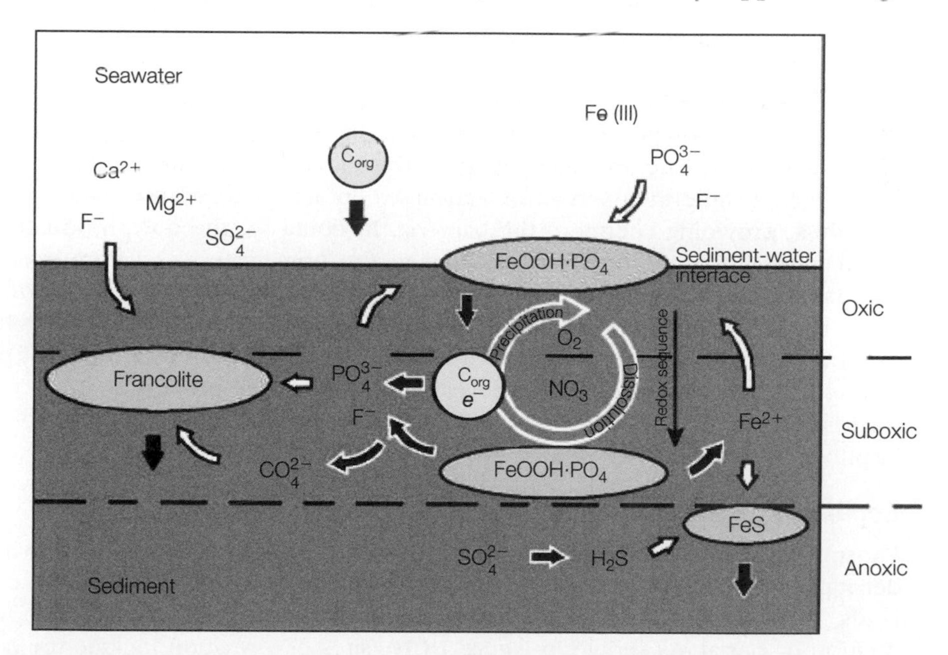

FIGURE 18.9

Schematic diagram of the coupled iron and phosphate cycles during early diagenesis in marine sediments. Light gray ovals and circles represent solid phases; black arrows are solid-phase fluxes. White outlined black arrows indicate reactions; white arrows are diffusion pathways. *Source*: From Ruttenberg, K. C. (2003). *Treatise on Geochemistry*, Elsevier Ltd. pp. 585–643.

dissolved iron into oxic seawater, leading the nearby deposition of metalliferous sediment. Thus, fluctuations in hydrothermal activity have the potential to alter the marine phosphorus cycle.

The iron-based redox cycle depicted in Figure 18.9 provides an effective preconcentrating step for phosphorus by trapping remineralized phosphate in oxic sediments. The conversion of phosphorus from POM to Fe(III)OOH to CFA is referred to as *sink switching*. Overall this process acts to convert phosphorus from unstable particulate phases (POM to Fe(III)OOH) into a stable particulate phase (CFA) that acts to permanently remove bioavailable phosphorus from the ocean. This is pretty important because most of the particulate phosphate delivered to the seafloor is remineralized. Without a trapping mechanism, the remineralized phosphate would diffuse back into the bottom waters of the ocean, greatly reducing the burial efficiency of phosphorus.

Recent research suggests that giant sulfur bacteria (*Thiomargarita namibiensis*) could be involved in the formation of phosphorites now depositing beneath coastal upwelling zones. These bacteria episodically release phosphate, creating supersaturated conditions that favor the precipitation of hydroxyapatite. They are adapted to survival under both oxic and anoxic conditions. They are very large cells, with diameters reaching as much as 1 mm! These giant bacteria dominate upwelling systems, forming thick mats at the sediment surface, and have been observed suspended in the oxic water column.

Under oxic conditions, the bacteria assimilate nitrate and phosphate. The phosphate is converted to polyphosphate and in doing so serves as an energy source. Both the nitrate and phosphate are stored in special vacuoles for use under anoxic conditions, during which time they serve as electron acceptors, enabling the oxidation of sulfide and, thus, providing energy to the bacteria. In doing so, the polyphosphate is broken down into phosphate, which is released to the pore waters and then precipitates as hydroxyapatite. Although hydroxyapatite has the potential to dissolve during periods when the bottom waters are oxic (due to lack of upwelling), this process is not significant because it is limited by the relatively slow rate of molecular diffusion of phosphate through the pore waters. The discovery of this bacterial process was not particularly surprising; its occurrence had already been surmised from microfossils commonly found in phosphorites.

Locations of Phosphorite Deposits

Phosphorite deposits are currently forming in areas of high organic productivity and low detrital input. These are typically coastal upwelling zones adjacent to arid continental lands. Phosphorites form at slow rates, so low detrital input is important to prevent dilution or burial. As shown in Figure 18.10, sites of formation include the continental margins of Peru, Chile, and southwest Africa.

Role of Phosphorite Deposition in the Crustal-Ocean-Atmosphere Factory

Much like the evaporites, layers or beds of phosphorites have been deposited over the eons. Some beds are 300 m thick with 20% w/w P_2O_5. These giant phosphorites

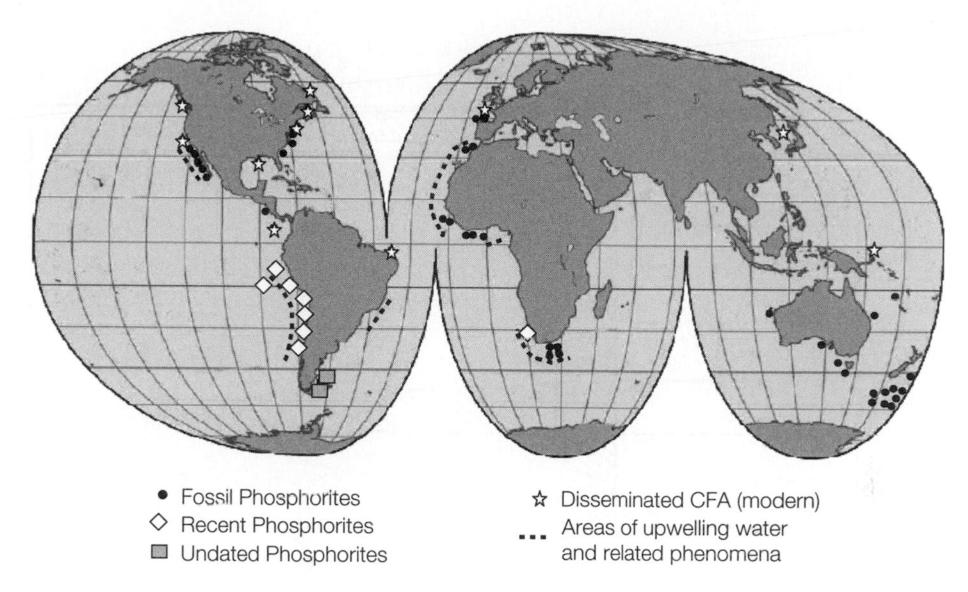

FIGURE 18.10

Locations of disseminated (nonphosphorite) authigenic CFA occurrence, as well as locations of phosphorites. Areas with substantial phosphorite deposits include the East China Sea between Korea and Japan, Ceara Rise, Saanich Inlet, eastern and western equatorial Pacific, California Borderland Basins, Gulf of St. Lawrence, Labrador Sea, Long Island Sound, Gulf of Mexico, North Atlantic continental platform, and Iberian margin in the northeastern Atlantic. *Source*: From Ruttenberg, K. C. (2003). *Treatise on Geochemistry*, Elsevier Ltd. pp. 585–643.

were deposited episodically over the Phanerozoic, particularly during the Miocene and Permian as shown in Figure 18.11. The fluctuations in deposition rates are thought to reflect changes in river fluxes of phosphate or changes in burial rates over time.

Because of the large amount of phosphorus trapped in phosphorites, their formation could have impacts on climate change. For example, during a glacial period, the declining sea level would expose phosphorites buried on the shelves and thereby increase the riverine phosphorus flux leading to enhanced primary production. This would cause a drop in atmospheric CO_2 levels and enhance global cooling.

Phosphorite deposition also has the potential to control atmospheric O_2 levels. For example, during periods of low atmospheric O_2, less Fe(III)OOH formation will occur due to a lowered O_2 content in the deep ocean, resulting in a lower burial efficiency of phosphorus. Thus, more phosphate will be available in seawater, thereby enhancing primary production and the generation of O_2. This negative feedback acts to buffer against changes in P_{O_2} preventing runaway ocean anoxia. By stabilizing atmospheric O_2 levels, this feedback probably had an influence on the evolution of life forms.

Current research is directed at quantifying the importance of dispersed sedimentary CFA. If this sink is large, it may play as important a role as phosphorites in the various feedbacks outlined earlier.

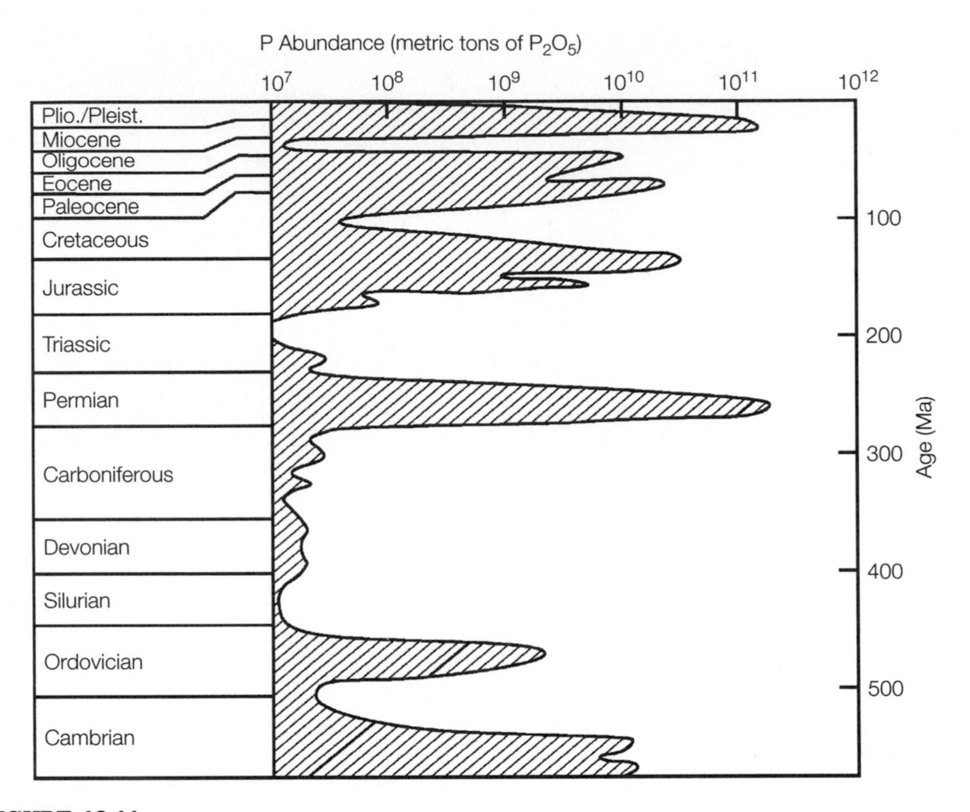

FIGURE 18.11

The abundance of phosphate in sedimentary rocks as a function of geologic age showing the episodicity of giant phosphorite deposition. *Source*: From Cook, P. J., and McElhinny, M. W. (1979) *Economic Geology* 74, 315–330.

Phosphorite Mining

Plate tectonics has caused the uplift of some of the ancient giant phosphorites. These deposits are our primary source of phosphorus.[1] Mining of phosphate in the United States began in 1867 in South Carolina, where the ore was first discovered in 1837. The oldest U.S. phosphate mining company, Charleston Mining and Manufacturing Company, began operations in 1866. The west bank of the Ashley River was considered to contain the richest, most accessible deposits of phosphate rock in the country. The ore occurred as a shallow, 3-foot thick stratum of phosphate rock. It was strip mined to provide a cheap and effective fertilizer. Because of financial desperation during Reconstruction, the phosphorite was strip mined from the grounds of former rice plantations, providing

[1] Other sources of minable phosphorous included huge guano deposits created by the droppings of seabirds on rocky coastlines. Intensive mining led to the depletion of these deposits by the early 1900s.

employment for thousands of newly freed slaves. Intensive mining led to the depletion of this deposit by the 1930s. Most of the active phosphorite mines in the United States are located in North Carolina and Florida.

In 2005, the International Fertilizer Industry Association estimated that worldwide demand for phosphate fertilizers would grow at an average rate of 2.3% per year over the next 5 years. This increase is due to population growth and an increasing trend on the part of developing nations to adopt a more Western-style, high-protein diet. As indicated in Table 18.3, humans are rapidly depleting the easily accessible, high-grade phosphate rock, and, thus, offshore deposits of phosphorite are likely to be exploited in the future.

18.3.2 **Barite**

Barite, $BaSO_4$, is a common component of marine sediments, averaging about 1% of the carbonate-free mass of deep-sea sediments. It is present as microcrystals or aggregates ranging in size from 0.5 to 5 mm. Most of the sedimentary barite is ultimately biogenic in origin with hydrogenous precipitation and diagenetic remobilization playing important reprocessing roles. Some sedimentary barite is also hydrothermal in origin. Recall from Chapter 9 that barium is a biointermediate element, so biogenic particles deliver this element to the seafloor. The exact nature of these biogenic particles is not known, but they are thought to be composed of barite produced by (1) direct biogenic precipitation by pelagic plankton and (2) hydrogenous precipitation of barium remineralized from detrital hard and soft parts. Because of this biogenic pump, about one third of the sinking flux of biogenic barite reaches the seafloor; the rest dissolves in transit. In the organic-rich sediment underlying regions of high biological productivity, barite can comprise as much as 5% of the carbonate-free fraction.

The hydrogenous precipitates are thought to form within anoxic microenvironments present in sinking aggregates of POM, BSi, and acantharian shells,[2] all of which are enriched in barium to variable degrees. Seawater is slightly undersaturated with respect to barite, so the semi-isolated volume of seawater within the aggregates affords a mechanism by which remineralized barium concentrations can rise to saturation levels, thereby enabling precipitation.

Pore waters tend to be supersaturated with respect to barite, especially in organic-rich sediments. This promotes the preservation of accumulating barite and the formation of diagenetic barite. Diagenetic barite forms from $Ba^{2+}(aq)$ supplied by upward diffusion from the sulfate-reducing zone, because barite dissolves as consequence of its sulfate getting reduced. As with diagenetically remobilized iron and manganese, barium tends to diffuse into the oxic zone, where it reprecipitates. Remineralization also occurs within anoxic microenvironments found in fecal pellets. In this setting, barite reprecipitates as a replacement material, leading to fossilization of the pellets. Such fossilized fecal pellets are called *coproliths*. Some of the sedimentary barite is deposited by benthic

[2] Acantharians are a type of radiolarian whose shells are composed of celestite ($SrSO_4$).

protozoans, called *xenophyophorans*. These organisms are widespread and abundant, but their relative importance in contributing to the sedimentary barite reservoir has not yet been quantified.

Since a large fraction of the biogenic barite formed in the water column reaches the sediment and its flux is well correlated with that of sinking POM, this mineral has potential as a paleoceanographic record of past changes in productivity. Unfortunately, complications arise from sedimentary production of barite by *xenophyophorans* and by post-diagenetic dissolution and remobilization. Another problem is that barite is also precipitated as a hydrogenous mineral from hydrothermal emissions. This leads to the formation of sediments with barite concentrations as high as 10% in the metalliferous sediments of the East Pacific Rise.

18.3.3 Oolitic Calcite and Aragonite Needles

Because warm surface seawater is usually supersaturated with respect to calcium carbonate, abiogenic precipitation of calcite and aragonite does occur, at least when supersaturations are very high. These conditions are limited to shallow water where temperatures can get sufficiently high, namely coastal tropical seas.

Nodules of calcite that are less than 2 mm in diameter are commonly found in and on the sediments of shallow tropical seas. Like manganese nodules, they are concretions that form around a nucleus. Because of their resemblance to fish eggs, they are called *calcitic ooids*. Some phosphorites also form ooids. After burial, continuing precipitation of calcium carbonate between the calcitic ooids can generate a pavement called a *calcitic oolite*. Formation of calcitic ooids requires: (1) a high degree of saturation with respect to calcite, (2) the presence of a precipitation nucleus, (3) sufficient agitation of water around the nodules, (4) a shallow water depth (subtidal to lower intertidal), and (5) maintenance of the ooids within the area favorable to their formation. These ooids are thought to form by rolling around like snowballs, being pushed by waves and currents. They are periodically buried. When they are reexposed, another layer of calcite is deposited, thereby forming concentric bands. Calcitic ooids are common in the shallow waters of the Bahamas, and the Arabian Gulf. Their presence in sediment cores is used as a paleoceanographic record of low sea-level stand.

Another calcium carbonate precipitation phenomenon common to the shallow tropical seas is called a *whiting*. These are white patches noticeable on satellite photography. The patches can be as much as 10 to 12 km long and cover 100 km^2 in surface area. They can extend from the sea surface to the sea floor. Whitings are common to the Bahama Banks, where individual events have an average duration of 2 d and cover 2% of the surface area of this region. The white appearance comes from aragonite needles with average dimensions of 0.25 μm in diameter and 2.5 μm in length. The current model for their formation involves periodic resuspension of the needles from the sediment. Once in the water column, the needles glow incrementally via hydrogenous precipitation from the supersaturated waters ($\Omega_{aragonite}$ = 2–6) until pelagic sedimentation returns the needles to the seafloor. Precipitation is probably autocatalyzed. The needles are thought to be resuspended into the water column 2 to 3 times per year, supporting leading a growth rate of 3% y^{-1}. Some aragonite needles are also produced biogenically by three genera of coralline algae, *Udotea, Halimeda*, and *Penicillus*. The detrital remnants of these biogenic needles are probably also resuspended during whitings.

18.3.4 **Glauconite and Green Clays**

Green clay minerals are a common component of marine sediments. Their presence imparts an overall green color to deposits that are thus called "green clays," "green muds," and "green sands." The coloration is due to the presence of Fe(II) and Fe(III). The relative proportions of these two oxidation states gives rise to clays that range in color from yellow-green to green to black-green. The green clays are thought to form diagenetically from reactions of iron(III) oxides with clay minerals, such as kaolinite and iron-montmorillonite, under reducing conditions found within microenvironments, such as in fecal pellets or the interiors of foram tests. Under these conditions, the partial reduction of Fe(III) remineralizes Fe(II), which reacts with the clay minerals to form glauconite and chamosite, both of which are classified as secondary clay minerals. Glauconites are mica-like and have high potassium concentrations (7% K_2O). Chamosite is structurally similar to chlorite.

Another type of green clay, bertherine, is formed in the sediments of river deltas. This black-green clay is relatively Fe(II) rich compared to the others. This is thought to be caused by the reducing conditions found in the organic-rich sediments characteristic

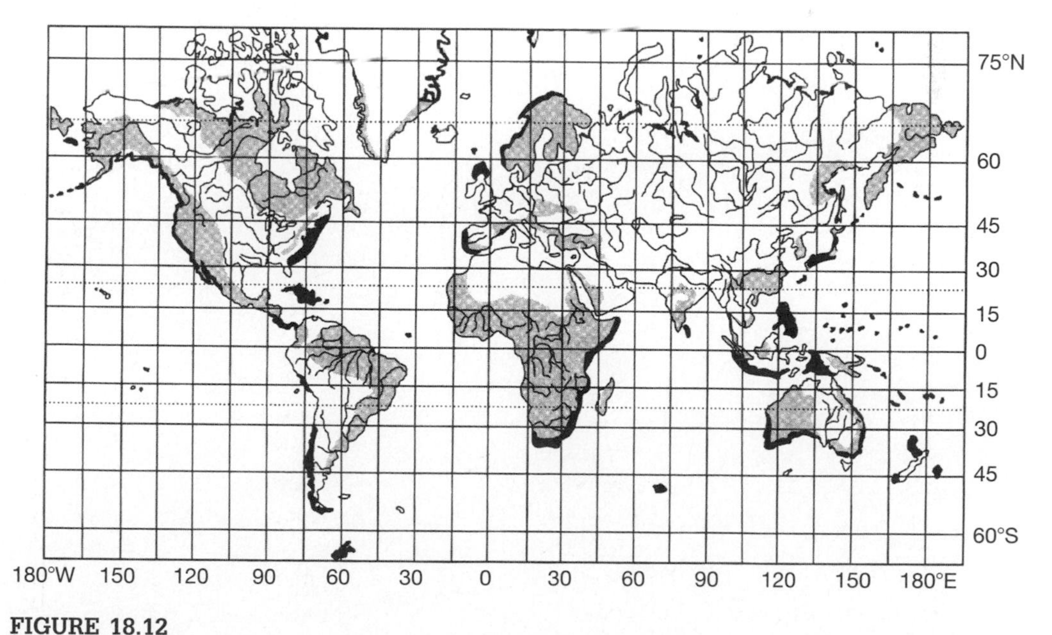

FIGURE 18.12

Geographic distribution of glauconite in marine sediments. *Source*: Galliher, E. W. (1935). *Bulletin of the American Association of Petroleum Geologists* 19, 1569–1601.

of these deposits. Like the other green clays, bertherine forms from postdepositional diagenetic reactions between fecal pellets, clay minerals, and iron(III) oxides.

Some green clay minerals occur as ovoids, probably having formed within fecal pellets or casts of forams. They also occur as films or stains on shells, sand grains, and phosphate nodules. They are found in highest density in sediments of the outer continental shelves and slopes where waters are shallow (20 to 700 m) and mildly suboxic. Slow sedimentation rates are necessary to prevent burial as green clays form at very slow rates. A notable example are the green muds found on the Blake Plateau.

As shown in Figure 18.12, glauconites are most abundant in nearshore sediments, reflecting a requirement for suboxic conditions that are supported by a high supply rate of POM. Glauconites are primarily found as ovoids. They are present in sediments of all ages and in sedimentary rocks, such as sandstones and carbonates. Hence, they appear to be a continuous part of the sedimentary record. Upon subaerial weathering, glauconite is converted into ochre, which is used as a paint.

CHAPTER 19
Metalliferous Sediments and Other Hydrothermal Deposits

All figures are available on the companion website in color (if applicable).

19.1 INTRODUCTION

Metalliferous sediments blanket the central rift valley and flanks of mid-ocean ridges and rises. They are also found along some convergent plate boundaries. The metals in these hydrogenous deposits are derived from hydrothermal fluids associated with underwater volcanic activity. Oceanographers had long surmised the occurrence of this activity from the geologic record and the geophysical necessity of accounting for heat dissipation from the inner Earth. The first direction observation of hydrothermal fluids was made in 1966, when oceanographers discovered a metal-rich brine accumulating above the spreading center located in the middle of the Red Sea. In 1977, improvements in bottom-water navigation, photography, and deep-sea submersibles led to the first direct observations of the venting of hydrothermal fluids at a mantle hot spot near the Galapagos Islands. Since then, oceanographers have obtained direct evidence of hydrothermal activity in all of the ocean basins. High-temperature venting has been observed to be occurring at mid-ocean spreading centers, in back-arc basins, and at microplate margins. Some of the venting occurs at lower temperatures along the flanks of mid-ocean ridges and at convergent plate boundaries. Some degree of geothermally driven convection is likely taking place through the sediments lying on continental margins.

The chemical reactions that occur in hydrothermal systems are largely the result of interactions between seawater and relatively young ocean crust. During these reactions, some elements are solubilized and released to seawater as ions or gases. Others are precipitated, forming minerals that end up as a component of new oceanic crust or the metalliferous sediments. For some elements, the resulting elemental fluxes rival those associated with river input, making hydrothermal activity a very important process in the crustal-ocean-atmosphere factory.

Metalliferous sediments are composed of metal-rich oxides and sulfides. At the ridge crests, the sediments can form huge mounds that are mostly metallic sulfides. Because of their large size—thousands to millions of tonnes—and diversity in metal enrichments—including copper, zinc, silver and gold—these deposits are termed *polymetallic massive sulfides* (PMS). Some of these deposits have been uplifted onto land. These deposits have been mined over all of human history for their metal ores. Efforts are now underway to attempt mining a PMS located off the coast of Papua New Guinea on the seafloor in water depths of 1600 m.

Hydrothermal activity plays a number of important roles in the crustal-ocean-atmosphere factory. First, it represents a sink for some elements and a source for others. These include some of the metals, alkalis, alkaline earths, phosphorus, and sulfur. Due to feedbacks with the carbon and sulfur cycles, hydrothermal activity has the potential to affect global climate. Second, hydrothermal activity is a key part of the rock cycle in which materials are exchanged between the mantle and the crust. Over very long time scales, the exchange of iron between the mantle and crust has likely controlled atmospheric O_2 levels. Third, hydrothermal systems represent important sites of intense biological activity. The energy fueling this is supplied by the reduced chemicals carried in the hydrothermal fluids. A similar ecosystem is supported by cold-water seeps because they also contain reduced chemicals. These seeps are found on passive and convergent continental margins. Some biologists hypothesize that the first life forms arose in the habitats provided by the hydrothermal vents and cold seeps. These habitats are thought to have served as refuges that sustained life during the various mass extinction events which have occurred throughout the Phanerozoic.

Hydrothermal activity is also of interest to geological and physical oceanographers. Since we have not yet drilled through the crust to permit direct observation of the mantle, hydrothermal systems provide the closest contact that scientists have yet had with magma. For physical oceanographers, some of the hydrothermal emissions, such as ^{3}He and Mn^{2+}, can be used as tracers of the rate and pathways of water-mass motion as already illustrated in Figure 11.19.

This chapter focuses on the geochemistry of hydrogenous metalliferous sediments and other hydrothermal deposits. As with the other hydrogenous sediments discussed in Chapters 17 and 18, research continues to reveal important connections with the marine biota. Most notably, microbes play an important role in hydrothermal chemistry, serving as a reaction catalyst and as the base of the chemoautolithotrophic food chain that supports the vent ecosystems. Color photographs and video of hydrothermal systems are available at websites maintained by NOAA (www.oceanexplorer.noaa.gov/ and www.pmel.noaa.gov/vents/), MBARI (http://www.mbari.org/volcanism), and the Ridge 2000 Program (http://www.ridge2000.org/).

19.2 THE PHYSICAL SETTING

Hydrothermal deposits are a by-product of chemical interactions between the ocean and lithosphere. The most notable of these interactions takes place at tectonic plate boundaries. At divergent plate boundaries, seafloor spreading causes rifting of the oceanic basement rock, thereby allowing magma to upwell from the mantle. As the magma rises, it cools and freezes into solid rock, i.e., new oceanic crust. In the present-day ocean, seafloor spreading has given rise to a 67,000-km-long continuous chain of volcanoes that

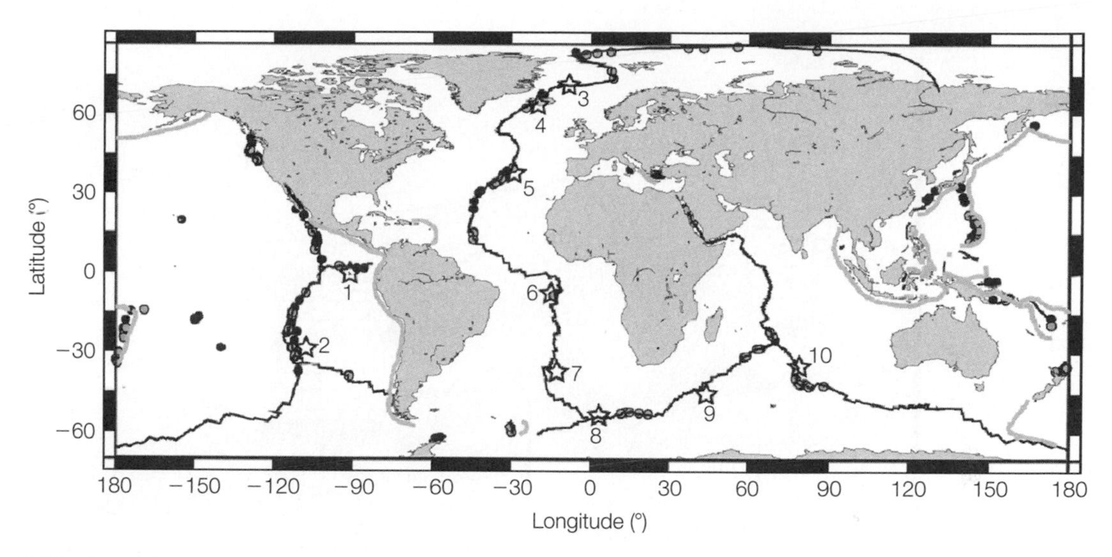

FIGURE 19.1

Seafloor spreading centers (solid black lines) and subduction zones (gray lines). Also shown are the sites of 144 known (black dots) and 133 inferred (gray dots) hydrothermal fields. Mantle hot spots (open stars) on or near (<500 km) the mid-ocean ridge include: (1) Galpagos, (2) Easter, (3) Jan Mayen, (4) Iceland, (5) Azores, (6) Ascension, (7) Tristan de Cunha, (8) Bouvet, (9) Crozet, and (10) Amsterdam–St. Paul. *Source*: From Baker, E. T., and C. R. German (2004). *Mid-Ocean Ridges: Hydrothermal Interactions between the Lithosphere and Oceans*, Geophysical Monograph Series 148, pp. 245–266.

runs through the middle of the Arctic, Atlantic, Indian, and Southern Oceans, forming a mid-ocean ridge (MOR). In the Pacific Ocean, the seafloor spreading center lies on eastern side of the basin and has a lower relief. Hence, it is referred to as a mid-ocean rise. Oceanic crust is destroyed at convergent plate boundaries by subduction of the basement rock back into the mantle. This process is thought to be driven by convection in which cooling of the oceanic crust increases its density causing it to sink. The present-day locations of seafloor spreading and subduction are shown in Figure 19.1.

19.2.1 Spreading Centers

At spreading centers, some of the upwelling magma freezes into place below the surface of the crust as an intrusive igneous rock. As shown in Figure 19.2, the intrusive rock can solidify either as a large globular mass, called a *pluton*, or as narrow ridge-parallel columns, called *dikes*. The latter form when magma flows into cracks and fractures that are generally 1 m wide by 10 km long and are hundreds of meters high. The magma that escapes out of the lithosphere erupts onto the seafloor as lava, which after cooling solidifies into extrusive igneous rocks called *pillow basalts*.

As the magma upwells, it is cooled by heat transfer to the surrounding crustal rocks or seawater. Two-thirds of this heat is dissipated conductively. The rest is lost via the

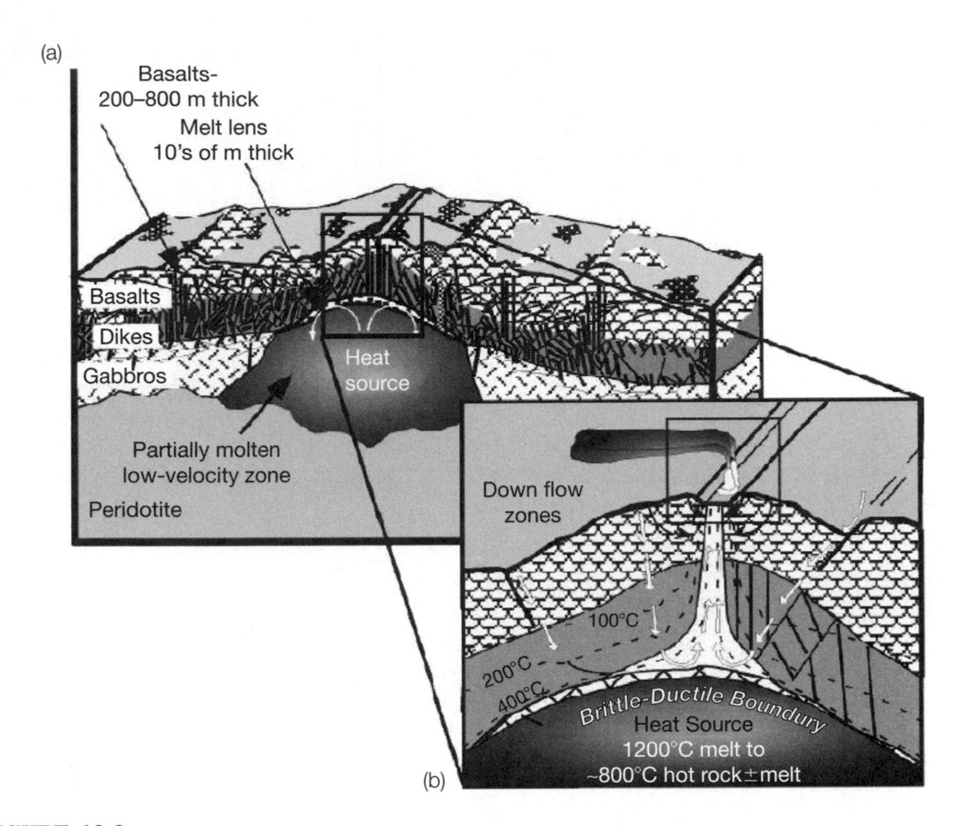

FIGURE 19.2

Schematic drawing showing mantle-crust relationships beneath ridge crests. (a) Crustal magma chambers fed from melt percolating through the underlying mantle section typically form at depths of 1 to 4 km below the seafloor. (b) Steep thermal gradients resulting from intrusion of 1175 ± 25°C basaltic melt into cool, water-saturated, and porous crustal rocks drive hydrothermal circulation beneath the spreading centers. High-temperature limbs of the resultant hydrothermal cells focus metal-rich, acidic fluids onto the seafloor. These form sulfide deposits upon mixing with cold, oxygenated seawater. *Source*: From Kelley, D. S., *et al*. (2002). *Annual Review of Earth and Planet and Science*, 30, 385–491.

convective circulation of seawater through the ridge systems. This hydrothermal circulation is driven by the conductive heating of seawater as it percolates through the hot fractured crust of the ridge systems. Heating lowers the density of the seawater, thereby increasing its buoyancy. The heated water rises until it escapes back into ocean. After sufficient mixing with cold seawater, the hydrothermal fluids become neutrally buoyant and cease to rise. Bottom-water currents then cause the diluted fluids to spread laterally as a plume.

Some of the hydrothermal emissions occur as focused and very hot discharges emitted through the annular orifices of rock chimneys that are tens of meters in height. The chimneys are mineral precipitates of Zn, Cu, Fe, Pb, Cd, and Ag sulfides. Other emissions occur as low-temperature diffuse flows that emanate through mounds of metalliferous sediments.

In both the high and low temperature settings, cold seawater is drawn down into the hydrothermal system to replace the discharged fluids. This process is termed *recharge*. The net effect is a convective flow pattern as shown by the arrows in Figure 19.2. Geochemists think that this convective flow extends as deep as 5 km below the seafloor, where magma temperatures are 1200°C, and several kilometers from the ridge crust. The annual fluid flux through the MOR is estimated to be 3.2 to 4.2×10^{13} kg/y. If this is spread equally and continuously along the 67,000 km of ridge crests, the flow rate is roughly equivalent to 2 kg per s per 100 m. At this rate, the whole ocean can be cycled through the ridge crest system every 33 to 44 million years.

Continued seafloor spreading pushes older oceanic crust away from the ridge crests, cutting it off from the magmatic heat source. Thus, as oceanic crust moves away from the ridge, it cools and subsides. Hydrothermal circulation occurs in this cooler crust, albeit at lower temperatures (Figure 19.3). Given the large volume of crust involved, this off-axis hydrothermal circulation is substantial. As shown in Figure 19.4, 70% of the heat flux from the crust occurs in the older rocks that lie on the ridge flanks. When off-axis circulation is included, the turnover time of the ocean with respect to hydrothermal circulation is quite short, with current estimates being on the order of 0.1 million years or less! Collectively, these systems are responsible for about 20% of Earth's total heat loss.

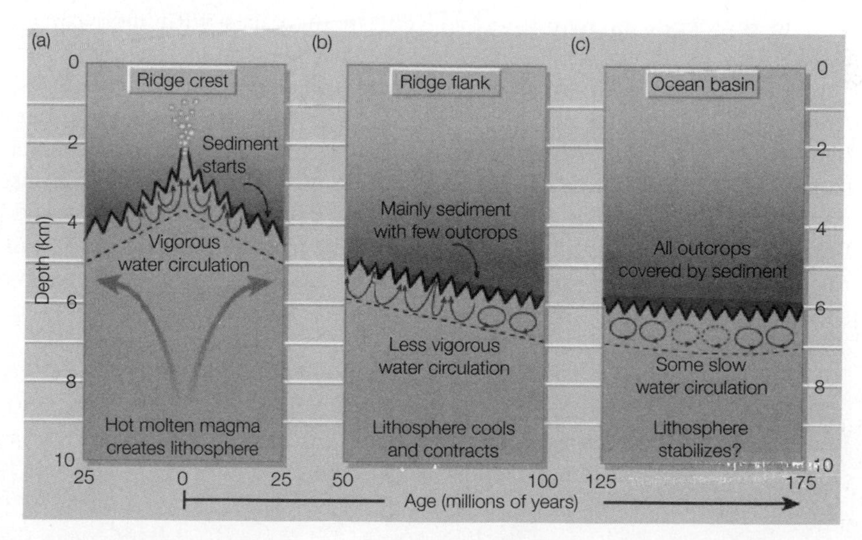

FIGURE 19.3

Changes in ocean-floor depth, sediment thickness, and hydrothermal circulation with increasing age of the lithosphere. (a) Heat loss is by advection, (b) heat loss is by advection and conduction, and (c) heat loss is by conduction alone. *Source*: From Sclater, J. G. (2003). *Nature* 421, 590–591.

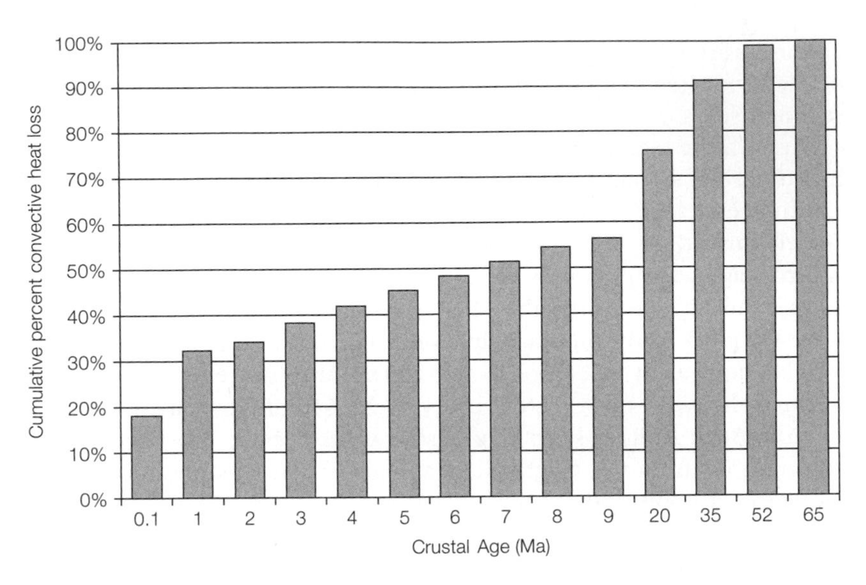

FIGURE 19.4

Cumulative percent convective heat loss. The effect of high-temperature hydrothermal emissions is included in the 0.1-million-year-old crustal age bin. Data from Wheat, C. G. (2003). *Geophysical Research Letters* 30, 1895.

The incidence of hydrothermal venting increases with increasing rate of spreading activity as shown in Figure 19.5. Spreading rates are fastest in the Pacific Ocean (2.4 to 15 cm/y) with lower rates in the Indian (6 cm/y), Atlantic (<2.0 cm/y), and Arctic (<1.0 cm/y) Oceans. At present only 20% of the ridges have been explored for hydrothermal venting, resulting in 144 confirmed vent sites and another 130 surmised from water-column chemical observations. Exploration efforts have focused on the eastern Pacific and northern Atlantic ridges. As a result, only two confirmed sightings exist for the entire 30,000 km span of ridge running from the equatorial Mid-Atlantic through the Indian Ocean to 38°S on the East Pacific Rise! Based on the observed relationship shown in Figure 19.5, the global population of active vent fields is estimated to be around 1000.

As shown in Figure 19.6, vent fields have been confirmed (or inferred) across depths ranging from 200 to 4300 m, with the most lying between 2200 and 2800 m. Very shallow systems are associated with volcanism that has built oceanic crust above sea level, such as near Iceland, the Hawaiian Islands, and the Azores.

19.2.2 **Subduction Zones**

About 80% of magmatic activity in the ocean occurs at the MOR, with another 10% each at subduction zones and mantle hot spots. Hydrothermal interactions take place at convergent plate boundaries where oceanic crust is being subducted back into the

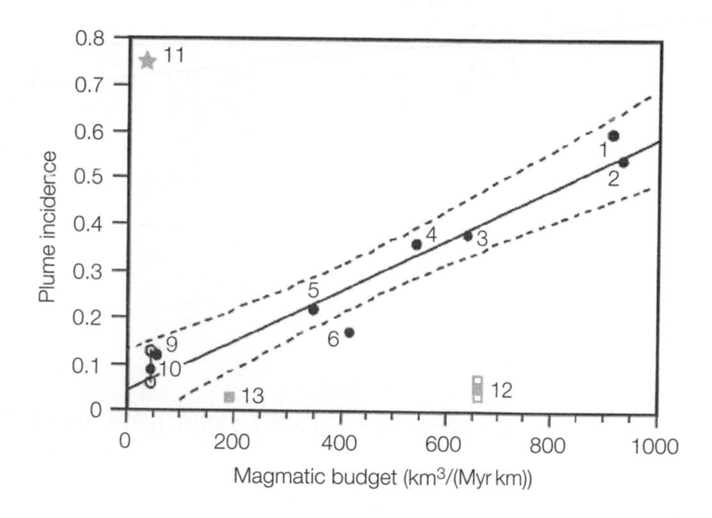

FIGURE 19.5

Plume incidence versus magmatic budget where the magmatic budget is the annual rate of delivery of a volume (km^3) of magma per length of ridge crest (km). It is a proxy for spreading rates. Plume incidence is the fraction of ridge crest length overlain by a significant hydrothermal plume. $r^2 = 0.93$, excluding ultra-slow and hotspot-affected ridges. Data points 1 through 6 are fast ridges (>55 mm/y), 7 through 10 are slow ridges (20–55 mm/y) and 11 is ultraslow (Gakkel <20 mm/y) ridges. 12 and 13 are mantle hot spots. *Source*: From Baker, E. T., and C. R. German (2004). *Mid-Ocean Ridges: Hydrothermal Interactions between the Lithosphere and Oceans*, Geophysical Monograph Series 148, pp. 245–266.

mantle. Most of the present-day subduction zones are located around the perimeter of the Pacific Ocean as a series of discontinuous submarine volcanic arcs and deep-sea trenches totaling about 3000 km in length. As the crust descends into the mantle, it carries the overlying sediment blanket with it. The resulting compaction and heating causes dewatering and melting that support fluid flows and volcanism. Seaward of the convergent plate boundary, fluid flows take the form of cold seeps. At the plate boundary, subduction produces hydrothermal emissions associated with the submarine arc volcanoes. Landward, back-arc spreading is common and also supports hydrothermal emissions.

19.2.3 **Mantle Hot Spots**

Some ridge sections are underlain by mantle melt anomalies, or hot spots, such as at the Azores and Galapagos Islands. These are marked by the stars in Figure 19.1 and data points 12 and 13 in Figure 19.5. Mantle upwelling beneath both these ridge sections has abnormally thickened the oceanic crust to at least about 10 km. Most of the 47 known hot spots lie more than 500 km from a ridge axis. The Hawaiian islands are a notable example.

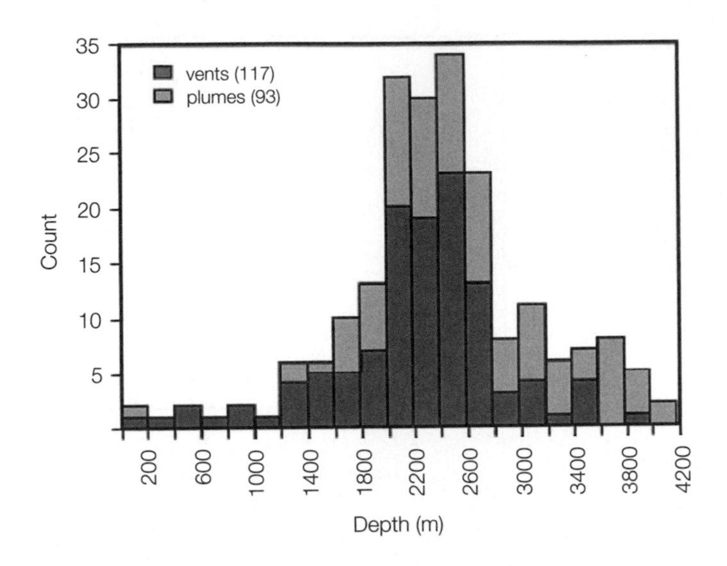

FIGURE 19.6

Frequency distribution of the depth of vent sites located along midocean ridge and back-arc basin spreading centers. Dark bars at the bottom of each column indicate depths determined from direct vent observations, and the lighter bars, depths inferred from plume sightings. *Source*: From Baker, E. T., and C. R. German (2004). *Mid-Ocean Ridges: Hydrothermal Interactions between the Lithosphere and Oceans*, Geophysical Monograph Series 148, pp. 245–266.

19.3 HYDROTHERMAL SYSTEM EVOLUTION

Resupply of magma is episodic at the spreading centers and subduction zones. This causes hydrothermal systems to vary over time in terms of volcanism and water circulation rates. As a result, the chemistry of the hydrothermal fluids and the biotic communities living around and in these systems can change over rather short time scales. Increasing effort is being directed at monitoring these changes.

As illustrated in Figure 19.7, hydrothermal venting seems to undergo an evolutionary sequence in which the discharge switches from high to low temperature as the magma chamber cools. Or, discharge can switch from low to high temperature as mineral precipitates seal subsurface conduits. In the latter case, hydrothermal emissions eventually cease upon complete sealing of the subsurface conduits. After hydrothermal discharge declines, the pillow basalts and other mineral precipitates react with seawater and become oxidized. These chemical changes undermine the structural strength of the minerals, causing the rock formations to crumble. This process is termed *mass wasting* and produces large mounds of metalliferous sediments.

The ridge crest is a dynamic setting in which volcanic activity creates new vents while old ones "die." On fast-spreading centers, hydrothermal circulation supports focused discharges through chimneys that have an average life span of a few decades.

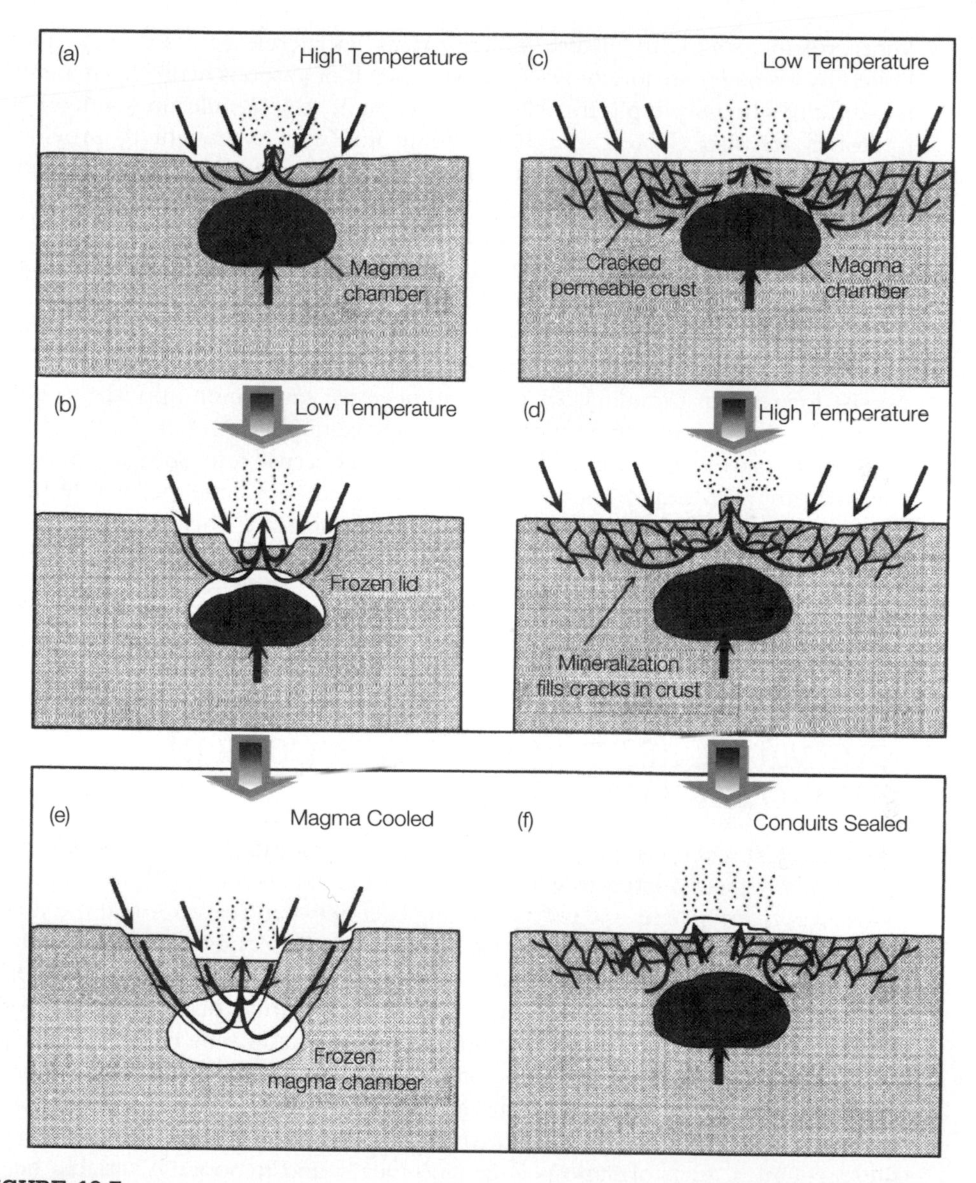

FIGURE 19.7

Mechanisms for vent evolution. Vents can evolve from (a) high to (b) low temperatures or from (c) low to (d) high temperatures. In the case of the former, cooling of the magma chamber is accompanied by an increase in depth of hydrothermal circulation. In the case of the latter, as crustal fractures are sealed off by hydrothermal mineral precipitates, the entrainment of fresh seawater is curtailed, so relatively undiluted hydrothermal fluids are discharged. In either case, final cooling results from either (e) total depletion of the heat source or (f) effective sealing of subsurface conduits or by a combination of the two processes. *Source*: After Haymon, R. M., and K. C. Macdonald (1985). *American Scientist* 73, 441–449.

Vent creation seems to involve an enormous initial release of gas-rich hot fluids followed by a smaller steadier flow. The initial belch of gaseous hydrothermal fluids creates megaplumes that are typically half a mile in height and 12 miles in diameter. Subsurface microbes are also ejected, creating a white floc that is thought to provide an initial food source promoting colonization of the newly formed vent. As discussed in a later section, the hydrothermal biotic community also undergoes a successional sequence in response to the shifting geochemical forces.

Continued seafloor spreading eventually pushes any given segment of oceanic crust away from the MOR, causing hydrothermal activity to decrease with increasing age. In general, venting off axis is diffuse and low in temperature, but can endure for very long periods. A recently discovered form of low-temperature off-axis venting is thought to persist for tens of thousands of years. Convective circulation appears to cease by the time the crust exceeds 65 million years in age (Figure 19.3).

Seafloor spreading eventually pushes oceanic crust into subduction zones where the hydrothermal sediments and rock are recycled back into the mantle. A small fraction of these deposits is uplifted, or *obducted*, onto land. These "rescued" deposits are termed *ophiolites*. Because of their metal enrichments, they serve as major ore bodies and have been mined for various precious metals, such as copper, for thousands of years.

19.4 CHEMICAL REACTIONS THAT OCCUR IN HYDROTHERMAL SYSTEMS

The chemical reactions that take place at the oceanic ridge crest as magma and igneous rock (mostly basalt) interact with seawater are summarized in Figure 19.8. These reactions are driven by extreme redox disequilibria between seawater and the mantle rocks. Vigorous chemical reactions take place as seawater descends into the system with some elements being solubilized from the rocks and others being precipitated. Other chemical changes occur in the hot reaction zone and during discharge. The net result of these reactions is a type of zone refining of elements in which chemicals stripped out of the mantle rocks are either precipitated as hydrogenous minerals or discharged into seawater. Similarly, some chemicals are stripped out of seawater to become part of the altered mantle rocks or part of the hydrogenous minerals. The exact nature of all these chemical transformations is geographically and temporally variable because they depend on water temperature, pressure, and circulation rates, as well as the porosity of the rock and its chemical composition. Most of these factors are related to spreading rates.

During this zone refining, the primary (igneous) rocks are transformed into secondary minerals. These include: (1) clay minerals, such as phillipsite, chlorite, montmorillonite (smectite), saponite, celadonite, and zeolite; (2) iron oxyhydroxides; (3) pyrite; (4) various carbonates; and (5) quartz. These minerals form rapidly, within 0.015 and 0.12 million years after creation of the oceanic crust at the MOR. During these alteration

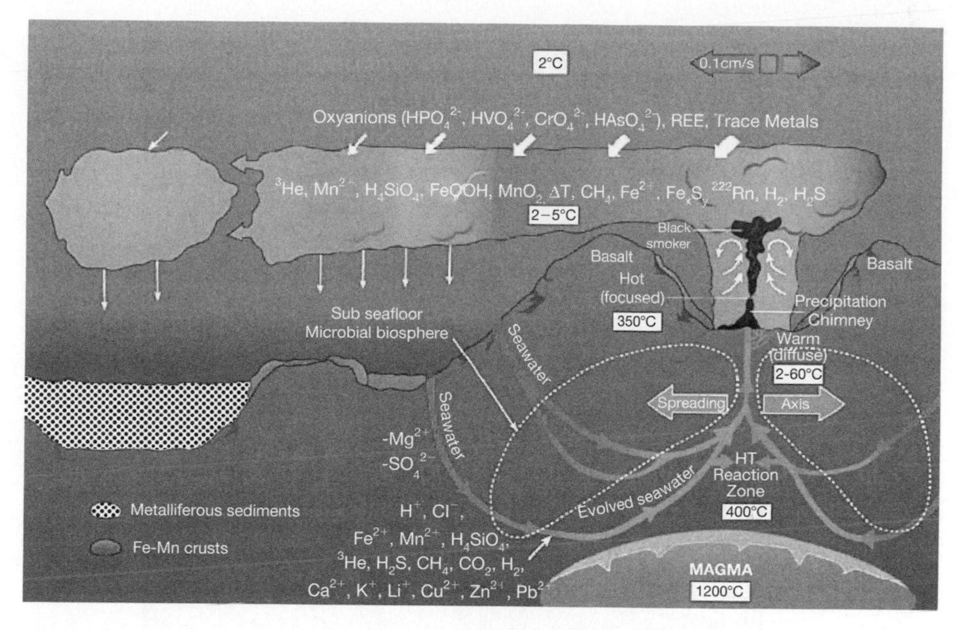

FIGURE 19.8

Chemical reactions in hydrothermal systems. High-temperature hydrothermal discharges can form plumes known as black smokers. Within the plumes, which sometimes travel thousands of miles from the vent, chemical reactions produce metal-rich particulates that settle on the ocean floor. REE, rare-earth elements, HT = high temperature. *Source*: After Garry Massoth, Pacific Marine Environmental Laboratory. NOAA Vents Program. http://www.pmel.noaa.gov/vents/chemistry/images/vents2.gif. (See companion website for color version.)

reactions, the crust also takes up large amounts of water and CO_2. Within 10 million years, mineral precipitation greatly reduces the porosity of rock in the oceanic crust, thereby limiting further chemical reaction to low-temperature processes. The details of the chemical reactions that occur during recharge, in the hot reaction zone, and during discharge are discussed next.

19.4.1 Seawater Recharge Reactions

As seawater begins its descent into the hydrothermal system, it increases in temperature and begins reacting with the hydrothermal sediments and rock as shown in Figure 19.9. The exact nature of this recharge process is not well known but is thought to occur over large areas as a diffuse flow. As temperature increases, the following reactions occur: (1) O_2 is removed by reaction with iron sulfides and iron-rich igneous rocks, and (2) alkalis, such as potassium, are incorporated into secondary clay minerals. At temperatures exceeding 150°C, (1) anhydrite ($CaSO_4$) precipitates because its solubility decreases with increasing temperature and (2) alkalis including potassium, rubidium, and lithium are solubilized from the crustal rocks. Since the hydrothermal

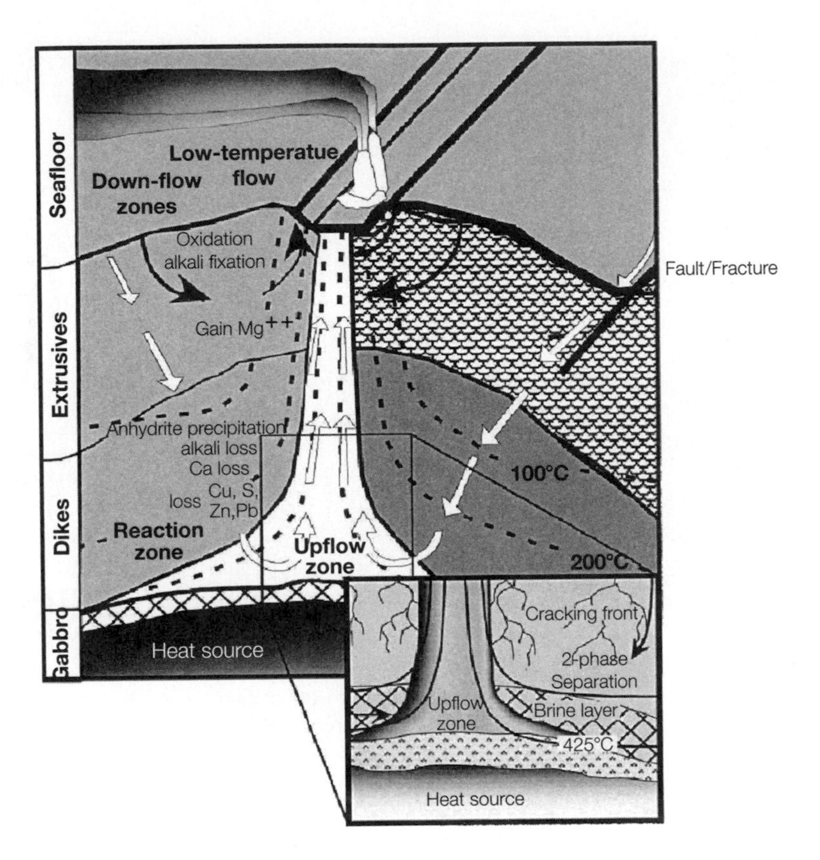

FIGURE 19.9

Schematic showing chemical reactions and mineral precipitation associated with down-welling recharge systems, low-temperature shallow circulation, and deep penetration by hydrothermal fluids into the reaction zone. *Source*: From Kelley, D. S., *et al*. (2002). *Annual Review of Earth and Planetary Sciences*, 30, 385–491.

fluids are acidic, the bicarbonate in seawater is converted to carbonic acid and CO_2. Mg is removed from seawater by incorporation into smectite (<200°C) and chlorite (>200°C).

19.4.2 High-Temperature Reactions in Upper Crust

The upper mantle of Earth is composed of rocks that have crystallized from magmas rich in iron and magnesium. As presented in Figure 14.2, crystallization conditions determine which igneous rock mineral is produced. Pyroxenes are the dominant mineral in basalt. Basaltic rocks also contain variable percentages of calcic plagioclase and olivine. They are extrusive and found in MOR systems as pillow basalts. Gabbros are the intrusive version of basaltic rock. Pyroxenes are silicate minerals whose crystal lattice contains variable degrees of magnesium, iron, and calcium, with lesser amounts of aluminum, sodium, lithium, manganese, or zinc. Another type of intrusive rock found in hydrothermal systems is peridotite, which is composed primarily of olivine. The empirical formula of this mineral is $(Mg, Fe)_2SiO_4$. All of these igneous rocks contain a relatively high amount of ferrous iron (Fe(II)) and, hence, are relatively reduced. This reducing power is transferred via high-temperature fluid-rock interactions from the igneous minerals into the form of reduced gases (H_2S, H_2, and CH_4). These gases then dissolve into the hydrothermal fluids, which are discharged through the vents, thereby making the reduced chemicals accessible to microbes living within and upon the hydrothermal systems.

High-temperature reactions occur several kilometers below the seafloor, where seawater comes into contact with magma and is heated to temperatures ranging from 340 to 465°C. (This seawater has already been somewhat altered from reaction in the recharge zone. For example, all O_2 has been removed.) Under these conditions, seawater undergoes a phase separation into a vapor and a brine phase. (This is discussed in more detail in the next section.) Magma contains reduced gases (H_2 and H_2S) and CO_2. These gases are released into the hydrothermal fluids via the physical process of degassing as the magma moves through the hydrothermal system.

Reaction conditions are thought to be such that chemical equilibrium is probably attained between the fluid phase and the magmatic rocks. One of the important high-temperature chemical reactions that basalt undergoes is

$$5Mg^{2+} + \underbrace{3SiO_2 + Al_2O_3}_{\text{basalt}} + 9H_2O \rightarrow \underbrace{Mg_5Al_2Si_3O_{10}(OH)_8}_{\text{chlorite}} + 10H^+ \qquad (19.1)$$

during which magnesium is removed from seawater. The reaction products are a secondary silicate mineral (chlorite) and acid. Some of the acid titrates HCO_3^-, generating CO_2, and some reacts with igneous rocks to solubilize metals (Fe, Mn, Zn, and Cu). Reduced gases, namely H_2S, H_2, and CH_4, are also produced. For example, H_2S is generated from the reduction of seawater sulfate via

$$\underbrace{Mg_{1.5}Fe_{0.5}Si_2O_6}_{\text{(Pyroxene)}} + 0.083H^+ + \underbrace{0.0416SO_4^{2-}}_{\text{from seawater}} + 0.5H_2O \rightarrow$$
$$\underbrace{Mg_{1.5}Si_2O_5OH}_{\text{(Talc)}} + \underbrace{0.167Fe_3O_4}_{\text{(Magnetite)}} + 0.0416H_2S \qquad (19.2)$$

(In Eq. 19.2, talc represents the magnesium hydroxide component of various secondary minerals such as chlorite and amphibole.) As a result, sulfate is stripped out of the hydrothermal fluids.

The high-temperature reactions make the hydrothermal fluids acidic, reducing, and metal rich. As shown in Table 19.1, dissolved metal concentrations are more than 1 million times greater than that of seawater. With all the sulfate reduced to sulfide and bicarbonate titrated to CO_2, chloride is the only anion left in solution. Although chloride appears to behave conservatively, hydrothermal fluids have been observed to have chlorinities that range from 6% to over 200% of that observed in seawater. This great range in chloride content is the product of phase separation caused by the high pressures and temperatures found at the vents. The chloride ion plays a very important role

Table 19.1 Ranges in Chemical Composition for All Known Vent Fluids.

Chemical Species	Units	Seawater	Overall Range	Slow[a] (0 to 2.5 cm y^{-1})	Intermediate[a] >2.5 to 6	Fast[a] >6 to 12	Ultrafast[a] >12	Sediment Covered[b]	Ultramafic Hosted[c]	Arc, Back-arc
T	°C	2	>2 to 405	40 to 369	13 to 382	8 to 403	16 to 405	100 to 315	40 to 364	278 to 334
pH	25°C, 1 atm	7.8	2.0 to 9.8	2.5 to 4.85	2.8 to 4.5	2.45 to >6.63	2.96 to 5.53	5.1 to 5.9	2.7 to 9.8	2.0 to 5.0
Alkalinity	meq kg^{-1}	2.4	~ 3.75 to 10.6	~ 3.4 to 0.31	~ 3.75 to 0.66	~ 2.69 to <2.27	~ 1.36 to 0.915	1.45 to 10.6		~ 0.20 to 3.51
Cl	mmol kg^{-1}	540	30.5 to 1245	357 to 675	176 to 1245	30.5 to 902	113 to 1090	412 to 668	515 to 756	255 to 790
SO_4	mmol kg^{-1}	28	<0 to <28	~ 3.5 to 1.9	~ 25 to 0.763	> to 8.76	~ 0.502 to 9.53	0	<12.9	0
H_2S	mmol kg^{-1}	0	0 to 110	0.5 to 5.9	0 to 19.5	0 to 110	0 to 35	1.10 to 5.98	0.064 to 1.0	2.0 to 13.1
Si	mmol kg^{-1}	0.03 to 0.18	<24	7.7 to 22	11 to 24	2.73 to 22.0	8.69 to 21.3	5.60 to 13.8	6.4 to 8.2	10.8 to 14.5
Li	µmol kg^{-1}	26	4.04 to 5800	238 to 1035	160 to 2350	4.04 to 1620	248 to 1200	370 to 1290	245 to 345	200 to 5800
Na	mmol kg^{-1}	464	10.6 to 983	312 to 584	148 to 924	10.6 to 983	109 to 886	315 to 560	479 to 553	210 to 590
K	mmol kg^{-1}	10.1	~ 1.17 to 79.0	17 to 28.8	6.98 to 58.7	~ 1.17 to 51	2.2 to 44.8	13.5 to 49.2	20.2 to 22	10.5 to 79.0
Rb	µmol kg^{-1}	1.3	0.156 to 360	9.4 to 40.4	22.9 to 59	0.156 to 31.1	0.39 to 6.8	22.5 to 105	28 to 37.1	8.8 to 360
Cs	nmol kg^{-1}	2	2.3 to 7,700	100 to 285	168 to 364	2.3 to 264		1000 to 7700	331 to 385	
Be	nmol kg^{-1}	0	10 to 91		10 to 37			12 to 91		
Mg	mmol kg^{-1}	52.2	0	0	0	0	0	0	<19	0
Ca	mmol kg^{-1}	10.2	~ 1.31 to 109	9.9 to 43	9.75 to 109	~ 1.31 to 106	4.02 to 65.5	26.6 to 81.0	21.0 to 67	6.5 to 89.0
Sr	µmol kg^{-1}	87	~ 29 to 387	42.9 to 133	0.0 to 348	~ 29 to 387	10.7 to 190	160 to 257	138 to 203	20 to 300
Ba	µmol kg^{-1}	0.14	1.64 to 100	<52.2	>8 to >46		1.64 to 18.6	>12	>45 to 79	5.9 to 100

Mn	μmol kg^{-1}	<0.001	10 to 7100	59 to 1000	140 to 4480	62.7 to 3300	20.6 to 2750	10 to 236	330 to 2350	12 to 7100
Fe	mmol kg^{-1}	<0.001	0.007 to 25.0	0.0241 to 5590	0.009 to 18.7	0.007 to 12.1	0.038 to 14.7	0 to 0.18	2.5 to 25.0	13 to 2500
Cu	μmol kg^{-1}	0.007	0 to 162	0 to 150	0.1 to 142	0.18 to 97.3	2.6 to 150	<0.02 to 1.1	27 to 162	0.003 to 34
Zn	μmol kg^{-1}	0.012	0 to 3000	0 to 400	2.2 to 600	13 to 411	1.9 to 740	0.1 to 40.0	29 to 185	7.6 to 3000
Co	μmol kg^{-1}	0.00003	<0.005 to 14.1	0.130 to 0.422	0.022 to 0.227			<0.005	11.8 to 14.1	
Ni	μmol kg^{-1}	0.012							2.2 to 3.6	
Ag	nmol kg^{-1}	0.02	<1 to 230		<1 to 38			<1 to 230	11 to 47	
Cd	nmol kg^{-1}	1.0	0 to 180	75 to 146	0 to 180			<10 to 46	63 to 178	
Pb	nmol kg^{-1}	0.01	<20 to 3900	221 to 376	183 to 360			<20 to 652	86 to 169	36 to 3900
B	μmol kg^{-1}	415	356 to 3410	356 to 480	465 to 1874	430 to 617	400 to 499	<2160		470 to 3410
Al	μmol kg^{-1}	0.02	0.1 to 18.7	1.03 to 13.9	4.0 to 5.2	0.1 to 18	9.3 to 18.7	0.9 to 7.9	1.9 to 4	4.9 to 17.0
Br	μmol kg^{-1}	840	29.0 to 1910	666 to 1066	250 to 1789	29.0 to 1370	216 to 1910	770 to 1180		306 to 1045
F	μmol kg^{-1}	68	<38.8	16.1 to 38.8	"0"					
CO_2	mmol kg^{-1}			3.56 to 39.9	<5.7	<200	8.4 to 22			14.4 to 200
CH_4	μmol kg^{-1}	0.0003		150 to 2150	<52		7 to 133		130 to 2200	
NH_4	mmol kg^{-1}	0	<15.6	<0.06	<0.65			5.6 to 15.6		
H_2	μmol kg^{-1}	0.0003	<38,000	1.1 to 727	<0.45	<38,000	40 to 1300		250 to 13,000	

[a] *These omit "sedimented covered" and "ultramafic hosted" vents.*

[b] *Includes: Guaymas, Escanaba, and Middle Valley vent fields.*

[c] *Includes Rainbow and Lost City vent fields.*

Source: After German, C. R., and K. L. Von Damm (2003). Treatise on Geochemistry, *Elsevier, pp. 181–222.*

in hydrothermal fluids because it forms chlorocomplexes, which enhances the solubility of the metal cations.

Silica is also solubilized from the igneous rocks during high-temperature reactions. This causes dissolved silica to become supersaturated, leading to the precipitation of quartz and other secondary silicates, such as albite, within the hydrothermal conduits. The formation of albite is an important reaction because it serves as a sink for Na^+ and K^+ and a source of Ca^{2+}:

$$(Na^+, K^+) + \underset{\text{anorthite (calcic plagioclase)}}{CaAl_2Si_2O_8} + 2SiO_2 + \frac{1}{2}H_2O + H^+ \longrightarrow Ca^{2+} + \underset{\text{albite}}{(K, Na)AlSi_3O_8} + \underset{\text{kaolinite}}{0.5Al_2Si_2O_5(OH)_4} \quad (19.3)$$

If bicarbonate is the source of acid driving this reaction, $CaCO_3$ and CO_2 are produced:

$$2(Na^+, K^+) + \underset{\text{anorthite (calcic plagioclase)}}{CaAl_2Si_2O_8} + 4SiO_2 + 2HCO_3^- \longrightarrow \underset{\text{calcite}}{CaCO_3} + \underset{\text{albite}}{2(K, Na)AlSi_3O_8} + CO_2 + H_2O \quad (19.4)$$

From this perspective, the high-temperature reactions are a potential carbonate buffering system.

In addition to the secondary silicate and oxide minerals, metal-rich minerals such as pyrite (FeS_2), chalcopyrite ($CuFeS_2$), and pyrrhotite ($Fe_{1-x}S$) are also formed by high-temperature reactions. These metal sulfide rocks catalyze the conversion of CO_2, CO, and H_2O into CH_4.

These generalizations regarding high-temperature fluid-rock reactions are based on field observations and laboratory simulations. Geochemists have attempted to run high-temperature reactions in the lab by reacting basalt with seawater at pressures up to 1000 bar and temperatures of 70 to 500°C at rock-to-water ratios from 1 to 62 for periods as long as 20 months. Their reaction products support the conclusions made from field observations.

19.4.3 Phase Separation: Changing Impact on Ion Speciation

In the hot reaction zone, seawater is heated to temperatures ranging from 340 to 465°C, which is considerably about the boiling point of pure water at 1 atm pressure. Although the high pressures in the deep sea considerably expand the thermal range at which water remains a liquid, the temperatures of the hydrothermal fluids are high enough that some type of phase separation does occur. This phase separation can take three forms listed in order of increasing pressure and temperature requirements: (1) a subcritical form akin to boiling that produces a vapor and a brine, (2) a supercritical form in which a high-chlorinity fluid condenses, and (3) an extreme supercritical form in which halite precipitates.

Phase separation greatly affects the chemistry of the hydrothermal emissions. More information on the fractionation of chemicals that occurs during phase separation is provided in the supplemental information for Chapter 19 available at **http://elsevierdirect.com/companions/9780120885305**.

19.4.4 Discharge via Venting from Mid Ocean Ridge Systems

Before the hydrothermal fluids can be discharged back into the ocean, they must rise through the rock system and any overlying sediments. As the fluids rise, they encounter and mix with descending seawater. The degree of this mixing is determined by plumbing within the hydrothermal system, such as the degree of fracturing, the porosity of the rock, and the water depth (overlying pressure). As the rising fluids cool and mix with seawater, chemical changes occur. Some of these chemical reactions cause mineral precipitation that eventually seals up the flow conduits, leading to a decline and eventual cessation of hydrothermal circulation.

Because of spatial and temporal variability in subsurface mixing, the temperature, flow rates, and chemical composition of hydrothermal discharges are similarly variable. For example, venting can range from highly focused, hot (200 to 400°C) turbid discharges emanating from mineral chimneys to diffuse, low-temperature (~25 to 60°C) flows rising from metalliferous sediment mounds (Figure 19.10). The ascending fluids continue to undergo rock-water interactions, producing secondary minerals and hydrogenous precipitates. Reduced chemicals in the fluids are also used as an energy source by chemoautolithotrophic microbes. The diversity in chemical composition of the venting fluids is illustrated in Table 19.1.

Vent lifetimes and abundances are related to magma supply and spreading rates. The fast-spreading ridges, such as those of the East Pacific Rise (EPR), support vents that can discharge for decades, as compared to vents on the slower spreading Mid-Atlantic Ridge (MAR), which can be active for centuries. Spreading rates also lead to differences in topography, with the fast ridges having a more narrow and shallow axial valley (50 to 80 m wide and 5 to 8 m deep) as compared to the slow ridges (Figure 19.11). Because of their shallow relief, "fast" ridges tend to discharge their hydrothermal fluids in a relatively narrow zone with plumes being readily dispersed off axis by local currents. The "slow" ridges of the MAR are characterized by deep, spacious (up to 15 km across), and generally enclosed rift valleys, commonly with several hundreds of meters of relief. Volcanic zones are not well defined, but deep and enduring faults are common. These facilitate discharge of hydrothermal circulation over a broad expanse of seafloor. Exploration of the ultraslow ridges, such as the Gakkel in the Arctic Ocean, has only recently been initiated. At this ridges, venting is thought to occur in a fashion similar to that found off-axis on the MAR as discussed in Chapter 19.4.5.

Hot Focused Flows

In locations where hydrothermal fluid exits directly into the bottom ocean with little subsurface dilution, large chimneys are deposited by the rapid precipitation of anhydrite

FIGURE 19.10

A typical hydrothermal vent system. Included are basal mounds of mineral precipitates, pillow basalts, a black smoker, and various vent organisms. *Source*: From Haymon, R. M. and K. C. Macdonald (1985). *American Scientist*, 73, 441–449.

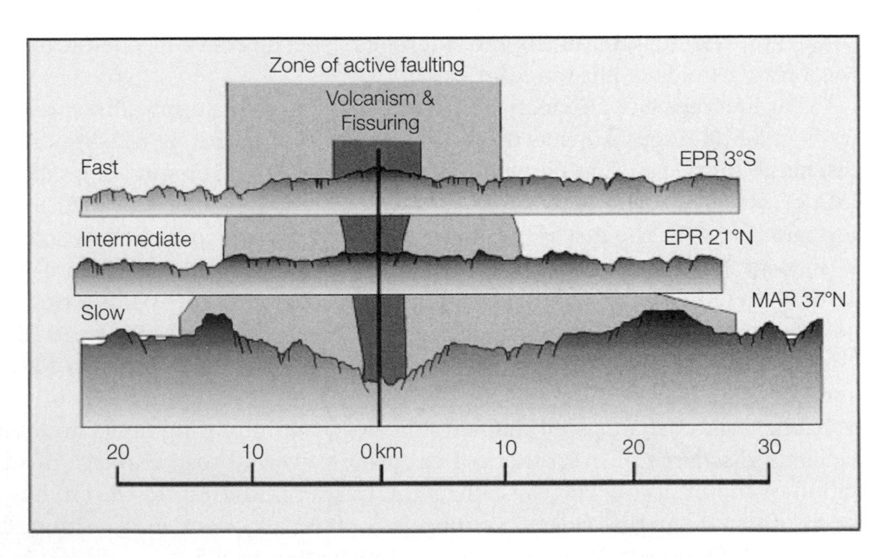

FIGURE 19.11

Cross-axis profiles of spreading centers. Topography, magma chamber depth, frequency of eruptive events, size of hydrothermal deposits, and vent fauna diversity and composition all show strong correlations with spreading rate. EPR, East Pacific Rise; MAR, Mid-Atlantic Ridge. *Source*: From Kelley, D. S., *et al*. (2002). *Annual Reviews of Earth and Planetary Sciences* 30, 385–491.

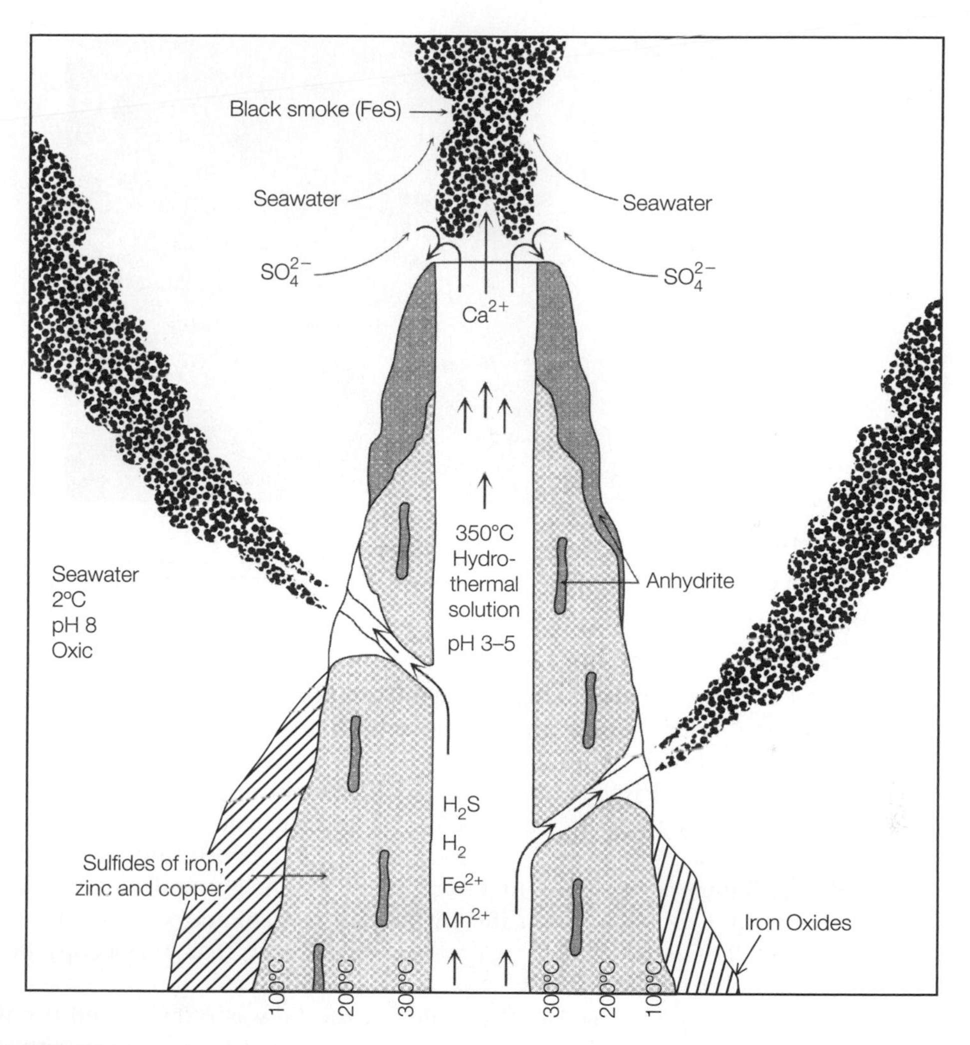

FIGURE 19.12

Hydrothermal chimney formation. *Source*: After Edmond, J. M. (1984). *Oceanus* 27, 15–19.

($CaSO_4$), barite ($BaSO_4$), and polymetallic sulfides. The precipitation reactions begin when the hot (200 to 400°C), acidic, sulfide- and metal-rich fluid mixes with the cold, oxic, alkaline seawater. The calcium from the hydrothermal fluid reacts with seawater sulfate, depositing anhydrite, which forms the leading edge of a "chimney," as shown in Figure 19.12.

The formation of these anhydrite walls prevents the hydrothermal fluids flowing through the chimney from mixing with seawater and provides a framework to enable precipitation of sulfide minerals. In some cases, the discharge of fluids is so rapid that the sulfide precipitates are emitted as clouds of black particles moving at a speed of

FIGURE 19.13

Photograph of a black smoker. Courtesy of the Woods Hole Oceanographic Institution, Woods Hole, MA.

meters per second. These discharges have the appearance of a jet of black smoke and are commonly referred to as "black smokers." They were first discovered in 1979 on the East Pacific Rise at 21°N. A photograph of one is shown in Figure 19.13.

Some of the discharged sulfide particles settle onto the chimney's exterior, where they are buried by the outward growth of anhydrite. Sulfide precipitation within the chimneys, causes copper, zinc, and iron sulfides to deposit and partially replace the anhydrite. Chimneys can build to several meters in height and their orifices range in diameter from 1 to 30 cm. Both the smoke and the chimneys are composed of polymetallic sulfide minerals, chiefly pyrrhotite (FeS), pyrite (FeS_2), chalcopyrite ($CuFeS_2$), and sphalerite or wurtzite (ZnS).

As the chimney interior fills with sulfides, the flow is reduced and the walls cool, causing anhydrite to dissolve. (The solubility of anhydrite increases with decreasing temperature.) Marine organisms colonize the cooling chimneys and the sulfides begin to oxidize. This destabilizes the chimneys and eventually they collapse, forming basal mounds that can reach heights of 2 m or more. The relatively small size of these mounds suggests that black smokers have short lifetimes, perhaps on the order of a decade. The continuing buildup and collapse of chimneys eventually creates a massive sulfide deposit.

In some locations, high-temperature fluids undergo considerable subsurface mixing with relatively "fresh" seawater. This leads to precipitation of the less soluble iron and copper sulfides within the conduits. The fluids discharging into the ocean generally have temperatures less than 400°C and are milky white because of zinc sulfide precipitates. These "white smokers" also build chimneys, some of which are as much as 13 m high. Because of their lower temperatures, white smokers are typically encrusted with worm

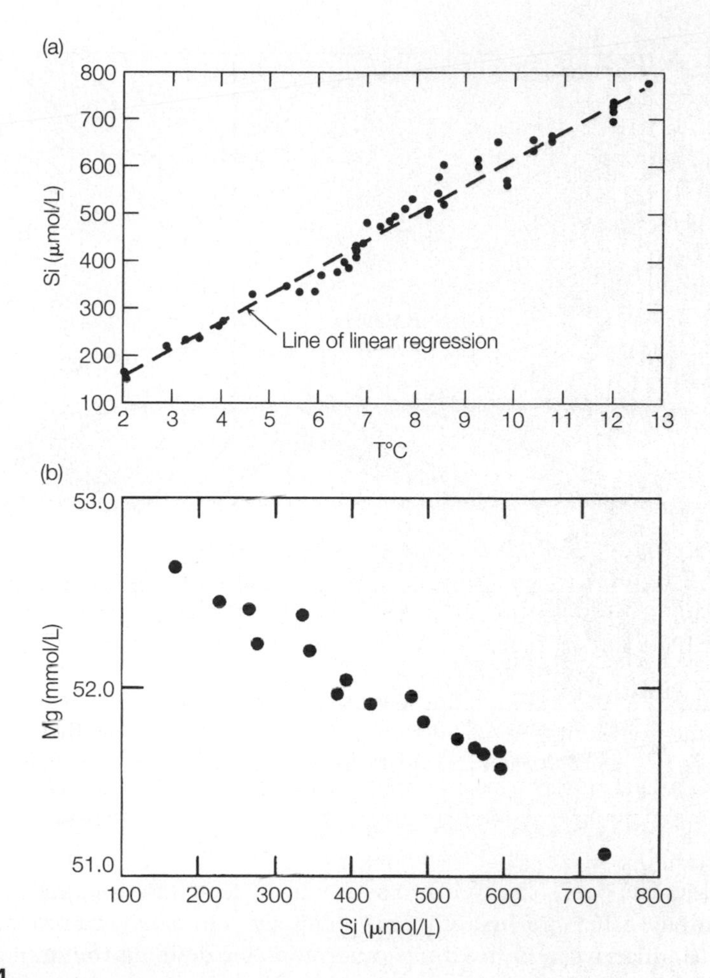

FIGURE 19.14

Chemical composition of hydrothermal fluids collected from the Galapagos vents: (a) Dissolved silicon concentrations versus temperature and (b) magnesium versus dissolved silicon concentrations. *Source*: From Edmond, J. M., *et al*. (1979). *Earth and Planetary Sciences Letters* 46, 1–18.

tubes and other organisms. Some white smokers are probably a later stage of venting that develops as cooling leads to fracturing of crustal rock as illustrated in Figures 19.7b and f.

Subsurface mixing can be tracked by following the conservative behavior of various chemicals in the hydrothermal fluids, as illustrated in Figure 19.14a for dissolved silica at the Galapagos vents. Silica is solubilized during rock-water interactions in the hot reaction zone. As this fluid rises and mixes with "fresh" seawater, its dissolved silicon concentration and temperature decrease in a linear fashion, reflecting conservative behavior.

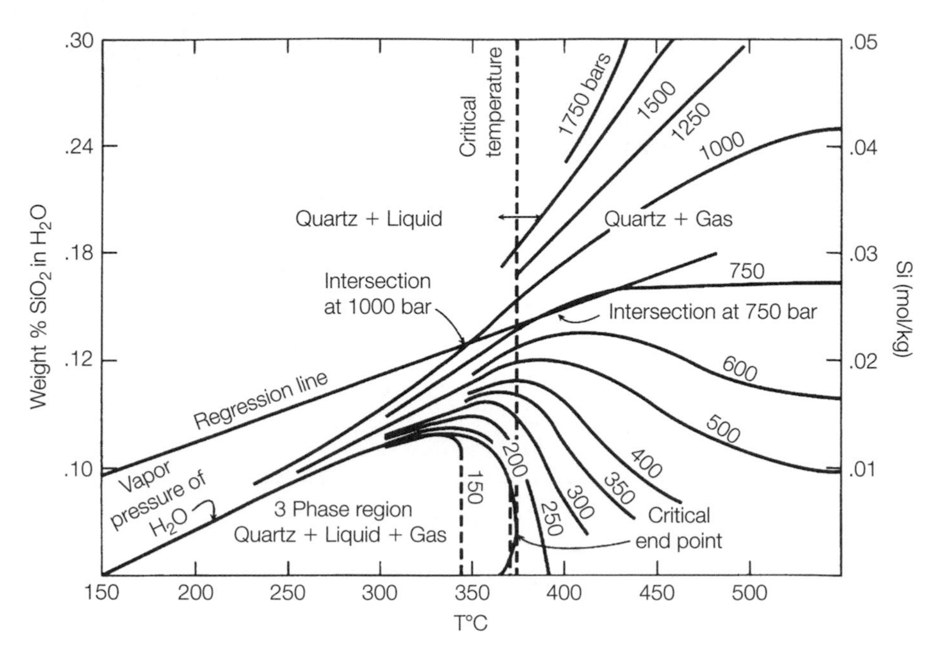

FIGURE 19.15

Solubility of quartz in vent water as a function of temperature (°C) and pressure (bars). The regression line from Figure 19.14a is also plotted and extrapolated to conditions where the solution becomes saturated with respect to quartz, e.g., 345°C at 1000 bar and 375°C at 750 bar. *Source*: From Edmond, J. M., *et al*. (1979). *Earth and Planetary Sciences Letters* 46, 1–18.

Quartz has been observed as a subsurface precipitate, suggesting that dissolved silica concentrations in the hydrothermal fluids are ultimately controlled by quartz solubility. Thus, quartz can be used a geothermometer defining the minimum temperature of the undiluted hydrothermal fluids responsible for quartz precipitation. This is done by extrapolating the line of linear regression shown in Figure 19.14a to the temperature and dissolved silica concentrations at which the hydrothermal fluid becomes saturated with respect to quartz. This is shown in Figure 19.15, neglecting the negligible effect of phase separation on dissolved silica-quartz saturation concentrations.

Since the saturation concentration is a function of pressure as well as temperature, the geothermometer temperature is dependent on pressures. At 1000 bar, which is roughly equivalent to a depth below the seafloor of 1.6 km (750 bar) in a 2.5-km water depth (250 bar), the regression line intersects the solubility curve of quartz at 345°C. Thus, the quartz geothermometer estimates a minimum temperature of 345°C for the undiluted hydrothermal fluid at the Galapagos vents.

Other chemicals behave conservatively during the low-temperature mixing process and, thus, can also be used as geothermometers. As shown in Figure 19.14b, the hydrothermal system is a sink for magnesium because of incorporation into silicate rocks (Eq. 19.1). Extrapolation of this trend to zero magnesium ion concentration also yields a temperature of approximately 345°C for the undiluted high-temperature hydrothermal fluid. The same has been observed for sulfate. For other chemicals, such as O_2, nitrate, copper, nickel, cadmium, chromium, uranium, and selenium, concentrations extrapolate to zero at temperatures between 30 to 40°C. This suggests that their removal occurs at fairly low temperatures following admixture of the hydrothermal fluid with descending "fresh" seawater, probably close to the surface of the crust. In other words, these chemicals exhibit nonconservative behavior during the subsurface mixing process.

Diffuse Low-Temperature Discharges

Warm water (~25 to 60°C) also emanates as a relatively dissipated flow (2 to 10 L/s) from the tops of mounds created by the crumbling of chimneys (Figure 19.10). These flows were first observed at the Galapagos vents (86°W) and are associated with a unique biotic community. Mounds at this location reach heights of 30 m. The low temperatures of these emissions are caused by extensive subsurface mixing of the high-temperature hydrothermal fluid with descending "fresh" seawater. Subsurface cooling causes all the sulfides to precipitate within the basalt conduits and sediments, so the exiting fluid produces no "smoke."

Plumes

Hydrothermal venting injects fluids into seawater as buoyant, jetlike plumes. These turbulent flows mix rapidly with seawater becoming diluted by factors of 10^4 to 10^5. This mixing eventually makes the plumes neutrally buoyant, after which they are transported laterally through the ocean basins as part of the intermediate and deepwater currents. Hydrothermal plumes have the potential to greatly affect seawater chemistry. From global estimates of hydrothermal fluid emissions and dilution ratios, a volume of seawater equivalent to the entire ocean can be entrained in the hydrothermal plumes every few thousand years.

Plume chemistry is highly variable. Volatile-rich emissions are associated with explosive megaplumes that initiate venting. Metal-rich emissions are characteristic of chronic venting from established black and white smokers. Megaplumes can rise as much as 800 m into the ocean, whereas chronic venting produces plumes that lie 100 to 200 m above the seafloor. Chronic plumes emitted in deep axial valleys tend to move laterally at rates of 1 to 5 cm/s along the ridge crests. Turbulent mixing caused by geothermal heating and rough bottom topography eventually disperses the plume. Longitudinal transport occurs at 5°N to 5°S due to strong equatorial currents. As illustrated in Figure 11.19, plumes can be traced as they move laterally through the ocean basins by their unique geochemical composition. Large-scale distributions of plumes are most easily mapped using towed in situ turbidity sensors, which are simple to use, inexpensive, and sensitive. A two-dimensional turbidity map of vent plumes over the Gakkel Ridge is shown in Figure 19.16. These maps are used to identify locations to explore for venting activity.

At the time of its injection into seawater, most of the particulate material in the plumes is composed of metal sulfides, anhydrite, barite, and quartz. As the plumes

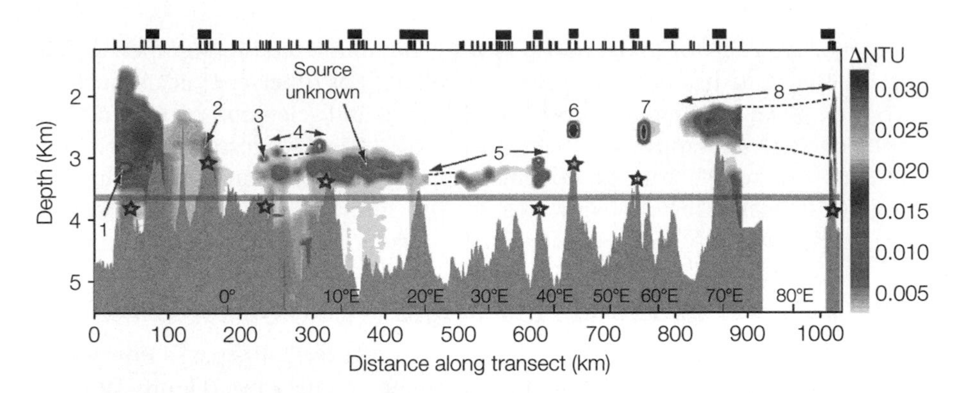

FIGURE 19.16

Along-axis transects of turbidity anomaly (units of NTU) at the Gakkel Ridge (4°W to 86°E) where ΔNTU is the difference between the NTU value at a particular depth and that for the profile background. Black bars show approximate location of volcanic centers. Numbers mark individual plumes, and the possible seafloor source location of each is shown by the underlying stars; source location of the extensive plume from 250 to 450 km along section, centered at ~3200 m, is unknown. Most of the NTU anomaly west of ~150 km has no thermal expression and is thus nonhydrothermal. Supplementary *x*-axis scale gives longitude along the transect. Gray line indicates approximate top of the bottom isopycnal layer. *Source*: From Baker, E. T., and C. R. German (2004). *Mid-Ocean Ridges: Hydrothermal Interactions between the Lithosphere and Oceans*, Geophysical Monograph Series 148, pp. 245–266.

travel, some of these particles sink back to the ocean floor, oxidize, and become a component of metalliferous sediment. Dissolved metals such as iron and manganese react with oxic seawater and are transformed into solids that efficiently co-precipitate most of the other dissolved metals emitted as part of the plumes. As a result, only Mn and Ca escape a rapid return to the seafloor. (As discussed in Chapter 18, manganese is the slowest to oxidize, so it is able to travel furthest before becoming transformed into a sedimentary particle.) The plume fallout forms a sedimentary halo around the hydrothermal systems, having a diameter of a few hundred kilometers.

The hydrothermal plumes tend to contain high concentrations of microbes, including archaea and viruses, along with zooplankton. The biotic communities in the plumes are sustained by chemoautolithotrophy using the reduced chemicals provided by the hydrothermal fluids. The lateral transport of the plumes carries the vent biota long distances. This transport is thought to be a mechanism by which vent biota are transported to new vents, especially for plumes that move axially along the ridge crests.

19.4.5 Off-Axis Discharges

Off-axis hydrothermal circulation is responsible for 70% of the convectively driven heat flow from hydrothermal systems. This circulation is thought to occur on the flanks of the mid-ocean ridges and rises at temperatures on the order of 20 to 54°C. The global fluid flux from this circulation is thought to be 30 to 300 times larger than that passing through the ridge crests and to be roughly equivalent to global flows from the world's rivers. Because of a lack of sampling, not much is known about this type of hydrothermal emission. Some insight was obtained in 2000 when a totally new type of hydrothermal activity was discovered on the seafloor above 15 km off axis from the MAR at 30°N in relatively shallow waters (750 to 900 m).

At this site, called the Lost City vent field, the venting water is <150°C with an average temperature of 90°C. It is enriched in the reduced gases, H_2 and CH_4, and has a very high pH (9 to 11). In contrast to high-temperature vents, these fluids are not metal enriched and no H_2S is present. They also have low CA and TDIC concentrations. As these fluids vent and mix with seawater, their high pH leads to the precipitation of aragonite as OH^- reacts with seawater bicarbonate to form carbonate. The OH^- also reacts with seawater Mg to precipitate brucite ($Mg(OH)_2$). These minerals form white vertical and lateral chimneys as tall as 60 m. Their walls are highly porous with networked channels in which seawater mixes with the clear vent fluids. The hydrothermal emissions occur as a diffuse flow through the channels, rather than through the annular orifices present in black and white smokers. Because of the low temperatures, no anhydrite precipitates. Continuing precipitation eventually clogs the networked channels. The nearby sediments are also lithified due to diffuse venting through the pore spaces. The entire system of venting carbonate towers at the Lost City field is about 300 m long.

The heat driving this hydrothermal system is not supplied by conduction from hot magma; rather, it is the product of the exothermic reaction of seawater with peridotite. Peridotite is an intrusive igneous rock that tends to be emplaced at deeper depths than basalts and gabbros (see Chapter 19.4.2). Faulting at slow spreading centers, such as the MAR, exposes peridotite rock to seawater. Like all igneous rocks, peridotite is highly reduced and reacts vigorously with oxic seawater. In this reaction, water is consumed, causing the rock to expand. This lowers its density, enabling uplift to shallow depths. Expansion also increases the degree of cracking in the rock, permitting more seawater to come in contact with the peridotite. Enough heat is produced to elevate rock temperatures to 260°C, thereby driving convective circulation through this hydrothermal system.

Reaction with heated seawater transforms the peridotite into a hydrous Mg-silicate mineral called serpentine. Hence, this process is called serpentinization, as illustrated for a common mineral in peridotite, olivine:

$$\underset{\text{olivine}}{2Mg_{1.8}Fe_{0.2}SiO_4} + 2.933H_2O \rightarrow \underset{\text{serpentine}}{Mg_{2.7}Fe_{0.3}Si_2O_5(OH)_4} + \underset{\text{brucite}}{0.9MgOH_2} + \underset{\text{magnetite}}{0.033Fe_3O_4} + 0.033H_2 \quad (19.5)$$

This reaction is the source of H_2 gas and OH^- (which is shown here as forming brucite) in the hydrothermal fluids. Peridotite is also composed of the mineral anorthite. As shown in Eq. 19.3, albitization of this mineral supplies Ca^{2+} to the hydrothermal fluids. This type of hydrothermal system is much more stable than the high-temperature ones venting on MOR. At the Lost City field, radiogenic carbon age-dating indicates that hydrothermal activity has persisted for at least 30,000 y.

At least some of the methane found in the vent fluids is thought to be of abiotic origin, having formed from the reaction of H_2 with CO_2 or $CaCO_3$. This reaction is catalyzed by magnetite and other iron-nickel sulfide minerals common to serpentized peridotites. Evolutionary biologists were particularly excited at the discovery of this low-temperature venting given that modern-day methanogens use Ni-based enzymes and ferredoxin. The latter is a ubiquitous enzyme, which is structurally similar to various hydrothermal iron-nickel sulfide minerals, such as Fe_5NiS_8. Recent experimental work has demonstrated that acetic acid can be generated from the reaction of CO and H_2S in the presence of an iron-nickel sulfide mineral and selenium. This evidence is interpreted as supporting the hypothesis that life evolved in similar settings.

19.4.6 Formation of Metalliferous Sediments

Seafloor spreading carries high-temperature vent fields away from the ridge crests. With increasing distance from the magma chamber, hydrothermal activity declines and eventually ends. Chemical changes continue to alter the hydrothermal deposits, including the chimneys, pillow basalts, and settled plume precipitates, as a result of reaction with seawater. In the case of the black smoker chimneys, reaction with O_2 converts the metal sulfides to oxides. As the system cools, anhydrite dissolves. Thus, chemical weathering leads to physical weathering, causing the rocks to be transformed into sediments that accumulate as mounds around former vent fields. Another type of mass wasting occurs as a result of the low-temperature chemical weathering of the pillow basalts, which eventually transforms the igneous rocks into clay minerals such as smectites, micas, and zeolites. Sediments are also contributed by pelagic sedimentation of hydrothermal plume precipitates. The sulfidic nature of these precipitates decreases with increasing distance from the ridge crest. All of the sulfidic minerals that accumulate in the sediments are subject to chemical weathering, primarily via oxidation reactions that appear to be facilitated by microbes such as iron oxidizing bacteria. These sediments are eventually buried by lithogenous and biogenous particles as the underlying crust moves away from the active spreading center (Figure W15.1).

The hydrothermally derived sediments have high concentrations of zinc, iron, copper, lead, cobalt, nickel, silver, chromium, titanium, and gold (see Table 18.1 listings for ridge and basal sediments). The global distribution pattern of these metalliferous sediments is shown in Figure 19.17. Concentrations are highest on the tops and flanks of mid-ocean ridges and rises. The Pacific Ocean has the greatest abundance of metalliferous sediments, with the largest metal enrichments associated with the East Pacific Rise, reflecting its rapid spreading rate. The southern lobe of metalliferous sediment in the South Pacific is a consequence of hydrothermal plume transport by currents. As noted in Chapter 18, the oxidation of manganese proceeds relatively slowly and involves co-precipitation of metals from seawater. Thus, considerable horizontal transport can occur before the MnO_2 particles settle to the seafloor, causing their deposition to be strongly controlled by the pattern of water circulation.

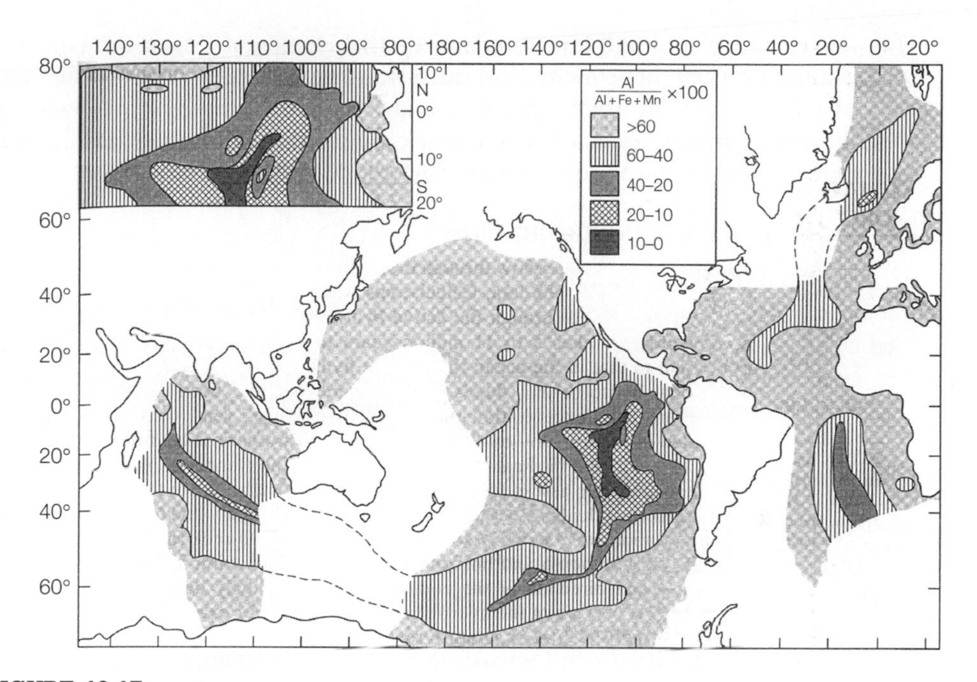

FIGURE 19.17

Distribution of metalliferous surface sediments in the ocean as indicated by the ratio Al/(Al + Fe + Mn). As these deposits are poor in aluminum relative to continental detritus, their abundance is inversely proportional to the value of the ratio. *Source*: From Bostrum, K. (1969). *Journal of Geophysical Research* 74, 3261–3270.

The low-temperature weathering of basalt is thought to extend 500 m below the seafloor due to gaps between the pillow formations, enabling contact with seawater. This weathering occurs at temperatures less than 100°C in a series of stages that are thought to be complete within a few million years after solidification of the basalt. During the first stage, the surface of the rock dissolves and a new mineral, called palagonite, is precipitated. A large amount of water and CO_2 is consumed, with some elements being solubilized (Si, Al, Mg, Ca, Na, and K). The CO_2 is consumed under alkaline conditions via reaction with Ca^{2+} to produce hydrogenous $CaCO_3$.[1]

The palagonite is thermodynamically unstable and, hence, reacts with seawater to form various clay minerals, including smectites (montmorillonite, nontronite, and saponite), micas (celadonite), and zeolites (phillipsite). This chemical weathering involves uptake of Si, Al, Mg, Ca, Na, and K and the release of water, reversing to some extent, the elemental effect of palagonitization. These mineral alterations tend to proceed progressively from the outer margin of the pillow basalts to their interior.

[1] These alkaline conditions are due in part to the consumption of acid during the low-temperature chemical weathering of plagioclase in a process similar to Eq. 19.4.

This makes the magnitude and direction of the elemental fluxes difficult to generalize because of their dependence on such variables as the water's temperature and circulation rate and the rock's porosity and chemical composition. Low-temperature weathering of basalt is also accompanied by precipitation of iron oxides and iron uptake into various clays (nontronite, saponite, and celadonite).

Economic Deposits including Ores

Hydrothermal sediments and ore bodies contain enough metals in high concentrations to be of economic interest. More information on the formation, location, and proposed mining of these deposits is provided in the supplemental information for Chapter 19 available at **http://elsevierdirect.com/companions/9780120885305**.

19.5 ROLE OF HYDROTHERMAL CHEMISTRY IN THE CRUSTAL-OCEAN-ATMOSPHERE FACTORY

Considerable interest has been focused on assessing the degree to which hydrothermal chemistry acts as a source or a sink of elements to the global ocean. This is a difficult task as it requires information on the geographic and temporal variability of hydrothermal fluid chemistry and production rates, including lower temperature reactions that occur off axis. Given that hydrothermal circulation is responsible for 20% of Earth's total heat loss and involves a large water flux, large impacts on ocean and crustal chemistry are likely. As discussed next, our current understanding suggests that the role of hydrothermal chemistry is large enough to control feedbacks for at least some elemental cycles.

19.5.1 Estimates of Global Hydrothermal Fluxes

Two approaches have been used to estimate the effect of hydrothermal chemistry on global elemental fluxes. One approach relies on comparisons of "fresh" to hydrothermally altered basalt, with changes in elemental composition being translated in elemental losses or gains. The other approach estimates these losses and gains by comparing the chemistry of hydrothermal fluid emissions to that of average seawater. The accuracy of neither method is reliable because of limited observational data and various simplifying assumptions required to enable flux calculations.

In the case of the hydrothermal fluid approach, the global flux of an element is estimated by multiplying its concentration-temperature gradient in an "average" vent by the global hydrothermal heat flux ($4.9 \pm 1.2 \times 10^{19}$ cal/y). For example, the data in Figure 19.14a for the Galapagos vents provide a silica concentration-temperature gradient of 0.0160 μmol/°C. This gradient is converted to units of μmol/kcal using the heat capacity of seawater for these environmental conditions. In the case of the silica profile in Figure 19.14a, the concentration-heat gradient is 62.5 μmol/kcal.

Multiplication of this gradient by the global heat flux generates a hydrothermal flux estimate of 3.1×10^{12} mol/y. An estimate of the hydrothermal magnesium flux can be obtained from the data presented in Figure 19.14b, by first converting the Mg-to-Si gradient (−0.00246 μmol Mg/μmol Si) into a concentration-temperature gradient (−0.00246 μmol Mg/μmol Si × 62.4 μmol Si/kcal = −154 μmol Mg/kcal).

Estimates of global hydrothermal fluxes are shown in Table 19.2 for the elements with the most comprehensive data sets. The lack of agreement between the estimates obtained from rock weathering as compared to those from fluid composition attests to the need for more data. For the hydrothermal fluids, global estimates are confounded by the increasingly large set of observations demonstrating that no two vents have exactly the same chemical composition. This composition reflects geographic differences in reaction temperature, water circulation rates, water-to-rock ratio (porosity), as well as in the depth and horizontal extent of the reaction zone all of which change over time due to the geological evolution of vent plumbing (Figure 19.7). This makes the magnitude and direction of the elemental fluxes highly dependent on the environmental setting of the hydrothermal reactions. For example, an element released from the high-temperature reaction zone can also be removed in one or more of the adjacent low-temperature reaction zones.

In general, the fluid flux estimates are biased by their reliance on data from high-temperature vents. These sites have relatively high circulation rates, water-to-rock ratios, and percentages of fresh oceanic crustal rocks. In comparison, very few data have been collected from ridge flanks and further off-axis vent fields. In the case of the rock data, only a small number of sites have been drilled deep enough to obtain useful information. No samples have yet been obtained from the reaction zone of a black smoker, which is the source of samples from which most of the hydrothermal fluid estimates are based. The rock approach to estimating fluxes is also complicated by the high spatial variability of crustal concentrations pre- and post-hydrothermal alteration.

Based on the fluid emission estimates, hydrothermal activity appears to be a major sink of magnesium and sulfur (as sulfate), rivaling or perhaps exceeding their riverine sources. The oceanic residence time of magnesium (13 million years) as estimated from river inputs is close to the turnover time of water through the hydrothermal system. This supports the hypothesis that uptake at hydrothermal systems is an important removal mechanism of this element from the ocean and, thus, acts to maintain its steady-state balance in the ocean. Hydrothermal emissions are a major source of iron and manganese along with lesser inputs of calcium, silicon, lithium, and rubidium relative to riverine fluxes.[2] For sodium and potassium, differences in the direction of the fluxes at the ridge crests as compared to the flanks make it impossible to determine whether the hydrothermal systems as a whole are a source or a sink of these elements. Because of the titration of alkalinity by the acidic hydrothermal fluids, a significant amount of alkalinity is also lost in the high-temperature systems. Finally, hydrothermal systems appear to be

[2] Although not shown in Table 19.2, hydrothermal fluids are also a significant source of copper and zinc.

Table 19.2 Comparison of Hydrothermal Fluxes to Riverine Fluxes.

	Fresh Mid-Ocean Ridge Basalt	Units	Bulk Rock Gains (+) and Losses (−) for Extrusives			Dikes	Gabbros	Average Crustal Gains/Losses	Total Flux from Crust (g/y)	Hydrothermal Fluxes from Submarine Vents (g/y)			River Fluxes (g/y)
			0–600 m	Error	600–1000 m					Flank	Axis Low	Axis High	
SiO_2	50.45	wt%	1.18	1.5	0.5	0.25	0.1	0.2370 g/100 g	−1.67E+14	1.08E+12	1.80E+13	6.61 E+13	3.85E+14
Al_2O_3	15.26	wt%	1.94	3					−9.65E+13				
FeO_0	10.43	wt%	−0.01	0.03						0.00E+00	1.65E+12	1.37E+13	1.65E+12
MnO	0.19	wt%	0	0.01	0.75	−0.035				3.69E+10	4.97E+11	3.97E+12	3.48E+11
MgO	7.58	wt%	−0.19	1	0	0	−0.2	−0.1569 g/100 g	1.11E+14	−2.18E+14	−4.03E+13	−1.05E+14	2.18E+14
CaO	11.3	wt%	2.14	1.5	0.3	0.04	−0.1	0.1335 g/100 g	−9.43E+13	2.64E+14	3.03E+11	1.23E+14	6.73E+14
Na_2O	2.679	wt%	0.15	0.2	0.15	0.15	0.03	0.06549 g/100 g	−4.63E+13	−3.84E+13			3.53E+14
K_2O	0.11	wt%	0.54	0.15	0.05	0	0	0.04845 g/100 g	−1.71E+13	−3.11E+13	6.12E+12	3.01E+13	1.22E+14
Rb	1.26	ppm	10.3	3	1	0	0	0.9268 mg/kg	−6.55E+09	−2.22E+09	1.37E+10	2.39E+10	3.16E+10
Cs	0.0141	ppm	0.183	0.050	0.03	0	0	0.01715 mg/kg	−1.21E+08				6.40E+08
CO_2	0.15	wt%	3.26	0.5	0.5	0.06	0.06	0.3552 g/100 g	−2.51E+14	−7.04E+12	−1.76E+12	−7.04E+12	1.41E+15
H_2O	0.2	wt%	2.81	1	2.09	1.5	0.11	0.4487 g/100 g	−3.17E+14				
S	960	ppm	−0.06		−0.06	−0.06	0.02	−0.0037 g/100 g	2.59E+12	−3.53E+13	−1.60E+13	−4.49E+13	2.85E+13
Li	4.5	ppm	2.8		2.8	−0.5	−2	−2 mg/kg	7.71E+09	−1.25E+10	4.86E+10	4.37E+11	9.72E+10
B	0.5	ppm	25.7	10	4	5.6	2.2	4.81 mg/kg	−3.40E+11	1.84E+11	6.49E+09	7.89E+10	5.84E+11
Sr	113	ppm	22	5	3	0.3	−1	1.4 mg/kg	−9.68E+10	2.19E+11	−9.00E+08	4.60E+09	2.02E+11
U	0.0711	ppm	0.3	0.05	0.1	0.01	0.007	0.03746 mg/kg	−2.65E+09				9.60E+09
							0.3		−3.60E+14	1.90E+14	−1.14E+13	1.02E+14	2.23E+15

Negative hydrothermal fluxes represent an uptake of the element or compound into the crust. In this table, $\times 10^{ab}$ *is abbreviated as "E + ab" ex.,* 1.08E + 12 = 1.08×10^{12}.
Source: After Staudigel, H. (2003). Treatise on Geochemistry, *Elsevier Ltd., pp. 511–535.*

a significant phosphorus sink, with off-axis processes playing a large role. The phosphorus appears to be removed by reaction with iron present in basalt, the overlying sediments, and the hydrothermal plumes. This removal of phosphorus appears to substantially exceed the river input and represents about one third of the total oceanic sink. As noted in Chapter 18, benthic remineralization supports a large flux of phosphorus back into the bottom waters.

19.5.2 Feedbacks between Elemental Cycles

Hydrothermal processes have the potential to contribute to feedbacks in the crustal-ocean-atmosphere factory that link various elemental cycles. For example, changes in spreading rates alter the Mg/Ca ratio in seawater because hydrothermal circulation removes Mg and supplies Ca to seawater. Thus, low Mg/Ca ratios tend to occur when spreading rates are high due to higher rates of release of Ca^{2+} and uptake of Mg^{2+}. This condition favors deposition of calcite, whereas periods of limited seafloor spreading lead to higher Mg/Ca ratios that favor aragonite deposition. These shifts have had important impacts on the evolution and survival of biocalcifying organisms. Changes in rates of seafloor spreading are thought to be related to quasiperiodic cycles of aggregation and dispersal of Earth's continental crust. Following a period of aggregation, buildup of heat beneath the thick continental crust leads to increased volcanism and rifting during which ocean basins open. Thus, periods of ocean basin opening are associated with high rates of seafloor spreading. These *supercontinent cycles* drive a periodic opening and closing of the ocean basins called *Wilson cycles*.

In addition to the hydrothermal controls, the elemental cycles of Ca and Mg are also linked to those of CO_2 and SiO_2 because the mobilization of Ca and Mg during terrestrial weathering from silicates, such as plagioclase, involves consumption of CO_2. In the case of calcium, the net chemical reaction is

$$\underset{\text{model calcium-rich rock such as wollastonite}}{CaSiO_3} + CO_2 \longrightarrow \underset{\text{biogenic calcite}}{CaCO_3} + \underset{\text{biogenic silica}}{SiO_2} \tag{19.6}$$

with the Ca and Si mobilized from rock weathering being removed in the present-day ocean via deposition of biogenic calcite and silicate. When these sediments are subducted back into the mantle, the reaction in Eq. 19.6 is driven in reverse, thereby resupplying CO_2 and calcium and magnesium-rich rocks, such as plagioclase. The elemental cycles of Na and K are similarly linked to each other and to CO_2 and SiO_2. In the case of sulfur, hydrothermal circulation acts to convert seawater sulfate to sulfide that serves to precipitate iron. This iron reacts with phosphorus, causing hydrothermal activity to serve as an important phosphorus sink. Because of the uptake and release of CO_2, HCO_3^-, and CH_4 from hydrothermal systems, even the elemental cycling of carbon and oxygen is interlinked.

Biogeochemists are working to construct numerical models that include all of these interlinked feedbacks to explain how the chemistry of seawater has changed over time in response to various forces, including tectonism, biological activity, ocean-atmosphere interactions, crustal weathering, and river runoff. To incorporate all of these linkages into a numerical multielemental model of seawater is very complex because most of the interactions are nonlinear (not directly proportional to reservoir concentrations) and involve a great number of reservoirs, such as the mantle, crust, ocean, land, and atmosphere. These models are discussed further in Chapter 21.

19.6 BIOLOGY OF HYDROTHERMAL SYSTEMS AND COLD SEEPS

Hydrothermal activity was first observed in 1977 at a series of vents that lie 2500 m below the sea surface, 280 km northeast of the Galapagos Islands. Geochemists were attempting to locate active vents by searching for temperature anomalies in the bottom waters of the East Pacific Rise. Instead, the actual discovery was made as a result of the unexpected and startling visual observations of large assemblages of clams, mussels, worms, and crabs clustered around fountains of shimmering water. The appearance of a typical vent community is shown in Figure 19.10. This type of community appears to be a ubiquitous feature of hydrothermal vents and even cold-water seeps.

The abundance and diversity of marine life at these vents is extraordinary. Prior to the discovery of these ecosystems, marine biologists had assumed that all life in the ocean ultimately depended on the photosynthetic production of POM by phytoplankton. Since this POM is inefficiently transferred to the deep-sea, the abundance of marine organisms decreases with increasing depth. The vent ecosystems are the sole exception to this because their food chain is based on nonphotosynthetic primary producers,[3] i.e., chemoautolithotrophs. The large chemical, redox, temperature, and geological gradients found in hydrothermal systems provide a diversity of ecological niches to support microbes and their consumers. Thus, it is not surprising that many novel bacteria and archaea have been discovered living in diverse settings such as on the surfaces of black and white smoker chimneys, in the metalliferous sediments, in the subsurface conduits within the hydrothermal system, and in the fluid plumes. Other chemoautolithotrophic-based communities have been discovered living in the off-axis hydrothermal systems, such as the Lost City vent field, and at cold seeps.

The reduced chemicals supplied by hydrothermal fluids are the source of energy fueling these ecosystems. As shown in Figure 19.18, the most energetic of these is sulfide oxidation, followed by iron oxidation, and methanotrophy, with sulfate reduction and methanogenesis generating significant amounts of energy only at higher temperatures. Since these metabolisms require different chemical reactants and temperatures, they are expected to be spatially and/or temporally separate. As noted in Chapter 7, the hyperthermophilic microbes are archaeans. Some of these microbes are able to grow temperatures well over 100°C but probably less than 150°C. Some are acidophiles capable of growing in waters with pH as low as 3.3.

[3] Technically, whale falls represent another exemption caused by the rapid delivery of a large amount of POM to the deep-sea floor that can sustain a prolific benthic community for a significant period of time.

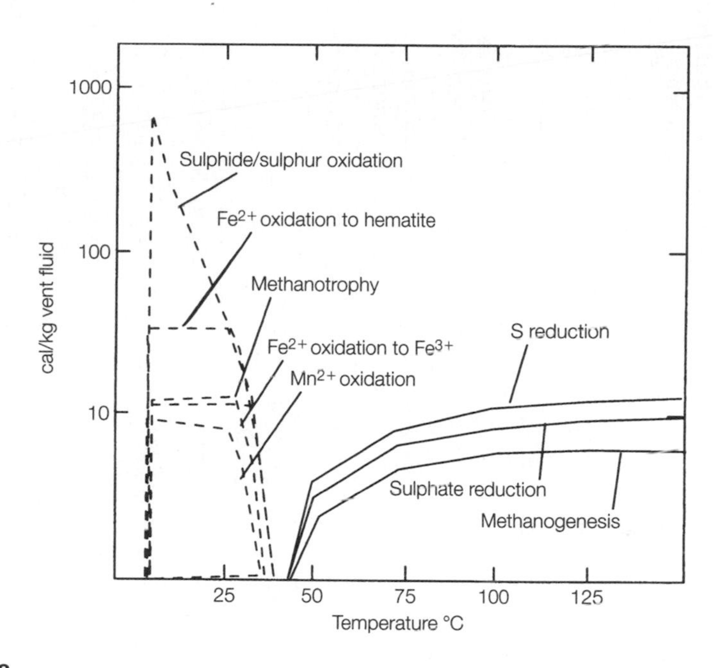

FIGURE 19.18

Theoretical estimates of energy available from hydrothermal vent fluids for various chemosynthesis reactions used by vent microbes. Oxidative reactions are shown as dashed lines and reductive reactions as solid lines. Iron reduction provides too little energy to appear on plot. H_2 oxidation was not calculated. *Source*: From Léveillé, R. J., and S. K. Juniper (2003). *Biogeochemistry of Marine Systems*, Blackwell Publications, pp. 238–292.

The chemical reactions involved in each of these metabolisms, along with others deemed likely to be occurring, are presented in Table 19.3. In many of these, microbes play the important role of facilitating mineral formation. For example, iron-oxidizing bacteria, including ***Leptothrix***, ***Gallionella***, ***Metallogenium***, and ***Leptospirillium***, have been observed to create iron and silica precipitates on cellular filaments, thereby promoting the formation of nontronite. Note that all of the aerobes require O_2 and, thus, are indirectly dependent on photosynthesis as a source of electron acceptors.

The hydrothermal-vent and cold-seep communities are dramatically different from the ecosystems typical of the abyssal plains. First, they are sites of high productivity supported by the abundant reduced chemicals in the hydrothermal fluids. Thus, these communities are independent of the skimpy flux of POM created in the surface waters that survives to settle on the seafloor. On the other hand, these communities have had to adapt to survival in hydrothermal systems that are ephemeral, disjunct, and characterized by extreme conditions, such as high temperatures, high concentrations of reduced metals and sulfur, and low pH. As a result, vent communities have high rates of endemism. Of the 712 recorded vent species, 71% are found in no other setting!

In terms of animals, most vent species are either vestimentiferan worms, annelids, mollusks, or arthropods, whereas the typical deep-sea biota is dominated by echinoderms,

Table 19.3 Known and Proposed Microbial Metabolic Pathways at Deep-Sea Hydrothermal Vents and Cold Seeps.

Pathway	Electron Donor	Electron Acceptor	Carbon Source	Reactions	Examples of Organisms and Environments
Sulphur oxidation	HS^-, S°, $S_2O_3^{2-}$, $S_4O_6^{2}$	O_2, possibly NO_3^-	CO_2	$S^{2-} + CO_2 + O_2 + H_2O \rightarrow SO_4^{-2} + [CH_2O]$ $H_2S + 1/2O_2 \rightarrow S^\circ + H_2O$ $2S^\circ + 2H_2O + 3O_2 \rightarrow 2SO_4^{2-} + 4H^+$	Mesophilic bacteria (e.g., *Beggiatoa*)
Hydrogen oxidation	H_2	O_2, possibly NO_3^-	CO_2	$4H_2 + O_2 + CO_2 \rightarrow [CH_2O] + 3H_2O$	Mesophilic bacteria detected by activity measurements
Metal oxidation	Fe^{2+}, Mn^{2+}	O_2, possibly NO_3^-	CO_2	$2Fe^{2+} + 0.5O_2 + 2H^+ \rightarrow 2Fe^{3+} + H_2O$	Mesophilic bacteria; putative *Gallionella, Leptothrix*
Methanotrophy/ methylotrophy	CH_4 and other C_1 compounds*	O_2	CH_4, CH_3OH, CO, CO_2	$CH_4 + O_2 \rightarrow [CH_2O]$ $CH_4 + 2H_2O \rightarrow CO_2 + 4H_2$ $2CH_4 + 2H_2O \rightarrow CH_3COOH + 4H_2$ $CH_4 + 2O_2 \rightarrow CO_2 + 2H_2O$	Nonthermophilic archaea
Nitrification	NH_4^+, NO_2^-	O_2	CO_2, organic compounds	$NH_3 + O_2 + CO_2 \rightarrow [CH_2O] + HNO_3$	Mesophilic bacteria
Heterotrophy	Organic compounds	O_2, NO_3^-	Organic compounds		Mesophilic and thermophilic bacteria
Sulphur reduction - chemolithotrophic	H_2	S°, $S_2O_3^{2-}$, SO_4^{2-}	CO_2	$H^+ + 4H_2 + SO_4^{2-} \rightarrow HS^- + 4H_2O$ $S + H_2 \rightarrow H_2S$ $CO_2 + SO_4^{2-} + 6H_2 \rightarrow [CH_2O] + S^{2-} + 5H_2O$	Mesophilic and thermophilic bacteria; hyperthermohilic archaea

Sulphur reduction - chemoheterotrophy	Organic compounds (e.g., acetate)	S°, SO_4^{2-}	Organic compounds	$CH_3COO^- + SO_4^{2-} \rightarrow 2HCO_3^- + HS^-$ $CH_3COOH + SO_4^{2} \rightarrow 2CO_2 + S^{2-} + 2H_2O$ $2CH_2O + 2H^+ + SO_4^{2-} \rightarrow H_2S + 2CO_2 + 2H_2O$	Mesophilic and thermophilic bacteria; hyperthermohilic archaea e.g., *Desulfovibrio*
Methanogenesis	H_2	CO_2	CO_2, possibly formate, acetate	$4H_2 + CO_2 \rightarrow CH_4 + 2H_2O$ (CO_2 reduction) $CH_3COOH \rightarrow CH_4 + CO_2$ (acetate fermentation)	Mesophilic to hyperthermophilic archaea at vents and seeps
Hydrogen oxidation Iron reduction	H_2 Organic acids	NO_3^- Fe^{3+} (oxyhydroxides)	CO_2 Organic acids		Identified from molecular analyses Mesophilic bacteria and hyperthermophilic archaea
Anaerobic methane oxidation (methanotrophy)	CH_4	SO_4^{2-}	CH_4	$CH_4 + SO_4^{2-} \rightarrow HCO_3^- + HS^- + H_2O$	Consortia of methanogenic archaea and SRBs at seeps and in hydrothermally active sediments
Acetogenesis	Fatty acids, alcohols, H_2	CO, fatty acids, alcohols, CO_2	CO, fatty acids, alcohols	$3CH_2O + H_2O \rightarrow H_3COOH + CO_2 + 2H_2$	Limited data from vents
Fermentation	Organic compounds	Organic compounds	Organic compounds	$CH_3COOH \rightarrow CH_4 + CO_2$	Mesophilic and thermophilic bacteria; hyperthermophilic archaea

Notes: *C_1 compounds (reduced single carbon compounds) include methane, formaldehyde, formate, methanol, and methylamine.*

Source: After Léveillé, R. J., and S. K. Juniper (2003) Biogeochemistry of Marine Systems, Blackwell Publications, pp. 238–292.

coelenterates, sponges, and arthropods. Most of the vent animals have protective covers, tubes, shells, and carapaces similar to adaptations found in the intertidal zone. Few colonial animals have been observed at the vents, although endo- and ectosymbionts are common. Overall the diversity of animals is relatively small because of limited food-web complexity reflecting a dependence on primary production by the chemoautolithotrophs. Due to the large environmental gradients within a vent field, pronounced species zonation occurs both vertically and concentrically. Other adaptations include rapid growth rates and high fecundity with larval planktonic forms easily dispersed by currents.

Many vent microbes are symbionts. Others form dense filamentous mats, such as the sulfide oxidizers *Beggiatoa* (oxic) and *Thioploca* (hypoxic). Similar mats probably develop subsurface within hydrothermal conduits. These subsurface mats are ejected into the water column during first stage of vent formation. The resulting explosive discharge of this biomass, which has the appearance of a white floc, has given rise to the term *snowblower vent*. The ejected microbes are thought to eventually settle back down onto the seafloor where they increase in number to form surficial mats that support the successional colonization of vent animals.

The existence of snowblower vents has motivated biogeochemists to search for subsurface biota in other sediments. Recent fieldwork indicates that a widespread subsurface chemoautolithotrophic community is probably living down to depths of 500 m. This microbial community appears to employ various metabolic strategies already reported as common components of diagenesis, i.e., carbon oxidation, ammonification, methanogenesis, anaerobic methane oxidation coupled with sulfate reduction, and manganese reduction, with some of the reduced reactants being supplied by upward transport of seawater that has chemically interacted with the basalt basement rocks.

Although the overall contribution of hydrothermal biomass production to global primary productivity is thought to be relatively small, it is locally significant and not seasonally variable. Thus, the vents represent an oasis in the biological desert of the abyssal plains. The biotic community assemblage at these vent fields varies geographically, with differences increasing with separation distance. Six biogeographic provinces have been distinguished with pronounced interbasin differences. This zonation is thought to be caused by barriers, such as ridge segmentation, transform valleys, fracture zones, and sills, that prevent the dissemination of larval forms. Genetic testing has demonstrated that none of the endemic animal species is older than 100 million years, attesting to the isolation of these communities. A few animal species are shared between the vents and cold seeps, but they are not phylogenetically related. Thus, although the cold seeps are thought to be the original source of the unique vent biota, most of the latter have undergone significant evolutionary change. Species dispersal is thought to occur via a stepping-stone approach using whale falls, seamounts, seeps, hot spots, and seamounts.

19.6.1 Ridge Crest Hydrothermal Systems

In high-temperature hydrothermal systems, sulfide-oxidizing bacteria are responsible for most of the primary production supporting the vent community. As shown in Eq. 19.7, the oxidizing agent is O_2, which is contributed by seawater transported to the vents by meridional overturning circulation:

$$CO_2 + S^{2-} + O_2 + H_2O \xrightarrow{\text{bacteria}} CH_2O + SO_4^{2-} \quad (19.7)$$

Thus, turbulent mixing of the hydrothermal fluids with oxic seawater is important to these microbes.

Many of the animals harbor endosymbiotic bacteria that are sulfide oxidizers, such as the giant tube worms (*Riftia pachyptila*), which reach lengths of 1 m. As shown in Figure 19.19, the tube worms are essentially a closed sac, having no mouth, digestive system, or other means of processing particulate food.

The worm absorbs O_2, sulfide, and CO_2 through a red gill-like respiratory plume that extends from its anterior tip. The red color is due to the presence of hemoglobin in the blood of these organisms. The chemicals are transported by the circulatory system to the trophosome that fills most of the worm's body cavity. The trophosome is composed of cells that contain dense colonies of endosymbiotic sulfur bacteria. These organisms convert the chemicals into organic matter that is then passed back into the blood, where it is aerobically respired by the worm to provide energy and nutrients. Scientists were initially amazed by this, because hydrogen sulfide had heretofore been observed to be toxic to animals at very low concentrations. Tube worms are able to tolerate sulfide because of the unique three-dimensional structure and high zinc content of their hemoglobin. The zinc reversibly binds the sulfide, keeping it from poisoning the sites on the hemoglobin that reversibly bind O_2. In this way, both sulfide and O_2 are concurrently transported within the blood and do not react with each other.

Similar sulfur-based endosymbioses have since been observed in sulfide-rich environments, such as mangrove swamps, petroleum seeps, sewage outfall zones, marshes, and cold seeps. The relationship between sulfur-metabolizing bacteria and the animal host tends to be species-specific; that is, each host species harbors a unique strain of bacteria. This suggests that endosymbioses with sulfur bacteria originated independently and repeatedly in diverse animal groups. The great abundance and diversity of the vent animals, as well as their relatively rapid growth rates and large sizes, indicate that these endosymbioses are a very successful adaptation.

Many of the smaller vent animals do not have symbionts. They obtain their reduced carbon and nutrients by filtering POM from seawater or by consuming other animals. The nitrogen needs of these organisms also appear to be met by the chemoautotrophic fixation of N_2 by an as-yet-unidentified microbe. Other microbial chemoautolithotrophs, such as methanotrophs and hydrogen bacteria, have been found in and near the vents. The methane and hydrogen gas that issue from the vents constitute their electron energy sources. As a result of the availability of reduced compounds, the concentration of microbes in the vent waters is four times greater than that in productive surface waters with thick mats of bacteria covering the surrounding metalliferous sediments.

Another potential source of energy supporting primary production at the ridge crests vents is radiation. Two types of radiation have been detected around hydrothermal systems: (1) visible light probably generated by chemical reactions, bubble formation, and

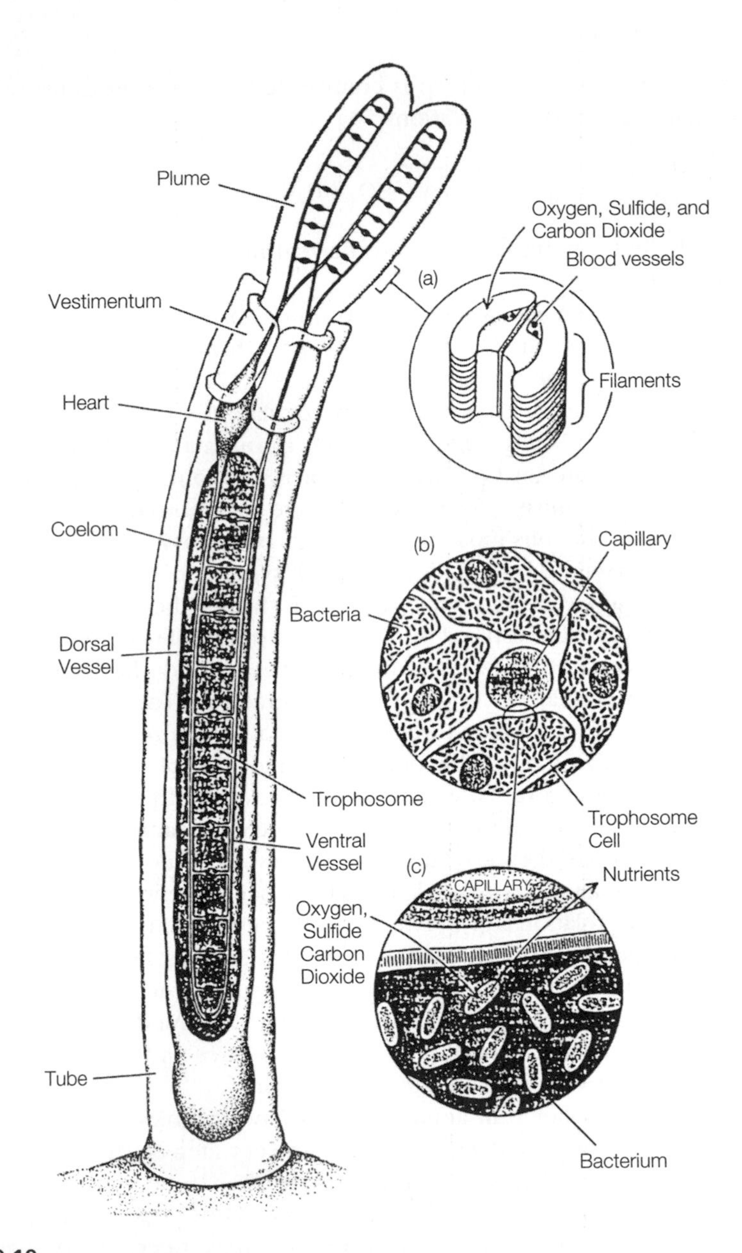

FIGURE 19.19

Internal structure of the tube worm *Riftia pachyptila*. (a) Oxygen, sulfide, and carbon dioxide are absorbed through the plume filaments and transported in the blood to the cells of the trophosome. (b) The chemicals are absorbed into these cells, which contain dense colonies of sulfur bacteria, where they are converted to organic compounds and (c) passed back into the circulatory system to act as an energy source for the worms. *Source*: From Childress, J. J., *et al*. (1987). *Scientific American*, 256, 114–121.

the formation and breaking of mineral crystal lattices, and (2) black-body radiation from the hot rocks. The latter generates infrared light. Some biologists have suggested that these forms of light energy are being used by photosynthetic bacteria analogous to the photosynthetic green sulfur bacteria living under low light conditions at 100 m in the euxinic Black Sea. But since vent bacteria are exposed to O_2, they must be a species different from that living in the anoxic waters of the Black Sea. Light-sensitive patches have been observed on species of shrimp and crabs that are otherwise eyeless, suggesting that some of the hydrothermal light is also being used by animals.

19.6.2 Off-Axis Hydrothermal Systems

Little is known about the biotic communities living around off-axis hydrothermal systems. The first discovery of a unique biotic community in such a setting was made in 2000 at the Lost City vent field (Chapter 19.4.5). Instead of sulfide and reduced metals, this community is exploiting CH_4 and H_2 under highly alkaline conditions. Because of the lower temperatures of the discharging vent fluids, the microbes are able to live within the porous chimney walls of the carbonate towers, causing the towers to contain 0.5% organic carbon by mass. The metabolism of the microbes at the Lost City vent field is diverse. Methanogens use H_2 from the vent fluids to create more methane, which is in turn used by the methanotrophs. Anaerobic methane-oxidizing archaea form a consortium with sulfate-reducing bacteria similar to that found in typical organic-rich sediments. These microbes exhibit a zonation that is temperature related as shown in Figure 19.20.

In comparison to the high-temperature hydrothermal vents, biomass at the Lost City vent field is much lower and the macrofauna are much smaller, typically less than 100 μm. These macrofauna are predominantly gastropods and polychaetes with 58% of the species being endemic, including nine new species of crustaceans!

19.6.3 Cold Seeps

In 1984, another type of chemoautolithotrophic-based vent community was discovered living at cold-water seeps located along the continental margins off Texas and western Florida in several hundred meters of water. Since then, 12 more cold seeps have been discovered on both passive and convergent continental margins in a variety of depths from intertidal to abyssal. The deepest one is located at 7326 m water depth in the Japan Trench. Others have been found in the North Sea, Gulf of Mexico, Arctic Ocean, Mediterranean Sea, the Barbados Islands, and around deep-sea trenches and fracture zones in the Pacific Ocean. This type of venting is very patchy with each vent field being only a few meters in radius.

The fluids at the cold seeps are low in O_2 and chemically enriched in CH_4, CO_2, N_2, H_2S, and, in some cases, oil. Hydrogenous minerals include calcite, aragonite, dolomite, and barite crusts. The reduced gases are mostly biogenic in origin, having been released during diagenesis of sedimentary organic matter. As shown in Figure 19.21, focused flows emerge along scarps, faults, and fracture zones. At some sites, flows are enhanced

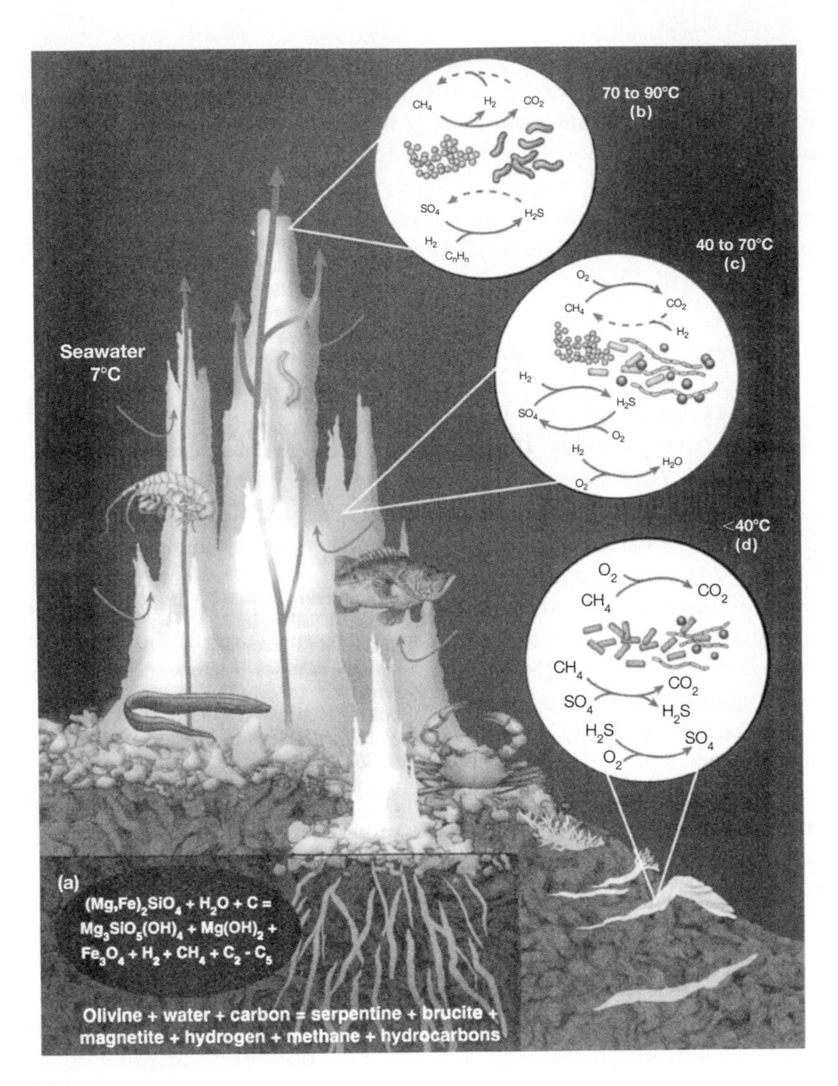

FIGURE 19.20

Microbial niches in serpentization-influenced environments at the Lost City hydrothermal vent field. (a) Subsurface serpentization reactions produce fluids of high pH, enriched in methane and H_2, as well as some hydrocarbons. (b) Within the warm interior of the carbonate chimneys, archaea (circles) form biofilms in which methane production occurs. Archaean methane oxidizers couple with bacterial sulfate reducers (wavy rods) to form a microbial consortium. (c) In the moderate-temperature endolithic environments, mixing of hydrothermal fluids and seawater supports a unique assemblage of archaea (rods) and proteobacteria (filaments and circles). The most likely metabolisms include methane oxidation and production from the oxidation of H_2. (d) Bacterial aerobic methanotrophs and sulfide oxidizers along with archaeans are living in the cooler lithified adjacent sediments. *Source*: From Boetius, A. (2005). *Science* 307, 1420–1422.

(a)

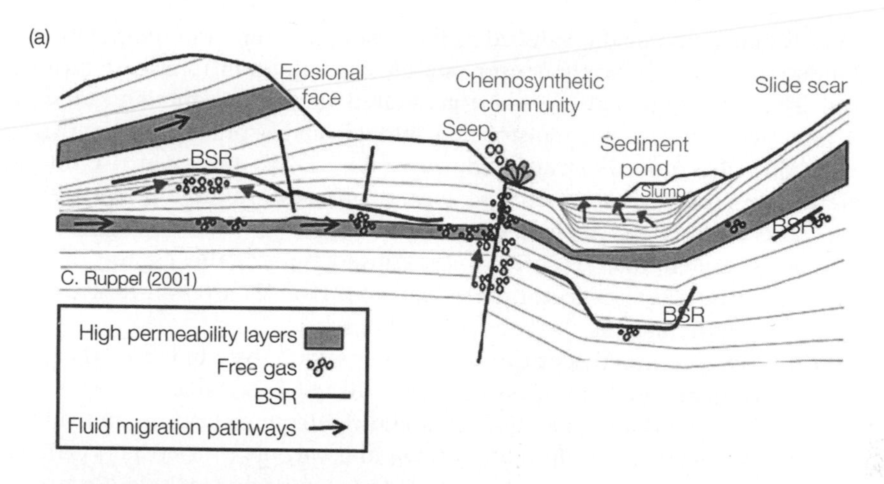

(b)

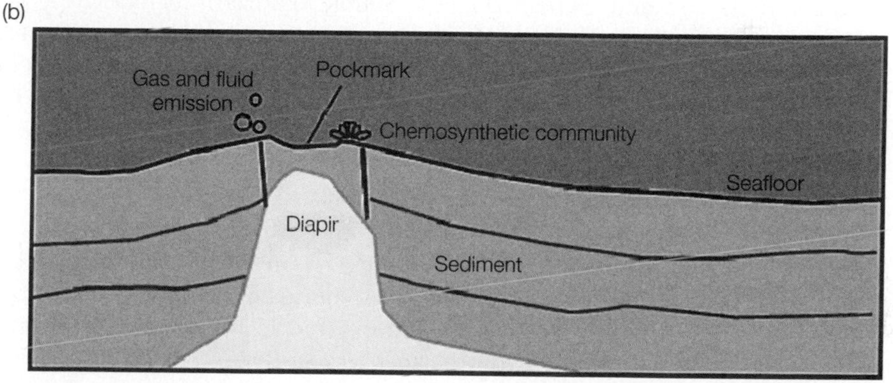

FIGURE 19.21

(a) Schematic diagram depicting the dynamics of a gas hydrate reservoir, including flow pathways for gas and other fluids. BSR denotes the bottom-simulating reflector, a seismic feature that often marks the base of gas hydrate in marine sediments. (b) Schematic diagram of a diapir rising up through the sediments. Faults that form around the diapir can lead to the development of flow systems for gas and fluids. Pockmarks, or small depressions in the seafloor, sometimes develop above these diapirs. *Source*: After Ruppel, C. (2005) Gas Hydrates Offshore Southeastern United States, NOAA Ocean Explorer/Explorations, http://www.oceanexplorer.noaa.gov/explorations/03windows/ background/hydrates/hydrates.html.

by the pumping action of bivalves living at the seeps. Sometimes, the seep fluids emerge explosively, creating mud volcanoes. At some of the seeps, temperatures and pressures are such that CH_4 freezes into a solid phase, called a methane hydrate. These hydrates can form large mounds on the seafloor. Microbial mats have been observed draped over these methane hydrate mounds, which also host burrowing worms.

Venting along convergent margins, venting is thought to be driven by subduction of accretionary wedges of organic-rich sediments into deep-sea trenches. Most of the seep flows are tectonically induced as increasing pressure and temperature converts the sediments into progressively denser phases, causing them to release structurally bound water. The heat and high pressure generated during subduction cause the sedimentary organic matter to be transformed into oil and asphalt along with H_2S and CH_4. Compaction also induces upward advective flows. The fluids can rise through interconnected pore spaces in sediments to form a diffuse flow or through discrete channels along faults to discharge as focused seeps.

On passive margins, seep flows are caused by: (1) the escape of petroleum and natural gas driven by salt tectonics (Figure 19.21b), (2) artesian flows, (3) catastrophic erosion of sediments, and (4) submarine slides.

The biotic communities at these cold seeps are most similar to those at the Pacific MOR, suggesting some kind of evolutionary linkage. Metabolic strategies are diverse and include aerobic methane oxidation, coupled methane oxidation, and sulfate reduction. The latter two lead to precipitation of calcium carbonate. Detrital POM fuels bacterial sulfate reduction and methanogenesis. Sulfide oxidizers include *Beggiotoa* mats and endosymbionts. Biogenic barite is remineralized as a result of sulfate reduction. The solubilized Ba^{2+} diffuses upward until it encounters seawater and then reprecipitates as barite. This produces a zone refining of barite. At sites where petroleum is present, this organic carbon provides an additional source of energy for the seep biota.

19.6.4 **Evolution of Life**

Life evolved soon after Earth's formation, before any continents were present, during a time when the oceans were chemically and thermally controlled by tectonic processes. Thus, it has been proposed that the life evolved in hot anaerobic submarine environments similar to present-day hydrothermal vent systems. This hypothesis is supported by the observation of structures, thought to be the remains of protocells, in rocks formed by hydrothermal processes 3.5 to 3.8 billion years ago. Thus, the first organisms on Earth were probably anaerobic hyperthermophiles. Hydrothermal vent habitats probably offered an additional benefit by providing a stable environment relatively isolated from the catastrophic effects of bolide impacts. In other words, submarine hydrothermal vents could have acted as refugia enabling survival of early life forms.

Two key abiotic reaction systems are provided in hydrothermal settings that appear to be stepping stones in the development of the first cells: (1) clay minerals, such as montmorillonite and (2) iron-nickel sulfides. As noted earlier, serpentization reactions generate iron-nickel sulfides similar to enzymes now found in methanogens. In the laboratory setting, iron-nickel sulfide minerals and selenium have been used to catalyze the production of key organic compounds from the reaction of CO and H_2S. The vent features created by serpentizing systems, such as observed at the Lost City vent field, are thought to be likely sites for the formation of the first protocells because the porous channels of the carbonate towers could have served to isolate bubbles of hydrothermal fluids from seawater. This isolation is proposed to have been achieved by the formation of gel-like iron precipitates within the channels. Such a precipitation would have been a consequence of the admixture of the alkaline hydrothermal fluids with early seawater, which is thought to have been iron-rich and acidic. In addition to creating a protocell by isolating a bubble of fluid, the mineral precipitate could also have played several other important roles, such as acting as (1) a scaffold to hold reactants in place, (2) a template that selected for specific reactants, and (3) a reaction catalyst promoting the formation of organic molecules. Similar roles have been suggested for montmorillonite, a clay mineral commonly found in hydrothermal systems.

These reaction systems do not explain additional key steps required for development of fully functional cell, i.e., a replication process based on either RNA or DNA and the transition to a cell membrane constructed wholly of organic compounds.

CHAPTER

Global Pattern of Sediment Distribution

20

All figures are available on the companion website in color (if applicable).

20.1 INTRODUCTION

In Chapters 14 through 19, each of the major sediment types has been discussed from the perspective of particle production, delivery to the seafloor, and preservation after burial. Marine sediments are generally classified on the basis of the most abundant particle type found in a given deposit. Therefore the accumulation rate of each contributing particle is critical to determining the type of sediment deposited, whether it be a biogenous ooze, abyssal clay, or metalliferous mud. In this chapter, some general "rules" are presented for predicting global distributions of the major sediment types. This prediction is then compared to the actual global distribution of modern-day surface sediments on an ocean-by-ocean basis. These patterns have changed over time in response to variations in the operation of the crustal-ocean-atmosphere factory. Oceanographers use downcore variations in sediment type as a record of these past conditions.

The four most important predictors of surface sediment compositions are: (1) proximity to particle source, (2) magnitude of particle source, (3) particle preservation potential, and (4) the relative accumulation rates of other particles. For a given particle type, accumulation rates decrease with increasing distance from the point where that particle enters the ocean. The major routes of lithogenous particle transport into the ocean are river runoff, winds, and ice. Biogenous and hydrogenous particles are generated in situ. The magnitude of a particle source can be predicted from knowledge of the processes controlling its production. For example, divergence in the Southern Ocean supplies nutrients to the surface waters. High nutrient concentrations support high levels of diatom production. This leads to the formation of siliceous oozes in the underlying sediments.

Some particles, particularly the biogenous ones, are prone to alteration as they settle onto the sediments and then undergo burial. The likelihood of particle preservation is generally enhanced in settings where the trip to the seafloor is short and burial rates are fast. The time a particle takes to settle onto the seafloor is determined by water depth and particle sinking rates. The latter is a function of particle shape and density. Seawater composition is an important determinant in preservation of some particle types, such as the biogenous hard parts. For these particles, conditions of mineral undersaturation can promote their dissolution enroute to the seafloor and after deposition in the sediments. For some particles, such as the calcareous hard parts and POM, losses in transit and prior to burial act to destroy a large fraction of the sedimenting flux. The degree to which particles are likely to be destroyed prior to burial can be predicted from knowledge of seawater chemistry, such as exemplified by the use of pH and alkalinity in mapping the depth of the saturation horizon for calcite.

While rapid burial enhances preservation, the type of sediment produced is determined by the relative particle composition of the deposit. For example, rapid burial of biogenic silicate by clay minerals helps protect the shells against dissolution, but the resulting deposit is classified as an abyssal clay, rather than a siliceous ooze, if the sediment is less than 30% by mass BSi. Thus, prediction of the sediment type likely to be found at a given location requires knowledge of the relative magnitudes of the accumulation rates of all particle types.

20.2 GENERAL MODEL OF SURFACE SEDIMENT DISTRIBUTIONS

A general model for predicting surface sediment distributions is presented in Figure 20.1 for an "average" ocean basin. This model can be partitioned into four settings: (1) continental margins, (2) abyssal plains, (3) polar seas, and (4) mid-ocean ridges and rises. General guidelines for predicting the sediment types found in each of these settings is provided next.

20.2.1 Continental Margins

Figure 20.1 indicates that most of the sediments on continental margins are lithogenous with some locally important biogenous deposits, namely coral reefs in tropical waters and siliceous oozes in coastal upwelling areas. At high latitudes, glacial marine deposits dominate. The surface area of the margins is small relative to the open ocean sediments. Nevertheless, these sediments play an important role in the crustal-ocean-atmosphere factory due to rapid sedimentation and burial rates. For example, most of the organic matter being buried in marine sediments environment is depositing on the continental margin.

The global distribution of continental shelves, including their widths, has fluctuated over geologic time in response to: (1) tectonism that changes the shape and elevation of the crustal plates, (2) isostatic readjustments in the elevation of land masses, and (3) climate shifts that alter the volume of the ocean. Changes in the extent and width of the shelves have affected the magnitude of various sedimentary sinks over time, with significant impacts on ocean chemistry. This topic is discussed further in Chapter 21.

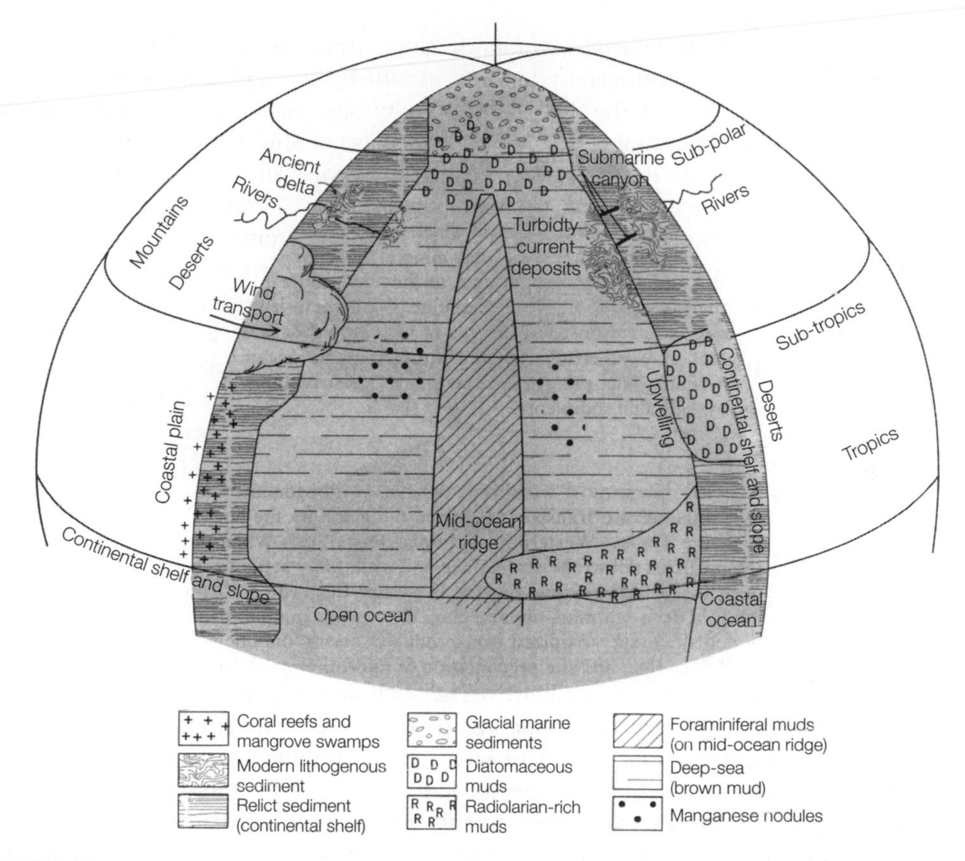

FIGURE 20.1

Schematic representation of the general distribution of sediment in a hypothetical "average" northern ocean basin. *Source*: From Gross, M. G. (1988). *Oceanography: A View of the Earth*, 4th ed., Prentice Hall, Inc., p. 94.

Lithogenous Coastal Sediments

The most common particle type found in the sediments of the continental margins are siliclastic materials produced by terrestrial weathering and delivered to the ocean via river runoff. These include clay minerals, quartz, feldspar, and mica. Sedimentation rates are spatially variable with approximately 70% of continental shelf covered by relict deposits that were formed at a time of lower sea level. The deposits now forming are mostly those located in the immediate vicinity of river mouths, close to the point of entry of the lithogenous materials. Most of the riverine input of lithogenous materials is presently depositing in estuaries as river deltas and the inner shelf. Where wave and tidal activity is low, terrigenous muds can form extensive deposits on the outer shelf and slope. Such areas are commonly referred to as *depocenters*. Aeolian transport is a minor source of terrigenous particles to actively forming coastal sediments.

Particle sizes in coastal sediments range from sands to silts to clays. The term "mud" is given to hemipelagic deposits in which the median grain is smaller than 5 μm and more than 25% of the mass is of terrigenous, volcanogenic, and/or neritic origin. Terrigenous muds contain 70% or more by mass of siliclastic materials.

Some of the shelf deposits are carried to the slope and rise by postdepositional transport processes, such as turbidity flows. As noted in Chapter 13, the resulting turbidites form the continental rise. Longshore and contour currents can resuspend and transport terrigenous sediments, but in general, most of these particles are deposited close to the latitude of their formation via terrestrial chemical weathering. Thus, the type of weathered siliclastic material characteristic of the sediments at a given latitude reflects the prevailing climatic conditions, such as temperature and rainfall, as these determine the intensity of the chemical weathering reactions. As described in Chapter 14, kaolinite is characteristic of tropical latitudes; chlorite is found in the sediments of high latitudes; and illite, being a general product of weathering, is found across all the climate zones as is quartz.

Biogenous Coastal Sediments

Pelagic marine organisms create particles that have the potential to settle onto the seafloor. The degree to which these particles are preserved in the sediment depends on the particle's chemical composition, the water depth, and chemistry of the surrounding seawater. The biogenous hard parts, such as shell and bone, are more likely to be preserved than the soft parts, such as tissues, fecal pellets, and molts. Though productivity is moderately high along most of the continental margins, biogenous oozes are rare in neritic sediments because of dilution by terrigenous particles from the nearby continents. The three exceptions to this are: (1) siliceous oozes found beneath coastal upwelling zones adjacent to land masses with few rivers, (2) coral reef communities, and (3) tropical latitudes where calcareous muds are common. The muds contain hard parts deposited by coralline algae and detrital remains of reef organisms.

Coral reef communities include a variety of organisms that create hard parts including coral animals, coralline algae, and sponges. Warm-water (>18°C) coral communities are most common in shallow waters surrounding tropical volcanic islands because the coral animals have algal endosymbionts that require sunlight. Reef communities are less common near continental land masses because the higher loads of terrigenous particles cloud the water and bury the reef. Cold-water coral communities (4–13°C) do not harbor symbiotic algae and generally live at depths between 200 and 1000 m. Both communities appear to be equally abundant (Figure 15.4). The amount of calcite produced by the neritic benthos is considerable and possibly equal to that of the pelagic plankton. Thus, changes in the extent and width of the continental shelves over time has the potential to alter the sedimentary sinks of carbon and calcium.

Diatomaceous oozes dominate sediments in coastal upwelling areas as these phytoplankton outcompete other algae when nutrient concentrations, including DSi, are high. These upwelling centers are found on the western margins of contents, such as off the coast of Peru and Africa at Walvis Bay (Figure 16.3). These coastal waters are adjacent to continental deserts, and, hence, river runoff of terrigenous particles is minimal.

Organic matter constituents a minor but important component of neritic sediments. The organic carbon content of these sediments ranges from 1 to 10% w/w. The most organic-rich sediments are located in salt-marsh estuaries and beneath coastal upwelling zones. In comparison, the organic carbon content of open-ocean sediments is generally less than 1%. The high organic content of neritic sediments is the result of four interdependent factors. First, the close proximity of the continents ensures a large nutrient flux that supports high levels of primary production. Shallow water depths prevent vertical segregation of the nutrients, enhancing bioavailability. Second, sinking POM reaches the seafloor fast enough to arrive in a relatively intact state because of the shallow water depths. Third, the large POM flux creates reducing conditions in the sediments, favoring preservation of the accumulating organic matter. Fourth, rapid sedimentation rates also enhance preservation by ensuring a quick burial. This combination of factors causes the continental margins to serve as the major present-day sink of sedimentary organic matter. As will be discussed in Chapter 26, some of this sedimentary organic matter has, over millions of years, been transformed into petroleum.

Hydrogenous Coastal Sediments

In some locations, hydrogenous sediments are found in the surface sediments of the continental margins. Most of these deposits are biogenic origin, i.e., green clays, glauconite, oolite, aragonite needles, and barite. Ancient phosphorites now outcrop in some relict sediments, such as those in Onslow Bay, NC. In some marginal seas, such as the Baltic and Black Seas, significant deposits of iron-manganese oxides are forming.

20.2.2 **Abyssal Plains**

Figure 20.1 illustrates that the major sediment type on the abyssal plains are the abyssal clays with some local exceptions, including equatorial radiolarian oozes, manganese nodules, and metalliferous sediments.

Abyssal Clays

The abyssal clays are composed primarily of clay-sized clay minerals, quartz, and feldspar transported to the surface ocean by aeolian transport. Since the winds that pick up these terrigenous particles travel in latitudinal bands (i.e., the Trades, Westerlies, and Polar Easterlies), the clays can be transported out over the ocean. When the winds weaken, the particles fall to the sea surface and eventually settle to the seafloor. Since the particles are small, they can take thousands of years to reach the seafloor. A minor fraction of the abyssal clays are of riverine origin, carried seaward by geostrophic currents. Despite slow sedimentation rates (millimeters per thousand years), clay minerals, feldspar, and quartz are the dominant particles composing the surface sediments of the abyssal plains that lie below the CCD. Since a sediment must contain at least 70% by mass lithogenous particles to be classified as an abyssal clay, lithogenous particles can still be the major particle type in a biogenous ooze.

Latitudinal patterns in clay mineral distributions are pronounced in abyssal plain sediments as illustrated in Figures 14.8 through 14.11. These latitudinal bands reflect the influence of climate on terrestrial weathering, which is the ultimate source of the clay minerals with the exception of montmorillonite. A significant amount of the montmorillonite is of hydrothermal origin. As in the neritic sediments, kaolinite concentrations are highest at low latitudes, due to transport by the Trades, and chlorite at high latitudes, due to transport by the Polar Easterlies. The effect of the Westerlies is seen on the distribution of detrital quartz, which reaches maximum concentrations in the abyssal plain sediments at mid-latitudes (Figure 14.12). Montmorillonite abundances are highest around mid-ocean ridges and rises.

Biogenous Oozes

On the abyssal plains, biogenous oozes are restricted to areas with high surface-water productivity, such as the equatorial waters where divergence causes upwelling, and where water depths lie above the CCD. The equatorial upwelling areas are characterized by high radiolarian productivity resulting in radiolarian oozes (Figure 16.9).

The model provided in Figure 20.1 is for an ocean basin whose abyssal plains all lie below the CCD. This most closely resembles the conditions in the North Pacific, whereas the rest of the ocean basins have a significant portion of their abyssal plains lying above the CCD, and, hence, contain some calcareous oozes. From a global perspective, calcareous oozes are more abundant than siliceous oozes. This is caused by two phenomena: (1) all seawater is undersaturated with respect to opal, whereas all surface waters and 20% of the deep waters are saturated with respect to calcite, and (2) siliceous plankton are dominant only in upwelling areas.

Iron-Manganese Nodules

Iron-manganese nodules and crusts form at very slow rates from metals injected into seawater as part of hydrothermal fluids. Because of very slow accretion rates, these hydrogenous deposits form only where the sedimentation rate of other particles types is very slow. These settings include: (1) mid-latitude abyssal plains midway between the MOR and the seaward edge of the continental margin, and (2) the current-swept tops of seamounts and guyots. As illustrated in Figure 18.2, these nodules and crusts can be dense enough to form pavements that cover the seafloor.

20.2.3 Polar Seas

At high latitudes (>40°S and >45°N), ice is an important transport agent. Glaciers flowing downslope to the coastline erode the land and can carry detrital particles ranging in size from clay to boulders. The glaciers that flow into the ocean calve to produce icebergs. When the ice melts, this glacial till settles to the seafloor. Glacial marine sediments are currently forming at latitudes above 60°. Relict glacial marine deposits were formed during ice ages when glaciers were more abundant and widespread.

The dominant clay mineral at high latitudes is chlorite. In addition to ice rafting, lithogenous materials are transported in the polar oceans by rivers and winds. Polar seas are also characterized by diatomaceous oozes due to the occurrence of upwelling supported by divergence at 60°N and 60°S.

20.2.4 Mid-ocean Ridges and Rises

As shown in Figure 20.1, foram oozes dominate the sediments of the mid-ocean ridges and rises because they lie at depths above the CCD. At the very top of the MORs, rapid production of new oceanic crust and hydrothermal venting cause metalliferous sediments to dominate. Where seafloor spreading rates are fast, such as at the East Pacific Rise, the oozes on the ridge flanks have high metal contents (Figure 19.17). Formation of these sediments ceases if volcanic activity and hydrothermal emissions stop. This can occur for two reasons: (1) infill of fracture zones (Figure 19.7) and (2) movement of new oceanic crust away from spreading centers. Once the production of metalliferous sediments ends, continuing pelagic sedimentation buries the deposit. As illustrated in Figure W15.1, oceanic crust subsides as it moves away from the mid-ocean spreading centers. This subsidence can carry the seafloor to depths below the CCD, thus terminating the formation of any calcareous oozes.

20.3 DISTRIBUTION PATTERN OF SEDIMENTS IN THE WORLD OCEAN

The actual distribution of surface sediments in the modern-day ocean is illustrated in Figure 20.2. These are somewhat different from the general model presented in Figure 20.1 due to water circulation patterns and geological features unique to each ocean basin.

As shown in Table 20.1, biogenous oozes are the most common type of pelagic oceanic sediment, with calcareous oozes dominating.

20.3.1 Atlantic Ocean Sediments

Foram oozes are the most abundant sediment type in the Atlantic Ocean, covering 65% of the seafloor. The Mid-Atlantic Ridge is blanketed by oozes that are greater than 80% carbonate by mass. Preservation of carbonate is favored in the Atlantic because most of its seafloor lies above the CCD. This is the result of (1) relatively shallow water depths and (2) the relative youth of its deep water. The Atlantic is a narrow ocean, so the MOR takes up a large percentage of its seafloor making the average water depth shallow relative to the other oceans. The deep water is relatively young as the Atlantic Ocean is the present-day site of deep-water formation (NADW and AABW) and, hence, has not accumulated as much remineralized CO_2 as compared to the Indian and Pacific Oceans.

Abyssal clays are found in greater abundance on the western side of the Atlantic Ocean than on the eastern side. This is due to bottom topography that restricts the flow of North Atlantic Deep Water and Antarctic Bottom Water to the western side of the basin. The lower temperature of the western waters causes the CCD to be somewhat shallower than on the east side of the basin as calcite solubility increases with decreasing

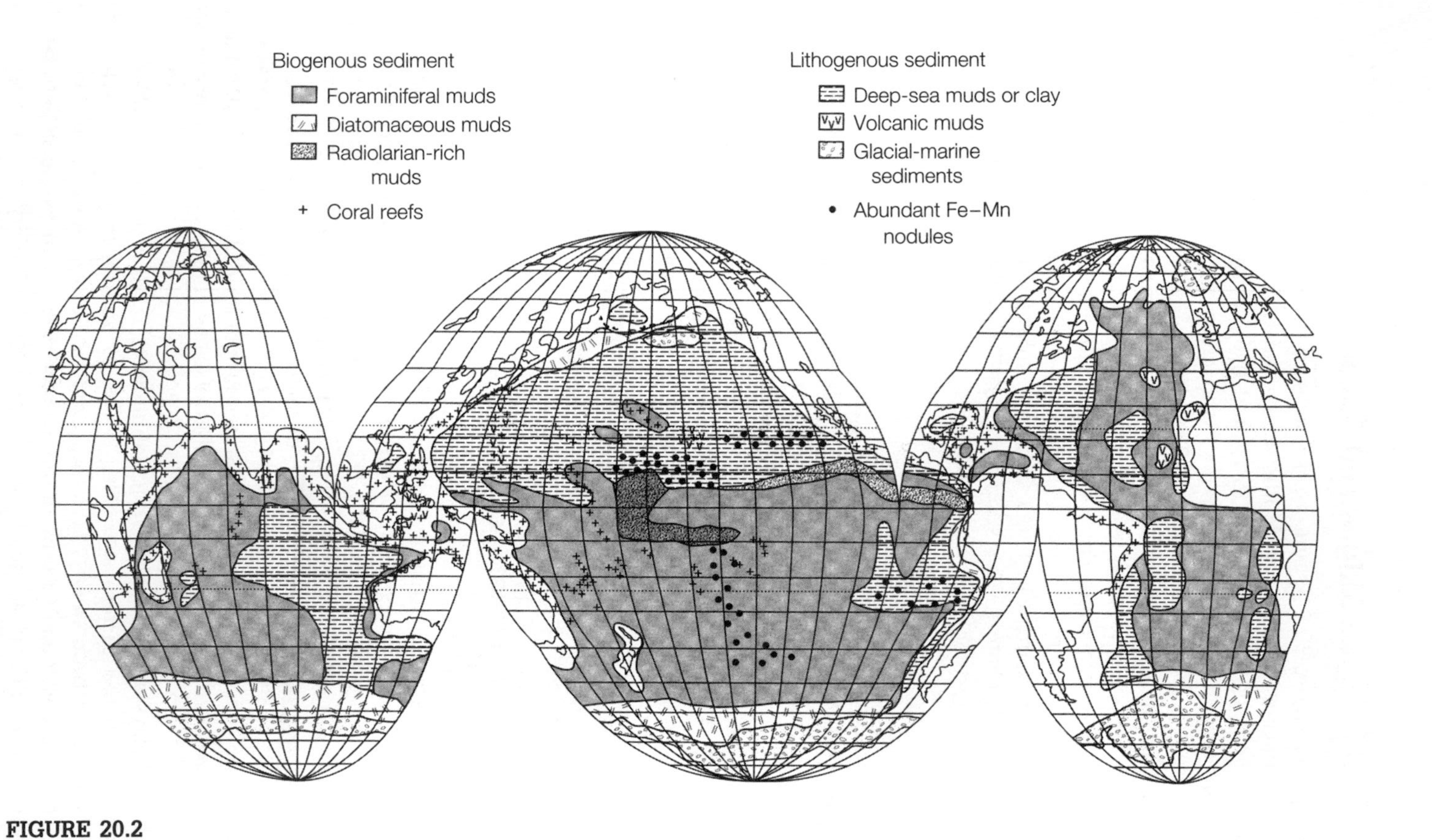

FIGURE 20.2
Distribution of deep-sea sediments. *Source:* From Davies, T. A., and D. S. Gorsline (1976). *Chemical Oceanography*, Vol. 5, Academic Press, pp. 1–80.

Table 20.1 Percent of Pelogic Sediment Coverage in the World Ocean.

Sediment type	Atlantic	Pacific	Indian	World
Calcareous ooze	65.1	36.2	54.3	47.1
Pteropod ooze	2.4	0.1		0.6
Diatom ooze	6.7	10.1	19.9	11.6
Radiolarian ooze		4.6	0.5	2.6
Red clays	25.8	49.1	25.3	38.1
Relative size of oceans (%)	23.0	53.4	23.6	100.0

Source: From Berger, W. H. (1976). Chemical Oceanography, *Volume 5, 2nd ed., Academic Press, pp. 265–388.*

temperature. In the absence of biogenic calcite, abyssal clay deposits dominate in the western basin, with most being of aeolian origin.

Significant amounts of manganese and phosphorite deposits are present on the top of the Blake Plateau, which lies at the foot of the continental margin off the southeastern United States at depths of 500 to 900 m. The Gulf Stream has eroded most of the unconsolidated sediments at this location, leaving only a carbonate platform, which has become covered with pavements of manganese and phosphorite covering an area of 5000 km^2.

On the eastern margin, a small deposit of siliceous ooze is located slightly south of the equator. This deposit is associated with the coastal upwelling area near Walvis Bay (23°S). The geographic spread of this deposit is limited because the seafloor in this area lies above the CCD, so calcite dilutes the BSi. This effect increases with increasing distance from the upwelling area.

20.3.2 Pacific Ocean Sediments

The distribution of sediment types in the Pacific Ocean is much different from that of the Atlantic. Except for the coastline of the northwest United States, the Pacific is ringed by deep-sea trenches and, hence, has relatively narrow continental shelves. The trenches effectively trap all the terrigenous particles carried to the sea by river runoff. The Pacific Ocean is much wider than the other oceans; thus the flux of windborne lithogenous particles is spread over a much greater area and produces a much lower mass flux, on an areal basis, to the seafloor. This makes other particles relatively important in determining the composition of the sediments in the Pacific ocean.

Because of the relative scarcity of lithogenous particles and fast seafloor spreading rates, metalliferous sediments are common around the East Pacific Rise and very high densities of manganese nodules are present on the abyssal plains, especially in the Southern Hemisphere. In these locations, the weathering products of volcanic detritus, such as montmorillonite, phillipsite, nontronite, and celadonite, are also found in great abundance.

In the South Pacific, the CCD is deep enough to permit the preservation of calcareous oozes except in the center of the basin, which as a result is covered by abyssal clays. The relatively rapid supply of hydrogenous sediments prevents the accumulation of calcareous oozes on the East Pacific Rise. In the North Pacific, abyssal clays dominate as this is the location where the CCD is shallowest. Aeolian transport is the source of the clay minerals that make up these deposits.

Upwelling in the Bering Sea, off Peru, and along the equator supports enough primary production to produce siliceous oozes in these locations. Radiolarian rather than foram oozes are found along the equator because these sediments lie below the CCD. They are also too far from land to have a significant input of lithogenous particles. The sediments in the Bering Sea are diatomaceous oozes. These phytoplankton are prolific in the North Sea as they dominate when temperatures are low and nutrient concentrations are high. The latter is supported by divergence at subpolar latitudes.

20.3.3 Indian Ocean Sediments

Like the Atlantic, the Indian Ocean is characterized by foram oozes along its mid-ocean ridge and most of its abyssal plains. Only in basins where the water depths exceed 5000 m are abyssal clays abundant.

20.3.4 Arctic Ocean Sediments

The continental shelves cover most of the seafloor in the Arctic Ocean, making this the shallowest ocean. Thus, most of the sediments are neritic. Because of light limitation, primary production is inhibited, so river runoff and ice rafting supply most of the particles to this ocean. As a result, lithogenous and glacial marine sediments are most common.

20.3.5 Southern Ocean Sediments

The neritic sediments surrounding the continent of Antarctica are dominated by glacial marine deposits. Divergence at 60°S causes oceanic upwelling, which transports cold, nutrient-rich waters to the euphotic zone. Since diatoms are the dominant phytoplankton, diatomaceous oozes encircle Antarctica between the latitudes 50° and 70°S.

CHAPTER 21

Why Seawater Is Salty but Not Too Salty

All figures are available on the companion website in color (if applicable).

21.1 INTRODUCTION

One of the most notable features of seawater is its high degree of saltiness. In previous chapters, we have discussed various sources of this salt, these being rivers, volcanic gases, and hydrothermal fluids. These elements have ended up in one of four places: (1) as dissolved ions in seawater, (2) as sedimentary minerals, (3) as hydrothermal minerals, and (4) as volatiles that reside in the atmosphere. The minerals are recycled via geologic uplift and subduction. Upon return to Earth's surface, these minerals are chemically weathered via acid attack by the atmospheric volatiles remobilizing the salts for return to the ocean in river runoff.

If the collective supply rate of a salt ion, such as Mg^{2+}, to the ocean is equal to its collective removal rate, its concentration remains constant over time. In other words, this ion is in a steady state within the oceanic reservoir. Until fairly recently, oceanographers thought that the major ion composition of seawater had been constant for the past 250 million years or so. Improvements in analytical chemistry have enabled geochemists to reconstruct the chemical composition of ancient seawater from fluid inclusions in bedded halites (evaporites). The results of this work have been interpreted as documenting that the major ion composition of seawater has undergone large changes over time, certainly during the Phanerozoic (past 500 million years), and probably prior, although the geochemical evidence is less substantive because of the increasing rarity of evaporites in older rocks. Significant shifts in major ion composition seem to coincide with other global changes, such as in climate, ocean anoxia (OAEs), and biological events, including mass extinctions and evolutionary explosions, like the development of planktonic calcifiers.

Biogeochemists are building increasingly more complicated numerical models to describe how the composition of seawater has changed over time. These models include positive and negative feedbacks within and among the elemental cycles of the major ions and with key biogenic elements, including carbon, phosphorus, oxygen, and iron. These models will be useful in predicting future changes that are likely to ensue as a result of the human activities which have altered some of the elemental cycling rates.

In Chapters 14 through 19, each of the sedimentary elemental sinks was discussed. In this chapter, we take an integrated view as to how feedbacks between the elemental cycles act to control the chemical composition of seawater. In doing so, we will see why: (1) the major ion ratios are geographically constant throughout the open ocean, (2) the average salinity of the ocean is presently 35‰, and (3) the salinity of the ocean has undergone significant changes over the past 500 million years.

21.2 THE CRUSTAL-OCEAN-ATMOSPHERE FACTORY

The global biogeochemical cycle was presented in Figure 1.1 as a crustal-ocean-atmosphere factory. From the largest scale perspective, the crustal-ocean-atmosphere factory involves material and energy transports between the mantle and Earth's surface. On the early Earth, the transfer of mantle materials via volcanic activity led to the formation of crustal rocks and atmospheric gases. The first crustal rocks were igneous and highly reduced. The initial volcanic gases released into the atmosphere are also thought to have been highly reduced, being comprised primarily of H_2 and CO. These substances are collectively termed *juvenile* or *primary* materials, reflecting their mantle origin.

These juvenile rocks and gases have since cycled through various reservoirs within the atmosphere, hydrosphere, biosphere, crust, and even back down into the mantle. For example, the chemical weathering of igneous silicates has produced clay minerals and ions, including Ca^{2+}. Much of this Ca^{2+} has since been transported by river runoff into the ocean and then sequestered in marine sediments as biogenic calcite. Clay minerals and calcite are examples of *secondary* minerals. Thus over time, the crustal-ocean-atmosphere factory has acted to transform primary materials in secondary ones, including minerals, solutes, and gases. Much of these secondary minerals are buried in oceanic sediments.

The pre-human natural rate at which rivers were carrying secondary minerals into the ocean was 1.4×10^{16} g/y.[1] Assuming that these solids have an average settled density of $1.6\,g/cm^3$,[2] this input would have filled the ocean basins (volume = $1.37 \times 10^{24}\,cm^3$ from Figure 2.1) within 157 million years. This has not happened because the sediments are recycled through the crust and the upper mantle (Figure 1.2).

Following burial, marine sediments are recycled via two pathways: a crustal route or a mantle route. In the crustal route, burial followed by diagenesis, catagenesis, and metagenesis transfers sediments into either sedimentary or metamorphic rocks. These rocks are eventually uplifted onto land by crustal motions associated with plate tectonics. In the mantle recycling route, sediments and sedimentary rocks are subducted at

[1] Pre-human suspended load carried by rivers as estimated by Syvitski, J. P. M., *et al.* (2005) *Science* 308, 376–380.

[2] The actual density of clay minerals is $2.7\,g/cm^3$, but these solids are surrounded by pore waters as they accumulate in the sediments. An average wet density of marine sediments is estimated at $1.6\,g/cm^3$ based on a porosity of about 65% (Figure 12.2).

convergent plate boundaries back into the mantle along with underlying oceanic crust. Some of the subducting materials are returned to Earth's surface by volcanic eruptions that occur near the convergent plate boundaries, such as at back-arc volcanoes. The rest of the subducting materials are driven down into the mantle, where they melt and mix with ambient magma. This mixture is then eventually erupted at spreading centers to form new oceanic crust. Subduction of secondary minerals has caused a long-term shift in the chemical composition of mantle magma such that its oxidation state has increased over time. For example, the primary components of volcanic gases are now H_2O and CO_2.

The sedimentary and metamorphic rocks uplifted onto land have become part of continents or oceanic islands. These rocks are now subject to chemical weathering. The dissolved and particulate weathering products are transported back to the ocean by river runoff. Once in the ocean, the weathering products are available for removal back into a marine sedimentary reservoir. At present, most mass flows on this planet involve transport of the secondary (recycled) materials rather than the chemical reworking of the primary (juvenile) minerals and gases. The nature of these transport and sediment formation processes has been covered in Chapters 14 through 19 from the perspective of the secondary minerals formed. We now reconsider these processes from the perspective of impacts on elemental segregation between the reservoirs of the crustal-ocean-atmosphere factory and the mantle.

21.3 SOURCES AND TRANSPORT PROCESSES

The major source of solutes and solids to the ocean is via river transport. The only major ion with a direct source associated with hydrothermal input seems to be calcium. The hydrothermal input of DSi is also significant. Volcanic gases are presently contributing a minor amount of HCl and sulfur gases (H_2S and SO_2). Each of these sources is discussed next with primary focus on how terrestrial chemical weathering provides most of the major ion input the oceans.

21.3.1 Terrestrial Weathering and River Transport

The chemical weathering of crustal rock was discussed in Chapter 14 from the perspective of clay mineral formation. It was shown that acid attack of igneous silicates produces dissolved ions and a weathered solid residue, called a clay mineral. Examples of these weathering reactions were shown in Table 14.1 using $CO_2 + H_2O$ as the acid (carbonic acid). Other minerals that undergo terrestrial weathering include the evaporites, biogenic carbonates, and sulfides. Their contributions to the major ion content of river water are shown in Table 21.1.

The sulfate and chloride minerals in evaporites (gypsum, anhydrite, halite) undergo congruent dissolution to produce $Ca^{2+}(aq)$, $SO_4^{2-}(aq)$, $Na^+(aq)$, and $Cl^-(aq)$. The dissolution of evaporite and biogenic carbonates (limestone, dolomite, and calcite) generates

Table 21.1 Sources of Major Elements in World River Water (in Percent of Concentrations).[e]

Element	Atmos. Cyclic Salt	Weathering			Pollution[b]
		Carbonates	Silicates	Evaporites[a]	
Ca^{2+}	0.1	65	18	8	9
HCO_3^-	<<1	61[c]	37[c]	0	2
Na^+	8	0	22	42	28
Cl^-	13	0	0	57	30
SO_4^{2-}	2[d]	0	0	22[d]	54
Mg^{2+}	2	36	54	<<1	8
K^+	1	0	87	5	7
H_4SiO_4	<<1	0	99+	0	0

[a]*Also includes NaCl from shales and thermal springs.*
[b]*Values taken from Meybeck (1979) except sulfate.*
[c]*For carbonates, 34% from calcite and dolomites and 27% from soil CO_2; for silicates, all 37% from soil CO_2; thus, total bicarbonate from soil (atmospheric) CO_2 = 27 + 37 = 64%.*
[d]*Other sources of river sulfate include natural biogenic emissions to atmosphere delivered to land in rain (3%), volcanism (8%) and pyrite weathering (11%).*
[e]*Actual concentrations include pollutive inputs and are provided in Table 21.2.*
Source: From Berner, E. K. and R. A. Berner (1996). Global Environment: Water, Air and Geochemical Cycles. *Prentice Hall, p. 376. Data source: Meybeck, M. (1979).* Revue de Geologie Dynamique et de Gèographie Physique, *21, 215–246.*

Ca^{2+}(aq), Mg^{2+}(aq), and HCO_3^-(aq). Silicate weathering is an incongruent process. The most important of these reactions involves the weathering of the feldspar minerals, orthoclase, albite, and anorthite. The dissolved products are K^+(aq), Na^+(aq), and Ca^{2+}(aq), and the solid products are the clay minerals, illite, kaolinite, and montmorillonite. The weathering of kaolinite to gibbsite and the partial dissolution of quartz and chert also produces some DSi.

Following burial in the sediment, terrestrial clay minerals are compacted into a sedimentary rock called shale. Marine shales deposited during times of ocean anoxia contain high concentrations of biogenic particulate organic matter and pyrite. They are called black shales. These rocks also tend to have high concentrations of trace metals due to biogenic enrichment during the formation of the organic matter (Table 11.5). When the shale is uplifted onto land, the organic matter and pyrite undergo oxidative weathering. In contrast, the clay minerals in the shale are relatively inert and, hence, do not readily undergo chemical weathering.

The terrestrial weathering of organic matter derived from shales and soils results in the oxidation of carbon, which generates CO_2. Dissolution of this CO_2 in water produces carbonic acid. This weak acid serves to enhance chemical weathering reactions by acting as a source of $H^+(aq)$. Pyrite also serves as a source of acid as its reduced sulfur is converted into sulfuric acid, via the following oxidative weathering reaction:

$$4FeS_2 + 15O_2 + 8H_2O \rightarrow 2Fe_2O_3 + 8H_2SO_4 \quad (21.1)$$

Because sulfuric acid is a strong acid, it is a more potent weathering agent than carbonic acid. By serving as a source of $H^+(aq)$, sulfuric acid is transformed into $SO_4^{2-}(aq)$. The oxidation of pyrite is responsible for 11% of the sulfate in river water, with pollution now contributing 54%. The latter is largely associated with the burning of sulfur-rich coal.

Chemical weathering requires water supplied by rain, ice melt, or groundwater. The dissolved and solid weathering products are transported downslope as this water makes its way to the ocean. The chemical composition of these waters is spatially and temporally variable because the intensity of chemical weathering depends on climate, topography, plant growth, and the chemical composition of the parent rock. The geographic variability in the chemical composition of river water is illustrated in Table 21.2. This table also lists the continental averages for water discharge and runoff ratio. The latter is the average runoff per unit area normalized to average rainfall. High runoff ratios promote the mobilization and transport of weathered materials. Asia has the largest water discharge, runoff ratio, and, as shown in Table 14.2, the largest sediment yield, reflecting the influence of the Himalayan mountains.

Considerable geographic variability exists in the distribution of the source rocks contributing salts to river and groundwaters. As shown in Table 21.3, most of the evaporites, which are the dominant natural source of Na^+ and Cl^- in river water, lie in marginal and endorheic (internal) seas. Some of these subsurface evaporite deposits dissolve into groundwaters, which eventually carry Na^+ and Cl^- into the ocean. Carbonates are the prevalent rock type between 15°N and 65°N. Precambrian-age crustal rocks and metamorphic minerals predominate between 25°S and 15°N and north of 55°N. Shales and sandstones represent on average 16% of the terrestrial surface lithology.

The overall effect of the terrestrial weathering reactions has been the addition of the major ions, DSi, and alkalinity to river water and the removal of O_2, and CO_2 from the atmosphere. Because the major ions are present in high concentrations in crustal rocks and are relatively soluble, they have become the most abundant solutes in seawater. Mass-wise, the annual flux of solids from river runoff (1.55×10^{16} g/y) in the pre-Anthropocene was about three times greater than that of the solutes (0.42×10^{16} g/y).[3] The aeolian dust flux (0.045×10^{16} g/y) to the ocean is about 30 times less than the river solids input.[4] Although most of the riverine solids are deposited on the continental margin, their input has a significant impact on seawater chemistry because most of these particles are clay minerals that have cations adsorbed to their surfaces. Some of these cations are desorbed

[3] This estimate of the suspended plus the bed load is from Syvitski, J. P. M., *et al.* (2005), *Science* 308, 376–380. The total input of dissolved solids in natural river water is estimated from the global mean river water TDS in Table 21.2 (99.6 mg/L) and the river runoff rate from Figure 2.1.

[4] *Source*: From Jickells, J. T., *et al.* (2005), *Science* 308, 67–71.

Table 21.2 Average Composition of River Water for the Different Continents.

By Continent	River Water Concentration (mg/L)[a]									Water Discharge (10^3 km^3/y)	Runoff Ratio[c]
	Ca^{2+}	Mg^{2+}	Na^+	K^+	Cl^-	SO_4^{2-}	HCO_3^-	SiO_2	TDS[b]		
Africa:											
Actual	5.7	2.2	4.4	1.4	4.1	4.2	26.9	12.0	60.5	3.41	0.28
Natural	5.3	2.2	3.8	1.4	3.4	3.2	26.7	12.0	57.8		
Asia:											
Actual	17.8	4.6	8.7	1.7	10.0	13.3	67.1	11.0	134.6	12.47	0.54
Natural	16.6	4.3	6.6	1.6	7.6	9.7	66.2	11.0	123.5		
S. America:											
Actual	6.3	1.4	3.3	1.0	4.1	3.8	24.4	10.3	54.6	11.04	0.41
Natural	6.3	1.4	3.3	1.0	4.1	3.5	24.4	10.3	54.3		
N. America:											
Actual	21.2	4.9	8.4	1.5	9.2	18.0	72.3	7.2	142.6	5.53	0.38
Natural	20.1	4.9	6.5	1.5	7.0	14.9	71.4	7.2	133.5		
Europe:											
Actual	31.7	6.7	16.5	1.8	20.0	35.5	86.0	6.8	212.8	2.56	0.42
Natural	24.2	5.2	3.2	1.1	4.7	15.1	80.1	6.8	140.3		
Oceania:											
Actual	15.2	3.8	7.6	1.1	6.8	7.7	65.6	16.3	125.3	2.40	—
Natural	15.0	3.8	7.0	1.1	5.9	6.5	65.1	16.3	120.3		
World average:											
Actual	14.7	3.7	7.2	1.4	8.3	11.5	53.0	10.4	110.1	37.4	0.46
Natural (unpolluted)	13.4	3.4	5.2	1.3	5.8	8.3 (5.3)[d]	52.0	10.4	99.6	37.4	0.46
Pollution	1.3	0.3	2.0	0.1	2.5	3.2 (6.2)[d]	1.0	0	10.5	—	—
World % pollutive	9%	8%	28%	7%	30%	28% (54%)[d]	2%	0%	—	—	—

[a]*Actual concentrations include pollution. Natural concentrations are corrected for pollution.*
[b]*TDS = total dissolved solids.*
[c]*Runoff ratio = (average runoff per unit area)/(average rainfall).*
[d]*Pollutive contribution as estimated by Berner and Berner (1996).*
Source: From Berner, E. K. and R. A. Berner (1996). Global Environment: Water, Air and Geochemical Cycles. *Prentice Hall, p. 376. Data source: Meybeck, M. (1979).* Revue de Geologie Dynamique et de Gèographie Physique, *21, 215–246.*

Table 21.3 Global Proportions of Surficial Lithologies for the Different Ocean Drainage Basins.[a]

Lithology	Arctic Ocean	Atlantic Ocean[b]	Indian Ocean	Pacific Ocean	Mediterranean + Black Sea	Endorheic (Internal)	Global
Water bodies	0.7	0.7	0.1	0.03	0.8	0.6	0.5
Polar ice and glaciers	0.08	...	...	0.08	...	...	0.02
Basic-ultrabasic plutonic rocks	0.34	*0.04*	0.38	0.27	0.25	0.2	0.2
Acid plutonic rocks	7.1	7.3	7.6	11.5	*3.0*	4.9	7.2
Basic volcanic rocks	4.3	3.1	6.4	**14.6**	4.8	4.3	5.8
Acid volcanic rocks	0.53	0.74	*0.21*	**3.5**	*0.04*	0.9	0.98
Precambrian basement	8.2	16.3	17.7	*4.4*	11.3	*3.9*	11.6
Metamorphic rocks	5.7	6.2	3.9	*1.9*	*0.21*	*1.8*	4.0
Basement + metamorphic	13.9	22.5	21.6	*6.3*	11.5	*5.7*	15.6
Complex lithology	6.9	3.1	2.6	9.6	3.5	9.5	5.4
Siliciclastic sedimentary rocks[c]	27	18.5	11.3	10.0	14.2	14.1	16.3
Mixed sedimentary rocks	*2.6*	5.9	13.2	12.7	*1.6*	9.7	7.8
Carbonate rocks—consolidated	13.1	9.1	11.2	6.7	**20.6**	8.1	10.4
Recent evaporites	...	...	0.21	...	0.12	**0.69**	0.12
Semi-to unconsolidated sedimentary rocks	5.3	11.1	7.2	7.2	14.0	15.6	10.1
Alluvial deposits	16.3	13.4	17.4	14.9	14.2	19.3	15.5
Loess	1.7	3.4	...	2.6	**6.6**	2.2	2.6
Dunes or shifting sand	*0.13*	1.15	*0.64*	*0.10*	**4.85**	**4.4**	1.6
Sum	100	100	100	100	100	100	100
Drainage area, M km^2	17.5	45.7	20.7	19.8	10.7	18.9	133.0

[a] *Different ocean basins denote nonglaciated areas (in percent of total area for each basin); boldface values are $\geq 2x$ the global average, italic values are $\leq 0.5 \times$ the global average). Three dots denote zero values.*
[b] *Atlantic Ocean does not include Mediterranean and Black Sea drainage basins.*
[c] *Includes shales, pyritic shales, and sandstones.*
Source: From Durr, H. D., and M. Meybeck (2005). Global Biogeochemical Cycles *19, GB4510.*

when the clays enter the ocean and, hence, are released into seawater. For the major ions, cation exchange is a significant source of calcium, constituting about 8% of the total river input.

The world's largest rivers lie in areas that are relatively inaccessible to scientists (Table 14.2). This has greatly limited our ability to chemically characterize "average" river water both geographically and temporally. Considerable effort is currently being directed at improving this data set because human impacts are significantly altering some of the elemental fluxes. As shown in Table 21.2, the overall effect of anthropogenic impacts has been to increase ion concentrations. This "salinization" is partly a result of agricultural activities and partly a result of salt ingestion by humans. The ingested salt is eventually released into sewage. The salt passes through sewage treatment facilities and is thereafter emitted into natural waterways. Notable exceptions include DSi and sediment. In the case of the former, fluxes are being reduced by enhanced diatom production in freshwaters caused by the runoff of fertilizers rich in nitrogen and phosphorus. The riverine flux of sediments has been reduced by about 10% from particle trapping by dams and reservoirs.

21.3.2 Volcanic Gases and Hydrothermal Emissions

The primary source of the anions in seawater are volcanic gases released during the eruption and cooling of magma. These gases are listed in order of relative abundance in Table 21.4. Spatial and temporal variability, combined with obvious sampling challenges, makes definition of an average volcanic gas concentration difficult. As noted in Section 21.2, the chemistry of the mantle has evolved over time such that volcanic gas composition has shifted from a reducing form (H_2/CO dominant) to a more oxidizing one (H_2O/CO_2 dominant).

Chlorine is released as HCl, which dissociates upon dissolution in water to generate $Cl^-(aq)$. Sulfur is released as either H_2S or SO_2. Both are transformed into $SO_4(aq)$ through chemical reactions involving oxidation by O_2 and dissociation/dissolution in water. The amounts of primary magmatic volatiles that have been degassed thus far are given in Table 21.5. About half of the chlorine has been retained in the ocean and the other half has been converted into evaporite minerals. In comparison, virtually

Table 21.4 Volcanic Gases in Order of Abundance.

(Most abundant ← ; ↑ Most abundant)			
Hydrogen	H_2O	H_2	
Carbon	CO_2	CO	CH_4
Sulfur	SO_2	H_2S	S_2
Halides	HCl	HF	
Inert gases	Ar	N_2	He

Table 21.5 Masses (in grams) of the Five Major Volatiles in the Upper Mantle, Crust, and Surface Reservoirs.

Species	Upper Mantle	Oceanic Crust[a]	Continental Crust[b]	Earth Surface[c]	Sediments[d]	Ocean[e]	Atmosphere[f]
H_2O	$(2 \text{ to } 5.5) \times 10^{23}$	6.2×10^{23}[g]		1.430×10^{24}	7.15×10^{22}	1.358×10^{24}	1.173×10^{19}
C	$(8.9 \text{ to } 16.6) \times 10^{22}$	9.200×10^{20}	2.576×10^{21}	7.784×10^{22}	7.780×10^{22}	3.850×10^{19}	7.850×10^{17}
N	1.11×10^{22}	?	2.240×10^{20}	4.890×10^{21}	1.000×10^{21}	2.263×10^{19}	3.867×10^{21}
Cl	5.55×10^{20}	?	1.680×10^{21}	4.311×10^{22}	1.735×10^{22}	2.576×10^{22}	
S	2.00×10^{23}	5.256×10^{21}	5.936×10^{21}	$(1.07 \text{ to } 1.42) \times 10^{22}$	$(9.5 \text{ to } 13) \times 10^{21}$	1.230×10^{21}	$0 \text{ to } 5.66 \times 10^{14}$

[a] *Mass of oceanic crust: 6.5 km × 3.61 × 10^8 km^2 × 2.7 g/cm^3 = 6.57 × 10^{24} g.*
[b] *Mass of continental crust: 30 km × 1.49 × 10^8 km^2 × 2.5 g/cm^3 = 11.2 × 10^{24} g.*
[c] *Sum of the masses in sediments, ocean, and atmosphere.*
[d] *Total sediment mass (continental and oceanic) 2.09 × 10^{24} g (Li, 2000). Sulfur content of sediments 0.45 to 0.62 wt %.*
[e] *From species concentrations in gram/liter. Table 3.7 and ocean volume 1.37 × 10^{21} L.*
[f] *Recent atmosphere.*
[g] *Water content of crystalline crustal rocks 3.5 wt %, range 2.5 to 4.5 wt %.*

Source: *From Mackenzie, F. T., and A. Lerman (2006).* Carbon in the Geobiosphere: Earth's Outer Shell. *Springer, and references therein. (Table 2.2 on p. 26).*

all of the carbon and most of the sulfur has become sequestered in the sediments. It is thought that the present-day rate of outgassing of HCl and H_2O is equal to their subduction fluxes. In other words, the mantle appears to be at a steady state with respect to chlorine and water so that volcanic activity is no longer increasing the crustal-ocean-atmosphere inventory of these volatiles.

Although global elemental fluxes from hydrothermal systems are still not well constrained, geochemists seem to agree that a significant amount of calcium and a minor amount of DSi are supplied from high-temperature venting. As shown in Tables 11.1 and 19.2, hydrothermal emissions are a significant source of trace elements, such as iron and manganese, and possibly chlorine (as chloride) and potassium.

21.4 STORAGE RESERVOIRS AND REMOVAL MECHANISMS

The major ions are removed from seawater by a variety of biogeochemical processes. These processes collectively operate at slower rates than those acting on the biolimited and particle-scavenged elements, such as phosphorus and iron. This causes the major ions to spend very long periods of time dissolved in seawater before being removed to the sediments.

For example, the average atom of potassium spends 10 million years dissolved in the ocean before becoming incorporated into the sediments. (Potassium is in steady state, so its oceanic residence time can be computed by dividing its input rate into the total amount in seawater.) This is plenty of time for ocean mixing, which occurs on time scales of a thousand years, to homogenize out any horizontal or vertical concentration gradients.

This is why the salinity of seawater is nearly the same throughout the open ocean, varying by only a few parts per thousand. (As per Figure 3.3, 75% of seawater has a salinity between 34 and 35‰.) The small degree of spatial variability is a consequence of geographic variations in the balance of evaporation versus precipitation in the surface waters. Recall that these surface waters are the source waters for intermediate and deep water masses. Since shifts in the relative rates of evaporation versus precipitation involve only addition or removal of water, the major ion ratios are unaltered. This is why the major ion ratios do not exhibit little if any spatial differences within the open ocean.

The major ions have two main escape routes from the ocean: (1) incorporation into sediments or pore water and (2) ejection into the atmosphere as seasalt spray. This spray is caused by bursting bubbles that produce small particles, called aerosols, that range in diameter from 0.1 to 1000 μm. The annual production rate of seasalt aerosols is large, on the order of 5×10^{12} kg/y, but virtually all of it is quickly returned when the spray falls back onto the sea surface. A small fraction (about 1%) is deposited on the coastal portions of land masses and carried back into the ocean by river runoff. As shown in Table 21.6, seasalts represent a significant fraction of dissolved solids in river runoff, especially for sodium and chloride. Due to the short timescale of this process, seasalt aerosol losses and inputs are considered by geochemists to be a short circuit in the crustal-ocean-atmosphere factory. The solutes transported by this process are collectively referred to as the "*cyclic salts*."

Table 21.6 Percentage of Cyclic Seawater Ions by Weight Relative to Their Total Weight in Rivers.

Element	Berner and Berner 1996[a]	Holland 1978[b]	Garrels and Mackenzie 1971[b]	Meybeck 1983[c]
Cl^-	13(18)	27	55	72
Na^+	8(11)	19	35	53
SO_4^{2-}	2(2)	39[d]	6	19
Mg^{2+}	2(2)	<3	7	15
K^+	1(1)	<14	15	14
Ca^{2+}	0.1(0.1)	1.3	0.7	2.5

[a] *Values are given as a percentage of "actual" (including pollution) world average river water and, in parenthesis, of "natural" world river water (corrected for pollution). [Cyclic Cl is set equal to 18%, which is the Amazon value given by Stallard R. E. and Edmond J. M. (1981).* Journal of Geophysical Research, *80, 9844–9858. Sea salt contributions to other ions are set in proportion to this based on the composition of average seawater.]*
[b] *Based on world average river water from Livingstone (1963), which is not corrected for pollution.*
[c] *Based on "natural" world average river water.*
[d] *Based on total atmospherically derived sulfur (natural + pollution).*
Source: From Berner, E. K., and R. A. Berner, (1996). Global Environment: Water, Air and Geochemical Cycles. *Prentice Hall, p. 201.*
See Berner and Berner (1996) for data sources: (1) Holland (1978), (2) Garrels and Mackenzie (1971), (3) Meybeck (1983), and (4) Livingstone (1963).

The major ions that are lost from the ocean by incorporation into the seafloor, whether as sediment or part of the oceanic crust, are removed for long periods of time as they are effectively entrained in the rock cycle (Figure 1.2). Some major ions are removed primarily into one sediment or mineral type, while others have multiple sinks. The chemical reactions responsible for these removals are discussed in the following sections (21.4.1 through 21.4.7). They include reactions controlled by a diverse set of processes including: (1) the actions of marine organisms, (2) the formation of evaporites, (3) the weathering of oceanic crust via hydrothermal processes, (4) cation exchange involving clay minerals, (5) burial of interstitial water, and (6) the reverse weathering of clay minerals. The quantitative contributions of the most important of these are given in Table 21.7 and indicate that the seawater concentrations of magnesium, sulfate, and sodium are close to, but not exactly in, steady state. Some of the imbalances between inputs and outputs are likely not significant, being within the uncertainty of current global flux estimates. Some of the imbalances are due to anthropogenic inputs, such as with sulfate. Still others, such as for Mg^{2+} and Ca^{2+}, are thought to reflect ongoing natural shifts in the processes that control the removals and inputs of these major ions.

Although the chemical removals of the major ions are slow, they collectively exert an important control on the salt composition of seawater. This is illustrated by comparing the major ion composition of seawater to that of river water as shown in Table 21.8. Although river water is the largest single source of the major ions to seawater, its ion

Table 21.7 Input-Output Balance for the Major Ions and Alkalinity.

		Cl^-		Na^+		Mg^{2+}		SO_4^{2-}		K^+		Ca^{2+}		Alk	
		J	**M**	**H**[c]	**M**	**H**	**M**	**H**	**M**	**H**	**M**	**H**	**M**	**H**	**M**
Ocean inventory ($\times 10^{18}$ mol)[a]		765		658		74		40		14		14		3	
Rates ($\times 10^{12}$ mol/y)[a]															
Rivers (natural)			6.1	2.4	8.5	6.1	5.2	4.2	3.2	1.4	1.2	15.0	12.5	37.6	31.9
Volcanic gas		0.2 to 0.6													
Hydrothermal systems	On axis[b]			−1.1	−0.9	−2.0	−3.1	−1.0	−1.7	0.6	−0.6	0.9	2.0	−0.1	−0.4
	Off axis					−0.4				−0.4					
Clays	Ion exchange			−1.5	−1.9	−0.3	−1.2			−0.3	−0.4	1.0	2.6		0.5
	Reverse Weathering									−0.8	−0.1				
Carbonate	Dolomite deposition					−1.7	−0.6					−1.7		−6.8	
	Carbonate deposition											−15.7	−17.0	−31.4	−35.0
Silicate	BSi deposition														
Sulfides	Sulfate reduction													4.0	
	Pyrite deposition							−2.0	−1.2						2.4

(Continued)

Table 21.7 *(Continued)*

		Cl^-		Na^+		Mg^{2+}		SO_4^{2-}		K^+		Ca^{2+}		Alk	
		J	**M**	**H**[c]	**M**	**H**	**M**	**H**	**M**	**H**	**M**	**H**	**M**	**H**	**M**
Salts	Atmospheric seasalt cycling, pore water burial, evaporite deposition		−6.1	[c]	−5.7		−0.3		−0.3		−0.06		−0.1		
	Atmospheric seasalt cycling	−1.1 to −4.5													
	Pore water burial	0.1													
	Evaporite deposition							−0.2				−0.2			
Total inputs – Total outputs			0.0	−0.2	0.0	1.7	0.0	1.0	0.0	0.5	0.0	−0.7	0.0	3.3	−0.6
Increment (+)/Decrement (−) estimated from data[d]				0.0		1.7		1.0		?		−0.7			
Conclusions from Holland (2005)				Imbalance within uncertainty of data		Nonsteady state supported by data		Nonsteady state supported by data		Outputs likely under-estimated		Nonsteady state supported by data		Imbalance within uncertainty of data	

[a] *Note that the Morel (1993) estimates are somewhat different from those in Table 21.8.*
[b] *On-axis entry includes high and low temperature reactions unless split out into On Axis and Off Axis.*
[c] *Evaporite and seasalt cycling effects are not included in this estimate. This was done by subtracting out molar concentrations equivalent to that of the chloride ion.*
[d] *Increment: seawater concentration is rising over time; decrement: seawater concentration is declining over time.*
H = *Holland, H. H. (2005).* American Journal of Science *305, 220–239.*
M = *Morel, F. M. M. and J. Herring (1993).* Principles and Applications of Aquatic Chemistry. *John Wiley & Sons.*
J = *Jarrard, R. D. (2003).* Geochemistry, Geophysics, Geosystems *4, 50.*

Table 21.8 Comparison of River and Seawater Composition.

Ion	River Water			Seawater				
	Concentration (μmol/L)	Runoff ($\times 10^{12}$ mol/y)	% Contribution by Mass to Total Dissolved Solids	Concentration[a] (μmol/L)	Inventory[b] ($\times 10^{18}$ mol)	% Contribution by Mass to Total Dissolved Solids	$\frac{[\text{Seawater}]}{[\text{Riverwater}]}$	Replacement Time (Million Years)
Na^+	226	10	6.0%	479,955	658	30.8%	2,122	65
Mg^{2+}	140	6	3.9%	54,050	74	3.7%	386	12
Ca^{2+}	334	15	15.5%	10,522	14	1.2%	31	1
K^+	33	1	1.5%	10,446	14	1.1%	314	10
Cl^-	164	7	6.7%	558,626	765	55.2%	3,415	104
SO_4^{2-}	55	2	6.1%	28,897	40	7.7%	524	16
HCO_3^-	852	38	60.2%	1,904	3	0.3%	2	0.068

[a] *Data from Table 3.7.*
[b] *Ocean Volume and Runoff Rates from Figure 2.1.*
Data from: Berner, E. K., and R. A. Berner, (1996). Global Environment: Water, Air and Geochemical Cycles. *Prentice Hall, p. 201.*

concentrations are lower than in seawater and its ion ratios are quite different. The increase in concentrations is caused by the evaporation of water, which has a much shorter residence time, a few thousand years, than any of the major ions. But clearly, something more than simple evaporation has transformed river water into seawater, because the dominant major ions in river water are bicarbonate and calcium, whereas in seawater, the dominant major ions are chloride and sodium.

What has happened to the bicarbonate and calcium delivered to the ocean by river runoff? As described later, these two ions are removed from seawater by calcareous plankton because a significant fraction of their hard parts are buried in the sediment. In contrast, the only sedimentary way out of the ocean for chloride is as burial in pore waters or precipitation of evaporites. The story with sodium is more complicated—removal also occurs via hydrothermal uptake and cation exchange. Because the major ions are removed from seawater by different pathways, they experience different degrees of retention in seawater and uptake into the sediments. Another level of fractionation occurs when the oceanic crust and its overlying sediments move through the rock cycle as some of the subducted material is remelted in the mantle and some is uplifted onto the continents.

21.4.1 Deposition of Biogenic Materials

Marine organisms control the production and destruction of biogenic calcite, pyrite, and silica, which are major elemental reservoirs of calcium, carbon, sulfur, and silicon, respectively. Although carbon and silicon are not major ions, they are mentioned herein because of their participation in geological reactions that act to link the crustal-ocean cycling of Ca, C, S, and Si. Calcium carbonate is the single largest sedimentary sink of the major ions and alkalinity. Most of the calcium and carbon that have entered the ocean are now sequestered in the crust as limestone or dolomite. These minerals are primarily biogenic in origin, although evaporitie deposition has been of greater importance in the past. As noted in Chapter 18.3.3, small amounts of hydrogenous calcium carbonate have also been deposited as oolites and aragonitic needle muds.

Most biogenic calcite is presently being deposited in pelagic sediments from the hard parts of foraminiferans, coccolithophorids, and pteropods. The rate of deposition of biogenic calcite on continental shelves is thought to increase during periods of high sea level due to an increase in habitat suitable for the growth of coral reefs and coralline algae. After burial in the sediments, biogenic calcite undergoes a diagenetic transformation in which some of its calcium is exchanged for magnesium. This converts calcite into the mineral dolomite and has the potential to remove a significant amount of magnesium from seawater leading to a decline in the Mg/Ca ratio of seawater. (Hydrothermal activity is another important sink for Mg that has the potential to alter the Mg/Ca ratio of seawater if significant changes in tectonic activity have occurred over geologic time.)

Prior to the advent of the coccolithophorids and planktonic foraminferans, 200 to 250 million years ago, all biogenic calcite precipitation must have been restricted to the shallow waters of the neritic zone. Thus, the evolution of the pelagic calcifiers ushered in an important innovation in the crustal-ocean-atmosphere factory—the creation of a pelagic sedimentary reservoir of biogenic calcite. This widespread reservoir has a far superior ability to buffer the ocean against changes in calcium carbonate saturation and, hence, short-term excursions in pH. Likewise, the evolution of diatoms around 200 mybp led to the development of the most important oceanic sink of silica, BSi. Both types of hard parts have also played a critical role in increasing the delivery of POM to the sediments and, hence, the burial of organic carbon. As mentioned in Chapter 15, this ballasting effect is a consequence of the less dense POM sinking faster when it is attached in some way to the denser, and, hence, more rapidly sinking, hard parts.

21.4.2 Pyrite Deposition

As listed in Table 21.7, approximately half the sulfur that enters the ocean is removed as pyrite (FeS_2). Some of this pyrite deposition is a by-product of biogenic sulfate reduction in anoxic sediments, with the rest occurring abiotically in hydrothermal systems. Production of the biogenic component is controlled by the redox state of the sediments and deep waters. Recall that the degree of oxygenation of the deep waters and sediments is controlled by the relative rates of deep-water circulation and the supply of sinking POM. As a result, the deposition rate of biogenic pyrite has varied over time, reaching peak values during ocean anoxic events (Figure W8.5). Similarly, variations in the rate of abiogenic sulfate reduction over geologic time have been caused by changing rates of tectonic activity. In both cases, pyrite deposition serves as a source of alkalinity via the following reaction:

$$\underset{\text{oxidized iron}}{2\text{“FeO”}} + 4SO_4^{2-} + \underset{\text{organic carbon}}{7C^o} + CO_2 + 4H_2O \longrightarrow \underset{\text{pyrite}}{2FeS_2} + \underset{\text{alkalinity}}{8HCO_3^-} \qquad (21.2)$$

which links the elemental control of sulfate to that of carbon. In this reaction, sulfate reduction is an intermediate reaction that generates acid, thereby solubilizing iron so that it is available to precipitate with $Fe^{2+}(aq)$. (In Eq. 21.2, “FeO” is used to denote that Fe(III) is present in the mineral phase. Examples of the latter include the iron (III) oxyhydroxides: magnetite and hematite. See the online appendix on the companion website for their empirical formulae.)

21.4.3 Deposition of Evaporites

Seawater is undersaturated with respect to the primary evaporite minerals, halite, gypsum, and anhydrite. While surface seawater is presently supersaturated with respect to calcium carbonate, kinetic barriers prevent its abiogenic precipitation except at very high degrees of supersaturation. To generate the saturated conditions required for evaporite deposition requires three factors: (1) climatic conditions that favor net evaporation, (2) shallow waters with a restricted connection to the ocean, and (3) long periods of stable climate and sea level. These three conditions are only episodically met, causing the rates of evaporite deposition to vary greatly over the ocean's history (Figure 17.8). In the present day, evaporite deposition is restricted to tidal mudflats in low latitudes (Table 17.1). Although actively forming evaporites are presently rare, only 0.004 to

0.02% of the ocean would have to be evaporitic to remove the annual river fluxes of sulfate and chloride.

The rate at which evaporites form is directly tied to plate tectonics because this process is a control on the length and width of the continental margins and marginal seas. This influence is random, as it is a product of crustal motions. Likewise, random motions act to uplift evaporites onto land, causing them to undergo chemical weathering, leading to their dissolution. These random controls preclude the establishment of equilibrium or any kind of feedback between the evaporite minerals and the ocean. Nevertheless, the relative constancy in evaporite mineralogy over geologic time puts severe limits on the extent to which the salinity and major ion composition of seawater could have varied. As a result, changes in the concentration of any of the major ions are unlikely to have exceeded a factor of 3. This is also supported by the fact that dissolution of all the existing sedimentary halite would serve to only double the chloride concentration in seawater.

Since evaporites are the sole repository of chloride and remove a significant amount of the sulfur input, their formation is the major anion sink. They are also a significant reservoir of sodium and magnesium. The magnesium in evaporites is in the form of dolomite, which is an authigenic carbonate. Despite its relatively large abundance in the geologic record and supersaturated state in most seawater, dolomites are presently forming at very slow rates. As a result, their mode of formation is not well understood and was probably quite different during periods of massive dolomite deposition. As noted in Chapter 21.4.1, most of this dolomite appears to be a secondary mineral formed when Mg^{2+} replaces some of the Ca^{2+} in biogenic calcite and aragonite. Direct precipitation from seawater appears to be inhibited by kinetic barriers.

21.4.4 Plate Tectonics and Hydrothermal Activity

Plate tectonics influences the major ion composition of seawater by controlling the rates of hydrothermal activity and the uplift of buried sedimentary and metamorphic rock. Plate tectonics also determines the length and width of the continental shelves. The global geochemical balances of magnesium and sulfate appear to be most affected by removal via hydrothermal processes. Magnesium is removed from seawater by high-temperature weathering at hydrothermal vents. As described in Chapter 21.7, changes in the rate of hydrothermal activity are thought to have altered the Mg/Ca ratio in seawater with high rates leading to greater Mg^{2+} removal. In the case of sulfur, abiogenic sulfate reduction in hydrothermal systems leads to the deposition of pyrite and other metallic sulfides. Alkalinity is also lost from seawater during albitization reactions in the high-temperature zone (Eq. 19.4).

An important series of linkages appears to exist among several of the reactions that occur in hydrothermal systems. These are thought to collectively control P_{CO_2} over time scales of millennia as compared to the feedbacks within the biogenic calcite system, which act over time scales of thousands of years. In this basalt-carbonate buffer system, hydrothermal activity produces minerals, such as calcium carbonate and albite, that, once uplifted onto land, serve as a sink for CO_2 during chemical weathering. A negative feedback is provided if this weathering rate increases with P_{CO_2} levels. Higher P_{CO_2} levels should promote faster reaction rates for two reasons: (1) atmospheric temperatures would be higher and (2) more reactant (H_2CO_3) would be available. Regeneration of the consumed CO_2 is thought to occur during formation of hydrothermal minerals.

Two likely hydrothermal processes that are thought to regenerate CO_2 include the one shown in Eq. 19.4 and the reaction sequence in Eqs. 21.3 through 21.5. This reaction sequence illustrates the intermediate steps involved in the high-temperature hydrothermal albitization and chloritization reactions (Eqs. 19.3 and Eq. 19.1, respectively). In the first stage of these reactions, Ca^{2+} and kaolinite are produced (Eq. 21.3). In the second stage, kaolinite is converted to either albite (albitization) or chlorite (chloritization). During this second stage, H^+ is generated. Some of the H^+ promotes further high-temperature acid-driven weathering of anorthite via Eq. 21.3. A portion of the acid is also consumed in thermogenic pyrite formation (Eq. 21.6). The Ca^{2+} released during production of kaolinite (Eq. 21.3) has two fates. Some precipitates in the hydrothermal system as abiogenic calcite (Eq. 21.7), thereby regenerating CO_2 and the remainder is released into seawater as Ca^{2+}(aq).

$$\underset{\text{anorthite}}{CaAl_2Si_2O_8} + 2H^+ + H_2O \rightarrow Ca^{2+} + \underset{\text{kaolinite}}{Al_2Si_2O_5(OH)_4} \quad (21.3)$$

$$\frac{1}{2}\underset{\text{kaolinite}}{Al_2Si_2O_5(OH)_4} + (Na^+,K^+) + 2SiO_2 \rightarrow \underset{\text{albite}}{(Na^+,K^+)\,AlSi_3O_8} + \frac{1}{2}H_2O + H^+ \quad (21.4)$$

$$Mg^{2+} + \frac{1}{5}\underset{\text{kaolinite}}{Al_2Si_2O_5(OH)_4} + \frac{1}{5}SiO_2 + \frac{7}{5}H_2O \rightarrow \frac{1}{5}\underset{\text{chlorite}}{Mg_5Al_2Si_3O_{10}(OH)_8} + 2H^+ \quad (21.5)$$

$$SO_4^{2-} + 2H^+ + \frac{11}{2}\underset{\text{iron silicate}}{Fe_2SiO_4} \rightarrow \frac{1}{2}\underset{\text{pyrite}}{FeS_2} + \frac{7}{2}\underset{\text{magnetite}}{Fe_3O_4} + \frac{11}{2}SiO_2 + H_2O \quad (21.6)$$

$$Ca^{2+} + \underset{\text{alkalinity}}{2HCO_3^-} \rightarrow \underset{\text{calcite}}{CaCO_3} + CO_2 + H_2O \quad (21.7)$$

Overall, these reactions cause the Ca^{2+} composition of the hydrothermal emissions to be inversely related to that of Mg^{2+}. Thus, as hydrothermal activity increases, the rate of supply of Ca^{2+} to seawater increases and the rate of Mg^{2+} removal increases. The Mg^{2+} concentration in seawater is further lowered by increased rates of dolomitization. This dolomitization takes place in biogenic carbonates deposited on continental shelves. The deposition of these carbonates increases with increasing hydrothermal activity because sea level rises due to an elevation in the topography of the MORs. At high sea levels, the continental edges are flooded, thereby increasing the width of the continental margins. This increases the area over which shallow biogenic carbonates are deposited.

A second mechanism by which CO_2 is regenerated as part of the crustal rock cycle is thought to occur under high pressures and temperatures such as found in subduction zones and under thick sedimentary prisms in the continental rise. This decarbonation process was briefly mentioned in Chapter 19.5.2. Its stoichiometry can be generalized as

$$\underset{\text{biogenic and abiogenic calcite and dolomite}}{\text{“Ca(Mg)CO}_3\text{”}} + \underset{\text{biogenic and abiogenic silica}}{SiO_2} \rightarrow \underset{\text{metamorphic silicate such as wollastonite}}{\text{“Ca(Mg)SiO}_3\text{”}} + \underset{\text{carbon dioxide gas}}{CO_2} \quad (21.8)$$

to accommodate natural variations in the magnesium content of calcite and dolomite. Other types of decarbonation reactions have been proposed to occur in these settings, such as

$$\underset{\text{dolomite}}{5CaMg(CO_3)_2} + \underset{\text{kaolinite}}{Al_2Si_2O_5(OH)_4} + \underset{\text{biogenic and abiogenic silica}}{SiO_2} + 2H_2O \rightarrow \underset{\text{chlorite}}{Mg_5Al_2Si_3O_{10}(OH)_8} + \underset{\text{calcite}}{5CaCO_3} + \underset{\text{carbon dioxide gas}}{5CO_2} \quad (21.9)$$

The resulting CO_2 gas is returned to the atmosphere by two means: (1) volcanic emissions associated with eruptions near subduction zones, i.e., back-arc volcanoes or (2) diffusion through the sediments of the continental rise into the ocean, followed by gas exchange across the air-sea interface. The combined production of CO_2 from these two settings is thought to exceed that from the high-temperature hydrothermal reaction zones.

The metamorphic silicates produced contain the calcium and magnesium that was buried in the sediments as calcite and dolomite. This calcium and magnesium is eventually returned to seawater by terrestrial weathering of the metamorphic minerals. This cycle is shown in Figure 21.1, which also depicts the fate of any sedimentary calcite (limestone) and dolomite that escape metamorphic transformation into silicates. These unreacted carbonates are eventually uplifted onto land and undergo chemical weathering, thereby returning their calcium and magnesium to the ocean. Note that some of the calcite and dolomite is of biogenic origin and some is abiogenic.

In the carbon cycling depicted in Figure 21.1, sedimentary silica plays an important role because is a reactant in the decarbonation reactions. Some of this sedimentary silica is biogenic in origin and some is abiogenic. Sources of the latter include detrital clay minerals and hydrothermal quartz. Terrestrial weathering of the metamorphic silicates returns DSi to the oceans. Thus, the crustal-ocean cycling of silica is linked to that of carbon, magnesium, and calcium.

The hydrothermal chemistry of methane also provides another buffering control on the global biogeochemical carbon cycle by serving as the site of reactions that act as sources and sinks of methane. Examples of source reactions are

$$\underset{\text{magnetite}}{2Fe_3O_4} + 3SiO_2 + 2H_2S + CO_2 \rightarrow \underset{\text{iron silicate}}{3Fe_2SiO_4} + CH_4 + 2SO_2 \quad (21.10)$$

$$\underset{\text{calcite}}{CaCO_3} + CH_4 + H_2S + 5H_2O \rightarrow \underset{\text{anhydrite}}{CaSO_4} + 2CO_2 + 8H_2 \quad (21.11)$$

These redox reactions are abiogenic, whereas the methane sinks are thought to be biogenic, such as the anaerobic oxidation of methane by archaea as observed at the Lost City vent fields (Figure 19.20). Microbial production of methane has also been observed at this site.

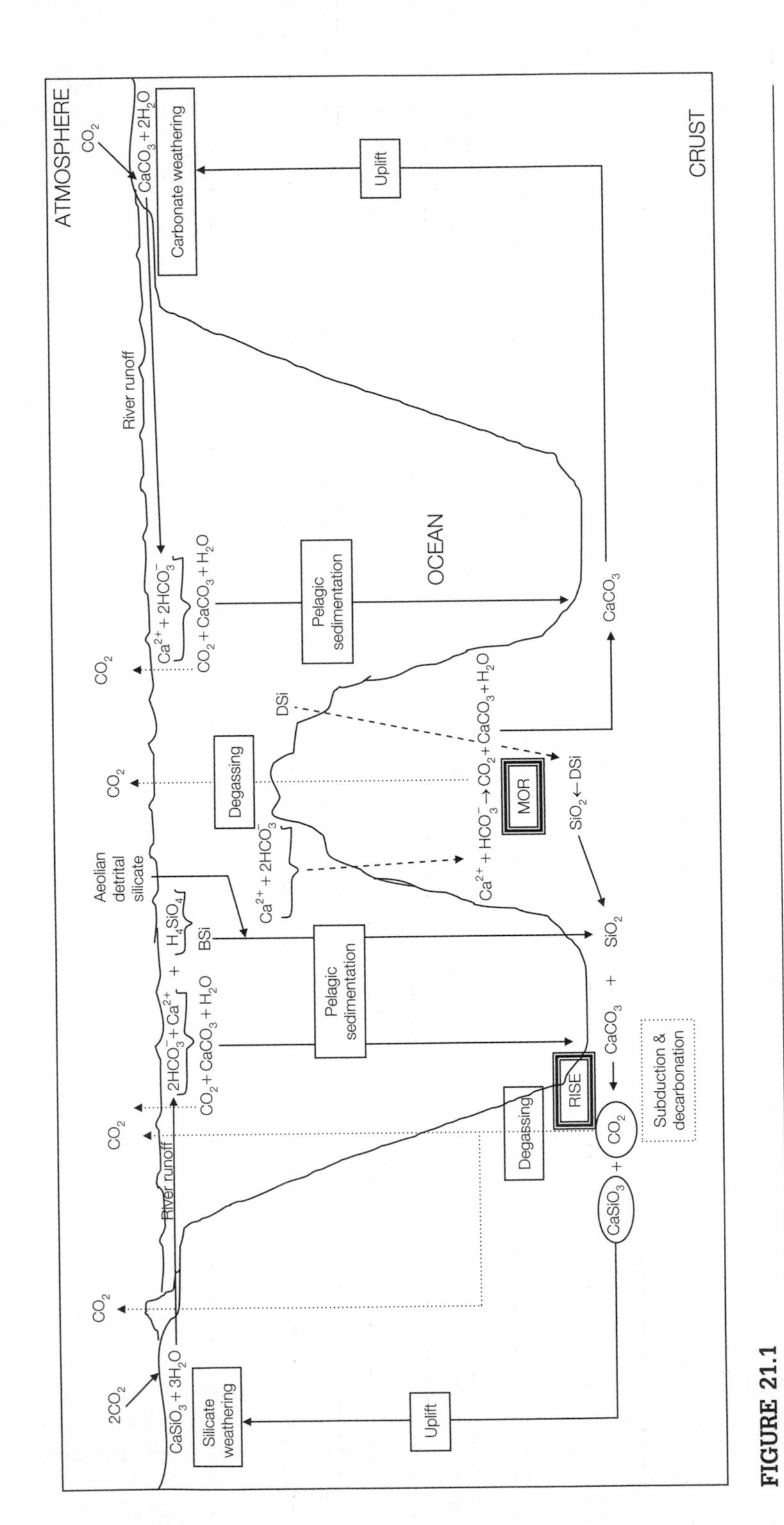

FIGURE 21.1
Linkages between the global carbon and crustal rock cycles.

21.4.5 Cation Exchange

The clay minerals carried by rivers into the ocean represent a net annual addition of 5.2×10^{15} mEq of cation exchange capacity. Most of these exchange sites are occupied by calcium. Within a few weeks to months following introduction into seawater, sodium, potassium, and magnesium displace most of the calcium. As shown in Table 21.7, this uptake removes a significant fraction of the river input of sodium, magnesium, and potassium.

The chemical transfer involved in cation exchange is presented in generic form in Eq. 14.1. The degree to which a cation is exchanged via this process depends on (1) the cation exchange capacity of the clay minerals, (2) the type of exchange sites on the clay, (3) the type of cation being adsorbed, (4) the concentration of the cation, (5) the ionic strength of the solution, and (6) the temperature of the solution. Ion exchange is governed by thermodynamics, as equilibrium is rapidly attained. Thus, an alteration in the cation concentration of seawater will lead to shifts in composition of the adsorbed ions. Because most of the clay minerals introduced into seawater are from river runoff and are deposited on the continental shelf, this cation sink is largely restricted to nearshore waters.

21.4.6 Reverse Weathering

Reverse weathering is thought to occur during halmyrolysis, diagenesis, and catagenesis as clay minerals react with the major ions, bicarbonate, and DSi in seawater. The nature of these reactions is not well known. Their general form is thought to be

$$\begin{aligned}&\text{Cation-poor aluminosilicates} + \text{DSi} + \text{bicarbonate} + \text{cations} \rightarrow \\ &\text{cation-rich aluminosilicates} + \text{carbon dioxide}\end{aligned} \quad (21.12)$$

or possibly

$$\begin{aligned}&\text{Clay mineral containing cation B} + \text{cation A} + \text{bicarbonate} \rightarrow \\ &\text{Clay mineral containing cation A} + \text{cation B} + \text{carbon dioxide} + \text{DSi}\end{aligned} \quad (21.13)$$

During this process, solutes mobilized by terrestrial weathering are returned to the aluminosilicate phase. In the present-day ocean, reverse weathering is thought to be an important sink for potassium. Evidence for rapid incorporation of potassium has been observed in deltaic sediments. In pelagic sediments, downcore gradients in porewater concentrations of potassium and magnesium suggest uptake of these ions that have been attributed to reverse weathering (Figure 21.2). Unfortunately, these reactions occur at very slow rates that are not amenable to laboratory study.

21.4.7 Burial and Subduction of Pore Waters

When sediment settles onto the seafloor, a considerable amount of sediment is trapped between the grains. As discussed in Chapter 12.2.2, pelagic sediments can initially have equal parts of pore water mixed with the solids. As burial progresses, compaction causes the upward vertical advection of pore water, thereby reducing the water content of the

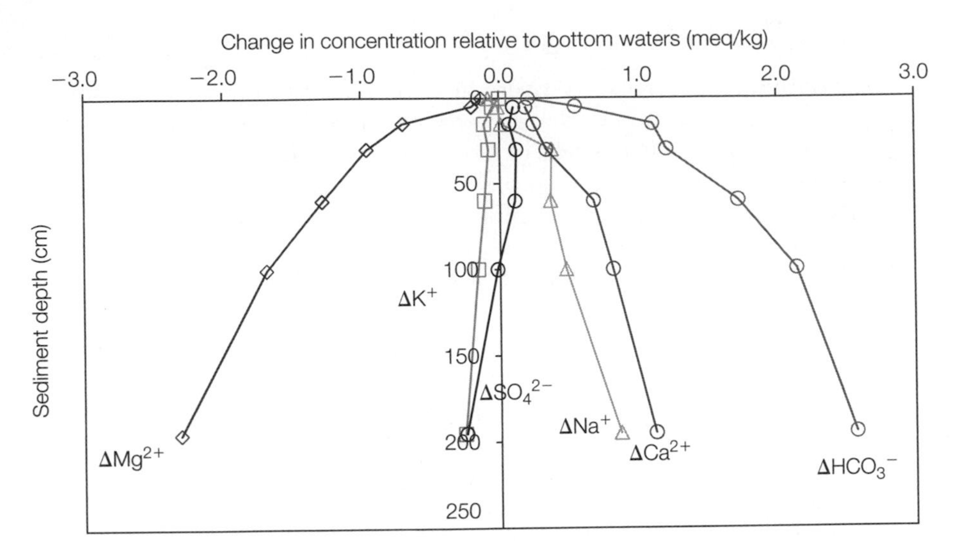

FIGURE 21.2

Change in concentrations relative to bottom waters for K^+, Mg^{2+}, Ca^{2+}, SO_4^{2-}, and HCO_3^- in interstitial waters from a core taken in the Brazil Basin in 4153 m water depth. *Source*: From Sayles, F. L. (1979). *Geochimica et Cosmochimica Acta* 43, 527–545.

sediments over time. Nevertheless, a substantial amount of seawater is retained in the sediments and represents a net loss of salts from the ocean. Estimating the importance of this sink is difficult because the pore-water content and salinity increase with increasing depth in the sediments in a highly variable fashion. As suggested by the estimates given in Table 21.9, the output of interstitial waters is probably a significant sink only for sodium and chlorine.

The oceanic crust is also a sink for seawater as its porosity increases with age because cooling and chemical weathering lead to the continuing development of fractures and fissures. Most of this trapped water is degassed back into ocean when the crust and its overlying sediment undergo subduction.

21.5 THE WHOLE PICTURE: INTEGRATED MODELS OF SEAWATER COMPOSITION

Marine chemists have taken increasingly more sophisticated approaches towards modeling seawater composition. The goal of these models is to understand the biogeochemical controls on seawater composition well enough that the effects of future perturbations can be predicted. As described next, the first modeling efforts were based on a series of reactions that were assumed to reach equilibrium; the next efforts took a steady-state approach as the composition of seawater was thought to have been relatively constant over time.

Table 21.9 Removal of Ions from the Ocean by Burial of Pore Water.

Ion	Annual Removal Rate ($\times 10^{12}$ mol/y)	% of Natural River Input
Cl^-	1.07	18%
Na^+	0.96	11%
Mg^{2+}	0.09	2%
SO_4^{2-}	0.06	2%
K^+	0.02	2%
Ca^{2+}	0.09	1%
Alkalinity	0.06	0.2%

Source: After Drever, J. I., et al. (1988). Chemical Cycles in the Evolution of the Earth, *John Wiley & Sons, p. 36.*

Improvements in analytical techniques have made possible reconstruction of ancient seawater composition from fluid inclusion trapped in marine halites. This has forced marine chemists to accept that the major ion composition has changed significantly—at least over the past 500 million years. Since marine halites older than 500 million years are rare, little is known about the major ion composition of seawater prior to the Phanerozoic eon. Thus, current modeling efforts are directed at simulating changes in seawater composition over the Phanerozoic.

An important generalization from these modeling efforts is that the biogeochemical processes controlling seawater composition are interlinked, such as illustrated in Figure 21.1. These linkages provide multiple opportunities for both positive and negative feedbacks. The time scales over which these interlinked controls act depend on the particular biogeochemical processes involved. For example, the feedbacks involved in calcite compensation (Chapter 15.6) stabilize seawater alkalinity and calcium over time scales of thousands of years, whereas the basalt-carbonate buffer acts over time scales of millennia.

21.5.1 Equilibrium Models

One of the first attempts at a model that explained the chemical composition of seawater was presented by Sillen in 1961. In this model, the major ion concentrations are assumed to be controlled by reactions that reach equilibrium. This control is suggested by the relatively homogeneous composition of seawater across the ocean basins and the long residence times of the major ions as compared to the much shorter turnover times of water and sinking particles. In Sillen's model, the cations are assumed to be supplied by igneous rocks. These reactants are represented in the equilibrium equations as SiO_2, Al_2O_3, NaOH, KOH, MgO, and CaO. The anions are assumed to be supplied as the magmatic volatiles: HCl, H_2O, and CO_2. In attaining equilibrium with seawater, these substances are assumed to have undergone reactions that have led their component elements to become redistributed among several gaseous, aqueous, and solid species commonly found in the oceans. The latter include sedimentary quartz, kaolinite, chlorite, illite, montmorillonite, calcite, and phillipsite. In this model, these solid phases act as the major sinks of Si, Al, Mg, K, Ca, and Na, respectively. Some of the equilibrium reactions are reverse weathering reactions (Eq. 21.12 and 21.13) that generate secondary clays and are presumably occurring in the sediments.

A unique solution for the equilibrium concentrations of each ion is obtained by fixing the temperature and chloride concentration. The resulting atmospheric level of CO_2 can also be calculated. An example of the numerical solution to this multicomponent equilibrium concentration calculation is shown in Table 21.10. The predicted major ion concentrations are close to the observed values. Nevertheless, this model is not widely accepted as realistic because little evidence has been found for the establishment of equilibria between seawater and the solid phases. In fact, concentration gradients in the bottom and pore waters suggest that equilibrium is not being attained (Figure 21.2). This model is also not able to predict chloride concentrations because the major sedimentary component (halite) is nowhere near saturation with respect to average seawater.

21.5.2 Steady-State Mass Balance Models

If the chemical composition of seawater has remained constant over time, a steady-state balance should exist in which the total supply rate of a particular ion is matched by its

Table 21.10 Results and Components in Kramer's Ocean Model.

Observed Concentration (M) at 25°C and 35‰	Equilibrium Concentration (M)	Aqueous Phase	Solids Controlling Concentrations
0.48	0.45	Na^+	Na-mont (cation exchange)
1.0×10^{-2}	9.7×10^{-3}	K^+	K-illite (cation exchange)
0.56	0.55 (defined)	Cl^-	0.55 (assumed)
2.9×10^{-2}	3.4×10^{-2}	SO_4^{2-}	$SrCO_3$, $SrSO_4$
1.1×10^{-2}	6.1×10^{-3}	Ca^{2+}	Phillipsite
5.4×10^{-2}	6.7×10^{-2}	Mg^{2+}	Chlorite
Variable	2.7×10^{-6}	PO_4^{3-}	OH-apatite
—	(1.7×10^{-3} atm)	CO_2	Calcite
7.0×10^{-5}	2.4×10^{-5}	F^-	F-CO_3-apatite
7.9×10^{-9}	4.7×10^{-9}	H^+	Given by electroneutrality

Source: *After Kramer, J. R. (1965)* Geochimica et Cosmochimica Acta *29, 921–945.*

total output rate. This balance does not require that the supply and output pathways be reversible, as is required by the equilibrium model. This is far more realistic for seawater. For example, ions supplied to the ocean by river runoff are not removed by any kind of riverine process. The geochemists who first tried constructing a steady-state model for the major ions in seawater had two goals in mind. First, they wanted to identify any unknown sources or sinks. Secondly, they wanted to evaluate the relative importance of the various ion supply and removal mechanisms in maintaining the steady-state composition of seawater. The results of such a calculation were first presented by Mackenzie and Garrels in 1966 and are reproduced in Table 21.11.

In this calculation, the major ion concentrations are assumed to remain constant over time as a result of processes that remove all of their input. The latter is assumed to be wholly from river runoff (the top row of Table 21.11). The time scale chosen for this model is 100 million years, which is somewhat larger than the longest residence time of the major ions. Thus, this time scale should be sufficient for all the removal processes to exert their full effect in maintaining a steady-state balance. The removal reactions are presented as numbered processes. For bookkeeping purposes, the effect of each type of mineral deposition is recorded as a deduction from the total amount of ions that must be removed. For example, reaction process 1 represents the deposition of pyrite, which results in removal of half of the sulfate input. The rest of the removal reactions represent the deposition of calcite, halite, dolomite, quartz, and anhydrite. After deducting the ions consumed in these processes, excess Na^+, Mg^{2+}, and K^+ remain. Thus, another series of sinks must be invoked to attain steady state. In this model, the missing sink reactions are assumed to be those involving reverse weathering.

To remove the excess Na^+, Mg^{2+}, and K^+ via reverse weathering requires the in situ production of montmorillonite, chlorite, and illite at rates of 20×10^{14}, 5.6×10^{14}, and 14×10^{14} g/y, respectively. This represents about 25% of the annual input of mineral solids to the ocean (river + aeolian). Hence, this model requires that a large fraction of the clay minerals in the ocean be transformed into secondary clays. Since reverse weathering acts on preexisting detritus, little change in the total mass of the clay minerals should occur. Given their structural similarity to detrital clays, the presence of these secondary clays should be difficult to detect. Not surprisingly, little direct evidence has been found to demonstrate their presence in significant amounts.

It is possible that a type of reverse weathering is occurring in unrecognized mineral phases or locations. For example, the incorporation of potassium into detrital clays could be producing green clays, like glauconite, rather than illite. As described in Chapter 18.3.4, glauconite is an amorphous iron-rich clay thought to form under reducing conditions in microenvironments of fecal pellets or the interiors of foram tests and diatom frustules. If most glauconite is authigenic, it could represent a sink for as much as 20% of the riverine flux of potassium, as well as 3% of the riverine flux of magnesium. In the case of K^+, uptake into the rapidly accumulating sediments of the Amazon River delta by incorporation into green clays has been identified as a significant global scale sink representing 7 to 10% of the river input of this major ion.

The need for invoking high rates of cation removal via reverse weathering was eliminated in the late 1970s when marine chemists realized that hydrothermal uptake and

Table 21.11 Mass Balance Calculation for the Removal of River-Derived Constituents from the Ocean.

Step no.	Reaction (balanced in terms of mmol of constituents used)	Constituent balance ($\times 10^{21}$ mmol)								HCO_3^-	CO_2	Products		
		SO_4^{2-}	Ca^{2+}	Cl^-	Na^+	Mg^{2+}	K^+	SiO_2	HCO_3^-	Consumed (−) Evolved (+)	Consumed (−) Evolved (+)	($\times 10^{21}$ mmol)		Total products formed (mol basis) (%)
Amount of material to be removed from ocean in 10^8 years		382	1220	715	900	554	189	710	3118					
1.	95.5 $FeAl_6Si_6O_{20}(OH)_4$ + 191 SO_4^{2-} + 47.8 CO_2 + 55.7 $C_6H_{12}O_6$ + 238.8 H_2O = 286.5 $Al_2Si_2O_5(OH)_4$ + 95.5 FeS_2 + 382 HCO_3^-	191	1220	715	900	554	189	710	3500	+382	−48	96 287	Pyrite Kaolinite	3 8
2.	191 Ca^{2+} + 191 SO_4^{2-} = 191 $CaSO_4$	0	1029	715	900	554	189	710	3500			191	$CaSO_4$	5
3.	52 Mg^{2+} + 104 HCO_3^- = 52 $MgCO_3$ + 52 CO_2 + 52 H_2O	0	1029	715	900	502	189	710	3396	−104	+52	52	$MgCO_3$ in magnesium calcite	2
4.	1029 Ca^{2+} + 2058 HCO_3^- = 1029 $CaCO_3$ + 1029 CO_2 + 1029 H_2O	0	0	715	900	502	189	710	1338	−2058	+1029	1029	Calcite and/or aragonite	29
5.	715 Na^+ + 715 Cl^- = 715 NaCl	0	0	0	185	502	189	710	1338			715	NaCl	20
6.	71 H_4SiO_4 = 71 $SiO_2(s)$ + 142 H_2O	0	0	0	185	502	189	639	1338			71	Free silica	2

7.	138 $Ca_{0.17}Al_{2.33}Si_{3.67}O_{10}(OH)_2$ + 46 Na^+ = 138 $Na_{0.33}Al_{2.33}Si_{3.67}O_{10}(OH)_2$ + 23.5 Ca^{2+}	0	24	0	139	502	189	639	1338			138	Sodic montmorillonite	4
8.	24 Ca^{2+} + 48 HCO_3^- = 24 $CaCO_3$ + 24 CO_2 + 24 H_2O	0	0	0	139	502	189	639	1200	−48	+24	24	Calcite and/or aragonite	1
9.	486.5 $Al_2Si_{2.4}O_{5.8}(OH)_4$ + 139 Na^+ + 361.4 SiO_2 + 139 HCO_3^- = 417 $Na_{0.33}Al_{2.33}Si_{3.67}O_{10}(OH)_2$ + 139 CO_2 + 625.5 H_2O	0	0	0	0	502	189	278	1151	−139	+139	417	Sodic montmorillonite	12
10.	100.4 $Al_2Si_{2.4}O_{5.8}(OH)_4$ + 502 Mg^{2+} + 60.2 SiO_2 + 1004 HCO_3^- = 100.4 $Mg_5Al_2Si_3O_{10}(OH)_8$ + 1004 CO_2 + 301.2 H_2O	0	0	0	0	0	189	218	147	−1004	+1004	100	Chlorite	3
11.	472.5 $Al_2Si_{2.4}O_{5.8}(OH)_4$ + 189 K^+ + 189 SiO_2 + 189 HCO_3^- = 378 $K_{0.5}Al_{2.5}Si_{3.5}O_{10}(OH)_2$ + 189 CO_2 + 661.5 H_2O	0	0	0	0	0	0	29	−42	−189	+189	378	Illite	11

Source: *After Mackenzie, F. T., and R. M. Garrels (1966).* American Journal of Science *264, 507–525.*

cation exchange were both significant in global elemental budgets. In the cases of magnesium and sulfur, current estimates of removal by hydrothermal systems (and for Mg, the additional effect of ion exchange) are large enough to also eliminate the need for invoking high rates of dolomite and pyrite deposition. A revised version of the Garrels and Mackenzie mode- ling approach is presented in Table 21.7 incorporating the recognition that magnesium, sulfate, and calcium do not currently appear to be in a steady state.

21.5.3 Dynamical Models

In the 1990s, marine chemists realized that the major ion composition of seawater has undergone significant shifts over the past 500 million years. Pale- oceanographers now agree that the marine cycles of magnesium, sulfate, and calcium do not appear to be currently at a steady state. These nonsteady-state conditions have been ascribed to changing rates of seafloor spreading and hydrothermal activity. As described in Chapter 21.4.4, these processes control the chemistry of hydrothermal emissions, sea level, and the length and width of the continental shelves. Figure 21.1 illustrates how these processes interlink with the carbon and silica cycles leading to controls on P_{CO_2}. Similar complicated linkages involve the deposition and weathering of pyrite, evaporites, apatite, iron oxides, and various clay minerals. These linkages create feedbacks, some positive and some negative, that are difficult to predict intuitively.

To adequately model how seawater's composition has changed over time requires simultaneously tracking the controls on the major ions (sodium, potassium, magnesium, calcium, chlorine, and sulfur) along with the ones acting on carbon, oxygen, iron, and phosphorus. Examples of such comprehensive models include MAGic (Mackenzie, Arvidson, Guidry Interactive Cycles), GEOCARB, BLAG (Berner, Lasaga, and Garrels), and COPSE (Carbon-Oxygen-Phosphorus-Sulfur-Evolution). In these models, fluxes between reservoirs; are represented as differential equations composed of terms representing fluxes generated by specific biogeochemical processes and time-dependent forcing functions. Note that primary production must be included because of its significant influence on the carbon, oxygen, calcium, phosphorus, and iron cycles.

Some of the forcing functions that have been included in these dynamical models are: (1) the size of the terrestrial organic matter reservoir; (2) weathering and runoff, which depend on rates of uplift, topographic relief, land area, lithology, terrestrial plant activity, and climate; (3) the area of continental shelves; (4) the evolution of planktonic calcifiers; and (5) subduction and seafloor spreading. The intensity of these forcing functions can change over time and, thus, must be included in the differential equations as time-dependent factors. The most ambitious of these models, MAGic, has 40 coupled nonlinear ordinary differential equations containing more than 100 fluxes and time-dependent forcing functions. Future efforts are planned to include additional fluxes and forcing functions, such as the mixing dynamics of the ocean. Although still considered to be crude, these models have greatly improved our understanding of how changes in the forcing functions have altered the major ion composition of seawater. Some of these insights are summarized in Chapter 21.7.

21.6 MEAN OCEANIC RESIDENCE TIMES AND REACTIVITY

Most of the major ion inputs to the ocean are derived from secondary sources, primarily from the terrestrial weathering of sedimentary and metamorphic rocks. Given the age of the Earth (4600 million years) and that an average Wilson cycle takes 250 million years, the crust has probably passed through the global rock cycle many times ($4600/250 \approx 18$ times). This leads geochemists to surmise that when considered on time scales greater than several Wilson cycles, the sedimentary reservoirs have likely achieved a steady state. Since element partitioning into the sediments is crucial in determining the major ion composition of seawater, the oceans would similarly be expected to achieve a steady state over long time scales. These hypotheses are supported by the fairly homogenous chemical composition of the various sedimentary rock types. The persistence of fundamentally similar life forms over hundreds of millions and even billions of years also suggests that the chemical composition of seawater has been fairly stable for long periods. The largest possible excursion in salinity that has occurred is thought to be no more than a factor of 2, this being the increase that would result from the complete dissolution of all known evaporites. On the other hand, over time scales less than a few Wilson cycles, excursions from steady state are possible. This appears to presently be the case for magnesium, sulfate, and calcium (Table 21.7).

On time scales over which the major ions attain a steady state, a residence time can be calculated using Eq. 1.1. Note that for all the major ions, the dominant input is from rivers making their residence times equivalent to their turnover times with respect to river runoff (Table 21.8). In the cases of sodium and chloride, geochemists usually subtract the seasalt component from the river fluxes to provide an estimate of cycling time through the entire crustal-ocean-atmosphere factory for these elements. Given the broad range of estimates for the percentage of seasalt contribution to river water (Table 21.6), these corrected residence times range from 150 to 900 million years. Recall that the long residence times of the major ions (>1 million years) give rise to their conservative behavior.

The residence times of the elements follow periodic trends that reflect their tendency to undergo reactions leading to their removal from seawater (Figure 11.5). In some cases, elements are so reactive that their residence times are much shorter than the mixing time of the ocean. For example, aluminum has a residence time of about 100 to 200 y. For such elements, a true residence time cannot be defined. Nevertheless, these "computed" residence times provide insight into an element's relative reactivity in the crustal-ocean-atmosphere factory. Geochemists call these computed values *mean oceanic residence times* (MORTs) to clarify that various assumptions are involved that might not be met, such as the ocean being well mixed with respect to an element.

The relationship between the MORT and the chemical reactivity of an element (A) can be defined as follows (this is similar to the presentation in Chapter 1.3.1). If $J_{A_{in}}$ and $J_{A_{out}}$ represent the total input and removal rates of element A, respectively, then

$$\frac{dM_A}{dt} = V\frac{dC_A}{dt} = J_{A_{in}} - J_{A_{out}} \tag{21.14}$$

where M_A equals the total amount of element A in the ocean, C_A is the mean concentration of A in seawater, and V is the volume of the ocean, which is assumed to be constant over time. If the removal rate of element A is assumed to be directly proportional to its concentration in seawater, then $J_{A_{out}} = k_A C_A$, where k_A is the net rate constant for removal. At steady state,

$$0 = V \frac{dC_A}{dt} = J_{A_{in}} - kC_A \tag{21.15}$$

or

$$\frac{1}{k} = \frac{C_A}{J_{A_{in}}} = \text{MORT}_A \tag{21.16}$$

Thus, for an element whose removal from seawater follows first-order reaction kinetics, its MORT is the inverse of its removal rate constant. This relationship predicts that reactive elements should have short residence times. As shown in Figure 21.3, the actual data do demonstrate a linear relationship ($r = 0.79$, $p = 0.00$), although a log-log plot is required to cover the several orders of magnitude diversity of MORT and concentrations exhibited by the solutes in seawater. A similar relationship exists between the MORT and the seawater-crustal rock partition coefficient (K_Y). The latter is defined as the ratio of the mean seawater concentration of an element to its mean concentration in crustal rocks. Elements with high partitioning coefficients would be expected to have low seawater concentrations. As shown in Figure 21.4, this is seen in the data and

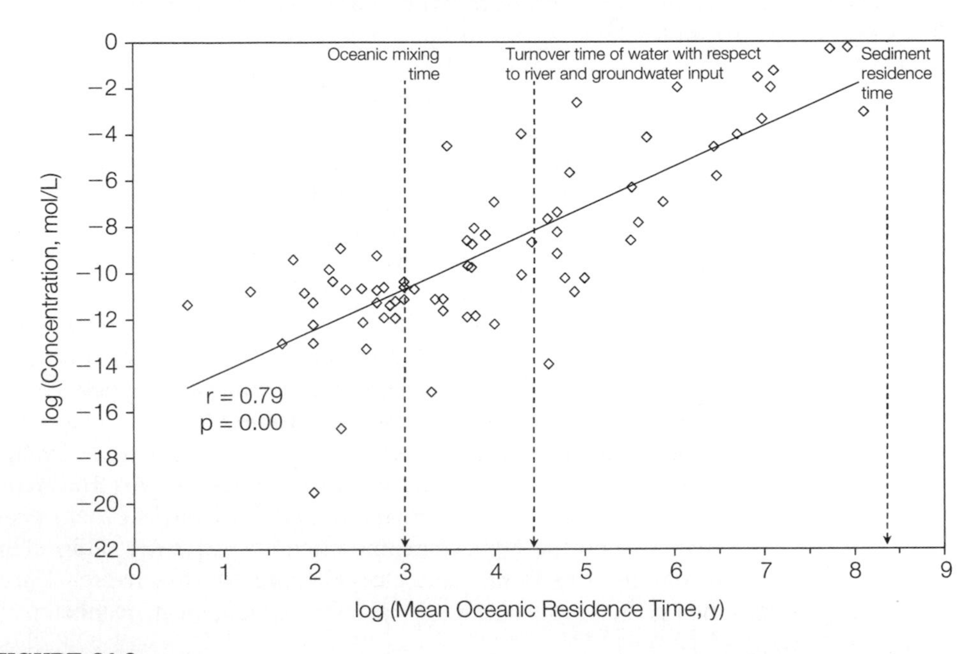

FIGURE 21.3

Mean seawater concentration versus mean oceanic residence time. Note that this is a log-log plot.

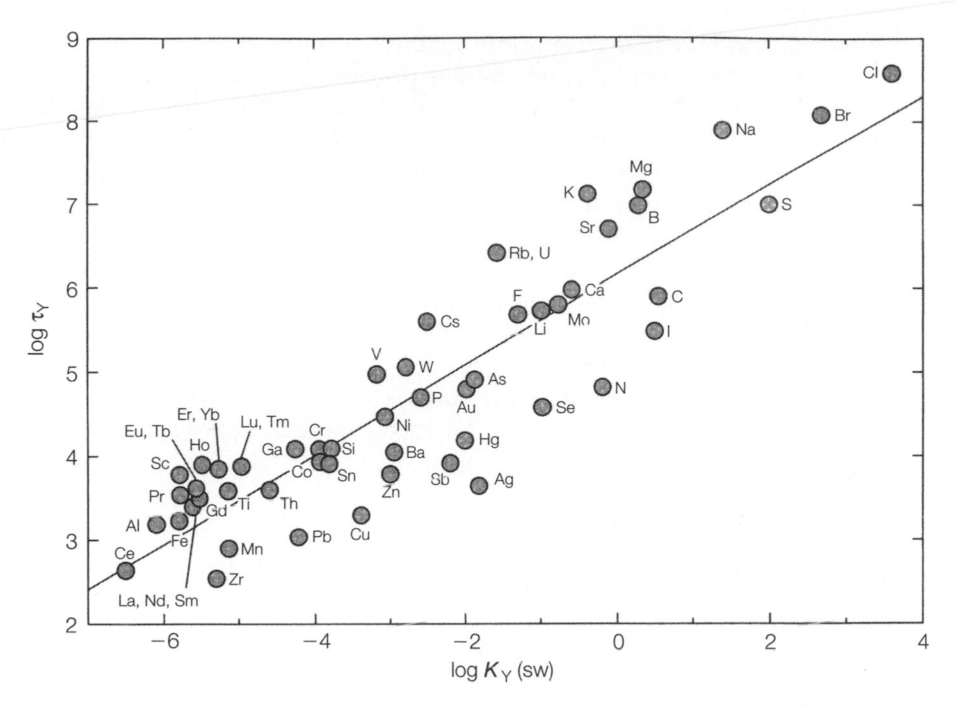

FIGURE 21.4

Seawater-crustal partition coefficient (K_Y(sw)) versus mean oceanic residence time (τ_Y). Note that this is a log-log plot. *Source*: From Turner, D. R., *et al*. (1980). *Marine Chemistry* 9, 211–218.

confirms the critical role of the sediments in determining the chemical composition of seawater.

An element's tendency towards partitioning into solid phases is related to its fundamental atomic properties. These properties follow periodic trends, giving rise to the trends shown in Figure 11.5 for the oceanic residence times of the elements. In the broadest sense, these trends reflect the relative tendency of an element towards electrostatic versus covalent interactions. Chemists have devised various measures of this tendency, such as an element's electronegativity and its ionization potential. The latter is a predictor of electrostatic interactions and is defined as the ratio, z^2/r, where z is the ionic charge of an ion and r is the ionic radius. In the case of the metals, z^2/r provides an estimate of an ion's propensity to form ionic bonds. For elements that are susceptible to covalent interactions, reactivity is best predicted by also considering their electronegativity, which is defined as the power of an atom in a molecule to attract electrons to itself. (Strictly considered, the electronegativity of an atom depends on its oxidation state and the energy levels of the valence electron(s) involved in the covalent interaction.)

General speciation behaviors, such as those summarized in Table 5.9, can also be predicted from an element's ionization potential and electronegativity. For example, the cations in seawater that have a low ionization potential and electronegativity tend to be unreactive and therefore are present primarily as their free (hydrated) ions. These include Na^+, K^+, Ca^{2+}, and Mg^{2+}. Because of their low reactivities, they have long residence times. The trace metals that have high ionization potentials and low-to-moderate electronegativities, such as iron, manganese, and aluminum, undergo hydrolysis. They are present primarily as hydroxide and oxide ion complexes. These molecules have limited solubility, causing these metals to have short residence times. Elements with a moderate-to-high degree of covalent character and low-to-moderate ionization potential tend to form soluble ion complexes, such as copper and cadmium with chloride and carbonate. Because of the solubility of their ion complexes, these elements tend to have intermediate residence times.

For elements that have multiple sources or sinks, a fractional residence time (τ_i), or turnover time, can be calculated for each supply or removal process. The residence time of the element (τ) is then given by

$$\frac{1}{\tau} = \sum_{i=1}^{i=n} \frac{1}{\tau_i} = \frac{1}{\tau_1} + \frac{1}{\tau_2} + \cdots \frac{1}{\tau_n} \tag{21.17}$$

This equation illustrates that the residence time will always be somewhat less than the process with the fastest turnover time. In the cases of sodium and chlorine, removal of the cyclic seasalt component significantly increases the computed MORTs for these elements.

21.7 EVOLUTION OF SEAWATER

A discussion of how the chemical composition of seawater is thought to have evolved since the Precambrian to the modern day is provided in the supplemental information for Chapter 21 available at **http://elsevierdirect.com/companions/9780120885305**.

21.8 WHY IS THE SALINITY OF SEAWATER AROUND 35‰?

As described in Chapter 21.7, a system of biogeochemical feedbacks act to stabilize the major ion composition of seawater. Some operate on short time-scale cycles, such as calcite compensation, and others operate over longer periods, such as the basalt-carbonate buffer. The linkages in the crustal-ocean-atmosphere factory that act on the major ions also influence atmospheric CO_2 levels and seawater's pH and alkalinity.

So why does seawater have an average salinity of 35‰ rather than some other value, such as 3.5 or 350‰? And why are sodium and chloride so much more abundant than

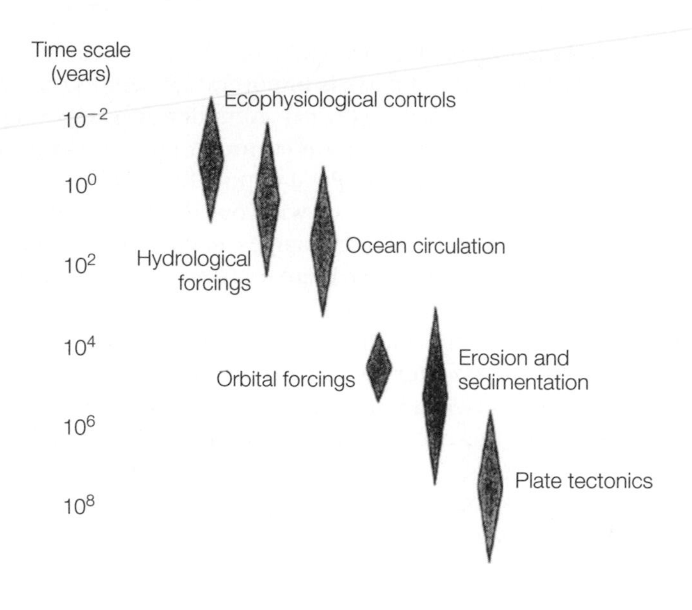

FIGURE 21.5

Examples of forcings of biogeochemical cycles and their characteristic time scales. *Source*: From Van Cappellen, P. (2003). *Biomineralization*, Reviews in MIneralogy and Geochemistry, 54, 357–381.

the other major ions? The answer lies partially in how the forcing functions, such as seafloor spreading and climate, vary over time. An important consideration is that these forcing functions also act over different time scales as shown in Figure 21.5. Note that these forcing functions have had large impacts on the marine biota, either by stimulating biological evolution or by causing mass extinction events. In both cases, changes in the abundance and diversity of the marine biota have altered marine cycling of many key elements.

The other reason why the average salinity of seawater is 35‰ lies in the fundamental chemistry of major ions. For example, the sevenfold increase in the Na^+/K^+ ratio between river water and seawater (Table 21.8) reflects the lower affinity of marine rocks for sodium as compared to potassium. In other words, the sodium sink is not as effective as the one for potassium. Thus, more sodium remains in seawater, with its upper limit, in theory, being controlled by the solubility of halite. Likewise, the Ca^{2+}/Mg^{2+} ratio in seawater is 12-fold lower than that of river water due to the highly effective removal of calcium through the formation of biogenic calcite.

Whereas the chemical nature of the major ions is fixed, the rates and nature of the forcing functions can change. An important consequence of such changes are alterations in the chemical composition of the sedimentary sources and sinks. For example, changes in the type of organisms, as a result of evolution and changing habitats, can lead to a shift in the chemical composition of biogenic particles. Similarly, a large enough population shift has the potential to alter seawater composition. For example, an increase in the rate of upwelling creates nutrient-rich conditions, which favor diatoms over coccolithophorids. Ballasting by hard parts is important in determining the efficiency of POM export to the sediments. Thus, biological shifts that lead to changes in the production of hard parts and/or their sinking rates, through changes in particle size, shape, and density, can result in changes in the POM export flux. This, in turn, affects atmospheric O_2 and CO_2 levels. In Chapter 25.4, we will consider further some of the global tectonic events that have caused major biotic changes, such as the development of hard-bodied life forms at the beginning of the Phanerozoic and the massive extinction events of the Permian.

While negative feedbacks within the crustal-ocean-atmosphere factory promote stability, their capacity to counter the effects of perturbations are not infinite. In other words, they can be overwhelmed by large-scale, long-term perturbations, leading to a permanent change in the chemical composition of seawater. As shown in Table 21.1, human activities are significantly increasing the rates of riverine input of the major ions, particularly for sodium, chloride, and sulfate. Our use of salt as a food additive and road deicer are responsible for the elevations of sodium and chloride. Sulfate levels are increasing due to burning of high-sulfur coal. On the other end of the spectrum, trapping of sediments behind dams is thought to be causing a decrease in the rate of delivery of silica to the oceans. Because of our imperfect understanding of the feedbacks in the crustal-ocean-atmosphere factory, we are presently unable to predict the effects of these increased rates on the rest of the system.

PART 4

Organic Biogeochemistry

CHAPTER

Marine Biogeochemistry: An Overview 22

All figures are available on the companion website in color (if applicable).

22.1 INTRODUCTION

Thus far, we have been using a simple model compound, $C_{106}(H_2O)_{106}(NH_3)_{16}$ PO_4, to represent an average molecule of organic matter. In reality, marine organic matter is composed of a great variety of molecules, ranging from low-molecular-weight hydrocarbons, such as methane, to high-molecular-weight complexes, called humic substances (molecular mass = 500 to 10,000). The marine biogeochemistry of specific organic molecules is the subject of Part IV of this text.

Organic compounds are molecules that contain carbon, with the exception of the simple oxides (CO, CO_2, HCO_3^-, H_2CO_3, CO_3^{2-}), elemental forms (diamonds and graphite), and carbonate minerals. In addition to exhibiting a wide range in molecular mass, organic compounds vary greatly in structure and elemental composition. Most contain oxygen and hydrogen. Nitrogen, sulfur, and phosphorus can also be present, but usually in lesser amounts. Some compounds form coordination complexes with metals, and others contain halides.

The ultimate source of a large fraction of the organic compounds found in seawater are the metabolic reactions conducted by marine organisms. As a result, their marine distributions are largely controlled by biological processes as depicted in Figure 8.1. Since equilibrium is rarely achieved in biologically mediated reactions, the distribution of organic compounds is controlled by kinetic, rather than thermodynamic, considerations. Abiotic reactions are important in some settings, such as surface waters where UV radiation is an energy source, deeply buried sediments, and hydrothermal systems.

Organic compounds constitute a relatively small elemental reservoir. Nevertheless, they have great influence on important parts of the crustal-ocean-atmosphere factory. This chapter provides an overview of the analytical and experimental approaches used in the study of marine organic biogeochemistry along with a description of the molecular structures of the most abundant classes of marine organic compounds. This information is used throughout Part IV to discuss the marine carbon, nitrogen, phosphorus, and sulfur cycles. The impacts of humans on these cycles is covered along with insights obtained from the paleoceanographic record of marine organic compounds buried in the sediments. At the end of Part IV in Chapters 27 and 28, two very important applications of organic biogeochemistry are discussed, namely the origin of petroleum hydrocarbons and the discovery of marine natural products.

22.2 THE IMPORTANCE OF ORGANIC COMPOUNDS IN THE MARINE ENVIRONMENT

The ultimate origin of most marine organic compounds is through metabolic synthesis conducted by marine organisms. While the organisms are living, these compounds can be released into seawater in solid and dissolved forms as a result of excretions or exudations. Examples of the solids include fecal pellets, cast-off feeding nets, and molts. Organic compounds are also released into seawater after the death of an organism, because of lysis of cell membranes. Once in seawater, most of these biomolecules are rapidly degraded by microbes.

A small fraction of the organic molecules in seawater is generated abiotically as a result of: (1) photochemical reactions in the surface waters, (2) high-temperature reactions in hydrothermal systems, and (3) spontaneous condensation and polymerization reactions among degraded biomolecules. The last process generates rather nonreactive, high-molecular-weight, complex compounds called humic substances. This process is thought to occur throughout the water column and in the sediments.

Organic compounds play several important roles in the crustal-ocean-atmosphere factory. Most importantly, they constitute the tissues of all organisms. In some cases, they also serve as structural support, i.e., the chitinous exoskeletons of crustacea. Organic compounds are the food source for heterotrophs. They also serve as communication and defense agents. In the case of animals, many are known to excrete pheromones into seawater to attract mates. Some phytoplankton, namely diatoms, excrete low-molecular-weight aldehydes to reduce the reproductive success of their grazers, copepods. As noted in Chapter 5.7.5, some plankton secrete organic compounds that complex with metals. Depending on the metal and the organic ligand, complexation can act to enhance bioavailability or reduce toxicity. Even bacteria use chemical signals. For example, some bacteria synchronize their metabolic and reproductive behaviors through excretion and detection of short-chain oligopeptides. Some bacteria release extracellular enzymes into seawater, such as proteases, to predigest food. Among this great diversity of marine biomolecules, many exhibit a high degree of physiological activity. Some of these have proven useful to humans, such as for drugs, cosmetics, and food additives. This subject is considered further in Chapter 28.

Dissolved organic matter adsorbs onto particle surfaces. As a result, all particulate matter in seawater eventually becomes coated with a layer of adsorbed organic compounds. This coating has a net negative charge and, hence, adsorbs cations. Since the particulate matter eventually sinks to the seafloor, these organic coatings collectively remove a significant amount of cations, primarily trace metals, from seawater. The coatings are also important as a food source to microorganisms. As a result, bacteria, protozoa, and fungi are commonly found growing on particles. Tunicates and sponges will colonize sufficiently large surfaces, such as the hulls of ships and piers. The progressive overgrowth of marine organisms on surfaces is termed *biofouling*. This is a very expensive problem for ships as biofouling increases drag on the hulls, thereby increasing the amount of fuel required to reach desirable speeds. To prevent this overgrowth, ship hulls have been painted with toxic compounds, such as tributyltin. Unfortunately, some of these chemicals dissolve in seawater and have lethal effects on nontarget organisms.

Organic matter also affects the mechanical properties of sediments. Sedimentary organic matter tends to increase cohesion between particles, thereby making the bulk sediment resistant to resuspension. It is also the sole food source for the heterotrophic members of the benthos. As described in Chapter 12.2.3, the feeding and burrowing activities of the benthos alter the physical, chemical, and geological characteristics of the sediment. For example, deposit feeders change the size distribution of particles by packaging of nonnutritive detritus into fecal pellets. Some types of bioturbation homogenize chemical gradients and increase the supply of O_2 to the sediments. The latter enhances the oxidation rate of sedimentary organic matter. Other types of bioturbation, such as the construction of burrows and feeding tubes, increases chemical and geological heterogeneity in the sediments (Figure 12.4).

Sedimentary organic matter is of particular interest to humans for three reasons. First, much of the economically useful deposits of petroleum are marine in origin, being the degradation products of sedimentary organic matter that has undergone diagenesis and catagenesis. Second, the steady-state balance of nutrients, O_2, and CO_2 in the crustal-ocean-atmosphere factory appears to be stabilized by feedbacks that involve the burial of organic matter in the sediments, primarily on the continental margins. Third, downcore variations in the quantity and types of sedimentary organic compounds provide a paleoceanographic record of changes in such phenomena as primary production, water circulation, and atmospheric chemistry. Some organic compounds are used to date sediments and to serve as stratigraphic markers.

22.3 CONCEPTUAL AND ANALYTICAL APPROACHES USED TO STUDY THE MARINE BIOGEOCHEMISTRY OF ORGANIC COMPOUNDS

Because of ongoing improvements in analytical technology, marine chemists can now detect and quantify a bewildering number of organic compounds. Nevertheless, many substances remain uncharacterized. Of the compounds that have been identified, most are present in low concentrations, usually less than 1 ppm. The analytical and experimental strategies that marine biogeochemists use to detect, quantify, and study marine organic matter are described in this section.

22.3.1 Analyses of Operationally Defined Fractions

Chemical analyses of marine organic matter usually starts with a segregation into the solid, dissolved, or colloidal phase. As described in Chapter 3.2, this segregation often

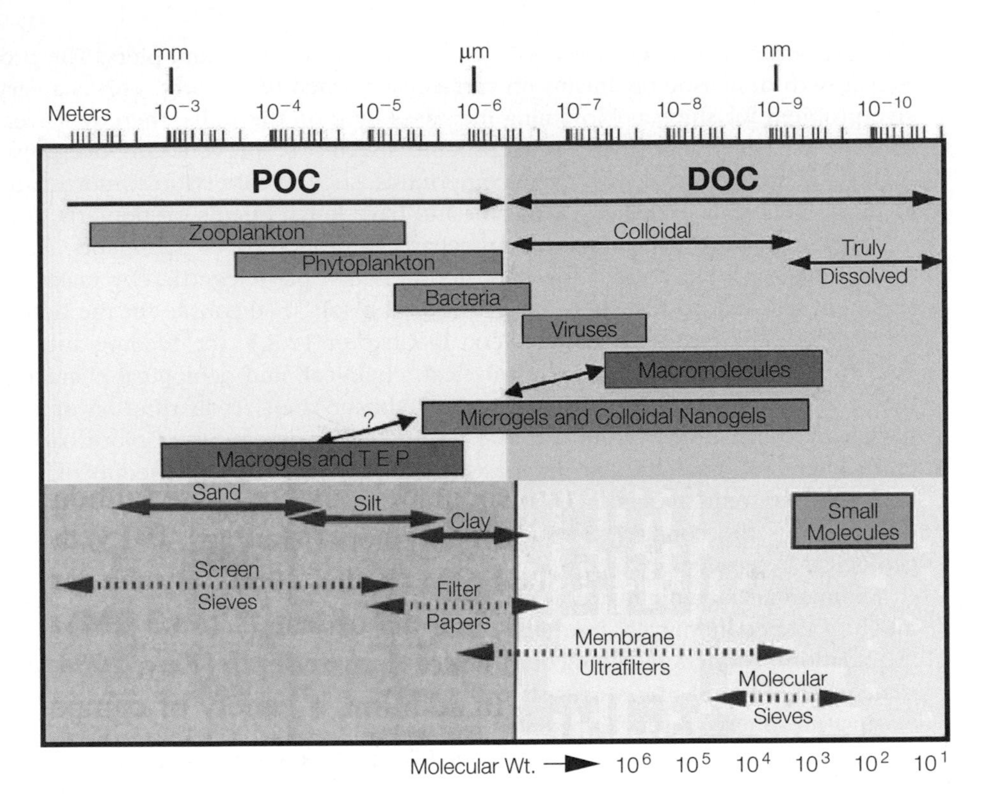

FIGURE 22.1

Size classes of organic compounds. *Source*: After Verdugo, P., *et al*. (2004). *Marine Chemistry* 92, 67–85.

involves some kind of mechanical filtration in which DOM is defined as the fraction that passes through the filter, and POM, as the fraction retained by the filter. An intermediate fraction, the colloids, is generally defined as ranging from 0.001 to 0.5 μm in diameter.[1] The upper end of this range has some functional importance as particles smaller than 1 μm are not prone to sink. Colloids are isolated from seawater by a process called cross-flow, or tangential-flow, filtration. Particles less than 1 nm in diameter are generally termed *true solutes*. This is an important size distinction as the *true solutes* are small enough to diffuse through cell membranes.

The size classes used in the study of organic compounds are summarized in Figure 22.1. Note that some of this information is also contained in Figures 3.2 (solids versus solutes), 5.3 (metal speciation), and 13.1 (sediments). Organic geochemists now distinguish another phase in seawater, called gels, which range in size from nanometers

[1] Some marine chemists define the upper end of the colloid size range as either 10 μm (Figure 3.2) or 0.1 μm (Figure 5.3).

to millimeters. They are extracellular networks of polymers that are highly interconnected through chemical and physical crosslinks. Water is trapped between the polymer molecules as are solutes and living organisms. This causes the region within the gels to represent a distinct geochemical microenvironment quite different from the ambient seawater. These gels can collide with each other and stick together to generate very large particles, called *transparent exopolymer particles* (TEPs). Small fragments of POM also tend to get stuck in TEPs. The end result of all this aggregation is production of a solid particle whose sinking rate is much faster than any of its contributing components.

The most commonly used acronyms for the operationally defined fractions of marine organic matter are presented in Table 22.1. Despite the widespread agreement on the size cutoff for true solutes, filters of varying pore size (0.2 to 1.0 μm) are still commonly used to isolate POM from DOM. Tangential flow filtration is used to subdivide DOM into a low-molecular-weight fraction (LMW, molecular mass <1000) and a high-molecular-weight fraction (HMW, molecular mass greater than 1000 and diameter

Table 22.1 Operationally Defined Fractions of Organic Matter.

Particulate	
POM	Particulate organic matter
POC	Particulate organic carbon
PON	Particulate organic nitrogen
POP	Particulate organic phosphorus
Dissolved	
DOM	Dissolved organic matter
DOC	Dissolved organic carbon
DON	Dissolved organic nitrogen
DOP	Dissolved organic phosphorus
CDOM	Colored DOM
UDOM	Ultrafiltered DOM
LMW DOM	Low molecular weight DOM
HMW DOM	High molecular weight DOM
Sedimentary	
%OC	Percent organic carbon
%N	Percent nitrogen
%P	Percent phosphorus
Particulate + Dissolved (Unfiltered)	
TOC	Total organic carbon
TN	Total nitrogen (PON + TDIN + DON)
TP	Total phosphorus (POP + TDIP + DOP)
Other	
TEP	Transparent exopolymer particles
UCM	Unresolved complex mixture

less than 1 nm).[2] Chemical characterization of these operationally defined fractions has provided much insight into the biogeochemistry of organic matter. The most basic characterization is determination of the elemental content of the organic matter. These are typically expressed as percentages, i.e., % OC in dry sediments, as a concentration, i.e., mg C/L seawater, or as a molar ratio, i.e., C/N.

Organic compounds can also be characterized by their spectroscopic properties, i.e., how they interact with various types of electromagnetic radiation. For example, the HMW fraction of DOM (called humic substances) absorbs infrared light at many frequencies and with varying intensity, as shown in Figure 22.2. The shape of the absorbance spectrum reflects the type of bonds present in the organic compounds. Though specific structures cannot be identified, differences in peak shape and position can be used to distinguish terrestrial from marine humic substances. Other investigative technologies that exploit various spectral properties of organic matter include UV-VIS absorbance, fluorescence, and nuclear magnetic resonance (NMR).

In addition to characterizing the elemental content, other approaches have been taken to obtain geochemical information from a complex mixture without explicitly identifying the structure of each component. For example, the source of an oil spill can be identified using a *gas chromatograph* to "fingerprint" the petroleum. In this instrument, the complex mixture is first heated, causing the compounds to volatilize. A stream of inert gas pushes the compounds onto a long, narrow column that is either packed or coated with special materials. Packed columns are 2 to 3 m long and are filled with diatomaceous earth impregnated with a high-boiling-point liquid. Alternatively, the liquid is applied as a thin film onto the inside wall of a very narrow (0.2 to 0.5 mm), very long (25 to 50 m) column. In either case, the liquid is termed the stationary phase.

The compounds dissolve in or are adsorbed by this stationary phase to a degree determined by their boiling points. Compounds that are not strongly retained by the column are rapidly pushed to the other end by the stream of inert gas. The compounds that are strongly retained by the column migrate at slower rates. Since the boiling point is inversely related to molecular weight, the smallest compounds tend to elute from the column first.

The time required for elution is called the retention time. As a substance elutes from the column, its presence is signaled by a detector that transforms some physical or chemical property of the substance into a measurable electrical signal. This detector records the information as a signal whose intensity is directly proportional to the compound's abundance. As illustrated in the top panel of Figure 22.3, the detector output is called a gas chromatogram, in which signal intensity is depicted as a peak.

Gas chromatographs are able to completely separate the components of simple mixtures, producing chromatograms with sharp, symmetric peaks, such as those shown in Figure 22.4b. The components of complex mixtures are not as well separated due to similarities in molecular structures that cause compound solubilities and boiling points

---

[2] Chemists have recently decided to switch from use of the term *molecular weight* to *molecular mass*, but the marine biogeochemists haven't quite caught up yet.

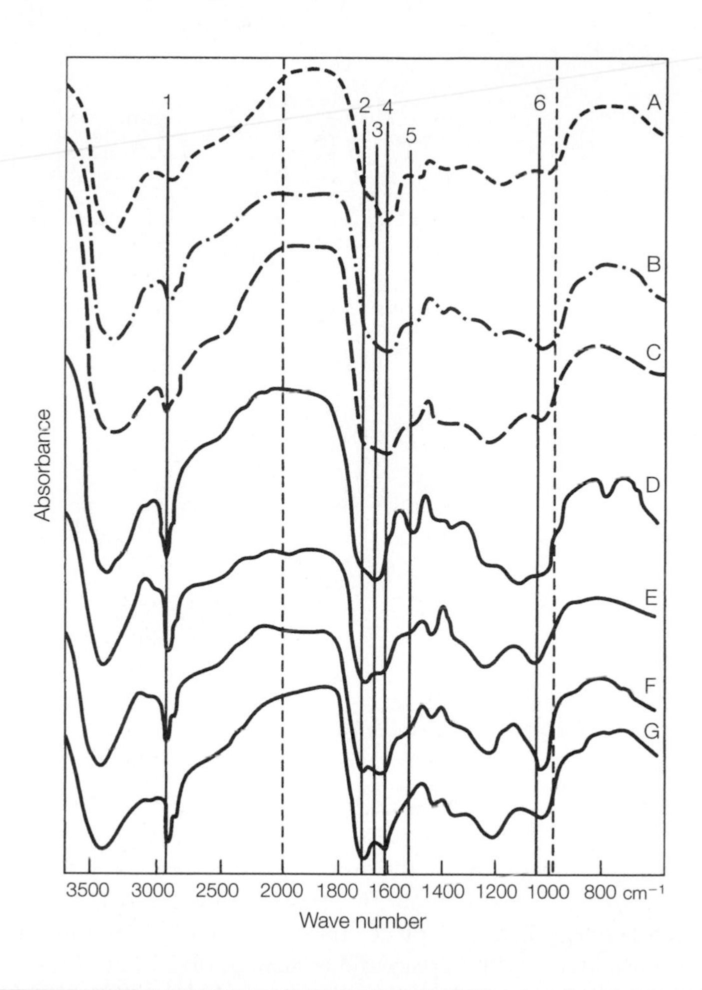

FIGURE 22.2

Comparison of IR spectra of humic acids from terrestrial soils (A–C) and marine sediments (D–G). Identification of IR bands: (1) aliphatic C–H; (2) C=O; (3 and 5) amides; (4) aromatic C=C; and (6) C–O in polysaccharides. *Source*: From Huc, A. Y., *et al*. (1974). Caractérisation Des Acides Humiques de Sédiments Marins Récents et Comparaison avec leurs Homologues Terrestres, Bulletin de L'ENSAIA De Nancy, 16, 59–75.

to be nearly identical. As a result, some of these compounds coelute and appear as large humps on the gas chromatograms, as illustrated in Figure 22.4a. Though the specific organic molecules present in a mixture are not always identifiable, gas chromatograms can be used as a fingerprint for matching purposes such as for source tracing of spilled oil. Because different fuel oils vary greatly in chemical composition (see Chapter 27.2 for an explanation), their gas chromatograms are unique. To trace the source of a spill, gas chromatograms of fuel oil from ships likely to have been the source are compared to that of the spilled oil. The one whose chromatogram most closely resembles that of the spilled oil is considered to be the most likely source.

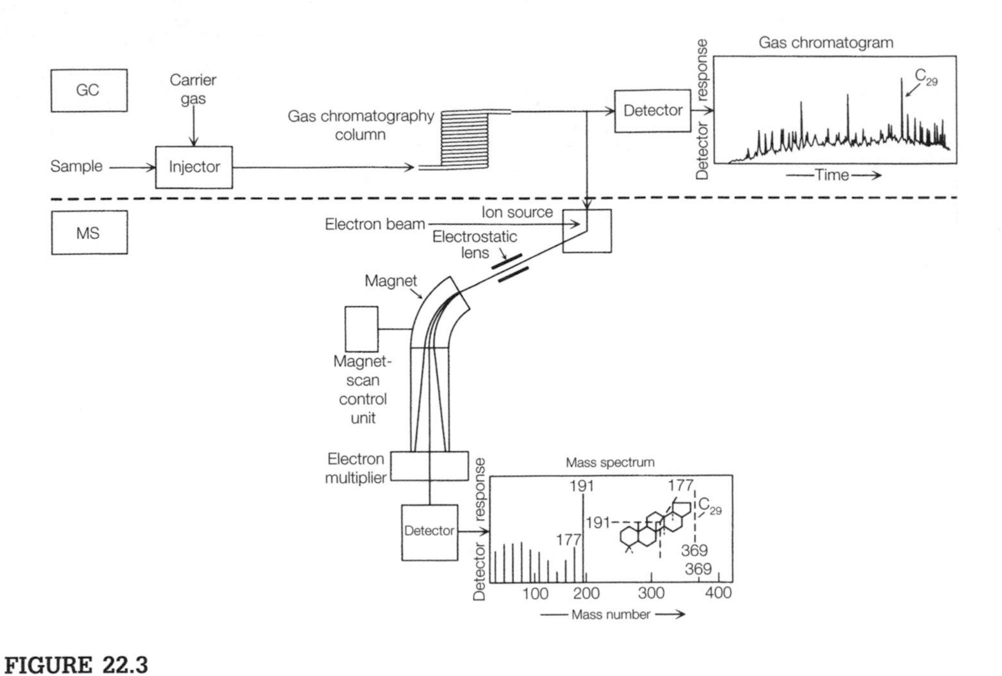

FIGURE 22.3

GC-MS analysis of organic compounds.

The molecular structures of each peak in a well-resolved gas chromatogram can be identified by *mass spectrometry*. This combination of technologies is called GC-MS and is illustrated schematically in the lower panel of Figure 22.3. As a compound elutes off the chromatographic column, it passes into the mass spectrometer. Bombardment with a high-energy beam of electrons creates molecular ions that can spontaneously break into fragments. The masses and relative abundances of the fragments are detected electronically and reported as a mass spectrum. An onboard computer compares this spectrum to those of known molecules to establish a best match and, hence, the likely identity of the compound. Once the compound has been identified, its presence in other gas chromatograms, obtained under similar operating conditions, can be determined from its retention time. Other types of chromatographic technologies are available, such as thin-layer chromatography (TLC) and high-pressure liquid chromatography (HPLC). The latter works at low temperatures and, thus, is used for separating compounds that are thermally unstable or that have low boiling points.

22.3.2 Biomarkers: Tracers of Organic Matter Source and History

The concentration distributions of individual organic compounds can provide information on biogeochemical processes. As shown in Figure 22.5, adenosine triphosphate (ATP) concentrations are highest at the sea surface. ATP is synthesized by all organisms.

(a)

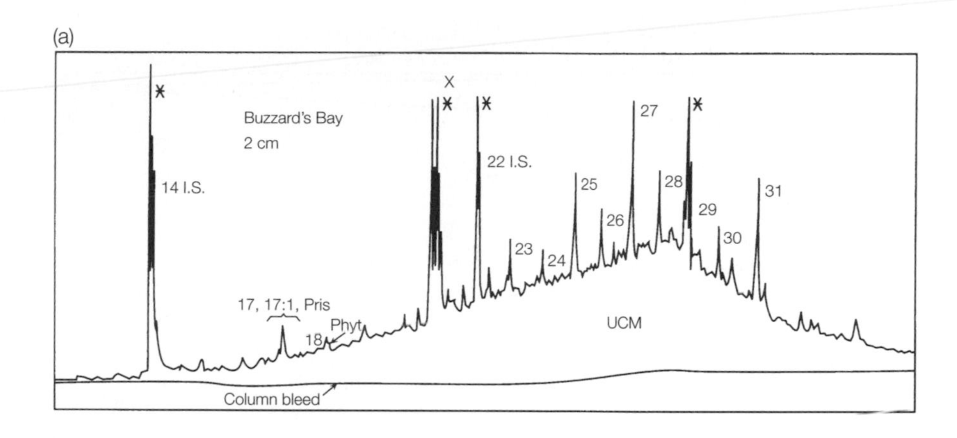

(b)

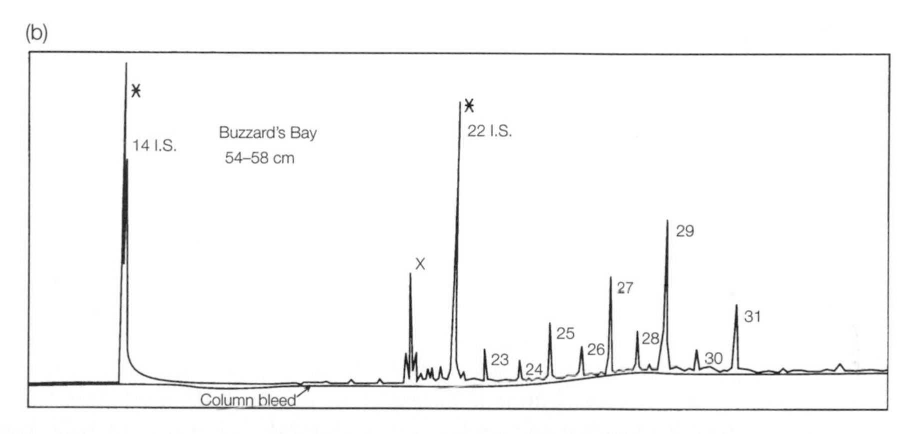

FIGURE 22.4

Gas chromatograms of anthropogenic petroleum hydrocarbons from two depths in a coastal marine sediment: (a) surface sediment and (b) 54 to 58 cm below the sediment surface. The degradation and migration of petroleum compounds is responsible for the decrease in mixture complexity with increasing depth. Some of the petroleum compounds in the surface sediments contribute to the hump, which is identified only as an unresolved complex mixture (UCM). Peaks identified by an asterisk were brought on scale by use of an auto attenuator. Peaks identified as I.S. are internal standards. X is used to designate groups of unsaturated compounds of unknown structure. *Source*: From Farrington J. W., *et al.* (1977). *Estuarine and Coastal Marine Science*, 5, 793–808.

In the case of plankton, cell lysis that occurs shortly after death causes ATP to be released into seawater. Like most biomolecules, ATP is rapidly degraded in seawater by microbes. Thus, high surface concentrations in Figure 22.5 reflect a rapid supply supported by the high rates of plankton production characteristic of the photic zone. Below the surface, concentrations decrease with increasing depth beneath the photic zone and, hence, distance from the biosynthetic source of the ATP.

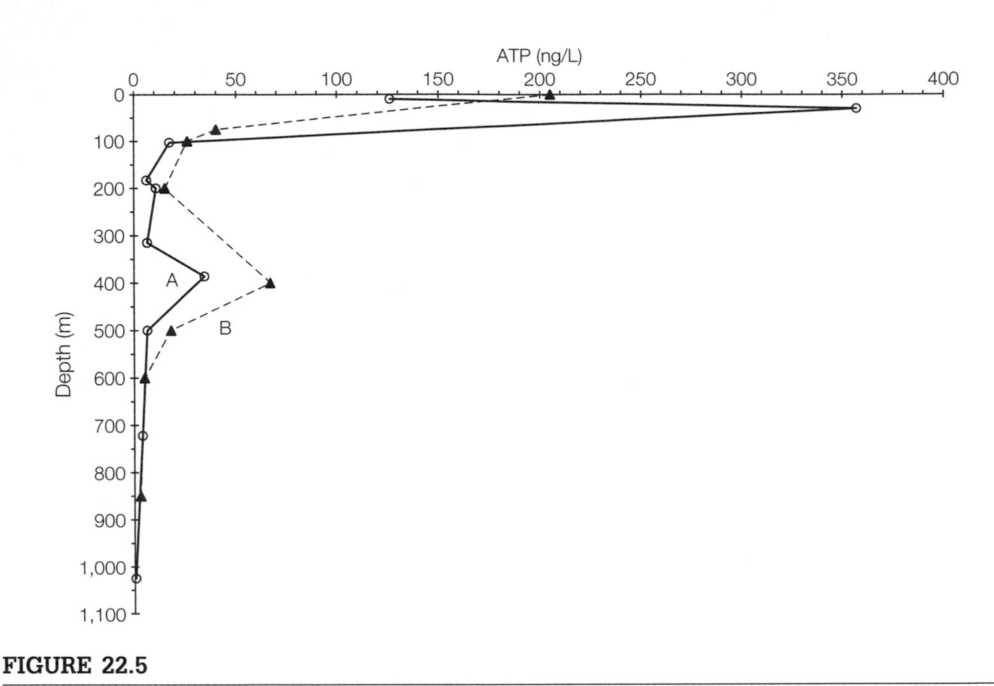

FIGURE 22.5

Depth profile of ATP concentrations at a site off the coast of California. Samples were collected in (A) May 1965 at 33°N 119°W and (B) October 1965 at 33°N 117°W. *Source*: From Holm-Hansen, O., and C. H. Booth (1966). *Limnology and Oceanography*, 11, 510–519.

A small ATP concentration maximum is located at 400 m, well below the bottom of the photic zone. This maximum is seasonal in nature with concentrations increasing from spring to fall. Possible causes for this mid-depth maximum include release of ATP from: (1) organisms that live at mid-depths, (2) detrital POM supplied by advective transport in a subsurface current, and (3) detrital POM sinking out of the surface water. Given the seasonal increase in concentration, the most likely mid-depth ATP source is sinking detrital POM. The depth of this subsurface concentration maximum reflects density stratification of the water column relative to the density of the sinking POM. This site is located at mid-latitudes where formation of a summer thermocline inhibits vertical mixing. Thus, POM generated through the spring and summer sinks through the highly stratified and increasingly dense waters, until it reaches a depth at which it is neutrally buoyant.

Some organic compounds are highly specific for their sources. For example, the sterol dinosterol is synthesized only by dinoflagellates. Thus, finding dinosterol in sinking POM provides unequivocal evidence for the contribution of dinoflagellate biomass to this particulate matter. Compounds like ATP and dinosterol that can be used as tracers of biological source and history are termed *biomarkers*. Three types of historical information can be gleaned from biomarkers. First, the presence of the biomarker compounds documents that the source organism was present. Second, some biomarkers reflect the health of an organism, which in turn reflects information on environmental conditions, such as temperature or exposure to pollutants. For example, elevated concentrations of cytochrome P4501A within an organism are an indicator of exposure to toxins. Third, biomarkers can undergo chemical change after release from the source organism, such as during diagenesis and catagenesis. The degradation products can be unique and, hence, continue to serve as source tracers. In some cases, the nature of the degradation products generated provides information on reaction conditions, including temperature, pH, redox state, and reaction duration. The latter provides a means of sediment dating.

The best biomarkers for source tracing are highly specific to one type of organism and are resistant to degradation. But even with minimal specificity, biomarkers can still be useful. For example, lignins are structural polymers synthesized by all woody terrestrial plants. Thus, detrital terrestrial POM transported by rivers into the ocean and buried in nearshore sediments contains lignin residues. These compounds are resistant to degradation and, hence, well preserved in shelf sediments. Their relative abundance in marine sediments can be used as a measure of the contribution of terrestrial organic matter to the total organic content of the deposit. In this setting, biomarkers are the most micro of microfossils.

The preferred use of biomarkers is a weight-of-evidence approach in which multiple compounds are used to confirm organic matter sources or history. Since modern analytical methods, such as GC-MS, can provide concentration data on hundreds of compounds within a sample, statistical techniques are now used to identify corroborating patterns in biomarker distributions. These patterns provide insight into biomarker source and history. This application of statistics, in which relationships among large numbers of measurements are used to gain insight into system processes and function, is termed *chemometrics*. An example of this approach is presented in Table 22.2, in which a statistical technique called principal component analysis was used to characterize organic matter sources (marine versus terrestrial) to the shelf sediments of the Arctic Ocean (See the footnotes in this table for details).

22.3.3 Isotopes as Tracers of the Source and Fate of Marine Organic Compounds

Isotopes of a particular element are atoms that have the same number of protons, but different numbers of neutrons. The naturally occurring stable isotopes of carbon, nitrogen, oxygen, and hydrogen have been used as source tracers. For example, the ratio of ^{13}C to ^{12}C is lower in terrestrial POM than in marine POM. Because these ratios are relatively constant with respect to their sources and not altered by diagenesis, the contribution of terrestrial organic matter to coastal marine sediments can be inferred from its relative carbon isotope composition. A similar approach is used with marine organisms to identify their food sources and trophic levels in food webs. Improvement in analytical technology now enable measurement of the naturally occurring isotopes in specific compounds. This has greatly improved source tracing capability.

The rates of biological processes can be measured by amending seawater or sediment samples with chemicals that possess high concentrations of a particular stable

Table 22.2 Estimated Percent Labile/Marine Content for Acyclic Hydrocarbons in Arctic Shelf Sediments.[a]

Biomarker	Carbon Range	Source[b]	Labile/Marine Content, (%)[c] Mean (Range)
n-Alkane	C_{17}–C_{21}	Algal, petroleum	8.4 (3.6–12)
	C_{22}–C_{30} even	Vascular plant. petroleum	18 (14–26)
	C_{23}–C_{30} odd	Vascular plant. petroleum	21 (16–31)
n-Alkanol	C_{14}–C_{18} even	Algal. zooplankton	34 (23–46)
	C_{15}–C_{19} odd	Bacteria?	20 (11–30)
	C_{20}–C_{30} even	Vascular plant. zooplankton	28 (13–40)
	C_{21}–C_{29} odd	Bacteria?	27 (12–47)
n-Alkanoic acids	C_{12}–C_{18} even	Algal, zooplankton. bacteria	85 (74–99)
	C_{13}–C_{17} odd	Bacteria?	89 (88–89)
	C_{20}–C_{28} even	Vascular plant. algal	58 (46–66)
	C_{21}–C_{27} odd	Bacteria?	60 (44–73)
Mono *n*-alkenoic acids (MUFA)	C_{14}–C_{20}	Algal, zooplankton. bacteria	89 (71–99)
Poly *n*-alkenoic acids (PUFA)	C_{16}–C_{20}	Algal, zooplankton. bacteria	77 (71–85)
x, ω *n*-Dicarboxylic acids	C_{20}–C_{26}	Vascular plant. bacteria	65 (48–75)
Branched alkanoic acids (*iso*-, anteiso-, branched)	C_{13}–C_{19}	Bacteria	75 (38–100)
Phytol		Algal	50

(Continued)

Table 22.2 *(Continued)*

Biomarker	Carbon Range	Source[b]	Labile/Marine Content, (%)[c] Mean (Range)
6.10.14-Trimethylpentadecan-2-one		Algal (from phytol)[d]	11
4.8.12-Trimethyltridecanoic acid		Zooplankton (phylol metabolism)[d]	53
4,8,12,16 Tetramethylheptadecan-4-olide		Unknown (see text)	62

[a] *Determined from principal component analysis and other chemometric techniques.*
[b] *Algal sources are ice algae, diatoms, cyanobacteria, dinoflagellates, and picoplankton; petroleum sources include eroded bitumens, oil seeps, etc.*
[c] *The best biomarkers for marine sources of organic matter have high percent labile/marine content and the best biomarkers for terrestrial organic matter have low percentages. Marine compounds tend to be more labile (reactive) than terrestrial organic matter.*
[d] *These isoprenoids are degradation products of phytol.*
Source: After Yunker, M., et al. (2006). Deep-Sea Research *II 52, 3478–3508.*

or radioactive isotope. For example, the rate of primary production is determined by measuring how much radiocarbon-labeled bicarbonate is converted into POC within a specified time interval. As in this example, the fate of the artificially introduced isotope is traced by periodically analyzing various fractions of the sample. At the end of the experiment, all of the tracer must be accounted for among the fractions measured, such as the water, biota, and sediment, or, if radioactive, through decay. Because the water and sediment samples must be removed from the ocean and incubated under controlled conditions, the experimental results can be questionable. As described in the next section, much effort has been expended in developing artificial systems simulating natural conditions to obtain more reliable results from manipulation experiments.

22.3.4 **Artificial Ecosystems**

Artificial ecosystems are used to simulate in situ environmental conditions. They are designed to function as much like the natural ocean as possible, while providing a system that can be experimentally manipulated and efficiently sampled. Some of the largest artificial ecosystems are the 14 MERL tanks (Marine Ecosystems Research Laboratory) operated by the University of Rhode Island on Narragansett Bay since 1975.

As illustrated in Figure 22.6, these tanks are approximately 16 ft high and 5 ft in diameter. They contain about 0.37 m of bay sediment and 13,000 L of water, which is slowly trickled in from the bay. The outflow rate is set so the water volume does not change over time. With a stirring device and the top of the tank open, the tanks function much like the neighboring bay. Typical experiments involve monitoring the

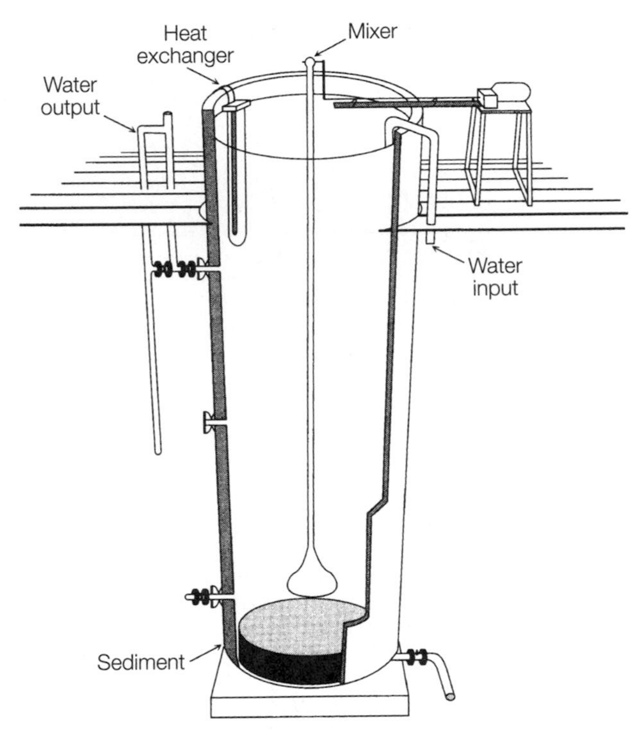

FIGURE 22.6

Cross-sectional view and dimensions of a MERL tank. *Source*: From Frithsen J. B., *et al.* (1985). *MERL Series*, *Report No. 3*, Marine Ecosystems Research Laboratory, University of Rhode Island.

effects of some perturbation, such as the introduction of nutrients, oil, or pesticides, on the water, sediment, and biota in the tank.

The term *mesocosm* is used to refer to systems with water volumes ranging from 1 to 10,000 m^3. Mesocosms provide an important link between simple small-scale laboratory experiments and field studies. In the case of plankton studies, the use of mesocosms eliminates the complicating effect of advection, thus enabling time-series sampling of the same population as it experiences natural variations in light intensity, photoperiod, and temperature. Mesocosms have been used to study plankton community and food web dynamics, aggregate formation and sinking rates, nutrient cycling, and the fate of pollutants. Systems of mesocosms, such as the 14 MERL tanks, are preferable as they allow for replication and control systems.

22.4 GENERAL CLASSES OF ORGANIC COMPOUNDS

Organisms obtain chemical energy by degrading organic compounds. This general process is termed *catabolism*. The chemical energy released can be used for locomotion, for reproduction, and to otherwise increase biomass. The general process of organic matter synthesis is termed *anabolism*. In its most general sense, anabolism involves the conversion of simple inorganic molecules into low-molecular-weight organic biomolecules that are then combined to form high-molecular-weight macromolecules.

An inventory of known biomacromolecules is provided in Table 22.3. Many of these play essential metabolic roles in enabling growth and reproduction, such as the carbohydrates, lipids, proteins, and polynucleotides. Others are components of cell walls and exoskeletons. Some organisms, such as bacteria, plankton, plants, and lower invertebrates, synthesize biomolecules, called *secondary metabolites*, that are used to control ecological relationships, including predator/prey, host/symbiont, mating/spawning, and competition for food or space.

All organisms synthesize carbohydrates, lipids, proteins, and polynucleotides, although the details of their molecular structures can be somewhat species specific. These basic classes of macromolecules have changed little over geologic time. The secondary metabolites are more species specific and have also changed little over geologic time. Many are resistant to degradation, and those provide excellent biomarkers that have been preserved in ancient marine sediments and petroleum deposits.

On the present-day Earth, most of the energy that is ultimately fueling the anabolic reactions is derived from sunlight. This energy is harvested by photosynthesizers who transform it into chemical energy that can be transferred up the food chain. This process is shown schematically in Figure 22.7. Catabolism is essentially the reverse of this process. A great number of reactions occur during anabolism and catabolism. As a result, organisms produce a myriad of organic molecules. The details of the metabolic reaction pathways tend to be species specific and in some cases produce compounds that can be used as biomarkers. In some metabolic processes, catabolism and anabolism proceed concurrently, such as in the Krebs (TCA) cycle (Figure 7.6b). These are referred to as *amphibolic* pathways.

The biomolecules are grouped into classes based on similarities in structure, most notably the presence of functional groups. The latter are small groups of atoms that undergo characteristic reactions. The reactivity of the functional groups greatly influences the overall chemistry of the host molecule. The most common of these functional groups are listed in Table 22.4. Organic molecules are often named after their functional groups. For example, a molecule with a hydroxyl group is classified as an alcohol and one with a carboxyl group is classified as a carboxylic acid. Most molecules contain more than one functional group. To deal with this, chemists have developed a highly rigorous system of nomenclature that is overseen by the International Union of Pure and Applied Chemistry (IUPAC). Biochemists also classify biochemicals into broad categories that are based on similarities in their molecular structures and metabolic roles. The major biochemical categories are described next and illustrated with representative molecular structures.

Table 22.3 Inventory of Presently Known Biomacromolecules, Their Occurrence in Extant Organisms, and Their Potential for Survival During Sedimentation and Diagenesis.

Biomacromolecules	Occurrence	"Preservation Potential"[a]
Starch	Vascular plants; some algae; bacteria	–
Glycogen	Animals	–
Fructans	Vascular plants; algae; bacteria	–
Laminarans	Mainly brown algae; some other algae and fungi	–
Poly-β-hydroxyalkanoates (PHA)	Eubacteria	–
Cellulose	Vascular plants; some fungi	–/+
Xylans	Vascular plants; some algae	–/+
Pectins	Vascular plants	–/+
Mannans	Vascular plants; fungi; algae	–/+
Galactans	Vascular plants; algae	–/+
Mucilages	Vascular plants (seeds)	+
Gums	Vascular plants	+
Alginic acids	Brown algae	–/+
Fungal glucans	Fungi	+
Dextrans	Eubacteria; fungi	+
Xanthans	Eubacteria	+
Chitin	Arthropods; copepods; crustacea; fungi; algae	+
Glycosaminoglycans	Mammals; some fish; Eubacteria	–/+
Proteins	All organisms	–/+
Extensin	Vascular plants; algae	–/+
Mureins	Eubacteria	+
Teichoic acids	Gram-positive Eubacteria	+
Teichuronic acids	Gram-positive Eubacteria	+
Lipoteichoic acids (LTA)	Gram-positive Eubacteria	+
Bacterial lipopolysaccharides (LPS)	Gram-negative Eubacteria	++
DNA, RNA	All organisms	–

(Continued)

Table 22.3 *(Continued)*

Biomacromolecules	Occurrence	"Preservation Potential"[a]
Glycolipids	Plants; algae; Eubacteria	+/ + +
Polyisoprenols (rubber and gutta)	Vascular plants	+
Polyprenols and dolichols	Vascular plants; bacteria; animals	+
Resinous polyterpenoids	Vascular plants	+/ + +
Cutins, suberins	Vascular plants	+/ + +
Lignins	Vascular plants	+ + + +
Tannins	Vascular plants; algae	+ + +/ + + + +
Sporopollenins	Vascular plants	+ + +
Algaenans	Algae	+ + + +
Cutans	Vascular plants	+ + + +
Suberans	Vascular plants	+ + + +
Cyanobacterial sheaths	Cyanobacteria	+

[a] *The preservation potential ranges from – (extensive degradation under depositional conditions) to + + + + (no degradation under any depositional conditions).*
Source: *From Tegelaar, E. W., et al. (1989).* Geochimica et Cosmochimica Acta, *3, 3103–3107.*

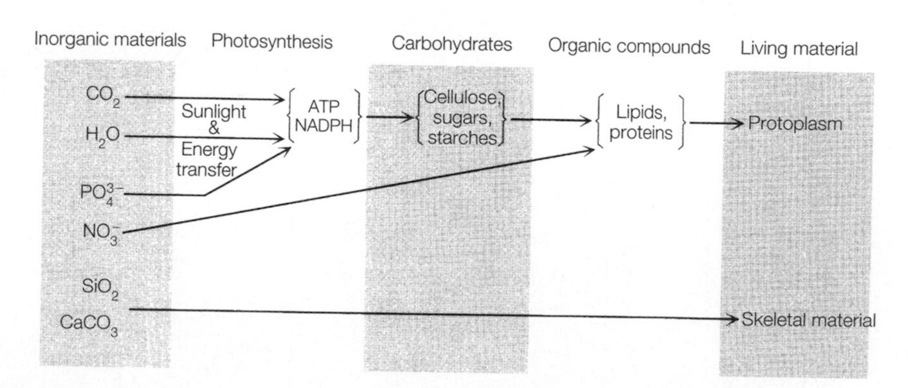

FIGURE 22.7

Various steps in the production of living material. *Source*: From Anikouchine, W. A., and R. W. Sternberg (1981). *The World Ocean: An Introduction to Oceanography*, Prentice Hall, Inc., p. 379.

Table 22.4 The Most Common Functional Groups Found in Naturally Occurring Organic Compounds.[a]

Chemical Class	Group	Formula	Graphical Formula	Prefix	Suffix	Example
Alkanoyl	Acyl[b]	-COR	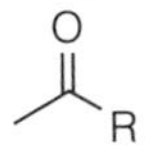	[b]	[b]	[b]
Alcohol	Hydroxyl	ROH	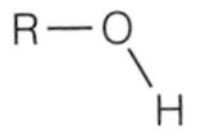	hydroxy-	-ol	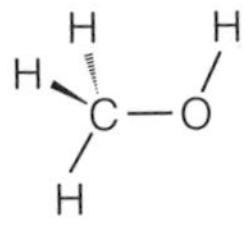 Methanol
Aldehyde	Aldehyde(carbonyl)	RCHO	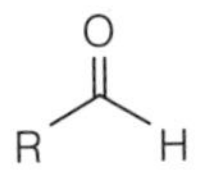	oxo-	-al	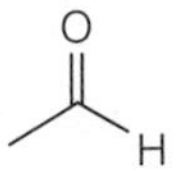 Acetaldehyde
Alkane	Alkyl	RH		alkyl-	-ane	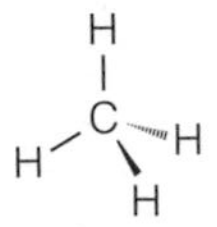 Methane
Alkene	Alkenyl	$R_2C{=}CR_2$	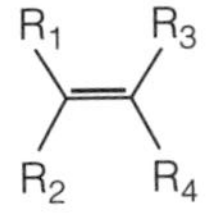	alkenyl-	-ene	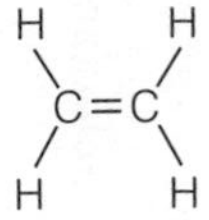 Ethylene

Amide	Carboxamide	$RCONR_2$	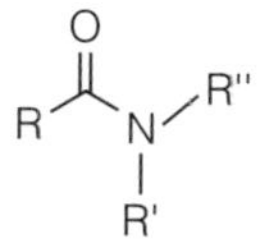 $R-C(=O)-N(R')-R''$	carboxamido-	-amide	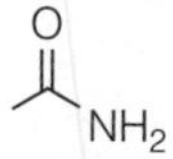 $CH_3C(=O)NH_2$ Acetamide
Amines	Primary amine	RNH_2	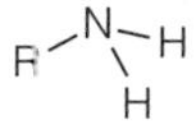 $R-NH_2$	amino-	-amine	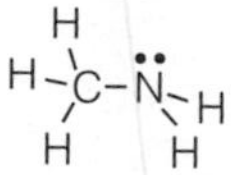 $H_3C-\ddot{N}H_2$ Methylamine
	Secondary amine	R_2NH	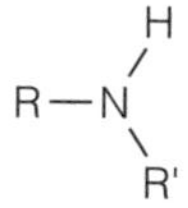 $R-N(H)-R'$	amino-	-amine	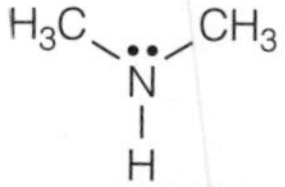 $H_3C-\ddot{N}(H)-CH_3$ Dimethylamine
	Tertiary amine	R_3N	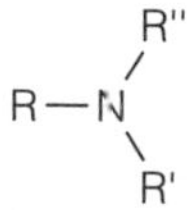 $R-N(R')-R''$	amino-	-amine	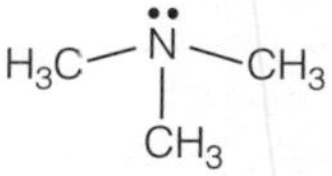 $H_3C-\ddot{N}(CH_3)-CH_3$ Trimethylamine
	Quaternary amine	R_4N+	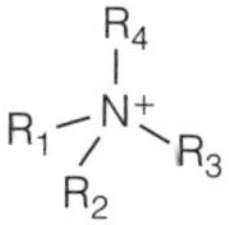 $R_1-N^+(R_2)(R_4)-R_3$	ammonio-	-ammonium	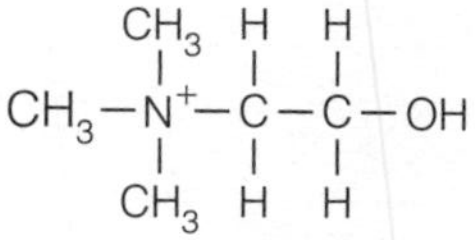 $CH_3-N^+(CH_3)_2-CH_2-CH_2-OH$ Choline

(Continued)

Table 22.4 *(Continued)*

Chemical Class	Group	Formula	Graphical Formula	Prefix	Suffix	Example
Carboxylic acid	Carboxyl	RCOOH	O ‖ R OH	carboxy-	-oic acid	Fatty Acids
Ether	Ether	ROR'	R O R'	alkoxy-	alkyl alkyl **ether**	
Ester	Ester	RCOOR'	O ‖ R OR'		-oate	
Haloalkane	Halo	RX	R — X	halo-	alkyl **halide**	
Ketone	Ketone	RCOR'	O ‖ R^1 R^2	keto-, oxo-	-one	
Benzene derivative	Phenyl	RC_6H_5	R — (benzene ring)	phenyl-	-benzene	
Phosphate	Phosphate	$ROP(=O)(OH)_2$	R O P(=O) OH OH	phospho-		
Thiol	Sulfhydryl	RSH	R — S — H	mercapto-, sulfanyl-	-thiol	H_3C — C(H)(H) — SH Ethanethiol *(Ethyl mercaptan)*
Disulfide	Disulfide	R_1-S-S-R_2				
Sulfide thioether	Sulfide thioether	R_1-S-R_2				

[a] *In the formulas, the symbols R and R' usually denotes an attached hydrogen, or a hydrocarbon side chain of any length, but may sometimes refer to any group of atoms.*

[b] *In organic chemistry, an acyl group is derived by the removal of a hydroxyl group from a carboxylic acid. The specific name of the functional group depends on the carboxylic acid involved, i.e., formic acid yields a formyl group, acetic acid yields an acetyl group, benzoic acid yields a benzoyl group.*

22.4.1 **Hydrocarbons**

The simplest biomolecules are the *hydrocarbons* which are composed solely of carbon and hydrogen atoms. In the marine environment, most hydrocarbons are degradation products resulting from the catabolism or diagenesis of biomacromolecules or they are terrestrial in origin, being derived from plants. The hydrocarbons are present as side chains in the biomacromolecules and, hence, are readily cleaved off during biotic and abiotic reactions. Hydrocarbon production through catagenic reactions in deeply buried organic-rich sediments has led to the formation of large petroleum deposits.

A large variety of hydrocarbons have been identified in marine sediments and petroleum. These chains of carbon atoms vary greatly in number and some are branched, as illustrated in Figure 22.8. Some have double or triple bonds. Hydrocarbons with double or triple bonds are said to be "unsaturated" with respect to hydrogen.

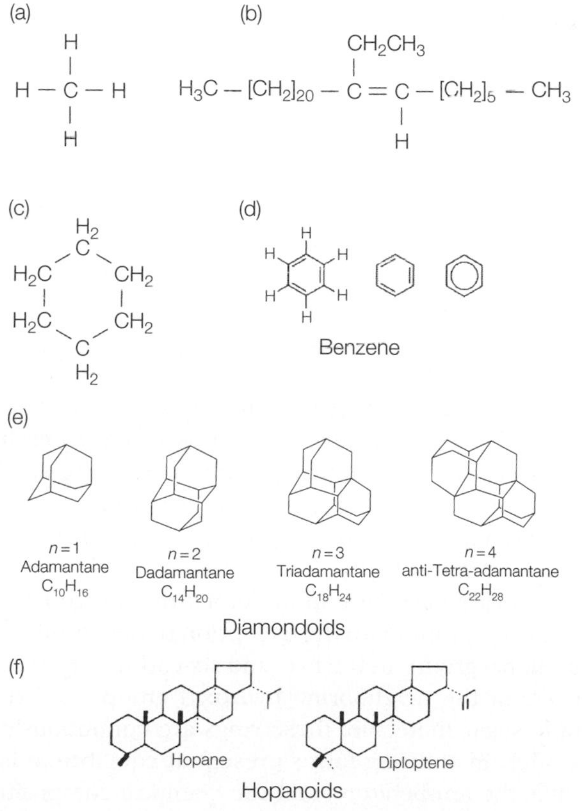

FIGURE 22.8

Structures of some hydrocarbons: (a) methane, (b) a branched and unsaturated C_{31} hydrocarbon, (c) cyclohexane, (d) benzene, (e) some diamondoids, and (f) some hopanoids.

Linear (unbranched) chains of carbon atoms are called *n*-alkanes if all the carbon bonds are saturated. Linear hydrocarbons that contain double or triple bonds are called *n*-alkenes or *n*-alkynes, respectively. Hydrocarbons can contain ring structures, such as cyclohexane. Like the linear chains, the rings can have one or more double bonds between the carbon atoms. In the case where the double bonds in the ring are alternating, the structure is termed an *aromatic hydrocarbon*. Benzene is the simplest of the aromatic hydrocarbons. The aromatic structure is often depicted as a circle within a hexagon created by the six linked carbons. This is meant to indicate that the electrons shared among the ring carbons are delocalized such that each carbon-to-carbon bond spends an equal amount of time as either a single or a double bond. Hydrocarbons that do not contain benzene rings are called *aliphatic hydrocarbons*.

For the short-chain *n*-alkanes, melting point increases with increasing molecular weight. Thus, low-molecular-weight *n*-alkanes ($<C_5$) are gases at room temperature, medium-length chains are liquids (C_5 to C_{35}), and long chains are solids ($>C_{35}$). In general, branching lowers the melting point of alkanes.

Most of the *n*-alkanes found in sediments and petroleum are derived from the degradation of bacterial cell membranes, algal algaenans, and vascular plant waxes (cutans). Other hydrocarbons found in petroleum deposition include the *polycyclic aromatic hydrocarbons* (PAHs), some of which are highly carcinogenic (Figure 28.16). PAHs are also created by burning of terrestrial biomass. Other petroleum hydrocarbons include the *diamondoids* and *hopanoids*, both of which are used as biomarkers, as they retain some of their original unique biological structures even after abiogenic alteration (Figures 22.8e and 22.8f). The diamondoids have cage-like structures similar to that of diamonds, which are thought to have potential for nanotechnological applications.

22.4.2 Simple Sugars and Complex Carbohydrates

Carbohydrates serve as short-term cellular reservoirs of chemical energy and as structural components. They are primarily composed of carbon, hydrogen, and oxygen with a molar ratio of hydrogen twice that of carbon and oxygen ($C_n(H_2O)_n$). The *simple carbohydrates*, or *monosaccharides*, are composed of single sugar molecules, of which glucose is an example. Its structure, along with those of the other common simple sugars, is given in Figure 22.9.

The simple sugars have carbon skeletons that range from C_3 to C_6. Each carbon is bound to a hydroxyl group with the exception of one terminal carbon, which contains an aldehyde functional group. In the five- and six-carbon sugars, the aldehyde spontaneously reacts with one of the neighboring hydroxyl groups to form a ring (Figure 22.10). In a single simple sugar molecule, these rings are continuously opening and closing. The proportions of chain and ring forms present at equilibrium is dependent on the type of sugar molecule, the temperature, and the chemical composition of the solution.

Complex carbohydrates are formed by the linking of many simple sugars through condensation reactions. In these reactions, a molecule of water is eliminated by extracting an acidic hydrogen from one sugar and the hydroxyl group from another. The resulting polymers are also referred to as *polysaccharides*. Some common ones are illustrated in

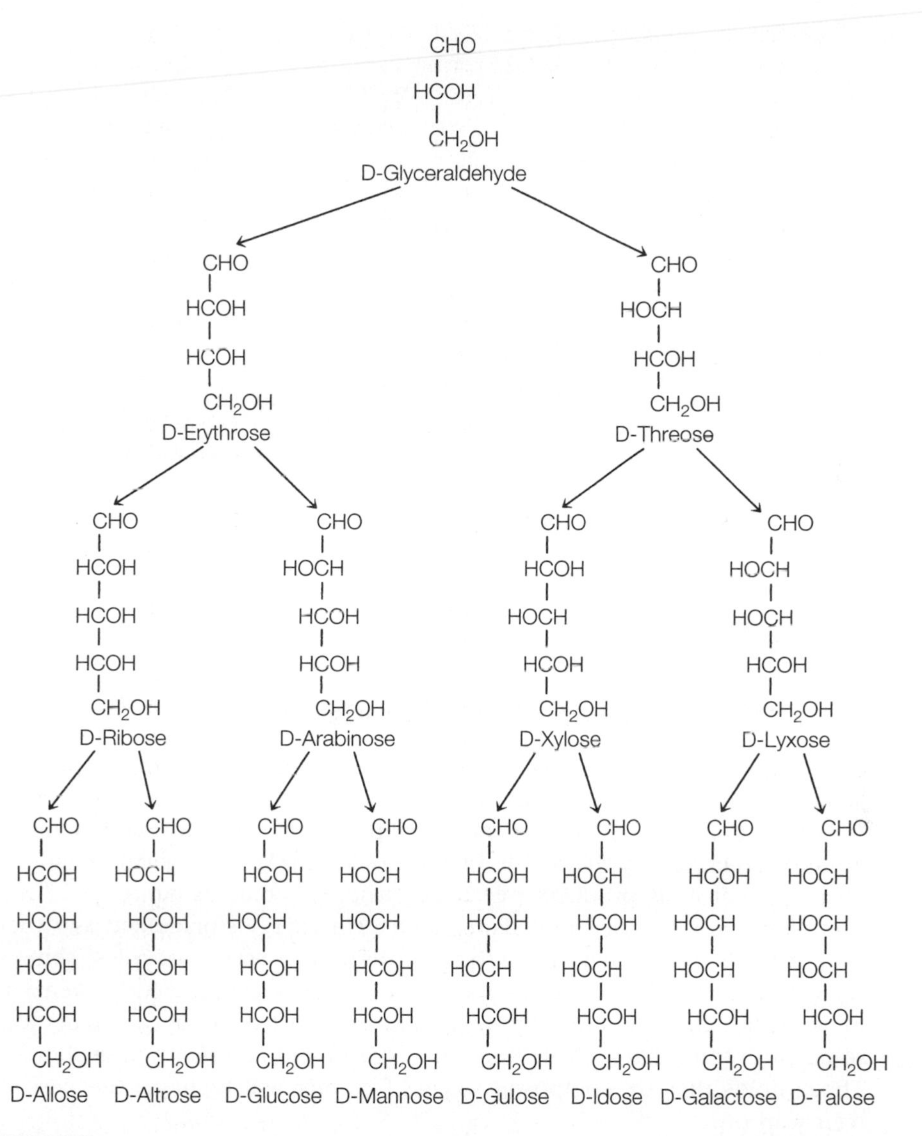

FIGURE 22.9

The simple sugars.

Figure 22.11. Cellulose is a structural and rather chemically inert component of terrestrial plants. Alginic acid is synthesized by marine algae and has medicinal properties. Some polysaccharides contain nitrogen in the form of amino sugars. The primary example of this is chitin, which composes the exoskeletons of crustacea. An amino sugar contains an amine in place of a hydroxyl group. In other polysaccharides, sugars combine with lipids and proteins. These form glycolipids and glycoproteins, respectively.

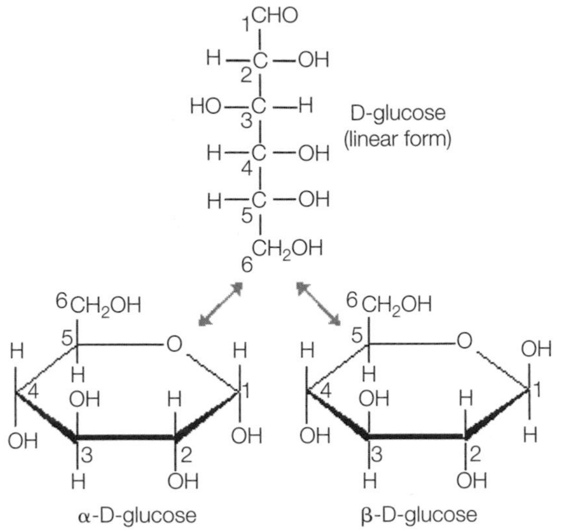

FIGURE 22.10

Glucose linear and ring forms.

22.4.3 Complex and Simple Lipids

Lipids are a diverse group of biomolecules that are soluble in nonpolar (organic) solvents and sparingly soluble in water. Note that this is an operational, rather than a structural definition. A further subdivision is based on the number of different compounds generated by hydrolysis of the lipid. Simple lipids yield at most two types of primary hydrolysis products per mole, whereas complex lipids yield three or more. *Complex lipids* are used in cells for long-term energy storage and as structural components of cell membranes. The *simple lipids* play a variety of roles; some are chemical regulators, such as the steroidal hormones. Others are pigments, coenzymes, and vitamins. Although carbon and hydrogen are the most common elements found in lipids, some structures have small amounts of oxygen, nitrogen, phosphorus, and sulfur. Those lipids that are composed solely of carbon and hydrogen are also classifiable as hydrocarbons.

Complex Lipids

The complex lipids contain various *fatty acids* attached to a "backbone" molecule. Fatty acids are hydrocarbons that have a terminal carboxylic acid group (R-COOH). About 40 are known to occur in nature. As shown in Table 22.5, the fatty acids vary in chain length, degree of unsaturation, and branching. Some of these features are unique to certain organisms. For example, branched chains are produced during the microbial degradation of organic detritus. Because the complex lipids contain fatty acids, they are commonly called *fats*. Some fats are liquids (*oils*) and some are solids, depending on

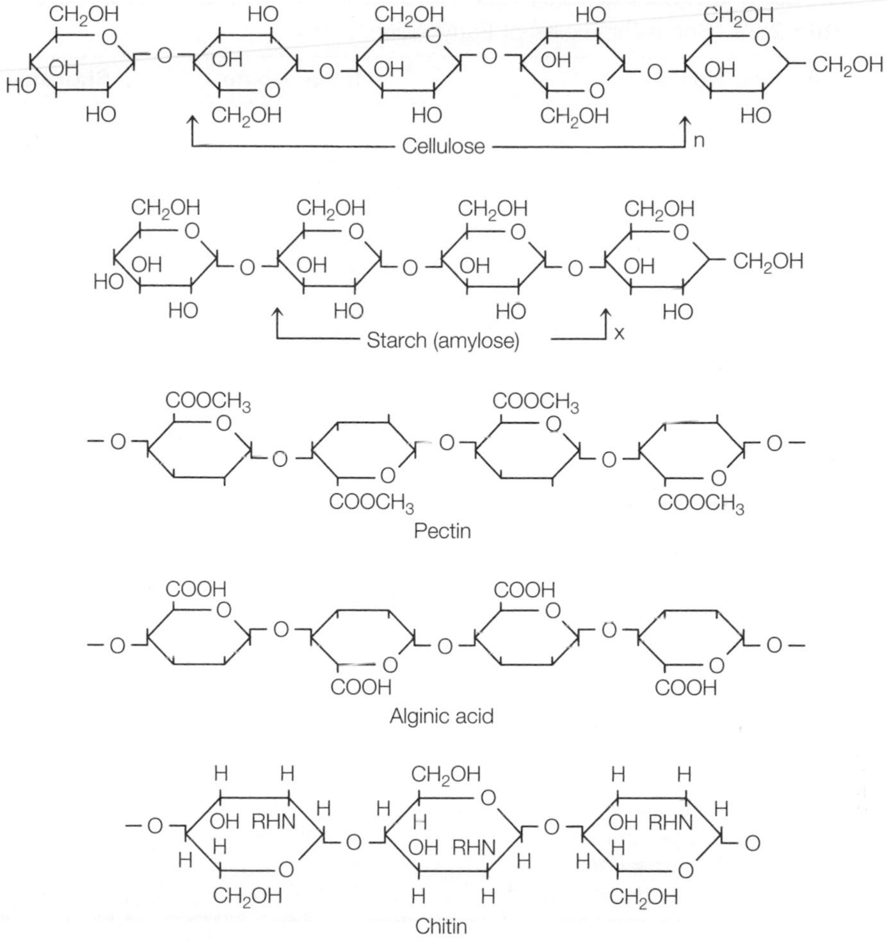

FIGURE 22.11

Some natural polysaccharides. For cellulose and starch, *n* and *x* reflect the different orientations of glucose, which distinguishes these two polymers.

their melting point and environmental conditions. In general, melting points increase with increasing molecular weight.

The molecules that serve as a backbone in the complex lipids are glycerol and sphingosine. As shown in Figure 22.12, glycerol serves as the backbone for the *triacylglycerols* and *glycerophospholipids*. Sphingosine serves as the backbone for the sphingolipids, some of which are phospholipids.

The triacylglycerols have three fatty acids esterified to the three hydroxyl groups of the glycerol backbone. Esterification is a chemical reaction during which a hydroxyl and a carboxyl group react to form an ester. As illustrated in Figure 22.13, one molecule

Table 22.5 Some Examples of Fatty Acids.

Systematic Name	Trivial Name	Shorthand[b]
Saturated fatty acids		
ethanoic	acetic	2:0
butanoic	butyric	4:0
hexanoic	caproic	6:0
octanoic	caprylic	8:0
decanoic	capric	10:0
dodecanoic	lauric	12:0
tetradecanoic	myristic	14:0
hexadecanoic	palmitic	16:0
octadecanoic	stearic	18:0
eicosanoic	arachidic	20:0
docosanoic	behenic	22:0
Monoenoic fatty acids		
cis-9-hexadecenoic	palmitoleic	16:1(n-7)
cis-6-octadecenoic	petroselinic	18:1(n-12)
cis-9-octadecenoic	oleic	18:1(n-9)
cis-11-octadecenoic	cis-vaccenic	18:1(n-7)
cis-13-docosenoic	erucic	22:1(n-9)
cis-15-tetracosenoic	nervonic	24:1(n-9)
Polyunsaturated fatty acids[a]		
9,12-octadecadienoic	linoleic	18:2(n-6)
6,9,12-octadecatrienoic	V-linolenic	18:3(n-6)
9,12,15-octadecatrienoic	α-linolenic	18:3(n-3)
5,8,11,14-eicosatetraenoic	arachidonic	20:4(n-6)
5,8,11,14,17-eicosapentaenoic	EPA	20:5(n-3)
4,7,10,13,16,19-docosahexaenoic	DHA	22:6(n-3)

[a] *All the double bonds are of the cis configuration.*
[b] *The first number represents the number of carbon atoms and the second, the number of double bonds. The number in parentheses, (n − x), denotes the location of the first double bond.*

of water is created for each ester bond formed. The triacylglycerols in animals tend to be present as solid fats or as liquid oils. The fatty acids in oils are more undersaturated than in fats. In other words, the melting point of the triacylglycerols decreases with degree of undersaturation. Other metabolic acylclycerols include the monoacylglycerols and diacylglycerols, which have only one or two fatty acids, respectively, esterified to the glycerol backbone.

The *phospholipids* are a major component of cell walls. Two types of phospholipids are found in nature: glycerophospholipids (phosphoglycerols) and some sphingolipids (phosphosphingolipids) (Figure 22.14). In the phosphoglycerols, two fatty acids are esterified at C_1 and C_2 with a phospho-X group at C_3. Examples of the small molecules that comprise the X groups are given in Figure 22.14. In the *sphingolipids*, the

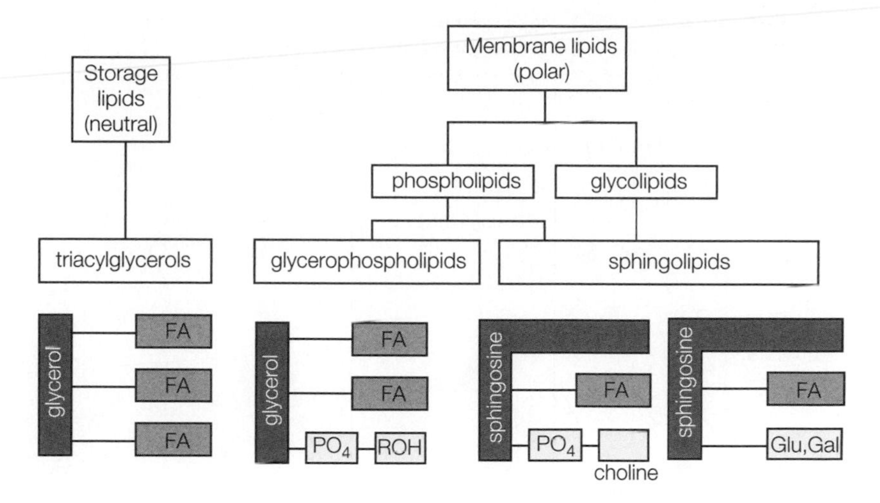

FIGURE 22.12

Complex lipid classification scheme. *Source*: After Lehninger, A. L., *et al.* (1993). Principles of Biochemistry, 2nd ed., W. H. Freeman. FA = fatty acid; Glu = glucose; Gal = galactose; ROH = alcohol.

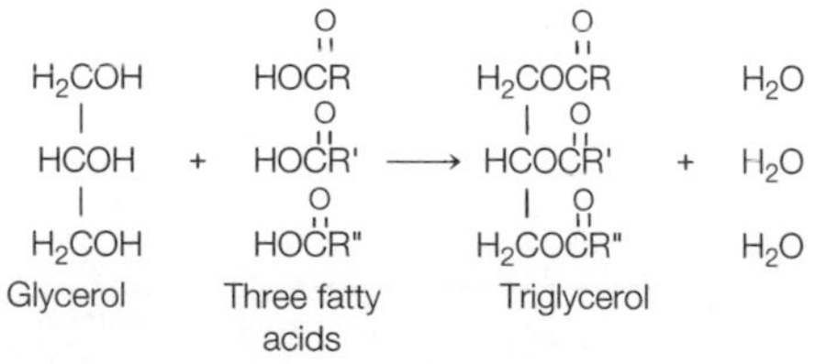

$$
\begin{array}{c} H_2COH \\ | \\ HCOH \\ | \\ H_2COH \\ \text{Glycerol} \end{array}
+
\begin{array}{c} HO\overset{O}{\overset{\|}{C}}R \\ HO\overset{O}{\overset{\|}{C}}R' \\ HO\overset{O}{\overset{\|}{C}}R'' \\ \text{Three fatty acids} \end{array}
\longrightarrow
\begin{array}{c} H_2CO\overset{O}{\overset{\|}{C}}R \\ | \\ HCO\overset{O}{\overset{\|}{C}}R' \\ | \\ H_2CO\overset{O}{\overset{\|}{C}}R'' \\ \text{Triglycerol} \end{array}
+
\begin{array}{c} H_2O \\ H_2O \\ H_2O \end{array}
$$

FIGURE 22.13

Formation of complex lipids through condensation reactions.

sphingosine backbone has a long alkyl group connected at C_1 and a primary amine at C_2. Various fatty acids bond to the amine group. A H atom or a phospho-X group is found at C_3. Some sphingolipids do not have a phosphate group. Instead, either a H atom (ceramide) or one or more sugar molecules (*glycosphingolipids*) are present at the C_3 carbon. A marine bacterium has been discovered that synthesizes a sphingolipid with a sulfonic acid group at the C_3 carbon.

The phosphate end of the phospholipids is relatively polar, making it water soluble or *hydrophilic*. In comparison, the fatty acid ends are relatively insoluble or *hydrophobic*. Molecules with hydrophilic "heads" and hydrophobic "tails" are termed *amphiphiles*. When their aqueous concentrations reach a critical level, they spontaneously aggregate to isolate their hydrophobic ends from water.

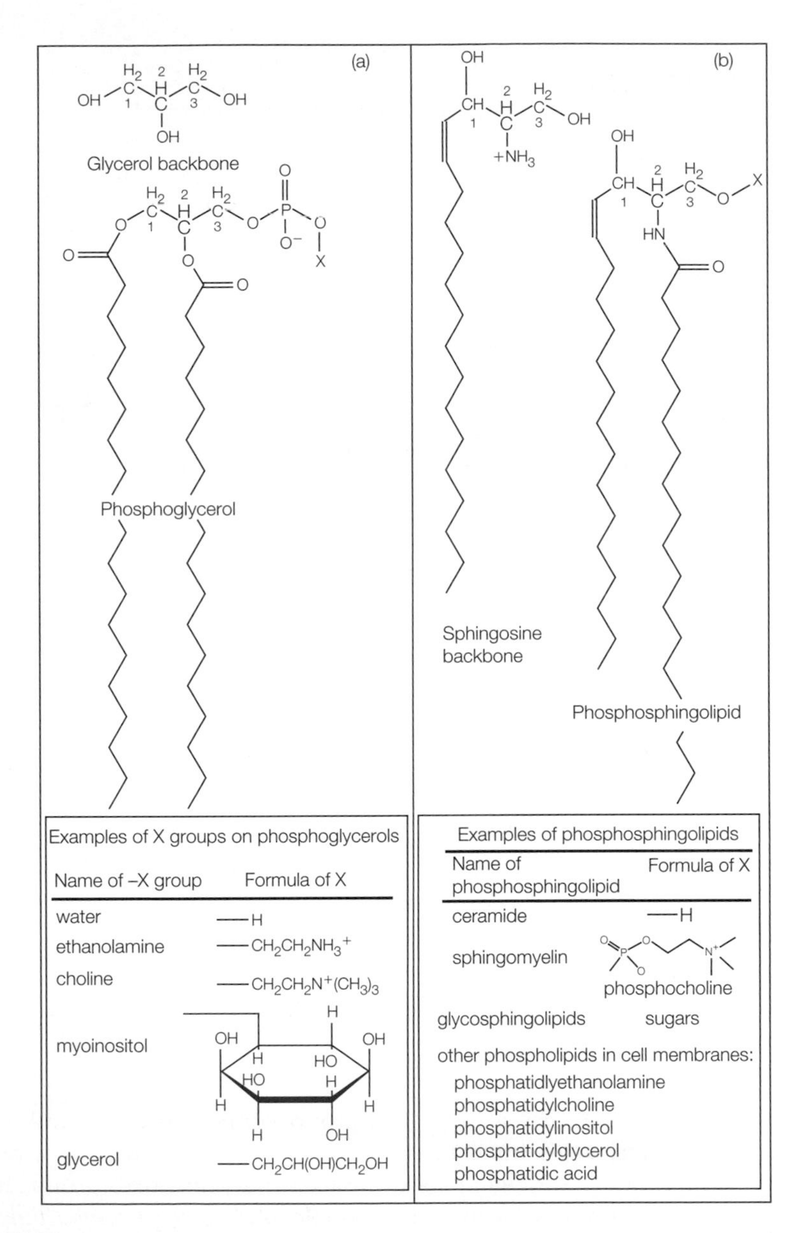

FIGURE 22.14

Phospholipids: (a) glycerol backbone and phosphoglycerols, (b) sphingosine backbone and phosphosphingolipids.

As illustrated in Figure 22.15a, amphiphiles with a single hydrophobic tail form *micelles.* Those with two tails, such as the phospholipids, form a vesicle with an aqueous interior (Figure 22.15b). This interior is isolated from the larger solution by a *bilayer* of phospholipid molecules in which one layer has its hydrophilic head dissolved in the interior and the other layer has its hydrophilic head dissolved in the larger solution. The hydrophobic ends of the molecules are thereby isolated from the aqueous environment. Amphiphiles spontaneously aggregate because of favorable thermodynamics; isolation of the hydrophobic tails from the aqueous solution is the stabler state. Cell membranes are functionally amphiphilic bilayers. Other biomolecules embedded or attached to inner surface of cell membranes include proteins (lipoproteins), sterols, glycoproteins, and glycolipids.

Simple Lipids

Simple lipids are a diverse group of compounds whose chemical structure is difficult to generalize. Some simple lipids are related to a base unit, called *isoprene*; others are not. For example, a *wax* is a long-chain alcohol, called a fatty alcohol, that has been esterified to a long-chain fatty acid. Waxes are common in plant leaves, feathers, and insect carapaces. In terrestrial organisms, they serve to control diffusion of gases, water, and aqueous solutes between the organisms and their environments. Waxes are also found in marine organisms. In fish, they are used as insulation and for buoyancy control and echo location. A sample structure is presented in Figure 22.16. In general, the fatty alcohol contains an even number of carbons ranging from 24 to 36 and the fatty acid chain has an even number of carbons, ranging from 16 to 36.

Alkenones are long-chain (C_{35}–C_{40}) unsaturated methyl or ethyl ketones produced by marine phytoplankton, for example by the coccolithophorid *Emiliania huxleyi* (Figure 22.17). Because the alkenones resist degradation, they are the most abundant extractable lipids in Quaternary marine sediments. These compounds are particularly useful as biomarkers because the degree of unsaturation in a given alkenone series has been shown to be a linear function of growth temperature. Thus, alkenones have been used to reconstruct past sea-surface temperatures on timescales ranging from interannual (El Niño) to millennial (glacial/interglacial). This temperature dependence reflects the phytoplankton's ability to regulate the melting point of its lipids in an effort to keep its cell membranes fluid. Recall that the melting point of lipids decreases with increasing unsaturation. Thus, phytoplankton growing at lower temperatures must increase the degree of unsaturation in their alkenones to keep their cell membranes fluid.

The *terpenes* are simple lipids whose base unit is isoprene. Oxygen-containing terpenes are called *terpenoids*, and terpenes with hydroxyl groups are called *terpenols*. Terpenes are further classified based on the number of isoprene units in the molecule as shown in Table 22.6. Examples of terpene molecular structures are presented in Figure 22.18.

Phytene is an example of a diterpene. It is found as the phytyl side chain in chlorophyll *a* and vitamin K. Haslene is an example of a sesterpene. It is an unsaturated and branched simple lipid synthesized by marine pennate diatoms. One of the largest families of terpenes

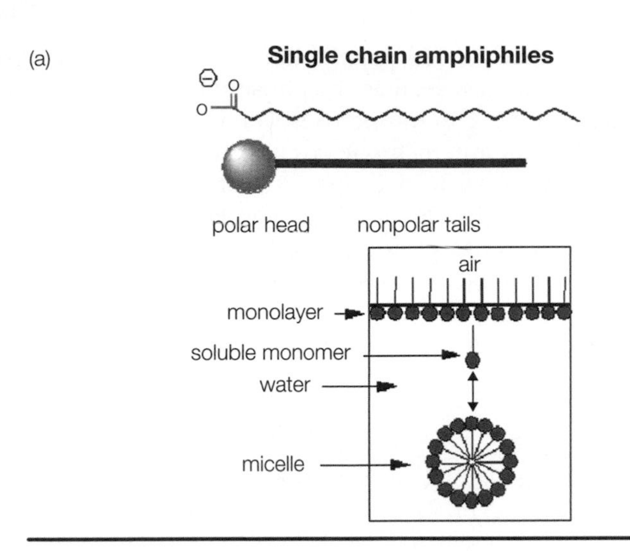

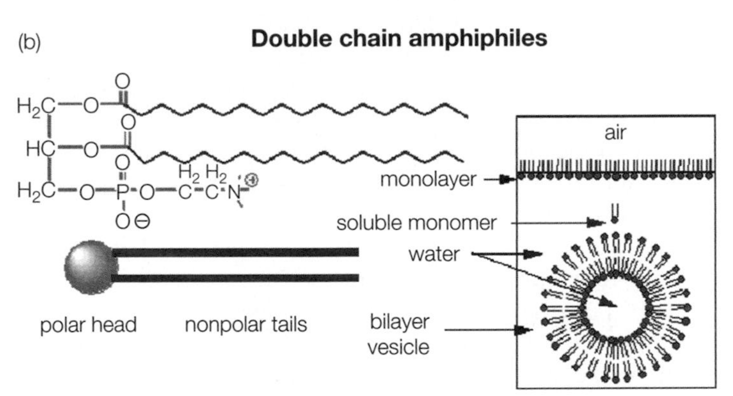

FIGURE 22.15

(a) A single-chain amphiphile and its micelle and (b) a double-chain amphiphile (e.g., phosphatidylcholine) forming a lipid bilayer. The resulting micelle is called a vesicle. *Source*: From Jakutowski, H. (2008). Biochemistry 331, St. John's University. http://employees.csbsju.edu/hjakubowski/classes/ch331/lipidstruct/ollipidwater.html.

are the triterpenoids, whose common precursor is the 20-carbon isoprenoid squalene.

FIGURE 22.16

A wax ester.

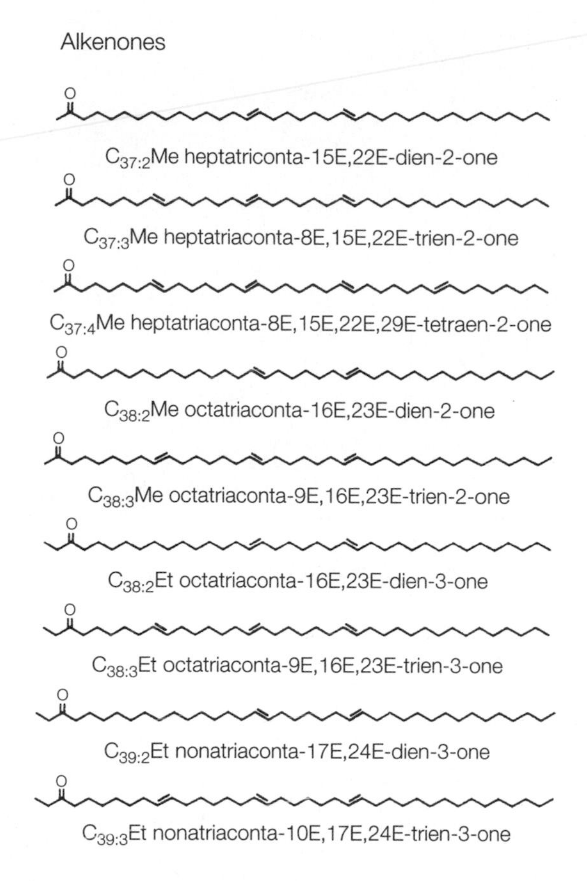

FIGURE 22.17

Examples of common long-chain alkenones found in marine sediments. Me = methyl; Et = ethyl.

Table 22.6 Terpene Classes.

	Terpenes	Isoprene units	Carbon atoms
1	Monoterpenes	2	10
2	Sesquiterpenes	3	15
3	Diterpenes	4	20
4	Sesterpenes	5	25
5	Triterpenes	6	30
6	Carotenoids	8	40
7	Rubber	>100	>500

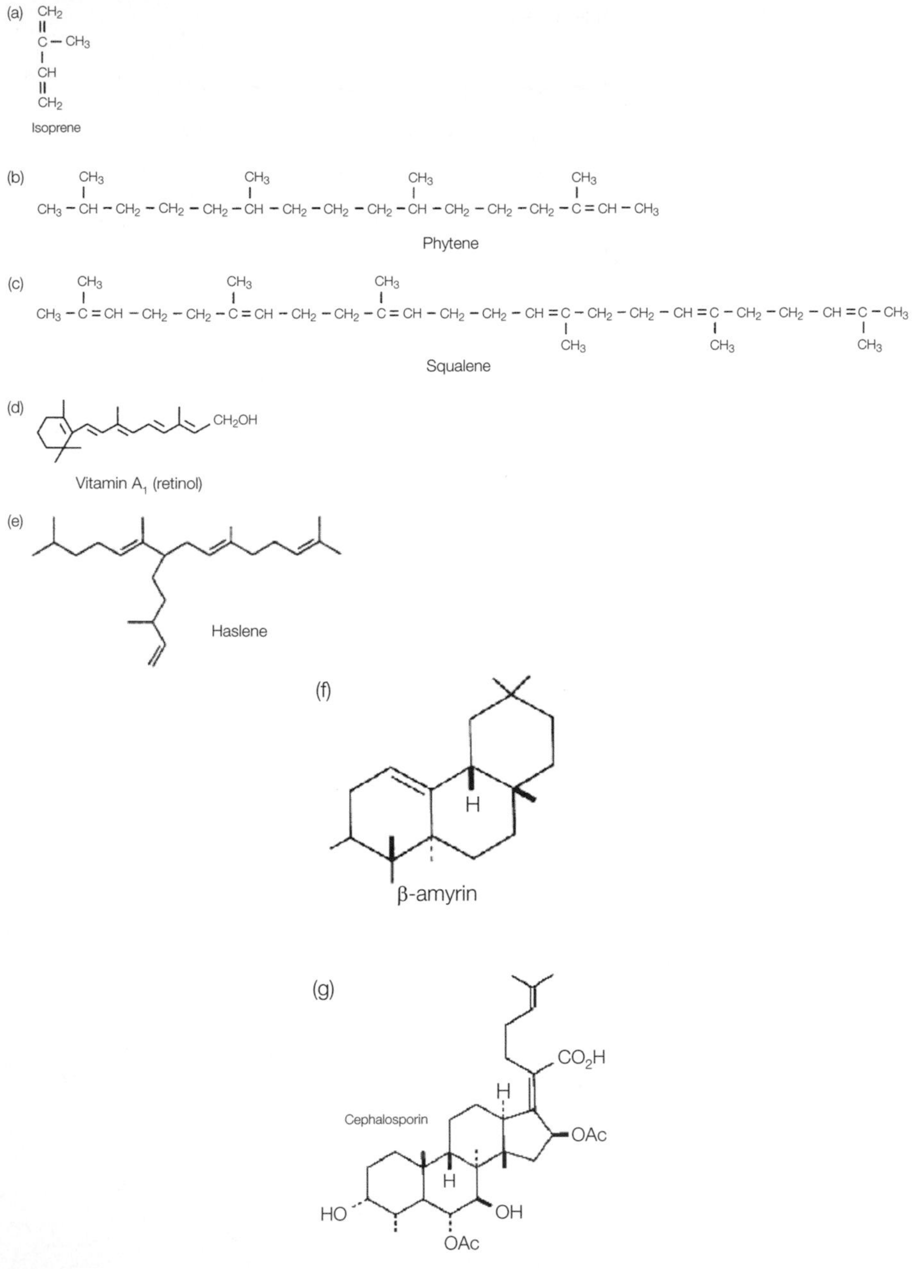

(a) $CH_2=C(CH_3)-CH=CH_2$

Isoprene

(b) $CH_3-CH(CH_3)-CH_2-CH_2-CH_2-CH(CH_3)-CH_2-CH_2-CH_2-CH(CH_3)-CH_2-CH_2-CH_2-C(CH_3)=CH-CH_3$

Phytene

(c) $CH_3-C(CH_3)=CH-CH_2-CH_2-C(CH_3)=CH-CH_2-CH_2-C(CH_3)=CH-CH_2-CH_2-CH=C(CH_3)-CH_2-CH_2-CH=C(CH_3)-CH_2-CH_2-CH=C(CH_3)-CH_3$

Squalene

(d) CH_2OH

Vitamin A_1 (retinol)

(e) Haslene

(f) H

β-amyrin

(g) CO_2H, H, OAc, H, HO, OH, OAc

Cephalosporin

FIGURE 22.18

Examples of terpenes: (a) isoprene, (b) phytene, which is a side chain of chlorophyll *a*, (c) squalene, (d) vitamin A, (e) haslene, (f) amyrin, and (g) cephalosporin.

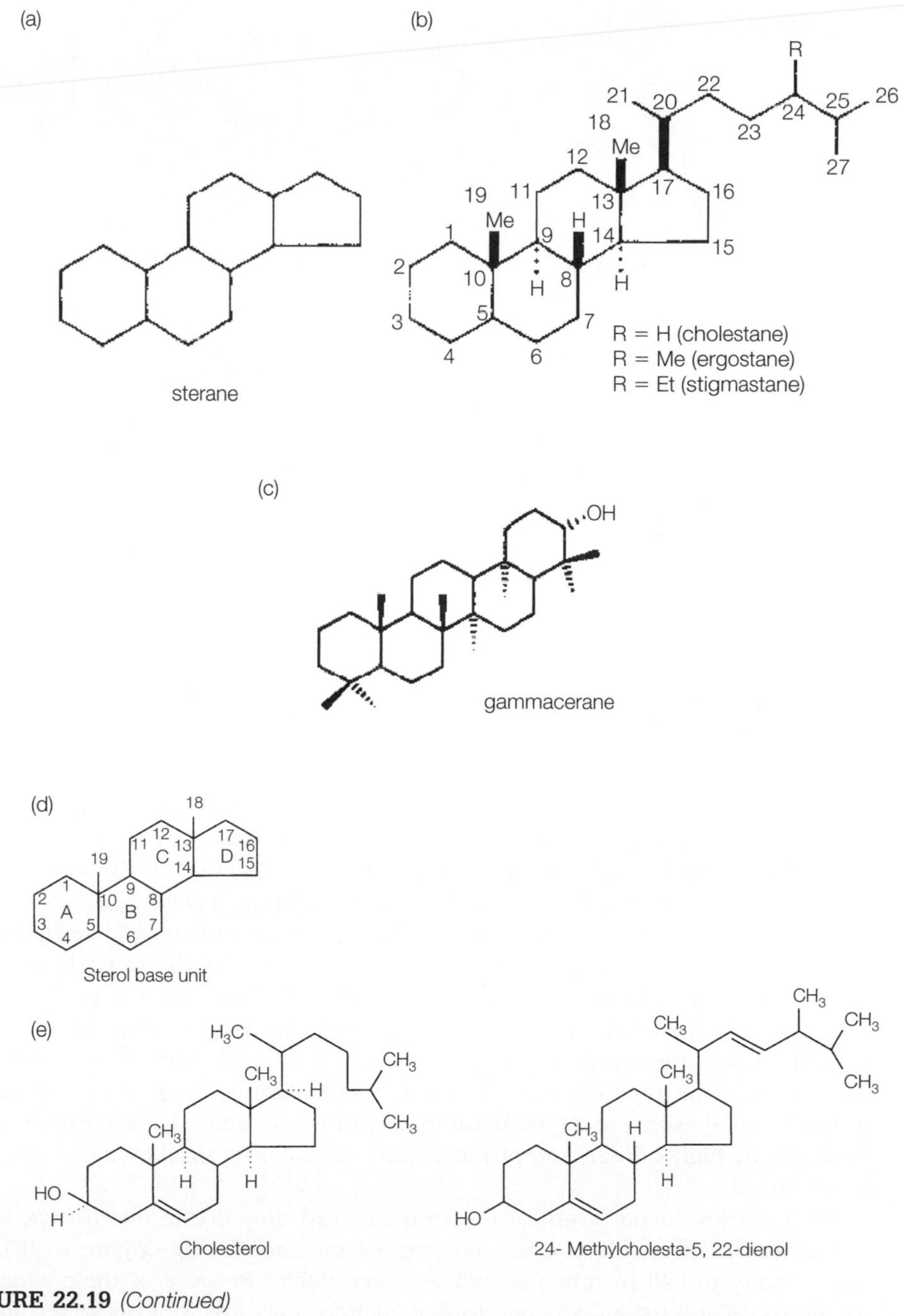

(a)

sterane

(b)

R = H (cholestane)
R = Me (ergostane)
R = Et (stigmastane)

(c)

gammacerane

(d)

Sterol base unit

(e)

Cholesterol

24- Methylcholesta-5, 22-dienol

FIGURE 22.19 *(Continued)*

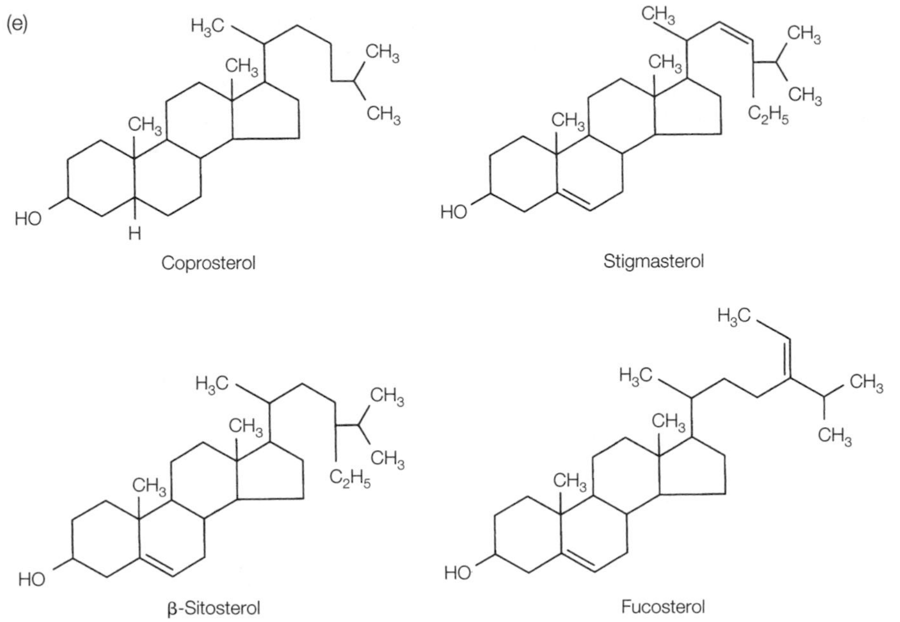

FIGURE 22.19

(a) Sterane base unit and (b) some examples, (c) gammacerane, and (d) sterol base unit, and (e) some examples. Me = methyl; Et = ethyl.

More than 1500 triterpenoids have been discovered based on 40 skeletal types. Some examples include amyrin (produced by terrestrial plants), cephalosporin (synthesized by marine bacteria), and the steranes. The skeletal structure of the last contains three six-membered carbon rings and one five-membered ring (Figure 22.19a).

Addition of functional groups to the steranes, such as ketones and hydroxyls, gives rise to the steroids. One particularly common category of the steroids are the ***sterols***, which contain a hydroxyl group at C_3 and a long alkyl chain at C_{17} (>8 carbons). Some examples of sterols are given in Figure 22.19e. They are synthesized by plants and animals. Some are components of cell membranes, and others are metabolic regulators. Cholesterol is the dominant sterol in zooplankton and red algae, whereas diatoms primarily synthesize the 24-methylsterols. β-Sitosterol and stigmasterol are synthesized primarily by higher-order land plants. Fucosterol is synthesized by brown algae. Vitamin D is a sterol.

Prokaryotes do not synthesize sterols. Instead they create ***hopanoids***, which have four six-membered carbon rings and one five-membered ring (Figure 22.8f). They provide rigidity to cell membranes and are very stable. Because of their widespread use by microbes and their resistance to degradation, they are well preserved in sediments and petroleum deposits, making them the most abundant natural products on Earth.

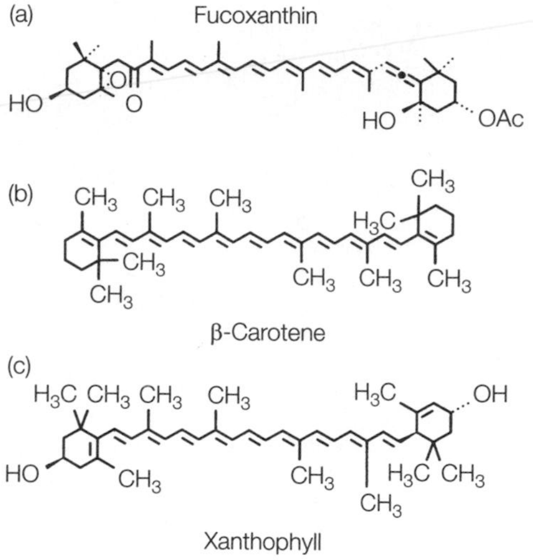

FIGURE 22.20

Structure of various terpenoid plant pigments: (a) fucoxanthin, (b) β-carotene, and (c) xanthophyll. OAc = acetate.

Because of their specificity for particular bacterial groups and environmental conditions, they make excellent biomarkers. Another steroidal biomarker is gammacerane (Figure 22.19b), which has six six-membered carbon rings. It is used as an indicator of water-column stratification in marine and lacustrine deposits.

The *carotenoids* are 40-carbon terpenoids. They serve as photosynthetic pigments in algae, photosynthetic bacteria, higher plants, and algae. They are also present in some nonphotosynthetic bacteria. The most abundant carotenoids are β-carotene, an orange pigment common in terrestrial and marine plants, and fucoxanthin, a yellow-brown pigment common in diatoms and dinoflagellates (Figure 22.20). Carotenoids make excellent biomarkers, especially in anoxic sediments where their diagenetic alteration produces unique residues. Carotene, which is a hydrocarbon, is also a precursor for biosynthesis of vitamin A.

22.4.4 Nucleotides and Nucleic Acids

Nucleotides are composed of an aromatic base, a five-carbon sugar, and one or more phosphate groups. The five aromatic bases that can be present are illustrated in Figure 22.21 and cause the nucleotides to be relatively nitrogen rich.

Nucleotides are also called mononucleotides. An example is given in Figure 22.22a. When linked together, they form polymers called *polynucleotides*. RNA and DNA, which store and transmit genetic information within cells, are examples of polynucleotides (Figures 22.22b and c). They are also commonly referred to as *nucleic*

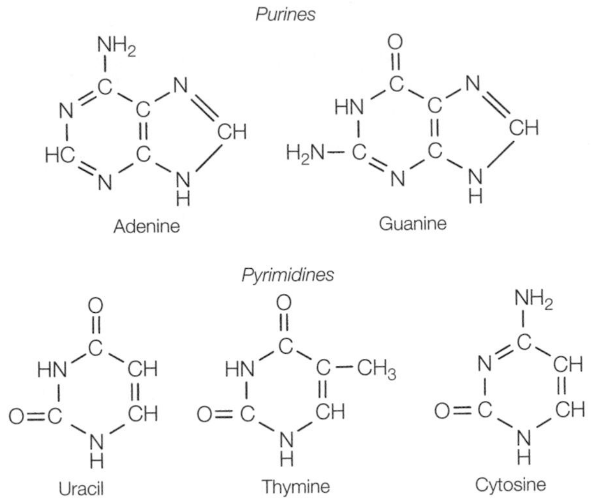

FIGURE 22.21

Aromatic bases present in nucleotides and nucleic acids.

acids. Other nucleotides include the energy-carrying molecules ATP and NADP (nicotinamide adenine dinucleotide phosphate). Extracellular DNA, RNA, and ATP are found in seawater and the sediments, having been released from tissues during bacterial decomposition.

22.4.5 Amino Acids and Proteins

Amino acids are the building blocks of proteins. These compounds all bear at least one amine group (R-NH_2) and a carboxyl group (R-COOH). The naturally occurring amino acids are listed in Table 22.7.

Amino acids become linked by reactions that occur between the amine group of one amino acid and the carboxyl group of another. As shown in Figure 22.23, this polymerization produces a molecule of water and, hence, is a condensation reaction. Naturally occurring polypeptides with molecular weights in excess of 10,000 daltons[3] are termed *proteins*. These biomolecules are ubiquitous in marine organisms and are not specific to particular species. Proteins are important components of enzymes as well as of structural parts and connective tissues.

Proteins are an important part of bacteria cell walls that surround their cell membranes. Most of the cell wall of gram-positive bacteria consists of interlinked layers of peptidoglycan. As illustrated in Figure 22.24, this polymer is composed of an alternating sequence of amino sugars, *N*-acetylglucosamine and *N*-acetylmuramic acid. Each peptidoglycan layer is connected, or crosslinked, to the other by a bridge composed of

[3] Protein chemists use units of daltons rather than molecular mass although the two are interchangeable.

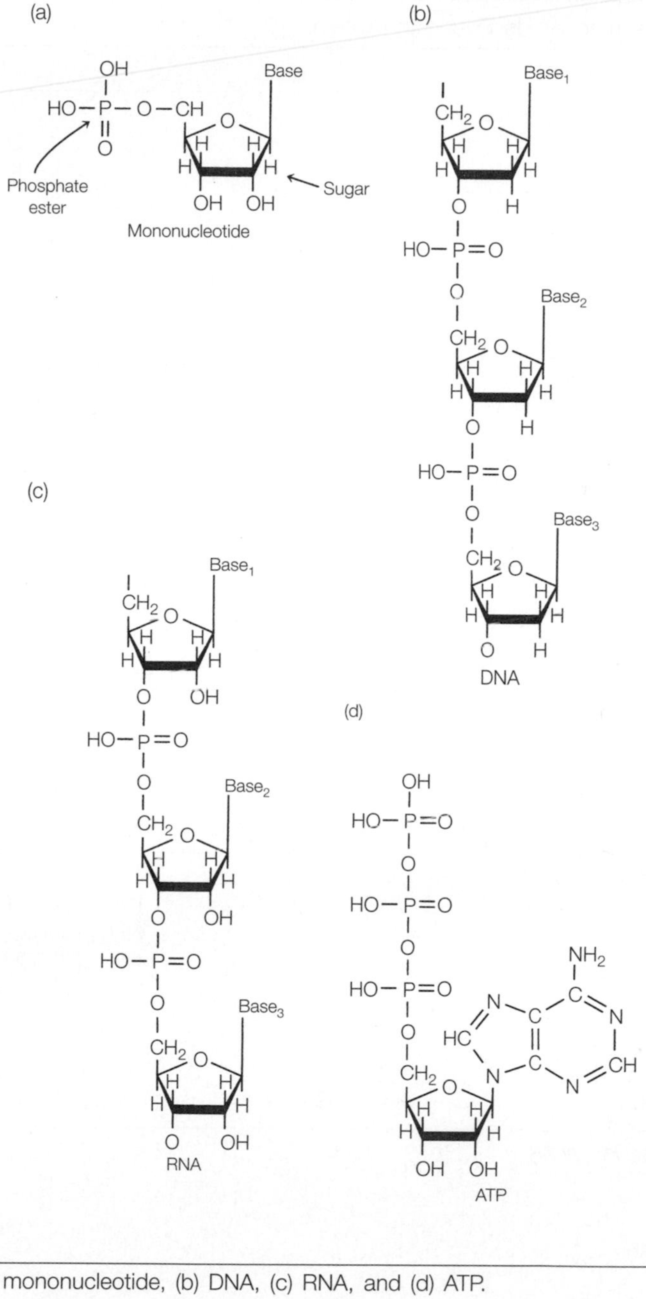

(a)

(b)

(c)

(d)

FIGURE 22.22

Structures of (a) a mononucleotide, (b) DNA, (c) RNA, and (d) ATP.

Table 22.7 Amino Acids Found in Proteins.

I. Aliphatic amino acids	
A. Monoaminomonocarboxylic acids	
Glycine	H H O N − C − C − OH H H
Alanine	H H O H − C − C − C − OH H N H H
Valine	$(CH_3)_2CHCH(NH_2)COOH$
Leucine	$(CH_3)_2CHCH_2CH(NH_2)COOH$
Isoleucine	$CH_3CH_2CH(CH_3)CH(NH_2)COOH$
Serine	$HOCH_2CH(NH_2)COOH$
Threonine	$CH_3CH(OH)CH(NH_2)COOH$
B. Sulfur-containing amino acids	
Cysteine	$HSCH_2CH(NH_2)COOH$
Methionine	$CH_3SCH_2CH_2CH(NH_2)COOH$
C. Monoaminodicarboxylic acids and their amides	
Aspartic acid	$HOOCCH_2CH(NH_2)COOH$
Asparagine	$NH_2COCH_2CH(NH_2)COOH$
Glutamic acid	$HOOCCH_2CH_2CH(NH_2)COOH$
Glutamine	$NH_2COCH_2CH_2CH(NH_2)COOH$
D. Basic amino acids	
Lysine	$NH_2CH_2CH_2CH_2CH_2CH(NH_2)COOH$
Hydroxylysine	$NH_2CHCH(OH)CH_2CH_2CH(NH_2)COOH$
Arginine	$NH_2C(NH)CH_2CH_2CH_2CH(NH_2)COOH$
Histidine	H N − C − $CH_2CH(NH_2)COOH$ H − C N − CH
II. Aromatic amino acids	
Phenylalanine	H H C = C HC C − $CH_2CH(NH_2)COOH$ C − C H H

(Continued)

Table 22.7 *(Continued)*

II. Aromatic amino acids	
Tyrosine	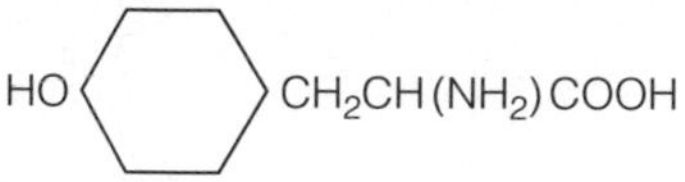 HO–C_6H_4–$CH_2CH(NH_2)COOH$
III. Heterocyclic amino acids	
Tryptophan	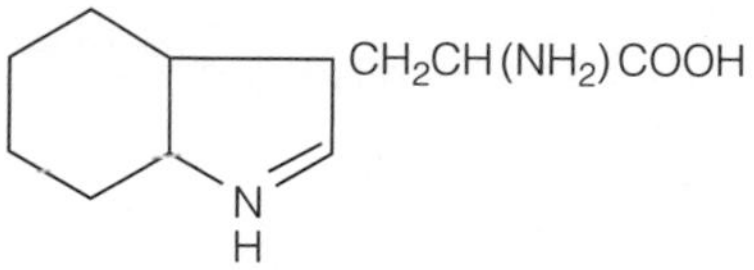 C_8H_5NH–$CH_2CH(NH_2)COOH$
Proline	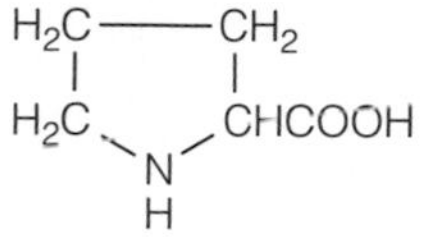 H_2C—CH_2, H_2C—N(H)—$CHCOOH$ (ring)
Hydroxyproline	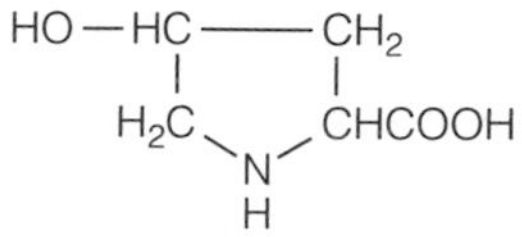 HO—HC—CH_2, H_2C—N(H)—$CHCOOH$ (ring)
Histidine (see above)	

Source: *From Horne, R. A. (1969).* Marine Chemistry, *John Wiley & Sons, Inc., p. 246.*

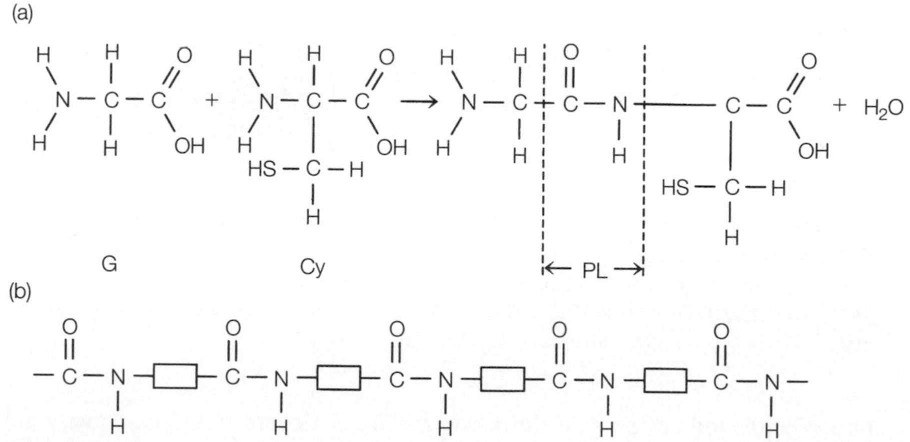

(a)

$$H_2N\text{-}CH_2\text{-}COOH + H_2N\text{-}CH(CH_2SH)\text{-}COOH \rightarrow H_2N\text{-}CH_2\text{-}C(=O)\text{-}NH\text{-}CH(CH_2SH)\text{-}COOH + H_2O$$

G Cy ← PL →

(b)

$$-C(=O)-NH-\square-C(=O)-NH-\square-C(=O)-NH-\square-C(=O)-NH-\square-C(=O)-NH-$$

FIGURE 22.23

(a) The condensation of two amino acids and (b) diagrammatic representation of a protein, showing the peptide linkage (PL) between the amino acids.

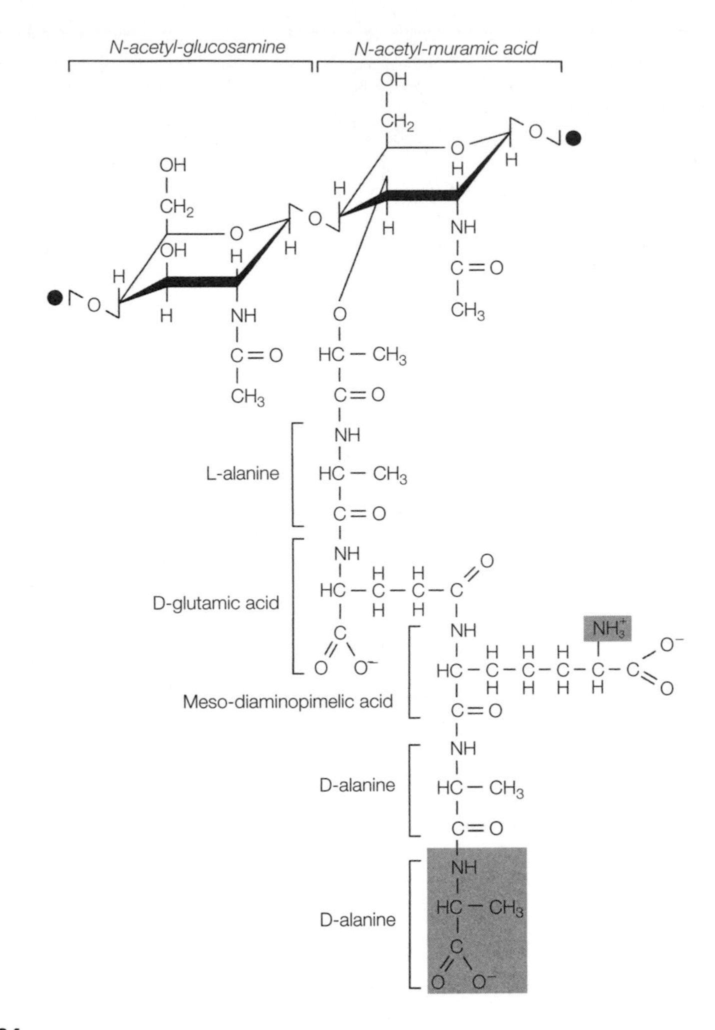

FIGURE 22.24

The peptidoglycan structure of bacterial cell walls. The shaded areas represent points of attachment of this macromolecule to the rest of the cell wall. The amino sugar units are joined end to end to form long, straight chains. The peptides form cross-links when the amino group of a meso-diaminopimelic acid in one chain replaces the terminal alanine in another chain. *Source*: From Koch, A. L. (1990). *American Scientist* 78, 327–341.

amino acids and amino acid derivatives. The particular amino acids vary among the different clades (i.e., bacterial "species"). The crosslinked peptidoglycan molecules form a network that covers the cell like a grid. Gram-negative bacteria also have a peptidoglycan layer in their cell wall, but it is much thinner and, hence, less interlinked than in gram-positive bacteria.

22.4.6 Pyrroles and Porphyrins

Pyrroles are components of important biomolecules, such as the pigments, chlorophyll, and phycobilin, as well as heme and vitamin B_{12}. The pyrrole group is a five-membered ring that contains one nitrogen atom (Figure 22.25a). The phycobilins are light-harvesting pigments (chromophores) found in cyanobacteria and the chloroplasts of red algae. They are efficient at absorbing red, orange, yellow, and green light at wavelengths not well absorbed by chlorophyll and, hence, serve as accessory pigments. They consist of an open chain of four pyrrole rings. The blue phycobilin, phycocyanobilin, is shown in Figure 22.25b.[4]

Chlorophylls are also composed of four pyrroles, but these are arranged in a ring, called a porphyrin (Figure 22.25c). The chlorophylls are green photosynthetic pigments found in plants and some bacteria (Figure 22.25d). Another important class of

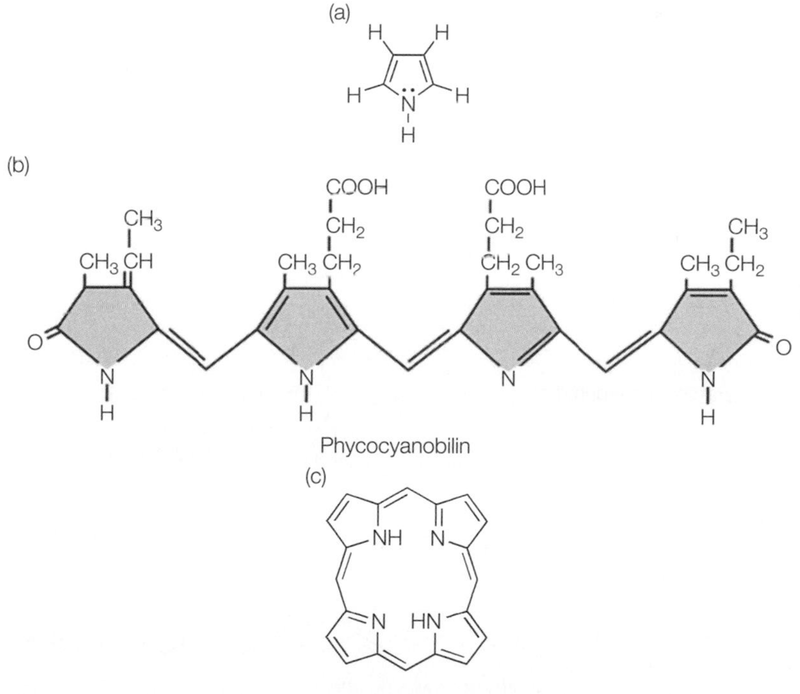

FIGURE 22.25 *(Continued)*

[4] In cyanobacteria and red algae, the phycobilins are covalently bound to proteins. The resulting structure is an example of a *phycobiliprotein*. In marine cyanobacteria and red algae, *phycoerythrins* are a common type of phycobiliprotein. The phycoerythins are distinguished by the large number of phycobilin groups that are present relative to the protein subunits.

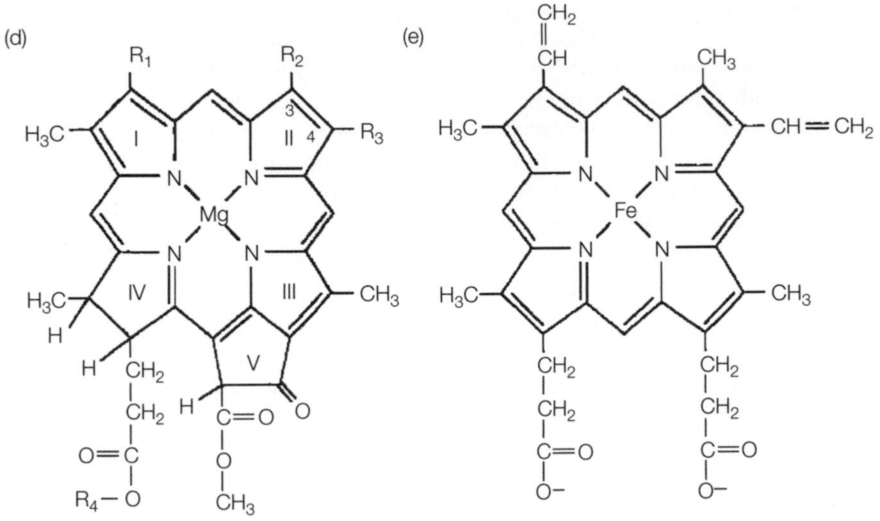

Chlorophy II

Iron–protoporphyrin IX

	R_1	R_2	R_3	R_4
Chlorophyll *a*	$-CH{=}CH_2$	$-CH_3$	$-CH_2-CH_3$	P
Chlorophyll *b*	$-CH{=}CH_2$	$-\overset{O}{\overset{\Vert}{C}}-H$	$-CH_2-CH_3$	P
Bacteriochlorophyll *a*	$-\overset{O}{\overset{\Vert}{C}}-CH_3$	$-CH_3$[a]	$-CH_2-CH_3$[a]	P or G
Bacteriochlorophyll *b*	$-\overset{O}{\overset{\Vert}{C}}-CH_3$	$-CH_3$[a]	$=CH-CH_3$[a]	P

[a] No double bond between positions C3 and C4.

P = — CH_2 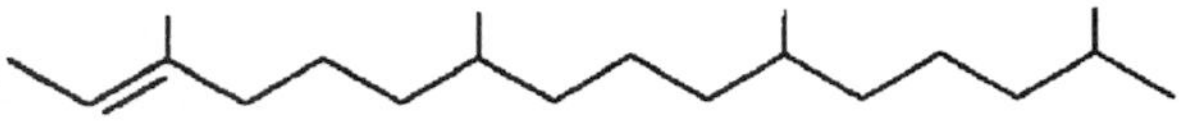

Phytyl side chain

G = — CH_2 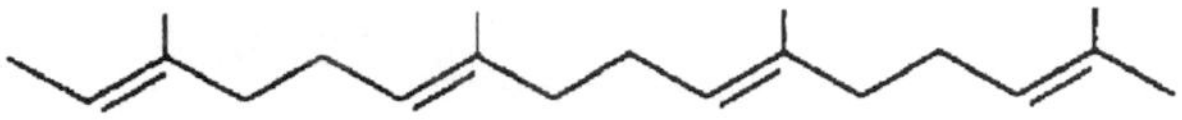

Geranylgeranyl side chain

FIGURE 22.25

(a) Pyrrole group, (b) phycocyanobilin, (c) porphyrin group, (d) chlorophylls, and (e) heme B.

biomolecules that contain porphyrins are the hemes. Figure 22.25e shows the molecular structure of heme B. Heme-containing biomolecules include the hemoglobins, myoglobins, and cytochromes. The hemoglobins and myoglobins, are responsible for oxygen transport and storage in living tissues. The cytochromes also serve as electron carriers in the electron transport chain (Table 7.11).

The porphyrins are notable for their ability to retain a metallic cation through the electron-donating power of the nitrogen atoms. In chlorophyll, this cation is magnesium. During the initial phases of diagenesis, chlorophyll tends to lose its central magnesium and its phytyl side chain (Figure 22.25c). The degraded porphyrin is then available to chelate trace metals from the surrounding sediments. This chelation causes the metals to be retained as the sediments are transformed into sedimentary rocks (shales) and petroleum.

In hemoglobin, the metallic cation bound by the porphyrin is either iron or zinc. Cobalt is bound by the tetrapyrrole structure found in vitamin B_{12}. Some experimental work suggests that vitamin B_{12} availability could be limiting phytoplankton growth in some marine environments. Phytoplankton do not synthesize this vitamin. Thus, they must assimilate it from seawater. Bacteria are thought to release vitamin B_{12} to seawater, thus providing phytoplankton with this essential biomolecule. Bacteria and some animals are also known to synthesize halogenated pyrrole compounds as shown later in Figure 22.29.

22.4.7 Low-Molecular-Weight Nitrogenous Compounds

Nitrogen is an important component of many biomolecules, such as proteins, amino acids, amino sugars, the aromatic bases in nucleotides, and the siderophores (Figure 5.11). Other nitrogen-containing compounds that have been found in seawater are excretory products such as creatinine, methylamine, urea, and uric acid (Figure 22.26).

22.4.8 Low-Molecular-Weight Carboxylic Acids

Low-molecular-weight carboxylic acids are produced and consumed during the Krebs cycle and glycolysis (see Figure 7.6), so these compounds are ubiquitous in marine organisms. Phytoplankton release a significant fraction of their fixed carbon into seawater as low-molecular-weight carboxylic acids, such as those shown in Figure 22.27.

22.4.9 Low-Molecular-Weight Phosphorus- and Sulfur-Containing Compounds

Organic compounds that contain phosphorus include ATP, NADP, and the phospholipids. The phosphate group is often temporarily attached to other organic compounds, such as pyruvate, as part of metabolic reactions, like glycolysis. As listed in Table 22.8, many of the phosphorus-containing biochemicals are found in seawater as a result of cell lysis and excretion.

Table 22.8 Low-Moleculor Weight Organic Phosphorus Compounds Indentified or Likely to Be Present in Seawater.

Compound	Chemical Formula (Molecular Weight)	P (% by Weight)	Molar C : N : P[b]
Monophosphate esters			
Ribose 5-phosphoric acid (R-5-P)	$C_5H_{11}O_8P$ (230.12)	13.5	5:_:1
Phospho(enol)pyruvic acid (PEP)	$C_3H_5O_6P$ (168)	18.5	3:_:1
Glyceraldehyde 3-phosphoric acid (G-3-P)	$C_3H_7O_6P$ (170.1)	18.2	3:_:1
Glycerophosphoric acid (gly-3-P)	$C_3H_9O_6P$ (172.1)	18.0	3:_:1
Creatine, phosphoric acid (CP)	$C_4H_{10}N_3O_5P$ (211.1)	14.7	4:3:1
Glucose, 6-phosphoric acid (glu-6-P)	$C_6H_{13}O_9P$ (260.14)	11.9	6:_:1
Ribulose l,5-bisphosphoric acid (RuBP)	$C_5H_6O_{11}P_2$ (304)	20.4	2.5:_:1
Fructose 1,6-diphosphoric acid (F-1,6-DP)	$C_6H_{14}O_{12}P_2$ (340.1)	18.2	3:_:1
Phosphoserine (PS)	$C_3H_8NO_6P$ (185.1)	16.7	3:1:1
Nucleotides and derivatives			
Adenosine 5'-triphosphoric acid (ATP)	$C_{10}H_{16}N_5O_{13}P_3$ (507.2)	18.3	3.3:1.7:1
Uridylic acid (UMP)	$C_9H_{13}N_2O_9P$ (324.19)	9.6	9:2:1
Uridine-diphosphate-glucose (UDPG)	$C_{15}H_{24}N_2O_{17}P_2$ (566.3)	10.9	7.5:1:1
Guanosine 5'-diphosphate-3'-diphosphate (ppGpp)	$C_{10}H_{17}N_5O_{17}P_4$ (603)	20.6	2.5:1.5:1
Pyridoxal 5-monophosphoric acid (PyMP)	$C_8H_{10}NO_6P$ (247.2)	12.5	8:1:1
Nicotinamide adenine dinucleotide phosphate (NADP)	$C_{22}H_{28}N_2O_{14}N_6P_2$ (662)	9.4	11:3:1
Ribonucleic acid (RNA)	Variable	~9.2	~9.5:4:1
Deoxyribonucleic acid (DNA)	Variable	~9.5	~10:4:1
Inositohexaphosphoric acid, or phytic acid (PA)	$C_6H_{18}O_{24}P_6$ (660.1)	28.2	1:_:1
Vitamins			
Thiamine pyrophosphate (vitamin B_1)	$C_{12}H_{19}N_4O_7P_2S$ (425)	14.6	6:2:1
Riboflavin 5'-phosphate (vitamin B_2-P)	$C_{17}H_{21}N_4O_9P$ (456.3)	6.8	17:4:1
Cyanocobalamin (vitamin B_{12})	$C_{63}H_{88}CoN_{14}O_{14}P$ (1355.42)	2.3	63:14:1
Phosphonates			
Methylphosphonic acid (MPn)	CH_5O_3P (96)	32.3	1:_:1
Phosphonoformic acid (FPn)	CH_3O_5P (126)	24.6	1:_:1
2-Aminoethylphosphonic acid (2-AEPn)	$C_2H_8NO_4P$ (141)	22.0	2:1:1
Other compounds/compound classes			
Marine fulvic acid (FA)[a]	Variable	0.4–0.8	80–100:_:1
Marine humic acid (HA)[a]	Variable	0.1–0.2	>300:_:1
Phospholipids (PL)	Variable	≤0.4	~40:1:1
Malathion (Mal)	$C_9H_{16}O_5PS$ (267)	11.6	9:_:1
Redfield phytoplankton	Variable	1–3	106:16:1

[a] *Marine HA and FA are operationally defined fractions, and their composition may be variable (values are from Nissenbaum, 1979). Phosphate associated with HA and FA may be organically bound. Alternatively, it may be inorganic orthophosphate linked to HA and/or FA through metal bridges (Laarkamp, 2000).*

[b] *Ratios represented by x:_:y are C:P ratios.*

Source: From Karl, D. M., and K. M. Björkman (2002). Biogeochemistry of Marine Dissolved Organic Matter, *Academic Press, pp. 249–349.*

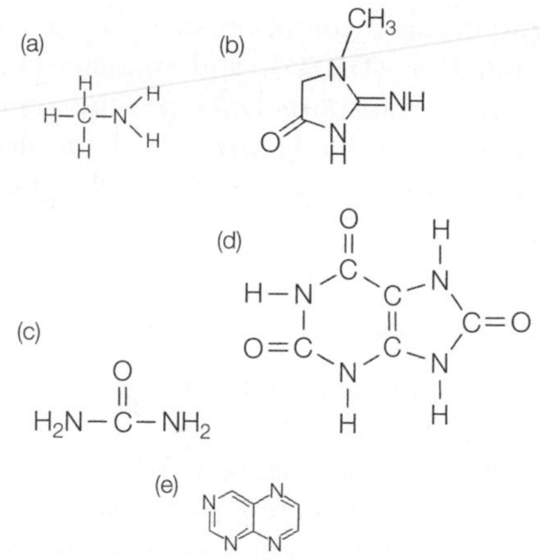

(a) (b) (c) (d) (e)

FIGURE 22.26

Low-molecular-weight nitrogenous compounds: (a) methylamine, (b) creatinine, (c) urea, (d) uric acid, and (e) pteridine.

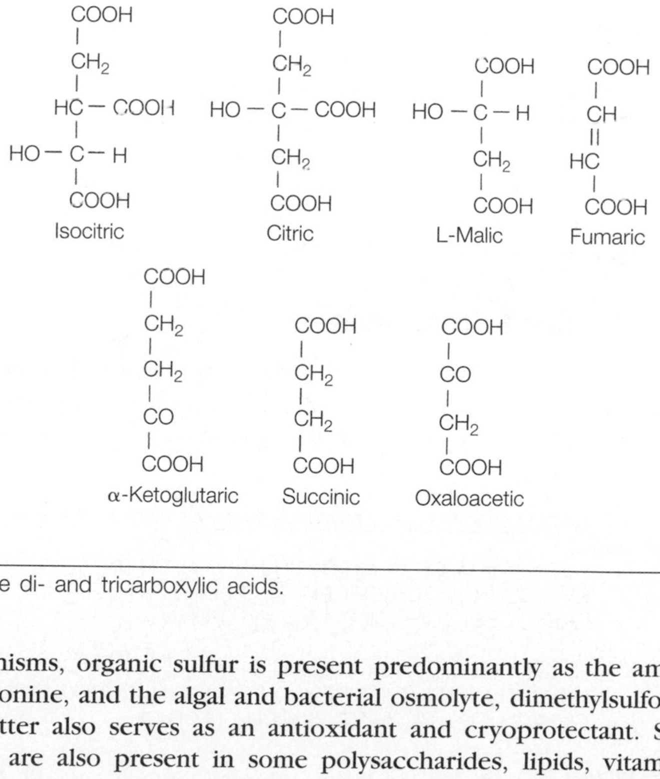

Isocitric Citric L-Malic Fumaric

α-Ketoglutaric Succinic Oxaloacetic

FIGURE 22.27

Structures of some di- and tricarboxylic acids.

Within organisms, organic sulfur is present predominantly as the amino acids cysteine and methionine, and the algal and bacterial osmolyte, dimethylsulfoniopropionate (DMSP). The latter also serves as an antioxidant and cryoprotectant. Small amounts of organosulfur are also present in some polysaccharides, lipids, vitamins, enzymes, and in the iron-sulfur protein ferrodoxin. Cell lysis and microbial degradation releases low-molecular-weight organic compounds, such as dimethyl sulfide (DMS), dimethyl disulfide, dimethylsulfoxide (DMSO), and methanethiol into seawater. Gas exchange across the air-sea interface transports DMS into the atmosphere where it plays an important role in climate regulation. See Figure W23.1 for more details. Because methanethiol forms complexes with DOM and trace metals, it has the potential to affect bioavailability of the micronutrients.

22.4.10 Organohalogens

A large number of organohalogen compounds have been found in seawater and marine organisms. This is not surprising given the high concentrations of chloride, bromine, and iodine in seawater. The organohalogens are toxic and are used as a chemical defense agent. Marine algae and seaweeds are a particularly prolific source of halomethanes and haloalkanes. Examples of these compounds are provided in Figure 22.28. These volatile compounds are thought to give a characteristic smell to sea air. Coastal salt marshes emit large quantities of methyl chloride and methyl bromide, representing a significant amount of the global atmospheric input. This is of particular interest because these chemicals react with stratospheric ozone, causing its destruction.

Marine animals, such as sponges and soft-bodied invertebrates, are well known to synthesize a large number of organohalogens or to harbor microbial symbionts that synthesize these compounds. Many have physiological activities that made them useful as marine natural products. Examples of these are provided in Chapter 28, namely furanone (S5), oroidin (S9), Tyrian purple (S20), spisulosine (S25), and salinosporamide A (S39).

Surprisingly, many halogenated compounds thought to be structurally similar to pollutants have been found to be naturally bioaccumulating in marine fish, birds, and mammals. These include the highly brominated bipyrroles (Figure 22.29). They are likely bacterial metabolites that are being passed up the food chain.

22.4.11 Humic Substances

Only 10% of the dissolved organic carbon in seawater has been characterized on a molecular level. About 5 to 23% of this DOC can be isolated from seawater by adsorption onto a nonionic macroporous resin followed by elution with an alkaline solution.

Volatile Organohalogens produced by marine algae and seaweed

Halomethanes: CH_3Cl, CH_3Br, CH_3I, CH_2Cl_2, CH_2Br_2, CH_2I_2, $CHCl_3$, $CHBr_3$, CHI_3, CCl_4, CBr_4, CH_2ClBr, CH_2ClI, CH_2Br_2, CH_2BrI, CH_2ClI, $CHBr_2Cl$, $CHBrI_2$, $CHBr_3$, $CHBrClI$, $CHBr_2I$, and $CHBrCl_2$

Haloalkanes: CH_3CH_2Br, CH_3CH_2I, $CH_3CH_2CH_2I$, $CH_3(CH_2)_3I$, $CH_3(CH_2)_4Br$, $CH_3(CH_2)_4I$, $(CH_3)_2CHI$, $CH_3CH_2CH(CH_3)I$, $(CH_3)_2CHCH_2I$, $BrCH_2CH_2Br$, $ClCH_2CH_2CH_2I$, $ClCH{=}CCl_2$, and $CH_3CH_2CH_2CH_2I$

FIGURE 22.28

Representative organohalogens found in seawater and marine organisms.

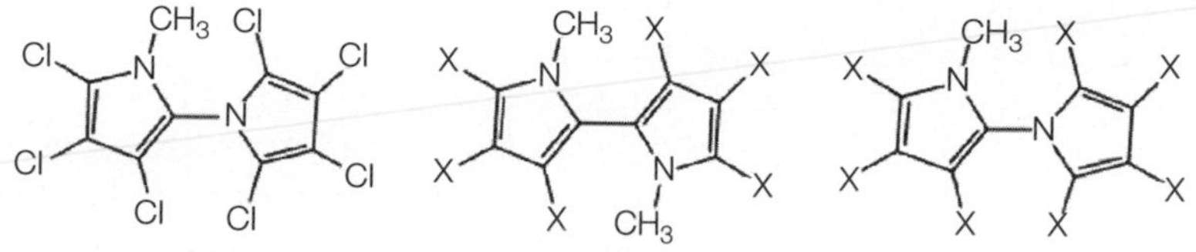

Q1
(MBP-Cl_7)

DMBP
X = Br/Cl
Br = 3-6; Cl = 0-3

MBP-Br_6Cl(X = Br_6Cl)
MBP-HBr_5Cl (X = HBr_5Cl)
MBP-Br_7 (X = Br_7)

FIGURE 22.29

Halogenated bipyrroles bioaccumulated in the marine food web. *Source*: From Teuton, E. L., *et al.* (2006). *Environmental Pollution* 144, 336–344.

This operationally defined fraction is referred to as ***humic substances***. These macromolecules range in molecular mass from 500 to 10,000. They impart a yellow-brown color to seawater. The humic substances are extremely variable in structure and elemental composition. They are thought to form abiotically from fragments of biomolecules generated during the microbial degradation of organic matter. Photochemical reactions at the sea surface play an important role in the formation and degradation of humic substances as these compounds are able to absorb UV energy. Humic substances are also found in solids, such as sediment, soil, and peat. They are relatively inert, making them difficult to study, so little is known of their structure or biogeochemistry. In Figure 22.2, IR spectroscopy has been used to distinguish marine humics from those of terrestrial origin, reflecting different biological sources of the reactant compounds that contribute to their formation.

CHAPTER

The Production and Destruction of Organic Compounds in the Sea 23

All figures are available on the companion website in color (if applicable).

23.1 INTRODUCTION

Though organic compounds constitute a relatively small reservoir in the crustal-ocean-atmosphere factory, they play a central role in the marine biogeochemical cycles of the biolimiting elements. Much of this stems from the role of organic matter as the essential food resource for all heterotrophs. Their consumption of organic matter leads to its transformation, either into more biomass or into various dissolved and particulate detrital materials. This processing controls the degree to which the deep ocean and its sediments can store carbon and nutrients over time scales that affect global climate. For example, biological processing that permits burial of organic carbon in marine sediments provides a sink for the greenhouse gases, such as CO_2 and CH_4.

The biological processing starts with the production of organic matter by autotrophs. The organic matter is passed to the heterotrophs, whose respiratory efficiency largely determines the fraction of organic matter that is transferred to the deep sea and sediments. In earlier chapters, the general function of this biological pump was described in relation to its role in sequestering nutrients and carbonates in the deep sea and sediments. In Chapter 12, we saw that the burial of organic matter in the sediments plays a controlling role in determining the redox levels in the ocean and sediments. In this chapter, we take a closer look at the molecular nature of the marine organic matter to better understand why some compounds are more reactive than others and, hence, differ in the degree to which they are retained in seawater and the sediments.

The fate of organic matter in the marine environment is largely related to its molecular structure, as this determines chemical reactivity. Compounds characterized by high concentrations in seawater and the sediments typically have slow loss rates relative to their production. These compounds tend to exhibit low chemical reactivity in the marine environment and can persist in seawater for thousands of years. Conversely, compounds with low concentrations typically have high loss rates relative to their production. Their high reactivity is generally due to rapid biotic uptake and transformation, leading to turnover times that are on the order of minutes to days. Thus, to understand the biogeochemistry of marine organic matter, marine scientists must measure transport and transformation rates as well concentrations.

Because of the central role of marine organisms in producing and consuming organic matter, this chapter focuses on the details of how temporal and geographic differences in plankton production lead to temporal and geographic differences in organic carbon storage in the deep sea and sediments. Some organic matter in the oceans is also of terrestrial origin, having been transported via river runoff and atmospheric processes. Recent research suggests that abiotic reactions in hydrothermal systems could also give rise to organic compounds emitted into the ocean.

In addition to serving as a food resource to heterotrophs, organic compounds play other important biological roles. As mentioned in Chapter 22.4, many marine organisms synthesize secondary metabolites that are used to manage ecological interactions. For example, some phytoplankton release a variety of dissolved organic compounds, some of which are toxic to their grazers, and others that complex with dissolved metals. This extracellular complexing serves to enhance the bioavailability of micronutrients and lower the bioavailability of toxic metals. In the surface waters, extracellular dissolved organic matter, mainly the high-molecular-weight compounds, absorbs solar radiation. This influences photochemical reactions and also serves to shield plankton from mutagenic UV radiation.

23.2 TECHNICAL LIMITATIONS IN THE STUDY OF MARINE ORGANIC MATTER AND VARIOUS WORK-AROUNDS

Marine chemists have had limited success in characterizing the molecular structure of organic matter, particularly for the dissolved compounds. Chemical analysis usually starts with the isolation of POM from DOM using a filter with a 0.2-μm pore size. This is generally followed by elemental analysis. More sophisticated approaches involve structural analysis, but this is usually limited to detection of functional groups or broad classes of compounds.

23.2.1 Elemental Analysis

From an elemental perspective, most of the mass of POM and DOM is carbon. Thus, DOC and POC concentrations are generally representative of the entire DOM and POM pools. Because DOC and POC concentrations are more easily measured than those of the other elements (nitrogen, phosphorus, oxygen, and hydrogen), far more data has been collected on their concentrations and reactivity as compared to that of DON, PON, DOP, or POP. DOM and POM concentrations are not measured for technical reasons and because the total mass of organic matter provides little insight into the biogeochemical processes responsible for its formation and destruction.

POC and DOC concentrations are generally reported in units of ppm C = mg organic carbon per kilogram of seawater. In the case of sedimentary organic matter, organic carbon concentrations are typically reported relative to the dry weight of the sediments, e.g., %C = gC per 100g dry sediment. Organic nitrogen and phosphorus concentrations are reported in similar units. Their concurrent measurement enables calculation of elemental ratios.

23.2.2 Analysis by Biomolecule Class

Characterizing the organic compounds in a sample by identifying and quantifying all the individual compounds is generally not possible because: (1) a very large number of organic compounds are present, (2) many of these compounds have large and complex molecular structures, and (3) the high salt level in seawater creates an intractable interference. As a result, marine chemists have adopted an operational approach to characterizing the molecular nature of organic compounds. This approach relies on the use of standard detection methods that were developed by biochemists for the measurement of broad classes of biomolecules, such as the sugars, amino acids, amino sugars, and lipids, in biological samples. In seawater, some of these biomolecules are present as dissolved, or free, species. They are also present as components of macromolecular dissolved and particulate organic matter. These component biomolecules are released from the macromolecules by sample pretreatment steps, such as acid hydrolysis and solvent extraction. By controlling the pretreatment conditions, subsets of the biomolecule classes can be quantified, i.e., free amino acids versus the combined amino acids (proteins) and free sugars versus the combined sugars (carbohydrates).

As shown in Table 23.1, relatively small fractions of the DOM and POM pools have been characterized on a molecular level. In the case of DOM, less than 10% of the surface ocean DOC and less than 5% of the deep ocean DOC have been structurally characterized. This suggests that most of the compounds in DOM have structures quite different from those of the broad classes of biomolecules. The story with POM is somewhat better. As exemplified from observations made in the Equatorial Pacific Ocean (Figure 23.1), 85% of the surface-water POC and 30% of the deepwater POC were characterizable. The higher degree of characterization of the surface POC is reflective of its relative "freshness," which makes it structurally similar to its major source, planktonic biomass produced in the photic zone. The characterizability of the POC decreases with depth below the photic zone with the least (20%) observed in the sediments. This suggests that as POC sinks through the water, it undergoes chemical transformations that make the molecular compounds in deepwater and sedimentary organic matter very different from those comprising the bulk of the photic-zone planktonic biomass.

23.2.3 Classification by Molecular Size: Ultrafiltered DOM

To aid in the characterization of the DOM pool, marine organic chemists have developed techniques for separating compounds into size fractions. Tangential flow ultrafiltration is used to isolate a high-molecular-weight (HMW) fraction from a low-molecular-weight (LMW) fraction. The size cutoff between these is approximately 1 nm, which equates

Table 23.1 Bulk and Molecular Composition of DOM in the Surface (<100 m) and Deep (>1000 m) Ocean.

Bulk Composition of DOM

	Surface Ocean	Deep Ocean	Redfield
Concentrations			
DOC (μM C)	60 to 90	35 to 45	
DON (μM N)	3.5 to 7.5	1.5 to 3.0	
DOP (μM P)	0.1 to 0.4	0.02 to 0.15	
Carbohydrates[a]	10 to 25	5 to 10	
Molar elemental ratios			
DOC: DON	9 to 18	9 to 18	7
DOC: DOP	180 to 570	300 to 600	106
DON: DOP	10 to 25	5 to 10	16

Molecular Composition of DOM

	Surface Ocean	Deep Ocean
Concentrations (nM C)		
Neutral Sugars[b]	200 to 800	20 to 170
Amino Acids[b]	200 to 500	80 to 160
Amino Sugars[b]	42 to 94	4 to 9
Lipids[c]	0.2 to 0.7	Not determined
% DOC as C		
Neutral Sugars[a]	2 to 6	0.5 to 2.0
Amino Acids[a]	1 to 3	0.8 to 1.8
Amino Sugars[a]	0.4 to 0.6	0.04 to 0.07
Lipids[b]	0.3 to 0.9	Not determined
% Identifiable DOC	3.7 to 10.5	1.3 to 3.9
% DON as N		
Amino Acids[a]	6 to 12	4 to 9
Amino Sugars[a]	0.8 to 1.7	0.2 to 0.4
% Identifiable DON	6.8 to 13.7	4.2 to 9.4

[a]*Detectable after strong acid hydrolysis of water sample.*
[b]*Detectable after mild acid hydrolysis of water sample.*
[c]*Detectable after solvent extraction of water sample.*
Data from high latitudes have been excluded, because these regions have well-mixed water columns that lead to a lack of vertical distinction between surface and deep-ocean DOM.
Source: After Benner, R. (2003). Biogeochemistry of Marine Dissolved Organic Matter, *Academic Press, pp. 59–90.*

to a molecular weight of about 1000 daltons. Most *ultrafiltered* DOM (UDOM) is LMW (75 to 80%). The majority of the colloidal DOM ends up in the HMW fraction. Because colloids compose 30 to 50% of the DOM pool, they represent a large reservoir of organic carbon, nitrogen, and phosphorus. Colloidal organic matter is of particular interest

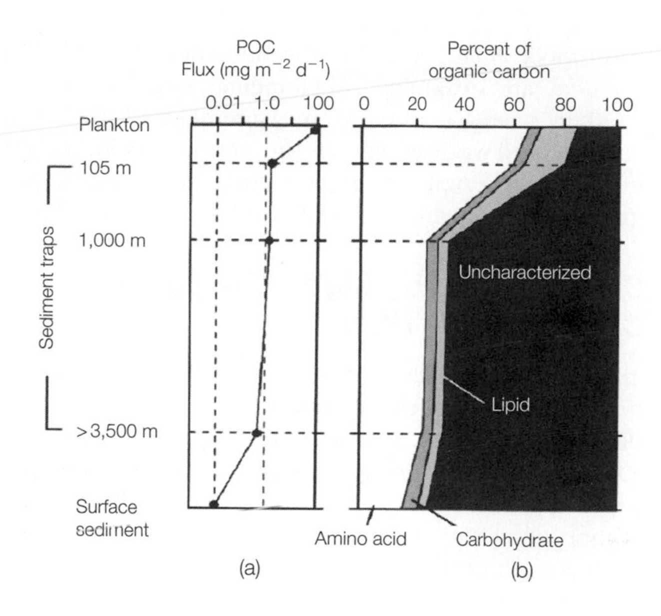

FIGURE 23.1

(a) POC depth profile and (b) percent organic carbon in various biochemical classes within plankton, sinking particles, and surface sediments in the equatorial Pacific Ocean. *Source*: From Eglinton, T. I., and D. J. Repeta (2003). *Treatise in Geochemistry*, Academic Press, pp. 145–180. Data sources provided therein.

because it is prone to a type of nonspecific surface interaction in which spontaneous aggregation leads to the formation of polymeric gels. These gels have a higher density than the colloids. They can sink and thereby their formation enhances the transport of organic matter into the deep sea. Because trace metals adsorb onto colloidal organic matter, formation of gels also serves to enhance the transport of trace metals into the deep sea.

23.2.4 Classification by Solubility: Humic Substances

Solid-phase extraction techniques are also used to isolate other operationally defined fractions of DOM, most notably the *humic substances*. During this process, a water sample is passed through a hydrophobic resin leading to the adsorption of a subset of DOM that is sparingly soluble in seawater. The sorbed DOM is eluted from the resin using solvents of varying pH. Because solubility is a function of molecular structure, the desorbed compounds are thereby fractionated into two classes, *humic acids* and *fulvic acids*. Fulvic acids are soluble at all pH values, whereas humic acids are insoluble at pH < 2. The humic substances that do not elute off the resin, and, hence, are insoluble at all pH values, are termed *humins*. The molecular weights of the humic substances range from 500 to 10,000 daltons. Thus, some are small enough to fall within the LMW category. Humic substances comprise approximately 5 to 15% of the DOC pool.

As with the bulk POM and DOM, the operationally defined fractions of UDOM and humic substances are quantified by elemental analysis and via broad molecular-class detection. Other strategies involve measurement of the natural isotopic composition, both stable and radioactive, of the various fractions. Efforts are underway to develop more sophisticated techniques, such as solid-state NMR and high-resolution mass spectrometry, for identification of specific bonds and functional groups.

23.2.5 Classification by Light Absorption: Chromophoric DOM

A substantial fraction of DOM is able to absorb light across the solar spectrum that reaches the sea surface. These compounds are responsible for imparting a characteristic yellow-brown tint to seawater. This organic matter is termed *chromophoric* DOM (CDOM) and is operationally quantified by the amount of light absorbed at specified wavelengths. CDOM is functionally important to the plankton as it limits the depth of the euphotic zone and absorbs harmful UV radiation. From a chemical perspective, CDOM acts a photochemical microreactor because it provides a setting in which sunlight can generate highly reactive oxygen free radicals. CDOM is also of analytical interest as it interferes with the in situ fluorescence measurements that are used for the quantification of chlorophyll. The molecular composition of CDOM is not known, nor has its actual concentration been quantified.

Finally, it is noteworthy that not all marine organisms are classifiable as POM. Viruses, small bacteria, and archaea can pass through 0.2-μm filter pores and, thus, are technically part of the DOM. Although these organisms are small (0.2 to 20 × 10^{-15} g C/cell for viruses and bacteria, respectively), their high numbers (10^9 to 10^{11} and 10^7 to 10^9 cells/L, respectively) cause their collective biomass in seawater to be similar to that of phytoplankton and zooplankton (<2 mm) (Table 23.2). The biomass of the archaea and the macrozooplankton (>2 mm) are currently unknown. Nonetheless, these two groups play very important biogeochemical roles as described later.

23.3 THE FATE OF TERRESTRIAL ORGANIC MATTER IN THE OCEAN

Most of the organic matter in seawater was created in situ by marine processes and is, hence, classified as *autochthonous*. Organic matter of nonmarine origin is classified as *allochthonous* and is primarily terrestrial detritus, transported by rivers or winds. The input of organic matter from rivers is small (0.4 Pg C/y) compared to primary productivity (40 to 50 Pg C/y). The aeolian input is unknown but thought to be significant.

The burial rate of organic matter in marine sediments is a factor of 2 smaller (0.15 to 0.19 Pg C/y) than the known terrestrial input. This requires that greater than half of the *terrestrial organic matter* be remineralized prior to burial. Marine chemists have been unable to establish how and where in the ocean this remineralization occurs. Similarly

Table 23.2 Distribution of Organic Carbon in the Oceans.[a]

Category	Gt C
DOC	700
Noncolloidal DOC	350 to 490
Colloidal DOC	210 to 350
POC	55
Detrital POC	50
Surface biota	3
Heterotrophic bacteria	0.35
Phytoplankton[b]	0.78
Viruses	0.20
Zooplankton (<2 mm)	0.26
Zooplankton (>2 mm)	Unknown

[a] *These estimates are somewhat different from those in Figures 25.1 and 25.2*
[b] *Phytoplankton biomass includes bacterial photoautotrophs.*
Sources: Le Quéré, C., et al. (2005). Global Change Biology *11, 2016–2040; and Suttle, C. A. (2005).* Nature *437, 356–361.*

perplexing is the fact that little or no detectable signatures of terrestrial DOC have been found in seawater beyond the immediate nearshore, although this pool comprises about half of the terrestrial organic input to the ocean with the remainder being POC. This requires that terrestrial DOC be very efficiently removed from seawater.

About 90% of all the organic matter buried in marine sediments is deposited on the continental margins with 30 to 50% being terrestrial in origin. As a result, terrestrial organic matter is estimated to contribute 25 to 50% of the organic carbon buried in marine sediments. This disproportionately large contribution suggests that terrestrial organic matter as a whole is relatively refractory (nonreactive) as compared with marine organic matter. Because the presence of terrestrial DOM in seawater is virtually nondetectable, it cannot be a major contributor to the sedimentary terrestrial organic matter pool. This leaves terrestrial POM as the remaining source of terrestrial organic matter to be buried in the sediments.

Some component of the terrestrial POM must be extremely nonreactive to enable a higher burial efficiency as compared to autochthonous POM. A possible candidate for this nonreactive terrestrial POM is *black carbon.* This material is a carbon-rich residue produced by biomass burning and fossil fuel combustion. Some black carbon also appears to be derived from graphite weathered from rocks. It is widely distributed in marine sediments and possibly carried to the open ocean via aeolian transport.

In Chapter 25, we will consider further the important role of the continental shelves in regulating organic matter burial. Because this type of organic matter burial is probably altered by changes in sea level, it provides a feedback in the crustal-ocean-atmosphere factory that acts on the biogenic gases involved in global climate (CO_2 and CH_4) and redox (O_2) control.

23.4 PRODUCTION AND CONSUMPTION OF ORGANIC COMPOUNDS BY MARINE ORGANISMS

23.4.1 Primary Producers

Most marine POM is generated in situ by primary producers, namely the phytoplankton, macroalgae, chemoautolithotrophic bacteria, and probably archaea. Most primary production is thought to be accomplished by photoautotrophic nanoplankton (2.0 to 20 μm in diameter) and picoplankton (0.2 to 2.0 μm in diameter). The picoplankton are composed of photoautotrophic bacteria whose light-gathering chromophores are located in their cell membranes rather than in a chloroplast as found in eukaryotic phytoplankton. Probably the most numerous members of the phytoplankton are photoautotrophic bacteria, *Synechococcus* and *Prochlorococcus.* These cyanobacteria use bacteriochlorophylls and phycoerythrins as pigments (Figure 22.25). The prevalence of bacteriochlorophylls and rhodopsins throughout the bacterioplankton suggest that microbial photoautotrophy is a larger component of primary production than currently estimated. For example, the marine bacterium *Pelagibacter ubique*, which is thought to be the most abundant organism in seawater (other than viruses and perhaps the archaea), contains rhodopsins. As described in Chapter 7.3.2, the latter are chromophores, suggesting that *Pelagibacter ubique* has some ability to harvest and use light energy to engage in photoassisted heterotrophy.

23.4.2 Elemental Composition of Biomolecules

The most abundant compound class found in phytoplankton and bacteria are the proteins. As shown in Table 23.3, proteins make up about half of their dry weight. In comparison to eukaryotic phytoplankton, bacteria are enriched in RNA and DNA. Because proteins and nucleic acids are relatively enriched in nitrogen as compared to carbohydrates and lipids (Table 23.4), bacterial biomass is enriched in nitrogen relative to eukaryotic phytoplankton.

23.4.3 Molecular Composition of Bacterial Cell Walls and Membranes

Bacteria are also notable for the chemically resistant biomolecules that make up their cell walls and membranes. These biopolymers include the peptidoglycans illustrated in Figure 22.24 found in the cell walls, along with lipopolysaccharides and porins, which are components of the cell membranes. The porins are proteins found in the cell membranes of gram-negative bacteria. Their three-dimensional structure creates pores large enough to enable passage of small (<600 daltons) hydrophilic compounds. Since bacteria do not have mouths or excretory organs, all food and waste materials must pass through the cell walls and membranes and, hence, must be solute sized. Hence, heterotrophic bacteria consume DOM. Most bacteria also assimilate inorganic nutrients.

Table 23.3 Average Composition of Phytoplankton and Bacteria by Compound Class.

Compound Class	% Dry Weight			
	Bacteria	Phytoplankton		
Protein	52	54.4	65	40
Polysaccharide	17	25.5	16	40
Lipid	9.4	16.1	19	15
RNA + DNA	19.2	4		5
Other	<3			
Reference	Southamer (1977)	Anderson (1991)	Hedges (2002)	Williams and Robertson (1991)

Sources: *Southamer, A. H. (1977).* Symposia of the Society for General Microbiology *27, 285–315; Anderson, L. A. (1995).* Deep Sea Research *I 42, 1675–1680; Hedges, J. I., et al. (2002).* Marine Chemistry *78, 47–63; and Williams, P. J. LeB, and J. E. Robertson (1991).* Journal of Plankton Research *13(suppl); 153–169.*

Table 23.4 Percent Elemental Composition of the Major Classes of Biomolecules.

	% Elemental Composition within a Biomolecular Class				
Biomolecule	**C**	**N**	**H**	**O**	**P**
Carbohydrate	44%		6%	49%	
Lipid	76%		12%	13%	
Protein	53%	16%	7%	23%	
Nucleic acid	37%	17%	4%	33%	10%

Source: *From Anderson, L. A. (1995).* Deep Sea Research *I 42, 1675–1680.*

Another major component of the cell membranes are the lipopolysaccharides, which are present as phospholipid bilayers. Following the death of bacteria, the biopolymers that constitute their cell walls and membranes become part of the detrital organic carbon pool. The great abundance of these biopolymers in seawater and the sediments is a reflection of their resistance to chemical degradation and the important role that bacterioplankton play in marine biomass production.

23.4.4 Consumption of Organic Matter: The Marine Food Web

The carbon fixed by marine autotrophs has several possible fates. Approximately one third is respired within the algal cells to provide energy for growth and maintenance.

Some is released into seawater as an extracellular exudate or as a result of viral lysis and from sloppy feeding by the grazers. Consumption by grazers entrains the organic matter in the marine food web as shown in Figure 23.2. Cells that are not consumed or lysed by a virus can sink or be otherwise carried out of the euphotic zone by currents.

Grazing by the Primary Consumers

The primary consumer (grazers) are size specialized. The smallest phytoplankton (1 to 20 μm) are consumed by the smallest members of the zooplankton, called the

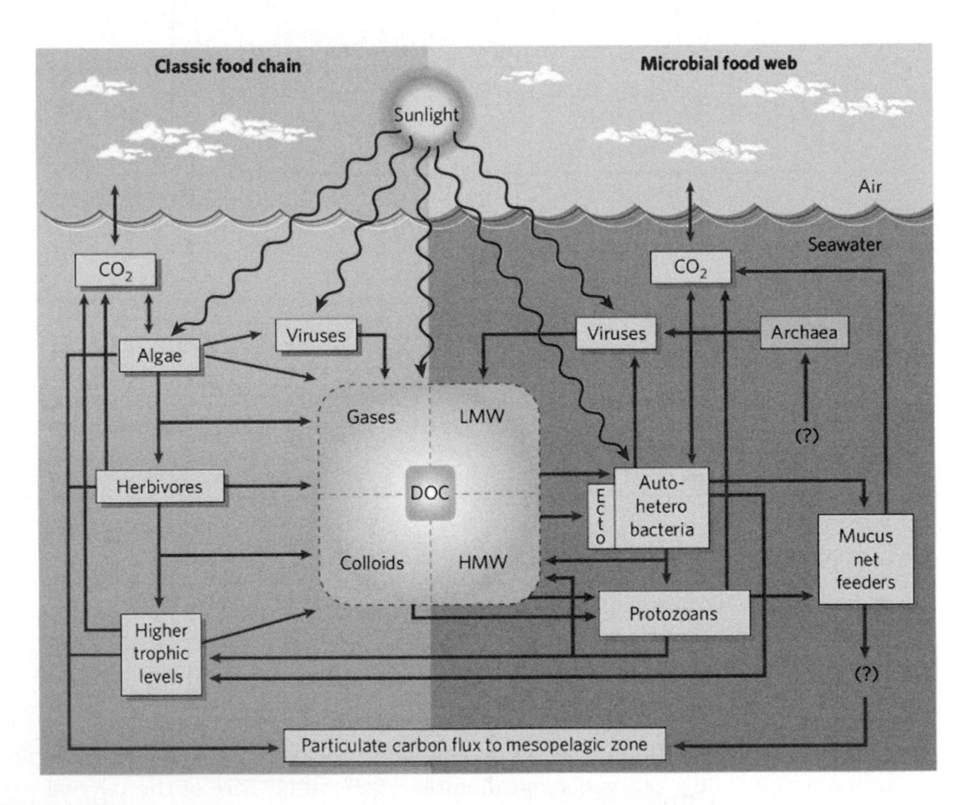

FIGURE 23.2

Schematic representation of the marine food web. Depicted on left is the part of the food web controlled by mesozooplankton grazing. On the right is the part of the food web controlled by the microbial loop and protozooplankton grazing. Bacteria use ectoenzymes (Ecto) to degrade high-molecular-weight (HMW) DOC down to a size that can pass through the bacterial cell membranes. Also shown are the roles of the viruses and archaea. The fraction of the biomass that sinks to the bottom of the euphotic zone is shown at the bottom of the diagram. This downward flux of particulate carbon (and energy) is thought to fuel most subeuphotic zone processes. Most of this flux is provided by the algae-herbivore grazer pathway (left side). *Source*: From DeLong, E. F., and D. M. Karl (2005). *Nature* 437, 336–342.

protozooplanton (5 to 200 μm). These members of the zooplankton are flagellated and ciliated protozoans, including the foraminiferans and radiolarians. The larger phytoplankton, such as the diatoms (20 to 200 μm), are consumed by mesozooplankton (200 to 2000 μm), which are dominated by the copepods, appendicularians, and amphipods. The largest members of the zooplankton, the macrozooplankton (>2000 μm), include the euphausiids, salps, and pteropods, which graze a wide spectrum of particle sizes. Each of these grazers plays an important and distinct biogeochemical role. By grazing on the pico and nanoplankton, the protozooplankton transfer microbial biomass into the marine food web. The mesozooplankton produce large and fast-sinking fecal pellets that enhance the export of POM into the deep sea. The copepods are the most abundant zooplankton in the ocean. The macrozooplankton produce even larger, and, hence, faster sinking, fecal pellets but tend to have patchy distributions as compared to the copepods.

Global estimates of grazing are still being formulated. The best guesses suggest that microzooplankton and mesozooplankton collectively remove the vast majority of the primary producers and heterotrophic prokaryotes through their grazing activities. Zooplankton growth efficiencies are on the order of 30%, so most of the consumed POM is lost to respiration and excretion.

Zooplankton are notable for their ability to excrete ammonium and other low-molecular-weight DON, such as urea (Figure 22.26c). This enhances the remineralization of nitrogen, thereby providing inorganic nutrients for the photoautotrophs. Zooplankton generate a wide variety of POM including their tissues, fecal pellets, molts, egg cases, undigested food particles, and discarded feeding nets. The appendicularians are particularly notable for generating very large feeding nets (up to 1 m across), which typically become clogged after about 24 h of use. The nets are then discarded and sink rapidly (800 m/d), carrying large amounts of detrital and living POM to the seafloor. Off the coast of Monterey Bay, CA, these mucous feeding nets and their associated filtered organic matter provide a annual flux of $7.6\,g\,C\,m^{-2}$ to the 3-km-deep surface sediments. Another impressive exporter of POM are the salps, a filter-feeding planktonic tunicate. In some locations, salps form swarms dense enough to filter 75% of the upper 50 m of the water column during an 8-h night-time period. The resulting fecal output from this depth zone over the 8-h period can be as high as 90 mg C per m^2!

DOM Release: The Viral Shunt, Algal Exudates, and Sloppy Feeding

Viral infections kill 2 to 77% of the marine bacteria in the surface waters on a daily basis, making their impact similar in scale to the grazing efforts of the protozooplankton. Viruses are also known to infect and kill eukaryotic phytoplankton, with 3% of the primary production estimated to be lost on an annual basis. Viruses kill cells by penetrating the host's cell membrane and replicating until their numbers are so large that the host's biovolume increases to the bursting point. The resulting lysis of the cell membrane causes all of the DOM contained within the cell to be dumped into seawater. From a biogeochemical perspective, cell lysis derails the transfer of carbon to higher trophic levels within the marine food web. This moves carbon into the DOC pool, enabling bacterial production at the expense of the eukaryotic photoautotrophs. Because viruses can acquire and transfer genes from and to their hosts, they functionally enable a type of sexual reproduction for the unicellular plankton. Viruses are also thought to have some ecological regulatory control as they are species specific for their hosts.

Sloppy feeding by grazers also leads to lysis of microbial and eukaryotic cell membranes with a similar outcome to viral lysis, i.e., conversion of POM into DOM as the cellular matrix is released into seawater.

DOM is also released into seawater by phytoplankton for reasons that are as yet unclear. On average, 13% of the phytoplankton carbon is released as exudates, some of which are low-molecular-weight compounds, such as free amino acids and tricarboxylic acids. Other exudates are high-molecular-weight compounds, such as the acylated heteropolysaccharides.[1] These macromolecules are relatively chemically resistant and appear to form a large portion of the HMW DOC pool.

Release rates of exudates from phytoplankton range from 0 to 80% of carbon fixed. These rates are dependent on species composition, physiological state, nutrient deficiency, temperature, and light limitation. Some evidence suggests that exudation is a mechanism for release of excess organic matter from cells when nutrient availability is too low to enable their usage as metabolic fuel.

In some cases, exudates enable the plankton to control their environment. For example, diatoms release compounds, called oxylipins, that induce natural abortions and growth reduction in the zooplankton that are their primary predators. Other exudates complex with trace metals, serving to reduce the bioavailability of toxics, such as copper, and enhance the bioavailability of micronutrients, such as iron. Examples of iron-binding extracellular DOM are the siderophores (Figure 5.11).

The Microbial Loop

Most of the DOM released into seawater is rapidly consumed by heterotrophic bacteria and probably archaea. Small amounts are likely consumed by some unicellular eukaryotes, such as heterotrophic flagellates and photoautotrophs, under light- and/or nutrient-limiting conditions. The bacterial production resulting from consumption of this DOM is equivalent to one third to one half of the primary production. This implies that bacterial uptake of DOM is substantial as microbial growth efficiencies are generally less than 20%. In other words, bacteria have to consume a lot of DOM to obtain enough chemical energy to create biomass. Although bacterial production is not too efficient, the resulting biomass does provide food for the protozooplankton and is, thus, a mechanism by which at least some of the DOM is transferred back into the marine food web. This is referred to as the *microbial loop*.

---

[1] Heteropolysaccharides are complex carbohydrates composed of more than one kind of simple sugar. See Table 22.4 for the structure of the acyl functional group.

The inefficiency of microbial heterotrophy does have a side benefit as it enhances nutrient remineralization rates. This serves to increase the availability of inorganic nitrogen and phosphorus for the photoautotrophs. The multiple roles of bacteria in the marine food web were shown in Figure 23.2, with the component of the food web controlled by the algal herbivores depicted on the left side and the microbial loop on the right. The viral shunt acts on both pathways.

23.4.5 Seasonal Fluctuations in Organic Matter Production and Export

In mid- and high latitudes, primary production varies seasonally in response to light and nutrient limitation. This in turn leads to seasonality in export of POM from the photic zone. In mid-latitudes, the annual cycle of plankton production and POM export is driven by a spring bloom and a secondary fall bloom. At high latitudes, summer-time blooms appear to be driven by iron delivery. For some phytoplankton, such as diatoms, export of POM is enhanced by bioflocculation. Small diatoms growing under upwelling or spring bloom conditions have been observed to exude mucus-like HMW DOM that causes them to coagulate into rapidly sinking aggregates.

Another type of seasonally driven export event is associated with larger diatoms (>50 μm) that grow under nutrient- and light-limited conditions at the base of the euphotic zone. These diatoms seem to undergo a mass settling event, called a *fall dump*, in response to destratification of the summer thermocline due to seasonal cooling and early winter storms. These diatoms sink rapidly and are relatively well preserved in the sediments.

Evidence for seasonal fluctuations in the export of sinking POM from the euphotic zone has been obtained from sediment trap studies. As shown in Figure 23.3, seasonal variations in the flux of sinking biogenic hard and soft parts are detectable even in the deep zone. Furthermore, interannual variations in POC and PIC production lead to similar shifts in the deepwater fluxes of these particles.

23.4.6 Abiotic Processes

A small fraction of POM is created via abiotic processes, all of which involve transformation of DOM into the particulate phase. As already noted, destabilization of colloidal DOM can lead to the formation of gels. Increasing salinity destabilizes colloids, so flocculation of DOM is common in estuaries.

DOM also tends to be concentrated at boundaries, such as the sediment-water and air-sea interfaces. The latter is also characterized by enrichments of POM, such as bacteria. Under calm conditions, the DOM and POM that collects at the air-sea interface forms a visible surface slick or microlayer. On windy days, this organic matter can be whipped up into an emulsion that has the appearance of a very sturdy foam. DOM can also be transferred into the POM pool by adsorbing onto organic particles.

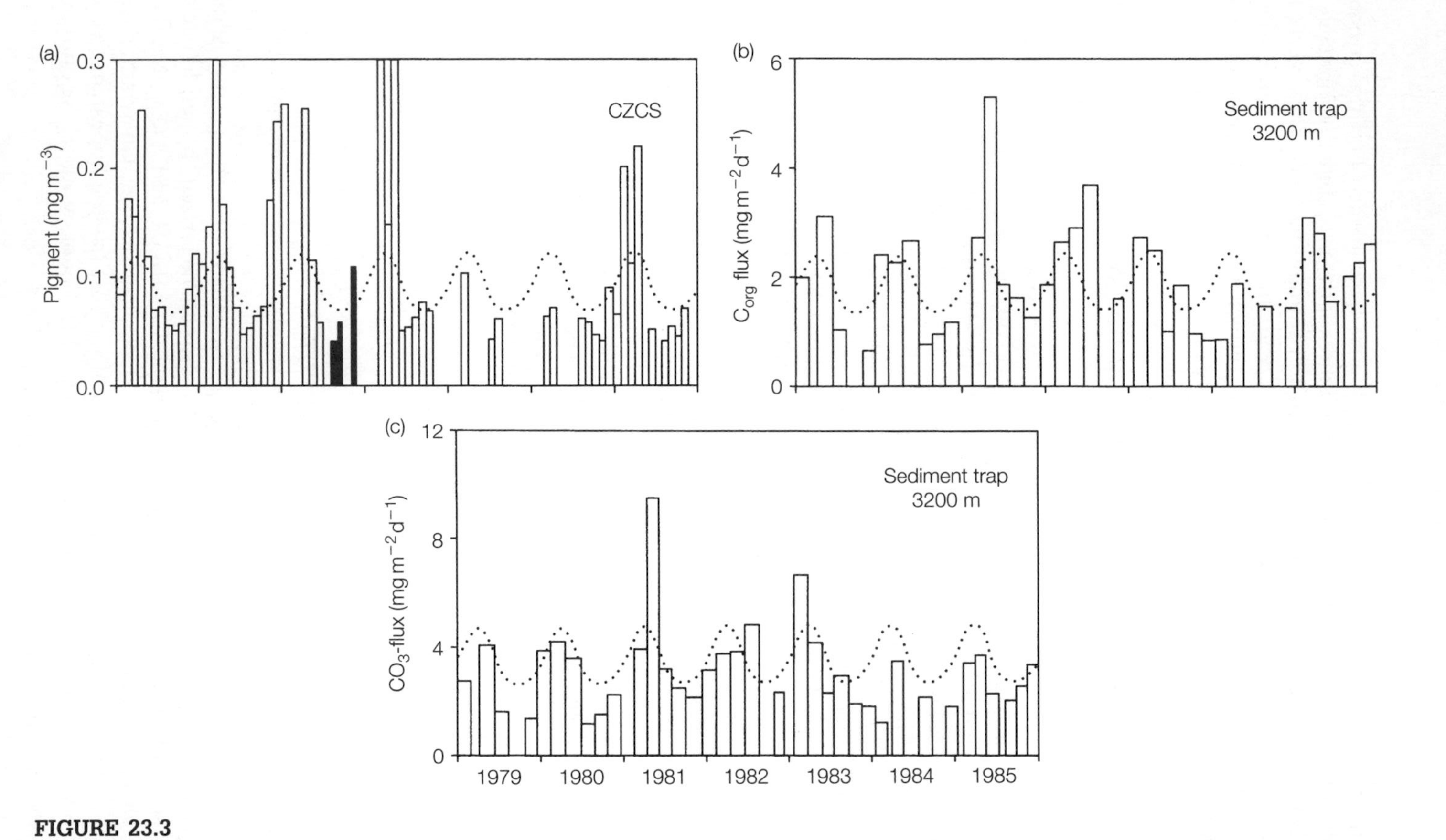

FIGURE 23.3

Variation with time in the fluxes of (a) pigment, as determined by the CZCS, coastal zone color scanner, and (b) particulate organic, and (c) inorganic carbon measured in a sediment trap deployed at 3200 m in the Sargasso Sea. Note the coincidence of variations in these fluxes with satellite-derived pigment concentrations shown in (a). The dotted line represents the average timing of the high and low particle mass fluxes at 3200 m. *Source*: From Deuser, W. G., *et al.* (1990). *Deep-Sea Research* 37, 1331–1343.

23.5 THE CHEMICAL AND PHYSICAL TRANSFORMATION OF DETRITAL PARTICULATE ORGANIC MATTER

Most of the marine biomass is produced in the surface waters. Following an organism's death, its biomass becomes part of the detrital POM pool along with molts, cast-off feeding nets, egg cases, fecal pellets, and marine snow. The depth profiles of detrital POC, PON, and POP exhibit vertical stratification similar to that of O_2 (Figure 23.4). Concentrations are high in the surface waters and decrease exponentially with depth to very low levels. This vertical stratification reflects production in the surface waters and consumption at depth. Most of the latter is due to heterotrophic bacteria that degrade detrital POM as it sinks through the water column.

As with the nutrients, only a small fraction of the POM generated in the surface waters survives the trip to the sediments. In terms of POC, about 18% of the annual primary production sinks below the euphotic zone with only 5% sedimenting onto the seafloor. Microbial degradation in the surface layers of the sediment destroys 93% of POC that reaches the seafloor. Thus, only 0.4% of the primary production is buried. Similar behavior is exhibited by PON and POP.

23.5.1 Bacterial Processing

Because bacteria do not have mouths or other organs of consumption, they must first degrade POM into fragments small enough to pass through their cell walls and membranes. This is achieved by the release into seawater of extracellular enzymes, called *ectoenzymes*. These enzymes are compound specific and include the proteases, lipases, phosphatases, nucleases, and glucosidases. The ectoenzymes break bonds between monomer pairs and thereby downsize biopolymers into successively smaller molecules. For example, cleavage of the bonds that link the amino acids within proteins first degrades polypeptides into oligopeptides and then into free amino acids. The resulting low-molecular-weight products are assimilated by the bacteria for production of either energy or biomass. In either case, the ensuing respiratory metabolic reactions generate CO_2 and remineralized nutrients.

The heterotrophic degradation of organic matter in pursuit of chemical energy and nutrients is termed *catabolism*. Redirection of the resulting chemical energy into anabolic pathways fuels the creation of biomass. The general catabolic pathways for the major classes of biomolecules are summarized in Table 23.5. In the case of the polynucleotides (nucleic acids), RNA and DNA are first fragmented by nucleases into nucleotides. Phosphatases remove the phosphate group(s), followed by the degradation of the residual ribonucleosides into ribose and a purine or pyrimidine base. The sugars and nitrogenous bases are finally degraded into CO_2 and inorganic nitrogen. The polysaccharides, such as cellulose and starch, are initially broken into oligosaccharides and then into monosaccharides. Their complete degradation yields CO_2. The degradation of lipids yields hydrocarbons, carbohydrates, and small carboxylic acids. These, in turn, are ultimately decomposed to CO_2. The exact nature of the metabolic intermediates and

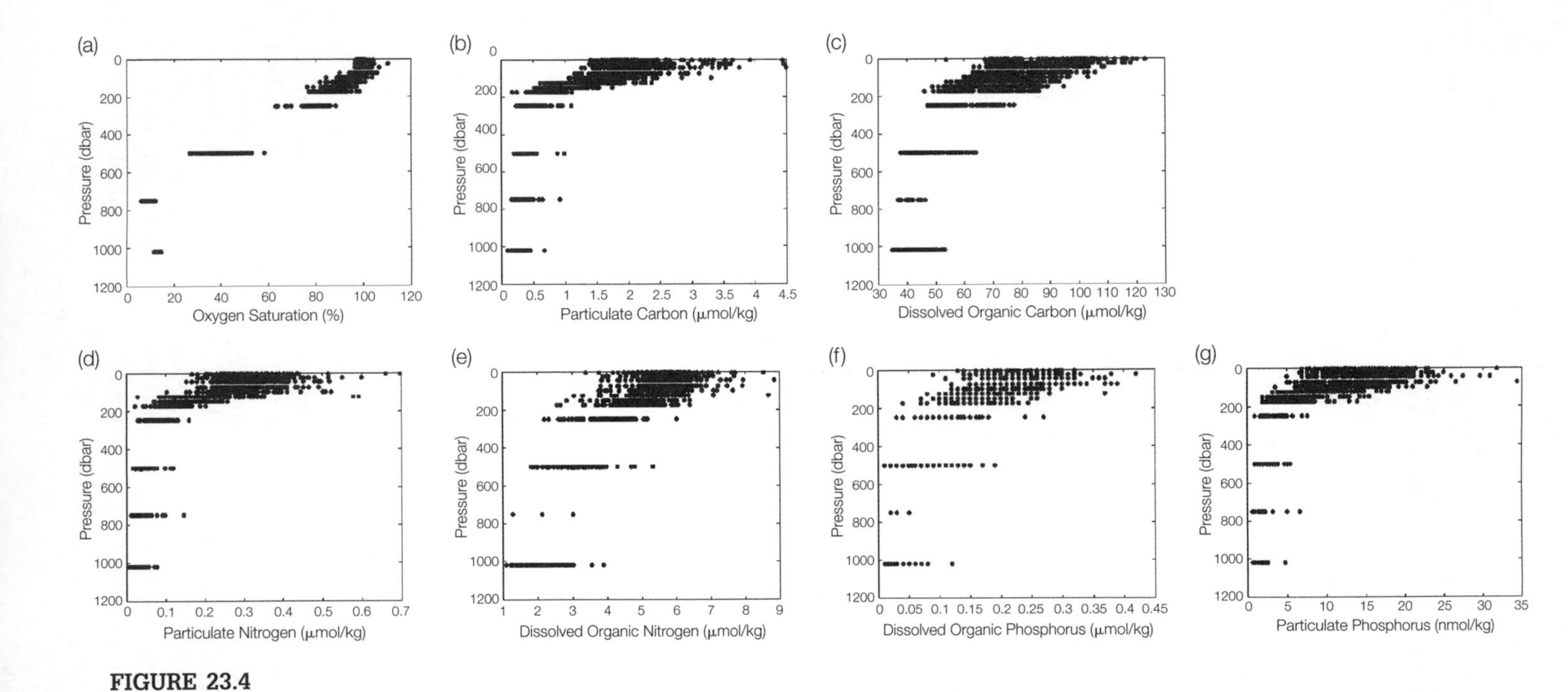

FIGURE 23.4

Depth Profiles where 1 dbar is approximately equal to 1 m water depth: (a) Percent Saturation of O_2, (b) Particulate Carbon, (c) DOC, (d) Particulate Nitrogen, (e) DON, (f) DOP and (g) Particulate Phosphorus at Station ALOHA (A Long-term Oligotrophic Habitat Assessment) located in 4750 m water depth, 100 km north of the island of Oahu. These profiles are part of the Hawaii Ocean Time Series (HOTS) in which vertical profiles have been measured monthly since 1989. Some of the variability exhibited at each depth represents seasonal effects since the station is located at mid-latitudes (22°45'N 158°00'W). Some of the variability is due to interannual effects, such as exerted by ENSOs (El Niño - Southern Oscillations) or global warming. *Source*: Hawaii Ocean Time-Series Data Organization & Graphical System (HOT-DOGS ©), Laboratory for Microbial Oceanography, University of Hawaii, http://hahana.soest.hawaii.edu/hot/hot-dogs/interface.html.

Table 23.5 Decomposition Products Produced from Organic Compounds Present in Detrital POM.

Life Substances	Decomposition Intermediates	Intermediates and Products Typically Found in Nonpolluted Natural Waters
Proteins	Polypeptides → $RCH(NH_2)COOH$ (amino acids) → { $RCOOH$, $RCH_2OHCOOH$, RCH_2CH, RCH_3, RCH_2NH_2 }	NH_4^+, CO_2, HS^-, CH_4, HPO_4^{2-}, peptides, amino acids, urea, phenols, indole, fatty acids, mercaptans
Polynucleotides	Nucleotides → purine and pyrimidine bases	
Lipids Fats Waxes Oils Hydrocarbons	RCH_2CH_2COOH (fatty acids) + $CH_2OHCHOHCH_2OH$ (glycerol) → { RCH_2OH, $RCOOH$, shorter chain acids, RCH_3, RH }	CO_2, CH_4, aliphatic acids, acetic, lactic, citric, glycolic, malic, palmitic, stearic, oleic acids, carbohydrates, hydrocarbons
Carbohydrates Cellulose Starch Hemicellulose	$C_x(H_2O)_y$ → { monosaccharides, oligosaccharides, chitin } → { hexoses, pentoses, glucosamine }	HPO_4^{2-}, CO_2, CH_4, glucose, fructose, galactose, arabinose, ribose, xylose
Lignin	$(C_2H_2O)_4$ → unsaturated aromatic alcohols → polyhydroxy carboxylic acids	
Porphyrins and Plant Pigments Chlorophyll Hemin Carotenes and Xanthophylls	Chlorin → Phaeophytin → hydrocarbons	Phytane, pristane, carotenoids, isoprenoids, alcohols, ketones, acids, porphyrins
Complex Substances Formed from Breakdown Intermediates, e.g.,	Phenols + quinones + amino compounds	→ Melanins, melanoidin, gelbstoffe
	Amino compounds + breakdown products of carbohydrates	→ Humic acids, fulvic acids, "tannic" substances

Source: After Stumm, W. S., and J. J. Morgan (1996) Aquatic Chemistry: Chemical Equilibria and Rates in Natural Waters, *3rd ed. Wiley-Interscience, p. 925.*

final products reflects the respiratory pathways of the microbe, with aerobic bacteria generating different organic residues than the various anaerobes (denitrifiers, sulfate reducers, fermenters, etc.).

23.5.2 Zooplankton and the Fecal Pellet Express

Zooplankton also exert a transformative effect on POM, but quite differently as compared to bacteria. Namely, they chemically and physically convert phytoplankton into rapidly sinking fecal pellets. Fecal pellet sinking rates range from 2 to 1000 km/y (Table 13.5). Although only a small fraction of the phytoplankton are repackaged in the form of a fecal pellet, this organic matter has a much higher likelihood of sinking into the deep sea and reaching the sediments than does an isolated detrital phytoplankton cell. This enhancement to the sinking flux of POM has been named "the fecal pellet express." It is responsible for a significant part of the mass flux to the sediments.

In addition to undigested organic matter, fecal pellets also contain fragments of calcareous and siliceous hard parts and clay-sized grains of clay minerals. Fecal pellets can harbor viable bacterial and phytoplankton cells. Because of their fast sinking rates, fecal pellets reach the seafloor relatively intact. Their organic matter represents the major source of food for the benthos.

23.5.3 Formation of Marine Snow

The other important contributor to the mass flux of POM is another type of rapidly sinking organic particulate matter termed *marine snow*. This amorphous material has a floc-like appearance. It is thought to form from the spontaneous aggregation of: (1) smaller particles of detrital POM; (2) small bacterial, protozoan, and phytoplankton cells; and (3) inorganic matter, including biogenic hard parts of clay minerals. Transparent exopolymer particles (TEPs) seem particularly important in stimulating the aggregation process. TEPs are produced by a wide variety of marine organisms such as suspension feeders (bivalves and tunicates) and phytoplankton growing under bloom conditions.

Because of its high organic content, the marine snow acts as a microhabitat that supports enhanced rates of heterotrophic microbial activity. The associated nutrient remineralization causes the seawater within and around the marine snow to be characterized by elevated nitrogen and phosphorus concentrations and low levels of O_2. The importance of these suboxic and anoxic microzones to the marine cycling of the biolimiting elements is unknown but potentially significant.

23.5.4 Ballasting of Sinking POM

The extent and types of transformations that POM undergoes as it sinks toward the seafloor depends on: (1) the types and relative abundances of the heterotrophs degrading the organic matter, (2) the original chemical composition of the sinking particles, and (3) their physical packaging, usually referred to as *ballasting*. The last enhances particle sinking rates and also appears to confer some protection against microbial attack.

For example, organic matter deeply embedded in the mineral matrix of biogenous hard parts would not be exposed to exoenzyme attack. This embedding could occur during deposition of the minerals or through adsorption of the organic matter from seawater. Most of the ballasting effect exerted on POM is conferred by calcareous and siliceous hard parts and by clay minerals.

23.5.5 Degradation of Sinking Particles: Horizontal, Vertical, and Temporal Trends

The depth profiles of POC, PON, and POP concentrations shown in Figure 23.4 were obtained by filtering particles from water samples. Below the euphotic zone, most of the particles are small (<20 μm) and, thus, sink slowly (≤1 m/d). The larger particles (fecal pellets and marine snow) are less numerous but collectively represent most of the POM flux to the sediments. The large particles are more representatively sampled with sediment traps (Figure 16.6). Because these devices sample over significant time periods—days to months—they are used to measure particle fluxes rather than concentrations. POC fluxes also decline exponentially with depth in the water column as shown in Figure 23.5. (These flux data were also used to generate Figure 23.1.) The steepest concentration gradients are located at two depths: (1) just below the euphotic zone and (2) at the sediment-water interface. Figure 23.5 shows that similar declines are also observed in the fluxes of the major classes of biomolecules with proteinaceous amino acids representing the vast majority of the molecularly characterizable POM at all depths.

In most general terms, the vertical declines in the flux of POC and constituent compound classes are driven by heterotrophic degradation. The nature of this process is not well understood. Although the fluxes of amino acids, carbohydrates, and lipids decline by several orders of magnitude over the water column, their proportions relative to each other are fairly well retained and quite similar to that of surface-water plankton. This is remarkable because microbes would be expected to selectively degrade molecules with high nutritive value, such as the nitrogen-rich amino acids. Also confounding is the increase in molecularly uncharacterizable compounds with increasing depth (Figure 23.1). Three explanations have been proposed to explain these observations: (1) As the characterizable POM degrades, it recombines to form uncharacterizable materials; and/or (2) the uncharacterizable material is always present but does not degrade because its molecular structure makes it relatively inert; and/or (3) mineral ballast physically shields some of the characterizable POM from microbial attack and, hence, nonselectively retains a fraction of this organic matter on the sinking particles.

The vertical trends in POM fluxes exhibit temporal and geographic variability. This was shown in Figure 23.3, in which seasonal shifts in surface productivity were seen to affect the subsurface particle fluxes even in deep waters. Other processes that can affect the sinking flux of POM include: (1) in situ production by mid-water microbes or zooplankton and (2) lateral transport of POM via advective currents. Both can produce mid-water maxima in the sinking organic matter fluxes. Geographic variability in these fluxes is common. As illustrated in Figure 23.6 for the central equatorial Pacific Ocean,

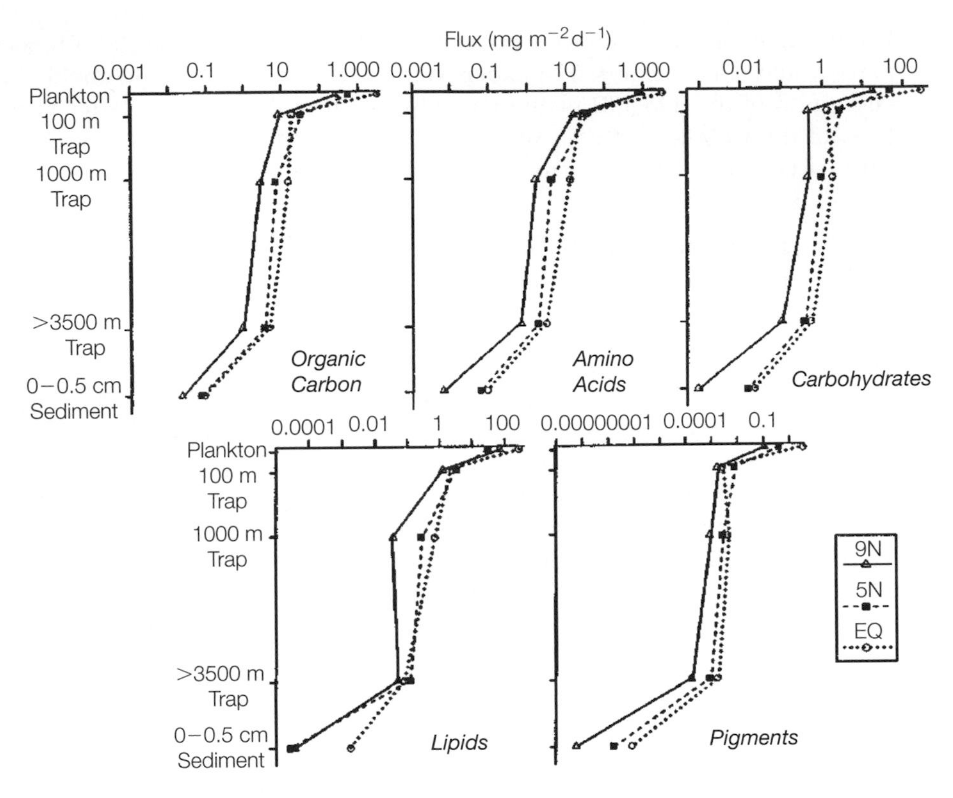

FIGURE 23.5

Fluxes (mg m^{-2} d^{-1}) of organic carbon and biochemical classes at 9°N, 5°N, and the equator. Compound class fluxes are in mg of compound, not mg C. The amino acids are present as peptides. Fluxes for net plankton are derived from primary production rates and measurements of biochemical content of net plankton. Fluxes into surface sediments were calculated using the sediment C_{org} content and accumulation rates and biochemical content measurements. *Source*: From Wakeham, S. G., *et al.* (1997). *Geochimica et Cosmochimica Acta* 61, 5363–5369.

latitudinal variations in primary production lead to geographic variations in sinking particle fluxes. These latitudinal variations demonstrate that regions with high primary production, like the equatorial Pacific, have elevated POC fluxes.

23.6 DISSOLVED ORGANIC COMPOUNDS

23.6.1 Importance of DOM

Table 23.2 indicates that DOM is the largest pool of organic matter in the ocean. From a biogeochemical perspective, marine DOM represents the largest pool of reactive carbon

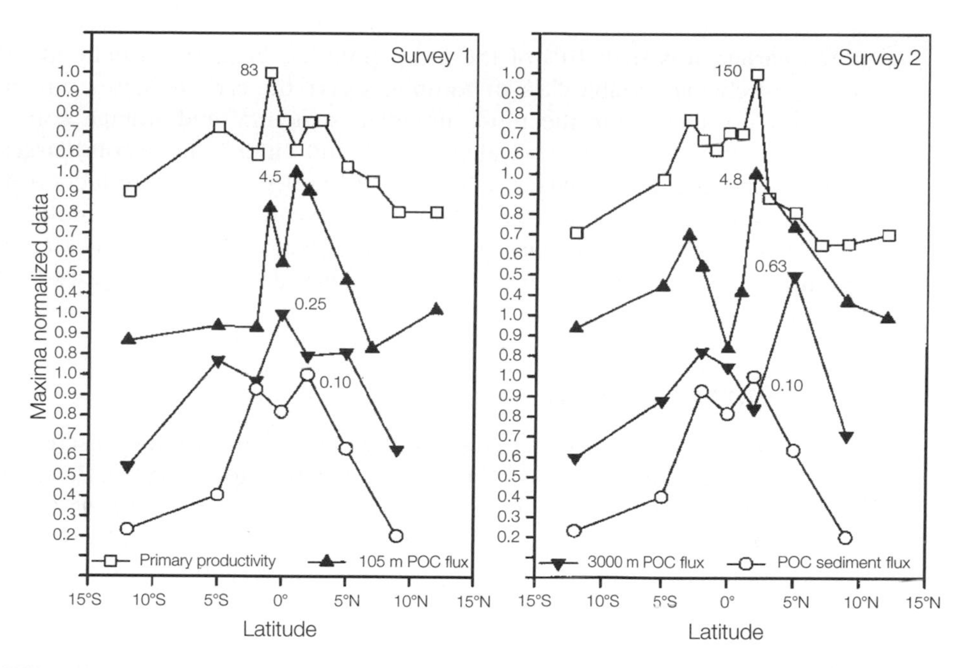

FIGURE 23.6

Relative primary productivity, POC fluxes at 105 and 3000 m, and POC sediment accumulation rates versus latitude in the central equatorial Pacific Ocean. Data are normalized to the maximum value in each transect. Survey 1 was conducted during February–March 1992 under El Niño conditions and Survey 2 from August to September 1992 under non–El Niño conditions at longitudes ranging from 135 to 140°W. Ordinate scale is reset to 1.0 at each maximum, and the absolute magnitude (mmol C m^{-2} d^{-1}) of each parameter is given next to its maximum. *Source*: From Hernes, P. J., *et al*. (2001). *Deep-Sea Research* I 48, 1999–2023.

on Earth. This reactivity has diverse impacts. Of primary interest is the potential for DOC to impact atmospheric CO_2 levels and, hence, global climate. This is considered likely as the DOC pool contains about the same amount of carbon as is present in atmospheric CO_2. From an ecological perspective, DOM represents the food source for heterotrophic bacteria and, hence, regulates their productivity and the microbial loop. The resulting bacterial respiration represents an important O_2 sink in the deep sea. Because of its ability to adsorb light, DOM alters the spectral properties of seawater and promotes photochemical reactions. DOM adsorbs onto particle surfaces, thereby altering the surface properties and sinking rates. DOM also complexes with trace metals, thereby altering their bioavailability. Finally, many marine organisms manage ecological relationships by secreting dissolved organic compounds, called secondary metabolites.

In addition to carbon, DOM also contains large amounts of nitrogen and phosphorus. As shown in Table 23.1, concentrations of DOC are on the order of tens of micromolar, a few micromolar for DON, and tenths of micromolar for DOP. The vertical concentration distributions typically exhibit stratification as exemplified by the profiles in Figure 23.4.

As noted earlier, less than 10% of the DOM pool has been structurally identified. The single largest characterizable class of compounds are the carbohydrates, which are likely degradation products from the most abundant structural and storage components of the phytoplankton, i.e., cellulose, alginic acid, and chitin. The second largest class of compounds are the amino sugars, which are likely degradation products from the cell walls of prokaryotes, as these are rich in peptidoglycans. Some lower molecular weight components of DOM are volatile, such as dimethylsulfide and, various, organobromine compounds, and, hence, degas into the atmosphere. Some of these gases likely play an important role in climate control.

23.6.2 Sources of DOM

The input of terrestrial DOM via rivers and aeolian transport was discussed in Chapter 23.3. Riverine concentrations of DOC range from 2 to 20 mg C/L. In contrast, little or no terrestrial DOM is detectable in seawater, leading to the current consensus that most is removed close to its point of input. In some estuaries, removal is associated with flocculation reactions promoted by the large increase in ionic strength that occurs when river water mixes with seawater. In other estuaries, DOC exhibits conservative behavior, leaving marine chemists with a mystery as to how and where DOC is removed.

Given the relatively small contribution of terrestrial organic matter to seawater, most of the DOM in the ocean is perforce of marine origin. Most of this DOM has as its ultimate source biomolecules released into seawater by excretions and exudations from microbes, plants, and animals. Equally important is the role of viral lysis in which cell membranes are ruptured, causing the DOM within the cells to be spilled into seawater. A similar effect can result from sloppy feeding by herbivorous zooplankton.

As noted in Chapter 23.4.4, phytoplankton "excrete" a significant amount of their fixed carbon in the form of low-molecular-weight compounds that are mostly carbohydrates. Although this process is variable, it is a ubiquitous and significant source of reactive (labile) DOC to surface seawater. Bacteria and zooplankton also release biomolecules into seawater and are, hence, another source of labile DOC. Although not restricted to the photic zone, they are most abundant in the surface waters, and, hence, this is where their contribution to the DOM is highest. The bacteria are notable for their role in converting a large part of the detrital POM pool into DOM by the actions of their ectoenzymes. Zooplankton are notable for their release of nitrogen-rich LMW compounds, particularly urea.

23.6.3 The Ecological Role of DOM

In addition to serving as the major food source to heterotrophic bacteria, DOM plays an important ecological role in enabling marine organisms to control various aspects of their environment including trophic interrelationships. This is accomplished by the secretion or exudation of specific molecules, called secondary metabolites. These are generally LMW compounds that tend to be species specific in their source and targets. Some act as toxins that repel or kill competitors or predators. As noted earlier, some diatoms release oxylipins to control the grazing activities of their predators. Some are used to enhance the bioavailability of micronutrients, e.g., the solubilization of iron by microbial siderophores. Others act to "neutralize" toxins, such as the complexation of copper by exudates from diatoms. Most of the metal-complexing abilities of DOM are conferred by amino and carboxylic acid groups. Other secondary metabolites, called pheromones, function as attractants and are used for mating purposes or to trigger metamorphosis of larvae. Bacteria and phytoplankton appear to use DOM to facilitate group responses to increased food resources or colonizable surfaces. This process is termed quorum sensing. The physiological activity of these secondary metabolites has been exploited by natural product chemists for development of novel pharmaceuticals, food additives, and other commercial chemicals. This subject is covered further in Chapter 28.

Phaeocystis *Blooms: TEP and DMSP Production*

Because of the importance of TEP in controlling metal availability and the conversion of DOM to marine snow, much effort is being directed at understanding the sources and biological uses of this DOM. Many marine organisms produce TEP. One genus of phytoplankton, *Phaeocystis*, can generate so much TEP when growing under bloom conditions that the resulting DOM can form a thick and smelly foam. This presents a nuisance when it washes up onto beaches. *Phaeocystis* blooms are also of great interest because they are a major source of dimethylsulfoniopropionate (DMSP). The microbial degradation of DMSP generates DMS. This volatile organic sulfur compound degases into the atmosphere where it plays an important role in climate regulation.

BOX 23.1 A further discussion of TEP and DMSP production is provided in the supplemental information for Chapter 23.6.3 that is available online at **http://elsevierdirect.com/companions/9780120885305**.

23.6.4 Molecular Composition

A small fraction of the DOM is composed of molecules that appear to be of direct biological origin, including carbohydrates, amino acids, and lipids. The concentration and molecular composition of this DOM is relatively well known and summarized in Table 23.1. The organic compound categories in this table are pooled by monomer type. Thus, the category of amino acids includes free amino acids and combined amino acids, with the latter ranging in size from oligopeptides to polypeptides. The carbohydrates are represented by the neutral sugars including the simple sugars and complex carbohydrates, with the latter ranging in size from oligosaccharides to polysaccharides. Phytoplankton are the likely source for most of the carbohydrate DOC. The amino sugars are likely derived from the microbial degradation of chitin and bacterial cell walls.

Recent improvements in analytical techniques have provided information on the chemical bonds present in UDOM, the molecularly uncharacterized fraction of the DOM pool. Most of this work has been done on the ultrafiltered, and, hence, HMW, fraction of the DOM pool. Spectral properties, such as NMR, are used to identify types of bonds, functional groups, and compound classes. Since the samples are pretreated to release molecular fragments that are amenable to study, the identified compounds likely represent components of much larger macromolecules. The following functional groups and compound classes have thus far been identified in the HMW DOC: (1) acylated heteropolysaccharides (APS) and (2) carboxyl-rich alicyclic molecules (CRAM), and in the HMW DON: (1) *N*-acetylaminopolysaccharides (*N*-AAPs) and (2) nonhydrolyzable amides.

The APS contain a large variety of neutral, amino, and acidic carbohydrates in which a high degree of branching and crosslinking exists among the monomeric sugars. Acetate is a common acyl group found in these molecules. APS are synthesized by algae and produced during microbial remineralization of POM. They appear to be relatively labile, exhibiting vertical stratification, with highest concentrations in the surface waters. This suggests rapid degradation, with high surface-water concentrations supported by production in the photic zone.

HMW DOM also contains a significant component of refractory organic matter. A ubiquitous and abundant part of this DOM are molecules with carboxylated and fused alicyclic rings (CRAM). Some model structures are shown in Figure 23.7. They have some structural similarity to terpenoids, such as the polycarboxylated fused-ring systems found in the sterols and hopanoids, suggesting they are degradation products of these biomolecules. The structural diversity found within CRAM and its substantial content of alicyclic rings and branching contribute to its resistance to biodegradation and refractory nature. The carboxyl groups probably confer a strong complexing ability for metals and calcium. The latter destabilizes colloidal DOM, leading to its aggregation and the subsequent formation of marine gels, thereby converting DOM to POM.

The likely sources of the *N*-acetylaminopolysaccharides (*N*-AAPs) found in DON are chitin and peptidoglycans from bacterial cells walls. The N-AAPs are similar to the APS in that they exhibit a high degree of vertical stratification, reflecting production in the surface waters and degradation with increasing depth. Most of the DON in the deep waters is composed of highly refractory amides. These amide-containing macromolecules are thought to be derived from proteins that have undergone abiotic transformations converting them to highly stable forms. More information on these abiotic reactions is provided in Chapters 23.6.5 and 23.6.6.

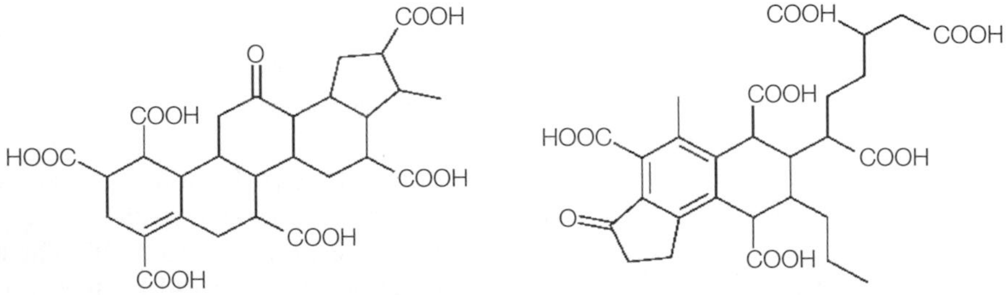

FIGURE 23.7

Model structures of a component of CRAM. *Source*: From Hertkorn, N., *et al.* (2006). *Geochimica et Cosmochimica Acta* 70, 2990–3010.

Unfortunately, most of the DOM in seawater is LMW (75 to 80%) and its chemical composition has not been as well studied as that of the HMW fraction. LMW DOM is thought to be composed primarily of biopolymers containing 10 or fewer monomers. Radiocarbon measurements indicate LMW is older than HMW DOM, suggesting that LMW is far less reactive than HMW DOM.

23.6.5 **Cycling of DOM**

DOM appears to participate in two cycles, one slow and one fast (Figure 23.8). In the fast cycle, labile LMW DOM is rapidly produced by phytoplankton through exudations, viral lysis, and sloppy feeding. These compounds are rapidly consumed by bacteria, including photosynthesizers that engage in mixotrophy, such as the cyanobacteria (Chapter 7.3.2). The most abundant of these labile LMW compounds are the oligosaccharides and oligopeptides, free sugars and amino acids, urea, DNA, RNA, ATP, and various lipids. Because of their rapid cycling, these compounds are present at low (nanomolar) concentrations and compose a small fraction of the LMW DOM.

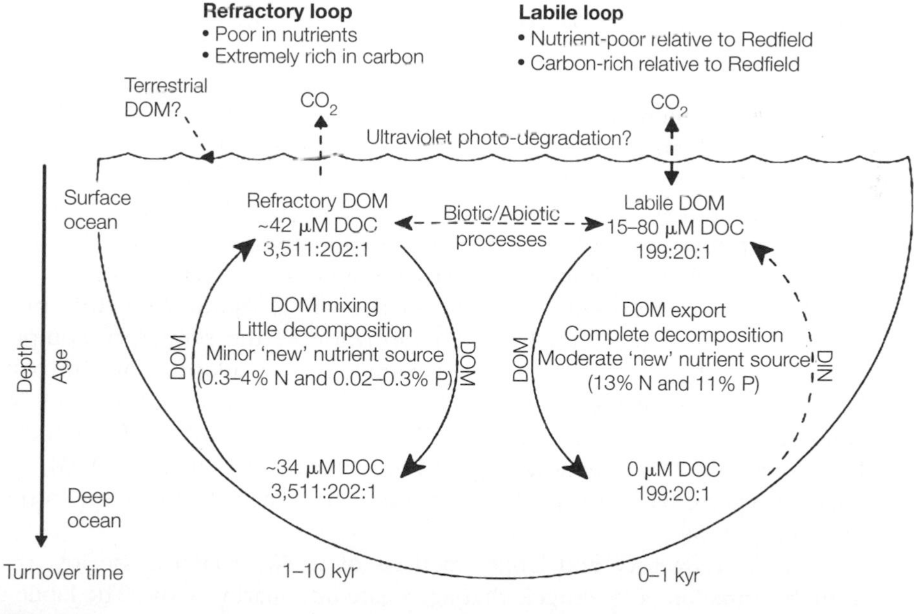

FIGURE 23.8

A simplified conceptualization of oceanic DOM pools based on a two-pool model showing carbon concentrations and elemental C:N:P ratios of the DOM pools. In reality, the reactivity of DOM probably spans a continuum that is presented herein as two pools, labile and refractory, for the purposes of simplification. Molar elemental ratios are in the format of C:N:P. *Source*: From Hopkinson, C. S., and J. J. Vallino (2005). *Nature* 433, 142–145.

The slow cycle involves degradation of the less reactive DOM. Likely sources of this DOM are the more resistant biomolecules contained in the cell walls and membranes of phytoplankton and bacteria. These biopolymers are degraded by the ectoenzymes of bacteria, with some of the products becoming part of the LMW pool and some part of the HMW pool. In both cases, the degradation products are more refractory than the source biomolecules. In the LMW pool, the refractory nature is thought to be a consequence of selective degradation of the reactive chemical bonds in the biomolecules. A useful index of this process is the degree to which nitrogen is depleted because C–N bonds are preferentially degraded over other carbon bond types. This causes the C/N ratio of DOM to increase with increasing degradation. As shown in Figure 23.8, the labile DOM pool has a C/N ratio (10:1) somewhat higher than that for plankton (6.6:1), whereas the refractory DOM has a much higher C/N ratio (17:1).

The refractory compounds in the HMW DOM pool seems to be generated through abiotic reactions that act to link degradation products into macromolecules. These new chemical bonds create molecular structures that enhance the overall refractory nature of the DOM. The chemical changes lead to increased crosslinking, aromaticity, cyclization, esterification, and nitrogen depletion. The general types of chemical reactions responsible are oxidations, polymerizations, and condensations. Considerable debate exists as to whether these reactions are wholly abiotic or whether they are, at least in part, microbially mediated.

The refractory macromolecules are termed *geopolymers* because their slow degradation rates cause them to persist in seawater for thousands of years. Similar geopolymers are found in sedimentary organic matter and in terrestrial DOM and soils. They were first studied by soil scientists in the 1930s, who developed their own operationally defined terminology, i.e., humic substances and its subdivisions: humic acids, fulvic acids, and humins (Chapter 23.2.4). These scientists also coined the term *humification* to describe the abiotic processes thought to be responsible for the formation of humic substances found in soils. As shown in Figure 23.9, this process closely resembles the one postulated for the formation of the refractory compounds in the HMW DOM pool. The soil science terminology was adopted by the petroleum geochemists to describe the chemical changes that take place during the catagenesis of sedimentary organic matter (Chapter 27.6). Prior to the advent of ultrafiltration, marine chemists adopted the methodology of the soil scientists to study DOM, particularly in estuaries where soil humics were assumed to be a major contributor to the DOM pool (Chapter 23.6.6).

Both the refractory and labile fractions of HMW DOM can be lost from seawater through formation of macrogels that aggregate into marine snow. The labile fraction that is known to participate in marine snow formation are the TEPs, such as mucopolysaccharides found in the mucus sheaths surrounding fecal pellets and plankton colonies. HMW DOM is also lost from seawater via: (1) adsorption onto sinking POM and minerals, (2) conversion into POM at the sea surface by turbulence associated with bursting bubbles, and (3) photochemical degradation.

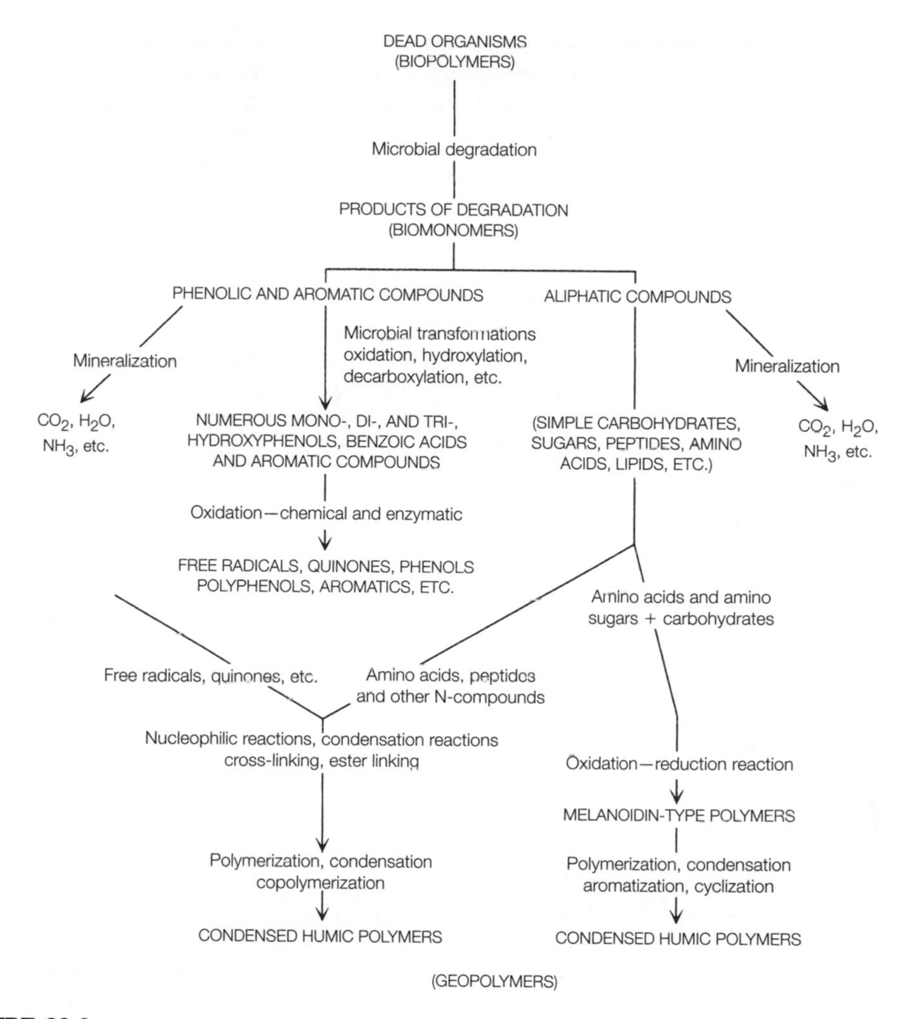

FIGURE 23.9

Major processes involved in the synthesis of humic compounds. *Source*: From Rashid, M. A. (1985). *Geochemistry of Marine Humic Compounds*. Springer-Verlag, p. 58.

23.6.6 Humic Substances

The most striking characteristic of the dissolved humic substances is their chromophoric nature. As part of the DOM, they impart a yellow-brown cast to marine and freshwaters and, hence, are part of the CDOM pool. Terrestrial humic substances compose a significant fraction of the riverine DOM entering the ocean. In seawater, humic substances compose 5 to 15% of the HMW DOM. Differences exist in the bulk properties of marine and terrestrial humic substances. These are summarized in Table 23.6. They have been used to trace the fate of terrestrial organic matter in the ocean.

Table 23.6 Certain Striking Differences in the Chemical Nature and Geochemical Behavior of Marine and Soil Humic Compounds.

Marine Humic Compounds	Soil Humic Compounds
General Composition	
Marine humic compounds are rich in aliphatic substances; the concentration of aliphatic carbon is almost twice that of aromatic carbon	A strong aromatic core is generally the characteristic feature of soil humic compounds; aromatic carbon is predominant in the molecule
Aromatic compounds are present but do not constitute a predominant part of the molecule; the polyphenolic and polyaromatic components are low in concentration and exhibit low degrees of substitution	Aliphatic side chains are attached to the aromatic core, but due to the participation of lignin degradation products, aromatic moieties are more abundant
Anaerobic conditions do not favor the development of highly condensed organic molecules	The molecules contain more polycyclic and polyphenolic constituents
	A high degree of oxidation leads to a high degree of condensation in humic molecules
Elemental Composition	
The carbon contents in fulvic and humic acids are 45–50% and 50–55%, respectively	The carbon content shows broader variations; because of the strong carbon lattice, the concentration of carbon is as high as 65%
Aquatic conditions favor greater hydrogen contents	Terrestrial humic compounds have generally low concentrations of hydrogen
Marine organisms are rich in nitrogen and therefore the humic compounds originating from them contain high amounts of nitrogen; some sedimentary humic compounds contain up to 10 times as much sulfur as the soil humic acids	The nitrogen content of soil humic compounds is generally low
Isotopic Composition	
The carbon isotopic composition of marine humic compounds is heavy; $\delta^{13}C$ values ranging from −20‰ to −23‰	Lighter carbon isotopic composition ($\delta^{13}C$ = −25 to −28‰) is characteristic of terrestrially derived humus

(Continued)

Table 23.6 *(Continued)*

Marine Humic Compounds	**Soil Humic Compounds**
	Isotopic Composition
$\delta^{15}N$ is generally about +9‰ and reflects nitrate as a nitrogen source	Terrestrial plants use atmospheric nitrogen fixed by bacteria; $\delta^{15}N$ is about +2‰
$\delta^{2}H$ values do not show much spread; the average is about −105‰	The $\delta^{2}H$ of soil humic compounds show much broader values ranging from −50‰ to −100‰
	Functional Groups
The total acidity is highly variable but in general it is lower than that of soil humic matter; acidity in marine humic compounds is largely a function of carboxyl groups; the phenolic hydroxyl content is generally low; the restricted aeration of bottom sediments favors greater accumulations of carbonyl groups	Soil humic compounds generally possess higher degrees of total acidity; the acidic characteristics arise from the participation of carboxyl as well as phenolic hydroxyl groups Because of the high degree of aeration, soil humic material contains more carboxyl groups and fewer carbonyl groups
Alcoholic hydroxyl contents are generally high	Alcoholic hydroxyl contents are low
Functional groups constitute about 20–30% of the humic molecules	About 25% of soil humic acids and about 50–60% of a fulvic acid molecule consist of functional groups

Isotopic compositions, e.g., $\delta^{13}C$, $\delta^{15}N$ and $\delta^{2}H$, are reported as values relative to a standard. Source: From Rashid, M. A. (1985). Geochemistry of Marine Humic Compounds, *Springer-Verlag, p. 105.*

The basic structure of humic substances involves a backbone composed of alkyl or aromatic units crosslinked mainly by oxygen and nitrogen groups. Major functional groups attached to the backbone are carboxylic acids, phenolic hydroxyls, alcoholic hydroxyls, ketones, and quinones. The molecular structure is variable as it is dependent on the collection of DOM available in seawater to undergo the various polymerization, condensation, and oxidation reactions and reaction conditions involved in humification, as well as the ambient physicochemical reaction conditions, such as temperature and light availability.

Recall from Chapter 23.2.4 that humic substances are isolated from seawater by adsorption on a hydrophobic resin followed by elution using solvents of varying pH. The desorbed compounds are fractionated into two classes, *humic acids* and *fulvic acids* based on their solubility behavior. A model structure for a humic acid is illustrated in Figure 23.10a in which fragments of biomolecules, such as sugars, oligosaccharides,

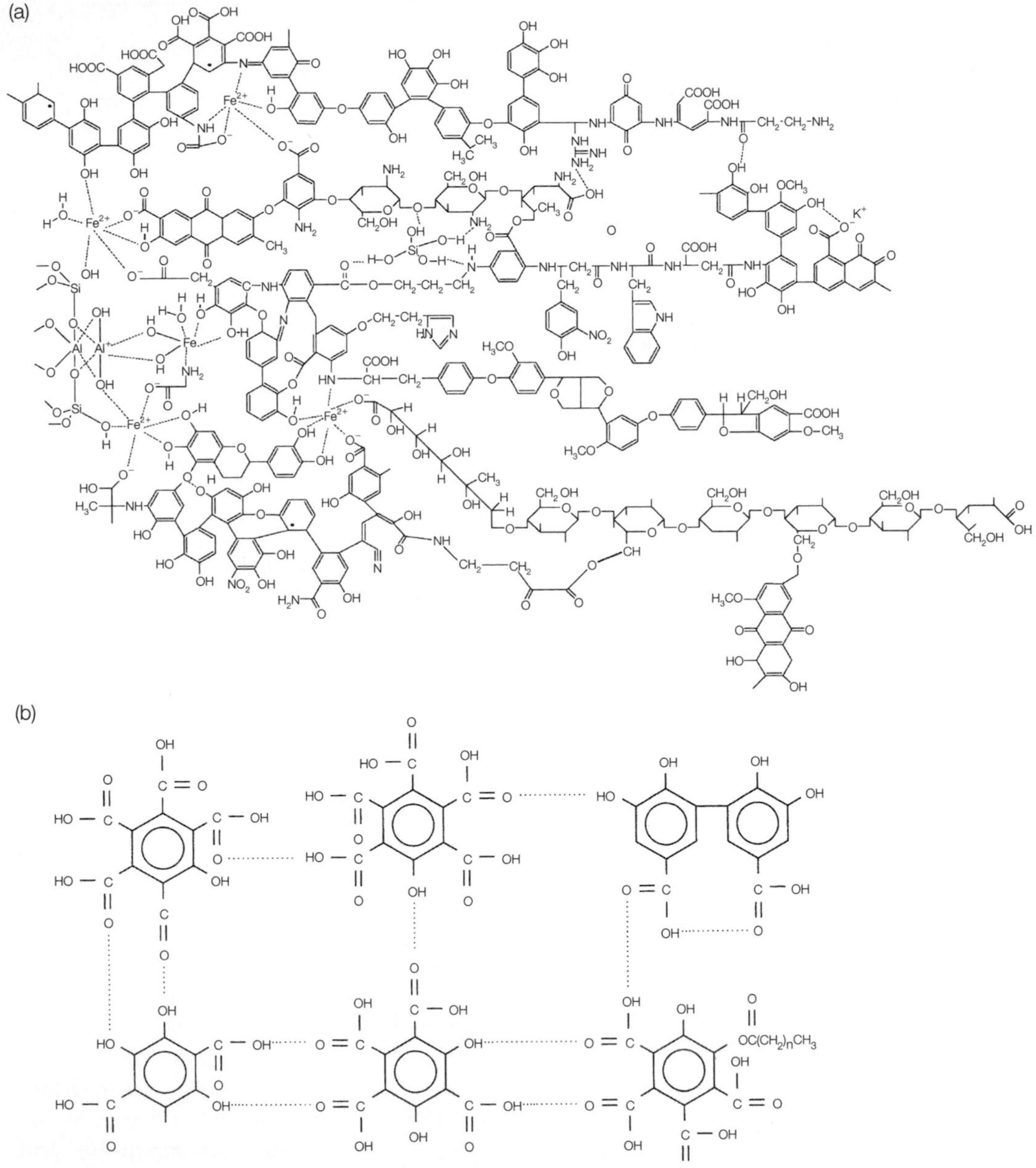

(a)

(b)

FIGURE 23.10

Model chemical structure of a soil. (a) Humic acid and (b) part of a fulvic acid. *Source*: From (*a*) Kleinhempel, D. (1970). Albrecht Thaer Archiv 14, 3–14; and (b) Rashid, M. A. (1985). *Geochemistry of Marine Humic Compounds*, Springer-Verlag, p. 75.

amino acids, and oligopeptides, are discernible. The carboxylic acid and amine groups provide opportunities for complexing with metal ions, such as iron. Adsorption onto a clay mineral crystal lattice is depicted on the left side.

A model structure for the fulvic acids is presented in Figure 23.10b. They tend to have a greater abundance of acidic functional groups than found in the humic acids. Most are phenolic acids (Figure 23.11b) whose source is thought to be degradation of lignin, an important structural polymer in vascular land plants (Figure 23.11a). The presence of lignin degradation products causes the terrestrial humic substances to have a relatively high degree of acidity and aromaticity, making them more susceptible to flocculation

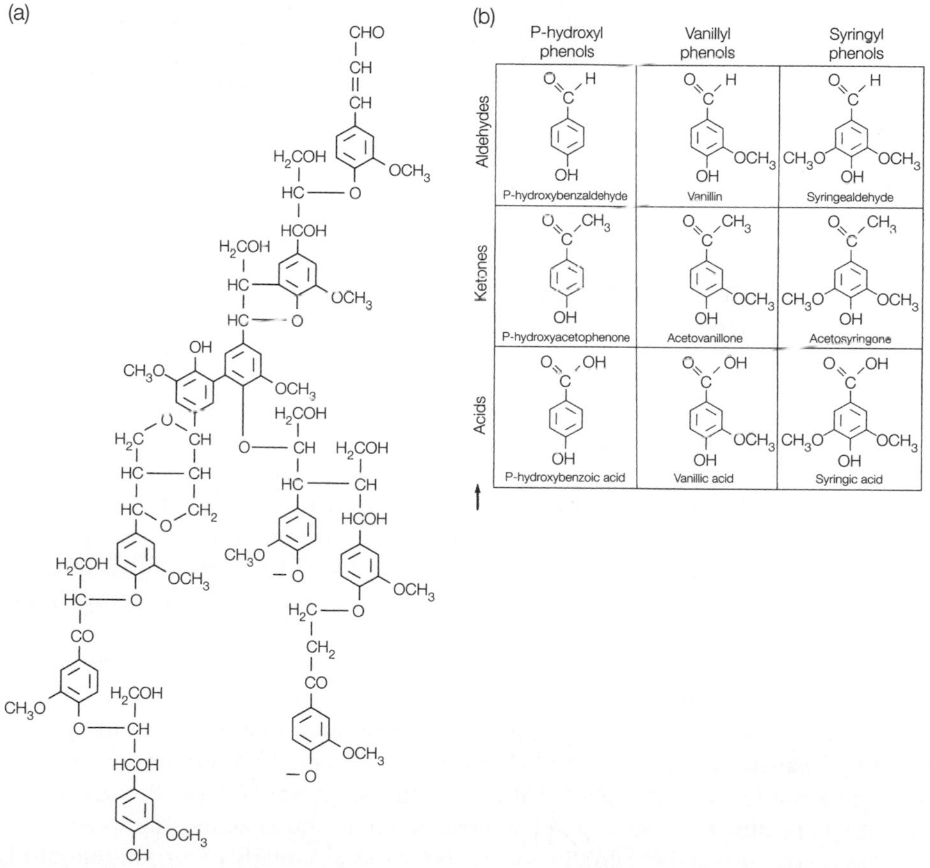

FIGURE 23.11

(a) The structure of part of a model lignin molecule. (b) Aromatic acids and phenols derived from lignin, which are used as indicators of terrestrial plant matter. *Source*: From Rashid, M. A. (1985). *Geochemistry of Marine Humic Compounds*, Springer-Verlag, p. 47; and Hedges, J. I. and P. L. Parker (1976). *Geochimica et Cosmochimica Acta* 40, 1019–1029.

when exposed to high concentrations of electrolytes. Since lignin is unique to land plants, the phenolic acid groups have been used as biomarkers for the presence of terrestrial organic matter. Their low abundances in open ocean sediments supports the hypothesis that most riverborne terrestrial organic matter is deposited in the nearshore.

Since most of the riverine DOM is comprised of humic substances, considerable attention has been focused on its fate in seawater. Little terrestrial DOM is detectable in seawater, suggesting the existence of an efficient removal process. This is surprising given the traditional view that humic substances are relatively refractory. Marine chemists are currently investigating the redox and photochemistry of humic substances to better understand its chemical fate in the oceans.

Some evidence exists that humic substances play an important role in the redox chemistry of iron and manganese by acting as an electron shuttle that facilitates the transfer of electrons from labile organic substrates to iron-manganese oxyhydroxides. This process seems to be microbially mediated and involves the dissimilatory reduction of the humic substances. The transfer of electrons into this electron shuttle is thought to provide an advantageous thermodynamic driving force that enables the iron-manganese oxyhydroxides to act as the ultimate electron sink. Denitrification appears to be similarly coupled to redox reactions involving humic substances.

23.6.7 Sea Surface Microlayer and the Role of Abiotic Photochemistry

Labile and refractory DOM undergo abiotic photochemical reactions in the photic zone, especially in the sea surface microlayer where physical processes concentrate DOM into thin films. Some of these reactions appear to be important in the formation of refractory DOM and others in its degradation. For example, DOM exuded by diatoms during plankton blooms has been observed to be transformed into humic substances within days of release into surface seawater. Laboratory experiments conducted in seawater have demonstrated that photolysis of labile LMW DOM promotes the chemical reactions involved in humification and produces chemical structures found in marine humic substances.

An example of a proposed reaction mechanism for the photolytic transformation of marine humic substances is shown in Figure 23.12 for a model LMW triglyceride (trilinolein). Most notable is the uptake of ammonium following photolysis. This provides an explanation as to why marine humic substances are more nitrogen-enriched than the terrestrial forms (Table 23.6). Although ammonia was used as the source of nitrogen in this experiment, other organic amines, such as amino acids, amino sugars, and nucleic bases (purines and pyrimidines), are likely to be similarly incorporated into DOM during humification.

The UV radiation adsorbed by chromophoric DOM stimulates production of free radical (singlet) O_2. This leads to high concentrations of the free radicals within and around the CDOM, creating local conditions of high reactivity. Thus, chromophoric DOM, which tends to be HMW, can be thought of as a photochemical microreactor in which free radical oxidations are promoted. Given its macromolecular nature, CDOM

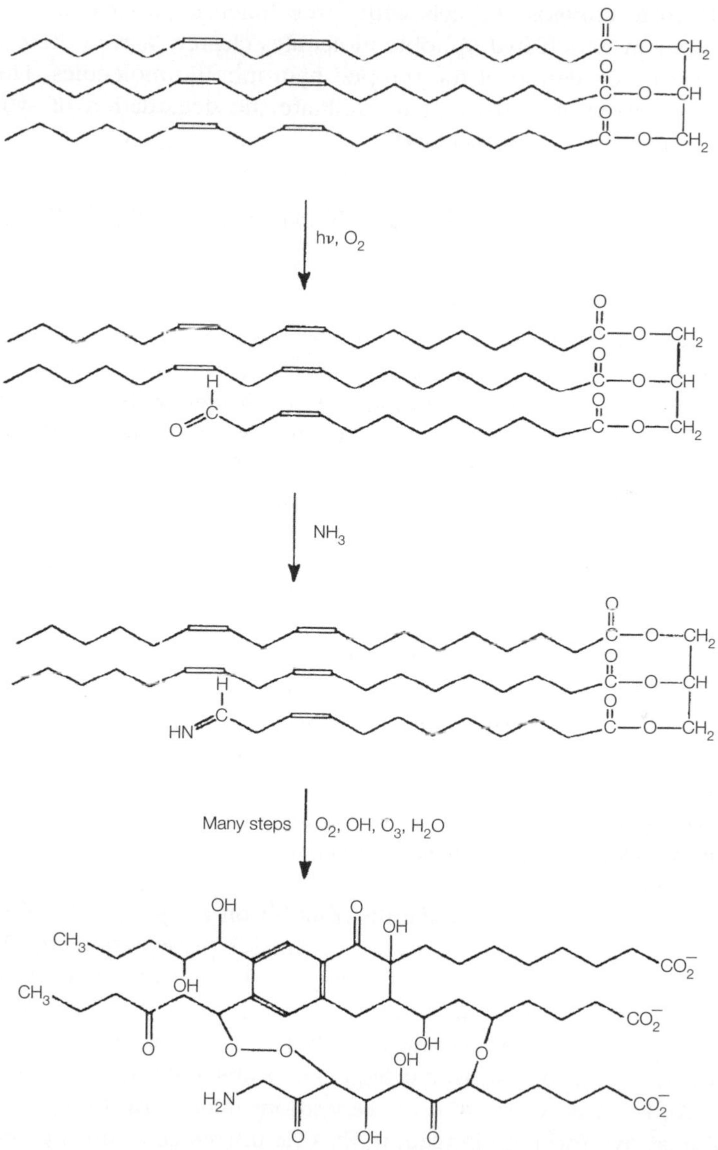

FIGURE 23.12

Proposed mechanisms for the incorporation of nitrogen into marine humic substances from photolyzed trilinolein (a model triglyceride). Photolysis of triglycerides generates aldehydes that are very reactive. The last two molecules in the series are representative of a family of related structures that contain covalently bound nitrogen. *Source*: From Kieber, R. J., *et al.* (1997). *Limnology and Oceanography* 42, 1454–1462.

tends to form colloids and gels with three-dimensional structures that possess interior microzones in which hydrophobic molecules cluster. Within these interiors, singlet O_2 promotes the oxidation of the trapped hydrophobic molecules. This process is of particular interest as it is thought to facilitate the degradation of synthetic hydrophobic organic compounds, such as DDT.

23.6.8 Export of DOM into Deep Ocean and Horizontal Segregation

DOM concentrations are highest in the surface water as this is where most organic matter is synthesized. Concentrations decrease exponentially through the thermocline (Figure 23.4). The deep waters are characterized by low and more uniform concentrations. As a rule, DOC concentrations are much higher than those of POC, especially in the deep sea. These observations collectively suggest that the deep-water DOC is composed of relatively refractory organic matter. This has been confirmed by radiocarbon dating in which the average "age" of deepwater DOC in the Atlantic has been found to be 4000 y and that of the Pacific Ocean, 6000 y. The radiocarbon ages of surface-water DOC are lower, on the order of 1000 to 2000 y, but clearly much older than would be expected if it were composed solely of labile DOC being rapidly cycled between the phytoplankton and bacteria.

Similar trends are seen in the C/N ratio of DOM. The C/N ratio of bulk deepwater DOM (18:1) is higher than bulk surface-water DOM (14:1), which is in turn higher than plankton (6.6:1). The radiocarbon and C/N observations gave rise to the two-pool model presented in Figure 23.8 in which surface-water DOM is viewed as an admixture of relatively young labile DOM and a much older refractory pool. The labile pool is effectively restricted to its site of formation in the surface waters, whereas the refractory pool is persistent enough to spread via advective currents throughout the surface and deep ocean.

The effect of this differential distribution on a typical vertical concentration profile is shown schematically in Figure 23.13. In this presentation, the DOM has been more realistically compartmentalized into four pools of varying reactivity ranging from the very labile, to the semilabile, to two types of refractory DOM, one which turns over on time scales of ocean mixing and the other that persists for several millennia. The latter is estimated from the horizontal segregation observed in deepwater DOM. As shown in Figure 23.14, deepwater DOC concentrations decline by 14 μM C along the pathway of meridional overturning circulation. DOC is introduced into the deep ocean by NADW, which forms seasonally from surface waters enriched in DOC. (The formation waters for AABW are low in DOC, and, hence, this water mass does not inject much DOC into the deep ocean.) The decline in DOC along the pathway of meridional overturning circulation is attributed to removal. Possible removal mechanisms include microbial respiration, adsorption onto sinking particles, and aggregation into marine snow.

Quite a bit of DOC is also injected into the subsurface ocean via isopycnal mixing in which water is transported down to the top of the thermocline. This subduction is associated with the seasonal formation of mode water, which occurs mostly on the

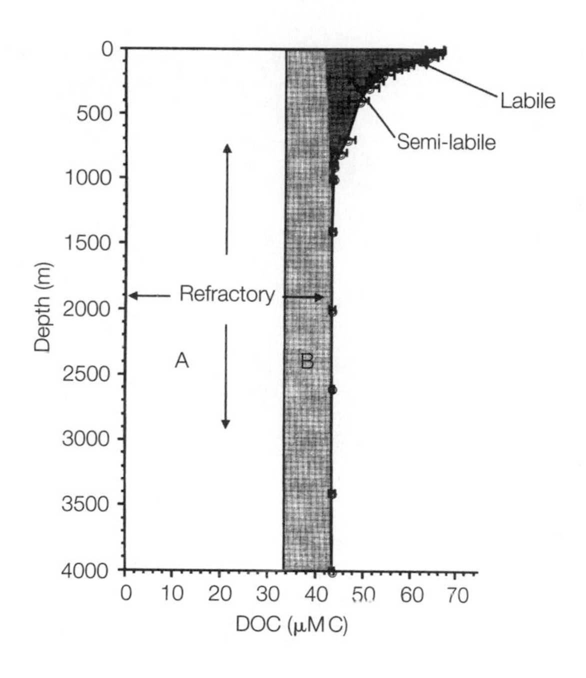

FIGURE 23.13

Conceptual cartoon of the various pools of refractory, semilabile, and labile DOC in the open ocean. *Source*: From Carlson, C. A. (2002). *Biogeochemistry of Marine Dissolved Organic Matter*, Academic Press, pp. 91–151.

western boundaries of the mid-ocean gyres between latitudes of 20 to 40° (Figure 4.16c). The DOC that subducts appears to form in areas of high productivity, such as equatorial upwelling zones. It is transported laterally via surface currents to the site of mode water formation. Measurements of the radiocarbon ages of DOM, fractionated by size and compound class, demonstrate that most of the labile DOM being subducted is rapidly decomposed and, hence, is not retained long enough to complete an oceanic mixing cycle. Some is decomposed into inorganic components and some into refractory DOM, primarily LMW in size and lipid in molecular character. The refractory pool degrades so slowly that it can complete one or more oceanic mixing cycles.

As per the model depicted in Figure 23.13, little or no labile DOM is present below the top of the thermocline. The refractory pool in the deep water experiences some degradation during a mixing cycle, leading to the observed horizontal segregation of DOC in the deep waters. The rest is stable enough to return to the surface waters. Twenty percent of the DOC would have to be removed during a 1000-y mixing cycle to maintain a steady-state age of 5000 y in the deepwater DOC. In other words, an average DOC carbon atom makes five oceanic mixing trips before being removed. The admixture of this refractory carbon with the labile surface water DOC is responsible for the relatively old age (1000 to 2000 y) of bulk surface-water DOC. Given the limited capacity for DOM removal in the deep waters, some degradation of the

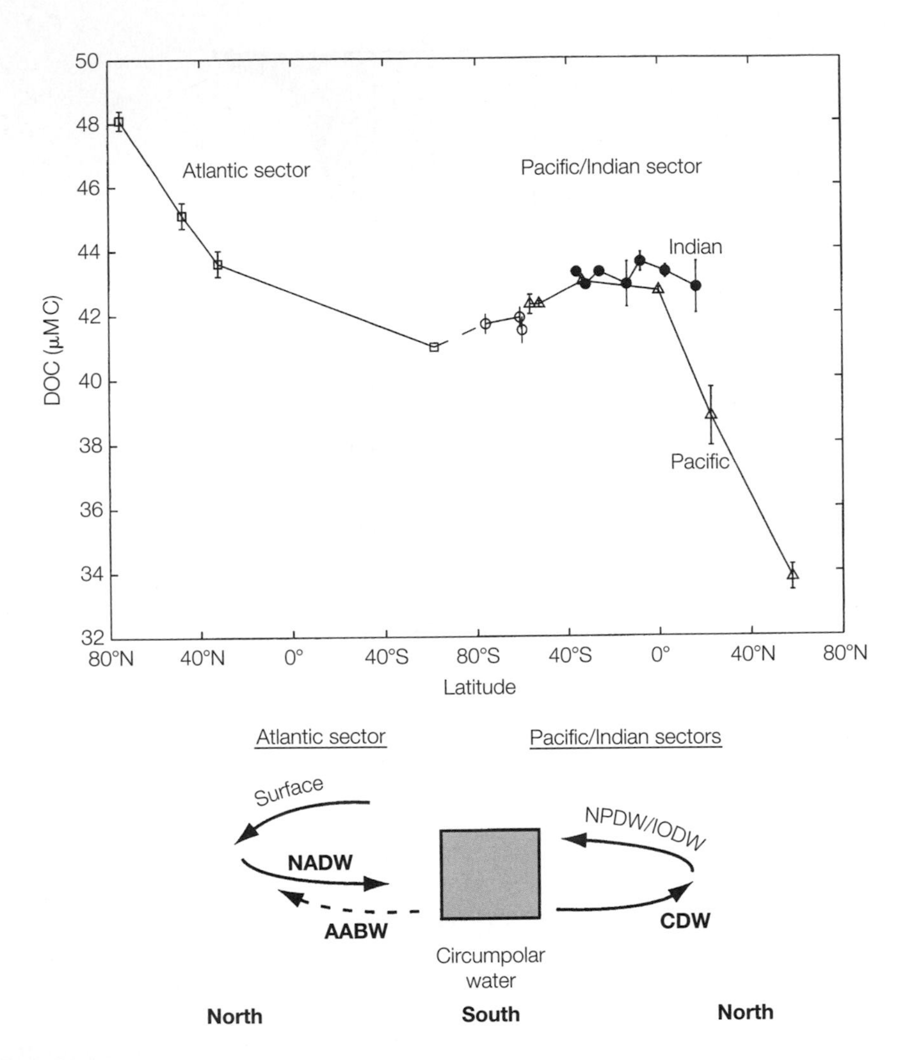

FIGURE 23.14

Distributions of DOC in the deep ocean. The *x*-axis is viewed in the context of the deep-ocean circulation, with formation in the North Atlantic, circulation around the Southern Ocean, and flow northward into the Indian and Pacific oceans. *Source*: From Hansell, D. A. (2002) *Biogeochemistry of Marine Dissolved Organic Matter*, Academic Press, pp. 685–715.

refractory compounds is posited to occur while it is in surface waters via photochemical reactions.

The export of DOC into the middle and bottom waters of the ocean represents 20% of the total annual export of carbon from the surface waters. Thus, DOC export represents a significant part of the biological pump contributing to the ocean's ability to extract and store atmospheric carbon. This export also has ramifications on nutrient and O_2 cycling. As shown in Figure 23.8, subsurface decomposition of the labile DOM has the potential to remineralize a significant amount of nutrients. Once returned to the surface waters, these nutrients support primary production. In contrast, the refractory DOM is nutrient poor. Upwelling of its remineralization products will not support much primary production and will instead act to remobilize carbon by returning it to the surface waters that are in contact with the atmosphere and UV radiation. If the deepwater DOC lost during horizontal segregation is being aerobically respired, it can account for 10% of the AOU of the deep waters and 30% in the main thermocline. Current accounting of the global carbon cycle suggests the deep ocean is functioning as a net heterotrophic system, with the subsurface DOC providing a significant amount of the necessary fuel.

All of these observations point to the important role of DOM in determining the ocean's ability to store carbon. The present-day DOC export flux from the surface waters appears to be pretty close, if not equal to, the net production of DOC. (The export flux composes only 17% of the *total* production due to rapid cycling of labile DOM in the surface waters.) This balance between net DOC production and its export is likely to shift over time as significant export can only occur when an ecosystem produces a large standing crop of DOC that is subducted out of the surface waters faster than it can be biologically consumed. Thus, timing is everything.

Once DOC has been exported into the subsurface waters and converted into refractory compounds, it enters a form that enables storage over time scales of millennia. If the DOC was instead retained in the labile surface-water pool, its concentration would likely fluctuate with primary production, both seasonally and interannually. Thus, the global DOC cycle has the potential to dampen rapid oscillations in the marine carbon cycle caused by seasonal and interannual shifts in primary production. Conversely, any alterations to the global marine cycle of DOM have the potential to change the ocean's ability to store carbon and, hence, could lead to global climate change. Two big unknowns in this aspect of marine DOM cycle are the fate of terrestrial DOM and the degree to which DOM produced during diagenesis acts as a source to the bottom waters via diffusion out of the sediments.

23.7 SEDIMENTARY ORGANIC COMPOUNDS: DIAGENESIS AND PRESERVATION

By serving as the collection site for sinking POM, the surface sediments represent a food-rich interface that supports the benthic food web. This food web is functionally similar to that in the water column, although some of the consumers are unique to the sediments, namely the meiofauna, the macrofauna, and the microflora that live in the guts of the macrofauna (Figure 23.15). Virtually all of the POC in the surface sediments passes through the guts of the benthic animals prior to final burial. The microflora in their guts aids in the metabolic breakdown of the ingested organic matter. This helps explain why degradation in the surface layers of the sediment destroys 93%

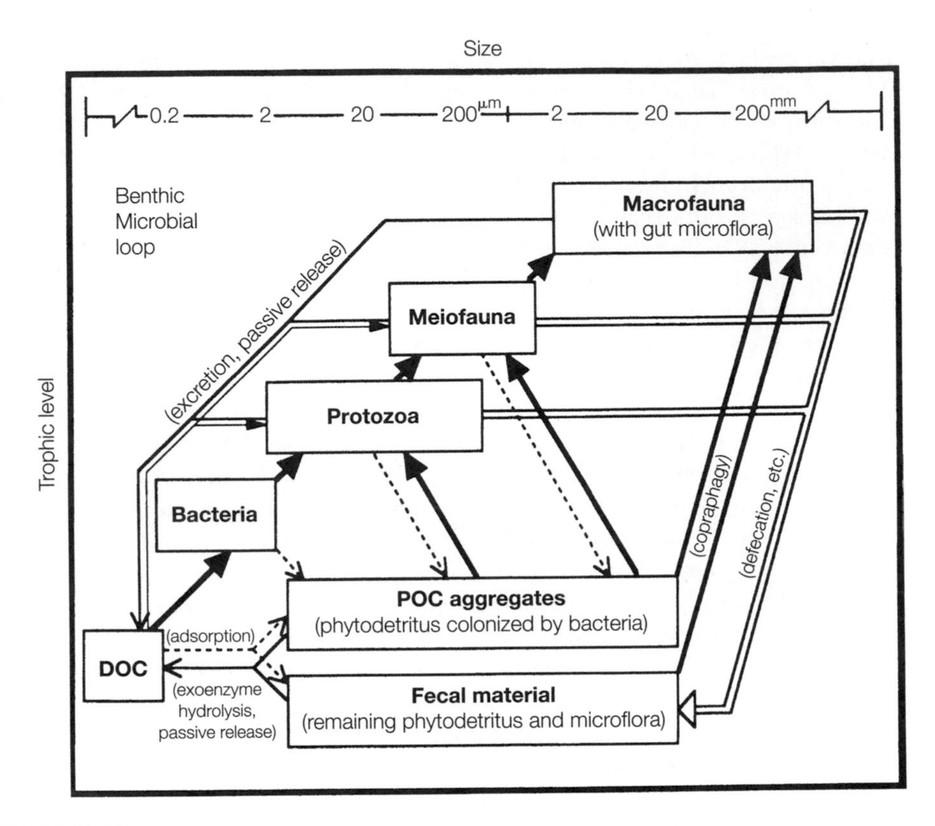

FIGURE 23.15

The benthic food web in marine sediments supported by sedimenting POM. Solid triangular arrowheads indicate energy flow to higher trophic levels; open arrowheads, loss of POC; line arrowheads, flow of DOC; dotted arrows, colonization. *Source*: From Deming, J. W., and J. A. Baross (1993). *Organic Geochemistry: Principles and Applications*, Plenum Press, pp. 119–139.

of the POC reaching the seafloor, leaving only 0.4% of the primary production to be buried.

Given the extensive degree of biological processing in the surface sediments, it is not surprising that most of the POC that manages to get buried in the sediments is highly altered. As shown in Figure 23.1, close to 80% of the sedimentary organic matter is molecularly uncharacterizable. This suggests that the POM that settles onto the sediments continues to undergo a progressive conversion into relatively refractory compounds. A conceptual model is shown in Figure 23.16 illustrating the reactions pathways by which the biomolecules in marine sediments are transformed into uncharacterizable forms. These pathways are further discussed through the rest of Chapter 23.7.

Note that a small fraction of the biomolecules delivered to the sediments as part of the sedimenting POM are not transformed. These biomolecules have molecular structures that are relatively resistant to degradation. (See Table 22.3 for a list of

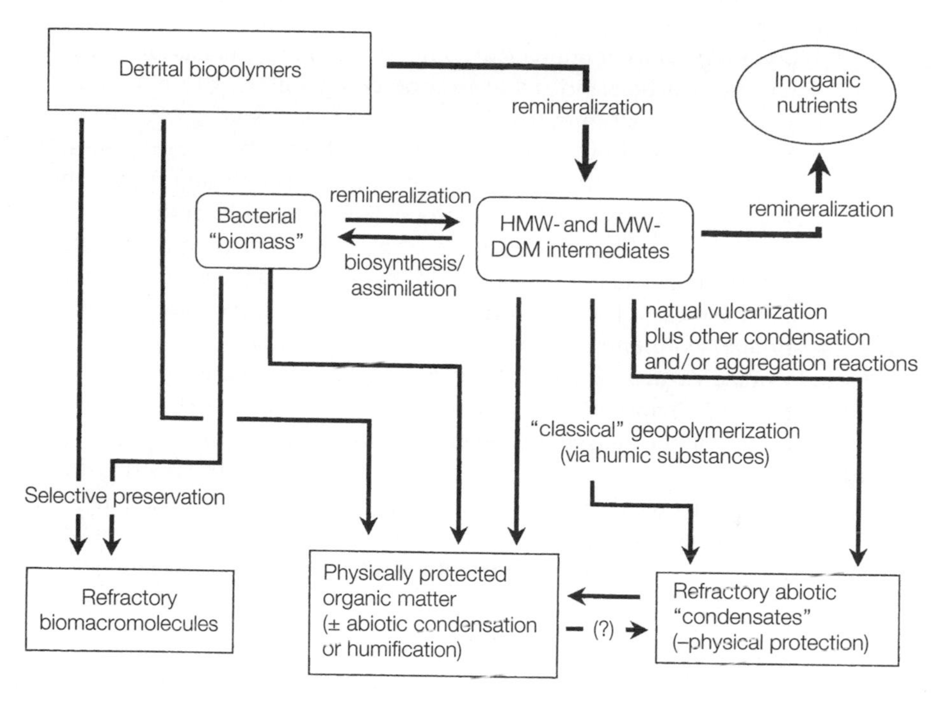

FIGURE 23.16

A conceptual model illustrating the pathways by which sedimenting POM is transformed into molecularly uncharacterizable organic matter. *Source*: From Burdige, D. J. (2006). *Geochemistry of Marine Sediments*. Princeton University Press.

biomacromolecules.) Figure 23.1 shows that of the small fraction of molecules in the sediments that are characterizable, most are amino acids, carbohydrates, and lipids. Within the last category are compounds that are species specific and, hence, make excellent biomarkers. Examples include the hopanoids, hopane and diploptene (Figure 22.8f); the terpenoids, haslene and amyrin (Figure 22.18); and the steranes (Figure 22.19).

23.7.1 Diagenetic and Catagenic Transformations

Because the vast majority of the sedimentary organic compounds are not amenable to direct study by current analytical techniques, organic geochemists rely on operational approaches to characterize them, such as measuring the %OC, %ON, OC/ON ratio and humin content.[2] Other strategies include molecular analysis of the small fraction of sedimentary organic compounds that are detectable. Some of these compounds make

[2] The elemental composition of the sediments is determined by first acidifying a sample to remove carbonate carbon. The sample is then combusted at high temperatures along with catalysts that convert the organic carbon to CO_2 and the organic nitrogen to N_2. The gases generated are quantified using a thermal conductivity sensor.

useful biomarkers as their sources are unique to the organisms that originally produced them. Their radiocarbon and stable isotope compositions can also provide information on sources and degradation rates.

Diagenetic changes in the surface layer are followed by catagenetic transformation of organic matter as sediments eventually become deeply buried below the seafloor. Current research indicates that viable microbes are present in sediments buried as deeply as 500 m below the seafloor, suggesting catagenesis is at least in part biologically mediated. The resulting organic residues are chemically distinct from that of the surface sediments. These residues are even more intractable to molecular characterization than are the surface sediments and are operationally defined as *kerogen*. Under certain temperatures and pressures, kerogen is eventually transformed into petroleum. This process is the subject of Chapter 27.

The nature of the diagenetic changes in the bulk of the organic sedimentary matter is still largely unknown due to our continuing inability to structurally characterize most of the molecules. Downcore trends in bulk properties are not consistent and, hence, difficult to generalize. At some sites, such as illustrated in Figure 23.17, %OC and %ON decrease with increasing depth, suggesting loss of organic matter via microbial

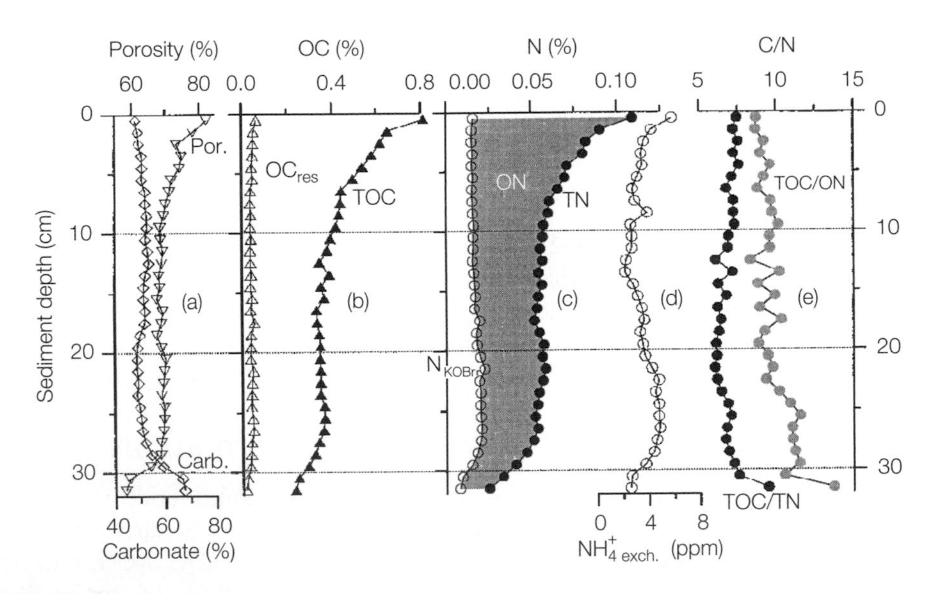

FIGURE 23.17

Depth distribution of (a) porosity and carbonate, (b) organic carbon fractionated into two pools: total organic carbon (TOC) and KOBr-resistant organic carbon (OC_{res}), (c) total nitrogen (TN), KOBr-resistant nitrogen (N_{KOBr}, mainly fixed ammonium), and by difference, organic nitrogen (ON, shaded area), (d) adsorbed ammonium (NH_{4exch}^{+}), and (e) TOC/TN and TOC/ON ratios. These sediments were collected from a mesotrophic site in the eastern subtropical Atlantic between the Canary Islands and the Moroccan coast in 1360 m water depth (28°53.4′N, 13°13.6′W). *Source*: From Freudenthal, T., *et al.* (2001). *Geochimica et Cosmochimica Acta* 65, 1795–1808.

degradation during diagenesis. Some of the nitrogen loss is due to ammonification[3] with the resulting ammonium becoming trapped in the mineral lattice of the association inorganic particles (Figure 23.17d). The increase in TOC/TON with depth indicates a preferential degradation of nitrogen-rich organic matter. At other sites, TOC/TON ratios do not change with depth, suggesting nonselective degradation, or they increase with depth, suggesting production of microbial biomass. Some biomarkers, such as the hopanoids, alkenones, and steranes, are common components of ancient marine sediments, providing further evidence for selective preservative of particular organic compounds.

23.7.2 Sedimentary Organic Matter as a Paleoceanographic Record

In some sediments, downcore variations in the bulk chemical composition are interpretable as records of temporal shifts in the elemental composition of the sinking POM. Such shifts are caused by changes in the production of sinking POM, which are in turn the result of fluctuations in the abundance and diversity of the overlying plankton community. In nearshore sediments, fluctuations in river runoff and lateral transport can lead to shifts in the supply rate of terrigenous organic matter. An example of a nearshore sediment core in which such fluctuations have been recorded is shown in Figure 23.18.

Periods of high sedimentation rate are associated with elevated organic carbon contents. The relative amounts of terrigenous and marine organic matter were estimated using biomarkers and the natural abundance ratios of the stable isotopes of carbon and nitrogen. Higher C/N ratios are also considered to be diagnostic for the presence of terrigenous organic matter. In this example, dating of the sediments established that glacial low sealevel stands were periods of enhanced sedimentation of terrigenous organic matter. Other lines of evidence confirm that diagenesis was also partly responsible for the overall decline in %C with depth, highlighting the limitations in using sediments are paleooceanographic records.

Note the large difference in core thickness between Figures 23.17 and 23.18. The sediment at the bottom of the latter core was about 100,000 years old! Collection of these kinds of core lengths is challenging as sampling must be done in such as way as not to disturb the sediment layers. Sites are carefully chosen to maximize the chances that sedimentation has been continuous over the time interval of interest. Also required is a high enough sedimentation rate to enable the formation of a layer thick enough to permit detailed subsampling. Geochemists also try to select for conditions where bioturbation has been minimal, i.e., anaerobic sediments.

---

[3] Ammonification is the microbial process by which an amine group is cleaved from POM and converted to ammonium ion.

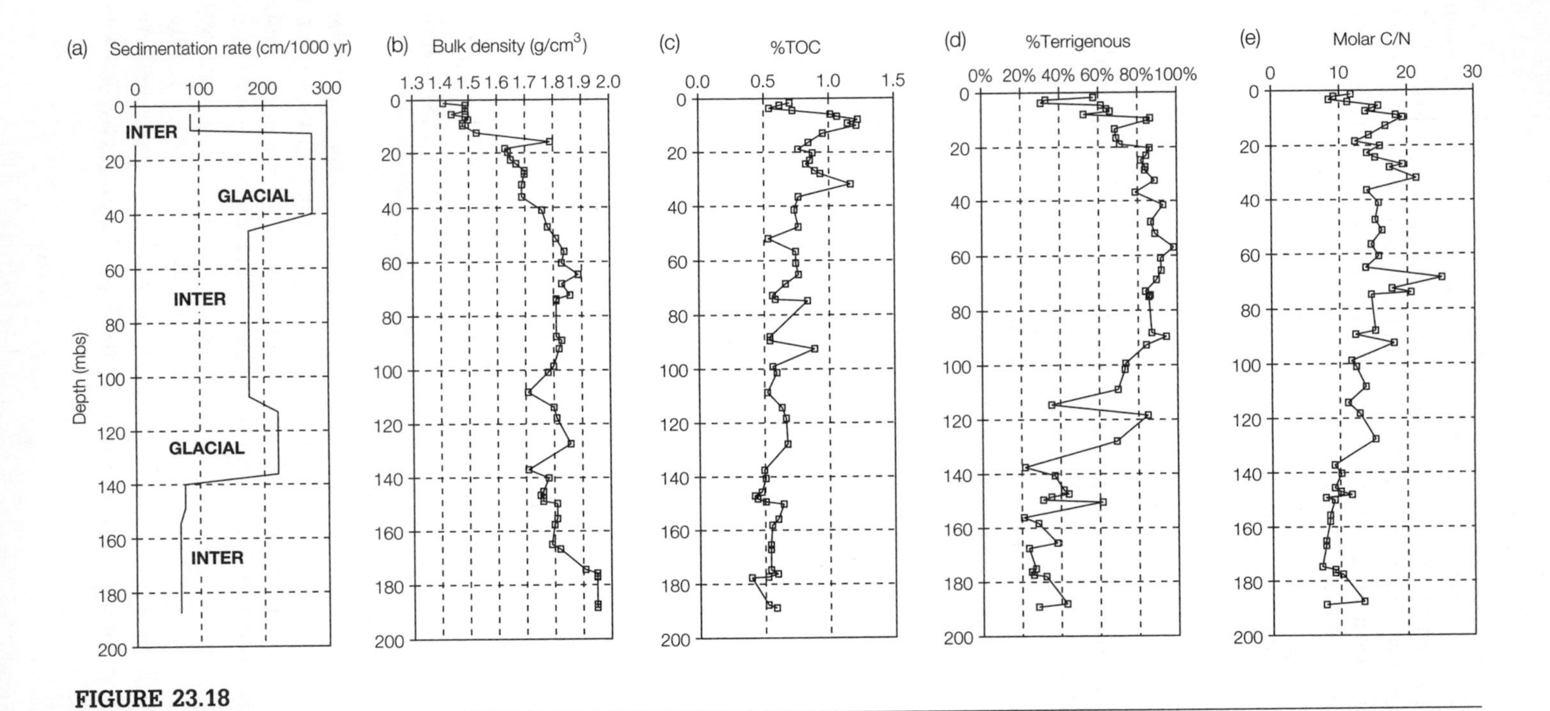

FIGURE 23.18

(a) Sedimentation rate, (b) bulk density, (c) % TOC, (d) % TOC from terrigenous sources, and (e) molar C/N ratio in sediment collected from 2259 m water depth in the Pigmy Basin in the northern Gulf of Mexico (27°11.61'N, 91°24.54'W). mbs = meters below seafloor. Data from Jasper, J. P., and R. B. Gagosian (1990). *Geochimica et Cosmochimica Acta* 54, 1117–1132.

23.7.3 Mechanisms of Preservation

Considerable attention is currently focused on determining what controls the global patterns of organic matter burial, as this is the ultimate storage reservoir of carbon in the crustal-ocean-atmosphere factory. As will be shown later, the total organic matter contents of the surface sediments do not always reflect the geographic patterns of primary production in the overlying surface waters. Some of the decoupling is due to variations in the efficiency with which sinking POM is exported from the photic zone, some from factors that control how much sinking POM reaches the seafloor, and some from the factors that influence burial rates. Export efficiencies are dependent on the plankton community composition and the aeolian dust flux as both influence the degree of ballasting in the sinking POM. Sedimentation is also dependent on sinking rates (again dependent on ballasting), the length of the water column (amount of time for microbial degradation), and possibly the O_2 content in the deep waters. Burial rates seem to be dependent on: (1) the chemical composition of the material reaching the seafloor; (2) sedimentation rates, as this determines the thickness of the oxic zone and, hence, the length of time that the sedimentary organic matter spends in contact with O_2; and (3) the types of anaerobic metabolisms occurring in the suboxic and anoxic layers. In the case of the last, some evidence suggests that degradation proceeds at slower rates under suboxic and anoxic conditions.

The inorganic matrix of the sediments seems to confer some protection against degradation onto organic matter, and most particularly onto marine organic matter. Sedimentary organic matter is so closely associated with the mineral fraction that the two cannot be physically separated. This causes the %OC in the sediments to be highly correlated with the surface area of the mineral particles. The particles with the highest surface areas are: (1) the three-layer montmorillonite clays, (2) the iron-manganese oxyhydroxides, and (3) diatom frustules.

Sedimentary organic matter can be chemically separated from its associated minerals, and when separated is labile. This supports the hypothesis that the inorganic matrix confers a protective effect to its associated organic matter. This protective effect does not seem to extend to terrestrial organic matter as evidenced by its low levels in all but very nearshore sediments. Two possible explanations for this differential protection are: (1) terrestrial organic matter is mineral free and (2) it does not form the kinds of mineral associations that provide the protective effect. Not surprisingly, the only types of terrestrial organic matter found in marine sediments are highly refractory ones, such as lignin decomposition products, soils, and black carbon.

The protective action of mineral association is less effective when the O_2 content of the bottom waters is low. The highest %OC levels in marine sediments occur in the surface layers at sites where anoxia is present, such as beneath upwelling zones. The high %OC level is undoubtedly reflective of a high sedimentation rate of POC, which is in itself protective by ensuring the rapid burial of organic matter, but some is due to slower degradation rates characteristic of low-O_2 environments. The latter has been confirmed through the observation of high degradation rates occurring in ancient organic-rich sediments that have become exposed to oxic bottom waters as a result of some geological event, such as erosion, uplift, or a turbidity current.

Other evidence supporting the role of O_2 in controlling the burial rate of sedimentary organic carbon is seen in the negative correlations of %OC with O_2 exposure times. The latter is defined as the depth of O_2 penetration divided by the sedimentation rate. (See Chapter 12.5.1 for a discussion of O_2 penetration depths.) O_2 exposure times integrate the effect of sediment accumulation rates and bottom-water O_2 levels on organic matter degradation. Nevertheless, marine chemists have not yet determined on a molecular level how the presence of O_2 enhances degradation rates.

Last, diagenetic and catagenic reactions may confer some protection against degradation of organic compounds. These reactions are thought to lead to structural rearrangements that enhance the chemical stability of various molecules. One likely candidate is the humification process and a second is sulfurization. The latter seems to occur in anoxic sediments and involves the formation of sulfide and polysulfide bridges that crosslink macromolecules. Lipids, and possibly carbohydrates, undergo such reactions. An important condition for sulfurization reactions is a low sedimentary iron content as high levels scavenge all the sulfide by precipitating pyrite. Sediments with sufficiently low iron contents are found in offshore OMZs and the distal ends of anoxic basins.

23.8 GLOBAL PATTERNS IN ORGANIC MATTER DISTRIBUTIONS

The organic matter distribution in seawater and sediments is heterogenous over time and space because its production and destruction are controlled by biogeochemical transformations ultimately driven by the atmospheric climate, ocean circulation, and global rock cycle. The latter involves plate tectonism, which determines continental erosion rates and, hence, the delivery of nutrients and micronutrients into the ocean. Plate tectonism creates the shape of the ocean basins and, thus, has a strong influence on the pattern of seawater circulation. Global oceanic circulation is also controlled by atmospheric climate and thereby regulates two critical determinants of primary production, the depth of the mixed layer and sea surface temperatures. The interaction of all of these factors leads to the spatial and seasonal trends in organic matter described below.

Long-term changes in climate, ocean circulation, and the rock cycle have led to large-scale changes in organic matter production. These changes have been recorded in ancient marine sediments as layers of varying organic content and molecular composition (Figure 23.18). The accompanying sedimentary fossil record shows that these changes record shifts in the marine plankton community structure, with some organisms becoming extinct and new ones evolving to occupy the vacant ecological niches. Similar shifts are currently being observed. At least some of them are thought to be related to global warming and are likely to affect the ocean's ability to store carbon.

23.8.1 Surface Production and Organic Matter Export from the Photic Zone

Estimating Chlorophyll from Satellite Imagery

The global standing stock of autotrophic biomass in the surface waters is estimated to be about 1 Gt C (Table 23.2). This estimate is based on a two-step conversion of satellite color imagery data of the global surface ocean. These color data are currently being provided by SeaWIFS (Sea-Viewing Wide Field of View Sensor) and the CZCS (Coastal Zone Color Scanner) along with other satellites described in Chapter 1.6.

In the first step, the light naturally radiated from the sea surface is used to obtain chlorophyll *a* concentrations. Most of this radiant light is reflected sunlight from which some wavelengths have been removed by molecular absorption, such as by chlorophyll *a*. Chlorophyll *a* absorbs light in the blue-violet (400 to 450 nm) and yellow-red (575 to 675 nm) wavelength ranges, leaving only green light to be reflected. By measuring the reflected light at carefully selected wavelengths in the 450 to 550 nm and infrared ranges, interferences from other substances that absorb and reflect light in this range, including CDOM and particles, can be minimized. In the second step, chlorophyll concentrations are converted to carbon concentrations.

The conversions conducted in both steps are currently based on empirical relationships that are more or less robust. For example, the relationship between the chlorophyll and carbon content in an "average" phytoplankton cell is dependent on factors that influence cell metabolism, including nutrient availability, temperature, and light. The temperature dependence of photosynthesis is associated with an enzyme-mediated step in the Calvin cycle (Figure 7.6a).

Some of the wavelengths collected as part of the satellite color dataset have been used to estimate POC concentrations. Figure 23.19 shows the geographic and seasonal trends in the surface water chlorophyll and POC. (Technically these are concentrations through the top 10 to 30 m of the water column as sunlight is reflected back through these depths.) The trends in chlorophyll and POC are generally similar, although chlorophyll composes only a small fraction of the POC pool and its abundance relative to POC is variable (0.1 to 1%). This similarity in spatial and temporal trends reflects the fundamental control of primary production in providing fixed carbon to the marine planktonic food web.

Three biogeographic zones have been defined to describe the spatial trends in the satellite-derived chlorophyll concentrations (C_{sat}): (1) oligotrophic = $C_{sat} < 0.1\ mg\ m^{-3}$, (2) mesotrophic $0.1 < C_{sat} < 1\ mg\ m^{-3}$, and (3) eutrophic = $C_{sat} > 1\ mg\ m^{-3}$. Most of the oligotrophic waters are located in the mid-ocean geostrophic gyres. Most of the eutrophic waters are in areas of divergence in the open ocean and coastal zone. Seasonal variability is strongest in mid- and high latitudes.

Estimating Net Primary Production

The entire phytoplankton biomass of the global oceans is consumed every 2 days to 2 weeks. Thus, the standing stock tells us little about how much organic

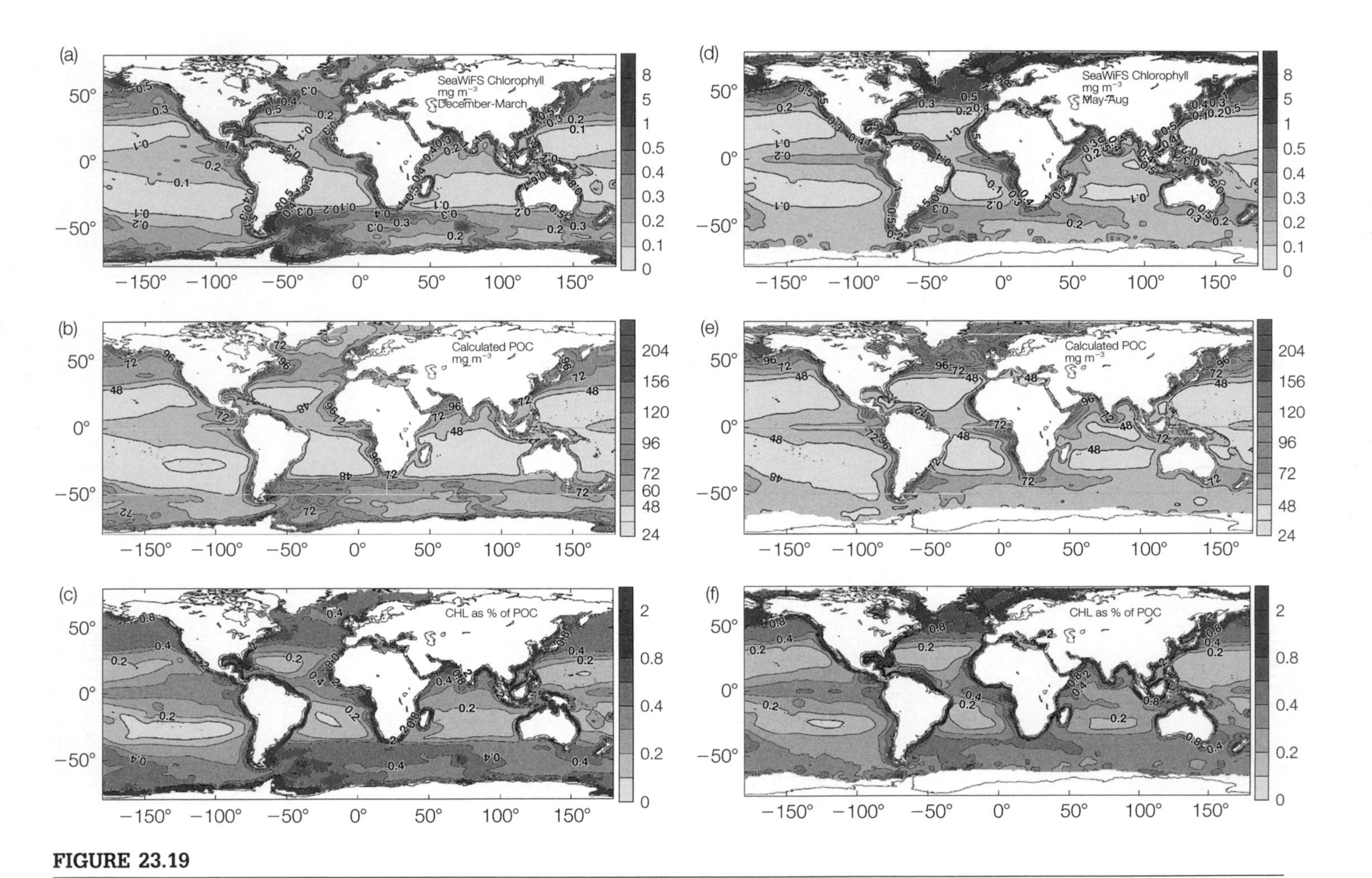

FIGURE 23.19

Average global distributions during 1997–2002 of wintertime (December–March): (a) SeaWIFS chlorophyll (mg m^{-3}); (b) average POC (mg m^{-3}); and (c) chlorophyll as a % of POC, and of summertime (May–August): (d) SeaWIFS chlorophyll (mg m^{-3}); (e) average POC (mg m^{-3}); and (f) chlorophyll as a % of POC. The concentrations are averaged over one attenuation depth, which ranged from 10 to 30 m. *Source*: After Gardner, W. D., *et al.* (2006). *Deep-Sea Research* II 53, 718–740. (See companion website for color version.)

matter is moving through the marine food web and ultimately exported to the seafloor. A better measure is provided by considering the rate at which autotrophically fixed carbon is made available to the first heterotrophic level. This rate is estimated from the *net primary production* (NPP), which is a measure of the primary production available for consumption. It is less than the gross primary production because some of the autotrophically fixed carbon is respired by the phytoplankton. Global annual NPP is estimated to be 50 Gt C/y. When this is compared to the biomass of the autotrophs (1 Gt C), it is clear that the surface-water pool of organic carbon must be rapidly cycling within the surface-water plankton community.

Numerous mathematical models have been developed to estimate global NPP. In these models, phytoplankton growth is estimated from a variety of controlling physical and biological factors, such as light, temperature, mixed layer depth, nutrient availability, and grazing. The results from one model are shown in Figure 23.20 and illustrate similar geographic and temporal trends to that observed in the chlorophyll data.

Some of the NPP models are based on the color imagery and some are not. In the latter, phytoplankton growth is estimated from coupled global circulation and biogeochemical models in which water motion controls nutrient availability. The water motion is controlled by climatic factors, such as temperature gradients and wind stress. The latest effort to compare model outputs was conducted with 31 different models and found that global estimates for a test year (1998) differed by as much as a factor of 2! The mean results from this model intercalibration experiment are shown in Table 23.7.

Most of the NPP occurs in mesotrophic waters mostly due to their large surface area. Note that the eutrophic waters make a disproportionately large contribution to the global NPP (11%) as compared to the area it occupies (3 to 5%). Similarly, the high-latitude regions (>70°N and >50°S) have a disproportionately lower contribution, reflecting the negative effect of colder temperatures on photosynthesis. These regions are also characterized by deep mixed layers that are not favorable to phytoplankton growth because vertical mixing causes the cells to spend time below the bottom of the photic zone where they cannot photosynthesize. Conversely, regions with very high surface temperatures are characterized by poor phytoplankton growth due to strong density stratification that prevents upwelling of nutrients. Regions where upwelling occurs at moderate to high temperatures (>20°S) have the highest open-ocean NPP. They include the equatorial regions and the zones of divergence located between 40 to 50° north and south.

Estimating POM Export from the Euphotic Zone

The type and amount of organic matter exported from the surface waters depends on the NPP and temperature. The general trend is for higher NPP to generate more export as more organic matter is available to sink below the photic zone. In the case of POM, this sinking occurs as particles settle to the seafloor. In the case of DOM, organic matter is transported into the deep sea by subduction of surface waters via mixing along isopycnals. Export is strongly controlled by temperature because the rate of bacterial heterotrophy increases with temperature, leading to a more efficient destruction of sinking organic matter.

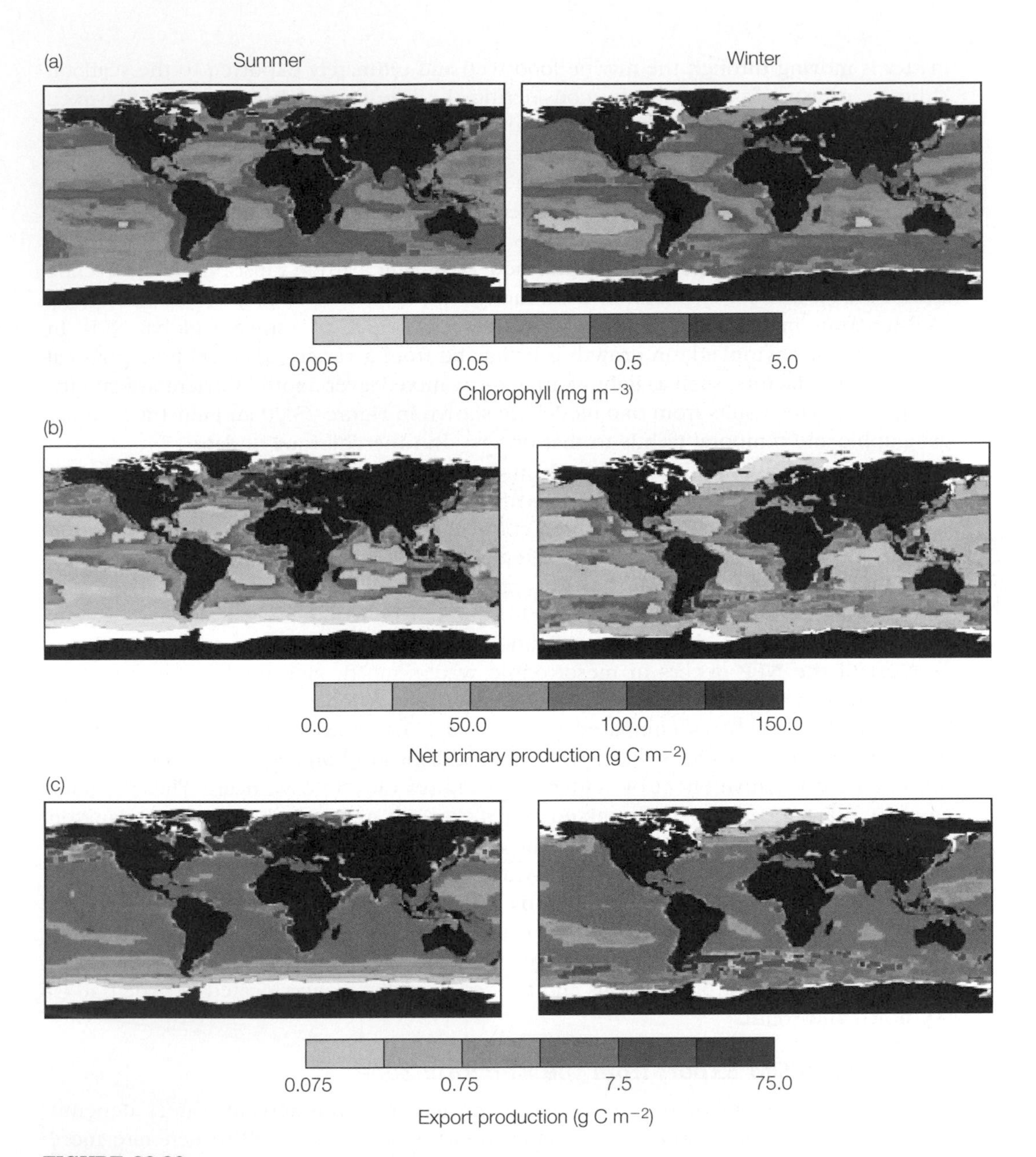

FIGURE 23.20

Composite global images for winter and summer of upper-ocean (a) chlorophyll concentrations derived from satellite-based observations of ocean color, (b) computed net primary production, and (c) POC export production. *Source*: After Falkowski, P. G. (2003). *Treatise in Geochemistry*, Elsevier, pp. 185–213, (See companion website for color version.)

POM export is also greatly affected by the types of plankton growing in the surface waters. The most efficiently exported POM is generated by diatoms and their grazers. This is due to the protective effect of siliceous hard parts on its associated organic matter and the relatively high density of diatom frustules, which causes them to have high sinking rates. In comparison, the phytoplankton that are small and do not have hard parts, such as the dinoflagellates, are poorly exported. The larger zooplankton, which are the herbivores, tend to generate the largest and densest fecal pellets and, hence, contribute significantly to the export flux. The calcareous plankton are also important contributors to the POM export flux.

As with the NPP, estimates of global export production have been inferred from mathematical models. The results from one effort are presented in Table 23.7 and Figure 23.20c. They indicate that the fraction of NPP exported from the surface waters (ef ratio) is low, being generally less than 30%. Geographic and seasonal patterns in POM export are similar to that of NPP. As noted earlier, seasonal shifts in particle fluxes have been detected in sediment trap samples (Figure 23.3). Thus, efforts are currently directed at including temporal dynamics into the POM export models along with details of phytoplankton community dynamics and zooplankton grazing.

Table 23.7 Ocean Net Annual Primary Production, Export Production, and ef Ratios.

	% Area	% Global NPP	Gt C/yr		ef ratio
			NPP	Export	
Basin					
Pacific	45%	44%	21	4.3	0.19
Atlantic	23%	27%	12.8	4.3	0.25
Indian	17%	21%	9.9	1.5	0.15
Southern	13%	5.5%	2.6	0.62	0.28
Arctic	12%	0.7%	0.33	0.15	0.56
Mediterranean	80%	0.95%	0.45	0.19	0.24
Global			47.1	11.1	0.21
Chlorophyll level					
Oligotrophic	26 to 32%	19%	9.2		
Mesotrophic	65 to 68%	70%	34.8		
Eutrophic	3 to 5%	11%	5.6		
Sea surface temperature range					
$<0°C$	2 to 4%	0.8%	0.52		
0 to 10°C	13 to 17%	10%	5.1		
10 to 20°C	~20%	25%	11.9		
$>20°C$	~60%	64%	32		

These estimates include a ~1 Gt C contribution from macroalgae. The ef ratios are a measure of the fraction of NPP exported from the surface waters. Data from Carr et al. (2006). Deep-Sea Research *II 53, 741–770; and Laws, E. A., et al. (2000).* Global Biogeochemical Cycles *14, 1231–1246.*

Interannual Variability

The first satellite color data were collected in 1979. Inspection of the first two decades of this color record suggests that significant interannual variations in chlorophyll concentration and NPP have occurred. Some are attributed to ocean circulation changes, such as ENSOs, and some to global warming. In the coastal zone, increases are likely being caused by enhanced runoff of nutrients from land. These changes are expected to be accompanied by shifts in plankton community structure. In Chapter 25, some observational evidence for such a shift is presented for the central Pacific Ocean. Interannual changes in the quantity and quality of NPP are likely to generate similar shifts in the POM export flux and, hence, the supply of organic particles to the seafloor. Evidence for past changes in this flux is provided by the down-core variations in organic matter content observed in the sediments (Figure 23.18).

23.8.2 **Sedimentary Organic Matter**

The ability of organic matter to become buried in the sediments provides for long-term storage of carbon, effectively removing it from the atmosphere for millennia. The

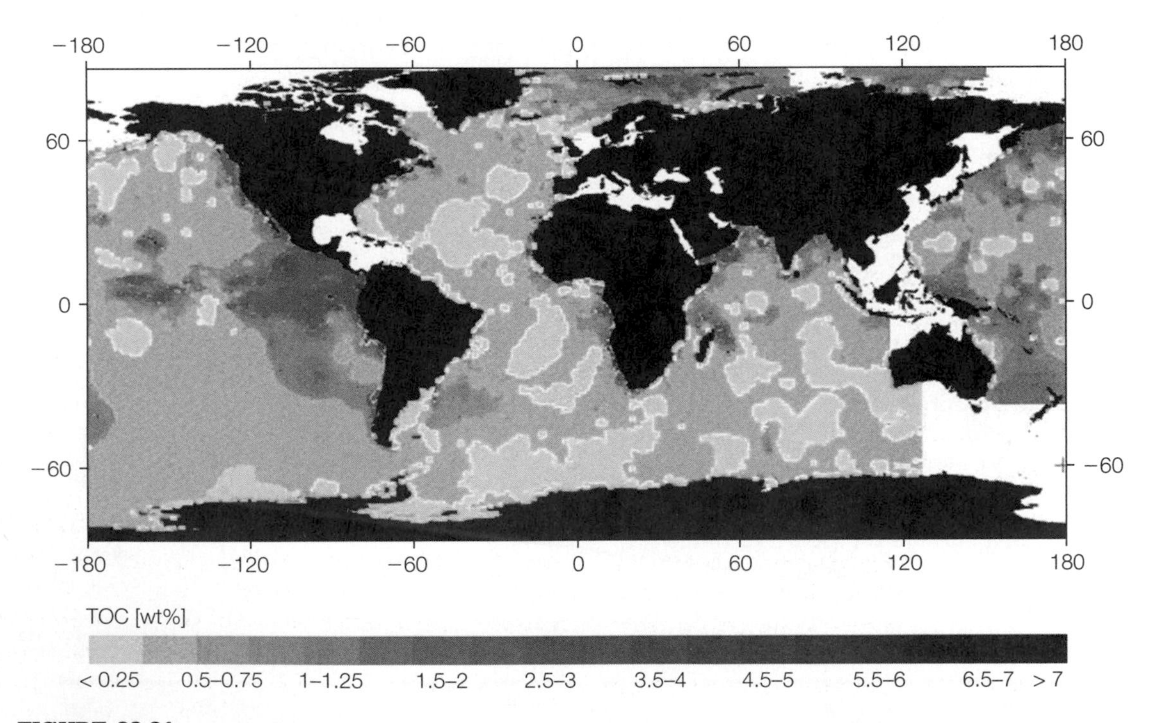

FIGURE 23.21

Global distribution pattern of the TOC content in surface sediments (<5 cm depth). *Source*: Redrawn from Seiter, K., *et al.* (2004). *Deep-Sea Research* I 51, 2001–2026.

geographic pattern of sedimentary organic deposition is shown in Figure 23.21 and indicates that most of the open ocean's surface sediments are characterized by very low organic carbon contents (<0.5%). This is largely due to the effective destruction of POM by heterotrophic bacteria as organic matter settles through the deep sea and undergoes diagenesis in the surface sediments. The only exception to this is seen in the equatorial upwelling zone of the Pacific Ocean, where siliceous oozes seem to confer some degree of preservative effect.

In the present-day ocean, the continental shelves are the location where most (90%) of the organic carbon is being buried. In this setting, organic carbon contents range from 1 to 10%. Most of organic matter accumulating in coastal sediments is produced locally by primary production, reflecting elevated levels of NPP nearshore supported by a high nutrient supply. Preservation of the exported POM is enhanced by the shorter length of the water column and rapid sedimentation rates that act to minimize CO_2 exposure times. Concentrations tend to increase with increasing proximity to the coastline. The notable exception is the very high organic content found in sediments lying below coastal upwelling areas. Resuspension and lateral transport can also create localized piles of organic-rich muds on the continental shelves.

CHAPTER

The Marine Nitrogen and Phosphorus Cycles 24

All figures are available on the companion website in color (if applicable).

24.1 INTRODUCTION

In this chapter, we consider the role of the ocean in the global nitrogen and phosphorus cycles. Because these two elements are strongly biolimiting, their cycles provide feedbacks in the crustal-ocean-atmosphere factory that affect important planetary characteristics including climate and atmospheric O_2 levels. Many earth scientists view the human alteration of the global nitrogen cycle as one of the most profound changes we have yet induced on our planet. Our largest impact has been in boosting the rate at which atmospheric N_2 is converted into reactive forms. By the end of the 20th century, human production of reactive nitrogen via N_2 fixation exceeded natural production rates. On the positive side, this increase in reactive nitrogen has provided food. At present, approximately 40% of the world's population is fed by crops sustained by the anthropogenic formation of reactive nitrogen. On the negative side, reactive nitrogen is now accumulating in the atmosphere and hydrosphere with deleterious consequences including eutrophication, hypoxia, harmful algal blooms, smog, acid rain, and loss of stratospheric ozone, all of which lead to habitat degradation and loss of species diversity.

Because of feedbacks within the crustal-ocean-atmosphere factory, our perturbation of the nitrogen cycle has the potential to influence global climate and, hence, oceanic circulation patterns and density stratification. In response to this growing problem, the International Council of Science organized the International Nitrogen Initiative in 2002 to optimize human uses of nitrogen such that sustainable food production is achieved without causing negative effects on human health and the environment.

Nitrogen is notable for its large number of naturally occurring oxidation states. Redox transformations between these oxidation states are an important feature of the marine nitrogen cycle. Most of these redox reactions are mediated by organisms. A large variety of metabolic pathways have evolved to exploit the redox chemistry of nitrogen. This complexity, compounded by great spatial and temporal variability, has made the marine nitrogen cycle difficult to study. Indeed, marine biogeochemists are still discovering new biological processes, such as the anammox reaction, which are likely of great importance to the crustal-ocean-atmosphere factory. As a result, our knowledge of global fluxes and reservoir sizes is still highly uncertain. The available data, which are presented in this chapter, suggest that either the marine biogeochemical cycle is not in a steady state or that some source of fixed nitrogen has been grossly underestimated. The consequences of departure from a steady state are not known, but likely affect the biogeochemical cycles of other biologically active elements such as carbon.

The story with the phosphorus cycle is similar, with humans having doubled the terrestrial inputs to the ocean, largely as a result of sewage discharges and fertilizer runoff. Phosphorus does not have the complicated redox chemistry of nitrogen, but important recent discoveries have suggested that this element, like nitrogen, is cycling through the ocean much faster than previously thought.

Various aspects of the marine chemistry of nitrogen and phosphorus have been discussed in earlier chapters. The energetics of the basic redox reactions were described in Chapter 7. The vertical and horizontal segregation of the dissolved inorganic species as a result of the biological pump was discussed in Chapters 8, 9, and 10 and diagenesis in Chapter 12. In this chapter, the role of the oceans in the global cycling of nitrogen and phosphorus is discussed in the context of changes being caused by human activities. Given the large role of the marine biota in transporting these elements, more detail is provided herein as to how these processes vary over time and space as a result of physical and biological controls.

24.2 NITROGEN SPECIES

The marine chemistry of nitrogen is largely controlled by redox reactions mediated by phytoplankton, bacteria, and probably archaea. As a result of these transformations, nitrogen is present in many oxidation states as listed in Table 24.1. The dissolved inorganic ions, NO_3^-, NO_2^-, and NH_4^+, are commonly referred to as DIN (dissolved inorganic nitrogen). Most of the nitrogen in the crustal-ocean-atmosphere factory is present as $N_2(g)$ and is relatively unreactive due to the great strength of the triple bond between the two nitrogen atoms. A few abiotic processes and microbes are able to convert N_2 into the more reactive compounds listed in Table 24.1. These compounds are collectively termed "fixed" or *reactive nitrogen* (Nr).

The oxidation number of nitrogen in all organic compounds is −III. Most of the dissolved nitrogen in seawater is in the form of DON except in the deep ocean where nitrate concentrations are very high. The mean surface water DON concentration is $6 \pm 2\,\mu M$ N and in deep waters it is $4 \pm 2\,\mu M$ N. PON generally represents only a small fraction of the fixed nitrogen pool.

As with DOC, a large fraction of the DON has yet to be molecularly characterized. Known major contributors to the DON pool include urea, amino acids, humic substances, nucleic acids (DNA and RNA), and the alkyl and quaternary amines. The last is thought to include amides of biological origin, including chitin, peptidoglycans from bacterial cell walls, *N*-acetylaminopolysaccharides (*N*-AAPs), and oligopeptides. Urea (Figure 22.26c) is a particularly large component of the DON pool. This low-molecular-weight species can reach concentrations of 13 μM N. It is released into

Table 24.1 Common Species of Marine Nitrogen.

Species	Molecular Formula	Oxidation Number of Nitrogen
Nitrate ion	NO_3^-	+V
Nitrogen dioxide gas	NO_2	+IV
Nitrite ion	NO_2^-	+III
Nitrous oxide gas	N_2O	+I
Nitric oxide gas	NO	+II
Nitrogen gas	N_2	0
Ammonia gas	NH_3	– III
Ammonium ion	NH_4^+	– III
Organic amine	RNH_2	– III

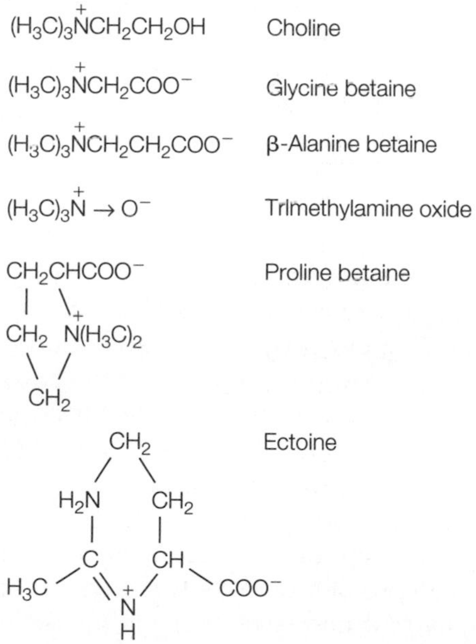

$(H_3C)_3\overset{+}{N}CH_2CH_2OH$ Choline

$(H_3C)_3\overset{+}{N}CH_2COO^-$ Glycine betaine

$(H_3C)_3\overset{+}{N}CH_2CH_2COO^-$ β-Alanine betaine

$(H_3C)_3\overset{+}{N} \rightarrow O^-$ Trimethylamine oxide

Proline betaine

Ectoine

FIGURE 24.1

Structure of quaternary amines and related compounds of ecological significance.

seawater by herbivorous zooplankton, cryptomonads (small flagellated green unicellular algae), bivalves, and fish.

Most marine organisms also contain alkyl and quaternary amines (Figure 24.1), which in some cases are present at concentrations close to that of the most abundant amino acids. The major function of these compounds is thought to be in osmoregulation so they are present almost entirely in solution as part of the intracellular matrix. These amines are probably most abundant in coastal marine organisms that are living in waters where salinities vary over time. These amines are released into seawater via excretion or cell lysis. Glycine betaine is of considerable interest because its production seems to be inversely related to that of DMSP and controlled by nitrogen availability. Given this linkage to DMSP, variations in the production of glycine betaine could potentially affect the CCN (cloud condensation nuclei) component of global climate (Figure W23.1).

24.3 THE GLOBAL NITROGEN CYCLE

The nitrogen flows between the land, sea, atmosphere, and sediments are illustrated in Figure 24.2. Three estimates are provided for different time intervals (circa 1860, early 1990s, and a projection for 2050) because human activities have significantly altered the rates of some of the critical transport pathways. The sizes of the major reservoirs are given in Table 24.2.

Most of the nitrogen in the ocean is in the form of N_2(g). This nitrogen is biologically inaccessible except to those few microbes, called nitrogen fixers, that can break N_2's strong triple bond. Most of the fixed nitrogen in the ocean is dissolved and present as deepwater nitrate and DON. The marine biota contain less than 0.05% of the marine Nr. Although the land biota contains a larger fraction of the terrestrial nitrogen (5%), the vast majority of the terrestrial nitrogen is present as part of the soil. Some of these reservoirs are also likely to be in a state of change due to anthropogenic impacts on the global nitrogen cycle.

The marine component of the global nitrogen cycle is quite distinct from that of the other biologically controlled elements, such as carbon, phosphorus, and sulfur, because of the strong controlling effect of in situ biological processes on the major input and output pathways. The major source of Nr to the ocean is via biological nitrogen fixation (87–156 $Tg\,N\,y^{-1}$) and the major loss is via denitrification (150–450 $Tg\,N\,y^{-1}$) and N_2O emission (4 $Tg\,N\,y^{-1}$). River input (48 $Tg\,N\,y^{-1}$ of DIN, DON, and PON) contributes a relatively small proportion that has little impact on the open ocean as most is lost to denitrification within the coastal zone. A small but significant input (33 $Tg\,N\,y^{-1}$) of Nr occurs via atmospheric deposition of the inorganic species (NO, NO_2, and NH_3). The rate of atmospheric deposition of DON is not known but is thought to be significant.

Compared to denitrification rates, burial in the sediments is a minor loss of nitrogen from the oceans. Current estimates suggest that shelf sediments are the major site of denitrification (14 $Tg\,N\,y^{-1}$) with a minor contribution from the sediments in the open ocean (0.8 $Tg\,N^{-1}$).

The globally integrated estimates of marine nitrogen fixation and denitrification have increased substantially over the past several decades in response to identification of previously unrecognized biological pathways by which these processes take place. As marine geochemists have refined their knowledge of nitrogen cycling rates, the estimated residence time of Nr in the ocean has gotten progressively shorter. Based on the

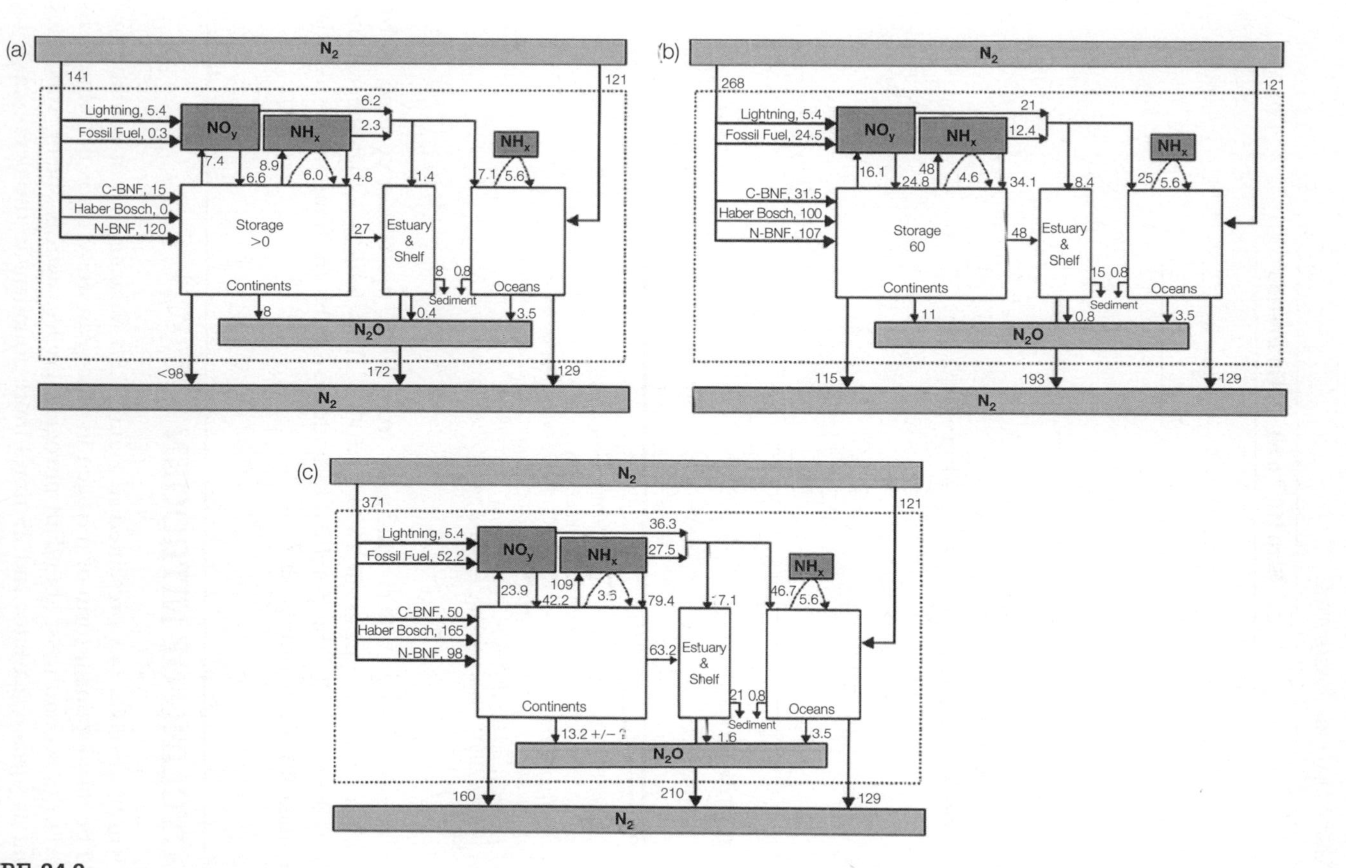

FIGURE 24.2

Components of the global nitrogen cycle for (a) 1860, (b) early 1990s, and (c) 2050, Tg N/yr. (1 Tg = 10^{12} g.) All shaded boxes represent reservoirs of nitrogen species in the atmosphere. Creation of Nr is depicted with bold arrows from the N_2 reservoir to the Nr reservoir (depicted by the dotted box). N–BNF is biological nitrogen fixation within natural ecosystems; C–BNF is biological nitrogen fixation within agroecosystems. Creation of N_2 from Nr via denitrification within the dotted box is also shown with bold arrows. All arrows that do not leave the dotted box represent inter-reservoir exchanges of Nr. The dashed arrows within the dotted box associated with NH_x represent natural emissions of NH_3 that are redeposited on fast time scales to the oceans and continents. *Source*: From Galloway, J. N., *et al.* (2004). *Biogeochemistry* 70, 153–226.

Table 24.2 Major Nitrogen Reservoirs.

Reservoir	Size (10^{15} g N)	Reference
Atmosphere		
N_2	3,950,000	a
N_2O	2	c
Terrestrial		
Biomass	10	a
Soil	190	a
Marine		
Biomass	0.5	a
Dissolved organic nitrogen	550	b
Detrital particulate organic nitrogen	3 to 24	b
NO_3^-	570	b
NH_4^+	7	b
NO_2^-	0.5	c
N_2O	0.2	c
N_2	22,000	b
Crustal		
Sedimentary organic nitrogen	999,600	a
Volcanic rock	1,000,000	c
Coal	200	c

Data sources:
[a]*MacKenzie, F. T. (2006).* Our Changing Planet: An Introduction to Earth System Science and Global Environmental Change, *2nd ed. Prentice Hall, p. 169.*
[b]*Capone, D. (2003). Biogeochemical Cycles N, BISC 419 Environmental Microbiology, University of Southern California, http://bioweb.usc.edu/courses/2003-spring/documents/bisc419-Ncycle_2002.pdf.*
[c]*Wada, E., and A. Hattori (1991).* Nitrogen in the Sea: Forms, Abundances, and Rate Processes. *CRC Press, p. 66.*

denitrification rates provided earlier, which are from 1995, the residence time of Nr is currently estimated as ranging from 1500 to 5000 y. This indicates that the marine nitrogen cycle is dynamic with the potential to undergo significant temporal and spatial shifts over geologically short time scales.

24.4 REDOX CYCLING OF NITROGEN

The thermodynamic predictions presented in Figure 7.8b indicate that at equilibrium, nitrate should be the dominant form of nitrogen in oxic seawater. This is not observed. Rather, nitrate is the second most abundant nitrogen species, with N_2 being 100 times more abundant in surface seawater and 25 times more abundant in the deep sea.

Chemical equilibrium is not attained for two reasons. First, N_2 is continually produced in the ocean by microbial denitrifiers. Second, the triple bond in N_2 is so strong that it cannot be broken without the intervention of a catalyst. In the marine environment, only microbial nitrogen fixers can produce the necessary enzyme, nitrogenase. During nitrogen fixation, N_2 is first reduced to NH_3. Some of the resulting ammonia is used within the cell to build biomolecules via transamination reactions, and some is released to seawater. Autotrophic N_2 fixation and production of organic matter is not restricted to phytoplankton. It also occurs in hydrothermal vent systems as a part of the microbial chemoautolithotrophy conducted by bacteria and archaea. The photoautotrophs that are not nitrogen fixers use the DIN in seawater (e.g., ammonia, nitrate, and nitrite) to meet their nitrogen needs.

The organic nitrogen synthesized by the autotrophs has several possible fates. Some is transferred up the food chain if the algal cells are consumed. This transfer of organic nitrogen is accompanied by losses due to the excretion and exudation of DIN and DON. PON is also lost via excretion and molting. Any cells that are not consumed are either killed by viral infection or sink into the deep sea, usually as part of a marine-snow type aggregate. Viral infection causes cell death via lysis. Cell lysis releases into seawater the dissolved nitrogen species contained in the cell matrix. It also fragments cell walls and membranes.

Detrital PON and DON are subject to microbial degradation, similar to that of POC and DOC, resulting in production of inorganic species. But unlike carbon, whose DIC species are all of the same oxidation state, the DIN species range in oxidation states from $-III(NH_4^+)$ to $+V(NO_3^-)$. These DIN species are generated through microbially mediated pathways that can be segregated into two groups, ones that occur in oxic seawater, and others that take place in suboxic and anoxic environments. Each of these pathways takes place in steps that are individually mediated by various microbes. Some of the nitrogen is used to build microbial biomolecules and some is transformed and released.

Nitrogen uptake that results in the formation of new biomolecules is termed an assimilation process, such as *assimilatory nitrogen reduction*. The processes that result in the release of DIN into seawater are referred to as dissimilations, such as *dissimilatory nitrogen reduction*. An example of the latter is denitrification, in which nitrate and nitrite obtained from seawater serve as electron acceptors to enable the oxidation of organic matter. This causes the nitrate and nitrite to be transformed into reduced species, such as N_2O and N_2, which are released back into seawater.

The biogeochemical cycling of nitrogen is very much controlled by redox reactions. This perspective is presented in Figure 24.3 for the redox reactions that take place in the water column and sediments. The major pathways of reduction are nitrogen fixation, assimilatory nitrogen reduction, and denitrification. The major oxidation processes are nitrification and anaerobic ammonium oxidation (anammox). Each of these is described next in further detail.

24.4.1 **Inorganic Nitrogen Assimilation**

Marine phytoplankton meet their nitrogen needs by importing DIN into their cells. Most phototrophs import fixed nitrogen, either ammonium, nitrate, or nitrite. This process is commonly called *nutrient assimilation*. Some phototrophs are able to break the triple bond in N_2 and, hence, can meet their nitrogen needs by importing N_2. This process is termed *biological nitrogen fixation* (BNF). Not all nitrogen fixers are photoautotrophs;

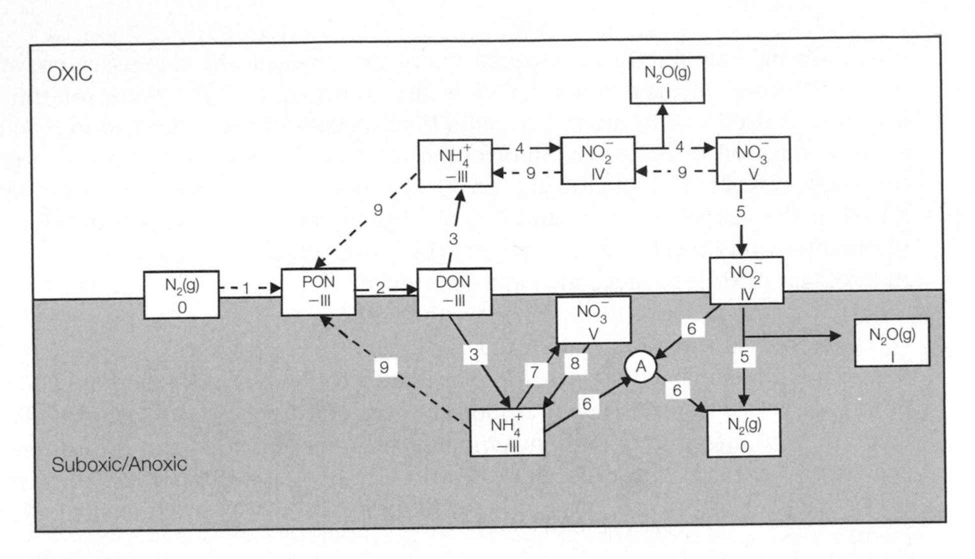

FIGURE 24.3

A simplified depiction of the marine nitrogen cycle illustrating redox and phase transitions mediated by microbes. The boxes contain the nitrogen species and its oxidation number. The arrows represent transformation reactions as follows: (1) nitrogen fixation, (2) solubilization, (3) ammonification, (4) nitrification, (5) denitrification, (6) anammox, (7) anaerobic nitrification mediated by manganese reduction, (8) dissimilatory nitrate reduction to ammonia (DNRA), (9) assimilatory nitrogen reduction. A = anammox microbes.

some are photoheterotrophs; and some are chemoheterotrophs. For this reason, a further discussion of this process is relegated to the next section. To further complicate matters, ammonium, nitrate, and nitrate are also assimilated by some chemoautolithotrophic bacteria, such as the nitrifiers, and by some heterotrophic bacteria. Nitrification is also discussed in a latter section.

The phototrophs that assimilate fixed nitrogen preferentially utilize ammonium over nitrate or nitrite. (Some phototrophs assimilate DON, such as urea, to meet their nitrogen needs, and, hence, are not strict autotrophs.) Ammonia is favored over the more oxidized forms of DIN as less redox energy is required to incorporate its nitrogen into biomolecules, primarily amino acids and the nucleotide bases.

During the process of nutrient assimilation, DIN is first actively transported across the cell membrane. This transport is mediated by species-specific enzymes called permeases that are present in the cell membrane. Once the inorganic nitrogen has crossed the cell membrane, it can participate in anabolic reactions. For example, ammonium helps build amino acids by first reacting with α-ketoglutaric acid to generate glutamic acid:

$$\mathrm{NH_4^+} + \underset{\alpha\text{-ketoglutaric acid}}{\mathrm{HOOCCO(CH_2)_2COOH}} + \mathrm{NADPH} \rightarrow$$

$$\mathrm{H_2O} + \underset{\text{glutamic acid}}{\mathrm{HOOCCH(NH_2)CH_2CH_2COOH}} + \mathrm{NADP^+} \quad (24.1)$$

The rest of the amino acids are synthesized by ***transamination reactions*** in which the amino group resident on one amino acid, such as glutamic acid, is transferred onto the ketone group of another molecule as shown in the following example:

$$\underset{\text{glutamic acid}}{HOOCCH(NH_2)CH_2CH_2COOH} + \underset{\text{pyruvic acid}}{CH_3COCOOH} \underset{\text{transaminase}}{\rightarrow}$$

$$\underset{\text{alanine}}{CH_3CH(NH_2)COOH} + \underset{\alpha\text{-ketoglutaric acid}}{HOOCCO(CH_2)_2COOH} \quad (24.2)$$

Note that in this transamination reaction, α-ketoglutaric acid is regenerated and, hence, is again available to transform more ammonium nitrogen into glutamic acid. α-Ketoglutaric acid is a metabolite produced via the Krebs cycle (Figure 7.6). Other nitrogenous biomolecules are similarly created from transaminations that ultimately stem from the starting reaction presented in Eq. 24.1. The nitrogenous biomolecules so produced include the nucleotides (building blocks of DNA/RNA and the energy carriers NAD/NADP, ADP/ATP, FAD), the amino sugars (building blocks of chitin and peptidoglycans in bacterial cell walls), photosynthetic pigments, vitamins, and other metabolites.

If ammonium concentrations in seawater are low, phytoplankton will assimilate nitrate and nitrite using chemical-specific permeases. Once inside the cell, these DIN species are transformed into ammonium via redox reactions in which nitrogen is reduced to the −III oxidation state:

$$NO_3^- + 2H^+ + 2e^- \underset{\text{nitrate reductase}}{\rightarrow} NO_2^- + H_2O \quad (24.3)$$

$$NO_2^- + 8H^+ + 6e^- \underset{\text{nitrite reductase}}{\rightarrow\rightarrow} NH_4^+ + 2H_2O \quad (24.4)$$

This process is commonly referred to as ***assimilatory nitrogen (nitrate or nitrite) reduction***. The electrons for these reductions are supplied by half-cell oxidations involving $NADPH/NADP^+$ and $NADH/NAD^+$ (Table 7.11). All of these reactions and membrane transport processes are mediated by enzymes that are specific to the DIN species. Considerable variation exists among the phytoplankton species in their ability to produce the necessary enzymes. Since marine phytoplankton are often nitrogen limited, the quantity and type of DIN available in the water column can greatly influence overall phytoplankton abundance and species diversity.

Heterotrophic bacteria also assimilate DIN. In the euphotic zone, their rates of DIN uptake can, in some cases, equal that of the phytoplankton. As with the phytoplankton, this nitrogen is used to synthesize biomolecules. (Recall from Chapter 23.4.4 that these bacteria can also assimilate DON.) Chemoautolithotrophic bacteria also assimilate DIN. For example, the proteobacterium *Thiomargarita namibiensis* is capable of reducing nitrate by using sulfide as the electron donor. This bacterium was discovered living in the sulfide-rich sediments on the Namibian shelf beneath the Benguela Upwelling zone. It also has the honor of being the largest known bacterium, reaching cell diameters of almost 1 mm!

24.4.2 Nitrogen Fixation

Biological nitrogen fixation is energetically costly, so this metabolism is found only in environments where Nr concentrations are low. As shown by the following stoichiometry, 8 electrons and 16 molecules of ATP are required to reduce one molecule of N_2.[1]

$$N_2 + 8H^+ + 8e^- + 16Mg\,ATP \underset{\text{nitrogenase}}{\longrightarrow} 2NH_3 + H_2 + 16MgADP + 16P_i \qquad (24.5)$$

With this reaction, nitrogen fixers are able to break the triple bond in N_2. The reduced nitrogen atoms are then incorporated, via anabolic reactions, into biomolecules. Since this is a very energy-expensive reaction, relatively few organisms can "fix" nitrogen, and they are all prokaryotes. The N_2 fixers are collectively referred to as *diazotrophs* (dinitrogen "eaters"). Although eukaryotes are not known to fix nitrogen, some marine diatoms, dinoflagellates, macroalgae, worms, coral, and zooplankton host nitrogen-fixing symbionts. Interrelated metabolic activities have also been observed among the microbial fixers and nonfixers living in shallow marine sediments, particularly in microbial mats.

Some diazotrophs are oxygenic photoautotrophs, such as *Trichodesmium* and the cyanobacteria, *Crocosphaera*. The purple sulfur bacteria are anoxygenic photoautotrophs. Chemoautotrophic N_2 fixers include hydrothermal archaea and the bacteria, *Thiobacillus* and *Beggiatoa*. Some N_2 fixers are heterotrophic, including the photoheterotrophic unicellular cyanobacterium, UCYN-A. The diazotrophs exhibit a wide variety of tolerance to O_2, with some being obligate anaerobes (e.g., *Desulfovibrio desulfuricans*), some facultative aerobes, and others obligate aerobes.

The enzyme required to facilitate nitrogen fixation is called nitrogenase. Iron is an essential cofactor and is required in large amounts, making iron availability a major factor limiting nitrogen fixation. Nitrogenase is inactivated by O_2. Diazotrophs exhibit a variety of strategies for protecting their nitrogenase from O_2. Some are strict anaerobes—these species are mostly found in anoxic shallow sediments, where many are photoautotrophs. Some species of cyanobacteria undergo cellular differentiation in response to nitrogen-limiting conditions. During this process, a small fraction (5 to 10%) of the vegetative cells switch to heterotrophy and form thick cell walls that prohibit O_2 exchange, thus creating an anoxic intracellular environment. These cells are called heterocysts and their conversion is irreversible. The nitrogen fixed within the heterocysts is passed to the vegetative cells by diffusion through cell wall pores. Since the cells must be in close contact, this strategy is most effective under bloom conditions.

Most free-living marine cyanobacteria are nonheterocystous. They protect their nitrogenase against O_2 exposure by conducting nitrogen fixation only at night when photosynthesis is not occurring. The most common nonheterocystous cyanobacterium, *Trichodesmium*, is also thought to be the most important marine nitrogen fixer. This microbe is an oxygenic photoautotroph found throughout the oligotrophic waters of all oceans. It is unusual in its ability to fix N_2 during the day by alternating between periods

---

[1] Magnesium plays an important role in stabilizing ATP, thereby affecting energy yields from the ATP-ADP half reaction as shown in Table 7.11.

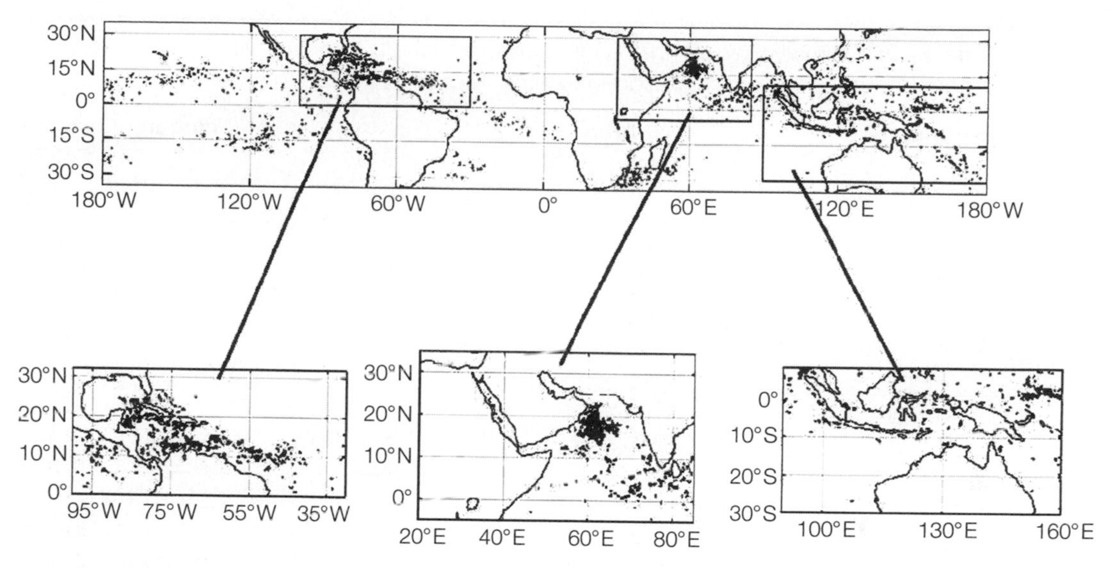

FIGURE 24.4

Positive *Trichodesmium* bloom retrievals using SEAWIFS imagery for a single 8-day composite (10–17 February 1998, 0.25° resolution). Shown are a (sub)-tropical view (35°N–35°S) and three regional extracts highlighting areas of activity: Gulf of Mexico/Caribbean, Indian Ocean/Arabian Sea, and the western Pacific/Indonesia. *Source*: From Westberry, T. K., *et al.* (2005). *Journal of Geophysical Research* 110, C06012.

of photosynthetic carbon fixation versus nitrogen fixation. Trichodesmium is most common in waters with temperatures above 20°C and forms blooms at temperatures above 25°C. At these high temperatures, the relative rates of O_2 and N_2 gas diffusion across its cell membrane and O_2 uptake via dark respiration are such that optimal nitrogen fixation rates are attained. At lower temperatures, the rates are not favorably balanced, causing *Trichodesmium* to be less abundant in temperate and cold seas. It is not known why free-living heterocystous diazotrophs are uncommon in the temperate and cold seas.

The unique optical properties of *Trichodesmium* have made it possible to map the distribution of this species using satellite imagery. These properties include a high CDOM content, unique pigments (such as phycoerythrin),[2] and backscatter caused by intracellular gas vacuoles. As shown in Figure 24.4, satellite imagery indicates that cell abundances are highest in the tropical ocean. Temporal surveys have confirmed that within the tropical ocean, the areas with the most persistently elevated cell concentrations are located in the western tropical Pacific, Caribbean, and Arabian Seas. A strong seasonal cycle in the Indian Ocean is thought to be related to the monsoons. Other nitrogen fixers form blooms. An example of this are the endosymbiotic diazotrophs, which form blooms when their host, such as the diatoms, are "blooming."

[2] See Chapter 22.4.6 for a structural description of this phycobiliprotein.

Table 24.3 Estimates of the Total Annual Contribution of Combined Nitrogen to the Global Nitrogen Cycle by Nitrogen Fixation in Benthic Marine Environments.[a]

Environment	Area ($km^2 \times 10^6$)	N_2 Fixation	
		($g\,m^{-2}\,y^{-1}$)	(Tg/y)
Depth			
>3000 m	272	0	0
2000–3000 m	31	0.0007	0.022
1000–2000 m	16	0.001	0.016
200–1000 m	16	0.01	0.16
0–200 m	27	0.1 ± 0.04	2.7
Bare estuary	1.08	0.4 ± 0.07	0.43
Sea grass	0.28	5.5	1.5
Coral reefs	0.11	25 ± 8.4	2.8
Salt marsh	0.26	24 ± 10.5	6.3
Mangroves	0.13	11	1.5
Total	363		15.4

[a] *Values for nitrogen fixation are annual averages ± one standard error.*
Source: From Capone, D. G., and E. J. Carpenter (1982). Science *217, 1140–1142.*

Benthic nitrogen fixation seems to be restricted to shallow water depths. This probably reflects a light requirement for the phototrophic nitrogen fixers and a need for large amounts of organic matter on the part of the heterotrophic nitrogen fixers. As shown in Table 24.3, most benthic nitrogen fixation is occurring in salt marshes, coral reefs, sea grass beds, mangroves, and sediments in water depths less than 200 m. Complex N_2-fixing microbial consortia with interrelated metabolic associations are commonly observed in shallow sediments. These associations usually involve some kind of photoautotrophic-chemoheterotrophic syntrophy that collectively serves to meet the reduced carbon and nitrogen needs of the community.

Most N_2 fixation is thought to be occurring in the open ocean at a global annual rate of 106 Tg N/y. Benthic N_2 fixation supplies another 15 Tg N/y. Thus the total rate of marine BNF is estimated to be 121 Tg N/y as shown in Figure 24.2, but marine chemists think it highly likely that this is a substantial underestimate. Locations where N_2 fixation is thought to be occurring for which no rate estimates have yet been made include: (1) hydrothermal systems, (2) anoxic zones in estuarine waters, and (3) anoxic microzones within sinking detrital POM. Application of new genetic mapping techniques also suggests the presence of many more N_2-fixing microbes than otherwise known. These probably include members of the archaea.

Because of the high iron demand by the nitrogenase enzyme, iron availability is thought to potentially be controlling where and when N_2 fixation takes place. In the surface waters, this supply is controlled by the aeolian dust flux. In deep waters, the advective transport of resuspended sediments from the continental margins is also thought to be important.

24.4.3 Solubilization and Ammonification

The nitrogen incorporated into organic form through the anabolic activities of the autotrophs and their consumers is returned to inorganic form via a series of microbially mediated reactions. The first steps in this process are solubilization and ammonification. Although nitrogen does not undergo a change in redox state, these processes involve degradation of organic compounds during which the carbon is oxidized. In the first stage, PON is decomposed into DON by fragmentation of biomolecules into compounds small enough to be classified as dissolved. This is exemplified in Figure 24.5, which illustrates the degradation of proteins via the breakage of carbon-nitrogen bonds between adjacent amino acid units. This generates oligopeptides with terminal amine groups. Continued degradation eventually results in the breakage of the carbon-nitrogen bond on this terminal amine, thereby releasing nitrogen into seawater as ammonia (NH_3). This ammonia reacts with H^+ or H_2O to form ammonium (NH_4^+). The entire process that leads to the generation of ammonium is called *ammonification*.

24.4.4 Ammonium Oxidation

Nitrification

In oxic seawater, chemoautolithotrophic bacteria of the family *Nitrobacteraceae* mediate the stepwise oxidation of ammonium to nitrite and then to nitrate. This process is

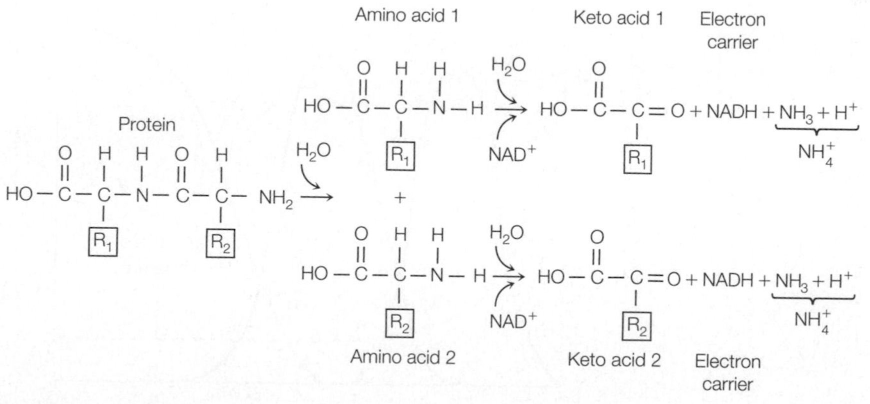

FIGURE 24.5

Ammonification. This process proceeds in steps: (1) The breakage of the peptide linkage between amino acids A and B, followed by (2) The deamination of the amino acids and formation of NH_4^+. The electron carrier (e.g., NAD^+, $NADP^+$) involved in such reactions depends on the type of protein undergoing degradation and the species of organism mediating the reaction. These reactions are catalyzed by enzymes.

called *nitrification*. The energy obtained from the oxidation of ammonium and nitrite is used to fuel metabolic processes including the fixation of CO_2 into organic carbon. The bacteria that oxidize nitrate include *Nitrosomonas*, *Nitrosospira*, *Nitrosococcus*, and *Nitrosolobus*. Those responsible for the oxidation of nitrite include *Nitrobacter*, *Nitrosospina, Nitrosospina*, *Nitrocystis*, and *Nitrococcus*. Some marine archaea are also thought to be nitrifiers.

Marine nitrifiers tend to be slow growers compared to heterotrophs. This reflects the relatively low energy yield from ammonia oxidation. The nitrifiers tend to be photoinhibited, making them less likely to compete with the photoheterotrophs for O_2. While O_2 is a requirement, some nitrifiers are capable of growing under suboxic conditions.

The stepwise nature of nitrification during the aerobic decomposition of detrital PON is illustrated in Figure 24.6. Initially, the degradation of PON produces ammonium, which stimulates the growth of the nitrate oxidizers. These bacteria transform the ammonium into nitrite, causing ammonium concentrations to decline and nitrite concentrations to rise. The elevated nitrite levels stimulate the growth of the nitrite oxidizers. These bacteria transform the nitrite into nitrate. Eventually all of the DIN is oxidized to nitrate. The residual pool of PON includes microbial biomass and any PON too inert to be degraded by aerobic marine bacteria.

In the water column, ammonification and nitrification can sometimes lead to the formation of subsurface ammonium and nitrite concentration maxima that are usually located toward the base of the euphotic zone in stratified water columns. This is called the *primary nitrite maximum*. Some of this nitrite is also contributed by releases

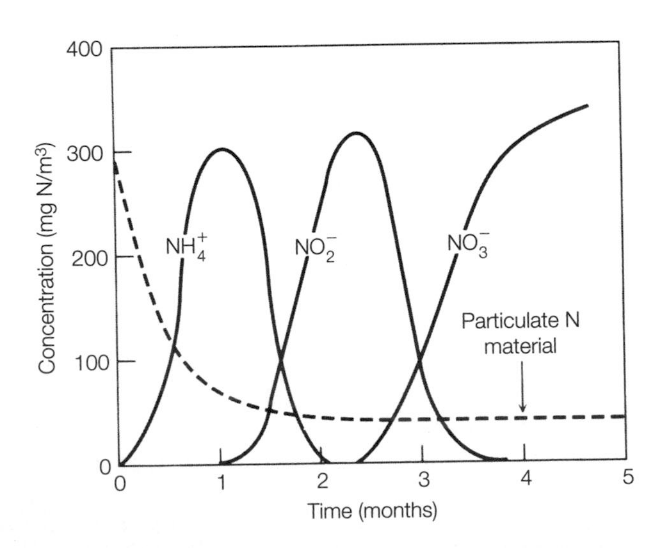

FIGURE 24.6

Production of nitrogenous material from the decomposition of phytoplankton in aerated waters stored in the dark. *Source*: From Horne, R. A. (1979). *Marine Chemistry*. John Wiley & Sons, Inc., p. 275.

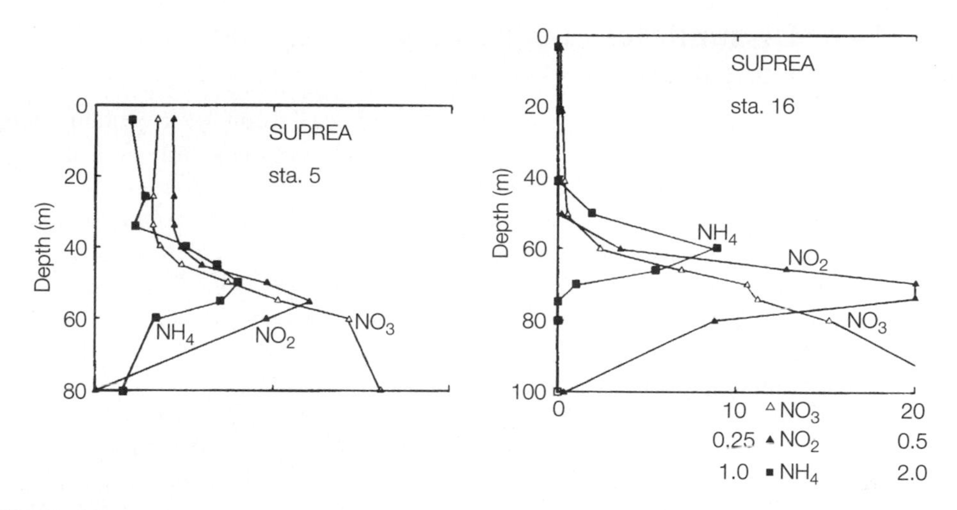

FIGURE 24.7

Representative euphotic zone depth profiles of DIN (μM N) in waters where ammonium was present. Shown are concentrations of nitrate, nitrite, and ammonium. The SUPREA cruise was conducted in August–September 1978 in the equatorial Atlantic Ocean off the Ivory Coast. *Source*: From Collos, Y., and G. Slawyk (1983). *Marine Biology Letters* 4, 295–308.

from phytoplankton growing under light-limited or iron-limited conditions. As shown in Figure 24.7, the primary nitrite maximum sometimes lies beneath an ammonium maximum.

Anaerobic Ammonium Oxidation

Ammonium is oxidized by microbes under anaerobic conditions in the water column and the sediments. Two general pathways for this process have been observed. In the anammox reaction, bacteria oxidize ammonium using nitrite as the electron acceptor. This is a type of dissimilatory nitrogen reduction by which DIN is converted to N_2. Recent research suggests that anammox could be more important than denitrification in removing fixed nitrogen from the ocean. Anammox has been observed to be occurring in the suboxic waters of oxygen minimum zones and in organic-rich sediments.

A second type of anaerobic ammonium oxidation has been observed in anoxic sediments and the Black Sea. It involves a redox coupling with manganese in which Mn(IV) acts as an electron acceptor enabling the oxidation of ammonium. This causes the manganese to be reduced to Mn^{2+} and the nitrogen to be oxidized to N_2. Further loss of fixed nitrogen results if the Mn^{2+} diffuses into the suboxic zone as it can then cause the reduction of nitrate to N_2. Since the Mn^{2+} is oxidized to particulate MnO_2, a catalytic loop can exist in which manganese is cycled between the anoxic and suboxic zones. (See Chapter 12.5.3 for a discussion of Mn^{2+}–MnO_2 cycling at the oxic-suboxic redox boundary.) The net effect is a conversion of ammonium and nitrate to N_2. Similar redox couplings with iron and iodine are thought to exist.

24.4.5 Dissimilatory Reductions under Suboxic and Anoxic Conditions

Under O_2-deficient conditions, marine microbes can facultatively obtain energy by oxidizing organic matter and other reduced species, such as sulfide and Mn^{2+}, using DIN as an electron acceptor. The reduced nitrogen is not incorporated into biomolecules. Rather it is released into seawater causing this process to be termed dissimilatory nitrogen reduction. A variety of metabolic strategies exist by which marine microbes have been observed to engage in dissimilatory nitrogen reduction. The known heterotrophic pathways are denitrification, anammox, and DNRA (dissimilatory reduction of nitrate to ammonium). The known chemoautolithotrophic pathways rely on sulfide and reduced manganese as electron donors.

Dissimilatory nitrogen reduction tends to be a sequential process in which the end products are the gases N_2 and N_2O. Conversion of DIN to N_2 and N_2O removes fixed nitrogen from the ocean. This is the major route by which nitrogen is removed from the sea as burial of fixed nitrogen in the sediments is minor (see Figure 24.2).

Dissimilatory nitrogen reduction requires oxygen-deficient conditions and, hence, is restricted to a narrow range of settings, such as the OMZs and organic-rich sediments. Human activities are increasing the extent of these conditions, and, thus, denitrification rates are thought to be rising. This could help remove some of the excess nitrogen that humans are introducing into the ocean if the excess nitrogen is exported to the areas where denitrification rates are rising. As with nitrogen fixation, the location and rates of fixed nitrogen loss are not well known. Part of the reason for this lack of quantitative knowledge is our incomplete understanding of microbial metabolic processes. For example, the first observations of anammox occurring in marine waters were made in 2003.

Denitrification

During denitrification, nitrate is reduced to $N_2(g)$ by heterotrophic bacteria. The stoichiometry for this process was presented in Eq. 8.11 and Table 12.1 for the water column and the sediments, respectively. Denitrification is thought to be performed by facultative aerobes that switch to nitrate respiration when O_2 concentrations are less than 5 μM (0.1 ml/L).

Like assimilatory nitrogen reduction, denitrification proceeds through a series of steps with nitrate first being reduced to nitrite, followed by reduction of nitrite to $N_2(g)$. Under some conditions, $N_2O(g)$ is also produced.

Some of the nitrogen that is denitrified is derived from the organic matter undergoing oxidation. In Table 12.1, two equations are provided to describe the stoichiometry of denitrification. In the first equation, all of the organic nitrogen present in the reactant organic matter is oxidized to $N_2(g)$. In the second equation, all of the organic nitrogen in the reactant organic matter is ammonified. The recent discovery of anammox in O_2-deficient waters demonstrates that a significant amount of the organic nitrogen is eventually being lost to $N_2(g)$. In other words, some of the ammonium produced during denitrification is used by anammox bacteria and thereby converted to $N_2(g)$.

In the water column, the O_2-deficient conditions supporting denitrification are found in the permanent OMZs of the open ocean and in seasonally O_2-deficient waters found in certain coastal zones. See Figure 8.4 for a map of the open-ocean OMZs. Recalling from Chapter 8, these zones are caused by: (1) high rates of surface production that supply sinking POM, (2) strong vertical density stratification that inhibits ventilation below the mixed layer, and (3) horizontal segregation of AOU. The largest of these OMZs is located in eastern tropical North Pacific and the eastern tropical South Pacific oceans. Coastal upwelling zones also support O_2-deficient conditions, i.e., the Peru and Benguela Upwelling Zones. Persistent O_2-deficient conditions are also present in several marginal seas, including the Arabian, Black, and Baltic Seas. Seasonally O_2-deficient conditions are found over continental shelves that lie beneath coastal waters with high productivity, such as in the Gulf of Mexico, the California borderland basins, and the West Indian shelf.

Depth profiles from the eastern tropical North Pacific (Figure 24.8) show the effects of nitrogen metabolism under O_2-deficient conditions. The thermocline is characterized by a sharp decline in O_2 concentrations that coincides with increasing nitrate and phosphate concentrations. The oxycline is produced by the respiration of sinking POM under vertically stagnant conditions. Below the oxycline, in depths where O_2 concentrations are suboxic, phosphate concentrations continue to increase, but at a slower rate. In contrast, nitrate concentrations decline and reach a mid-water minimum that coincides with a nitrite maximum. The latter is referred to as the *secondary nitrite maximum*. (At this site the primary nitrite maximum is located at 50 m.)

At this site in the eastern tropical North Pacific, denitrification is responsible for the midwater loss of nitrate and production of nitrite. The size of the secondary nitrite maximum is dependent on the relative rates of its production from NO_3^- and its loss via dissimilatory reduction to N_2. The amount of nitrate lost to denitrification is shown as the difference between the measured nitrate and the calculated nitrate. The latter was estimated by multiplying the observed phosphate concentrations by the average nitrate-to-phosphate ratio in the three deepest samples ($11.9 \pm 1.6\ \mu$mol N/L). Note that the zone of denitrification is restricted to mid-depths, i.e., the depths of the OMZ at this site.

In suboxic waters, a secondary ammonium maximum can also be present. It typically lies just above the secondary nitrite maximum. This secondary maximum is supported by high rates of ammonification. Because the waters are suboxic, nitrification rates are slow permitting the buildup of ammonium.

As shown by the estimates presented in Table 24.4, roughly one third of the denitrification occurring in the oceans is thought to take place in OMZs and the rest in the coastal regions, particularly the shelf and estuarine sediments because of the higher abundance of sedimentary organic matter in shallow depths. The areal limitation of denitrification is notable—pelagic denitrification is restricted to less than 0.1% of the volume of the ocean, and most seems to be occurring in the eastern tropical Pacific Ocean and in the Arabian Sea in conjunction with the West Indian Shelf (Figure 24.9). Thus changes in the biogeochemical functioning of these relatively small areas have the potential to exert a global effect on the nitrogen cycle. This is

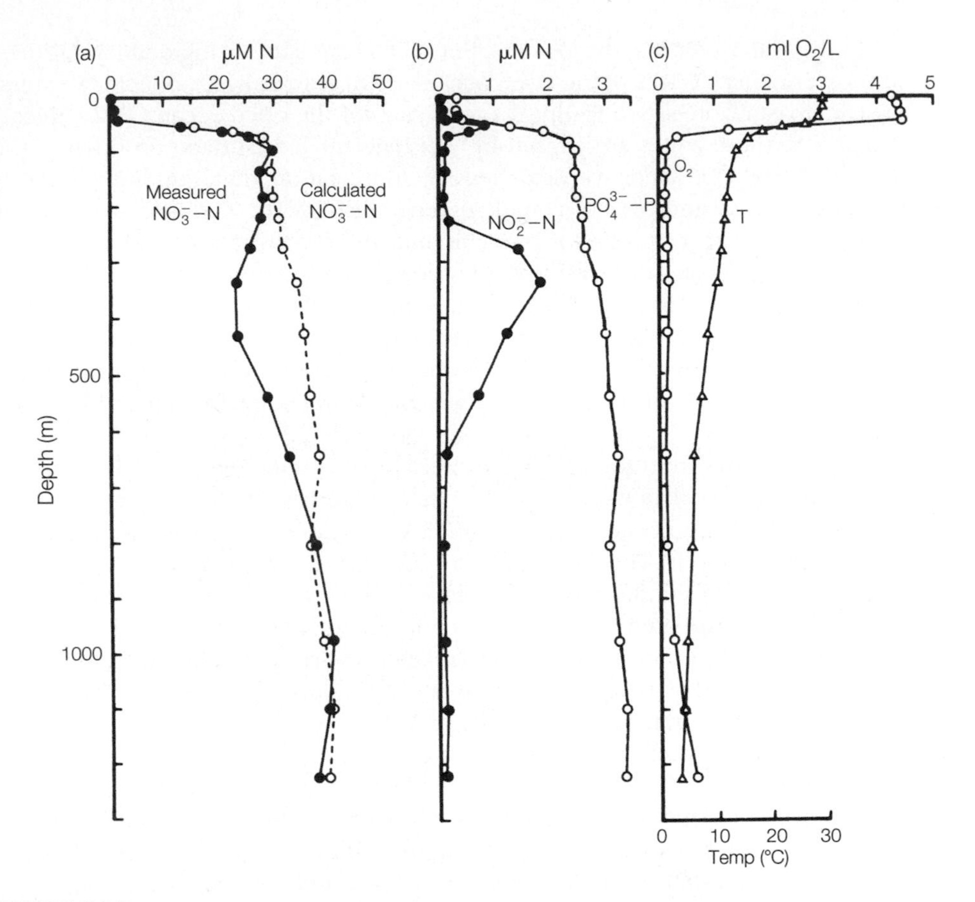

FIGURE 24.8

Vertical concentration profiles of (a) nitrate, (b) nitrite and phosphate, and (c) O_2 and temperature in the eastern tropical North Pacific in August 1962 (15°N 100°W). The calculated nitrate concentrations were estimated by multiplying the observed phosphate concentrations by the average nitrate to phosphate ratio in the three deepest samples (11.9 ± 1.6 μmol N/L). *Source*: From Thomas, W. H. (1966). *Deep-Sea Research* 13, 1109–1114.

particularly worrisome because global climate change is likely to alter circulation and, hence, the physical processes that control vertical and horizontal segregation of O_2 in the OMZs.

Enhanced nutrient loading from land also appears to be increasing the areal extent, duration, and intensity of hypoxia and anoxia in the coastal zone and marginal seas. While the perennial open-ocean systems occupy a far larger volume of seawater than the seasonal coastal systems, O_2 deficiencies are more intense in the latter, often leading to anoxia. Thus, considerable attention is currently focused on reducing nutrient loading to nearshore waters.

Table 24.4 Rates of Denitrification (Tg N/yr).

Location	Galloway et al. (2004)[a]	Seitzinger (2006)[b]	Capone (2003)[c]
Water column			
East Tropical North Pacific	22 to 40	81	10 to 230
East Tropical South Pacific	26 to 40	26	19 to 25
East Tropical Atlantic	0 to 5		5
Indian Ocean	33 to 65	33	30
Total pelagic	81 to 150	140	64 to 290
Benthic			
Estuaries		8	10 to 35
Pelagic sediments	8.4 to 17		
Shelf	57 to 287	250	50 to 75
Total benthic	65 to 304	258	60 to 110
Marine total	146 to 454	398	124 to 400

[a]*Galloway, J.N., et al. (2004).* Biogeochemistry, *70, 153–226.*
[b]*Seitzinger, S., et al. (2006).* Ecological Applications, *16, 2064–2090.*
[c]*Capone, D. (2003) Biogeochemical Cycles N, BISC 419 Environmental Microbiology, University of Southern California, http://bioweb.usc.edu/courses/2003-spring/documents/bisc419-Ncycle_2002.pdf*

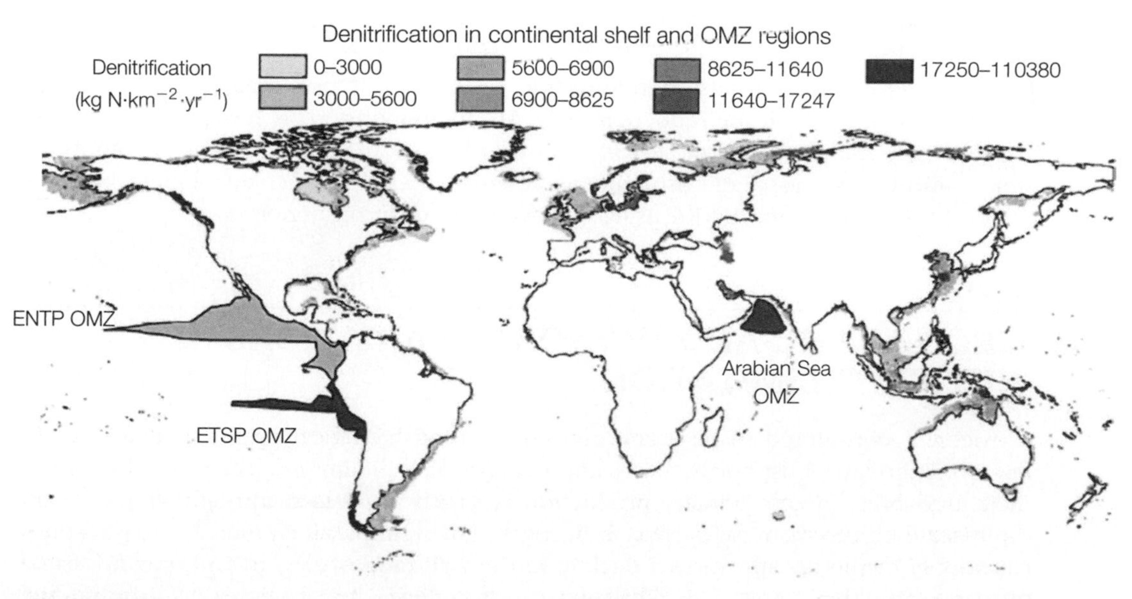

FIGURE 24.9

Denitrification in continental shelves and OMZs (oxygen minimum zones). *Source*: From Seitzinger, S., *et al.* (2006). *Ecological Applications* 16, 2064–2090. (See companion website for color version.)

A notable example of a coastal system in which intense denitrification occurs seasonally is the west Indian shelf and inshore waters. In this setting, denitrification has been observed to completely remove all nitrate and nitrite, enabling sulfate reduction to proceed. Water column depth profiles documenting the spatial and temporal development of these conditions are provided in the supplemental information for Chapter 24.4.5 that is available at **http://elsevierdirect.com/companions/9780123885305**.

Anammox

Before the recent discovery of anammox, most of the loss of fixed nitrogen from the ocean was assumed to be occurring via denitrification. We now know that some, if not most, of this loss is truly due to anammox. Anammox occurs in coastal sediments, anoxic marginal seas, and in OMZs. The bacteria responsible for anammox are similar to the ones found in wastewater bioreactors (*Planctomycetales*).

The prevalence of anammox has important consequences for the ocean because it results in a more efficient loss of nitrogen than via denitrification alone. Global biogeochemical models that attempt to predict changes in plankton production based on the availability of nitrogen and phosphorus will have to be adjusted to include the effect of anammox. Since anammox removes fixed nitrogen, it acts to lower the N/P ratio of the DIN and DIP in seawater, thereby leading to, or intensifying, conditions of nitrogen limitation for the phytoplankton.

Dissimilatory Reduction of Nitrate to Ammonium

Under anaerobic conditions, heterotrophic bacteria will use organic matter as an electron donor to drive the reduction of nitrate to ammonia. This process is called *dissimilatory reduction of nitrate to ammonium (DNRA)* and appears to be an important sink of nitrate in coastal and estuarine sediments. Similarly, some sedimentary bacteria are probably performing DNRA using sulfide as the electron donor.

24.5 GEOGRAPHIC AND TEMPORAL VARIATIONS IN NUTRIENT LIMITATION

Nitrogen is considered to be a key nutrient limiting biological production of organic matter. Although phosphorus, silica, and iron are also biolimiting, the mode by which their availability affects primary production is largely expressed through alterations in the marine biogeochemical cycling of nitrogen. For example, if an increase in phosphorus supply to the ocean causes a decline in the N/P ratio of DIN to DIP, conditions of nitrogen limitation result. This limitation can be relieved by an increase in the rate of BNF. Such an increase can only be achieved if the rate of iron supply to the ocean is sufficient. Recall from earlier chapters that iron is delivered to the open ocean primarily via aeolian dust with production being dependent on climate. An increased rate of BNF will eventually lead to a draw-down in DIP concentrations to levels that induce conditions of phosphorus limitation. This example illustrates how several nutrients can concurrently act to limit primary production. This phenomenon has been termed *resource colimitation*.

In addition to nutrients, various physical factors are important in controlling primary production, such as light availability, water temperature, the depth of the mixed layer, and local vertical mixing rates. Temporal shifts in these physical controls cause shifts in primary production. For example, mid- and high-latitude waters experience seasonal changes in primary production. Interannual changes appear to be occurring over time scales that range from decades to millennia. Information on decadal variability is being supplied by remote sensing data and long-term monitoring at selected sites such as Station ALOHA (Figure 23.4). Paleoceanographic records buried in marine sediments reflect shifts coincident with glacial-interglacial transitions suggesting a cause-and-effect relationship. All of these phenomena are discussed next, with a concluding proposition that the marine biogeochemical cycling of nitrogen, phosphorus, silica, iron, and probably other trace metal micronutrients, such as zinc, interacts to form a set of feedbacks that stabilize the N/P ratios of seawater and plankton at the Redfield-Richards value. These feedbacks are built upon the contributions of marine plankton species, each of which provides a unique biogeochemical service. Thus, in stabilizing the nutrient levels in seawater, the marine planktonic community also confers on itself a degree of ecological stability.

24.5.1 **Seasonal Variations**

Spatial patterns exist in the way that plankton communities respond to seasonal changes in physical conditions and nutrient availability. These patterns reflect local controls on physical and biogeochemical resources. For example, light limitation increases with increasing latitude due to an increasing angle of incidence in insolation (Figure 4.1). Nutrient limitation increases with decreasing latitude due to increasing density stratification and, hence, vertical segregation. In recognition of these spatial patterns, some marine scientists have suggested that the ocean can be classified into a set of biogeochemical provinces, referred to as *domains*, in which climate is the primary defining parameter. Climate is recognized as the master forcing function because of its controlling influence in determining water temperature, the depth of the photic zone, and water motion. The last controls nutrient availability by determining: (1) the depth of the mixed layer, (2) rates of horizontal advection, and (3) the rate of vertical resupply of exported nutrients via turbulent mixing and vertical advection.

The four major biogeochemical zones, thus defined, are the polar, easterly, and trade domains. A fourth zone is defined to encompass the coastal regions. The major functional characteristics of these domains are presented in Table 24.5.

The depth of the mixed layer is important for two reasons. First, phytoplankton can be carried out of the photic zone and, hence, halt net primary production if the mixed layer is deeper than the photic zone. Second, the bottom of the mixed layer marks the upper limit to which density stratification in the thermocline inhibits upward vertical transport of nutrients. If the photic zone extends into the thermocline, phytoplankton

Table 24.5 Marine Biogeochemical Domains.

Domain	Latitude	Mixed layer depth control	Plankton bloom development	Notes
Polar	>60°	Seasonality of sea-ice cover leading to formation of an extensive brackish surface layer from ice melt each spring. Also upwelling around edges of ice sheets and floes.	Growth is initially light-limited by sun angle and/or ice cover, but as irradiance increases in spring or as ice cover dissipates, a bloom may be rapidly induced because the shallow pycnocline is already shallower than the algal critical depth.	Area of extensive continental shelves.
Westerlies	30 to 60°	Balance between solar heating and local stirring effect of the mid-latitude westerlies.	■ Spring bloom is initiated as increasing insolation and reduced wind stress causes shoaling of the pycnocline above the critical depth for net algal growth. ■ The spring bloom becomes nutrient-limited when the initial charge of inorganic nitrate (or, in some cases, silicate) in the mixed layer becomes exhausted. ■ Summer time conditions of nutrient limitation result from grazing and vertical segregation of nutrients. ■ Fall bloom caused by recharge of nutrients from early winter storms and relaxation of consumption.	Initial pre-bloom conditions for nutrients at the end of winter are related to the depth of winter mixing.

Trades	30°N to 30°S	Basinwide geostrophic adjustments in the zonal current systems in response to changes in trade wind stress. Also local wind mixing.	■ Trade wind intensification during boreal summer (May–October) tilts the Atlantic thermocline (but not the Pacific or Indian Ocean thermoclines) to maintain geostrophic balance, so that uplift occurs in the east and induces a seasonal bloom throughout the eastern basin. ■ Intensified trades at the same season are associated with equatorial divergence upwelling east of 10°W, and intensification of the Guinea, Angola, and the Costa Rica thermal domes.	Seasonality is generally weak. Significant control exerted by grazers.
Coastal	all	Controls by the mixed layer are modified by seasonality of winds causing coastal upwelling, tidal mixing, formation of fronts and river discharge.		Includes marginal seas.

Source: After Longhurst, A. (1998). Progress in Oceanography *36, 77–167.*

obtain access to nutrients. Nitrogen exported from the surface waters is returned in the form of nitrate. The returning nitrogen is transported vertically back into the surface waters by turbulent mixing and vertical advection. The source of the upwelled nitrate is remineralization and nitrification of detrital PON that has sunk into the subsurface waters. Some nitrate is also supplied by the nitrification of ammonium excreted by zooplankton living below the euphotic zone. Within the surface waters, nutrient regeneration supplies a mixture of ammonium and nitrate because of the inhibiting effect of light on nitrification. Nutrient regeneration in the surface water is substantial and usually comprises the majority of the planktonic uptake. As shown in Table 24.6, this is particularly the case in oligotrophic waters where nutrient recycling in the surface water supports greater than 90% of the primary production. As described in Chapter 24.4.2, BNF represents a third source of fixed nitrogen to the euphotic zone.

At mid-latitudes (Westerlies domain), seasonal changes in light availability, mixed layer depth, and temperature support two plankton blooms, one in the spring and a lesser one in the fall (Figure 24.10). In the winter, phytoplankton growth is light limited. (The carbon fixation reaction is also slower at lower temperatures.) Thus as heterotrophic microbes remineralize detrital POM, DIN concentrations rise.

With the approach of spring, insolation increases, relieving light limitation and decreasing the depth of the mixed layer (Figure 4.7). With light and nutrients now abundant, a phytoplankton bloom develops. The increased plankton biomass stimulates an increase in the abundance of grazers, i.e., protozoans and the herbivorous zooplankton. The ensuing increase in grazing causes phytoplankton numbers to peak in mid-spring. Regeneration of DIN by microbes and excretion by zooplankton enables the plants to grow despite intense grazing pressure. Thus during this period, nitrogen is rapidly cycled between the consumers and producers. This recycling is not wholly efficient. A significant amount is lost from the euphotic zone by the sinking of detrital PON. As

Table 24.6 Percent Annual Primary Production Supported by Recycling of Nitrogen in Surface Waters.

Biome	Depth of Euphotic Zone (m)	Productivity (g C/m^2/y)	Productivity ($\times 10^{15}$ g C/y)	% Recycled Production
Oligotrophic	100	26	3.8	94%
Transitional	50	51	4.2	87%
Equatorial and Subpolar Divergence	30 to 50	73	6.3	82%
Inshore	10 to 20	124	4.8	70%
Neritic (Shelf)	10 to 20	365	3.9	54%

"Oligotrophic" denotes waters of the central parts of the subtropical gyres. "Transitional" denotes waters that lie between the subtropical and subpolar zones and also includes the extremities of the equatorial divergence zones. Source: After Eppley, R. W. and B. J. Peterson (1979). Nature *282, 677–680.*

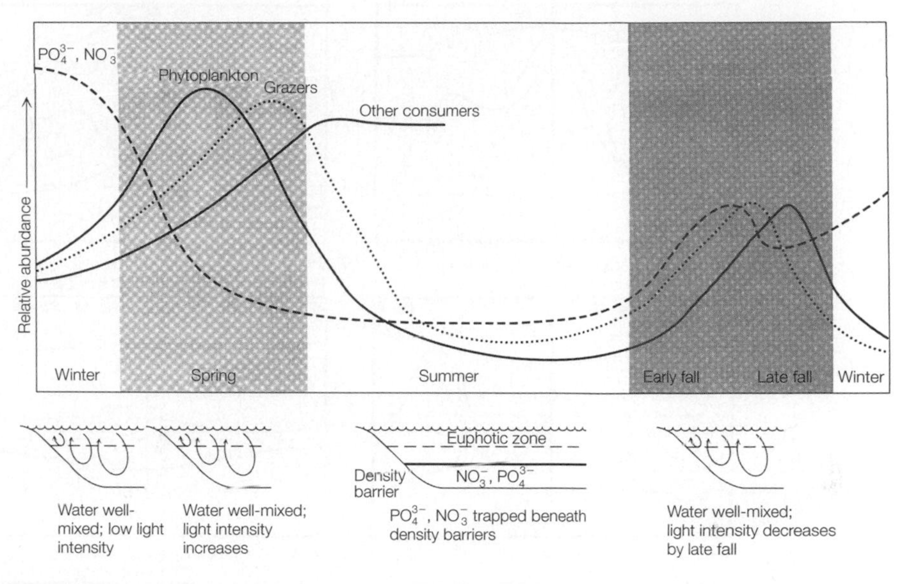

FIGURE 24.10

Seasonal shifts at mid-latitudes in the standing stocks of nutrients, phytoplankton, and the heterotrophic consumer community of bacteria, protozoa, and zooplankton. Also shown are seasonal changes in density stratification of the mixed layer. *Source*: From Black, J. A. (1986). *Oceans and Coasts*, Wm. C. Brown Publishers, p. 143.

the summer progresses, this leak eventually strips the surface waters of DIN, causing plankton growth rates and numbers to decline. As a result, the latter part of the summer is characterized by relatively low productivity.

The detrital PON exported from the euphotic zone is remineralized in the thermocline and deep zone. The return of these remineralized nutrients to the photic zone is inhibited by the strong density stratification associated with formation of a summer thermocline (Figure 4.7). In the fall, declining atmospheric temperatures and early winter storms destroy the summer thermocline. This causes some of the exported remineralized nitrogen to be mixed back up into the euphotic zone. Sunlight is still plentiful, so the injection of this nitrogen fuels a second, though smaller, plankton bloom. This bloom is also attributable to a decrease in grazing pressure. As winter approaches, light levels decline, causing a decrease in plant growth rates. Microbial heterotrophy continues, serving to regenerate DIN from the detrital pool of PON contained in the mixed layer. The regenerated nutrients accumulate until spring conditions return.

The interplay of physical controls is less complicated in the Polar and Trade (tropical) domains. As shown in Figure 24.11a, only one phytoplankton bloom occurs in the Polar domain, but is larger in amplitude than at mid-latitudes (Westerlies). Phytoplankton growth in the subpolar region is prolific because uniformly cold atmospheric temperatures suppress density stratification of the water column. Abundant winds ensure that

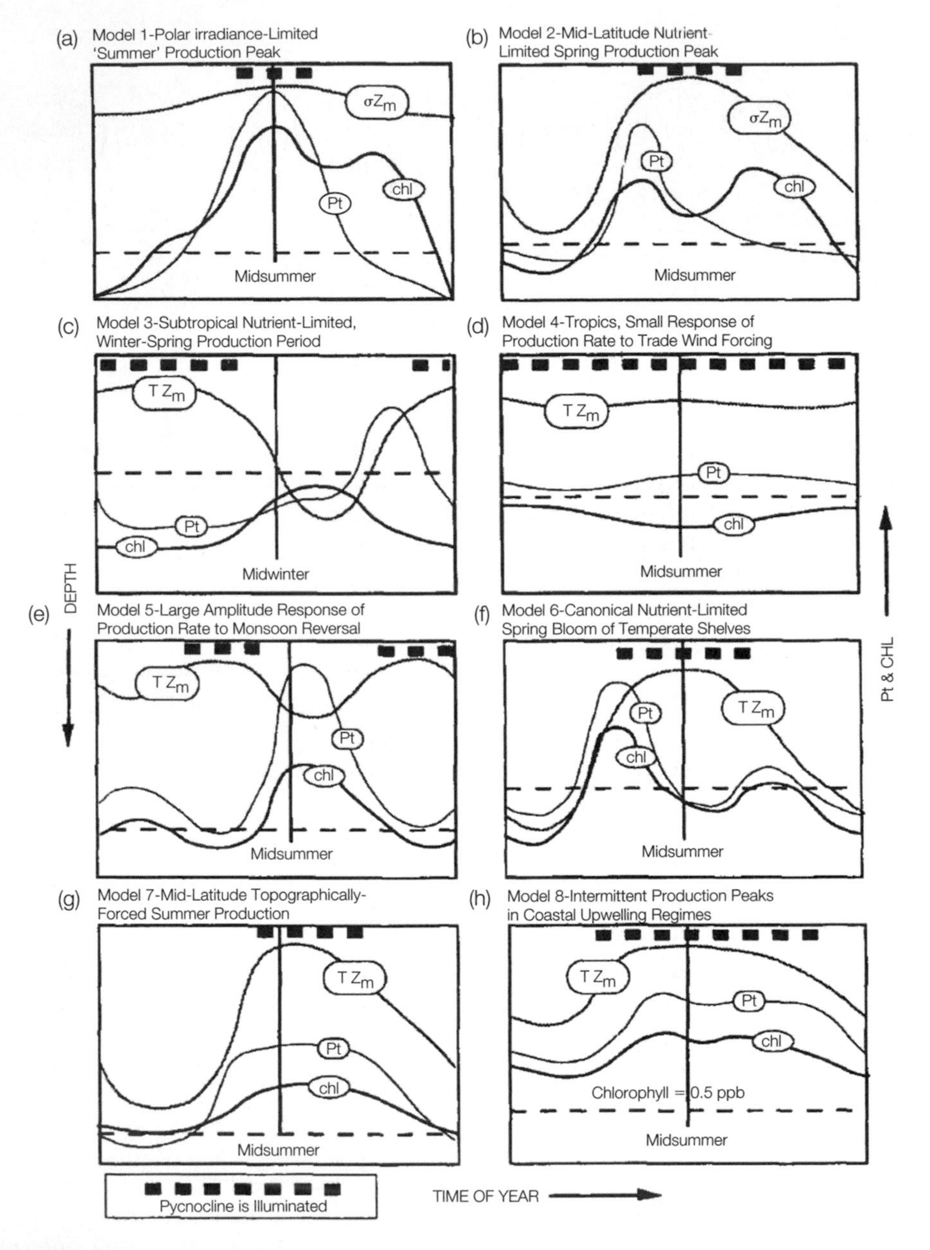

FIGURE 24.11

Cartoons showing the general features of the seasonal cycles of pigment values (chl), primary production rate (Pt), and mixed layer depth (σZ_m or TZ_m) (as assessed by either a temperature or density criterion where regionally appropriate) for the four major biogeochemical domains. The dashed line represents a comparative chlorophyll baseline of 0.5 ppb. The period over which the photic zone extends into the pycnocline is shown at the top of each panel as a heavy dashed line. (a) Polar, (b) Westerlies equatorward, (c) Westerlies poleward, (d) Trades, and (e–h) Coastal domains. *Source*: From Longhurst, A. (1995). *Progress in Oceanography* 36, 77–167.

the water column is well mixed and, hence, prevent vertical segregation of DIN. When insolation increases in early spring, a large bloom occurs and is sustained until light levels diminish at the end of the summer. Despite a limited growth period, the subpolar plankton are responsible for a significant amount of the global marine primary production. Plankton growing at high latitudes have biochemical mechanisms for coping with low light levels. These mechanisms involve iron and, hence, lead to an increased cellular demand for the micronutrient. As shown in Figure 24.12, nitrate, phosphate, and silicate are present in polar and subpolar surface waters, whereas iron concentrations are low. This suggests that iron is persistently biolimiting in these locations as well as in equatorial waters.

Tropical and equatorial waters (Trades Domain) are characterized by uniformly low productivity (Figure 24.11d) because of permanent nutrient limitation. This condition is the result of a strongly stratified water column supported by high and constant insolation. As a result, the nutrients transported out of the euphotic zone in the form of sinking detrital POM are not locally returned by seasonal convective overturn or by turbulent mixing and vertical advection. Because the surface waters have very low DIN concentrations, the energy-expensive process of nitrogen fixation is ecologically favored at these latitudes. As noted earlier, warm waters and high iron concentrations are required to support N_2 fixation, both of which are found in extensive areas of the subtropical Atlantic and Indian Oceans. As shown in Figure 24.12, iron concentrations are very low in the equatorial waters and appear to limit growth in this region. These low concentrations reflect low rates of aeolian deposition (Figure 11.4.)

24.5.2 Decadal Variations

Since 1989, marine scientists have been collecting nutrient and carbon data from a site in the oligotrophic subtropical North Pacific Ocean called Station Aloha. Some of these data were presented in Figure 23.4 as vertical profiles. Over the past two decades, interannual shifts in the N/P ratio of dissolved inorganic nutrients, suspended POM in the surface waters, and exported particles has been observed as shown in Figure 24.13. These shifts suggest that the plankton community are alternating between states of nitrogen and phosphorus limitation. This alternation is thought to form a cycle that starts when low N-to-P ratios (<16:1) are present in the dissolved nutrient pool in the near-surface waters (step 1 in Figure 24.13). This condition favors N_2-fixing microorganisms that generate POM with high (>16:1) N/P ratios (step 2). Export and remineralization of this POM causes an increase in the N-to-P ratio of the subsurface dissolved nutrient pool (steps 3 and 4). Turbulent diffusion and vertical advection eventually return these nutrients to the surface waters, providing relief from nitrogen-limiting conditions and, hence, discouraging further N_2 fixation (step 5). The cycle starts anew if some perturbation, such as an increase in denitrification and anammox rates, causes a differential loss of nitrogen relative to phosphorus (step 6). During periods dominated by N_2 fixation, nitrogen, phosphorus, and iron are likely to be colimiting.

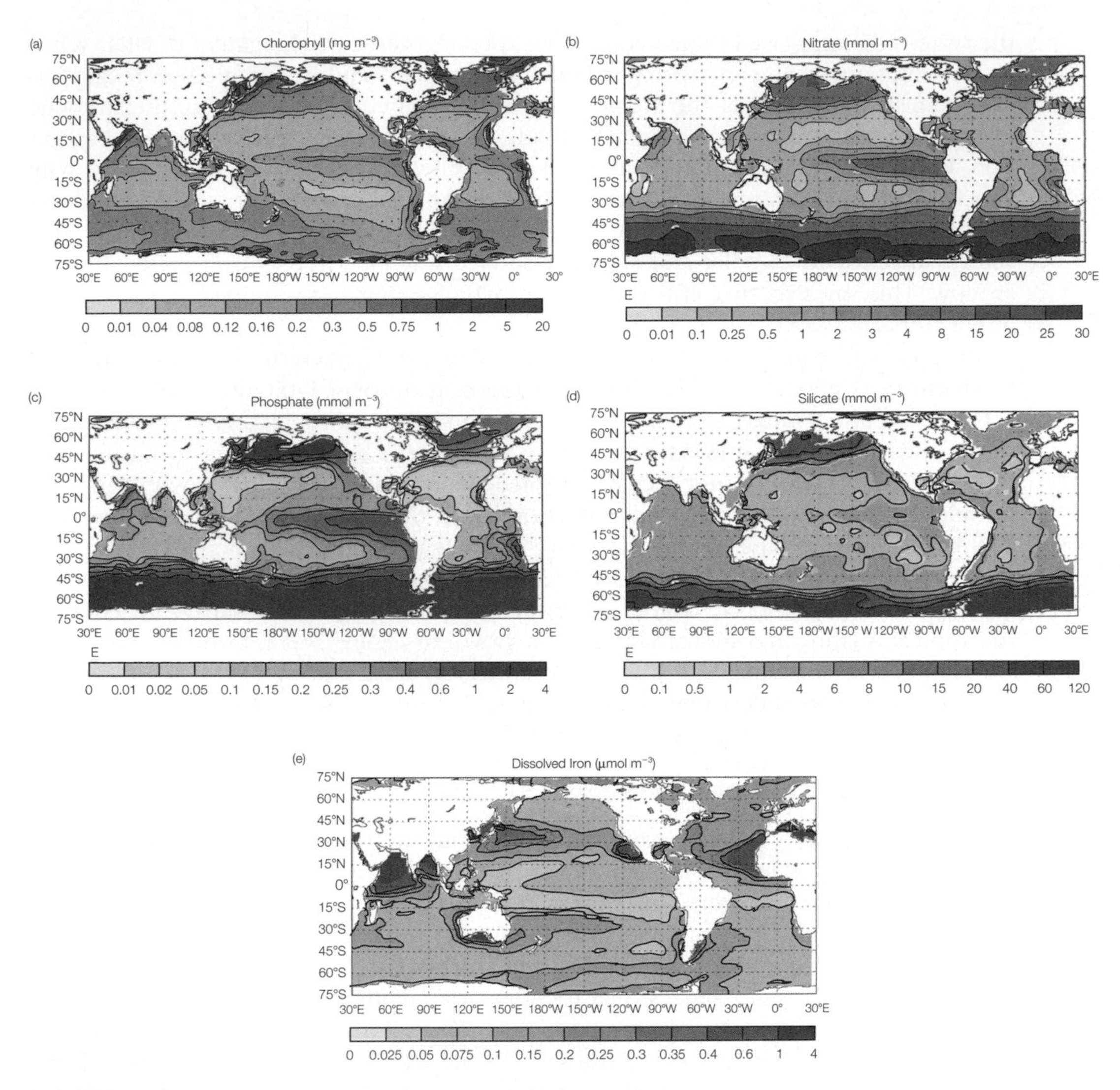

FIGURE 24.12

Annual mean surface seawater concentrations of (a) chlorophyll (b) nitrate, (c) phosphate, (d) silicate, and (e) iron. Data from Conkright, M., *et al.* (2002). *World Ocean Atlas 2001: Vol. 4 Nutrients*. Vol. NOAA Atlas NESDIS 52. U.S. Government Printing Office. Plots from M. Vichi, *et al.* (2007). *Journal of Marine Systems* 64, 110–134. (See companion website for color version.)

These data suggest that biogeochemical domains undergo a systems switch on time scales required for return of exported nutrients to the sea surface. In oligotrophic waters, this return requires decades to centuries as the strong density stratification at these sites forces the return to proceed through the meridional overturning circulation.

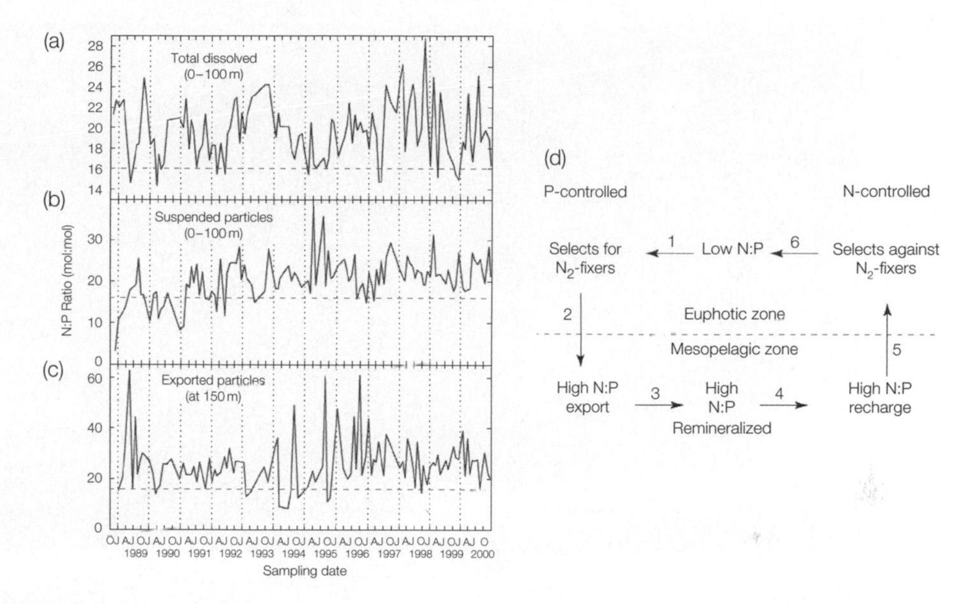

FIGURE 24.13

Station ALOHA (see Figure 23.4 for location information). Three-point running mean observations of N/P molar ratios in (a) total dissolved inorganic plus organic pool, (b) total suspended particulate matter in the upper 0–100 m, (c) in exported particulate matter at 150 m depth, and (d) cycling in nutrient limitation (described in text). *Source*: From Karl, D. M. (2002). *Trends in Microbiology* 10(9), 410–418.

These sustained shifts in biogeochemical function suggest that the species composition of the plankton also undergoes sustained shifts, switching from a N_2 fixer dominated community to one dominated by nonfixers.

24.5.3 **Millennial Variations**

Considerable interest exists in assessing the ocean's role in modulating P_{CO_2} and, hence, global climate. In Chapter 25, we will consider further how shifts in marine primary production can lead to changes in the air-sea exchange of CO_2 and storage of carbon in the ocean. Since nutrient availability controls primary production, it could be a master variable in global climate control. A series of negative feedbacks has been postulated to act as a stabilizing force in the nitrogen cycle, thereby selecting for a narrow range of plankton species assemblages. This feedback is illustrated in Figure 24.14, starting with present-day conditions in which subsurface waters are upwelling with N/P ratios less than 16 (Figure 8.3). In regions of upwelling, primary production is high and large amounts of POM are exported. Remineralization of this POM in the subsurface waters generates O_2-deficient conditions that support denitrification and anammox. This causes a net loss of fixed nitrogen from the ocean, thereby lowering the N/P ratio of

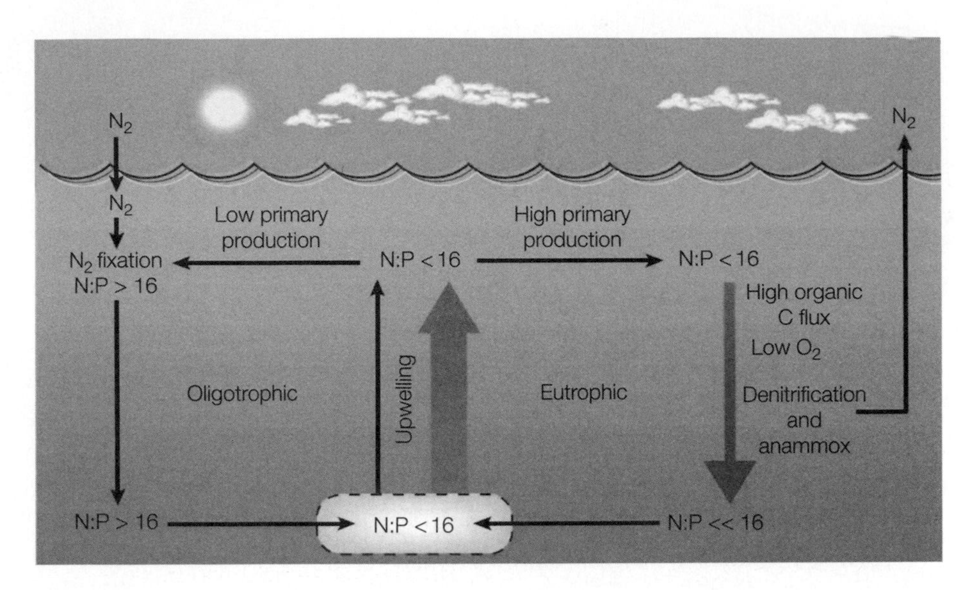

FIGURE 24.14

The global ocean balance between N_2 fixation and the loss of fixed nitrogen through anammox and denitrification. *Source*: From Arrigo, K. R. (2005). *Nature* 437, 349–351.

the subsurface waters. Eventual return of these nitrogen-depleted waters to the surface favors N_2 fixers that can overcome their nitrogen limitation, thereby generating additional POM. Because this POM has an N/P greater than 16, its export and remineralization in the deep ocean returns the N/P ratio in the subsurface waters to the long-term average of 16.

These feedbacks rely on large biologically mediated fluxes of nitrogen in and out of the marine Nr pool. Most of the terrestrial Nr entering the coastal waters is either trapped in the nearshore sediments or lost to denitrification and, hence, is not well linked to nitrogen cycling in the open ocean. Because of the high recycling efficiency of nitrogen in the open ocean, the burial rate of Nr in pelagic sediments is small. This makes estimates of the mean oceanic residence time of Nr highly dependent on the relatively fast rates of BNF and denitrification. Assuming a steady state exists, these rates suggest a residence time of only a few thousand years for Nr. Deviations from a steady state on these time scales are considered highly likely as the sites where most of the Nr input is occurring, i.e., N_2 fixation in tropical oligotrophic waters, are geographically distant from the locales where most of the Nr loss takes place, i.e., denitrification in OMZs and nearshore sediments. Thus, any coupling between the input and output processes is subject to a lag time determined by the rate at which water movement can transport Nr through the oceans. Considerable debate is now focused on assessing the time scales over which imbalances between the Nr inputs and outputs could act to alter fluxes in the marine carbon cycle.

24.6 THE GLOBAL PHOSPHORUS CYCLE

The negative feedbacks illustrated in Figure 24.14 act through adjustments to the N/P ratio of the dissolved inorganic nutrients. Since deficits and surpluses in Nr are thought to be compensated through shifts in the rates of BNF and denitrification, large changes in primary production must be driven by changes in the phosphorus cycle. In contrast to Nr, the residence time of phosphorus is on the order of 10,000 to 20,000 y, reflecting much slower input and loss rates. As shown in Figure 24.15, the sole source of phosphorus to the ocean is via river runoff and the sole loss is through burial in the sediments. (A small amount of phosphorus is delivered via dust deposition.) Thus, a change in the total amount of phosphorus in the ocean can arise only from shifts in the rates of these physical processes.

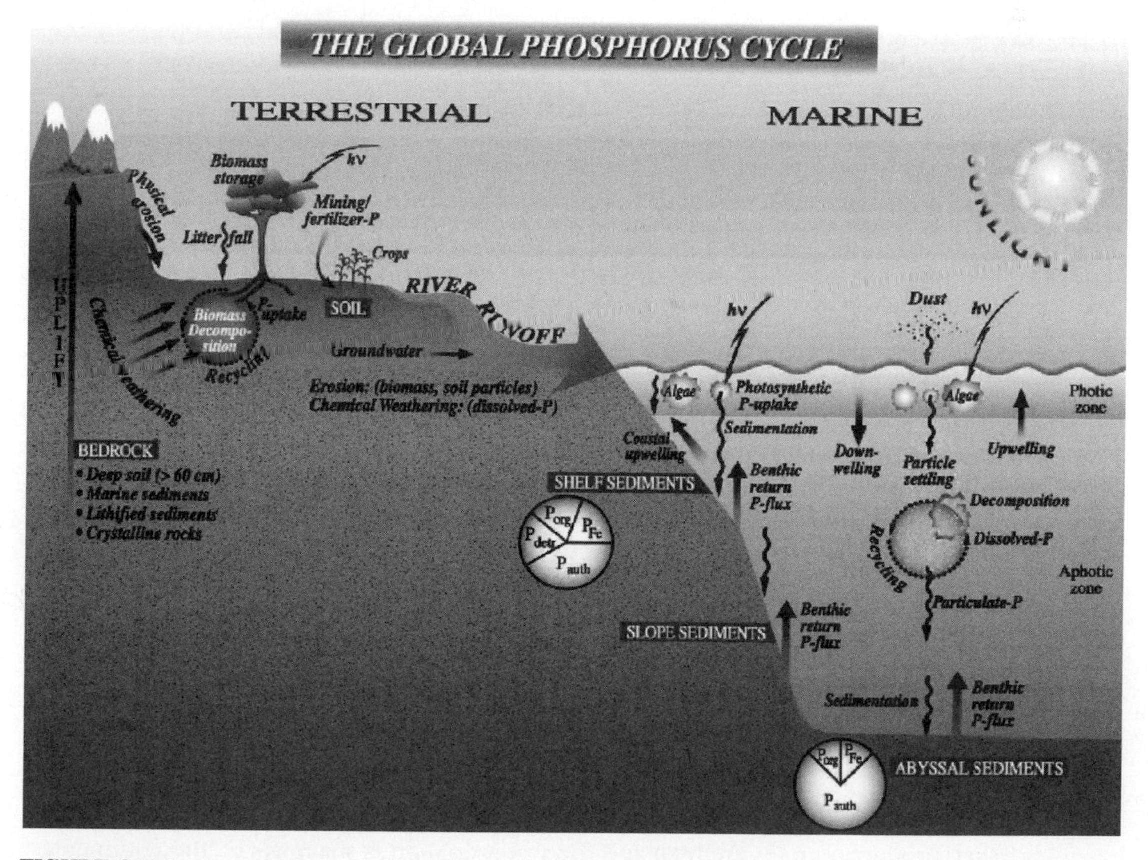

FIGURE 24.15

Marine phosphorus cycle. *Source*: From Ruttenberg, K. C. (2003). *Treatise on Geochemistry*, Academic Press, pp. 585–683.

Table 24.7 Global Phosphorus Reservoirs (Tg P).

Reservoir	Coastal	Open Ocean		Total
		Surface	Deep water	
Atmosphere				
Dust[a]				0.0279
Terrestrial				
Soil P (<60 cm deep)[a]				96,100 to 199,950
Biota[a]				2,601 to 3,001
Marine				
Fish[b]	3	21	21	
Biota[a]				50 to 138
Water Column: Dissolved P[b]	155	1,457	86,490	88,102
Water Column: Particulate Organic P[b]	71	1,116	16,430	17,617
Crustal				
Sediments (marine sediments, crustal rocks and soils >60 cm deep)[a]				837,000,000 to 4,030,000,000
Minable P[a]				10,013 to 19,995

[a]*Ruttenberg, K. C. (2003).* Treatise on Geochemistry, *Academic Press, pp. 585–683.*
[b]*Slomp, C. P., and P. Van Cappellen (2006).* Biogeosciences Discussions *3, 1587–1629.*

The major reservoirs in the global phosphorus cycle are presented in Table 24.7. In contrast to nitrogen, little phosphorus is present in the atmosphere. The ocean contains roughly the same mass of phosphorus as found on land, whereas the oceanic reservoir of Nr is much larger than the terrestrial one. Two sets of estimates are given in Table 24.7, illustrating that a consensus has yet to be reached on reservoir sizes largely because of the difficulty in characterizing the molecular forms of particulate phosphorus in sinking detritus and the sediments.

24.7 SEDIMENTARY NITROGEN AND PHOSPHORUS TRANSFORMATIONS

As with organic carbon, the most nitrogen-rich sediments are deposited where overlying primary production rates are high and the water column is shallow, i.e., the continental margins. The organic nitrogen concentrations in the near shore are 1% or less and decline to $< 0.1\%$ on the abyssal plains (Figure 24.16).

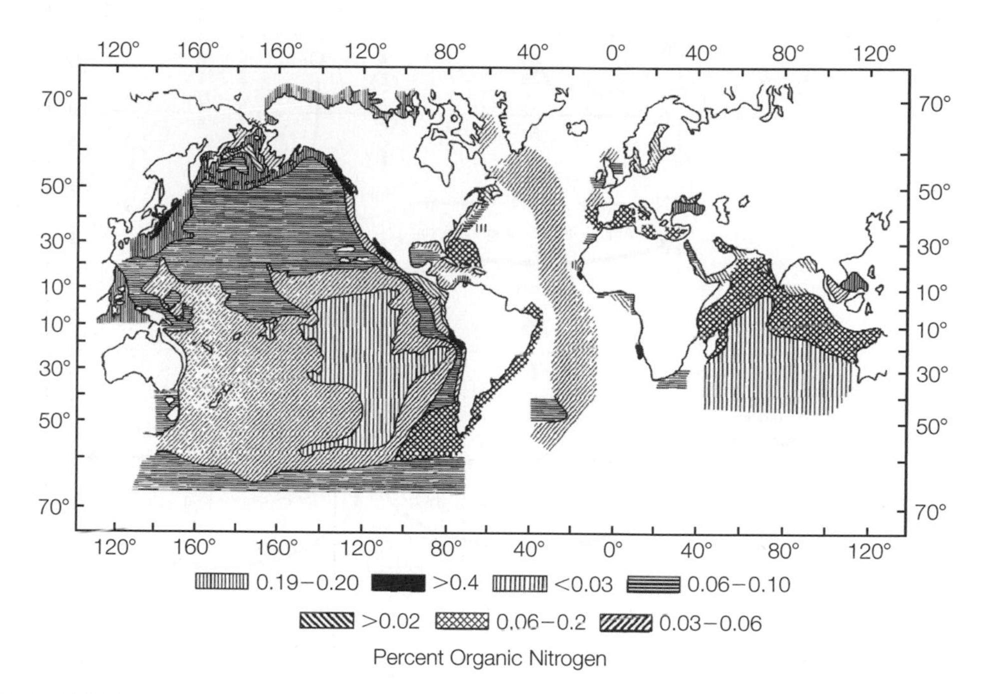

FIGURE 24.16

Distribution of organic nitrogen in the surficial sediments of the world ocean. *Source*: From Premuzic, E. T., *et al*. (1982). *Organic Geochemistry* 4, 63–77.

As with carbon, postdepositional remineralization leads to the loss of a large fraction of the sedimentary organic nitrogen (>90%). In the case of nitrogen, the remobilized nitrogen is subject to redox reactions. These reactions were first presented in Table 12.1. They are summarized in Figure 24.17, illustrating a depth segregation reflecting control by ambient redox conditions. Remobilized Nr can reenter the ocean via diffusion and porewater advection into the bottom waters. In deep-sea sediments, upward diffusion of remineralized ammonium generally leads to its oxidation to nitrate. Thus, most of the Nr supplied to the bottom waters of the open ocean by sedimentary diagenesis is in the form of either nitrate or DON. As described in Chapter 23, only a small fraction of the organic matter in the sediments and pore waters has been molecularly characterized. In the case of nitrogen, most of the characterized organic structures are amino acids and amino sugars (Figure 23.1 and Table 23.1).

In coastal sediments where organic carbon concentrations are high, the redox boundary is at or near the sediment-water interface. Under these conditions, denitrification acts as a sink for nitrate. In some settings, the rate of sedimentary denitrification is fast enough to drive a diffusive flux of nitrate from the bottom waters into the sediments. Remineralization of organic matter under suboxic and anoxic conditions releases

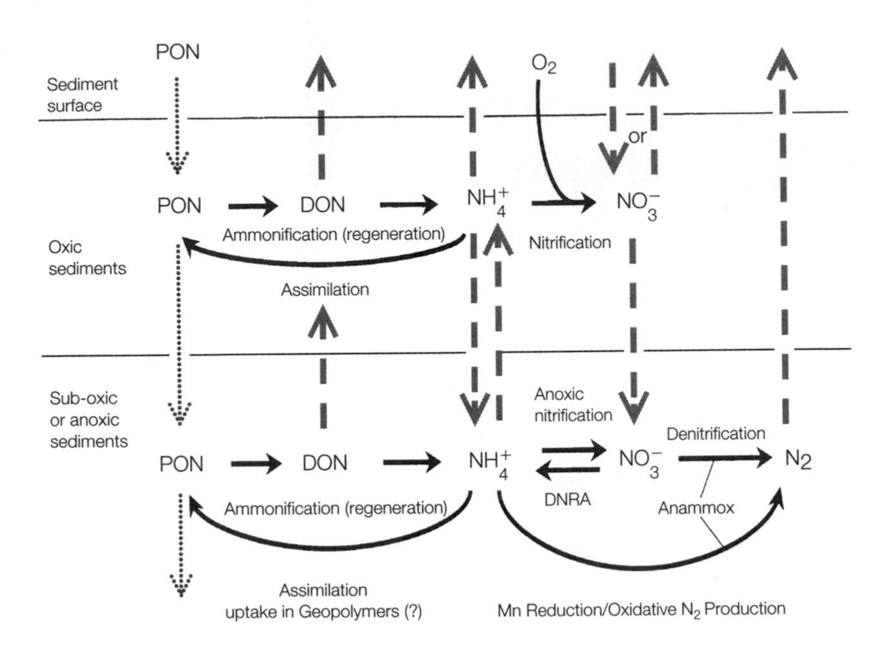

FIGURE 24.17

A conceptual model of sedimentary nitrogen cycling. Dashed arrows represent pore water diffusion and advection. Dotted arrows represent sedimentation. *Source*: After Burdige, D.J. (2006). *Geochemistry of Marine Sediments*. Princeton University Press, p. 453.

ammonium into the pore waters. Some of this ammonium absorbs onto clay minerals and some is converted to N_2 by the anammox process. (Evidence for the former process is provided in Figure 23.17.) Nevertheless, ammonium production rates in coastal sediments can be large enough to support a significant flux into the bottom waters. As shown in Table 24.8, this flux can support a large fraction of the local primary production.

Nitrification and denitrification can be tightly coupled in marine sediments because the nitrate generated from nitrification can fuel denitrification. This occurs if the production of nitrate in the overlying oxic layer is sufficient to support a downward diffusive flux into the suboxic layer where denitrification is occurring. This coupling can also be enhanced by the irrigating activities of microfauna. In shallow waters, rates of nitrification and denitrification are also influenced by the activities of benthic algae as the phototrophs are a source of O_2 and organic matter. They also elevate pore-water pH and compete with the nitrifiers for ammonium. Seasonal shifts in this coupling occur at locations where primary production experiences seasonal swings.

The cycling of phosphorus in marine sediments is quite different from that of nitrogen in that its remineralized form, HPO_4^{2-}, is readily sequestered in several mineral phases, namely carbonate fluoroapatite (CFA) and iron-rich clays and oxyhydroxides.

Table 24.8 Benthic Fluxes of Nitrogen Species and Their Relative Contribution to Local Primary Production.

Environments	N flux across sediment–water interface (mg N m^{-2} d^{-1})			Denitrification		N requirements of primary production	
	NH_4^+	NO_3^-	Total	Rate (mg N m^{-2} d^{-1})	Percentage total N mineralization	Rate (g N m^{-2} y^{-1})	Percentage supplied by benthic mineralization
La Jolla Bight (CA, USA)	12	2	14	–	–	77	7
Loch Turnaig (UK)	13	0	13	–	–	29	17
Buzzards Bay (MA, USA)	18	1	19	–	–	10	70
Cap Blanc (Africa)	78	58	136	–	–	128	39
Belgian coastal zone	25	32	57	17	23	27	78
Southern Bight of the North Sea	13	17	30	5	15	28	38
Narraganset Bay (RI, USA)	34	4	38	33	16	–	–
South River (NC, USA)	38	0.5	38.5	–	–	48	29
Neuse River estuary (NC, USA)	76	1	77	–	–	107	26
Georgia Bight (USA)	55	3	59	–	–	132	16
Great Belt (USA)	10	7	17	49	16	17	36
W. Kattegat (DK)	13	5	18	69	22	15	42
E. Kattegat (DK)	20	5	25	28	8	17	53
Limfjord (DK)	25	13	38	188	26	26	55

Source: From Billen, G., and C. Lancelot (1988). Nitrogen Cycling in Coastal Marine Environments, Scope 33, *John Wiley & Sons, pp. 341–379.*

As described in Chapter 18.3.1, this process is a type of sink switching that has only recently been recognized as highly important to the phosphorus cycle. Phosphorus is delivered to the sediments as part of POM. Not all of the associated phosphorus is present in organic form; a significant percentage appears to be inorganic. This inorganic phosphorus is thought to be present in phytoplankton as: (1) a solute in their intracellular fluids and (2) some kind of mineral that is closely associated with BSi.

Most of the organic phosphorus in the sediments is present as phosphate esters that are found in nucleic acids and phospholipids. The sediments contain the largest pool of extracellular DNA on the planet and constitute a significant reservoir of phosphorus. A small amount of organic phosphorus is present as phosphonates (Table 22.8). The rate at which phosphorus derived from remineralization of sedimentary organic matter is converted into mineral form is not well known. Current estimates are provided in Table 24.10 and suggest that the residence time of phosphorus in the ocean is significantly shorter than traditional estimates, i.e., 10,000 to 20,000 y versus 30,000 to 50,000 y.

24.8 ARE THE MARINE NITROGEN AND PHOSPHORUS CYCLES IN A STEADY STATE?

Efforts to estimate global transport rates for the marine nitrogen and phosphorus cycles are still underway. As shown in Tables 24.9 and 24.10, considerable disagreement exists among biogeochemists as to the magnitude of most of the major rates.

In the case of nitrogen, a consensus does exist that the current estimates of marine BNF are probably far too low. Marine scientists hypothesize that other microbes, in addition to *Trichodesmium*, are providing a significant amount of pelagic nitrogen fixation. Genomic techniques are being used in an effort to identify these "missing" N_2 fixers. Estimates of the loss of Nr from the ocean are also suspect given the recent discoveries of anammox and other anaerobic ammonium oxidation pathways. Other poorly known transports include the atmospheric deposition rate of organic nitrogen and the rate at which nutrients are exported from coastal waters into the open ocean. In the case of phosphorus, geochemists are just starting to understand how sink switching in the sediments acts to sequester remineralized phosphate (Figure 18.9). And they are still unsure as to the molecular form of sedimenting particulate phosphorus.

Spatial and temporal variability make global transport rates difficult to estimate. This is especially the case for river runoff and atmospheric deposition. Other complications arise from the necessity of having to simultaneously track the numerous chemical species that both nitrogen and phosphorus are present in and to distinguish the reactive forms from the unreactive ones. One approach to dealing with these issues is to engage in high-resolution sampling through space, time, and chemical

Table 24.9 Recent Estimates of Rates in the Marine Nitrogen Cycle (Tg N/y).

Process	Gruber (2004)[a]	Galloway et al. (2004)[b]	Seitzinger (2005)[c] & (2006)[d]	Capone (2003)[e]
Sources				
Pelagic N_2 fixation	120 ± 50	121.0	87 to 156[d]	
Benthic N_2 fixation	15 ± 10			
River input (DIN)	35 ± 10	47.8	24.8°	
River input (DON)			11.5°	
River input (PON)	45 ± 10		29.6°	
Atmospheric Deposition	50 ± 20	39		
Total Sources	265 ± 55	207.8		
Sinks				
Organic N export from Fishing and Guano deposition				
Benthic dentrification	180 ± 50	322.0	258[d]	60 to 110
Water column dentrification	65 ± 10		140[d]	60 to 290
Sediment burial	25 ± 10	15.8		
N_2O loss to atmosphere	4 ± 2	4.3		
Total Sinks	275 ± 55	342.1		

[a]Gruber, N. (2004). The Ocean Carbon Cycle and Climate, *NATO ASI series. Kluwer Academic, pp. 97–148;*
[b]Galloway, J. N., et al. (2004). Biogeochemistry *70, 153–226;*
[c]Seitzinger, S. P., et al. (2005). Global Biogeochemical Cycles *19, GB4S01;*
[d]Seitzinger, S., et al. (2006). Ecological Applications *16, 2064–2090;*
[e]*Capone, D. (2003). Biogeochemical Cycles N, BISC 419 Environmental Microbiology, University of Southern California, http://bioweb.usc.edu/courses/2003-spring/documents/bisc419-Ncycle_2002.pdf.*

speciation. Another approach has been to develop models that can extrapolate from limited observational data. This has recently been done to refine estimates of river inputs.[3]

[3] For an example, see http://marine.rutgers.edu/globalnews/mission.htm for the Global Nutrient Export from Watersheds model that was developed by UNESCO's Intergovernmental Oceanographic Commission in 2005.

Table 24.10 Recent Estimates of Current Rates in the Marine Phosphorus Cycle (Tg P/yr).

Process	Data Sources: Ruttenberg (2003)[a]	Slomp and Van Cappellen (2006)[b]				Seitzinger (2005)[c]
Sources						
River input (dissolved)	0.99 to 1.80	2.79				1.76
River input (particulate)	18.29 to 20.15					9.03
Atmospheric Deposition	0.62 to 1.55					
Total Inputs	19.90 to 23.50					10.79
Total Reactive P inputs	7.60 to 9.33	2.79				
Sinks (by location)						
Sediment burial - shelf	4.65					
Sediment burial - Abyssal Plains	4.03					
Total P Burial	8.68					
Sinks (by phosphorous species)		Coastal	Open Ocean		Total	
			Surface	Bottom		
Particulate Organic P Burial		0.50	0.22		0.71	
CFA Burial		0.99	0.43		1.43	
FeP burial		0.50	0.22		0.71	
Total Reactive P burial	5.49 to 7.50	1.98	0.87		2.85	
Internal Cycling						
Primary Production		176	1054		1230	
Remineralization of Particulate Organic P		165	905	145	1215	
Export from Nearshore to Offshore						
Soluble Reactive Phosphorus		18.3			18	
Particulate Organic Phosphorus		9.4			9.4	
Fish Organic and Inorganic						
Production		1.8	10.5	10.5[d]	12	
Dissolution		1.7	0.0	10.5	12	
Downwelling from Surface to Deep Water			121		121	
Sinking Particle Export from Surface to Deep Water			146		146	
Upwelling from Deep to Surface Water		25.1	251		276	

[a] *Ruttenberg, K. C. (2003).* Treatise on Geochemistry, *Academic Press, 585–683.*
[b] *Slomp, and Van Cappellen (2006).* Biogeosciences Discuss., *3, 1587–1629.*
[c] *Seitzinger, S. P., et al. (2005).* Global Biogeochemical Cycles, *19, GB4S01, 5.*
[d] *Fish export from surface to deep water.*

Given all these uncertainties, it is not currently possible to determine whether the nitrogen and/or phosphorus cycles are in a steady state. Indeed, anthropogenic inputs of both are now so large that the maintenance of a steady state seems unlikely. As noted earlier, "natural" deviations from a steady state in the nitrogen cycle are also deemed likely given the large spatial separation between the locales where denitrification and BNF take place.

Climate change is a likely cause of natural imbalances between Nr input and loss. For example, climate change alters the hydrological cycle and, hence, the rate of nutrient delivery from land and oceanic circulation. An increase in surface-water nutrient concentrations is likely to lead to enhanced POM export, followed by subsurface suboxic conditions, and finally, increased denitrification rates. Conversely, a decrease in surface-water nutrient concentrations, due to either a declining nutrient supply from land or strengthened density stratification, promotes conditions favorable for BNF. Biogeochemists think they are seeing evidence of such impacts at Station ALOHA in the subtropical North Pacific (Figure 24.13), where a shift toward more chronic phosphorus limitation is coinciding with a period of shallow stratification and greater oligotrophy. Whether this condition is widespread in the Pacific Ocean is not known, nor is it known whether a similar shift in denitrification is occurring.

Fluctuations in the marine nitrogen cycle also have the potential to affect global climate. For example, the ocean is currently a net source of $N_2O(g)$ to the atmosphere. This gas is a by-product of both nitrification and denitrification (Figure 24.3). Although N_2O degassing represents a minor loss of Nr from the ocean, it is a significant source of a very potent greenhouse gas. As shown in Figure 24.18, most of the N_2O flux to the atmosphere occurs in coastal and open-ocean upwelling areas. Thus, any changes in the factors that control N_2O production in these areas, such as oceanic circulation rates and local primary production, have the potential to exert a disproportionate effect on global climate.

24.9 ANTHROPOGENIC PERTURBATIONS OF THE MARINE NITROGEN AND PHOSPHORUS CYCLES

Humans are having a significant impact on the global phosphorus and nitrogen cycles. In the case of nitrogen, much of this perturbation derives from our newfound ability to fix N_2 via the Haber-Bosch process. The drive to develop an artificial source of fixed nitrogen arose toward the end of the 19th century, when the growing human population increased the demand for nitrogenous fertilizers beyond what could be provided by natural sources, these being manure, guano found on islands with large bird rookeries, and some isolated nitrate salt deposits in the Chilean desert. In 1913, chemical engineers developed a high-temperature combustion reaction, called the Haber-Bosch process, in which N_2 and H_2 are combined to form NH_3. The resulting ammonia is converted to a form suitable for fertilizer or other use. This is a very energy-intensive reaction

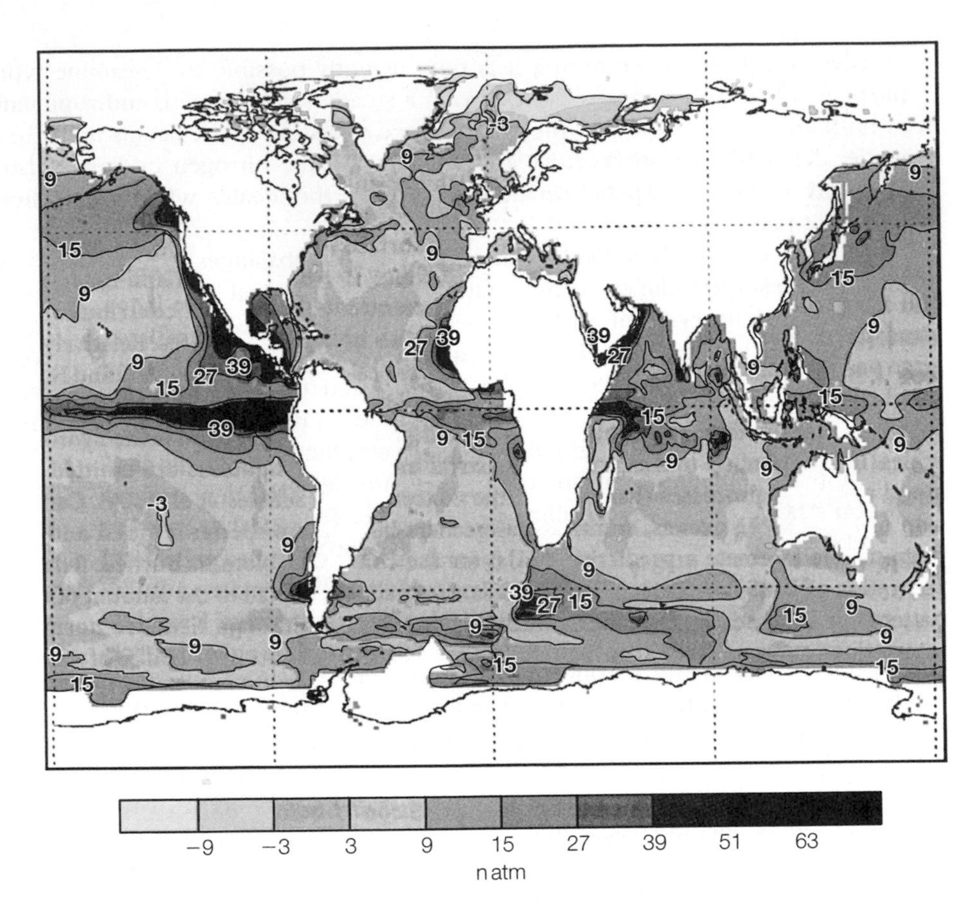

FIGURE 24.18

Annual composite surface ocean–atmosphere difference in partial pressure of N_2O (in n atm = 10^{-9} atm). Positive values reflect ocean to atmosphere fluxes. *Source*: After Nevison, C. D., *et al.* (1995). *Journal of Geophysical Research* 100, 15809–15820. (See companion website for color version.)

and chemical engineers are still working to develop better catalysts to improve its efficiency.

The problem with this fixed nitrogen is that its use as a fertilizer requires land application. While some of the nitrogen is retained by the plants, much is carried off the land as stormwater runoff. This increases the concentration of DIN in groundwater and riverwater. Drainage of these waters into the coastal ocean supplies nutrients that stimulate plankton growth. Remineralization of the plankton biomass can lead to development of hypoxic and anoxic conditions in coastal waters.

Secondary processes provide further routes by which anthropogenically fixed nitrogen enters the ocean. These include: (1) volatilization of nitrogen from urine generated by livestock fed on crops grown with fertilizers, (2) discharge of sewage from humans fed on crops and livestock, and (3) increased soil erosion from agricultural activities. Since most of the human population is concentrated along the coastlines, a significant fraction of human sewage is disposed of via discharge into the oceans, either directly or via rivers. For phosphorus, sewage discharge is the major anthropogenic input route. Although the United States has banned the introduction of untreated sewage into the coastal waters, this practice continues in other counties. In 2006, the United Nations identified untreated sewage discharge as one of the most serious and growing threats to the marine environment.

Humans are also fixing nitrogen somewhat unintentionally by farming crops, such as legumes and rice, that harbor microbial nitrogen fixers. Some N_2 fixation also occurs during the high-temperature combustion of fossil fuels. This combustion process also generates oxidized gases, N_2O, NO, and NO_2, thereby mobilizing nitrogen that had been sequestered in the form of petroleum and coal. Nitrogen is also mobilized from trees and soil via deforestation and biomass burning. The nitrogen oxide gases (N_2O, NO, and NO_2) released by fossil fuel and biomass burning contribute to other environmental problems such as smog, acid rain, and loss of ozone in the stratosphere.

Another important anthropogenic perturbation of the nitrogen cycle has been an increase in the production of N_2O as a result of enhanced rates of marine denitrification. These enhanced rates are being caused by anthropogenic nutrient loading in estuarine and coastal waters. The increased nutrient input is supporting higher rates of primary production and, hence, higher rates of POM input. Since N_2O is a potent greenhouse gas, an increase in its production contributes to climate change.

As shown in Table 24.11, the anthropogenic rate of N_2 fixation had exceeded the natural global rate by the mid-1990s, largely as a result of the Haber-Bosch process. Increases in the rates of other transports in the crustal-ocean-atmosphere factory have occurred in response to this increased generation of Nr. Between 1860 and the mid-1990s, global food production rose by a factor of 7 and helped support a threefold increase in global population. By switching from biomass to petroleum as our major energy source, global energy production increased 90-fold. The net effect of these rate shifts was a ninefold increase in the rate at which humans fixed nitrogen. This resulted in a fourfold increase in the atmospheric deposition of inorganic nitrogen to the oceans. River transfer rates increased about twofold. By 2050, the anthropogenic rate of N_2 fixation is expected to rise to a level that is double the natural rate.

Atmospheric deposition of new Nr to the oceans appears to be increasing at a faster rate than is transfer via river discharge (Figure 24.19). In 1860, atmospheric deposition conveyed only 34% of the terrestrial input to the ocean. By the mid-1990s, atmospheric deposition constituted 44% and by 2050, it is projected to supply almost 52%. This has important consequences as most of the Nr entering the oceans via river flow is thought to be removed via denitrification in coastal and shelf regions and, hence, has little or no impact on the open ocean. In contrast, atmospheric deposition has the potential to cause more geographically widespread impacts. This is shown in Figure 24.20 which also illustrates how aeolian input has impacted even the open ocean. Between

Table 24.11 Changes in Rates of Natural and Human Sources of Fixed Nitrogen (1 Tg = 10^{12} g).

	1860	Early-1990s	2050
Nr creation			
Natural			
Lightning	5.4	5.4	5.4
BNF-terrestrial	120	107	98
BNF-marine	121	121	121
Subtotal	246	233	224
Anthropogenic			
Haber-Bosch	0	100	165
BNF-cultivation	15	31.5	50
Fossil fuel combustion	0.3	24.5	52.2
Subtotal	15	156	267
Total	262	389	492

Source: After Galloway, J. N., et al. (2004). Biogeochemistry *70, 153–226.*

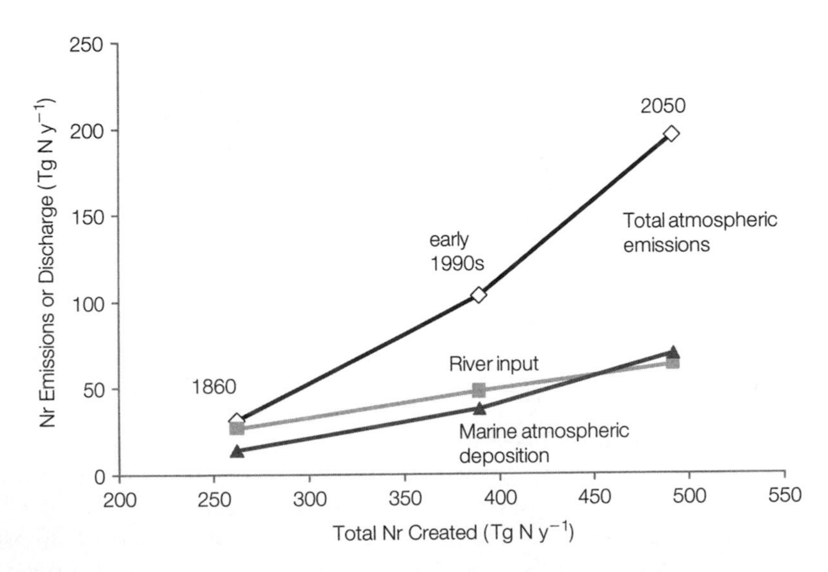

FIGURE 24.19

Increase in fixed nitrogen and Nr emission rates over time due to anthropogenic inputs. Data from Galloway, J. N., *et al.* (2004). *Biogeochemistry* 70, 153–226.

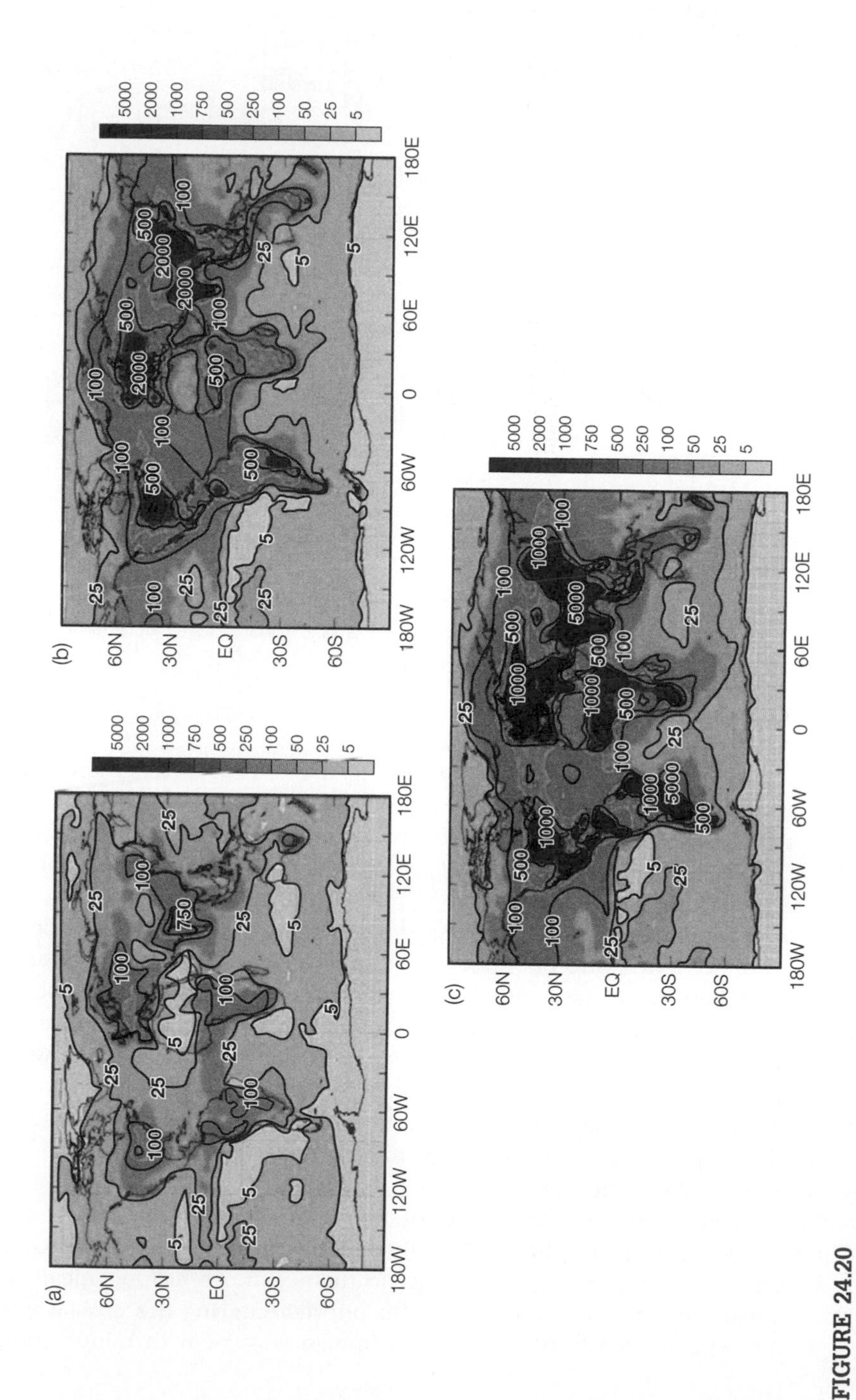

FIGURE 24.20

Spatial patterns of atmospheric nitrogen deposition in (a) 1860, (b) early 1990s, and (c) 2050. *Source:* After Galloway, J. N., *et al.* (2004). *Biogeochemistry* 70, 153–226. (See companion website for color version.)

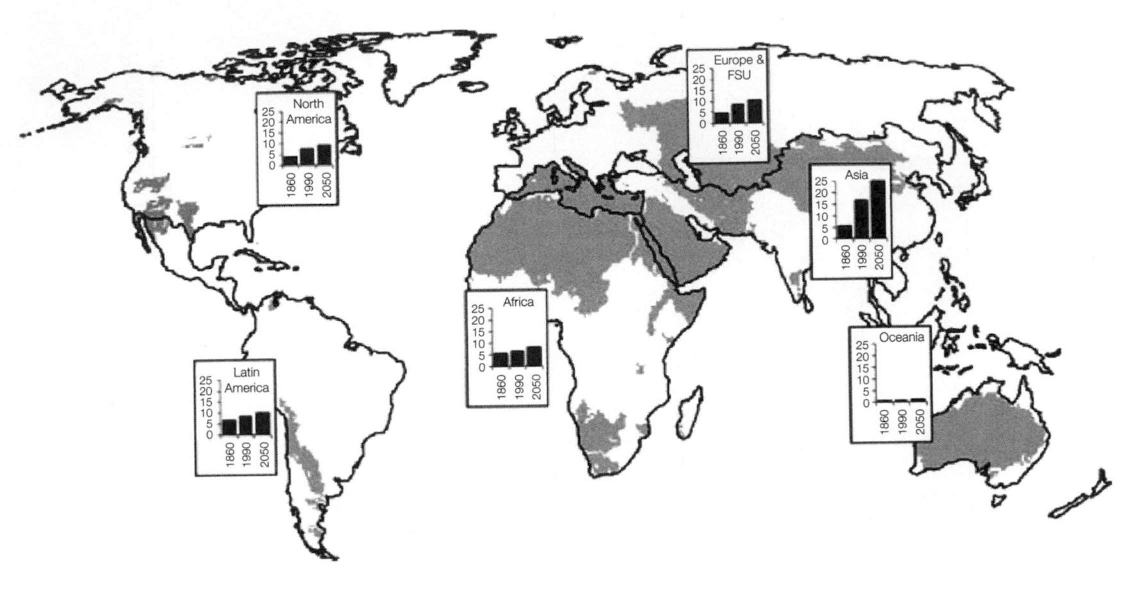

FIGURE 24.21

Riverine Nr export to the coastal zone (Tg N/y) in the past (1860, left bar), present (1990, center bar) and future (2050, right bar). Dry and inland watershed regions that do not transmit to coastal areas are shown in gray. *Source*: From Galloway, J. N., *et al*. (2004). *Biogeochemistry* 70, 153–226.

1860 and the mid-1990s, areal deposition rates in the coastal waters increased by more than a factor of 10 and in open ocean regions by at least a factor of 2, particularly in the North Atlantic. The aeolian deposition pattern now reflects transport via the Westerlies from North America to Europe and via the Trade Winds from Africa to South America.

Another important feature of the anthropogenic release of nitrogen is that Nr export via rivers is concentrated into hot spots. As shown in Figure 24.21, the riverine export is presently highest from Asia, although its rate in 1860 was comparable to that of other regions. Export rates from the other continents have increased, although not as dramatically. Projections for 2050 suggest that the riverine export rates will increase significantly.

The anthropogenic impacts on the rate of phosphorus transfer into the ocean are summarized in Figure 24.22. Phosphorite mining, and its ensuing use as a fertilizer, has more than doubled the rate of phosphorus inputs to the environment, leading to a threefold increase in the river runoff of this element into the ocean. Fortunately, only the dissolved inorganic phosphate (DIP) is biologically reactive and constitutes a minor fraction of the total phosphorus input to the ocean. Anthropogenically mobilized DIP now constitutes 65% of the reactive phosphorus entering the coastal waters. The largest anthropogenic source is discharge of human sewage with minor contributions

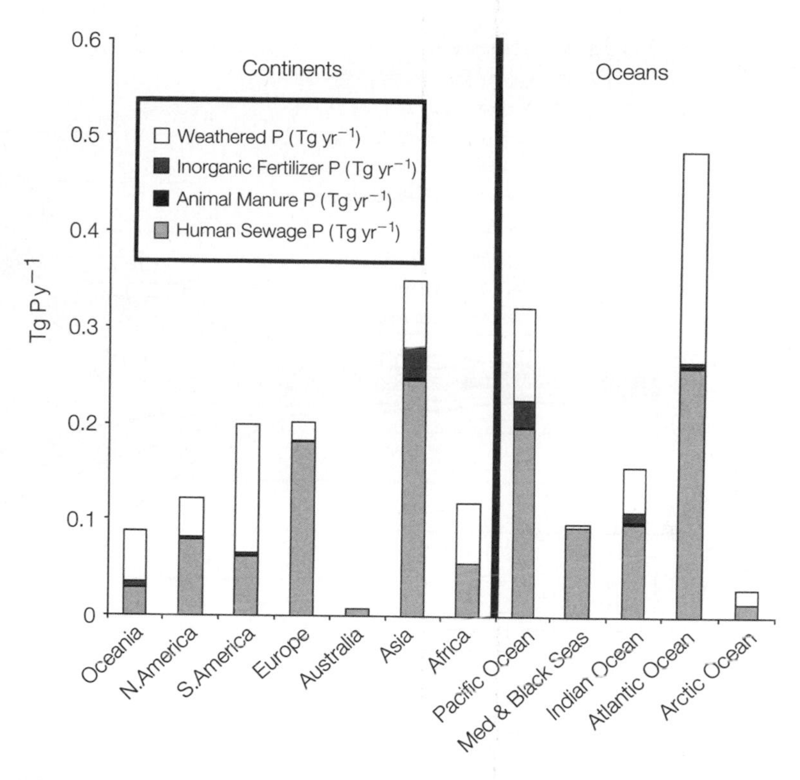

FIGURE 24.22

River export of DIP (Tg P y^{-1}) from continents and to ocean basins as estimated from the Global Nutrient Export from Watersheds model. The "weathered P" represents the natural input. *Source*: From Harrison, J. A., *et al*. (2005). *Global Biogeochemical Cycles* 19, GB4S03.

from inorganic fertilizers and animal manure. Asia is the largest continental exporter of both DIP and Nr.

Projected population increases suggest that anthropogenic nutrient loading to the coastal ocean will continue to increase. Models have been developed to predict the fate of this additional nutrient input. The results from one such model are presented in Figure 24.23 and illustrate a considerable difference in the fates of nitrogen and phosphorus. First, most of the riverine flux of phosphorus is in particulate inorganic form, which is not biologically reactive and, thus, is less likely to exert an impact on the ocean. From a biological standpoint, most of the coastal ocean is nitrogen limited. Thus, continued increases in the input of Nr is expected to further stimulate primary production. Although Nr has an additional pathway into the coastal ocean, namely atmospheric deposition, this input should be somewhat countered by increased denitrification rates. The latter are expected to result from increased hypoxia caused by enhanced primary production rates. Nevertheless, the increased river inputs are projected to increase the

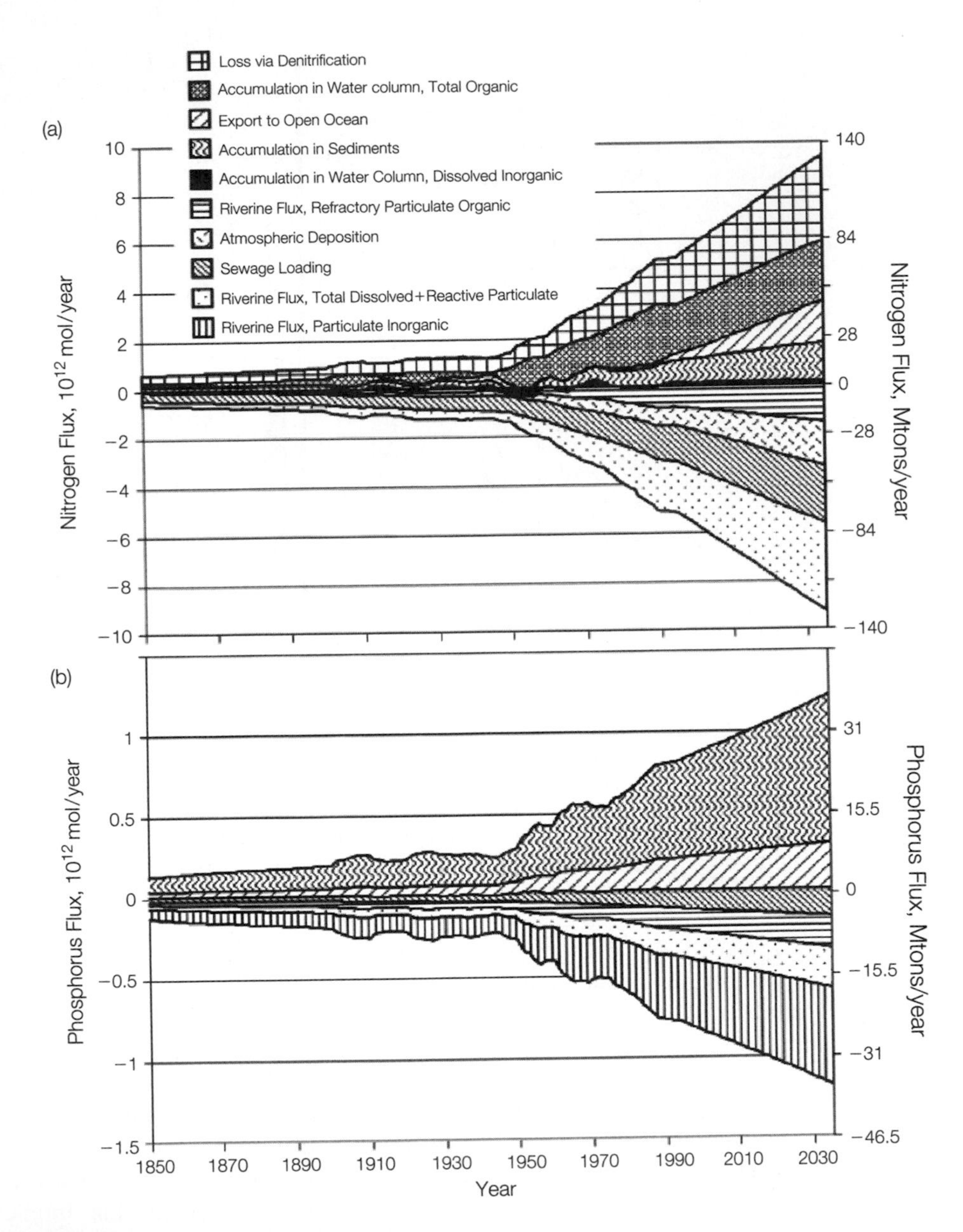

FIGURE 24.23

The model-calculated partitioning of the human-induced perturbation fluxes in the global coastal margin of (a) nitrogen and (b) phosphorus for the period from 1850 to the present (2000) and projected to 2035 under the business-as-usual scenario, in units of 10^{12} mol/y and Mtons/y. The anthropogenic sources are plotted on the (−) side and the resulting accumulations and enhanced export fluxes are plotted on the (+) side. *Source*: From Mackenzie, F. T., *et al*. (2002). *Chemical Geology* 190, 13–32.

storage of both nitrogen and phosphorus in the coastal sediments and increase their export to the open ocean. In the case of nitrogen, water column concentrations are also expected to increase.

In Chapter 28, we will consider further the effects of unintentional fertilization of coastal waters. These include development of hypoxia and blooms of harmful algae. Unfortunately, the size and persistence of coastal hypoxic zones and harmful algal blooms seems to be increasing worldwide.

CHAPTER 25

The Marine Carbon Cycle and Global Climate Change

All figures are available on the companion website in color (if applicable).

In this chapter, we consider the ocean's role in the global carbon cycle. This is a matter of great interest as the ocean has a large capacity to absorb atmospheric CO_2 and, hence, mitigate the effect of human emissions of this greenhouse gas. Burning of fossil fuels and land use change have driven atmospheric levels higher than they have been in the past 650,000 y. As a greenhouse gas, an increase in atmospheric CO_2 levels has the potential to cause global warming. In 2006, the Intergovernmental Panel on Climate Change reported a 90% likelihood that the global warming observed since 1750 is largely due to human activities. This human-driven climate change is superimposed over natural variations, the most important being a warming trend that has followed the end of the last Ice Age 18,000 years ago.

At present, most of the anthropogenic climate change has been driven by rising emissions of CO_2. Other anthropogenic greenhouse gases are becoming equally important, including methane, water, N_2O, halocarbons, and ozone. The ocean appears to have acted as the single largest sink of anthropogenic CO_2 emissions. About one third of the anthropogenic carbon released to the atmosphere from fossil fuel emissions and land use change has been taken up by the oceans and a similar fraction by the terrestrial biosphere. A bit less than half has remained in the atmosphere, where it acts as a greenhouse gas causing global warming.

Global climate change is arguably the single most important environmental threat that humans have faced. Continued warming is likely to result in severe environmental impacts, from rising sea level, to changes in the hydrological cycle, to shifts in floral and faunal distributions, including a predicted increase in the geographic spread of pathogens and parasites. Thus, considerable attention has been focused on understanding how the crustal-ocean-atmosphere factory responds to perturbations in the global carbon cycle and climate. Research efforts have focused in four areas: (1) quantifying how the modern carbon cycle is behaving through direct observational measurements; (2) understanding past changes in the carbon cycle by studying the sedimentary, rock, and fossil records; (3) performing controlled laboratory and field experiments to observe how key cycling rates respond to various perturbations;

and (4) developing mathematical models to predict future conditions under specified carbon-loading scenarios.

How the global carbon cycle responds to human perturbations depends on the time scale being considered. For example, the ocean can take up some CO_2 quickly, over time scales of years, through simple dissolution in surface seawater. Over time scales of millennia, meridional overturning circulation enhances this uptake through water mass exchange, driving the anthropogenic CO_2 into the deep sea and bringing fresh seawater to the ocean's surface to engage in further CO_2 uptake. An even larger amount of carbon can be removed via burial in marine sediments as organic matter or carbonate, but this is a slow process and hence requires tens of thousands of years to achieve a significant removal.

Another important characteristic of the carbon cycle is that it has undergone several rather abrupt changes during periods of Earth's history. These changes have been associated with mass extinction events that led to major shifts in the evolutionary trajectory of life on Earth. Many of these events are associated with large shifts in the size of the atmospheric CO_2 reservoir. This reservoir plays a crucial role in carbon cycling as it interacts with the hydrosphere and geosphere. Thus, it serves as a unifying linkage made particularly effective by its fast mixing rate and very rapid turnover time (a few years). As a result, the atmospheric CO_2 reservoir is extremely responsive to changes in the global carbon cycle. In this chapter, we consider the ocean's role in controlling the atmospheric CO_2 reservoir. We will first use the paleoceanographic record to provide insight into carbon cycling over geologic history, followed by a description of the ocean's current response to anthropogenic carbon emissions, and ending with predictions of future trends.

25.1 THE GLOBAL CARBON CYCLE

Current estimates of the carbon reservoir sizes and transfer rates in the crustal-ocean-atmosphere factory are shown in Figure 25.1 for the preindustrial state. As discussed later, the carbon cycle has evolved over geologic time with large oscillations marking times of significant global climate change, such as the Earth's entrance into and exit from Ice Ages. Estimates for reservoir sizes and transfer rates under glacial conditions are provided in parentheses.

The crust is the largest carbon reservoir in the crustal-ocean-atmosphere factory (8×10^7 Pg C including the sediments). Most of this carbon is in the form of inorganic minerals, predominantly limestone, with the rest being organic matter, predominantly contained in shale and secondarily in fossil fuel deposits (coal, oil, and natural gas). The oceanic reservoir (4×10^4 Pg C) and the terrestrial reservoir (2 to 3×10^3 Pg C) are both far smaller than the crustal reservoir. The smallest reservoir is found in the atmospheric, primarily as CO_2 (preindustrial 6×10^2 Pg C, now 8×10^2 Pg C and rising). The flux estimates in Figure 25.1 have been constrained by an assumption that the preindustrial atmospheric and oceanic reservoirs were in steady state over intermediate time scales (millennia).

The atmospheric carbon reservoir is largely controlled by biotic processes. To obtain a better sense of this, it is helpful to consider the turnover time of carbon in the

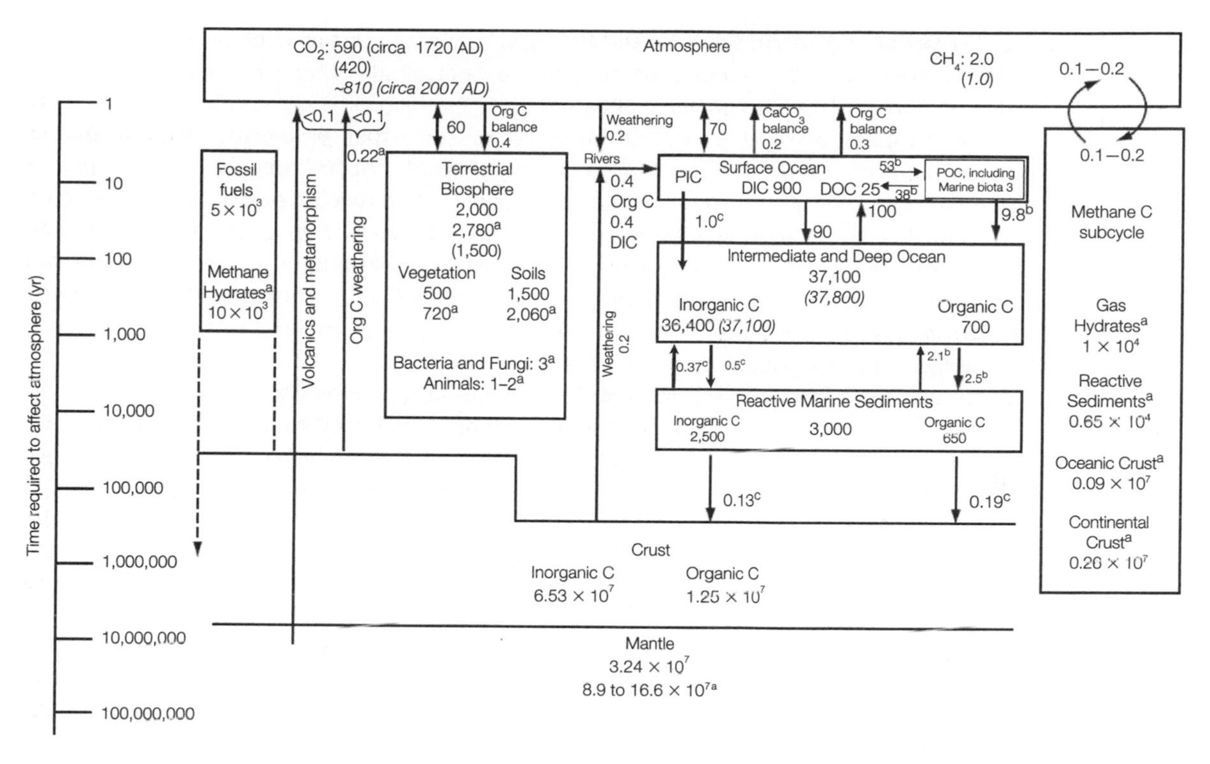

FIGURE 25.1

The crustal-ocean-atmospheric cycle of carbon. Reservoirs are in boxes (Pg C = 10^{15} g C) and fluxes are arrows (Pg C/y) for preindustrial conditions. Values for glacial periods, where available, are shown in parentheses. Values for the glacial and preindustrial (1750 AD) atmospheric carbon reservoir are based on assumed P_{CO_2} levels of 200 and 278 ppmv, respectively. The vertical bar on the left shows the approximate time (in years) necessary for the different reservoirs to affect the atmosphere. Atmospheric "balance" fluxes indicate the small net atmospheric exchange required to maintain a steady state with respect to sedimentation of organic carbon and calcium carbonate. Oceanic inorganic carbon fluxes are strictly pelagic; neritic values are not shown due to uncertainty in estimates–see Figure 25.2 for an estimate from NOAA. The terrestrial biosphere and oceanic reservoir values are rounded to the nearest 100 Pg C and the total reactive marine sediments to the nearest 1000 Pg C. All other reservoirs are rounded to the number of significant figures shown. All fluxes are rounded to one or two significant figures. *Source*: After Sunquist, E. T., and K. Visser (2003). *Treatise on Geochemistry*, Academic Press, pp. 425–472. Additional data from: [a]Mackenzie, F. T., and A. Lerman (2006). *Carbon in the Geobiosphere*. Springer-Verlag, p. 9; [b]Sarmiento, J. L., and N. Gruber, *Ocean Biogeochemical Cycles*. Princeton University Press, p. 360; and [c]Dunne, J. P., *et al.* (in preparation). *Global Biogeochemical Cycles*.

atmosphere as a result of its uptake by photosynthesis and release by respiration. On average, these processes were balanced (or nearly balanced)[1] in the preindustrial carbon cycle and, hence, are represented as two-headed arrows in Figure 25.1. The sum of

[1] Interannual deviations from a long-term steady-state balance between respiration and photosynthesis are likely and thought to be caused by large-scale phenomena, such as ENSO events.

the terrestrial and marine fluxes was 130 Pg C/y, yielding a turnover time of 4.5 y. The turnover time of atmospheric carbon with respect to all other processes was on the order of 1000 y [(590 Pg C)/(0.5 to 0.6 Pg C/y)].

The biomass of land plants is much larger than that of marine phytoplankton. Because land plants tend to grow more slowly than phytoplankton, the rates of marine and terrestrial production are approximately equal. Most detrital organic matter on land is oxidized, leading to very low soil organic matter production rates. As a result, the soil carbon reservoir is quite small. In contrast, a very large amount of carbon has been buried in the marine sediments, primarily as mineral carbonates and secondarily as organic matter. Both are biogenic in origin, having been created in surface seawater by various types of plankton, some of whom are calcifiers.

The carbon fueling the marine plankton is ultimately derived from atmospheric CO_2 that has dissolved into surface seawater. Sinking and burial of detrital biogenic carbon has led to the formation of a very large crustal limestone reservoir and a smaller amount of organic deposits. Since the dissolved CO_2 consumed by marine plants is readily replaced by gas exchange across the air-sea interface, the burial of biogenic carbon in marine sediments represents a large carbon sink. Over long time scales, on the order of 500 million y, the carbon buried in the crust is eventually recycled back to CO_2. This recycling involves the rock cycle, in which marine sediments are transformed into sedimentary rocks that undergo metamorphosis and subduction (Figure 1.2). During subduction, the rocks are decarbonated and the resulting CO_2 is emitted back into the atmosphere as a magmatic gas that issues out of volcanoes and hydrothermal vents (Figure 21.1). Decarbonation also takes place in sediments buried beneath the thick deposits of the continental rise (Chapter 21.4.4).

The cycling of carbon through the geologic reservoirs (sediment and crust) is often referred to as the *geologic* or *deep carbon cycle* to distinguish it from the far more rapid cycling that occurs as carbon is transferred among the atmospheric, biospheric, and hydrospheric reservoirs. The overall function of the geologic carbon cycle can be summarized in the following set of equations, which starts with the formation of biogenic hard parts followed by their burial in the sediments:

$$Ca^{2+} + 2HCO_3^- \rightarrow CaCO_3 + CO_2 + H_2O \tag{25.1}$$

$$H_4SiO_4 \rightarrow SiO_2 + 2H_2O \tag{25.2}$$

The net effect is

$$Ca^{2+} + 2HCO_3^- + H_4SiO_4 \rightarrow CaCO_3 + SiO_2 + CO_2 + 3H_2O \tag{25.3}$$

Decarbonation reactions produce a calcium silicate mineral:

$$CaCO_3 + SiO_2 \rightarrow CO_2 + CaSiO_3 \tag{25.4}$$

This reaction is similar to Eq. 21.8. Wollastonite is an example of a calcium silicate mineral produced by the metamorphosis of detrital biogenic calcium carbonate and BSi.

As the rock cycle continues, the calcium silicate minerals are eventually uplifted onto land where they undergo chemical weathering. This reaction involves acid hydrolysis driven by carbonic acid. The latter is derived from the dissolution of the magmatic CO_2 in rainwater:

$$2CO_2 + CaSiO_3 + 3H_2O \rightarrow Ca^{2+} + 2HCO_3^- + H_4SiO_4 \quad (25.5)$$

Combining Eqs. 25.3 and 25.5 yields the net effect of the deep carbon cycle,

$$CO_2 + CaSiO_3 \rightleftarrows CaCO_3 + SiO_2 \quad (25.6)$$

This is a very sketchy depiction of the deep carbon cycle because it illustrates only the behaviors of calcium and silica. In reality, a wide variety of other cations are present in the silicate minerals, such as in the plagioclase feldspars (Table 13.2). Furthermore, not all of the limestone is converted into silicate minerals; some remains as limestone. Uplift of the limestone onto land, followed by chemical and biological weathering, is another sink for atmospheric CO_2, via

$$CO_2 + CaCO_3 + H_2O \rightarrow Ca^{2+} + 2HCO_3^- \quad (25.7)$$

The products of chemical weathering, Ca^{2+}, H_4SiO_4, and $2HCO_3^-$, are transported by river runoff into the ocean, where they are then available to be returned to biogenic form by marine plankton. (Marine plankton have an enzyme, carbonic anhydrase, that converts bicarbonate to CO_2.)

The biogenic soft parts that become buried in the sediments are also transformed into sedimentary rocks, predominantly shale. Geologic uplift followed by chemical weathering leads to the oxidation of the organic carbon, i.e.,

$$\text{“}CH_2O\text{”} + O_2 \rightarrow CO_2 + H_2O \quad (25.8)$$

This generates CO_2, thus tying the carbon cycle to the global oxygen cycle (Chapter 8) and, hence, to the global sulfur cycle as the burial and weathering of pyrite has a large impact on P_{O_2}. The carbon and sulfur cycles are also linked directly through the pyrite deposition process depicted in Eq. 21.2. Because an important part of the deep carbon cycle relies on rock weathering, cycling rates are thought to be ultimately controlled by the rate of crustal uplift and by atmospheric temperatures. In the case of the latter, higher temperatures lead to faster rates of chemical weathering.

Although the amount of carbon stored in the fossil fuel reservoir is a very small component of the total crustal reservoir, it is large in comparison to the atmospheric reservoir. By burning fossil fuels, we have greatly accelerated the rate at which this carbon is released back into the atmosphere. Once in the atmosphere, this remobilized carbon is then able to interact with the biospheric and hydrospheric reservoirs.

Humans started using fossil fuels in a big way in the 1850s. Between 1800 and 1994, 244 Pg C was released as a result of fossil fuel emissions and cement production.[2]

[2] Cement production is a source of CO_2 because it involves the precipitation of calcium carbonate from calcium and bicarbonate, i.e., $Ca^{2+} + 2HCO_3^- \rightarrow CaCO_3 + CO_2 + H_2O$.

If all of this carbon had remained in the atmosphere, it would have increased the size of this reservoir by about 40%. Fortunately (unless you take the perspective of the planktonic calcifiers), a large measure of this carbon dissolved into the ocean, so that the atmospheric reservoir increased by only 28%.

25.1.1 The Oceanic Carbon Cycle

A more detailed depiction of the oceanic carbon cycle is presented in Figure 25.2, in which internal cycling rates in the coastal and open ocean are shown. The ocean

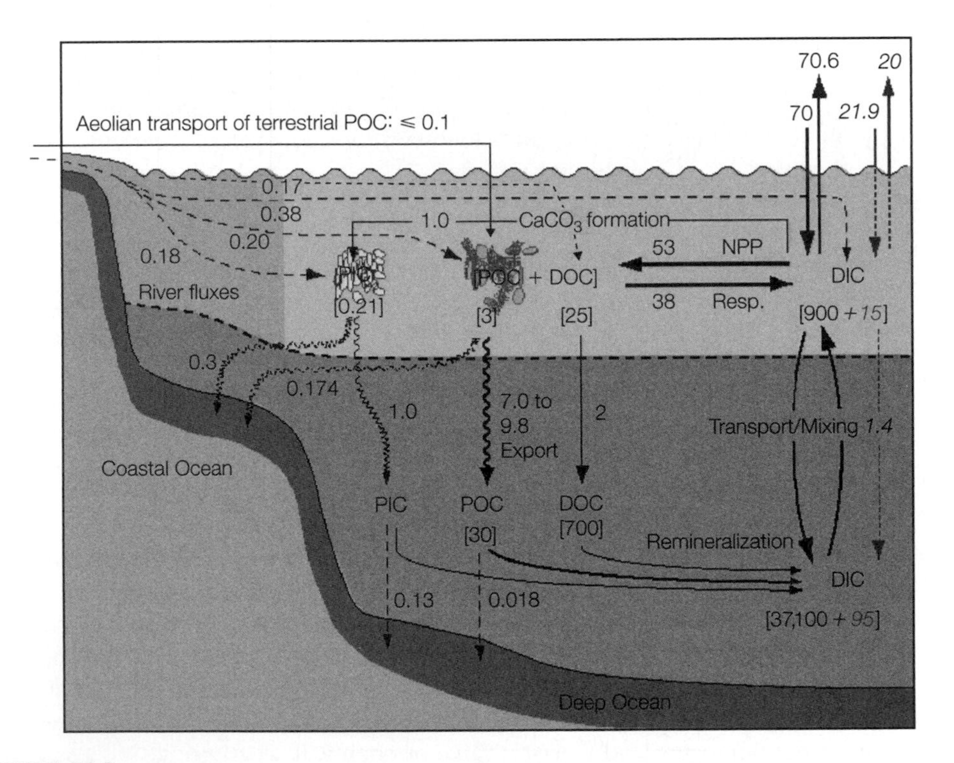

FIGURE 25.2

Oceanic carbon cycle showing the preindustrial (black) and anthropogenic (gray) ocean-atmosphere and land-atmosphere exchange fluxes in units of Pg C/y. Anthropogenic fluxes are shown in italics and by the dashed arrows. Reservoir sizes (Pg C) are in brackets. Anthropogenic fluxes are average values for the 1980s and 1990s. This figure includes an estimate for PIC burial on the shelves, unlike Figure 25.1. Some values differ from those in Figure 25.1, reflecting scientific uncertainty. *Source*: After NOAA PMEL Carbon Group, Global Carbon Cycle, *http://www.pmel.noaa.gov/co2/gcc.html*, accessed Sept. 2008. Additional data sources as per Figure 25.1 plus: (1) riverine organic fluxes from Seitzinger, S. P., *et al.* (2005) *Global Biogeochemical Cycles*, 19, GB4S01, (2) detrital POC flux from Mackenzie, F. T., and A. Lerman (2006). *Carbon in the Geobiosphere*. Springer-Verlag, p. 9, and (3) aeolian POC flux from Hedges, J. I. and R. G. Keil (1995) *Marine Chemistry* 49. 81–115, and Chester, R. (2003) *Marine Geochemistry*, 2nd ed., Blackwell Publishing, p. 68 and 219.

represents the single largest biologically active reservoir in the global carbon cycle. It contains 50 times more carbon than either the atmosphere or the terrestrial biosphere. Thus, it is less affected by changes in the rates or reservoir sizes of the biologically driven components of the carbon cycle. Conversely, it is large enough to exert a controlling influence on the other reservoirs. As a result, many scientists hypothesize that the ocean carbon cycle controls the atmospheric CO_2 reservoir and, hence, global climate on the intermediate time scales, i.e., millennia. Based on the rates and reservoir sizes in Figure 25.2, the residence time of carbon in the preindustrial ocean was on the order of 30,000 to 40,000 y [(38,758 Pg C)/(1.13 to 1.22 Pg C/y)].

The oceanic carbon inventory was presented in Table 15.3. Most of the carbon is inorganic (98%), predominantly in the form of bicarbonate (87%), and is located in the intermediate and deep waters. Of the 2% that is organic, the majority is DOC (see Table 23.2). At present, the ocean is acting as a net sink for atmospheric CO_2. In the modern-day carbon cycle, the sole oceanic sink for carbon is burial in the sediments in the form of detrital biogenic PIC and POC.

Most of the POC burial (95%) occurs on the continental margin. As shown in Table 25.1, burial in deltaic sediments, although geographically restricted, is very important, accounting for 44% of the organic carbon burial. The geographic distribution of PIC burial is not as well known because of lack of information on burial rates in continental margin sediments. Some scientists think that the burial rates in pelagic and shelf sediments are similar. A significant lateral transport of resuspended shelf sedimentary carbon into the pelagic ocean and its sediments is also thought to be occurring, but has not yet been quantified. This transport likely occurs sporadically when swift currents

Table 25.1 Organic Carbon Burial Rates in Different Marine Sediment Regimes (Pg C/y).

Sediment Type	Location	Carbon Burial Rate (Pg C/y)	% of Total Burial
Deltaic	Delta	0.070	44%
Shelf and Upper Slope	[a]	0.068	43%
High Productivity Zones	Slope	0.007	4%
High Productivity Zones	Pelagic	0.003	2%
Shallow-water Carbonates	Shelf	0.006	4%
Low Productivity Zones	Pelagic	0.005	3%
Anoxic Basins	Shelf	0.001	1%
Total		0.160	100%

[a] *Outside of the high productivity zones on the continental slopes.*
Source: Data from Hedges, J. I., and R. G. Keil (1995). Marine Chemistry *49, 81–115.*

contact the seafloor, creating "benthic storms" strong enough to resuspend settled material. If the resuspended particles are entrained in deepwater currents, such as the deep Western Boundary Currents, they have the potential to be transported long distances. Based on the sediment burial estimates provided in Figure 25.2, PIC burial rates are twice that of POC burial rates (0.43 versus 0.19 Pg C/y, respectively).

The delivery of PIC and POC to the sediments depends on biological production in the surface waters. The input of terrestrial carbon is relatively minor. The relative proportions of PIC and POC generated in the surface waters reflect the species composition of the plankton, i.e., calcifiers versus noncalcifiers. The success of detrital PIC and POC in surviving the trip to the seafloor is also dependent on the species composition of the plankton. As discussed in Chapter 23.5.4, the presence of biogenic hard parts confers protection on POC, inhibiting its remineralization.

Figure 25.1 indicates that a substantial amount of carbon is trapped in marine sediments as frozen methane. This gas is produced in the surface sediments by methanogens from metabolism of detrital organic matter and abiotically in more deeply buried sediments. At sufficiently low temperatures, the methane becomes encased in a cage of water molecules, forming a solid gas hydrate. As shown in Figure 25.3, the stability

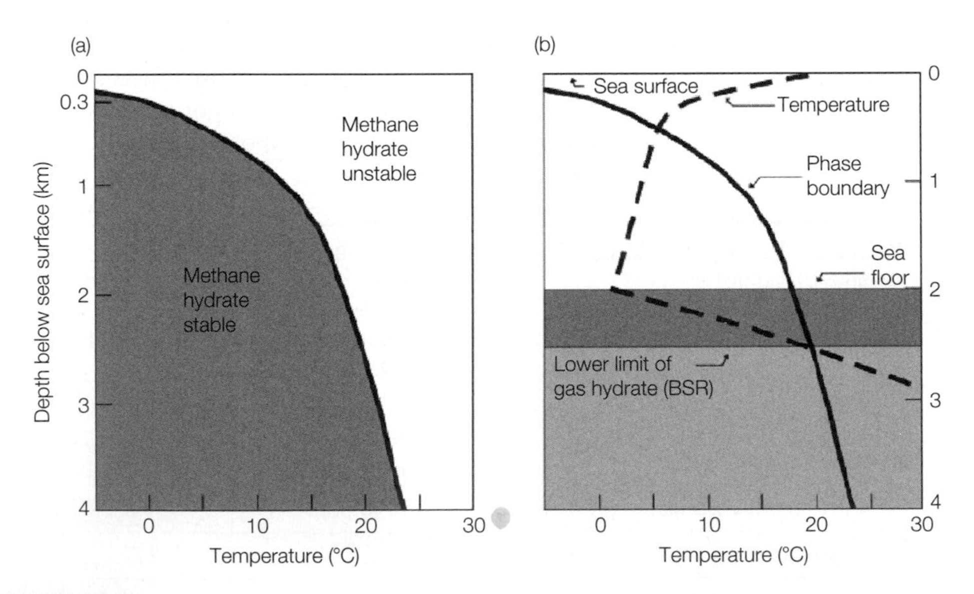

FIGURE 25.3

(a) Phase boundary and stability fields for methane hydrate. The vertical axis represents the effect of hydrostatic pressure, increasing downward, and plotted as equivalent depth (km). (b) The same phase boundary as shown in (a) with addition of a typical marine temperature profile through the water column and seafloor (dashed line). Also inserted is the position of the seafloor (2 km below the sea surface). The dark gray area denotes the vertical extent of methane hydrate under the specified conditions. BSR = bottom simulated reflector. *Source*: From Sundquist, E. T., and K. Visser (2003). *Treatise in Geochemistry*, Academic Press, pp. 425–472.

of methane hydrates is also pressure dependent. If these sediments are warmed, their hydrates will melt, thereby releasing methane into the ocean where it can be degassed into the atmosphere. At subpolar latitudes, a substantial amount of methane is also frozen in continental permafrost. This methane is also subject to mobilization as a result of global warming.

25.1.2 The Atmospheric Reservoir and Its Greenhouse Problem

The relatively small amount of carbon present in the atmosphere is primarily in the form of CO_2 with minor amounts of CH_4. As shown in Figure 25.4, the size of the atmospheric CO_2 reservoir oscillates seasonally. Concentrations are lowest in the summer, reflecting an increase in net uptake into plant biomass. During the winter, plant growth rates decline, while animals and microbes continue to respire organic matter. This results in a wintertime net emission of CO_2 that increases atmospheric P_{CO_2} levels. The degree of seasonal oscillation varies latitudinally (Figure 24.5b). The largest amplitude oscillations occur at the latitudes that experience the largest seasonal shifts in terrestrial primary productivity (i.e., northern hemispheric mid-latitudes). The latitudinal source effect is preserved in the atmospheric P_{CO_2} distribution because wind mixing is mostly driven by the Trades, Westerlies, and Polar Easterlies, thereby inhibiting cross-latitudinal transport. Note that the phase of the oscillation is reversed in the southern hemisphere as compared to the north, corresponding to the timing of the growing seasons. In the tropics, little seasonality is seen in atmospheric P_{CO_2} levels. At these latitudes, seasonal changes in moisture affect terrestrial photosynthesis and respiration rates almost equally. This causes seasonal shifts in the rates of these two processes to remain largely in phase, with the result being little or no net flux of CO_2.

Despite these seasonal oscillations and latitudinal variations, a long-term increase in the atmospheric CO_2 level is clearly evident. This is largely the result of CO_2 emissions from fossil-fuel burning, cement production, and changes in land-use patterns, primarily deforestation. As shown in Figure 25.5a, the increasing trend tracks the rate of population increase. Note that the growth rate of world population accelerated in the late 20th century, leading to a shortened doubling time of atmospheric P_{CO_2}. A similar increase in the rate of CO_2 emissions occurred around 1950.

Interestingly the interannual rate of rise of P_{CO_2} has not been constant, typically varying by 5 Pg C/y (Figure 25.5b). Most of the interannual variability is due to climate-forced changes in the terrestrial biosphere with a minor contribution coming from variability in ocean-atmosphere gas exchange (although the latter is largely out of phase with terrestrial flux). Most of the land-atmosphere flux variability occurs in the tropics where ENSO-related climate anomalies alter the hydrological cycle and atmospheric energy balance. This leads to the high correlation of the Southern Oscillation Index (a measure of ENSO intensity) with interannual P_{CO_2} variations shown in Figure 25.5b. Likewise, the amplitude of the seasonal oscillations in P_{CO_2} has been increasing since the mid-1990s, rising by 20% at Mauna Loa and 40% at Point Barrow, Alaska. This implies that the rate of processing of carbon may be increasing, perhaps as a result of

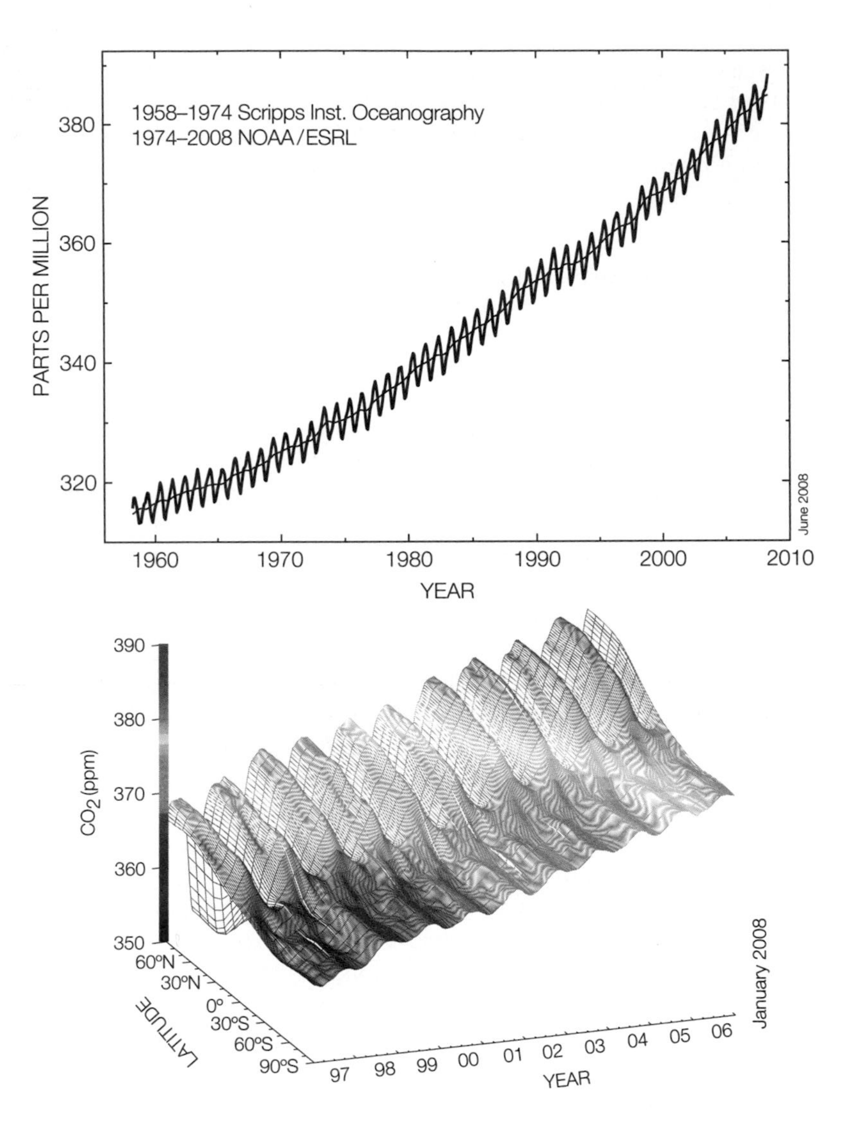

FIGURE 25.4

Seasonal and interannual trends in atmospheric carbon dioxide concentrations reported as mole fraction in dry air. (a) Monthly mean values at Mauna Loa Observatory, Hawaii. Data are also presented as 6-month running average to eliminate the seasonal effects and (b) three-dimensional representation of latitudinal distributions of monthly mean values. *Source*: After P. Tans and T. Conway, NOAA/ESRL Global Monitoring Division (www.esrl.noaa.gov/gmd/ccgg/trends). (See companion website for color version.)

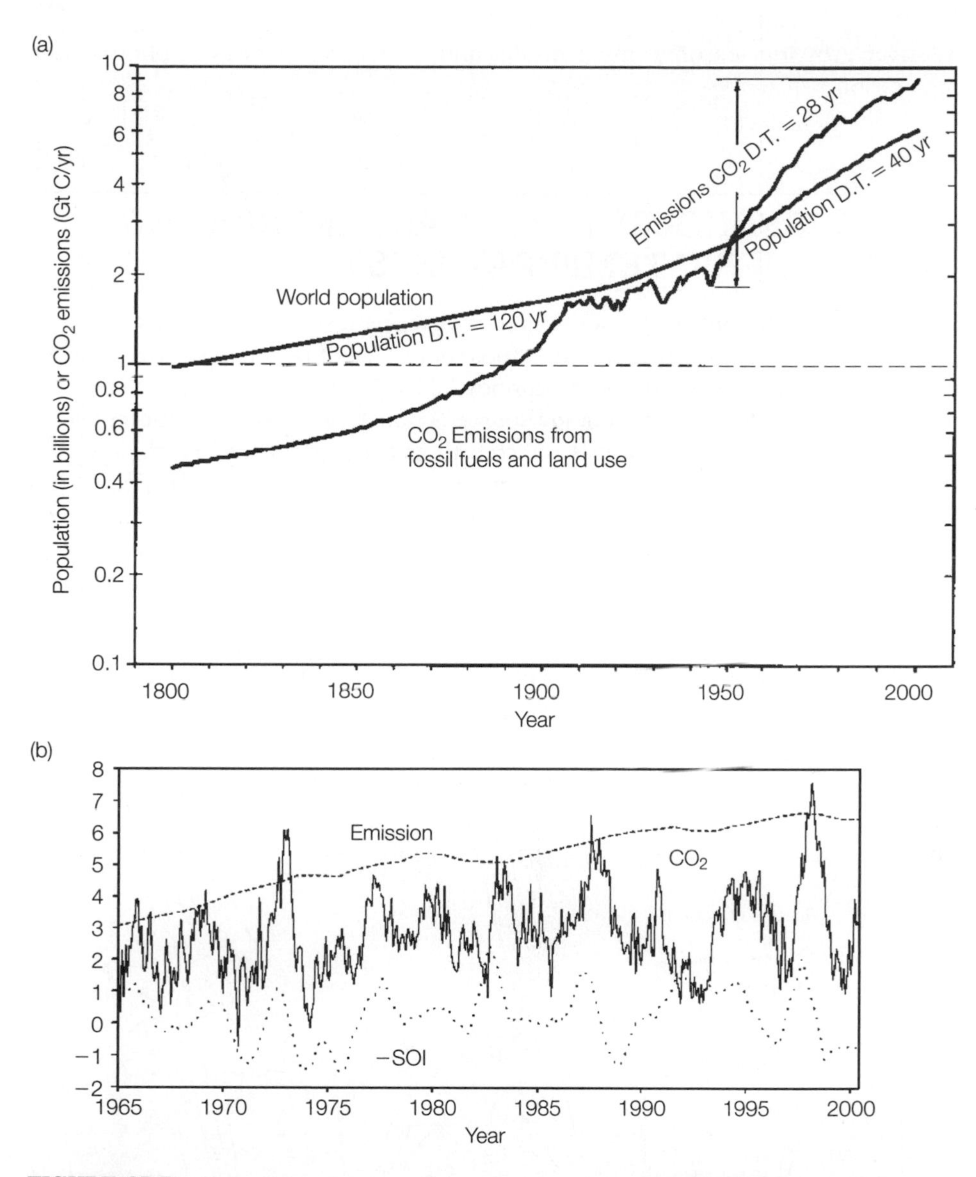

FIGURE 25.5

(a) Fossil fuel emissions and population increases over time. *Source*: From Mackenzie, F. T., and A. Lerman (2006). *Carbon in the Geobiosphere—Earth's Outer Shell*. Springer, p. 320.
(b) Anthropogenic CO_2 emission and atmospheric CO_2 growth rate (monthly from January 1965 to June 2000 with seasonal cycle removed) at Mauna Loa, Hawaii, in Pg C/y. Also plotted is the negative Southern Oscillation Index (−SOI, in mbar), which is an indicator of the tropical ENSO phenomenon. *Source*: From Zeng, N., *et al.* (2005). *Global Biogeochemical Cycles* 19, GB1016.

a longer growing season at mid-latitudes and, hence, faster rates of photosynthesis and respiration.

25.2 THE GREENHOUSE EFFECT AND INCREASING ATMOSPHERIC TEMPERATURES

To understand the linkage between greenhouse gas levels and climate control, we must consider the energy budget in the atmosphere. This budget is shown in Figure 25.6. (The heat flux from the inner Earth supported by radioactive decay is so small compared to the solar heat flux that this energy source is not shown.) The retention of solar energy by the greenhouse gases water is referred to as the *greenhouse effect*.

In the situation where Earth's atmosphere is at steady state with respect to insolation, the amount of energy (shown in W/m^2) coming into the atmosphere is balanced by the amount exiting to outer space. Some of the energy is reflected back into space, some is adsorbed by the greenhouse gases, and some is absorbed by Earth itself. (A very small fraction is harvested by the phototrophs.) Because of a process called *black-body radiation*, the solar energy absorbed by the Earth is transformed into longer wavelengths and reradiated back into the atmosphere, where it can either exit to outer space or become absorbed by the greenhouse gases.

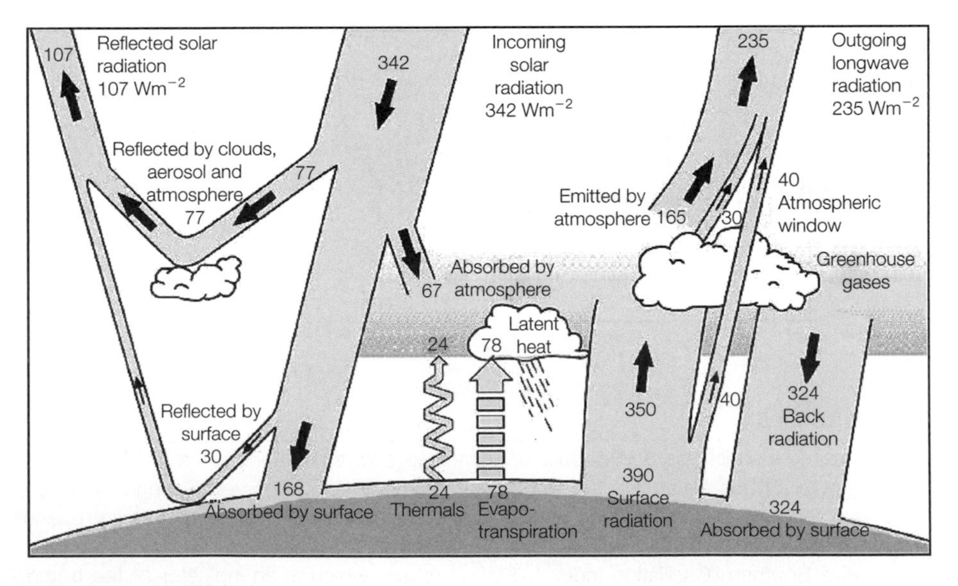

FIGURE 25.6

The earth's annual global mean energy budget. Units are W/m^2. *Source*: From Kiehl, J. T., and K. E. Trenberth (1997). *Bulletin of the American Meteorological Society* 78, 197–208.

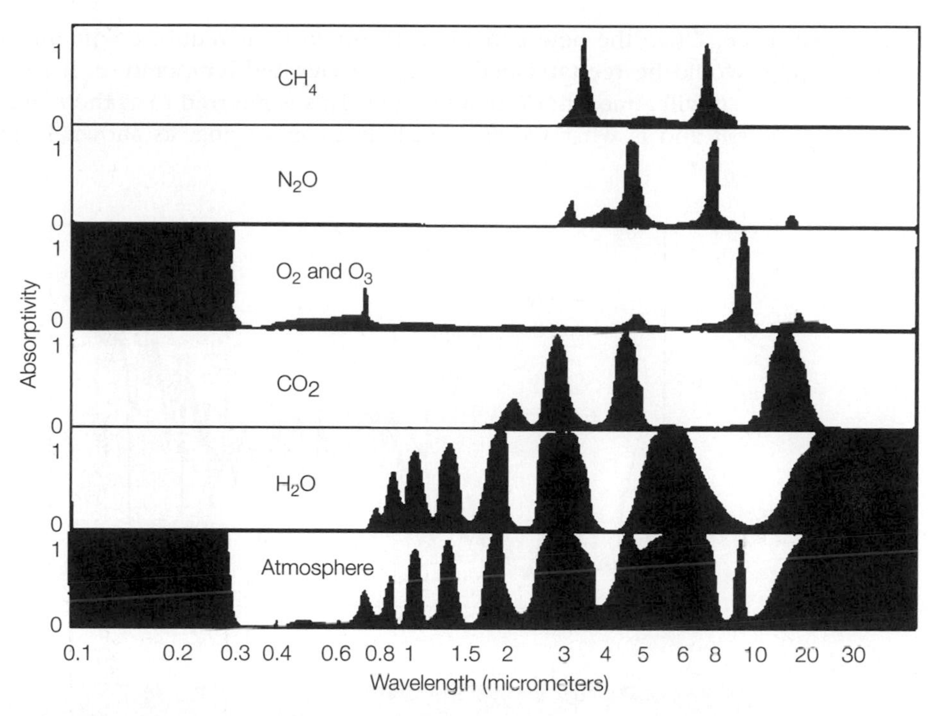

FIGURE 25.7

Absorptivity of various atmospheric gases and the atmosphere as a whole as a function of the wavelength of radiation. An absorptivity of 0 means no absorption; a value of 1 means complete absorption. The dominant absorbers of infrared (long-wave) radiation are water vapor and carbon dioxide. O_2 and O_3 are responsible for absorbing shorter wavelengths (UV-VIS). Incoming solar radiation ranges in wavelength from 0.25 to 3 μm and outgoing black-body radiation ranges in wavelength from 3 to 40 μm. *Source*: From Nese, J. M., and L. M. Grenci (2001). *A World of Weather: Fundamentals of Meteorology*, Kendall Hunt.

The wavelengths over which the greenhouse gases absorb the incoming and outgoing radiation are shown in Figure 25.7. The most important greenhouse gas is H_2O, which has absorption bands that fall across a broad range of wavelengths. In only a few narrow ranges of wavelengths is radiative energy not absorbed by one or more of the greenhouse gases that include CO_2, CH_4, O_3, NO, and various halocarbons. As a result, roughly 60% of insolation is absorbed by a greenhouse gas. When radiation is absorbed by a greenhouse gas, it is transformed into kinetic energy and then dissipated as thermal energy. This heat is radiated in all directions and warms the atmosphere.

On an Earth without greenhouse gases, no solar energy would be transformed into thermal energy and Earth's surface would be −4°C. At a constant level of greenhouse gases, a steady-state atmospheric temperature is achieved. If gas levels rise, the rate at which solar energy is absorbed by the atmosphere would temporarily exceed its loss to outer space. Once the new gases had absorbed their requisite amount of radiation, steady state would be reestablished, but at an elevated temperature. If gas levels continue to rise, so will atmospheric temperature. This is referred to as the *runaway greenhouse problem* and is what we are currently experiencing, as shown in Figure 25.8.

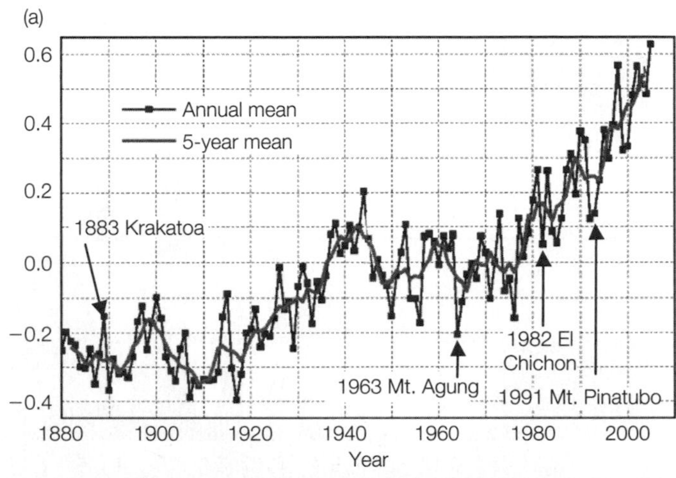

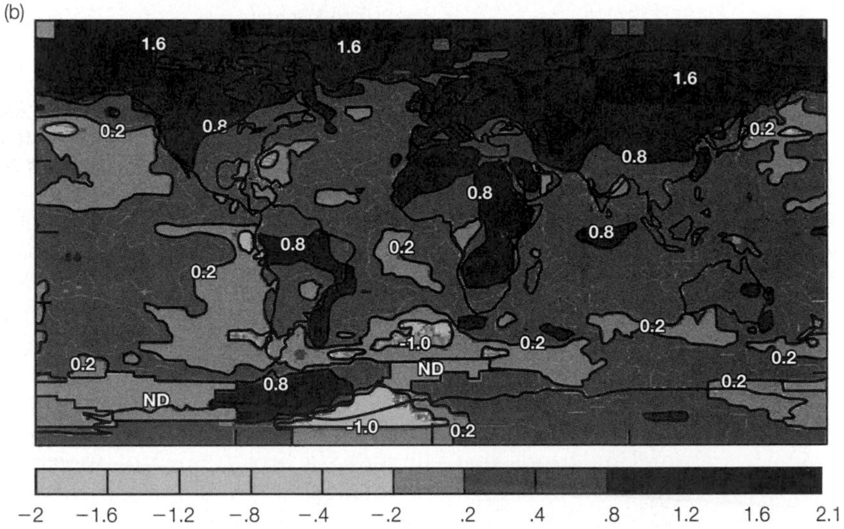

FIGURE 25.8

Increases in global surface temperatures. (a) Time trends reflecting annual and 5-year means. Also shown are dates of large volcanic eruptions and (b) geographic trends based on mean values for the period 2001 to 2005 AD. Temperature anomalies are the temperature differences for a given year or location relative to a base mean value for that year or location. The base period is 1951–1980 AD. *Source*: After Hansen, J., *et al.* (2006). *Proceedings of the National Academy of Sciences of the United States of America* 103, 14288–14293. (See companion website for color version.)

Table 25.2 Global Warming Potentials (Mass Basis) for Major Greenhouse Gases over Different Time Periods.

Gas	Atmospheric Lifetime (y)	Global Warming Potential by Time Horizon		
		20 y	100 y	500 y
CO_2		1	1	1
CH_4	12	62	23	7
N_2O	114	275	296	256
CFCs[a]	1.4 to 260	40 to 9,400	12 to 12,000	4 to 10,000
PFCs[b]	2,600 to 50,000	3,900 to 8,000	5,700 to 11,900	8,900 to 18,000
SF_6	3,200	15,100	22,200	32,400

[a]*CFCs, chlorofluorocarbon gases;*
[b]*PFCs, perfluorocarbon gases.*
Source: From Robertson, G. P. (2004). Scope 62, The Global Carbon Cycle, *Island Press pp. 493–506.*

Note that global mean temperatures are now within 1°C of the highest temperatures that have occurred over the past 1 million years.

Some greenhouse gases cause more warming than others. The degree to which a greenhouse gas can exert a warming effect on the atmosphere is called its *global warming potential* (GWP). Gases with long residence times and high radiative efficiencies have high global warming potentials. The Intergovernmental Panel on Climate Change (IPCC) has developed a GWP index relative to CO_2. As shown in Table 25.2, the gases with anthropogenic sources, methane, nitrous oxide, and the chlorofluorocarbons, have GWPs greater than 1 and, hence, exert more global warming impact on a mass basis than CO_2. Their concentrations are rising at a high enough rate that their contribution to the runaway greenhouse problem is significant. Even at the current rate of growth in CO_2 emissions, the other GHGs are predicted to eventually exceed the contribution of CO_2 to global warming due to their long atmospheric residence times.

Note that the heat budget model presented in Figure 25.6 does not take into account latitudinal effects. Neither the poles nor the tropics are in radiative balance. In the tropics, solar radiational input exceeds long-wavelength (infrared) radiation loss. The opposite occurs at the polar latitudes. The transport of heat energy from the tropics to the poles by wind and ocean currents alleviates this imbalance. Another important component of the budget presented in Figure 25.6 is reflection of radiation, both incoming solar radiation and the outgoing long-wave radiation. This reflection occurs off ice, clouds, and aerosol particles and is called Earth's *albedo*. On an ice-covered Earth, the albedo is very high and, hence, most of the incoming solar energy is reflected, providing a positive feedback that enhances cooling. Natural sources of reflective aerosols include forest fires, biogenic sulfur emissions, and volcanic eruptions.

25.3 HOW THE MODERN OCEAN TAKES UP CARBON—THE THREE PUMPS

Carbon enters the ocean through gas exchange across the air-sea interface, river input, aeolian deposition of terrestrial POC, and hydrothermal emissions. The ultimate sinks for this carbon are burial in the sediments or emission back into the atmosphere. In preindustrial times, the ocean was a net source of carbon to the atmosphere. The ocean is now serving as a sink of atmospheric carbon in response to anthropogenic loading of CO_2. About half of the anthropogenic carbon emitted into the atmosphere has been taken up by the ocean, partially mitigating the enhanced greenhouse effect caused by increasing P_{CO_2}.

Scientists are working hard to understand how the ocean takes up carbon so they can forecast how well this sink will operate in the future. Some of the questions they are seeking to answer are: How much carbon can the ocean take up and how fast? What are the ultimate controls on the ocean's carbon uptake rate? Are other human activities affecting the ocean's carbon uptake system and, hence, affecting its ability to mitigate the anthropogenic CO_2 problem? Do natural feedback systems exist that will enhance or decrease the uptake system as we continue to load CO_2 into the atmosphere? To answer these questions, we first look at the three ways in which the modern-day ocean assimilates carbon. These are called the *Solubility or Gas Exchange Pump*, the *Soft Tissue Pump*, and the *Carbonate Pump*. Then, we will consider how past changes in the marine carbon cycle, as gleaned from paleoceanographic records, provide insight into the controlling mechanisms. Finally, we will consider how the ocean is responding to atmospheric CO_2 loading and end with model projections for the most likely future conditions.

25.3.1 The Solubility or Gas Exchange Pump

In Chapter 6, we saw that gases are more soluble at lower temperatures. As a consequence, CO_2 is twice as soluble in the cold surface waters found at polar latitudes as compared with the warm surface waters located at the equator. These cold CO_2-rich waters sink into the deep sea to form the intermediate, deep and bottom water masses. The sinking of these water masses pumps CO_2 below the thermocline. The impact of this pump is seen in Figure 25.9, which shows that anthropogenic CO_2 has penetrated far below the thermocline at the sites of formation of NADW, AABW, and AAIW. At present, half of the anthropogenic CO_2 that has invaded the ocean is still located at depths less than 400 m reflecting the limited rates and geographic scope of subthermocline water-mass formation. As further time elapses, scientists expect the anthropogenic CO_2 to spread laterally as it continues to be carried along by the subthermocline water masses.

Because this pump is driven by physical processes associated with gas solubility and water circulation, it is termed the *solubility or gas exchange pump*. The efficiency of this carbon pump is controlled by three factors: (1) the physical movement of seawater,

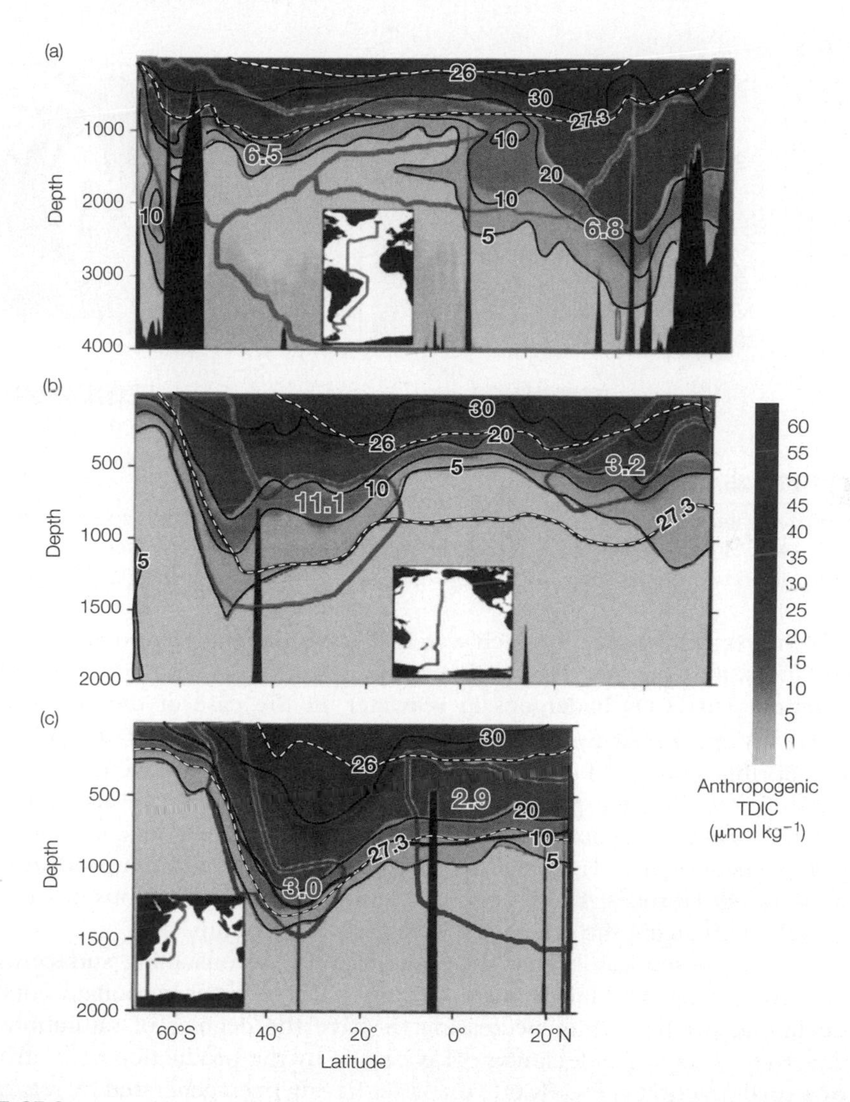

FIGURE 25.9

Representative sections of anthropogenic TDIC (μmol/kg) from (a) the Atlantic, (b) Pacific, and (c) Indian oceans. Note that the depth scale for (a) is twice that of the other figures, reflecting the deep penetration of anthropogenic TDIC in the North Atlantic. Potential density contours are represented by the black dashed lines. Gray outlined regions and numbers therein define the intermediate and deepwater masses and their total inventory of anthropogenic CO_2 (Pg C). In the southern latitudes, the intermediate water mass is Antarctic Intermediate Water. In the northern latitudes, the demarcated water masses are North Atlantic Deep Water in (a), North Pacific Intermediate Water in (b), and Red Sea/Persian Gulf Intermediate Water in (c). The cruise tracks are shown in the map insets. *Source*: After Sabine, C. L., *et al.* (2004). *Science* 305, 367–371. (See companion website for color version.)

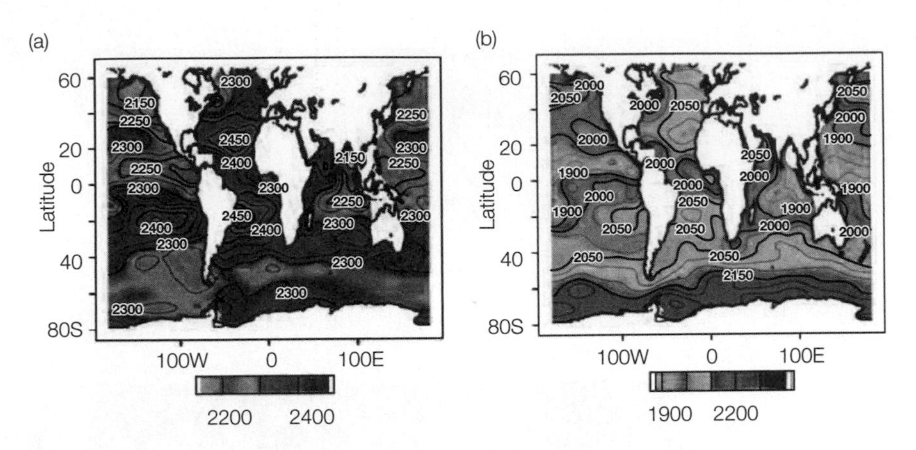

FIGURE 25.10

Surface water concentrations of (a) total alkalinity (μmol/kg) and (b) dissolved inorganic carbon (μmol/kg). *Source*: After Key, R. M., *et al.* (2004). *Global Biogeochemical Cycles* 18, GB3011. (See companion website for color version.)

which dictates the rate at which waters from below the thermocline exchange with the surface waters; (2) the effect of temperature on P_{CO_2} solubility; and (3) the chemical reactions that CO_2 undergoes in seawater. In the case of the last, solubility of CO_2 is greatly enhanced by reaction with carbonate ($CO_3^{2-} + H_2O + CO_2 \rightarrow 2HCO_3^-$). The equilibrium constant for this reaction is temperature dependent, with increasing temperature favoring the products. The availability of carbonate alkalinity (CO_3^{2-}) in the surface waters is probably as important as temperature in controlling the solubility pump. As shown in Figure 25.10, surface alkalinities and temperatures are highest at mid- to low latitudes. These conditions promote high saturations of calcium carbonate, thereby enhancing the efficiency of the solubility pump.

In the present-day ocean, the invading CO_2 is consuming surface-water carbonate and converting it to bicarbonate. Because surface-water carbonate concentrations are declining, pH has begun decreasing, as have the degrees of saturation of calcite and aragonite. Part of the declining pH is caused by the production of H^+ from dissociation of a small fraction of the HCO_3^-, the latter having been generated by reaction of CO_2 and H_2O with the CO_3^{2-}. If equilibrium reactions control the speciation of inorganic carbon, the fraction of HCO_3^- that undergoes dissociation should increase as the amount of CO_2 taken up by seawater increases. In other words, a type of CO_2 buffering takes place in which CO_2 uptake leads to a decline in seawater's capacity to take up additional CO_2. Evidence for such a decline in the solubility pump has already been observed.

Over time scales of millennia, the return flow of meridional overturning circulation will carry a fresh supply of carbonate into the mixed layer, eventually providing a second level of buffering against alkalinity loss and pH decline. But if anthropogenic CO_2 continues to invade the ocean, this second level of buffering will eventually become exhausted once anthropogenic CO_2 penetrates through the entire ocean and depletes all seawater of its carbonate alkalinity. At this point, a third level of buffering will kick in, as acidic waters come into contact with calcareous sediments, leading to their dissolution. On still longer time scales (tens of thousands of years), the deep carbon cycle will eventually provide a fourth level of buffering of atmospheric P_{CO_2} through chemical weathering of uplifted carbonate and silicate rocks.

25.3.2 The Biological Pumps

Carbon is also pumped into the deep sea as a result of the sinking of biogenic hard and soft parts, i.e., detrital PIC and POC. This is a two-stage pumping action. In the first stage, detrital particulates sink out of the surface waters. Although some of these particles are remineralized, the resolubilized carbon is trapped below the thermocline and, hence, is effectively removed from interaction with the atmosphere—at least on time scales that are less than a few thousand years. The particles that survive to become buried in the sediments represent a second stage of pumping in which carbon is effectively removed from interaction with the atmosphere over much longer time scales.

The first stage of the biological pump is responsible for the vertical segregation of TDIC (Figure 15.10b). In the preindustrial ocean, most of this vertical segregation was caused by the soft tissue pump. This pump is driven by the formation of POC by surface-water plankton. The rate of POC export from the surface waters depends on the rates of biological productivity and the degree to which detrital POC can be exported below the thermocline. As discussed in Chapter 24, plankton production is controlled by complex interactions among physicochemical factors, such as nutrient limitation, light availability and mixed layer depth, and by biotic forces, namely grazing. The effects of the former have led to the classification of the surface ocean into biogeographic provinces (Table 24.5).

The degree to which surface-water production is exported into the deep sea relies in large measure upon species composition of the plankton community. For example, POC generated by the phytoplankton that deposit hard parts, namely the diatoms and coccolithophorids, is more effectively exported than is the POC generated by the soft-bodied plankton. The hard parts act as ballast and as protection against microbial attack. Seasonal diatom blooms also lead to enhanced export as cells aggregate into large rapidly sinking clumps. Another effectively exported type of POC are the large fecal pellets created by the mesozooplankton. Members of the macrozooplankton, such as the salps and larvaceans, are also highly effective at creating rapidly sinking organic-rich particles.

The kinds of plankton and their relative abundance vary over time and space in response to changes in physicochemical and biotic controls. The marine food web in the context of its effect on carbon cycling is illustrated in Figure 23.2. Complex interactions within this marine food web collectively act to direct species composition within planktonic communities. On one hand, these interactions can be viewed as biotic relationships, such as grazing and competition for nutrients. Alternatively, the various species can be viewed as functional types capable of exerting unique biogeochemical effects that influence the rest of the planktonic community. For example, conditions of nitrogen limitation favor the dominance of N_2 fixers. These organisms act to channel new nitrogen into the community. As shown in Figure 24.14, this leads to a ripple effect on the biogeochemistry of the deep waters, which are eventually returned to the sea surface by the return flow of meridional overturning circulation. These temporal shifts in nutrient availability are thought to cause alternating ecosystem states. At least some of this alternation is likely controlled by phosphorus and/or micronutrient availability, particularly iron. A classification scheme that groups plankton into biogeochemical functional types is presented in Table 25.3.

The degree to which POC is remineralized as it sinks below the thermocline is dependent, at least in part, on dissolved O_2 levels. A greater portion of the POC flux seems to survive the trip to the seafloor and become buried in the sediments when PO_2 is low. Thus, the rate of meridional overturning circulation, which controls the rate of oxygenation of the deep waters, plays an important role in determining the efficiency of the second stage of the soft-tissue pump. Some of the POC remineralized during export is transformed into DOC. Given the long average age of deep and surface-water DOC (Chapter 23.6.8), conversion of carbon into this refractory form seems to be rather effective in removing it from interaction with the atmosphere, at least over time scales of a few thousand years or so.

One of the important planktonic functional types are the marine calcifiers. Approximately half of the PIC is generated by coccolithophorids and the rest by foraminiferans. Production of PIC by the planktonic calcifiers drives the carbonate pump. As shown in Figure 25.2, the PIC flux is a small contributor as most of the biological pump is supported by export of POC (1 versus 10 Pg C/y, respectively). Although PIC is a small component of the overall export, it is better preserved than the POC and, hence, constitutes 41% of the carbon buried in the sediments. In the crust (Figure 25.1), it comprises 84% of the carbon.

In recognition of the unique and interlinked effects of the various plankton species in determining carbon export, marine scientists have begun to incorporate ecosystem dynamics into the global models used to simulate global climate. These global models generally include various submodels, such as one that simulates the important interactions between ocean circulation and climate. Submodels are now being developed to cover ecosystem interactions that link the carbon cycle to global climate. In the largest of these models, linkages enable the physical submodels of climate and circulation to act on the biogeochemical-ecosystem submodels and vice versa. Geological submodels are also being developed to cover connections between terrestrial rock weathering and climate as this linkage controls the input of phosphorus, silica, iron, and alkalinity to the ocean, as well as serving as atmospheric O_2 and CO_2 sinks. To validate the ecosystem models, marine scientists are working out methods by which SeaWIFS ocean color spectral data can be used to map functional plankton types. An example is shown in Figure 25.11, tracking monthly trends in coccolithophorids, cyanobacteria, and diatoms.

25.3.3 CO_2 Fluxes across the Air-Sea Interface

Surface-water P_{CO_2} varies geographically and seasonally, ranging from 150 to 550 μ atm. The P_{CO_2} of the atmosphere is now around 380 ppm, making some surface water greatly undersaturated with respect to P_{CO_2} and some greatly supersaturated. These

Table 25.3 Plankton Functional Types.

Functional Type	Common Cell Size (μm)	Main Members	Functional Role/Impact	Notes
	<0.2	Viruses	Shunt that prevents transfer of organic carbon and nutrients to higher trophic levels	Infects prokaryotes and eukaryotes; sometimes very specific to host. Thought to be responsible for lateral gene transfer
Picoheterotrophs	0.3–1.0	Heterotrophic bacteria and archaea	Remineralization of POC and DOC	
Picoautotrophs	0.7–2.0	Picoeukaryotes and non-N_2-fixing photosynthetic bacteria	Widespread significant component of primary production especially in oligotrophic and HNLC[a] areas	*Synechococcus* and *Prochlorococcus*. Might be really underestimated given recent reports of widespread chromophores including bacteriochlorophylls, rhodopsins, and phycoerythrins in bacteria
Phytoplankton N_2 fixers	0.5–2.0	*Trichodesmium* and N_2-fixing unicellular prokaryotes	N_2 fixers can relieve N limitation if sufficient Fe is present, making primary production essentially P limited	
Phytoplankton DMS producers	<20	*Phaeocystis* and small autotrophic flagellates	Produce dimethylsulfoniopropionate (DMSP) and convert it into DMS using an extracellular enzyme (DMSPlyase). Thus, they affect the atmospheric sulfur cycle. Have high P requirement	Particularly abundant in coastal areas, where they are often observed in colonies. Calcifiers are also important for the DMS cycle

(Continued)

Table 25.3 *(Continued)*

Functional Type	Common Cell Size (μm)	Main Members	Functional Role/Impact	Notes
Phytoplankton calcifiers	5–10	Coccolithophorids	Produce half the PIC flux, the rest being generated by forams	Produce the densest ballasts observed in sinking particles destroyed in regions where the mixing depth of the ocean is below the euphotic zone. Their calcification rate is reduced at low pH. Zn limited
Phytoplankton silicifiers	5–200	Diatoms	Dominant microphytoplankton. Comprises most of the primary production and biomass in spring blooms at temperate and polar regions. They require more Fe and P than most of the smaller nano- and pico-phytoplankton	Contribute to carbon export far more effectively than smaller plankton through direct sinking of single cells, through key grazing pathways, and through mass sedimentation events at the end of the spring blooms when nutrients are depleted. Silicifiers require and deplete Si. They respond to enhanced Fe input in HNLC[a] regions as long as Si is available. They produce little DMSP compared with most other phytoplankton
Mixed phytoplankton	2–200	Autotrophic dinoflagellates and *Chrysophyceae*	No direct impact on the cycles of S, Si, or $CaCO_3$	Background biomass of phytoplankton. Do not bloom in the open ocean and have low seasonality

Protozooplankton	5–200	Ciliates, heterotrophic flagellates	Unicellular heterotrophs that dampen bloom formation of small phytoplankton. Includes the smaller foramiferans that generate PIC	Graze preferentially on small phytoplankton (1–20 μm), such as the pico- and nanophytoplankton. Their growth rates are similar to that of phytoplankton in the pico- and nano-size range, and their ingestion rates are closely coupled to the production rates of their prey
Mesozooplankton	200–2000	Copepods, appendicularians, amphipods	Produce large and fast-sinking fecal pellets. Are an important food for fish. Includes larger foraminiferans that generate PIC	Graze preferentially on larger plankton (20–200 μm), such as proto-zooplankton and phytoplankton silicifiers. Their grazing and reproductive rates are slower than that of protozooplankton
Macrozooplankton	>2000	Euphausiids, salps, pteropods	Produce large fecal pellets, which sink much faster than those of mesozooplankton	Graze across a wide spectrum of sizes. Can achieve very high biomass locally, but tend to have a patchy distribution

[a]*High nutrient low chlorophyll regions include the Southern Ocean, equatorial and subarctic North Pacific, and some strong upwelling areas such as off Peru and California.*

Source: After LeQuere, C., et al. (2005). Global Change Biology *11, 2016–2040.*

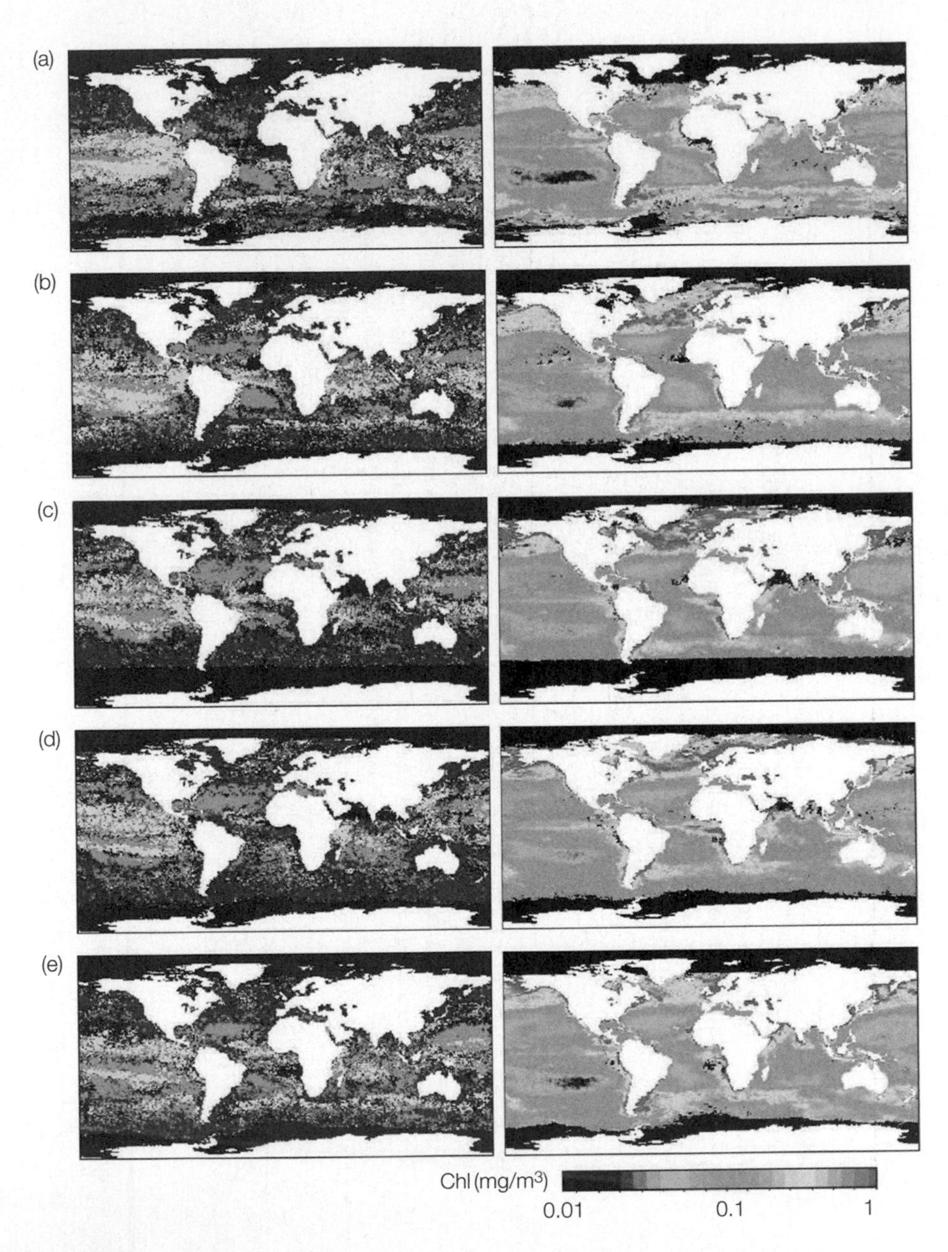

FIGURE 25.11

Left panels: Monthly mean maps of the phytoplankton assemblages for (a) January, (b) April, (c) June, (d) August, and (e) October 2001 (haptophytes (coccolithophorids) in blue, *Prochlorococcus* (cyanobacteria) in green, *Synechococcus*-like cyanobacteria in yellow, and diatoms in red). Right panels: Chlorophyll *a* monthly maps (mg/m^3) obtained from SeaWIFs data. *Source*: After Alvain, S., *et al.* (2005). *Deep-Sea Research* I 52, 1989–2004. (See companion website for color version.)

disequilibria drive net fluxes of P_{CO_2} across the air-sea interface. Since no practical technique has been developed to directly measure this flux, marine scientists generate estimates from data-based computations. The most common practice has been the use of Fick's first law (Eq. 6.16) in which the flux is computed from the air-sea P_{CO_2} gradient. CO_2 concentrations are measured shipboard using an infrared gas analyzer.

Other techniques for estimating the air-sea CO_2 flux rely on tracers of gas exchange, such as the relative abundance of the naturally occurring radioactive isotope of Rn (^{222}Rn) compared to its parent isotopic source, ^{226}Ra. Atmospheric O_2/N_2 ratios have also been used to infer P_{CO_2} fluxes. Efforts are now being directed at obtaining these fluxes computationally from the Ocean General Circulation Model coupled to a biogeochemical-ecosystem model. These coupled models are fed various data such as climatological parameters or geochemical tracer concentrations like radiocarbon. The model output is validated against other observational data, such as nutrients, chlorophyll, and P_{CO_2} levels. The advantage to this approach, over the use of Fick's first law, is that no diffusivity coefficient (D_A) is required. As discussed in Chapter 6.4.1, semiempirical approaches must be employed to parameterize D_A for each gas including the effects of wind stress, temperature, and salinity.

Fluxes generated from the application of Fick's first law to observed air-sea gradients in P_{CO_2} are shown in Figure 25.12. This map of the mean annual net air-sea flux for

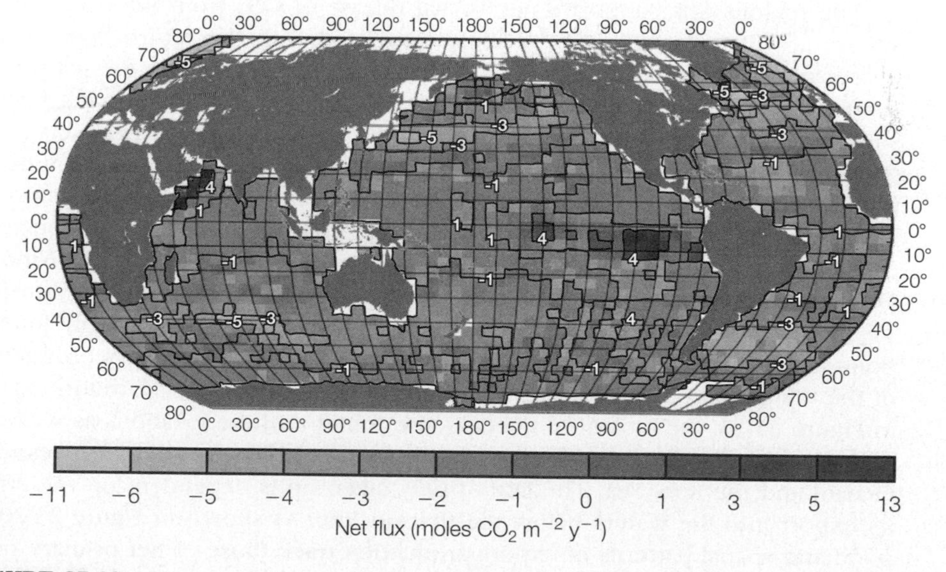

FIGURE 25.12

Mean annual net air-sea flux for CO_2 (mol CO_2 m^{-2} y^{-1}) for reference year 1995. Areas that are a source of atmospheric CO_2 are red-yellow. Areas that are a sink are blue-purple. *Source*: After Takahashi, T., *et al.* (2002). *Deep-Sea Research* II 49, 1601–1622. (See companion website for color version.)

CO_2 is based on 940,000 measurements of surface-water P_{CO_2} made between 1956 and 2000. Some areas of the surface ocean, like the equatorial waters, act as a net source of P_{CO_2} to the atmosphere and some, like the North Atlantic, act a net sink. Overall the ocean is presently acting as a net sink, taking up CO_2 from the atmosphere at a rate of about 2 Pg C/y. This is a significant departure from preindustrial conditions during which the ocean is thought to have acted as a small net source (0.6 Pg C/y; see Figure 25.2).

The regions that support a net annual uptake of CO_2 are the sites where NADW, AAIW, and North Pacific Intermediate Water are formed. Gaseous disequilibrium is supported by the continuous movement of warm waters poleward by the geostrophic currents. The resulting drawdown of P_{CO_2} is partly due to enhanced solubility from cooling and to high wind speeds. It is also partly due to biological uptake supported by the high nutrient concentrations found at the subpolar latitudes. The North Atlantic as a whole is a site of net uptake because mode water formation occurs over a substantial part of its area (Figure 4.16c). High wind speeds over these low-P_{CO_2} waters lead to increased CO_2 uptake rates. The Southern Ocean also appears to be currently acting as a small net sink with the direction of CO_2 flux highly dependent on the intensity of the circumpolar Westerlies. Only a small net sink exists because the enhancing effect of high wind speeds on CO_2 uptake is countered by Ekman-driven upwelling of CO_2-rich waters to the sea surface. The wind-driven upwelling is also countering the strong biological pump that exists in the Southern Ocean (Figure 10.10).

The regions that support a net annual release of CO_2 from the ocean to the atmosphere are the sites where upwelling, associated with the return flow of meridional overturning circulation, brings CO_2-enriched deep waters back to the sea surface. Once these cold, alkaline waters enter the mixed layer, they warm. Warming decreases the solubility of $CaCO_3$, driving the reaction in Eq. 25.1 toward the products and thereby generating CO_2. Warming also decreases the solubility of CO_2. Thus, warming decreases the CO_2 uptake capacity of seawater. The major area of CO_2 degassing is presently the equatorial zone of the Pacific Ocean with lesser contributions from the equatorial zones of the Atlantic and Indian Oceans. When upwelling ceases during El Niño/Southern Oscillation (ENSO) events, the Pacific Equatorial Zone becomes a net P_{CO_2} sink.

The P_{CO_2} fluxes can also vary on a seasonal basis. This is primarily found in locations subject to large seasonal swings in temperature and/or primary productivity. Most of the temperature effects occur at the sites where water masses are formed. As shown in Figure 25.13, the biological effect is also found in these locations as well as in areas with intense upwelling such as the eastern Equatorial Pacific, the northwestern Arabian Ocean, and the Ross Sea. The fate of CO_2 taken up is dependent on the efficiency of its export into the waters below the thermocline. As shown in Figure 23.20, the temporal and spatial patterns of export production track those of net primary production, with highest values seen in subpolar waters and coastal upwelling zones. Considerable attention is currently being directed toward creating coupled physical-biogeochemical-ecosystem models that can predict export fluxes of carbon to the deep sea.

If the preindustrial atmosphere was in steady state, net transports through the atmosphere and oceanic currents would have been required to balance out regional

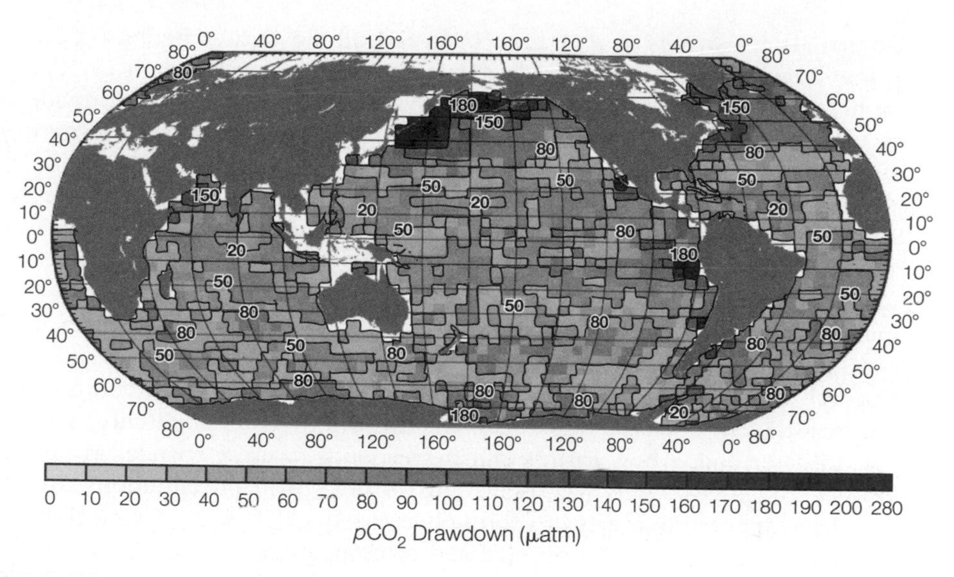

FIGURE 25.13

P_{CO_2} drawdown based on seasonal differences in P_{CO_2} concentrations in reference year 1995, corrected for temperature effects. This drawdown reflects the degree of biological utilization of CO_2. Values exceeding 150 μatm (yellow-orange-red) are observed in the northwestern subarctic Pacific and Atlantic, the eastern equatorial Pacific, the northwestern Arabian Sea, and the Ross Sea. These high-biology areas are associated with intense upwelling of subsurface waters with high CO_2 and nutrient concentrations. Small areas near the coasts of Mexico, Chile, and Argentina also show strong biological drawdown effects. Data are insufficient to characterize seasonal changes in the southern Weddell Sea and the coastal upwelling areas off West Africa. *Source*: After Takahashi, T., *et al.* (2002). *Deep-Sea Research* II 49, 1601–1622. (See companion website for color version.)

differences in net P_{CO_2} fluxes. In the case of the Pacific Ocean, the northern and southern basins currently appear to have approximately the same magnitude of uptake. In contrast, the Atlantic Ocean is strongly imbalanced because of a large uptake rate in the northern basin. If such a regional disparity existed during preindustrial times, a south to north transport of P_{CO_2} must have taken place through the atmosphere with a return flow as part of the NADW deepwater current. In the present day, anthropogenic loading of P_{CO_2} is concentrated in the northern hemisphere as this is where most of the land masses and, hence, people are located. Thus, anthropogenic activities are likely to have altered the hemispheric carbon exchanges.

25.4 PAST VARIATIONS IN CLIMATE AND ATMOSPHERIC CARBON DIOXIDE LEVELS

The geologic carbon cycle was described in Chapter 25.1 as providing the negative feedback loop responsible for stabilizing P_{CO_2} over time scales of 10^5 to 10^7 y. Nevertheless, large transient swings in P_{CO_2} have occurred during various periods of Earth's history. A few have led to long-term reorganizations of the global carbon cycle. Paleoceanographers are particularly interested in studying these events as they provide some clues as to what changes we can expect as a result of our anthropogenic perturbations of atmospheric CO_2 levels.

Over the past 1.8 million years, Earth has cycled in and out of Ice Ages during which large continental ice sheets have waxed and waned. These changes in ice volume and regional temperatures are well correlated with oscillations in the greenhouse gases, CO_2, CH_4, and N_2O. As in earlier times, high GHG levels have coincided with warm and interglacial conditions. Conversely, cold and glacial conditions have coincided with low GHG levels. The role of the GHGs in climate change is still a matter of hot debate and research effort. While the astronomical Milankovitch cycles that control insolation are the major agent of long-term climate change, the GHGs undoubtedly play an important role in determining how Earth's climate responds to these changes in insolation.

Most scientists agree that oscillations in the GHG concentrations over Earth's history have affected past climates through a variety of physical, chemical, biological, and geological feedbacks, some positive and some negative. One of the key features of the atmospheric P_{CO_2}, P_{CH_4}, and temperature oscillations is their slow decline into glacial conditions followed by a rapid return to high levels during the interglacials. This suggests that a threshold is reached that triggers a rapid shift from glacial to interglacial conditions.

The last Ice Age terminated 10,000 to 15,000 ybp and has been followed by a period of general warming termed the Holocene Epoch. Prior to 1700 AD, atmospheric CO_2 levels during the Holocene varied by no more than 10 ppmv, fluctuating between 275 ppm and 285 ppm. The increase in P_{CO_2} that has taken place from 1700 to 2000 AD has been about 85 ppm, equivalent to 175 Pg C, or 30% of the preindustrial level. Atmospheric CO_2 concentrations reconstructed from ice cores extending back 650,000 y have established that in the Late Quaternary, the typical glacial-interglacial difference has been 100 ppm, and was accompanied by temperature swings of about 10°C. The present-day P_{CO_2}, which exceeds 380 ppm, represents a large departure from that of the past 650,000 y (Figure 25.14 and Figure W25.3). Most of this increase is attributed to human activities as our known emissions are more than adequate to explain the observed atmospheric increase (Figure 25.5). As shown in Figures 25.4 and 25.14, the time frame over which atmospheric P_{CO_2} has risen rapidly coincides with the period over which anthropogenic emission rates have increased. Figure 25.4b also documents that the changes in P_{CO_2} have been largest in the northern hemisphere, matching the geographic source of our emissions. During this same period, the ^{14}C content of the atmospheric CO_2 has decreased, further supporting a fossil-fuel source.

Figure 25.14 documents that the concentrations of other GHGs have also increased over the past century. Methane levels have roughly doubled. Although still in the ppb range, methane's higher GWP makes its climate-forcing effect equivalent to 30% of that currently contributed by CO_2. Fortunately methane is prone to oxidation, making its atmospheric lifetime relatively short (8 to 10 y). The increase in its concentration over

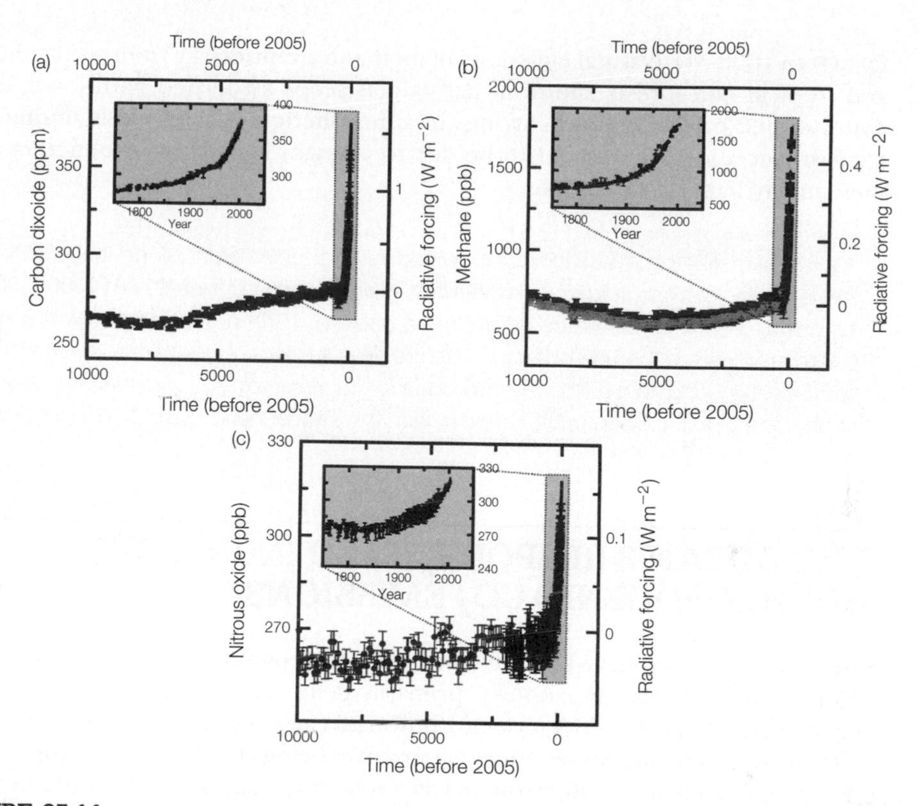

FIGURE 25.14

Atmospheric concentrations of (a) carbon dioxide, (b) methane, and (c) nitrous oxide over the past 10,000 years (large panels) and since 1750 (inset panels). The corresponding radiative forcings are shown on the right-hand axes of the large panels. *Source*: From IPCC Working Group I (2007). *Climate Change 2007: The Physical Science Basis, Contribution of Working Group I to the Fourth Assessment Report of the IPCC*. Cambridge University Press.

time follows a temporal trend similar to that of CO_2 until 1985 AD. After 1985 AD, the annual growth rate of methane has declined such that atmospheric levels have now stabilized.[3] More than 50% of present-day global methane emissions are anthropogenic in origin, coming from fossil-fuel production (escape of natural gas from leaky pipelines), livestock (cattle), rice cultivation, and waste handling (including animal waste, domestic sewage, and landfills). Biomass burning has also been an important source, particularly

[3] This stabilization is thought to have been caused by the drying of wetlands, leading to decreased natural emissions and thereby countering some of the anthropogenic input. In 2007 AD, P_{CH_4} rose for the first time since 1998, jumping from an annual global mean of 1775 ppb to 1783 ppb. Likely sources are increased anthropogenic emissions from rapidly industrializing Asia and melting permafrost in the Arctic.

from 0 to 1000 AD. Natural emissions of methane are primarily from wetlands, peatlands, and tropical rain forests and from natural gas seeps associated with fossil-fuel deposits and the melt of gas hydrates. Swings in atmospheric methane levels during the end of the last glaciation are thought to be due to changes in wetland production rather than methane-hydrate destabilization.

Additional material on this subject is provided in the supplemental information for Chapter 25.4 that is available online at **http://elsevierdirect.com/companions/9780120885305**. Key topics covered are the role of tectonism in the geologic carbon cycle and how the evolution of pelagic calcifiers in the Phanerozoic led to the development of feedbacks, some stabilizing and some destabilizing, that act on the atmospheric CO_2 reservoir. Also included is a short summary of how the global carbon cycle interacts with the atmospheric O_2 and sulfur cycles.

25.5 THE OCEAN'S RESPONSE(S) TO INCREASED ANTHROPOGENIC CO_2 EMISSIONS

Since 1850, 275 Pg C have been released from the combustion of fossil fuels and another 155 Pg C from changes in land use, primarily deforestation. The observed rise in P_{CO_2} is only 175 Pg C requiring operation of a powerful carbon sink(s), of which the ocean is the most likely candidate. Much effort is currently being directed at developing global carbon cycle models to elucidate the ocean's role in CO_2 uptake during the Anthropocene. The results from one such model are shown in Figure 25.15.

25.5.1 Past to Present Emissions

The consensus from these modeling efforts and the observational paleoceanographic data is that during preindustrial times, the ocean as a whole acted as a small net sink for atmospheric CO_2. This represents a global average that encompasses considerable latitudinal variations. As shown in Figure 25.15, these latitudinal variations arise from the combined effects of physical and biological phenomena on carbon uptake and release. For example, high outgassing rates in the equatorial waters are caused by physical effects, such as decreased CO_2 solubility at higher temperatures. This causes a strong degassing at the equator because this is the site at which upwelling returns CO_2-rich waters to the sea surface. Strong biological uptake at these latitudes counters some of the physical loss. This illustrates that the physical controls on the air-sea exchange of CO_2 tend to act in opposition to the biological effects. The physical controls also tend to be the dominant controlling agent. The major exception to the latter is the Southern Ocean, where the biological effect dominated during preindustrial times. But in this region, the biological pump was very inefficient and led to net loss of CO_2 from the ocean. As described in Chapter 25.3.3, the regions of high CO_2 uptake are centered around the sites where subsurface water masses, i.e., mode water and intermediate and deep water, are created.

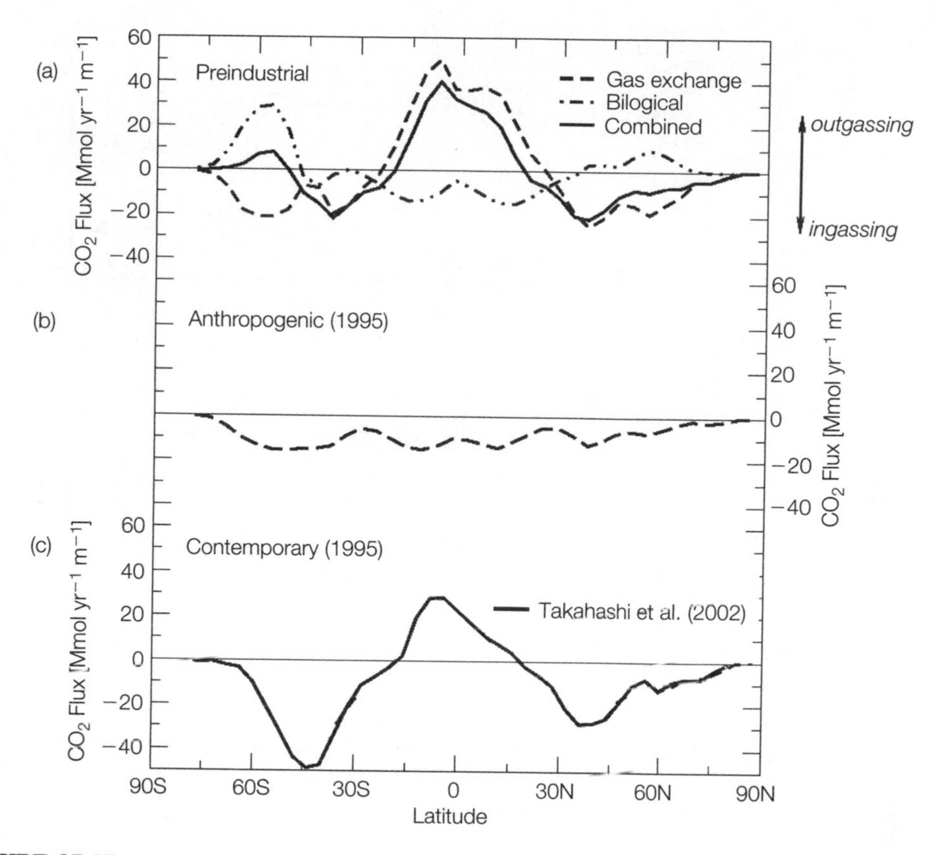

FIGURE 25.15

Zonally integrated global sea-air exchange fluxes for CO_2 (positive fluxes denote outgassing from the ocean). (a) Effects of the gas exchange pump and the biological pumps (preindustrial). (b) Anthropogenic gas flux circa 1995 and (c) observed present day total sea-air flux. *Source*: From Sarmiento, J. L., and N. Gruber (2006). *Ocean Biogeochemical Cycles*. Princeton University Press, p. 353. See Figure 25.12 for Takahashi et al. (2002) citation.

The present-day carbon transports are zonally similar to the preindustrial pattern (Figure 25.15c) except that the sites of outgassing have reduced in strength and those of ingassing have increased. The overall effect has been to change the ocean from a small net atmospheric carbon source to a large net sink. A timeline for this switchover is presented in Figure 25.16, which shows that the carbon emissions from fossil-fuel burning, cement production and land-use changes have been variously transferred into the atmosphere, ocean, and presumably the terrestrial biosphere. An accounting of this carbon transfer by rate and mass is presented in Table 25.4. This mass balance shows that during the period of 1800 to 1994, only half of the CO_2 released by humans remained in the atmosphere—30% was taken up by oceans and 20% presumably by the terrestrial biosphere.

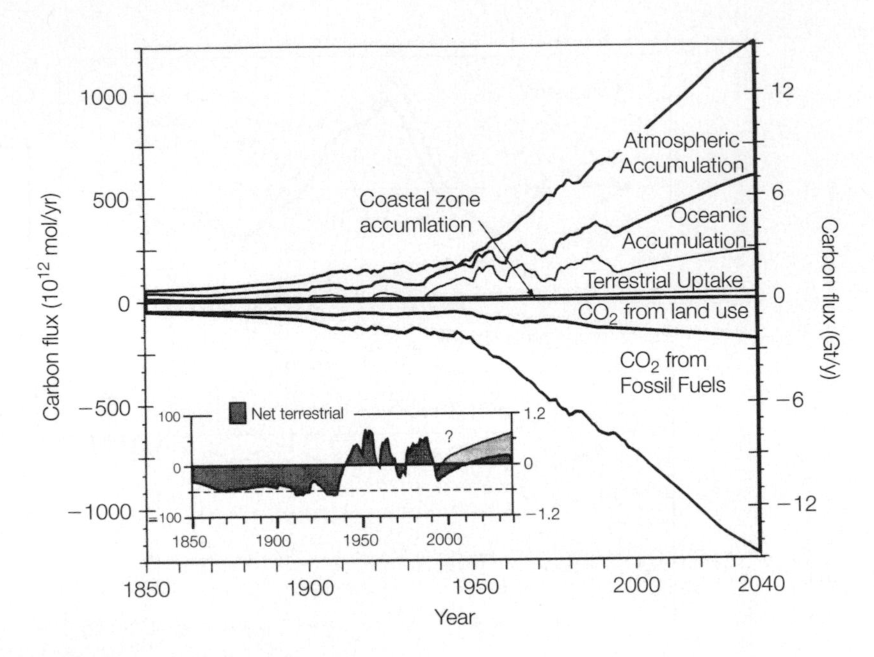

FIGURE 25.16

Partitioning of the human-induced perturbation fluxes of carbon for the period 1850 to 2000 and projected to the year 2040 under a business-as-usual scenario in units of 10^{12} mol C/y and Gt C/y (1 Gt = 1 Pg). The anthropogenic sources are plotted below the zero line on the (–) side. The resulting accumulations and enhanced export fluxes are plotted above the zero line on the (+) side. The insert shows the net terrestrial organic carbon flux with a range of projections from 2000 onward that depend on whether or not changes in land use will result in more or less emissions of CO_2 to the atmosphere. This diagram illustrates only the anthropogenically derived fluxes. Not shown are the large natural exchanges of carbon. *Source*: From Mackenzie, F. T. and A. Lerman (2006). *Carbon in the Geobiosphere*. Springer-Verlag (Figure 11.4).

Prior to the 1960s, a large fraction of the CO_2 released by humans was a consequence of land-use change. This made combined effect of terrestrial processes (biomass production versus land-use losses) a net CO_2 source to the atmosphere. Since the 1960s, fossil-fuel emissions have increased such that they are now the dominant CO_2 source.[4] Between 1980 to 1999, the terrestrial biospheric sink appears to have increased slightly, causing the combined effect of terrestrial processes to switch from being a net source to being a small sink. Because the percentage of CO_2 remaining in the atmosphere has remained constant, a small decline in ocean uptake must have occurred. This effect is an expected outcome arising from the loss of CO_2 buffering capacity that follows continued CO_2 uptake (Chapter 25.3.1).

[4] This includes emissions from uncontrolled fires in coal mines, which is a significant source of CO_2.

Table 25.4 Carbon Budgets (a) During the Decade of the 1990s, (b) Cumulative Effects from 1850 to 2000, and (c) Terrestrial Net Balance over the Anthropocene.

(a) Balance of CO_2 uptake and release rates during the 1990s (Pg C/yr)

Atmospheric CO_2 increase		Fossil Fuel Emissions		Land Use Emissions		Ocean Uptake		Presumed terrestrial sink
3.2 ± 0.2	=	6.3 ± 0.4	+	2.2 ± 0.8	–	2.4 ± 0.7	–	2.9 ± 1.1

(b) Cumulative effects over the anthropocene (amounts in Pg C)

Time Period	Fossil Fuel Emissions	=	Atmospheric Increase	+	Oceanic Uptake	+	Terrestrial Net Balance[a]
1800 to 1994	244 ± 20	=	165 ± 4	+	118 ± 19	+	–39 ± 28 (release)
1980 to 1999	117 ± 5	=	65 ± 1	+	37 ± 8	+	15 ± 9 (uptake)

(c) Terrestrial net balance over the anthropocene (amounts in Pg C)

Time Period	Terrestrial Net Balance[a]	=	Accumulation in Undisturbed Ecosystems	–	Release from Land-Use Change
1800 to 1994	–39 ± 28 (release)	=	61 to 141	–	100 to 180
1980 to 1999	15 ± 9 (uptake)	=	39 ± 18	–	24 ± 12

Source: From Woods Hole Research Center (2007). The Missing Carbon Sink, http://www.whrc.org/carbon/missingc.htm; *and Sabine, C. L, et al. (2004).* Science *305, 367–371.* [a]*Terrestrial net release is the difference between a presumed terrestrial sink and known land-use change emissions.*

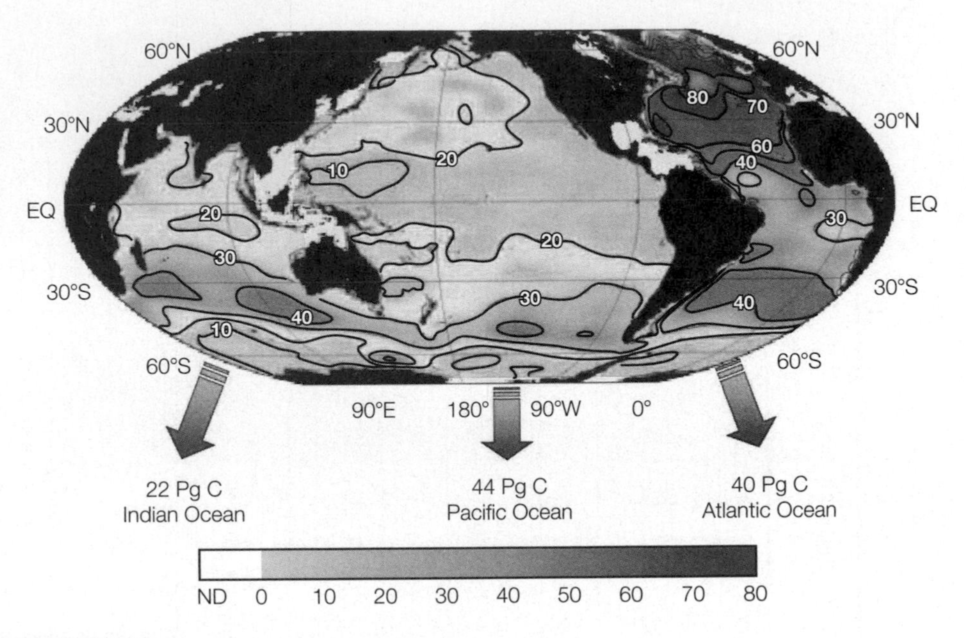

FIGURE 25.17

Vertical column inventory of anthropogenic CO_2 in the ocean (mol/m^2). High inventories are associated with deepwater formation in the North Atlantic and intermediate and mode water formation between 30°S and 50°S. Total inventory of shaded regions is 106 ± 17 Pg C. *Source*: From Sabine, C. L., *et al.* (2004). *Science* 305, 367–371. (See companion website for color version.)

The penetration of anthropogenic carbon into the ocean was shown in Figure 25.9 in the form of longitudinal cross sections for each ocean. By 1994, 106 ± 17 Pg C had been injected into the open ocean. Including marginal seas and the Arctic Ocean brings the total to 118 ± 19 Pg C. About one third of the open-ocean injection was achieved by uptake into subsurface water masses, although only 7% has penetrated beneath 1500 m. The strong influence of uptake at the sites of water-mass formation is shown in Figure 25.17, which maps the vertical column inventory of anthropogenic CO_2. AAIW is the largest of the water mass sinks (20 Pg C), followed by NADW (7 Pg C), North Pacific Intermediate Water (3.2 Pg C), and Indian Intermediate Water (3 Pg C). Similar penetration has been observed for other tracers, such as radiocarbon and the chlorofluorocarbons (CFCs).

25.5.2 Response of the Solubility Pump to Future Emissions

The future response of the ocean to increasing injection of anthropogenic carbon into the atmosphere can be predicted by considering how the physical and biological pumps are most likely to respond. Initially, the physical factors will lead to a decrease in the ocean's uptake rate. This is a result of four factors. First, as discussed in Chapter 25.3.1, the CO_2 buffer capacity of seawater decreases as its P_{CO_2} increases due to shifts in carbonate-bicarbonate ion equilibria, i.e., the Revelle factor becomes less favorable for CO_2 uptake. This makes surface seawater more resistant to taking up carbon. Second, as surface seawater warms, gas solubility is decreased. Third, warming increases stratification, which reduces vertical mixing and the rate of formation of subsurface water. This traps CO_2 in the mixed layer and prevents its injection into the deep sea. Fourth, anthropogenic CO_2 emission rates are increasing over time.

The initial decrease in ocean uptake rate will eventually be reversed once enough time has passed (millennia) for meridional overturning circulation to recycle the subsurface waters. This will serve to inject CO_2 below the mixed layer. Once the surface waters have been sufficiently acidified and are transported into the deep sea, they will eventually start dissolving sedimentary calcium carbonate. The resupply of alkalinity via this route will provide sufficient enhancement to ocean carbon uptake capacity to enable removal of 90% of the anthropogenic CO_2 emissions. This is expected to occur over time scales of 10^4 to 10^6 years.

25.5.3 Response of the Biological Pumps to Future Emissions

Less well constrained are the likely impacts to the biological pumps. Even if primary production is increased, the net effect on the global carbon cycle will depend on whether carbon export and burial in the sediments increases. Mechanisms that can lead to increased net primary production include: (1) increased nutrient loading to surface waters caused by changes in meridional overturning circulation and (2) enhanced utilization of nutrients due to increased iron inputs, changes in plankton species, and faster rates of photosynthesis at higher temperatures. Mechanisms that can lead to increased export production are: (1) changes in the Richards-Redfield ratio of detrital POM caused by shifts in plankton species and ballasting and (2) increases in the ratio of POC to PIC caused by a shift in plankton species composition away from coccolithophorid dominance. Quantitative estimates of the relative importance of these mechanisms await results from the global models that are under development and which include feedback linkages between biogeochemical processes, climate, and ocean circulation.

As the ocean acidifies, production of biogenic $CaCO_3$ is expected to decline because of a decline in $CaCO_3$ mineral saturation in surface seawater. Laboratory experiments have demonstrated that nearly all marine calcifiers are negatively impacted by declining carbonate ion concentrations even when mineral saturation remains above 1. On a more positive note, the loss of calcifiers will increase ocean CO_2 uptake rates for two reasons: (1) calcification produces CO_2 (Eq. 25.1), so less calcification leads to less marine production of CO_2 and, hence, less consumption of surface seawater's buffer capacity, leaving more capacity for the uptake of anthropogenic CO_2; and (2) as old shells dissolve, carbonate ion concentrations will rise, increasing seawater's alkalinity and ability to consume more CO_2. On the other hand, biogenic calcite represents an important ballasting material and its loss is expected to reduce the net export of POC into the deep sea.

If the current rate of increase in anthropogenic CO_2 emissions continues unabated, the pH of the surface ocean is predicted to decline by more than 0.4 units by the end of the century as P_{CO_2} rises to 800 ppm. At this point, surface-water DIC concentrations will have increased by 12% and carbonate ion concentrations decreased by 60%. Such changes have probably not occurred for more than 20 million years. Regions of aragonite undersaturation are expected to develop first in the Southern Ocean, followed by the subarctic Pacific Ocean as shown in Figure 15.13. As noted earlier, the degree to which seawater is supersaturated has a profound effect on calcification rates for individual species and communities in both planktonic and benthic habitats. This is especially true for the biocalcifiers that deposit aragonite as this mineral is more soluble than calcite. Aragonite is deposited by reef-building corals, planktonic pteropods and heteropod mollusks (Chapter 15.2). Thermal stress will also negatively affect coral communities once water temperatures exceed 31°C.

Projections of the combined effects of acidification and thermal stress in restricting coral habitat is shown in Figure 25.18. Once P_{CO_2} reaches 425 ppmv, which is shown as occurring in 2060–2069, no coral habitat will remain. Figure 25.2 shows that $CaCO_3$ burial in neritic waters is a large sedimentary sink. Loss of this sink is likely to have significant consequences on the ocean's overall carbon uptake ability. The only long-term data sets available to examine the changing $CaCO_3$ mineral saturation of seawater are from the North Pacific (Hawaiian Ocean Time Series, 1990 to present) combined with the GEOSECS cruise work of the 1970s. Over the past 30 years, calcite supersaturation in the North Pacific surface seawater has declined from 6.4 to 5.5.

25.5.4 Response of the Coastal Ocean and Estuaries to Future Emissions

Another poorly constrained component of the ocean's response to continued anthropogenic CO_2 loading is the behavior of the shallow coastal ocean and estuaries. The carbon dynamics in these regions are also being altered by increased terrestrial runoff of nutrients. This has led to increased primary production with an as yet unknown effect on carbon export to the underlying sediments and open ocean. The latter is referred to as the *continental shelf pump*. The results from a recent modeling effort directed at predicting how carbon cycling in the coastal ocean will be affected by climate change, nutrient loading, and increased P_{CO_2} is shown in Figure 25.19. After 2000, carbonate ion concentrations are predicted to have declined sharply, leading to a reduction in the saturation state of the $CaCO_3$ minerals and eventually the termination of biocalcification. A similar effect should be occurring in the sediments. Increased production of POM supported by nutrient loading should be increasing the sedimentation of POM. The increased input of POM to the sediments should be causing a decline in the porewater saturation of the $CaCO_3$ minerals, leading to increased dissolution rates.

Based on the results of this modeling effort, the global coastal ocean appears to have acted as a net source of CO_2 to the atmosphere over the past 300 y due to the combined production of CO_2 from biocalcification and an overall state of net heterotrophy. This role is expected to switch to that of a net CO_2 sink in response to rising atmospheric CO_2 levels and increasing nutrient runoff from land. The latter is stimulating new production and causing an overall state of net autotrophy. Recent observational

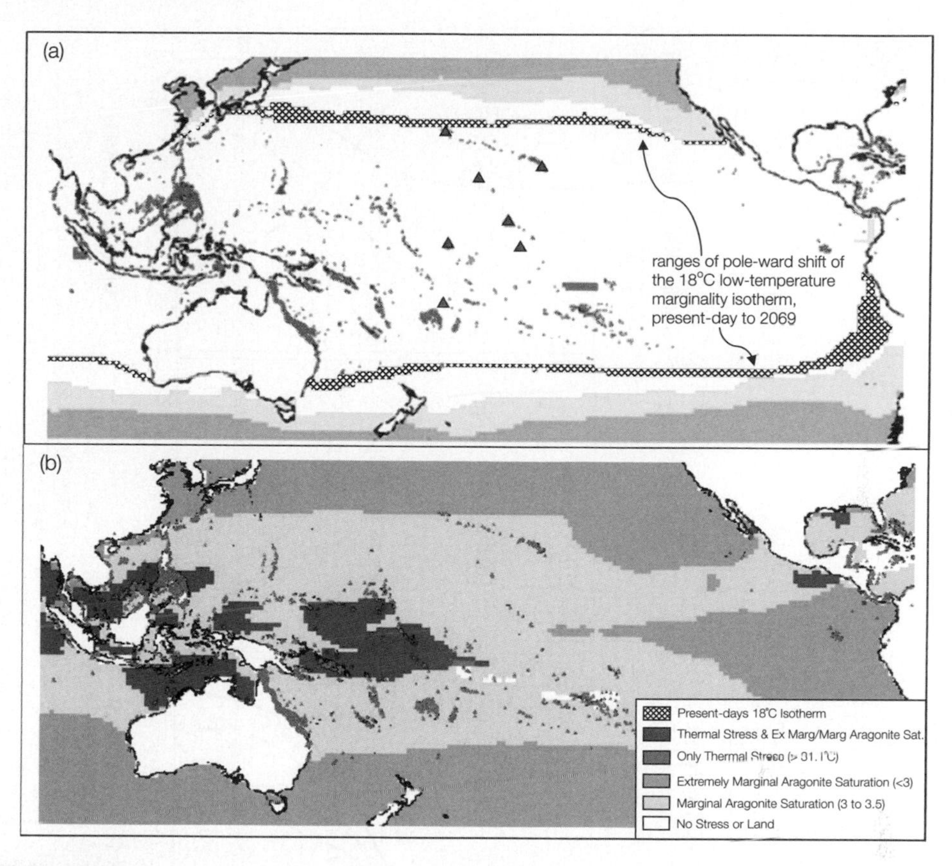

FIGURE 25.18

Global distribution of warm-water coral reef habitat and causes for loss of this habitat. Dots mark documented reef locations. (a) Preindustrial sea-surface temperature and Ω_{arag} marginal ranges, showing locations of U.S. National Wildlife Refuges and Marine Sanctuaries (triangles) and range of poleward shift of the 18°C low temperature marginality isotherm (hatched area) from present to 2069. (b) Projected locations of temperature and Ω_{arag} marginal ranges, 2060 to 2069; P_{CO_2} = 517 ppmv. Ω_{arag} = aragonite saturation state. See Eq. 15.8 for a definition of Ω. *Source*: After Guinotte, J. M., *et al.* (2003). *Coral Reefs* 22, 551–558. (See companion website for color version.)

data suggest that the global coastal ocean, including estuaries, is currently a small net source of CO_2 (0.1 Pg C/y). Without the estuaries, the coastal region is a stronger sink (0.3 Pg C/y) demonstrating that the nearshore systems (estuaries, mangroves, marshes, coral reefs, and upwelling area) are significant sources despite their relatively small area. These regions are particularly at risk for change due to anthropogenic impacts. Any future alterations in their function have the potential to significantly affect the role that the shallow coastal oceans will play in responding to anthropogenic CO_2 emissions.

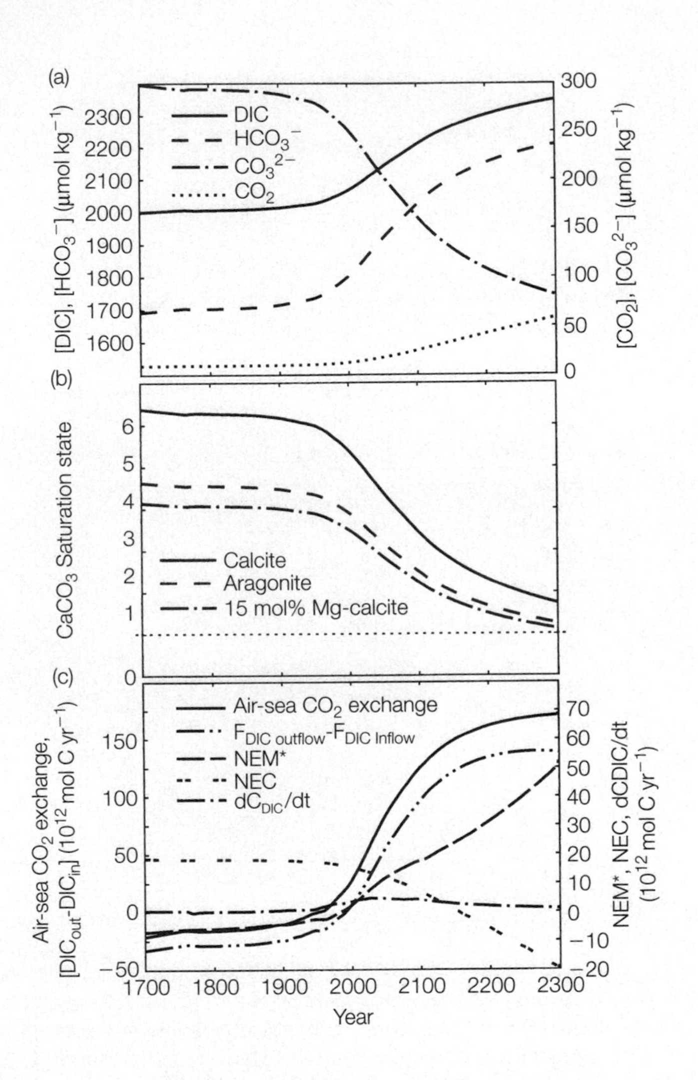

FIGURE 25.19

Surface water dissolved inorganic carbon chemistry between 1700 and 2300 AD predicted at 25°C and 35 psu. (a) Inorganic carbon species concentrations, (b) saturation state of calcite, aragonite, and 15 mol% magnesian calcite, and (c) air-sea CO_2 exchange. NEM = net ecosystem metabolism (gross primary production – respiration), NEC = net ecosystem calcification, dC_{DIC}/dt = atmospheric CO_2 concentration forcing, and $F_{DIC_{outflow}} - F_{DIC_{inflow}}$ = (export to the open ocean – input from rivers and upwelling). *Source*: From Andersson, A. J., *et al.* (2006). *Global Biogeochemical Cycles* 20, GB1S92.

25.5.5 Climate Feedbacks and Tipping Points

Much of the concern over continued global climate change rests on the potential for positive feedbacks to accelerate ongoing increases in temperature, leading to: (1) faster rates of rise in sea level, (2) faster transport rates in the global hydrological cycle, and (3) faster changes to the terrestrial and marine biospheres. The crustal-ocean-atmosphere factory also appears to have some negative feedbacks, but if these require very long time scales, the undesirable effects of climate change may be too far gone to be retrievable, resulting in species extinctions. The planetary climate feedbacks can be classified into four categories: (1) albedo effects, (2) shifts in the hydrological cycle including atmospheric water content and ocean circulation, (3) changes in the biosphere that affect carbon retention and GHG emissions, and (4) release of methane stored in methane hydrates, wetlands, etc. Some of the positive feedbacks are likely governed by thresholds above which rapid change proceeds. Synergy between positive feedbacks is also likely, suggesting that their cumulative effects will follow a nonlinear enhanced rate of change.

Albedo Changes

Albedo changes are a consequence of climate feedbacks that alter the formation and loss of clouds, aerosols, snow, and ice. Clouds are the biggest uncertainty in climate modeling as their net effect, warming versus cooling, is dependent on their altitude, size, and composition. Cooling is a result of insolation reflection. Heating is a consequence of water's GHG properties. An important control in cloud formation is the presence of cloud condensation nuclei (CCN). Various types of atmospheric aerosols can act as CCN, including anthropogenic dust, volcanic ash, sulfate particles from volcanic gases and planktonic DMS, sea salt, aerosols from forest fires, and factory smoke. Some of the aerosols, such as the black carbon (soot) produced by biomass and fossil-fuel burning, act like GHGs, absorbing incoming solar radiation and reradiating it as heat. In some regions, such as China and India where black carbon emissions are high, the resulting heating is leading to local changes in atmospheric circulation and rainfall patterns. But on a global scale, the net effect of aerosols is cooling. As shown in Figure 25.8a, significant interruptions in the global warming trend occurred immediately following major volcanic eruptions. In a sad irony, reduction of air pollution could serve to enhance global warming due to the removal of reflective aerosols.

Changes in the Hydrological Cycle and Ocean Circulation

The warming climate is likely to induce changes in the hydrological cycle that will lead to further climate change. Increased heating should increase the rate of evaporation and, hence, the amount of water vapor, which is a GHG. The IPCC's *Fourth Assessment Report*, published in 2007, finds that "the average atmospheric water vapor content has increased since at least the 1980s over land and ocean as well as in the upper troposphere."

High atmospheric P_{CO_2} leads to faster evapotranspiration rates, providing another mechanism by which water vapor levels increase. This is singularly important as water vapor changes are now recognized by the IPCC as the largest feedback affecting climate sensitivity.

Other important feedbacks in the hydrological cycle include a possible shutdown or slowdown in meridional overturning circulation due to reduced formation of NADW.

This is a matter of considerable controversy given the current lack of consensus on the importance of winds and tidally driven internal waves in driving meridional overturning circulation. Nevertheless, evidence for changes in circulation have been construed from a freshening of low-latitude surface seawater in the Atlantic Ocean and a slowdown in NADW formation between 1998 and 2004.

Cyclical phenomena, such as ENSO events, the North Atlantic Oscillation, the North Pacific Oscillation, and a phenomenon in the Southern Ocean called the "Southern Annular Mode," have important impacts on the solubility pump. This was illustrated for the ENSO events in Figure 25.5b. These oscillations are all sensitive to global climate change. In the case of the Southern Ocean, the impact of climate change on circulation is complicated by the zonation associated with the polar and subpolar frontal boundaries (Figure 10.10). As a result, some parts of the Southern Ocean are expected to respond differently than others to changes in P_{CO_2} increases and meridional overturning circulation rates.

Finally, the volume of ocean water is increasing due to the loss of continental and sea ice as well as thermal expansion. Approximately one fourth of the resulting sea level rise has been due to thermal expansion. Because liquid water has a higher heat capacity than ice, an increase in the amount of liquid water in the ocean should increase its heat storage capacity. Efficient storage of this heat in the ocean requires that it be mixed below the surface layer, and such mixing takes time, especially since global warming will likely cause a slowdown in the rate of meridional overturning circulation.

Changes in Biospheric Carbon Processing

Changes in climate and meridional overturning circulation are likely to affect plankton species distributions. As discussed in Chapter 25.5.3, such changes have the potential to alter the biological carbon pumps. The associated feedbacks are also likely to be affected due to synergistic interactions with other pollution problems including nutrient loading.

Of particular concern are the impacts of seawater acidification on biocalcification and the burial rates of sedimentary carbon. Carbonate ion concentrations in the surface waters have already declined by 16%. Thus, it is not surprising that the abundance of tropical/subtropic planktonic foraminiferan species appears to have declined since the 1960s. This information was obtained by studying the rapidly accumulating sediments of the Santa Barbara Basins off the coast of California.

Changes in climate and the hydrological cycle are likely to affect terrestrial flora, most importantly forests, as these are the putative carbon sink responsible for about 20% of the uptake of anthropogenic carbon. Loss of these forests could reduce or eliminate this important sink. The expansion of insects and pathogens as a result of global warming is another positive feedback that has the potential to impair this carbon sink. Furthermore, as sea level rises, wetlands and mangrove habitat will shift inland at rates that may be too fast to accommodate their biotic adaptation. This could alter the role of the nearshore systems in carbon storage.

Changes in Emission Rates of Natural GHGs

Although atmospheric methane concentrations appear to have stabilized over the past few decades, melting of gas hydrates in permafrost and shallow marine sediments have the potential to rapidly release large quantities of this potent greenhouse gas. As noted in the IPCC's Fourth Assessment Report, "Temperatures at the top of the permafrost layer have generally increased since the 1980s in the Arctic (by up to 3°C). The maximum area covered by seasonally frozen ground has decreased by about 7% in the Northern Hemisphere since 1900, with a decrease in spring of up to 15%."

25.5.6 **Climatic Predictions for the Next Century**

Earth is now absorbing $0.85 \pm 0.15\,W/m^2$ more energy from the Sun than it is emitting to space. The ocean has absorbed 84% of the total heating experienced since the 1950s, thereby moderating the effect of global warming on atmospheric temperatures. This heat retention has led to a detectable warming of the surface ocean, providing some of the most convincing evidence for global warming. As shown in Figure 25.20, the rate and depth of penetration of warming has been highest where NADW, NPIW, and AAIW are formed. This illustrates the important role of advection in driving heat uptake. Ocean-to-ocean differences are occurring. For example, the Indian Ocean is not absorbing as much heat as the Atlantic, probably due to solar dimming from the high aerosol production on the Indian subcontinent.

Paleoclimate information retrieved from marine sediments has been used to reconstruct sea surface temperatures in the western Equatorial Pacific over the past 13,500 y as shown in Figure 25.21. This evidence documents that Earth's temperature has now climbed to a level that has not been seen since 12,000 years ago and is within 1°C of the maximum reached over the past million years. During the past 30 years, the mean surface temperature at this location has been warming at a rate of 0.2°C per decade. This trend is found in most of the other oceans, suggesting that global atmospheric temperatures are now at or near the highest level of the Holocene.

As temperatures are expected to continue to increase, we will soon reach levels last seen 3 million years ago during an interglacial in which sea level was approximately

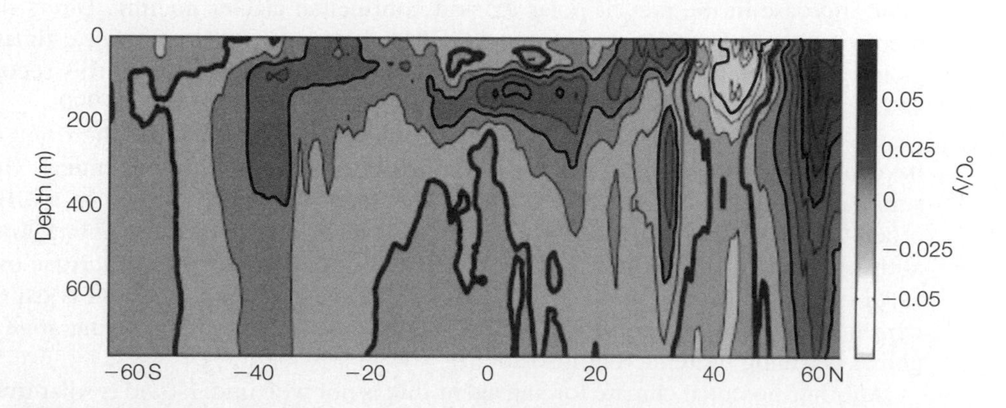

FIGURE 25.20

Zonally averaged, 10-y temperature trend in °C/y for 1993 to 2003 calculated using a least squares fit from in situ data. *Source*: After Willis, J. K., *et al.* (2004). *Journal of Geophysical Research* 109, C12036. (See companion website for color version.)

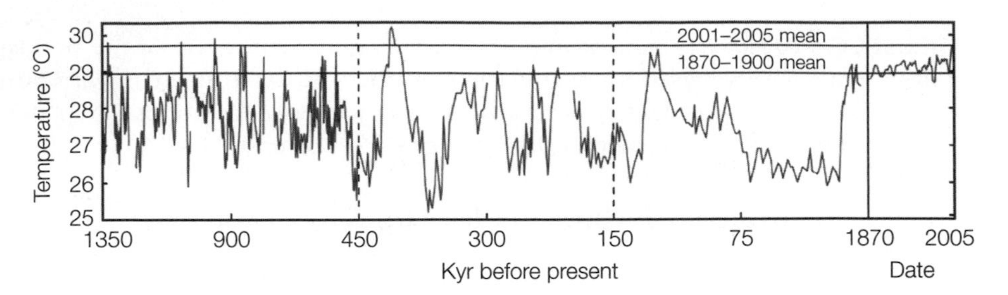

FIGURE 25.21

Modern sea surface temperatures in the Western Equatorial Pacific (0.3°N, 159.4°E) as compared with paleoclimate proxy data for the past 1.35 million years. Modern data are the 5-y running mean, while the paleoclimate data have a resolution on the order of 1000 y. *Source*: From Hansen, J., M., *et al.* (2006). *Proceedings of the National Academy of Science of the United States of America* 103, 14288–14293.

25 m higher than the present day. In support of this, the globally averaged rate of sea level rise also appears to have increased since the 1950s. The 130-y record from 1870 to 2000 yields a rate of rising sea level equal to 1.44 mm/y. In 1900 to 2000, this rate increased to 1.7 ± 0.3 mm/y, yielding an acceleration rate of 0.013 ± 0.006 mm/y. This rate of rise was partially slowed by solar dimming caused by volcanic aerosols ejected by the 1963 eruption of Mt. Agung, the 1982 eruption of El Chichon, and the 1991 eruption of Mt. Pinatubo. Figure 25.8a illustrated a similar slowdown in the rate of atmospheric temperature rise.

Other lines of evidence that document the effects of global climate change are presented in Table 25.5. Over the past 30 y, climate zones have been moving poleward at a rate of 40 km per decade. Recent large-scale surveys of global flora and fauna have found habitats shifting poleward at a rate of 6 km per decade. Particularly alarming has been a rapid increase in the rate of polar ice and continental glacier melting. Direct measurement of tropospheric temperatures extend back to the 1880s enabling estimates of a global mean temperature (Figure 25.8a). The eight warmest years in this record have all occurred since 1998, with the 14 warmest having occurred since 1990.

Computer models of global climate that include ocean and biosphere interactions have been used to identify the most important drivers of global warming. These are summarized in Figure 25.22, reinforcing the important role of CO_2 as a GHG. The other GHGs, ozone, CH_4, N_2O, and the CFCs, presently contribute about 20% of the anthropogenic climate forcing. Their concentrations are expected to increase over time as all but the CFCs are by-products of fossil-fuel combustion whose use is expected to continue to increase. (Recall that anthropogenic aerosols are countering some of this global warming by reflecting insolation.)

Another potential climate forcing agent that is not well understood is solar irradiance. (This was discussed at the end of Chapter 25.4.4.) Direct measurements have been underway only since 1978 and provide no evidence for any upward trend over the past 25 or so years, although the existence of the 11-y sunspot cycle is apparent. Variations in

Table 25.5 Evidence of Climate Change.

Ocean Physics

- Beginning in the 1960s, average temperatures of surface waters in all of the world's major oceans began to rise, caused by penetration of anthropogenic heat.
- Slowing of meridional overturning circulation in the North Atlantic from 1998 to 2004. Gulf Stream was deflected southward. Deep southerly flows declined by 50%.

Ocean Chemistry

- Surface salinity decreasing at high latitudes and increasing at low latitudes caused by melting of Arctic sea ice and Greenland Ice Sheet.
- pH has declined and alkalinity risen. Carbonate ion concentrations in surface ocean have been reduced by 16%.
- Ocean has switched from being a small net carbon source to a large sink.
- Changes in nutrient supply rates.[a]

Ocean Biology

- Coral bleaching and decrease in spatial extent of optimal coral habitat.
- Increase in frequency of harmful algal blooms such as *Phaeocystis* in Baltic Sea.
- "Blooms" of giant jellyfish off Japan and China. Thought to be caused in part by shifts in coastal currents associated with changes in river flows.
- Widespread ecosystem changes throughout the North Pacific representing a plankton regime shift.
- Warming of the California Current leading to a change in planktonic foraminiferan species.

Atmosphere

- Continuing temperature increase in troposphere.
- Possible increase in intensity and frequency of hurricanes in Pacific and Indian Oceans.

Land

- Increased frequency of extreme weather events, such as flooding and droughts.
- Increased wildfires caused by drying of soil.
- Increased disaster losses from lightening strikes.
- Amazon River flows are declining.
- Permafrost in subpolar latitudes is melting.
- Increased rate of river runoff since beginning of the century. Caused by changes in precipitation patterns, haze, vegetation cover, and decreased evapotranspiration. When P_{CO_2} is high, terrestrial plants are more efficient in their use of soil moisture, so their stomata do not open as much or for as long. This causes a drop in evapotranspiration rates.
- Changes in geographic distribution of species including migration of treelines.
- Changes in timing of biological processes, such as seasonal migrations and reproduction cycles.

Cryosphere

- Loss of sea ice with rate of melting increasing over time. Within this century, the Arctic Ocean is predicted to be ice free in the summer.
- Polar ice sheets are being lost faster than expected. A possible explanation is that ice sheets are hydroplaning on their melt waters.
- Retreat of alpine glaciers.

[a] *Climate-induced changes in nutrient supply rates are a consequence of alterations in the hydrological cycle and in meridional overturning circulation.*

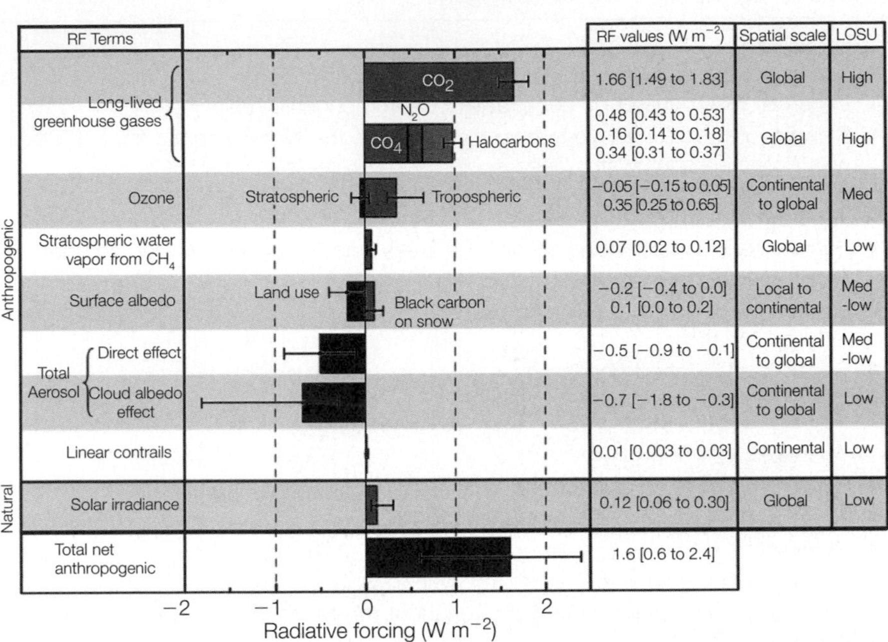

FIGURE 25.22

Global-average radiative forcing (RF) estimates and ranges in 2005 for anthropogenic carbon dioxide, methane, nitrous oxide, and other important agents and mechanisms, together with the typical geographical extent (spatial scale) of the forcing and the assessed level of scientific understanding (LOSU). The net anthropogenic radiative forcing and its range are also shown. Volcanic aerosols contribute an additional natural forcing, but are not included in this figure due to their episodic nature. The range for linear contrails does not include other possible effects of aviation on cloudiness. *Source*: After IPCC Working Group I (2007). *Climate Change 2007: The Physical Science Basis, Contribution of Working Group I to the Fourth Assessment Report of the IPCC*. Cambridge University Press, http://www.ipcc.ch/ipccreports/ar4-wg1.htm.

solar radiation incident at Earth's surface (insolation) indicate a decline from 1960 until 1990, corresponding to a global dimming of 4 to 6%. Part of the dimming is likely related to a period of lowered sunspot activity (1960 to 1975). But since the 1990s, widespread brightening has occurred. This is thought to be at least partly due to the clearing of aerosols generated by the 1991 eruption of Mt. Pinatubo and to the implementation of stricter air quality regulations in Europe and North America.

Computer models have been developed to predict the effects of the forcing functions shown in Figure 25.22 on future climate change and sea level rise under a variety of "what-if" scenarios. A recent set of projections based on several likely GHG emission scenarios is presented in Figure 25.23. In 1990, the IPCC presented a first set of projections, estimating that between 1990 and 2005, global average temperatures would

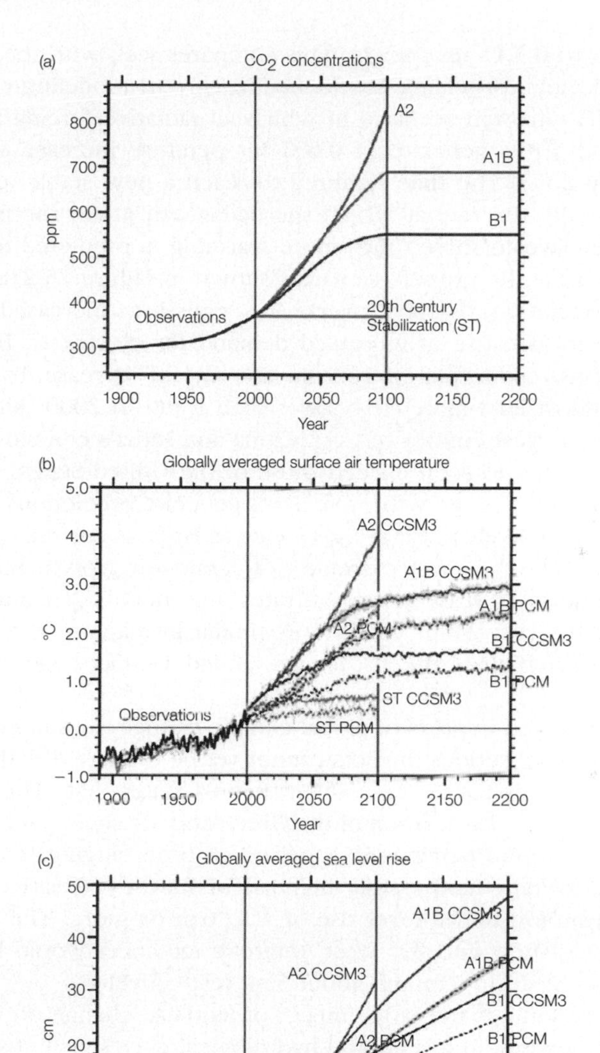

FIGURE 25.23

Time series predictions for various greenhouse gas emission scenarios (A2, A1B, B1, and 20th century stabilization) and actual observations: (a) CO_2 concentrations; (b) globally averaged surface air temperatures; and (c) sea level rise from thermal expansion only. A much larger sea level rise will result if continental ice sheets are lost. In the case of (b) and (c), several models were used to obtain the scenario estimates, i.e., PCM, (Parallel Climate Model) and CCSM3 (Community Climate System Model version 3). *Source*: After Meehl, G. A., *et al.* (2005). *Science* 307, 1769–1772.

rise by 0.15 to 0.3°C per decade. This compares well with the observed value of about 0.2°C per decade, providing confidence in current modeling efforts.

The GHG emission scenario in which all radiative forcing agents are held constant at year 2000 levels generates a 0.6°C temperature increase and a sea level rise of 5 to 10 cm by 2050. The time required to reach a new stable temperature and sea level is controlled by the rate at which the ocean can attain thermal equilibrium with the atmosphere. Two to three times more warming is predicted to occur by 2050 if GHG emissions follow the growth scenarios shown in Figure 25.23. This is probably a more realistic expectation. In 2005, overall U.S. emissions increased 0.8% from the previous year, primarily because of increased demand for electricity. Between 1990 and 2005, U.S. emissions rose a total of 16%. At this rate of increase, by 2020, the United States will emit almost 20% more GHG gases than it did in 2000. Although the United States has been the largest emitter of CO_2, China and India's emissions are growing at a rapid rate and expected to soon exceed those of the United States.

The GHG emission growth projections generate predictions that the resulting climate change will very likely to be at least twice as large as any change expected from natural variability. Under the most extreme CO_2 emission growth scenario, where emissions rates continue to increase at current rates, the models generate a 2 to 6°C temperature increase by the year 2400. This shift is similar in magnitude to the amount of warming that has occurred since the last ice age ended 18,000 y ago, but will take place much faster.

Other expected impacts of global climate change are summarized in Table 25.6. The severity of several serious impacts cannot yet be estimated with much confidence. Most notably, the IPCC's Fourth Assessment Report states that "There is medium confidence that at least partial deglaciation of the Greenland ice sheet, and possibly the West Antarctic ice sheet, would occur over a period of time ranging from centuries to millennia for a global average temperature increase of 1 to 4°C (relative to 1990 to 2000), causing a contribution to sea level rise of 4 to 6 m or more. The complete melting of the Greenland ice sheet and the West Antarctic ice sheet would lead to a contribution to sea-level rise of up to 7 m and about 5 m, respectively."

Also worrisome is the large impact of land-use change on the complex interactions that link the carbon, nutrient, and hydrological cycles. Scientists estimate that between one third to one half of our planet's land surfaces have been transformed by human development. Our ability to predict the long-term effects of these land-use changes is still rudimentary. For example, in 2006, researchers discovered that terrestrial plants are likely a significant source of methane. Furthermore, scientists are still not able to identify what plants and land regions are responsible for the terrestrial storage of carbon inferred from the mass balance presented in Table 25.4.

25.6 USING THE OCEAN TO SOLVE THE CO_2 PROBLEM

Climate change projections suggest that we are already committed to a temperature increase of 0.6°C just by holding GHG emissions constant at the rates that existed during

Table 25.6 Key impacts of changing climate as a function of increasing global average temperature change. The solid black lines link impacts and dotted arrows indicate impacts continuing with increasing temperature. Entries are placed so that the left-hand side of the text indicates the approximate onset of a given impact. Quantitative entries for water stress and flooding represent the additional impacts of climate change relative to the conditions projected across the range of emission scenarios shown in Figure 25.23. Adaptation to climate change is not included in these estimations. Confidence levels for all statements are high.

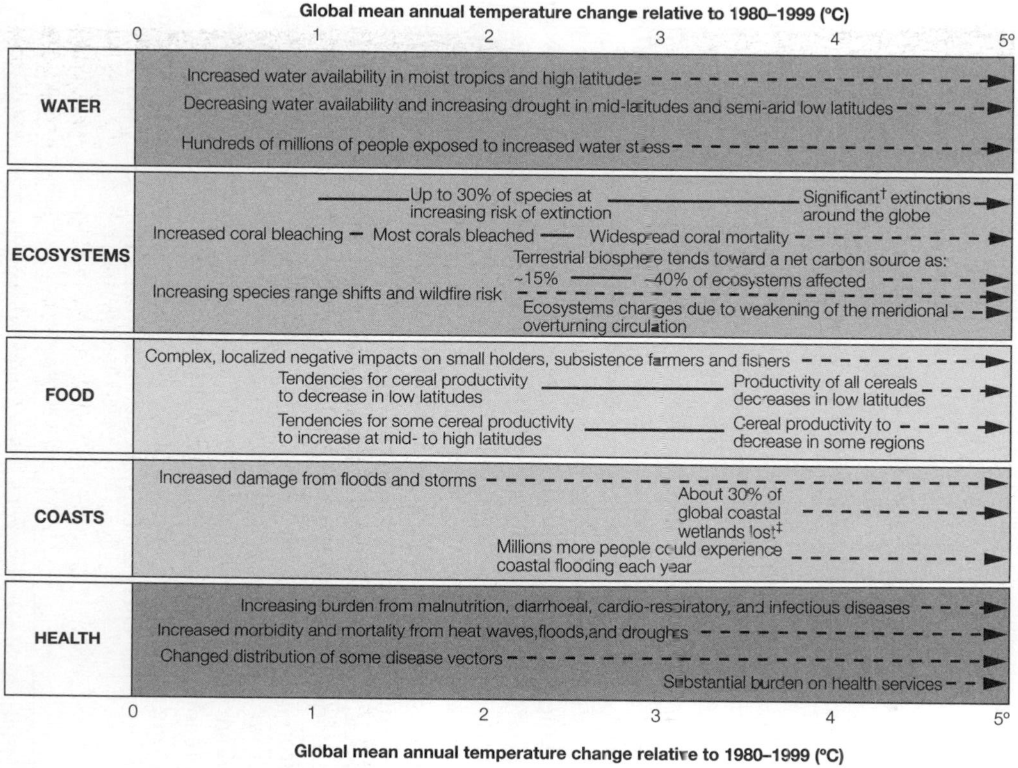

Global mean annual temperature change relative to 1980–1999 (°C): 0 – 1 – 2 – 3 – 4 – 5°

WATER	Increased water availability in moist tropics and high latitudes - - - →
	Decreasing water availability and increasing drought in mid-latitudes and semi-arid low latitudes - - - →
	Hundreds of millions of people exposed to increased water stress - - - →
ECOSYSTEMS	—— Up to 30% of species at increasing risk of extinction —— Significant[†] extinctions around the globe →
	Increased coral bleaching — Most corals bleached —— Widespread coral mortality - - - →
	Terrestrial biosphere tends toward a net carbon source as: ~15% —— ~40% of ecosystems affected - - - →
	Increasing species range shifts and wildfire risk - - - →
	Ecosystems changes due to weakening of the meridional overturning circulation - - →
FOOD	Complex, localized negative impacts on small holders, subsistence farmers and fishers - - - →
	Tendencies for cereal productivity to decrease in low latitudes —— Productivity of all cereals decreases in low latitudes - - →
	Tendencies for some cereal productivity to increase at mid- to high latitudes —— Cereal productivity to decrease in some regions - - →
COASTS	Increased damage from floods and storms - - - →
	About 30% of global coastal wetlands lost[‡] - - - →
	Millions more people could experience coastal flooding each year - - - →
HEALTH	Increasing burden from malnutrition, diarrhoeal, cardio-respiratory, and infectious diseases - - →
	Increased morbidity and mortality from heat waves, floods, and droughts - - - →
	Changed distribution of some disease vectors - - - →
	Substantial burden on health services - - →

Global mean annual temperature change relative to 1980–1999 (°C): 0 – 1 – 2 – 3 – 4 – 5°

[†] Significant is defined here as more than 40%.
[‡] Based on average rate of sea level rise of 4.2 mm/year from 2000 to 2080.

Source: From IPCC (2007). Climate Change 2007: Impacts, Adaptation and Vulnerability. Contribution of Working Group II to the Fourth Assessment Report of the Intergovernmental Panel on Climate Change, Cambridge University Press, pp. 7–22.

2000 AD. Higher temperature increases are likely given current growth in emission rates of most of the GHGs. Therefore, attention has focused on mitigating this rise by reducing GHG emissions. The ocean provides several strategies to enable this as illustrated in Figure 25.24.

Injection of compressed CO_2 into the deep ocean has already been tested. The goal of this approach is to emplace the CO_2 into waters with low temperatures, ensuring the formation of relatively immobile gas hydrates. This strategy has the potential to sequester thousands of gigatonnes of carbon, but likely environmental impacts include: (1) a change in the pH in the seawater near the emplaced gas hydrates, (2) benthic kills, (3) other ecosystem impacts, and (4) release back to the atmosphere as an eventual consequence of meridional overturning circulation.

Also in the testing phase is mineral carbonation in which CO_2 is reacted with a calcium or magnesium oxide to form the carbonate salt,

$$CO_2 + MgO \rightarrow MgCO_3 \quad (25.9)$$

$$CO_2 + CaO \rightarrow CaCO_3 \quad (25.10)$$

The salts would then be disposed of by emplacement in the seabed at depths above the CCD.

Another strategy that is already in use is storage in geological formations, both onshore and offshore. In the United States, the first offshore injection of CO_2 into the seabed begun in 2004 in the Gulf of Mexico at a rate of 177 tonnes of CO_2 per day. CO_2 is trapped in geological formations via: (1) reaction with minerals, (2) dissolution into liquids, and (3) physical retention by structural and stratigraphic barriers. Trapping in geological formations is expected to be able to sequester 2000 Gt CO_2. For any of these approaches to be economically realistic, the CO_2 source would have to be a large, stationary emitter (>0.1Mt CO_2 per year). Potential sources include power plants, cement production facilities, petroleum refineries, and the iron and steel production industry. In 2005, power plants and industrial emissions of CO_2 in the United States amounted to 3.8 Gt.

Several other technologies are actively being explored as mechanisms for enhancing oceanic removal of anthropogenic CO_2 emissions. Research and development in this arena is a direct response to the development of several national and international carbon emission trading systems in which companies can profit from the sequestration of anthropogenic CO_2 emissions. On the international level, the carbon trading market stems from a provision of the Clean Development Mechanism of the Kyoto Protocol. Technologies under development rely on two strategies: (1) stimulating the growth of plankton and (2) speeding up chemical weathering of coastal volcanic rocks (Eq. 14.2) by spraying them with a strong acid, i.e., HCl generated electrochemically from seawater using solar energy.[5] This accelerated weathering generates HCO_3^- (Eq. 14.2). By

---

[5] House *et al.* (2008) *Environmental Science and Technology*, 42(24), 8464–8470.

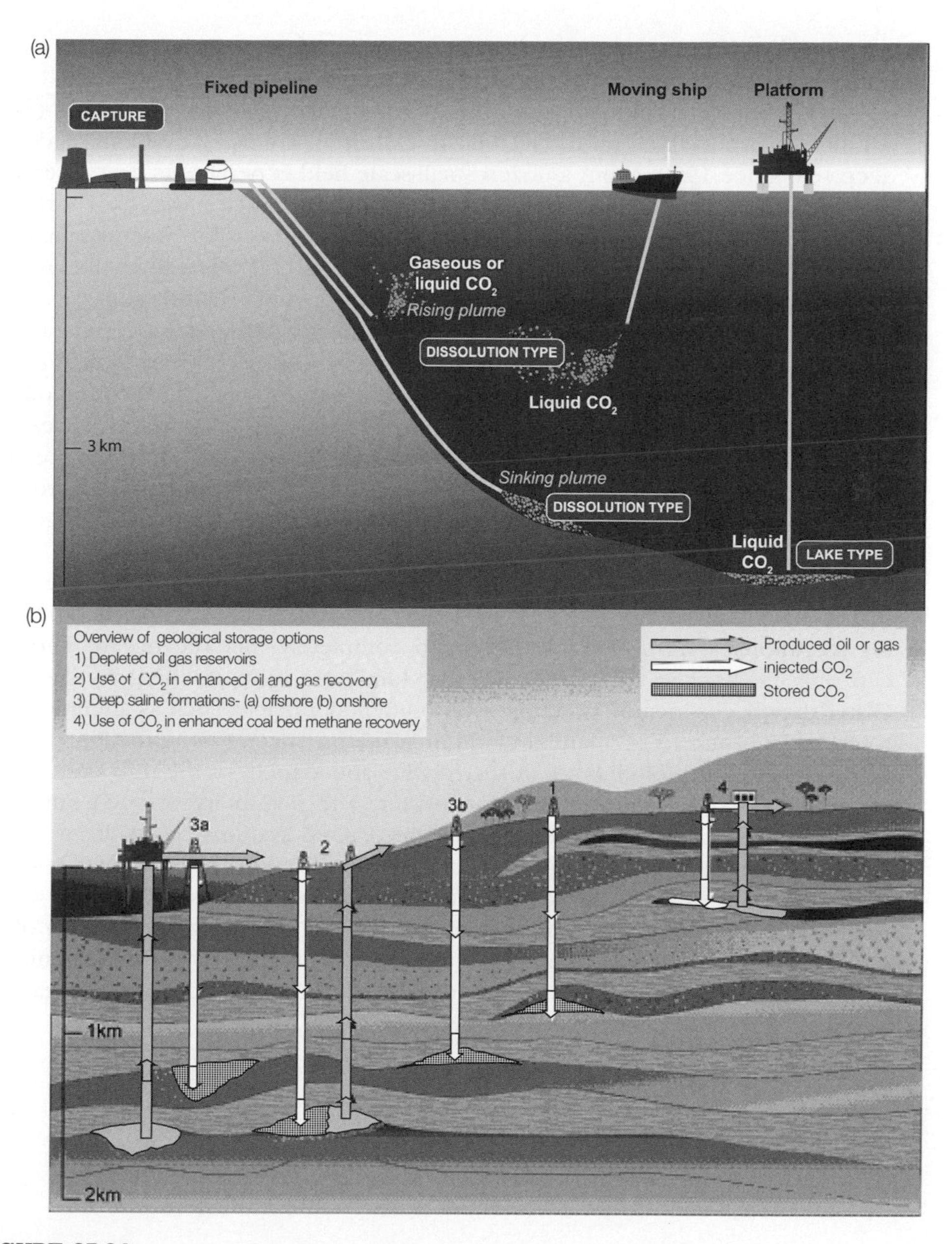

FIGURE 25.24

Options for CO_2 sequestration: (a) oceanic and (b) geological. *Source*: After Metz, B., *et al.* (2005). *IPCC Special Report on Carbon Dioxide Capture and Storage*. Cambridge University Press. (See companion website for color version.)

releasing this bicarbonate into seawater, a net increase in alkalinity would be achieved, thereby increasing the ocean's ability to take up atmospheric CO_2.

The fertilization approach to oceanic CO_2 uptake relies on the assumption that the additional carbon fixed into POC will be exported, via pelagic sedimentation, into the deep sea. Since 1993, about a dozen small-scale field experiments have been conducted in which iron has been intentionally dumped into the surface waters of various regions. While primary production was stimulated, little evidence has been obtained for POC export out of the surface waters. Nonetheless, several companies are attempting to commercialize this process (such as Climos, http://www.climos.com/) and, thus, sell carbon credits. Field trials are also underway to test whether dumping urea into surface waters will significantly stimulate phytoplankton growth (Ocean Nourishment Corp., http://www.oceannourishment.com/). Still another approach to fertilization uses flexible plastic tubes (10 m diameter and 100 to 1000 m long) as one-way valves that create upwelling using wave action to force nutrient-rich deepwater into the mixed layer. One company, Atmocean, Inc. (http://www.atmocean.com/), is using this approach to stimulate the growth of salps. As discussed in Chapters 23.4.4, these animals generate large fecal pellets that are more likely to export POC into the deep ocean.

A drawback to all of these POC-export-based strategies is that CO_2 remineralized in the deep sea will degas back into the atmosphere once enough time has passed for meridional overturning circulation to completely mix the ocean. Therefore, CO_2 can only be kept in the ocean over the long-term by continuing fertilization efforts. Other drawbacks include potential environmental impacts, such as alterations in ecosystem function due to permanent shifts in plankton species composition. The latter is a likely consequence of changes in the relative abundances of the nutrients and micronutrients, and to the acidification of seawater. The possibility of such impacts led the International Maritime Organization of the United Nations to issue an official statement of concern in 2007 regarding ocean iron fertilization (http://climos.com/imo/IMO/IMO_LC_LP_Statement_of_Concern_June2007.pdf). A similar statement of concern was issued in 2008 by the Scientific Committee on Oceanic Research (SCOR) of the International Council for Science and the Joint Group of Experts on the Scientific Aspects of Marine Environmental Protection (GESAMP) of the United Nations regarding ocean nutrient fertilization (http://www.scor-int.org/SCOR-GESAMP.pdf).

CHAPTER 26
The Origin of Petroleum in the Marine Environment

Please visit **http://elsevierdirect.com/companions/9780120885305** for the complete version of Chapter 26.

Much of the most valuable petroleum deposits were formed tens of millions of years ago in shallow highly productive marine settings. Our understanding of how petroleum formed and where to find it represents one of the most successful applications of paleoceanography. The most commonly used research tools are geochemical biomarkers and sedimentary stratigraphy. As described in Chapter 22.3.2, biomarkers provide information on organic matter sources and transformations. From this information, we know that the organic matter comprising petroleum is largely of biotic origin, with marine plankton being a major contributing source. The degree to which this organic matter became buried in marine sediments and transformed into petroleum was largely controlled by the physical characteristics of its depositional environment, such as water depth and circulation. Information on ancient marine depositional environments has been gleaned from sedimentary stratigraphy by studying grain size, mineralogy, microfossils, and stable isotope composition.

As shown in Figure 25.1, the fossil fuel reservoir of carbon is much smaller than that of limestone. This reservoir is composed of oil, natural gas, and coal.[1] Nevertheless, the rapid return of this carbon to the atmosphere as a result of fossil-fuel burning represents a significant perturbation to the crustal-ocean-atmosphere factory because it is causing atmospheric CO_2 levels to rise. Despite its negative effects on global climate, fossil fuels are the major source of energy fueling the world economy. Petroleum is also an important feedstock for the manufacture of petrochemicals used in the creation of synthetic rubber, synthetic fibers, drugs, detergents, and plastics. Components of petroleum also serve as lubricating oils and solvents. Because of these valuable uses, many have suggested that we are wasting a valuable feedstock resource by continuing to burn it for energy.

[1] Coal was formed in shallow aquatic environment and natural gas in both the marine and aquatic settings.

Considerable effort continues to be directed at determining how and where petroleum has formed. As the easily recoverable deposits of petroleum have been depleted, interest is being increasingly directed at finding and recovering petroleum from sediments located at great depth below the sea surface and sea floor.

Chapter 26, which is available online, describes how petroleum forms in the marine environment. This story requires knowledge of the timing and locations of past conditions under which organic matter production and preservation were favored. This knowledge, in turn, relies on an understanding of the role of marine organisms in the production and destruction of organic matter and how the crustal-ocean-atmosphere factory has controlled water circulation as the latter determines the rates of O_2 and nutrient delivery. Therefore, Chapter 26, in describing the formation of petroleum in the marine environment, provides a synthesis and practical application of most of the topics previously covered in this text.

CHAPTER

Organic Products from the Sea: Pharmaceuticals, Nutraceuticals, Food Additives, and Cosmoceuticals

27

Please visit **http://elsevierdirect.com/companions/9780120885305** for the complete version of Chapter 27.

Humans have long used the ocean as a source of food and minerals. The inorganic resources obtained from the sea are primarily salt, sand, and gravel. In coastal regions where oysters grow, their shells have been used to formulate a type of concrete called tabby. Diatomaceous earth is used as an insecticide, a filtration medium, and an abrasive. Although most of the organic materials harvested from the ocean serve as seafood, an increasing amount are being used for medicinal purposes, as food additives, in cosmetics, and as pesticides.

Since the 1970s, much effort has been directed at searching the ocean and its organisms for novel biomolecules. This interest was stimulated by the recognition that marine organisms are likely to have developed unique biosynthetic pathways to generate compounds that help them survive the environmental conditions found only in the oceans. Although many unique compounds have been identified, few new products have yet to be brought to market. Recent advances in biotechnology, such as genomics and bioinformatics, are now being used to overcome problems associated with the commercial development of new marine products. As a result, several marine biomolecules are now in clinical trials for use in treating cancer, infections, Alzheimer's disease, and asthma. Others are being tested for use as antifouling agents. These advances are supported by an interdisciplinary approach involving the expertise of marine biologists, chemical ecologists, synthetic organic chemists, and pharmacologists, along with experts in genomics, biotechnology, and mariculture.

In Chapter 27, which is available online, a description of the marine organic products in current use is provided along with a consideration of how the structures of these molecules confer predictable physiological activities. This is followed by a discussion of the strategic approaches used to discover and develop marine organic products along with the reasons why new products are so hard to bring to market. Finally, some examples of products now in the development pipeline are presented. These products are being targeted for use as biomedicines, nutraceuticals, cosmoceuticals, antifouling agents, pesticides, research probes, and biosensors, and in mariculture and environmental applications.[1] The latter include in situ pollutant degradation and removal of toxic heavy metals.

[1] A marine nutraceutical is defined as "a marine-derived substance that can be used as a dietary supplement or a food ingredient that provides a medicinal or health benefit beyond basic nutrition" [Barrow, C., and F. Shahidi (2008). Marine Nutraceuticals and Functional Foods. CRC Press, 494 pp.] Cosmoceuticals are cosmetic products with drug-like benefits conferred by ingredients such as vitamins, phytochemicals, enzymes, antioxidants, and essential oils.

PART 5

Marine Pollution

CHAPTER 28
Marine Pollution: The Oceans as a Waste Space

All figures are available on the companion website in color (if applicable).

28.1 INTRODUCTION

In the preceding chapters, we have discussed the ocean's pivotal role in the crustal-ocean-atmosphere factory. For example, the ocean serves as a receptacle for chemical flows originating from land. We have seen that the ocean's ability to either store these chemicals or bury them in the sediments is a crucial component of the global biogeochemical cycles that influence climate and, hence, the hydrological cycle and ocean circulation. These and other linkages support feedbacks that act on biological diversity and abundance, terrestrial erosion, and atmospheric composition.

The effect of humans on this complex web of interactions is a relatively new phenomenon—at least on geological time scales. Some scientists think our first effects commenced with fishing and land-use changes associated with our aboriginal origins as hunters, gatherers, and eventually farmers. Many scientists now view humans as the major agent of change on planet Earth. Through our activities, we have already transformed about half of the land surface, causing mass deforestation, loss of wetlands, and changes in riverine flows of water and sediments. We have altered the chemical composition of the atmosphere, most notably through additions of greenhouse gases in large enough amounts to augment global warming. Overfishing, probably the first large-scale effect that humans have had on the oceans, has severely reduced the abundance and diversity of the top trophic levels of the marine food web. This has led to alterations in the structure of the lower levels of the food web, which have in turn made the oceans less resilient to the impacts of pollutants and prone to overgrowth of algae and microbes.

In this chapter, we consider the different types of pollutants entering the ocean, their transport pathways, and their effects. Most pollutants in the ocean originate from land-based sources; the rest come from ocean dumping and shipping activities. Discharge from land-based sources is a direct consequence of the concentration of human population in coastal areas. Sixty percent of us now live within 100 km of the coastline. Most large cities (65% of those with more than 2.5 million people) are also located along the coast, primarily on estuaries. Pollutants from land-based sources are transported into the ocean via river flows, submarine groundwater discharge, and atmospheric processes.

The last include wet and dry deposition of particles and solutes and gas exchange across the air-sea interface. Because of proximity to source, coastal waters tend to be more polluted than the open ocean. A notable exception is the worldwide acidification of surface waters caused by CO_2 emissions. Of all the coastal waters, estuaries tend to be the most impacted. This is due to high rates of pollutant loading and to natural processes that act to concentrate these pollutants in the local sediments and biota. This is most unfortunate as estuarine waters support the world's largest fisheries and are where recreation is concentrated.

Land-use change in the coastal zone has accelerated the rate of pollutant loading for three reasons. First, removal of vegetation mobilizes materials, such as sediment. Second, the emplacement of impervious surfaces, such as roads and roofs, enhances pollutant transport as part of stormwater runoff. Third, loss of natural habitats, such as wetlands, eliminates important ecosystem services, such as pollutant uptake and degradation.

Water quality, habitat, and biotic assessments conducted periodically by the United Nations and in the United States by the USEPA and NOAA have repeatedly demonstrated significant and widespread environmental degradation throughout most of the world's oceans, with greatest impacts in the coastal regions. Some of this degradation is now having negative impacts on human health due to increasing exposure to pathogens and biotoxins and from consumption of tainted seafood.

Current efforts at understanding and managing marine pollution focus on the use of an ecosystem approach that recognizes the necessity of functional habitats in sustaining marine life and human health. This approach is complicated by the immutable nature of change in marine ecosystems arising from natural biological evolution and inevitable "catastrophes," such as various plate tectonic and astronomical events, like the eruption of massive flood basalts and meteorite impacts. To further complicate matters, an important part of our pollution loading is now biological in nature. These include: (1) impacts from invasive species, (2) pathogens whose habitat ranges are being expanded as a consequence of climate change, and (3) stimulation of biotoxin production by "natural" algae and bacteria as a consequence of nutrient and dust loading. Another alarming trend is the increasing load of solid pollution, primarily plastics, that is accumulating into giant floating rafts of litter. A mat of debris the size of Texas now exists in the Northern Pacific gyre and has been named the "Eastern Garbage Patch."

We are now at a crossroads in planetary history. For a long time, we have been able to use the ocean simultaneously as a resource, for food, minerals, recreation and transportation, and for waste disposal. Since the ocean is vast, the effects of these conflicting uses were not evident until fairly recently. But our inputs and their impacts have now grown so large that we are imperiling the ocean's ability to continue to supply us with food, recreational opportunities, and safe transportation. We are also altering parts of the crustal-ocean-atmosphere factory that regulate climate and, hence, ocean circulation. Climate change is likely to intensify some of the impacts of pollution on marine biota, most notably by creating abiotic stresses such as reduced O_2 solubility.

Notable progress has been made over the past three decades in reducing some forms of marine pollution, such as oil, radioactive substances, heavy metals (with the exception of mercury), and some of the persistent organic pollutants, such as DDT and

PCBs. The United States stopped discharging untreated sewage into the ocean in 1992, although we continue to emit treated effluents and polluted stormwater runoff. Unfortunately, this is not happening in developing countries where discharge of untreated sewage is the norm. All countries are emitting stormwater runoff containing high concentrations of nutrients acquired from the wash-off of excess fertilizers. Discharge of sewage and nutrient-enriched runoff, along with habitat loss, are now considered to be the largest pollution threats to the coastal oceans as they collectively result in a cascade of impacts including cultural eutrophication, hypoxia, and harmful algal blooms.

Coastal population is expected to double in the next 30 years, suggesting that pollutant loads and habitat loss will continue to increase unless we choose to develop and live in a dramatically different, and more sustainable way. The good news is that more and more effort is being directed at effecting such a change. As marine scientists, we have a moral responsibility to increase and apply our understanding of marine biogeochemistry to help this happen.

28.2 WHY HAS THE OCEAN BEEN USED FOR WASTE DISPOSAL?

Knowing what we know now, it is hard to understand why the ocean was ever, and still is, used as a waste-disposal site. The rationale has been the following—that the ocean is vast enough to accommodate our wastes without undergoing an unacceptable amount of change. This view has been based on the assumption that any potentially toxic wastes would either be degraded or diluted to innocuous levels with oceanic currents eventually transporting the residues far from our coastlines. Although this may have worked in the past, the assimilative capacity of the coastal ocean now appears to have been exceeded for many pollutants. This is probably a result of three effects: (1) a cumulative buildup of pollutants from continual loading over time, (2) an ever-increasing rate of input due to population growth coupled with increases in per-capita resource usage, and (3) attainment of threshold concentrations at which synergistic effects are now operational.

28.3 WHAT IS MARINE POLLUTION?

Progress in controlling marine pollution has been impeded by the difficulty of defining it. Two international advisory groups (the United Nations Group of Experts on the Scientific Aspects of Marine Pollution and the International Commission for Exploration of the Sea) have adopted the following definition:

> *"Pollution of the marine environment" means the introduction by man, directly or indirectly, of substances or energy into the marine environment, including estuaries, which results or is likely to result in such deleterious effects as harm to living*

resources and marine life, hazards to human health, hindrance to marine activities, including fishing and other legitimate uses of the sea, impairment of quality for use of sea water and reduction of amenities. (Article 1.4 in the United Nations Convention on the Law of the Sea (UNCLOS 1982))

A naturally occurring substance is classified as a *pollutant* if its concentration is "above the natural background level for the area and for the organism." These pollutants are termed *contaminants*. Examples include nutrients (nitrogen and phosphorus), trace metals, sediments (eroded soils and dredge spoils), and organic matter, such as from sewage. Other pollutants were never in the ocean until humans put them there. These are primarily organic compounds, such as polychlorinated biphenyls (PCBs), most pesticides, like DDT, and plastics. Not all pollutants are chemical; some are biotic, such as pathogens and invasive species, and some are physical, such as heat and noise. The most worrisome of the marine pollutants are listed in Table 28.1.

28.4 WHY POLLUTION IS HARD TO MEASURE

Another major factor impeding progress in the control of marine pollution is the great difficulty associated with measuring pollutant levels and impacts. First, our direct observational data of chemical conditions in the ocean is largely limited to the past three decades due to the relatively recent development of the necessary analytical techniques and commitment to long-term monitoring. The atmospheric CO_2 record shown in Figure 25.4 is one of the longest, extending back only into the late 1950s. Variability on short time and spatial scales makes dense sampling a requisite for obtaining representative measures of pollutant concentrations. Satellite and in situ sensors are now being used to overcome the high costs of dense sampling, but these innovations are recent. For example, the SeaWIFS satellite was launched in 1990s. As a result of our short observational database, the "natural" levels of most contaminants are unknown, preventing a determination of what constitutes an elevated concentration. A similar problem exists in the biological and physical realms; we have limited knowledge of what the ocean was like prior to human impacts. This is why so much effort is now being directed at developing paleoceanographic tools for reconstruction of past biotic and abiotic conditions.

A second reason why marine pollution is difficult to measure lies in the large number of pollutants being introduced and the myriad of pathways by which they enter the oceans. In 2006, 82,000 different chemicals were in commercial use, with approximately 1000 new ones being added annually. Most have the potential to reach the ocean. In some cases, groups of these compounds can be measured in one analysis, such as the PCBs (polychlorinated biphenyls) and PAHs (polycyclic aromatic hydrocarbons). Nevertheless, a large number of analyses are required to measure all the pollutants that can potentially be present in seawater.

The major pathways of pollutant input to the ocean are listed in Table 28.2. These can be classified into two categories: emissions from discrete sources, called

Table 28.1 Forms of Marine Pollution.

Forms	Sources	Effects
Contaminants		
Biostimulants (organic wastes, nutrients, dust)	Sewage, industrial and mariculture wastes; runoff from farms and urban areas; airborne nitrogen gases from combustion of fossil fuels and volatization of ammonia from livestock operations; dust from devegetated lands affected by drought	Cultural eutrophication and hypoxia. Changes in plankton community structure including enhancement of nitrogen fixation and harmful algal blooms.
Petroleum hydrocarbons (oil and combustion by-products such as PAHs)	Runoff and atmospheric deposition from land activities; shipping and tanker operations; accidental spills, coastal and offshore oil and gas production; natural seepage	Toxic effects including birth defects, cancer, and systemic poisoning. Tar balls degrade beach habitat.
Sediments	Erosion from farming, forestry, mining, and development; river diversions; coastal dredging and mining	Reduces water clarity and changes bottom habitats; clogs fish gills; carries toxins and nutrients.
Trace metals (arsenic, cadmium, chromium, copper, nickel, lead, mercury, zinc)	Industrial and municipal wastewaters; runoff from urban areas and landfill; erosion of contaminated soils and sediments; atmospheric deposition	Toxic effects including birth defects, reproductive failure, cancer, and systemic poisoning.
Chlorinated organic compounds (dioxins, other halocarbons)	Combustion of municipal wastes, paper processing, cleaning solvents	Toxic effects including birth defects, reproductive failure, cancer, and systemic poisoning.
Greenhouse gases	Burning of fossil fuels	Declining pH of surface waters from CO_2 uptake. Change in climate leading to change in water temperatures and circulation.
Physical Pollutants		
Thermal	Cooling water from power plants and industry	Lethal to some sedentary species, displaces others.

(Continued)

Table 28.1 *(Continued)*

Forms	Sources	Effects
Noise	Vessel propulsion, sonar, seismic prospecting, low-frequency sound use in defense and research	May disturb marine mammals and other organisms that use sound for communication.
Biological Pollutants		
Harmful algal blooms	Stimulation of the growth of algae and bacteria that synthesize biotoxins	Biotoxins are generally neurotoxic.
Invasive species	Ships and ballast water, fishery stocking, aquarists	Displaces native species, introduces new diseases.
Pathogens from humans and domesticated animals	Sewage, urban runoff, livestock, pets	Introduces new diseases. Contaminates swimming waters and shellfish.
Artificial Pollutants		
Organic compounds (chlorinated pesticides, phosphorus-based pesticides, freons)	Industrial and municipal wastewaters; runoff from urban areas and landfill; erosion of contaminated soils and sediments; atmospheric deposition	Toxic effects including birth defects, reproductive failure, cancer, and systemic poisoning.
Organometallic compounds (tributyltin, tetraethyl lead)	Anti-fouling paints, gasoline additive	Toxic effects including birth defects, reproductive failure, cancer, and systemic poisoning.
Plastics and other debris	Runoff from urban areas and landfill; dumping and loss from cargo, military, and cruise ships; loss of fishing nets	Entangles marine life or is ingested; degrades beaches, wetlands, and nearshore habitats.
Radioactive substances	Atmospheric fallout from bomb testing; loss from sunken submarines; emissions in cooling waters of nuclear power plants	Bioaccumulation into seafood poses health risk to humans.

Source: *After Boesch, D. F., et al. (2001). Pew Oceans Commission, p. 2.*

Table 28.2 Pathways of Pollutant Entry into the Ocean.

Land-based
Direct outfalls into coastal waters (pipes)
Flows of contaminated river water
Flows of contaminated groundwater
Sheet flow of polluted stormwater runoff
Deposition of aeolian particles and gas dissolution
Ocean-based
Offshore dumping of dredge spoils, sewage sludge and industrial wastes
Offshore industrial activities and accidents
Mariculture
Bilge pumping
Deposition of aeolian particles and gas dissolution

point-source inputs, and emissions that are widely dispersed, called *nonpoint-source inputs*. Point sources include pipes and smoke stacks that discharge emissions from industrial, power-generating, and sewage treatment facilities. They are more easily monitored and regulated than the nonpoint source inputs. Most of the latter stem from decentralized activities that deposit pollutants onto the land surface. Rainfall mobilizes these pollutants, causing them to either infiltrate into groundwaters or wash-off the land as polluted runoff. The polluted overland flows eventually enter the river systems that terminate in the coastal oceans. Atmospheric processes provide another nonpoint source transport route as they include the deposition of widely dispersed particles and solutes and the uptake of gases across the air-sea interface.

Eighty percent of marine pollution has a land-based source. The practices responsible for pollution at sea include intentional dumping, primarily of dredge spoils. Most of this material is sediment removed periodically from harbors. Approximately two thirds of the dredge spoils are deposited in estuaries, with the rest split between the coastal and open ocean. Since river-borne pollutants tend to become trapped in estuarine sediments, dredging can remobilize them and create toxic conditions in the water column both at the dredging and spoil deposition sites. Abandoned fishing nets, cargo blown or washed overboard, and garbage dumping all contribute to an enormous load of debris, mostly plastics, that is released from ships into the ocean. Other significant sea-based sources of pollution include discharges from ships, including leaching of antifouling paints and air pollutants, and similar discharges from oil-drilling platforms.

Assessing the impact of a pollutant is difficult because its effects can be complex and take a long time to develop. This is the result of the multitude of pathways that a pollutant can move through once it has entered the marine environment, as illustrated in Figure 28.1. With so many pollutants present, synergistic effects often occur in which the pollutants interact such that their combined impact are greater than a simple addition of their individual effects.

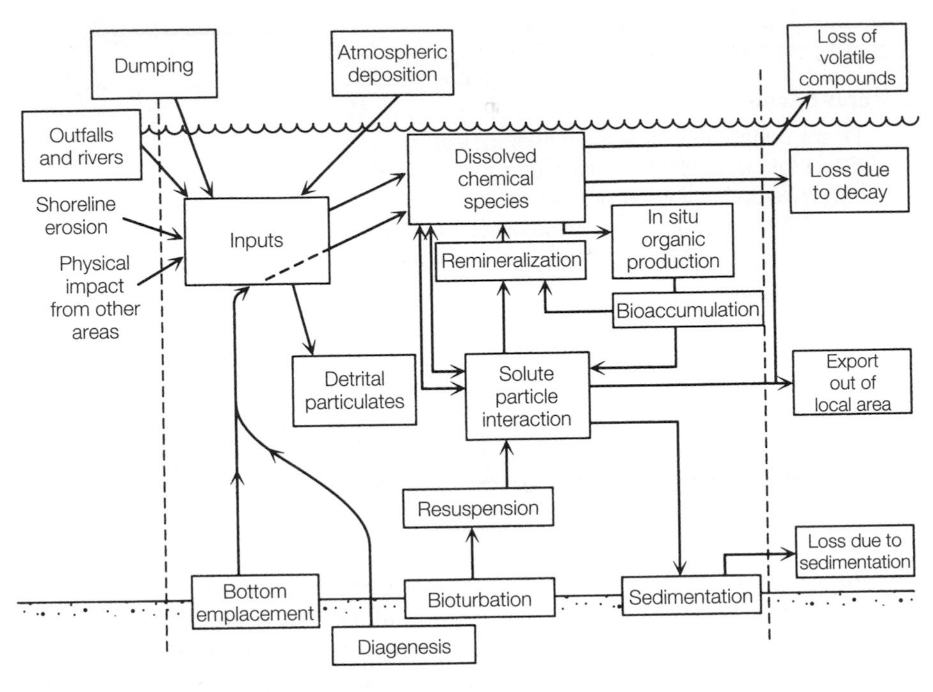

FIGURE 28.1

Important biogeochemical processes affecting the fate of chemical pollutants in marine ecosystems. *Source*: From Farrington, J. W., and J. Westall (1986). *The Role of Oceans as a Waste Disposal Option*, NATO ASI Series No. 172. D. Reidel Publ., pp. 361–425.

Marine organisms tend to concentrate pollutants in their tissues by a process termed *bioaccumulation*. This pollutant enrichment is caused by either: (1) the passive adsorption of pollutants from seawater or (2) active uptake followed by retention in tissues or hard parts as a result of nonexcretion. The degree of enrichment is variable and depends on factors such as (1) the chemical nature of the pollutant, (2) the type of organism, (3) its physiological state, and (4) environmental conditions such as water temperature, salinity, and pH. Environmental conditions affect reaction rates and the chemical speciation of the pollutant. Enrichment factors are greatest for metals, some of which are as high as 10^9. (In Table 11.5 and Eq. 11.1, bioaccumulation of trace metals was quantified as an enrichment factor.) Although some pollutants are eventually excreted or degraded, the rates of these processes tend to be slow. Thus, the consumption of tainted tissues causes many pollutants to be passed up the food chain. In the case where bioaccumulation occurs during each transfer to higher trophic levels, the concentration of the pollutant increases. This process is termed *biomagnification* and causes organisms at the top of the food chain to have high pollutant concentrations. Note that not all pollutants undergo biomagnification. Furthermore, the body burden of a pollutant is often not dispersed homogeneously. Some pollutants are strongly concentrated in fats, whereas others are deposited the hard parts or in detoxifying organs, such as the liver.

Because biomagnification and other transport processes take time, the harmful effect of many compounds may not become evident for decades. This makes direct causal relationships between specific pollutants and environmental change difficult to establish. Substantiating such relationships is further complicated by the complex network of positive and negative feedbacks that occur among most parts of the crustal-ocean-atmosphere factory.

Finally, the effects of pollution are difficult to establish because of the requirement that a "significant change" must result. Definition of what constitutes a significant change is essentially subjective. The easiest approach, and, hence, the most commonly used, has been to focus on lethal effects resulting from acute and chronic exposures, such as an oil spill or hypoxia. These endpoints have commonly been used to establish regulatory *water-quality criteria*, such as in the U.S. Clean Water Act. They are typically based on laboratory bioassays that employ test organisms specially bred for this use. For example, an LD_{50} (LD for lethal dosage) is the concentration that causes a 50% mortality in the test organisms during a designated exposure period.

Chronic exposures to sublethal concentrations of pollutants induce effects that are harder to detect and quantify. These effects can extend over several levels of biological organization. As shown in Table 28.3, they can involve physiological, behavioral, and ecological effects, as well as increased susceptibility to environmental stresses, such as disease and climate change. Current efforts at developing methods for detecting sublethal effects are focused on the detection of toxicological biomarkers within organisms. These chemicals are produced in response to a stressor, which can be chemical, biological, or physical in nature. They include altered enzyme activities and metabolites associated with an immune system response. Other approaches involve measurement of pollutant levels in the blood or tissues, enumeration of lesions and tumors, and correlation with parasite loads and disease state. Efforts are underway to identify genetic markers that are sensitive to pollutant exposures and other environmental stressors. Marine organisms being explored as candidates for this genomic approach include sessile animals, such as oysters.

28.5 LAND-USE IMPACTS ON THE COASTAL ZONE

Because the human population is concentrated along the coastline, the waters of the coastal zone are strongly influenced by our activities. These impacts take the form of geological effects in which humans have altered the rate of sediment delivery to the coasts, removed wetlands, dredged deepwater ports, and hardened the shoreline with seawalls, groins, and breakwaters. The most sweeping biological effects are a result of the destruction of wetlands, sea grasses, mangroves, and overfishing. In the lower 49 states of the United States, half of the coastal wetlands have been lost. From a hydrological and chemical standpoint, the most important alterations to river flows have been caused by damming and discharge of pollutants into waterways, most notably nutrients. Over the past few decades, some progress has been made in cleaning up the rivers and harbor areas of cities in developed countries. But this has largely been

Table 28.3 Summary of Response to Pollutant Stress.

Level	Adaptive Response	Destructive Response	Result at Next Level
Biochemical—cellular	Detoxification		Adaptation of organism
		Membrane disruption	Reduction in condition of organism
		Energy imbalance	
Organismal	Disease defense		Regulation and adaptation of populations
	Adjustment in rate functions		
	Avoidance		
		Metabolic changes	Reduction in performance of populations
		Behavior aberrations	
		Increased incidence of disease	
		Reduction in growth and reproduction rates	
Population	Adaptation of organism to stress		No change at community level
	No change in population dynamics		
		Changes in population dynamics	Effects on coexisting organisms and communities
Community	Adaptation of populations to stress		No change in community diversity or stability
			Ecosystem adaptation
		Changes in species composition and diversity	Deterioration of community
		Reduction in energy flow	Change in ecosystem structure and function

Source: *From Capuzzo, J. M. (1981).* Oceanus *24(1), 25–33.*

swamped by the increasing impacts from the developing countries due to a concurrent increase in their populations and standard of living. Even in the United States, the most recent rating of the overall health of the coastal zone is "fair" as shown in Figure 28.2, with water quality ranging between good and fair.

Human activities have already significantly converted or modified the natural land cover of half of Earth's terrestrial surface. Virtually no region remains untouched by human impacts of some kind. The major land conversions we have accomplished over the past 10,000 years or so include: (1) land clearing, mostly through deforestation and

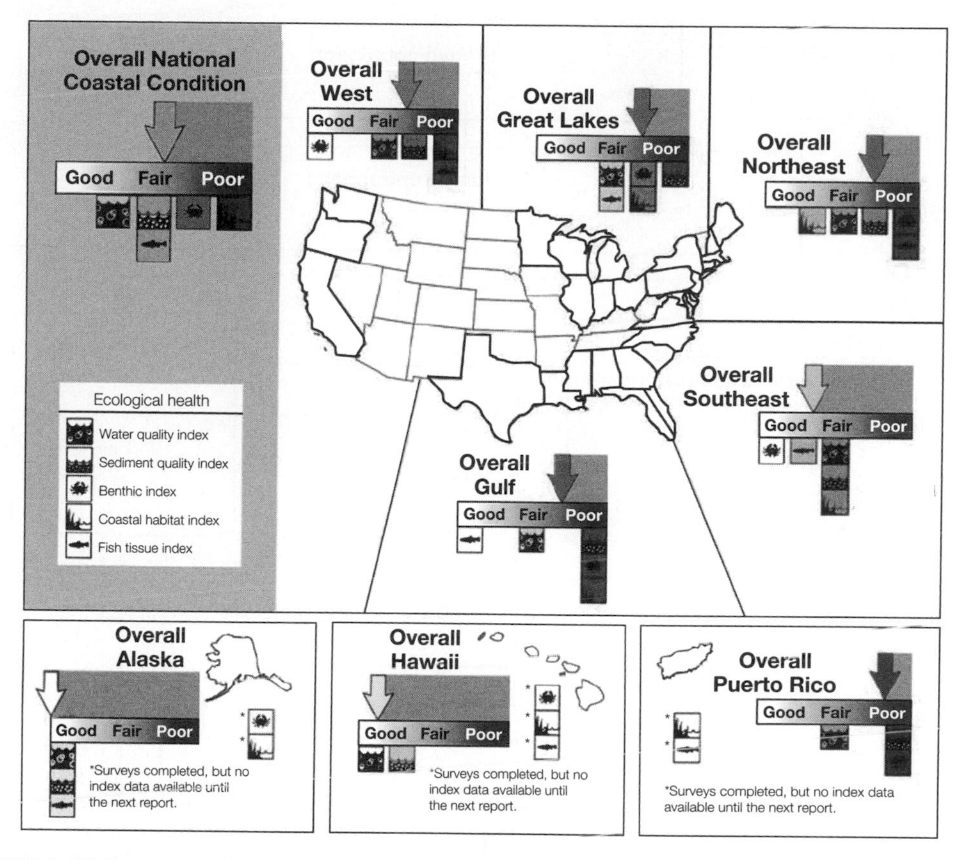

FIGURE 28.2

Overall national and regional coastal condition, primarily between 2001 and 2002. *Source*: After U.S. EPA (2007). Draft National Coastal Condition Report III. United States Environmental Protection Agency. Office of Research and Development and Office of Water. (See companion website for color version.)

biomass burning, (2) agricultural activities of annual crops and trees, (3) domesticated animal husbandry requiring land conversion to pasture for grazing, and (4) installation of impervious surfaces, including asphalt, concrete, and buildings. These land use changes have complicated synergistic impacts on crustal-ocean-atmosphere factory as illustrated in Figure 28.3. Most of the impacts are focused in the coastal zone. This is most unfortunate as this region contains some of the most productive ecosystems on Earth and their natural physical and biogeochemical processes cause them to trap pollutants.

Of all the marine ecosystems, temperate estuaries seem to be undergoing the most degradation. Evidence for this is seen in: (1) increased sedimentation and turbidity, (2) more extended and expansive episodes of hypoxia or anoxia, (3) loss of seagrasses and dominant suspension feeders, primarily bivalves, and (4) a higher frequency and duration of nuisance and toxic algal blooms, jellyfish infestations, and fish kills. Overall, these

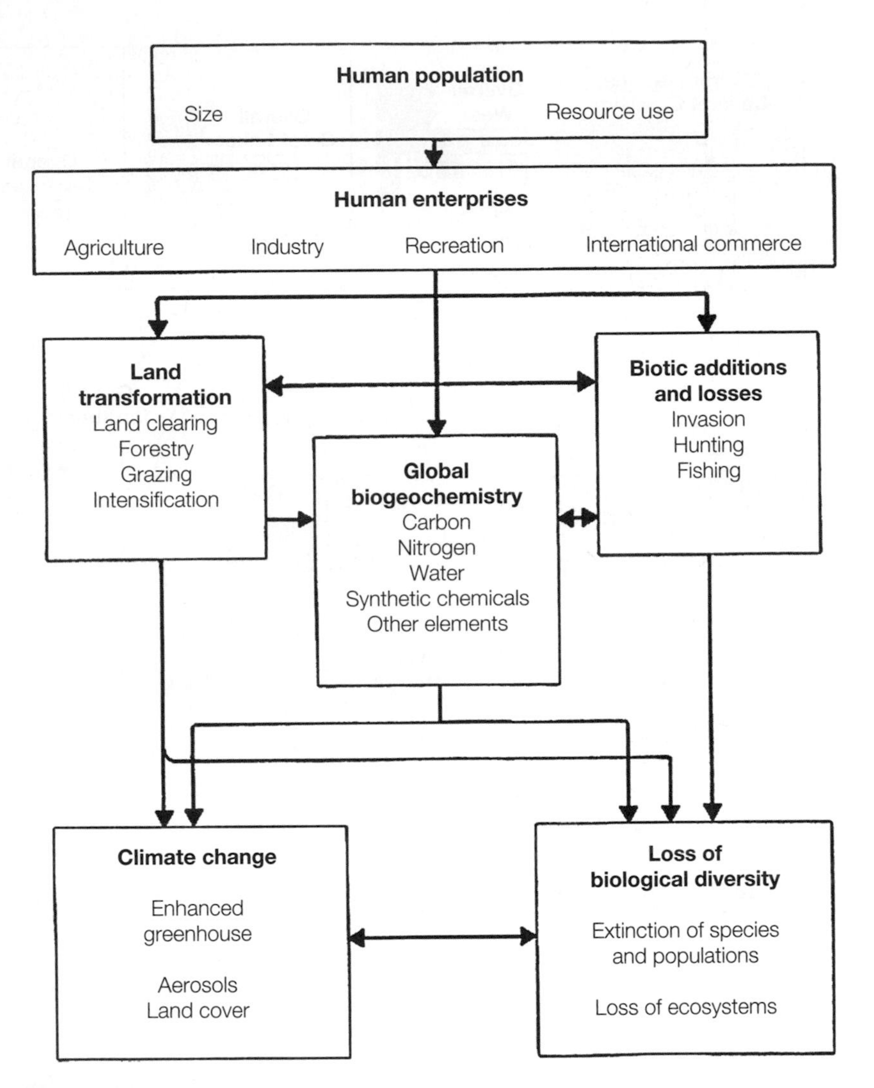

FIGURE 28.3

The scope, degree, and interactive nature of human alteration of the Earth system. *Source*: From Vitousek, P. M., *et al*. (1997). *Science* 277, 494–499.

ecosystems have shifted from ones dominated by benthic primary production to those dominated by planktonic primary production. This is thought to largely be a consequence of cultural eutrophication caused by nutrient loading from land-based discharges although some of this shift is undoubtedly due to the loss of filter-feeding bivalves through overfishing and disease.

Pollutant trapping in estuaries is the result of two related phenomena: patterns of water circulation and patterns of sediment deposition and resuspension. As shown in Figure 28.4, estuarine circulation is characterized by a subsurface inflow of dense saline

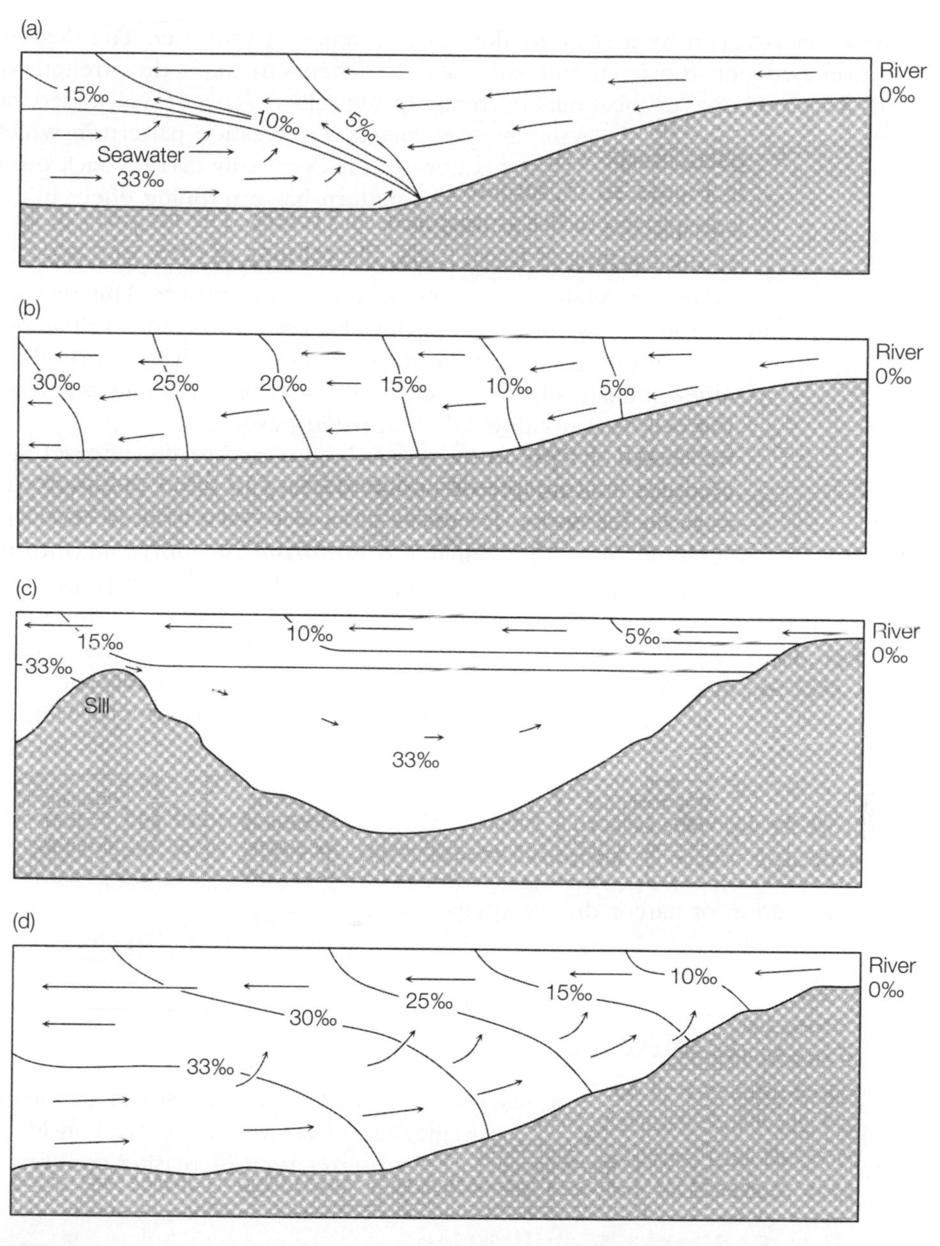

FIGURE 28.4

Salinity gradients and direction of water flow in (a) salt-wedge, (b) partially mixed, (c) fjord-type, and (d) well-mixed estuaries. Salinity is given in parts per mil.

seawater overlain by a seaward flow of low-density freshwater. The degree of stratification between the fresh and salt water is dependent upon the strength of the river discharge relative to local tidal currents. In the salt-wedge and well-mixed cases, turbulent mixing between the water masses creates a circulation pattern in which seawater becomes entrained in the surface waters and is eventually carried back out to sea. For the biolimiting elements, this circulation pattern has a trapping effect initiated by the creation of biogenic particles in the surface waters. These particles sink into the salty bottom waters where remineralization resolubilizes the biolimiting elements. Turbulent mixing returns the resolubilized nutrients to the surface waters. This recycling process supports high productivity in estuaries, but also serves to trap nutrient pollution discharged into estuaries in the form of fertilizer runoff and sewage. A similar trapping is observed for other pollutants that bioaccumulate, such as heavy metals and hydrophobic organic compounds, most notably the chlorinated pesticides.

In salt-wedge and partially mixed estuaries, pollutants are also trapped by adsorption onto fine-grained particles. These particles are largely terrestrial soils that have been carried by stormwater runoff into the estuaries. These fine-grained particles tend to collect in the water column in a zone called the *estuarine turbidity maximum* (ETM) in which sediment concentrations range from 10^2 to 10^4 mg/L. ETMs are thought to be the result of four physical phenomena: (1) bottom resuspension, (2) inhibition of turbulent mixing by stratification, (3) tidal asymmetries, and (4) the convergence of bottom residual flows. ETMs are typically found in association with a deposit of mud on the seabed. The locations of the ETM and its associated mud deposit can shift over time in response to changes in river flow and tidal amplitude as illustrated in Figure 28.5. These fluctuations produce a cycle of sedimentation, temporary storage, and resuspension that acts to concentrate the fine-grained particles, and by association, the sorbed pollutants. Dredging of these fine-grained sediments risks resuspension of the sorbed pollutants and can lead to significant water-quality problems. This is why federal permits are required for relocation of harbor dredge spoils.

28.6 CONTAMINANTS

In the following section, we discuss each of the major classes of contaminants listed in Table 28.1. On a global basis, the greatest negative impacts are considered to result from sewage and nutrient loading, although litter is an increasing concern.

28.6.1 Sediment Mobilization

Humans have arguably become the dominant geomorphic agent shaping the terrestrial landscape. This is largely the result of land-use changes associated with construction, mining, deforestation, and agriculture. Terrestrial erosion caused by humans is estimated to be occurring at a rate that is 10 times faster than the sum of all natural processes. As shown in Figure 28.6, our net effect began to exceed that of natural processes towards

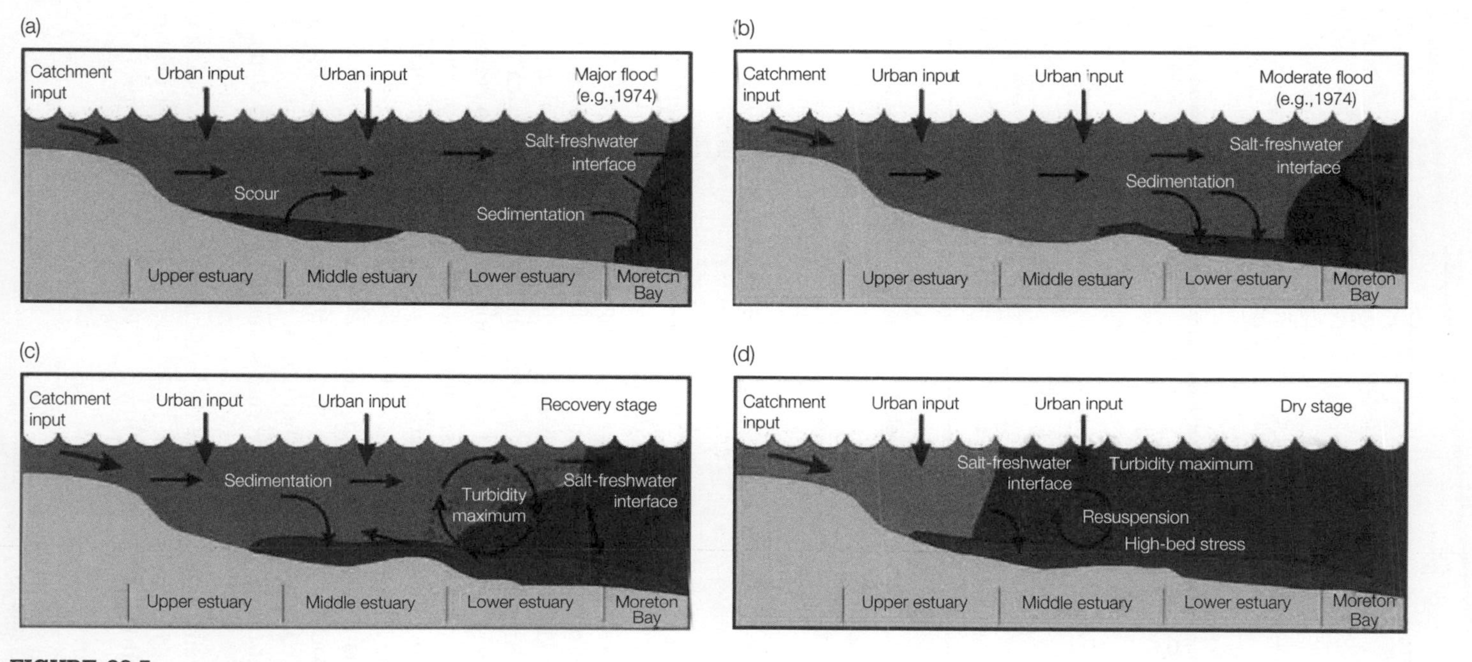

FIGURE 28.5

Conceptual model of some processes maintaining turbidity in the Brisbane River Estuary. (a) Major Flood, (b) Moderate Flood, (c) Recovery Stage, and (d) Dry Stage. Note that the interface between marine and freshwater migrates in response to flooding events. *Source*: From Dennsion, W. C., and E. G. Abal, (1999). Moreton Bay Study: A scientific basis for the Healthy Waterways Campaign, South East Queensland Water Quality Management Strategy, p. 245. http://www.ozestuaries.org/indicators/turbidity.jsp. (Figure 2).

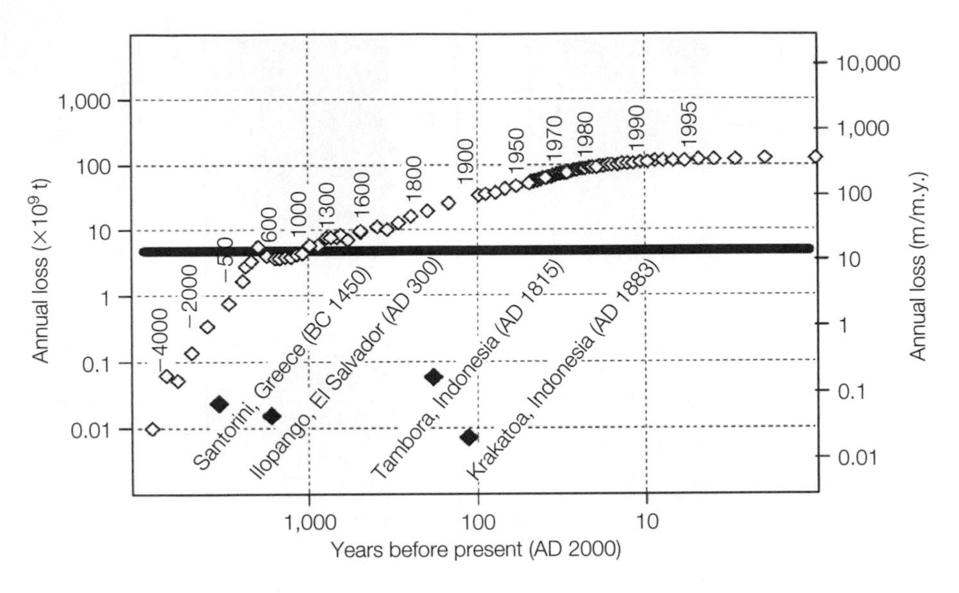

FIGURE 28.6

Historical rates of anthropogenic erosion (open diamonds). Solid black diamonds are volumes of several large volcanic eruptions (dates in parentheses); heavy black line is the mean deep-time denudation rate of 24 m/10^6 y. *Source*: From Wilkinson, B. H. (2005). *Geology* 33(3), 161–164.

the end of the first millennium AD. Agricultural soil loss represents about 70% of the human impact on the terrestrial denudation rate—most of the rest is from construction. On a per capita basis, this downscales to 21 tonnes per person per year with 6 tonnes coming from construction and 15 tonnes from farming. As discussed in Chapter 14.5.1, most soils are produced from weathering of rocks located at high elevation and on steep slopes, i.e. in Southeastern Asia, Oceania and South America (Table 14.2). The weathering products are transported downslope, with most depositing in depressions and floodplains to create fertile soils. Anthropogenic impacts on denudation rates are concentrated in the low-lying areas largely as a result of soil loss associated with agricultural activities. The anthropogenic soil and rock mobilization rate (~ 120×10^9 tonnes/y) is much larger than the current total sediment flux carried by rivers into the ocean (18 to 20×10^9 tonnes/y) indicating that most of the mobilized materials are being retained on land.[1] As discussed in Chapter 14.5.1, the construction of large dams has led to a significant trapping of the riverine sediment load, estimated at 20 to 30% of the total particle flux.

Dams have been constructed over the millennia to prevent flooding and to serve as reservoirs for drinking, irrigation, and hydropower. During the last century, the number of dams increased dramatically as shown in Figure 28.7. Prior to 1950, most of the

[1] Suspended plus bed load carried by rivers during the Anthropocene as estimated by Syvitski, J. P. M., *et al.* (2005). *Science* 308, 376–380.

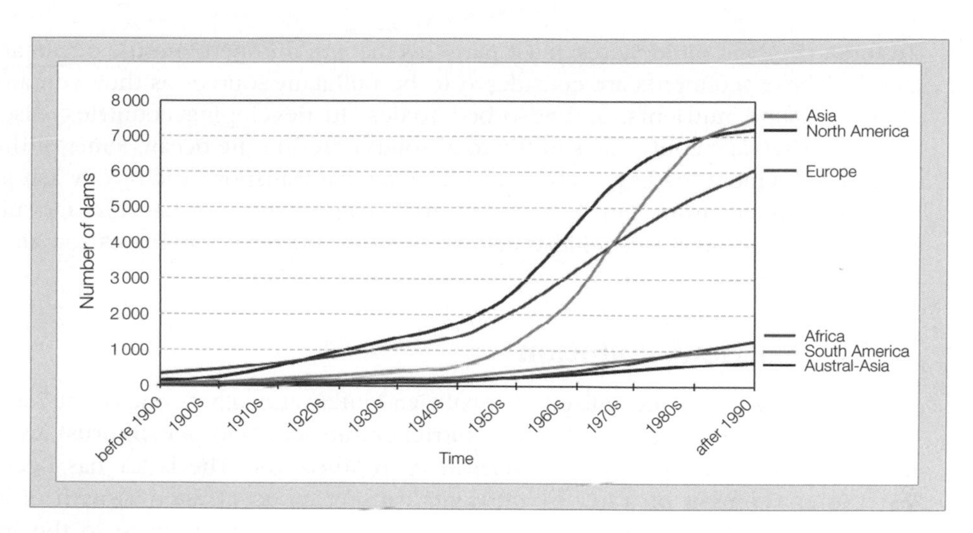

FIGURE 28.7

Dams constructed over time by region (1900–2000). *Note*: Information excludes the time trend of dams in China. *Source*: After International Commission on Large Dams (ICOLD), 1998 as published on p. 9 in World Commission on Dams (2000). *Dams and Development: A New Framework for Decision Making*. Earthscan Publications Ltd., http://www.dams.org/docs/report/wcdch1.pdf.

large dams were located in industrialized countries. Between 1949 and the end of the 20th century, the number of large dams increased from 5000 to over 45,000, with two thirds now located in developing countries. The countries with the greatest number of large dams are China (22,000), the United States (6390), India (4000), Spain (1000), and Japan (1200). The increase of dams in China is particularly notable, as only 22 existed in 1949. Dam construction is now occurring primarily in the developing countries. In the developed countries, the current trend is toward dam removal in an effort to restore hydrology to improve fisheries and ecological processes.

From a temporal perspective, the initial impact of humans on sediment delivery to the ocean was to cause a large increase in sediment fluxes. High loads of particles in river water discharging into the ocean have several negative consequences. These include: (1) smothering of the benthos, (2) clogging of gills and filter-feeding organs, and (3) increasing contaminant concentrations, such as those of particulate organic matter and nutrients. In countries where industrialization lead to dam construction, sediment loads have now declined below natural levels. This is problematic as riverine sediments are needed to resupply coastal regions, such as beaches and marshes, thereby replacing particles lost to erosion and sea-level rise. From a geographic perspective, riverine sediment loads are still high in developing countries due to intensive land-use change in response to increasing population and economic development, although this will likely change over time with continued dam construction.

As noted in the previous sections, coastal sediments are also being relocated through dredging that is being performed to maintain deepwater harbors and to renourish beaches. In the United States, most particles that are dumped into the ocean are dredge spoils. These sediments are considered to be pollutant sources as they contain organic matter, bound nutrients, and adsorbed toxics. In developing countries, discharge of untreated sewage contributes to the total solids entering the ocean. Some of the anthropogenic sediment loading is also a result of aeolian transport, such as fly ash generated by waste incinerators. Finally, desertification caused by unsustainable agriculture and climate change increases wind transport of soils and hence increases the aeolian load of particles to the ocean.

28.6.2 Nutrient Pollution

Nutrient pollution is recognized nationally and internationally as one of the most serious threats to ocean and human health. Nutrients (nitrogen and phosphorus) are classified as pollutants when they cause *cultural eutrophication*. The latter has been defined as "the enrichment of water by nutrients causing an accelerated growth of algae and higher forms of plant life that produces an undesirable disturbance to the balance of organisms present in the water and to the quality of the water concerned."[2] Undesirable disturbances which have thus far been observed include: (1) changes in food web structures, (2) proliferation of nuisance and harmful organisms, and (3) hypoxia. The geographic scope and frequency of these disturbances have increased greatly since the 1950s. The most drastic change has been in the number of coastal sites now subject to periods of hypoxia, increasing from fewer than nine prior to the 1960s to an estimated 200 in 2007.

The effects of nutrient pollution are a consequence of two phenomena: (1) increased terrestrial fluxes of nitrogen and phosphorus leading to their increased availability in coastal waters and (2) shifts in the relative abundances of the macronutrients and micronutrients, particularly iron. The most important macronutrient ratio shift has involved DSi, whose availability in coastal waters has been declining through the period over which nitrogen and phosphorus inputs have been rising. The reason for this is that unlike nitrogen and phosphorus, DSi has only a natural source—weathering of silicate rocks. Diatom growth in rivers transforms this DSi into particulate form (BSi), which is readily retained in the sediments trapped behind dams. In contrast, anthropogenic loading of phosphorus and nitrogen has led to increases in their fluxes to coastal waters (Figures 24.18, 24.19, and 24.20). The resulting decline in the relative abundance of DSi in coastal waters now favors the proliferation of cyanobacteria over diatoms. Increased iron inputs also appear to be playing a role in favoring overgrowth of the planktonic species that form red tides by stimulating N_2-fixing microbes. Red tides are an example of a ***harmful algal bloom*** (HAB).

Nutrient pollution has contributed to other notable disturbances in the structure of marine food webs. These include: (1) bioinvasions of macroalgae and microbial mats

---

[2] *Source*: From OSPAR (1998). OSPAR Agreement 1998-18. OSPAR Strategy to Combat Eutrophication. The OSPAR Commission comprises countries bordering the northeastern Atlantic Ocean.

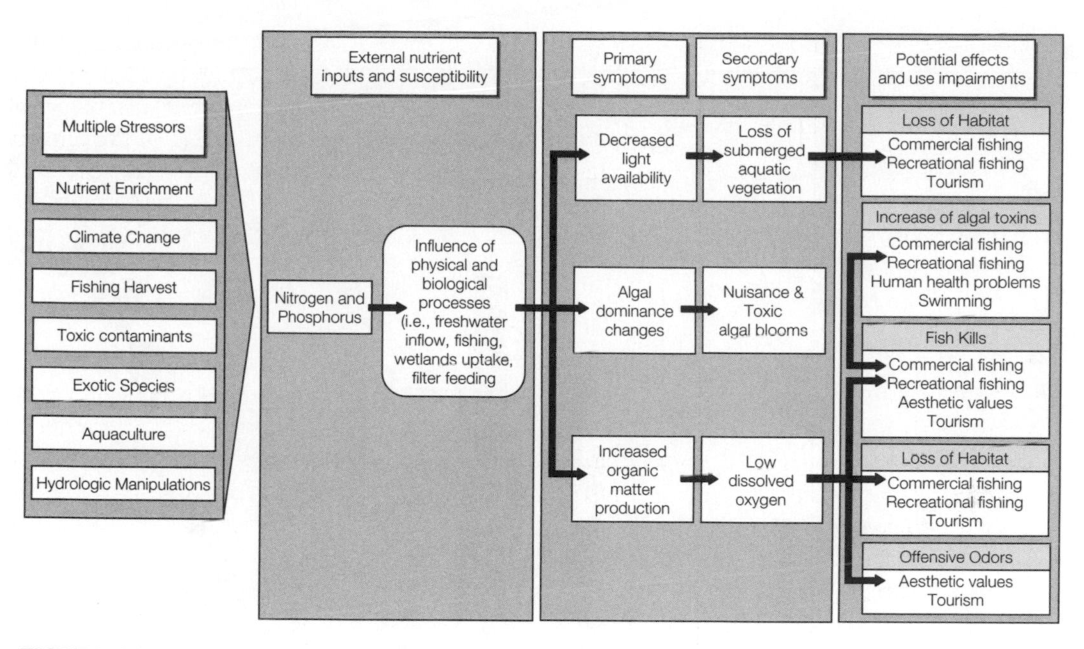

FIGURE 28.8

Disturbances to ocean and human health resulting from cultural eutrophication. *Source*: After Committee on Environment and Natural Resources (2003). *An Assessment of Coastal Hypoxia and Eutrophication in U.S. Waters*. National Science and Technology Council Committee on Environment and Natural Resources, p. 11.

that have supplanted seagrass and kelp beds, (2) coral reef die-offs, (3) infestations of jellyfish, and (4) blooms of mucilaginous phytoplankton, such as *Phaeocystis* (as described in Chapter 23.6.3). Overall, coastal zones have seen a decline in planktonic grazing and commercial fish production. The records of primary productivity obtained from satellite imagery are still too short to allow accurate evaluation of long-term trends, although preliminary interpretations suggest that increases are occurring. As shown in Figure 28.8, the impacts of nutrient pollution are being enhanced by other ecosystem stressors including: (1) habitat destruction, particularly of salt marshes and mangroves; (2) removal of filter feeding grazers, such as oysters; (3) shifts in climate that have altered water temperature and circulation patterns; and (4) introduction of exotic invasive species.

The degree to which cultural eutrophication has affected coastal waters is assessed using various rating systems, such as the one shown in Figure 28.9b. In the United States, estuarine eutrophication status is classified into high, moderate (or medium), and low levels. Eutrophication rating systems are variously based on: (1) threshold levels of nitrogen, phosphorus, chlorophyll and dissolved oxygen concentrations; (2) turbidity as measured by Secchi depth; (3) declines in submerged aquatic vegetation; and (4) prevalence of algal blooms, including planktonic and epiphytic forms. As shown in

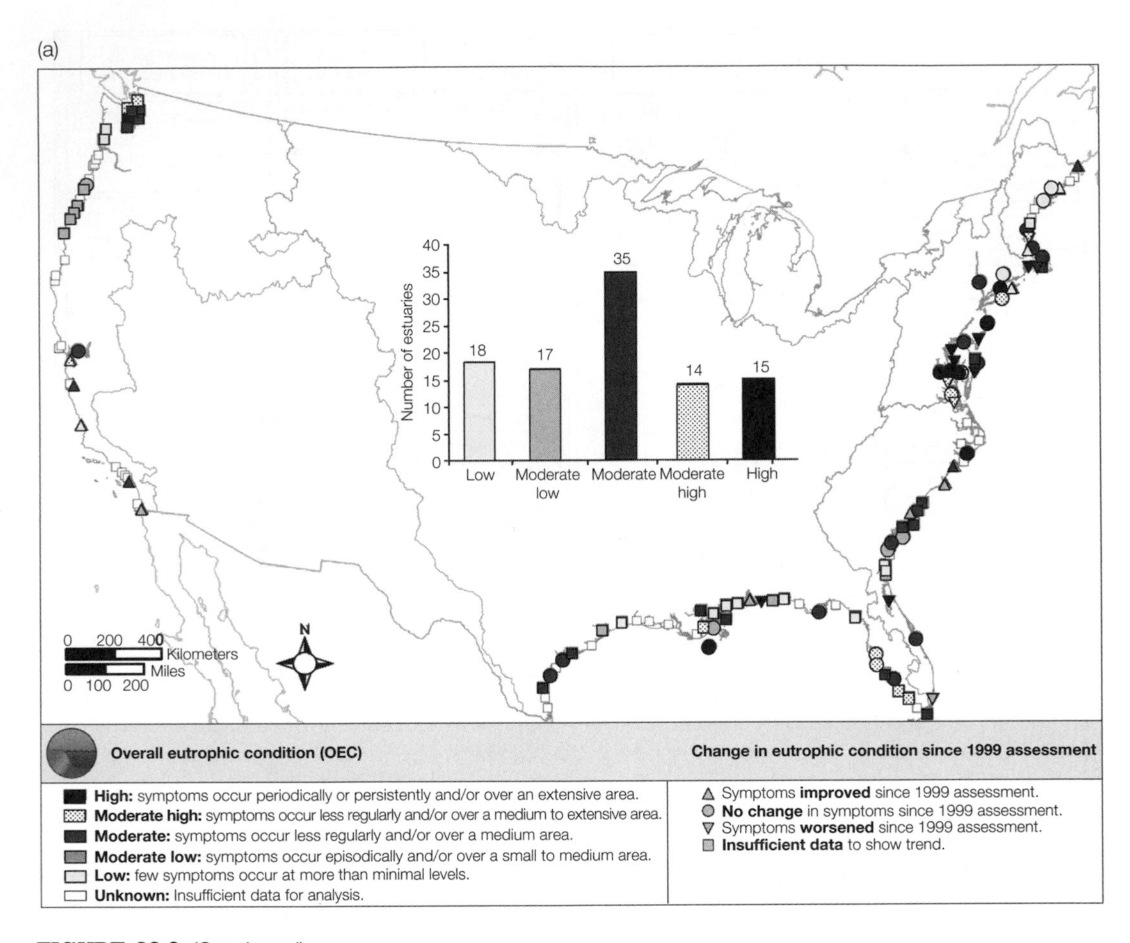

FIGURE 28.9 *(Continued)*

Figure 28.9a, twenty-nine estuaries in the United States are now classified as having a moderate high to high level of eutrophication and 35 fall into the moderate level.

The ultimate sources of nitrogen and phosphorus causing cultural eutrophication are industrial N_2 fixation, fossil-fuel burning, and the mining of phosphorite. The nitrogen and phosphorus used as fertilizer have three possible fates: they either enter the food chain, become part of the soil, or are washed off the land by stormwater runoff.[3] The nutrients that enter the food chain eventually end up as either animal wastes or dead biomass. Animal wastes include human sewage, livestock manure, and pet feces. Sanitary

[3] In the case of nitrogen, some is transferred to the atmosphere as a consequence of denitrification, which takes place in soils, sewage treatment plants, and natural waters. Some nitrogen is also volatilized as ammonia from animal feedlots.

(b)

Symptom	Parameters	Expression: Low	Expression: Moderate	Expression: High
Chlorophyll a (phytoplankton) Typical high concentration ($\mu g\ L^{-1}$) in an annual cycle determined as the 90th percentile value.	*Spatial coverage:* High >50%; Moderate 25–50%; Low 10–25%; Very low 0–10% *Frequency:* Episodic; Periodic; Persistent *Concentration:* High >20 $\mu g\ L^{-1}$; Medium 5–20 $\mu g\ L^{-1}$; Low 0–5 $\mu g\ L^{-1}$	*Low symptom expression:* *Conc. / Coverage / Frequency* low / any / any medium / mod.–v. low / episodic high / low–v. low / episodic	*Moderate symptom expression:* *Conc. / Coverage / Frequency* medium / high / episodic medium / moderate / periodic high / low–v. low / periodic high / moderate / episodic	*High symptom expression:* *Conc. / Coverage / Frequency* medium / high / periodic medium / mod.–high / periodic high / high / episodic
Macroalgae Causes a detrimental impact on any natural resource.	*Frequency of problem:* Episodic (occasional/random) Periodic (seasonal,annual, predictable) Persistent (always/continuous)	No macroalgal bloom problems have been observed.	Episodic macroalgal bloom problems have been observed.	Periodic or persistent macroalgal bloom problems have been observed.
Dissolved oxygen Typical low concentration (determined as the 10th percentile value) in an annual cycle.	*Spatial coverage:* High >50%; Moderate 25–50%; Low 10–25%; Very low 0–10% *Frequency:* Episodic; Periodic; Persistent *State:* Anoxia 0 mg L^{-1}; Hypoxia 0–2 mg L^{-1}; Biol. stress 2–5 mg L^{-1}	*Low symptom expression:* *State / Coverage / Frequency* anoxia / mod.–low / episodic anoxia / very low / periodic hypoxia / low–v. low / periodic hypoxia / moderate / episodic stress / any / episodic stress / mod.–v. low / periodic	*Moderate symptom expression:* *State / Coverage / Frequency* anoxia / high / episodic anoxia / low / periodic hypoxia / moderate / periodic hypoxia / high / episodic stress / high / periodic	*High symptom expression:* *State / Coverage / Frequency* anoxia / moderate-high / periodic hypoxia / high / periodic
Submerged aquatic vegetation A change in SAV spatial area observed since 1990.	*Magnitude of change:* High >50%; Moderate 25–50%; Low 10–25%; Very low 0–10%	The magnitude of SAV loss is low to very low.	The magnitude of SAV loss is moderate.	The magnitude of SAV loss is high.
Nuisance/toxic blooms Causes detrimental impact on any natural resources.	*Duration:* Persisten, seasonal, months, variable, weeks, days, weeks to seasonal, weeks to months, or days to weeks *Frequency:* Episodic, periodic, or persistent	Blooms are either a) short in duration (days) and periodic in frequency; or b) moderate in duration (days to weeks) and episodic in frequency.	Blooms are either a) moderate in duration (days to weeks) and periodic in frequency; or b) long in duration (weeks to months) and episodic in frequency.	Blooms are long in duration (weeks, months, seasonal) and periodic in frequency.

FIGURE 28.9

(a) National assessment of eutrophication conditions in the United States and (b) eutrophication rating system used to construct this assessment. *Source*: After Bricker, S., *et al.* (2007). NOAA Coastal Ocean Program Decision Analysis Series No. 26. National Centers for Coastal Ocean Science, p. 33 and p. 17. (See companion website for color version.)

sewer systems usually merge sewage with other domestic wastes, such as wash waters, which contain phosphate-based detergents, and industrial wastes. A large percentage of these municipal wastewaters are discharged untreated into the ocean. The United Nations reported in 2004 that 80 to 90% of the wastewaters from central and eastern Europe, Latin America, east Asia, southern Asia, the southeast Pacific, and west and central Africa are being discharged untreated.

Even where sewage is supposed to be treated prior to discharge, such as in the industrialized countries of North America and western Europe, untreated effluents are still periodically emitted. These emissions are the result of breaks in sewer lines, pump failures, and combined sewer overflows (CSOs). CSOs occur in sanitary sewer systems that, intentionally or not, merge wastewaters with stormwater runoff. The stormwater flows are usually too large for the treatment plants to handle, leading to the discharge of untreated wastewaters following rain events. CSOs are especially common in large cities, such as New York City, Seattle, and San Francisco.

In some coastal regions, a substantial amount of nitrogen is delivered via atmospheric deposition of NO_x from fossil fuel burning and as ammonia that has volatilized from urine generated in feedlots (Figure 24.18). Both nitrogen and phosphorus are also carried to ocean as a component of eroded soils. Finally, mariculture is now a large enough enterprise in some regions to be a significant source of nutrient pollution.

Nitrogen pollution has received far more attention than that of phosphorus for two reasons. First, it has been considered as the nutrient-limiting primary production in estuaries and coastal waters. Second, its loading into the coastal zone has been far greater than that of phosphorus (Figure 24.21). It is also more efficiently exported into the ocean due in part to formation of iron phosphate minerals in anoxic estuarine sediments.

The impact of anthropogenic nutrient emissions in the coastal zone is heightened by its chemical speciation. Pollutant nitrogen and phosphorus are delivered to the coastal waters primarily in inorganic form, whereas most of the natural riverine dissolved nitrogen and phosphorus are components of organic compounds, i.e., DON and DOP. Thus, the pollutant nutrients are delivered to the coastal waters in a chemical form that can be directly assimilated by coastal plankton, whereas the organically bound (natural) forms must first be remineralized.

Human activities have approximately doubled the oceanic inputs of nitrogen and phosphorus. In the case of nitrogen, atmospheric deposition is becoming an increasingly more important transport pathway (Figure 24.17). Globally, most of the nutrient input is directly linked to agricultural activity, e.g., fertilizer runoff and volatilization of animal wastes. Around large coastal cities, sewage discharge is also significant. Projections based on population growth and demographic trends suggest that these nutrient inputs will continue to increase. As shown in Figure 24.19, most of the increase in riverine nitrogen export is expected to occur in Asia. Projections for increases in phosphorus loading are somewhat lower than those for nitrogen due to reductions achieved through restrictions in the use of phosphate-based detergents.

Much effort is currently being directed at predicting the fate of these additional nutrients in the coastal zone. Model projections shown in Figure 24.21 indicate an increase in the burial of organic matter on the continental shelf. In other words, the additional nutrients are expected to stimulate primary production with the resulting biomass being exported to the sediments. At present, coastal waters are generally net heterotrophic, so anthropogenic nutrient loading is predicted to cause a system shift to net autotrophy. With more organic matter in the shelf sediments, an increase in sedimentary denitrification is also predicted, thereby providing a mechanism to reduce the nitrogen inventory in the coastal zone. A large unknown in these models is the degree to which shifts in open-ocean circulation affect coastal circulation patterns, most notably coastal upwelling. Some of these shifts are natural and occur over decadal time scales, including ENSO and the North Atlantic and Pacific Oscillations. Others, which are as yet unknown, are anticipated as a consequence of long-term global climate change. Another unknown is the degree to which sedimentary organic matter can be transported from the continental margin to the deep ocean.

28.6.3 Hypoxia

Marginal Seas and Bays

Worldwide cultural eutrophication is most commonly seen in marginal seas and bays. In Europe, observations since the 1950s have documented increased cultural eutrophication in the North, Baltic, Adriatic, Irish, Mediterranean, Black, and Kattegat Seas. In the United States, cultural eutrophication hot spots include the Gulf of Mexico, Chesapeake Bay, Long Island Sound, Narragansett Bay, Florida Bay, and Tampa Bay. In most of these locales, cultural eutrophication has progressed such that hypoxia is now a seasonal or persistent feature.

The largest hypoxic zone in the United States is located in the Gulf of Mexico along the Texas-Louisiana coastline (Figure 28.10). Although some component of this hypoxic zone may have been natural, its area has tripled in size since the 1980s. This growth is attributed to increased nutrient inputs from agricultural runoff transported into the Gulf of Mexico via the Mississippi River. This river's drainage basin is the third largest in the world, exceeded in size only by the watersheds of the Amazon River and Congo River. It drains 41% of the 48 contiguous states of the United States, including all or parts of 31 states and two Canadian provinces. A similar increase in the prevalence of seasonal hypoxia has been observed since the 1930s in Chesapeake Bay. This increase is also attributed to nutrient loading derived from agriculture with atmospheric emissions playing an important role. Another key factor has been the loss of a large oyster fishery through overharvesting, disease, habitat loss, and sedimentation. These filter feeders played an important role in removing algal biomass with subsequent conversion into oyster tissues that exert less oxygen demand to the bay's waters as compared with the detrital remains of plankton.

The development of hypoxia involves a sequence of events triggered by the presence of certain chemical, physical, and biological factors (Figure 28.11). The key biological event is the consumption of oxygen by aerobic heterotrophs, mainly bacteria. The source of this organic matter can be sewage or eroded soils, or it can be algal biomass whose production was stimulated by nutrient pollution. To create waters that are undersaturated with respect to O_2 requires that the biological uptake rate exceed resupply. Resupply routes include photosynthesis and uptake of atmospheric O_2 across the air-sea interface.

The most common scenario in which O_2 uptake exceeds its resupply is in vertically stratified water columns. Oxygen deficiencies develop in the subsurface waters as sinking POM is remineralized. Density stratification prevents vertical mixing and hence the transport of O_2 into the subsurface waters. Aerobic respiration of sedimentary organic matter can also contribute to the subsurface water oxygen deficit and thereby release nutrients and micronutrients, such as iron, back into the water column. Eventual return of these nutrients into the euphotic zone restimulates algal growth.

Density stratification in coastal waters can result from increased freshwater flows from land due to heavy rainfall, and from seasonal surface warming. Changes in winds and currents alter upwelling conditions that can also affect stratification, while concurrently affecting nutrient resupply to the surface waters. These changing environmental

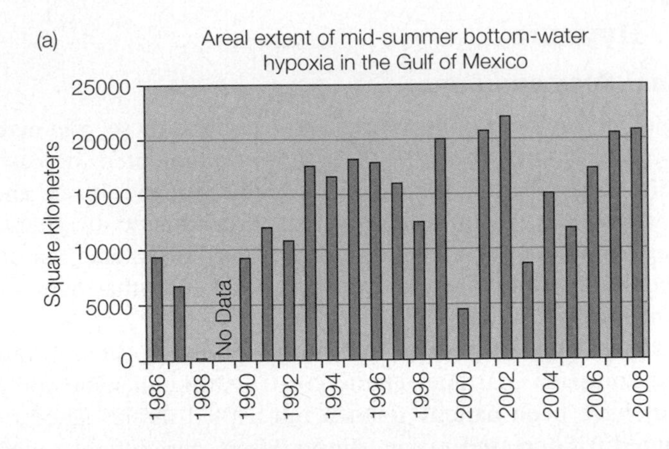

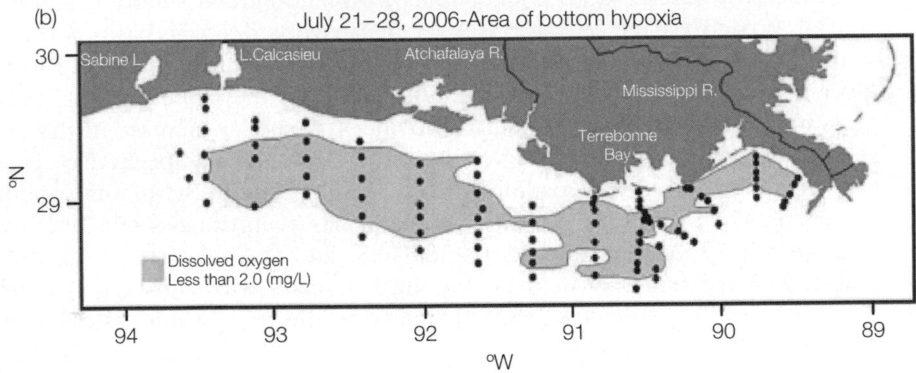

FIGURE 28.10

(a) The estimated area of bottom water dissolved oxygen less than 2 mg/L during mid-summer in the northern Gulf of Mexico. Hypoxia occurs in mid-to-late July. The map is constructed from measurements conducted at 80 to 90 stations within a 5-to-6 day period. *Source*: N. N. Rabalais and R. E. Turner (August 12, 2008). Results, Gulf of Mexico O_2, LUMCON, http://www.gulfhypoxia.net/research/shelfwidecruises/area.asp. and (b) U.S. EPA (2008) Gulf of Mexico Program, Hypoxia Release and Map, July 28, 2006, http://www.epa.gov/gmpo/nutrient/hypoxia_pressrelease.html.

conditions cause shifts in the plankton community toward species that are less readily consumed by grazers. This increases the export efficiency of POM export and burial.

Hypoxic conditions are generally restricted to subsurface waters and lead to profound changes in the biotic community structure, particularly in the benthos. This is a consequence of low O_2 concentrations, low pH, and high concentrations of ammonium and even hydrogen sulfide. The latter three are the waste products of aerobic and anaerobic microbial metabolism and, hence, accumulate over time as organic matter is degraded. These chemically harsh conditions tend to favor opportunistic species and diminish species diversity, leading to an increased dominance of microbes. While fish

(a)

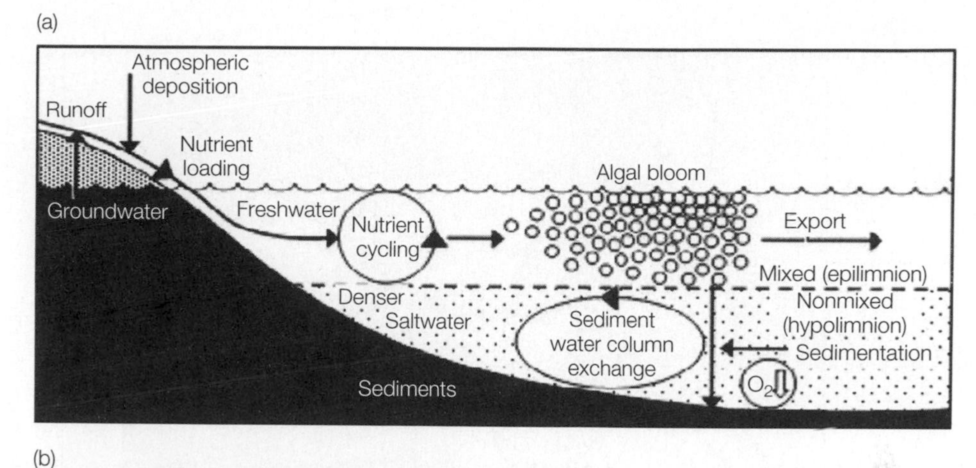

(b)

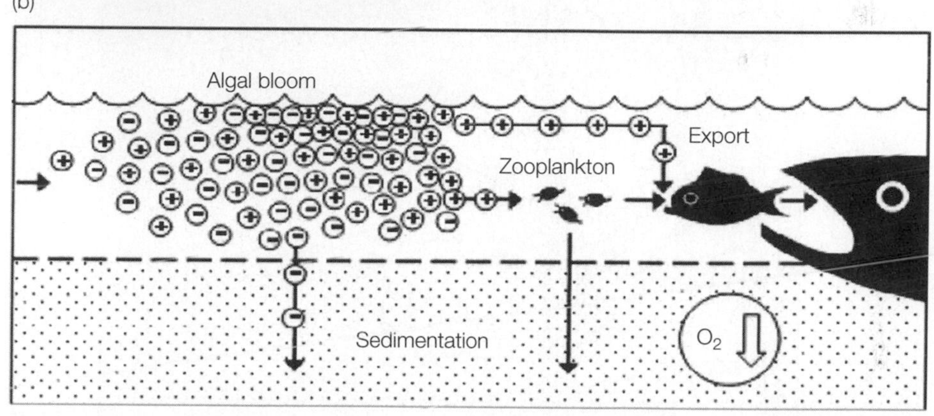

FIGURE 28.11

(a) Conceptual figure showing the linkage between external nutrient loading, internal nutrient cycling, nutrient-enhanced algal bloom formation, and hypoxia under salinity-stratified estuarine conditions. (b) Differential impact on hypoxia of phytoplankton species that are readily consumed (labeled +) versus species that are not (–). Species that are not consumed form a larger share of sedimented organic matter and, therefore, represent a larger burden on the hypoxia potential in the estuary. *Source*: From Paerl, H. W. (2003). *Fish Physiology, Toxicology, and Water Quality*, Proc. 7th International Symposium. EPA/600/R-04/049/, p. 41.

have some ability to relocate to more oxic waters, benthic invertebrates experience mass fatalities. The largest impacts to fisheries thus far has been on shrimp and bivalves.

Worldwide more than half of the major estuaries now exhibit hypoxia either seasonally or episodically (Figure 28.12). Estuaries and marginal seas are especially prone to hypoxia because water mixing and flushing, as driven by wind and tides, are geographically restricted by land masses and shallow water depths. An increase in the number of sites experiencing hypoxia began in the 1960s, with some locales having reached the organic matter overloading threshold earlier, such as the Chesapeake Bay in the 1930s,

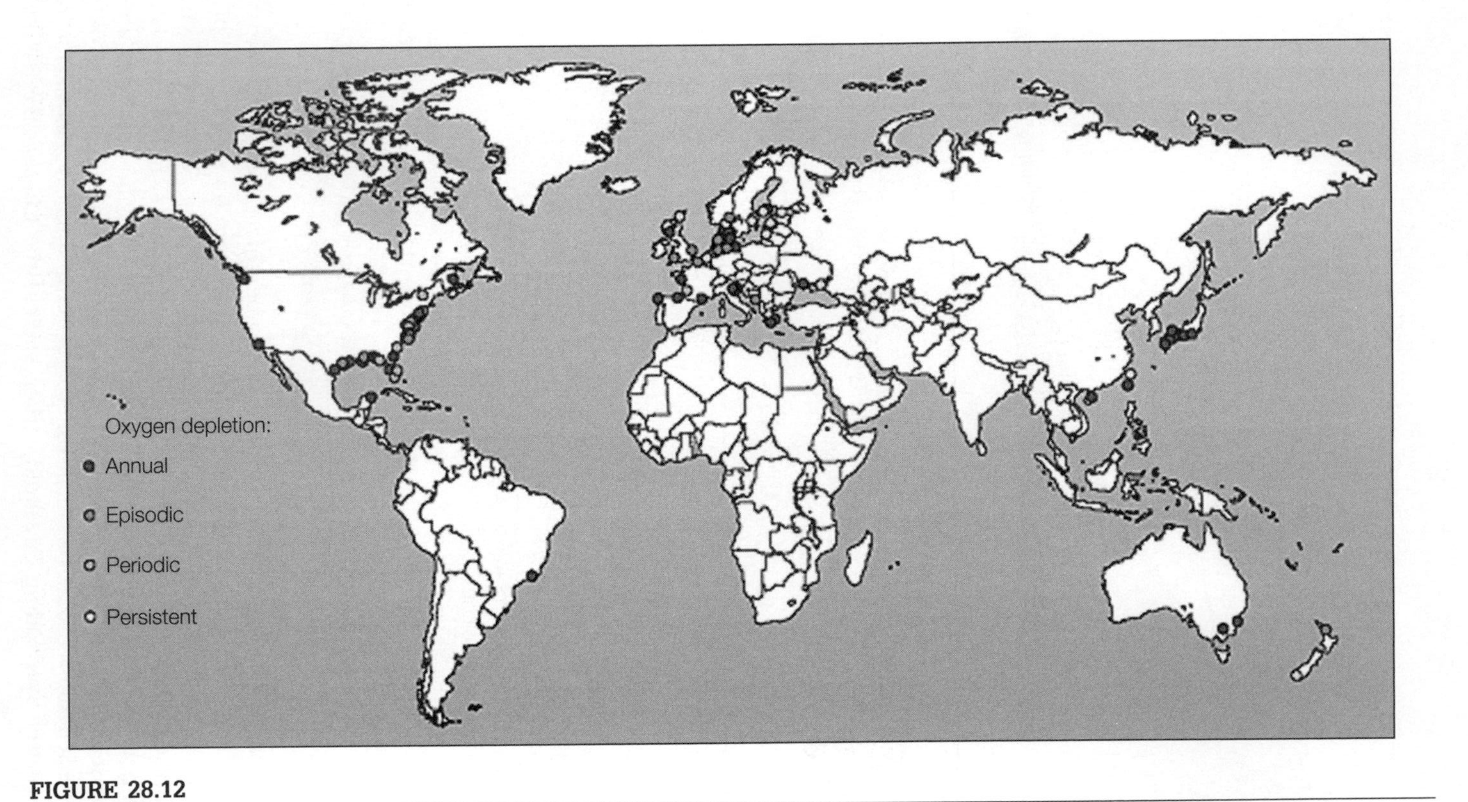

FIGURE 28.12

Global distribution of coastal areas experiencing hypoxic conditions circa 2003. *Source*: After UNEP (2003). *Global Environment Outlook Year Book 2003*, Box 4, United Nations Environmental Programme. (See companion website for color version.)

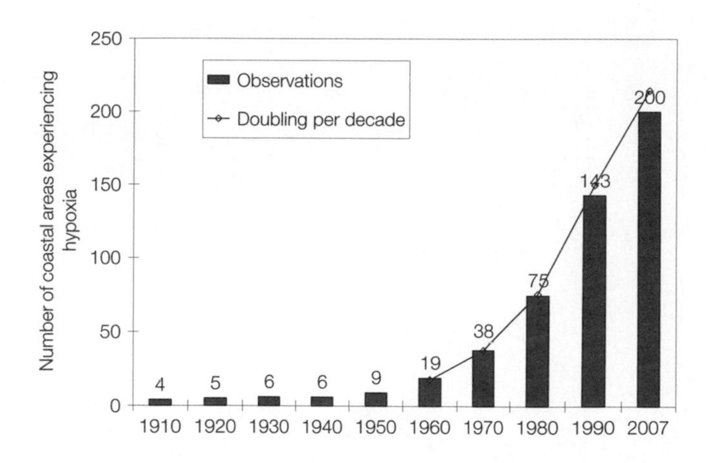

FIGURE 28.13

Areas experiencing hypoxia. Counts are accumulated by decade. Less than 6% of areas in which hypoxia has occurred have later experienced increases in dissolved oxygen concentrations. 2007 data are based on press release reports from UNEP. Other data from Diaz, R. J. (2006). NOAA Great Lakes Seminar Series: 2006 Past Seminars. ftp://ftp.glerl.noaa.gov/webcast/2006/diaz/20061019.pdf.

the Baltic Sea in the 1950s, and the East River in New York City in the 1920s. Since the 1960s, the number of sites reporting hypoxia has doubled each decade (Figure 28.13). About half of the areas exhibiting hypoxia do so seasonally, usually during the summer and fall, reflecting the impact of high temperatures on oxygen solubility, primary production, and respiration rates. About 20% of the sites experience hypoxia episodically. Generally, sites that initially experience episodic hypoxia eventually transition over to seasonal or persistent occurrences. About 20% of the sites now experience hypoxia several times during a year or are persistently hypoxic. Areas where hypoxic or anoxic conditions are now persistent include the Gulf of Finland and central areas of the Baltic, Black, and Caspian Seas.

A small percentage of the coastal systems experiencing hypoxia have either stabilized or improved their oxygen conditions. This has been achieved through intensive regulation of nutrient or carbon inputs. These include the main stem of Chesapeake Bay, Raritan Bay, and the Hudson, East, and Delaware Rivers. Temporary improvements have been observed in systems as a result of favorable changes in hydrology or nutrient inputs, i.e., changes not intentionally orchestrated by humans. These include the Gulf of Mexico (Figure 28.10a), the Black Sea, the Baltic Sea, and Gulf of Finland.

Continental Shelves and Open Ocean

Hypoxia has also been observed on the continental shelves considerably offshore. Most has been ascribed to natural coastal upwelling, such as off Peru and western Africa. But in some settings, such as coastal Oregon and New Jersey, controversy exists as to the role of anthropogenic forcing. For New Jersey, the most dramatic hypoxic episode

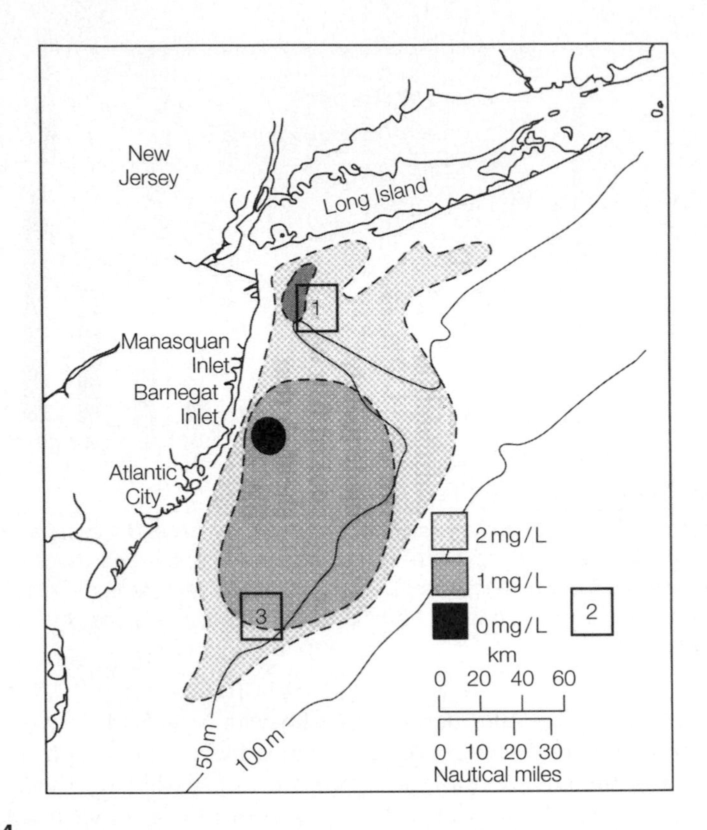

FIGURE 28.14

Dissolved O_2 concentrations (in mg O_2/L) in the bottom waters of the Mid-Atlantic bight during the summer of 1976. The numbered boxes mark the locations of active and historical dump sites. The 12-mile dump site (1) was used for the disposal of NYC's municipal wastes until 1981. The city then switched to the 106-mile dumpsite (2), which was to used until 1991. The Philadelphia sludge site (3) was used until 1981. *Source*: From Steimle, F. W., and C. J. Sindermann, (1978). *Marine Fisheries Review* 40, 17–26.

was observed in 1976. As illustrated in Figure 28.14, bottom waters were hypoxic (<2 ppm O_2) in a large area that included several offshore dumpsites. The so-called 12-mile dumpsite was being used for the disposal of NYC's municipal wastes. In response to the development of hypoxia in 1976, the offshore dumpsite was moved 106 miles offshore where water depths are 2500 m. This site was used until 1992, after which all sewage sludge dumping ceased. This site contains the largest amount of sludge that has ever been dumped into the "deep" ocean. Although 8 million tons of sewage sludge was discharged annually at this site between 1986 and 1992, little evidence for impact on oxygen concentrations has been observed. Furthermore, holistic modeling efforts involving biogeochemical and physical processes have demonstrated that the hypoxic event observed in 1976 could have resulted strictly from natural causes in which a perfect storm of supportive environmental conditions were present. These included:

(1) a warm winter with large runoff, (2) a low frequency of spring storm events, (3) a deep summer thermocline, (4) persistent southerly winds with few reversals, (5) a large autochthonous carbon load, and (5) low grazing pressure by zooplankton.

While sludge dumping has ended, the discharge of organic wastes has not. Major sources are now nonpoint-source runoff, CSOs, permitted discharges of treated effluents, and dry-weather failures in pumps, collection lines, and treatment plants. In 1988, the New Jersey coastline experienced a famous wash-up of medical wastes that was traced to litter carried into the ocean by stormwater runoff. In 1997, NYC spilled 13 million gal of raw sewage into the Mid-Atlantic Bight as a result of dry-weather failures in the pipes and pumps of the city's sanitary sewer system. By 2000, this had been cut 60% through improved surveillance and maintenance. The United States is not the only country with dry-weather pump failures. In April 2007, the Seafield waste treatment works in Edinburgh experienced a pump failure and had to discharge about 100 million liters (26 million gallons) of filtered, but otherwise untreated, sewage into the sea (Firth of Forth) over a period of 3 days to avoid having it back up into streets and homes.

In addition to dry-weather failures, stormwater runoff is a major problem as most cities have combined sewer overflows. About 700 point sources emit CSOs from New Jersey and NYC into New York Harbor. In NYC, a CSO event occurs once a week on average, resulting in the discharge of approximately 500 million gal of raw sewage. Efforts to address this are being directed at construction of large (tens of millions of gallons) holding tanks.

Most coastal U.S. cities emit treated effluents into the ocean. One of the largest is the Deer Island Treatment Plant, which serves the metropolitan area of Boston. As shown in Figure 28.15, the outfall from this facility runs through a deep rock tunnel extending 9.5 miles east. The effluent is emitted into Massachusetts Bay from more than 400 ports that serve to diffuse and dilute the discharge. Prior to 2000, when this system came on line, effluent was being discharged into Boston Harbor. Secondary treatment was first implemented in 1997, along with source reduction of toxics.

In at least two offshore areas, changes in coastal currents and climate are thought to be contributing to an intensification of hypoxia in natural upwelling zones. The seasonal hypoxic zone found off the Oregon coast has been expanding in temporal and geographic scope and in the degree of O_2 deficiency. This has been attributed to wind shifts that are altering the timing and strength of upwelling. A similar effect is being observed in three settings in the Arabian Sea. First, along the Somalian and Omani coasts, enhanced upwelling has caused the average summer time plankton biomass to increase by a factor greater than 3.5. The wind shift responsible for the enhanced upwelling is thought to be caused by changes in the monsoons driven by a reduction in the ice cover on the Himalayan mountains. Second, in the open ocean of the Arabian Sea, an intensification of the permanent OMZ is attributed to increased organic matter export from the euphotic zone, also supported by enhanced primary production. The latter is partially due to blooms of a green dinoflagellate, *Noctiluca milaris*. This change in plankton growth dynamics is thought to be supported by an increased dust flux of iron. Third, on the eastern side of the Arabian Sea, subsurface O_2 deficiency and

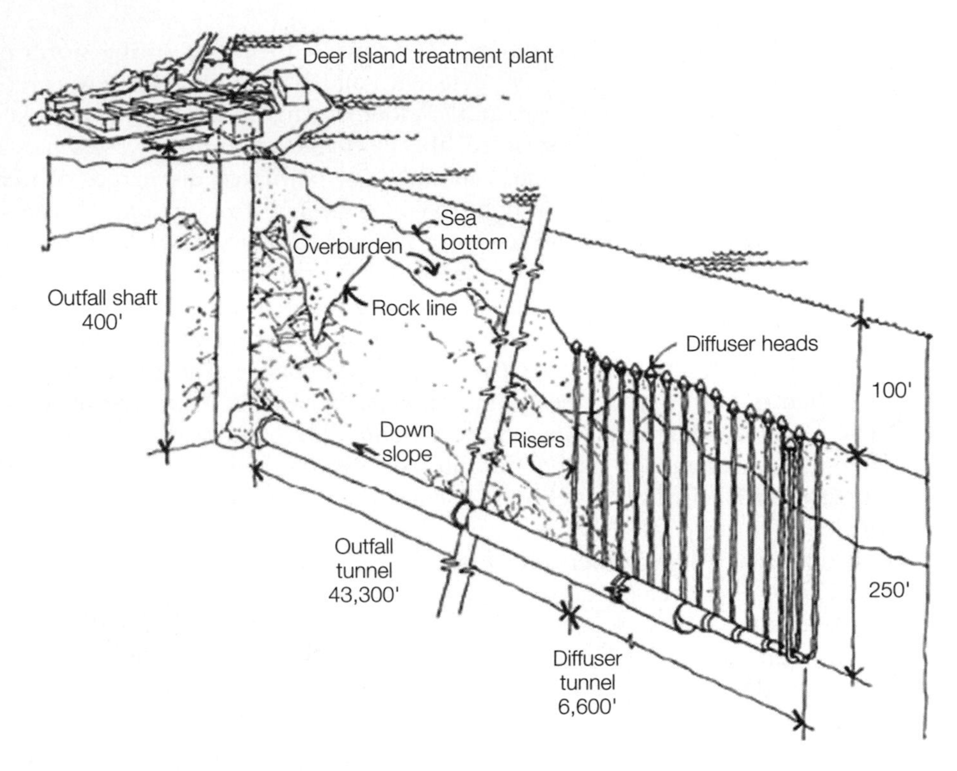

FIGURE 28.15

Boston Harbor outfall pipe at the Deer Island wastewater treatment plant. *Source*: From Massachusetts Water Resources Authority, the Massachusetts Bay Outfall, http://www.mwra.com/harbor/graphic/diffuser_linedrawing.git, Accessed July 2008.

denitrification are also increasing. These increases are attributed to nutrient loading from India via riverine runoff and atmospheric deposition. As illustrated in Figure W24.1, some of the bottom waters in this region are now experiencing anoxia accompanied by subsurface accumulations of H_2S.

Although the time series data for open-ocean O_2 concentrations are limited, evidence is beginning to emerge suggesting declines have been occurring over decadal time scales. For example, in the subarctic Pacific Ocean (Ocean Station Papua located at 50°N, 145°W), O_2 concentrations have decreased between 17% and 30% (20–40 μmol O_2 kg^{-1}) within the depth range of 125–300 m over the period between 1956 to 2006.[4] At this rate, hypoxic concentrations will be attained within a few decades. These waters are a significant feedstock to the upwelling zone that lies off the coast of the northwestern U.S. As noted above (and see Figure 8.4), this region is already characterized by a subsurface OMZ due to high productivity in the overlying surface waters. Thus any decrease in the

[4] Whitney, F. A., *et al.* (2007). *Progress in Oceanography*, 75, 179–199.

O_2 levels of the waters feeding this upwelling area are likely to contribute to an expansion in the intensity and vertical extent of its hypoxic zone. An important consequence would be a loss of oxygenated habitat in the underlying shelf and slope ecosystems resulting in a widespread mortality of benthic species.

Expansion of the vertical extent of the OMZ has already been observed in the eastern tropical Atlantic and equatorial Pacific Ocean where O_2 concentrations have been declining at rates of 0.09 to 0.34 μmol O_2 kg^{-1} y^{-1} over the past 50 years within the depths (150 to 800 m) in which hypoxic waters (< 90 μmol O_2 kg^{-1}) are present.[5] These open-ocean subsurface declines in O_2 are hypothesized to be a consequence of global climate change, resulting from three effects: (1) rising water temperatures lead to lower O_2 concentrations in newly forming water masses due to the inverse relationship between gas solubility and temperature, (2) respiration rates increase with increasing environmental temperatures, and (3) global circulation models predict that warming will slow the rate at which the deep sea is ventilated because of alterations in meridional overturning circulation and various turbulent mixing processes.

28.6.4 Harmful Algal Blooms

Harmful algal blooms (HABs) seem to be increasing in geographic scope and frequency. The term HABs is used very loosely to refer to any algal growths that exert harmful biological effects. These effects include: (1) poisoning from toxins, (2) mechanical damage such as the clogging or laceration of fish gills, and (3) various negative consequences resulting from the overproduction of biomass, such as hypoxia.

From the human perspective, HABs are problematic because they cause: (1) risks to human health, (2) loss of natural or cultured seafood resources, (3) impairment of tourism and recreational activities, and (4) damage to noncommercial marine resources and wildlife. Exposure pathways include: (1) consumption of toxic shellfish that have accumulated phytoplankton toxins filtered from the water, (2) consumption of tropical fish that have accumulated phytoplankton toxins (ciguatera), (3) inhalation of aerosolized toxins ejected from the sea surface, and (4) skin contact resulting in irritations due to allergy-like reactions. Harmful health effects from acute exposures have been relatively well studied. Less well known are the health effects resulting from chronic exposures to low toxin levels. This is of particular concern with regards to marine mammals and seabirds.

Nutrient pollution is thought to be a major causative agent in the apparent increase of HAB scope and frequency observed since the 1970s. Other candidate causes include: (1) alteration of food webs by overfishing, (2) modifications to water flow, (3) increased mariculture operations, (4) species dispersal through currents, storms, and other natural mechanisms influenced by climate change, and (5) transport and dispersal of HAB species via ballast water or shellfish seeding activities. Some of the reported increase is likely due to improved detection and monitoring.

---

[5] Stramma, L., *et al.* (2008). *Science*, 320, 655–658.

A further review of this subject is provided in the supplemental information for Chapter 26.6.4 that is available at **http://elsevierdirect.com/companions/9780120885305**.

28.6.5 Microorganisms

Pathogen pollution is the introduction of disease-causing agents as a direct or indirect consequence of human activities. The incidence of disease resulting from pathogen pollution appears to be increasing in both humans and marine organisms. At least some of this increase is due to global climate change, which has expanded the range of many species. In marine mammals, an increased incidence of infectious disease has been attributed, in part, to a weakened immune system caused by exposure to chemical pollutants, such as heavy metals and persistent organic pollutants.

Pathogenic microbes include bacteria, protozoans, and viruses. Parasitic worms are another common disease agent. The infective agents whose source is external to the ocean are termed *pollutogens*. Pathways of transport of pollutogens into the ocean include: (1) discharge of treated and untreated sewage including bilge pumping, (2) stormwater runoff, (3) groundwater seeps, (4) aeolian dust, and (5) off-loading of ballast water. Humans are exposed to pollutogens through skin contact and ingestion of contaminated seafood. Consumption of shellfish is particularly problematic as these filter feeders concentrate pathogens in their tissues. Since shellfish tend to live in estuaries, their exposure to pollutogens is high, especially following rain events, because stormwater runoff mobilizes contaminants from domesticated animal feces, broken sanitary sewer system pipes, and dysfunctional septic tanks. These same sources can contaminate groundwater and thereby eventually seep into the coastal waters. While sewage treatment can reduce pathogen levels, not all microbes are removed, so even treated effluents are a potential pollutogen source. Because of the extensive use of antibiotics, pathogenic bacteria are becoming increasingly antibiotic resistant and prevalent in treated effluents.

A further review of this subject is provided in the supplemental information for Chapter 26.6.5 that is available at **http://elsevierdirect.com/companions/9780120885305**.

28.6.6 Petroleum

The natural processes responsible for the formation of petroleum in marine sediments are described in Chapter 27. The original source of the organic compounds that make up petroleum is the biomass of marine plankton. Earthquakes and other natural geologic phenomena can cause failures in stratigraphic and structural traps, enabling oil and gas to seep naturally from reservoir beds into the ocean. As shown in Table 28.4, natural seeps are the single largest source of petroleum into the marine environment when viewed globally on an annual basis. This input accounts for about half of the total annual petroleum release, although considerable uncertainty is inherent in all of the estimates in Table 28.4.

Table 28.4 Average Annual Releases (1990–1999) of Petroleum by Source (in Thousands of Tonnes).

Sources	North America				Worldwide		
	Best Est.	Regions[a]	Min.	Max.	Best Est.	Min.	Max.
Natural seeps	160	160	80	240	600	200	2000
Extraction of petroleum	3.0	3.0	2.3	4.3	38	20	62
Platforms	0.16	0.15	0.15	0.18	0.86	0.29	1.4
Atmospheric deposition	0.12	0.12	0.07	0.45	1.3	0.38	2.6
Produced waters	2.7	2.7	2.1	3.7	36	19	58
Transportation of petroleum	9.1	7.4	7.4	11	150[+]	120	260
Pipeline spills	1.9	1.7	1.7	2.1	12	6.1	37
Tank vessel spills	5.3	4.0	4.0	6.4	100	93	130
Operational discharges (cargo washings)	na[b]	na	na	na	36	18	72
Coastal facility spills	1.9	1.7	1.7	2.2	4.9	2.4	15
Atmospheric deposition	0.01	0.01	trace[c]	0.02	0.4	0.2	1
Consumption of petroleum	84	83	19	2000	480[++]	130	6000
Land-based (river and runoff)	54	54	2.6	1900	140	6.8	5000
Recreational marine vessel	5.6	5.6	2.2	9	nd[d]	nd	nd

(Continued)

Table 28.4 *(Continued)*

Sources	North America				Worldwide		
	Best Est.	Regions[a]	Min.	Max.	Best Est.	Min.	Max.
Spills (nontank vessels)	1.2	0.91	1.1	1.4	7.1	6.5	8.8
Operational discharges (vessels ≥ 100 GT)	0.10	0.10	0.03	0.30	270	90	810
Operational discharges (vessels < 100 GT)	0.12	0.12	0.03	0.30	nd[e]	nd	nd
Atmospheric deposition	21	21	9.1	81	52	23	200
Jettisoned aircraft fuel	1.5	1.5	1.0	4.4	7.5	5.0	22
Total (without natural seeps)	260	250	110	2300	1300	470	8300

[a]*"Regions" refers to 17 zones or regions of North American waters for which estimates were prepared.*
[b]*Cargo washing is not allowed in U.S. waters, but is not restricted in international waters. Thus, it was assumed that this practice does not occur frequently in U.S. waters.*
[c]*Estimated loads of less than 10 tonnes per year reported as "trace."*
[d]*Worldwide populations of recreational vessels were not available.*
[e]*Insufficient data were available to develop estimates for this class of vessels.*

Notes:

I. Totals may not equal sum of components due to independent rounding.

2. Generally assumes an average specific volume of oil at 294 gallons per tonne (7 barrels per tonne). Where the specific commodity is known, the following values are applied when converting from volume to weight:

Gasoline: 333 gallons per tonne
Light distillate: 285 gallons per tonne
Heavy distillate: 256 gallons per tonne
Crude oil: 272 gallons per tonne

3. Numbers reported to no more than 2 significant figures using rules: http://web.mit.edu/10.001/Web/Course_Notes/Statistics_Notes/Significant_Figures.html.
[+]*Also included as a component in operational discharges.*
[++]*Significant petroleum hydrocarbon inputs into the oceans related to consumption of petroleum include river and urban runoff, oil spills from cargo ships, operational discharges from commercial vessels and recreational craft, and atmospheric deposition of petroleum hydrocarbons.*

Source: From National Research Council, National Academy of Sciences (2004). Oil in the Sea III: Inputs, Fates, and Effects. *National Academies Press, p. 69.*

In the 1990s, the largest anthropogenic contributors were (1) operational discharges from large vessels and tankers (24%), (2) land-based emissions via rivers and stormwater runoff (11%), and (3) catastrophic spills associated with nonroutine tanker operations, such as collisions, explosions, and running aground (8%). Other sources of catastrophic spills include oil well and pipeline accidents. Atmospheric deposition accounts for 4% of all inputs. Overall, each of the anthropogenic inputs has decreased significantly since the mid-1980s.

Anthropogenic inputs of petroleum can be separated into two categories: chronic and acute releases. Chronic discharges maintain a constant elevated level of contamination, whereas acute discharges support very high concentrations for short periods of time. Most acute releases are concentrated and, hence, are point-source in nature, such as catastrophic spills from tankers. An example of a routine tanker operation that is a chronic source of oil is bilge pumping. After a tanker offloads its oil, the bilge is filled with seawater to stabilize the ship for its trip back to home port. Just before arrival, the oily ballast water is pumped back into the ocean. Other operational discharges from large ships are associated with refueling and leaks from inefficient engines.

The effects of operational discharges are most evident in coastal waters. Satellite imagery is now being used to assess oil spill coverage by detecting oil slicks. As shown in Figure 28.16, a large fraction of the Mediterranean Sea is affected by oil spills. In the open ocean, oil slicks of anthropogenic original are most common in the North Atlantic and North Pacific. Although bilge pumping is now banned in most coastal waters, noncompliance is common. Other chronic sources of petroleum are diffuse in nature. These include stormwater runoff and groundwater seeps contaminated by (1) washoff of petroleum residues from roads, (2) disposal of motor oil in storm drains, and (3) release of municipal

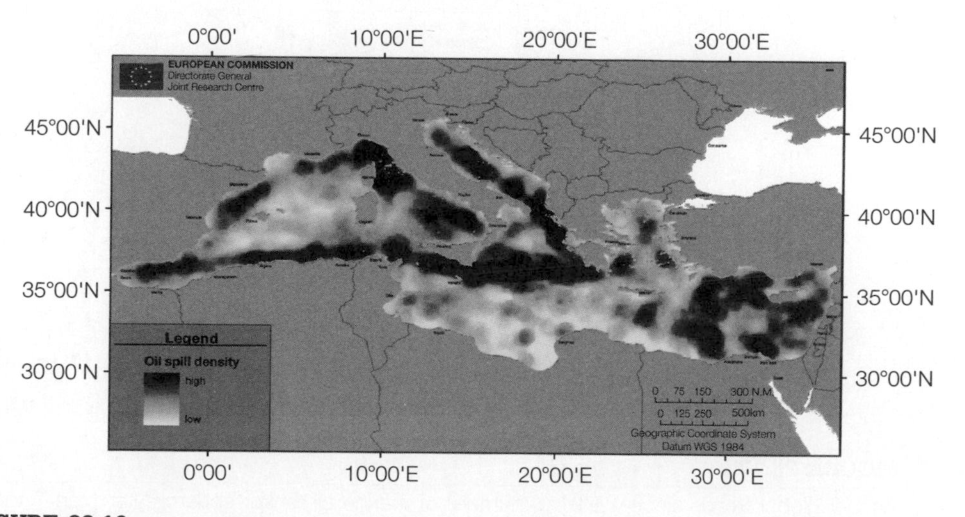

FIGURE 28.16

Oil spill density for the Mediterranean Sea for the period 1999 to 2004. *Source*: After Ferraro, G., *et al*. (2007). *Marine Pollution Bulletin* 54, 403–422.

and refinery waste waters into rivers and coastal waters. This source is expected to increase due to the projected growth of population in coastal areas.

The number of catastrophic oil spills has been decreasing in number and volume since the 1980s (Figure 28.17). The largest oil spills on record are listed in Table 28.5a. Most

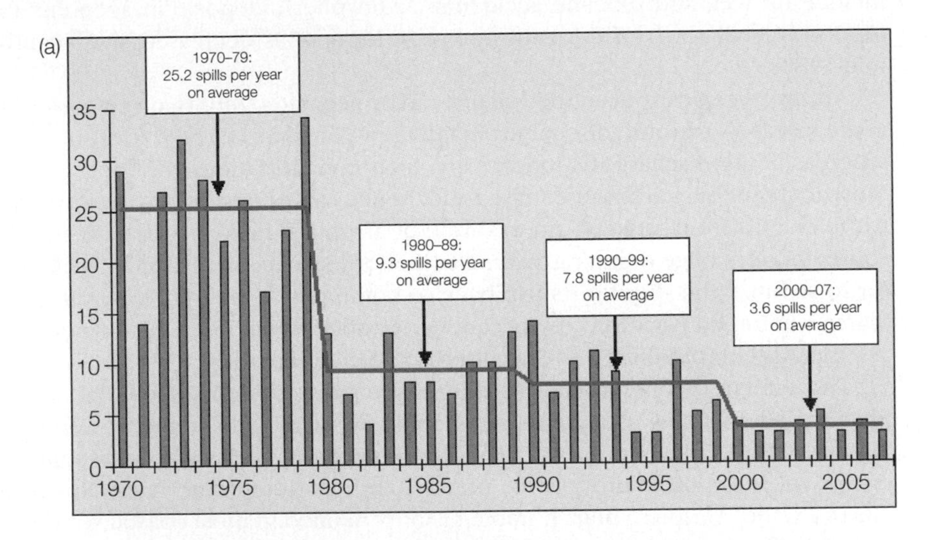

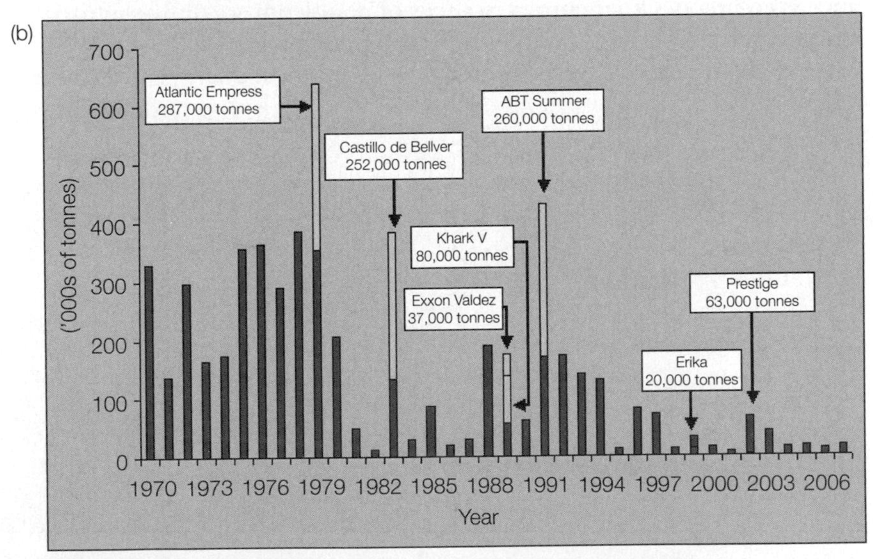

FIGURE 28.17

Annual global trends since 1970: (a) number of marine oil tanker spills greater than 7 tonnes; (b) volume of oil spilled into the oil in thousands of tonnes from those spills. *Source*: From International Tanker Owners Pollution Federation Limited (2007). Statistics http://www.itopf.com/information-services/data-and-statistics/statistics/#major.

Table 28.5a Oil Spills over 100,000 Tonnes (30 Million U.S. Gallons).

Spill/Tanker	Location	Date	Tons Crude Oil
Gulf War oil spill	Persian Gulf, off coasts of Kuwait and Saudi Arabia	1991 January 23	1,770,000
Ixtoc I exploratory oil well	Gulf of Mexico, Bahia Del Campeche	1979 June 3–1980 March 23	454,000–480,000
Atlantic Empress collision with *Aegean Captain*	Caribbean Sea, 32 km northeast of Trinidad and Tobago	1979 July 19	287,000
Fergana Valley oil well[a]	Uzbekistan, Fergana Valley	1992 March 2	285,000
Nowruz platform oil well	Persian Gulf, Nowruz field	1983 February	260,000
Tanker *ABT Summer*	Atlantic Ocean, 700 nautical miles off Angola	1991 May 28	260,000
Tanker *Castillo de Bellver*	Saldanha Bay, South Africa	1983 August 6	252,000
Tanker *Amoco Cadiz*	Atlantic Ocean, Brittany, France	1978 March 16	223,000
Tanker *Haven*	Mediterranean Sea, near Genoa, Italy	1991 April 11	144,000
Tanker *Odyssey*	Atlantic Ocean, 700 nautical miles off Nova Scotia, Canada	1988 November 10	132,000
Tanker *Sea Star*	Gulf of Oman	1972 December 19	115,000
Tanker *Torrey Canyon*	Atlantic Ocean, between Isles of Scilly and the western coast of Cornwall, UK	1967 March 18	121,000
Tanker *Irenes Serenade*	Mediterranean Sea, Navarino Bay, Greece	1980 February 23	100,000
Tanker *Urquiola*	Atlantic Ocean, La Coruna Harbor, Spain	1976 May 12	101,000
Tanker *Texaco Denmark*	North Sea, off Belgium	1971 December 7	102,000
Tanker *Hawaiian Patriot*	Pacific Ocean, 593 km west of Kauai Island, Hawaii	1977 February 23	101,000
Storage tanks[a]	Kuwait, Shuaybah	1981 August 20	101,000
Kharyaga-Usinsk Pipeline[a]	Russia, Usinsk	1994 October 25	100,000

[a]*Not marine*

One tonne of crude oil is roughly equal to 308 U.S. gallons, or 7.33 barrels.

Source: After International Tanker Owners Pollution Federation Limited (2007). Statistics, http://www.itopf.com/information-services/data-and-statistics/statistics.

Table 28.5b Major Oil Spills in Persian Gulf and Gulf of Oman.

Date	Spill	County	Location	Million Gallons
26 January 1991	Terminals, tankers, burning of oil wells	Kuwait	Off coast in Persian Gulf and in Saudi Arabia	240.0
4 February 1983	Platform No. 3 well	Iran	Persian Gulf, Nowruz Field	80.0
19 December 1972	Tanker *Sea Star*	Oman	Gulf of Oman	37.9
20 August 1981	Storage tanks	Kuwait	Shuaybah	31.2
25 May 1978	Well and pipeline No. 126	Iran	Ahvazin	28.0
6 December 1985	Tanker *Nova*	Iran	Persian Gulf, 140 km south of Kharg Island	21.4
9 December 1983	Tanker *Pericles GC*	Qatar	Persian Gulf, 30 km east-northeast of Doha	14.0
26 August 1979	Ore/bulk/oil carrier *Patianna*	UAE	Persian Gulf, 11 km off Dubai	11.2

of these are the result of marine tanker accidents. The most notable exceptions are the two largest events, the Ixtoc I oil well blowout in the Gulf of Mexico, and the massive crude oil release resulting from the Persian Gulf War of 1991. The latter was three times as large as the Ixtoc disaster, which itself was close to twice the size of the next largest spill.

The release in the Persian Gulf is considered to be an act of environmental and economic warfare. About 5 million barrels was released by the Iraqi military from offshore terminals and tankers a few days after Coalition forces began bombing in response to Iraq's invasion of Kuwait. This was followed a month or so later, at the start of the ground war, by ignition of 650 Kuwaiti oil wells and uncapping of another 82 wells. The former was estimated to have burned 3 to 4 million barrels of oil a day with smoke fallout affecting the nearby water and land. The releases from the 82 uncapped wells created huge basins of oil. Although much of the petroleum evaporated or dispersed, a significant fraction washed ashore, destroying over 400 mi of coastal habitat and wildlife, including more than 30,000 seabirds. Follow-up surveys in 1992 and 1993 found that stranded oil had penetrated up to 40 cm into the sediment. Persistent pavements had formed in the upper intertidal zone, effectively sealing subsurface oil in place. Because Kuwaiti crude oil forms a very stable emulsion, most that washed ashore stranded in the intertidal zone, leaving the subtidal region relatively unimpacted. A similar volume of crude oil was released in the Persian Gulf and neighboring Gulf of Oman over the period of 1978 to 1985 as a result of four tanker and two oil-well accidents (Table 28.5b).

A particularly notable chronic spill exists in the Greenpoint neighborhood, which lies along Newton Creek in Brooklyn, New York. Petroleum refining has been conducted in this area since 1866. By 1870, more than 50 refineries were located along the banks of

Newtown Creek. Over the years, millions of gallons of oil and petroleum by-products were spilled on land. A significant fraction of these spills seeped into Newton Creek. When first discovered in the 1970s, the oil that had accumulated underground was estimated at 17 million gal and appeared to be spread over 100 acres. About 9.3 million gal has been recovered and the spill area reduced to 55 acres.

Petroleum pollution in the marine environment has a variety of negative impacts on ocean and human health. The nature of these impacts depends on the chemical composition and concentration of the petroleum along with the length of exposure. As discussed in Chapter 27, the chemical composition of crude oil is highly variable as it reflects the original organic carbon source and environmental conditions of diagenesis, catagenesis, and migration. Petroleum discharged into the marine environment continues to undergo physical and chemical changes as illustrated in Figure 28.18. These changes are collectively referred to as *weathering*. The ultimate fate of the petroleum depends on the solubility, volatility, and chemical reactivity of its component hydrocarbons as well as environmental conditions, including: (1) sea state, (2) wind speed,

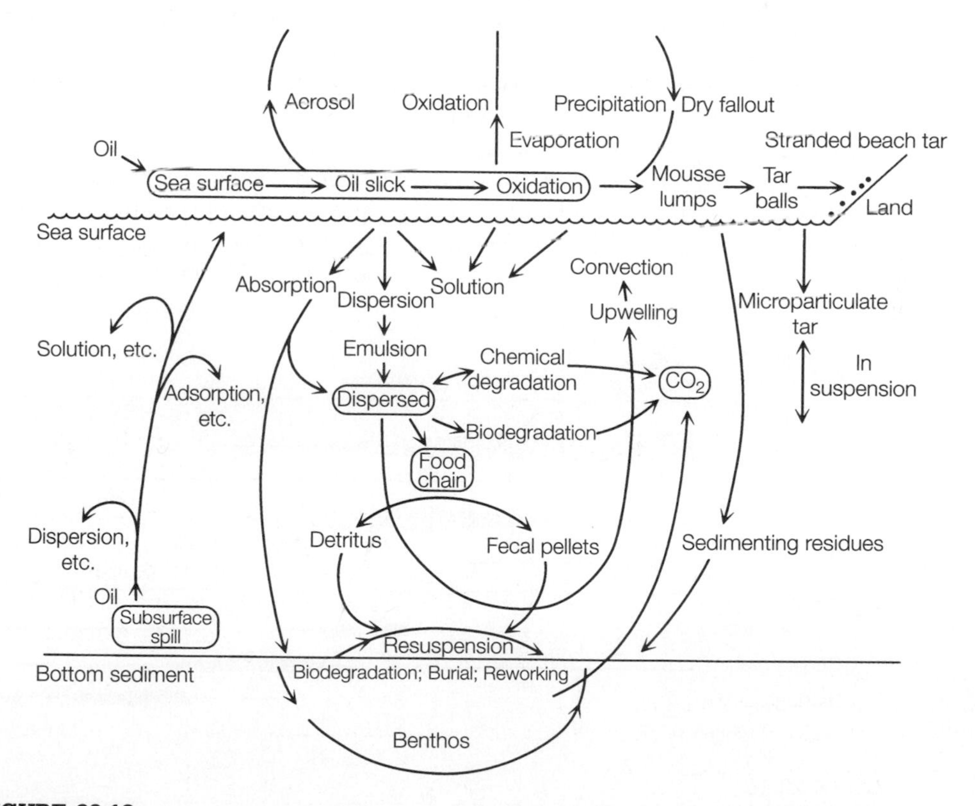

FIGURE 28.18

The weathering of a crude oil slick at sea. *Source*: From Preston, M. R. (1988). *Chemical Oceanography*, Vol. 9, Academic Press, pp. 53–96.

(3) temperature, (4) water depth, (5) geology of the seafloor and shoreline, and (6) local biological activity.

The density of crude oil is on the order of 0.85 g/cm^3, so if the sea surface is calm, an oil spill will initially form a slick. The slick is subject to physical processes, such as advection and turbulence, causing it to move vertically and/or horizontally. Advection tends to lead to dispersal or, if land is nearby, shoreline stranding. Turbulence promotes the formation of emulsions, called "chocolate mousse," which can be transformed via weathering into tarballs. The lower-molecular-weight compounds tend to evaporate or dissolve. Some fractions of petroleum have solubilities in seawater on the order of tens of milligrams per liter. Some are also photochemically oxidized.

As shown in Figure 28.19, the chemical changes involved in weathering are largely complete within 1 d after the oil spill, leaving a residue that is far different in composition from the original petroleum discharge. On longer time scales, the altered petroleum is transported either horizontally by currents or vertically by the sinking of tar balls. Some petroleum is also carried to the seafloor as a result of adsorption onto sinking particles or incorporation into POM following biological uptake. Tarballs can be transported great distances, due to their inert nature and low density, before reaching the seafloor or becoming stranded on shorelines. This has caused them to become a common feature on beaches worldwide. Petroleum that reaches the sediments tends to persist for decades due to slow biodegradation rates. Thus, the toxic impacts on the benthos can also persist for decades following deposition of petroleum onto the seafloor.

Exposure to oil pollution can cause sublethal effects on marine biota, such as reduced growth rate and reproductive failure. The type and severity of the effect depends on: (1) the concentration of petroleum, (2) length of exposure, (3) persistence and bioavailability of specific hydrocarbons, and (4) the ability of organisms to accumulate and metabolize

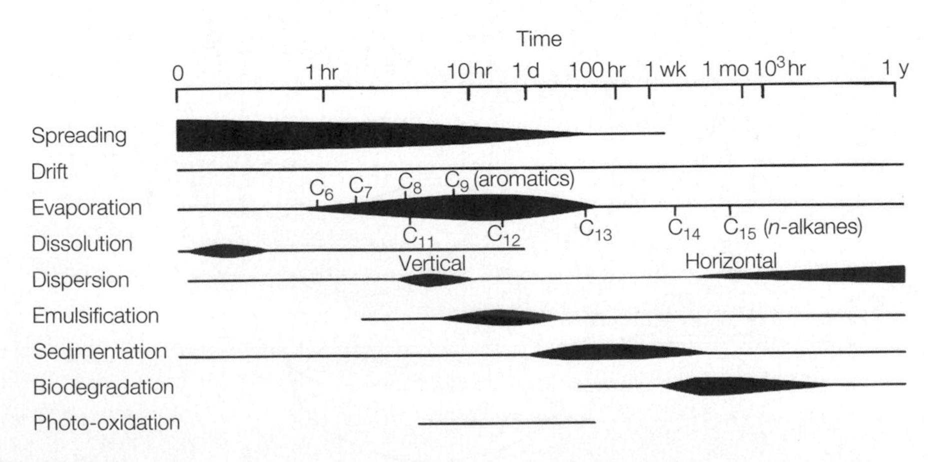

FIGURE 28.19

Time course of factors affecting oil spilled on the sea. *Source*: From Whittle, K. J., *et al.* (1982). Philosophical Transactions of the Royal Society of London, Series B, Biological Sciences 297, 193–218.

the petroleum, and (5) the degree to which the specific hydrocarbons interfere with normal metabolic processes critical to survival and reproduction. For example, some hydrocarbons have narcotic effects on nerve transmission. In general, biological effects of oil pollution are greatest on immature marine animals, i.e., egg, larvae, and juveniles. Species that release epipelagic eggs are at particularly high risk from surface oil slicks. In areas where biological stress is also present, such as from overfishing or other pollutants, petroleum contamination adds further stress. From a physical standpoint, oil can damage habitats by smothering the benthos and by coating skin, fur, and feathers. On seabirds, a coating of oil can lead to death from hypothermia or drowning because the petroleum alters the insulating properties of feathers and impairs swimming ability. For humans, oil on beaches detracts from recreational uses, such as sunbathing and surf fishing, and contaminates seafood.

Most of our knowledge of the biological effects of petroleum pollution is based on studying the acute effects of major spills or heavily contaminated sediments. These effects tend to be directly lethal. Acute toxicity has been found to be largely related to water solubility, with cumulative toxicity reflecting the sum of the effects of each individual hydrocarbon. Relatively little is known about the effects of chronic exposures at lower concentrations, especially in the presence of other stressors, such as heavy metals. Effects of chronic inputs from land-based sources on populations, communities, and ecosystem structure and function are also not sufficiently known.

Particular attention has been focused on the toxic effects of aromatic hydrocarbons because these chemicals have proven highly carcinogenic to humans and marine life. Of greatest concern are the PAHs, which are toxic to the benthos at the ppb level. The most common compounds are shown in Figure 28.20; their structures are based on fused aromatic rings. These high-molecular-weight compounds are very nonpolar and, hence, have low solubilities. Once in seawater, they tend to adsorb onto particles and become incorporated in the sediments. The toxicity of PAHs is enhanced by photochemical reaction with UV radiation. Photo-activated toxicity is especially problematic in shallow-water sediments, such as found in estuaries.

PAHs are a common, although low-level, constituent of crude oil. They are also formed during the burning of organic matter, including fossil fuels, wood, and meat. PAHs are highly concentrated in asphalts and petroleum-based road sealants, both of which leach toxicants into stormwater runoff. Although PAHs are naturally generated during forest fires, volcanic eruptions, and anaerobic diagenesis, the anthropogenic source is now dominant. High levels of contamination are common in the sediments of urban harbors particularly near CSOs. This presents a problem for harbor dredging as the spoils must treated as hazardous wastes. Because PAHs are produced during burning, they are a common constituent of aeolian particles and have been transported worldwide, leading to widespread contamination. For example, PAHs are now found in polar ice cores.

Because PAHs are nonpolar compounds, they are hydrophobic and, hence, highly retained in the fatty tissues of organisms. This is most unfortunate given their high toxicity, which in addition to cancer, has been shown to include reproductive failures and impairments to immune systems. Fish and mammals tend to metabolize and excrete PAHs fairly rapidly, so these compounds are less prone to bioaccumulation or biomagnification

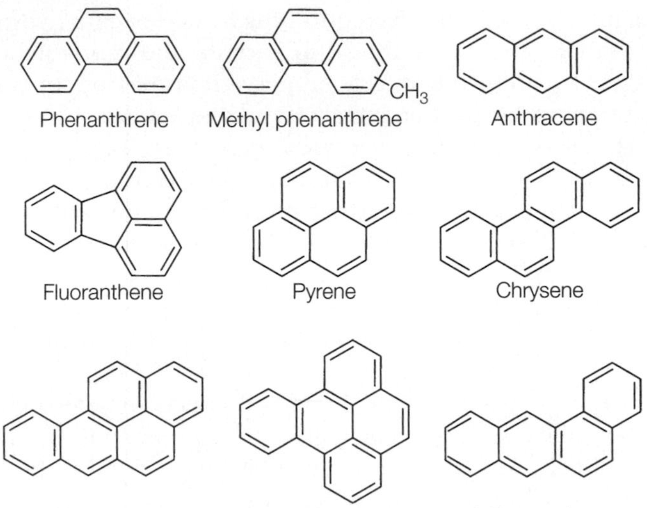

CH_3

Phenanthrene Methyl phenanthrene Anthracene

Fluoranthene Pyrene Chrysene

3, 4-Benzopyrene 1, 2-Benzopyrene 1, 2-Benzanthracene

FIGURE 28.20

Structures of polynuclear aromatic hydrocarbons found in natural waters.

than other nonpolar organic pollutants such as the PCBs and chlorinated pesticides. The toxicity of the PAHs is intensified by exposure to UV radiation, such as is found in shallow benthic environments.

28.6.7 Radioactivity

Radioactivity is a form of energy emitted by radioactive elements (radioisotopes or radionuclides). Radionuclides can present a health threat to humans and marine organisms because of the ionizing ability of the emitted radiation. The major pathway by which marine organisms and humans are exposed to radionuclides comes from consumption of fish and shellfish due to bioaccumulation of ^{210}Po and ^{210}Pb. Most of the ^{210}Po and ^{210}Pb in the ocean is natural, but, human activities have increased their inputs to the coastal ocean.

The anthropogenic sources of radionuclides to the marine environment can be divided into two categories: (1) emissions of artificially generated isotopes and (2) mobilization of naturally occurring isotopes. Emissions of the artificially generated isotopes result from activities of the nuclear industries. They include weapons testing, accidents, dumping, and discharges from power and fuel reprocessing plants, and well as disposal of medical wastes. Humans mobilize the naturally occurring isotopes during mineral mining, oil drilling, and fossil-fuel burning.

The global dispersion and deposition of debris from atmospheric nuclear weapons is by far the largest source of artificial radioactivity to the terrestrial and marine environment. Nevertheless, this caused only a slight increase in the total global inventory of radioactivity, with the major contributor being tritium (^3H). Evidence for significant radionuclide exposures to humans and marine life seems to be limited to a few areas where inputs have been geographically concentrated, such as around the discharge outfalls serving nuclear fuel reprocessing plants.

The fate of radionuclides in the marine environment is similar to that of the stable isotopes, being dependent on chemical speciation, including redox state, solubility, and tendency to form complex ions. For example, $^{238,239,240}Pu$ and ^{241}Am are particle reactive, whereas ^{137}Cs, ^{99}Tc, and ^{129}I are not. The particle reactive radionuclides are strongly concentrated in marine sediments. Some, such as ^{210}Po, also have high bioaccumulation factors in fish and shellfish. The artificial radionuclides that exhibit conservative behavior, such as ^{3}H, ^{90}Sr, and ^{137}Cs, have been used by oceanographers as tracers of ocean currents because most of their injection occurred over a very short period as a result of above-ground nuclear-bomb testing.

The scope and scale of pollution from radionuclides has been greatly reduced due to: (1) the cessation of aboveground nuclear bomb testing, (2) an international ban on the dumping of nuclear wastes at sea, and (3) better control of discharges from power and fuel reprocessing plants.

A further review of this subject is provided in the supplemental information for Chapter 26.6.7 that is available at **http://elsevierdirect.com/companions/9780120885305**.

28.6.8 **Trace Metals**

Mining and fossil-fuel burning have greatly increased the rate at which some trace metals are being transferred from Earth's crust to the atmosphere, hydrosphere, and biosphere. Estimates of natural and anthropogenic fluxes for the most affected of the metals (Cd, Hg, Pb, Cu, Zn, As, Cr, Sn, and Ni) are presented in Table 11.2. Distinguishing anthropogenic impacts from natural inputs is complicated as some of our effects are essentially perturbations of natural processes, such as accelerating the rate of weathering and forest fires. Humans are also delivering trace metals to seawater via dumping, outfall discharges, and the use of organometallic paints that prevent marine biofouling. Indirect anthropogenic inputs are occurring as part of river flows, groundwater seeps, and atmospheric transport.

Most of the trace metals carried by rivers are converted into particulate form as a result of biogeochemical processing in estuaries and coastal waters. This leads to their deposition in nearshore sediments. Thus, atmospheric transport is the major source of metal pollution to the open ocean (Table 11.2). The major source of these metals is fossil-fuel combustion followed by mining and the smelting of mineral ores and scrap metal. Aeolian transport involves both the wet and dry deposition of aerosols. Some metals are also present in the gas phase, i.e., mercury (Hg^0 and methylated forms), lead (as tetraalkyl lead), and arsenic (AsH_3 and methylated forms). Cadmium is also considered to be a semivolatile metal. The gaseous species tend to dissolved in rain water or adsorb to settling particles. They can also undergo gas exchange across the air-sea interface.

Spatial and Temporal Trends

Winds have carried the aerosolized and gaseous metals worldwide. Their eventual deposition has left a record in ice cores, lake sediments, peat, and coastal marine sediments of large increases coinciding with historical events, such as the development of ore smelting techniques, the Industrial Revolution, and the adoption of fossil fuel as a primary energy resource. Many of the metals are bioaccumulated and biomagnified, causing an increased body burden in many marine organisms.

Anthropogenic metal inputs to the oceans from atmosphere transport show strong regional variations reflecting proximity to, and intensity of, the pollutant source. Thus, trace metal pollution is highest in the North Atlantic, followed by the North Indian and North Pacific. The southern hemisphere contamination is lower than that in the North, following the same ocean pattern as in the north, i.e., South Atlantic greater than South Indian greater than South Pacific. As shown in Table 11.2, some of the metals had anthropogenic inputs close or equal to those from natural sources in the mid-1980s. Most of these emissions were from coal-fired power plants, and for lead, from the use of leaded gasoline in automobiles. By the mid-1990s, environmental controls implemented in North America and Europe had caused these atmospheric emissions to drop substantially for all metals, except for V and Ni. The ongoing increase in these metals is attributed to a worldwide increase in oil combustion for electricity production and heating. Regionally, Asia is the largest atmospheric source of anthropogenic metal emissions. Because of limited environmental controls and population growth, emissions in this region are continuing to increase over time.

Speciation and Toxicity

The trace metals listed in Table 11.2 (with the inclusion of Sn) are of particular concern as they are toxic at low concentrations. For historical reasons, these elements are commonly referred to as the *heavy metals*. The degree to which the heavy metals cause toxic effects is dependent on their concentration, chemical speciation, and other environmental conditions, such as temperature. As illustrated in Table 28.6, the type and physiological state of the target organism are also important as these factors determine the degree to which internal metabolic processes can detoxify or eliminate the pollutant.

The chemical speciation of metals is of particular importance in determining its biological effect. The physical and chemical factors influencing chemical speciation were discussed in Chapter 5 and are summarized in Table 28.7. In general, methylated forms tend to be most toxic as they are more biologically reactive than inorganic forms. The methylated metals species are also the species most greatly biomagnified. Methylation is thought to be the result of anaerobic bacterial activity. Many methylated metals are also quite volatile.

Bioconcentration and Biomagnification

A few of the heavy metals, such as Zn and Ni, are also micronutrients. As shown in Figure 28.21, these elements have a stimulatory effect on biological activity when present at low concentrations. They become toxic at concentrations above an optimal threshold

Table 28.6 Factors Influencing the Toxicity of Heavy Metals in Solution.

Factor			
Form of metal in water	Inorganic	Soluble	Ion
	Organic		Complex ion
			Chelate ion
			Molecule
		Particulate	Colloidal
			Precipitated
			Adsorbed
Presence of other metals or poisons	Synergy		
	No interaction		
	Antagonism		
Environmental factors (influences on the physiology of organisms and possible form of metal in water)	Temperature		
	pH		
	Dissolved oxygen concentration		
	Light intensity		
	Salinity		
Condition of organism	Stage in life history (egg, larva, etc.)		
	Changes in life cycle (e.g., molting, reproduction)		
	Age and size		
	Sex		
	Starvation		
	Activity		
	Additional protection (e.g., shell)		
	Adaptation to metals		
Behavioral response			

Source: *From Bryan, G. W. (1976).* Marine Pollution, *Academic Press, pp. 185–302.*

level. In contrast, the nonessential heavy metals, such as Hg and Pb, have no stimulatory effect at any concentration. Some have very low threshold concentrations above which toxic effects are induced. Most heavy metals tend to have very high enrichment factors (Table 11.5 and W28.11) and slow clearance rates. As a result, biological uptake is a significant sink for these metals and, hence, can provide some measure of water purification.

Various terms have been defined to quantify the degree to which chemicals accumulate in the tissues and hard parts of marine and aquatic organisms. As illustrated in Figure 28.22, the bioconcentration factor (BCF) is used to quantify uptake from the surrounding medium (seawater). The biomagnification factor (BMF) is a measure of the uptake resulting from dietary intake. Biomagnification is important primarily in seabirds

Table 28.7 Some Factors Affecting Chemical Speciation and Thus Toxicity of Trace Metals in Estuarine and Marine Organisms.

Variables	Explanation
pH, alkalinity, organic and inorganic ligands	Changes metal species distribution and free ion concentration. Influences formation of hydroxo, carbonato, and other complexes. Changes absorbability of metal ions and cellular uptake rates.
Density of organisms	Reduces available total dissolved metal concentrations and changes metal species distribution because of adsorption on cell surfaces and/or by complexation with exudates from organisms.
Concentration of particles and colloids	Metals are sequestered by particulate oxides of iron and manganese. Organic colloids are particularly effective at binding metals.
Redox potential	Affects oxidation state of metal. Biomethylation often occurs more readily at low redox potentials.

Source: *From Stumm, W. S. and J. J. Morgan (1981).* Aquatic Chemistry, *2nd ed. John Wiley & Sons, p. 702.*

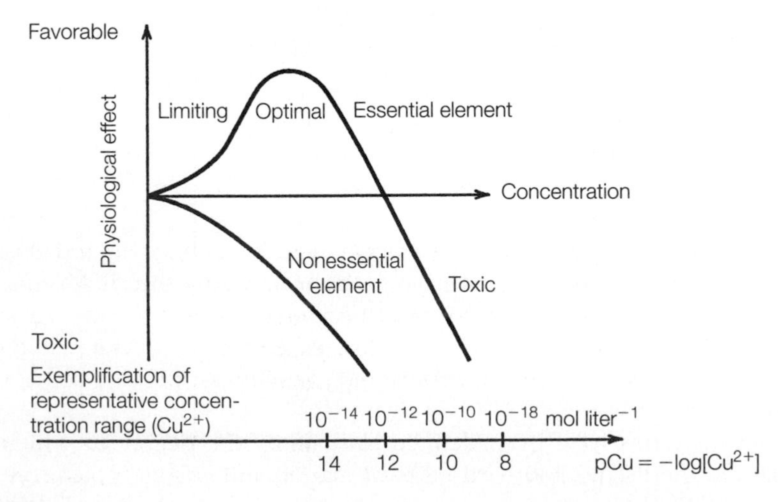

FIGURE 28.21

Relation between the concentration of an element and its physiological effect. *Source*: From Stumm, W., and J. J. Morgan (1996). *Aquatic Chemistry, Chemical Equilibria and Rates in Natural Waters*, 3rd ed. John Wiley & Sons, Inc., p. 633.

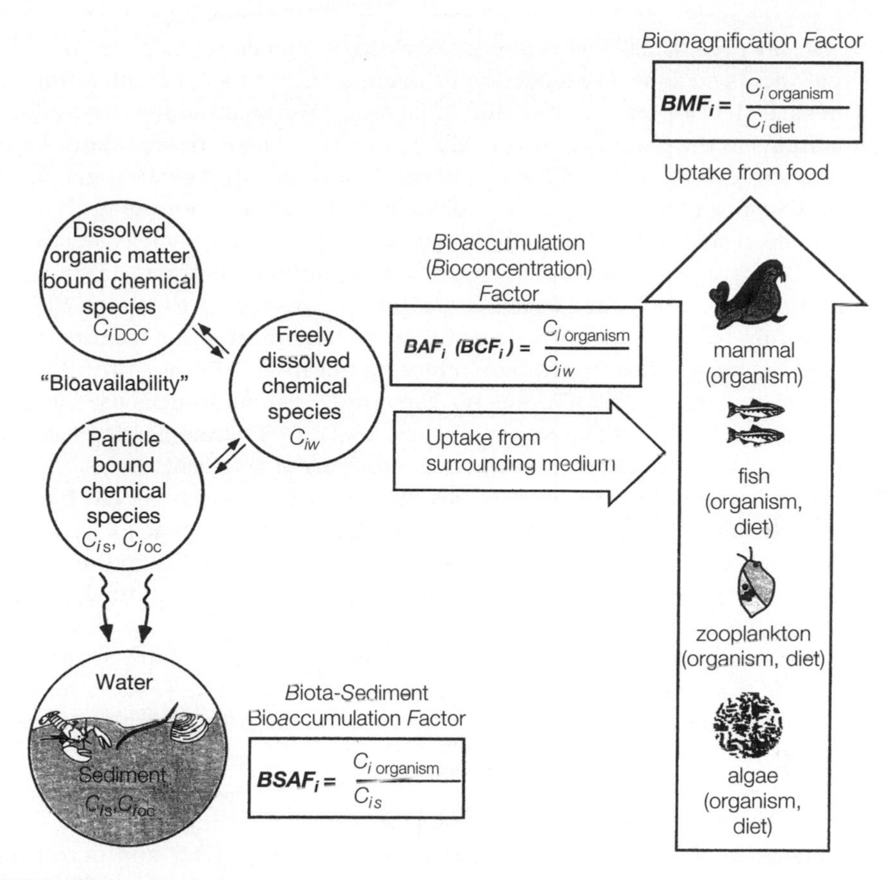

FIGURE 28.22

Terms and parameters frequently used to describe accumulation of chemicals in aquatic organisms. *Source*: From Schwarzenbach, R. P., *et al.* (2003). *Environmental Organic Chemistry*, 2nd ed. Wiley Interscience, p. 345.

and marine mammals, reflecting their high trophic level in marine food webs. The bioaccumulation factor (BAF) describes the total accumulation by all possible routes, including bioconcentration and biomagnification. The degree of bioconcentration and biomagnification tends to be linearly related to metal concentrations in seawater and diet, respectively. This enables definition of BCFs, BMF, and BAFs as concentration ratios. As shown in Tables 11.5 and W28.11, the values of these ratios are dependent on the metal being enriched and the target organism.

The tendency of a marine organism to bioaccumulate a toxic metal depends in part on its lifestyle, as this determines the degree to which it is exposed to elevated concentrations in seawater or the sediments. For example, benthic animals that burrow will be exposed to sediment pore waters rather than to the overlying seawater. Animals that use mucous feeding nets, such as larvaceans, will be more prone to sorb metals because of the large surface area of their nets.

On an organismal level, responses to toxic metals include death and more subtle impairments such as: (1) induction of tolerance/resistance, (2) inhibition of growth, (3) abnormal development (especially in larvae), (4) impairment of reproduction, and (5) alterations in the hatching rate of eggs. Some, like Hg, are nerve toxins. Ingestion of contaminated seafood and shellfish can impair human health. The most graphic examples are mercury poisoning (Minamata Bay disease) and cadmium poisoning (Itai-Itai disease).

Shellfish strongly bioaccumulate a variety of pollutants, such as metals and pesticides. Thus, their tissue concentrations provide a long-term record of the average pollutant concentrations to which they have been exposed. This makes them a useful environmental monitoring tool, commonly referred to as a "bioindicator" or "sentinel" species. In the United States, shellfish-based biomonitoring began in 1976 with the Mussel Watch program. The current version of this program provides data sources used in formulating the National Coastal Condition Report (Figure 28.2). This biomonitoring has documented an overall long-term decline in metal concentrations of shellfish.

On a worldwide basis, toxic concentrations of the heavy metals have thus far been limited to industrialized harbors. The only metals that appear to have accumulated to toxic levels on a regional scale are mercury, cadmium, and lead in the Arctic Ocean. This concentration of mercury and lead has been facilitated by a natural process, called the *grasshopper effect*, which acts to transport volatile compounds poleward. This transport plays a major role in redistributing the volatile organic pollutants, such as the PCBs, and, hence, is discussed at further length in Chapter 26.7. The process responsible for the cadmium enrichment in the Arctic appears to involve low-altitude transport of the fine particles that compose Arctic haze.

Trace Metals as Anthromarkers in Coastal Sediments

Discharge of municipal and industrial wastes through river and ocean outfalls injects metals directly into natural waters. As shown in Table 28.8, these discharges have high concentrations of the toxic trace metals. Some of these metals sorb onto particles and, hence, accumulate in the sediments. Some of the discharged metals are bioaccumulated. The efficient removal of trace metals is a characteristic of estuaries as exemplified in Figure 5.1 for iron. In contrast to the anthropogenic emissions, most of the natural riverine load of metals is present in a biologically unavailable form, i.e., as part of the crystal lattice of weathered aluminosilicate minerals. Some is also reversibly sorbed to these clays.

Desorption of natural and pollutant metals from particulate matter can occur, especially during transport through a salt gradient such as found in estuaries. As the particles enter increasingly saline water, the major cations displace the less abundant trace metals from the exchange sites on the clays. The degree to which a heavy metal undergoes ion exchange is related to its ionic charge density (z/r). This causes the order of desorption for the heavy metals to be Hg > Cu > Zn > Pb > Cr > As. This pool of desorbed metals is important because it is biologically available.

The solubilized metals form complexes with organic and inorganic anions. The chemical speciation of these complexes changes as the metal moves seaward through the estuary due to increasing salinity. These shifts can be predicted from equilibrium speciation calculations as described in Chapter 5.7. Two examples are shown in Figure 28.23 for

Table 28.8 Comparison of the Dissolved Concentrations of Trace Metals (nM) in the Ocean Outfall Seawater off the Tanshui Coast and in Other Heavily Polluted Marine Environments around the World.

Location	Al	Cd	Cr(VI)	Cu	Fe	Mn	Pb	Zn
Tanshui coast	66.9–131.4	0.03–0.14	4.35–6.86	0.37–1.74	24.3–93.0	8.5–24.2	0.09–0.43	10.3–74.9
The San Francisco Bay	ND	ND	ND	9–73	8–1680	ND	ND	3.6–28
The Hudson estuary	ND	ND	ND	15–97	15–720	8–1100	ND	45–460
The Euripos Straits, Greece	ND	0.36–3.38	ND	7.1–325	ND	6.2–56	0.7–13	39–1835
The Humber estuary	ND	ND	ND	31–157	ND	6.2–56.1	ND	54–306
The Seine estuary	ND	ND	ND	26–40	ND	20–450	ND	90–200
The Scheldt estuary	ND	0.09–1.42	ND	6.3–25.2	ND	ND	0.14–1.45	7.7–191
Port Jackson estuary, Australia	ND	0.05–0.89	ND	14.7–40	ND	6–1838	ND	50–148
North Atlantic[a]	—	0.04–0.22	1.7–2.3	0.8–1.6	0.45–2.7	0.2–0.5	0.02–0.10	3.0–4.6

ND, no data.
[a]*Schmidt, D., and W. Gerwinski, (1992). A baseline study for trace metals in the open Atlantic Ocean. International Council for the Exploration of the Sea.* ICES CM 1992/E18, *15pp.*
For other data sources, see Fang, T.-H., et al. (2006). Marine Environmental Research *61, 224–243.*

mercury and copper. At low salinities, mercury speciation is dominated by complexation with humic substances and by chloro-complexes at high salinity. Copper exhibits a similar shift, with humic complexation at low salinities and inorganic complexation with hydroxide and carbonate at high salinities. Part of these shifts reflect increased concentrations of the inorganic ions at high salinity and part, to the removal of humic substances via flocculation reactions promoted by increasing ionic strength. Other colloidal materials, such as metal oxyhydroxides, tends to coprecipitate as ionic strength increases. Most flocculation is complete by the time estuarine mixing has elevated salinities to 15‰. Bioconcentration associated with high biological productivity in estuaries also serves to transform trace metals into particulate form. Thus, multiple processes conspire to trap trace metals in estuarine sediments.

Because of this trapping effect, estuarine sediments can provide excellent chronological records of contamination histories. An example is provided in Figure 28.24 for sediment collected from two sites located in two watersheds in the Elizabeth River, VA, USA. In general, sedimentary trace metals tend to be closely related to the transport and deposition of fine-grained and organic-rich sediments. The downcore profiles presented

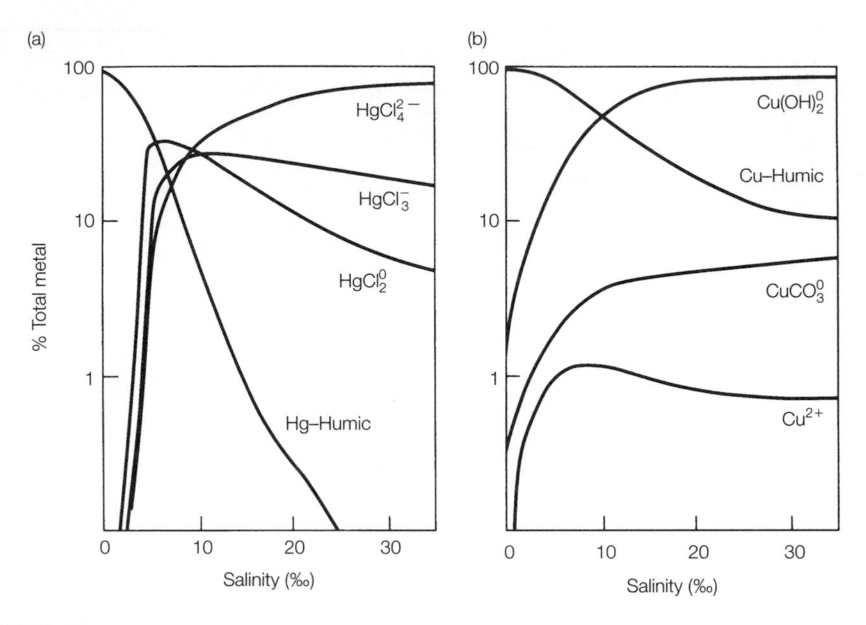

FIGURE 28.23

Calculated equilibrium speciation of (a) mercury and (b) copper during estuarine mixing of hypothetical river water with seawater. Hum, humic substance. Note logarithmic scale on *y*-axis. *Source*: From Mantoura, R. F. C., *et al.* (1978). *Estuarine and Coastal Marine Science* 6, 387–408.

in Figure 28.24 have been grouped to illustrate common characteristics. At both sites, Cd, Pb, and Zn are highly concentrated in layers deposited between 1940 and 1970. (Dating was performed using the ^{137}Cs and ^{210}Pb records.) The metal data are presented as an enrichment factor in which concentrations are normalized to average crustal abundances:

$$EF = ([Me_s]/[Fe_s])/([Me_{cr}]/[Fe_{cr}]) \quad (28.1)$$

where the subscripts s and cr denote the concentrations at the experimental stations and crustal abundances, respectively. An EF greater than 1 indicates that the trace metal concentration is higher than predicted from the presence of natural crustal materials and, hence, is attributable to anthropogenic inputs. The differences between the two cores (PC-1 and WB-2) reflects differences in land uses between the watersheds. PC-1 was collected from a highly industrialized harbor identified by the U.S. EPA as a toxic hot spot. The high enrichments of Cd, Cu, Pb, and Zn are attributed to fossil-fuel operations along the banks of the river and from automotive sources. Chromium, Co, and Ni are thought to be associated primarily with the mineral matrix of the sediment and, thus, do not exhibit significant enrichments. Although the metal enrichments have decreased in recent years (as a result of the implementation of environmental controls), surface sediment concentrations of Cd, Cu, Pb, and Zn are still two to five times higher than those at the bottom of the cores.

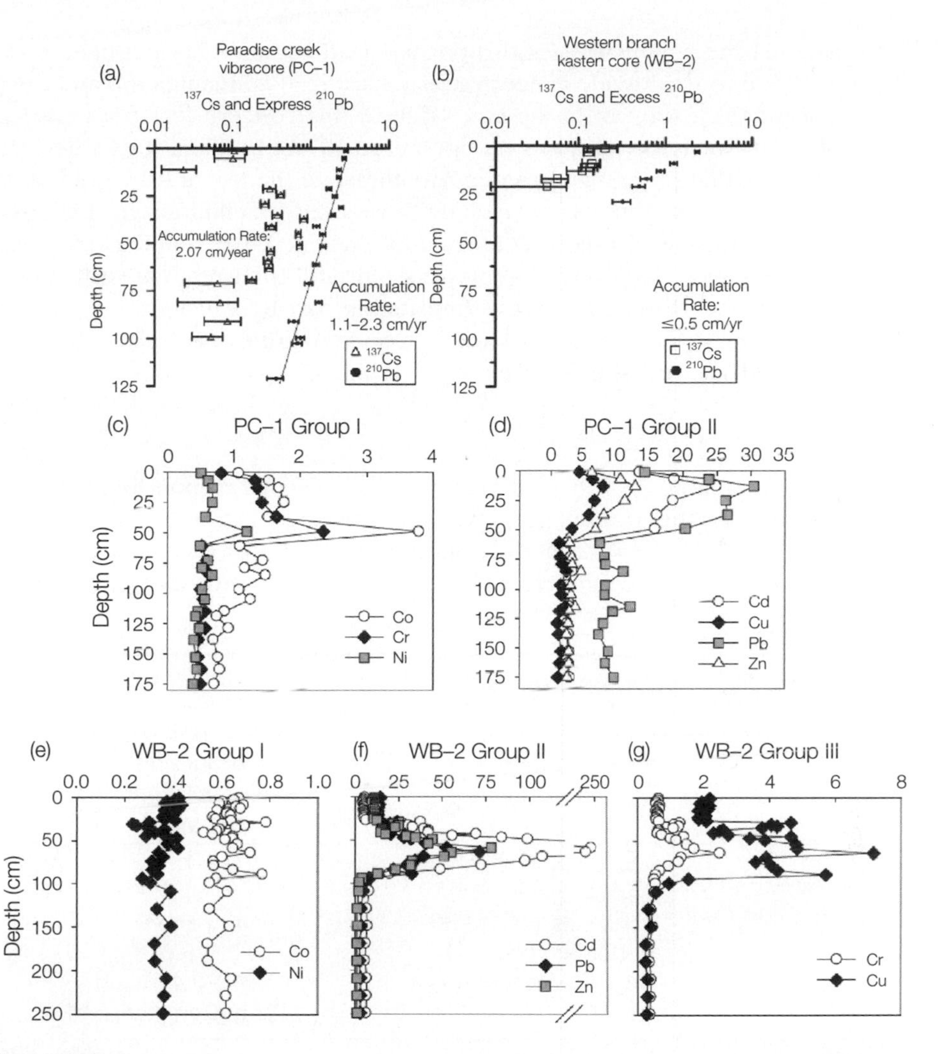

FIGURE 28.24

Metal profiles for two sediment cores from the Elizabeth River, VA, USA. Land use along the shores adjacent to collection site PC-1 (Paradise Creek) is primarily industrial and includes oil terminals, shipyard installations, coal transfer facilities, petroleum distribution and shipment operations, and wood treatment facilities. It has been identified as a toxic hot spot by the U.S. EPA. Land-use adjacent to WB-2 (Western Branch) is primarily residential. Excess ^{210}Pb and ^{137}Cs profiles for (a) PC-1 and (b) WB-2 profiles. These were used to determine accumulation rates (1.1 to 2.3 cm/y at PC-1 and ≤0.5 cm/y at WB-2). Trace metal enrichment factor profiles (see Eq. 28.1 in text) are presented in profiles (c–g) in groups determined by the depth and shape of their concentration peaks. *Source*: From Conrad, C. F., *et al.* (2007). *Marine Pollution Bulletin* 54, 385–395.

The marine chemistry of anthropogenic lead and mercury has been particularly well studied due to worldwide concern over widespread contamination and the potential for significant toxic effects. In the case of lead, this concern led to a phase-out of leaded gasoline in North America and Europe in the 1970s. In 2003, the United Nations issued a mandate calling for all governments to eliminate the use of lead in gasoline, paint, and other manufactured items, particularly those used by children. In the case of mercury, the United Nations has recently (circa 2007) deemed that voluntary international efforts have been ineffective in controlling emissions and is currently investigating adoption of legally binding treaties. This was stimulated in part by a recognition that contamination in freshwater fish is widespread and that some marine species also have body burdens that exceed safe ingestion levels.

Lead

As shown in Table 28.9, human activities are presently responsible for most of the lead transported to the atmosphere, rivers, and oceans. The majority of this lead is mobilized by the burning of leaded gasoline, the refining of ores, and the burning of coal. More than 90% of environmental Pb is a result of past anthropogenic activities. Although

Table 28.9 Global Lead Emission from Natural and Anthropogenic Sources.

Source		Production (in 1000 tons per year)
Natural	Windborne soil particles	0.3–7.5
	Seasalt spray	0–2.8
	Volcanoes	0.5–6.0
	Wild forest fires	0.1–3.8
	Biogenic processes	0–3.4
	Total	**0.9–23.5**
Anthropogenic	Fuel combustion	
	Coal	1.8–14.6
	Oil	0.9–3.9
	Gasoline	248
	Wood	1.2–3.0
	Nonferrous metal industry	
	Primary	30.0–68.2
	Secondary	0.1–1.4
	Other industries and use	5.1–33.8
	Waste incineration	1.6–3.1
	Total	**288.7–376.0**

Source: From Nriagu, J. O. (1989). Nature *338, 47–49; Nriagu, J. O., and J. M. Pacyna (1988).* Nature *333, 134–139.*

leaded gasoline is no longer in use in North America and Europe, historical discharges have created a global background concentration that is now recirculating through the soils, atmosphere, hydrosphere, and biosphere. As noted earlier, Pb is a nonessential element that is toxic at low exposure levels. It is strongly bioaccumulated with some marine species exhibiting very high body burdens, particularly animals at the top of the food web and those living in the Arctic region.

The aerosol and gaseous lead that has settled onto polar ice eventually becomes buried. Thus, ice cores provide an excellent chronological record of lead inputs as shown in Figure 28.25. Humans began smelting lead ores dates back to at least 5000 BC and can be detected in the ice core record. Lead production increased significantly at the start

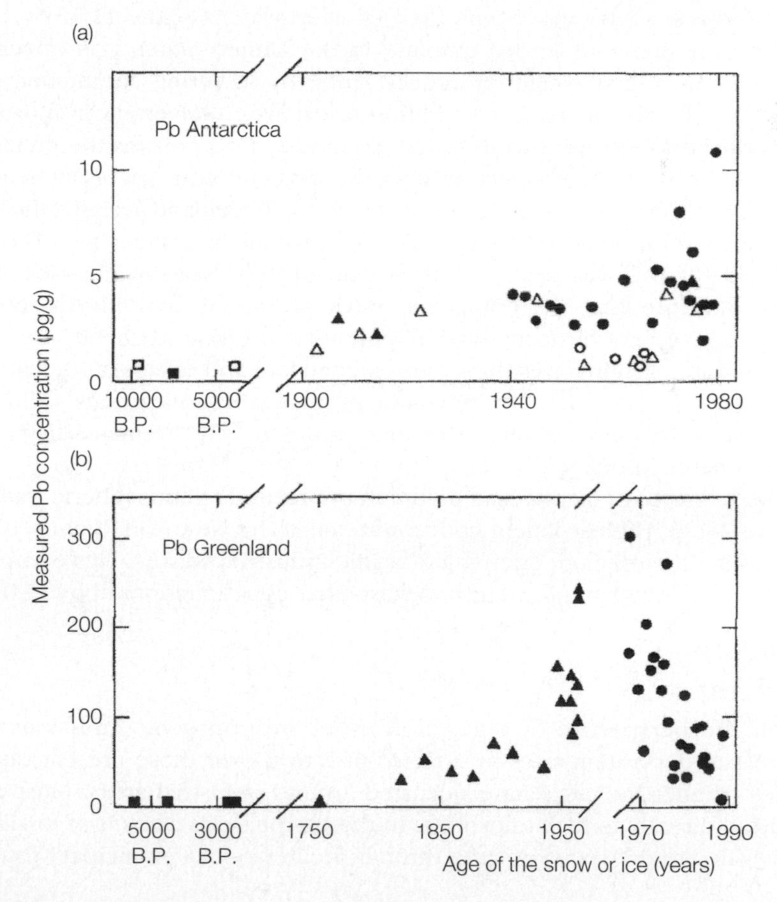

FIGURE 28.25

(a) Changes in lead concentrations (pg/g) in Antarctic snow from 1900 to 1980. (b) Changes in lead concentrations in Greenland snow from 1750 to 1989. Data are compiled from various ice cores represented by different symbols. *Source*: From Boutron, C. F., *et al.* (1994). *Geochimica et Cosmochimica Acta* 58(15), 3217–3225.

of the Industrial Revolution, with a far steeper increase in the 1950s marking the advent of leaded gasoline. In Greenland, anthropogenic input has led to a 100-fold increase in ice concentrations (from 1 to 250 pg Pb/g by the mid-1960s) and in Antarctica, a ten-fold increase (from 0.5 to 5 pg Pb/g by the 1970s), demonstrating that tropospheric transport had enabled Pb to reach the most remote areas of Earth. A sharp decline in Pb occurred from the early 1970s to the late 1980s following the rapid reduction in use of lead alkyl additives in gasoline, especially in the United States. Similar ice core concentration gradients have been observed for Cd, As, and Cu, reflecting adoption of pollution controls on coal-fired power plants and automobiles. The source of the remaining lead input is now being traced using stable isotope ratios ($^{206}Pb/^{207}Pb$). These measurements indicated that Eurasia now accounts for most of the total Pb deposition in the Arctic. Declining concentrations have also been observed in seawater (Figure 11.16).

The phase-out of leaded gasoline in the United States was driven by the passage of regulations that sought to reduce smog by lowering automobile emissions of CO and NO_x. To do this required addition of catalytic converters to motor vehicles. These devices cannot be used with leaded gasoline as lead poisons the metal catalysts. Unfortunately, some of the metals used in the catalytic converters, namely Pt, Pd, and Rh, are now showing concentration increases in Greenland ice. By the late 1990s, their concentrations were 40- to 120-fold higher than in ancient ice. The largest increase, 120-fold for Rh, was second only to that formerly seen in Pb. The most recent snow layers indicate that the deposition rates of Pt, Pd, and Rh are continuing to rise. The toxicity and bioaccumulation potential of these elements are largely unknown. Another concern involves the environmental fate and effects of the anti-knock agent that replaced lead, i.e., methylcyclopentadienyl manganese tricarbonyl (MMT). The use of this compound has also declined over time as design improvements have reduced or eliminated engine knocking.

An example of marine lead pollution not related to atmospheric transport is shown in Figure 28.26. High sediment concentrations in the New York Bight sediments have been caused by the offshore dumping of acidic industrial wastes. This dumping was halted in 1988. Extensive benthic testing suggests that concentrations above 110 ppm are toxic.[6]

Mercury

Sources

From the perspective of a global average, anthropogenic emissions of mercury have increased deposition rates by a factor of 2 to 4 over those present during preindustrial times. Higher increases have occurred around industrial areas. Most of the mercury is emitted directly to the atmosphere in the gas phase as a result of fossil-fuel burning, primarily in coal-fired power plants, through incineration of municipal wastes, and via mining.

---

[6] This value is the median determined from a meta-data analysis (Long, E. R. and L. G. Morgan [1991]. *The Potential for Biological Effects of Sediment-Sorbed Contaminants Testing in the National Status and Trends Program*, NOAA Technical Memorandum NOS OMA 52, 175 pp.) The lower threshold at which 10% of the data demonstrated toxicity was 35 ppm.

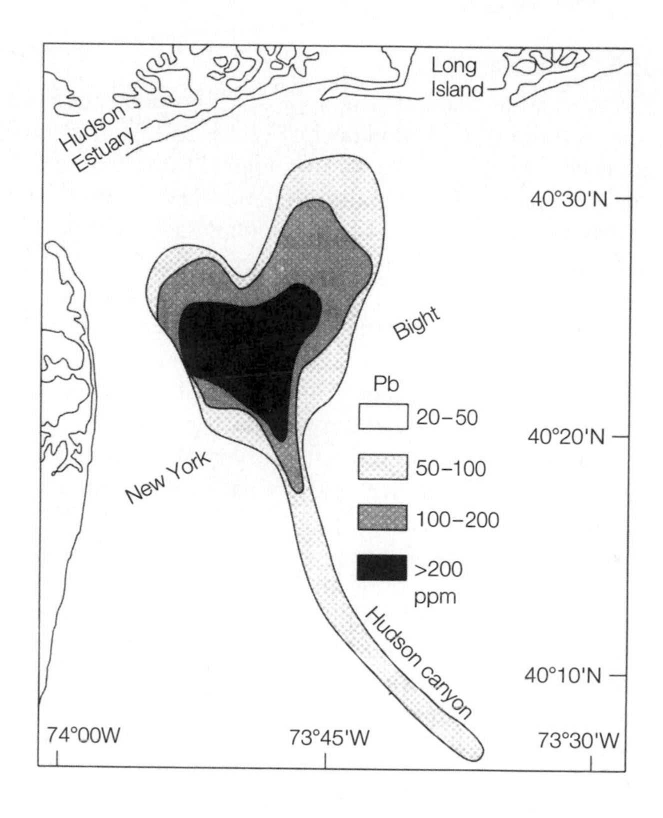

FIGURE 28.26

Distribution of lead (ppm) in bottom sediments from the New York bight. See Figure 28.14 for locations of dumpsites. *Source*: From Carmody, D. J., *et al.* (1973). *Marine Pollution Bulletin*, 4, 132–135.

Municipal wastes acquire mercury from disposal of fluorescent lamps, switches, button batteries, and thermometers. Other industrial processes that use very large amounts of mercury are chloralkali plants and gold-mining operations. The latter is an emerging concern, as loss of mercury via associated discharge into rivers presently constitutes 10% of the annual anthropogenic input and is growing. Over the past 30 years, implementation of environmental regulations has led to a decrease in mercury emissions from developed countries, but this has been largely offset by an increasing emission rate from developing countries, particularly those in East Asia.

Some mercury has also been inadvertently dumped at sea. In 2003, after years of searching, the Norwegian Navy found a German U-boat that had sunk in 1945 during World War II about 56 km from their coast in 140-m water depth. This ship was transporting 65 tonnes of mercury to Japan. About 4 kg has escaped and contaminated an area covering 30,000 m^2, with elevated concentrations found in the sediments and benthos. A debate is now underway as to whether to try to recover the mercury or to seal it in place with layers of concrete, sand, and gravel.

Speciation/Cycling

Mercury is very volatile. Natural atmospheric emissions emanate from crustal degassing, such as from volcanoes and hydrothermal vents, and from weathering of minerals, such as cinnabar (HgS). Mercury undergoes biological transformations during which it bioaccumulates and biomagnifies. Unfortunately, mercury is a nonessential metal that exhibits toxicity at fairly low exposure levels, particularly as monomethylmercury (MMHg). The chemical speciation of mercury includes inorganic and organic forms, namely elemental mercury (Hg^0), ionic mercury (Hg^+, Hg^{2+}), and methylated mercury [CH_3Hg^+, $(CH_3)_2Hg$].

The mercury cycle in the oceans is highly complex, involving transformations among gas, solute, and particulate phases, and reactions including oxidations, reductions, and biomethylations. Atmospheric mercury is also subject to phase and redox transformations, particularly near the sea surface, such as the oxidation of Hg^0 to the more reactive ionic forms. Most of the atmospheric mercury is gaseous with a turnover time of about 1 to 2 y. Its deposition occurs via wet and dry pathways as much as 1000 to 2000 km downwind of its source. This has enabled global dispersion of both natural and anthropogenic emissions. Deposition varies geographically and seasonally with high fluxes in the tropics due to rainfall (wet deposition) and to enhanced rates of oxidation of atmospheric Hg^0 into more reactive forms. Deposition is also high in the western North Atlantic and Pacific Oceans, reflecting their proximity to industrial sources in the eastern United States and Asia. Mercury is subject to the Grasshopper Effect which has led to elevated concentrations in polar regions.

As shown in Figure 28.27, anthropogenic inputs are increasing the oceanic inventory of mercury as a result of atmospheric deposition, and to a much lesser degree, river runoff. Once the mercury is deposited onto the sea surface, it is subject to a wide variety of processes, including reduction to elemental mercury, biomethylation, and particle scavenging. Regeneration from sinking particles produces a subsurface mercury concentration maximum composed of MMHg and dimethylmercury (DMHg). Methylated mercury is lipophilic and, hence, readily bioaccumulated. The subsurface enrichment in MMHg and DMHg may account for the relatively high mercury content of mesopelagic fish, such as tuna and swordfish. Scavenging of MMHg onto sinking particles results in mercury export to the sediments. In contrast, DMHg tends to degas into the atmosphere. Biologically mediated reduction of Hg(II) to Hg^0 also facilitates degassing of mercury from the sea surface. This reduction appears to be driven by photochemical reactions, particularly in the presence of humic substances. While anthropogenic input has elevated the concentration of mercury in surface waters, the net effect of the marine mercury cycle has led most of ocean's anthropogenic mercury to be transferred into the deep sea.

Given its high toxicity, considerable attention has been focused on understanding MMHg cycling. Most mercury methylation is microbially mediated and, hence, influenced by environmental factors such as temperature, pH, redox potential, activity and structure of the bacterial community, and the presence of inorganic and organic complexing agents. Biomethylation is commonly found in settings that are warm, shallow, organic-rich, acidic, and anoxic, such as estuarine and shelf sediments where bacterial activity is also high. Mercury has a very large binding constant for DOM, probably as a result of complexation with thiol groups in the organic compounds. This complexation prevents the

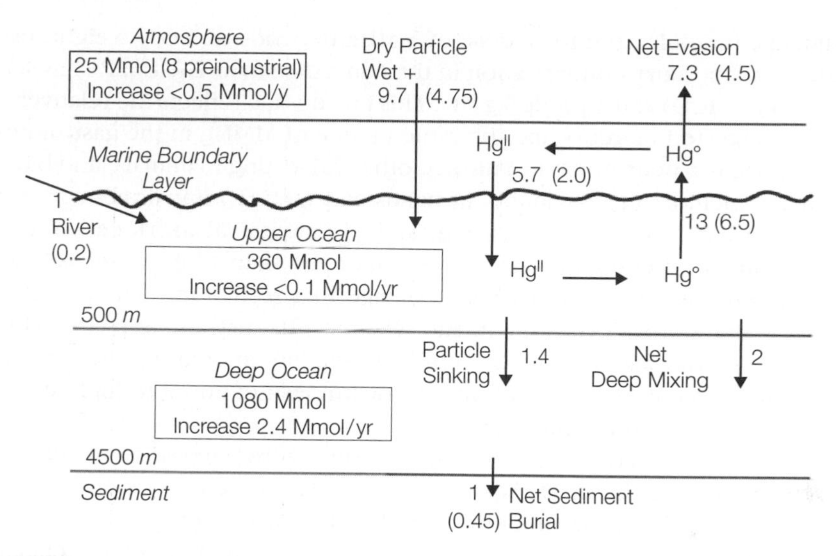

FIGURE 28.27

Marine mercury cycle. All fluxes are in units of Mmol/y. Preindustrial fluxes are in parentheses. The marine boundary layer is the lower part of the troposphere in which mercury transformations associated with the air-sea interface occur. *Source*: After Mason, R. P., and G.-R. Sheu (2002). *Global Biogeochemical Cycles* 16, GB001440.

precipitation of mercury as $HgS(s)$ that otherwise would be thermodynamically favored in anoxic sediments.

Complexation of Hg with DOM provides a potential mechanism for abiotic methylation, although this has not been definitively shown to occur. DOM can act to both inhibit and stimulate biomethylation. Inhibition arises from a reduction in bioavailability caused by the complexation of Hg with DOM. The stimulation arises from the usage of DOM as an energy and carbon source by the biomethylating bacteria. Considerable evidence suggests that sulfate-reducing bacteria are responsible for biomethylating mercury. The net effect is a high production rate of MMHg in estuarine and shelf sediments. In shelf sediments, this production rate is sufficient to support a diffusive flux of MMHg into the coastal ocean that is equivalent to the amount needed to account for the elevated MMHg concentrations observed in coastal fish. The common saltwater grass, *Spartina*, also appears to facilitate the export of MMHg to coastal waters through uptake from the sediments via their roots, followed by release as a chloride salt. Other heavy metals are similarly depurated from estuarine sediments. This uptake mechanism seems to be associated with the process by which *Spartina* obtains water from seawater, i.e., by releasing salts from its leaves.

Bioaccumulation and Fish

MMHg is lipophilic, causing it to bioaccumulate and biomagnify through the marine food web. About 85 to 95% of the total mercury in fish is MMHg, most of which is the result of biomagnification. Marine fish consumption accounts for most of the mercury burden in humans. In adults, the toxic dose of MMHg depends on body weight, fish consumption rates, and mercury concentration in the consumed fish. The USEPA has set the daily reference dose (RfD) at 0.1 μg Hg/kg bw. This low dosage reflects the relatively slow clearance rate of ingested mercury and the rapid uptake of MMHg in the gastrointestinal tract.

MMHg is a neurotoxin to humans, other fish-eating mammals, and birds. Symptoms in humans include: (1) numbness in hands and feet; (2) impairment of peripheral vision, coordination, speech, and hearing; and (3) mental disturbances. Young children and developing fetuses are more sensitive to mercury than adults. Maternal mercury is passed to the fetus where it causes brain damage that results in mental retardation, delays in speech and motor development, and seizures. Mercury also appears to be a contributing factor in cardiovascular disease in men. In wildlife, mercury poisoning causes neurotoxicity, impairment of reproduction, and death. Fish exposed to high levels of MMHg also experience impaired reproduction.

The most dramatic case of mercury poisoning from ingestion of contaminated seafood occurred in Japan in the 1950s in Minamata Bay. Mercury was emitted into the bay as a component of industrial effluents by a chemical processing plant that used the metal as a catalyst in the production of acetaldehyde. Once in the bay, the mercury was bioconcentrated by the phytoplankton and passed up the food chain until it was consumed by the bay's residents and their pet cats. As shown in Table 28.10, the fish and shellfish had mercury concentrations well above any safe ingestion level (considered by the U.S. EPA to be 1 mg methylmercury per kg wet weight of fish tissue). To make matters worse, the people of Minamata Bay had a very high rate of fish ingestion. Consumption of contaminated

Table 28.10 Mercury Concentrations in Samples from Minamata, Japan, and from Nearby Areas Devoid of Mercury Pollution.

Material	mg/kg Dry Weight	
	Minamata Bay	**Unpolluted Area**
Clams from the beach	11–39	1.70–6.00
Fish	10–55	0.01–1.70
Cats (from Minamata with mercury poisoning)		
Liver	40–145	0.64–6.60
Kidney	12–36	0.05–0.82
Brain	8–18	0.05–0.13
Man (from Minamata with mercury poisoning)		
Liver	22–70	0.07–0.84
Kidney	22–144	0.25–10.70
Brain	2–25	0.05–1.50
Hair	281–705	0.14–7.50

Source: *From Tokuomi, H. (1969).* Revue Internationale Océanographie Médicale *13–14, 5–35.*

seafood led to large mercury enrichments in the internal organs of the humans and cats. The problem was first recognized when local inhabitants developed severe neurological problems, including muscle spasms, loss of equilibrium, and impaired mental state. Birth defects and death were common. Similar symptoms were seen in the cats. The horrible consequences of this poisoning resulted in strict controls on the discharge of mercury in industrial effluents.

Although discharge of mercury in industrial effluents has been greatly restricted, other sources, such as emissions from coal-fired power plants in Asia, are increasing. Mercury emitted in the past by anthropogenic practices and by natural phenomena, such as volcanic eruptions, continues to undergo cycling through the atmosphere, biota, and hydrosphere. As a result, fish contamination is common and frequently high enough to require consumption advisories. Nationwide, mercury is the most common toxicant leading to fish consumption advisories. In 2006, 92% of the U.S. Atlantic coast was under a fish consumption advisory for mercury, as was 100% of the Gulf coast. Hawaii has a statewide advisory for mercury in several marine species. No Pacific coast state has issued a statewide advisory for its coastal waters.

While MMHg is well recognized as a toxicant at very low exposure levels, disagreement remains as to a "safe" dosage. The U.S. FDA allows up to 1 ppm Hg in commercial seafood and claims that consumers can safely eat up to 7 ounces of fish per week at this concentration. But according to the U.S. EPA's RfD, for a 60-kg person, fish containing 1 ppm MMHg should only be eaten at a rate of 1.5 ounces per week. The most recent fact sheet published by the U.S. EPA (EPA-823-F-01-011) lists 0.3 ppm as the toxic criterion for human health consumption of organisms.

As shown in Table 28.11, mercury levels in seafood vary by species and by geography. Because older fish have bioaccumulated more mercury, concentrations tend to be higher in larger fish within a species. Even controlling for all these factors, considerable variability exists in mercury concentrations, making determination of safe seafood consumption rates difficult. Ingestion of large ocean fish, such as tuna, shark, and swordfish, is the leading source of methylmercury exposure for the general U.S. population. Tuna alone accounts for approximately 40% of the total intake. Tunas from the Mediterranean Sea have higher concentrations (0.87 mg/kg) than those from the Atlantic or Pacific Oceans (0.47 and 0.17 mg/kg, respectively). This is caused by natural emissions from local cinnabar (HgS) deposits. Evidence of elevated mercury levels due to frequent consumption of fish with high levels of mercury can be found in people worldwide. In some situations, this is recognized as a public health problem. For example, in the United States between 1999 and 2002, blood Hg levels measured in women of childbearing years were high enough to suggest that approximately 6% of this group (3.8 million individuals) were exceeding the daily RfD. This has the potential to affect the health of the women as well as their children.

Although the rate of mercury input to the ocean's surface waters has increased as a result of anthropogenic activities, it is not clear that the relatively high MMHg concentrations now seen in the larger fish, e.g., tuna, shark, and swordfish, are the result of pollution. Measurements of mercury levels in preserved fish collected over the past 100 years have proven inconclusive due to small sample sizes and contamination effects.

Table 28.11 Total Mercury Concentrations in Marine Fish (mg/kg)[a] Aggregated by Geographic Region.

Species	FDA		Imports		Atlantic		Pacific	
	(mean ± SD)	No.[b]	(mean ± SD)	No.	(mean ± SD)	No.	(mean ± SD)	No.
Anchovies	0.04	40	0.06 ± 0.01	53	No landings		0.04 ± 0.01	40
Herring	0.04	38	0.13 ± 0.03	14	0.14 ± 0.06	15	0.04 ± 0.02	131
Sardine	0.02	22	0.03 ± 0.003	35	No landings		No landings	
Shad	0.07	59	0.07 ± 0.01	59	0.02 ± 0.02	40	0.07 ± 0.01	59
Bluefish	0.34 ± 0.13	52	None consumed		**0.45 ± 0.33**	288	No landings	
Clamsa	ND	6	0.06 ± 0.01	3	0.01 ± 0.002	4	0.01 ± 0.002	2
Cod	0.10 ± 0.08	39	0.07 ± 0.01	19	0.06 ± 0.02	21	0.11 ± 0.03	28
Crabs	0.06 ± 0.11	63	0.10 ± 0.02	27	0.26 ± 0.44	369	0.15 ± 0.07	56
Croaker	0.07 ± 0.04	50	None consumed		0.07 ± 0.08	315	0.12 ± 0.10	45
Haddock	0.03 ± 0.02	4	0.06 ± 0.01	31	0.03 ± 0.02	4	No landings	
Hake and whiting	0.01 ± 0.02	11	0.13 ± 0.01	88	0.07 ± 0.02	22	0.01 ± 0.02	11
Monkfish	0.18	81	0.13 ± 0.01	25	0.18 ± 0.04	81	No landings	
Flounder	0.05 ± 0.05	23	0.05 ± 0.07	55	0.08 ± 0.04	60	0.07 ± 0.07	58
Plaice	0.05 ± 0.05	23	0.05 ± 0.02	33	0.05 ± 0.02	33	No landings	
Sole	0.05 ± 0.05	23	0.10 ± 0.10	64	No landings		0.06 ± 0.02	518
Grouper	**0.47 ± 0.29**	43	0.34 ± 0.07	17	0.36 ± 0.14	100	**0.47 ± 0.29**	43
Sea bass	0.22 ± 0.23	47	0.19 ± 0.12	29	0.14 ± 0.04	14	0.22 ± 0.23	47
Rockfish	0.22 ± 0.23	47	None consumed		No landings		0.29 ± 0.22	314
Halibut	0.25 ± 0.23	46	0.23 ± 0.05	11	0.25 ± 0.23	46	0.28 ± 0.09	11
Scorpionfish	0.29	78	0.11 ± 0.003	7	No landings		0.22 ± 0.05	79
Lobster	0.17 ± 0.09	16	0.10 ± 0.005	13	0.28 ± 0.15	106	0.17 ± 0.09	16

Mackerel, all	0.15	432	0.15 ± 0.10	432	0.22 ± 0.16	877	0.09 ± 0.06	30
Marlin	**0.49 ± 0.24**	16	**0.49 ± 0.24**	16	No landings		**0.57 ± 0.41**	39
Mussels	NA	NA	0.03 ± 0.009	80	0.08 ± 0.09	729	0.03 ± 0.02	330
Oysters	ND	34	0.01 ± 0.01	27	0.07 ± 0.09	2,082	0.06 ± 0.03	63
Ocean perch	ND	6	0.09 ± 0.02	53	0.08 ± 0.02	50	0.08 ± 0.02	50
Orange roughy	0.54	26	**0.55 ± 0.11**	32	No landings		No landings	
Pollock	0.06	37	0.03 ± 0.002	12	0.02 ± 0.01	115	0.06 ± 0.03	37
Sablefish	0.22	102	0.22 ± 0.04	102	No landings		0.22 ± 0.04	103
Salmon, fresh	0.01	34	0.04 ± 0.01	69	0.13 ± 0.17	11	0.04 ± 0.01	289
Salmon, canned	ND	34	0.04 ± 0.01	32	No landings		0.04 ± 0.01	289
Scallops	0.05	66	0.06 ± 0.02	21	0.01 ± 0.003	12	0.04 ± 0.001	3
Sea trout	0.25	27	None consumed		0.21 ± 0.15	1,220	No landings	
Shrimp	ND	24	0.03 ± 0.01	106	0.04 ± 0.05	171	0.03 ± 0.01	44
Skate	0.14	56	None consumed		0.14 ± 0.03	56	0.14 ± 0.03	56
Snapper	0.19 ± 0.12	25	0.21 ± 0.15	324	0.28 ± 0.43	363	0.25 ± 0.09	17
Porgy	NA	NA	None consumed		0.08 ± 0.07	14	No landings	
Sheepshead	0.13	59	None consumed		0.18 ± 0.20	268	No landings	
Squid	0.07	200	0.07 ± 0.01	200	No supply		No supply	
Shark	**0.99 ± 0.63**	351	**0.99 ± 0.63**	351	**0.75 ± 0.70**	585	**0.80 ± 0.37**	35
Swordfish	**0.98 ± 0.51**	618	**1.03 ± 0.54**	689	**0.98 ± 0.51**	618	**0.98 ± 0.51**	618
Tilefish	**1.45**	60	None consumed		**1.45 ± 0.29**	60	No landings	

(Continued)

Table 28.11 *(Continued)*

Species	FDA		Imports		Atlantic		Pacific	
	(mean ± SD)	No.[b]	(mean ± SD)	No.	(mean ± SD)	No.	(mean ± SD)	No.
Tuna, canned albacore	0.35	179	0.37 ± 0.12	318	0.37 ± 0.12	318	0.37 ± 0.12	318
Tuna, canned light	0.12	131	0.11 ± 0.10	199	0.11 ± 0.10	199	0.11 ± 0.10	199
Tuna, fresh and frozen	0.38	131	**0.48 ± 0.24**	422	0.28 ± 0.12	496	0.24 ± 0.10	555
Whitefish	0.07 ± 0.05	25	0.07 ± 0.01	25	No landings		No landings	
Whale meat for sale in Korean and Japanese markets								
North Pacific minke whale	0.01–**0.54**	53						
Cuvier's beaked whale	0.43	1						
Harbour porpoise	0.42–**0.54**	15						
Finless porpoise	0.11–**1.81**	32						
Common dolphin	0.44–**1.89**	21						
Pacific white-side dolphin	1.04–**1.61**	4						
Risso's dolphin	**1.85**	18						
Blainville's beaked whale	**1.71–9.21**	2						
Stejneger's beaked whale	**2.01–3.30**	1						
False killer whale	**1.39–81.0**	13						
Bottle-nose Dolphin	0.59–**98.9**	40						
Killer Whale	**1.06–13.3**	7						

[a]*Wet weight.* [b]*Number of samples analyzed.*
Values in bold exceed the USEPA's safety threshold of 0.3 ppm.
NA = not analyzed
ND = not detectable
Source: After (upper table) Sunderland, E. M. (2007). Environmental Health Perspectives *115, 235–242. Whale data from Endo, T., et al. (2007).* Marine Pollution Bulletin *54, 669–677.*

Because of the complex cycling of MMHg in the ocean, carnivorous fish at the top of the food chain could be naturally enriched in mercury. But mercury levels in seabird feathers show significant increases over time, suggesting that anthropogenic emissions have similarly caused an increase in fish tissues.

Global Climate Change

Global climate change is expected to lead to a worsening of mercury contamination in the ocean and seafood. This is an expected consequence for the following reasons: (1) enhancement of biomethylation as warmer temperatures stimulate bacterial activity, (2) increased emissions from northern peatlands due to melting of permafrost and increased fires (peatlands are naturally enriched in mercury), (3) alteration of food webs including changes in the age distribution of species, (4) increased nonpoint-source runoff into rivers whose watersheds are experiencing land-use change, and (5) changes in aeolian transport associated with changing winds and dust fluxes. An example of the latter was seen in late January 2006, when satellites detected a mass of thick smog pushed by westerly winds from China to the Korean Peninsula. Atmospheric concentrations in the smog were 10 times higher than normal. This dust storm exported 5 to 15 tonnes of mercury from China to Korea. Some fraction of that mercury was naturally associated with the crustal materials composing the dust. But China is recognized as the major global mercury emitter, generating about 600 t/y of gaseous mercury from industrial sources (coal burning) and another 150 t/y from biomass/biofuel combustion. About half the mercury pollution in the air in Seoul, Korea, comes from China.

28.7 SYNTHETIC ORGANIC COMPOUNDS

Humans are creating new organic compounds at a rate of about 1000 per year with about 70,000 now in commercial use. Some 5000 of these are recognized as high-production-volume chemicals. Some of these have proven toxic to marine life and now pose health threats to humans who ingest seafood. The toxic effects include endocrine disruption, carcinogenicity, and impairment of immune system function. The most problematic ones have slow degradation rates and large bioconcentration factors causing them to accumulate in high concentrations, particularly in the biota that occupy top trophic levels. These pollutants are not readily metabolized and, hence, have slow clearance times. For many of them, the major route by which they are removed from the body is maternal transfer to the fetus.

For the semivolatile compounds, such as the halogenated hydrocarbons, worldwide dispersal has been facilitated by the grasshopper effect. As illustrated in Figure 28.28, these compounds migrate poleward in steps initiated by volatilization from the land during warmer seasons followed by atmospheric transport to higher (cooler) latitudes where condensation brings them back to Earth's surface. Seasonal warming revolatilizes the compounds, enabling them to move to even higher latitudes. The more volatile the compound, the farther poleward it is transported. The net effect is a global distillation of the semivolatile compounds far from their point of release into the environment, leading to their concentration in polar and mountainous regions. Many of these pollutants are lipophilic and, hence,

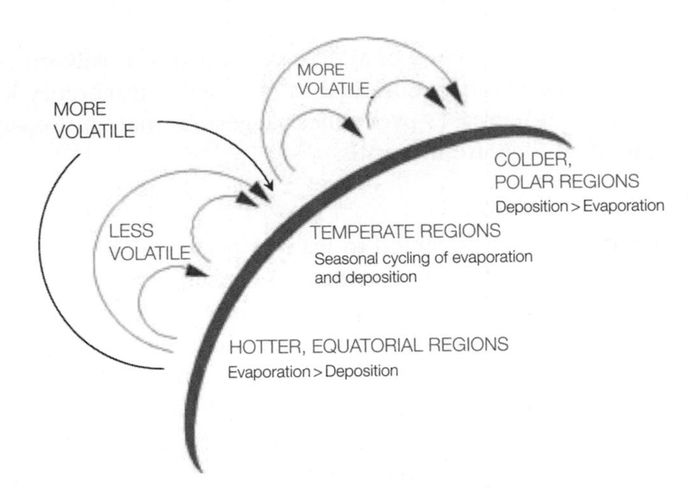

FIGURE 28.28

The Grasshopper Effect (also known as Global Distillation) leading to the long-range transport of semi-volatile persistent organic pollutants. The more volatile POPs are able to take more "hops" and thus can undergo longer range transport than the less volatile ones.

are effectively concentrated in fatty tissues. Unfortunately polar animals tend to have high fat levels because blubber is an effective insulator. This causes the polar animals to have high BAFs, and, hence, high body burdens, of the lipophilic pollutants.

The artificial organic compounds are introduced into seawater via atmospheric input, river runoff, and groundwater seeps. They are a common component of untreated and treated municipal and industrial wastes. Seawater concentrations tend to be low, ranging from 0.1 to 1000 ng/L as these compounds are hydrophobic and, hence, sparingly soluble. Thus, they tend to adsorb to particles and bioaccumulate. Some compounds are lipophilic and others bind preferentially to proteins. Toxic effects thresholds can be very low. Some of the compounds are biomagnified through the marine food web. Contamination of marine life is now global, due in part to dispersal via the grasshopper effect. Concentration ranges are provided in Table 28.12, for DDT, PCBs, and PAHs in various biota.[7] Note that the EU minimal risk level (MRL) for chronic exposure to ingested PCB is 20 ng/kg per day. Thus, a 30-kg person should eat no more than about an ounce per day of seafood with PCB concentrations greater than 20 ng/g wet weight. The MRL for DDT is 500 ng/kg/d.

28.7.1 Persistent Organic Pollutants

Particular concern is currently focused on the organic pollutants that are highly toxicity, have large BAFs and slow environmental degradation rates. Reflecting their physicochem-

[7] Technically PAHs are a contaminant but are generally considered as a serious organic pollutant due to their large anthropogenic source, high toxicity, and lipophilic nature.

Table 28.12 Concentration Ranges of Some Organic Contaminants in Marine Organisms (ng/g wet weight).

Organisms	Total DDT	Total PCB	Total PAH	Region	Reference[a]
Edible clams	0.6–3.4	1.6–15.4	2.1–24.5	Europe	Binelli and Provini (2003)
Crustacean (crabs, shrimp)	18–24	6.1–14	98–180	Kara Sea	Sericano et al. (2001)
Deep sea fish (muscle)	7.4–12.6	13.8–24.0	0.2–0.6	Britain	Sole et al. (2001)
Seabird (herring gull)	0.2–18.8	0.4–340	0–333	Arctic	Shore et al. (1999)
Ringed seal blubber	34.8–904	501–6010	65.8–140.7	Northern Alaska	Kucklick et al. (2002) and Holsbeek et al. (1999)
Humpback dolphin blubber	9111–59,542	69–19,979	2762–3275	South China	Leung et al. (2005)

[a]*For references, see: Sarkar A., et al. (2006)* Ecotoxicology *15, 333–340.*

ical behavior, these pollutants are collectively referred to as *persistent organic pollutants (POPs)* or *persistent bioaccumulative toxics (PBTs)*. (The latter term also is broad enough to include the heavy metals.) The United Nations has prioritized global control of the twelve most problematic POPs. These compounds fall into three groups: (1) pesticides (aldrin, chlordane, DDT, dieldrin, endrin, heptachlor, hexachlorobenzene, mirex, toxaphene), (2) industrial chemicals (hexachlorobenzene and polychlorinated biphenyls [PCBs]), and (3) unintended industrial by-products (dioxins and furans). Most of these are organochlorines. In 2001, the United Nations adopted the International Convention on the Control of Harmful Anti-fouling Systems on Ships, which calls for a worldwide ban on use of tributyltin (TBT) in antifouling paints.

Substantial evidence suggests that current environmental levels of the POPs are exerting adverse effects on marine mammal populations. These include: (1) depression of the immune system and the subsequent triggering of infectious diseases, (2) reproductive impairment, (3) lesions of the adrenal glands and other organs, (4) cancers, (5) alterations in skeletal growth and ontogenetic development, and (6) induction of bone lesions. Similar problems are now seen in sea birds. High body burdens of POPs are also found in people who consume large amounts of marine fish, shellfish, and marine mammals. The indigenous Arctic peoples have particularly elevated concentrations of PCBs and pesticides. In East Greenland, 100% of the population have blood concentrations that exceed levels of concern and 30% have concentrations at which a change of diet is advised.

In recognition of the threat posed by these compounds, the United Nations ratified the Stockholm Convention on Persistent Organic Pollutants in 2004. This agreement seeks to end the release and use of 12 of the most problematic POPs with limited use of DDT being permitted to control malaria. Although production has been halted for many of these POPs, considerable stockpiles still exist (500,000 tonnes of obsolete or unused pesticides now exist and would cost $2.5 billion U.S. to destroy).

The UN intends to expand the list of banned POPs pending further review of new environmental ecotoxicological data. Other compounds of concern are: (1) chlorinated pesticides including lindane (hexachlorohexane [HCH]), chlordanes, and endosulfan; (2) organometallics, namely TBT; (3) polybrominated diphenyl ethers (PBDEs) and hexabromocyclododecanes (HBCDs); (4) PAHs (which are technically contaminants); (5) low-molecular-weight volatile organic compounds (VOCs), including freons; (6) alkyl and aryl phthalate esters; (7) polychlorinated napthalenes (PCN); and (8) per- and polyfluorinated alkyl substances (PFAS).

An emerging concern has been the degree to which POPs cause reproductive failure and developmental disturbances by disrupting endocrine system function. Toxicants that are endocrine disruptors are defined by the U.S. EPA as "an exogenous chemical substance or mixture that alters the structure or function(s) of the endocrine system and causes adverse effects at the level of the organism, its progeny, populations, or sub-populations of organisms." Examples of endocrine disrupting effects include: (1) eggshell thinning in top predator birds, (2) imposex in mollusks, (3) reproductive disorders in seals, and (4) sex/hermaphroditism in fish, reptiles, and mammals. In human populations, endocrine disruptors have been implicated in increased incidences of cancers (particularly breast and testicular cancer) and disorders of the male reproductive system such as declines in sperm counts, cryptorchidism, and hypospadias.

Chemicals that have demonstrated endocrine disrupting effects at environmental levels are listed in Table 28.13. Note that each of the POPs covered by the Stockholm Convention is on this list. The lipophilic POPs tend to be exported to the fetus during gestation and can lead to the development of reproductive impairments in the offspring prior to and after birth. The most well-documented cause and effect linkage between an endocrine disruptor and reproductive failure is the induction of imposex in dog whelk mollusks by TBT. Imposex is the occurrence of induced male sex characteristics in female gastropods. The effects threshold is 1 to 2 ng/L, with complete sterilization resulting at concentrations greater than 3 to 5 ng/L. Imposex has now been observed worldwide in gastropods living in urbanized marinas and ports, reflecting the large-scale usage of TBT-based antifouling paints.

Marine mammals tend to have elevated body burdens of POPs and heavy metals due to their high blubber content, long lifetimes, and tendency to biomagnify. Large geographic differences in body burdens have been observed. As shown in Figure 28.29 for DDT and PCBs, concentrations are particularly high in the Mediterranean Sea and the western coast of the United States. High concentrations of mercury and TBT have also been observed (Tables 28.11 and 28.16). A series of mass mortality events involving harbor seals, turtles, and dolphins during 1986 to 1992 led to the hypothesis that these toxicants were causing an impairment of immune function, making the animals susceptible to fatal viral infections. Hence, attention has been directed at quantifying immune system activities by measuring concentrations of toxicological biomarkers such as cytochrome P450. These biomarkers are a component of natural detoxification pathways, suggesting that marine

Table 28.13 Chemicals in Common Use Known to Have Endocrine-Disrupting Effects.

Herbicides			
2,4-D	Alachlor	Atrazine	Nitrofen
2,4,5-T	Amitrole	Metribuzine	Trifluralin
Fungicides			
Benomyl	Mancozeb	Tributyltin	Ziram
Hexachlorobenzene	Metiram-complex	Zineb	
Insecticides			
α-HCH	DDT (metabolites)	Methomyl	Parathion
Carbaryl	Endosulfan	Methoxychlor	Synthetic pyrethroids
Chlordane	Heptochlor	Mirex	Toxaphene
Docofol	Lindane	Oxychlordane	Transnonachlor
Dieldrin			
Nematocides			
Aldicarb	DBCP		
Industrial chemicals			
Cadmium	Mercury	Pentachlorophenols	Phthalates
Dioxin	Polybrominated biphenyls	Penta- to Nonylphenols	Styrenes
Lead	Polychlorinated biphenyls		

Source: *From Colborn, T., et al. (1993).* Environmental Health Perspectives *101, 378–384.*

mammals have always been exposed to structurally similar, but natural, toxicants. For some of the organochlorines and organobromines, natural marine sources may exist, such as for some of the PBDEs.

Pesticides and Herbicides

Unsaturated carbon rings and halides confer toxicity to organic compounds making them excellent biocides. Synthetic organic chemists have used these substituents to develop potent pesticides, such as those illustrated in Figure 28.30. Another large class of pesticides are based on the toxic effect of phosphorus bound to sulfur (Figure 28.30k and l). All of these pesticides are particularly effective for two reasons: (1) they are not species specific and, hence, kill a wide spectrum of insects, and (2) they are relatively stable in the environment and, hence, remain active for a long time. Unfortunately, these characteristics also make pesticides toxic to nontarget organisms, such as birds and mammals. Mirex is also used as a flame retardant.

The chlorinated and phosphate pesticides are applied to plants and soils. Due to their stability, many persist long enough to be transported by stormwater and groundwater runoff into the ocean. Many of these compounds are semivolatile and, hence, prone to transport via the grasshopper effect. Once in the ocean, these compounds are bioconcentrated and biomagnified. They also adsorb onto sinking particles. Along the way, some are degraded by metabolic processes. The most common degradation products of dichlorodiphenyltrichloroethane (DDT) are dichlorodiphenyldichloroethane

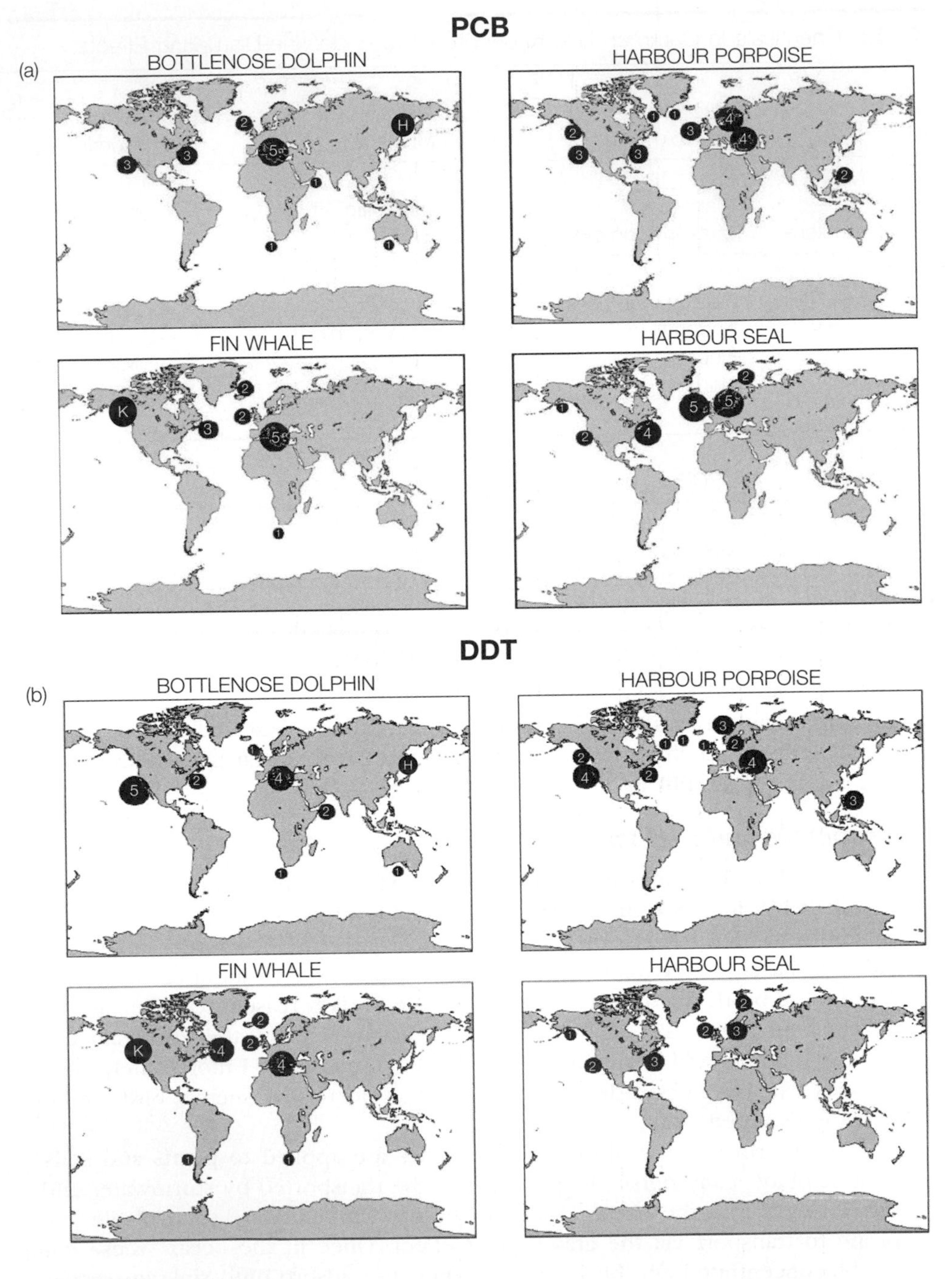

FIGURE 28.29 *(Continued)*

LEGEND	Concentration Rating (mg xenobiotic/kg blubber)				
Mammal	1	2	3	4	5
Bottlenose dolphin	<10	10-30	30-100	100-500	>500
Harbour porpoise	<5	5-12	12-25	25-50	>50
Fin whale	<1	1-2.5	2.5-5	5-10	>10
Harbour seal	<2	2-5	5-10	10-20	>20

(H) Humpback dolphin (Ratings 1–3); (K) Killer whale (Ratings 3–5)

FIGURE 28.29

Marine mammal blubber concentrations of (a) PCBs and (b) DDT. *Source*: From Aguilar, A., *et al.* (2002). *Marine Environmental Research* 53, 425–452. Additional data from: Leung, C. C. M., *et al.* (2005). *Marine Pollution Bulletin* 50, 1713–1744; Hayteas, D. L. and D. A. Duffield, (2000). *Marine Pollution Bulletin* 40, 558–561; and Ross, P. S., *et al.* (2000). *Marine Pollution Bulletin* 40, 504–515.

Table 28.14 Estimated Residence Times for Some Halogenated Hydrocarbons in the Euphotic Zone of Oligotrophic, Mesotrophic, and Eutrophic Waters.

Station Area Type	Residence Time (*y*)		
	$\sum$HCH	$\sum$DDT	PCBs
Oligotrophic	5.1–10	0.19–0.37	0.38–0.76
Mesotrophic	4.3–6.4	0.08–0.12	0.082–0.12
Eutrophic	2.0–3.4	0.031–0.052	0.070–0.12

$\sum$HCH = [α – HCH] + [β – HCH] + [γ – HCH], $\sum$DDT = [*p*,*p*′ – DDE] + [*p*,*p*′ – DDT] + [*o*,*p* – DDT] *The PCB congeners are listed in Figure 28.32. Source: From Tanabe, S., R. Tatsukawa (1983).* Journal of the Oceanographical Society of Japan *39, 53–62.*

(DDD) and dichlorodiphenyldichloroethylene (DDE). Since the removal of pesticides from seawater is enhanced by the presence of large amounts of particles and biological activity, residence times are longest in the open ocean and shortest in estuarine waters (Table 28.14).

The most dramatic demonstration of biomagnification and bioaccumulation was seen in DDT. As illustrated in Figure 28.31, DDT is passed up the food chain leading to high concentrations in fish-eating birds. In the late 1960s, a dramatic decline in the reproductive success of pelicans was observed along the coast of Southern California. As shown in Table 28.15, this decline was accompanied by very high concentrations of DDT in their prey (sardines). The cause for this decline was later traced to the effect of DDT on the birds, which caused them to lay very thin-shelled eggs that were crushed during gestation. The decline in the pelican population was probably worsened by a concurrent

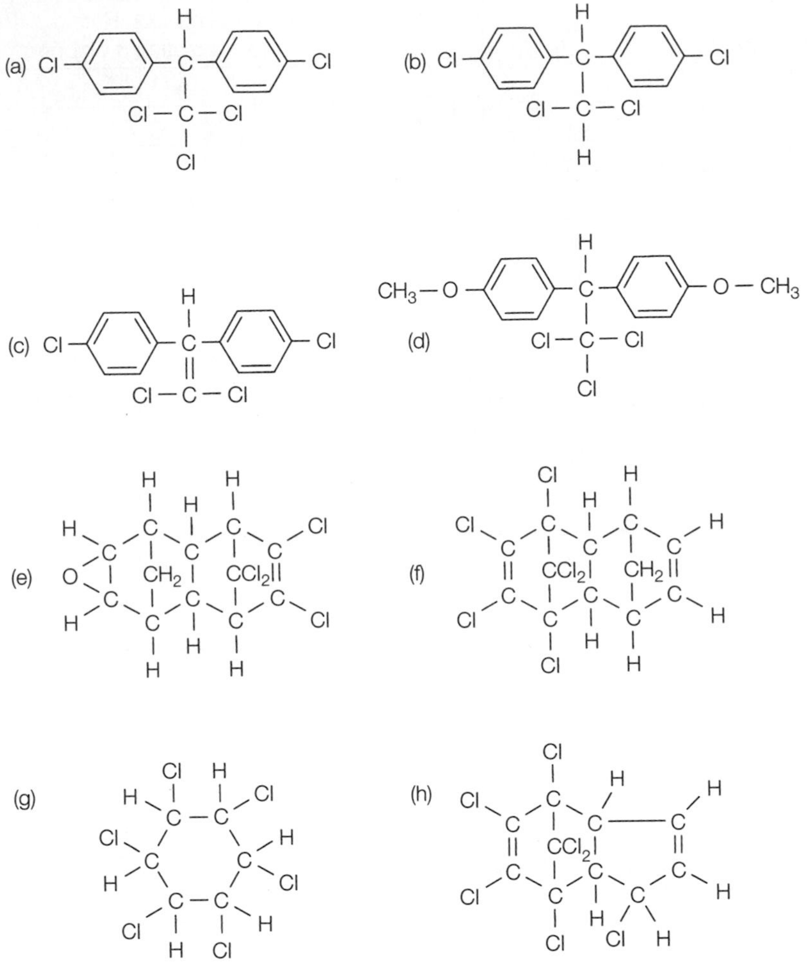

FIGURE 28.30 *(Continued)*

enrichment in PCBs and a concurrent decrease in sardine abundance. No evidence was found to attribute the latter effect to the DDT enrichment in the sardines themselves. Partly in response to the reproductive failure of this pelican population, the use of DDT in the United States was severely curtailed in 1970 and banned in 1972. Pelican numbers rebounded in the early 1970s as environmental levels of the pesticide rapidly declined.

Although the use of DDT has been banned in the United States, a large stockpile remains and other countries continue its use. In the early 1980s, DDT levels were observed to be increasing on land and in the coastal waters of the United States. This DDT is thought to be derived from either the illegal use of the pesticide or from the import of foodstuffs grown with its use. Animals can also be important transport mechanisms for

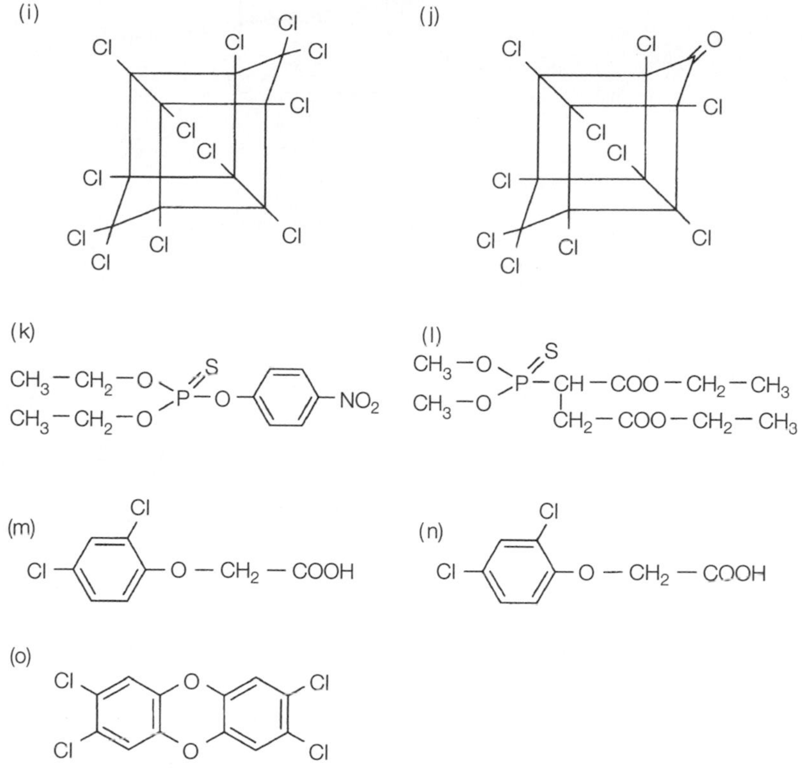

(i) (j) (k) (l) (m) (n) (o)

FIGURE 28.30

The structural formulae of some common pesticides and *herbicides: (a) p, p′-DDT, (b) p, p′-DDD, (c) p, p′-DDE, (d) methoxychlor, (e) dieldrin and its stereoisomer, endrin, (f) aldrin and its stereoisomer, isodrin, (g) lindane (HCH), (h) heptachlor, (i) mirex, (j) kepone, (k) parathion, (l) malathion, (m) *2,4-D, and (n) *2,4,5-T.

POPs. For example, seabirds in the Arctic appear to be conveying Hg, HCBs, and DDT to their rookeries via guano deposition. Similarly, salmon contaminate upland streams following their death after spawning as their body burden of toxicants is eventually released to the terrestrial ecosystem.

In the case of coastal California, DDT levels stopped declining in the 1980s. This has been attributed to a continuing input of sediments that were contaminated by stormwater runoff decades ago. From 1947 to 1971, the Montrose Chemical Company released about 2000 tonnes of DDT into county sewers that eventually discharged into the ocean. Most of this DDT ended up in the sediments, with the most severe contamination covering several square miles of seafloor in San Redondo Bay. This deposit is estimated to contain 100 tonnes of DDT. The U.S. EPA is now considering sealing this deposit off by covering it with a thick layer of sand.

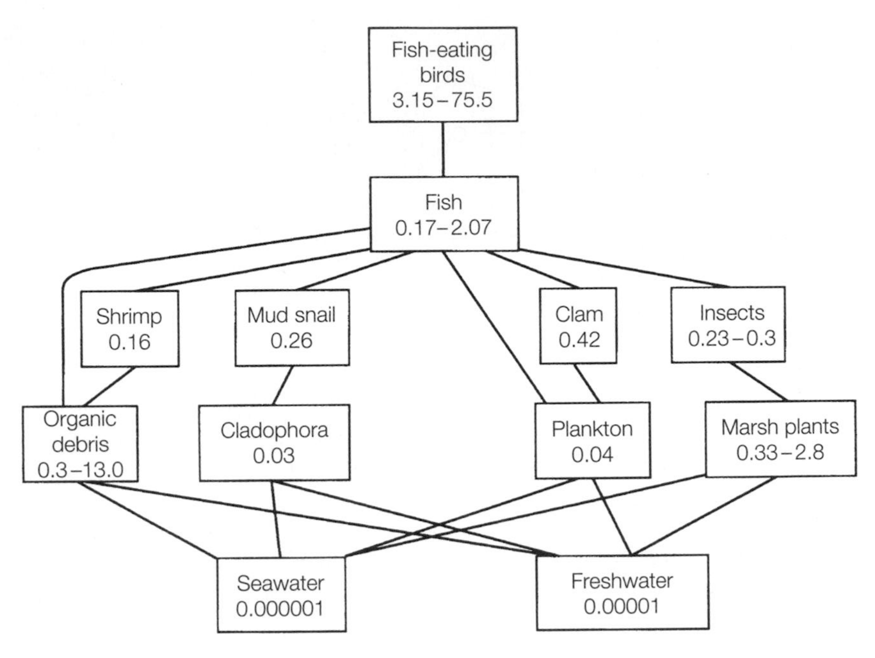

FIGURE 28.31

The bioaccumulation and biomagnification of DDT (ppm) in a marine and aquatic food web. *Source*: From Edwards, C. A. (1973). *Persistent Pesticides in the Environment*, 2nd ed. CRC Press, p. 80.

Table 28.15 $\sum$DDT Concentrations ([DDT] + [DDD] + [DDE]) in Sardines and Intact Pelican Eggs from the Coastal Islands of Southern California during 1969 to 1974.

Year	Number of Nests	Number of Young Birds	DDT in Intact Eggs (mg/kg Fat)	DDT in Sardines (mg/kg Wet Weight)	Abundance of Sardines (1000 Shoals per Sea Area)
1969	1125	4	907	4.27	140
1970	727	5		1.40	70
1971	650	42		1.34	80
1972	511	207	221	1.12	195
1973	597	134	183	0.29	275
1974	1286	1185	97	0.15	355

Also listed are numbers of pelican nests, young birds, and sardines living in the vicinity of Anacapa and Coronado Norte islands off the coast of Southern California. Source: From Anderson, D. W., et al. (1975). Science *190, 806–808.*

Polychlorinated Benzenes, Biphenyls, Dibenzodioxins, and Dibenzofurans

PCBs are the most abundant of the chlorinated aromatic contaminants now circulating in the marine ecosystem. They are synthetic organic compounds that were used as dielectrics in transformers and large capacitors. Because they are very stable at high temperatures, they serve as flame retardants and were added to hydraulic, drilling and heat-exchange fluids, inks, oil, paints, self-copying paper, plastics, and insecticides. Their commercial use began in the 1920s. Because of their toxic effects, which include endocrine disruption, immunosuppression and carcinogenicity, production was terminated in the United States in 1977 and worldwide in 1989. Before these restrictions were introduced, about 2 million tonnes of PCBs had been manufactured.

While some of the PCBs have been degraded or incinerated, the remainder have been released to the environment or is still in use or storage as part of a formerly manufactured product. Continuing emissions are due to evaporation from existing products, incineration of municipal and industrial wastes, building fires, and bursting transformers. One particularly egregious emission has been volatilization from waste oils used to control dust on roadways. PCBs are strongly influenced by the grasshopper effect and, thus, are undergoing a continued enrichment in polar regions. There is a substantial disparity between the quantities of PCBs reaching the northern and southern hemispheres, with the North Atlantic receiving a proportionally larger input.

Former emissions included discharge into natural waters of industrial effluents. From 1947 to 1977, the General Electric Co. discharged 1.3 million pounds (590,000 kg) of PCBs into the Hudson River from two capacitor manufacturing plants located north of Albany. The spread of PCBs throughout the river and its food chain has required the institution of fish consumption advisories. About 200 mi of the river are designated as a Superfund site. PCB wastes were also discharged directly into New Bedford Harbor by two electrical component manufacturing plants causing such a high degree of sediment contamination that this site was added to the Superfund list in 1983. Cleanups are presently underway at both sites with costs at each estimated to be on the order of hundreds of millions of dollars.

The base unit for PCBs is the biphenyl group shown in Figure 28.32a. A wide variety of PCBs have been produced, generally as mixtures by various substitutions of chlorine onto this base unit (Figure 28.32b). These derivatives are termed congeners. Of the 209 PCB congeners, 12 are recognized as especially toxic. Toxicity is related to the chlorine content and is highest in the congeners in which the chlorine atoms are coplanar with the rest of the molecule.

Because of their hydrophilic nature, PCBs are sparingly soluble in seawater and readily removed by adsorption onto sediments (Table 28.14). They are lipophilic, causing them to bioconcentrate. Due to their great stability, PCBs are not readily metabolized and, hence, undergo biomagnification, leading to very high body burdens in marine mammals, especially those living near industrialized harbors and in polar waters (Table 28.12 and Figure 28.29). Females are generally less contaminated than males because they transfer some of their PCB load to the developing fetus during pregnancy. The highest

(a)

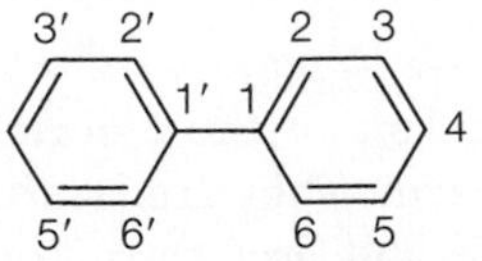

3′ 2′ 2 3
4′ 1′ 1 4
5′ 6′ 6 5

(b)

IUPAC No.	*Chlorine Substitution*
8	2, 4′
28	2, 4, 4′
29	2, 4, 5
44	2, 2′, 3, 5′
49	2, 2′, 4, 5′
52	2, 2′, 5, 5′
60	2, 3, 4, 4′
70	2, 3′, 4′, 5
86	2, 2′, 3, 4, 5
87	2, 2′, 3, 4, 5′
95	2, 2′, 3, 5′, 6
101	2, 2′, 4, 5, 5′
105	2, 3, 3′, 4, 4′
110	2, 3, 3′, 4′, 6
118	2, 3′, 4, 4′, 5
128	2, 2′, 3, 3′, 4, 4′
129	2, 2′, 3, 3′, 4, 5
137	2, 2′, 3, 3′, 6, 6′
138	2, 2′, 3, 4, 4′, 5
143	2, 2′, 3, 4, 5, 6′
153	2, 2′, 4, 4′, 5, 5′
156	2, 3, 3′, 4, 4′, 5
180	2, 2′, 3, 4, 4′, 5, 5′

(c)

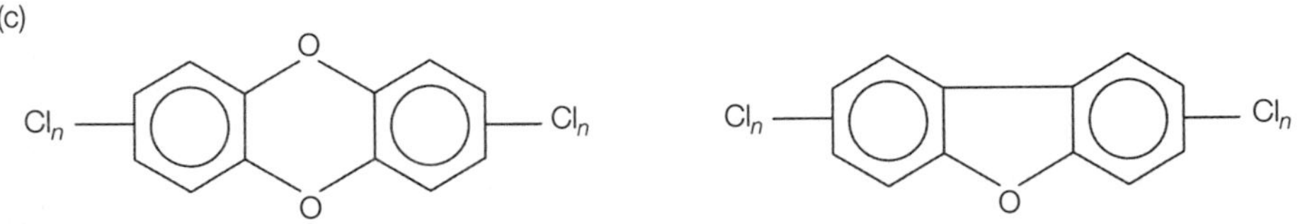

Cl_n O O Cl_n

PCDDs (I)

Cl_n O Cl_n

PCDFs (II)

FIGURE 28.32 *(Continued)*

(d)

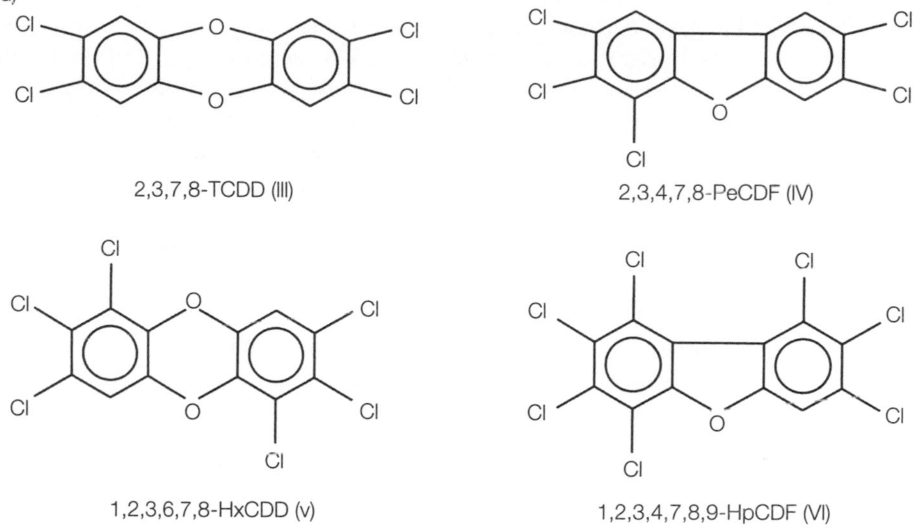

2,3,7,8-TCDD (III)

2,3,4,7,8-PeCDF (IV)

1,2,3,6,7,8-HxCDD (v)

1,2,3,4,7,8,9-HpCDF (VI)

FIGURE 28.32

(a) Base unit of PCBs, (b) molecular formulae of PCB congeners, (c) base unit of PCDDs and TCDFs, and (d) some congeners of PCDDs (left) and PCDFs (right).

PCB contamination has been observed in the toothed whales (*Odontoceti*) that consume fish, seabird, and other marine mammals. Sperm whales (*Physeter macrocephalus*), which feed on deepwater squid, have intermediate loads, whereas the predatory dolphins, which feed closer to the surface, carry higher loads. The highest loads have been observed in killer whales (*Orcinus orca*), whose diets consists of salmon and seal. In contrast, the baleen whales (*Mysticeti*) carry much lower loads as their diet is composed of zooplankton. Even whales living in the middle of the ocean far from land have detectable levels of PCBs and DDT.

In mammals, PCBs are potent endocrine disruptors, inducing pathological changes in reproductive organs and cycles. They also depress the immune system and are thought to have contributed to the mass mortality events involving harbor seals, turtles, and dolphins that occurred between 1986 to 1992 in the North Atlantic. Fish consumption is the major route of human exposure to PCBs. Some marine-fish- and mammal-eating populations, namely the Arctic Inuit and people living on the Faroe Islands, have body burdens high enough to pose a health threat to themselves and to their offspring.

The most toxic POPs are the polychlorinated dibenzo-p-dioxins (PCDD) and polychlorinated dibenzofurans (PCDF) (often simply referred to as "dioxins" and "furans," respectively, and collectively as PCDD/F). As shown in Figures 28.32c and d, a large variety of congeners exists due to variable substitution of chlorine atoms on the base units. PCDD/Fs are strongly concentrated in fish tissues with enrichment factors on the order of 10^4. The PCDD/Fs are potent endocrine disruptors. They also cause liver damage, weight loss, atrophy of the thymus gland, and immunosuppression. At higher concentrations and long exposure, they are likely carcinogens. One congener, 2,3,7,8-tetrachlorodipenzol-*p*-dioxin (TCDD), is so toxic that the U.S. EPA has established an ambient water quality standard of 0.014 pg/L to ensure safe levels in seafood. PCDD/Fs are by-products of industrial processes and waste incineration as are polychlorinated benzenes.

In 1996–1997, the major sources of dioxin in North America were municipal waste incinerators (25%), backyard trash burning (22%), cement kilns burning hazardous waste (18%), medical waste incinerators (11%), secondary copper smelters (8%), and iron sintering plants (7%). Lesser amounts are produced during the bleaching of wood stocks in paper mills and the manufacture of the herbicide, 2,4,5-T, and the wood preservative, 2,4,5-trichlorophenol. It is also generated during the environmental degradation of DDT, DDE, and lindane (HCH). Examination of downcore concentration profiles in lake and marine sediments suggests that the natural production of PCDD/Fs from forest fires is a relatively minor source compared to anthropogenic inputs.

PCDD/Fs are transported to the ocean primarily via effluent discharges, stormwater runoff and by wet and dry deposition of aerosols. Because of their high molecular weight and low volatility, most PCDD/F contamination is local, with adsorption onto sediments being the primary sink, although transport to the Arctic is occurring.

Hexachlorobenzene (HCB) was used as a pesticide to protect against fungus and in the manufacture of fireworks, ammunition, and synthetic rubber. This polychlorinated benzene now is found in coastal marine sediments. It also bioaccumulates in marine organisms.

28.7.2 Other Semivolatile Synthetic Organic Compounds of Concern

Other artificial organic compounds that have a widespread distribution in marine biota, sediments, and terrestrial settings, including human tissues include: (1) the polybrominated diphenyl ethers (PBDEs), (2) hexabromocyclododecanes (HBCDs), (3) per- and polyfluorinated alkyl substances (PFAS), (4) polychlorinated napthenes (PCNs), and (5) the alkyl and aryl phthalate esters. They are all extremely resistant to physical, chemical, or biological degradation, making them highly persistent. They also appear to be bioaccumulative endocrine disruptors. Some are ampiphilic, binding to proteins, and, hence, are heptocarcinogens.

The PBDEs (decaBDE, octaBDE, and pentaBDE) and are used as flame retardants in plastics, electronic equipment, printed circuit boards, vehicles, furniture, textiles, carpets, and building materials. Global demand has increased rapidly since the 1970s with 70,000 tonnes produced in 2001. Their flame retardant activity relies on decomposition at high temperatures, leading to the release of bromine atoms. This slows the chemical reactions that drive O_2-dependent fires. HBCDs are a flame retardant added to extruded and expanded polystyrene that is used as thermal insulation in buildings.

The PFAS are used in the production of stain-repelling agents, and fluoropolymers, such as Teflon, pesticides, lubricants, paints, medicines, and fire-fighting foams. PCNs are used in cable insulation, wood preservatives, engine oil additives, electroplating masking compounds, and in dye production. PCNs are also present in PCB formulations as impurities. Although their production and use were banned in the United States and Europe in the 1980s, PCNs continue to circulate around the environment due to slow degradation rates and tendency to bioaccumulate. The alkyl and aryl phthalate esters, such as nonylphenol, are used as additives in polyvinyl chloride (PVC) and various other plastic resins to impart flexibility. The phthalates are also used as plasticizers in building materials, home furnishings, transportation, clothing, and to a limited extent in food (packaging) and medical products.

28.7.3 Volatile Organic Compounds

Low-molecular-weight organic compounds tend to have high vapor pressures and, hence, are also termed *volatile organic compounds* or VOCs. As shown in Figure 28.33, many are chlorinated. The primary use of VOCs, such as carbon tetrachloride and benzene, is as industrial solvents. Trichloroethylene is commonly used as dry-cleaning fluid. As a result of poor waste-disposal procedures, most of the groundwaters in the United States now contain some of these compounds. This is a matter of great concern as most VOCs are carcinogenic. VOCs are introduced into the ocean via river runoff and gas exchange at the air-sea interface. The latter process is not well understood as some VOCs are thought to have a natural marine sources. The ocean appears to be a significant global sink for the synthetic ones, such as the freons.

The freons are chlorofluorocarbons (CFCs) and hydrochlorofluorocarbons (HCFCs) used as refrigerants, flame retardants, and aerosol propellants. In 1989, the United Nations banned the use of several of the freons as part of its treaty, the Montreal Protocol on Substances that Deplete the Ozone Layer. Loss of stratospheric ozone is of concern as it absorbs ultraviolet rays, thereby shielding Earth's surface from this highly mutagenic radiation. It is somewhat ironic that anthropogenic input of ozone to the lower atmosphere (troposphere) is occurring as a result of fossil-fuel combustion. In the troposphere, ozone reacts with other air pollutants and sunlight to form photochemical smog and, hence, poses a significant health threat to humans. The chlorofluorocarbons are also greenhouse gases.

28.7.4 Organometallic Paints

Organometallic derivatives tend to be more toxic than their parent inorganic metal, with the alkyl derivatives being more toxic than the aryl ones. Tetraalkyl lead is an example. Other organometallics in widespread use are the arsenicals (methylarsenates) and tins (tributyltin [TBT] and triphenyltin [TPT]). Sodium and disodium methylarsenate are used as herbicides. Inorganic arsenic is also being released into the environment, most notably through the use of chromated copper arsenate (CCA) as a wood preservative. Phytoplankton assimilate and biomethylate inorganic arsenic. Methylated arsenic is passed up the marine food chain, but fortunately is not as toxic as the inorganic species. In comparison, TBT and TPT exert toxic effects at very low concentrations (1–2 ng/L).

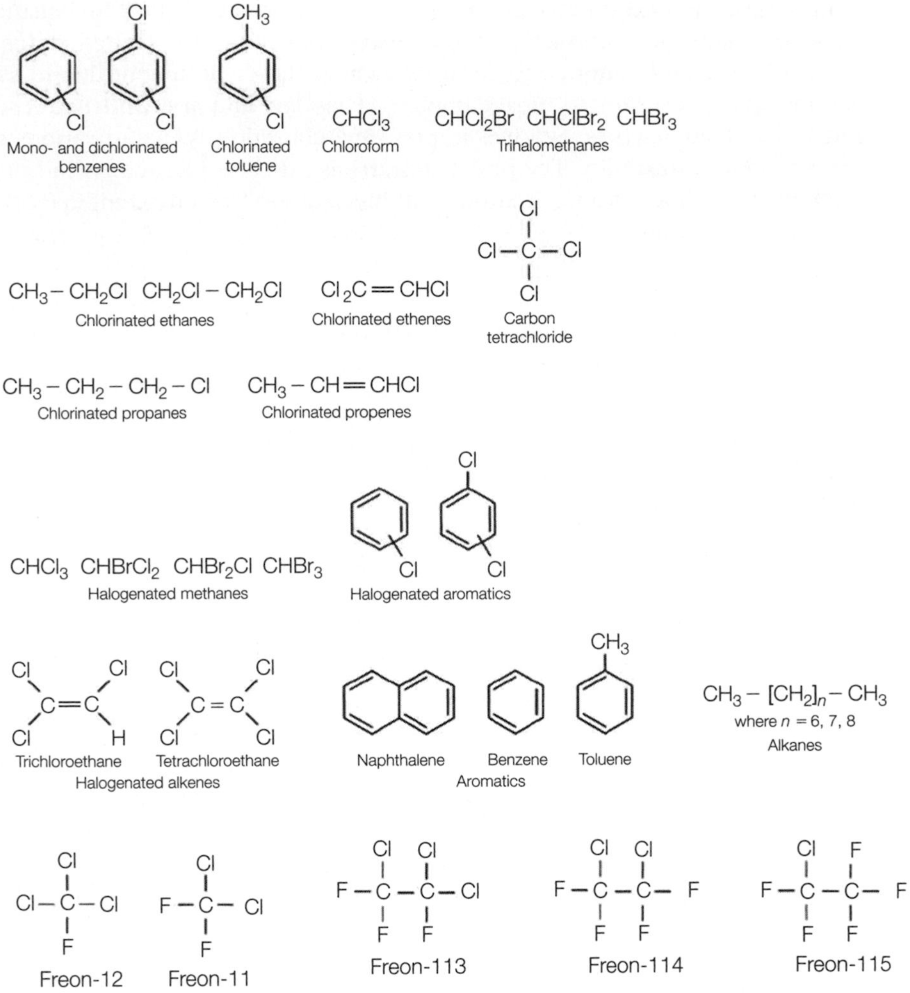

Mono- and dichlorinated benzenes — Chlorinated toluene — $CHCl_3$ Chloroform — $CHCl_2Br$ $CHClBr_2$ $CHBr_3$ Trihalomethanes

CH_3-CH_2Cl CH_2Cl-CH_2Cl Chlorinated ethanes — $Cl_2C=CHCl$ Chlorinated ethenes — CCl_4 Carbon tetrachloride

$CH_3-CH_2-CH_2-Cl$ Chlorinated propanes — $CH_3-CH=CHCl$ Chlorinated propenes

$CHCl_3$ $CHBrCl_2$ $CHBr_2Cl$ $CHBr_3$ Halogenated methanes — Halogenated aromatics

Trichloroethane — Tetrachloroethane (Halogenated alkenes) — Naphthalene, Benzene, Toluene (Aromatics) — $CH_3-[CH_2]_n-CH_3$ where $n = 6, 7, 8$ Alkanes

Freon-12 — Freon-11 — Freon-113 — Freon-114 — Freon-115

FIGURE 28.33

Structures of some volatile organic compounds commonly found in natural waters.

TBT and TPT have been used in antifouling boat paints since the early 1970s. By the 1980s, 140,000 tonnes of TBT-based antifouling paints were consumed annually in the United States. In the late 1990s, the global consumption of TBT-based paint was estimated to be around 7.2 million tonnes per year, covering 70% of seagoing vessels. Its popularity was due to its effectiveness and relative long lifetime (4 to 7 y).

Over time, TBT and TPT are solubilized from the paint. In sea water, TBT exists as a cation ($(C_4H_9)_3Sn^+$) complexed mostly with hydroxide. The most important sink for TBT is adsorption onto sediments where degradation causes it to have a half-life of a few years.

Aqueous and sedimentary TBT and TPT cause chronic and acute effects in algae, zooplankton, crustacea, mollusks, fish, and animals. These effects have been local in nature, occurring mostly in harbors near industrialized lands. TBT is bioaccumulated in many species, which is unfortunate as it is a potent endocrine disruptor. The enrichment factor in mussels, snails, and oysters ranges from 10,000 to 60,000. As mentioned in Section 28.7.1, TBT induces imposex in marine gastropods.

Because of slow clearance times, TBT levels in mussels and oysters are useful for monitoring environmental levels. TBT concentration data from the Mussel Watch program indicates that concentrations are highest in harbors and marinas, but have been decreasing since TBT monitoring started in 1989, reflecting the declining use of this biocide. While TBT does not appear to be biomagnified, TPT does, with BMFs approaching those of the PCBs. Butyl tins have been found in sea otters, cetaceans, seals, and water birds. Concentrations tend to be vary among the organs, with highest levels found in the livers of mammals as the butyltins tend to bind to proteins rather than lipids. Liver concentrations for cetaceans from various waters are presented in Table 28.16. Very high concentrations were measured in bottle-nose dolphins that were found stranded along southeastern U.S. Atlantic and Gulf coasts during 1989–1994, with the highest liver concentration reaching 11 mg/g wet wt. This high level of contamination is thought to have contributed to this mass mortality event.

Given the toxic effects of TBT and TPT, efforts are now underway to halt their use in antifouling paints. The United States and European Union banned their use on boats smaller than 25 m circa 1990. In 2001, the United Nations adopted the International Convention on the Control of Harmful Anti-fouling Systems on Ships, which calls for a worldwide ban on use of TBT in antifouling paints and is slated to come into force in 2008. As noted above, areas where the use of butyltin has been restricted have experienced a decline in butyltin concentrations. A similar decline has been observed in the incidence of imposex.

Concern is now directed at the environmental impacts of antifouling agents that are replacing TBT. For example, silicone-based paints contain polydimethylsiloxanes (PDMS). Although these compounds are not reactive and do not bioaccumulate, they have the potential to exert negative effects through physical-mechanical means, such as by trapping and suffocation of organisms.

28.7.5 **Marine Litter**

Marine litter is considered to be any persistent, manufactured, or processed solid material discarded, disposed of, or abandoned in the marine and coastal environment. Marine litter has become widely dispersed and is now found throughout all the world's oceans, either floating on the sea surface, stranded on the shoreline, or lying on the seabed. In 1997, the U.S. National Academy of Sciences estimated the global annual input of marine litter into the oceans to be 6.4 million tonnes with 8 million items being introduced per day. As shown in Table 28.17, marine litter is comprised of plastics, wood, glass, and metal fragments. Because of their low density and durability, plastics represent about 90% of the floating debris. Plastics are organic polymers, such as polyethylene, polypropylene,

Table 28.16 Comparison of Residue Levels (ng/g wet weight)[a] of Butyltin Compounds in the Liver of Cetaceans from Various Waters.

Species	Location	MBT	DBT	TBT	$\sum$BT[b]
Killer whale	Rausu, Japan	40 (<3.0–84)	210 (8.0–650)	20 (29–62)	250 (13–770)
Finless porpoise	Japan coasts	1300 (130–3000)	2900 (790–6100)	700 (200–1100)	4900 (1100–10,000)
Short-finned pilot dolphin	Ayukawa, Japan	510 (340–650)	1200 (770–1600)	350 (260–520)	2100 (1500–2600)
Risso's dolphin	Taiji, Japan	430 (35–1300)	2400 (270–4400)	820 (210–1200)	3700 (550–6000)
Dall's porpoise	Sanriku, Japan	97 (14–150)	430 (210–1200)	230 (120–540)	760 (350–1700)
Killer whale	Taiji, Japan	710 (480–1100)	1600 (1500–1900)	180 (150–220)	2500 (2200–2700)
Striped dolphin	England and Wales	< 4	160 (84–230)	80 (77–82)	240 (160–310)
Common dolphin	England and Wales	< 4	160 (130–200)	61 (53–68)	220 (190–260)
Bottlenose dolphin	Italy	180 (150–200)	1200 (800–1600)	330 (250–400)	1700 (1200–2200)
Bottlenose dolphin	Florida, U.S.	640 (100–2300)	1900 (290–8300)	200 (5.8-770)	2700 (420–11,000)
Harbour porpoise	Black Sea, Turkey	23 (10–34)	130 (100–164)	26 (16–35)	180 (140–220)
Finless porpoise	Dongshan, China	130 (99–220)	670 (570–880)	70 (57–130)	890 (730–1200)
Long-snouted spinner dolphin	Sulu Sea, Philippines	3.1 (<3 – 3.1)	32 (23–41)	21 (19–23)	56 (42–67)
Indo-Pacific hump-backed dolphin	Bay of Bengal, India	22 (11–46)	47 (22–71)	54 (34–100)	120 (67–200)
Bottlenose dolphin	Bay of Bengal, India	26 (11–29)	44 (15–84)	35 (16–55)	110 (53–170)

[a]*Mean values with ranges in parentheses.*
[b]$\sum BT = [MBT] + [DBT] + [TBT]$.
Source: From Kajiwara, N., et al. (2006). Marine Pollution Bulletin *52, 1066–1076.*

Table 28.17 Some Types of Marine Litter.

Plastics	
Fishing gear	Nets, floats, ropes, traps, and fishing lines
Cargo-associated wastes	Plastic strapping and sheeting; debris from the offshore petroleum industry; and pellets of polyethylene, polypropylene, and polystyrene
Sewage-associated wastes	Condoms, tampon applicators, sanitary napkins, and disposable diaper shields
Domestic	Bags, bottles, lids, six-pack rings, and balloons
Rubber and metal fragments	
Glass bottles	
Wood and paper	
Metal drums containing synthetic wastes	
Munitions	

styrofoam, Saran, and nylon. The molecular structures of the base units of some of these plastic polymers are illustrated in Figure 28.34.

A significant portion of the litter enters the ocean as a result of land-based activities, namely: (1) discharge of untreated municipal sewage and stormwater into rivers or coastal waters, (2) similar discharges from industrial facilities, (3) loss from municipal and industrial landfills sited near rivers and the shoreline, and (4) coastal tourism. Surveys have demonstrated that tourism is responsible for about 60% of the litter found on beaches. Water and winds are both important in transferring litter from land into the sea. In addition to litter that is blown off land, atmospheric transport carries balloons out over the ocean where they are eventually deposited on the sea surface and mistaken as prey by sea turtles. Sea-based sources of litter include intentional dumping and unintentional releases. The latter is associated with losses from vessels during bad weather, equipment failure, and operational activities. For example, containerized cargo is frequently washed off the deck of ships during storms. Fishing nets lost at sea continue to catch fish, resulting in a process called "ghostfishing." Operational losses also take place from the offshore petroleum and mariculture industries.

After entering the marine environment, litter can (1) be blown around, (2) remain floating on the water surface, (3) drift in the water column, (4) get entangled on shallow, tidal bottoms, or (5) sink to the deeper seabed. About 70% lands on the seabed, 15% on beaches and the shoreline, and 15% remains floating in the water column. Concentrations of floatables are highest in the vicinity of shipping lanes, around fishing areas, and in oceanic convergence zones. Geostrophic currents are thought to have a focusing effect, giving rise to the so-called Pacific Trash Vortex, which is composed of a western Garbage Patch, located south of Japan, in the Northwestern Pacific and an eastern Garbage Patch, located between San Francisco and Hawaii, in the Northeastern Pacific

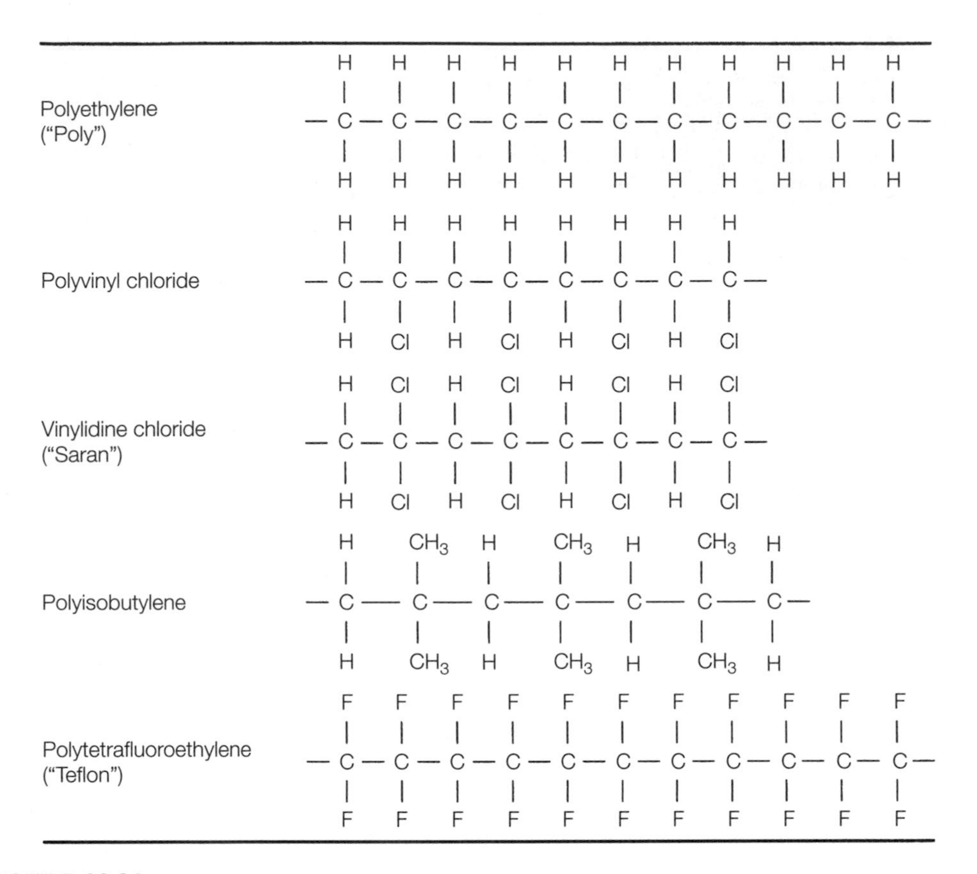

FIGURE 28.34

Molecular structure of some plastic polymers. *Source*: From Horne, R. A. (1969). *The Chemistry of Our Environment*. John Wiley & Sons, Inc., p. 441.

Ocean. The former is estimated to have a surface area the size of Texas. The focusing effect of the Kuroshio Current on the density of plastics collected in plankton net tows is shown in Figure 28.35. In some locations, particle densities are greater than that of the local zooplankton. High concentrations have also been reported on the seabed near industrialized areas such as the Mediterranean and the North Seas. In 2000, the volume of litter estimated to be residing on the seabed of the North Sea was 600,000 m^3.

The most common litter item are small pieces of plastics whose lengths are on the order of a few millimeters. An important component are thermoplastic resin pellets and beads that are raw materials, mostly polypropylene and polyethylene, intended for manufacture into commercial items. Loss during ship transport and stormwater runoff are major sources of the pellets and beads to the ocean. Densities of 3500 per km^2 have been reported floating on the surface in the Sargasso Sea. On the beaches of New Zealand located near industrialized areas, concentrations as high as 100,000 per km^2 are now being observed. The pellets and beads are carried by currents until they are either

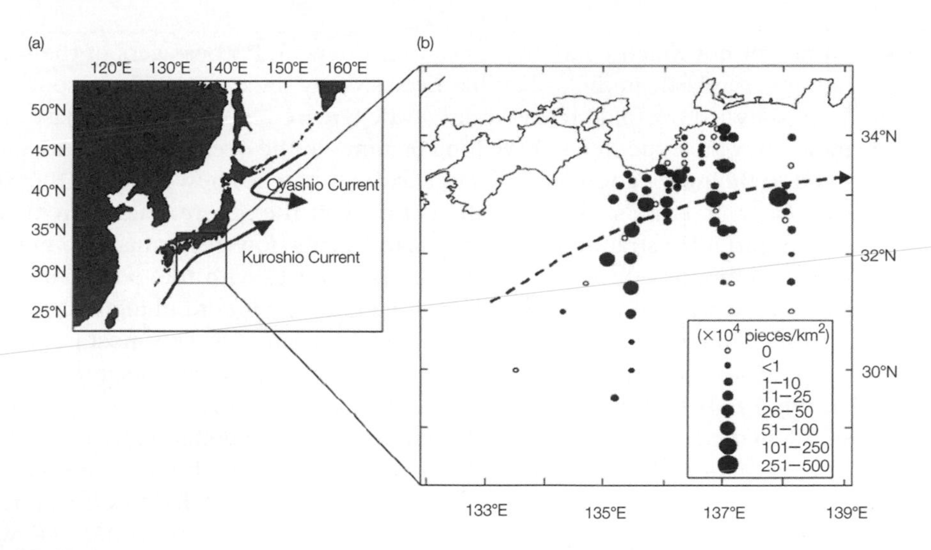

FIGURE 28.35

Locations of stations sampled for plastic in 2000–2001. (a) Distribution and abundance (pieces/km^2) of total plastics in the Kuroshio Current area. (b) Samples were collected by a surface tow using a neuston net (mouth opening 50 × 50 cm; side length 3 m; mesh size 330 μm). The numerically dominant size class (62%) was 1–3 mm. Broken line in (b) denotes the Kuroshio flow path during the study period. *Source*: From Yamashita, R., and A. Tanimura (2007). *Marine Pollution Bulletin* 54, 464–488.

degraded photochemically, consumed by marine organisms, or stranded on beaches. The pellets and beads have been found embedded in tarballs, forming concretions that coat some Caribbean beaches. They have also been shown to have elevated concentrations of lipophilic pollutants, such as the PCBs, PAHs, DDT, DDE, and nonylphenol, that have presumably been absorbed from seawater. Concentration factors up to 10^6 have been observed, making these pellets a potential source of toxicants to the marine organisms that ingest them.

Marine litter poses threats to the health and safety of human and marine life. Threats to humans include: (1) fouling of beaches, marinas, and harbors, (2) damage to vessels and gear, (3) damage to cooling water intakes in power stations and desalinization plants, (4) clogging of waterways that contributes to flooding, and (5) transport of invasive species. Beach litter presents an aesthetic, recreational, and health hazard. Frequent beach cleaning can have harmful effects on native ecosystems. Threats to marine life from litter include: (1) ingestion and entanglement of seabirds, mammals, and turtles; (2) smothering of the seabed, affecting corals and sea grasses; and (3) injuries to cattle grazing in coastal pastures. Exposure to toxins is an eventual likely secondary effect of litter associated with military debris. Millions of tonnes of munitions have been dumped in metal drums that contain explosives and biowarfare agents. As these drums corrode, their contents will be released into the ocean.

Although not much data exist, oceanographers and other users of the sea agree that the amount of plastic in the oceans has risen sharply since the 1950s. In some places, studies have shown a ten-fold increase per decade (Figure 28.36). Because plastics are slow to degrade, they are expected to have long lifetimes in the ocean. This has undoubtedly contributed to their collection into the large garbage patches now found in the North Pacific Ocean. Oceanic islands that lie in the pathway of the geostrophic currents are particularly prone to litter stranding. The Hawaiian Islands now accumulate so much trash that a large, continuous effort must be maintained to deal with this pollutant. This includes beach clean ups and recovery of relict fishing gear that is continuing to catch fish, turtles, and marine mammals. In the case of the Hawaiian Islands, litter is most prone to stranding during periods when the subtropical convergence zone migrates south.

In the 1970s the United Nations enacted controls to reduce litter by adopting two treaties: (1) the Convention on the Prevention of Marine Pollution by Dumping of Wastes and Other Matter, aka the London Convention, and (2) the International Convention for the Prevention of Pollution from Ships, aka MARPOL. The London Convention has been updated by the London Protocol, which came into force in 2006. MARPOL was updated in 1978 with new provisions coming into force in 2006.

Under the updated London Convention, now banned is dumping of all industrial and radioactive wastes as well as incineration at sea of industrial wastes and sewage sludge. Still permitted is the dumping of: (1) dredged material, (2) sewage sludge, (3) fish wastes or material resulting from industrial fish-processing operations, (4) vessels and platforms

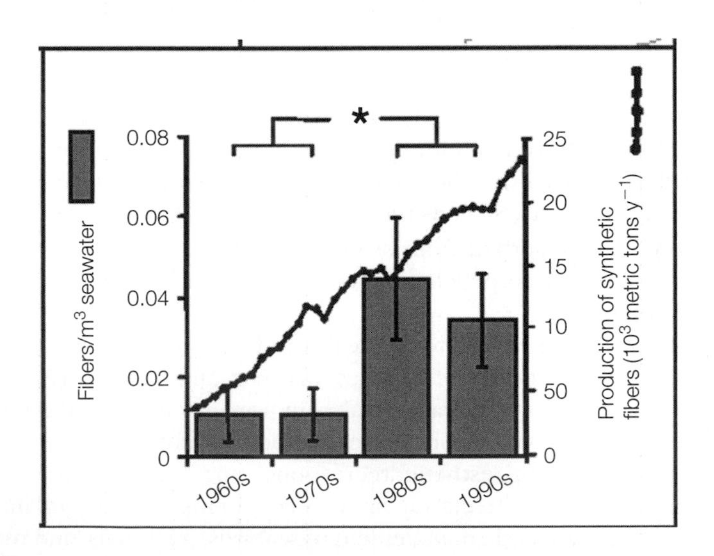

FIGURE 28.36

Microscopic plastic collected as part of continuous plankton recorder samples showing a significant (*) increase in abundance when samples from the 1960s and 1970s were compared to those from the 1980s and 1990s. (Samples were collected at 10 m depth through a 127 mm^2 aperture onto a scrolling 280 μm-mesh silkscreen.) Approximate global production of synthetic fibers is overlain for comparison. *Source*: From Thompson, R. C., *et al.* (2004). *Science* 304, 838.

It is illegal for any vessel to dump plastic trash anywhere in the ocean or navigable waters of the United States. Annex V of the Marpol Treaty is a new International Law for a cleaner, safer marine environment. Each violation of these requirements may result in civil penalty up to $25,000, a fine up to $50,000, and imprisonment up to 5 years.
U.S. lakes, rivers, bays, sounds, and 3 miles from shore
ILLEGAL TO DUMP
Plastic Garbage
Paper Metal
Rags Crockery
Glass Dunnage
Food
3 to 12 miles
ILLEGAL TO DUMP
Plastic
Dunnage (lining & packing materials that float) also if not ground to less than one inch:
Paper Crockery
Rags Metal
Glass Food
12 to 25 miles
ILLEGAL TO DUMP
Plastic
Dunnage (lining & packing materials that float)
Outside 25 miles
ILLEGAL TO DUMP
Plastic
State and local regulations may further restrict the disposal of garbage

FIGURE 28.37

MARPOL regulations on ocean dumping. *Source*: From O'Hara, K. J., *et al.* (1988). *A Citizen's Guide to Plastics in the Ocean: More Than a Litter Problem*. Center for Marine Conservation, p. 37.

or other man-made structures at sea, (5) inert organic geologic materials (mining wastes), (6) organic material of natural origin, and (7) bulky items primarily composed of iron, steel, and concrete and similarly unharmful materials for which the concern is physical impact. The last is limited to those circumstances where such wastes are generated at locations, such as small islands with isolated communities, that have no practicable access to disposal options other than dumping. Also permitted are CO_2 streams associated with carbon sequestration processes.

The current version of MARPOL contains regulations designed to prevent loss of hazardous materials from ships including cargo, oil, sewage, garbage, and air pollution. As illustrated in Figure 28.37, Annex V of MARPOL specifies the distances from land at which different types of litter can be disposed of and bans the dumping at sea of all forms of plastic.

28.8 THE FUTURE OF THE OCEAN AS A WASTE SPACE

In the 1970s and 1980s, some marine scientists proposed that the oceans continue to be used, albeit with careful planning and control, as an intentional dumpsite. This proposition was motivated by the recognition that human population was continuing to grow, with densities increasing in coastal areas, ensuring an increase in waste production. During this period, a crisis in waste management was already brewing, as landfills were filling up, and various legal deadlines halting dumping at sea were approaching. Since no one wants to live near a landfill, we were running out of safe and secure places to put trash. In 1987, a barge traveled up and down the eastern seaboard of the United States for 6 months in a futile attempt to find a place to dump the garbage it was carrying. Another telling event occurred in 1996, when a 300-foot-high mountain of garbage that had been dumped into a municipal landfill slid down the rocky shoreline at O Portino, Spain, toward the Atlantic Ocean. This avalanche was caused by the formation of methane and liquefaction of the wastes.

The marine scientists advocating a responsible use of the ocean as a waste space, pointed out that "there is no free lunch"—that even with recycling and conservation, we were still going to need acceptable sites for waste disposal. They asserted that the ocean could provide these sites as it, like the land, has some finite capacity to assimilate various compounds without causing an unacceptable impact on living organisms or nonliving resources. They contended that waste disposal, if carefully managed, could be kept within these limits. This left at issue how to define an unacceptable impact. The scientists advocated a definition that reflected socioeconomic pressures, thereby taking into account the relative harm to human life posed by land-based disposal, such as contamination of drinking water.

Despite scientific evidence suggesting that intentional and controlled dumping was not leading to significant environmental degradation, political leaders adopted a precautionary approach, i.e., "If you don't know, don't dump." They felt we had so much to lose and such a poor understanding of the fate, transport, and impacts of dumped wastes, that it was best to abandon any effort to continue intentionally using the ocean as a waste disposal option. They also questioned whether a regulatory management and surveillance system could be implemented to guide legalized dumping. Thus, since the 1980s, the United Nations has continued strengthening the London and MARPOL conventions, with similar efforts occurring in the United States through reauthorization of the Ocean Dumping and Clean Water Acts. It is notable that a recent amendment to the London Convention permits burial of CO_2 in the seabed, vindicating the contention that socioeconomic forces will ultimately require the oceans to play some role in human waste disposal.

While dumping has been variously controlled, pollutant fluxes from other transport pathways have increased to the point where they are now larger than any effect that could have been induced by continued dumping. These continuing and growing sources include: (1) discharge of treated/untreated municipal effluents, particularly sewage as fully 60% of the human population still lacks access to basic sanitation; (2) stormwater runoff, particularly of agricultural chemicals, including nutrients, herbicides, and pesticides; (3) atmospheric transport of nitrogen, POPs, and volatile metals (Hg, Pb, Cd); and (4) uptake of gaseous CO_2 leading to the acidification of surface seawater. New synthetic organic compounds are being identified in seawater, sediments and marine organisms. An emerging concern is the widespread distribution of pharmaceuticals and personal care products (PPCPs) in fresh and marine waters due to discharge from sewage treatment plants.

Since the 1990s, an increasing awareness has developed that ocean health is intimately tied to human health. Environmental degradation of the ocean threatens its use as a food and recreational resource. Loss of species means loss of potential natural products yet to be discovered. Changes that promote expansion of pollutogens and natural pathogens, such as cholera, present safety problems for all who come in contact with seawater. Changes in ocean circulation caused by global climate change also impact sea transportation, mineral mining operations, and coastal land uses. Much of this environmental degradation is a consequence of population growth, land-use change, and introduction of invasive species. From a biotic standpoint, the most profound impacts have been habitat destruction and alteration of biotic community structures. In the ocean, impacts to the marine biota have the potential to alter the biological pumps which play key roles in the biogeochemical cycles involved in climate control.

Humans have so altered material fluxes in the crustal-ocean-atmosphere factory that scientists deem our effects have launched us into a new geological epoch, dubbed the Anthropocene, which is discernable in marine sediments as a geochemically distinct stratum. Some of these changes have overwhelmed stabilizing feedbacks in the global biogeochemical cycles, leading to catastrophic ecological changes, such as the loss of coral reefs and aragonitic plankton. To reverse this trend requires adoption of pollution control strategies that are global scale while tailored to local ecosystems. They must also be practical and cost effective. Identification of these strategies should rely on a science-based evaluation of management options. The best framework for conducting these evaluations is predictive modeling. The relevant time scales for these models depends on the threat being managed, ranging from emergency responses to oil spills to control of greenhouse gases.

Formulation of reliable predictive models is now an area of intense scientific effort and has given rise to a new field called *ecological forecasting*. These models rely on paleoceanographic data, much like retrospective epidemiological studies, to explore the effects of past changes on the function of the crustal-ocean-atmosphere factory. In the marine system, focused efforts are underway to better understand: (1) the role of the coastal ocean as an elemental sink and source; (2) how the plankton community structure, nutrient availability, and ocean circulation affect the biological pumps; (3) the temporal and spatial variability of hydrothermal sources and sinks; and (4) how to quantify gas exchange rates across the air-sea interface and the role of sea-surface microlayer as a concentrating agent. Many of these require a better understanding of physical processes, such as turbulence and water-mass renewal, as well as geological processes, such as particle transport, weathering processes and plate tectonics.

Other critical areas in which more knowledge is needed are in elucidating the role of polar zones in the global biogeochemical cycles, especially as they are likely to be the regions most affected by global climate change. Finally, our knowledge of microbial life in the oceans, sediments, and even crustal rocks, is still limited such that we're continuing to identify new life forms, new metabolisms, and new habitats. We are just beginning to understand that these microbes (and other members of the plankton) are capable of interspecies interactions that resemble those of a superorganism, enabling their continued joint evolution through genetic exchanges.

Effective implementation of pollution control strategies requires assessment of progress, which in turn, requires monitoring. We know that we need to keep watch across many time and spatial scales. Fortunately modern technology is making this increasingly more possible through satellite imagery and in situ sensors—if we have the will to provide funding and a trained corps of scientists to operate the equipment and manage the data. The latter is critical as the large amounts of data now being collected need to be archived in a way that makes them usable to a diverse group of stakeholders. The Global Earth Observation System of Systems (GEOSS) described in Section 1.7 is leading the way in organizing national efforts into a coherent international program. The Global Ocean Observing System (GOOS) is focusing on the ocean. The Land-Ocean Interactions in the Coastal Zone Project of the International Geosphere-Biosphere Programme (LOICZ) is doing the same for coastal zones, recognizing estuaries and their rivers as areas of particular concern. The UN's Joint Group of Experts on the Scientific Aspects of Marine Protection (GESAMP) is now producing global summary assessments and management recommendations. In the United States, efforts are underway to fully implement an Ocean Action Plan that was developed in 2004 and established a new cabinet-level Committee on Ocean Policy. These efforts all recognize that environmental management is best practiced from an ecosystem and watershed perspective.

One thing that we can be sure about is that we don't presently know all that we need to know to develop perfect management strategies. Thus monitoring and assessment data should be used to refine management approaches; this process is termed *adaptive management*. To identify future options for effective environmental management and to implement and assess them requires an enormous international effort, involving a large-scale commitment of time and resources. Obviously the need for new marine biogeochemists has never been greater or more important. In the words of the U.S. National Science Foundation (2003):

> As our technological and research capacity increases, we face both the promise of understanding the environment and our relationship to it, and the responsibility of making wise decisions about managing the complex relationships among people, ecosystems, and planetary processes.[8]

and the 45th Vice President of the United States, Al Gore, Jr. (1993–2001),

> Put your knowledge into action.

[8] Pfirman, S., and the NSF Advisory Committee for Environmental Research and Education (2003). Complex Environmental Systems: Synthesis for Earth, Life, and Society in the 21st Century, A report summarizing a 10-year outlook in environmental research and education for the National Science Foundation, 68 pp.

APPENDIX

1 The Periodic Table of the Elements

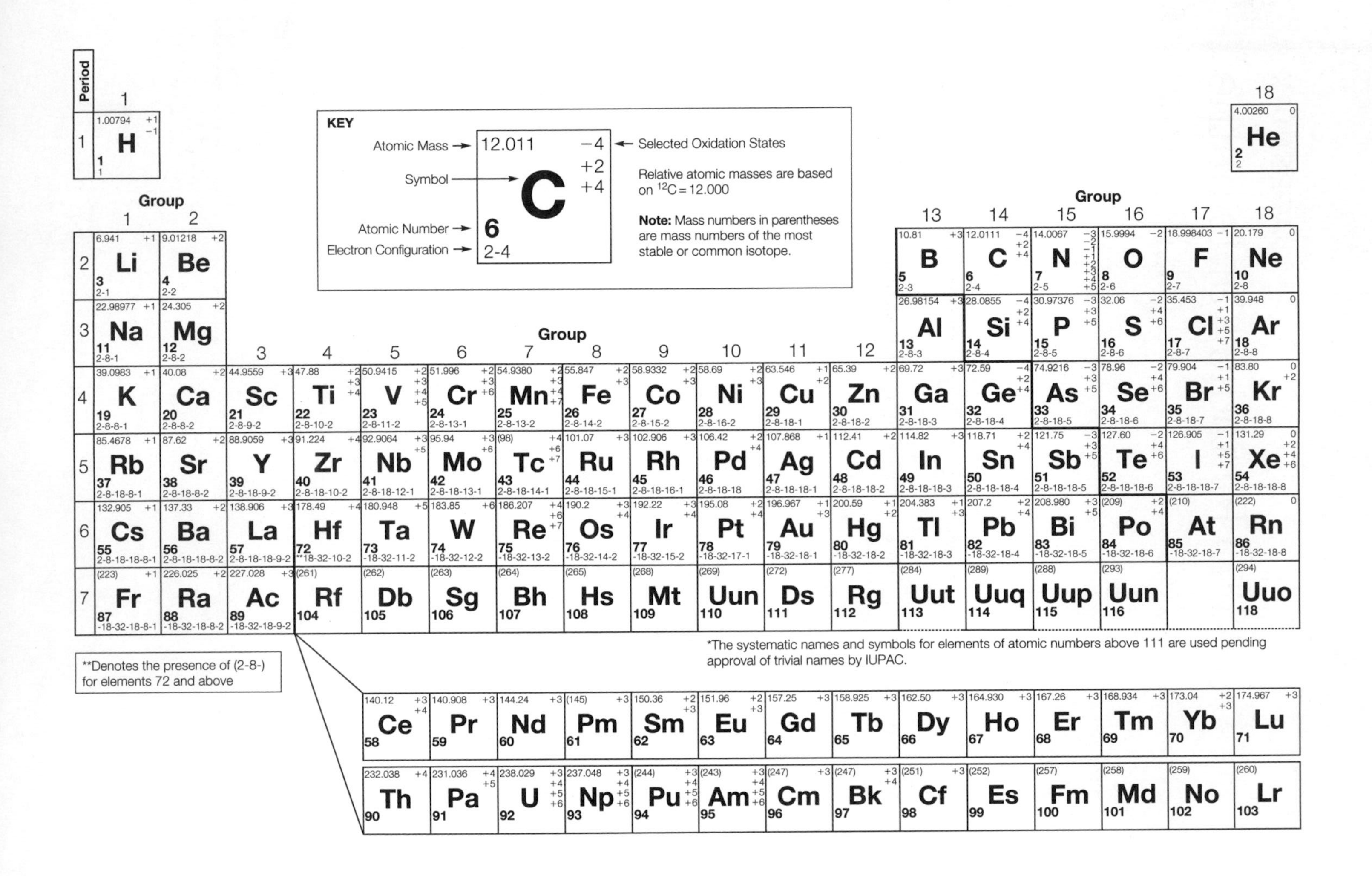
KEY
Atomic Mass → 12.011
Symbol → C
Atomic Number → 6
Electron Configuration → 2-4
−4 +2 +4 ← Selected Oxidation States
Relative atomic masses are based on 12C = 12.000
Note: Mass numbers in parentheses are mass numbers of the most stable or common isotope.
*The systematic names and symbols for elements of atomic numbers above 111 are used pending approval of trivial names by IUPAC.
**Denotes the presence of (2-8-) for elements 72 and above

APPENDIX 2

Common Names and Chemical Formulae

Table A2.1

Most Abundant Elements Present in Biological Tissues	
C	Carbon
H	Hydrogen
N	Nitrogen
O	Oxygen
P	Phosphorus
S	Sulfur
Major Ions in Seawater	
Na^+	Sodium
Cl^-	Chloride
Ca^{2+}	Calcium
Mg^{2+}	Magnesium
K^+	Potassium
SO_4^{2-}	Sulfate
Carbonate System	
CO_2	Carbon dioxide
H_2CO_3	Carbonic acid
HCO_3^-	Bicarbonate
CO_3^{2-}	Carbonate
$CaCO_3$	Calcium carbonate
$C(H_2O)$	Organic matter
TDIC or ΣCO_2 or C_T	Total dissolved inorganic carbon
CA	Carbonate alkalinity
TA	Total alkalinity

(Continued)

Table A2.1 *(Continued)*

Major Gases in Seawater	
N_2	Nitrogen
O_2	Oxygen
Ar	Argon
Kr	Krypton
CO_2	Carbon dioxide
CH_4	Methane
H_2	Hydrogen
H_2S	Hydrogen sulfide
DMS	Dimethylsulfide
CO	Carbon monoxide
N_2O	Nitrous oxide
Inorganic Nitrogen Compounds	
N_2	Nitrogen gas
NO_3^-	Nitrate
NO_2^-	Nitrite
N_2O	Nitrous oxide
NH_4^+	Ammonium
NH_3	Ammonia
Inorganic Phosphorus Compounds	
PO_4^{3-}	Phosphate
HPO_4^{2-}	Hydrogen phosphate
$H_2PO_4^-$	Dihydrogen phosphate
Inorganic Silicon Compounds	
H_4SiO_4	Silicic acid
SiO_2	Silica
BSi	Biogenic silica
DSi	Dissolved silica

APPENDIX 3

Metric Units and Equivalents

Table A3.1

Metric Prefixes

atto (a)	= 0.000000000000000001	$= 10^{-18}$
femto (f)	= 0.000000000000001	$= 10^{-15}$
pico (p)	= 0.000000000001	$= 10^{-12}$
nano (n)	= 0.000000001	$= 10^{-9}$
micro (μ)	= 0.000001	$= 10^{-6}$
milli (m)	= 0.001	$= 10^{-3}$
centi (c)	= 0.01	$= 10^{-2}$
eci (d)	= 0.1	$= 10^{-1}$
deca (da)	= 10	$= 10^{1}$
hecto (h)	= 100	$= 10^{2}$
kilo (k)	= 1000	$= 10^{3}$
mega (M)	= 1,000,000	$= 10^{6}$
giga (G)	= 1,000,000,000	$= 10^{9}$
tera (T)	= 1,000,000,000,000	$= 10^{12}$
peta (P)	= 1,000,000,000,000,000	$= 10^{15}$

Table A3.2

Units of Length

1 ångström (Å)	= 0.0001 micron
1 nanometer (nm)	$= 10^{-9}$ meter
1 micron (μm)	= 0.001 millimeter (or 10^{-3} mm)
1 millimeter (mm)	= 1000 microns
	0.001 meter
1 centimeter (cm)	= 10 millimeters
	0.394 inch
1 decimeter (dm)	= 0.1 meter

(Continued)

Table A3.2 *(Continued)*

Units of Length	
1 meter (m)	= 100 centimeters
	3.28 feet
1 kilometer (km)	= 1000 meters
	3280 feet
	0.62 statute mile
	0.54 nautical mile
1 inch	= 2.54 centimeters
1 foot	= 0.3048 meter
1 yard	= 3 feet
	0.91 meter
1 fathom	= 6 feet
	1.83 meters
1 statute mile	= 5280 feet
	1.6 kilometers
	0.87 nautical mile
1 nautical mile	= 6076 feet
	1.85 kilometers
	1.15 statute miles

Table A3.3

Units of Area	
1 square centimeter (cm^2)	= 100 square millimeters
	0.155 square inch
1 square meter (m^2)	= 10^4 square centimeters
	10.8 square feet
1 acre (a)	= 100 square meters
1 square kilometer (km^2)	= 10^6 square meters
	247.1 acres
	0.386 square statute mile
	0.292 square nautical mile
1 hectare (ha)	= 10,000 square meters

Table A3.4

Units of Speed	
1 knot	= 1 nautical mile/hour
	1.15 statute miles/hour
	1.85 kilometers/hour
Velocity of sound in water of salinity 35‰	= 1507 meters/second

Table A3.5

Units of Volume

1 milliliter (mL)	= 0.001 liter
	1 cm^3 (or 1 cc)
1 liter (L)	= 1000 cm^3
	1.06 liquid quarts
1 gallon	= 3.785 dm^3 = 3.785 liters
1 cubic meter (m^3)	= 1000 liters

Table A3.6

Units of Mass

1 milligram (mg)	= 0.001 gram
1 kilogram (kg)	= 1000 grams
	2.2 pounds
1 tonne (or ton) (t)	= 1 metric ton
	10^6 grams
	1 Mg
1 pound (lb)	= 453.6 grams
1 long ton (UK)	= 1.016 Mg
1 short ton (U.S.)	= 0.907 Mg

Table A3.7

Units of Pressure

1 atmosphere	= 760 torr
	12.5 psi
	14.7 lbs per square inch

Table A3.8

Units of Temperature

°C = 5/9(°F − 32)
K = 273.15 + °C

Table A3.9

Units of Energy

1 cal = 4.184 joule
1 kcal = 1000 cal

Table A3.10

Units of Radioactivity

1 Bq (Bequerel)	= 1 dps (disintegrations per second)
	= 0.0167 dpm (disintegrations per minute)
10^{15} Bq	= 10^3 TBq = 1 PBq

Table A3.11

Units of Concentration (also see Table 5.1)

M (molarity)	= moles of solute per liter of solution
m (molality)	= moles of solute per kilogram of solvent
μg-atom/L	= μM = mg-atoms/m^3
μg/L	= mg/m^3
% (parts per hundred)	= 1 g/100 g (= dg/100 mL)*
ppt (parts per thousand) or (parts per mille)[a]	= g/1000 g (= g/L)*
ppm (parts per million)	= 10^{-3} g/1000 g (= mg/L)*
ppb (parts per billion)	= 10^{-6} g/1000 g (= μg/L)*
dpm	= disintegrations per minute

[a]*mille (or mil) is Latin for 1000.*
**for solution with density of 1g/cm^3, i.e., pure water at 4°C.*

APPENDIX 4

Symbols, Constants, and Formulae

Table A4.1

Symbols	
a	Ionic activity
A	Radioactive activity
λ	Radioactive decay constant
γ	Ionic activity coefficient
K	Equilibrium constant
I	Ionic strength
G	Gibbs free energy
H	Enthalpy
ρ	Density
n	Moles
d	Finite increment
∂	Infinitesimally small increment
z	Vertical dimension
x, y	Horizontal dimensions
C	Conservative ionic solute
S	Nonconservative ionic solute
t	Time
S‰	Salinity
Cl‰	Chlorinity
P	Pressure
V	Volume

Table A4.2 Constants and Formulae.

I. Ideal Gas Law	$PV = nRT$ $R = 0.0820578$ L atm °K^{-1} mol^{-1} $T(K) = 273.15 + °C$ 1 mole of an ideal gas occupies 22.4141 L at STP
II. Acidity	$pH = -\log\{H^+\}$ $pOH = -\log\{OH^-\}$ $pH + pOH = 14$ at 25°C and $I = 0\,m$
III. Redox	$F = 96.485$ coulombs mol electrons^{-1} $= 23.066$ k cal volt^{-1} mol electrons^{-1} $R = 1.987 \times 10^{-3}$ kcal °K^{-1} mol electrons^{-1}
IV. Equilibrium	For $A + B \rightleftharpoons AB$, $K = \frac{[AB]}{[A][B]}$
V. Logarithms	$\log a + \log b = \log(ab)$ $\log a - \log b = \log(a/b)$ $\log 10^a = a$ $\ln x = 2.303 \log x$
VI. O_2 Conversions	1 ml O_2/L = 11.2 × mg-at O_2/L 1 mg O_2/L = 16.0 × mg-at O_2/L 1 ppm O_2 = 0.032 × μmol O_2/kg 1 mL O_2/L = 0.700 × ppm O_2 × seawater density (kg/L) 1 mL O_2/L = 0.0224 × μmol O_2/kg$_2$ × seawater density (kg/L)
VII. Atmosphere	Mass of dry air = 5.148×10^{21} g Moles of dry air = 1.777×10^{20} mol Mean dry molar mass of air = 28.97 g/mol

APPENDIX 5

Geologic Time Scales

1999 GEOLOGIC TIME SCALE

CENOZOIC

PERIOD		EPOCH		AGE
QUATERNARY		HOLOCENE		
		PLEISTOCENE		CALABRIAN
TERTIARY	NEOGENE	PLIOCENE	L	PIACENZIAN
			E	ZANCLEAN
		MIOCENE	L	MESSINIAN
				TORTONIAN
			M	SERRAVALLIAN
				LANGHIAN
			E	BURDIGALIAN
				AQUITANIAN
	PALEOGENE	OLIGOCENE	L	CHATTIAN
			E	RUPELIAN
		EOCENE	L	PRIABONIAN
			M	BARTONIAN
				LUTETIAN
			E	YPRESIAN
		PALEOCENE	L	THANETIAN
				SELANDIAN
			E	DANIAN

AGE (Ma): 5, 10, 15, 20, 25, 30, 35, 40, 45, 50, 55, 60, 65

MESOZOIC

PERIOD	EPOCH		AGE
CRETACEOUS	LATE		MAASTRICHTIAN
			CAMPANIAN
			SANTONIAN
			CONIACIAN
			TURONIAN
			CENOMANIAN
	EARLY		ALBIAN
			APTIAN
		NEOCOMIAN	BARREMIAN
			HAUTERIVIAN
			VALANGINIAN
			BERRIASIAN
JURASSIC	LATE		TITHONIAN
			KIMMERIDGIAN
			OXFORDIAN
	MIDDLE		CALLOVIAN
			BATHONIAN
			BAJOCIAN
			AALENIAN
	EARLY		TOARCIAN
			PLIENSBACHIAN
			SINEMURIAN
			HETTANGIAN
TRIASSIC	LATE		RHAETIAN
			NORIAN
			CARNIAN
	MIDDLE		LADINIAN
			ANISIAN
	EARLY		OLENEKIAN
			INDUAN

AGE (Ma): 70, 80, 90, 100, 110, 120, 130, 140, 150, 160, 170, 180, 190, 200, 210, 220, 230, 240

PALEOZOIC

PERIOD		EPOCH	AGE	
PERMIAN		L	TATARIAN	
			UFIMIAN-KAZANIAN	
		E	KUNGURIAN	
			ARTINSKIAN	
			SAKMARIAN	
			ASSELIAN	
CARBONIFEROUS	PENNSYLVANIAN	L	GZELIAN	S.
			KASIMOVIAN	
			MOSCOVIAN	W.
			BASHKIRIAN	N.
	MISSISSIPPIAN	E	SERPUKHOVIAN	
			VISEAN	
			TOURNAISIAN	
DEVONIAN		L	FAMENNIAN	
			FRASNIAN	
		M	GIVETIAN	
			EIFELIAN	
			EMSIAN	
			PRAGHIAN	
			LOCKHOVIAN	
SILURIAN		L	PRIDOLIAN	
			LUDLOVIAN	
		E	WENLOCKIAN	
			LLANDOVERIAN	
ORDOVICIAN		L	ASHGILLIAN	
			CARADOCIAN	
		M	LLANDEILIAN	
			LLANVIRNIAN	
		E	ARENIGIAN	
			TREMADOCIAN	
CAMBRIAN		D	SUNWAPTAN	
			STEPTOEAN	
		C	MARJUMAN	
			DELAMARAN	
		B	DYERAN	
			MONTEZUMAN	
		A		

AGE (Ma): 260, 280, 300, 320, 340, 360, 380, 400, 420, 440, 460, 480, 500, 520, 540

PRECAMBRIAN

EON	ERA
PROTEROZOIC	LATE
	MIDDLE
	EARLY
ARCHEAN	LATE
	MIDDLE
	EARLY

AGE (Ma): 750, 1000, 1250, 1500, 1750, 2000, 2250, 2500, 2750, 3000, 3250, 3500, 3750

FIGURE A5.1

Source: From Palmer, A. R. and J. Geissman (1999). *1999 Geologic Time Scale*. The Geological Society of America.

Glossary

For definitions of units, chemical composition of minerals, chemical formulae of molecules and time spans of geologic periods, see the Appendices.

AABW, Antarctic bottom water The densest water mass in the open ocean.

Abiogenic Formed without the intervention or support of organisms.

Abiotic Without life.

Absorb To become incorporated in.

Abyssal Of the deep sea.

Acantharia Pelagic protozoans that deposit hard parts composed of strontium sulfate.

Accrete The process by which inorganic bodies grow larger by the addition of fresh material to the outer surface.

Acidic Having a pH less than 7 (i.e., the hydrogen ion concentration is greater than the hydroxide ion concentration).

Activity (1) The effective concentration of a solute. Solute activity is significantly lower than total concentration in solutions of high ionic strength. (2) Rate of radioactive decay, usually measured in disintegrations per minute (dpm).

Adsorb To adhere to the exterior surface.

Advection The large-scale mass transport of matter.

Aeolian Of the wind.

Aerobic respiration The biological process during which animals and some bacteria oxidize organic carbon to carbon dioxide to yield cellular energy. The oxidizing agent is O_2.

Aerosols Minute liquid and solid particles that are suspended in the atmosphere. Sizes range from 0.001 to 100 microns in diameter.

Air-Sea interface The boundary between the surface ocean and atmosphere.

Albedo The fraction of incident sunlight that is reflected

Aliphatic hydrocarbons Hydrocarbons that do not contain benzene rings.

Alkaline Having a pH greater than 7 (i.e., the hydrogen ion concentration is less than that of the hydroxide ion concentration).

Alkalinity The concentration of negative charge in a solution that can be titrated by a strong acid. In seawater, HCO_3^- and CO_3^{2-} contribute most of this negative charge. The units of measurement are meq/L.

Allochthonous Formed outside the ocean. Opposite of autochthonous.

Alpha (α) particles Helium nuclei (^4_2He). A common by-product of the radioactive decay of primordial radionuclides.

Aluminosilicates Minerals composed primarily of the elements aluminum, oxygen, and silicon.

Amino acids Biochemicals found in all organisms. They are the building blocks of proteins. Each contains at least one amine and one carboxylic acid group.

Ammonification The microbial process by which dissolved organic nitrogen is converted to ammonium.

Anabolism Metabolic reactions in which the resultant molecules are larger than those of the reactants.

Anaerobes Organisms that live in the absence of O_2.

Analog A molecule with a fundamentally similar structure but subtly different, such as in the chemical composition of a side chain.

Anhydrite A mineral composed of calcium sulfate. It is a common hydrogenous mineral deposited in hydrothermal systems.

Anion A negatively charged ion.

Annamox Anaerobic ammonium oxidation. This process is mediated by bacteria.

Anoxic Without O_2.

Anthropogenic Caused by humans.

AOU, apparent oxygen utilization The difference in O_2 concentration between that in a deep-water sample and its NAEC. It is a measure of the amount of O_2 consumed via the respiration of organic matter since a deep-water mass was last at the sea surface.

Aragonite A mineral form of calcium carbonate. It is more soluble than calcite. Some marine organisms, such as pteropods, deposit shells composed of aragonite. This mineral is also a common component of evaporites.

Archaea Unicellular microorganisms that are structurally similar to prokaryotes but include some metabolic characteristics of eukaryotes. The archaea are genetically distinct from the other two domains. Most extremophiles are members of the archaea.

Arctic Throughflow Surface currents that move seawater between the Arctic Ocean and the North Pacific and Atlantic Oceans.

Aromatic hydrocarbons Hydrocarbons that contain benzene rings.

Aromaticity The number of benzene groups in an organic molecule.

Artificial radionuclides Radionuclides produced by humans via the explosion of atomic bombs and in nuclear reactors.

Asphalt A solid residue left after distillation has removed all other compounds from petroleum. The residue is composed of asphaltenes.

Asphaltenes A class of compounds that are part of the high-molecular-weight component of petroleum. They are relatively inert, have high boiling points, and have high degrees of unsaturation.

Assimilation The uptake of nutrients and other dissolved materials by phytoplankton.

Assimilatory nitrate reduction The reduction of nitrate to organic nitrogen compounds that constitute the tissues of marine organisms. Plankton and some bacteria assimilate nitrogen via this process.

Asynchrony Not in phase; out of sync.

ATP, adenosine triphosphate A biomolecule that acts to transport energy within cells by enabling electron transfer.

Authigenic Created by inorganic processes in the ocean.

Autochthonous Formed within the ocean. Opposite of allochthonous.

Autotrophs Organisms that can use inorganic substances as their energy source.

Bacteria See **Prokaryotes**.

Barite A mineral form of barium sulfate that is most common beneath surface waters of high productivity. The barium is primarily biogenic in origin.

Basalts The igneous extrusive rocks that constitute the oceanic crust.

Basic Of a solution that has a greater concentration of hydroxide (OH^-) ions as compared to hydrogen (H^+) ions.

Benthic Of the organisms that live in or on the seafloor.

Bioaccumulation An enrichment of a particular chemical caused by either passive adsorption from seawater or active uptake followed by retention in living tissues or hard parts as a result of nonexcretion.

Bioassay A technique of concentration measurement that relies on the measured response of an organism, or some part of an organism, to an analyte.

Bioavailability In a form that is readily consumed or assimilated by organisms.

Bioflocculation Flocculation caused by living organisms (e.g., the formation of large clumps of phytoplankton during a bloom).

Biofouling The sequential colonization of marine surfaces by microbes and higher order organisms.

Biogenic Of marine organisms.

Biogenic silica A mineral form of silica that is amorphous in structure and deposited by marine organisms such as diatoms and radiolaria. Also called opal or opaline silica.

Biogenous ooze Sediments that contain more than 30% by mass hard or soft parts that were synthesized by marine organisms.

Biogenous sediments Sediments that are composed of hard or soft parts, such as shells and tissues that were synthesized by marine organisms.

Biogeochemical cycle The transport of materials on the earth as a result of biological, chemical, and geological processes.

Biogeochemistry The science that studies the biological, chemical, and geological aspects of environmental processes.

Biointermediate element An element whose marine distributions are controlled by both physical and biogeochemical processes.

Bioirrigation The movement of water through marine sediments caused by benthic animals.

Biolimiting element An element whose marine distributions are controlled primarily by biogeochemical processes. Such elements are characterized by low concentrations in the surface waters and tend to limit the growth of phytoplankton.

Biological nitrogen fixation Nitrogen fixation, the conversion of N_2 to fixed forms, such as ammonium, by the actions of bacteria.

Biological pump The transport of chemicals in the ocean as a result of incorporation into biological hard and soft parts, followed by pelagic sedimentation into the deep sea.

Biomagnification The concentration of a chemical with increasing trophic level. This is caused by bioaccumulation that occurs during each transfer of the chemical to higher trophic levels. Biomagnification causes organisms at the top of the food chain to have the highest pollutant concentrations.

Biomarkers Organic compounds that are synthesized by specific organisms and thus can be used as an indicator of the current or past presence of those organisms.

Biomass The quantity of living particulate organic matter.

Biomolecules Organic compounds synthesized by organisms.

Biopolymers Organic molecules synthesized by organisms. They are composed of repeating units.

Biota All of the organisms in an ecosystem.

Bioturbation The physical mixing of sediments caused by the burrowing and feeding activities of benthic organisms.

Biounlimited element An element whose marine distributions are controlled primarily by physical processes and are relatively uninfluenced by biogeochemical phenomena. These elements demonstrate conservative behavior.

Bitumen The collective mass of organic matter represented by the petroleum and asphalt in a geologic deposit.

Blackbody Radiation The process by which solar energy absorbed by the Earth is transformed into longer wavelengths and reradiated back into the atmosphere. In physics, a blackbody is a perfect adsorber of electromagnetic radiation that can be released at other wavelengths with no loss of total energy.

Black carbon Various carbonaceous products of the incomplete combustion of organic matter. Includes chars, charcoals and soots.

Box model A conceptual representation of a geochemical cycle in which reservoirs are depicted as boxes and transports by arrows. No details as to the processes that occur in the boxes are provided.

Bunsen solubility coefficient (α_A) The term that relates the concentration of a gas in seawater to its partial pressure in the atmosphere. It is dependent on temperature and salinity.

Caballing The process of densification in which the nonlinear equation of state of seawater causes the product of conservative mixing of two water masses to be denser than either of the two mixing end members.

Calcite compensation See **Carbonate compensation**.

Calcite compensation depth See **Calcium carbonate compensation depth**.

Calcitic ooids Nodules of calcite that are less than 2 mm in diameter. They are commonly found in and on the sediments of shallow tropical seas.

Calcitic oolite A pavement of calcitic ooids formed by continuing precipitation of calcium carbonate between the calcitic ooids.

Calcium carbonate compensation depth (CCD) The depth below which calcium carbonate is not found in marine sediments due to its dissolution.

Calorie A measure of heat energy. One (1) calorie is the amount of heat energy required to raise 1 g of pure water 1°C from 14.5 to 15.5°C.

Cap rock A relatively impermeable rock that retards the migration of petroleum through marine sediments. Many are composed of ancient evaporites, also called diapirs.

Carbohydrates A class of biopolymers whose building blocks are composed of simple sugars such as glucose and fructose. These compounds contain only carbon, hydrogen, and oxygen.

Carbon fixation reactions Biochemical reactions conducted by plants and some bacteria in which inorganic carbon is incorporated into organic molecules.

Carbonate alkalinity The concentration of negative charge in seawater contributed by bicarbonate (HCO_3^-) and carbonate (CO_3^{2-}). It is usually reported in units of meq/L.

Carbonate compensation The ocean's response to P_{CO_2} perturbations through shifts in its carbonate chemistry. These shifts require changes in the carbonate ion concentration that change the depth of the calcium carbonate compensation depth and hence lead to changes in the burial rate of carbon as biogenic calcium carbonate.

Carotenoids Organic compounds that are 40-carbon terpenoids. They serve as photosynthetic pigments in algae, photosynthetic bacteria, and higher plants. They are also present in some nonphotosynthetic bacteria.

Catabolism The biochemical process by which organic compounds are degraded. This provides cellular energy.

Catagenesis Geochemical reactions that occur in the sediments following burial on time scales greater than 1000 years and at temperatures of 50 to 150°C.

Cation A positively charged ion.

Cation exchange capacity See **CEC**.

Cation exchange The displacement of one cation for another on the surface of a negatively charged solid, such as a clay mineral.

CCD, calcium carbonate compensation depth The depth below which calcium carbonate is not found in marine sediments due to its dissolution.

CEC, cation exchange capacity A measure of the amount of cations that will adsorb to the negatively charged surface of a clay mineral. It is usually measured in units of meq of charge per 100 g of clay mineral. This adsorption is reversible.

Cell potential (E_h) A measure of how far a redox reaction is from equilibrium. It is reported in units of volts. The higher the E_h, the greater the driving force for reaction.

Chelation The complexation of a metal cation by an organic ligand. Some ligands have multiple binding sites. Those with two sites are called bidentate ligands.

Chemoautolithotrophs Organisms that use inorganic chemicals as their source of electrons, energy and carbon.

Chemometrics The science of relating measurements made on a chemical system or process to the state of the system via application of mathematical or statistical methods.

Chlorite A magnesium-rich clay mineral produced by terrestrial weathering under polar and subpolar conditions.

Chromophore The region of an organic or organometallic molecule that is capable of absorbing visible light, causing the molecule to appear colored. In biological molecules, chromophores serve to capture or detect light energy.

Circumpolar Surrounding either the north or south poles.

Clathrate hydrates Solid cages of water that form around small gas molecules such as methane, hydrogen, or carbon dioxide under conditions of high pressure and low temperature such as found on the deep sea floor and within the sediments.

Clay (1) A grain whose diameter is less than 4 μm. Most are inorganic silicates. (2) A sedimentary deposit that is composed of more than 70 percent by mass clay-sized grains.

Clay mineral A layered aluminosilicate, such as kaolinite, illite, chlorite, and montmorillonite. Most are formed by chemical weathering of rocks on land.

Coastal upwelling The upward advection of water from the base of the mixed layer toward the sea surface caused by Ekman Transport. This water motion brings nutrient-rich water to the sea surface.

Coccolithophorids Phytoplankton that deposit calcareous plates called coccoliths.

Coccoliths Calcareous plates deposited by phytoplankton called coccolithophorids.

Colligative property Any property of a solution that depends only on the number of solute particles, rather than on the nature of the particles.

Colloid A form of matter that ranges in length from 0.001 to 10 μm. This is in between that of a solute and a solid. Some scientists recognize 100 μm as the upper size limit.

Compaction Compression in marine sediments caused by over pressure. Leads to a reduction in sediment porosity.

Complex carbohydrates Polymers whose building blocks are simple sugars.

Complex lipids Lipids whose hydrolysis produces several other biomolecules, such as fatty acids, simple sugars, glycerol, sphingosine, etc.

Complexation A covalent chemical bond between a cation and a ligand that ranges in behavior from polar to nonpolar.

Compressibility The degree to which the volume of a given quantity of material can be reduced as a result of an increase in pressure.

Condensation The phase change during which a gas is transformed into a liquid.

Conduction (1) The transport of heat via molecular processes. (2) The transport of electrons causing an electrical current to flow.

Conductivity A measure of the salt content of seawater, which relies on the presence of ions that conduct electricity.

Congener Chemicals that are related, such as elements within the same periodic group, or PCBs whose number and location of chlorine substituents varies.

Conservative Chemical behavior that is controlled by physical processes. In other words, physical transport is much faster than any chemical processes that can either supply or remove the chemical. All of the major ions exhibit conservative behavior. The concentrations of conservative species are directly proportional to each other and to salinity.

Conservative ions Ions that exhibit conservative behavior. These include the major ions in seawater.

Contaminant Naturally occurring substances whose concentrations have been increased as a result of human activity.

Continental crust The thickened part of the crust that comprises the continents and consists primarily of granitic rock.

Continental margin The zone separating the emergent continents from the deep-sea bottom; it generally consists of a continental shelf, slope, and rise.

Continental rise The large sedimentary deposit that lies at the foot of the continental slope.

Continental shelf pump Export processes that serve to transport carbon from the coastal margins to the open ocean.

Continental shelf The sea floor adjacent to a continent, extending from the low-water line to the change in slope, usually at 180 m water depth, where the continental shelf and slope join.

Continental slope A declivity that extends from the outer edge of the continental shelf to the continental rise. The angle is approximately 4 to 5°.

Contour currents Water currents that flow along the foot of the continental slope.

Convection The transport of heat as a result of the physical movement of a carrier, such as air, water, or magma. Convection occurs spontaneously due to density differences.

Coordination complex A molecule composed of one or more metal atoms, each of which is covalently bonded to more than one ligand or electron donor.

Coprecipitate To precipitate together. Due to their chemical similarity, most metals tend to precipitate together into solids that are thus composed of a large and variable amount of these elements (e.g., polymetallic sulfides and oxyhydroxides).

Coriolis effect The apparent force acting on moving particles that results from Earth's rotation. The Coriolis Effect causes freely moving bodies to be deflected to the right of their direction of motion in the northern hemisphere and to the left in the southern hemisphere.

Cosmic dust Particles with an interstellar origin that range in size from a few molecules to 0.1 mm in diameter.

Cosmic rays Protons and α particles that enter Earth's atmosphere from outer space. When they collide with atoms or molecules of atmospheric gas, high-energy neutrons can be given off. These neutrons can then undergo nuclear reactions with other atmospheric gas atoms and molecules.

Cosmogenous From outer space.

Covalent bond A chemical bond in which the bonding electrons are shared between the bonded atoms. If the sharing is equal, the bond is termed nonpolar covalent. If one atom is more electronegative, the electrons are not equally shared. See Polar covalent bond.

Cracking reactions Chemical reactions that occur during catagenesis and metagenesis in marine sediments and sedimentary rocks. During this process, petroleum compounds are formed as hydrocarbons are broken off heteroatomic macromolecules.

Crude oil The liquid portion of petroleum. It contains various fractions, such as kerosene, gasoline, and napthenes that can be isolated by distillation.

Crustal Of Earth's crust, which is the outer shell of the planet. It is composed of sedimentary, metamorphic, and igneous rocks and ranges in thickness from 5 to 35 km.

Crustal-ocean-atmosphere factory The conceptual model that describes the material flows between the crust, ocean, atmosphere, and upper mantle.

Crystal lattice The regular, repeating framework created by atoms, ions, or molecules constituting a crystalline solid.

Cultural eutrophication Overgrowth of algae in natural waters caused by runoff of fertilizers from land-based applications.

Cyanobacteria These bacteria are photosynthetic and hence are sometimes referred to as blue-green algae.

Cyclic salts Salt ions that undergo rapid cycling between the ocean, atmosphere and land. This cycle involves ejection into the atmosphere via bursting bubbles and return via either dry deposition onto the sea surface or onto land followed by runoff back into the ocean.

Cytotoxic Toxic to cells.

Daughters The radionuclides produced by the decay of primordial radionuclides.

Debouches Exits from an orifice.

Decay constant The constant that describes the rate at which a radionuclide decays. It is equal to $0.693/\tau$, where τ is the half-life of the radionuclide.

Deep water or zone That portion of the water column from the base of the permanent thermocline or pycnocline to the ocean floor.

Degas The process by which gas is lost from a solution.

Denitrification The conversion of fixed nitrogen, typically nitrate, into N_2 gas. This is achieved by heterotrophic bacteria that use nitrate as an electron acceptor under suboxic and anoxic conditions.

Density Mass per unit volume of a substance, usually expressed in units of g/cm^3.

Density stratification Gradients in the density of seawater caused by the presence of different water masses. In a stable density configuration, density increases with increasing depth.

Deposit An accumulation of crustal solids on the seafloor.

Desertification The process by which fertile land is turned into a desert. This is usually caused by removal of vegetation and/or climate change.

Desorb The reverse of adsorb; to be released from a surface.

Detrital (1) Nonliving; (2) from land.

Diagenesis Geochemical reactions that occur in the sediments following burial on time scales less than 1000 years and at temperatures ranging from 0 to 150°C.

Diagenetic remobilization The solubilization of materials from sedimentary particles after their accumulation on the seafloor.

Diamondoids Hydrocarbons with cage-like molecular structures that resemble the crystal structure of diamonds.

Diapir An ancient evaporite that has become buried in marine sediments. The overlying pressure causes the rock to flow like toothpaste out of a tube, thereby forming vertical pillar and domal structures.

Diatoms Microscopic phytoplankton that deposit siliceous tests.

Diazotrophs Nitrogen-fixing bacteria. In the marine environment, some nitrogen fixers live as symbionts in phytoplankton, particularly diatoms.

DIC, dissolved inorganic carbon CO_2, H_2CO_3, CO_3^{2-}, and HCO_3^-.

Diel Daily.

Diffusion The transfer of matter or heat as a result of molecular motion. This motion causes net transport from regions of high concentration (or heat) to regions of lower concentration. In the absence of gradients, this motion is random and causes no net transport. Also see **Eddy diffusion**.

Dinoflagellates Phytoplankton that have flagella. These organisms do not synthesize hard parts, although some have exoskeletons composed of chitin.

Disequilibria Not in thermodynamic equillbrium and thus liable to spontaneous reactions.

Disintegrations per minute (dpm) The unit measurement of radioactive decay rates.

Disproportionation A reaction in which a molecule reacts with itself.

Dissimilatory nitrate reduction Denitrification.

Dissolved silica At the pH and ionic strength of seawater, the dominant dissolved species of silicon is orthosilicic acid (H_4SiO_4(aq) or $Si(OH)_4$(aq)). About 5% is in the form of $H_3SiO_4^-$(aq).

Divalent Having two units of electrical charge.

Divergence Horizontal flow of water from a common center or zone that results in upwelling. This occurs in the open ocean at the equator and 60°S as result of surface currents driven by the Trade winds and Westerlies, respectively.

DOC, dissolved organic carbon The fraction of the organic carbon pool in seawater that is dissolved.

DOM, dissolved organic matter The fraction of the organic matter pool in seawater that is dissolved.

Domain (1) A biological grouping above the kingdom level, i.e. Prokaryotes, Eukaryotes, and Archaea. (2) Biogeochemical provinces in which climate is the primary defining parameter.

DON, dissolved organic nitrogen The fraction of the organic nitrogen pool in seawater that is dissolved.

Downwelling The downward advection of water that leads to water-mass formation.

D_z Vertical diffusivity coefficient used to numerically model the physical process of turbulence over depth scales in the water column.

Ebullition The movement of solutes through pore waters via transport in rising gas bubbles.

Ecological forecasting The use of models to predict the effects of perturbation in the form of impacts at the ecosystem level.

Eddy diffusion Transport as a result of turbulent mixing within and among water bodies.

Eddy pumping Vertical water displacement caused by internal waves.

Ejecta Small fragments spewed out of volcanoes during an eruption. Most are pieces of ash or glassy fragments of basalt.

Ekman transport The advection of water in the mixed layer caused by the winds and the Coriolis Effect. The latter causes net water transport to occur at a direction that is at a 90° angle relative to the wind direction.

Electrolytes Ionic solutes that conduct electricity within a solution.

Electron A negatively charged subatomic particle that constitutes an insignificant amount of the mass of an atom.

Electron activity (pε) {e^-}; A measure of how far a redox reaction is from equilibrium. The larger the pε, the greater the driving force toward spontaneous reaction.

Electron carriers Biomolecules, such as NAD and NADP that act to transport energy within cells by enabling electron transfers.

Electronegativity The degree to which an atom can attract electrons when it is bonded to other atoms.

Electroneutrality Having no net electrical charge.

Eleotrostriction The drawing together of water molecules in the presence of ionic solutes as a result of specific and nonspecific interactions.

Elute To remove a solute as a result of transport in an advecting solution. This usually involves desorption from a solid support.

Embayment An indentation in the coastline.

Empirical Determined through observation and measurement rather than from theoretical principles.

Empirical formulae The molar combining ratio of a compound that has a variable size, such as a crystalline solid.

Endosymbiotic Of a close biological association in which an organism lives inside its host.

Enrichment factor (E.F.) The degree to which a marine organism is enriched in a particular chemical with respect to the seawater concentration, e.g., E.F. = [metal concentration in biogenic material]/[metal concentration in seawater].

Equation of state of seawater The semiempirical equation that relates the density of seawater to its salinity, temperature, and pressure.

Equilibrium Constant (K) An expression that relates the equilibrium concentrations (or pressures) of reactive species to one another. Its value is dependent on temperature, pressure, and the ionic strength of the solution in which the reaction is occurring.

Equilibrium The chemical state in which the concentrations of reactive species do not change over time. Ongoing chemical reactions are reversible.

Estuarine Of estuaries, which are coastal areas where freshwater mixes with salt water. Most are semienclosed bodies subject to tidal motions, such as the seaward end of a river valley.

Eukaryotes Multicellular organisms and protists. All eukaryotic cells contain a nucleus.

Euphotic Of the depth zone through which light penetrates. The bottom of the euphotic zone is defined as the depth at which less than 1 percent of the incident solar radiation remains, the rest having been either absorbed or reflected.

Eutrophic Of waters that have a great abundance of marine life, usually due to high nutrient levels.

Eutrophication The overgrowth of algae in marine and fresh waters caused by an overabundance of nutrients. Once the algae die and their remains settle below the sea surface, microbially mediated decomposition of their biomass leads to O_2 depletion and, potentially to fish and benthic kills.

Evaporation The phase change during which a liquid is transformed into the gaseous state.

Evaporite A suite of minerals formed as a result of the evaporation of seawater.

Exothermic Releases heat and is therefore the result of a spontaneous reaction.

Extracellular Occurs outside the cell.

Extremophiles Microbes that can grow under very adverse environment conditions. Classifications include: thermophiles (temperature $> 40^\circ C$), acidophiles (pH < 2), alkalophiles (pH > 11), halophiles (salt $> 20\%$ w/v), barophiles (pressure > 100 atm), and psycrophiles (temperature $< 20^\circ C$).

Exudates Materials that are slowly discharged from various pores and orifices of organisms.

Facies The physical characteristics of a rock, which usually reflect its conditions of origin.

Fats Complex lipids; triglycerides.

Fatty acids Biochemicals that are composed of hydrocarbons attached to a terminal carboxylic acid group. They vary in the number of carbons and degree of multiple bonding and branching in the hydrocarbon.

Fecal pellet Solid organic excrement generally ovoid in form and less than 1 mm in length. Produced primarily by marine invertebrates.

Feedback loop A set of geochemical processes that influence each other. In a negative feedback, an alteration in the rate of one process is at least partially compensated for by changes in the rates of the other interconnected processes. In a positive feedback, an alteration in the rate of one process is amplified by accompanying changes in the rates of the other interconnected processes.

Fick's first law States that a net diffusive flux occurs spontaneously, moving solutes from regions of high concentration to regions of low concentration, with a magnitude that is proportional to the concentration gradient $([\partial C]/\partial z)$. In one spatial dimension (z), Flux $= -D_z([\partial C]/\partial z)$.

Fick's second Law A mathematical description of the effect of diffusion and turbulent mixing on the rate of change in the concentration of a solute, S, $\partial[S]/\partial t = -D_z(\partial^2[S]/\partial z^2)$.

Filter feeding The filtering or trapping of edible particles from seawater. This feeding mode is typical of many zooplankton and other marine organisms of limited mobility.

Floc Clotlike masses composed of small particles.

Flocculation The process by which small particles aggregate into clumps.

Fluidize To cause material to flow by increasing the motion of individual particles.

Flux The transport of matter or energy through a given surface area or volume in a given unit of time.

Foraminifera Protozoans that deposit hard parts, called forams, which are composed of calcite.

Formation constant An equilibrium constant for a reaction in which an ion complex is formed.

Fossil fuel Refined petroleum products that are burned in automobiles and factories to provide energy.

Fractional residence time Turnover time.

Free energy change (ΔG) The total energy of a chemical system that would have to be expended or absorbed to reach the equilibrium state.

Free ions Ions that are completely hydrated.

Free water Water that is not contained within the crust.

Freezing The phase change during which a liquid is transformed into a solid.

Frustules The hard parts deposited by diatoms.

Fulvic acids Humic substances that are soluble at all pHs.

Gaia hypothesis The hypothesis that living organisms and inorganic material on Earth are part of a dynamic homeostatic system.

Gas chromotagraph A scientific instrument used to identify and quantify organic compounds of intermediate boiling points.

Gas exchange pump See **Solubility pump**.

Gasoline The fraction of petroleum that has a boiling point less than 180°C. Most of the compounds are low-molecular-weight, straight-chain hydrocarbons.

Geomorphological Of the general configuration of Earth's surface.

Geopolymers Complex, refractory, high-molecular-weight compounds formed from the polymerization of biomolecules during diagenesis and catagensis.

Geostrophic current The advection of water resulting from the balance between gravity, wind stress, and the Coriolis Effect.

Geothermometers Minerals whose chemical composition can be used to determine their temperature of crystallization.

Gibbs free energy (G) The total chemical energy of a system.

Glacioeustatic Of changes in sea level caused by variations in continental ice loading.

Glauconite A hydrogenous mineral commonly found in carbonate sediments of continental shelves located in the tropics. This iron-rich, greenish, hydrous silicate contains fossilized biogenic detritus such as fecal pellets and siliceous tests. It is found as nodules and encrustations. In some locations, concentrations are high enough to give the sediments a greenish cast and hence are referred to as green muds.

Global warming potential The degree to which a greenhouse gas can exert a warming effect on the atmosphere. Gases with long residence times and high radiative efficiencies have high global warming potentials.

Glycerophospholipids Phosphoglycerides. Glycerol-based phospholipids.

Glycolysis The chemical reactions during which glucose is catabolized to generate cellular energy.

Graded bedding A type of sediment stratification in which each stratum displays a gradation in grain size from coarse below to fine above.

Gradient The rate of decrease (or increase) of one quantity with respect to another.

Granite A crystalline intrusive igneous rock consisting of alkaline feldspar and quartz.

Grasshopper effect The migration of volatile and semivolatile pollutants poleward in steps initiated by volatilization from the land during warmer seasons followed by atmospheric transport to higher (cooler) latitudes where condensation brings them back to Earth's surface. Seasonal warming revolatilizes the compounds, enabling them to move to even higher latitudes. The more volatile the compound, the farther poleward it is transported. The net effect is a global distillation of the semivolatile compounds far from their point of release into the environment, leading to their concentration in polar and mountainous regions.

Gravimetric Of the chemical analyses that involve mass measurements.

Green tides Proliferations of the macroalgal *Chlorophyta* (*Ulva* spp).

Greenhouse effect The warming of Earth's atmosphere as a result of the retention of solar radiation. This retention is possible because insolation absorbed by the land and ocean is radiated back to the atmosphere as IR energy. This energy is absorbed by atmospheric gases and then radiated from them as heat.

Greenhouse problem Continued warming of earth's atmosphere caused by the buildup of heat-retaining gases released by anthropogenic activities.

Groundwater That part of the subsurface water that is in the soil or flows through and around terrestrial crustal rocks.

Gypsum A mineral form of calcium sulfate. Gypsum is a common component of evaporites, and hydrothermal deposits.

Gyres A set of four interlocking geostrophic currents that move water in each ocean basin. Northern hemisphere gyres move surface waters clockwise, while southern hemisphere gyres move water counterclockwise.

Half-cell reaction A conceptual representation of electron transfer in which the number of electrons gained by a molecule or atom is indicated. For example, the half-cell reduction of Mn^{4+} to Mn^{2+} is written as: $Mn^{4+} + 2e^- \rightarrow Mn^{2+}$.

Half-life The time required for some physical or chemical process to remove half of the original amount of a substance.

Halite A mineral form of sodium chloride. Halite is a common component of evaporites.

Halmyrolysis The processes that alter the chemical composition of terrestrial clay minerals during their first few months of exposure to seawater.

Halocline A region in which a strong salinity gradient exists.

Halogenated Having one or more atoms of a halide (e.g., chloride, bromide, or iodide).

Harmful algal blooms Proliferation of phytoplankton that have toxic effects on other marine organisms and humans. The harmful effects include, but are not limited to, poisoning from biotoxins and mechanical abrasion.

Heat capacity The amount of heat required to raise the temperature of a substance by a given amount. If given on a per unit mass basis, this is called the specific heat capacity or specific heat.

Heavy metal Metals whose atomic weights exceed 20 amu.

Hemipelagic sediments Sediments that lie in water depths of 200 to 3000 m (roughly encompassing the continental slope and upper part of the rise).

Henry's Law The solubility of a gas in an aqueous solution is directly proportional to its atmospheric partial pressure. The constant of proportionality is a function of temperature and the ionic strength of the solution.

Heterotrophs Organisms that require organic matter as their carbon source.

Hopanoids Hydrocarbons that have four six-membered carbon rings and one five-membered ring. They are synthesized by prokaryotes and are well preserved in sediments due to their low reactivity.

Horizontal segregation The horizontal gradient in biogenic materials, such as nutrients and O_2, that is established by the interaction between the biogeochemical cycling of particulate organic matter and meridional overturning circulation.

Humic acids Humic substances that are insoluble at acidic pHs.

Humic substances High-molecular-weight organic compounds that are variable in composition, have complex structures, and are relatively inert. They comprise a large fraction of the DOM. Found in soils, sediment, fresh, and seawater.

Humification The formation of humic substances via abiotic reactions, including condensations and polymerizations.

Hydration The binding of water molecules by adsorption onto solid surfaces or through electrostatic attractions to ions or molecules.

Hydrocarbons Organic compounds composed entirely of carbon and hydrogen atoms.

Hydrogen bond The relatively weak intermolecular interaction that occurs between oppositely charged ends of adjacent water molecules.

Hydrogenous Same as Authigenic.

Hydrographic Pertaining to water.

Hydrological cycle The global water cycle involving the transport of this substance between the atmosphere, hydrosphere, and lithosphere.

Hydrolysis A chemical reaction involving water.

Hydrosphere The water portion of the earth, as distinguished from the solid (lithosphere) and gaseous (atmosphere) parts. This includes water in lakes, ponds, streams, rivers, glaciers, icebergs, the ocean, pore waters, and that which is trapped in crustal rocks.

Hydrothermal Of the hot-water systems that are present at active mid-ocean spreading centers.

Hydroxyl A chemical group composed of an oxygen atom bound to a hydrogen atom (i.e., $-OH$).

Hypersaline Water with a salinity in excess of that at which halite will spontaneously precipitate.

Hypoxic Waters with dissolved oxygen concentrations less than 2 to 3 ppm (2 mL/L).

IAP, ion activity product If the product of the ion activities exceeds the K_{sp} of a mineral, it will spontaneously precipitate.

Ice rafting The transport of lithogenous material by icebergs. The rafted material is rock eroded from the continents as the icebergs flowed seaward.

Ice The solid form of water.

Iceberg A large mass of detached land ice that either floats in the sea or is stranded in shallow water.

ideal gas law PV = nRT.

Igneous rock A rock formed by the solidification of magma.

Illite A potassium-rich clay mineral.

In situ Occurring or formed in place.

Indonesian throughflow Surface currents that move seawater from the southwestern Pacific Ocean into the Indian Ocean.

Inert Unreactive.

Inorganic carbon pump The transport of carbon into the deep sea as a result of the sinking of detrital calcium carbonate formed by cococcolithophorids and foraminiferans.

Insolation Solar radiation received at Earth's surface.

Intercalibration A standardization process in which scientists compare measurements of a single substance as produced by different equipment or methods at different times and locations.

Interface A surface separating two substances of different properties, such as different densities, salinities, or temperatures (e.g., the air-sea interface or the sediment-water interface).

Interstitial water Water trapped in the sediments in the pore spaces between the particles.

Ion An atom that is electrically charged due to a surplus or deficiency of electrons. The former is termed an anion and the latter a cation.

Ion complex Coordination complex. A molecule comprised of a cation (metal) and a ligand held together by a bond with significant covalent character.

Ion exchange See Cation exchange.

Ion pair A weak electrostatic attraction that occurs between solutes in a highly concentrated solution. This is an example of a specific interaction.

Ion pairing Electrostatic attractions that arise between cations and anions in solutions of high ionic strength. These attractions are not as strong as ionic bonds. Nevertheless, their presence can influence the physical and chemical behavior of the solution.

Ionic bond A chemical bond in which electron(s) have been transferred from the less electronegative atom to the more electronegative atom. The difference in these electronegativities is somewhat greater than that in a polar covalent bond.

Ionic solid A solid held together by ionic bonds between monatomic ions or complex ions e.g., NaCl(s).

Ionic strength The total concentration of positive and negative charges in a solution contributed by ionic solutes, usually given in units of molality.

IR, infrared radiation This type of electromagnetic radiation has a wavelength that ranges from 0.75 to 500 μm.

Iron-manganese oxides Hydrogenous precipitates composed of iron and manganese oxides. Most of the metals are hydrothermal in origin.

Isomorphic substitution The replacement of some of the aluminum and silicon in aluminosilicate minerals by cations of similar ionic charge and radius. This usually occurs as a result of chemical weathering.

Isopycnal A line on a chart or map that connects points of constant seawater density.

Isostacy The condition of gravitational equilibrium that is attained by lithospheric units as a result of adjustments in their positions as they float in the asthenosphere.

Isotope Nuclides that have the same number of protons in their nucleus, and hence belong to the same element, but have different numbers of neutrons.

Iteration Repetition.

Juvenile Of material that has directly issued from Earth's interior, i.e., it has not been recycled through the sediments.

K_{sp} The equilibrium constant describing the solubility of minerals.

Kaolinite A clay mineral produced by intense weathering under tropical conditions. As a result, it is depleted in all cations except for silicon and aluminum.

Kerogen The complex mixture of solid organic compounds formed from the diagenesis and catagenesis of soils and marine sediments.

Kinetic energy Energy of motion. This energy causes particles to move with a velocity which increases with temperature and decreases with increasing mass.

Labile Reactive.

Lagoonal Of lagoons, which are semi-isolated bodies of seawater trapped between coral reefs and volcanic islands or between the mainland and barrier islands. Seawater in lagoons tends to be hypersaline.

Latent heat of vaporization The amount of heat required to transform a specified mass of liquid water into steam or the amount of heat that must be removed to transform a specified mass of steam into liquid water.

Latent heat of fusion The amount of heat required to transform a specified mass of ice into liquid water or the amount of heat that must be removed to transform a specified mass of liquid water into ice.

Lava See **Volcanism**.

LD_{50} The lethal dose at which a chemical causes the mortality of 50 percent of the test organisms during a specified period of exposure.

Ligand An electron donor. Ligands form ion complexes with cations (i.e., electron acceptors).

Lignin A large polymeric macromolecule synthesized only by woody plants. The degradation of lignin is a unique source of phenolic acids.

Lipids A class of organic compounds composed of carbon, hydrogen, and oxygen atoms. Complex lipids contain fatty acids attached to a backbone molecule such as glycerol. Simple lipids, such as carotene, are polymers of terpene. Lipids are used in organisms for energy storage.

Liquid water Water in its liquid state.

Lithogenous Of the continental or oceanic crust.

Lithosphere The outer, solid portion of Earth, including the crust and upper mantle.

Lysis The breakage of cell membranes.

Lysocline The depth at which shell dissolution starts to have a detectable impact on the calcium carbonate content of the surface sediments.

Macroalgal Of large marine plants such as seaweed and marsh grasses.

Macrofauna Benthic animals that are larger than 0.5 mm.

Magma Mobile, usually molten rock material generated within the earth. It can intrude into crevices and fissures as it upwells through the crust. Alternatively, it can upwell through fissures to be extruded onto the crust's surface. Solidification produces igneous rock.

Major ions The six most abundant solutes in seawater: Na^+, Mg^{2+}, Ca^{2+}, K^+, Cl^- and SO_4^{2-}. These ions are present in constant proportions to each other and to salinity.

Mantle The bulk of the Earth. It is the layer that lies between the crust and core.

Marcet's Principle See the Rule of Constant Proportions.

Marginal sea A semienclosed body of water adjacent to, widely open to, and connected with the ocean at the water surface but bounded at depth by submarine ridges.

Mariculture Human efforts at the cultivation and propagation of seafood.

Marine pollutant Any substance introduced into the ocean by humans that alters any natural feature of the marine environment.

Marine snow Large, loosely aggregated solids composed of biogenous and lithogenous particles. The organic material is often colonized by microbes.

Mass balance equation An equation that accounts for the total amount of material in a given system.

Mass spectrometer (1) A scientific instrument used to identify organic compounds. (1) A scientific instrument used to measure the relative abundance of stable isotopes in a sample, e.g. isotope ratio mass spectrometer (IRMS).

Melting The phase transition during which a solid is transformed into a liquid.

Meridional overturning circulation (MOC) Deep water circulation driven by thermohaline processes, tides and winds.

Mesocosm Artificial ecosystems with water volumes ranging from 1 to 10,000 m^3.

Mesophiles Microbes that grow best at moderate temperatures (15 to 40°C).

Mesotrophic Seawater that has biological productivity intermediate between that of eutrophic and oligotrophic areas.

Metabolites Compounds produced or required by the reactions that occur within cells.

Metagenesis Abiotic chemical reactions that take place in deeply buried sediments at temperatures over 200°C. Considered to be a type of very low-grade metamorphism

Metal sulfides Hydrogenous sulfides, such as FeS_2 and CuS_2. Metal sulfides are common components of metalliferous sediments; most of the constituents are hydrothermal in origin.

Metalliferous Metal rich.

Metalliferous sediments Metal-rich sediments. Most are found around active spreading centers because hydrothermal activity is the primary source of the metals.

Metamorphic rock Rocks formed in the solid state as a result of changes in temperature, pressure, or chemical environment.

Meteorological Of changes in the temperature, humidity, or air motion of the atmosphere.

Methanogenesis The anaerobic microbial process that produces methane.

Methanogens Archaeans that conduct methanogenesis.

Methanotrophs Microbes that use methane as an electron donor.

Methylotrophs Microbes that use single-carbon compounds, such as methanol and formate, as electron donors.

Microlayer The thin layer present at the sea surface, also referred to as a "sea slick." This layer tends to have high solute and particle concentrations.

Micronutrient Elements that are required by organisms in smaller amounts than nitrogen and phosphorus. Most are trace metals.

Microtektites Small tektites formed by the impacts of meteorites on Earth's surface.

Microzone A small volume of water or solid matter in which the redox environment is considerably different from that in the surrounding sediment or seawater.

Migration The movement of petroleum through marine sediments and sedimentary rock as a result of overlying pressure.

Milankovitch cycles Changes in global ice volume that reach maxima every 23,000, 41,000, and 100,000 years. They are thought to be related to changes in astronomical alignments that have similar periods.

Mineral An inorganic substance that occurs naturally in the Earth and has distinctive physical properties. Its chemical composition can be expressed as an empirical formula that shows the molar combining ratios of the constituent elements.

Mixed layer Near-surface waters down to the pycnocline that are isohaline and isothermal as a result of wind mixing.

Molecular diffusion The random motion of solutes in a solution. In the absence of external forces, solutes spontaneously undergo net diffusion from regions of higher concentration to lower concentration. This continues until a homogeneous distribution of the solute is achieved.

Mollusks Soft, unsegmented animals usually protected by a calcareous shell and having a muscular foot for locomotion; includes snails, clams, and octopuses.

Mononucleotide Building blocks of polynucleotides. Composed of a phosphate group linked to a five carbon sugar which is linked to a purine or pyrimidine base.

Montmorillonite An iron-rich clay mineral that has a very high cation exchange capacity. Unlike the other clay minerals, a significant amount of sedimentary montmorillonite is hydrothermal in origin.

Mud A sedimentary deposit composed of 70 percent or more by mass silt and clay-sized grains; also called deep-sea muds.

Mutagenic A form of energy or chemical that can alter the molecular structure of the genes.

NAD, nicotine adenine dinucleotide A biochemical that functions as a cellular electron carrier.

NADP, nicotine adenine dinucleotide phosphate A biochemical that functions as a cellular electron carrier.

NADW, North Atlantic Deep Water This is a deep water mass formed in the North Atlantic.

NAEC, normal atmospheric equilibrium concentration The gas concentration that a water mass would attain if it reached equilibrium with the atmosphere. The NAEC of a gas is a function of water temperature and salinity as well as the partial pressure of the gas in the atmosphere.

Napthenes Cycloparaffins, such as cyclohexane. They are a common component of gasoline.

Natural gases Hydrocarbons that have fewer than 5 carbons, i.e., methane, ethane, propane, and butane.

Natural products Secondary metabolites that are generally used by organisms to control ecological relationships.

Nepheloid layer Deep and bottom waters that have high concentrations of resuspended sediment.

Neritic Of the coastal ocean (i.e., over the continental margin).

Nernst equation The mathematical equation that relates the cell potential of a redox reaction to the temperature and concentrations of the reacting chemicals, e.g., $E_{cell} = E^{\circ}_{cell} - ((RT/nF)\ln Q)$.

Neutron An electrically neutral particle found in the nucleus of atoms.

Nitrification The microbial oxidation of ammonium to nitrite and then to nitrate by O_2. In the ocean, this process is mediated by nitrifying bacteria.

Nitrogen fixation The process by which some bacteria and phytoplankton are able to convert N_2 into organic nitrogen. The energy required is large because a triple bond must be broken.

Nodules Mineral precipitates, such as iron-manganese oxides and glauconites, that form as roundish lumps.

Nonconservative Chemical behavior that is largely controlled by biogeochemical reactions. The concentrations of nonconservative substances are not directly proportional to salinity.

Nonideal Physicochemical behavior that does not conform to ideal thermodynamic predictions.

Nonpelagic Sediments that accumulate at rates in excess of 1 cm/1000 y.

Non-point-source inputs Pollutants introduced into the ocean from widely dispersed sources, such as stormwater runoff.

Nonpolar covalent bond See Covalent bond.

Nonspecific interactions Solute-solute or solute-solvent interactions that occur within solutions without causing the formation or breakage of a chemical bond. In seawater, most nonspecific effects are electrostatic in nature.

Normal atmospheric equilibrium concentration See NAEC.

Nucleic acids Biopolymers whose monomers are nucleotides. They are one of the building blocks of DNA and RNA.

Nucleotides See mononucleotides.

Nucleus The central, positively charged part of an atom, composed primarily of protons and neutrons.

Nuclide A species of atom uniquely identified by the number of protons and neutrons in its nucleus.

Nutrient An inorganic or organic solute necessary for the nutrition of primary producers.

Nutrient regeneration The process whereby particulate organic nitrogen and phosphorous are transformed into dissolved inorganic species, such as nitrate and phosphate. Microorganisms are largely responsible for this process.

Oceanic crust The mass of basaltic material, approximately 5–7 km thick, which underlies the ocean basins.

Oceanic Of the water or sediment beyond the continental margins.

Oils Liquid fats.

Oligotrophic Waters with very low biological productivity, such as those of the mid-ocean gyres.

Oolite A hydrogenous precipitate commonly found in carbonate sediments of continental shelves located in the tropics. They are composed primarily of calcium carbonate and are thought to be an abiogenic precipitate formed from warm seawater supersaturated with respect to calcite and aragonite.

Ooze A sediment that contains greater than 30 percent detrital biogenic hard parts by mass.

Opal (or opaline silica) An amorphous silicate formed through the polymerization of silicic acid molecules. Though most is biogenic in origin, some forms as a result of diagenesis.

Open ocean That part of the ocean seaward of the continental margin.

Operational definition A property defined by the manner in which it is measured. Opposite of a theoretical or conceptual definition.

Ophiolites Ancient pieces of oceanic crust that have been thrust up onto the continents as a result of geologic uplift.

Organometallic Compounds that contain both organic structures and metals.

Orthosilicic acid The dominant dissolved species of silicon in seawater (written as $H_4SiO_4(aq)$ or $Si(OH)_4(aq)$).

Osmolyte A neutral solute that reacts minimally with the contents of the cell. It serves to protect cells from drying out and/or in responding to salinity changes.

Osmotic pressure The pressure exerted across a semipermeable membrane separating two solutions that differ in their concentrations of a dialyzable solute.

Outfalls Discrete locations where large quantities of water and/or human wastes are introduced into rivers or the ocean.

Oxic Waters that contain O_2 concentrations close to that of their NAEC.

Oxidation number Oxidation state. A measure of the degree of oxidation of an atom in a substance.

Oxidation The loss of electrons.

Oxidizing agent A chemical that gets reduced, thereby causing some other reactant to become oxidized.

Oxycline The depth over which O_2 concentrations decrease rapidly. This usually coincides with the thermocline.

Oxygen free radical An atom or molecule of oxygen that contains extra electrons and hence is negatively charged. These species are very strong oxidizing agents.

Oxygen minimum zone (OMZ) Subsurface undersaturations in dissolved oxygen gas found within the thermocline.

Oxyhydroxides Amorphous precipitates of oxides and hydroxides that form in alkaline solutions such as seawater. These precipitates usually contain a variety of cations, such as trace metals.

p_ – log of a quantity, such as pH, pε, pK.

Paleoceanography The study of the ancient oceans that seeks to reconstruct past environmental conditions by examining chemical and fossil records left in the sediments and polar ice cores.

Paleocene eocene thermal maximum (PETM) A hyperthermal period, 55.8 mybp, during which average global temperatures increased by ~6°C for a period of 20,000 years.

Paraffins Saturated straight-chain, aliphatic hydrocarbons that have five to ten carbons. A common component of gasoline.

Partial pressure The pressure exerted by a particular gas in a liquid or gaseous solution.

Particle ballasting Packaging of particles which enhances their sinking rates and also appears to confer some protection against microbial attack.

Particulate matter Any solid.

pε Electron activity, i.e., $\{e^-\}$.

pelagic Of, relating to, or living or occurring in the open sea.

Pelagic sedimentation Sedimentation that occurs at rates less than 1 cm/1000 y. This is characteristic of sediment on the abyssal plains and mid-ocean ridges.

Persistent bioaccumulative toxics (PBTs) Pollutants that have slow degradation rates and high bioaccumulation factors.

Persistent organic pollutants (POPs) Organic pollutants with slow degradation rates, such as PCBs.

pH The negative log of the hydrogen ion activity. Solutions with pH less than 7 are acidic and those with pH greater than 7 are alkaline. A solution is neutral if its pH equals 7.

Phospholipids Lipids that contain phosphate groups.

Phosphorite A hydrogenous mineral that forms in surficial sediments underlying surface waters of high biological productivity. It is composed primarily of calcium phosphate that is biogenic in origin.

Phosphorylation The biochemical reactions in which a phosphate group is attached to a biomolecule. The attachment and removal of such groups is used as a means of energy transfer in cells.

Photic Same as Euphotic.

Photoautolithotrophs Organisms that use solar radiation as their energy source and inorganic substances as their carbon and electron sources.

Photodissociation The decomposition of a molecule as a result of the input of solar energy.

Phycoerythrins A phycobiliprotein synthesized by cyanobacteria and red algae.

Physicochemical Of a process that involves either physical and/or chemical change.

Phytoplankton Single-celled plants that have weak swimming ability.

Pillow basalts Large mounds of basalt that form when lava is extruded onto the seafloor at active spreading centers.

Piston velocity The rate at which supersaturated gases are moved from the surface ocean into the atmosphere by molecular diffusion. Transfer velocity.

Plankton Organisms that drift passively or are weak swimmers. Includes mostly microscopic plants (phytoplankton) and protozoans, as well as larval animals and small filter feeders (zooplankton).

Plate tectonics The theory that posits crustal plate movement is caused by seafloor spreading and subduction.

Platform carbonates Massive concretions of calcium carbonate in neritic sediments deposited by corals, corraline algae and other organisms that deposit calcareous hard parts.

POC, particulate organic carbon The fraction of the organic carbon pool that is not dissolved in seawater. In practical terms, this includes all organic compounds that do not pass through a filter with a 0.45 μm pore size.

Point-source inputs Inputs of pollutants into the ocean from discrete sources, such as sewage outfall pipes.

Polar covalent bond Chemical bonds in which electrons are not equally shared due to the greater electronegativity of one of the atoms. As a result, the more electronegative atom acquires a small net negative charge relative to the less electronegative one. The difference in electronegativities is somewhat smaller than that in an ionic bond.

Pollutogens Infective agents whose source is external to the ocean.

Polycyclic aromatic hydrocarbons Aromatic hydrocarbons composed of fused benzene rings. Many are highly carcinogenic.

Polymetallic Containing a variety of different metals.

Polymetallic sulfides Hydrogenous metal sulfides that form from the precipitation of metals and sulfides that are hydrothermal in origin. The metals (Fe, Cu, Co, Zn, and Mn) tend to coprecipitate to form heterogeneous sulfide deposits.

Polynuclear Of an organometallic molecule that contains more than one atom of metal.

Polynucleotides Nucleic acids. Biopolymers whose building blocks are nucleotides (mononucleotides).

Polyprotic Of an acid that contains more than one proton that it can donate to a base.

Polysaccharide A biopolymer composed of two or more simple sugars, such as glucose. They are used for energy storage in organisms.

POM, particulate organic matter The fraction of the organic matter pool that is not dissolved in seawater. In practical terms, this includes all organic compounds that do not pass through a filter with a 0.45-micron pore size.

PON, particulate organic nitrogen The fraction of the organic nitrogen pool that is not dissolved in seawater. In practical terms, this includes all organic nitrogen compounds that do not pass through a filter with a 0.45-micron pore size.

Ponding The slumping of abyssal sediments into depressions. This process tends to smooth out irregular topographic features on the seafloor of the deep ocean.

Poorly sorted sediments Sediments that contain a wide range of different size class particles, such as clays, sands, and pebbles.

Pore waters Same as Interstitial waters.

Porosity A measure of the open space between grains in a sedimentary deposit.

Postdepositional Of a change that occurs in a sediment following its accumulation on the seafloor.

Potash Mineral form of potassium carbonate.

Pourbaix diagrams Graphs that show redox speciation as a function of pH and either pε or E_h.

Precipitation (1) Rainfall. (2) Formation of solids from dissolved materials.

Preformed Refers to the fraction of a solute in a water mass whose origins are not related to remineralization of detrital particles in situ.

Pressure solution The dissolution of deeply buried solids in marine sediments caused by the increase of mineral solubility as a result of increasing pressure.

Primary nitrite maximum Subsurface concentration maximum in nitrite found toward the base of the euphotic zone in stratified water columns. Mainly caused by ammonification and nitrification.

Primary production The amount of organic matter synthesized by organisms from inorganic substances in a unit volume or area of water.

Primary productivity The rate at which organic matter is synthesized by organisms from inorganic substances per unit volume or area of seawater.

Primary See Juvenile.

Primordial radionuclides Long-lived radionuclides that were present at Earth's formation. The most abundant are ^{235}U, ^{238}U, and ^{232}Th. These isotopes decay to radioactive daughters and hence form a series of decay steps ending in a stable nuclide of lead.

Prokaryotes Single-celled organisms that lack a nucleus, includes bacteria, archaeans, and viruses.

Proteins Biopolymers composed of amino acids and hence are nitrogen-rich organic compounds.

Protists Eukaryotic microorganisms that are neither animal, fungi, plant, or archaean. Unicellular forms include the amoeboid protozoans and algae, such as the foraminferans and radiolarians, and dinoflagellates and diatoms, respectively. Some algae are either multicellular or colonial, such as the red algae and freshwater *Volvox*, respectively.

Proton A nuclear particle that contains one unit of positive electrical charge.

Protozoans Heterotrophic single-celled eukaryotes. Benthic and pelagic forms are present in the ocean.

Pteropods Free-swimming gastropods in which the foot is modified into fins; both shelled and nonshelled forms exist.

Pycnocline The vertical region which density increases rapidly with depth. This usually coincides with the thermocline.

Radioactive decay law The mathematical description of how the amount of radioactive material diminishes over time as a result of radioactive decay. $A = A_o e^{-kt}$.

Radioactive decay See Radioactivity.

Radioactivity The spontaneous breakdown of the nucleus of an atom that results in the emission of radiant energy either as particles or waves.

Radiocarbon The radioactive isotope of C, ^{14}C, which has a half-life of approximately 5730 y.

Radioisotope See Radionuclide.

Radiolaria A group of amoeboid protozoans that deposit siliceous tests.

Radionuclide A radioactive isotope.

Rain rate The rate at which particulate matter settles to the seafloor. Since some of the particles can dissolve prior to permanent burial, the rain rate can be less than the accumulation rate.

Reactive nitrogen (Nr) The chemical forms of nitrogen other than N_2 gas. Fixed nitrogen.

Recycling efficiency The degree to which a biolimiting element is remineralized prior to its removal from the ocean via sedimentation.

Red tide A red or red-brown discoloration of surface waters caused by high concentrations of certain microorganisms, particularly dinoflagellates. Most common in coastal waters.

Redfield-Richards Ratio The average molar elemental ratio of C to N to P (106:16:1) that is present in the marine plankton as sampled by a net tow.

Redox reactions Chemical reactions that involve changes in the oxidation numbers of some of the participating species.

Reducing agent A chemical that is oxidized, thereby causing some other reactant(s) to become reduced.

Reduction Gain in electrons.

Refractory Inert; unreactive.

Relict sediment Sediments that are no longer forming. Most of the sediments on the continental margins are presently relict. Some are even eroding.

Remineralization The dissolution of hard parts or the degradation of POM that leads to solubilization of nutrients and micronutrients.

Reprecipitate The formation of solids from solutes whose origins were some dissolution or remineralization process.

Reservoir beds Marine sediments that are porous enough for large amounts of petroleum to migrate into.

Reservoir The biogeochemical form and/or location of a chemical in the crustal-ocean-atmosphere factory.

Residence time The average amount of time that a chemical species spends in the ocean or sediment, assuming the species is at a steady state.

Resolubilization Same as Remineralization.

Reverse weathering Chemical reactions that are theorized to occur in the sediments. In these reactions, seawater is thought to react with clay minerals and bicarbonate producing secondary clays and consuming alkalinity and some cations. This process is approximately the reverse of chemical weathering on land that produces clay minerals.

River runoff The transport of water, solutes, and lithogenous particles from the continents to the ocean as a result of riverine input.

Riverine Of rivers.

Rule of Constant Proportions The relative abundances of the six major cations (Na^+, K^+, Ca^{2+}, Mg^{2+}, Cl^-, and SO_4^{2-}) is constant regardless of the salinity of seawater.

Sabkha Intertidal mudflats in which modern-day evaporites form. Most are located in the Middle East, the northwestern coast of Australia and lagoons along the coast of Texas.

Salinas Coastal salt lakes in which modern-day evaporites form. Most are located in the Middle East and Western Australia.

Salinity A measure of the salt content of seawater now expressed in practical salinity units related to measurement via conductivity.

Salinometer The device used to measure the salinity of seawater by determining its electrical conductivity.

Salt A molecule composed of one or more cations and anions held together by ionic bonds, e.g., $KC1$ and Na_2SO_4.

Salting out A decrease in the solubility of a gas or solute with increasing ionic strength of the solution.

Saltmarsh A relatively flat area of the shore where fine-grained sediment is deposited and salt-tolerant grasses grow. One of the most biologically productive ecosystems on Earth's surface.

Sand Particles that range in diameter from 1/16 to 1 mm.

Sapropel Kerogen that is of marine or aquatic origin.

Saturated (1) Of a solution containing the equilibrium concentrations of solutes dictated by the solubility of a particular solid. As a result, the mass of the solid in solution will remain constant over time. (2) Of a solution that contains the equilibrium concentration of a gas. This concentration is determined by the temperature and ionic strength (salinity) of the solution and the partial pressure of the gas in the atmosphere.

Saturation horizon The depth range over which seawater is saturated with respect to calcium carbonate, i.e., $D = 1$. At depths below the saturation horizon ($D < 1$), calcium carbonate will spontaneously dissolve if exposed to the seawater for a sufficient period of time.

Scavenging The removal of dissolved materials (such as trace metals) from seawater by adsorption onto the surfaces of sinking particles.

Scavenging turnover time The amount of time required to remove the entire inventory of a solute from a specified region of seawater via scavenging.

Sea state A description of the ocean surface with respect to the height of the wind waves.

Secondary Of the material that has been recycled through the sediments.

Secondary nitrite maximum Subsurface concentration maximum in nitrite found in the oxygen minimum zone. Mainly caused by denitrification.

Sediment Particulate inorganic or organic particles that accumulate in loose, unconsolidated form.

Sedimentary rock A rock resulting from the consolidation of loose sediment.

Sedimentation The process by which sediments accumulate on the seafloor.

Sessile Of benthic organisms that live attached to the seafloor.

Seuss Effect The lowering of the $^{14}C/C$ in atmospheric CO_2 as a result of the burning of fossil fuels.

SHE, standard hydrogen electrode The electrode used as a standard against which all other half-cell potentials are measured. The following reaction occurs at the platinum electrode when immersed in an acidic solution and connected to the other half of an electrochemical cell: $2H^+(aq) + 2e^- \rightarrow H_2(g)$. The half- cell potential of this reaction at 25°C, 1 atm and 1 m concentrations of all solutes is agreed, by convention, to be 0 V.

Sigma-T A measure of the density of seawater (i.e., $\sigma_t = (\rho - 1) \times 1000$). It is a function of temperature, salinity, and hydrostatic pressure.

Silica Silicon oxide ($SiO_2(s)$), which can occur in crystalline form, e.g., quartz, or noncrystalline form, e.g., opal.

Silicic acid Orthosilicic acid, H_4SiO_4.

Silico-flagellates Members of the plankton that deposit hard parts of silica and have flagella.

Sill Shallow portion of the seafloor that partially restricts water flow, usually at the mouth of a marginal sea or estuary.

Silt Particles that range in diameter from 1/256 to 1/32 mm.

Simple lipids Lipids whose hydrolysis produces a relative small set of degradation products. Most are terpenes.

Sink A reservoir that is the recipient of material transport.

Snowball Earth Global glaciation events during which all, or nearly all, of Earth's surface, including the surface ocean, was covered by ice.

Solubility A measure of the maximum quantity of a substance that can dissolve in a given volume of solvent under a specified set of conditions.

Solubility product (K_{sp}) The equilibrium constant that describes the dissolution of a solid in a solvent.

Solubility pump The transport of carbon into the deep sea via physical processes associated with gas solubility and water circulation.

Solubilized Dissolved.

Solute A substance that is dissolved in a solvent. For a solute to be dissolved, it must be hydrated, i.e., completely surrounded by water molecules.

Solution A state in which a solute is dissolved in a solvent.

Solvent The most abundant substance in a solution.

Source The reservoir that is the origin of material undergoing transport, e.g., water is transported from lakes to the ocean by river runoff. Thus the lakes are the source and the ocean the sink.

Spatial Of the geographic dimension.

Speciation The chemical form(s) that an atom is in.

Specific activity The radioactive decay rate of a specified amount of sample, usually expressed as disintegrations per liter (dpm/L) or disintegrations per gram (dpm/g).

Specific interactions Chemical processes that involve the formation or breakage of intramolecular bonds. This behavior can be described by equilibrium constants and expressions.

Spherules Small spheres.

Spicules Minute, needle-like spines produced by some of the microorganisms that deposit calcareous and/or siliceous hard parts.

Stability constant The equilibrium constant for a reaction which describes the decomposition of a molecule, i.e., the reverse of a formation constant.

Steady state The absence of change over time in the abundance of a chemical caused by equal rates of its production and removal. Chemical equilibrium is a special case of a steady state in which the processes producing and removing the chemicals are reversible.

Steam The gaseous phase of water.

Sterols Biomolecules that are aliphatic hydrocarbons, whose base unit is composed of three rings containing six carbons and one ring containing five carbons.

Stoichiometry The relative abundance of atoms in a molecule or in a reaction, given in terms of molar quantities.

Stormwater Sheet flow of rainwater that flows across the land surface until it either percolates into the soil, flows into rivers and the coastal ocean, or evaporates.

STP, standard temperature and pressure For gases, the temperature is 273.15°K and the pressure is 1 atm.

Stratographic trap Regions of the seabed where steep gradients in porosity retard the migration of petroleum and hence cause large pools of oil, gas and tar to accumulate.

Stratosphere That part of Earth's atmosphere that lies between the troposphere and the upper layer (ionosphere).

Stromatolites Domal concretions formed on the seafloor, primarily shallow waters, by microorganisms. These deposits represent the oldest unequivocal fossilized remains of life on Earth.

Structural trap Geological structures, such as diapirs, that retard the migration of petroleum through marine sediments and sedimentary rock.

Subaqueous Below the air-sea interface.

Suboxic O_2 concentrations significantly below that of the NAEC, i.e., less than 0.2 mL/L.

Subpolar Latitudes ranging from 40 to 60°.

Substituents Atoms or groups of atoms that are attached to a molecular framework or backbone.

Substrate The base upon which an organism lives and grows. Can refer to physical habitat or the organic growth medium.

Subterranean estuaries Coastal aquifers in which saline pore waters mix with terrestrial groundwaters.

Subtidal At depths below the lowest reach of the low tide.

Supersaturated (1) Of a solution containing concentrations of solutes that exceed equilibrium levels for a particular solid. As a result, the ions will spontaneously precipitate to form that solid. (2) Of a solution that contains greater than the equilibrium concentration of a gas. If the solution is in contact with the atmosphere, a net flux of gas out of the solution will occur.

Supratidal That part of the shoreline that lies above the highest reach of the high tides. This zone receives seawater only as a result of storm surges and wind transport.

Surface tension The force that exists among water molecules at the air-liquid interface. This force is the result of hydrogen bonding between water molecules.

Surfactant A chemical that has soap-like behavior (i.e., it is able to cause the dissolution of nonpolar solutes in a polar solvent).

Surficial Of the surface of a solid.

Symbiosis A close relationship between two species upon which the survival of at least one member is dependent.

Taxonomic Of the system used to classify organisms (e.g., kingdom, phylum, class, species, etc.).

Tectonism Earthquake, volcanic, and other crustal motions associated with the process of plate tectonics.

Tektites Glassy rocks that range up to a few centimeters in length. Formation is thought to have been a result of meteorite impacts that liquefied and ejected molten crustal rocks. As the ejecta fell back to Earth, they cooled and froze into teardrop and dumbbell shapes. Small tektites are called microtektites.

Temporal Of time.

Terpenes Organic compounds comprised of repeating units of isoprene. Terpenes are simple lipids.

Terrigenous Of planet earth.

Tests Microscopic shells of marine plankton.

Terpenoids Oxygen-containing terpenes.

Thermocline The depth range over which temperature decreases rapidly with depth. At mid latitudes this zone usually spans depths ranging from 300 to 1000 m.

Thermodynamic equilibrium See Equilibrium.

Thermohaline circulation Deep-water circulation caused by density differences created in the surface waters of polar regions. Cooling increases the density of the surface waters, which sink and then advect horizontally throughout the deep ocean. The water is returned to the sea surface by eddy diffusion.

Tillites The lithified remains of unsorted glacial marine sediments. Contains large fragments of unweathered lithogenous particles.

Titrimetry The analytical technique of concentration measurement that involves reaction of a solution containing an analyte with a known quantity of standard (e.g., titration).

Total alkalinity (T.A.) Alkalinity contributed by the carbonate species and other titratable charge, such as borate, minus the in-situ hydrogen ion concentration.

Total dissolved inorganic concentration (TDIC and $\sum CO_2$) Concentration of carbon contributed by all dissolved inorganic forms, i.e., carbon dioxide, carbonic acid, carbonate, and bicarbonate.

Trace element An element whose dissolved concentration is between 50 and 0.05 μmol/kg. Most are metals and are referred to as trace metals.

Transamination The reaction in which an amine group is transferred from an amino acid to a carboxylic acid, thereby creating a new amino acid.

Transparent exopolymer particles (TEPs) Detrital particulate organic matter whose origin is secretions and exudates from marine organisms.

Tritium The radioactive isotope of H, ^3H, which has a half-life of approximately 12.5 y.

Trophic level The relative position occupied by an organism in a food chain or web.

Troposphere The portion of the atmosphere closest to Earth's surface in which temperature generally decreases with increasing altitude, clouds form, and convection is active.

T-S diagram The x-y graph of the temperature and salinity of water samples used to identify source water masses and types.

Turbidite The sedimentary deposit created by turbidity currents. The latter are underwater mudslides common to the continental slope and rise.

Turbidity current An underwater mudslide common to the continental slope. The particles deposit at the foot of the slope to form the continental rise. The resulting deposit is often referred to as a turbidite.

Turbulence The random motion of water molecules.

Turbulent mixing coefficient Constant of porportionality used in Fick's First and Second Laws to predict changes in solute concentration over time or fluxes from solute gradients that arise from turbulent mixing. The constant is a function of length scales and the degree of density stratification and hence is termed a "coefficient".

Turnover time The time required for a specific process or set of processes to remove all of a substance from a reservoir.

Unconsolidated Composed of small particles or grains, e.g., a sandy beach.

Undersaturated (1) Of a solution containing concentrations of solutes that are less than those required to reach equilibrium with a particular solid phase. As a result, that solid phase will spontaneously dissolve. (2) Of a solution that contains less than the equilibrium concentration of a gas. If the solution is in contact with the atmosphere, a net flux of gas into the solution will occur.

Universal solvent Water, which is so named due to its ability to dissolve at least a small amount of all substances.

Unsorted sediments Sediments that contain unconsolidated grains exhibiting a wide range of particle sizes. Same as poorly sorted sediments.

USEPA, United States Environmental Protection Agency The federal agency charged with protecting air, water, and soil quality for the protection of human health and aquatic life. This agency sets water quality criteria.

UV, ultraviolet radiation This form of electromagnetic radiation has wavelengths that range from 10 to 400 nm.

Vertical segregation The vertical gradient in biogenic materials, such as nutrients and O_2, that is established by the interaction between the biogeochemical cycling of particulate organic matter and the vertical density stratification of the water column. Strongest at mid and low latitudes.

Volatiles Compounds with such high vapor pressures under environmental conditions that a significant fraction is present in the gas phase.

Volcanism Volcanic eruptions during which magma from the mantle is extruded into or onto Earth's crust. Once the magma has escaped from the mantle, it is termed lava.

Water mass Water parcels of common original that exhibit a range of temperature and salinity values. The variability in temperature and salinity within a particular water mass is due to spatial and temperature variations in the processes responsible for their formation, i.e., cooling, evaporation, sea ice formation, etc.

Water type Water parcels defined by a single salinity and temperature.

Water-quality criteria Legal limits that define the permissible levels of pollutants in marine and fresh waters.

Weathering Destruction or partial destruction of rock by thermal, chemical, biological, and/or mechanical processes.

Well-sorted sediments Sediments composed of one size class of particles, e.g., a deep-sea clay.

Zeolites Authigenic clay minerals whose production is associated with geochemical processes occurring at spreading centers, including volcanism and chemical weathering of ocean crust by seawater.

Zooplankton The animal and protozoan members of the plankton.

Index

C

D

E

J

K

L

M

N

O

P

T

U

V

W

X

Y

Z